GENETICS

OF RELATED INTEREST
FROM THE BENJAMIN/CUMMINGS SERIES IN THE LIFE SCIENCES

GENERAL BIOLOGY

N. A. Campbell
Biology, Fourth Edition (1996)

N. A. Campbell, L. G. Mitchell, and J. B. Reece
Biology: Concepts and Connections, Second Edition (1997)

J. Dickey
Laboratory Investigations for Biology (1995)

J. Hagen, D. Allchin, and F. Singer
Doing Biology (1996)

A. Lawson and B. D. Smith
Studying for Biology (1995)

J. G. Morgan and M. E. B. Carter
Investigating Biology: A Laboratory Manual for Biology,
Second Edition (1996)

J. Pechenik
A Short Guide to Writing Biology, Third Edition (1997)

R. A. Wallace
Biology: The World of Life, Seventh Edition (1997)

R. A. Wallace, G. P. Sanders, and R. J. Ferl
Biology: The Science of Life, Fourth Edition (1996)

BIOCHEMISTRY

R. F. Boyer
Modern Experimental Biochemistry, Second Edition (1993)

C. K. Mathews and K. E. van Holde
Biochemistry, Second Edition (1996)

G. L. Sackheim
Chemistry for Biology Students, Fifth Edition (1995)

CELL BIOLOGY

W. M. Becker, J. B. Reece, and M. F. Poenie
The World of the Cell, Third Edition (1996)

R. J. King
Cancer Biology (1996)

L. J. Kleinsmith and V. M. Kish
Principles of Cell and Molecular Biology, Second Edition (1995)

GENETICS

D. S. Falconer and T. F. C. Mackay
Introduction to Quantitative Genetics, Fourth Edition (1995)

A. Radford, S. Baumberg, and D. Cove
A Primer of Genetics (1995)

P. J. Russell
Fundamentals of Genetics (1994)

MOLECULAR BIOLOGY

M. V. Bloom, G. A. Freyer, and D. A. Micklos
Laboratory DNA Science (1996)

L. E. Hood, I. L. Weissman, W. B. Wood, and J. H. Wilson
Immunology, Second Edition (1984)

J. D. Watson, N. H. Hopkins, J. W. Roberts, J. A. Steitz, and A. M. Weiner
Molecular Biology of the Gene, Fourth Edition (1987)

GENETICS

Fifth Edition

Peter J. Russell

REED COLLEGE

An imprint of Addison Wesley Longman, Inc.

Menlo Park, California • Reading, Massachusetts • New York • Harlow, England
Don Mills, Ontario • Sydney • Mexico City • Madrid • Amsterdam

Sponsoring Editor: Nina Horne
Senior Sponsoring Developmental Editor: Susan Weisberg
Editorial Assistants: Erika Buck, Amy Dhillon
Production Editor: Lisa Weber
Art Supervisor: Carol Ann Smallwood
Artists: J/B Woolsey Associates, Inc., Karl Miyajima
Photo Editor: Kathleen Cameron
Composition and Film Buyer: Vivian McDougal
Senior Permissions Editor: Ariane de Pree-Kajfez
Copyeditors: Sylvia Stein Wright, Lisa Weber
Indexer: Barbara Littlewood
Cover Designer: Yvo Riezebos
Cover Photo: © Dr. Gopal Murti/Science Photo Library/Photo Researchers, Inc.

This title published by Benjamin/Cummings, an imprint of Addison-Wesley Longman, Inc.

The cover photo is a colorized TEM (transmission electron micrograph) of replicating DNA from a human HeLa cancer cell.

Library of Congress Cataloging-in-Publication Data
Russell, Peter J.
    Genetics / Peter J. Russell. — 5th ed.
        p.   cm.
    Includes bibliographical references and index.
    ISBN 0-321-00038-2
    1. Genetics.   I. Title.
QH430.R87   1998
576.5 — dc21   97-29688
                                                              CIP

ISBN 0-321-00038-2

  2 3 4 5 6 7 8 9 10—RNV—01 00 99 98

The Benjamin/Cummings Publishing Company, Inc.
2725 Sand Hill Road
Menlo Park, CA 94025

BRIEF CONTENTS

DETAILED CONTENTS

4

EXTENSIONS OF MENDELIAN GENETIC ANALYSIS 97

5

GENETIC MAPPING IN EUKARYOTES I 133

6

GENETIC MAPPING IN EUKARYOTES II 168

13

TRANSCRIPTION, RNA MOLECULES, AND RNA PROCESSING   379

14

THE GENETIC CODE AND THE TRANSLATION OF THE GENETIC MESSAGE   418

15

RECOMBINANT DNA TECHNOLOGY AND THE MANIPULATION OF DNA   446

16

REGULATION OF GENE EXPRESSION IN BACTERIA AND BACTERIOPHAGES 501

17

REGULATION OF GENE EXPRESSION AND DEVELOPMENT IN EUKARYOTES 535

18

GENETICS OF CANCER 585

PREFACE

OVERVIEW OF THE TEXT

It is now more than forty years since the mysteries of DNA were first unravelled, and in that time genetics has become one of the most exciting and ground-breaking of the sciences. A continual flood of discoveries not only expands our understanding about heredity but affects our daily lives in areas ranging from disease therapy to courtroom evidence.

Genetics, Fifth Edition, reflects this dynamic nature of the field. It has been extensively updated to include many significant advances made since the last edition, particularly those advances resulting from the use of molecular experimental approaches. Like previous editions, this new edition emphasizes an experimental, inquiry-based approach, with solid treatment of many research experiments that have contributed to our knowledge of genetics. In this way, students are exposed to the "process of science," learning about the formulation and study of scientific questions in a way that will be of value in their study of genetics and in all areas of science.

A consistent effort has been made to present the important experiments without including excessive facts and detail that could obscure the central concepts of genetics. *Genetics*, Fifth Edition, is ideally suited for students who have had some background in biology and chemistry and who are interested in learning the concepts of genetics from a research-oriented perspective. Great care has been taken to keep the text accessible to students by making it easy to read, with a consistent level of coverage and a logical progression of ideas. In addition, increased attention to editorial development in this edition ensures a high degree of accuracy throughout the text.

Genetics, Fifth Edition, retains the overall approach, organization, and pedagogical features (such as Principal Points, Keynotes, Summaries, and Analytical Approaches for Solving Genetics Problems) that have made previous editions valued learning tools for students of genetics. As in the past, problem solving is a major feature of the book, and this edition includes more and updated end-of-chapter Questions and Problems, which have been consistently praised by reviewers of all editions. It also retains a format that allows in-structors to use the chapters out of sequence to accommodate various teaching approaches.

The fifth edition includes the following improvements:

- A new chapter, Genetics of Cancer (Chapter 18), discusses the relationship of the cell cycle to cancer, mutational models for cancer, genes and cancer, the multistep nature of cancer, and chemicals and radiation as carcinogens. Cancer is of obvious importance to the human population, and this chapter imparts important knowledge to students about the nature of the disease and the various roles that genes play in its development.

- All the molecular aspects of genetics are updated, so that the book continues to reflect our current understanding of genes at the molecular level. New material is presented, for example, on the structure of eukaryotic chromosomes (Chapter 11), the control of the cell cycle by cyclins and cyclin-dependent DNA kinases (Chapter 12), the use of mutations to define telomerase function (Chapter 12), promoter elements and initiation of transcription in eukaryotes (Chapter 13), 3' end generation and polyadenylation in eukaryotes (Chapter 13), alternative splicing in the production of mRNAs (Chapter 17), control of mRNA degradation (Chapter 17), modern molecular screens for the isolation of mutants (Chapter 19), and investigation of genetic relationships by mitochondrial DNA analysis (Chapter 21).

- Human examples continue to be used extensively throughout the text, and they have been updated particularly with regard to new molecular understandings of various human genetic diseases.

- The coverage of mapping genes in human chromosomes (Chapter 6) now involves a modern presentation of constructing physical and genetic maps of the human genome.

- The coverage of recombinant DNA technology has been updated and expanded to reflect the rapid developments in that area. New material is presented, for example, on the use of PCR (polymerase chain reaction) in the analysis of diseases, the current status of the human genome project, techniques for transforming genes into plants, and the use of transposable elements for transforming *Drosophila*.

- The coverage of developmental genetics has been updated to include current information on *Drosophila* development and on the relationship between homeotic genes of *Drosophila* and mammals. The recent experiment demonstrating that the nucleus of an adult sheep cell could direct the development of a normal lamb from an enucleated egg cell is described.

- Approximately 100 new questions and problems have been added throughout the book. Many test the ability of students to think in an investigative way, and a number integrate concepts presented in more than one chapter.

- Many new references have been added to the Suggested Readings for each chapter.

ORGANIZATION AND COVERAGE

The three major areas of genetics—transmission genetics, molecular genetics, and population and quantitative genetics—are covered in 23 chapters. Chapter 1 is an introductory chapter designed to summarize the main branches of genetics, to explain what geneticists do and what their areas of research encompass, and to introduce the main properties of genes and the main experimental approaches used in genetics research today. The next seven chapters deal with transmission of the genetic material. Chapters 2 and 3 present the basic principles of genetics in relation to Mendel's laws. Chapter 2 is focused on Mendel's contributions to our understanding of the principles of heredity, while Chapter 3 covers mitosis and meiosis in the context of both animal and plant life cycles, the experimental evidence for the relationship between genes and chromosomes, and methods of sex determination. Mendelian genetics in humans is introduced in Chapter 2 with a focus on pedigree analysis and autosomal traits. The topic is continued in Chapter 3 with respect to sex-linked genes.

The exceptions to and extensions of Mendelian analysis (such as the existence of multiple alleles, the modification of dominance relationships, gene interactions and modified Mendelian ratios, essential genes and lethal alleles, and the relationship between genotype and phenotype) are described in Chapter 4. In Chapters 5 and 6, genetic mapping in eukaryotes is presented. In Chapter 5, we describe how the order of and distance between the genes on eukaryotic chromosomes are determined in genetic experiments designed to quantify the crossovers that occur during meiosis. Chapter 6 focuses on tetrad analysis, primarily in fungal systems; on mapping eukaryotic genes through mitotic analysis; and on the construction of genetic and physical maps in humans. With the understanding of the relationship between genes and chromosomes obtained from Chapters 2 through 6, chromosomal mutations—changes in normal chromosome structure or chromosome number—are discussed in Chapter 7. Chromosomal mutations in eukaryotes and human disease syndromes that result from chromosomal mutations, including triplet repeat mutations, are emphasized. In Chapter 8, we discuss the ways of mapping genes in bacteriophages and in bacteria, which take advantage of the processes of transformation, conjugation, and transduction. Fine structure analysis of bacteriophage genes concludes this chapter.

Chapters 9 through 15 comprise the "molecular core" of *Genetics*, Fifth Edition, detailing the current level of our knowledge about the molecular aspects of genetics. In Chapter 9, we examine some aspects of gene function, such as the genetic control of the structure and function of proteins and enzymes, and the role of genes in directing and controlling biochemical pathways. A number of examples of human genetic diseases that result from enzyme deficiencies are described to reinforce the concepts. The discussion of gene function in Chapter 9 enables students to understand the important concept that genes specify proteins and enzymes, setting them up for the following chapters in which gene structure and expression is discussed.

In Chapter 10, we cover the structure of DNA, presenting the classical experiments that revealed DNA and RNA to be genetic material and that established the double helix model as the structure of DNA. The details of DNA structure and organization in prokaryotic and eukaryotic chromosomes are set out in Chapter 11. We cover DNA replication in prokaryotes and eukaryotes, and recombination between DNA molecules in Chapter 12.

After thoroughly explaining the nature of the gene and its relationship to chromosome structure, in Chapter 13, we discuss the first step in the expression of a gene—transcription. First, we describe the general process of transcription and then present the currently understood details of the transcription of messenger RNA, transfer RNA, and ribosomal RNA genes, and the processing of the initial transcripts to the mature RNAs for both prokaryotes and eukaryotes. In Chapter 14, we describe the structure of proteins, the evidence for the nature of the genetic code, and a detailed expression of our current knowledge of translation in both prokaryotes and eukaryotes. A brief discussion of how eukaryotic proteins are sorted to the compartments in which they function is also included in Chapter 14. In Chapter 15, we discuss recombinant DNA technology and other molecular techniques that are now essential tools of most areas of modern genetics. There are descriptions of the use of recombinant DNA technology to clone and characterize genes and to manipulate DNA, followed by a discussion of the applications of recombinant DNA technology in the analysis of biological processes, the diagnosis of human diseases, the isolation of human genes, the Human Genome Project (with the goal of mapping and sequencing the complete genomes of humans and other selected organisms), forensics (DNA typing), gene therapy, the development of commercial products, and the genetic engineering of plants.

The next two chapters focus on regulation of gene expression in prokaryotes (Chapter 16) and eukaryotes (Chapter 17). In Chapter 16, we discuss the operon as a unit of gene regulation, the current molecular details in the regulation of gene expression in bacterial operons, and regulation of genes in bacteriophages. In Chapter 17, we explain how eukaryotic gene expression is regulated, stressing molecular changes that accompany gene regulation, short-term gene regulation in simple and complex eukaryotes, gene regulation in development and differentiation, and immunogenetics.

Next, in Chapter 18—the new chapter on the genetics of cancer—we discuss the relationship of the cell cycle to cancer, and the various types of genes which, when mutated, play a role in the development of cancer. We also discuss the fact that cancer usually requires a number of independent mutational events in order for it to develop. Lastly, we consider the induction of cancer by chemicals and radiation (carcinogens).

In Chapters 19 and 20, we describe some of the ways in which genetic material can change or be changed. Chapter 19 covers the processes of gene mutation, the procedures that screen for potential mutagens and carcinogens (the Ames test), some of the mechanisms that repair damage to DNA, and some of the procedures that are used to screen for particular types of mutants. Chapter 20 presents the structures and movements of transposable genetic elements in prokaryotes and eukaryotes.

In Chapter 21, we address the organization and genetics of extranuclear genomes of mitochondria and chloroplasts. We cover the current molecular information about the organization of genes within the extranuclear genomes, and the classical genetic experiments that are used to study the inheritance of extranuclear genes. New to this chapter are discussions of RNA editing and genomic imprinting.

In Chapters 22 and 23, we describe the genetics of populations and quantitative genetics, respectively. In Chapter 22, "Population Genetics," we present the basic principles in population genetics, extending our studies of heredity from the individual organism to a population of organisms. This chapter includes an integrated discussion of the developing area of conservation genetics. In Chapter 23, "Quantitative Genetics," we consider the heredity of traits in groups of individuals that are determined by many genes simultaneously. In this chapter we also discuss heritability: the relative extent to which a characteristic is determined by genes or by the environment. Both Chapters 22 and 23 include discussions of the application of molecular tools to these areas of genetics. Chapter 22, for example, includes a section on measuring genetic variation with RFLPs and DNA sequencing, and a discussion of molecular evolution.

PEDAGOGICAL FEATURES

Because the field of genetics is complex, making the study of it potentially difficult, we have incorporated a number of special pedagogical features to assist students and to enhance their understanding and appreciation of genetic principles. These features have proved to be very effective in previous editions of this text:

- Each chapter opens with an outline of its contents and a section called "Principal Points." Principal Points are short summaries that alert students to the key concepts they will encounter in the material to come.
- Throughout each chapter, strategically placed "Keynote" summaries emphasize important ideas and critical points.
- Important terms and concepts—highlighted in bold—are clearly defined where they are introduced in the text. For easy reference, they are also compiled in a Glossary at the back of the book.
- Some chapters include boxes covering special topics related to chapter coverage. Some of these boxed topics are: *Genetic Terminology* (Chapter 2); *Equilibrium Density Gradient Centrifugation* (Chapter 12); *Labeling of DNA* (Chapter 15); *Hardy, Weinberg, and the History of Their Contribution to Population Genetics* (Chapter 22); and *Analysis of Genetic Variation with Protein Electrophoresis* (Chapter 22).

- Chapter summaries close each chapter, further reinforcing the major points that have been discussed.
- With the exception of the introductory Chapter 1, all chapters conclude with a section entitled "Analytical Approaches for Solving Genetics Problems." Genetics principles have always been best taught with a problem-solving approach. However, beginning students often do not acquire the necessary experience with basic concepts that would enable them to attack assigned problems methodically. In the "analytical approaches" sections (pioneered in earlier editions of this text), typical genetic problems are talked through in step-by-step detail to help students understand how to tackle a genetics problem by applying fundamental principles.
- The problem sets that close the chapters include approximately 650 questions and problems designed to give students further practice in solving genetics problems. The problems for each chapter represent a range of topics and difficulty. The answers to questions indicated by an asterisk (*) can be found at the back of the book, and answers to all questions are available in a separate supplement, the *Study Guide and Solutions Manual*.
- Comprehensive and up-to-date suggested readings for each chapter are listed at the back of the book.
- Special care has been taken to provide the most useful Index—extensive, accurate, and well cross-referenced.

SUPPLEMENTS

A *Study Guide and Solutions Manual* to accompany this text has been prepared by Bruce Chase of the University of Nebraska. In addition to detailed solutions for all the problems in the text, the *Guide* contains the following features for each chapter: a review of important terms and concepts; an "Analytical Approaches for Solving Genetics Problems" section, which provides guidance and tips on solving problems and avoiding common pitfalls; and additional questions for practice and review.

The study guide includes a variety of essays, multiple choice and matching questions, Key Terms and Concepts, Chapter Outline, Approach to Analytical Thinking and Comprehensive Solutions.

The printed Test Bank, prepared by Holly Ahern of Adirondack Community College, includes approximately 900 multiple-choice, fill-in-the-blanks, and short-answer questions. The entire set of questions is also available in Macintosh and Windows test-generating software.

A set of 146 full-color transparencies complements the text.

The full complement of supplemental teaching materials is available to qualified instructors.

ACKNOWLEDGMENTS

I would like to thank Robert Kaplan (Reed College) for updating the chapters on "Population Genetics" and "Quantitative Genetics" for the fourth edition. These chapters continue to present the important concepts for these areas of genetics in this new edition.

Also making an invaluable contribution to this text are John and Bette Woolsey of J/B Woolsey Associates, Inc., who generated the new figures for this edition in the same style as the critically acclaimed full-color art program they developed and executed for my *Genetics*, Third and Fourth Editions. I also thank artist Karl Miyajima for assistance in art production and art supervisor Carol Ann Smallwood for coordinating the art program for this edition.

I would also like to thank all of the reviewers involved in this edition:

Mike Bentley, University of Calgary
Anna Berkovitz, Purdue University
Roymarie Bollister, University of California, Santa Barbara
Aaron Cassill, University of Texas, San Antonio
Bruce Chase, University of Nebraska, Omaha
Denise Clark, University of New Brunswick
Janice Clark, University of North Dakota
Qingquan Quentin Fang, Georgia Southern University
David Fromson, California State University, Fullerton
Clay Fuqua, Trinity University
Jack R. Girton, Iowa State University
Dr. Elliott Goldstein, Arizona State University
David Hicks, Manchester College
J. Spencer Johnston, Texas A & M University
Lewis Kleinsmith, University of Michigan
Sally Mackenzie, Purdue University
John C. Osterman, University of Nebraska, Lincoln
Gregory C. Phillips, New Mexico State University
Kevin Piers, Red Deer College
David Sadava, Claremont Colleges
Peter Wejksnora, University of Wisconsin, Milwaukee
Vernon L. Wranosky, Colby Community College

For this edition, I would also like to thank Bruce Chase (University of Nebraska, Omaha), who provided excellent new end-of-chapter problems; thanks also to Janis Shampay (Reed College) and Anna Berkovitz (Purdue University), who acted as accuracy checkers.

I am grateful to the literary executor of the late Sir Ronald A. Fisher, F.R.S., to Dr. Frank Yates, F.R.S., and to Longman Group Ltd. London, for permission to reprint Table IV from their book *Statistical Table for Biological, Agricultural and Medical Research* (Sixth Edition, 1974).

Finally, I wish to thank those at Benjamin/Cummings who helped to make *Genetics*, Fifth Edition, a physical reality. In particular, I thank Susan Weisberg, senior sponsoring developmental editor, for her excellent and perceptive comments on the new edition as it was being written. Unquestionably she contributed greatly to the quality of the Fifth Edition. I thank Lisa Weber, production editor, for her smooth handling of the book production process; Kathleen Cameron, photo editor; Ariane de Pree-Kajfez, senior permissions editor; Sylvia Stein Wright and Lisa Weber, copyeditors; Yvo Riezebos, cover designer; Erika Buck and Amy Dhillon, editorial assistants; and Vivian McDougal, composition and film buyer. I also thank Barbara Littlewood for her continued efforts in maintaining an excellent, comprehensive index for this book.

Peter J. Russell

1 GENETICS:
An Introduction

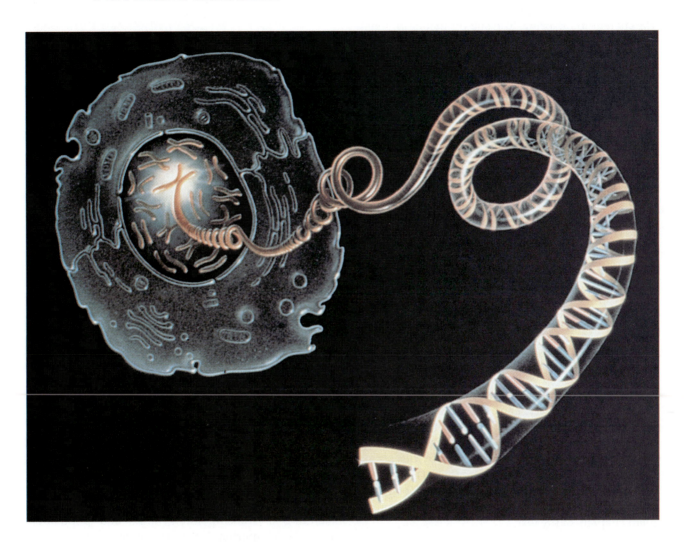

*W*elcome to the fascinating and exciting subject of **genetics**, the science of heredity. Genetics is primarily concerned with the understanding of biological properties that are transmitted from parent to offspring. The subject matter of genetics includes the phenomenology of heredity, the molecular nature of the genetic material, how genes control metabolism and development, and the distribution and behavior of genes in populations.

Genetics is of central importance to biology because genes are the principal determinants of all life processes, from cell structure and function to reproduction of the organism. Unquestionably, genetics is directly relevant to all living things. Learning what genes are, how genes are transmitted from generation to generation, how genes are expressed, and how gene expression is regulated is the focus of this book. The subject of genetics is expanding so rapidly, though, that it is simply not possible to describe everything that is known about genetics between these covers. The important principles and concepts are presented carefully and thoroughly; those students who want to go further should consult the references at the end of the text.

In this particular chapter we present an introduction to the subject of genetics to acquaint you, the student, with what is to be found in the rest of the book. We assume that you are beginning your excursion into genetics with at least an awareness of the general features of gene transmission and gene expression that were presented in your introductory biology course. Here we introduce the subject by discussing the main branches of genetics, geneticists and their areas of research, and the main properties of genes. These topics are discussed more fully in the subsequent chapters of the text.

THE BRANCHES OF GENETICS

Geneticists—scientists who study the processes of genetics—often divide genetics into three main branches (Figure 1.1). The branch dealing with the transmission of genes from generation to generation and how genes recombine is **transmission genetics** (sometimes called classical genetics). Analyzing the transmission of traits in a human pedigree or in crosses of experimental organisms are examples of transmission genetics studies. The branch dealing with the structure and function of genes at the molecular level is called **molecular genetics**. Analyzing the molecular events involved in the expression of genes is an example of a molecular genetics study. Finally, the branch dealing with the distribution and behavior of genes

~ **FIGURE 1.1**

The three main branches of genetics: transmission genetics, molecular genetics, and population genetics.

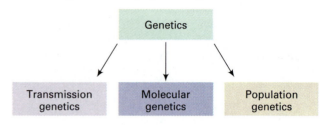

(usually in mathematical terms) in populations is called **population genetics**. Analyzing the frequency of a disease-causing gene in the human population is an example of a population genetics study. Although the subdivisions help us think about genes from different perspectives, there is no sharp boundary between them.

RELATIONSHIP OF GENETICS TO OTHER AREAS OF BIOLOGY

Through the products they encode, genes play a role in determining all aspects of the life of an organism. Since an understanding of gene transmission, gene structure, and gene expression is important for many biological fields of study, genetics is considered to be a central subject in biology. For example, an understanding of transmission genetics and population genetics is important for studies in, for example, population biology, ecology, evolution, and animal behavior. Similarly, an understanding of molecular genetics is important, for example, for studies in neurobiology, cell biology, developmental biology, animal physiology, plant physiology, and immunology.

GENETICISTS AND GENETICS RESEARCH

The material presented in this book is the result of an incredible amount of research done by geneticists working in many areas of biology. In this section we discuss geneticists as scientists, the nature of research, and the types of organisms that have been the fruitful objects of geneticists' research.

What Are Geneticists?

Geneticists use the methods of science to investigate all aspects of genes: how they are segregated from generation to generation, how they are expressed, and how they behave in populations. As researchers, geneticists typically use the **hypothetico-deductive method of investigation** (Figure 1.2). This consists of making *observations*, forming *hypotheses* to explain the observations, making experimental *predictions* based on the hypotheses, and finally *testing* the predictions. The last step produces new observations and so a cycle is set up leading to a refinement of the hypotheses and perhaps eventually to the establishment of a law or an accepted principle. Given that the questions being posed in genetics research span a broad area, a particular genetics researcher may well not carry out all of the steps. That is, geneticists may specialize by primarily formulating hypotheses, for example, or by primarily testing predictions experimentally.

As is true of all other areas of scientific research, the exact path a research project will follow cannot be predicted precisely. In part, it is the unpredictability of the research that makes it exciting and motivates the scientists engaged in the research. It is important to realize, for example, that the discoveries that have revolutionized genetics were not planned—they developed out of research in which basic genetic principles were being examined. Barbara McClintock's

discovery of jumping genes is an excellent example. From careful genetic studies of patches of color on corn kernels she theorized the existence of segments of DNA that could move (transpose) from place to place in the genome. Many years later, these DNA segments—called *transposons* or *transposable elements*—were isolated and characterized in detail. We know now that transposons are ubiquitous, playing a role not only in the evolution of species, but also in some human disease.

What Is Basic and Applied Research?

Genetics research, and research in general, may either be basic or applied. In **basic research**, experiments are done in order to gain an understanding of fundamental phenomena, whether or not that understanding leads to immediate applications. In **applied research**, experiments are done with an eye toward overcoming specific problems in society or exploiting discoveries. Most of the facts we will discuss in this book became known as the result of basic research. For example, we know how the expression of many prokaryotic and eukaryotic genes is regulated through basic research on model organisms such as the bacterium *Escherichia coli* (*E. coli*), *Saccharomyces cerevisiae* (yeast), and *Drosophila melanogaster* (the fruit fly). This knowledge is part of our overall understanding of phenomena in the area of molecular genetics. As the knowledge obtained from basic research has been and continues to be accumulated, it has been useful largely to fuel more basic research to make our understanding as complete as possible about the phenomena being studied.

Applied research is done with different goals in mind. Unquestionably, applied genetics research has been very important in many areas of significance to humankind. In agriculture, applied genetics has contributed significantly to improvements in animals bred for food (such as reducing the amount of fat in beef and pork) and in crop plants (such as increasing the amount of protein in soybeans). In medicine, a number of diseases are caused by genetic defects, and great strides are being made in understanding the molecular bases of some of those diseases. Drawing from the knowledge gained from basic research, applied genetics research involves, for example, developing rapid diagnostic tests for genetic diseases and producing new pharmaceuticals for treating diseases.

Since practical applications of science naturally intrigue the populace more than advances in basic knowledge, the public has become more educated about genetics as hopes for eradicating genetic dis-

~ **FIGURE 1.2**

The hypothetico-deductive method of investigation. Observations are made, then hypotheses are proposed to explain the observations. Experimental predictions are generated based on the hypotheses and the predictions are tested, generating new observations. The cycle continues, with each cycle leading to a refinement of the hypotheses.

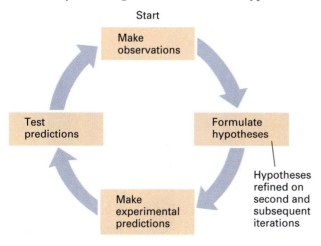

eases are raised. On the other hand, the public is not always accepting of new advances, as evidenced by many people's resistance to the release of genetically engineered organisms into the environment (e.g., bacteria to confer greater frost resistance to strawberries and other plants; the marketing of genetically engineered crops), and the prospects of gene therapy in humans. These advances and controversies frequently bring genetics into the public eye through presentations on television and in newspapers and magazines. These presentations offer excellent opportunities for the student and the layperson to learn about genetics and its relevance to present and future society.

It is important to realize that there is no sharp dividing line between basic and applied research. Indeed, both areas of research use similar techniques and depend on the accumulated body of genetic information for building hypotheses. For example, **recombinant DNA technology**—experimental procedures that allow molecular biologists to splice a DNA fragment from one organism into DNA from another organism and to clone (make many identical copies of) the new recombinant DNA molecule—has had a profound effect on both basic and applied research. That is, recombinant DNA techniques are used in all areas of basic genetics research to investigate genetic phenomena. In applied research, many biotechnology companies owe their existence to recombinant DNA technology as they seek to clone and manipulate genes for the synthesis of commercial products, the improvement of plant crops and animals, the devel-

opment of diagnostic tools for genetic diseases, the development of new or more effective pharmaceuticals, and so on. In the area of plant breeding, for example, recombinant DNA technology has expanded the introduction of traits such as disease resistance from noncultivated species into cultivated species. Such crop improvement traditionally has been done by genetic crosses. Figure 1.3a shows the Flavr Savr tomato, a fruit that has been genetically engineered to spoil at a much slower rate than normal. As a result, tomatoes can be picked later, when they have developed more flavor. In the area of animal breeding, recombinant DNA technology is being employed in the beef, dairy cattle, and poultry industries to improve yields and to develop better strains. In medicine the results are equally impressive. Recombinant DNA technology is being used in the production of many antibiotics, hormones, and other medically important agents such as clotting factor, human insulin (marketed under the name Humulin: Figure 1.3b), and in the diagnosis and treatment of a number of human genetic diseases. In forensics, *DNA typing* (also called *DNA fingerprinting* or *DNA profiling*) is being used to obtain evidence in paternity cases and for solving crimes. In short, the science of genetics is currently in an exciting and dramatic growth phase. By understanding the basic principles of genetics as presented to you in this text, you will have a greater capacity to understand and perhaps contribute to new and exciting applications, for there is still much to discover.

~ FIGURE 1.3

Examples of products in the marketplace that were developed as a result of recombinant DNA technology. (a) The Flavr Savr tomato. This tomato has been engineered so that its rate of spoilage is much lower than usual, so it can be picked at a later stage of ripeness. As a result, it has more flavor for the consumer. (b) Humulin, human insulin for insulin-dependent diabetics.

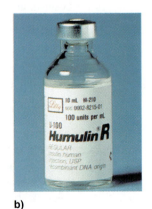

a)

b)

What Organisms Are Suitable for Genetic Experimentation?

Geneticists use experimental organisms for their research. The principles of heredity were first established by *Gregor Mendel*'s experiments in the nineteenth century with the garden pea. Since Mendel's time, a large number of different organisms have been used for genetics experiments. There are several qualities that make an organism well suited for genetic experimentation:

1. If an experimenter wants to test a genetic hypothesis, the genetic history of the organism involved must be well known. Therefore the genetic background of the parents used in the experimental crosses must be known.
2. The organism must have a relatively short life cycle so that a large number of generations occur within a relatively short time. In this way data over many generations can be obtained rapidly. Fruit flies, for example, produce offspring in 10 to 14 days.
3. A large number of offspring must be produced from a mating, since much genetic information can only be obtained if there are numerous progeny to study.
4. The organism should be easy to handle. Hundreds of fruit flies can easily be kept in half-pint milk bottles for experimental purposes, but hundreds of elephants would certainly be more difficult and much more expensive to maintain!
5. Most importantly, there must be genetic variation among the individuals in the population. If there are no discernible differences among the individual organisms under study, it is impossible to study the inheritance of traits. The more marked the differences, the easier the genetic analysis. For example, fruit flies may have one of many different eye colors, humans may be able to taste or not taste certain chemicals, and bacteria may or may not have particular nutritional requirements.

Given the above ideal qualities, what organisms have been used extensively? Both eukaryotic and prokaryotic organisms and their viruses have been used.

EUKARYOTES. Eukaryotes (meaning "true nucleus") are organisms that have cells in which the genetic material (DNA) is located in the **nucleus**, a discrete structure within the cell that is bounded by a nuclear membrane. Eukaryotes can be unicellular or multicellular. Many eukaryotic organisms are used in genetic research, each chosen for its usefulness for studying the scientific questions of interest to the research scientist. Thus, as the directions and emphases of genetics research have changed with new knowledge, so the experimental organisms have changed. In contemporary genetics, for example, a great deal of genetics research is done with six organisms (Figure 1.4a–f): *Saccharomyces cerevisiae* (yeast); *Drosophila melanogaster* (fruit fly); *Caenorhabditis elegans* (nematode); *Arabidopsis thaliana* (a small weed of the mustard family); *Mus musculus* (mouse); and *Homo sapiens* (humans). Over the years, research with the following seven organisms has also contributed significantly to our understanding of genetics (Figure 1.4g–m): *Neurospora crassa* (orange bread mold); *Tetrahymena* (a protozoan); *Paramecium* (a protozoan); *Chlamydomonas reinhardtii* (a green alga); *Pisum sativum* (garden pea); *Zea mays* (corn); *Gallus* (chicken). Of these, *Tetrahymena*, *Paramecium*, *Chlamydomonas*, and *Saccharomyces* are unicellular organisms, and the rest are multicellular. Humans are included in this list, not because they fulfill all the criteria for an organism well suited for genetic experimentation, but because ultimately we wish to understand as much as we can about human genes and their expression. This understanding is necessary for us to make progress in combating genetic diseases, as well as for the inherent fundamental knowledge about our species' development and evolution.

What are the features of eukaryotic cells of importance to genetics? You learned about many of them in your introductory biology course. Figure 1.5a is a cutaway diagram of a generalized eukaryotic (animal) cell, and Figure 1.5b is a cutaway diagram of a generalized higher plant cell. Surrounding the cytoplasm of both animal and plant cells is a membrane called the *plasma membrane*. Plant cells, but not animal cells, have a rigid cell wall outside of the plasma membrane. The nucleus of eukaryotic cells contains DNA, the genetic material, and is separated from the rest of the cell, the cytoplasm and associated organelles, by the double membrane of the nuclear envelope. The membrane is selectively permeable and has pores about 20 to 80 nm (nm = nanometer = 10^{-9} meters) in diameter, which make it possible for materials to move between the nucleus and the cytoplasm. The DNA is complexed with proteins and is organized into a number of linear structures called *chromosomes*. Within the nucleus is the **nucleolus**, the organelle in which ribosomes are assembled.

The cytoplasm of eukaryotic cells contains a vast amount of different materials and organelles. Of special interest for geneticists are the *centrioles, endoplasmic reticulum* (ER), *ribosomes, Golgi apparatus, mitochondria,* and *chloroplasts.* Centrioles (also called basal bodies) are found in nearly all animal cells, but not

~ FIGURE 1.4

Eukaryotic organisms that have contributed significantly to our knowledge of genetics principles: (a) *Saccharomyces cerevisiae* (a budding yeast); (b) *Drosophila melanogaster* (fruit fly); (c) *Caenorhabditis elegans* (a nematode); (d) *Arabidopsis thaliana* (Thale cress, a member of the mustard family); (e) *Mus musculus* (mouse); (f) *Homo sapiens* (humans); (g) *Neurospora crassa* (orange bread mold); (h) *Tetrahymena* (a protozoan); (i) *Paramecium* (a protozoan); (j) *Chlamydomonas reinhardtii* (a green alga); (k) *Pisum sativum* (a garden pea); (l) *Zea mays* (corn); (m) *Gallus* (chicken).

a)

b)

c)

d)

e)

f)

g)

h)

i)

j)

k)

l)

m)

~ FIGURE 1.5

Eukaryotic cell: Cutaway diagrams of (a) A generalized animal cell, and (b) A generalized higher plant cell, showing the main organizational features and the principal organelles.

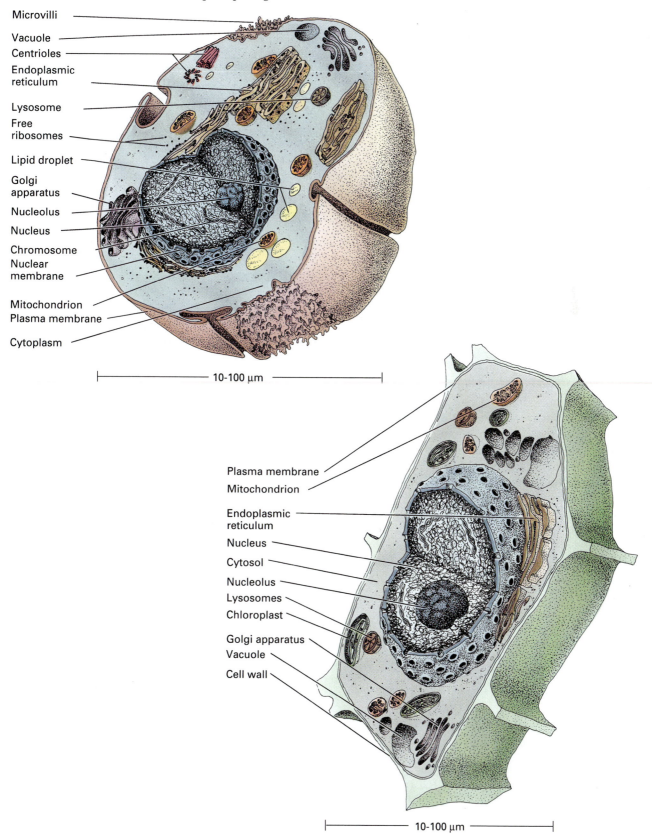

usually in plant cells (except in cells of lower plants) (Figure 1.6). A centriole is a small, cylindrical organelle about 0.2 μm (μm = micrometer = 10^{-6} meter) wide and 0.4 μm long. It consists of a ring of nine groups of three fused microtubules (a microtubule is a specialized protein filament). In animal cells, a pair of centrioles is at the center of the centrosome, a region of undifferentiated cytoplasm that organizes the spindle fibers that function in mitosis and meiosis, processes that are discussed in Chapter 3.

The endoplasmic reticulum (ER) is a double-membrane system that runs through the cell. There are two types of ER: rough and smooth. The former has ribosomes attached to it, giving it a rough appearance, and the latter does not. Ribosomes bound to the rough endoplasmic reticulum synthesize proteins to be secreted by the cell or to be localized in the cell membrane or particular vacuoles within the cell. The synthesis of proteins other than those that are distributed via the ER is performed by ribosomes that are free in the cytoplasm.

Eukaryotic cells also contain *mitochondria* (singular: *mitochondrion*), one of which is illustrated in Figure 1.7. These large organelles are surrounded by a double membrane, the inner one of which is highly convoluted. Mitochondria play an extremely important role in processing energy for the cell. They also contain genetic material in the form of a circular, double-stranded molecule of DNA. This DNA encodes some of the proteins that function in the mitochondrion, as well as some of the components of the mitochondrial protein synthesis machinery.

Finally, many plant cells contain chloroplasts, the large, double-membraned, chlorophyll-containing organelles involved in photosynthesis (Figure 1.8). Chloroplasts also contain genetic material and, as in mitochondria, its form is a circular, double-stranded DNA molecule. This DNA encodes some of the proteins that function in the chloroplast, as well as some of the components of the chloroplast protein synthesis machinery.

~ FIGURE 1.7

Cutaway diagram of a mitochondrion. Energy-processing mechanisms involve interrelationships between the organelle's intermembranal space, the inner membrane, and the matrix.

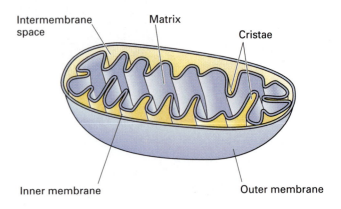

PROKARYOTES. Prokaryotes (meaning "prenuclear"), unlike eukaryotes, do not have a nuclear membrane surrounding their genetic material (DNA); this is the major distinguishing feature of prokaryotes. Included in this group are all the **bacteria**. Bacteria are spherical, rod-shaped, or spiral-shaped organisms; most are single-celled, although a few are multicellular and filamentous. The shape of the bacterium is maintained by a rigid cell wall located out-

~ FIGURE 1.8

Cutaway diagram of a chloroplast. The organelle's energy-harvesting mechanisms involve interrelationships between the intermembranal space, the stroma, the thylakoids stacked in grana, and the area within each thylakoid.

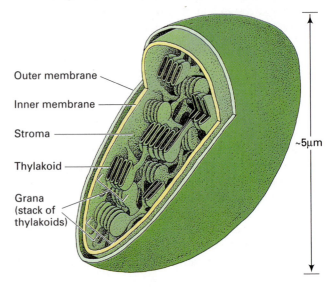

~ FIGURE 1.6

Schematic of the centriole: (a) End view; (b) Side view.

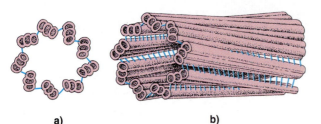

a)   b)

~ FIGURE 1.9

Colorized scanning electron micrograph of *Escherichia coli,* a rod-shaped bacterium common in human intestines.

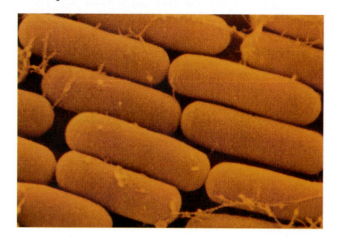

side the cell membrane. The bacteria are divided into two distantly related groups, the *eubacteria* and the *archaebacteria.* Eubacteria are the common varieties found in living organisms (naturally or by infection), soil, and water. Archaebacteria are found in much more inhospitable conditions, such as hot springs, salt marshes, methane-rich marshes, or in the ocean depths. Bacteria generally vary in size from about 100 nm in diameter to 10 μm in diameter and 60 μm long; one species, the surgeonfish symbiont *Epulopiscium fishelsoni,* is as large as 60 × 800 μm, fully a million times larger than *E. coli!* In most cases, the bacteria studied in genetics are eubacteria. The most intensely studied bacterium in genetics is *Escherichia coli* (Figure 1.9), a rod-shaped bacterium common in human intestines. Studies of this bacterium have resulted in significant advances in our understanding of the regulation of gene expression and in the development of molecular biology. Today *E. coli* is used extensively in recombinant DNA experiments.

THE DEVELOPMENT OF THE FIELD OF GENETICS

Genetics has shown us that many of the differences between organisms are the result of differences in the genes they carry, differences that have resulted from the evolutionary processes of **mutation** (a change in the genetic material), **recombination** (exchange of genetic material between chromosomes), and **selection** (the favoring of particular combinations of genes in a given environment). While breeding experiments with domesticated animals and cultivated plants

were done for centuries, the principles of heredity were first written about by Gregor Mendel in the 1860s. The importance of his findings were not recognized in his lifetime and it was only around 1900, when Mendel's principles of heredity were rediscovered, that the development of the subject of genetics truly began. Figure 1.10 presents a time line of *some* of the important historical landmarks in genetics, which will be discussed in more detail later in the text. In the context of this time line of genetics discoveries, let us now discuss briefly what we know about genes at the molecular level, and some of the modern ways of studying genes.

What Is the Nature of the Genetic Material?

The genetic material of all living organisms, both eukaryotes and prokaryotes, is **DNA (deoxyribonucleic acid).** DNA is also the genetic material of many viruses that infect prokaryotes and eukaryotes. A number of other viruses have **RNA (ribonucleic acid)** as genetic material.

From light microscopy studies, electron microscopy studies, biochemical studies, and molecular studies, we have learned about the organization of the genetic material in prokaryotes and eukaryotes. In eukaryotes, the nucleus contains most of the genetic material of the cell. In the nucleus, the genetic material is complexed with protein and is organized into a number of linear structures called **chromosomes** (Figure 1.11). *Chromosome* means "colored body" and is so named because these threadlike structures are visible under the light microscope only after they are stained with dyes. Each eukaryotic chromosome has a single molecule of DNA going from one end to the other. In prokaryotes, the DNA is organized into a single circular chromosome and has relatively few proteins associated with it.

Lastly, as already has been mentioned, both mitochondria and chloroplasts contain genetic material. In both cases the genetic material is in the form of double-stranded DNA, associated with few, if any, structural proteins. Knowledge of the genes contained in both mitochondrial and chloroplast genomes has come from studies of mutants and, more recently, from modern molecular approaches involving the sequencing of entire mitochondrial and chloroplast genomes.

How Is the Genetic Material Transmitted from Generation to Generation?

We have a good understanding of how the genetic material is transmitted from generation to generation for both eukaryotes and prokaryotes. As we have

~ FIGURE 1.10

Time line of some of the important historical landmarks in genetics.

1856-1863	**Gregor Mendel** Conducted his famous pea experiments concerning gene segregation
1859	**Charles Darwin** Published *On the Origin of the Species*; modern theory of evolution is identified with Darwin
1866	**Gregor Mendel** Published a research paper on his work establishing the basic principles of heredity
1871	**Fredrich Miescher** Isolated nuclein from nucleus; nuclein is now known to be DNA
1882-1885	**E. Strasburger, Walther Flemming** Showed that nuclei contained chromosomes
1900	**Hugo de Vries, Carl Correns, Erich von Tschermak-Seysenegg** Independently produced results confirming Mendel's principles of heredity
1902	**Archibald Garrod** Identified the first human genetic disease
	Walter Sutton, Theodor Boveri Proposed the chromosome theory of heredity
1903	**William E. Castle** First to recognize the relationship between allele and genotype frequencies (see 1908, Hardy and Weinberg)
1905	**William Bateson** Called the science of heredity "genetics"
	W. Bateson, R. C. Punnett Demonstrated linkage between genes
1908	**Godfrey H. Hardy, Wilhelm Weinberg** Formulated the Hardy-Weinberg principle relating mathematically the genotypic frequencies to the frequencies of alleles in randomly mating populations
	H. Nilsson-Ehle Obtained experimental proof for multigene inheritance as the basis for continous traits
1909	**W. Johannsen** Introduced the word "gene"
1910	**Edward M. East** Elucidated the role of sexual reproduction in evolution
	Thomas Hunt Morgan Found the first sex-linked gene, *white*, an eye color gene in *Drosophila melanogaster*
1911	**Thomas Hunt Morgan** Proposed that genetic linkage was the result of the genes involved being on the same chromosome
1913	**Alfred Sturtevant** Devised the principle for constructing a genetic linkage map
1922	**Ronald A. Fisher** Published a substantial paper presenting a quantitative examination of the evolutionary consequences of Mendelian inheritance

1927	**Herman J. Muller** Showed that X rays can induce mutations
1928	**Frederick Griffith** Discovered genetic transformation of a bacterium and called the agent responsible the "transforming principle"
1930	**Ronald A. Fisher** His comprehensive theory of evolution, combining Mendelian inheritance and Darwinian selection was published as "The Genetical Theory of Natural Selection"
	Sewall Wright Developed his own genetical theory for natural selection, and laid the important theoretical foundation for genetic drift, the random change in gene frequency due to chance
1931	**Harriet Creighton, Barbara McClintock** Showed that genetic recombination in maize results from a physical exchange of homologous chromosomes
	Curt Stern Showed that genetic recombination in *Drosophila* results from physical exchange of homologous chromosomes
1941	**George Beadle, Edward Tatum** Proposed the one gene–one enzyme hypothesis
1944	**Oswald Avery, Colin MacLeod, Maclyn McCarty** Showed that Griffith's transforming principle (see 1928) was DNA
1946	**Joshua Lederberg, Edward Tatum** Discovered conjugation in bacteria
1950	**Barbara McClintock** Reported results of maize experiments indicating movable genes, now called transposable elements
1952	**Alfred Hershey, Martha Chase** Showed that the genetic material of bacteriophage T2 is DNA
1953	**James Watson, Francis Crick** Proposed double helical model for DNA
1957	**Heinz Fraenkel-Conrat, B. Singer** Showed that the genetic material of tobacco mosaic virus was RNA
1958	**Matthew Meselson, Franklin Stahl** Proved the semiconservative model for DNA replication
	Arthur Kornberg Isolated DNA polymerase I from *E. coli*
1959	**Severo Ochoa** Discovered the first RNA polymerase
	Sidney Brenner, François Jacob, Matthew Meselson Discovered messenger RNA (mRNA)

1961	**François Jacob, Jacques Monod**
	Put forward the operon model for the regulation of gene expression in bacteria
1966	**Marshall Nirenberg, H. Gobind Khorana**
	Worked out the complete genetic code
1972	**Paul Berg**
	Constructed the first recombinant DNA molecule *in vitro*
1973	**Herb Boyer, Stanley Cohen**
	First to use a plasmid to clone DNA
1975	**Edward M. Southern**
	Developed a method for transferring DNA fragments separated in a gel to a filter, preserving the relative positioning of the fragments. This is one of the most valuable techniques for identifying cloned genes.
1977	**Walter Gilbert, Frederick Sanger**
	Devised methods for sequencing DNA
	Phillip Sharp and others
	Discovered introns in eukaryotic genes
	Frederick Sanger
	Obtained the complete nucleotide sequence of a virus, bacteriophage ΦX174
1983	**Thomas Cech, Sidney Altman**
	Discovered self-splicing of an intron RNA
1986	**Kary Mullis and others**
	Developed the polymerase chain reaction (PCR), a technique for amplification of selected DNA segments without cloning
1989	**L.–C. Tsui and John Riordan, and Francis Collins's group**
	Identified and cloned the human gene responsible for cystic fibrosis
1990	**James Watson and many other scientists**
	Launched the Human Genome Project to map and sequence the complete genomes of a number of genetically important organisms, including humans
1993	**Huntington Disease Collaborative Research Group**
	Discovered molecular basis for Huntington disease, a human genetic trait
1994	**M. Skolnick and other scientists**
	Cloned the first breast cancer gene *(BRCA1)*
1996	**Many scientists in several international research groups**
	Published the first complete DNA sequence of a eukaryotic organism, the yeast *Saccharomyces cerevisiae*
	J. Craig Venter and many other scientists in several U.S. research groups
	Published the complete DNA sequence of a member of the Archaea, *Methanococcus jannaschii*. The sequence data confirm that the Archaea are a third major branch of life distinct from prokaryotes and eukaryotes.
	National Institutes of Health
	Reported that almost 150 clinical trials have been approved for the transfer of genes into humans as part of long-term goals to treat genetic diseases by gene therapy

~ **FIGURE 1.11**

Colorized scanning electron micrograph of human chromosomes.

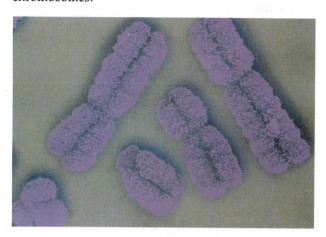

already mentioned, Gregor Mendel, the founder of modern genetics, was the first to understand the principles of gene segregation. Mendel performed a series of careful breeding experiments with the garden pea. In brief, he picked strains of peas that differed in particular characteristics (also called *traits*). These differences were clearly distinguishable, making identification and scoring of the **phenotypes** (the measurable attributes of the organism) unambiguous. For example, the pea seeds were either smooth or wrinkled, and the flowers were either purple or white (Figure 1.12). Then he made genetic crosses, counted the progeny, and interpreted the results. (This basic experimental design is still used in studies of gene transmission today.) From this kind of data, Mendel was able to conclude that phenotypic traits were controlled by *factors*, later called *Mendelian factors* and now called *genes*.

~ **FIGURE 1.12**

Example of easily distinguishable alternative traits: purple-flowered (left) vs. white-flowered (right) pea plants.

Mendel considered the factors that controlled the phenotypes he studied in abstract terms. He correctly deduced that the factors segregated randomly into the gametes and that the two factors controlling one trait assorted independently from the two factors controlling another trait. It was almost two decades after his death (in 1884) that the material basis of gene segregation from generation to generation was shown. In 1902, Walter Sutton and Theodor Boveri proposed the chromosome theory of heredity, which posited that genes are on chromosomes and that the segregation patterns of genes can be explained entirely by the segregation of chromosomes from generation to generation. This, of course, is generally accepted today, with gene segregation paralleling the segregation of chromosomes in meiosis.

What Are Genetic Maps?

Since 1902, much effort has been made to construct **genetic maps** (Figure 1.13) for the commonly used experimental organisms in genetics. Genetic maps are like road maps that show the relative locations of towns along a road; that is, they show the arrangements of genes along the chromosomes and the genetic distances between the genes. The goal of this work has been to obtain an understanding of the organization of genes along the chromosomes; for example, to inform us whether or not genes with related functions are on the same chromosome and, if they are, whether or not they are close to each other. Genetic maps are now proving useful in efforts to clone and sequence particular genes of interest.

What Are Genes at the Molecular Level?

The complete genetic makeup of an organism is its genotype. Genes control all aspects of the life of an organism, encoding the products that are responsible for development, reproduction, and so forth. All of the measurable attributes that an organism has are its *phenotype*. The term *phenotype* is also used to describe the attributes of a particular trait. The genotype alone is not responsible for the phenotype; rather, the genotype interacts with the environment—the external environment and the internal environment—to produce the phenotype. Thus, two individuals with identical genotypes (identical twins, for example) will not necessarily be exactly identical in phenotype. Everyone knows that identical twins are not truly identical; there are always some differences, even when they have been exposed to virtually identical environments. The differences are even greater if the twins are reared apart instead of together, because the environmental influences are much different. Chapter

~ **FIGURE 1.13**

Example of a genetic map: some of the genes on chromosome 2 of the fruit fly, *Drosophila melanogaster*. (From *Principles of Genetics* by Robert H. Tamarin. Copyright © 1996. Reproduced with permission of The McGraw-Hill Companies.)

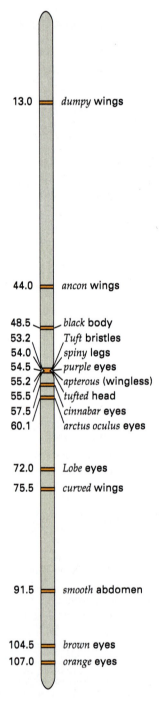

13.0	*dumpy* wings
44.0	*ancon* wings
48.5	*black* body
53.2	*Tuft* bristles
54.0	*spiny* legs
54.5	*purple* eyes
55.2	*apterous* (wingless)
55.5	*tufted* head
57.5	*cinnabar* eyes
60.1	*arctus oculus* eyes
72.0	*Lobe* eyes
75.5	*curved* wings
91.5	*smooth* abdomen
104.5	*brown* eyes
107.0	*orange* eyes

4 discusses environmental influences on gene expression, and the chapters of the book on molecular genetics discuss the expression of genes at the molecular level. Ultimately, it is the understanding of what

genes code for and how genes are regulated that will help us comprehend the complete pathway from genotype to phenotype.

Several landmark discoveries—listed in Figure 1.10—paved the way to our present intensive investigations in molecular genetics, and we will discuss some of the discoveries briefly here. They are discussed in more detail in later chapters.

GENES ENCODE POLYPEPTIDES. In 1941, George Beadle and Edward Tatum, working with mutations of the fungus *Neurospora crassa* that imposed nutritional requirements on the organism, showed that there was a firm relationship between genes and enzymes (special subset of proteins). That is, cells function through a myriad of biochemical pathways, each step of which is catalyzed by an enzyme (or more than one enzyme). Their studies of mutants elegantly showed that each of the steps in the biochemical pathways they studied was under genetic control. Since the steps are catalyzed by enzymes, they proposed the **one gene–one enzyme hypothesis**. This hypothesis has later been modified to the *one gene–one polypeptide hypothesis*, since not all proteins are enzymes and not all proteins consist of only one polypeptide.

DNA AND GENE EXPRESSION. In 1953, James Watson and Francis Crick, using chemical and physical evidence obtained by Erwin Chargaff, and by Rosalind Franklin and Maurice Wilkins, proposed the double-helix model for DNA (Figure 1.14). DNA is made up of two chains (also called strands); each chain consists of building blocks called **nucleotides**, each of which consists of the sugar, deoxyribose, a phosphate group, and a base. The arrangement of the nucleotides in the chains results in sugar-phosphate backbones on the outside of the helix with the bases oriented toward the center of the helix. Specific pairing of the bases by hydrogen bonds is the force holding the two chains of DNA together.

Once the structure of the genetic material was established, the doors were opened to many exciting discoveries in molecular genetics. We know from biochemical studies how DNA is replicated, and we know from studies of mutants many of the genes involved in the DNA replication process. Most importantly, we know that a gene at the molecular level is a sequence of nucleotide pairs in DNA.[1] All

---

[1]This is true for prokaryotes and eukaryotes. Viruses may have single-stranded or double-stranded DNA or single-stranded or double-stranded RNA as their genetic material; thus virus genes here are sequences of nucleotides for single-stranded genomes, and sequences of nucleotide pairs for double-stranded genomes. The nucleotides are DNA nucleotides or RNA nucleotides depending upon the nature of the genetic material.

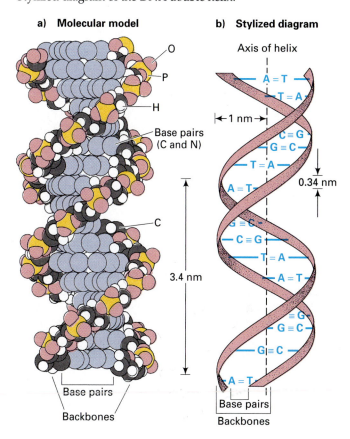

~ FIGURE 1.14

DNA: (a) Three-dimensional molecular model of DNA; (b) Stylized diagram of the DNA double helix.

genes are transcribed into RNA molecules. In **transcription**, the DNA separates locally into single strands and an enzyme makes an RNA copy of one of the strands of the DNA molecule (Figure 1.15). The enzyme that catalyzes the synthesis of a new RNA molecule using the DNA nucleotide pair sequence as a template is **RNA polymerase**. In prokaryotes three classes of RNA are made: **messenger RNA (mRNA)**, **transfer RNA (tRNA)**, and **ribosomal RNA (rRNA)**. Eukaryotes encode these three classes of RNA and in addition **small nuclear RNA (snRNA)**. The RNA molecules are essential for cell function in both prokaryotes and eukaryotes. The mRNAs specify the amino acid sequences of proteins which are important structural and functional components of cells. The base-pair information that potentially specifies the amino acid sequence of a protein is called the **genetic code**. Each amino acid is specified by a three-nucleotide sequence of the mRNA; this sequence is called a codon. Other sequences of the DNA specify where the RNA copy is to stop.

~ FIGURE 1.15

Transcription process. The DNA separates locally into single strands and RNA polymerase makes an RNA copy of one of the DNA strands.

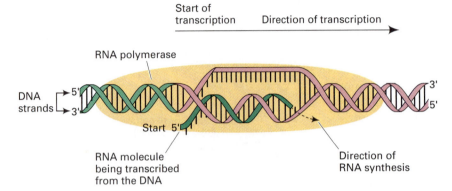

The process by which the base sequence information in mRNA is converted into an amino acid sequence in proteins is called **translation**. Translation occurs on **ribosomes** (Figure 1.16), large complexes of rRNA molecules and proteins.

How Is Gene Expression Controlled?

At any one time, only some of the genes in a particular genome are active and, in complex multicellular organisms, only a specific set of genes are active in each tissue and organ. How is all of this accomplished? We do not have anywhere near a complete understanding yet, but at the general level, this is the result of a finely tuned array of gene regulation signals determining which genes are active and which are inactive.

Our comprehension of the regulation of gene expression began in bacteria, when François Jacob

and Jacques Monod in 1961 proposed a model to explain the regulation of expression of genes that encode enzymes needed for the metabolism of lactose (a sugar). In their model, a genetic switch is involved. When the switch is set one way, transcription of the genes encoding the enzymes is blocked, and when the switch is set in the other way, transcription of the genes can take place. This model has been shown to be a general one for controlling many gene systems in bacteria and their viruses.

Genetic switches are also used to regulate gene expression in eukaryotes, but those switches are different from and typically more complex than those identified in bacteria. Moreover, many mechanisms are eukaryote-specific. Unquestionably, although much has been learned about gene regulation in eukaryotes, much remains to be learned.

What Experimental Approaches Are Commonly Used in Genetics Research Today?

Since the turn of the century, genetics has been a powerful tool for studying biological processes. An important approach used by many geneticists has been to isolate mutants affecting a particular biological process and then, by comparing the mutants with normal strains, to obtain an understanding of the process. Such research has gone in many directions; for example, analyzing heredity in populations, analyzing evolutionary processes, identifying the genes that control the steps in a process, mapping the genes involved, determining the products of the genes, and analyzing the molecular features of the genes includ-

~ FIGURE 1.16

A ribosome, the organelle on which translation of mRNA (= protein synthesis) takes place. Two views of three-dimensional models of the *E. coli* ribosome are shown. The large subunit is red and the small subunit is yellow.

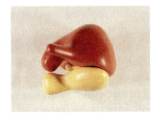

ing the regulation of the genes' expression. Research in genetics was revolutionized in 1972, when Paul Berg constructed the first recombinant DNA molecule *in vitro*, and in 1973, when Herbert Boyer and Stanley Cohen cloned a recombinant DNA molecule for the first time. Now, recombinant DNA technology and other molecular technologies spawned from it are used in every area of genetics.

Since so much of our present knowledge of the molecular aspects of genetics has come from the use of recombinant DNA techniques, let us consider some of those techniques in brief. The construction of recombinant DNA molecules is possible with **restriction enzymes** and **cloning vectors**. Restriction enzymes are enzymes that recognize specific nucleotide pair sequences in DNA and, in many cases, cut both strands of the DNA specifically within the sequence at symmetrical positions on each side of the midpoint of the sequence (Figure 1.17); such cuts are called *staggered cuts*. Since any DNA cut with the same enzyme will be cut within the same nucleotide pair sequence, the ends of all of the fragments produced will be the same—they are called *sticky ends*. For an enzyme that produces a staggered cut, the ends of different fragments can bond with each other by base pairing. This is a fundamental principle of constructing a recombinant DNA molecule.

A cloning vector is a DNA molecule capable of replication in a host organism and into which a piece of DNA to be studied can be specifically inserted at known positions. The recombinant DNA molecule is introduced into a host such as *E. coli*, yeast, an animal cell, or a plant cell. Reproduction of the host cell results in the replication of the recombinant DNA molecule, thereby producing many identical copies. Thus, to clone genes, a piece of DNA is taken from an organism, cut with a restriction enzyme, and spliced into a cloning vector that has been cut with the same restriction enzyme (Figure 1.18). There are many reasons for doing this. For example, suppose we want to study the gene for a particular human protein in order to determine its DNA sequence, and how its expression is regulated. Each human cell contains only two copies of that gene, making it an extremely difficult—almost impossible—task to isolate enough copies of the gene for analysis. By contrast, an essentially unlimited number of copies of the gene can be produced by cloning. And, as we will see throughout the book, there have been many, many discoveries made in genetics using recombinant DNA approaches. There have been many commercial applications of recombinant DNA technology also, including the development of pesticide-resistant crop plants, the

~ **FIGURE 1.17**

Example of how a restriction enzyme cuts DNA. Within the DNA segment shown is the nucleotide pair sequence recognized by the restriction enzyme *Eco*RI ("echo-R-one"). This *Eco*RI site is symmetrical about the mid-point of the sequence. Cleavage of the DNA with *Eco*RI produces two fragments, each with a single-stranded sticky end.

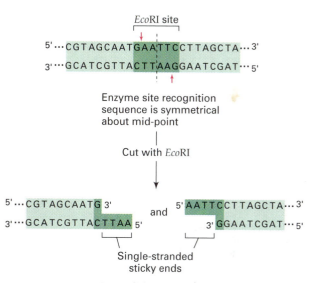

development of tomatoes that do not spoil as quickly, the development of diagnostic tests for human genetic diseases and for bacterial and viral infections, and the production of vaccines and pharmaceuticals.

The advent of recombinant DNA technology has initiated a revolution in biology that is still taking place. We can anticipate an ever-increasing number of exciting discoveries in the future that will further our knowledge of basic biological functions and that will lead to improvements in the quality of human life. Already the complete genomic DNA sequences have been determined for a handful of organisms and in the relatively near future we will have determined the sequences for many more organisms, including humans. As scientists analyze the data, we can expect major contributions to our biological knowledge. For example, we will know about each and every gene in the human genome—where they are in the genome, their sequences, and their regulatory sequences. Such knowledge will undoubtedly lead to a better understanding of human genetic diseases, and contribute to their cures. With such discoveries in sight, this is an exciting time to be a student of genetics.

~ FIGURE 1.18

Basic scheme for the construction of a recombinant DNA molecule using a plasmid cloning vector. In the example, the plasmid and the DNA to be cloned are both cut with the restriction enzyme *Eco*RI.

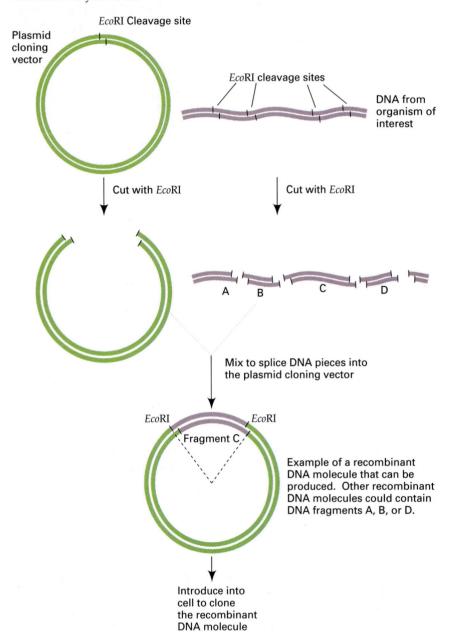

SUMMARY

In this chapter we have introduced the subject of genetics very briefly to acquaint you with its scope, and we will address the main areas in the remainder of the book. It is an extremely fascinating subject in which significant discoveries are being made daily, particularly in the area of molecular genetics. This is an exciting, stimulating time for geneticists, and we hope to communicate some of this excitement to you

as you use this text. In the rest of the chapters we hope to help you as much as we can with the inclusion of learning aids such as principal points (brief summaries at the starts of the chapters of the key points of the chapters), keynotes (summaries of the concepts strategically placed in the chapters), solved problems (Analytical Approaches for Solving Genetics Problems), and questions and problems that have solutions either at the end of the text or in the study guide.

2 MENDELIAN GENETICS

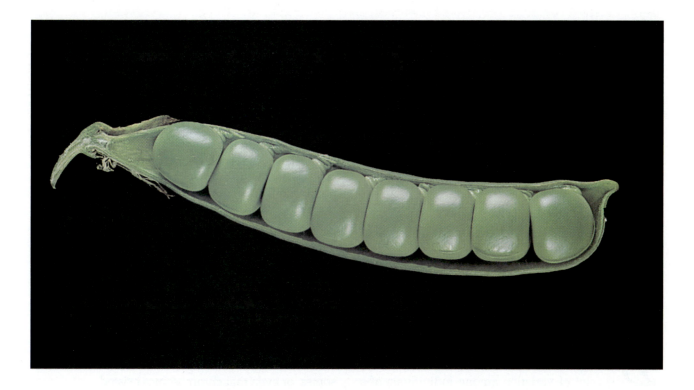

PRINCIPAL POINTS

~ The genotype is the genetic makeup of an organism, while the phenotype is the observable manifestation of genetic traits.

~ The genes give the potential for the development of characteristics; this potential is affected by interactions with other genes and with the environment.

~ Mendel's first law, the principle of segregation, states that the two members of a gene pair (alleles) segregate from each other in the formation of gametes.

~ To determine the genotype of an unknown individual, usually expressing the dominant phenotype, a cross is made between that individual and a known homozygous recessive individual. This cross is called a testcross.

~ Mendel's second law, the principle of independent assortment, states that members of different allele pairs are transmitted independently of one another during the production of gametes.

~ Mendelian principles apply to humans as well as to peas and all other eukaryotes. The study of the inheritance of genetic traits in humans is complicated by the fact that no controlled crosses can be done. Instead, human geneticists analyze genetic traits by pedigree analysis; that is, by examining the occurrences of the trait in family trees of individuals who clearly exhibit the trait.

*B*y simple observation it is evident that there is a lot of variation among individuals of a given species. Among dogs, for example, are many breeds, including Bernese mountain dogs, Dalmatians, pointers, dachshunds, Pomeranians, and so on (Figure 2.1). All dogs belong to the same species, *Canis familiaris*, yet each breed is clearly distinguishable by its size, shape, color, and behavior. Similarly, differences among individual humans include eye color, height, skin color, and hair color, even though all humans belong to the species *Homo sapiens*. Within a species individual differences are seen at each generation, yet the species remains clearly identifiable. The differences among individuals within and between species are mainly the result of differences in the DNA sequences that constitute their genes. It is largely the genes that determine the structure, function, and development of the cell and organism. Thus the genetic information coded in DNA is responsible for species and individual variation.

The understanding of how genes are transmitted from parent to offspring began with the work of Gregor Johann Mendel (1822–1884), an Augustinian monk. In this chapter we will learn the basic principles of the transmission of genes by examining Mendel's work. Throughout this chapter we must remember that the segregation of genes is directly related to the behavior of chromosomes. However, although Mendel analyzed the patterns of segregation of hereditary traits, he did not know that genes

control the traits, or that genes are located in chromosomes, or even that chromosomes existed.

GENOTYPE AND PHENOTYPE

Before we begin our study of Mendel's work, we must distinguish between the nature of the genetic material each individual organism possesses and the physical characteristics that result from the expression of genes.

The characteristics of an individual that are transmitted from one generation to another are sometimes called **hereditary traits** (Mendel called them **characters**). These traits are under the control of DNA segments called **genes** (Mendel called them *factors*). The genetic constitution of an organism is called its **genotype**, and the observable properties of an organism, produced by the genotype in association with the environment, is called a **phenotype**.

The genes that every individual carries give only the potential for the development of a particular phenotypic characteristic. The extent to which that potential is realized depends not only on interactions with other genes and their products but also on environmental influences and random developmental events (Figure 2.2). A person's height, for example, is controlled by many genes, the expression of which can be significantly affected by internal and external environmental influences. Notable among these influences are nutrition (an external environmental influence) and the effects of hormones during puberty (an

~ FIGURE 2.1

Variation among dogs: (a) Bernese mountain dog; (b) Dalmatian; (c) Pointer; (d) Dachshund; (e) Pomeranian.

a)

b)

c)

d)

e)

~ FIGURE 2.2

Influences on the physical manifestation (phenotype) of the genetic blueprint (genotype): interactions with other genes and their products (such as hormones) and with the environment (such as nutrition).

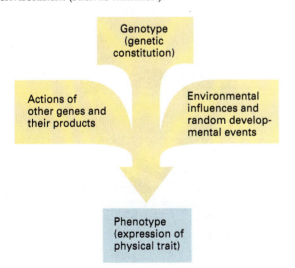

internal environmental influence). While we can say that a child is "tall like her father," there is no simple genetic explanation for that statement. She may be tall because excellent nutritional habits are interacting with the genetic potential for height.

Because of the effects of environment, then, individuals may have identical genotypes (as in identical twins) but different phenotypes. Studies have shown, for instance, that identical twins raised apart and exposed to different environmental effects tend to differ more in their appearance than identical twins raised together in the same environment (Figure 2.3). Similarly, individuals of a species may have virtually identical phenotypes but very different genotypes. These examples show that: (1) genes are a starting position for determining the structure and function of an organism, and (2) the pathway to the mature phenotypic state is highly complex and involves a myriad of interacting biochemical pathways.

It is important to understand for our future discussions that, although the phenotype is the product of interaction between gene(s) and environment, the

~ FIGURE 2.3

Identical twins (a) reared apart, and (b) together.

 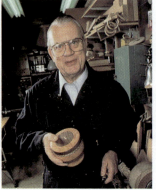

a)

b)

contribution of the environment varies. In some cases, the environmental influence is great, but in others, the environmental contribution is nonexistent. We will develop the relationship between genotype and phenotype in more detail as the text proceeds, but for our current task of examining Mendel's experiments, the important aspects mentioned above are sufficient.

ᛕEYNOTE

The genotype is the genetic constitution of an organism. The phenotype is the observable manifestation of the genetic traits. The genes give the potential for the development of characteristics; this potential is often affected by interactions with other genes and with the environment. Thus individuals with the same genotype can have different phenotypes, and individuals with the same phenotypes may have different genotypes.

MENDEL'S EXPERIMENTAL DESIGN

The first person to gain some understanding of the principles of heredity was Gregor Johann Mendel (Figure 2.4), whose work is considered to be the foundation of modern genetics. Mendel was born in 1822 in the village of Heinzendorf in northern Moravia (then a part of Austria). In 1843 he was admitted as a novice to the Augustinian Monastery in Brünn,

Austria (now Brno in the Czech Republic), where he later carried out his experiments. In 1847 he became a priest. From 1851 to 1853 Mendel attended the University of Vienna, where he studied a variety of sciences and mathematics. This education—particularly in physics—gave him an excellent background for designing experiments and analyzing experimental data.

Mendel returned to the monastery in 1853 and began teaching natural science and physics. In 1854

~ FIGURE 2.4

Gregor Johann Mendel, the founder of the science of genetics.

he began a series of breeding experiments with the garden pea *Pisum sativum* in an attempt to learn something about the mechanisms of heredity. As a result of his creativity, Mendel discovered some fundamental principles of genetics.

From the results of crossbreeding pea plants that exhibited differences in characteristics such as height, flower color, and seed shape, Mendel developed a simple theory to explain the transmission of hereditary characteristics or traits from generation to generation. (Note that Mendel had no knowledge of mitosis and meiosis. Now, of course, we know genes segregate according to chromosome behavior.) Although Mendel reported his conclusions in 1865, their significance was not fully realized until the late 1800s and early 1900s—several years after his death in 1884. In view of his significant contributions, Mendel is regarded as the father of genetics.

Mendel was not the first to perform breeding experiments. In fact, many individuals had bred animals and plants for many centuries. However, none of them was able to establish the key principles of genetics as Mendel did. Why was Mendel successful in doing this? First, he chose an experimental organism that was easy to grow and crossbreed. Recall our Chapter 1 discussion of some of the ideal criteria for experimental organisms to be used for genetics studies (p. 5). The garden pea fits these criteria very well. Pea varieties are easy to obtain and cultivate, they take up relatively little space, their generation time is relatively short, and many offspring are produced in each generation. Second, Mendel planned and performed his genetics experiments very carefully. In fact, his general experimental designs and methods of analysis are still used in genetic studies today. For example, he followed only one or at most a few pairs of alternative characteristics in each experiment, thereby making the analysis relatively uncomplicated. Third, he kept excellent records of his experiments, particularly of the different kinds of progeny produced and the numbers of each kind. These quantitative data were extremely important tools with which Mendel worked to formulate the principles of gene segregation. In short, Mendel was an early user of the hypothetico-deductive method of research (see Chapter 1, p. 3). That is, his experimental approach was effective because he developed a simple interpretation of the ratios of progeny he obtained from his crosses and then carried out direct and convincing experiments to test this hypothesis. Finally, he studied seven traits that were transmitted independently of each other.

Let us consider the organism of Mendel's genetics studies, *Pisum sativum* (the garden pea), in more detail. Figure 2.5 presents a cross section of a flower of the

~ **FIGURE 2.5**

Procedure for crossing pea plants.

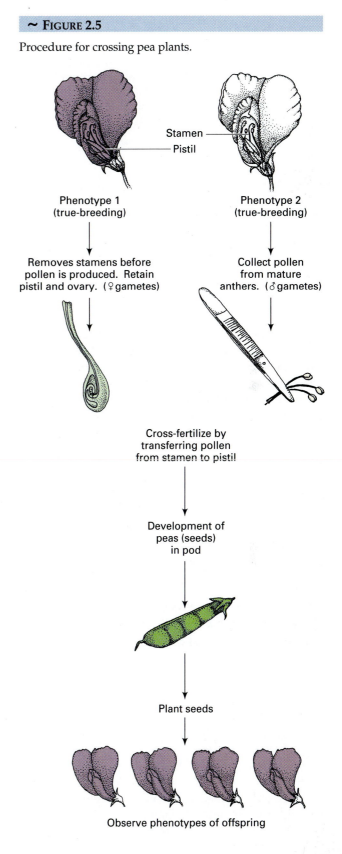

Stamen
Pistil

Phenotype 1
(true-breeding)

Phenotype 2
(true-breeding)

Removes stamens before pollen is produced. Retain pistil and ovary. (♀ gametes)

Collect pollen from mature anthers. (♂ gametes)

Cross-fertilize by transferring pollen from stamen to pistil

Development of peas (seeds) in pod

Plant seeds

Observe phenotypes of offspring

~ **FIGURE 2.6**

General procedure for making a genetic cross in animals. Male and female diploid individuals with different known phenotypes produce haploid gametes by meiosis. Fusion of male and female gametes produces diploid zygotes that develop by mitotic divisions. The phenotypes of the diploid offspring of the cross can be analyzed to determine their hereditary basis.

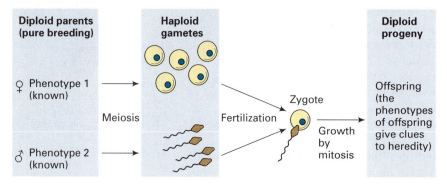

garden pea, showing the stamens (male reproductive organs) and the pistils (female reproductive organs). The pea normally reproduces by **self-fertilization**. That is, the stamen produces pollen (microspore of a flowering plant that germinates to form the male [♂] gametophyte), which lands on the pistil (containing the female [♀] gametes) within the same flower and fertilizes the plant. This process is also called **selfing**. Fortunately, it is a relatively simple procedure to prevent self-fertilization of the pea by removing the stamens from a developing flower bud before they produce any mature pollen. Then pollen taken from the stamens of another flower can be dusted onto the stigma of the pistil of the emasculated one to pollinate it.

Cross-fertilization, or more simply a **cross**, is the term used for the fusion of male gametes (pollen) from one individual and female gametes (eggs) from another. Once cross-fertilization has occurred, the zygote develops in the seeds (peas), which are then planted. The phenotypes of the plants that grow from the seeds are then analyzed. (Figure 2.6 outlines the general procedure of a genetic cross in animals.)

Mendel knew enough about the reproductive cycle of the garden pea to plan the crosses carefully. For his experiments he obtained 34 strains of pea plants that differed in a number of traits. He allowed each strain to self-fertilize for many generations to ensure that the traits he wanted to study were inherited and to remove from consideration those strains that produced progeny with traits different from the parental type. This preliminary work ensured that he only worked with pea strains in which the trait under investigation remained unchanged from parent to offspring for many generations. Such strains are called **true-breeding** or **pure-breeding strains**.

Next, Mendel selected seven traits to study in breeding experiments. Each trait had two easily distinguishable, alternative appearances (phenotypes),

as shown in Figure 2.7. These traits affect the appearance of most parts of the pea plant, including:

1. flower and seed coat color (grey versus white seed coats, and purple versus white flowers—note that a single gene controls these particular color properties of both seed coats and flowers)
2. seed color (yellow versus green)
3. seed shape (smooth versus wrinkled)
4. pod color (green versus yellow)
5. pod shape (inflated versus pinched)
6. stem height (tall versus short)
7. flower position (axial versus terminal)

MONOHYBRID CROSSES AND MENDEL'S PRINCIPLE OF SEGREGATION

Before discussing Mendel's experiments, we will clarify the terminology encountered in breeding experiments. The parental generation is called the **P generation**. The progeny of the P mating is called the **first filial generation**, or **F_1**. The subsequent generation produced by breeding together the F_1 offspring is the **F_2 generation**. Interbreeding the offspring of each generation results in F_3, F_4, F_5 generations, and so on.

Mendel first performed crosses between true-breeding strains of peas that differed in a single trait. Such crosses are called **monohybrid crosses**. For example, he pollinated pea plants that gave rise only to smooth seeds with pollen from a true-breeding variety that produced only wrinkled seeds.[1] As

---

[1]Seeds are the diploid progeny of sexual reproduction. If a phenotype concerns a part of the mature plant, such as stem size or flower color, the seeds must be germinated before that phenotype can be observed.

~ FIGURE 2.7

Seven character pairs in the garden pea that Mendel studied in his breeding experiments.

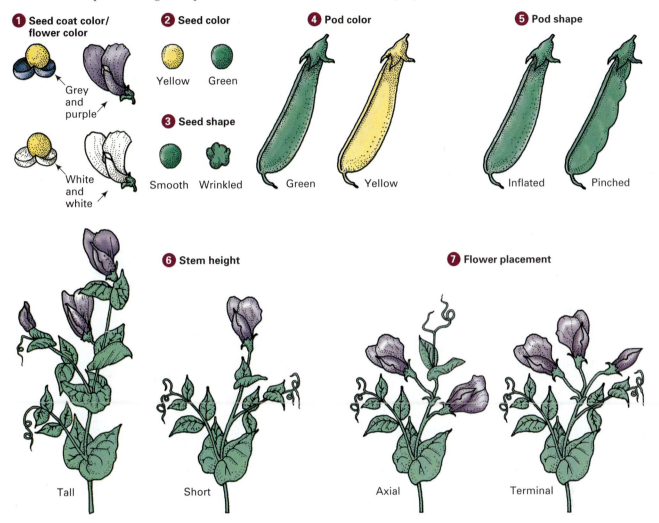

1 Seed coat color/ flower color

Grey and purple

White and white

2 Seed color

Yellow   Green

3 Seed shape

Smooth   Wrinkled

4 Pod color

Green   Yellow

5 Pod shape

Inflated   Pinched

6 Stem height

Tall   Short

7 Flower placement

Axial   Terminal

Figure 2.8 shows, the outcome of this monohybrid cross was all smooth seeds. The same result was obtained when the parental types were reversed; that is, when the pollen from a smooth-seeded plant was used to pollinate a pea plant that gave wrinkled seeds. (Matings that are done both ways, smooth female [♀] × wrinkled male [♂] and wrinkled female [♀] × smooth male [♂], are called **reciprocal crosses.** Conventionally, the female is given first in crosses of plants.) If the results of reciprocal crosses are the same, the interpretation is that the trait is not dependent on the sex of the organism.

The significant point of this cross was that all the F$_1$ progeny seeds of the smooth × wrinkled reciprocal crosses were smooth; that is, they exactly resembled only one of the parents in this character rather than being a blend of both parental phenotypes. That all offspring of true-breeding parents are alike is sometimes referred to as the *principle of uniformity in F$_1$*.

~ FIGURE 2.8

Results of one of Mendel's breeding crosses. In the parental generation he crossed a true-breeding pea strain that produced smooth seeds with one that produced wrinkled seeds. All the F$_1$ progeny seeds were smooth.

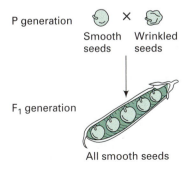

P generation

Smooth   Wrinkled
seeds   seeds

F$_1$ generation

All smooth seeds

Next, Mendel planted the seeds and allowed the F_1 plants to self-fertilize (= to self) to produce the F_2 seed. Both smooth and wrinkled seeds appeared in the F_2 generation, and both types could be found within the same pod. Typical of his analytical approach to the experiments, Mendel counted the number of each type. He found that 5,474 were smooth and 1,850 were wrinkled (Figure 2.9). The calculated ratio of smooth to wrinkled seeds was 2.96:1, which is very close to a 3:1 ratio.

Using the same quantitative approach, Mendel analyzed the behavior of the six other pairs of traits. Qualitatively and quantitatively, the same results were obtained (Table 2.1). From the seven sets of crosses he made the following general conclusions about his data:

1. The results of reciprocal crosses were always the same.
2. All F_1 progeny resembled one of the parental strains.
3. In the F_2 generation the parental trait that had disappeared in the F_1 generation reappeared. Further, the trait seen in the F_1 was always found in the F_2 at about three times the frequency of the other trait.

How can a trait present in the P generation disappear in the F_1 and then reappear in the F_2? Mendel observed that, while the F_1 resembled only one of the parents in their phenotype, they did not breed true, a fact that distinguishes the F_1 from the parent they resembled. Moreover, the F_1 could produce F_2 some of

~ **FIGURE 2.9**

The F_2 progeny of the cross shown in Figure 2.8. When the plants grown from the F_1 seeds were self-pollinated, both smooth and wrinkled F_2 progeny seeds were produced. Commonly, both seed types were found in the same pod. In Mendel's actual experiments he counted 5,474 smooth and 1,850 wrinkled F_2 progeny seeds for a ratio of 2.96:1.

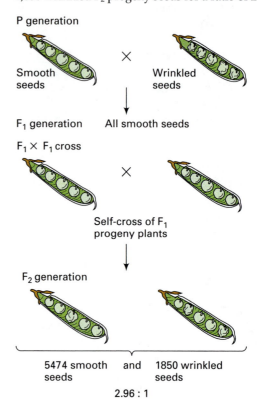

P generation

Smooth seeds   ✕   Wrinkled seeds

F_1 generation   All smooth seeds

$F_1 \times F_1$ cross

✕

Self-cross of F_1 progeny plants

F_2 generation

5474 smooth seeds   and   1850 wrinkled seeds

2.96 : 1

~ **TABLE 2.1**

Mendel's Results in Crosses Between Plants Differing in One of Seven Characters

| CHARACTER[a] | F_1 | F_2 (NUMBER) | | | F_2 (RATIO) |
		DOMINANT	RECESSIVE	TOTAL	DOMINANT: RECESSIVE
Seeds: smooth versus wrinkled	All smooth	5,474	1,850	7,324	2.96:1
Seeds: yellow versus green	All yellow	6,022	2,001	8,023	3.01:1
Seed coats: grey versus white[b]	All grey ⎫	705	224	929	3.15:1
Flowers: purple versus white	All purple ⎬				
Flowers: axial versus terminal	All axial	651	207	858	3.14:1
Pods: inflated versus pinched	All inflated	882	299	1,181	2.94:1
Pods: green versus yellow	All green	428	152	580	2.82:1
Stem: tall versus short	All tall	787	277	1,064	2.84:1
Total or average		**14,949**	**5,010**	**19,959**	**2.98:1**

[a] The dominant trait is always written first.

[b] A single gene controls both the seed coat and the flower color trait.

which had the parental phenotype that had disappeared in the F_1. Mendel concluded that the alternative traits in the crosses—for example, smoothness or wrinkledness of the seeds—were determined by **particulate factors**. He reasoned that these factors, which were transmitted from parents to progeny through the gametes, carried hereditary information. We now know these factors by another name, *genes*.

Since Mendel was examining pairs of traits (e.g., wrinkled/smooth), each factor was considered to exist in alternative forms (which we now call **alleles**), each of which specified one of the traits. For the gene that controls the shape of the pea seed, for example, there is one form, or allele, that results in the production of a smooth seed and another allele that results in a wrinkled seed.

Mendel reasoned further that a true-breeding strain of peas must contain a pair of identical factors. Since the F_2 exhibited both traits while the F_1 exhibited only one of those traits, then each F_1 individual must have contained both factors, one for each of the alternative traits. In other words, crossing two different true-breeding strains brings together in the F_1 one factor from each strain: the eggs contain one factor from one strain and the pollen contain one factor from the other strain. Further, since only one of the traits was seen in the F_1 generation, the expression of the "missing" trait must somehow have been masked by the visible trait; this masking is called *dominance*. For the smooth × wrinkled example the F_1 seeds were all smooth. Thus, the allele for smoothness is masking or **dominant** to the allele for wrinkledness. Conversely, wrinkled is said to be **recessive** to smooth because the factor for wrinkled is masked (Figure 2.10). Similar conclusions can be made for the other six pairs of traits. The dominant and the recessive forms for each pair of traits are indicated in Table 2.1.

A simple way to visualize the crosses is to use symbols for the alleles, as did Mendel. For the smooth × wrinkled cross we can give the symbol S to the allele for smoothness and the symbol s to the allele for wrinkledness. The letter used is based on the dominant phenotype and the convention in this case is that the dominant allele is given the uppercase (capital) letter and the recessive allele the lowercase (small) letter. (This convention was used for many years, particularly in plant genetics. Now it is more conventional to assign the letter based on the recessive phenotype. We will use this convention later.) Using these symbols, we can diagram the cross as shown in Figure 2.11; the production of the F_1 is shown in Figure 2.11a, and the F_2 in Figure 2.11b. (In both Figures 2.10 and 2.11 the genes are shown on chromosomes. Keep in mind that the segregation of genes from generation to generation follows the behavior of chromosomes, although Mendel had no knowledge of the existence of chromosomes.) Since each parent is true breeding and diploid (i.e., has two sets of chromosomes), each must contain two copies of the same allele. Thus the genotype of the parental plant grown from the smooth seeds is SS and that of the wrinkled parent is ss. True-breeding individuals that contain two copies of the same specific allele of a particular gene are said to be **homozygous** for that gene.

When plants produce gametes by meiosis (see Chapter 3), each gamete contains only one copy of the gene (i.e., one allele, as indicated previously); the plants from smooth seeds produce S-bearing gametes, and the plants from wrinkled seeds produce s-bearing gametes. When the gametes fuse during the fertilization process, the resulting zygote has one S allele and one s allele, a genotype of Ss. Plants that have two different alleles of a particular gene are said to be **heterozygous**. Because of the dominance of the smooth S allele, only smooth seeds develop from the F_1 zygotes.

The plants grown from the F_1 seeds differ from the smooth parent in that they produce two types of gametes in equal numbers: S-bearing and s-bearing. All the possible fusions of F_1 gametes are shown in the matrix in Figure 2.11, called a **Punnett square** after its originator, R. Punnett. These fusions give rise to the zygotes that produce the F_2 generation.

In the F_2 generation three types of genotypes are produced: SS, Ss, and ss. As a result of the random fusing of gametes, the relative proportion of these zygotes is 1:2:1, respectively. However, since the S factor is dominant to the s factor, both the SS and Ss seeds are smooth, and the F_2 generation seeds exhibit a phenotypic ratio of 3 smooth : 1 wrinkled seeds. Further crosses are needed to test the genotypes of the smooth-seeded progeny. The results were the same

~ **FIGURE 2.10**

Dominant and recessive alleles of a gene for seed shape in peas.

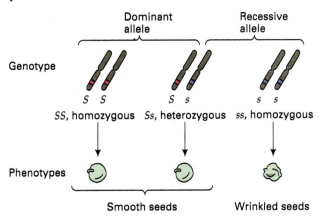

~ FIGURE 2.11

The same cross as in Figures 2.8 and 2.9. Here, genetic symbols illustrate the principle of segregation of Mendelian factors. (a) Production of F_1 generation; (b) Production of F_2 generation.

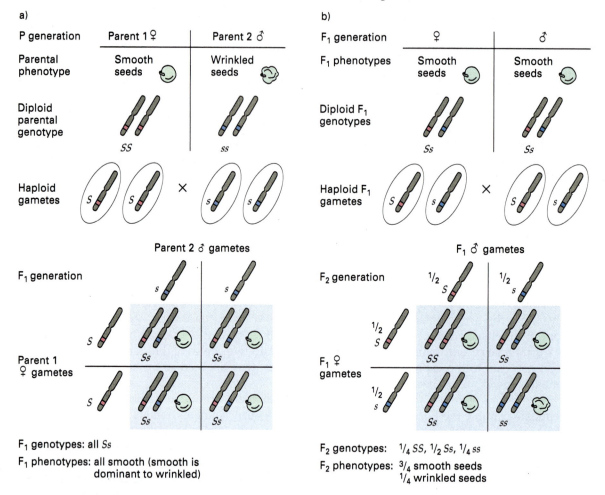

F_1 genotypes: all Ss
F_1 phenotypes: all smooth (smooth is dominant to wrinkled)

F_2 genotypes: $1/4 SS$, $1/2 Ss$, $1/4 ss$
F_2 phenotypes: $3/4$ smooth seeds
                    $1/4$ wrinkled seeds

for crosses involving the other six of Mendel's character pairs.

The Principle of Segregation

From the sort of data we have discussed, Mendel proposed his **first law**, the **principle of segregation**, which states that *the two members of a gene pair (alleles) segregate (separate) from each other in the formation of gametes.* As a result, half the gametes carry one allele and the other half carry the other allele. In other words, each gamete carries only one allele of each gene. The progeny are produced by the random combination of gametes from the two parents.

In proposing the principle of segregation, Mendel had clearly differentiated between the factors (genes) that determined the traits (the genotype) and the traits themselves (the phenotype). From a modern perspective we know that genes are on chromosomes and the specific location of a gene on a chromosome is called a **locus** (or **gene locus**; plural *loci*). Further, Mendel's first law means that at the gene level the members of a pair of alleles of a gene on a pair of chromosomes segregate during meiosis so that any offspring receives only one member of a pair from each parent. Thus **gene segregation** parallels the separation of homologous pairs of chromosomes at anaphase I in meiosis (discussed in more detail in Chapter 3).

Box 2.1 presents a summary of the genetics concepts and terms we have learned so far in this chapter. A thorough familiarity with these terms is essential to your study of genetics.

Box 2.1
Genetic Terminology

Gamete: a mature reproductive cell that is specialized for sexual fusion. Each gamete is haploid and fuses with a cell of similar origin but of opposite sex to produce a diploid zygote.

Cross: a mating between two individuals, leading to the fusion of gametes.

Zygote: the cell produced by the fusion of male and female gametes.

Gene (Mendelian factor): the determinant of a characteristic of an organism. Gene symbols are underlined or italicized.

Locus (gene locus): the specific place on a chromosome where a gene is located.

Alleles: alternative forms of a gene. For example, S and s alleles represent the smoothness and wrinkledness of the pea seed. (Like gene symbols, allele symbols are underlined or italicized.)

Genotype: the genetic constitution of an organism. A diploid organism (i.e., one with two copies of each chromosome) that has both alleles the same for a given gene locus is said to be **homozygous** for that allele. Homozygotes produce only one gametic type with respect to that locus. For example, from our pea example, true-breeding smooth individuals have the genotype SS, and true-breeding wrinkled individuals have the genotype ss; both are homozygous. The smooth parent is **homozygous dominant**; the wrinkled parent is **homozygous recessive**.

Diploid organisms that have two different alleles at a specific gene locus are said to be **heterozygous**. So F_1 hybrid plants from the cross of SS and ss parents have one S allele and one s allele. Individuals heterozygous for two allelic forms of a gene produce two kinds of gametes (S and s).

Phenotype: the physical manifestation of a genetic trait that results from a specific genotype and its interaction with the environment. In our example the S allele was dominant to the s allele, so in the heterozygous condition the seed is smooth. Therefore both the homozygous dominant SS and the heterozygous Ss seeds have the same phenotype (smooth), even though they differ in genotype.

𝒦EYNOTE

Mendel's first law, the principle of segregation, states that the two members of a gene pair (alleles) segregate (separate) from each other in the formation of gametes; half the gametes carry one allele, and the other half carry the other allele.

Representing Crosses with a Branch Diagram

The use of a Punnett square to consider the pairing of all possible gamete types from the two parents in a monohybrid cross (shown in Figure 2.11) is a relatively simple way to learn how to predict the relative frequencies of genotypes and phenotypes in the next generation. In this section, however, we will describe an alternative method, one which you are encouraged to master: the branch diagram. (Box 2.2 discusses some elementary principles of probability that will help you understand this approach.) In order to use the branch diagram approach, it is necessary to know the dominance/recessiveness relationship of the allele pair so that the progeny phenotypic classes can be determined. Figure 2.12 illustrates the application of the branch diagram analysis of the $F_1 \times F_1$ self of the smooth × wrinkled cross diagrammed in Figure 2.11.

The F_1 seeds from the cross in Figure 2.11 have the genotype Ss. In the meiotic process an equal number of S and s gametes are expected to be produced, so we can say that half the gametes are S and the other half are s (Figure 2.12). Thus 1/2 is the predicted frequency of each of these two types. But just as tossing a coin many times does not always give exactly half heads and half tails, the two gametes may not be produced in an exactly 1:1 ratio. However, the more chances (e.g., tosses), the closer you will come to the true frequency.

From the rules of probability, the expected frequencies of the three possible genotypes in the F_2 generation can be predicted. To produce an SS plant, an S egg must pair with an S pollen grain. The frequency of S eggs in the population of eggs is 1/2, and

Box 2.2
Elementary Principles of Probability

A **probability** is the ratio of the number of times a particular event is expected to occur to the number of trials during which the event could have happened. For example, the probability of picking a heart from a deck of 52 cards, 13 of which are hearts, is $p(\text{heart}) = 13/52 = 1/4$. That is, we would expect, on the average, to pick a heart from a deck of cards once in every four trials.

Probabilities and the *laws of chance* are involved in the transmission of genes. As a simple example, let us consider a couple and the chance that their child will be a boy or a girl. Assume that an exactly equal number of boys and girls are born (which is not precisely true, but we can assume it to be so for the sake of discussion). The probability that the child will be a boy is 1/2, or 0.5. Similarly, the probability that the child will be a girl is 1/2.

Now a rule of probability can be introduced: the **product rule**. The product rule states that *the probability of two independent events occurring simultaneously is the product of each of their individ-*ual probabilities. Thus the probability that both children in families with two children will be girls is 1/4. That is, the probability of the first child being a girl is 1/2, the probability of the second being a girl is also 1/2, and by the product rule the probability of the first and second being girls is $1/2 \times 1/2 = 1/4$. Similarly, the probability of having three boys in a row is $1/2 \times 1/2 \times 1/2 = 1/8$.

Another rule of probability is the **sum rule**, which states that *the probability of either one of two independent mutually exclusive events occurring is the sum of their individual probabilities.* For example, if two dice are thrown, what is the probability of getting two sixes or two ones? The individual probabilities are calculated as follows: The probability of getting two sixes is found by using the product rule. The probability of getting one six, $p(\text{one six})$, is 1/6, since there are six faces to a die. Therefore the probability of getting two sixes, $p(\text{two sixes})$, when two dice are thrown is $1/6 \times 1/6 = 1/36$. Similarly, $p(\text{two ones}) = 1/36$. To roll two sixes *or* two ones involves independent events, so the sum rule is used. The sum of the individual probabilities is $1/36 + 1/36 = 2/36 = 1/18$. To return to our family example, the probability of having two boys or two girls is $1/4 + 1/4 = 1/2$.

~ FIGURE 2.12

Calculating the ratios of phenotypes in the F_2 generation of the Figure 2.11 cross by using the branch diagram approach.

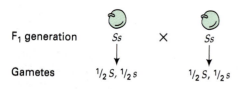

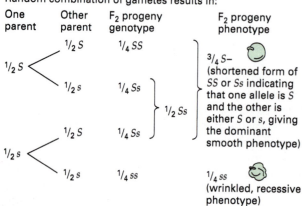

the frequency of S pollen grains in the pollen population is 1/2. Therefore the expected proportion of SS smooth plants in the F_2 is $1/2 \times 1/2 = 1/4$. Similarly, the expected proportion of ss wrinkled progeny in the F_2 is $1/2 \times 1/2 = 1/4$.

What about the Ss progeny? Again, the frequency of S in one gametic type is 1/2, and the frequency of s in the other gametic type is also 1/2. However, there are two ways in which Ss progeny can be obtained. The first involves the fusion of an S egg with s pollen, and the second is a fusion of an s egg with S pollen. Using the product rule (Box 2.2), the probability of each of these events occurring is $1/2 \times 1/2 = 1/4$. By using a sum rule (Box 2.2), the probability of *one or the other occurring* is the sum of the individual probabilities, or $1/4 + 1/4 = 1/2$.

Given the rules of probability, then, the prediction is that one-fourth of the F_2 progeny will be SS, half will be Ss, and one-fourth will be ss, exactly as was found with the method shown in Figure 2.11. Either method—the Punnett square or the branch diagram—may be used with any cross.

Confirming the Principle of Segregation: The Use of Testcrosses

When formulating his principle of segregation, Mendel did a number of tests to ensure the validity of his results. He continued the self-fertilizations at each generation up to the F_6 and found that in every generation some of the preceding generation produced progeny, some with the dominant character and the remainder with the recessive character. He concluded, therefore, that the principle of segregation was valid no matter how many generations were carried out.

Another important test concerned the F_2 plants. As shown in Figure 2.11, a ratio of 1:2:1 occurs for the genotypes SS, Ss, and ss for the smooth × wrinkled example. Phenotypically, the ratio of smooth to wrinkled is 3:1. At the time of his experiments, the presence of segregating factors that were responsible for the smooth and wrinkled phenotypes was only an hypothesis. To test his factor hypothesis, Mendel allowed the F_2 plants to self-pollinate. As he expected, the plants produced from wrinkled seeds bred true, supporting his conclusion that they were pure for the s factor (gene).

Selfing the plants derived from the F_2 smooth seeds produced two different types of progeny. One-third of the smooth F_2 seeds produced all smooth-seeded progeny, whereas the other two-thirds produced both smooth and wrinkled seeds in each pod in a ratio of 3 smooth : 1 wrinkled (Figure 2.13). For the plants that produced both seed types in the progeny, the actual ratio of smooth:wrinkled seeds was 3:1; that is, the same ratio as seen for the F_2 progeny. These results completely support the principle of segregation of genes. The random combination of gametes that form the zygotes of the original F_2 produces two genotypes that give rise to the smooth phenotype (e.g.,

Figures 2.11 and 2.12); the relative proportion of the two genotypes SS and Ss is 1:2. The SS seeds give rise to true-breeding plants, whereas the Ss seeds give rise to plants that behave exactly like the F_1 plants when they are self-pollinated in that they produce a 3:1 ratio of smooth:wrinkled progeny. *Mendel explained these results by proposing that each plant had two factors, while each gamete had only one. He also proposed that the random combination of the gametes generated the progeny in the proportions he found. Mendel obtained the same results in all seven sets of crosses.*

The self-fertilization test of the F_2 progeny proved a useful way of confirming the genotype of a plant with a given phenotype. A more common test to find out the genotype of an organism is to perform a **testcross**, a cross of an individual of unknown genotype, usually expressing the dominant phenotype, with a homozygous recessive individual in order to determine the genotype of the unknown individual.

Consider again the cross shown in Figure 2.11. We can predict the outcome of a testcross of the F_2 progeny showing the dominant, smooth-seed phenotype. If the F_2 individuals are homozygous SS, the result of a testcross with an ss plant will be all smooth seeds. As Figure 2.14a shows, the Parent 1 smooth SS plants produce only S gametes. Parent 2 is homozygous recessive wrinkled, ss, so it produces only s gametes. Therefore, all zygotes are Ss, and all the resulting seeds are smooth in phenotype. In actual practice, then, if a plant showing the dominant trait is testcrossed, and only the dominant phenotype is seen among the progeny, the plant must have been homozygous for the dominant allele. On the other hand, if heterozygous Ss F_2 plants are testcrossed with a homozygous ss plant, a 1:1 ratio of dominant to recessive phenotypes is expected. As Figure 2.14b shows, the Parent 1 smooth Ss produces both S and s gametes in equal proportion, while the homozygous ss Parent 2 produces only s gametes. As a result, half the progeny of the testcross are Ss heterozygotes and have a smooth phenotype because of the dominance of the S allele, and the other half are ss homozygotes and have a wrinkled phenotype. In actual practice, then, if a plant showing the dominant trait is testcrossed, and the progeny exhibit a 1:1 ratio of dominant to recessive phenotypes, the plant must have been heterozygous.

In summary, testcrosses of the F_2 progeny from Mendel's crosses that showed the dominant phenotype indicated that there was a 1:2 ratio of homozygous dominant:heterozygous genotypes in the F_2 progeny. That is, one-third of the F_2 progeny with the dominant phenotype gave only progeny with the dominant phenotype in crosses with the homozygous recessive and therefore were homozygous for the

~ FIGURE 2.13

Determining the genotypes of the F_2 smooth progeny of Figure 2.11 by selfing the plants grown from the smooth seeds.

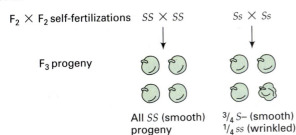

$F_2 × F_2$ self-fertilizations $SS × SS$ $Ss × Ss$

F_3 progeny

All SS (smooth) progeny

$^3/_4$ $S–$ (smooth)
$^1/_4$ ss (wrinkled)
(i.e., both kinds of progeny)

~ FIGURE 2.14

Determining the genotypes of the F$_2$ generation smooth seeds (Parent 1) of Figure 2.11 by testcrossing plants grown from the seed with a homozygous recessive wrinkled *(ss)* strain (Parent 2). (a) If Parent 1 is *SS*, all progeny seeds are smooth in phenotype. (b) If Parent 1 is *Ss*, 1/2 of the progeny seeds are smooth and 1/2 are wrinkled.

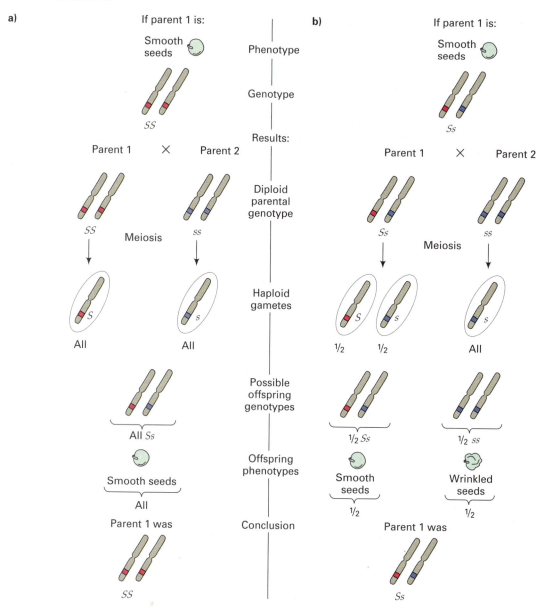

dominant allele. The other two-thirds of the F$_2$ progeny with the dominant phenotype gave progeny with a 1:1 ratio of dominant phenotype progeny to recessive phenotype progeny, and therefore were heterozygous.

KEYNOTE

A testcross is a cross of an individual of unknown genotype, usually expressing the dominant phenotype, with a known homozygous recessive individual in order to determine the genotype of the unknown individual. The phenotypes of the progeny of the testcross indicate the genotype of the individual tested. If the progeny all show the dominant phenotype, the individual was homozygous dominant. If there is an approximately 1:1 ratio of progeny with dominant and recessive phenotypes, the unknown individual was heterozygous.

DIHYBRID CROSSES AND THE MENDELIAN PRINCIPLE OF INDEPENDENT ASSORTMENT

The Principle of Independent Assortment

Mendel also analyzed a number of crosses in which two pairs of traits were simultaneously involved. In each case he obtained the same results. From these experiments he proposed his **second law**, the **principle of independent assortment**, which states that the factors (genes) for different traits assort independently of one another. In modern terms, this means that genes on different chromosomes behave independently in the production of gametes. This principle was unknown to Mendel and will be considered when linkage and crossing-over are discussed in Chapter 5.

Consider an example involving smooth (*S*)/wrinkled (*s*) and yellow (*Y*)/green (*y*) seed traits (yellow is dominant to green). Mendel made crosses between true-breeding smooth-yellow plants (*SS YY*) and wrinkled-green plants (*ss yy*), with the results shown in Figure 2.15. All the F$_1$ seeds from this cross were smooth and yellow, as predicted from the results of the monohybrid crosses. As Figure 2.15a shows, the smooth-yellow parent produces only *S Y* gametes, which give rise to *Ss Yy* zygotes upon fusion with the *s y* gametes from the wrinkled-green parent. Because of the dominance of the smooth and the yellow traits, all F$_1$ seeds are smooth and yellow.

The F$_1$ are heterozygous for two pairs of alleles at two different loci. Such individuals are called dihybrids, and a cross between two of these dihybrids of the same type is called a **dihybrid cross**.

When Mendel self-pollinated the dihybrid F$_1$ plants to give rise to the F$_2$ generation (Figure 2.15b p. 32) he considered two possible outcomes. One was that the genes for the traits from the original parents would be transmitted together to the progeny. In this case a phenotypic ratio of 3:1 smooth-yellow:wrinkled-green would be predicted.

The other possibility was that the traits would be inherited independently of one another. In this case, the dihybrid F$_1$ would produce four types of gametes: *S Y*, *S y*, *s Y*, and *s y*. Because of the independence of the two pairs of genes, each gametic type is predicted to occur with equal frequency. In F$_1$ × F$_1$ crosses, the four types of gametes would be expected to fuse randomly in all possible combinations to give rise to the zygotes and, hence, the progeny seeds. All the possible gametic fusions are represented in the Punnett square in Figure 2.15b. In a dihybrid cross there are 16 possible gametic fusions. The result is nine different

~ FIGURE 2.15a

The principle of independent assortment in a dihybrid cross. This cross, actually done by Mendel, involves the smooth/wrinkled and yellow/green character pairs of the garden pea. (a) Production of the F$_1$ generation. (Note that, compared with previous figures of this kind, only one box is shown in the F$_1$, instead of four. This is because only one class of gametes exists for Parent 2 and only one class for Parent 1. Previously we showed two gametes from each parent, even though those gametes were identical.)

F$_1$ genotypes: all *Ss Yy*
F$_1$ phenotypes: all smooth-yellow seeds

genotypes but, because of dominance, only four phenotypes are predicted:

1 *SS YY*, 2 *Ss YY*, 2 *SS Yy*, 4 *Ss Yy* = 9 smooth-yellow
1 *SS yy*, 2 *Ss yy*   = 3 smooth-green
1 *ss YY*, 2 *ss Yy*   = 3 wrinkled-yellow
1 *ss yy*   = 1 wrinkled-green

According to the rules of probability, if pairs of characters are inherited independently in a dihybrid cross, then the F$_2$ from an F$_1$ × F$_1$ cross will give a 9:3:3:1 ratio of the four possible phenotypic classes. This ratio is the result of the independent assortment of the two gene pairs into the gametes and of the random fusion of those gametes.

This prediction was met in all dihybrid crosses that Mendel performed. In every case the F$_2$ ratio was close to 9:3:3:1. For our example he counted 315 smooth-yellow, 108 smooth-green, 101 wrinkled-

~ FIGURE 2.15b

(b) The F₂ genotypes and 9:3:3:1 phenotypic ratio of smooth-yellow:smooth-green:wrinkled-yellow:wrinkled-green are derived by using the Punnett square.

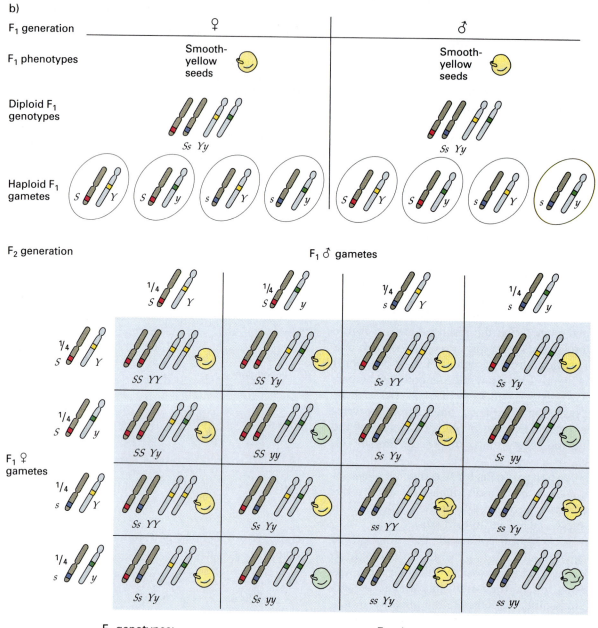

F₂ genotypes:

$^1/_{16}$ (SS YY) + $^2/_{16}$ (Ss YY) + $^2/_{16}$ (SS Yy) + $^4/_{16}$ (Ss Yy) = $^9/_{16}$ smooth-yellow seeds
$^1/_{16}$ (SS yy) + $^2/_{16}$ (Ss yy) = $^3/_{16}$ smooth-green seeds
$^1/_{16}$ (ss YY) + $^2/_{16}$ (ss Yy) = $^3/_{16}$ wrinkled-yellow seeds
$^1/_{16}$ (ss yy) = $^1/_{16}$ wrinkled-green seeds

F₂ phenotypes:

yellow, and 32 wrinkled-green seeds—very close to the predicted ratio. To Mendel this result meant that the factors (genes) determining the specific, different character pairs he was analyzing were transmitted independently.

$\mathcal{K}$EYNOTE

Mendel's second law, the principle of independent assortment, states that genes for different traits assort independently of one another in the production of gametes.

Branch Diagram of Dihybrid Crosses

As perhaps is already apparent, it is quite tedious and slow to construct a Punnett square of gamete combinations and then to count up the numbers of each phenotypic class from all the genotypes produced. This chore is not too difficult for a dihybrid cross, but for more than two gene pairs it becomes more complex. It is easier to get into the habit of calculating the expected ratios of phenotypic or genotypic classes by using a branch diagram, which is basically to consider the traits one at a time and use the laws of probability to determine the likelihood of a specific outcome. *With practice it should be possible to simply calculate the probabilities of outcomes of various crosses just by using the laws of probability without any need for drawing out the branch diagram. It cannot be overemphasized that this requires the diligent working of problems.*

Using the same example, in which the two gene pairs assort independently into the gametes, we will consider each gene pair in turn. Earlier we showed that an F_1 self of an *Ss* heterozygote gave rise to progeny of which three-fourths were smooth and one-fourth were wrinkled. Genotypically, the former class had at least one dominant *S* allele; that is, they were *SS* or *Ss*. A convenient way to signify this situation is to use a dash to indicate an allele that has no effect on the phenotype. Thus *S–* means that phenotypically the seeds are smooth and genotypically they are either *SS* or *Ss*.

Now consider the F_2 produced from a selfing of *Yy* heterozygotes; again, a 3:1 ratio is seen, with three-fourths of the seeds being yellow and one-fourth being green. Since this segregation occurs independently of the segregation of the smooth/wrinkled pair, we can consider all possible combinations of the

phenotypic classes in the dihybrid cross. For example, the expected proportion of F_2 seeds that are smooth and yellow is the product of the probability that an F_2 seed will be smooth and the probability that it will be yellow, or $3/4 \times 3/4 = 9/16$. Similarly, the expected proportion of F_2 progeny that are wrinkled and yellow is $3/4 \times 1/4 = 3/16$. Extending this calculation to all possible phenotypes, as shown in Figure 2.16, we obtain the ratio of 9 *S– Y–* (smooth, yellow) : 3 *S– yy* (smooth, green) : 3 *ss Y–* (wrinkled, yellow) : 1 *ss yy* (wrinkled, green).

The testcross can be used to check the genotypes of F_1 progeny and F_2 progeny from a dihybrid cross. In the example the F_1 is a double heterozygote, *Ss Yy*. This F_1 produces four types of gametes in equal proportions, as was the case in Figure 2.15: *S Y, S y, s Y,* and *s y*. In a testcross with a doubly homozygous recessive plant, in this case *ss yy*, the phenotypic ratio of the progeny is a direct reflection of the ratio of gametic types produced by the F_1 parent. In a testcross like this one, then, there will be a 1:1:1:1 ratio in the offspring of *Ss Yy : Ss yy : ss Yy : ss yy* genotypes, which means a 1:1:1:1 ratio of smooth-yellow : smooth-green : wrinkled-yellow : wrinkled-green phenotypes. This 1:1:1:1 phenotypic ratio is diagnostic of testcrosses in which the "unknown" parent is a double heterozygote.

In the F_2 of a dihybrid cross there were nine different genotypic classes but only four phenotypic classes. The genotypes can be ascertained by testcrossing as

~ **FIGURE 2.16**

Using the branch diagram approach to calculate the F_2 phenotypic ratio of the Figure 2.15 cross.

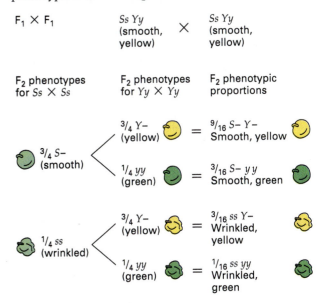

we have shown. Table 2.2 lists the expected ratios of progeny phenotypes from such testcrosses. No two patterns are the same, so here the testcross is truly a diagnostic approach to confirm genotypic type.

Trihybrid Crosses

Mendel also confirmed his laws for three characters segregating in other garden pea crosses. Such crosses are called **trihybrid crosses**. Here the proportions of F_2 genotypes and phenotypes are predicted with precisely the same logic used before, considering each character pair independently. Figure 2.17 shows a branch diagram derivation of the F_2 phenotypic classes for a trihybrid cross. The independently assorting character pairs in the cross are smooth versus wrinkled seeds, yellow versus green seeds, and purple versus white flowers. There are 64 combinations of 8 maternal and 8 paternal gametes. Combination of these gametes gives rise to 27 different genotypes and 8 different phenotypes in the F_2 generation. The phenotypic ratio in the F_2 is 27:9:9:3:9:3:3:1.

Now that enough examples have been considered, we can make some generalizations about phenotypic and genotypic classes. In each of the examples discussed, the F_1 is heterozygous for each gene involved in the cross, and the F_2 is generated by selfing (where that is possible) or by allowing the F_1 progeny to interbreed. In monohybrid crosses there are two phenotypic classes in the F_2, in dihybrid crosses there are four, and in trihybrid crosses there are eight. The general rule is that there are 2^n phenotypic classes in the F_2, where n is the number of independently assorting, heterozygous gene pairs (Table 2.3). (This rule holds *only* when a true dominant-recessive relation holds for each of the gene pairs.)

~ **TABLE 2.3**

Number of Phenotypic and Genotypic Classes Expected from Self-Crosses of Heterozygotes in Which All Genes Show Complete Dominance

NUMBER OF SEGREGATING GENE PAIRS	NUMBER OF PHENOTYPIC CLASSES	NUMBER OF GENOTYPIC CLASSES
1^a	2	3
2	4	9
3	8	27
4	16	81
n	2^n	3^n

[a] For example, from $Aa \times Aa$, two phenotypic classes are expected, with genotypic classes of AA, Aa, and aa.

Furthermore, we saw that there are 3 genotypic classes in the F_2 of monohybrid crosses, 9 in dihybrid crosses, and 27 in trihybrid crosses. A simple rule is that the number of genotypic classes is 3^n, where n is the number of heterozygous gene pairs.

Incidentally, the phenotypic rule (2^n) can also be used to predict the number of classes that will come from a multiple heterozygous F_1 used in a testcross. Here the number of genotypes in the next generation will be the same as the number of phenotypes. For example, from $Aa\ Bb \times aa\ bb$ there are four progeny genotypes (2^n, where n is 2)—$Aa\ Bb$, $Aa\ bb$, $aa\ Bb$, and $aa\ bb$—and four phenotypes—both dominant phenotypes, the A dominant phenotype and b recessive phenotype, the a recessive phenotype and B dominant phenotype, and both recessive phenotypes.

"REDISCOVERY" OF MENDEL'S PRINCIPLES

Mendel published his treatise on heredity in 1866 in *Verhandlungen des Naturforschenden Vereines* in Brünn, but it received little attention from the scientific community until the "rediscovery" of his principles in the early 1900s. In 1985 Iris and Laurence Sandler proposed a possible reason. They contend that it may have been impossible for the scientific community from 1865 to 1900 to understand the significance of Mendel's work because it did not fit into that community's conception of the relationship of heredity to other sciences. To Mendel's contemporaries, heredity included not only those ideas that are today understood as genetic, but also those that are considered developmental. In other words, their concept of heredity included what we now know as genetics and embryology. More pertinently, they also viewed heredity as simply a particular

~ **TABLE 2.2**

Proportions of Phenotypic Classes Expected from Testcrosses of Strains with Various Genotypes for Two Gene Pairs

TESTCROSSES	PROPORTION OF PHENOTYPIC CLASSES			
	$A-\ B-$	$A-\ bb$	$aa\ B-$	$aa\ bb$
$AA\ BB \times aa\ bb$	1	0	0	0
$Aa\ BB \times aa\ bb$	1/2	0	1/2	0
$AA\ Bb \times aa\ bb$	1/2	1/2	0	0
$Aa\ Bb \times aa\ bb$	1/4	1/4	1/4	1/4
$AA\ bb \times aa\ bb$	0	1	0	0
$Aa\ bb \times aa\ bb$	0	1/2	0	1/2
$aa\ BB \times aa\ bb$	0	0	1	0
$aa\ Bb \times aa\ bb$	0	0	1/2	1/2
$aa\ bb \times aa\ bb$	0	0	0	1

~ FIGURE 2.17

Branch diagram derivation of the relative frequencies of the eight phenotypic classes in the F_2 of a trihybrid cross.

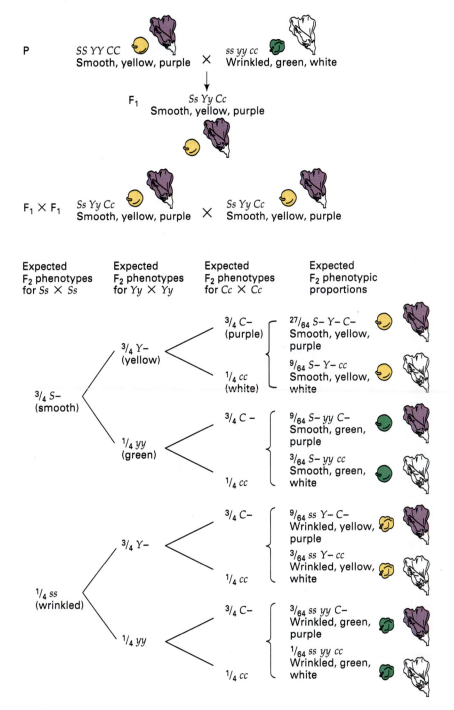

moment in development, and not as a distinct process requiring special analysis. By 1900 conceptions had changed enough that the significance of Mendel's work was more apparent.

At about the turn of the century three researchers working independently on breeding experiments came to the same conclusions as had Mendel. The three men were Carl Correns, Hugo de Vries, and Erich von Tschermak. Correns concentrated mostly on maize (corn) and peas, de Vries worked with a number of different plant species, and von Tschermak studied peas.

The first demonstration that Mendelism applied to animals came in 1902 from the work of William Bateson, who experimented with fowl. Bateson also coined the terms *genetics, zygote,* F_1, F_2, and **allelomorph** (literally, "alternative form," meaning one of an array of different forms of a gene). The last was shortened by others to *allele.* The term *gene* as a replacement for Mendelian factor was introduced by W. L. Johannsen in 1909. In the years after Bateson's work, a number of investigators showed the general applicability of Mendelian principles to all sexually reproducing eukaryotic organisms.

STATISTICAL ANALYSIS OF GENETIC DATA: THE CHI-SQUARE TEST

Data from genetic crosses are quantitative. A geneticist typically uses statistical analysis to interpret a set of data from crossing experiments in order to understand the significance of any deviation between observed results and the results predicted from the hypothesis being tested. That is, the observed phenotypic ratios among progeny rarely match expected or predicted ratios due to chance factors inherent in biological phenomena. The hypothesis is presented as a *null hypothesis* that states that there is no real difference between the observed data and the predicted data. Every data set will show some difference from what was predicted. Thus, appropriate statistical analysis is used to determine whether or not the difference is due to chance and, hence, whether the null hypothesis should be rejected. If the null hypothesis is rejected, a new hypothesis must be developed to explain the data. A relatively simple statistical analysis used to test null hypotheses is called the **chi-square (χ^2) test.** The chi-square test is essentially a *goodness-of-fit test.* In the genetic crosses we have discussed so far, the progeny seemed to fit particular ratios (such as 1:1, 3:1, and 9:3:3:1), and this is where a null hypothesis can be posed and where the chi-square test can tell us if the data are consistent with that hypothesis.

To illustrate the use of the chi-square test, we will analyze theoretical progeny data from a testcross of a smooth-yellow double heterozygote (*Ss Yy*) with a wrinkled-green homozygote (*ss yy*) (see p. 33 and Table 2.2). (Further applications of the chi-square test are given in Chapter 5.) The progeny data are:

154 smooth, yellow
124 smooth, green
144 wrinkled, yellow
146 wrinkled, green

Total   568

We hypothesize that the two genes assort independently and use the chi-square test to test the hypothesis, as shown in Table 2.4.

If the two genes assort independently, then a testcross should result in a 1:1:1:1 ratio of the four phenotypic classes. First, in column 1 we list the four phenotypes expected in the progeny of the cross. Then we list the observed (*o*) numbers for each phenotype, using actual numbers and not percentages or proportions (column 2). Next, we calculate the expected (*e*) number for each phenotypic class, given the total number of progeny (568) and the hypothesis under evaluation (in this case 1:1:1:1; column 3). Thus we list $1/4 \times 568 = 142$, and so on. Now we subtract the expected number (*e*) from the observed number (*o*) for each class to find differences, called the deviation value (*d*). The sum of the *d* values is always zero (column 4).

In column 5 the deviation squared (d^2) is computed by multiplying each deviation value in column 4 by itself. In column 6 the deviation squared is then divided by the expected number (*e*). The chi-square value, χ^2 (item 7 in the table), is the total of the four values in column 6. The more the observed data deviate from the data expected on the basis of the hypothesis being tested, the higher χ^2 will be. In our example χ^2 is 3.43. The general formula is

$$\chi^2 = \Sigma \frac{d^2}{e}, \text{ where } d^2 = (o - e)^2$$

The last value in the table, item 8, is the degrees of freedom (df) for the set of data. The degrees of freedom in a test involving *n* classes are usually equal to $n - 1$. For example, if we have four classes (as in our example), then once three classes have been assigned, there is only one class left; in other words, there are three degrees of freedom.

~ TABLE 2.4

Chi-Square Test Example

(1) PHENOTYPES	(2) OBSERVED NUMBER (*o*)	(3) EXPECTED NUMBER (*e*)	(4) *d* (= *o* − *e*)	(5) d^2	(6) d^2/e
Smooth, yellow	154	142	+12	144	1.01
Smooth, green	124	142	-18	324	2.28
Wrinkled, yellow	144	142	+2	4	0.03
Wrinkled, green	146	142	+4	16	0.11
Total	568	568	0		3.43

(7) $\chi^2 = 3.43$ (8) Degrees of freedom (df) = 3

The χ^2 value and the degrees of freedom are next used to determine the probability (P) that the deviation of the observed values from the expected values is due to chance. The P value for a set of data is obtained from tables of χ^2 values for various degrees of freedom. Table 2.5 presents part of a table of chi-square probabilities. For our example, $\chi^2 = 3.43$ with three degrees of freedom, the P value is between 0.30 and 0.50. This is interpreted to mean that in 30 to 50 out of 100 trials (i.e., 30 to 50 percent of the time) we could expect χ^2 values of this magnitude or greater due to chance with the hypothesis being true. We can reasonably regard this deviation as simply being a sampling, or chance, error. We must be cautious how we use this result, however. That is, a result like this does not tell us that the hypothesis is *correct*; it only shows us that the experimental data provided no statistically compelling argument against it.

As a general rule, if the probability of obtaining the observed χ^2 values is greater than 5 in 100 (5 percent of the time: $P > 0.05$), the deviation between expected and observed is not considered statistically significant and the hypothesis being tested is not thrown out.

Let us consider that, in another chi-square analysis of a different set of data, we obtained $\chi^2 = 15.85$ with three degrees of freedom. By looking up the value in Table 2.5, we see that the P value is less than 0.01 and greater than 0.001 ($0.001 < P < 0.01$). Thus from 0.1 to 1 times out of 100 (0.1 to 1 percent of the time) we could expect χ^2 values of this magnitude or greater due to chance with your hypothesis being true. This P value, being less than 0.05, indicates that the results are not statistically consistent with the 1:1:1:1 hypothesis being tested because of the poor fit.

~ TABLE 2.5

Chi-Square Probabilities

					PROBABILITIES					
df	0.95	0.90	0.70	0.50	0.30	0.20	0.10	0.05	0.01	0.001
1	0.004	0.016	0.15	0.46	1.07	1.64	2.71	3.84	6.64	10.83
2	0.10	0.21	0.71	1.39	2.41	3.22	4.61	5.99	9.21	13.82
3	0.35	0.58	1.42	2.37	3.67	4.64	6.25	7.82	11.35	16.27
4	0.71	1.06	2.20	3.36	4.88	5.99	7.78	9.49	13.28	18.47
5	1.15	1.61	3.00	4.35	6.06	7.29	9.24	11.07	15.09	20.52
6	1.64	2.20	3.83	5.35	7.23	8.56	10.65	12.59	16.81	22.46
7	2.17	2.83	4.67	6.35	8.38	9.80	12.02	14.07	18.48	24.32
8	2.73	3.49	5.53	7.34	9.52	11.03	13.36	15.51	20.09	26.13
9	3.33	4.17	6.39	8.34	10.66	12.24	14.68	16.92	21.67	27.88
10	3.94	4.87	7.27	9.34	11.78	13.44	15.99	18.31	23.21	29.59
11	4.58	5.58	8.15	10.34	12.90	14.63	17.28	19.68	24.73	31.26
12	5.23	6.30	9.03	11.34	14.01	15.81	18.55	21.03	26.22	32.91
13	5.89	7.04	9.93	12.34	15.12	16.99	19.81	22.36	27.69	34.53
14	6.57	7.79	10.82	13.34	16.22	18.15	21.06	23.69	29.14	36.12
15	7.26	8.55	11.72	14.34	17.32	19.31	22.31	25.00	30.58	37.70
20	10.85	12.44	16.27	19.34	22.78	25.04	28.41	31.41	37.57	45.32
25	14.61	16.47	20.87	24.34	28.17	30.68	34.38	37.65	44.31	52.62
30	18.49	20.60	25.51	29.34	33.53	36.25	40.26	43.77	50.89	59.70
50	34.76	37.69	44.31	49.34	54.72	58.16	63.17	67.51	76.15	86.66

← Fail to reject | Reject →
at 0.05 level

Source: From Table IV in *Statistical Tables for Biological, Agricultural, and Medical Research* by Fisher and Yates, 6th ed., 1974. Reprinted by permission of Addison Wesley Longman Ltd.

MENDELIAN GENETICS IN HUMANS

After the rediscovery of Mendel's laws in 1900, geneticists found that the inheritance of genes follows the same principles in all sexually reproducing eukaryotes, including humans. W. Farabee in 1905 was the first to document a genetic trait in humans, brachydactyly (abnormally broad and short fingers: Figure 2.18). By analyzing the trait in human families, Farabee learned that brachydactyly is inherited. The pattern of transmission of the abnormality over several generations led to the conclusion that the trait is a simple dominant trait. In this section we will introduce some of the methods used to determine the mechanism of hereditary transmission in humans, and we will learn about some inherited human traits.

Pedigree Analysis

The study of human genetics is complicated because controlled matings of humans cannot be made. This means that geneticists must study crosses that happen by *chance* rather than by design. The inheritance patterns of human traits are usually established by examining the way the trait occurs in the family trees of individuals who clearly exhibit the trait. The family tree investigation is called **pedigree analysis**, and it involves the careful compilation of phenotypic records of the family over several generations. The "affected" individual through whom the pedigree is discovered is called the **proband** (= **propositus** if male, **proposita** if female). The more information there is, the more likely the investigator will be able to make some conclusions about the mechanism of inheritance of the gene (or genes) responsible for the trait being studied.

One of the modern applications of pedigree analysis is **genetic counseling**. A geneticist makes predictions about the probabilities of particular traits (deleterious or not) occurring among a couple's children. In most cases the couple comes to the counselor because

~ FIGURE 2.18

Photograph of (a) normal hands alongside (b) hands with brachydactyly.

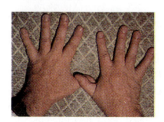
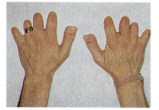

a)   b)

~ FIGURE 2.19

Symbols used in human pedigree analysis.

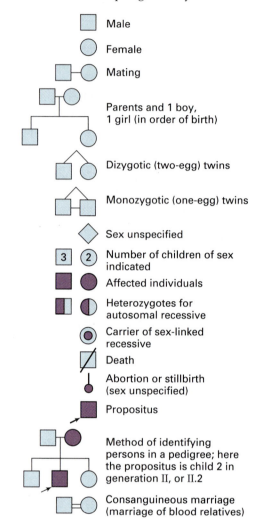

there is some possibility of the existence of an undesirable heritable trait in one or both families. As we might expect, pedigree analysis is most useful for traits that are the result of a single gene difference.

Pedigree analysis has its own set of symbols; Figure 2.19 summarizes the basic symbols that will be used here and elsewhere in the text. (The terms *autosomal* and *sex-linked* used in the figure are explained in Chapter 3; they are included here for completeness.) Figure 2.20 presents a hypothetical pedigree to show how the symbols are assigned to the family tree.

The trait presented in Figure 2.20 is determined by a recessive mutant allele *a*. (Note that recessive mutant alleles may be rare or common in a population.) Generations are numbered with Roman numerals, while individuals are numbered with Arabic numerals; this makes it easy to refer to particular people in the pedigree. The trait in this pedigree results from homozygosity for the allele, brought about by cousins mating. Since cousins share a fair proportion

~ **FIGURE 2.20**

A human pedigree, illustrating the use of pedigree symbols.

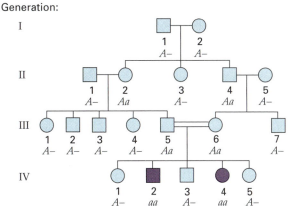

Generation:

of their genes, a number of alleles will become homozygous in their offspring; in this case one mutant recessive allele became homozygous and resulted in an identifiable genetic trait.

Gene symbols are included in the pedigree to show the deductive reasoning possible with such pedigrees; normally, such symbols would not be present and the researcher would have to analyze the pedigrees without that information. For example, the following reasoning could take place: The trait appears first in generation IV. Since neither parent (the two cousins) had the trait, while two children were produced with the trait (IV-2 and IV-4), the simplest hypothesis is that the trait is caused by a recessive mutant allele. Thus, IV-2 and IV-4 would both have the genotype *aa*, and their parents (III-5 and III-6) both must have the genotype *Aa*. All other individuals who did not have the trait must have at least one *A* allele; that is, they must be *A–* (i.e., *AA* or *Aa*). Since III-5 and III-6 are both heterozygotes, then at least one of each of their parents must have carried an *a* allele. Further, since the trait appeared only after cousins had children, the simplest assumption is that the *a* allele was inherited from individuals with bloodlines shared by III-5 and III-6. This means that II-2 and II-4 are likely both *Aa* and that one of I-1 and I-2 is *Aa* (perhaps both, unless the allele is rare).

Examples of Human Genetic Traits

RECESSIVE TRAITS. A large number of human traits are known to be caused by homozygosity for mutant alleles that are recessive to the normal allele. Such recessive mutant alleles produce mutant phenotypes because of a *loss of function* of the gene product resulting from the mutation involved.

Many serious abnormalities or diseases result from homozygosity for recessive mutant alleles. Two individuals expressing the recessive trait of albinism

(deficient pigmentation) are shown in Figure 2.21a and a pedigree for this trait is shown in Figure 2.21b. Individuals with albinism do not produce the melanin pigment that protects the skin from harmful ultraviolet radiation. As a consequence, persons with albinism have considerable skin and eye sensitivity to sunlight. Frequencies of recessive mutant alleles are usually higher than frequencies of dominant mutant alleles because heterozygotes for the recessive mutant allele are not at a significant selective disadvantage. Nonetheless, individuals homozygous for recessive mutant alleles are usually rare. In the United States approximately 1 in 17,000 of the white population and 1 in 28,000 of the African American population have albinism. Among the Irish, about 1 in 10,000 have albinism.

The following lists some general characteristics of recessive inheritance for a relatively rare trait (see Figure 2.21):

1. Most affected individuals have two normal parents, both of whom are heterozygous. The trait appears in the F_1 since a quarter of the progeny are expected to be homozygous for the recessive allele.

~ **FIGURE 2.21**

(a) Persons with albinism: the blues musicians Johnny (left) and Edgar Winter (right). (b) A pedigree showing the transmission of the autosomal recessive trait of albinism.

a)

b) Pedigree

Generation:

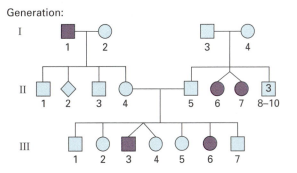

If the trait is rare or relatively rare, an individual expressing the trait is likely to mate with a homozygous normal individual; thus the next generation is represented by heterozygotes who do not express the trait. That is, recessive traits often "skip" generations. In the pedigree, for example, II-6 and II-7 must both be *aa* and this means both parents (I-3 and I-4) must be *Aa* heterozygotes. I-1 is also *aa* and so II-4 must be *Aa*. Since II-4 and II-5 produce some *aa* children, II-5 also must be *Aa*.

2. Matings between two normal heterozygotes should produce an approximately 3:1 ratio of normal progeny to progeny exhibiting the recessive trait. However, in the analysis of human populations (families), it is difficult to obtain enough numbers to make the data statistically significant, especially if a biochemical test is necessary to confirm the presence of the trait because, in such cases, only the living members of a family can be surveyed. For recessive genes that have less deleterious effects, the allele can reach significant frequencies in the population. An example of such a recessive trait is attached earlobes. There are significant numbers of heterozygotes and homozygous recessives for this trait in the population. As a result, there is a strong possibility of *Aa × aa* matings, and half the progeny will have the trait.

3. When both parents are affected, all their progeny will usually exhibit the trait.

DOMINANT TRAITS.   There are many known dominant human traits. Figure 2.22a illustrates one such trait called woolly hair, in which an individual's hair is very tightly kinked, is very brittle, and breaks

~ **FIGURE 2.22**

(a) Members of a Norwegian family, some exhibiting the trait of woolly hair; (b) Part of a pedigree showing the transmission of the autosomal dominant trait of woolly hair.

a)

b) Generation:

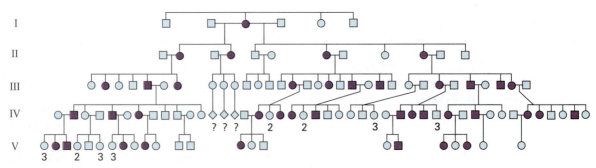

before it can grow very long. The best examples of pedigrees for this trait come from Norwegian families; one of these pedigrees is presented in Figure 2.22b. Since it is a fairly rare trait and since not all children of an affected parent show the trait, it can be assumed that most woolly-haired individuals are heterozygous for the dominant allele involved.

Dominant mutant alleles are expressed in a heterozygote when they are in combination with what is usually called the **wild-type allele**—the allele that predominates (is present in the highest frequency) in the population found in the "wild." Because many dominant mutant alleles that give rise to recognizable traits are rare, it is extremely unusual to find individuals homozygous for the dominant allele. Thus an affected person in a pedigree is likely to be a heterozygote, and most pairings that involve the mutant allele are between a heterozygote and a homozygous recessive (wild type). Most dominant mutant genes that are clinically significant (that is, they cause medical problems) fall into this category.

The following are some general characteristics of a dominant trait (refer to Figure 2.22b):

1. Every affected person in the pedigree must have at least one affected parent.
2. The trait usually will not skip generations.
3. An affected heterozygous individual will, on average, transmit the mutant gene to half of his or her progeny. Suppose the dominant mutant allele is designated A, and its wild-type allele is a. Then most crosses will be $Aa \times aa$. From basic Mendelian principles half the progeny will be aa (wild type) and the other half will be Aa and show the trait.

$\mathcal{K}$EYNOTE

Mendelian principles apply to humans as well as to peas and all other eukaryotes. The study of the human inheritance of genetic traits is complicated by the fact that no controlled crosses can be done. Instead, human geneticists analyze genetic traits by pedigree analysis, that is, by examining the occurrences of the trait in family trees of individuals who clearly exhibit the trait. Many recessively inherited and dominantly inherited genetic traits have been identified as a result of pedigree analysis.

SUMMARY

In this chapter we discussed fundamental principles of gene function and gene segregation. Genes in this chapter are defined as DNA segments that control the biological characteristics that are transmitted from one generation to another; that is, the hereditary traits. An organism's genetic constitution is called its genotype, while the physical manifestation of a genetic trait is called the phenotype. An organism's genes only give the potential for the development of that organism's characteristics. That potential is influenced during development by interactions with other genes as well as with the environment. Thus, individuals with the same genotype can have different phenotypes, and individuals with the same phenotype may have different genotypes.

The first person to obtain some understanding of the principles of heredity—that is, the inheritance of certain traits—was Gregor Mendel. From his breeding experiments with garden peas, Mendel proposed two basic principles of genetics. In modern terms, the principle of segregation states that the two members of a single gene pair (the alleles) segregate from each other in the formation of gametes. For each gene with two alleles, half of the gametes carry one allele, and the other half carry the other allele. The principle of independent assortment, proposed on the basis of experiments involving more than one gene, states that genes for different traits behave independently in the production of gametes. Both principles are recognized by characteristic phenotypic ratios—called Mendelian ratios—in particular crosses. For the principle of segregation, in a monohybrid cross between two true-breeding parents, one exhibiting a dominant phenotype and the other a recessive phenotype, the F_2 phenotypic ratio will be 3:1 for the dominant:recessive phenotypes. For the principle of independent assortment, in a dihybrid cross, the F_2 phenotypic ratio will be 9:3:3:1 for the four phenotypic classes.

The gene segregation patterns can be studied more definitively by determining the genotypes for each phenotypic class. This is done using a testcross, in which an individual of unknown genotype is crossed with a homozygous recessive to determine the genotype of the unknown individual. For example, in a monohybrid cross resulting in an F_2 3:1 phenotypic ratio, the dominant class can be shown to consist of 1 homozygous dominant : 2 heterozygotes by using a testcross.

After the rediscovery of Mendel's laws in 1900, geneticists found that Mendelian principles of gene segregation apply to all other eukaryotes, including humans. The study of the inheritance of genetic traits in humans is complicated by the fact that no controlled crosses can be done. Instead, human geneticists analyze genetic traits by pedigree analysis, that is, by examining the occurrences of the trait in family trees of individuals who clearly exhibit the trait. Many recessively inherited and dominantly inherited genetic traits have been identified as a result of pedigree analysis.

ANALYTICAL APPROACHES FOR SOLVING GENETICS PROBLEMS

The most practical way to reinforce Mendelian principles is to solve genetics problems. In this and all following chapters we will discuss how to approach genetics problems by presenting examples of such problems and discussing their answers. These problems present familiar and unfamiliar examples and pose questions designed to get problem solvers to think analytically.

Q2.1 A purple-flowered pea plant is crossed with a white-flowered pea plant. All the F_1 plants produced purple flowers. When the F_1 plants are allowed to self-pollinate, 401 of the F_2s have purple flowers and 131 have white flowers. What are the genotypes of the parental and F_1 generation plants?

A2.1 The ratio of plant phenotypes in the F_2 is very close to the 3:1 ratio expected of a monohybrid cross. More specifically, this ratio is expected to result from an $F_1 \times F_1$ cross in which both are identically heterozygous for a specific gene pair. In addition, since the two parents differed in phenotype and only one phenotypic class appeared in the F_1, it is likely that both parental plants were true breeding. Further, since the F_1 phenotype exactly resembled one of the parental phenotypes, we can say that purple is dominant to white flowers. Assigning the symbol P to the gene that determines purpleness of flowers and the symbol p to the alternative form of the gene that determines whiteness, we can write the genotypes:

P generation: PP, for the purple-flowered plant
                pp, for the white-flowered plant
F_1 generation: Pp, which, because of dominance, is purple-flowered

We could further deduce that the F_2 plants have an approximately 1:2:1 ratio of $PP:Pp:pp$ by performing testcrosses.

Q2.2 Consider three gene pairs Aa, Bb, and Cc, each of which affects a different character. In each case the uppercase letter signifies the dominant allele and the lowercase letter the recessive allele. These three gene pairs assort independently of each other. Calculate the probability of obtaining:

a. an $Aa\ BB\ Cc$ zygote from a cross of individuals that are $Aa\ Bb\ Cc \times Aa\ Bb\ Cc$;

b. an $Aa\ BB\ cc$ zygote from a cross of individuals that are $aa\ BB\ cc \times AA\ bb\ CC$;

c. an $A\ B\ C$ phenotype from a cross of individuals that are $Aa\ Bb\ CC \times Aa\ Bb\ cc$;

d. an $a\ b\ c$ phenotype from a cross of individuals that are $Aa\ Bb\ Cc \times aa\ Bb\ cc$.

A2.2 Again, we must break down the question into simple parts in order to apply basic Mendelian principles. The key is that the genes assort independently, so we must multiply the probabilities of the individual occurrences to obtain the answers.

a. First, we must consider the Aa gene pair. The cross is $Aa \times Aa$, so the probability of the zygote being Aa is 2/4 since the expected distribution of genotypes is 1 AA : 2 Aa : 1 aa, as we have discussed. Then the probability of BB from $Bb \times Bb$ is 1/4, and that of Cc from $Cc \times Cc$ is 2/4, following the same sort of logic. Using the product rule (see Box 2.2), the probability of an $Aa\ BB\ Cc$ zygote is $1/2 \times 1/4 \times 1/2 = 1/16$.

b. Similar logic is needed here, although we must be sure of the genotypes of the parental types since they differ from one gene pair to another. For the Aa pair the probability of getting Aa from $AA \times aa$ has to be 1. Next, the probability of getting BB from $BB \times bb$ is 0, so on these grounds alone we cannot get the zygote asked for from the cross given.

c. This question and the next ask for the probability of getting a particular phenotype, so we must start thinking of dominance. Again, we break up the question and consider each character pair in turn. The probability of an A phenotype from $Aa \times Aa$ is 3/4, from basic Mendelian principles. Similarly, the probability of a B phenotype from $Bb \times Bb$ is 3/4. Lastly, the probability of a C phenotype from $CC \times cc$ is 1. Overall, the probability of an $A\ B\ C$ phenotype is $3/4 \times 3/4 \times 1 = 9/16$.

d. An $a\ b\ c$ phenotype from $Aa\ Bb\ Cc \times aa\ Bb\ cc$ is $1/2 \times 1/4 \times 1/2 = 1/16$.

Q2.3 In chickens the white plumage of the leghorn breed is dominant over colored plumage, feathered shanks are dominant over clean shanks, and pea comb is dominant over single comb. Each of the gene pairs segregates independently. If a homozygous white, feathered, pea-combed chicken is crossed with a homozygous colored, clean, single-comb chicken, and the F_1s are allowed to interbreed, what proportion of the birds in the F_2 will produce only white, feathered, pea-combed progeny if mated to colored, clean-shanked, single-combed birds?

A2.3 This example is typical of a question that presents the unfamiliar in an attempt to get at the familiar. The best approach to such questions is to reduce them to their simple parts and, wherever possible, to assign gene symbols for each character. We are told which character is dominant for each of the three gene pairs, so we can use W for white and w for colored, F for feathered and f for clean shanks, and P for pea comb and p for single comb. The cross involves true-breeding strains and can be written as follows:

P generation: $WW\,FF\,PP \times ww\,ff\,pp$
F_1 generation: $Ww\,Ff\,Pp$

Now the question asks the proportion of the birds in the F_2 that will produce only white, feathered, pea-combed progeny if mated to colored, clean-shanked, single-combed birds. The latter are homozygous recessive for all three genes: that is, $ww\,ff\,pp$, as in the parental generation. For the result asked for, the F_2 birds must be white, feathered, and pea-combed, and they must be homozygous for the dominant alleles of the respective genes in order to produce only progeny with the dominant phenotype. What we are seeking, then, is the proportion of the F_2 chickens that are $WW\,FF\,PP$ in genotype. We know that each gene pair segregates independently, so the answer can be calculated by using simple probability rules. We consider each gene pair in turn. For the white/colored case the $F_1 \times F_1$ is $Ww \times Ww$, and we know from Mendelian principles that the relative proportion of F_2 genotypes will be $1\ WW : 2\ Ww : 1\ ww$. Therefore the proportion of the F_2s that will be WW is $1/4$. The same relationship holds for the other two pairs of genes. Since the segregation of the three gene pairs is independent, we must multiply the probabilities of each occurrence to calculate the probability for $WW\,FF\,PP$ individuals. The answer is $1/4 \times 1/4 \times 1/4 = 1/64$.

QUESTIONS AND PROBLEMS

* Solutions to Questions and Problems marked with an asterisk are found at the end of this text, starting on page S-1. Solutions to all problems are found in the Student Study Guide and Solutions Manual that accompanies this text.

*2.1 In tomatoes, red fruit color is dominant to yellow. Suppose a tomato plant homozygous for red is crossed with one homozygous for yellow. Determine the appearance of (a) the F_1; (b) the F_2; (c) the offspring of a cross of the F_1 back to the red parent; (d) the offspring of a cross of the F_1 back to the yellow parent.

2.2 In maize, a dominant allele A is necessary for seed color as opposed to colorless (a). Another gene has a recessive allele wx that results in waxy starch as opposed to normal starch (Wx). The two genes segregate independently. Give phenotypes and relative frequencies for offspring resulting when a plant of genetic constitution $Aa\ WxWx$ is testcrossed.

*2.3 F_2 plants segregate 3/4 colored : 1/4 colorless. If a colored plant is picked at random and selfed, what is the probability that more than one type will segregate among a large number of its progeny?

*2.4 In guinea pigs rough coat (R) is dominant over smooth coat (r). A rough-coated guinea pig is bred to a smooth one, giving eight rough and seven smooth progeny in the F_1.

a. What are the genotypes of the parents and their offspring?

b. If one of the rough F_1 animals is mated to its rough parent, what progeny would you expect?

2.5 In cattle the polled (hornless) condition (P) is dominant over the horned (p) phenotype. A particular polled bull is bred to three cows. Cow A, which is horned, produces a horned calf; a polled cow B produces a horned calf; and horned cow C produces a polled calf. What are the genotypes of the bull and the three cows, and what phenotypic ratios do you expect in the offspring of these three matings?

*2.6 In the Jimsonweed, purple flowers are dominant to white. When a particular purple-flowered Jimsonweed is self-fertilized, there are 28 purple-flowered and 10 white-flowered progeny. What proportion of the purple-flowered progeny will breed true?

*2.7 Two black female mice are crossed with the same brown male. In a number of litters female X produced 9 blacks and 7 browns and female Y produced 14 blacks. What is the mechanism of inheritance of black and brown coat color in mice? What are the genotypes of the parents?

2.8 Bean plants may differ in their symptoms when infected with a virus. Some show local lesions that do not seriously harm the plant. Other plants show general systemic infection. The following genetic analysis was made:

P   local lesions × systemic lesions
F_1 all local lesions
F_2 785 local lesions : 269 systemic lesions

What is probably the genetic basis of this difference in beans? Assign gene symbols to all the genotypes occurring in the above experiment. Design a testcross to verify your assumptions.

2.9   A normal *Drosophila* has both brown and scarlet pigment granules in the eyes, which appear red as a result. Brown (*bw*) is a recessive allele on chromosome 2 which, in homozygous condition, results in the absent of scarlet granules (so that the eyes are brown). Scarlet (*st*) is a recessive on chromosome 3 which, when homozygous, results in scarlet eyes due to the absence of brown pigment. Any fly homozygous for recessive brown and recessive scarlet alleles produces no eye pigment and has white eyes. The following results are obtained from crosses:

P   brown-eyed fly × scarlet-eyed fly
F_1 red eyes (both brown and scarlet pigment present)
F_2 9/16 red : 3/16 scarlet : 3/16 brown : 1/16 white

a.   Assign genotypes to the P and F_1 generations.
b.   Design a testcross to verify the F_1 genotype, and predict the results.

*2.10   Grey seed color (*G*) in garden peas is dominant to white seed color (*g*). In the following crosses, the indicated parents with known phenotypes but unknown genotypes produced the listed progeny. Give the possible genotype(s) of each female parent based on the segregation data.

PARENTS	PROGENY		FEMALE PARENT
FEMALE × MALE	GREY	WHITE	GENOTYPE
grey × white	81	82	?
grey × grey	118	39	?
grey × white	74	0	?
grey × grey	90	0	?

*2.11   Fur color in babbits, a furry little animal and popular pet, is determined by a pair of alleles, *B* and *b*. *BB* and *Bb* babbits are black, and *bb* babbits are white.

A farmer wants to breed babbits for sale. True-breeding white (*bb*) female babbits breed poorly. The farmer purchases a pair of black babbits, and these mate and produce six black and two white offspring. The farmer immediately sells his white babbits, then he comes to consult you for a breeding strategy to produce more white babbits.

a.   If he performed random crosses between pairs of F_1 babbits, what proportion of the F_2 progeny would be white?
b.   If he crossed an F_1 male to the parental female, what is the probability that this cross will produce white progeny?
c.   What would be the farmer's best strategy to maximize the production of white babbits?

2.12   In Jimsonweed, purple flower (*P*) is dominant to white (*p*), and spiny pods (*S*) are dominant to smooth (*s*). In a cross between a Jimsonweed homozygous for white flowers and spiny pods and one homozygous for purple flowers and smooth pods, determine the phenotype of (a) the F_1; (b) the F_2; (c) the progeny of a cross of the F_1 back to the white, spiny parent; (d) the progeny of a cross of the F_1 back to the purple, smooth parent.

*2.13   Cleopatra is a normally a very refined cat. When she finds even a small amount of catnip, however, she purrs madly, rolls around in the catnip, becomes exceedingly playful, and appears intoxicated. Cleopatra and Antony, who walks past catnip with an air of indifference, have produced five kittens who respond to catnip just like Cleopatra. When the kittens mature, two of them mate and produce four kittens that respond to catnip and one that does not. When another of Cleopatra's daughters mates with Augustus (a non-relative), who behaves just like Antony, three catnip-sensitive and two catnip-insensitive kittens are produced. Propose an hypothesis for the inheritance of catnip sensitivity that explains these data.

2.14   Using the information in Problem 2.12, what progeny would you expect from the following Jimsonweed crosses? You are encouraged to use the branch diagram approach.

a.   *PP ss* × *pp SS*
b.   *Pp SS* × *pp ss*
c.   *Pp Ss* × *Pp SS*
d.   *Pp Ss* × *Pp Ss*
e.   *Pp Ss* × *Pp ss*
f.   *Pp Ss* × *pp ss*

*2.15   In summer squash white fruit (*W*) is dominant over yellow (*w*), and disk-shaped fruit (*D*) is dominant over sphere-shaped fruit (*d*). In the following problems the appearances of the parents and their progeny are given. Determine the genotypes of the parents in each case.

a.   White, disk × yellow, sphere gives 1/2 white, disk and 1/2 white, sphere.
b.   White, sphere × white, sphere gives 3/4 white, sphere and 1/4 yellow, sphere.
c.   Yellow, disk × white, sphere gives all white, disk progeny.
d.   White, disk × yellow, sphere gives 1/4 white, disk; 1/4 white, sphere; 1/4 yellow, disk; and 1/4 yellow, sphere.
e.   White, disk × white, sphere gives 3/8 white, disk; 3/8 white, sphere; 1/8 yellow, disk; and 1/8 yellow, sphere.

*2.16   Genes *a*, *b*, and *c* assort independently and are recessive to their respective alleles *A*, *B*, and *C*. Two triply heterozygous (*Aa Bb Cc*) individuals are crossed.

a.   What is the probability that a given offspring will be phenotypically *ABC*, that is, exhibit all three dominant traits?

b. What is the probability that a given offspring will be genotypically homozygous for all three dominant alleles?

2.17 In garden peas, tall stem *(T)* is dominant over short stem *(t)*, green pods *(G)* are dominant over yellow pods *(g)*, and smooth seeds *(S)* are dominant over wrinkled seeds *(s)*. Suppose a homozygous short, green, wrinkled pea plant is crossed with a homozygous tall, yellow, smooth one.

a. What will be the appearance of the F_1?

b. What will be the appearance of the F_2?

c. What will be the appearance of the offspring of a cross of the F_1 back to its short, green, wrinkled parent?

d. What will be the appearance of the offspring of a cross of the F_1 back to its tall, yellow, smooth parent?

2.18 *C/c*, *O/o*, and *I/i* are three independently segregating pairs of alleles in chickens. *C* and *O* are dominant alleles, both of which are necessary for pigmentation. *I* is a dominant inhibitor of pigmentation. Individuals of genotype *cc*, or *oo*, or *Ii*, or *II* are white, regardless of what other genes they possess.

Assume that White Leghorns are *CC OO II*, White Wyandottes are *cc OO ii*, and White Silkies are *CC oo ii*. What types of offspring (white or pigmented) are possible, and what is the probability of each, from the following crosses?

a. White Silkie × White Wyandotte

b. White Leghorn × White Wyandotte

c. (Wyandotte—Silkie F_1) × White Silkie

2.19 Two homozygous strains of corn are hybridized. They are distinguished by six different pairs of genes, all of which assort independently and produce an independent phenotypic effect. The F_1 hybrid is selfed to give an F_2.

a. What is the number of possible genotypes in the F_2?

b. How many of these genotypes will be homozygous at all six gene loci?

c. If all gene pairs act in a dominant-recessive fashion, what proportion of the F_2 will be homozygous for all dominants?

d. What proportion of the F_2 will show all dominant phenotypes?

*2.20 The coat color of mice is controlled by several genes. The agouti pattern, characterized by a yellow band of pigment near the tip of the hairs, is produced by the dominant allele *A*; homozygous *aa* mice do not have the band and are nonagouti. The dominant allele *B* determines black hairs, and the recessive allele *b* determines brown. Homozygous $c^h c^h$ individuals allow pigments to be deposited only at the extremities (e.g., feet, nose, and ears) in a pattern called Himalayan. The genotype *C–* allows pigment to be distributed over the entire body.

a. If a true-breeding black mouse is crossed with a true-breeding brown, agouti, Himalayan mouse, what will be the phenotypes of the F_1 and F_2?

b. What proportion of the black agouti F_2 will be of genotype *Aa BB* Cc^h?

c. What proportion of the Himalayan mice in the F_2 are expected to show brown pigment?

d. What proportion of all agoutis in the F_2 are expected to show black pigment?

2.21 In cocker spaniels, solid coat color is dominant over spotted coat. Suppose a true-breeding, solid-colored dog is crossed with a spotted dog, and the F_1 dogs are interbred.

a. What is the probability that the first puppy born will have a spotted coat?

b. What is the probability that if four puppies are born, all of them will have a solid coat?

2.22 In the F_2 of his cross of red-flowered × white-flowered *Pisum*, Mendel obtained 705 plants with red flowers and 224 with white.

a. Is this result consistent with his hypothesis of factor segregation, from which a 3:1 ratio would be predicted?

b. In how many similar experiments would a deviation as great as or greater than this one be expected? (Calculate χ^2 and obtain the approximate value of P from the table.)

2.23 In tomatoes, cut leaf and potato leaf are alternative characters with cut *(C)* dominant to potato *(c)*. Purple stem and green stem are another pair of alternative characters with purple *(P)* dominant to green *(p)*. A true-breeding cut, green tomato plant is crossed with a true-breeding potato, purple plant and the F_1 plants are allowed to interbreed. The 320 F_2 plants were phenotypically 189 cut, purple; 67 cut, green; 50 potato, purple; and 14 potato, green. Propose an hypothesis to explain the data, and use the χ^2 test to test the hypothesis.

*2.24 The simple case of just two mating types (male and female) is by no means the only sexual system known. The ciliated protozoan *Paramecium bursaria* has a system of four mating types, controlled by two genes (*A* and *B*). Each gene has a dominant and a recessive allele.

The four mating types are expressed under the following scheme:

GENOTYPE	MATING TYPE
AA BB	*A*
Aa BB	*A*
AA Bb	*A*
Aa Bb	*A*
AA bb	*D*
Aa bb	*D*
aa BB	*B*
aa Bb	*B*
aa bb	*C*

It is clear, therefore, that some of the mating types result from more than one possible genotype. We have four strains of known mating type — "A," "B," "C," and "D" — but unknown genotype. The following crosses were made with the indicated results:

	MATING TYPE OF PROGENY			
CROSS	A	B	C	D
"A" × "B"	24	21	14	18
"A" × "C"	56	76	55	41
"A" × "D"	44	11	19	33
"B" × "C"	0	40	38	0
"B" × "D"	6	8	14	10
"C" × "D"	0	0	45	45

Assign genotypes to "A," "B," "C," and "D."

*2.25 In bees, males (drones) develop from unfertilized eggs and are haploid. Females (workers and queens) are diploid and come from fertilized eggs. *W* (black eyes) is dominant over *w* (white eyes). Workers of genotype *RR* or *Rr* use wax to seal crevices in the hive; *rr* workers use resin instead. A *Ww Rr* queen founds a colony after being fertilized by a black-eyed drone bearing the *r* allele.

a. What will be the appearance and behavior of workers in the new hive, with their relative frequencies?

b. Give the genotypes of male offspring, with relative frequencies.

c. Fertilization normally takes place in the air during a "nuptial flight," and any bee unable to fly would effectively be rendered sterile. Suppose a recessive mutation, *c*, occurs spontaneously in a sperm that fertilizes a normal egg, and that the effect of the mutant gene is to cripple the wings of any adult not bearing the normal allele *C*. The fertilized egg develops into a normal queen named Madonna. What is the probability that wingless males will be *found in a hive* founded two generations later by one of Madonna's granddaughters?

d. By one of Madonna's great-great-grandaughters?

*2.26 Consider the pedigree below. The allele responsible for the trait (*a*) is recessive to the normal allele (*A*).

a. What is the genotype of the mother?

b. What is the genotype of the father?

c. What are the genotypes of the children?

d. Given the mechanism of inheritance involved, does the ratio of children with the trait to children without the trait match what would be expected?

Generation

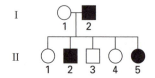

2.27 For pedigrees A and B below, indicate whether the trait involved in each case could be (a) recessive or (b) dominant. Explain your answer.

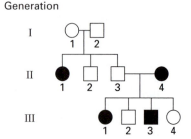

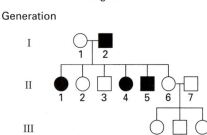

2.28 After a few years of marriage, a woman comes to believe that among all of the reasonable relatives in her and her husband's families, her husband, her mother-in-law and her father have so many similarities in their unreasonableness that they must share a mutation. A friend taking a course in genetics assures her that it is unlikely that this trait has a genetic basis and, even if it did, all of her children would be reasonable. Diagram and analyze the relevant pedigree to evaluate whether this advice is accurate.

*2.29 Gaucher's disease is caused by a chronic enzyme deficiency that is more common among Ashkenazi Jews than in the general population. A Jewish man has a sister afflicted with the disease. His parents, grandparents, and three siblings are not affected. Discussions with relatives in his wife's family reveal that the disease is not likely to be present in his wife's family, although some relatives recall that the brother of his wife's paternal grandmother suffered from a very similar disease. Diagram and analyze the relevant pedigree to determine (a) the genetic basis for inheriting this trait; (b) the highest probability that if this couple has a child, the child will be affected (i.e., what is the chance of the "worst case" scenario occurring?).

3 CHROMOSOMAL BASIS OF INHERITANCE, SEX DETERMINATION, AND SEX LINKAGE

MITOSIS AND MEIOSIS
    Chromosome Complement of Eukaryotes
    Asexual and Sexual Reproduction
    Mitosis
    Genetic Significance of Mitosis
    Meiosis
    Genetic Significance of Meiosis
    Locations of Meiosis in the Life Cycle

CHROMOSOME THEORY OF INHERITANCE
    Sex Chromosomes
    Sex Linkage
    Nondisjunction of X Chromosomes

SEX DETERMINATION
    Genotypic Sex Determination Systems
    Environmental Sex Determination Systems

ANALYSIS OF SEX-LINKED TRAITS IN HUMANS
    X-Linked Recessive Inheritance
    X-Linked Dominant Inheritance
    Y-Linked Inheritance

PRINCIPAL POINTS

~ Diploid eukaryotic cells have two haploid sets of chromosomes, one set coming from each parent. The members of a pair of chromosomes are called homologous chromosomes. Haploid eukaryotic cells have only one set of chromosomes.

~ Mitosis is the process of nuclear division in eukaryotic cells represented by M in the cell cycle (i.e., G_1, S, G_2 and M). Mitosis results in the production of daughter nuclei that contain identical chromosome numbers and are genetically identical to one another and to the parent nucleus from which they arose. Prior to mitosis, the chromosomes replicate, producing duplicates of each. Mitosis usually is followed by cytokinesis, the actual division of the cell into two genetically identical progeny cells.

~ Meiosis occurs in all sexually reproducing eukaryotes. It is a process in which a specialized diploid cell (or cell nucleus) with two sets of chromosomes is transformed through one round of DNA replication and two rounds of nuclear division into four haploid cells (or four nuclei), each with one set of chromosomes. In the first of two divisions, pairing (synapsis) of homologous chromosomes occurs. The meiotic process, in combination with fertilization, results in the conservation of the number of chromosomes from generation to generation.

~ Meiosis generates genetic variability through the processes by which maternal and paternal chromosomes are (a) reassorted in progeny nuclei and (b) by crossing-over between members of a homologous pair of chromosomes.

~ The chromosome theory of inheritance states that the chromosomes are the carriers of the genes.

~ A sex chromosome is a chromosome in eukaryotic organisms that is represented differently in the two sexes. In many organisms with sex chromosomes, the female has two X chromosomes (she is XX) while the male has one X and one Y chromosome (he is XY).

~ Sex linkage is the physical association of genes with the sex chromosomes of eukaryotes. Such genes are referred to as sex-linked genes.

~ The correlation between gene segregation patterns and the patterns of chromosome behavior in meiosis supports the chromosome theory of inheritance.

~ In many eukaryotic organisms, sex determination is related to the sex chromosomes. In humans and other mammals, for example, the presence of a Y chromosome specifies maleness, while its absence results in femaleness. Several other sex-determination mechanisms are known in eukaryotes.

~ In humans, the gene responsible for a trait can be inherited in one of five main ways: autosomal recessive, autosomal dominant, X-linked recessive, X-linked dominant, or Y-linked.

Soon after the beginning of the twentieth century, scientists realized that Mendelian laws applied to a large variety of eukaryotic organisms and that the principles of Mendelian analysis could be used to predict the outcome of crosses in these organisms. On Mendel's foundation early geneticists began to build genetic hypotheses that could be tested by appropriate crosses and began to investigate the nature of Mendelian factors. We now know that Mendelian factors are genes and that genes are located on chromosomes; this is called the **chromosome theory of inheritance.** The formulation of the chromosome theory of inheritance was very important to the development of genetics. In particular, it allowed researchers to relate the segregation patterns of genes in crosses with the behavior of physical entities that are visible under the microscope.

In this chapter we focus on the behavior of genes and chromosomes. We start by learning about the transmission of chromosomes from cell division to cell division, and from generation to generation by the processes of mitosis and meiosis, respectively. We then consider the evidence for the chromosomal theory of inheritance. In so doing we will learn about the segregation of genes located on the sex chromosomes. Next, we will learn about various mechanisms of sex determination, and lastly we will discuss sex-linked traits in humans.

MITOSIS AND MEIOSIS

The chromosome theory of inheritance was developed as a result of the efforts of cytologists, who examined the behavior of chromosomes, and geneticists, who examined the behavior of genes. Cytologists established how chromosomes behave during the two types of nuclear division in eukaryotes, **mitosis** and **meiosis**, which are the subjects of this section.

Chromosome Complement of Eukaryotes

The genetic material of eukaryotes is distributed among multiple chromosomes; the number of chromosomes, with rare possible exceptions, is characteristic of the species. Many eukaryotes have two copies of each type of chromosome in their nuclei, so their chromosome complement is said to be **diploid**, or 2N. Diploid eukaryotes are produced by the fusion of two **gametes**, one from the female parent and one from the male parent. The fusion produces a diploid **zygote**, which then undergoes embryological development. Each gamete has only one set of chromosomes and is said to be **haploid** (N). The complete complement of genetic information in a haploid chromosome set is called the **genome**. Two examples of diploid organisms are humans, with 46 chromosomes (23 pairs), and *Drosophila melanogaster*, with 8 chromosomes (4 pairs). By contrast, the bread mold *Neurospora crassa*, with 7 chromosomes, is haploid.

In diploids, the members of a chromosome pair that contain the same genes and that pair at meiosis are called **homologous chromosomes;** each member of a pair is called a **homolog.** Homologous chromosomes are usually identical with respect to their visible structure; one chromosome is inherited from each parent. Chromosomes that contain different genes and that do not pair during meiosis are called **nonhomologous chromosomes.** Figure 3.1 illustrates the chromosomal organization of haploid and diploid organisms.

In animals and in some plants there are differences in the chromosome complement of male and female cells. One sex has a matched pair of **sex chromosomes**, chromosomes related to the sex of the organism; the other sex has an unmatched pair or a single sex chromosome. For example, human females have two X chromosomes (XX), while human males have one X and one Y (XY). Chromosomes other than sex chromosomes are called **autosomes**.

$\mathcal{K}$EYNOTE

Diploid eukaryotic cells have two haploid sets of chromosomes, one set coming from each parent. The members of a pair of chromosomes, one from each parent, are called homologous chromosomes. Haploid eukaryotic cells have only one set of chromosomes.

~ **FIGURE 3.1**

Chromosomal organization of haploid and diploid organisms.

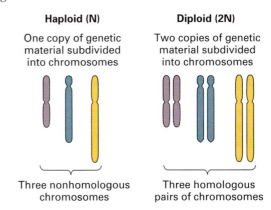

When viewed under the microscope, chromosomes differ in size and morphology within and between species. Each chromosome has a specialized region somewhere along its length that is often seen as a constriction under the microscope. This constriction, called a **centromere,** is important in the activities of the chromosomes during cellular division and can be located in one of four general positions in the chromosome (Figure 3.2). A **metacentric chromosome** has the centromere in approximately the center of the chromosome so that it appears to have two approximately equal arms. **Submetacentric chromosomes** have one arm longer than the other, **acrocentric chromosomes** have one arm with a stalk and often with a "bulb" (called a *satellite*) on it, and **telocentric chromosomes** have only one arm, since the centromere is at the end. Chromosomes also vary in relative size. Chromosomes of mice, for example, are all similar in length, whereas those of humans show a wide range of relative lengths. Chromosome length and cen-

~ **FIGURE 3.2**

General classification of eukaryotic chromosomes into metacentric, submetacentric, acrocentric, and telocentric types, based on the position of the centromere. (The chromosomes are shown as they appear following duplication in mitosis or meiosis.)

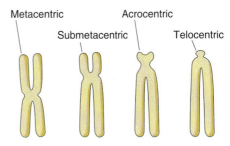

tromere position are constant for each chromosome and help in the identification of individual chromosomes.

Asexual and Sexual Reproduction

Eukaryotes can reproduce by asexual or sexual reproduction. In **asexual reproduction** a new individual develops from either a single cell or from a group of cells (vegetative reproduction) in the absence of any sexual process. Asexual reproduction is found in both unicellular and multicellular eukaryotes. Single-celled eukaryotes (such as yeast) grow, double their genetic material, and generate two progeny cells, each of which contains an exact copy of the genetic material found in the parental cell. This process repeats as long as there are sufficient nutrients in the growth medium. In multicellular organisms asexual reproduction is sometimes referred to as **vegetative reproduction**. Multicellular fungi, for example, can be propagated vegetatively by taking a small piece of the tissue and transferring it to new medium. Many higher plants, such as roses and fruit trees, are commercially maintained by cuttings, which is an asexual means of propagation.

    Sexual reproduction is the fusion of two haploid gametes (sex cells) to produce a single diploid zygote cell. From this zygote a new multicellular individual develops. Sexual reproduction involves the alternation of diploid and haploid phases. A key effect of sexual reproduction is that it achieves genetic recombination; that is, it generates gene combinations in the offspring that are distinct from those in the parents. With the exception of self-fertilizing organisms (such as many plants), the two gametes are from different parents. Note that it is during the production of gametes or spores that genetic recombination takes place. Figure 3.3 shows the cycle of growth and sexual reproduction in animals; here the haploid gametes are sperm and eggs. Note that the number of chromosomes is kept constant from generation to generation.

    Sexually reproducing organisms have two sorts of cells: somatic (body) cells and germ (sex) cells. Somatic cells are haploid or diploid, depending on the eukaryote; lower eukaryotes such as yeast are often haploid, while higher eukaryotes typically are diploid. All somatic cells reproduce by a process called mitosis. The products of meiosis, which are always haploid, may be gametes (animals) or spores (plants). (In plants gametes are produced by mitosis.) Figure 3.4 illustrates the differences between asexual and sexual reproduction.

Mitosis

In both unicellular and multicellular eukaryotes, cellular reproduction is a cyclical process of growth,

---

~ FIGURE 3.3

Cycle of growth and reproduction in a sexually reproducing animal. Sexual reproduction involves the alternation of diploid (somatic) and haploid (gametic) phases.

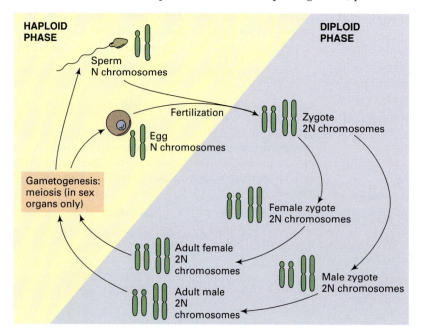

~ FIGURE 3.4

The differences in chromosome complement of organisms undergoing either asexual or sexual reproduction.

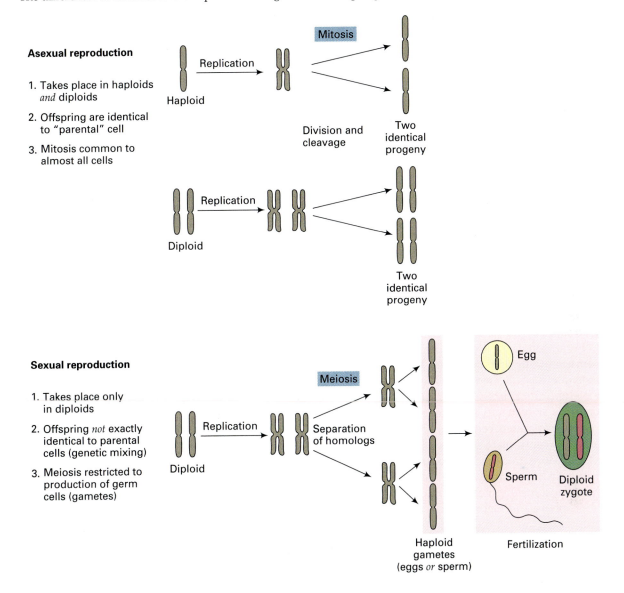

Asexual reproduction

1. Takes place in haploids *and* diploids

2. Offspring are identical to "parental" cell

3. Mitosis common to almost all cells

Haploid — Replication — Mitosis — Division and cleavage — Two identical progeny

Diploid — Replication — Two identical progeny

Sexual reproduction

1. Takes place only in diploids

2. Offspring *not* exactly identical to parental cells (genetic mixing)

3. Meiosis restricted to production of germ cells (gametes)

Diploid — Replication — Meiosis — Separation of homologs — Haploid gametes (eggs *or* sperm) — Egg — Sperm — Fertilization — Diploid zygote

---

mitosis (*nuclear division* or *karyokinesis*) and (usually) **cell division** (*cytokinesis*). In most cases, these two processes are linked with mitosis rapidly followed by cell division. The cycle of growth, mitosis and cell division is called the **cell cycle**. In proliferating somatic cells the cell cycle consists of two phases (Figure 3.5): the mitotic (or dividing) phase (M) and an interphase between divisions. Interphase consists of three stages, G_1 (gap 1), S, and G_2 (gap 2). During G_1 (presynthesis stage), the cell prepares for DNA and chromosome replication, which take place in the S stage. In G_2 (the postsynthesis stage), the cell prepares for cell division, or the M phase. That is, the precise replication of chromosomes takes place in

interphase and then mitosis occurs, resulting in the distribution of a complete chromosome set to each of the two progeny nuclei. Most of the cell cycle is spent in the G_1 stage, although the relative time spent in each of the four stages varies greatly among cell types. Further, in a given organism, variation in the length of the cell cycle depends primarily on the length of G_1, with the duration of S, G_2, and M being approximately the same in all cell types. For example, cancer cells and early fetal cells of humans spend minutes in G_1, while some differentiated adult cells (such as nerve cells) spend years in G_1. Finally, some cells exit the cell cycle and enter a quiescent, nondividing state called G_0.

~ FIGURE 3.5

Eukaryotic cell cycle. This cycle assumes a period of 24 hours in culture, although there is great variation between cell types and organisms *in vivo*.

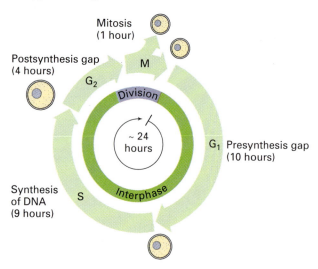

During interphase of the cell division cycle, the individual chromosomes are extended and are difficult to see under the light microscope. The DNA of each chromosome is replicated, and the product of chromosome duplication is two exact copies, called **sister chromatids,** which are held together by the replicated but unseparated centromere. (Because of the latter, only one centromere structure is seen under the microscope.) More precisely, a **chromatid** is one of the two visibly distinct, longitudinal subunits of all replicated chromosomes, which become visible between early prophase and metaphase of mitosis (and between prophase I and the second metaphase of meiosis, discussed later). After mitotic anaphase, when the centromeres separate, they become known as *daughter chromosomes.*

Mitosis occurs in both haploid and diploid cells. It is a continuous process, but for purposes of discussion it is usually divided into four cytologically distinguishable stages called *prophase, metaphase, anaphase,* and *telophase.* In Figure 3.6, photographs show the typical chromosome morphology in interphase and in the four stages of mitosis in plant (onion root tip) and animal (whitefish blastula) cells. Figure 3.7 (p. 54) shows the four stages in simplified diagrams.

PROPHASE. At the beginning of **prophase** (see Figures 3.6b and 3.7) the chromatids are very elongated and cannot be seen under the light microscope. In preparation for mitosis they begin to coil tightly so that they appear shorter and fatter under the micro-

scope. By late prophase each chromosome, which was duplicated during the preceding S phase in interphase, is seen to consist of two sister chromatids.

During prophase, the mitotic spindle (spindle apparatus) assembles outside the nucleus. Each of the spindle fibers in the bipolar mitotic spindle is approximately 25 nm in diameter and consists of microtubules made of special proteins called *tubulins.* In most animal cells, the foci for spindle assembly are the centrioles (see Figure 1.6 on p. 8). (Higher-plant cells usually do not have centrioles, but they do have a mitotic spindle.) Prior to the S phase, the cell's pair of centrioles have replicated and each new centriole pair becomes the focus for a radial array of microtubules called the *aster.* Early in prophase the two asters are adjacent to one another close to the nuclear membrane. By late prophase the two asters have moved far apart along the outside of the nucleus and are spanned by the microtubular spindle fibers.

Near the end of prophase the nuclear envelope breaks down and the nucleolus or nucleoli disappear, allowing the spindle to enter the nuclear area. Specialized structures called **kinetochores** form on either face of the centromere of each chromosome and become attached to special microtubules called *kinetochore microtubules* (Figure 3.8, p. 55). These microtubules radiate in opposite directions from each side of each chromosome and interact with the spindle microtubules.

METAPHASE. Metaphase (see Figures 3.6c and 3.7) begins when the nuclear envelope has completely disappeared. During metaphase the chromosomes become arranged so that their centromeres become aligned in one plane halfway between the two spindle poles and with the long axes of the chromosomes at 90 degrees to the spindle axis. The kinetochore microtubules are responsible for this chromosome alignment event. The plane where the chromosomes become aligned is called the **metaphase plate**. Figure 3.9 (p. 55) shows electron micrographs of human chromosomes at this stage of the cell cycle. Note the highly condensed state of the sister chromatids. The electron micrograph (EM) in Figure 3.9c shows a human chromosome from which much of the protein has been removed. In the center of the chromosome is a dense framework of protein called a *scaffold,* which retains the form of the chromosome. The scaffold is surrounded by a halo of DNA filaments that have uncoiled and spread outward.

ANAPHASE. Anaphase (see Figures 3.6d and Figure 3.7) begins when the two sister chromatids separate at the centromere giving rise to two daughter chromosomes. Once the paired kinetochores on

~ FIGURE 3.6

Interphase and the stages of mitosis in onion root tip (left) and whitefish blastula (right): (a) Interphase; (b) Prophase; (c) Metaphase; (d) Anaphase; (e) Telophase.

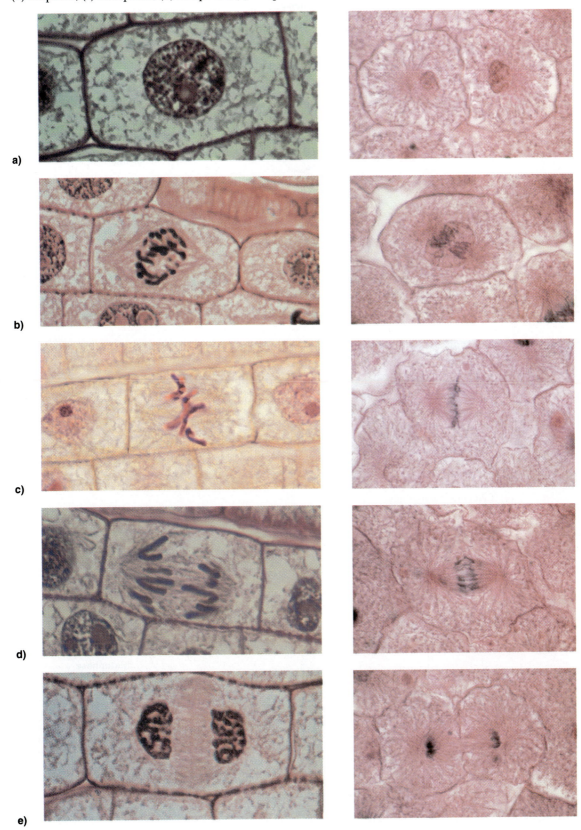

~ FIGURE 3.7

Interphase and mitosis in an animal cell.

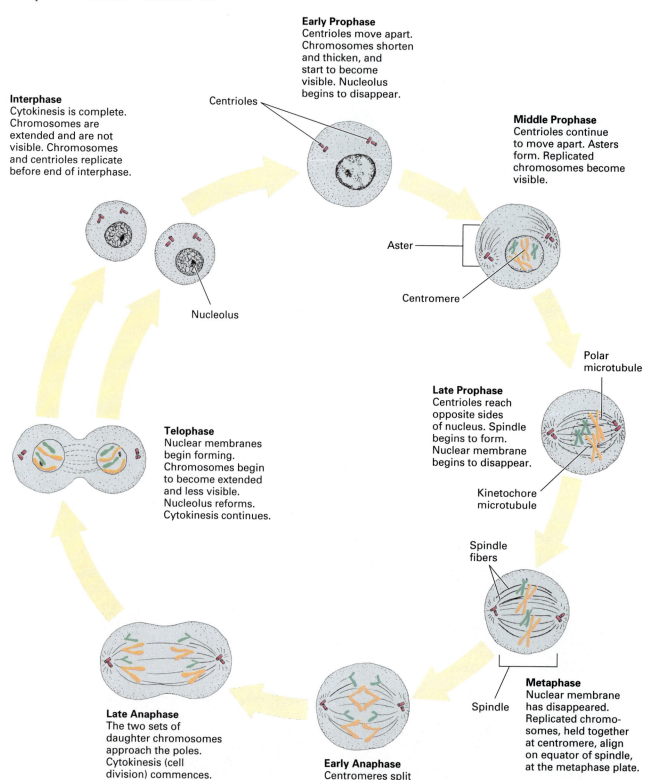

Early Prophase
Centrioles move apart.
Chromosomes shorten
and thicken, and
start to become
visible. Nucleolus
begins to disappear.

Interphase
Cytokinesis is complete.
Chromosomes are
extended and are not
visible. Chromosomes
and centrioles replicate
before end of interphase.

Centrioles

Middle Prophase
Centrioles continue
to move apart. Asters
form. Replicated
chromosomes become
visible.

Nucleolus

Aster

Centromere

Polar
microtubule

Late Prophase
Centrioles reach
opposite sides
of nucleus. Spindle
begins to form.
Nuclear membrane
begins to disappear.

Telophase
Nuclear membranes
begin forming.
Chromosomes begin
to become extended
and less visible.
Nucleolus reforms.
Cytokinesis continues.

Kinetochore
microtubule

Spindle
fibers

Late Anaphase
The two sets of
daughter chromosomes
approach the poles.
Cytokinesis (cell
division) commences.

Early Anaphase
Centromeres split
and daughter chromo-
somes begin migration
to opposite poles.

Spindle

Metaphase
Nuclear membrane
has disappeared.
Replicated chromo-
somes, held together
at centromere, align
on equator of spindle,
at the metaphase plate.

~ FIGURE 3.8

Kinetochores and kinetochore microtubules.

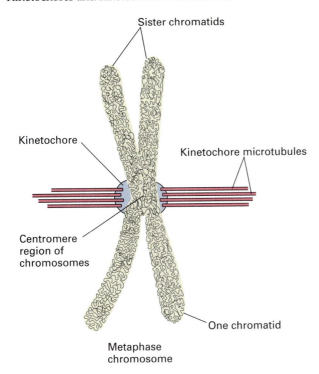

Sister chromatids

Kinetochore

Kinetochore microtubules

Centromere region of chromosomes

One chromatid

Metaphase chromosome

each chromosome separate, the sister chromatid pairs undergo **disjunction** (separation), and the daughter chromosomes (the former sister chromatids) move toward the poles. In anaphase the daughter chromosomes migrate toward the opposite poles of the cell as they are pulled by the contracting spindle fibers attached to their centromeres. Now the chromosomes assume characteristic shapes related to the location of the centromere along the chromosome's length. For example, a metacentric chromosome will appear as a V as the two roughly equal-length chromosome arms trail the centromere in its migration toward the pole. Similarly, a submetacentric chromosome will appear as a J with a long and short arm. Figure 3.10 gives an interpretation of the mitotic apparatus in anaphase in an animal cell. Cytokinesis (cell division) usually begins in the latter stages of anaphase.

TELOPHASE. During **telophase** (see Figures 3.6e and 3.7), the migration of daughter chromosomes to the two poles is completed. The two sets of progeny chromosomes are assembled into two groups at opposite ends of the cell. The chromosomes begin to uncoil and assume the extended state characteristic of interphase. A nuclear envelope forms around each chromosome group, the spindle microtubules disappear, and the nucleolus or nucleoli reform. At this point, nuclear division is complete: the cell has two nuclei. Usually, cell division continues in telophase, giving rise to two cells, each with one nucleus.

~ FIGURE 3.9

Human metaphase chromosome. (a) Transmission electron micrograph of an intact chromosome; (b) Colorized scanning electron micrograph of an intact chromosome; (c) Transmission electron micrograph of an intact chromosome from which much of the protein has been removed; the DNA filaments have uncoiled.

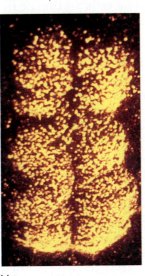

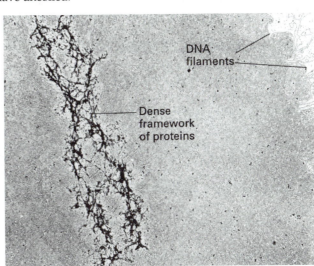

DNA filaments

Dense framework of proteins

a)

b)

c)

~ FIGURE 3.10

Mitotic apparatus at anaphase in an animal cell. A centriole pair is located at each pole, with spindle microtubules spanning the cell. The separation of the sister chromatids depends on spindle microtubules attached to the centromere of each chromatid. The daughter chromosomes shown in the figure are metacentric.

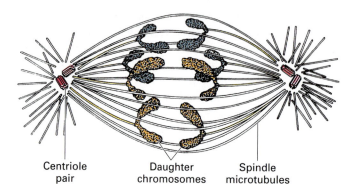

Centriole   Daughter   Spindle
pair   chromosomes   microtubules

CYTOKINESIS.   *Cytokinesis* refers to division of the cytoplasm; usually it follows the nuclear division stage of mitosis and is completed by the end of telophase. Cytokinesis compartmentalizes the two new nuclei into separate daughter cells, completing the mitosis/cell division process. In animal cells cytokinesis occurs by the formation of a constriction in the middle of the cell which contracts until two daughter cells are produced (Figure 3.11a). However, most plant cells do not divide by the formation of a constriction. Instead, a new cell membrane and cell wall are assembled between the two new nuclei to form a *cell plate* (Figure 3.11b). Cell wall material coats each side of the cell plate, and the result is two progeny cells. Figure 3.11c is a scanning electron micrograph of an animal cell in tissue culture shortly after its division.

Genetic Significance of Mitosis

Mitosis maintains a constant amount of genetic material from generation to generation. Mitosis occurs in haploid or diploid cells after DNA and chromosome duplication have taken place. It is a highly ordered process in which one copy of each duplicated chromosome is segregated into each of the two daughter cells. Thus, for a haploid (N) cell, chromosome duplication produces a cell with two sets of chromosomes. Mitosis then results in two progeny haploid cells, each with one complete set of chromosomes (= a genome). For a diploid (2N) cell, which has two sets of chromosomes (= two genomes), chromosome duplication produces a cell in which each chromosome set has doubled its content. Mitosis then results in two genetically identical progeny diploid cells, each with two sets of chromosomes (= two genomes).

Thus, there is an equal distribution of genetic material and no loss of genetic material.

KEYNOTE

Mitosis is the process of nuclear division in eukaryotes. It is one part of the cell cycle (i.e., G₁, S, G₂ and M) and results in the production of daughter nuclei that contain identical chromosome numbers and that are genetically identical to one another and to the parent nucleus from which they arose. Prior to mitosis, the chromosomes duplicate. Mitosis is usually followed by cytokinesis. Both haploid and diploid cells proliferate by mitosis.

Meiosis

Meiosis is the term applied to the two successive divisions of a diploid nucleus following only one DNA replication cycle. That original diploid nucleus contains one maternally derived and one paternally derived haploid set of chromosomes. Meiosis occurs only at a special point in the organism's life cycle. Unlike mitosis, meiosis is not responsible for cell proliferation. Meiosis results in the formation either of haploid gametes (**gametogenesis**) or of meiospores (in *sporogenesis*). (A meiospore undergoes mitosis to produce a gamete-bearing, multicellular stage called the gametophyte.) During meiosis, homologous chromosomes replicate, then pair, then undergo two divisions. As a consequence, each of the four cells resulting from the two meiotic divisions receives one

~ FIGURE 3.11

Cytokinesis (cell division). (a) Diagram of cytokinesis in an animal cell; (b) Diagram of cytokinesis in a higher plant cell; (c) Scanning electron micrograph of an animal cell in tissue culture shortly after its division.

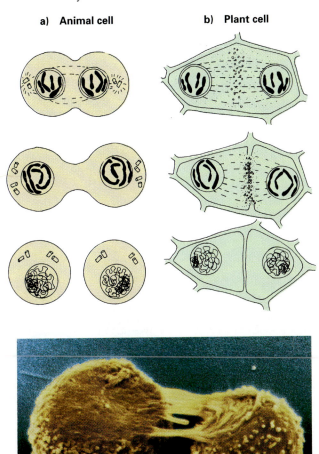

a) Animal cell   **b) Plant cell**

c)

chromosome of each chromosome set (i.e., one complete haploid genome). The two nuclear divisions of a normal meiosis are called meiosis I and meiosis II. Diagrams of these divisions are presented in Figure 3.12. The first meiotic division results in reduction in the number of chromosomes from diploid to haploid (reductional division), and the second division results in separation of the sister chromatids (equational division). In most cases the divisions are accompanied by cytokinesis, so the result of the meiosis of a single diploid cell is four haploid cells.

MEIOSIS I: THE FIRST MEIOTIC DIVISION.

Meiosis I, in which the chromosome number is reduced from diploid to haploid, consists of four cyto-

logically distinguishable stages: *prophase I, metaphase I, anaphase I,* and *telophase I.*

Prophase I.   As prophase I begins, the chromosomes have already replicated. In prophase I the chromosomes become shorter and thicker; they pair off, crossing-over occurs, the spindle apparatus forms, and the nuclear membrane and nucleolus/nucleoli disappear (see Figure 3.12). Prophase I is divided into a number of stages. Except for the behavior of homologous pairs of chromosomes, and crossing-over, prophase I of meiosis is very similar to the mitotic prophase.

In the classical view of meiosis, during prophase I homologous chromosomes first undergo **synapsis** (the formation of a very tight association of homologous chromosomes), and then **crossing-over** (the reciprocal physical exchange of chromosome segments at corresponding positions along pairs of homologous chromosomes) takes place. However, some very recent data have changed that view. These data were obtained by separate studies involving detailed time-course analyses of synchronized populations of meiotic cells to order the specific events in meiosis, and from the study of a gene identified as being required specifically for synapsis. In this new view of meiosis, *pairing*—the loose alignment of homologous chromosomes along their lengths—is mechanistically and temporally distinct from synapsis—the very tight association of the homologous chromosomes along their lengths. Crossing-over is initiated after pairing occurs and before synapsis has occurred. This new view is reflected in the following discussion.

In **leptonema** (early prophase—the leptotene stage) the chromosomes have begun to coil. Once a cell enters leptonema, it is committed to meiotic process. A key event in leptonema is *pairing* of homologous chromosomes; that is, the loose alignment of homologous regions of the chromosomes. Under the microscope, the threadlike chromosomes seen at this time are the sister chromatids paired together at the centromere. Once pairing has been completed, a most significant event begins: crossing-over, the reciprocal exchange of chromosome segments at corresponding positions along pairs of homologous chromosomes. If there are genetic differences between the homologs, crossing-over can produce new gene combinations in a chromatid. There is usually no loss or addition of genetic material to either chromosome, since crossing-over involves reciprocal exchanges. A chromosome that emerges from meiosis with a combination of genes that differs from the combination with which it started is called a **recombinant chromosome.** Therefore crossing-over is a mechanism that can give rise to **genetic recombination,** a concept we will examine more fully in later chapters.

~ FIGURE 3.12

Diagrams of the stages of meiosis in an animal cell.

Early prophase I

Chromosomes, already replicated, become visible.

Middle prophase I

Homologous chromosomes shorten and thicken. The chromosomes synapse and crossing-over occurs.

Late prophase I

Results of crossing-over become visible as chiasmata. Nuclear membrane begins to disappear. Spindle apparatus begins to form.

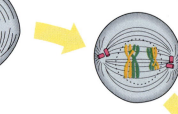

Metaphase I

Assembly of spindle is completed. Each chromosome pair (bivalent) aligns across the equatorial plane of the spindle.

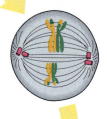

Telophase II

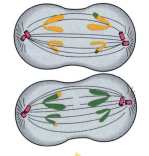

4 gametes

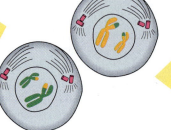

Anaphase I

Homologous chromosome pairs (dyads) separate and migrate toward opposite poles.

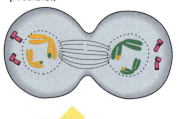

Anaphase II

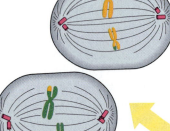

Telophase I

Chromosomes (each with two sister chromatids) complete migration to the poles and new nuclear membranes may form. (Other sorting patterns are possible.)

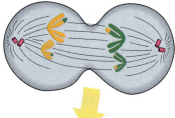

Metaphase II

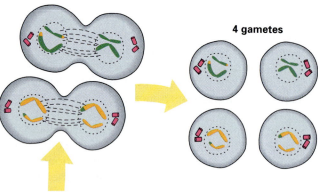

Prophase II

Cytokinesis

In most species, cytokinesis occurs to produce two cells. Chromosomes do not replicate before meiosis II.

In **zygonema** (early/mid-prophase—the zygotene stage) a key event is synapsis, defined as the formation of an intimate association of homologous chromosomes brought about by the formation of a zipper-like structure along the length of the chromatids called the **synaptonemal complex** (Figure 3.13). Because of the replication that occurred earlier, each synapsed set of homologous chromosomes consists of four chromatids and is referred to as a **bivalent**. The chromosomes are maximally condensed prior to the start of synaptonemal complex formation.

Following zygonema is **pachynema** (mid-prophase—the pachytene stage). The formation of the synaptonemal complex is completed in pachynema; the structure is disassembled at the end of pachynema. The chromosomes have started to extend already by this time.

The next stage in prophase I is **diplonema** (mid/late prophase—the diplotene stage). Here the chromosomes begin to move apart. The result of crossing-over becomes visible during diplonema as a cross-shaped structure called a **chiasma** (plural, *chiasmata*: Figure 3.14a). Since all four chromatids may be involved in crossing-over events along the length of the homologs, the chiasma pattern at this stage may be very complex.

Diplonema is followed rapidly in most organisms by the remaining stages of meiosis. However, in many animals the oocytes (egg cells) can remain in diplonema for very long periods. For example, in the human female, oocytes go through meiosis I up to diplonema by the seventh month of fetal development and then remain arrested in this stage for many years. At the onset of puberty and until menopause, one oocyte per menstrual cycle completes meiosis I and is ovulated. If the oocyte is fertilized by a sperm as it passes down the fallopian tube, it quickly completes meiosis II and a functional zygote is then produced by fertilization of the resulting haploid egg.

Diakinesis (late prophase) follows diplonema. During this stage, the chiasmata often *terminalize*; that is, they move down the chromatids to the ends (Figure 3.14b). As a result, the chromatids at this stage appear to be attached near the tips. That is, the chromosomes are sliding past each other so that, in a sense, the chiasmata delay the separation of the chromatids. In addition, the nucleolus and nuclear envelope break down. The chromosomes can be most easily counted at this stage of meiosis.

The pairing, crossing-over, and synapsis phenomena described for prophase I applies to homologous chromosomes; namely, the autosomes. The sex chromosomes are not homologous chromosomes. However, the mammalian Y chromosome has been shown to have a small portion (a terminal piece of the short arm) that is homologous to a portion of the X chromosome; the rest of the Y chromosome contains Y-specific DNA sequences. The homologous region has been called the *pseudoautosomal region*. In male meiosis, recombination due to crossing-over between

~ FIGURE 3.13

Electron micrograph of the synaptonemal complex formed between two homologous chromatid pairs during the zygonema stage of prophase I.

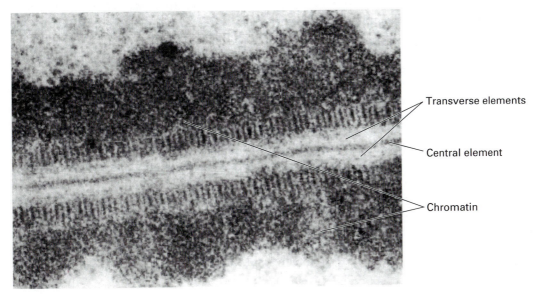

Transverse elements

Central element

Chromatin

~ FIGURE 3.14

(a) Appearance of chiasmata, the visible evidence of crossing-over, in diplonema;
(b) Terminalization of a chiasmata during diakinesis.

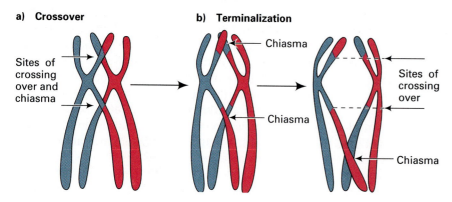

a) Crossover

b) Terminalization

Sites of crossing over and chiasma

Chiasma

Chiasma

Sites of crossing over

Chiasma

the pseudoautosomal regions of the X and Y chromosomes takes place; in fact, this recombination is required for correct segregation of X and Y chromosomes in the subsequent stages of meiosis. The term pseudoautosomal comes, then, from the fact that genes located in the pseudoautosomal region segregate in genetic crosses like autosomal genes, yet they are on the sex chromosomes.

Metaphase I. By the beginning of metaphase I (see Figure 3.12), the nuclear envelope has completely broken down, and the bivalents become aligned on the equatorial plane of the cell. The spindle apparatus is completely formed now, and the microtubules are attached to the centromeres of the homologs. Note particularly that it is the pairs of homologs (the bivalents) that are found at the metaphase plate. In contrast, in mitosis of most (but not all) organisms, replicated homologous chromosomes (sister chromatid pairs) align independently at the metaphase plate.

Anaphase I. In anaphase I (see Figure 3.12) the chromosomes in each bivalent separate, so the chromosomes of each of the homologous pairs disjoin and they migrate toward opposite poles, the areas in which new nuclei will form. (At this stage, each of the separated chromosomes is called a *dyad*.) This migration assumes that: (1) maternally derived and paternally derived centromeres are segregated randomly at each pole (except for the parts of chromosomes exchanged during the crossing-over process); and (2) at each pole there is a haploid complement of replicated centromeres with associated chromosomes. It is important to remember that *at this time the segregated*

sister chromatid pairs remain attached at their respective centromeres. In other words, a key difference between mitosis and meiosis I is that sister chromatids remain joined after metaphase in meiosis I, whereas in mitosis they separate.

Telophase I. The length of telophase I (see Figure 3.12) varies considerably among species. Nonetheless, in telophase I the dyads complete their migration to opposite poles of the cell, and (in most cases) new nuclear envelopes form around each haploid grouping. In most species, cytokinesis follows, producing two haploid cells. Thus, meiosis I, which begins with a diploid cell that contains one maternally derived and one paternally derived set of chromosomes, ends with two nuclei, each of which is haploid and contains one mixed-parental set of replicated chromosomes. After cytokinesis, each of the two progeny cells has a nucleus with a *haploid set of replicated chromosomes.*

MEIOSIS II: THE SECOND MEIOTIC DIVISION. The second meiotic division is very similar to a mitotic division. Figure 3.12 presents diagrams of prophase II, metaphase II, anaphase II, and telophase II of meiosis II.

Prophase II (see Figure 3.12) is a stage of chromosome contraction.

In **metaphase II** (see Figure 3.12) each of the two daughter cells organizes a spindle apparatus that attaches to the centromeres which still connect the sister chromatids. The centromeres line up on the equator of the second-division spindles.

During **anaphase II** (see Figure 3.12) the centromeres split and the chromatids are pulled to the opposite poles of the spindle: One sister chromatid of each pair goes to one pole, while the other goes to the opposite pole. The separated chromatids are now referred to as chromosomes in their own right.

In the last stage, **telophase II** (see Figure 3.12), a nuclear envelope forms around each set of chromosomes, and cytokinesis takes place. After telophase II, the chromosomes become more extended and again are invisible under the light microscope.

The end products of the two meiotic divisions are four haploid cells from one original diploid cell. Each of the four progeny cells has one chromosome from each homologous pair of chromosomes. Remember, however, that these chromosomes are not exact copies of the original chromosomes because of the crossing-over that occurs between chromosomes during pachynema of meiosis I. Figure 3.15 compares mitosis and meiosis.

Genetic Significance of Meiosis

Meiosis has three significant results:

1. Meiosis generates haploid cells with half the number of chromosomes found in the diploid cell that entered the process because two division cycles follow only one cycle of DNA replication (S period). Fusion of the haploid nuclei (fertilization) restores the diploid number. Therefore through a cycle of meiosis and fertilization, the chromosome number is maintained in sexually reproducing organisms.
2. In metaphase I of meiosis, each maternally derived and paternally derived chromosome has an equal chance of aligning on one or the other side of the equatorial metaphase plate. As a result, each nucleus generated by meiosis will have some combination of maternal and paternal chromosomes.

    The number of possible chromosome combinations in the haploid nuclei resulting from meiosis is large, especially when the number of chromosomes in an organism is large. Consider a hypothetical organism with two pairs of chromosomes in a diploid cell entering meiosis. Figure 3.16 shows the two possible combinations of maternal and paternal chromosomes that can occur at the metaphase plate.

    The general formula states that the number of possible chromosome combinations in the nuclei resulting from meiosis is 2^n, where n is the number of chromosome pairs. In *Drosophila*, which has four pairs of chromosomes, the number of possible arrangements is 2^4, or 16; in humans, which have 23 chromosome pairs, over 8 million metaphase arrangements are possible. Therefore since there are many gene differences between the maternally derived and paternally derived chromosomes, the nuclei produced by meiosis will be genetically quite different from the parental cell and from one another.
3. The crossing-over between maternal and paternal chromatid pairs during meiosis I generates still more variation in the final combinations. Crossing-over occurs during every meiosis, and because the sites of crossing-over vary from one meiosis to another, the number of different kinds of progeny nuclei produced by the process is extremely large. Given the genetic features of meiosis, this process is of critical importance for understanding the behavior of genes, as will be seen in the following chapters.

It is important to note that the events that occur in meiosis are the bases for the segregation and independent assortment of genes according to Mendel's Laws, as discussed in Chapter 2.

KEYNOTE

Meiosis occurs in all sexually reproducing eukaryotes. It is a process by which a specialized diploid (2N) cell or cell nucleus with two sets of chromosomes is transformed, through one round of chromosome replication and two rounds of nuclear division, into four haploid (N) cells or nuclei each with one set of chromosomes. In the first of two divisions, pairing, crossing-over, and synapsis of homologous chromosomes occur. The meiotic process, in combination with fertilization, results in the conservation of the number of chromosomes from generation to generation. It also generates genetic variability through the various ways in which maternal and paternal chromosomes are combined in the progeny nuclei and by crossing-over (the physical exchange of genes between maternally or paternally derived homologs).

~ **FIGURE 3.15**

Comparison of mitosis and meiosis in a diploid cell.

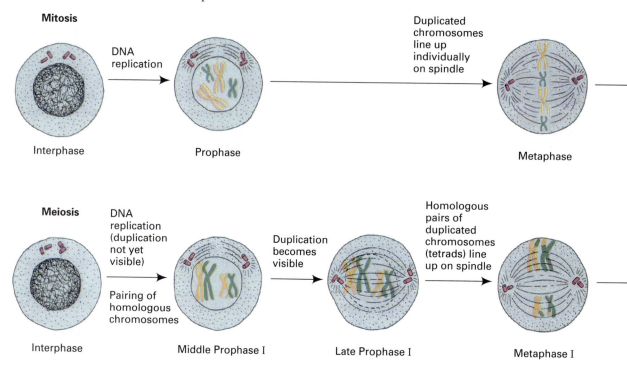

~ **FIGURE 3.16**

The two possible arrangements of two pairs of homologous chromosomes on the metaphase plate of the first meiotic division. Paternal chromosomes are shown in yellow and green; maternal chromosomes in purple and tan.

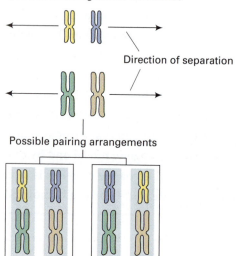

Locations of Meiosis in the Life Cycle

MEIOSIS IN ANIMALS.   Most multicellular animals are diploid through most of their life cycle. In such animals meiosis produces haploid gametes, which produce a diploid zygote when their nuclei fuse in the fertilization process. The zygote then divides mitotically to produce the new diploid organism. Thus the gametes are the only haploid stages of the life cycle. Gametes are only formed in specialized cells. In the male the gamete is the sperm, produced through a process called **spermatogenesis**. The female gamete is the egg, produced by **oogenesis**. Spermatogenesis and oogenesis are illustrated in Figure 3.17 (p. 64).

In male animals the **sperm cells** (also called **spermatozoa**) are produced by the testes. The testes contain the primordial germ cells (*primary spermatogonia*), which, through mitotic division, produce *secondary spermatogonia*. Spermatogonia transform into *primary spermatocytes (meiocytes)*, each of which undergoes

~ FIGURE 3.15 continued

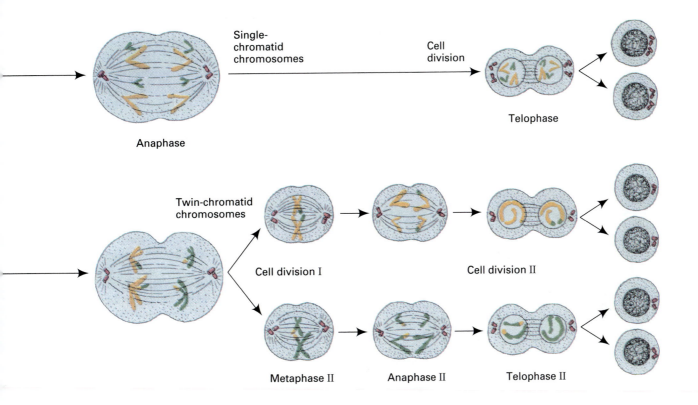

meiosis I and gives rise to two *secondary spermatocytes.* Each spermatocyte undergoes meiosis II. As a result of these two divisions, four haploid *spermatids* arise, which eventually differentiate into the mature male gametes, the spermatozoa.

In female animals the ovary contains the primordial germ cells *(primary oogonia),* which, by mitosis, give rise to *secondary oogonia.* These cells transform into *primary oocytes,* which grow until the end of oogenesis. The diploid, primary oocyte goes through meiosis I and unequal cytokinesis to give two cells: a large one called the **secondary oocyte** and a very small one called the *first polar body.* In the second meiotic division the secondary oocyte produces two haploid cells. One is a very small cell and is called a *second polar body;* the other is a large cell that rapidly matures into the mature egg cell, or **ovum.** The first polar body may or may not divide during the second meiotic division. The polar bodies have no function in most species and are discarded. Only the ovum is a viable gamete. Thus in the female animal only one

mature gamete (the ovum) is produced by meiosis of a diploid cell. In humans, all oocytes are formed in the fetus, and one oocyte completes meiosis I each month in the adult female, but does not progress further through meiosis unless stimulated to do so by fertilization with a sperm.

MEIOSIS IN PLANTS. The life cycle of sexually reproducing plants typically has two phases, the **gametophyte** or haploid stage in which gametes are produced, and the **sporophyte** or diploid stage in which haploid spores are produced by meiosis.

In angiosperms, the flowering plants, the flower is the structure in which sexual reproduction occurs. Figure 3.18 shows a generalized flower containing both male and female reproductive organs, the **stamens** and **pistils,** respectively. Each stamen consists of a single stalk, the filament, on the top of which is an anther. From the anther are released the pollen grains, which are immature male gametophytes. The pistil contains the female gametophytes and typically

~ FIGURE 3.17

Spermatogenesis and oogenesis in an animal cell.

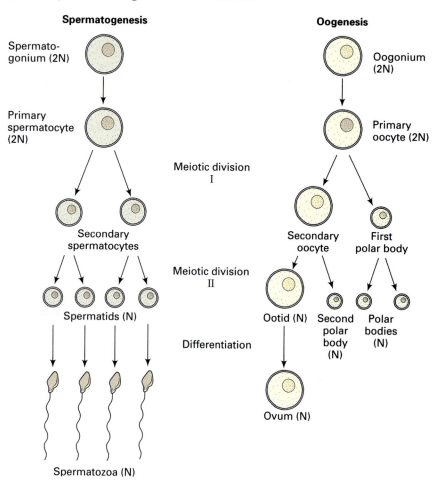

~ FIGURE 3.18

Generalized structure of a flower.

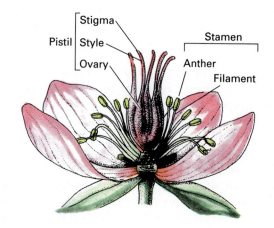

consists of the stigma, a sticky surface specialized to receive the pollen; the style, a thin stalk down which a pollen tube grows from the adhered pollen grain; and, at the base of the structure, the ovary, within which are the ovules. Each ovule encloses a female gametophyte (the *embryo sac*) with a single egg cell. When the egg cell is fertilized, the ovule develops into a seed.

Meiosis occurs in specific ways in the female and male parts of the flower. Meiosis in the female part of the flower occurs as follows (Figure 3.19): Within the ovary of the flower, each ovule contains a large 2N cell called the megaspore mother cell. Meiosis in the megaspore mother cell involves two divisions as usual, resulting in four haploid cells. Three of the cells disintegrate leaving a single large cell with one haploid nucleus. This cell is called the *megaspore* and its production is called **megasporogenesis**.

~ **FIGURE 3.19**

Megasporogenesis: Meiosis in the female part of the flower and the production of the embryo sac.

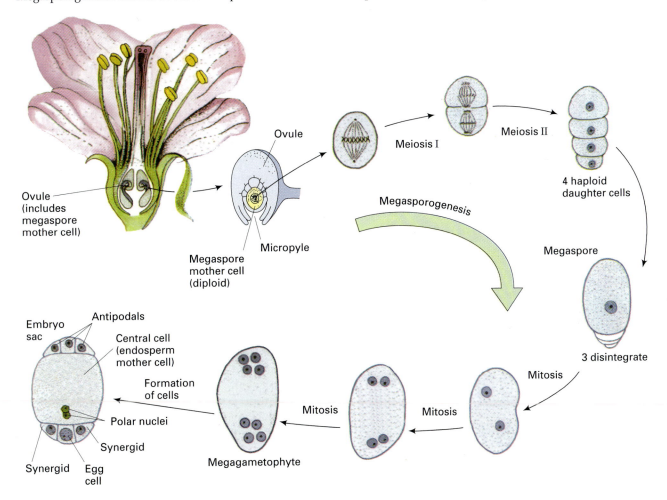

Next, the megaspore undergoes three successive nuclear mitotic divisions to produce a *megagametophyte* cell with eight identical haploid nuclei in a common cytoplasm (Figure 3.19). Plasma membranes and partial cell walls form around three of the nuclei at the micropylar end—the "bottom"—of the megagametophyte. Two of the cells are the *synergids*, and the third becomes the *egg cell* of the female gametophyte. Another two of the original eight nuclei—the *polar nuclei*—assume a position near the three cells just described. These two nuclei are in a common cytoplasm of a very large central cell taking up most of the volume of the female gametophyte. The three nuclei remaining become cells as plasma membranes and walls are formed; these cells—the *antipodals*—take up a position at the opposite end of the gametophyte from the egg cell and the two synergids. The

entire seven-celled megagametophyte structure is called the *embryo sac*.

In the anther, meiosis occurs by similar, although less complex, events to produce the pollen grains, or male gametophytes (Figure 3.20). This series of events is called **microsporogenesis**. Anthers contain four pollen sacs within which are many diploid microspore mother cells, each of which undergoes meiosis to produce four haploid microspores. Each microspore then divides once mitotically, producing a pollen grain consisting of a generative cell and a tube cell surrounded by a tough wall.

During pollination in flowering plants, the pollen grains are deposited on the stigma and each "germinates," producing a pollen tube (Figure 3.21, p. 67). There is no common path and mechanism for the growth of the pollen tube. In many plants the pollen

~ **FIGURE 3.20**

Microsporogenesis: Meiosis in the male part of the flower and the production of a pollen grain.

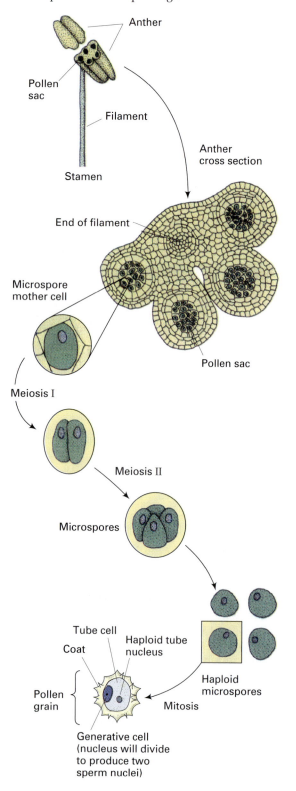

Anther

Pollen sac

Filament

Anther cross section

Stamen

End of filament

Microspore mother cell

Pollen sac

Meiosis I

Meiosis II

Microspores

Tube cell

Coat

Haploid tube nucleus

Pollen grain

Haploid microspores

Mitosis

Generative cell (nucleus will divide to produce two sperm nuclei)

tube grows down through the stigma and into the style, producing enzymes to digest the tissues ahead of its growth. In many species, the haploid generative cell nucleus divides by mitosis before the pollen grain germinates to produce two sperm nuclei (the gametes); the tube cell nucleus does not divide. In other species, the generative nucleus divides as the pollen tube grows down the style. In either case, the three nuclei stay close to the growing tip of the pollen tube. The pollen tube eventually penetrates the tissues of the ovule at a very small pore called the *micropyle*, allowing the two sperm nuclei to enter the embryo sac. One sperm nucleus fertilizes the haploid egg cell, producing the diploid zygote. The other sperm nucleus fuses with the two haploid polar nuclei in the central cell (endosperm mother cell), producing a triploid nucleus, thereby completing fertilization. Until recently, it was thought that this **double fertilization** occurs only in flowering plants. There is now evidence that double fertilization occurs in some gymnosperms (nonflowering seed plants).

After fertilization, the diploid zygote, the triploid endosperm cell, and the other tissues begin cell division. In this structure, the embryo is derived from the zygote, and the stored food within the seed derives from the triploid endosperm cell.

Among living organisms only plants produce gametes from special bodies called gametophytes. Thus plants have two distinct reproductive phases, called the **alternation of generations** (Figure 3.22). Meiosis and fertilization constitute the transition between these stages. The haploid gametophyte generation begins with spores that are produced by meiosis. In flowering plants the spores are the cells that ultimately become pollen and embryo sac. Fertilization initiates the diploid sporophyte generation. The diploid sporophyte generation produces specialized haploid cells called spores, which, in turn, produce gametophytes. Thus the sporophyte is the second alternate generation.

Figure 3.22 indicates the relationship of each phase of the alternation of generations to meiosis. The spores of ferns and mosses, and the microspores and megaspores of higher plants are all produced by meiosis. Thus in all green plants the alternation of generations involves an alternation between stages of haploid cells and diploid cells: The gametophyte cells are haploid, and the sporophyte cells are diploid.

CHROMOSOME THEORY OF INHERITANCE

Around the end of the nineteenth century and the beginning of the twentieth century, cytologists had

~ FIGURE 3.21

Pollination and fertilization in a flowering plant.

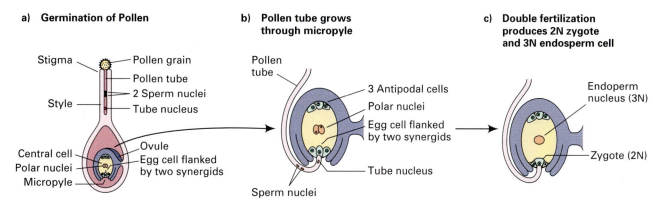

a) **Germination of Pollen**

Stigma — Pollen grain
Pollen tube
Style — 2 Sperm nuclei
Tube nucleus
Central cell — Ovule
Polar nuclei — Egg cell flanked
Micropyle — by two synergids

b) **Pollen tube grows through micropyle**

Pollen tube
3 Antipodal cells
Polar nuclei
Egg cell flanked by two synergids
Tube nucleus
Sperm nuclei

c) **Double fertilization produces 2N zygote and 3N endosperm cell**

Endoperm nucleus (3N)
Zygote (2N)

~ FIGURE 3.22

Alternation of gametophyte and sporophyte generations in green flowering plants.

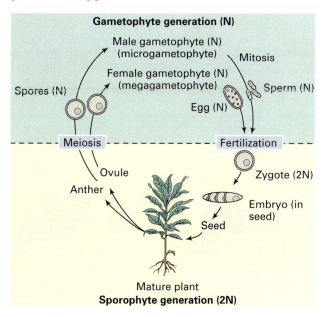

Gametophyte generation (N)

Male gametophyte (N) (microgametophyte)
Female gametophyte (N) (megagametophyte)
Mitosis
Spores (N)
Sperm (N)
Egg (N)
Meiosis — — — — — — — Fertilization
Ovule
Anther
Zygote (2N)
Seed
Embryo (in seed)
Mature plant
Sporophyte generation (2N)

Sutton and Theodor Boveri independently recognized that the transmission of chromosomes from one generation to the next closely paralleled the pattern of transmission of genes from one generation to the next. To explain this correlation, they proposed the **chromosome theory of inheritance**. This theory states that the chromosomes are the carriers of the genes. In this section, we will consider some of the evidence obtained by cytologists and geneticists to support this theory.

KEYNOTE

The chromosome theory of inheritance states that the chromosomes are the carriers of the genes. The first clear formulation of the chromosome theory was made in 1902 by Sutton and by Boveri, who independently recognized that the transmission of chromosomes from one generation to the next closely paralleled the pattern of transmission of genes from one generation to the next.

Sex Chromosomes

The support for the chromosome theory of inheritance came from experiments that related the hereditary behavior of particular genes to the transmission of the **sex chromosome**, the chromosome in (many) eukaryotic organisms that is represented differently

established that, within a given species the total number of chromosomes is constant in all cells, while the chromosome number varies considerably between species (Table 3.1). And, as we have just discussed, the processes of mitosis and meiosis had been described.

In 1900 Mendel's work on the nature of hereditary factors was rediscovered. Within a short time Mendel's experiments became known and appreciated throughout the scientific world. In 1902 Walter

~ Table 3.1

Chromosome Number in Various Organisms[a]

Organism	Total Chromosome Number
Human	46
Chimpanzee	48
Rhesus monkey	42
Dog	78
Cat	72
Mouse	40
Horse	64
Mallard	80
Chicken	78
Alligator	32
Cobra	38
Bullfrog	26
Toad	36
Goldfish	94
Starfish	36
Fruit fly (*Drosophila melanogaster*)	8
Housefly	12
Mosquito	6
Australian ant (*Myrecia pilosula*)	♂ 1, ♀ 2
Nematode	♂ 11, ♀ 12
Neurospora (haploid)	7
Sphagnum moss (haploid)	23
Field horsetail	216
Giant sequoia	22
Tobacco	48
Cotton	52
Kidney bean	22
Broad bean	12
Onion	16
Potato	48
Tomato	24
Bread wheat	42
Rice	24
Baker's yeast (*Saccharomyces cerevisiae*)	34

[a]Except as noted, all chromosome numbers are for diploid cells.

in the two sexes. For many animals the sex chromosome composition of the individual is directly related to the sex of the individual. The sex chromosomes typically are designated the **X chromosome** and the **Y chromosome**. In humans and the fruit fly, *Drosophila melanogaster*, for example, the female has two X chro-

mosomes (i.e., she is XX with respect to the sex chromosomes) while the male has one X chromosome and one Y chromosome (i.e., he is XY). Figure 3.23a shows the appearance of male and female *Drosophila*, and Figure 3.23b shows the chromosome complements of the two sexes. Because the male produces two kinds of gametes with respect to sex chromosome content (X or Y), and because the female produces only one gametic type (X), the male is called the **heterogametic sex** and the female is called the **homogametic sex**. In *Drosophila* the X and Y chromosomes are similar in size but their shapes are different. (Note: In some organisms the male is homogametic and the female is heterogametic. This reversal will be discussed later in the chapter.)

The pattern of transmission of X and Y chromosomes from generation to generation is straightforward (Figure 3.24). In this figure the X is represented by a straight structure much like a slash mark, and

~ Figure 3.23

(a) Female (left) and male (right) *Drosophila melanogaster* (fruit fly), an organism used extensively in genetics experiments. Top: Adult flies; Bottom: Drawings of ventral abdominal surface to show differences in genitalia; (b) Chromosomes of *Drosophila melanogaster* diagrammed to show their morphological differences. A female (left) has four pairs of chromosomes in her somatic cells, including a pair of X chromosomes. The only difference in the male is an XY pair of sex chromosomes instead of two Xs.

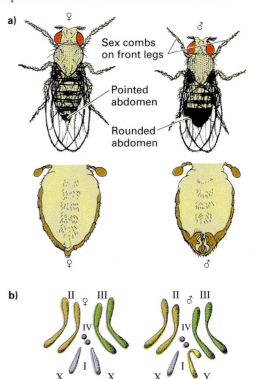

~ FIGURE 3.24

Inheritance pattern of X and Y chromosomes in organisms where the female is XX and the male is XY: (a) Production of the F_1 generation; (b) Production of the F_2 generation.

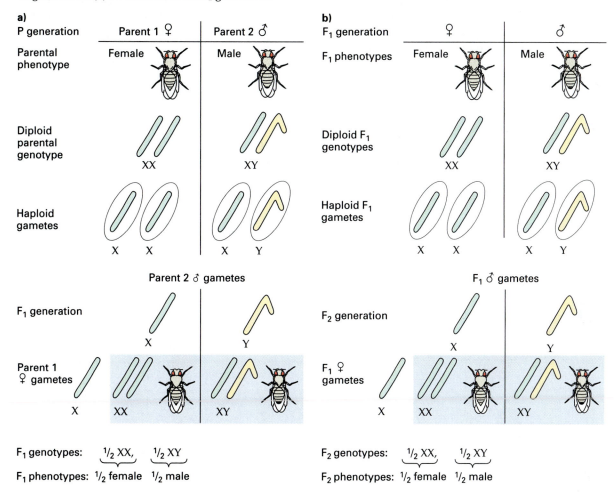

the Y by a similar structure topped by a hook to the right. The mother produces X-bearing gametes, while the male produces both X-bearing and Y-bearing gametes. Random fusion of these gametes produces an F_1 generation with 1/2 XX (female) and 1/2 XY (male) flies.

KEYNOTE

A sex chromosome is a chromosome in eukaryotic organisms that is represented differently in the two sexes. In many of the organisms encountered in genetic studies, one sex possesses a pair of identical chromosomes (the X chromosomes). The opposite sex possesses a pair of visibly different chromosomes: one is an X chromosome, and the other, structurally and functionally different, is called the Y chromosome. Commonly, the XX sex is female, and the XY sex is male. The XX and XY sexes are called the homogametic and heterogametic sexes, respectively.

Sex Linkage

Evidence to support the chromosome theory of heredity was obtained by using *Drosophila* (fruit flies) in genetic experiments. In 1910 Thomas Hunt Morgan reported his results of crossing experiments with the fruit fly.

Morgan found a male fly in one of his true-breeding stocks that had white eyes instead of the brick red eyes characteristic of the **wild type**. The term *wild type* refers to a strain, organism, or gene that is most prevalent in the "wild" population of the organism with respect to genotype and phenotype. For example, a *Drosophila* strain with all wild-type genes will have brick-red eyes. Variants of a wild-type strain arise from mutational changes of the wild-type genes that produce **mutant alleles**; the result is strains with mutant characteristics. Mutant alleles may be recessive or dominant to the wild-type allele; for example, the mutant allele that causes white eyes in *Drosophila* is recessive to the wild-type (red-eye) allele.

When Morgan crossed the white-eyed male with a red-eyed female from the same stock, he found that all the F_1 flies were red-eyed and concluded that the white-eyed trait was recessive. Next, he allowed the F_1 progeny to interbreed and counted the phenotypic classes in the F_2 generation; there were 3,470 red-eyed

and 782 white-eyed flies. The number of individuals with the recessive phenotype was too small to fit the Mendelian 3:1 ratio. In addition, *Morgan noticed that all the white-eyed flies were male.* (Later, he determined that the lower-than-expected number of flies with the recessive phenotype was the result of lower viability of white-eyed flies.) Morgan's cross is diagrammed in Figure 3.25. The *Drosophila* gene symbolism used is different from that we used for Mendel's crosses and is described in Box 3.1. The *Drosophila* gene symbolism should be understood before proceeding with this discussion. As we go through the discussions, note that the mother-son inheritance pattern presented in Figure 3.25 is the result of the segregation of genes located on a sex chromosome.

~ **FIGURE 3.25**

The X-linked inheritance of red eyes and white eyes in *Drosophila melanogaster*. The symbols w and w^+ indicate the white- and red-eyed alleles, respectively. (a) A red-eyed female is crossed with a white-eyed male; (b) The F_1 flies are interbred to produce the F_2s.

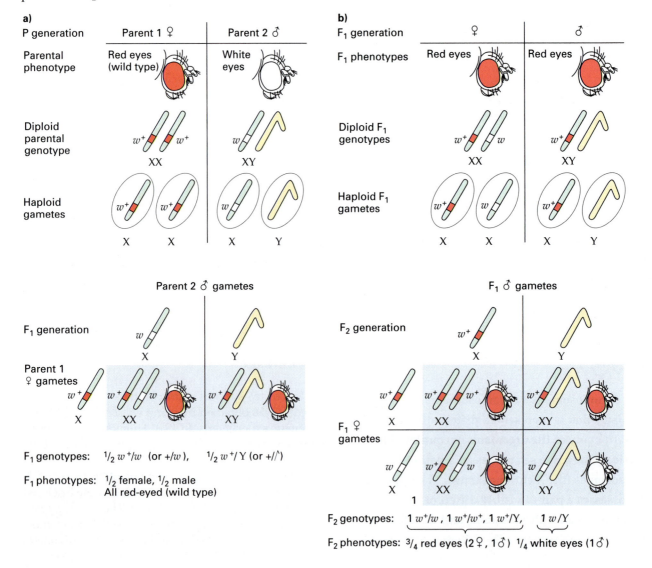

Box 3.1
Genetic Symbols Revisited

Unfortunately, there is no one system of gene symbols used by geneticists; the gene symbols used for *Drosophila* are different from those we saw used for peas in Chapter 2. The *Drosophila* symbolism is commonly, but not exclusively, used in genetics today. In this system the symbol + indicates a wild-type allele of a gene. A lowercase letter designates mutant alleles of a gene that are *recessive* to the wild-type allele, and an uppercase letter is used for alleles that are *dominant* to the wild-type allele. *The letters are chosen on the basis of the phenotype of the organism expressing the mutant allele.* For example, a variant strain of *Drosophila* has bright orange eyes instead of the usual brick red. The mutant allele involved is recessive to the wild-type brick red allele, and because the bright orange eye color is close to vermilion in tint, the allele is designated v and is called the vermilion allele. The wild-type allele of v is v^+, but when there is no chance of confusing it with other genes in the cross, it is often shortened to +. In the "Mendelian" terminology used up to now, the recessive mutant allele would be v, and its wild-type allele would be V.

A conventional way to represent the chromo-somes (instead of the way we have been using in the figures) is to use the slash (/). Thus v^+/v or $+/v$ indicates that there are two homologous chromosomes, one with the wild-type allele (v^+ or +) and the other with the recessive allele (v). The Y chromosome is usually symbolized as a Y or a bent slash ($\wedge$). Thus Morgan's cross of a true-breeding red-eyed female fly with a white-eyed male could be written $w^+/w^+ \times w/Y$ or $+/+ \times w/\wedge$.

The same rules apply when the alleles involved are dominant to the wild-type allele. For instance, some *Drosophila* mutants, called *Curly*, have wings that curl up at the end rather than the normal straight wings. The symbol for this mutant allele is Cy, and the wild-type allele is Cy^+, or + in the shorthand version. Thus, a heterozygote would be Cy^+/Cy or $+/Cy$.

In the rest of the book, both the A/a ("Mendelian") and a^+/a (*Drosophila*) symbolisms will be used, so it is important that you be able to work with both. Since it is easier to verbalize the "Mendelian" symbols (e.g. big A, small a), many of our examples will follow that symbolism, even though the *Drosophila* symbolism in many ways is more informative. That is, with the *Drosophila* system, the wild-type and mutant alleles are readily apparent since the wild-type allele is indicated by a +. The "Mendelian" system is commonly used in animal and plant breeding. A good reason for this is that, after many years (sometimes centuries) of breeding, it is no longer apparent what the "normal" (wild-type) gene is.

Morgan proposed that the gene for the eye color variant is located on the X chromosome. The condition of X-linked genes in males is said to be **hemizygous** since the gene is represented only one time in this case because they have no corresponding allele on the Y. For example, the white-eyed *Drosophila* males have an X chromosome with a white allele and no other allele of that gene in their genome: These males are said to be hemizygous for the white allele. Since the white allele of the gene is recessive, the original white-eyed male must have had the recessive allele for white eyes (designated w; see Box 3.1) on his X chromosome. The red-eyed female came from a true-breeding stock, so both of her X chromosomes must have carried the dominant allele for red eyes, w^+.

The F_1 flies are produced in the following way (see Figure 3.25a): The males receive their only X chromosome from their mother and hence have the w^+ allele and are red-eyed. The females receive a dominant w^+ allele from their mother and a recessive w allele from their father. As a result of this inheritance pattern, the F_1 females are also red-eyed.

To produce the F_2 flies, Morgan interbred F_1 red-eyed females and red-eyed males (see Figure 3.25b). In the F_2 the males that received an X chromosome with the w allele from their mother are white-eyed; those that received an X chromosome with the w^+ allele are red-eyed. The gene transmission shown in this cross from a male parent to a female offspring ("child") to a male grandchild is called **crisscross inheritance**.

Morgan also crossed a true-breeding white-eyed female (homozygous for the w allele) with a red-eyed male (hemizygous for the w^+ allele) (Figure 3.26). (This cross is the *reciprocal cross* of Morgan's first cross performed—white male × red female—shown in Figure 3.25.) All the F_1 females receive a w^+-bearing X

~ **FIGURE 3.26**

Reciprocal cross of that shown in Figure 3.25. (a) A homozygous white-eyed female is crossed with a red-eyed (wild-type) male; (b) The F_1 flies are interbred to produce the F_2s. The results of this cross differ from those in Figure 3.25 because of the way sex chromosomes segregate in crosses.

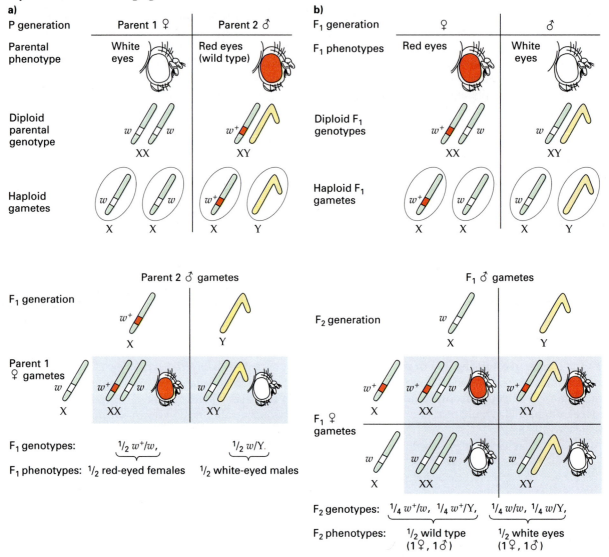

from their father and a w-bearing X from their mother (Figure 3.26a). Consequently, they are heterozygous w^+/w and have red eyes. All the F_1 males receive a w-bearing X from their mother and a Y from their father, and so they have white eyes (Figure 3.26a). This result is distinct from that of the cross in Figure 3.25. Further, these results are different from the normal results of a reciprocal cross because of the inheritance pattern of the X chromosome.

Interbreeding of the F_1 flies (Figure 3.26b) involves a $w/$ male and a w^+/w female giving approximately equal number of male and female red and white-eyed flies in the F_2. This ratio differs from the results obtained in the first cross, where an approximately 3:1 ratio of red-eyed:white-eyed flies

was obtained and where none of the females and approximately half the males exhibited the white-eyed phenotype. The difference in phenotypic ratios in the two sets of crosses reflects the transmission patterns of sex chromosomes and the genes they contain.

Morgan's crosses of *Drosophila* involved eye color characteristics that we now know are coded for by genes found on the X chromosome. These characteristics and the genes that give rise to them are referred to as **sex-linked**, or, more correctly, **X-linked** because the gene locus is part of the X chromosome. *X-linked inheritance* is the term used for the pattern of hereditary transmission of X-linked genes. When the results of reciprocal crosses are not the same, and different ratios are seen for the two sexes of the offspring, sex-linked characteristics may well be involved. By comparison, the results of reciprocal crosses are always the same when they involve genes located on the **autosomes** (chromosomes other than the sex chromosomes). Most importantly, Morgan's results strongly supported the hypothesis that genes were located on chromosomes. Morgan found many other examples of genes on the X chromosome in *Drosophila* and in other organisms, thereby showing that his observations were not confined to a single species. Later in this chapter we will discuss the analysis of X-linked traits in humans.

KEYNOTE

Sex linkage is the linkage of genes with the sex chromosomes of eukaryotes. Such genes, as well as the phenotypic characteristics these genes control, are referred to as sex-linked. Those genes only on the X chromosome are called X-linked. Morgan's pioneering work with the inheritance of sex-linked genes of *Drosophila* strongly supported, but did not prove, the chromosome theory of inheritance.

Nondisjunction of X Chromosomes

Proof for the chromosome theory of inheritance came from the work of Morgan's student, Calvin Bridges. Morgan's work showed that from a cross of a white-eyed female (w/w ♀) with a red-eyed male (w^+/Y ♂), all the F_1 males should be white-eyed, and all the females should be red-eyed. Bridges found that there are rare exceptions to this result: About 1 in 2,000 of the F_1 flies from such a cross are either white-eyed females or red-eyed males.

To explain these data, Bridges hypothesized that a problem had occurred with chromosome segregation in meiosis. Normally, homologous chromosomes (in meiosis I) or sister chromatids (in meiosis II or mitosis) move to opposite poles at anaphase (see p. 60). When this fails to take place, chromosome **nondisjunction** results. Nondisjunction can involve either autosomes or the sex chromosomes. For the crosses being analyzed, occasionally the two X chromosomes failed to separate, so eggs were produced either with two X chromosomes or with no X chromosomes instead of the usual one. This particular example of nondisjunction is called **X chromosome nondisjunction** (Figure 3.27). When it occurs in an individual with a normal set of chromosomes, it is

~ **FIGURE 3.27**

Nondisjunction in meiosis involving the X chromosome. (Nondisjunction of autosomal chromosomes and chromosomes in mitosis occurs in the same way.) (a) Normal X chromosome segregation in meiosis; (b) Nondisjunction of X chromosomes in meiosis I; (c) Nondisjunction of X chromosomes in meiosis II.

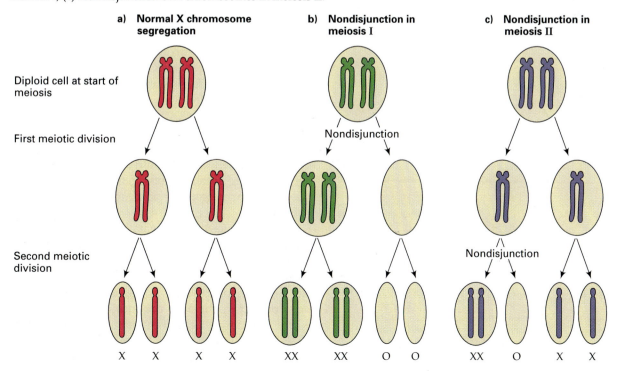

called **primary nondisjunction**. Normal disjunction of the X chromosomes is illustrated in Figure 3.27a, and nondisjunction of the X chromosomes in meiosis I and meiosis II is shown in Figures 3.27b and 3.27c, respectively.

Nondisjunction of the X chromosomes can explain the exceptional flies in Bridges's first experimental cross. When primary nondisjunction occurs in the w/w female (Figure 3.28), two classes of exceptional eggs result with equal and low frequency: those with two X chromosomes and those with no X chro-

~ **FIGURE 3.28**

Rare primary nondisjunction during meiosis in a white-eyed female *Drosophila melanogaster*, and results of a cross with a normal red-eyed male. Note the decreased viability of XXX and YO progeny.

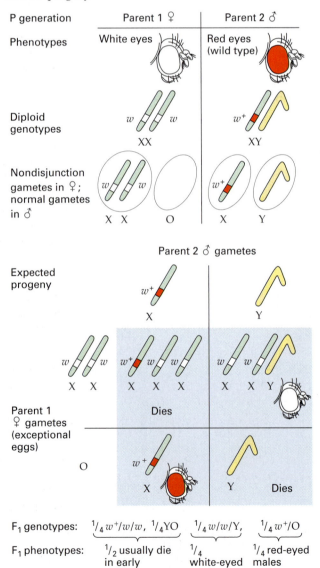

P generation	Parent 1 ♀	Parent 2 ♂
Phenotypes	White eyes	Red eyes (wild type)
Diploid genotypes	w w XX	w^+ XY
Nondisjunction gametes in ♀; normal gametes in ♂	w w X X O	w^+ X Y

Expected progeny

| | | Parent 2 ♂ gametes | |
		w^+ X	Y
Parent 1 ♀ gametes (exceptional eggs)	w w X X	w^+ w w X X X Dies	w w X X Y
	O	w^+ X	Y Dies

F₁ genotypes:	$\frac{1}{4}w^+/w/w$, $\frac{1}{4}$YO	$\frac{1}{4}w/w/$Y,	$\frac{1}{4}w^+/$O
F₁ phenotypes:	$\frac{1}{2}$ usually die in early development	$\frac{1}{4}$ white-eyed females	$\frac{1}{4}$ red-eyed males (sterile)

mosomes. The XY male is $w^+/$Y, and produces equal numbers of w^+- and Y-bearing sperm. When these eggs are fertilized by the two types of sperm, the result is four types of zygotes. Two types usually do not survive: the YO class, which has a Y chromosome but no X chromosome, and the triplo-X (XXX) class, which has three X chromosomes and, as illustrated in Figure 3.28, the genotype $w/w/w^+$. The YO die because they lack the X chromosome and its genes that code for essential cell functions. The XXX often die because the flies apparently cannot function with the extra dose of X chromosome genes.

The surviving classes are the red-eyed XO males (in *Drosophila* the XO pattern produces a sterile male), with no Y chromosome and a w^+ allele on the X, and the white-eyed XXY females (in *Drosophila*, XXY produces a fertile female), with a w allele on each X. The males are red-eyed because they received their X chromosome from their fathers, and the females have white eyes because their two X chromosomes came from their mothers. This result is unusual, since sons normally get their X from their mothers, and daughters get one X from each parent.

Bridges's hypothesis was confirmed by examining the chromosome composition of the exceptional flies: the white-eyed females were XXY and the red-eyed males were XO. (**Aneuploidy** is the term used for the abnormal condition—as is the case here—in which one or more whole chromosomes of a normal set of chromosomes either are missing, or are present in more than the usual number of copies.)

Bridges tested his hypothesis further by crossing the exceptional white-eyed XXY females with normal red-eyed XY males (Figure 3.29). The XXY female is homozygous for the w allele on her two Xs. The male has the w^+ allele on his X. Both parents have no equivalent eye color allele on their Y chromosomes. The two X chromosomes of the XXY parent were expected to segregate into different gametes: one X gamete and one XY gamete. Fusion with an X-bearing sperm from the male would give XX and XXY progeny, respectively, both of which would be heterozygous w^+/w and therefore would have red eyes. Again, flies with unexpected phenotypes resulted from this cross: A low percentage of the male progeny had red eyes and a similarly low percentage of the female progeny had white eyes.

To account for these unusual phenotypes, Bridges hypothesized that segregation of chromosomes in the meiosis of an XXY female can occur in two ways. In normal disjunction, the two X chromosomes separate and migrate to opposite poles, with one of them accompanied by the Y, to produce equal numbers of X- and XY-bearing eggs. This pattern is the one that the Xs should follow during meiosis (Figure 3.29a).

~ FIGURE 3.29

Results of a cross between the exceptional white-eyed XXY female of Figure 3.28 with a normal red-eyed XY male. Again, XXX and YY progeny usually die. (a) Normal disjunction of the X chromosomes in the XXY female. (b) Secondary nondisjunction of the homologous X chromosomes in meiosis I of the XXY female.

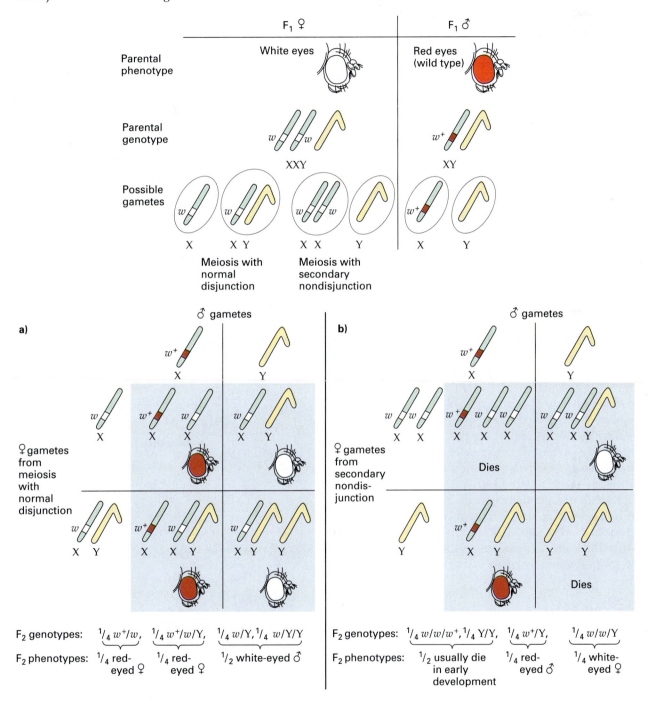

In the second pattern, which takes place only about 4 percent of the time, there is nondisjunction of the Xs (Figure 3.29b). Bridges called this segregation **secondary nondisjunction** because it occurred in the progenies of females that were produced by primary nondisjunction. Secondary nondisjunction results in the two Xs migrating together to one pole and the Y migrating to the other; the eggs are XX and Y. When these eggs are fertilized by the two classes of sperm (X and Y), the two surviving classes are the exceptional red-eyed (XY) males and white-eyed (XXY) females. As indicated above, the other two classes, the XXX and YO, die, usually early in development. Bridges verified his secondary nondisjunction hypothesis by microscopic examination of the chromosomes of the flies collected from the cross.

Thus, Bridges's experiments showed that the odd pattern of inheritance always went hand in hand with the specific aneuploid types (XO and XXY). Such a correlation could not be due to some accidental parallelism, proving without any doubt that a specific phenotype was associated with a specific chromosome complement.

In sum, gene segregation patterns follow the patterns of chromosome behavior in meiosis. In Figure 3.30 this parallel is illustrated for a diploid cell with a homologous pair of metacentric chromosomes and a homologous pair of acrocentric chromosomes. The cell is genotypically *Aa Bb* with the *A/a* gene pair on the metacentric homologs and the *B/b* gene pair on the telocentric homologs. As the figure shows, the two homologous pairs of chromosomes align on the metaphase plate independently, giving rise to two different segregation patterns for the two gene pairs. Since each of the two alignments, and hence segregation patterns, is equally likely, meiosis results in cells that exhibit equal frequencies of the genotypes *AB*, *ab*, *Ab*, and *aB*. Genotypes *AB* and *ab* result from one chromosome alignment, and genotypes *Ab* and *aB* result from the other alignment. In terms of Mendel's laws, we can see how the principle of segregation (two members of a gene pair segregate from each other in the formation of gametes) applies to the segregation pattern of one homologous pair of chromosomes and the associated gene pair (e.g., the metacentric chromosomes and gene pair *A/a* in Figure 3.30), while the principle of independent assortment (genes for different traits assort independently of one another during the production of gametes) applies to the segregation pattern of both homologous pairs of chromosomes and the associated two gene pairs in Figure 3.30.

KEYNOTE

Bridges observed an unexpected inheritance pattern of an X-linked mutant gene in *Drosophila*. He correlated this pattern directly with a rare event during meiosis, called nondisjunction, in which members of a homologous pair of chromosomes do not segregate to the opposite poles. The correlation between gene segregation patterns and the patterns of chromosome behavior in meiosis supported the chromosome theory of inheritance.

~ **FIGURE 3.30**

Illustration of the parallel behavior between Mendelian genes and chromosomes in meiosis. In the hypothetical *Aa Bb* diploid cell there is a homologous pair of metacentric chromosomes, which carry the *A/a* gene pair, and a homologous pair of telocentric chromosomes, which carry the *B/b* gene pair. The independent alignment of the two homologous pairs of chromosomes at metaphase I results in equal frequencies of the four meiotic products, *AB*, *ab*, *Ab*, and *aB*, illustrating Mendel's principle of independent assortment.

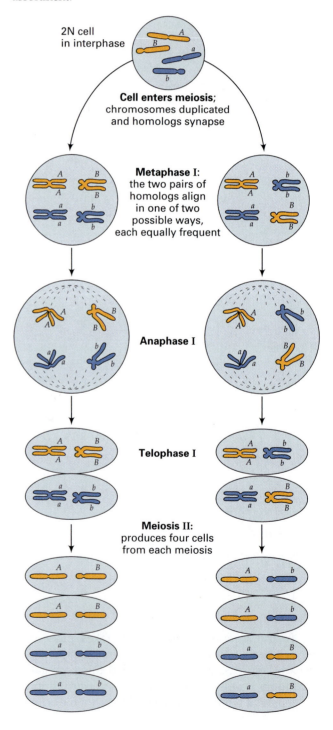

SEX DETERMINATION

In this section some of the mechanisms for sex determination will be discussed. In *genotypic sex determination systems*, sex is governed by the genotype of the zygote or spores; in *environmental sex determination systems*, sex is governed by internal and external environmental conditions.

Genotypic Sex Determination Systems

Genotypic sex determination, in which the sex chromosomes play a decisive role in the inheritance and determination of sex, may occur in one of two ways. In the **Y chromosome mechanism of sex determination** (seen in humans), the Y chromosome of the heterogametic sex is active in determining the sex of an individual. Individuals carrying the Y chromosome are genetically male, while individuals lacking the Y chromosome are genetically female. In the **X chromosome-autosome balance system** (seen in *Drosophila* and the nematode, *Caenorhabditis elegans*), the main factor in sex determination is the *ratio between the number of X chromosomes and the number of sets of autosomes*. In this system the Y chromosome has no effect on sex determination, but is required for male fertility.

SEX DETERMINATION IN MAMMALS. In humans and other placental mammals the mode of sex determination is the Y-chromosome mechanism of sex determination. If there is no Y chromosome present, the gonadal primordia develop into ovaries.

Evidence for the Y-Chromosome Mechanism of Sex Determination. The early evidence for the Y-chromosome basis of sex determination mechanism in mammals came from studies in which meiotic nondisjunction produced an abnormal sex chromosome complement. While sex determination in these cases follows directly from the sex chromosomes present, those individuals with unusual chromosome complements display many unusual characteristics.

Nondisjunction, for example, can produce exceptional XO individuals who have an X but no Y chromosome. In humans, XO individuals with the normal two sets of autosomes are female and sterile, and they exhibit **Turner syndrome**. A Turner syndrome individual is shown in Figure 3.31a, and her set of metaphase chromosomes is shown in Figure 3.31b. (A complete set of metaphase chromosomes in a cell is called its **karyotype**; this is described fully in Chapter 11.) Note that there is only one sex chromosome—an X chromosome. These aneuploid females have a genomic complement of 45,X, indicating that they have a total of 45 chromosomes (sex chromosomes + autosomes), in contrast to the normal 46, and that the sex chromosome complement consists of one X chromosome. Turner syndrome individuals occur with a

~ FIGURE 3.31

Turner syndrome (XO): (a) Individual; (b) Karyotype.

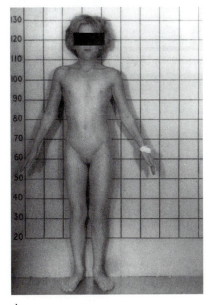
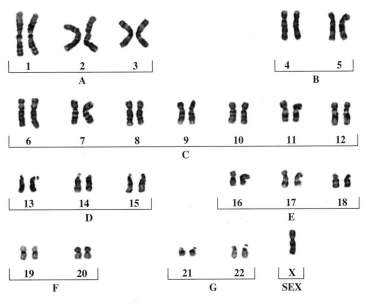

a)

b)

frequency of 1 in 10,000 female births. It is estimated that up to 99 percent of all 45,X embryos die before birth. Surviving Turner syndrome individuals have few noticeable major defects until puberty, when they fail to develop secondary sexual characteristics. They tend to be shorter than average, and they have web-like necks, poorly developed breasts, and immature internal sexual organs. They have a reduced ability to interpret spatial relationships and are usually infertile. All of these defects in XO individuals indicate that two X chromosomes are needed for normal development in females.

Nondisjunction can also result in the generation of XXY humans, which are male and have **Klinefelter syndrome**. Figure 3.32a shows a Klinefelter syndrome individual, and Figure 3.32b shows a karyotype for such an individual. Note the presence of two X chromosomes and one Y chromosome. About 1 in 1,000 males born have Klinefelter syndrome. These 47,XXY males have underdeveloped testes, and are often taller than the average male. Some degree of breast development is seen in about 50 percent of affected individuals, and some show subnormal intelligence. Individuals with similar phenotypes are also found with higher numbers of X and/or Y chromosomes, e.g., 48,XXXY, 49,XXXXY and 48,XXYY. The defects in Klinefelter individuals indicate that one X

and one Y chromosome are needed for normal development in males.

Some individuals have one X and two Y chromosomes; they have *XYY syndrome*. These 47,XYY individuals are male because of the presence of the Y. The XYY karyotype results from nondisjunction of the Y chromosome in meiosis. About 1 in 1,000 males born have XYY syndrome. They tend to be taller than average and occasionally there are adverse effects on fertility.

About 1 in 1,000 females born have three X chromosomes instead of the normal two. These 47,XXX (triplo-X) females are mostly completely normal, although they are slightly less fertile and a small number have less than average intelligence.

Table 3.2 summarizes the consequences of exceptional X and Y chromosome complements in humans. In each case the normal two sets of autosomes are associated with the sex chromosomes.

Dosage Compensation Mechanism for Extra X Chromosomes. Mammals can tolerate abnormalities in the complement of sex chromosomes quite well, whereas, with rare exceptions, they cannot tolerate any variation in the number of autosomes. That is, mammals with an unusual number of autosomes usually die. This is because there is a **dosage compen-**

~ **FIGURE 3.32**

Klinefelter syndrome (XXY): (a) Individual; (b) Karyotype.

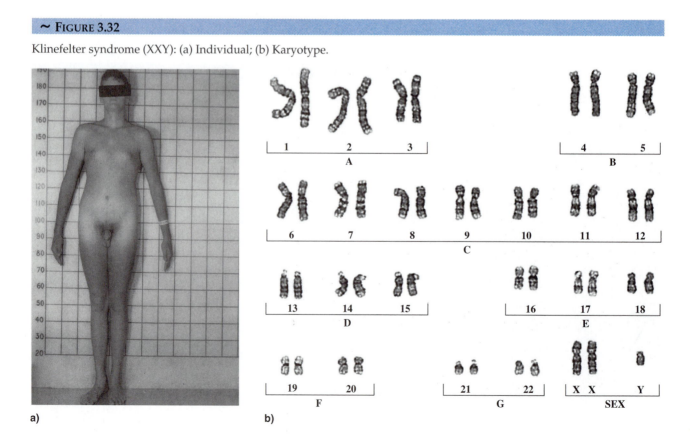

a)   b)

~ **TABLE 3.2**

~ **TABLE 3.2**

Consequences of Various Numbers of X- and Y-Chromosome Abnormalities in Humans, Showing Role of the Y in Sex Determination

CHROMOSOME CONSTITUTION[a]	DESIGNATION OF INDIVIDUAL	EXPECTED NUMBER OF BARR BODIES
46,XX	Normal ♀	1
46,XY	Normal ♂	0
45,X	Turner syndrome ♀	0
47,XXX	Triplo-X ♀	2
47,XXY	Klinefelter syndrome ♂	1
48,XXXY	Klinefelter syndrome ♂	2
48,XXYY	Klinefelter syndrome ♂	1
47,XYY	XYY syndrome ♂	0

[a]The first number indicates the total number of chromosomes in the nucleus, and the Xs and Ys indicate the sex chromosome complement.

sation mechanism in mammals that compensates for X chromosomes in excess of one and no such mechanism exists for extra autosomes. If we examine the nuclei of normal XX females we can see a highly condensed mass of chromatin—named the **Barr body** after its discoverer, Murray Barr—not found in the nuclei of normal XY male cells. That is, the somatic cells of XX individuals have one Barr body, and the somatic cells of XY individuals have no Barr bodies (Figure 3.33). In 1961 this concept was expanded by Mary Lyon and Lillian Russell. In what is now called the *Lyon hypothesis*, the following was proposed:

1. The Barr body is a highly condensed and (mostly) genetically inactive X chromosome (it has become "lyonized"; the process is called **lyonization**).
2. The inactivation occurs at about the sixteenth day following fertilization.
3. The X chromosome to be inactivated is randomly chosen from the maternal and paternal X chromosomes in a process that is independent from cell to cell. (Once a maternal or paternal X chromosome is inactivated in a cell, all descendants of that cell inherit the inactivation pattern.)

Because of X-chromosome inactivation, mammalian females heterozygous for X-linked traits are effectively genetic mosaics; that is, some cells show the phenotypes of one X chromosome, while the other cells show the phenotypes of the other X chromosome.

When lyonization operates in cells with extra X chromosomes, all but one of the X chromosomes typically become inactivated to produce Barr bodies. Thus, a general formula for the number of Barr bodies is the number of X chromosomes minus one. This dosage compensation process therefore minimizes the effects of multiple copies of X-linked genes. In normal individuals, lyonization results in males and females each with one active X chromosome per cell.

The number of Barr bodies associated with the normal and abnormal human X chromosome constitutions we have discussed are given in Table 3.2.

~ **FIGURE 3.33**

Barr bodies: (a) Nucleus of normal human female cells (XX), showing Barr bodies; (b) Nucleus of normal human male cells (XY), showing no Barr bodies.

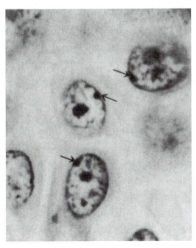

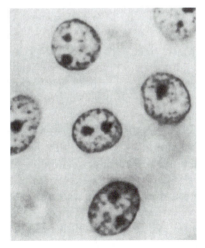

a)   b)

Normal 46,XX females have one Barr body, while normal 46,XY males have no Barr bodies. Turner syndrome females (45,X) have no Barr bodies; 47,XXY Klinefelter syndrome males have one Barr body, triplo-X (47,XXX) females have two Barr bodies, and 47,XYY syndrome males have no Barr bodies.

The process of X-inactivation begins at the *X-inactivation center* (*Xic* in mice/*XIC* in humans) and spreads in both directions. The *Xic/XIC* must contain elements both for counting chromosomes and for initiating X inactivation. The first gene identified within the human *XIC* region had properties that made it likely to have a role in X inactivation. This *XIST* gene (*Xist* in mice)(for X inactive specific transcripts) is expressed from the inactive X rather than from the active X; this is the opposite of other X-linked genes. Other genes are known that escape inactivation, but these are on *both* the active and inactive X chromosomes. The *XIST/Xist* gene encodes a 15 kb (1 kb = 1,000 bases; so, 15,000 bases) RNA molecule that is not translated into protein.

Recently, 450 kb (450,000 base pairs) of the mouse X chromosome containing *Xic* and *Xist* was cloned using recombinant DNA technology (see Chapter 15) and was introduced into autosomes of male mouse cells in tissue culture. Organisms or cells that have extra genes introduced into their genomes by genetic manipulation are called **transgenic organisms or cells,** and the introduced gene is called a **transgene.** In normal male cells with one X chromosome, X-chromosome inactivation does not occur. However, in the transgenic mouse cells with *Xic* and *Xist* added to an autosome, chromosome inactivation was turned on, with either the X or the autosome becoming inactivated in a random fashion. This means that the genetically modified male cells showed properties typical of X-chromosome inactivation in normal female cells. Thus, the 450 kb of DNA with *Xic* and *Xist* must contain all the sequences that are necessary and sufficient for chromosome counting and for the initiation of X-chromosome inactivation.

The Gene on the Y Chromosome for Male Sex Determination. Since the Y chromosome determines maleness in placental mammals, it was hypothesized that the Y uniquely carries an important gene or genes that encodes a product that sets the switch toward male sexual differentiation. This product is called **testis-determining factor** and the corresponding hypothesized genes were called *TDF* (testis-*d*etermining *f*actor gene) in humans and *Tdy* (testis-*d*etermining gene on the *Y*) in mice. Testis-determining factor causes the gonadal primordia to differentiate into testes instead of ovaries. This is the central event in sex determination of many mammals; all other dif-

ferences between the sexes are secondary effects resulting from hormone action or from the action of factors produced by the gonads. Therefore, sex determination is equivalent to testis determination.

A gene called *SRY* (sex-determining region Y) in humans and *Sry* in mice has been identified and molecularly cloned; these are most likely the *TDF* and *Tdy* genes. How was this gene found? In the human population, there are males who are XX (instead of XY), and females who are XY (instead of XX). These unusual individuals are said to exhibit *sex reversal*. It was hypothesized that individuals showing sex reversal would have changes in the sex chromosomes affecting the testis-determining factor. Indeed, in XX males a small fragment from near the tip of the small arm of the Y chromosome had broken off during the production of gametes and become attached to one of their X chromosomes. Among XY females, a number had deletions of the same region of the Y chromosome. This suggested that the testis-determining factor gene was in that small segment of the Y. More careful molecular analysis of the DNA of XX males and XY females resulted in the identification of a male-specific gene sequence near the end of the small arm. That is, this DNA is present in XY males and in the unusual "XX" males, and it is absent in XX females and in the unusual "XY" females. This gene is the *SRY* gene. An equivalent gene, *Sry*, has been identified in mice. The *SRY* and *Sry* genes have many of the properties expected of a testis-determining gene. For example, the mouse *Sry* gene is expressed *only* at the time and place expected for the testis-determining factor; that is, in the undifferentiated genital ridges of the embryo just before the formation of testes. Further, in an experiment involving recombinant DNA technology, a DNA clone containing the *Sry* gene was introduced into XX mouse embryos by microinjection. This experiment showed that the DNA fragment carrying *Sry* is sufficient to induce testis differentiation and subsequent normal male secondary sexual development; that is, a full phenotypic sex reversal in an XX transgenic adult. The proteins encoded by *SRY* and *Sry* are probably transcription factors, meaning that they bind to DNA and regulate expression of genes, most likely those involved in testis differentiation.

SEX DETERMINATION IN *DROSOPHILA*. *Drosophila melanogaster* has four pairs of chromosomes: one pair of sex chromosomes and three pairs of autosomes. In this organism, the homogametic sex is the female (XX) and the heterogametic sex is the male (XY). However, an XXY fly is female and an XO fly is male, indicating that the sex of the fly is not the consequence of the presence or absence of a Y chromosome. In fact, the sex of the fly is determined by the ratio of the number of X chromosomes (X) to the number of *sets* of autosomes (A). Since *Drosophila* is diploid, there are two sets of autosomes in a wild-type fly, although abnormal numbers of sets can be produced as a result of nondisjunction. *Drosophila* exemplifies the *X chromosome–autosome balance system of sex determination.*

Table 3.3 presents some chromosome complements and the sex of the resulting flies. In a normal female there are two Xs and two sets of autosomes; hence the X:A ratio is 1.00. A normal male has a ratio of 0.50. If the X:A ratio is greater than or equal to 1.00, the fly will be female; if the X:A ratio is less than or equal to 0.50 the fly will be male. If the ratio is between 0.50 and 1.00, the fly is neither male nor female; it is an intersex. Intersex flies are variable in appearance, generally having complex mixtures of male and female attributes for the internal sex organs and external genitalia. Such flies are sterile.

Sex determination in the nematode *Caenorhabditis elegans* (Figure 3.34) also occurs by the X chromosome–autosome balance system. *C. elegans* has become a popular tool for developmental geneticists because an adult has only about a thousand cells and the lineages of every one of those cells has been carefully defined from egg to adult. *C. elegans* has two sexual types: hermaphrodites and males. Most individuals are **her-**

maphroditic; that is, they have both sex organs, the ovaries and testes. They make sperm when they are larvae and store that sperm as development continues. In the adult, the ovary produces eggs that are fertilized by the stored sperm as the eggs migrate to the uterus. Self-fertilization in this way almost always produces more hermaphrodites. However, 0.2 percent of the time males are produced from self-fertilization. These males can fertilize hermaphrodites if the two mate, and such matings result in about equal numbers of hermaphrodite and male progeny because their sperm has a competitive advantage over the sperm stored in the hermaphrodite. Genetically, hermaphrodites are XX and males are XO; that is, an X chromosome–autosome ratio of 1.00 results in hermaphrodites and a ratio of 0.50 results in males.

SEX CHROMOSOMES IN OTHER ORGANISMS. Not all organisms have an X-Y sex chromosome makeup like that found in mammals and in *Drosophila*. In birds, butterflies, moths, and some fish, the sex chromosome composition is the opposite of that in mammals: The male in these organisms is the homogametic sex and the female is the heterogametic sex. To prevent confusion with the X and Y chromosome convention, we designate the sex chromosomes in these organisms as Z and W. Thus the males are ZZ and the females are ZW. Genes on the Z chromosome behave just like X-linked genes in the earlier examples except that hemizygosity is found only in females. All the daughters of a male homozygous for a Z-linked recessive gene express the recessive trait, and so on. Figure 3.35 (p. 83) presents an example, the inheritance of the sex-linked (Z-linked), dominant barred plumage (*B*) in poultry.

Higher plants exhibit quite a variety of sexual situations. Some species (the ginkgo, for example) have plants of separate sexes, with male plants producing flowers that contain only stamens and female plants producing flowers that contain only pistils. These species are called **dioecious**. Other species have both male and female sex organs on the same plant. If the sex organs are in the same flower, the plant is hermaphroditic (e.g., the rose and the buttercup), and the flower is said to be a *perfect flower*. If the sex organs are in different flowers on the same plant, it is said to be **monoecious** (= hermaphroditic; e.g., corn), and the flower is said to be an *imperfect flower*.

Some dioecious plants have sex chromosomes that differ between the sexes, and a large proportion of these plants have an X-Y system. Such plants typically have an X chromosome–autosome balance system of sex determination like that in *Drosophila*. However, many other sex determination systems are seen in dioecious plants.

~ TABLE 3.3

Sex Balance Theory of Sex Determination in *Drosophila melanogaster*

SEX CHROMOSOME COMPLEMENT	AUTOSOME COMPLEMENT (A)	X:A RATIO[a]	SEX OF FLIES
XX	AA	1.00	♀
XY	AA	0.50	♂
XXX	AA	1.50	Metafemale (sterile)
XXY	AA	1.00	♀
XXX	AAAA	0.75	Intersex (sterile)
XX	AAA	0.67	Intersex (sterile)
X	AA	0.50	♂ (sterile)

[a]If the X chromosome:autosome ratio is greater than or equal to 1.00 (X:A ≥ 1.00), the fly will be a female. If the X chromosome:autosome ratio is less than or equal to 0.50 (X:A ≤ 0.50), the fly will be male. Between these two ratios, the fly will be an intersex.

~ **FIGURE 3.34**

The nematode, *Caenorhabditis elegans*: (a) Photograph and (b) Drawing of a hermaphrodite (XX); (c) Photograph and (d) Drawing of a male (XO).

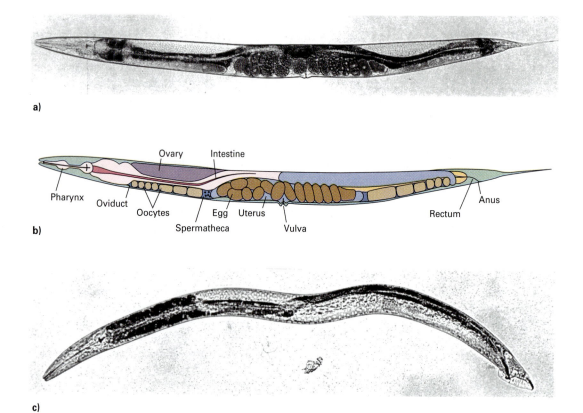

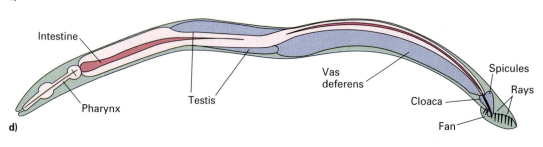

Many species, particularly eukaryotic microorganisms, do not have sex chromosomes but instead rely on a *genic system* for the determination of sex, that is, a system in which the sexes are specified by simple allelic differences at a small number of gene loci. For example, the yeast *Saccharomyces cerevisiae* is a haploid eukaryote that has two "sexes"—*a* and α— referred to as **mating types**. The mating types are morphologically indistinguishable, but crosses can occur only between individuals of opposite type. These mating types are controlled by the *MATa* and *MAT*α alleles, respectively, of a single gene.

Environmental Sex Determination Systems

In **environmental sex determination**, the environment plays a major role in sex determination. These types of sex determination mechanisms are much rarer than the ones we have discussed up to now. In the marine worm *Bonellia* (Figure 3.36, p. 84), for example, the free-swimming larval forms are sexually undifferentiated. If an individual settles down alone, it becomes female. If a larva attaches to the body of an adult female, the larva will differentiate into a male. Thus sex differentiation in *Bonellia* is not determined

~ **FIGURE 3.35**

Sex-linked inheritance in chickens. The barred-feather (*B*) phenotype is caused by a gene that is dominant to the allele for nonbarred (*b*) plumage. (a) A barred-feather female is crossed with a nonbarred male; (b) An F₁ nonbarred female is crossed with an F₁ barred male.

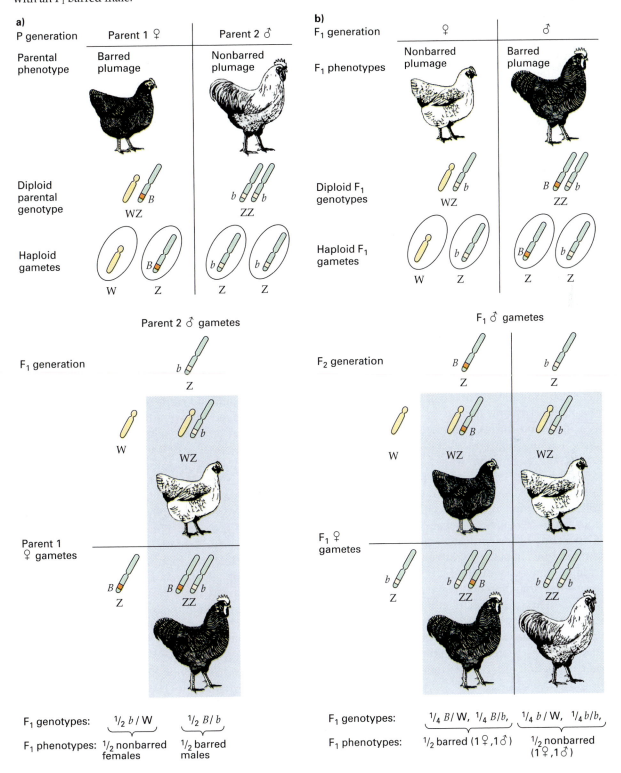

F₁ genotypes: ½ *b* / W   ½ *B* / *b*

F₁ phenotypes: ½ nonbarred females   ½ barred males

F₁ genotypes: ¼ *B* / W, ¼ *B* / *b*,   ¼ *b* / W, ¼ *b* / *b*,

F₁ phenotypes: ½ barred (1♀, 1♂)   ½ nonbarred (1♀, 1♂)

~ FIGURE 3.36

Drawing of a female and male *Bonellia,* a marine worm.

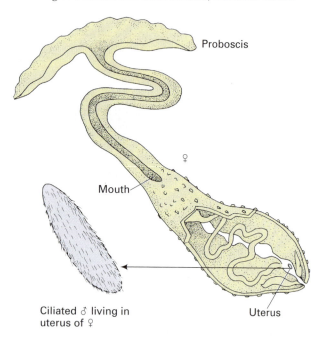

at fertilization by some genetic component but is probably directed by environmental factors related either to the association or lack of association with other members of the species.

There are also examples where temperature plays a major role in determining the sex of the progeny. For example, in certain turtles, eggs incubated above 32°C produce females, eggs incubated below 28°C produce males, and eggs in between these temperatures produce a mixture of males and females. That there is no common system involved in this temperature control of sex is shown by the snapping turtles, eggs of which produce females *either* at 20°C or below *or* at 30°C or above, and produce predominantly males at temperatures in between. Interestingly, it has been speculated that this method of temperature-dependent sex determination may have been used by dinosaurs. If so, then a slight change in environmental temperature might have changed the conditions so that only males or females hatched from eggs, therefore eliminating the possibility of breeding and accelerating their extinction. Of course, there is no firm evidence for this idea.

𝒦EYNOTE

Many eukaryotic organisms have sex chromosomes that are represented differentially in the two sexes; in humans and many other mammals the male is XY and the female is XX. In other eukaryotes with sex chromosomes, the male is ZZ and the female is ZW. In many cases sex determination is related to the sex chromosomes. For humans and many other mammals, for instance, the presence of the Y chromosome confers maleness, and its absence results in femaleness. *Drosophila* and *Caenorhabditis* have an X chromosome–autosome balance system of sex determination: The sex of the individual is related to the ratio of the number of X chromosomes to the number of sets of autosomes. Several other sex-determining systems are known in the eukaryotes, including genic systems, found particularly in the lower eukaryotes, and phenotypic (environmental) systems.

ANALYSIS OF SEX-LINKED TRAITS IN HUMANS

In this section we will discuss examples of the analysis of X-linked and Y-linked traits in humans. Recall that, in Chapter 2, we introduced the analysis of recessive and dominant traits in humans. Those traits were not sex-linked but were the result of alleles carried on autosomes; that is, chromosomes other than the sex chromosomes. For the analysis of all pedigrees, whether the trait is autosomal or X-linked, it is important to realize that collecting reliable human pedigree data is a difficult task. Since one has to rely on the family's recollections in many cases, the accuracy of the pedigree data is open to question. Also, there may not be enough affected people to allow an unambiguous determination of the mechanism of inheritance involved, especially when the trait is rare and when the family is small. Further, the expression of a trait may vary, resulting in some individuals erroneously being classified as normal. Lastly, the same mutant phenotype could result from mutations in more than one gene, making it possible that different pedigrees will indicate, correctly, that different mechanisms of inheritance are involved for the "same" trait.

X-Linked Recessive Inheritance

A trait due to a recessive mutant allele carried on the X chromosome is called an **X-linked recessive trait.** At least 100 human traits are known for which the gene has been traced to the X chromosome. Most of the traits involve X-linked recessive alleles. The best known X-linked recessive pedigree is that of hemophilia A in Queen Victoria's family (Figure 3.37). Hemophilia (the bleeder's disease) is a serious ailment in which the blood lacks a clotting factor. A cut or even a serious bruise can be fatal to a hemophiliac. In Queen Victoria's pedigree the first instance of hemophilia was in one of her sons, so she was either a carrier for this trait, or a mutation had occurred in her germ cells.

In X-linked recessive traits the female usually must be homozygous for the recessive allele in order to express the mutant trait. The trait is expressed in the male who possesses but one copy of the mutant allele on the X chromosome. Therefore, affected males

normally transmit the mutant gene to all their daughters but to none of their sons. The instance of a father-to-son inheritance of a rare trait in a pedigree would tend to rule out X-linked recessive inheritance.

Other characteristics of X-linked recessive inheritance are as follows (refer to Figure 3.37):

1. For X-linked recessive mutant alleles many more males than females should exhibit the trait, owing to the different number of X chromosomes in the two sexes.
2. All sons of a homozygous mutant mother should show the trait, since males receive their only X chromosome from their mothers.
3. The sons of heterozygous (carrier) mothers should show an approximately 1:1 ratio of nor-

~ **FIGURE 3.37**

(a) Painting of Queen Victoria as a young woman. (b) Pedigree of Queen Victoria (III.2) and her descendants, showing the inheritance of hemophilia. [Refer to Figure 2.19, p. 38, for an explanation of symbols used in pedigrees. In the pedigree shown here, in many cases the marriage partner has been omitted as a shorthand way to save space. Those marriage partners were normal with respect to the trait.] Since Queen Victoria was heterozygous for the sex-linked recessive hemophilia allele with no cases in her ancestors, the trait may have arisen as a mutation in one of her parents' germ cells (the cells that give rise to the gametes).

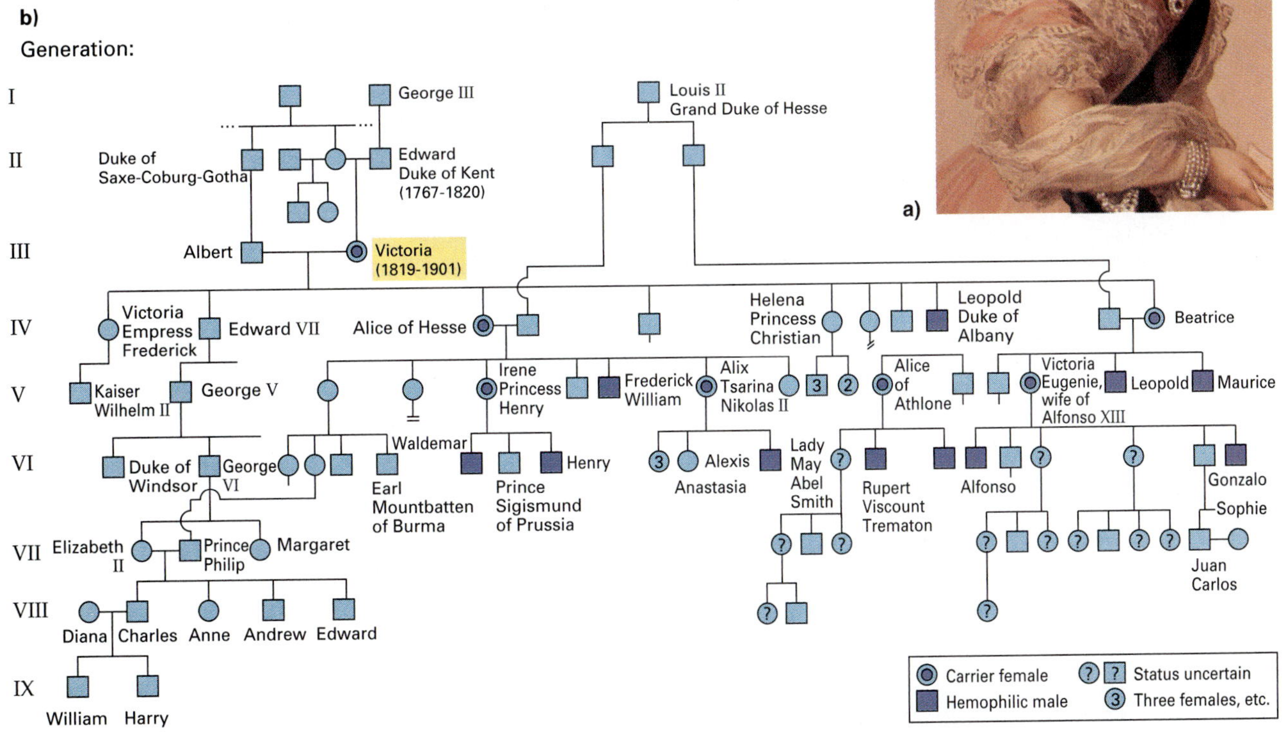

mal individuals to individuals expressing the trait. That is, $a^+/a \times a^+/Y$ gives half a^+/Y and half a/Y sons.

4. From a mating of a carrier female with a normal male all daughters will be normal, but half will be carriers. That is, $a^+/a \times a^+/Y$ gives half a^+/a^+ and half a^+/a females. In turn, half the sons of these carrier females will exhibit the trait.

5. A male expressing the trait, when mated with a homozygous normal female, will produce all normal children, but all the female progeny will be carriers. That is, $a^+/a^+ \times a/Y$ gives a^+/a females and a^+/Y (normal) males.

X-Linked Dominant Inheritance

A trait due to a dominant mutant allele carried on the X chromosome is called an **X-linked dominant trait**.

Only a few X-linked dominant traits have been identified.

An example of an X-linked dominant trait that causes faulty tooth enamel and dental discoloration is shown in Figure 3.38a and a pedigree for this trait is shown in Figure 3.38b. Note that all the daughters and none of the sons of an affected father (III.1) are affected, and that heterozygous mothers (IV.3) transmit the trait to half their sons and half their daughters. Webbing to the tips of the toes in a family in South Dakota (studied in the 1930s) is also an X-linked dominant mutant trait, as is a severe bleeding anomaly called constitutional thrombopathy. In the latter (also studied in the 1930s), bleeding is not due to the absence of a clotting factor (as in hemophilia) but to interference in the formation of blood platelets, which are needed for blood clotting.

The X-linked dominant traits follow the same sort of inheritance rules as the X-linked recessives, except

~ FIGURE 3.38

(a) A person with the X-linked dominant trait of faulty enamel; (b) A pedigree showing the transmission of the faulty enamel trait. This pedigree illustrates a shorthand convention which omits parents who do not exhibit the trait. Thus it is a given that the female in generation I paired with a male who did not exhibit the trait.

a)

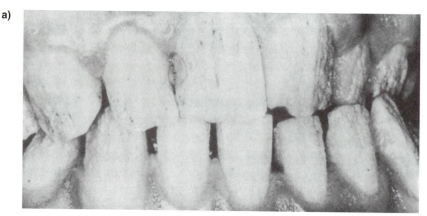

b) **Pedigree**

Generation:

that heterozygous females express the trait. In general, X-linked dominant traits tend to be milder in the female than in the male. Also, since females have twice the number of X chromosomes as males, X-linked dominant traits are more frequent in females than in males. If the trait is a rare one, females with the trait are likely to be heterozygous. These females will pass the trait on to 1/2 of their male progeny and to 1/2 of their female progeny. A male with an X-linked dominant trait will pass the trait on to all of his daughters and to none of his sons.

Y-Linked Inheritance

A trait due to a mutant gene carried on the Y chromosome but with no counterpart on the X is called a **Y-linked**, or **holandric** ("wholly male"), **trait**. Such traits should be readily recognizable, since every son of an affected male should have the trait and no females should ever express it. Several traits with Y-linked inheritance have been suggested. In most cases the genetic evidence for such inheritance is poor or nonexistent. In fact, there is no clear evidence for genetic loci on the Y chromosome other than those in the pseudoautosomal region (see p. 59) and the *SRY* gene.

A possible example of Y-linked inheritance is the hairy ears, or hairy pinnae, trait in which bristly hairs of atypical length grow from the ears (Figure 3.39). This trait is common in parts of India, although some other populations also exhibit it. While this trait shows father-to-son inheritance, there is no doubt that it is a complex phenotype, and many of the collected pedigrees can be interpreted in other ways, such as autosomal inheritance. The trait could also be the result of the interaction of a gene with the male hormone testosterone, similar to the appearance of hair on the face and chest. Another putative Y-linked trait is bilateral radio-ulnar synostosis. In this trait the radius and ulna bones of both arms are fused.

~ FIGURE 3.39

Individuals exhibiting the possibly Y-linked trait of hairy ears.

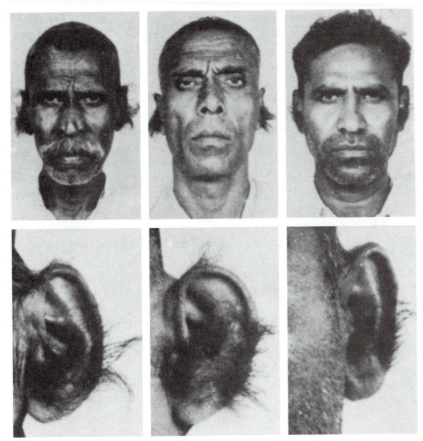

KEYNOTE

Controlled matings cannot be made in humans, so analysis of the inheritance of genes in humans must rely on pedigree analysis. This technique involves the careful study of the phenotypic records of the family extending over several generations. The data obtained from pedigree analysis enable geneticists to make judgments, with varying degrees of confidence, about whether a mutant gene is inherited as an autosomal recessive, an autosomal dominant, an X-linked recessive, an X-linked dominant, or a Y-linked allele.

SUMMARY

Mitosis and Meiosis

The genetic material of eukaryotes consists of double-stranded DNA distributed among several linear chromosomes. In these chromosomes, the DNA is complexed with many proteins. Typically, eukaryotic cells contain either one or two sets of chromosomes: the former are haploid cells and the latter are diploid cells. Proliferation of both haploid and diploid cells occurs by mitosis. Mitosis involves one round of DNA replication followed by one round of nuclear division, often accompanied by cell division. Thus, mitosis results in the production of daughter nuclei that contain identical chromosome numbers and that are genetically identical to one another and to the parent nucleus from which they arose.

In all sexually reproducing organisms, meiosis occurs at a particular stage in the life cycle. Meiosis is the process by which a diploid cell (never a haploid cell) or cell nucleus undergoes one round of DNA replication and two rounds of nuclear division to produce four specialized haploid cells or nuclei. The products of meiosis are gametes or meiospores. Unlike mitosis, meiosis generates genetic variability in two main ways: (1) through the various ways in which maternal and paternal chromosomes are combined in progeny nuclei; and (2) through crossing-over between maternally derived and paternally derived homologs to produce recombinant chromosomes with some maternal and some paternal genes.

Chromosome Theory of Inheritance

In Chapter 2, genes were considered as abstract entities that control hereditary characteristics. Prior to the beginning of this century, genes were called Mendelian factors. As a result of the efforts of cytologists working in the nineteenth century, information was being accumulated about cell structure and cell division. The fields of genetics and cytology came together in 1902 when Sutton and Boveri independently hypothesized that genes are on chromosomes, "bodies" within the cell nucleus. The chromosome theory of heredity states that the chromosome transmission from one generation to the next closely parallels the patterns of gene transmission from one generation to the next.

Support for the chromosome theory of inheritance came from experiments that related the hereditary behavior of particular genes to the transmission of the sex chromosome. The sex chromosome in eukaryotic organisms is the chromosome that is represented differently in the two sexes. In most organisms with sex chromosomes, the female has two X chromosomes, while the male has one X and one Y chromosome. The Y chromosome is structurally and genetically different from the X chromosome. The association of genes with the sex-determining chromosomes of eukaryotes is called sex linkage. Such genes, and the phenotypes they control, are called sex-linked.

Sex Determination Mechanisms

In this chapter, a number of sex determination mechanisms were discussed. In many cases sex determination is related to the sex chromosomes. In humans, for example, the presence of a Y chromosome specifies maleness, while its absence results in femaleness. The recent identification and characterization of a gene on the Y chromosome, *SRY/Sry*, the product of which directs male sexual differentiation, opens the way to a detailed analysis of the genetic control of sex determination in mammals. Several other sex determination systems are known in the eukaryotes, including X chromosome–autosome balance systems (in which sex is determined by the ratio of the number of X chromosomes to the number of sets of autosomes), genic systems (in which a simple allelic difference determines the sex of an individual), and phenotypic systems (in which sex is determined by environmental cues).

Analysis of Sex-Linked Traits in Humans

Also discussed in this chapter were sex-linked traits in humans. In Chapter 2 we discussed the fact that the inheritance patterns of traits in humans are usually studied by charting the family trees of individuals exhibiting the trait, a method called pedigree analysis. In this chapter we considered examples of X- and Y-linked human traits to illustrate the features of those mechanisms of inheritance in pedigrees. Note that evidence supporting Y-linked inheritance in most cases is questionable. Autosomal traits were discussed in Chapter 2. It is important to realize that collecting reliable human pedigree data is a difficult task. In many cases the accuracy of record keeping within the families involved is open to question. Also,

particularly with small families, there may not be enough affected people to allow an unambiguous determination of the inheritance mechanism involved, especially when rare traits are involved. Moreover, the degree to which a trait is expressed may vary, so that some individuals may be erroneously classified as normal. It is also possible for the same mutant phenotype to be produced by mutations in different genes, and therefore different pedigrees may, correctly, indicate different mechanisms of inheritance of the "same" trait.

ANALYTICAL APPROACHES FOR SOLVING GENETICS PROBLEMS

The concepts introduced in this chapter may be reinforced by solving genetics problems. The types of problems are similar to those introduced in Chapter 2. When sex linkage is involved, remember that one sex has two kinds of sex chromosomes, whereas the other sex has only one; this feature alters the inheritance patterns slightly. Most sample problems presented in this section center on interpreting data and predicting the outcome of particular crosses.

Q3.1 A female from a pure-breeding strain of *Drosophila* with vermilion-colored eyes is crossed with a male from a pure-breeding wild-type, red-eyed strain. All the F_1 males have vermilion-colored eyes, and all the females have wild-type red eyes. What conclusions can you draw about the mode of inheritance of the vermilion trait, and how could you test them?

A3.1 The observation is the classic one that suggests a sex-linked trait is involved. Since none of the F_1 daughters have the trait and all the F_1 males do, the trait is presumably X-linked recessive. The results fit this hypothesis, because the F_1 males receive the X chromosome from their homozygous v/v mother, with the v gene on the X chromosome. Furthermore, the F_1 females are v^+/v since they receive a v^+-bearing X chromosome from the wild-type male parent and a v-bearing X chromosome from the female parent. If the trait were autosomal recessive, all the F_1 flies would have had wild-type eyes. If it were autosomal dominant, both the F_1 males and females would have had vermilion-colored eyes. If the trait were X-linked dominant, all the F_1 flies would have had vermilion eyes.

The easiest way to verify this hypothesis is to let the F_1 flies interbreed. This cross is v^+/v ♀ × v/Y ♂, and the expectation is that there will be a 1:1 ratio of wild-type: vermilion eyes in both sexes in the F_2. That is, half the

females are v^+/v and half are v/v; half the males are v^+/Y and half are v/Y. This ratio is certainly not the 3:1 ratio that would result from an $F_1 \times F_1$ cross for an autosomal gene.

Q3.2 In humans, hemophilia A or B is caused by an X-linked recessive gene. A woman who is a nonbleeder had a father who was a hemophiliac. She marries a nonbleeder, and they plan to have children. Calculate the probability of hemophilia in the female and male offspring.

A3.2 Since hemophilia is an X-linked trait, and since her father was a hemophiliac, the woman must be heterozygous for this recessive gene. If we assign the symbol h to this recessive mutation and h^+ to the wild-type (nonbleeder) allele, she must be h^+/h. The man she marries is normal with regard to blood clotting and hence must be hemizygous for h^+, that is, h^+/Y. All their daughters receive an X chromosome from the father, and so each must have an h^+ gene. In fact, half the daughters are h^+/h^+ and the other half are h^+/h. Since the wild-type allele is dominant, none of the daughters are hemophiliacs. However, all the sons of the marriage receive their X chromosome from their mother. Therefore they have a probability of 1/2 that they will receive the chromosome carrying the h allele, which means they will be hemophiliacs. Thus the probability of hemophilia among daughters of this marriage is 0, and among sons it is 1/2.

Q3.3 Tribbles are hypothetical animals that have an X-Y sex determination mechanism like that of humans. The trait bald (*b*) is X-linked and recessive to furry (*b^+*), and the trait long leg (*l*) is autosomal and recessive to short leg (*l^+*). You make reciprocal crosses between true-breeding bald, long-legged tribbles and true-breeding furry, short-legged tribbles. Do you expect a 9:3:3:1 ratio in the F_2 of either or both of these crosses? Explain your answer.

A3.3 This question focused on the fundamentals of X chromosome and autosome segregation during a genetic cross and tests whether or not you have grasped the principles involved in gene segregation. Figure 3.A diagrams the two crosses involved, and we can discuss the answer by referring to it.

Let us first consider the cross of a wild-type female tribble (b^+/b^+, l^+/l^+) with a male double-mutant tribble (b/Y, l/l). Part (a) of the figure diagrams this cross. These F_1s are all normal—that is, with furry bodies and short legs—because for the autosomal character both sexes are heterozygous, and for the X-linked character the female is heterozygous and the male is hemizygous for the b^+ allele donated by the normal mother. With the production of the F_2 progeny the best approach is to treat the X-linked and autosomal traits separately. For the X-linked trait, random combination of the gametes produced gives a 1:1:1:1 genotypic ratio of b^+/b^+ (furry female) :

b^+/b (furry female) : b^+/Y (furry male) : b/Y (bald male) progeny. Collecting by phenotypes, we see that all the females are furry; half the males are furry and half are bald. For the autosomal leg trait the $F_1 \times F_1$ is a cross of two heterozygotes, so we expect a 3:1 phenotypic ratio of short-legged:long-legged tribbles in the F_2. Since autosome segregation is independent of the inheritance of the X chromosome, we can multiply the probabilities of the occurrence of the X-linked and autosomal traits to calculate their relative frequencies. The calculations are presented in part (a) of the figure, from which we see that the ratio of the four possible phenotypic classes differs in females and males.

The first cross, then, has a 9:3:3:1 ratio of the four possible phenotypes in the F_2. However, note that the ratio in each sex is not 9:3:3:1, owing to the inheritance pattern of the X chromosome. This result contrasts markedly with the pattern of two autosomal genes segregating independently, where the 9:3:3:1 ratio is found for both sexes.

FIGURE 3.A

a) Furry, short-legged (wild-type) ♀ ✕ bald, long-legged ♂

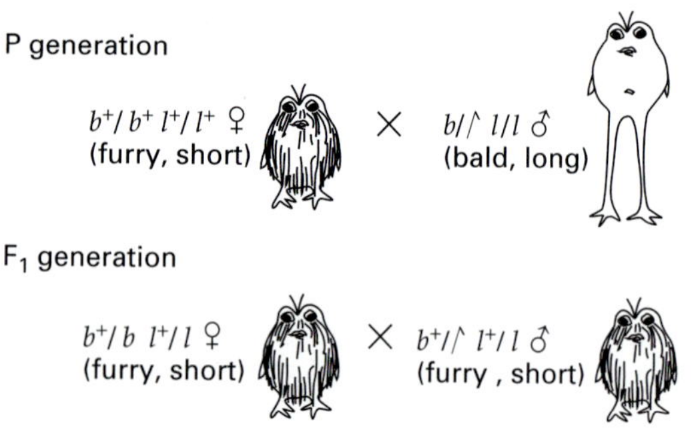

P generation

b^+/b^+ l^+/l^+ ♀
(furry, short)  ✕ $b/\curlywedge$ l/l ♂
(bald, long)

F_1 generation

b^+/b l^+/l ♀
(furry, short) 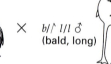 ✕ $b^+/\curlywedge$ l^+/l ♂
(furry , short)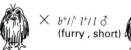

F_2 generation

Sex-linked phenotypes and genotypes	Autosomal phenotypes and genotypes

Genotypic results:

$\frac{1}{2}$ b^+ ($\frac{1}{2}$ b^+/b^+, $\frac{1}{2}$ b^+/b; furry) ♀ $\left\{\begin{array}{l}\frac{3}{4}\ l^+\ (l^+/l^+\ \text{and}\ l^+/l;\ \text{short})\\ \frac{1}{4}\ l\ (l/l\ ;\ \text{long})\end{array}\right.$

$\frac{1}{4}$ b^+ ($b^+/\curlywedge$; furry) ♂ $\left\{\begin{array}{l}\frac{3}{4}\ l^+\ (\text{short})\\ \frac{1}{4}\ l\ (\text{long})\end{array}\right.$

$\frac{1}{4}$ b ($b/\curlywedge$; bald) ♂ $\left\{\begin{array}{l}\frac{3}{4}\ l^+\ (\text{short})\\ \frac{1}{4}\ l\ (\text{long})\end{array}\right.$

Phenotypic ratios:

	Furry-short		Furry-long		Bald-short		Bald-long
	b^+l^+		b^+l		$b l^+$		$b l$
♀	6	:	2	:	0	:	0
♂	3	:	1	:	3	:	1
Total	9	:	3	:	3	:	1

b) Bald, long-legged ♂ ✕ furry, short-legged (wild-type) ♀

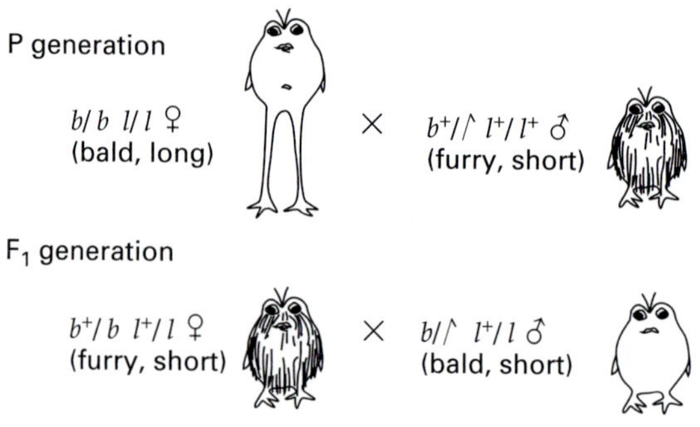

P generation

b/b l/l ♀
(bald, long) ✕ $b^+/\curlywedge$ l^+/l^+ ♂
(furry, short)

F_1 generation

b^+/b l^+/l ♀
(furry, short) ✕ $b/\curlywedge$ l^+/l ♂
(bald, short)

F_2 generation

Sex-linked phenotypes and genotypes	Autosomal phenotypes and genotypes

Genotypic results:

$\frac{1}{4}$ b^+ (b^+/b^+; furry) ♀ $\left\{\begin{array}{l}\frac{3}{4}\ l^+\ (l^+/l^+\ \text{and}\ l^+/l;\ \text{short})\\ \frac{1}{4}\ l\ (l/l\ ;\ \text{long})\end{array}\right.$

$\frac{1}{4}$ b (b/b; bald) ♀ $\left\{\begin{array}{l}\frac{3}{4}\ l^+\ (\text{short})\\ \frac{1}{4}\ l\ (\text{long})\end{array}\right.$

$\frac{1}{4}$ b^+ ($b^+/\curlywedge$; furry) ♂ $\left\{\begin{array}{l}\frac{3}{4}\ l^+\ (\text{short})\\ \frac{1}{4}\ l\ (\text{long})\end{array}\right.$

$\frac{1}{4}$ b ($b/\curlywedge$; bald) ♂ $\left\{\begin{array}{l}\frac{3}{4}\ l^+\ (\text{short})\\ \frac{1}{4}\ l\ (\text{long})\end{array}\right.$

Phenotypic ratios:

	Furry-short		Furry-long		Bald-short		Bald-long
	b^+l^+		b^+l		$b l^+$		$b l$
♀	3	:	1	:	3	:	1
♂	3	:	1	:	3	:	1
Total	6	:	2	:	6	:	2

The second cross (a reciprocal cross) is diagrammed in part (b) of the figure. Since the parental female in this cross is homozygous for the sex-linked trait, all the F_1 males are bald. Genotypically, the F_1 males and females differ from those in the first cross with respect to the sex chromosome but are just the same with respect to the autosome. Again, considering the X chromosome first as we go to the F_2, we find a 1:1:1:1 genotypic ratio of furry females : bald females : furry males : bald males. In this case, then, half of both males and females are furry and half are bald, in contrast to the results of the first cross in which no bald females were produced in the F_2. For the autosomal trait we expect a 3:1 ratio of short:long in the F_2, as before. Putting the two traits together, we get the calculations presented in part (b) of the figure. (Note: We use the total 6:2:6:2 here rather than 3:1:3:1 because the numbers add to 16, as does $9 + 3 + 3 + 1$.) So in this case we do not get a 9:3:3:1 ratio; moreover, the ratio is the same in both sexes.

This question has forced us to think through the segregation of two types of chromosomes and has shown that we must be careful about predicting the outcomes of crosses in which sex chromosomes are involved. Nonetheless, the basic principles for the analysis were no different from those used before: Reduce the questions to their basic parts and then put the puzzle together step by step.

QUESTIONS AND PROBLEMS

In 3.1 through 3.3, select the correct answer.

*3.1 Interphase is a period corresponding to the cell cycle phases of
a. mitosis.
b. S.
c. $G_1 + S + G_2$.
d. $G_1 + S + G_2 + M$.

3.2 Chromatids joined together by a centromere are called
a. sister chromatids.
b. homologs.
c. alleles.
d. bivalents (tetrads).

*3.3 Mitosis and meiosis always differ in regard to the presence of
a. chromatids.
b. homologs.
c. bivalents.
d. centromeres.
e. spindles.

3.4 State whether each of the following statements is true or false. Explain your choice.

a. The chromosomes in a somatic cell of any organism are all morphologically alike.
b. During mitosis the chromosomes divide and the resulting sister chromatids separate at anaphase, ending up in two nuclei, each of which has the same number of chromosomes as the parental cell.
c. At zygonema, any chromosome may synapse with any other chromosome in the same cell.

3.5 For each mitotic event described below, write the name of the event in the blank provided in front of the description. Then put the events in the correct order (sequence); start by placing a 1 next to the description of interphase, and continue through 6, which should correspond to the last event in the sequence.

NAME OF EVENT		ORDER OF EVENT
_____	The cytoplasm divides and the cell contents are separated into two separate cells.	_____
_____	Chromosomes become aligned along the equatorial plane of the cell.	_____
_____	Chromosome replication occurs.	_____
_____	The migration of the daughter chromosomes to the two poles is complete.	_____
_____	Replicated chromosomes begin to condense and become visible under the microscope.	_____
_____	Sister chromatids begin to separate and migrate toward opposite poles of the cell.	_____

*3.6 Decide whether the answer to these statements is *yes* or *no*. Then explain the reasons for your decision.
a. Can meiosis occur in haploid species?
b. Can meiosis occur in a haploid individual?

3.7 The general life cycle of a eukaryotic organism has the sequence
a. 1N → meiosis → 2N → fertilization → 1N.
b. 2N → meiosis → 1N → fertilization → 2N.
c. 1N → mitosis → 2N → fertilization → 1N
d. 2N → mitosis → 1N → fertilization → 2N.

*3.8 Which statement is true?
a. Gametes are 2N; zygotes are 1N.
b. Gametes and zygotes are 2N.
c. The number of chromosomes can be the same in gamete cells and in somatic cells.
d. The zygotic and the somatic chromosome numbers cannot be the same.
e. Haploid organisms have haploid zygotes.

3.9 All of the following happen in prophase I of meiosis *except*

a. chromosome condensation.
b. pairing of homologs.
c. chiasma formation.
d. terminalization.
e. segregation.

*3.10 Give the name of the stages of mitosis or meiosis at which the following events occur:

a. Chromosomes are located in a plane at the center of the spindle.
b. The chromosomes move away from the spindle equator to the poles.

3.11 Given the diploid, meiotic mother cell shown below, diagram the chromosomes as they would appear

a. in late pachynema;
b. in a nucleus at prophase of the second meiotic division;
c. in the first polar body resulting from oogenesis in an animal.

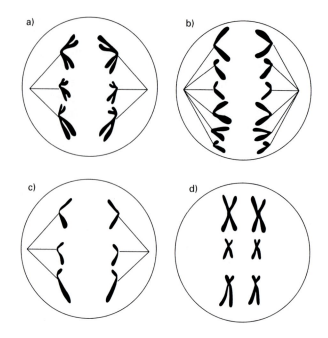

Mother cell

a) Middle prophase I

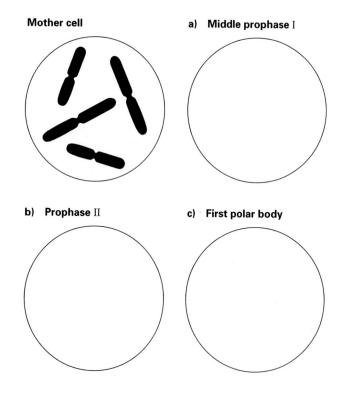

b) Prophase II   **c) First polar body**

3.12 The cells in the following figure were all taken from the same individual (a mammal). Identify the cell division events happening in each cell, and explain your reasoning. What is the sex of the individual? What is the diploid chromosome number?

3.13 Does mitosis or meiosis have greater significance in genetics (i.e., the study of heredity)? Explain your choice.

*3.14 Consider a diploid organism that has three pairs of chromosomes. Assume that the organism receives chromosomes A, B, and C from the female parent and A′, B′, and C′ from the male parent. To answer the following questions, assume that no crossing-over occurs.

a. What proportion of the gametes of this organism would be expected to contain all the chromosomes of maternal origin?
b. What proportion of the gametes would be expected to contain some chromosomes from both maternal and paternal origin?

*3.15 Normal diploid cells of a theoretical mammal are examined cytologically at the mitotic metaphase stage for their chromosome complement. One short chromosome, two medium-length chromosomes, and three long chromosomes are present. Explain how the cells might have such a set of chromosomes.

3.16 Is the following statement true or false? Explain your decision. "Meiotic chromosomes may be seen after appropriate staining in nuclei from rapidly dividing skin cells."

*3.17 Is the following statement true or false? Explain your decision. "All of the sperm from one human male are genetically identical."

3.18 The horse has a diploid set of 64 chromosomes, and a donkey has a diploid set of 62 chromosomes. Mules (viable but usually sterile progeny) are produced when a male donkey is mated to a female horse. How many chromosomes will a mule cell contain?

*3.19 The red fox has 17 pairs of large, long chromosomes. The arctic fox has 26 pairs of shorter, smaller chromosomes.

a. What do you expect to be the chromosome number in somatic tissues of a hybrid between these two foxes?

b. The first meiotic division in the hybrid fox shows a mixture of paired and single chromosomes. Why do you suppose this occurs? Can you suggest a possible relationship between this fact and the observed sterility of the hybrid?

*3.20 At the time of synapsis, preceding the reduction division in meiosis, the homologous chromosomes align in pairs and one member of each pair passes to each of the daughter nuclei. In an animal with five pairs of chromosomes, assume that chromosomes 1, 2, 3, 4, and 5 have come from the father, and 1′, 2′, 3′, 4′, and 5′ have come from the mother. Assuming no crossing over, in what proportion of the germ cells of this animal will all the paternal chromosomes be present together?

*3.21 In a male *Homo sapiens*, from which grandparent could each sex chromosome have been derived? (Indicate yes or no for each option.)

	MOTHER'S		FATHER'S	
	MOTHER	FATHER	MOTHER	FATHER
X chromosome	———	———	———	———
Y chromosome	———	———	———	———

3.22 Depict each of the following crosses, first using Mendelian and then using *Drosophila* notation. Give the genotype and phenotype of the F_1 progeny that can be produced.

a. In humans, a mating between two individuals, each heterozygous for the recessive trait phenylketonuria whose locus is on chromosome 12.

b. In humans, a mating between a female heterozygous for both phenylketonuria and X-linked colorblindness and a male with normal color vision and heterozygous for phenylketonuria.

c. In *Drosophila*, a mating between a female with white eyes, curled wings, and normal long bristles and a male that has normal red eyes, normal straight wings, and short, stubble bristles. In these individuals, curled wings result from a heterozygous condition at a gene whose locus is on chromosome 2, while the short, stubble bristles result from a heterozygous condition at a gene whose locus is on chromosome 3.

d. In *Drosophila*, a mating between a female from a true-breeding line that has eyes of normal size that are white, black bodies (a recessive trait on chromosome 2), and tiny bristles (a recessive trait called *spineless* on chromosome 3) and a male from a true-breeding line that has normal red eyes, normal gray bodies, normal long bristles but reduced eye size (a dominant trait called *eyeless* on chromosome 4).

3.23 In *Drosophila*, white eyes are a sex-linked character. The mutant allele for white eyes (w) is recessive to the wild-type allele for brick-red eye color (w^+).

a. A white-eyed female is crossed with a red-eyed male. An F_1 female from this cross is mated with her father, and an F_1 male is mated with his mother. What will be the eye color of the offspring of these last two crosses?

b. A white-eyed female is crossed with a red-eyed male, and the F_2 from this cross is interbred. What will be the eye color of the F_3?

*3.24 One form of color blindness (c) in humans is caused by a sex-linked recessive mutant gene. A woman with normal color vision (c^+) and whose father was color-blind marries a man of normal vision whose father was also color-blind. What proportion of their offspring will be color-blind? (Give your answer separately for males and females.)

3.25 In humans, red-green color blindness is due to an X-linked recessive gene. A color-blind daughter is born to a woman with normal color vision and a father who is color-blind. What is the mother's genotype with respect to the alleles concerned?

*3.26 In humans, red-green color blindness is recessive and X-linked, while albinism is recessive and autosomal. What types of children can be produced as the result of marriages between two homozygous parents, a normal-visioned albino woman and a color-blind, normally pigmented man?

*3.27 In *Drosophila*, vestigial (partially formed) wings (vg) are recessive to normal long wings (vg^+), and the gene for this trait is autosomal. The gene for the white-eye trait is on the X chromosome. Suppose a homozygous white-eyed, long-winged female fly is crossed with a homozygous red-eyed, vestigial-winged male.

a. What will be the appearance of the F_1?

b. What will be the appearance of the F_2?

c. What will be the appearance of the offspring of a cross of the F_1 back to each parent?

3.28 In *Drosophila*, two red-eyed, long-winged flies are bred together and produce the offspring given in the following table:

	FEMALES	MALES
red-eyed, long-winged	3/4	3/8
red-eyed, vestigial-winged	1/4	1/8
white-eyed, long-winged	—	3/8
white-eyed, vestigial-winged	—	1/8

What are the genotypes of the parents?

3.29 In poultry a dominant sex-linked gene (*B*) produces barred feathers, and the recessive allele (*b*), when homozygous, produces nonbarred feathers. Suppose a nonbarred cock is crossed with a barred hen.

a. What will be the appearance of the F_1 birds?

b. If an F_1 female is mated with her father, what will be the appearance of the offspring?

c. If an F_1 male is mated with his mother, what will be the appearance of the offspring?

3.30 In chickens, barred plumage (*B*) is dominant over non-barred (solid color)(*b*); the locus for this plumage phenotype is located on the sex chromosomes. (In birds, the female is the heterogametic sex.) The phenotypes can be distinguished in newly hatched chicks. Commercial chicken breeders in England have used this difference to separate male and female chicks, otherwise a difficult task. In order to accomplish this, what must be the genotype of: (a) the female parent; and (b) the male parent?

*3.31 A man (A) suffering from defective tooth enamel, which results in brown-colored teeth, marries a normal woman. All their daughters have brown teeth, but the sons are normal. The sons of man A marry normal women, and all their children are normal. The daughters of man A marry normal men, and 50 percent of their children have brown teeth. Explain these facts.

3.32 In humans, differences in the ability to taste phenylthiourea are due to a pair of autosomal alleles. Inability to taste is recessive to ability to taste. A child who is a nontaster is born to a couple who can both taste the substance. What is the probability that their next child will be a taster?

*3.33 Cystic fibrosis is inherited as an autosomal recessive. Two parents without cystic fibrosis have two children with cystic fibrosis and three children without. They come to you for genetic counseling.

a. What is the numerical probability that their next child will have cystic fibrosis?

b. Their unaffected children are concerned about being heterozygous. What is the numerical probability that a given unaffected child in the family is heterozygous?

3.34 Huntington's disease is a human disease inherited as a Mendelian autosomal dominant. The disease results in choreic (uncontrolled) movements, progressive mental deterioration, and eventually death. The disease affects the carriers of the trait any time between 15 and 65 years of age. The American folk singer Woody Guthrie died of Huntington's disease, as did one of his parents. Marjorie Mazia, Woody's wife, had no history of this disease in her family. The Guthries had three children. What is the probability that a particular Guthrie child will die of Huntington's disease?

3.35 Suppose gene *A* is on the X chromosome, and genes *B*, *C*, and *D* are on three different autosomes. Thus *A–* signifies the dominant phenotype in the male or female. An equivalent situation holds for *B–*, *C–*, and *D–*. The cross *AA BB CC DD* females × *aY bb cc dd* males is made.

a. What is the probability of obtaining an *A–* individual in the F_1?

b. What is the probability of obtaining an *a* male in the F_1?

c. What is the probability of obtaining an *A– B– C– D–* female in the F_1?

d. How many different F_1 genotypes will there be?

e. What proportion of F_2s will be heterozygous for the four genes?

f. Determine the probabilities of obtaining each of the following types in the F_2: (1) *A– bb CC dd* (female); (2) *aY BB Cc Dd* (male); (3) *AY bb CC dd* (male); (4) *aa bb Cc Dd* (female).

*3.36 As a famous mad scientist, you have cleverly devised a method to isolate *Drosophila* ova that have undergone primary nondisjunction of the sex chromosomes. In one experiment you used females homozygous for the sex-linked recessive mutation causing white eyes (*w*) as your source of nondisjunction ova. The ova were collected and fertilized with sperm from red-eyed males. The progeny of this "engineered" cross were then backcrossed separately to the two parental strains. What classes of progeny (genotype and phenotype) would you expect to result from these backcrosses? (The genotype of the original parents may be denoted as *ww* for the females and *w+/Y* for the males.)

3.37 In *Drosophila*, the bobbed gene (*bb+*) is located on the X chromosome: *bb* mutants have shorter, thicker bristles than wild-type flies. Unlike most X-linked genes, however, the Y chromosome also carries a bobbed gene. The mutant allele *bb* is recessive to *bb+*. If a wild-type F_1 female that resulted from primary nondisjunction in oogenesis in a cross of bobbed female with a wild-type male is mated to a bobbed male, what will be the phenotypes and their frequencies in the offspring? List males and females separately in your answer. (Hint: Refer to the chapter for information about the frequency of nondisjunction in *Drosophila*.)

3.38 A Turner syndrome individual would be expected to have the following number of Barr bodies in the majority of cells:

a. 0   c. 2

b. 1   d. 3

3.39 An XXY Klinefelter syndrome individual would be expected to have the following number of Barr bodies in the majority of cells:

a. 0   c. 2

b. 1   d. 3

*3.40 In human genetics, the pedigree is used for analysis of inheritance patterns. The female is represented by a circle and the male by a square. The following figure presents three, two-generation family pedigrees for a trait in humans. Normal individuals are represented by unshaded symbols and people with the trait by shaded symbols. For each pedigree (A, B, and C), state, by answering yes or no in the appropriate blank space, whether transmission of the trait can be accounted for on the basis of each of the listed simple modes of inheritance:

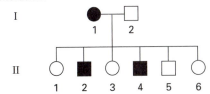

	PEDIGREE A	PEDIGREE B	PEDIGREE C
Autosomal recessive	_____	_____	_____
Autosomal dominant	_____	_____	_____
X-linked recessive	_____	_____	_____
X-linked dominant	_____	_____	_____

3.41 Looking at the following pedigree, in which shaded symbols represent a "trait," which of the progeny (as designated by numbers) eliminate X-linked recessiveness as a mode of inheritance for the trait?
a. I.1 and I.2
b. II.4
c. II.5
d. II.2 and II.4

*3.42 When constructing human pedigrees, geneticists often refer to particular persons by a number. The generations are labeled by roman numerals and the individuals in each generation by Arabic numerals. For example, in the pedigree in the figure at the top of the next column, the female with the asterisk would be I.2. Use this means to designate specific individuals in the pedigree. Determine the probable inheritance mode for the trait shown in the affected individuals (the shaded symbols) by answering the following questions. Assume the condition is caused by a single gene.

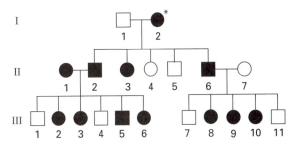

a. Y-linked inheritance can be excluded at a glance. What two other mechanisms of inheritance can be definitely excluded? Why can these be excluded?
b. Of the remaining mechanisms of inheritance, which is the most likely? Why?

3.43 A three-generation pedigree for a particular human trait is shown in the following figure:

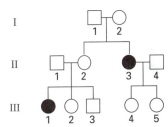

a. What is the mechanism of inheritance for the trait?
b. Which persons in the pedigree are known to be heterozygous for the trait?
c. What is the probability that III.2 is a carrier (heterozygous)?
d. If III.3 and III.4 marry, what is the probability that their first child will have the trait?

3.44 For the more complex pedigrees shown in the Figure 3B (p. 96), indicate the probable mechanism of inheritance: autosomal recessive, autosomal dominant, X-linked recessive, X-linked dominant, Y-linked.

*3.45 If a rare genetic disease is inherited on the basis of an X-linked dominant gene, one would expect to find the following:
a. Affected fathers have 100 percent affected sons.
b. Affected mothers have 100 percent affected daughters.
c. Affected fathers have 100 percent affected daughters.
d. Affected mothers have 100 percent affected sons.

3.46 If a genetic disease is inherited on the basis of an autosomal dominant gene, one would expect to find the following:
a. Affected fathers have only affected children.
b. Affected mothers never have affected sons.
c. If both parents are affected, all of their offspring have the disease.
d. If a child has the disease, one of his or her grandparents also had the disease.

FIGURE 3.B

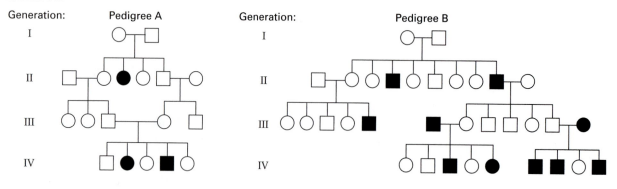

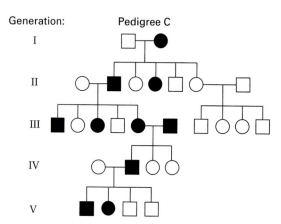

*3.47 If a genetic disease is inherited as an autosomal recessive, one would expect to find the following:

a. Two affected individuals never have an unaffected child.

b. Two affected individuals have affected male offspring but no affected female children.

c. If a child has the disease, one of his or her grandparents will have had it.

d. In a marriage between an affected individual and an unaffected one, all the children are unaffected.

3.48 Which of the following statements is *not* true for a disease that is inherited as a rare X-linked dominant?

a. All daughters of an affected male will inherit the disease.

b. Sons will inherit the disease only if their mothers have the disease.

c. Both affected males and affected females will pass the trait to half the children.

d. Daughters will inherit the disease only if their father has the disease.

*3.49 Women who were known to be carriers of the X-linked, recessive, hemophilia gene were studied in order to determine the amount of time required for the blood clotting reaction. It was found that the time required for

clotting was extremely variable from individual to individual. The values obtained ranged from normal clotting time at one extreme, all the way to clinical hemophilia at the other extreme. What is the most probable explanation for these findings?

3.50 Hurler syndrome is a genetically transmitted disorder of mucopolysaccharide metabolism characterized by short stature, mental retardation and various bony malformations. There are two specific types described with extensive pedigrees in the medical genetics literature. They are:

Type I—recessive autosomal
Type II—recessive X-linked

You are a consultant in a hospital ward with several Hurler syndrome patients who have asked you for advice concerning their relatives' offspring. Being aware that both types are extremely rare, and that afflicted individuals virtually never reproduce, what counsel would you give to a Type I Hurler syndrome female patient (whose normal brother's daughter is planning marriage) concerning the offspring of the proposed marriage? In your answer, state the probabilities of affected offspring and sex differences where relevant.

4 EXTENSIONS OF MENDELIAN GENETIC ANALYSIS

PRINCIPAL POINTS

~ Many allelic forms of a gene can exist. This phenomenon is called multiple allelism. However, any given diploid individual can possess only two different alleles of a given gene.

~ With complete dominance, the same phenotype results whether the dominant allele is heterozygous or homozygous. In incomplete dominance the phenotype of the heterozygote is intermediate between those of the two homozygotes, whereas in codominance the heterozygote exhibits the phenotypes of both homozygotes.

~ In many cases, different genes interact to determine phenotypic characteristics. In epistasis, for example, modified Mendelian ratios occur because of interactions of nonallelic genes: the phenotypic expression of one gene depends upon the genotype of another gene locus.

~ Alleles of certain genes may be fatal to the individual. The existence of such lethal alleles of a gene indicates that the product usually produced by the non-lethal allele is essential for the function of the organism; the gene is called an essential gene.

~ Penetrance is the frequency within the population with which a dominant or homozygous recessive gene manifests itself in the phenotype of an individual. Expressivity is the kind or degree of phenotypic manifestation of a penetrant gene or genotype in a particular individual.

~ The zygote's genetic constitution only specifies the organism's potential to develop and function. As the organism develops and differentiates, many things can influence gene expression. One such influence is the organism's environment, both internal and external. Examples of the former may include age and sex of the individual. Examples of the latter include nutrition, light, chemicals, temperature, and infectious agents.

~ Variation in most of the genetic traits considered in the earlier discussion of Mendelian principles is determined predominantly by differences in genotype; that is, phenotypic differences result from genotypic differences. For many traits, however, the phenotypes are influenced by both genes and the environment. The debate over the relative contribution of genes and environment to the phenotype has been termed the nature versus nurture controversy.

Mendel's principles apply to all eukaryotic organisms and form the foundation for predicting the outcome of crosses in which segregation and independent assortment might be occurring. As more and more geneticists did experiments, though, they found that there are exceptions and extensions to Mendel's principles. Several of these cases will be discussed in this chapter. We will examine examples of genes that have many different alleles rather than the simpler two-allele genes discussed so far, cases in which one allele is not completely dominant to another allele at the gene locus, and situations in which products of different genes interact to produce modified Mendelian ratios. We will also look at essential genes, and at the effects of the environment on gene expression. This discussion will give us a broader knowledge of genetic analysis, particularly in terms of how genes relate to the phenotypes of an organism.

MULTIPLE ALLELES

So far in our genetic analyses we have considered only pairs of alleles which control characters, such as smooth versus wrinkled seeds in peas, red versus white eyes in *Drosophila*, and unattached versus attached earlobes in humans. The allele that predominates in populations of the organism in the wild is the wild-type allele, and the alternative allele is the variant or mutant allele. In a population of individuals, however, there can be many alleles of a given gene (one wild type and the rest mutant), not just two. Such genes are said to have **multiple alleles**, and the alleles are said to constitute a *multiple allelic series* (Figure 4.1). Although a gene may have multiple alleles in a given population of individuals, a *single diploid individual can have a maximum of two of these alleles*, one on each of the two homologous chromosomes carrying the gene locus.

ABO Blood Groups

An example of multiple alleles of a gene is found in the human ABO blood group series, which was discovered by Karl Landsteiner in the early 1900s. Since certain ABO blood groups are incompatible, these alleles are of particular importance when blood transfusions are contemplated. (There are many blood group series other than ABO; these also can cause

~ FIGURE 4.1

Allelic forms of a gene.

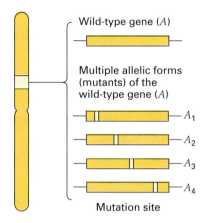

Wild-type gene (A)

Multiple allelic forms
(mutants) of the
wild-type gene (A)

A_1

A_2

A_3

A_4

Mutation site

problems in blood transfusions and need to be checked through the process of cross-matching.)

Four blood group phenotypes occur in the ABO system; O, A, B, and AB; Table 4.1 gives their possible genotypes. The six genotypes that give rise to the four phenotypes represent various combinations of three ABO blood group alleles, I^A, I^B, and i. People homozygous for the recessive i allele are of blood group O. Both I^A and I^B are dominant to i. Thus you will express blood group A if you are either I^A/I^A or I^A/i, and you will express blood group B if you are either I^B/I^B or I^B/i. Heterozygous I^A/I^B individuals express blood group AB; that is, essentially both blood groups A and B simultaneously (see discussion of codominance in this chapter on p. 104).

The genetics of this system follows basic Mendelian principles. An individual who expresses blood group O, for example, must be i/i in genotype. The parents of this person could both be O ($i/i \times i/i$), they could both be A ($I^A/i \times I^A/i$, to produce one-fourth i/i progeny), they could both be B ($I^B/i \times I^B/i$), or one could be A and one could be B ($I^A/i \times I^B/i$). Simply put, each parent would have to be either homozygous i or heterozygous, with i as one of the two alleles.

~ TABLE 4.1

ABO Blood Groups in Humans, Determined by the Alleles I^A, I^B, and i

PHENOTYPE (BLOOD GROUP)	GENOTYPE
O	i/i
A	I^A/I^A or I^A/i
B	I^B/I^B or I^B/i
AB	I^A/I^B

Blood-typing (the determination of an individual's blood group) and the analysis of the blood group inheritance are sometimes used in legal medicine in cases of disputed paternity or maternity or in cases of an inadvertent baby switch in a hospital. In such cases genetic data cannot prove the identity of the parent. Genetic analysis on the basis of blood group can only be used to show that an individual is *not* the parent of a particular child; for example, a child of phenotype AB (genotype I^A/I^B) could not be the child of a parent of phenotype O (genotype i/i). (Note: Blood-type data alone are usually not sufficient for a legal decision in many cases, according to the laws in most states. And, these days, more modern tests may be used, including DNA typing (see Chapter 15).

When giving blood transfusions, the blood groups of donors and recipients must be carefully matched because the blood group alleles specify molecular groups, called *cellular antigens*, that are attached to the outside of the red blood cells. An **antigen**—<u>anti</u>body <u>gen</u>erating substance—is any molecule that is recognized as foreign by an individual and therefore stimulates the production of specific protein molecules called **antibodies**, which bind to the antigen. An antibody is a protein molecule that recognizes and binds to the foreign substance (antigen) introduced into the organism. A given individual has a large number of antigens on cells and tissues, many of which are foreign to another individual; hence the concern over the blood type in blood transfusions and tissue type in organ transplants. However, the antigens are usually not recognized as foreign by the individual expressing them (autoimmune diseases are an exception). The I^A allele specifies the A antigen, and in people with blood group A (I^A/I^A or I^A/i) the blood serum contains naturally occurring antibodies against the B antigen (called anti-B antibodies), but none against the A antigen. Antibodies against the B antigen will agglutinate, or clump, any red blood cells which have the B antigen on them. Since clumped cells cannot move through the fine capillaries, agglutination may lead to organ failure and possibly death. Conversely, people of the B blood group (I^B/I^B or I^B/i) have the B antigen on their red blood cells and have serum containing naturally occurring anti-A antibodies, but no anti-B antibodies. For the AB blood type (I^A/I^B), both A and B antigens are on the blood cells, so neither anti-A nor anti-B antibodies occur in the serum of people with this blood type. In people with blood type O (i/i), the blood cells have neither A nor B antigen, and therefore the serum contains both anti-A and anti-B antibodies. These antigen-antibody relationships are shown in Figure 4.2. Agglutination (clumping) of the

~ FIGURE 4.2

Antigenic reactions that characterize the human ABO blood groups. Blood serum from each of the four blood groups was mixed with blood cells from the four types, in all possible combinations. In some cases, such as a mix of B serum with A cells, the cells become clumped.

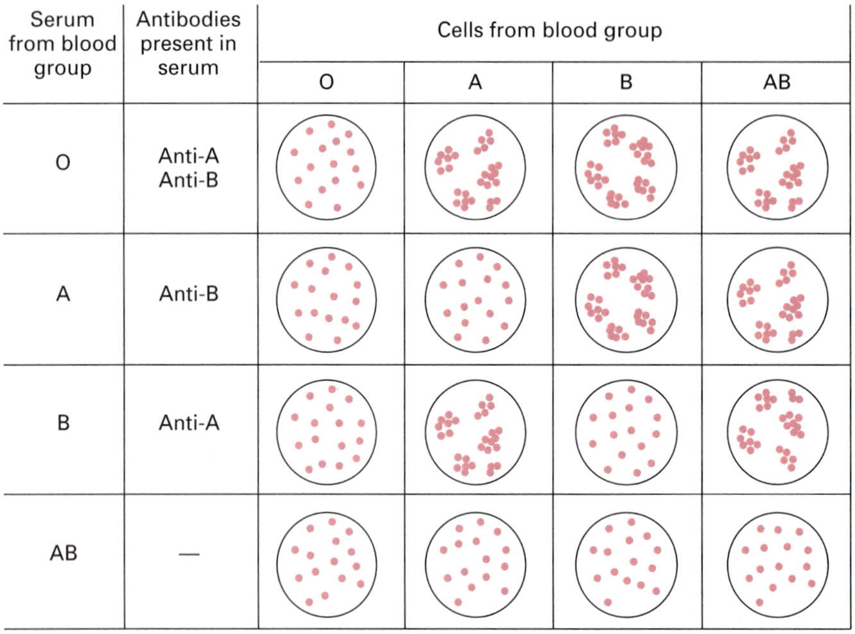

Serum from blood group	Antibodies present in serum	Cells from blood group			
		O	A	B	AB
O	Anti-A Anti-B				
A	Anti-B				
B	Anti-A				
AB	—				

red blood cells is seen in each case where an antibody interacts with the antigen for which it is specific.

What transfusions are safe, then, between people with different blood groups in the ABO system?

1. Individuals in blood group A produce the A antigen, so their blood can be transfused only into recipients who do *not* have the anti-A antibody, that is, people of blood groups A and AB.
2. Those in blood group B produce the B antigen, so their blood can be transfused only into recipients who do *not* have the anti-B antibody, that is, people of blood groups B and AB.
3. AB individuals produce both the A and B antigens, so their blood can be transfused only into recipients who do *not* have either the anti-A antibody or the anti-B antibody, that is, people of blood group AB.
4. O individuals produce neither A nor B antigens, so their blood can be transfused into any recipient, that is, people of blood groups A, B, AB, and O.

You will note from this presentation that people of blood group AB can receive transfusions of blood from people of any of the four blood groups. Thus, blood group AB individuals are called *universal recipients*. Similarly, blood group O blood can be used as donor blood for any recipient since it elicits no

reaction. Thus, blood group O individuals are called *universal donors*.

Drosophila Eye Color

Another example of multiple alleles concerns the white (*w*) locus of *Drosophila*. Recall alleles are alternative forms of a gene, each of which may affect a character differently. Recall from Chapter 3 that the w^+ allele results in red eyes, and the *w* allele, when homozygous or hemizygous, results in white eyes.

It was Morgan's work with a white-eyed variant of *Drosophila* that indicated the presence of genes on the X chromosome. Soon after that discovery he found evidence for other distinct genes on the *Drosophila* X chromosome (such as the bar-eye shape and the vermilion eye color genes). Morgan experimented with strains that had different eye color genes. When he crossed a white-eyed female with a vermilion-eyed male, the results were unexpected: the F_1 females were all red-eyed. The simplest explanation is that the white and vermilion eye color traits are specified by *two different genes*. Using genetic symbols, the cross was $w\,v^+/w\,v^+\ ♀ \times w^+\,v/\!/\!\!\wedge\ ♂$.[1] The F_1

[1]As a reminder, the / represents the homologous chromosomes on which the alleles are found, and the ∕ represents the Y chromosome (see Box 3.1, p. 71).

females from this cross are doubly heterozygous $w\,v^+/w^+\,v$ and therefore show the dominant effect resulting from the presence of the wild-type allele for each allelic pair; namely, the brick-red eyes characteristic of wild-type flies.

In 1912 Morgan obtained data for X-linked eye color genes that could not be explained so simply. The eye color variants were *white* and *eosin* (reddish orange). Like *white*, *eosin* is X-linked recessive to the wild-type eye color. However, when a female from an eosin-eyed strain is crossed with a male from a white-eyed strain, all F_1 females have eosin eyes. In 1913, Alfred Sturtevant observed that (1) red (wild-type) eye color is dominant to *eosin* and to *white*, and (2) *eosin* is recessive to wild type but dominant to *white*.

He concluded *eosin* and *white* are both mutant alleles of a single gene. In other words, there are multiple alleles of the white-eye gene.

Sturtevant's concept becomes clearer if we assign symbols to the alleles. A lowercase letter (or letters) designates the gene, and superscripts designate the different alleles, as in the following example: w^+ is the wild-type allele of the white-eye gene, w is the recessive white-eye allele, and w^e is the eosin allele. Figure 4.3a uses this notation to track the cross of an eosin-eyed female with a white-eyed male. The F_1 females are w^e/w and have eosin eyes because w^e is dominant over w. When these F_1 females are crossed with red-eyed males (Figure 4.3b), all the female progeny are heterozygous and red-eyed since they contain the w^+ allele; they are

~ FIGURE 4.3

Results of crosses of *Drosophila melanogaster* involving two mutant alleles of the same locus, white (w) and white-eosin (w^e). (a) white-eosin-eyed (w^e/w^e) ♀ × white eyed (w/Y) ♂. (b) F_1 (w^e/w) ♀ × red-eyed (wild type) (w^+/Y) ♂.

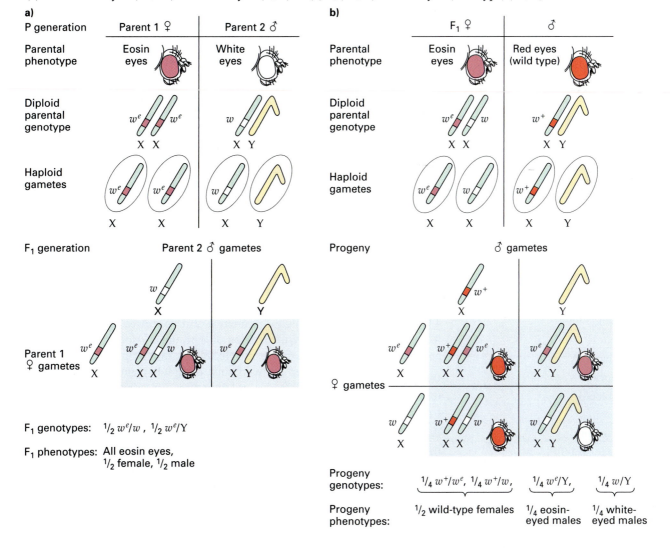

either w^+/w^e or w^+/w. Half the male progeny are eosin-eyed (w^e/Y), and the other half are white-eyed (w/Y).

Multiple allelic series exist for all types of genes, not just X-linked ones. But let us expand on the example of the white-eyed gene of *Drosophila* a little. There are many known alleles of the white-eye gene that are distinguishable because they produce different eye colors ranging from white to near-wild type when the alleles are homozygous or hemizygous. Eye color phenotype is related to the amount of pigment deposited in the eye cells. It is possible to extract those pigments from the eyes and to quantify the amount of pigment present in wild type and in strains with different mutant alleles of the white locus (Table 4.2). As expected, the original mutant allele w has the least amount of pigment. Between white and wild, there is a wide range of pigment amounts. The phenotypic expressions of different alleles of the same gene reflect the varying extent to which the biological activity of the white gene product has been altered.

As you can see from Table 4.2, the number of allelic forms of a gene is not restricted to the three we have discussed; indeed, many hundreds of alleles are known for some genes. The number of possible genotypes in a multiple allelic series depends on the number of alleles involved (Table 4.3). With one allele, only one genotype is possible. With two alleles, A^1 and A^2, three genotypes are possible: A^1/A^1 and A^2/A^2 homozygotes and the A^1/A^2 heterozygote. The general formula is for n alleles, $n(n + 1)/2$ genotypes are possible, of which n are homozygotes and $n(n - 1)/2$ are heterozygotes.

~ TABLE 4.2

Eye Pigment Quantification for *Drosophila* White Alleles

GENOTYPES	RELATIVE AMOUNT OF TOTAL PIGMENT
w^+/w^+ (wild type)	1.0000
w/w (white)	0.0044
w^t/w^t (tinged)	0.0062
w^a/w^a (apricot)	0.0197
w^{bl}/w^{bl} (blood)	0.0310
w^e/w^e (eosin)	0.0324
w^{ch}/w^{ch} (cherry)	0.0410
w^{a3}/w^{a3} (apricot-3)	0.0632
w^w/w^w (wine)	0.0650
w^{co}/w^{co} (coral)	0.0798
w^{sat}/w^{sat} (satsuma)	0.1404
w^{col}/w^{col} (colored)	0.1636

~ TABLE 4.3

Genotype Number of Multiple Alleles

NUMBER OF ALLELES	KINDS OF GENOTYPES	KINDS OF HOMOZYGOTES	KINDS OF HETEROZYGOTES
1	1	1	0
2	3	2	1
3	6	3	3
4	10	4	6
5	15	5	10
n	$n(n + 1)/2$	n	$n(n - 1)/2$

KEYNOTE

Many allelic forms of a gene can exist in a population. When they do, the gene is said to show multiple allelism, and the alleles involved constitute a multiple allelic series. However, any given diploid individual can possess only two different alleles of a given gene. Multiple alleles obey the same rule of transmission as alleles of which there are only two types, although the dominance relationships among multiple alleles vary from one group to another.

MODIFICATIONS OF DOMINANCE RELATIONSHIPS

In most of the genetic examples discussed so far—the ABO locus is an exception—one allele is dominant to the other, so that the phenotype of the heterozygote is the same as the homozygous dominant. This phenomenon is called **complete dominance**. With **complete recessiveness**, the recessive allele is phenotypically expressed only when the organism is homozygous. Complete dominance and complete recessiveness are the two extremes of a range of dominance relationships. For example, all the allelic pairs that Mendel studied showed such complete dominance/complete recessiveness relationships. However, many allelic pairs do not show this dominance relationship.

Incomplete Dominance

When one allele is not completely dominant to another allele, it is said to show **incomplete**, or **partial**, **dominance**. With incomplete dominance the heterozygote's phenotype is between that of individuals homozygous for either individual allele involved.

Plumage color in chickens presents a good example of incomplete dominance. Crosses between a true-breeding black strain (homozygous $C^B C^B$) and a true-breeding white strain (homozygous $C^W C^W$) give F_1 birds with bluish-grey plumage (Figure 4.4a), called Andalusian blues by chicken breeders. (Note that the symbols are designed to give equal weight to the two alleles since neither dominates the phenotype. In this particular example, the C signifies color, while "B" and "W" indicate black and white, respectively.) An Andalusian blue cannot be true breeding because it is heterozygous, so in Andalusian × Andalusian crosses (Figure 4.4b) the two alleles will segregate in the offspring and produce black, Andalusian blue (bluish-grey), and white fowl in a ratio of 1:2:1. The most efficient way to produce Andalusian blues is to cross black × white, since all progeny of this mating are Andalusian blues.

Another example of incomplete dominance is the Palomino horse, which has a golden-yellow body color and a mane and tail that are almost white (Figure 4.5a). Palominos do not breed true. When they are interbred, the progeny are 1/4 cremellos (extremely light-colored: Figure 4.5b), 1/2 Palominos, and 1/4 light chestnuts (Figure 4.5c). The 1:2:1 ratio resulting from interbreeding is characteristic of incomplete dominance. The interpretation is that there are two alleles, C and C^{cr}, involved in producing Palominos. That is, the genotype C/C, in combination with other genes, results in horses with a light chestnut (also called sorrel) color. The genotype C/C^{cr}, in combination with the same genes as above, results in

---

~ **FIGURE 4.4**

Incomplete dominance in Andalusian fowls. (a) A cross between a white and a black bird produces F_1s of intermediate grey color, called Andalusian blues. (b) The F_2 generation shows the 1:2:1 phenotype ratio characteristic of incomplete dominance.

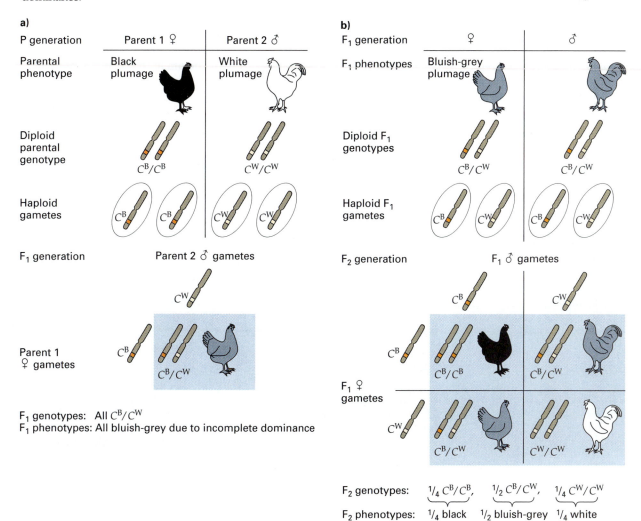

a dilution of the normal reddish-brown of light chestnuts to yellow or cream to give the Palomino color. The genotype C^{cr}/C^{cr} brings about an extreme dilution of the phenotype to give the cremello color.

In the plant kingdom there are many examples of incomplete dominance, such as flower color in the snapdragon (Figure 4.6). There are two alleles, C^R and C^W. A cross of a red-flowered variety (C^RC^R; i.e., homozygous for the Colored-Red allele) with a white-flowered variety (C^WC^W; i.e., homozygous for the Colored-White allele) produces F_1 plants with pink flowers (Figure 4.6a). The F_1 plants are C^RC^W heterozygotes. Interbreeding the F_1 plants produces an F_2 with a 1:2:1 ratio of red:pink:white (Figure 4.6b). The red and white flowers are determined by homozygosity for the respective color alleles, the pink flowers by heterozygosity for the two alleles. Thus, plants with pink flowers do not breed true.

Codominance

Codominance is a modification of the dominance relationship that is related to incomplete dominance. In codominance the heterozygote exhibits the phenotypes of both homozygotes. Because the phenotypes of both homozygotes are expressed, codominance differs from incomplete dominance, in which the heterozygote exhibits a phenotype intermediate between the two homozygotes.

The ABO blood series discussed earlier in this chapter provides a good example of codominance. Heterozygous I^A/I^B individuals are blood group AB because both the A antigen (product of the I^A allele) and the B antigen (product of the I^B allele) are produced. Thus, the I^A and I^B alleles are codominant.

The human M-N blood group system is another example of codominance. In terms of transfusion compatibility, this system is of less clinical importance than the ABO system. In the M-N system three blood types occur: M, N, and MN. They are determined by the genotypes L^M/L^M, L^M/L^N, and L^N/L^N, respectively. As in the ABO system, the M-N alleles result in the formation of antigens on the red blood cell surface. The heterozygote in this case has both the M and the N antigens and shows the phenotypes of both homozygotes. Inheritance patterns for the M-N blood groups are illustrated in Table 4.4.

What is the explanation of incomplete dominance and codominance at the molecular level? A general interpretation is that, in codominance (for example, the ABO blood groups), products result from both alleles in a heterozygote so that frequently (but not always) both homozygote phenotypes are observed (e.g., L^M/L^N individuals express both M and N blood group antigens). In incomplete dominance, in a heterozygote, only one allele is expressed to produce a product. In a homozygote for that allele, then, there are two doses of the gene product and full phenotypic expression results (e.g., red flowers in snapdragons). In a homozygote for the allele that is not expressed, a phenotype characteristic of no gene expression results (e.g., white flowers in snapdragons). In a heterozygote, the single allele expressed results in only enough product for an intermediate phenotype (e.g., pink flowers in snapdragons).

~ **FIGURE 4.5**

Incomplete dominance in horses. (a) Palomino horse, genotype C/C^{cr}; (b) Cremello horse, genotype C^{cr}/C^{cr}; and (c) Light chestnut (sorrel) horse, genotype C/C.

a)   b)   c)

~ FIGURE 4.6

Incomplete dominance in snapdragons. (a) A cross of true-breeding red-flowered variety with a white-flowered variety produces F_1s with pink flowers. (b) A 1:2:1 ratio of red:pink:white is seen in the F_2.

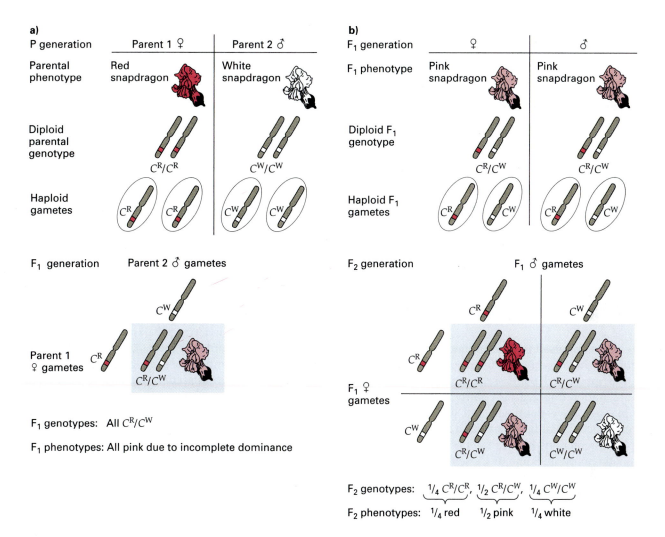

~ TABLE 4.4

Inheritance Patterns for the M-N Blood Types

PARENTAL PHENOTYPES	PARENTAL GENOTYPES	PROGENY PHENOTYPES	PROGENY GENOTYPES
M × M	$L^M/L^M \times L^M/L^M$	All M	All L^M/L^M
N × N	$L^N/L^N \times L^N/L^N$	All N	All L^N/L^N
N × M	$L^N/L^N \times L^M/L^M$	All MN	All L^M/L^N
MN × M	$L^M/L^N \times L^M/L^M$	1/2 MN:1/2 M	1/2 L^M/L^N:1/2 L^M/L^M
MN × N	$L^M/L^N \times L^N/L^N$	1/2 MN:1/2 N	1/2 L^M/L^N:1/2 L^N/L^N
MN × MN	$L^M/L^N \times L^M/L^N$	1/4 MM:1/2 MN:1/4 NN	1/4 L^M/L^M:1/2 L^M/L^N:1/4 L^N/L^N

KEYNOTE

With complete dominance the same phenotype results whether the dominant allele is heterozygous or homozygous. With complete recessiveness the allele is phenotypically expressed only when the genotype is homozygous recessive; the recessive allele has no effect on the phenotype of the heterozygote. Complete dominance and complete recessiveness are two extremes between which all transitional degrees of dominance are possible. In incomplete dominance, the phenotype of the heterozygote is intermediate between those of the two homozygotes, while in codominance the heterozygote exhibits the phenotypes of both homozygotes.

GENE INTERACTIONS AND MODIFIED MENDELIAN RATIOS

No gene acts by itself in determining an individual's phenotype; the phenotype is the result of highly complex and integrated patterns of molecular reactions that are under direct gene control. All the genetic examples we have discussed and will discuss have discrete biochemical bases and in a number of cases complex interactions between genes can be detected by genetic analysis. Some of these examples will be discussed in this section.

In Chapter 2 we discussed Mendel's principle of independent assortment, which states that genes on different chromosomes behave independently in the production of gametes. Let's assume that there are two independently assorting gene pairs, each with two alleles; A and a, and B and b. The outcome of a cross between individuals, each of whom is doubly heterozygous ($A/a\,B/b \times A/a\,B/b$), will be nine genotypes in the following proportions:

$$1/16\;A/A\;B/B$$
$$2/16\;A/A\;B/b$$
$$1/16\;A/A\;b/b$$
$$2/16\;A/a\;B/B$$
$$4/16\;A/a\;B/b$$
$$2/16\;A/a\;b/b$$
$$1/16\;a/a\;B/B$$
$$2/16\;a/a\;B/b$$
$$1/16\;a/a\;b/b.$$

If the phenotypes determined by the two allelic pairs are distinct—for example, smooth versus wrinkled peas, long versus short stems—and there is complete dominance, then we get the familiar dihybrid phenotypic ratio of 9:3:3:1 (see Figure 2.15b, p. 32). That is, 9/16 of the progeny show both dominant phenotypes, 3/16 show the first gene pair's dominant phenotype and the other gene pair's recessive phenotype, 3/16 show the first gene pair's recessive phenotype and the other pair's dominant phenotype, and 1/16 show both recessive phenotypes. Any alteration in this standard 9:3:3:1 ratio indicates that the phenotype is the product of the interaction of two or more genes. As we discuss each modified Mendelian ratio, we will refer to this distribution of genotypes.

For the 9:3:3:1 phenotypic ratio the genotypes for these phenotypes can be represented in a shorthand way as, respectively, $A/-\,B/-$, $A/-\,b/b$, $a/a\,B/-$, $a/a\,b/b$. The dash indicates that the phenotype is the same, whether the gene is homozygous dominant or heterozygous (e.g., $A/-$ means either A/A or A/a). This system cannot be used when incomplete dominance or codominance is involved.

The following sections discuss the main processes that result in modified Mendelian ratios. The first section describes examples of interactions between nonallelic genes that control the same general phenotypic attribute. In the second section, we discuss examples of interactions of nonallelic genes in which one allele masks the expression of alleles of another gene. This second type of interaction is called *epistasis*. (In both cases the discussions are confined to dihybrid crosses in which the two pairs of alleles assort independently for a clear, uncomplicated presentation of the topics. In the "real world" there are many more complex examples of gene interactions, involving more than two pairs of alleles and/or genes that do not assort independently.) For many of the examples we will discuss, *hypothetical* molecular explanations are presented. *It is important to note that these presentations are theoretical in nature.* They are included because the processes of experimental science typically involve proposing of hypotheses based on theories or models, and doing experiments designed to test the hypotheses (see Chapter 1, p. 3), and thus it is appropriate to consider models that are compatible with the modified Mendelian ratios being discussed.

Gene Interactions That Produce New Phenotypes

In all the previous examples of dihybrid crosses, the two genes have acted independently in terms of phenotype. For example, the allelic pair for round/wrinkled peas had no effect on the allelic pair for long/short stem. If, however, the two allelic pairs affect the same phenotypic characteristic, there is a chance for gene product interaction to give novel phenotypes, and the result can be modified phenotypic ratios, depending on the particular interaction between the products of the nonallelic genes.

COMB SHAPE IN CHICKENS. A classic example of such gene interactions is comb shape in chickens. New comb shape phenotypes are a consequence of interactions between two allelic pairs. Figure 4.7 shows the four comb phenotypes that result from the interaction of the alleles of two gene loci. Each of these four types can be bred true if the alleles involved are homozygous.

Crosses made between true-breeding rose-combed and single-combed varieties showed that rose was completely dominant over single. When the F_1 rose-combed birds were bred together, there was a clear segregation into 3 rose : 1 single in the F_2. Similarly, pea comb was found to be completely dominant over single, with a 3 pea : 1 single ratio in the F_2. When true-breeding rose and pea varieties were crossed, however, the result was new and interesting (Figure 4.8). Instead of showing either rose or pea combs, all birds in the F_1 showed a new comb form, different from either the rose or the pea comb (Figure

~ **FIGURE 4.7**

Four distinct comb shape phenotypes in chickens, resulting from all possible combinations of a dominant and a recessive allele at each of two gene loci: (a) rose comb $(R/- p/p)$; (b) walnut comb $(R/- P/-)$; (c) pea comb $(r/r P/-)$; (d) single comb $(r/r p/p)$.

a) Rose comb

b) Walnut comb

c) Pea comb

d) Single comb

4.8a). This new form was called walnut comb, since it resembles half a walnut meat.

When the F_1 walnut-combed birds were bred together, another fascinating result was observed in the F_2 (Figure 4.8b): Not only did walnut-, rose-, and pea-combed birds appear, but so did single-combed birds. These four comb types occurred in a ratio of 9 walnut : 3 rose : 3 pea : 1 single. Such a ratio is characteristic in F_2 progeny from two parents each heterozygous for two genes. The doubly dominant class in the F_2 was walnut, while the proportion of the single-combed birds indicated that this class contained both recessive alleles.

The overall explanation of the results is as follows (see Figure 4.8b): The walnut comb depends on the presence of two dominant alleles, R and P, both located at two independently assorting gene loci. In $R/- p/p$ birds, a rose comb results, in $r/r P/-$ birds, a pea comb results and in $r/r p/p$ birds, a single comb results.

Thus it is the interaction of two dominant alleles, each of which individually produces a different phenotype, that produces a new phenotype. No modification of typical Mendelian ratios is involved. The molecular basis for the four comb types is not known. At a very general level we can propose that the single-comb phenotype results from the activities of a number of genes other than the R and P genes. In other words, $r/r p/p$ birds do not produce any functional gene product that influences the comb phenotype beyond the basic single appearance. The dominant R allele might produce a gene product that interacts with the products of genes controlling the single-comb phenotype to produce a rose-shaped comb. Similarly, the dominant P allele might produce a gene product that interacts with the products of the single-comb genes to produce a pea-shaped comb. When the products of both the R and P alleles are present, they interact to produce another comb variation, the walnut comb.

FRUIT SHAPE IN SUMMER SQUASH (THE 9:6:1 RATIO). In the comb shape example, each dominant allele alone produces a different phenotype (rose and pea), but both dominant alleles together produce a new phenotype (walnut). And when there is homozygosity for the recessive alleles of both loci, yet another comb phenotype results—single. Similarly, fruit shape in summer squashes shows complete dominance at both gene pairs, and interaction between both dominants results in a new phenotype. In this case, though, each dominant allele alone produces the same phenotype, so this example differs from the comb shape example discussed above.

Two of the many varieties of summer squash have long fruit (Figure 4.9a) and sphere-shaped fruit

~ FIGURE 4.8

Complete dominance in the fowl. The genetic crosses show the interaction of genes for comb shape in fowl. (a) The cross of a true-breeding rose-combed bird with a true-breeding pea-combed bird gives all walnut-combed offspring in the F_1. (b) When the F_1 birds are interbred, a 9:3:3:1 ratio of walnut:rose:pea:single occurs in the F_2.

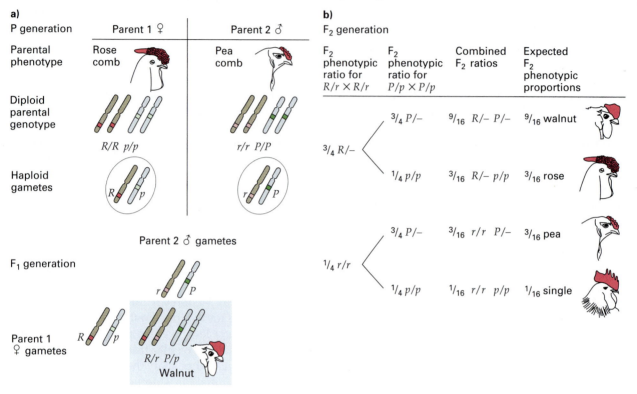

a)

P generation

Parental phenotype — Rose comb | Pea comb

Diploid parental genotype

$R/R \; p/p$ | $r/r \; P/P$

Haploid gametes — $R \; p$ | $r \; P$

Parent 2 ♂ gametes

F_1 generation — $r \; P$

Parent 1 ♀ gametes — $R \; p$

$R/r \; P/p$ Walnut

F_1 genotypes: All $R/r \; P/p$

F_1 phenotypes: All walnut comb

b)

F_2 generation

F_2 phenotypic ratio for $R/r \times R/r$	F_2 phenotypic ratio for $P/p \times P/p$	Combined F_2 ratios	Expected F_2 phenotypic proportions
$3/4 \; R/-$	$3/4 \; P/-$	$9/16 \; R/- \; P/-$	$9/16$ walnut
	$1/4 \; p/p$	$3/16 \; R/- \; p/p$	$3/16$ rose
$1/4 \; r/r$	$3/4 \; P/-$	$3/16 \; r/r \; P/-$	$3/16$ pea
	$1/4 \; p/p$	$1/16 \; r/r \; p/p$	$1/16$ single

(Figure 4.9b). The long-fruit varieties are always true breeding. However, in some crosses between different varieties of true-breeding sphere-shaped plants, the F_1 fruit is disk-shaped (Figure 4.9c). In such instances, the F_2 fruit show approximately 9/16 disk-shaped, 6/16 sphere-shaped, and 1/16 long-shaped (Figure 4.10); this is a modification of the typical Mendelian ratio. The explanation is as follows: Either dominant allele alone ($A/- \; b/b$ or $a/a \; B/-$) specifies spherical fruit, while the two nonallelic dominant alleles ($A/- \; B/-$) interact together to produce a new phenotype, namely, disk-shaped fruit. The doubly homozygous recessive ($a/a \; b/b$) gives a long fruit shape. Thus in the cross above the sphere-shaped parentals are $A/A \; b/b$ and $a/a \; B/B$. The F_1s are disk-shaped and doubly heterozygous $A/a \; B/b$. The F_2 disk-shaped fruits are $A/- \; B/-$, the spherical fruits are $A/- \; b/b$ or $a/a \; B/-$, and the long-shaped fruits are $a/a \; b/b$, giving the 9:6:1 ratio.

Again, the precise molecular bases for the different shapes of squash fruit are not known. In the

~ FIGURE 4.9

Three fruit shapes of summer squash and the genotypes that produce them: (a) long; (b) sphere; (c) disk.

a)   Long
$a/a \; b/b$

b)   Sphere
$A/- \; b/b$ or
$a/a \; B/-$

c)   Disk
$A/- \; B/-$

~ **FIGURE 4.10**

Generation of an F_2 9:6:1 ratio for fruit shape in summer squash.

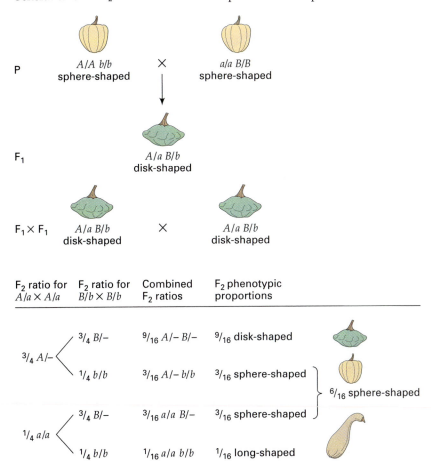

a/a b/b squash, in the absence of *A* and *B* allele products, the functions of other genes determine the long fruit shape. If either the *A* or the *B* allele product, but not both, is present, the basic long fruit shape is modified into a sphere shape. The disk fruit shape presumably occurs through a modification of the sphere shape due to the interaction of A and B products.

Epistasis

Epistasis is a form of gene interaction in which one gene interferes with the phenotypic expression of another nonallelic gene so that the phenotype is governed by the former gene and not by the latter gene when both genes are present in the genotype. Unlike the chicken comb shape example we have just discussed, no new phenotypes are produced by this type of gene interaction. A gene that masks another gene's expression is said to be epistatic, and a gene whose expression is masked by a nonallelic gene is said to be

hypostatic. If we think about the F_2 genotypes *A/– B/–, A/– b/b, a/a B/–,* and *a/a b/b*, epistasis may be caused by the presence of homozygous recessives of one gene pair, so that *a/a* masks the effect of the *B* allele. Or epistasis may result from the presence of one dominant allele in a gene pair. For example, the *A* allele might mask the effect of the *B* allele. Moreover, the epistatic effect need not be just in one direction, as in the examples we just mentioned. Epistasis can occur in both directions between two gene pairs. All these possibilities produce quite a number of modifications of the 9:3:3:1 ratio in a dihybrid cross.

COAT COLOR IN RODENTS (THE 9:3:4 RATIO). In one type of epistasis, recessive epistasis, *a/a B/–* and *a/a b/b* individuals have the same phenotype, so the overall phenotypic ratio in the F_2 is 9:3:4 rather than 9:3:3:1. An example is coat color in rodents. The ancestral coat color of mice is the greyish color seen in ordinary wild mice, due to the presence of black and yellow pigments in the fur. Individual hairs are mostly black with narrow yellow bands near the tip (Figure 4.11). This coloration, the agouti pattern, has a camouflage function and is found in many wild rodents, including the wild rabbit, the guinea pig, the grey squirrel, and wild mice.

Several other coat colors are seen in domesticated rodents. The most familiar example is the albino, in which the complete absence of pigment in the fur and in the irises of the eyes causes a white coat and pink eyes. Albinos are true breeding, and this variation behaves as a complete recessive to any other color. Another variant has black coat color as the result of the absence of the yellow pigment found in the agouti pattern. Black also is recessive to agouti.

When true-breeding agouti mice are crossed with albinos, the F_1 progeny are all agouti. When these F_1 agoutis are interbred, the F_2 progeny consist of approximately 9/16 agouti animals, 3/16 black, and 4/16 albino, as shown in Figure 4.12. This pattern occurs because the parents differ in a gene necessary for the development of any color, which the black mice have (they are $C/-$) but the albinos do not (they are c/c), and in a gene for the agouti pattern ($A/-$ for agouti; a/a for nonagouti), which results in a banding of the black hairs with yellow. (Note: The symbols here are the actual ones used for the genes involved in rodent coat color. Do not confuse the a and c loci here with the a and b loci referred to in our continuing discussion of modified ratios that began on p. 106.) Phenotypically, $A/- C/-$ are agouti, $a/a C/-$ are black, and $A/- c/c$ and $a/a c/c$ are albino, giving a 9:3:4 phenotypic ratio of agouti: black:albino. Thus, in this example, there is epistasis of c/c over $A/-$. In other words, white hairs are produced in c/c mice, regardless of the genotype at the other locus.

It is now known that three gene loci are involved in the phenotypes of rodent coat color. At one locus the dominant C allele specifies a product that is necessary for the production of any pigment in the coat; the recessive c allele, when homozygous, prevents pigment formation regardless of the genotypes of other coat-color genes and hence the mice are albino. At a second locus the dominant allele A specifies a product that determines the agouti factor. Its recessive allele a in the homozygous state produces nonagouti mice in which color is determined by the other color genes present. The dominant allele B of the third locus specifies a product that governs the synthesis of black pigment. The recessive allele b, when homozygous, results in brown pigment. This latter locus is relevant to the example only insofar as the basic hair color is black. All the mice involved in Figure 4.12 must have had at least one B allele, otherwise some brown mice would have been seen (e.g., $A/- C/- b/b$ are brown).

FLOWER COLOR IN SWEET PEAS (THE 9:7 RATIO). A number of true-breeding varieties occur in the

~ **FIGURE 4.11**

Pigment patterns in rodent fur: (a) agouti; (b) albino; and (c) black.

a) Agouti
$A/- C/-$

b) Albino
$A/- c/c$ or
$a/a c/c$

c) Black
$a/a C/-$

~ **FIGURE 4.12**

Generation of an F_2 9:3:4 ratio for coat color in rodents.

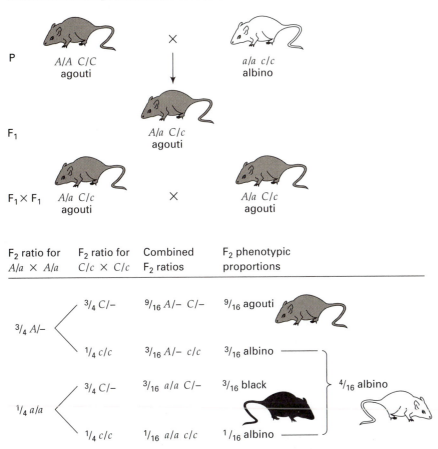

sweet pea, most of which have descended from a purple-flowered wild sweet pea of Sicily. Purple flower color is dominant to white and gives a typical 3:1 ratio in the F_2. White-flowered varieties of sweet peas breed true, and crosses between different white varieties usually produce white-flowered progeny. In some cases, however, crosses of two true-breeding white varieties give only purple-flowered F_1 plants. When these F_1 hybrids are self-fertilized, they produce an F_2 generation consisting of about 9/16 purple-flowered sweet peas and 7/16 white-flowered, as shown in Figure 4.13. The 9:7 ratio is a modification of the 9:3:3:1 ratio. Even though they are not all homozygous for the alleles in question, all the F_2 white-flowered plants breed true when self-fertilized. One ninth of the purple-flowered F_2 plants—the $C/C\ P/P$ genotypes—breed true.

These results may be explained by considering the interaction of two nonallelic genes. In contrast to the example of comb shape in chickens, no new traits appear in the F_1 or F_2. The 9/16 purple-flowered F_2

plants suggests that colored flowers appear only when two independent dominant alleles are present together and that the color purple results from some interaction between them. White flower color would then be due to the absence of either or both of these alleles. Thus gene pair C/c specifies whether or not the flower can be colored, and gene pair P/p specifies whether or not purple flower color will result. An interaction of two genes to give rise to a specific product is a form of epistasis called duplicate recessive *epistasis* or *complementary gene action*.

Figure 4.14 (p. 113) presents a purely theoretical pathway for the production of purple pigment. In this pathway a colorless precursor compound is converted through several steps (via compounds 1, 2, and 3) to a purple end product. Each step is controlled by a functional gene product. To explain the F_2 ratio in the sweet pea example, we can propose an hypothesis that gene C controls the conversion of white compound 1 to white compound 2, and that gene P controls the conversion of compound 2 to compound 3

~ **FIGURE 4.13**

Generation of an F_2 9:7 ratio for flower color in sweet peas.

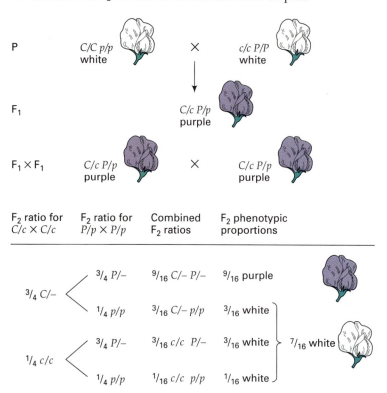

F_2 ratio for $C/c \times C/c$	F_2 ratio for $P/p \times P/p$	Combined F_2 ratios	F_2 phenotypic proportions	
$^3/_4\ C/-$	$^3/_4\ P/-$	$^9/_{16}\ C/-\ P/-$	$^9/_{16}$ purple	
	$^1/_4\ p/p$	$^3/_{16}\ C/-\ p/p$	$^3/_{16}$ white	
$^1/_4\ c/c$	$^3/_4\ P/-$	$^3/_{16}\ c/c\ P/-$	$^3/_{16}$ white	$^7/_{16}$ white
	$^1/_4\ p/p$	$^1/_{16}\ c/c\ p/p$	$^1/_{16}$ white	

(Figure 4.14a). Therefore, homozygosity for the recessive allele of either or both of the C and P genes will result in a block in the pathway, and only white pigment will accumulate. That is, $C/-p/p$, $c/c\ P/-$, and $c/c\ p/p$ genotypes will all be white (Figure 4.14b). The only plants that produce purple flowers will be those in which the two steps are completed so that the rest of the pathway leading to the colored pigment can be carried out. This situation occurs only in $C/-P/-$ plants. (One purely theoretical pathway proposed for this sweet pea example is applicable to other 9:7 ratios, with modification, of course, for the particular phenotypes encountered.)

In the cross described, the two white parentals were $C/C\ p/p$ and $c/c\ P/P$ and the F_1 plants were purple and doubly heterozygous. Interbreeding the F_1 gives a 9:7 ratio of purple:white in the F_2. The true-breeding purple F_2 plants were $C/C\ P/P$. In this case, recessive epistasis occurs in both directions between two gene pairs. The consequence of this gene interaction is that the same phenotype (white) is exhibited whenever one or the other gene pair is homozygous recessive.

FRUIT COLOR IN SUMMER SQUASH (THE 12:3:1 RATIO).
Summer squash has three common fruit colors: white, yellow, and green. In crosses of white and yellow and of white and green, white is always expressed. In crosses of yellow and green, yellow is expressed. Yellow thus is recessive to white but dominant to green.

Consider two gene pairs, W/w and Y/y. In squashes that are $W/-$ in genotype, the fruit is white no matter what genotype is at the other locus. In w/w plants, (1) the fruit will be yellow if a dominant allele of the other locus is present, and (2) green if it is absent. In other words, $W/-Y/-$ and $W/-y/y$ plants have white fruits, $w/w\ Y/-$ plants have yellow fruits, and $w/w\ y/y$ plants have green fruits. The F_2 progeny of an F_1 self of doubly heterozygous individuals shows a 12:3:1 ratio of white:yellow:green fruits in the plants (Figure 4.15, p. 114). The 12:3:1 ratio is a modification of the 9:3:3:1 ratio. This example shows complete dominance at both gene pairs, with one gene (the white one here), when dominant, epistatic to the other gene; that is, the example illustrates dominant epistasis.

A theoretical biochemical pathway to explain the 12:3:1 ratio of squash color is shown in Figure 4.16 (p. 115). The hypothesis is that a white substance is converted to a yellow end product via a green intermediate. Both steps are controlled by genes, the green

~ FIGURE 4.14

One purely theoretical pathway for the production of (a) purple pigment and (b) white pigment in sweet peas to explain the 9 purple : 7 white ratio in the F₂ of a dihybrid self.

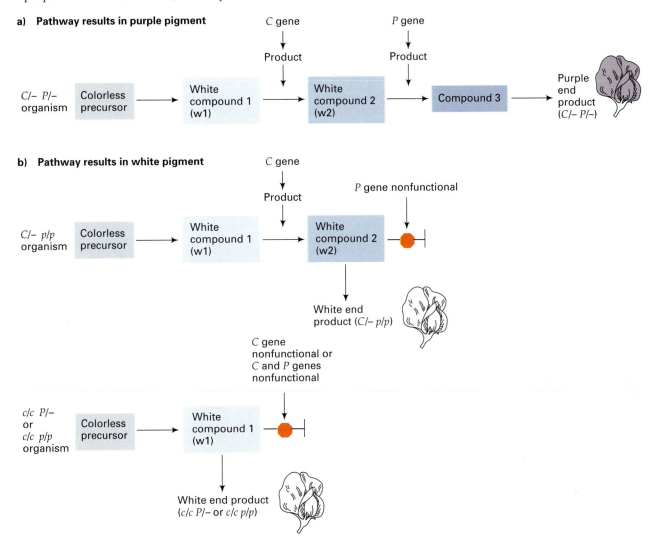

to yellow step being controlled by the *Y* gene. To explain the F₂ phenotypic ratio, we will hypothesize that the dominant *Y* allele is needed for the conversion of the green substance to yellow, and if the dominant *W* allele specifies a product that inhibits the white-to-green conversion step. Thus all plants that have at least one *W* allele will be white-fruited, no matter which alleles are present at the *Y* locus, since the *W* gene product prevents the making of green substance (Figure 4.16a). As a result, 12/16 of the F₂ squashes are white-fruited and have the genotypes *W/– Y/–* and *W/– y/y*, and 3/16 of the F₂ are yellow-fruited with the genotype *w/w Y/–* (Figure 4.16b). In this case green substance is made since there is no inhibition of the white-to-green step, and functional *Y* allele

product catalyzes the conversion of green substance to yellow substance. Lastly, 1/16 of the F₂s are green-fruited and are the *w/w y/y* individuals (Figure 4.16c). Again, green substance is produced because there is no inhibition of the white-to-green step, but in the absence of a dominant *Y* allele, the green substance cannot be converted to yellow substance.

In sum, many types of phenotypic modifications are possible as a result of interactions between the products of different gene pairs. Geneticists detect such interactions when they observe deviations from the expected phenotypic ratios in crosses. We have discussed some examples in which two nonallelic genes assort independently and in which complete dominance is exhibited in each gene pair. The ratios

~ FIGURE 4.15

Generation of an F$_2$ 12:3:1 ratio for fruit color in summer squash.

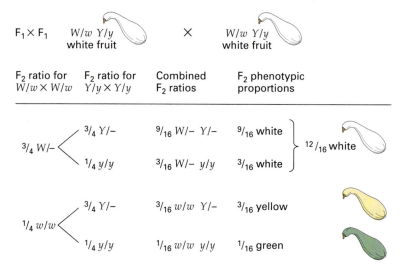

~ TABLE 4.5

Examples of Epistatic F$_2$ Phenotypic Ratios from an $A/a\ B/b \times A/a\ B/b$ in Which Complete Dominance Is Shown for Each Gene Pair (From *Science of Genetics*, 6th ed. by George W. Burns and Paul J. Bottino. Copyright © 1989. Reprinted by permission of Prentice-Hall, Inc., Upper Saddle River, NJ.)

		A/A B/B	A/A B/b	A/a B/B	A/a B/b	A/A b/b	A/a b/b	a/a B/B	a/a B/b	a/a b/b
More than four phenotypic classes	A and B both incompletely dominant	1	2	2	4	1	2	1	2	1
	A incompletely B completely dominant	3		6		1	2	3		1
Four phenotypic classes	A and B both completely dominant (classic ratio)	9				3		3		1
Fewer than four phenotypic classes	a/a epistatic to B and b; recessive epistasis	9				3		4		
	A epistatic to B and b; dominant epistasis	12						3		1
	A epistatic to B and b; b/b epistatic to A and a; dominant and recessive epistasis	13[a]						3		
	a/a epistatic to B and b; b/b epistatic to A and a; duplicate recessive epistasis	9					7			
	A epistatic to B and b; B epistatic to A and a; duplicate dominant epistasis	15								1
	Duplicate interaction	9				6				1

[a]The 13 is composed of the 12 classes immediately above plus the one $a/a\ b/b$ from the last column.

~ **FIGURE 4.16**

Hypothetical pathway for the purpose of illustration to explain the F_2 ratio of 12 white : 3 yellow : 1 green color in squashes. (a) Production of white color; (b) Production of yellow color; (c) Production of green color.

a) White fruit-producing pathway

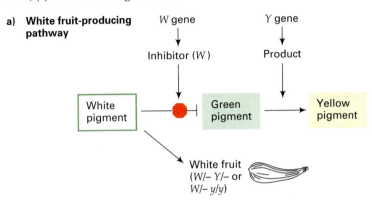

b) Yellow fruit-producing pathway

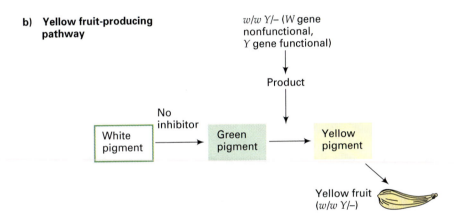

c) Green fruit-producing pathway

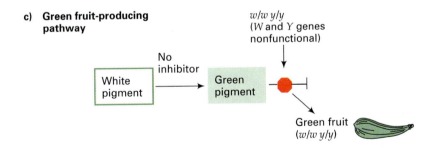

we discussed would necessarily be modified further if the genes did not assort independently and/or if incomplete dominance or codominance prevailed. Table 4.5 shows examples of epistatic F_2 phenotypic ratios from an $A/a\ B/b \times A/a\ B/b$ cross.

$\mathcal{K}$EYNOTE

In many instances, nonallelic genes interact to determine phenotypic characteristics. In some cases interaction between genes results in new phenotypes without modification of typical Mendelian ratios. In another type of gene interaction, called epistasis, interaction between genes causes modifications of Mendelian ratios because one gene interferes with the phenotypic expression of another nonallelic gene (or genes). The phenotype is controlled largely by the former gene and not the latter when both genes occur together in the genotype. The analysis of epistasis is complicated further when one or both gene pairs involve incomplete dominance or codominance, or when gene pairs do not assort independently.

ESSENTIAL GENES AND LETHAL ALLELES

For a few years after the rediscovery of Mendel's principles, geneticists believed that mutations only changed the appearance of a living organism, but then they discovered that a mutant allele could cause the death of an organism. In a sense this mutation is still a change in phenotype, with the new phenotype being lethality. An allele that results in the death of an organism is called a **lethal allele,** and the gene involved is called an essential gene. **Essential genes** are genes which, when they are mutated, can result in a lethal phenotype. If the mutation is due to a **dominant lethal allele,** both homozygotes and heterozygotes for that allele will show the lethal phenotype. If the mutation is due to a **recessive lethal allele,** only homozygotes for that allele will be lethal.

In 1905, Lucien Cuenot was interested in the inheritance of yellow body color in mice. As we have discussed, wild-type mice have an agouti color resulting from a banding of black and yellow pigments in the hairs (see p. 110). Cuenot found that yellow mice never bred true; that is, when yellows were bred to nonyellows, the progeny showed an approximately 1:1 ratio of yellow:nonyellow mice. (The nonyellow color depends upon other coat color genes.) This ratio was expected from the mating of a heterozygote with a recessive, and suggested that yellow mice are heterozygous. When the yellow heterozygotes were interbred, a phenotypic ratio of about 2 yellow : 1 nonyellow was observed instead of the predicted 3:1 ratio.

Cuenot's data were correctly interpreted in 1910 by W. Castle and C. Little, who proposed that the yellow homozygotes were aborted *in utero*; in other words, the yellow allele has a *dominant* effect with regard to coat color, but acts as a *recessive* allele with respect to the lethality phenotype, since only homozygotes die. We now know that homozygotes for the yellow allele die at the embryo stage.

The yellow allele has been shown to be an allele at the agouti locus (*a*) and has been given the symbol A^Y. The yellow × yellow cross is shown in Figure 4.17. Genotypically, the cross is $A^Y/A \times A^Y/A$. We expect a genotypic ratio of $1/4\ A^Y/A^Y : 2/4\ A^Y/A : 1/4\ A/A$ among the progeny. The $1/4\ A^Y/A^Y$ mice die before

birth, giving a birth ratio of $2/3\ A^Y/A$ (yellow) : $1/3$ A/A (nonyellow). Because the A^Y allele causes lethality in the homozygous state, it is called a *recessive lethal*. Characteristically, when two heterozygotes are crossed, recessive lethal alleles are recognized by a 2:1 ratio of progeny types.

More recently, the agouti locus has been molecularly cloned, permitting analysis of the lethal yellow allele. In wild-type agouti mice, the agouti gene is expressed in skin samples taken a few days after birth, at the time when the yellow band in the hair is

~ FIGURE 4.17

Inheritance of a lethal allele A^Y in mice. A mating of yellow by yellow gives 1/4 nonyellow mice, 1/2 yellow mice, and 1/4 dead embryos. The viable yellow mice are heterozygous A^Y/A, and the dead individuals are homozygous A^Y/A^Y.

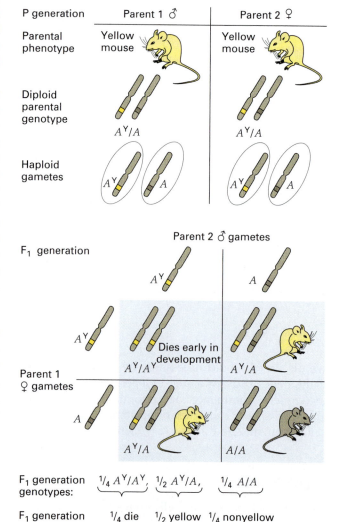

F1 generation genotypes: $1/4\ A^Y/A^Y$, $1/2\ A^Y/A$, $1/4\ A/A$

F1 generation phenotypes: 1/4 die   1/2 yellow   1/4 nonyellow
Of the viable progeny:
2/3 yellow, 1/3 nonyellow

being produced, in skin during regeneration of hair after plucking, and in no other tissues and at no other time. In heterozygous lethal yellow (A^Y/A) mice, the agouti-yellow allele is expressed at high levels in all tissues and at all developmental stages, indicating that tissue-specific regulation of expression has been lost. Moreover, the RNA transcribed from the agouti-yellow allele is more than 50 percent larger than that from the wild-type agouti allele. The explanation is that the A^Y allele has resulted from deletion of a large DNA segment between the agouti locus and an upstream gene called *Raly*, such that the *Raly* promoter and first part of that gene are now fused to the agouti gene. The *Raly* promoter now controls the expression of the attached agouti gene. The RNA transcripts are longer because of the added part of the *Raly* gene, and the expression in all tissues is due to regulatory signals in the *Raly* promoter. The embryonic lethality of yellow homozygotes is probably due to the absence of *Raly* gene activity, rather than an effect of a defective agouti gene.

Essential genes are found in all diploid organisms. In humans there are many known recessive lethal alleles. One example is *Tay-Sachs disease* (see p. 275). Homozygotes appear normal at birth, but before about one year of age they show symptoms of central nervous system deterioration. Progressive mental retardation, blindness, and loss of neuromuscular control follow. Afflicted children usually die at three to four years of age. The genetic defect in Tay-Sachs results in an enzyme deficiency that prevents proper nerve function.

There are X-linked lethal mutations as well as autosomal lethal mutations, and dominant lethal mutations as well as recessive lethals. In humans, for example, the genetic disease hemophilia is caused by an X-linked recessive allele. Untreated, hemophilia is lethal. Dominant lethals exert their effect in heterozygotes, so the organism usually dies at conception or at a fairly young age. Dominant lethals cannot be studied genetically unless death occurs after the organism has reached reproductive age. For example, the symptoms of the autosomal dominant trait *Huntington disease*—involuntary movements and progressive central nervous system degeneration—may not begin until the affected individuals reach their early thirties; as a result, parents may unknowingly pass on the gene to their offspring. Death occurs usually when the afflicted persons are in their forties or fifties. The well-known American folksinger Woody Guthrie died from Huntington disease. (The gene that causes Huntington disease has been molecularly cloned. See Chapter 15 for a discussion of the mutation responsible for the disease.)

KEYNOTE

An allele that is fatal to the individual is called a lethal allele. Recessive lethal and dominant lethal alleles exist, and they can be sex-linked or autosomal. The existence of lethal alleles of a gene indicates that the gene's normal product is essential for the function of the organism; the gene is called an essential gene.

THE ENVIRONMENT AND GENE EXPRESSION

A gene is a segment of DNA, and an organism's set of genes (its genome) is found in its chromosomes. The *development* of a multicellular organism from a zygote is a process of *regulated growth* and *differentiation* that results from the interaction of the genome with the internal cellular environment and with the external environment. Development is a tightly controlled, programmed series of phenotypic changes that, under normal environmental conditions, is essentially irreversible. Four major processes interact with one another to constitute the complex process of development: (1) replication of the genetic material, (2) growth, (3) differentiation of the various cell types, and (4) the aggregation of differentiated cells into defined tissues and organs.

Think of development as a series of intertwined, complex biochemical pathways. Any of these pathways is susceptible to environmental influences if the products of the genes controlling the pathways are affected by the internal or external environment. Though genes influence development, no gene by itself totally determines a particular phenotype. That is, a phenotype is the result of tightly interwoven interactions among many factors during development. So, even if individual organisms have the same genetic constitution, the same phenotypes may not necessarily result. This phenomenon is most readily studied in experimental organisms where the genotype is unequivocally known. The extent to which the gene manifests its effects under varying environmental conditions can then be seen. We will consider some examples in the following section.

Penetrance and Expressivity

In some cases not all individuals who are known to have a particular genotype show the phenotype specified by that gene. The frequency with which a dominant or homozygous recessive gene manifests itself in individuals in a population is called the **penetrance** of the gene. Penetrance depends on both the genotype (e.g., the presence of epistatic or other genes) and the environment. Figure 4.18 illustrates the concept of penetrance. In some cases not all individuals who are known to have a particular genotype show the phenotype specified by that gene. Penetrance is complete (100 percent) when all the homozygous recessives show one phenotype, when all the homozygous dominants show another phenotype, and when all the heterozygotes are alike. For example, if all individuals carrying a dominant mutant allele show the mutant phenotype, the allele is exhibiting complete penetrance. Many genes show complete penetrance: The seven gene pairs in Mendel's experiments and the alleles in the human ABO blood group system are two examples. If less than 100 percent of the individuals with a particular genotype exhibit the phenotype expected, penetrance is incomplete. For example, an organism may be genotypically $A/-$ or a/a but may not display the phenotype typically associated with that genotype. If 80 percent, say, of the individuals carrying a particular gene show the corresponding phenotype, we say that there is 80 percent penetrance.

There are many examples of incomplete penetrance in humans. For example, brachydactyly, an autosomal

~ FIGURE 4.18

Complete and incomplete penetrance.

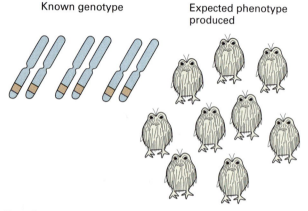

Known genotype

Expected phenotype produced

Complete penetrance

Identical genotypes

Expected phenotype produced (100%)

Incomplete penetrance

Identical genotypes

Expected phenotype produced (<100%)

~ FIGURE 4.19

Joseph Merrick, the Elephant Man.

dominant trait that causes shortened index fingers and toes, shows 50 to 80 percent penetrance. Another human gene that shows incomplete penetrance is the autosomal dominant gene that causes neurofibromatosis. Individuals with neurofibromatosis develop tumorlike growths (neurofibromas) over the body. This genetic disease also shows 50 to 80 percent penetrance. Joseph Merrick (1862–1890) had neurofibromatosis; he was called the Elephant Man because of the disfiguring growths that he developed (Figure 4.19).

We can also determine the degree to which a gene influences a phenotype. **Expressivity** refers to the degree to which a penetrant gene or genotype is phenotypically expressed. Expressivity may be described either qualitatively or quantitatively (Figure 4.20); for example, expressivity may be referred to as severe, intermediate, or slight. Like penetrance, expressivity depends on the genotype and the internal and external environments, and it may be constant or variable. An example of variation in expression is found in the human condition osteogenesis imperfecta. The three

~ FIGURE 4.20

Expressivity of a gene: severe, intermediate, and slight.

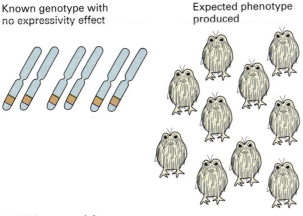

Known genotype with no expressivity effect

Expected phenotype produced

Variable expressivity:

Identical, known genotype

A range of phenotypes produced

Severe   Intermediate   Slight

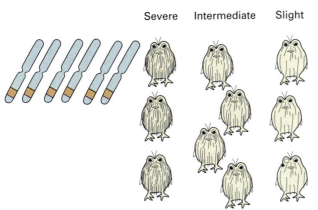

~ FIGURE 4.21

Variable expressivity in individuals with neurofibromatosis. Top: Café-au-lait spot. Middle: Café-au-lait spot and freckling. Bottom: Large number of cutaneous neurofibromas (tumorlike growths).

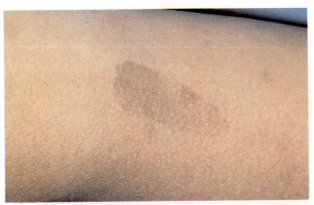

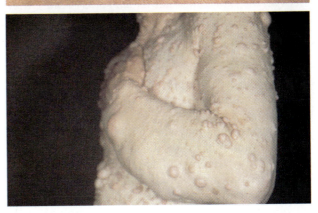

main features of this disease are blue sclerae (the whites of the eyes), very fragile bones, and deafness. This condition is inherited as an autosomal dominant with almost 100 percent penetrance; that is, 100 percent of individuals with the genotype have the condition. However, the trait shows variable expressivity: A person with the gene may have any one or any combination of the three traits. Moreover, the fragility of the bones for those who exhibit this condition is also very variable. Variable expressivity is also seen in neurofibromatosis (see text above and Figure 4.21). In its

~ FIGURE 4.22

Wild-type *Drosophila* eye (top) and eye of a fly that is homozygous for the recessive mutant gene *eyeless* (*ey*) (bottom).

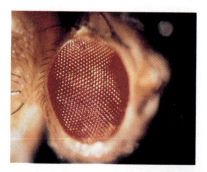

mildest form, individuals have only a few pigmented areas on the skin (called *café-au-lait* spots because they are the color of coffee with milk). In more severe cases, one or more other symptoms may be seen, including neurofibromas of various sizes, high blood pressure, speech impediments, headaches, large head, short stature, tumors of the eye, brain or spinal cord, curvature of the spine, and others. Therefore, we see that in medical genetics it is important to recognize that a gene may vary widely in its expression, a qualification that makes the task of genetic counseling that much more difficult. Another example of expressivity occurs in *Drosophila* homozygous for the recessive mutant gene, *eyeless*. Such flies may have phenotypes varying from no eyes to completely normal eyes, although the common condition is an eye significantly smaller than normal (Figure 4.22).

KEYNOTE

Penetrance is the frequency within the population with which a dominant or homozygous recessive allele manifests itself in the phenotype of an individual. Expressivity is the type or degree of phenotypic manifestation of a penetrant allele or genotype in a particular individual.

Effects of the Internal Environment

Complex biochemical responses occur in cells and in the organism as a result of certain stimuli from the internal environment. Not enough is known about cell biochemistry as it relates to development and adult function, though, to describe all of these environmental effects at the molecular level. Two of these internal stimuli are age and sex.

AGE OF ONSET.   The age of the organism reflects internal environmental changes that can affect gene function. All genes do not continually function; instead over time, there is programmed activation and deactivation of genes as the organism develops and functions. As indicated previously, the Huntington disease gene is one whose effects are not manifested until later in the organism's existence. Somehow, the gene responds to the age of the individual, and the phenotypic effects, devastating as they are, commence. Numerous other age-dependent genetic traits occur in humans; pattern baldness (i.e., the bald spot creeping forward from the crown of the head) appears in males between 20 and 30 years of age, and Duchenne severe muscular dystrophy appears in children between 2 and 5 years of age. In most cases the nature of the age dependency is not understood.

SEX.   The expression of particular genes may be influenced by the sex of the individual. In the case of sex-linked genes, as mentioned earlier, differences in the phenotypes of the two sexes are related to different complements of genes on the sex chromosomes. However, in some cases genes that are autosomal, i.e., not located on the sex chromosomes, affect a particular character that appears in one sex but not the other. Traits of this kind are called **sex-limited traits**.

One such trait is the feathering pattern of chickens (Figure 4.23). A common difference between male and female fowls is in the structure of many of the feathers. Cocks often have long, narrow, pointed feathers on their hackles and saddle, pointed feathers on the cape, back, and wing, and long, curving, pointed sickle feathers on the tail. This pattern is called cock feathering. Hens usually show the contrasting feather characteristics of short, broad, blunt, and straight feathers. This pattern is called hen feathering. If heterozygous (h^+/h) hen-feathered males and females are interbred, all progeny females are hen-

~ FIGURE 4.23

Sex-limited inheritance of feathering in chickens. From a cross between heterozygous F_1 parents the F_2 progenies show a 3:1 ratio of hen-feathered:cock-feathered male birds, while all females are hen-feathered. The gene controlling the phenotype is autosomal.

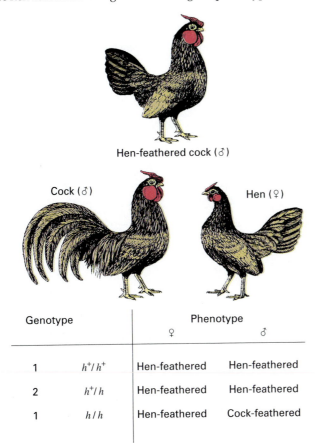

Hen-feathered cock (♂)

Cock (♂)   Hen (♀)

Genotype		Phenotype	
		♀	♂
1	h^+/h^+	Hen-feathered	Hen-feathered
2	h^+/h	Hen-feathered	Hen-feathered
1	h/h	Hen-feathered	Cock-feathered

feathered, whereas progeny males exhibit a 3:1 ratio of hen-feathered:cock-feathered.

The explanation is that the trait is controlled by an autosomal gene, hen feathering (h^+) being dominant to cock feathering (h). In females, h^+/h^+, h^+/h, and h/h give rise to hen feathering. In males, the h^+/h^+ and h^+/h genotypes are hen-feathered while the h/h chickens are cock-feathered. So, from a $h^+/h \times h^+/h$ cross, 1/4 of the male progeny will be h/h and exhibit cock-feathering, while the other 3/4 of the male progeny will exhibit hen-feathering. In other words, in the homozygous condition (h/h) the cock-feathered trait is expressed in males but not in females (male-limited), showing that sex, as an internal environmental parameter, can affect gene expression. Several experiments have indicated that biochemical differences in sex hormones between the two sexes in association with the genetic constitution of the skin (in which the feathers originate) determine the state of feathering in the fowl.

Other examples of sex-limited traits are milk yield in dairy cattle (where the genes involved obviously operate in females but not in males), the appearance of horns in certain species of sheep (where males with genes for horns have horns and females with genes for horns do not have horns), the formation of breasts and ovaries in female humans (and other mammals), the distribution of facial hair in humans, and the ability to produce eggs or sperms in animals.

A slightly different situation is found in **sex-influenced traits** which, like sex-limited traits, are often controlled by autosomal genes. Such traits appear in both sexes, but either the frequency of occurrence in the two sexes is different or the relationship between genotype and phenotype is different.

An example of a sex-influenced trait is pattern baldness in humans (Figure 4.24). One pair of alleles of an autosomal gene is involved in this trait, b^+ and b. The b/b genotype specifies pattern baldness in *both*

~ FIGURE 4.24

Sex-influenced inheritance of pattern baldness in humans. The b allele is recessive in one sex and dominant in the other. (a) Cross of a nonbald female with a bald male who is homozygous b/b; (b) A cross between two F_1 heterozygotes produces a 3:1 ratio of bald:nonbald in males and 1:3 ratio of bald:nonbald in females.

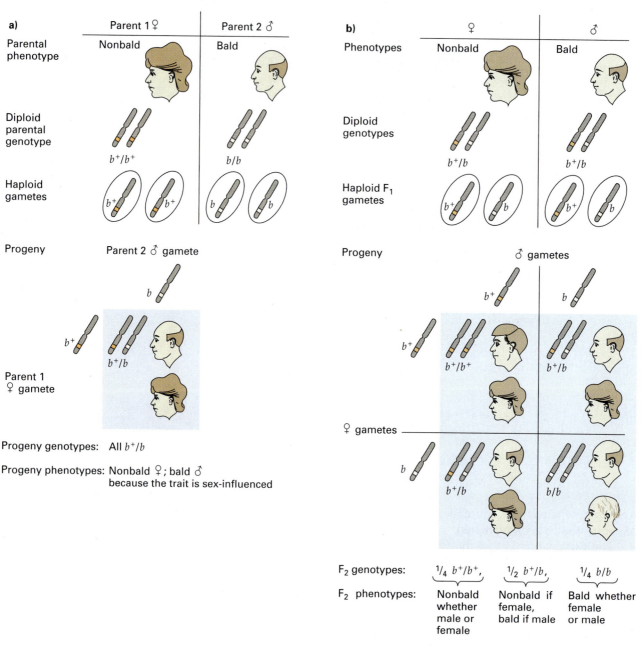

males and females, and the b^+/b^+ genotype gives a nonbald phenotype in both sexes. The difference lies in the heterozygote: In males b^+/b leads to the bald phenotype, and in females it leads to the nonbald phenotype. In other words, the b allele acts as a dominant in males but as a recessive in females. The expression of the b allele is influenced by the sex hormones of the individual; that is, the male hormone testosterone is responsible for the expression of the pattern baldness allele b when it is present in one

dose. The sex-influenced pattern of inheritance and gene expression explains why pattern baldness is far more frequent among men than among women. From a large sampling of the progeny of matings between two heterozygotes, 3/4 of the daughters are nonbald and 1/4 are bald, and 3/4 of the sons are bald and 1/4 are nonbald. Lastly, we must realize that baldness is not a straightforward trait to study. One reason is that there is variable expressivity in the baldness phenotype: as is apparent by inspection of the adult population, baldness may occur early in life or late, it may appear first on the crown (e.g., Prince Charles) or on the forehead, and the degree of baldness varies from minimal to extreme. In fact, a number of genes can affect the presence of hair on the head, including the pattern baldness gene. The final phenotype, then, is mostly the result of the interaction between the environment and the particular set of those genes present. For example, while *b/b* females show pattern baldness, the onset of baldness in these women occurs much later in life than is the case in men because of the influence of hormones in the female internal environment.

Other examples of sex-influenced traits are horn type in Dorset Horn sheep, in which the ram's horns are heavier and more coiled than those of the ewe (Figure 4.25); cleft lip (incomplete fusion of the upper lip) in humans, in which there is a 3:2 ratio of the trait in males:females; some gouts in humans (8:1 ratio in males:females); rheumatoid arthritis in humans (1:3 ratio in males:females); and lupus in humans (an autoimmune disease—1:9 ratio in males:females).

Effects of the External Environment

Although many factors in the external environment influence gene expression (temperature, nutrition, light, chemicals, infectious agents, etc.), we will only discuss two significant factors here, temperature and chemicals.

TEMPERATURE. Biochemical reactions within the cell are catalyzed by enzymes. Normally, enzymes are relatively unaffected by temperature changes within a reasonable range. However, some alleles of an enzyme-coding gene may give rise to an enzyme that is temperature-sensitive; that is, it may function normally at one temperature, but be nonfunctional at another temperature. An example of a temperature effect on gene expression is fur color in Himalayan rabbits (Figure 4.26). Certain genotypes of this white rabbit cause dark fur to develop at the extremities (ears, nose, and paws), where the local surface temperature is lower (Figure 4.26a). Since all body cells develop from a single zygote, this distinct fur pattern cannot be the result of a genotypic difference of the cells in those areas. It can be hypothesized that the fur pattern results from external environmental influences. (A similar situation applies to Siamese cats.)

This hypothesis can be tested by rearing Himalayan rabbits under different temperature conditions. When a rabbit is reared at a temperature above 30°C, all its fur, including that of the ears, nose, and paws, is white (Figure 4.26b). If a rabbit is raised at a temperature of approximately 25°C, the typical Himalayan phenotype results (Figure 4.26c). Lastly, if a rabbit is raised at 25°C while its left rear flank is artificially cooled to a temperature below 25°C, the rabbit develops the Himalayan coat phenotype, and exhibits an additional patch of dark fur on the cooled flank area (Figure 4.26d).

CHEMICALS. Certain chemicals in the internal and external environment can have significant effects on an organism, as the following two examples show.

Phenylketonuria. Since most genes produce protein products, ultimately all the interactions we have discussed occur at the chemical level as those proteins function in cell metabolism. The human disease phenylketonuria (PKU), for example, is an autosomal recessive trait. In individuals homozygous for the

---

~ FIGURE 4.25

Sex-influenced horns in Dorset Horn sheep. (a) The ram's horns are much heavier and more coiled than (b) the ewe's.

a)   b)

(a) Himalayan rabbit. (b) White extremities result when Himalayan rabbits are reared at above 30°C. (c) Normal Himalayan pattern when reared at 25°C. (d) Normal Himalayan pattern when rabbit reared at 25°C, with a dark patch on the side where flank was cooled to below 25°C.

a)

b)   White extremities, reared at >30°C

c)   Normal Himalayan pattern, reared at 25°C

d)   Himalayan pattern with dark patch on flank, reared at 25°C, flank cooled to below 25°C

recessive allele, a variety of symptoms appear, most notably mental retardation at an early age. The cause of the phenotypes is a defective gene product in a biochemical pathway that is needed for the metabolism of phenylalanine, an amino acid. (An amino acid is a protein building block.) The degree to which the symptoms of PKU manifest themselves depends upon diet. Problem foods include protein containing phenylalanine, such as the protein of mother's milk, which may be considered to be an internal environment effect. PKU can be diagnosed soon after birth by a simple blood test and then treated by restricting the amount of phenylalanine in the diet. More discussion of PKU is in Chapter 9, p. 272.

Phenocopies Induced by Chemicals. Changes in the chemical composition of the environment can also influence the expression of one or more genes. The most sensitive time period is early development, since relatively small changes at that time can result in great changes later. When administered to a developing embryo, certain drugs, chemicals, and viruses produce effects that mimic those of known specific mutant alleles. The abnormal individual resulting from this treatment is called a **phenocopy** (phenotypic copies). A phenocopy is defined as a nonhereditary, phenotypic modification (caused by special environmental conditions) that mimics a similar phenotype caused by a gene mutation. In other words, although the individual expresses a mutant phenotype, the genotype is normal. The agent that produces a phenocopy is called a *phenocopying agent*. There are many examples of phenocopies, and, in some instances, studies of phenocopies have provided useful information about the actual molecular defects caused by the mutant counterpart. Remember that the abnormalities seen in these situations occur in a genotypically normal individual without any alteration of the genetic constitution.

In humans, for example, cataracts, deafness, and heart defects are sometimes produced when an individual is homozygous for rare recessive alleles; these disorders may also result if the mother is infected with rubella (German measles) virus during the first 12 weeks of pregnancy. Another human trait for which there is a phenocopy is *phocomelia*, a suppression of the development of the long bones of the limbs, which is caused by a rare dominant allele with variable expressivity. Between 1959 and 1961 similar phenotypes were produced by the sedative thalidomide when it was taken by expectant mothers during the thirty-fifth to fiftieth day of gestation. The drug was removed temporarily from the market when its devastating effects were discovered.

KEYNOTE

The phenotypic expression of a gene depends on several factors, including its dominance relationships, the genetic constitution of the rest of the genome (e.g., the presence of epistatic or modifier genes), and the influences of the internal and external environments. In some cases special environmental conditions can cause a phenocopy, a nonhereditary and phenotypic modification that mimics a similar phenotype caused by a gene mutation.

Nature Versus Nurture

We are left with the nature-nurture question; that is, what is the relative contribution of genes and environment to the phenotype? (Nature-nurture is discussed more in Chapter 23.) Up to this point, variation in most of the traits we have examined has been determined largely by differences in genotype; that is, phenotypic differences have reflected genetic differences. However, earlier in this section we learned that the phenotypes of many traits are influenced by both genes and environment. Let us consider some other human examples.

Human height, or stature, is definitely influenced by genes. On the average, tall parents tend to have tall offspring and short parents tend to have short offspring. There are also a number of genetic forms of dwarfism in humans. Achondroplasia is a type of dwarfism in which the bones of the arms and legs are shortened, but the trunk and head are of normal size; achondroplasia is due to the presence of a single dominant allele. But genes alone do not determine human height; environment also plays a role. For example, human height has increased about 1 inch per generation over the past 100 years. This increase is not the result of changes in the genotype but is probably due to better diet and health care, both of which are known to affect human stature. Thus, genes and the environment interact in determining human height.

For a trait such as height, genes set certain limits or potential for the phenotype. The phenotype an individual develops within these limits depends on the environment. The extent to which the phenotype varies with the environment is termed the **norm of reaction**. For some genotypes, the norm of reaction is small; the phenotype produced by a genotype is the same in different environments. For other genotypes, the norm of reaction is large and the phenotype produced by the genotype varies greatly in different environments.

Many human behavioral traits are the result of interaction between genes and the external environment. For example, alcoholism is a major health problem in the United States—about 10 million Americans are problem drinkers and 6 million are severely addicted to alcohol. Numerous studies have shown that alcoholism is influenced by genes. For example, sons of alcoholic fathers who are separated from their biological parents at birth and adopted into foster homes with nonalcoholic parents are four times more likely to become alcoholic than sons adopted at birth whose biological fathers were not alcoholic. Thus, genes affect alcoholism. However, there is no gene that forces a person to drink alcohol. That is, one cannot become an alcoholic unless one is exposed to an environment in which alcohol is available and drinking is encouraged. What genes do is make certain people more or less susceptible to alcohol abuse; they increase or decrease the risk of developing alcoholism. How genes influence our susceptibility to alcohol abuse is not yet clear. They may affect the way we metabolize alcohol, which might affect how much we drink. Or, genes may influence certain of our personality traits, which make us more or less likely to drink heavily. The important point is that a behavioral trait such as alcoholism may be influenced by genes, but the genes alone do not produce the phenotype.

Nowhere has the role of genes and environment been more controversial than in the study of human intelligence. In the past, there was a tendency to think of human intelligence as either genetically preprogrammed or produced entirely by the environment. The clash of these opposing views was termed the nature-nurture controversy. Today, geneticists recognize that neither of these extreme views is correct; human intelligence is the product of both genes and environment. That genes influence human intelligence is clearly evidenced by genetic conditions that produce mental retardation. For example, PKU (see pp. 123–124) is an autosomal recessive disease that results in mental retardation if untreated; Down syndrome (see Chapter 7, pp. 209–211), caused by an extra copy of chromosome 21, produces mental retardation. Numerous studies also indicate that genes influence differences in IQ among nonretarded people. (IQ, or Intelligence Quotient, is a standardized measure of mental age compared to chronological age; it is relatively stable over time.) For example, adoption studies show that the IQ of adopted children is closer to that of their biological parents than to the IQ of their foster parents. However, it is also clear that IQ is influenced by environment. Identical twins frequently differ in IQ, which can only be explained by environmental differences. Family size, diet, and culture are environmental factors that are known to affect IQ. Thus, IQ results from the interaction of genes and environment. Consequently, if two people (other than identical twins) differ in IQ, it is impossible to attribute that difference solely to either genes or environment, because both interact in determining the phenotype. It is also important to recognize that, although we cannot change our genes, we can alter the environment and thus can affect a phenotypic trait like intelligence.

KEYNOTE

Variation in most of the genetic traits considered in the discussion of Mendelian principles is determined predominantly by differences in genotype; that is, phenotypic differences resulted from genotypic differences. For many traits, however, the phenotypes are influenced by both genes and the environment. The debate over the relative contribution of genes and environment to the phenotype has been termed the nature versus nurture controversy.

SUMMARY

A variety of exceptions to and extensions of Mendel's principles were discussed in this chapter. These include the following:

1. Multiple alleles: A gene may have many allelic forms, and these alleles are called multiple alleles of a gene. Any given diploid individual can have only two different alleles of a given multiple allelic series.

2. Modified dominance relationships: In complete dominance, the same phenotype results whether an allele is heterozygous or homozygous. In incomplete dominance, the phenotype of the heterozygote is intermediate between those of the two homozygotes.

In codominance, the heterozygote exhibits the phenotypes of both homozygotes.

3. Gene interactions and modified Mendelian ratios: In many cases, nonallelic genes do not function independently in determining phenotypic characteristics. In epistasis, modified Mendelian ratios occur because of interactions of nonallelic genes: the phenotypic expression of one gene depends upon the genotype of another gene locus. In other interactions a new phenotype is produced.

4. Essential genes and lethal alleles: Alleles of certain genes result in the lack of production of a necessary functional gene product and this gives rise to a lethal phenotype. Such lethal alleles may be recessive or dominant. The existence of lethal alleles of a gene indicates that the normal product of the gene is essential for the organism to function.

5. It is not always the case that a gene's phenotypes are simply wild type or mutant. The phenomenon of penetrance describes the condition in which not all individuals who are known to have a particular allele show the phenotype specified by that allele. That is, penetrance is the frequency with which a dominant or homozygous recessive allele manifests itself in individuals in the population. The related phenomenon of expressivity describes the degree to which a penetrant gene or genotype is phenotypically expressed in an individual. Both penetrance and expressivity depend on the genotype and the external environment.

6. An organism's potential to develop and function is specified by the zygote's genetic constitution. As an organism develops and differentiates, gene expression is influenced by a number of factors, including the aforementioned dominance relationships, genetic constitution of the rest of the genome, and the influences of the internal and external environment. That is, the phenotypes of many traits are influenced by both genes and environment. The debate over the relative contribution of genes and environment to the phenotype has been termed the nature versus nurture controversy.

ANALYTICAL APPROACHES FOR SOLVING GENETICS PROBLEMS

Q4.1 In snapdragons red flower color (C^R) is incompletely dominant to white flower color (C^W); the heterozygote has pink flowers. Also, normal broad leaves (L^B) are incompletely dominant to narrow, grasslike leaves (L^N); the heterozygote has an intermediate leaf

breadth. If a red-flowered, narrow-leaved snapdragon is crossed with a white-flowered, broad-leaved one, what will be the phenotypes of the F_1 and F_2 generations, and what will be the frequencies of the different classes?

A4.1 This basic question on gene segregation includes the issue of incomplete dominance. In the case of incomplete dominance, remember that the genotype can be directly determined from the phenotype. Therefore we do not need to ask whether or not a strain is true breeding because all phenotypes have a different (and, therefore, known) genotype.

The best approach here is to assign genotypes to the parental snapdragons. Let $C^R/C^R L^N/L^N$ represent the red, narrow plant, and $C^W/C^W L^B/L^B$ represent the white, broad plant. The F_1 plants from this cross will all be double heterozygotes, $C^R/C^W L^B/L^N$. Owing to the incomplete dominance, these plants are pink-flowered and have leaves of intermediate breadth. Interbreeding the F_1s gives the F_2 generation, but it does not have the usual 9:3:3:1 ratio. Instead, there is a different phenotype for each genotype. These genotypes and phenotypes and their relative frequencies are shown in Figure 4.A.

Q4.2 In snapdragons, red flower color is incompletely dominant to white, with the heterozygote being pink; normal flowers are completely dominant to peloric-shaped ones; and tallness is completely dominant to dwarfness. The three gene pairs segregate independently. If a homozygous red, tall, normal-flowered plant is crossed with a homozygous white, dwarf, peloric-flowered one, what proportion of the F_2 will resemble the F_1 in appearance?

A4.2 Let us assign symbols: C^R = red and C^W = white; N = normal flowers and n = peloric; T = tall and t = dwarf. Then the initial cross becomes $C^R/C^R T/T N/N \times C^W/C^W t/t n/n$. From this cross we see that all the F_1 plants are triple heterozygotes with the genotype $C^R/C^W T/t N/n$ and the phenotype pink, tall, normal-flowered. Interbreeding the F_1 generation will produce 27 different genotypes in the F_2; this answer follows from the rule that the number of genotypes is 3^n, where

FIGURE 4.A

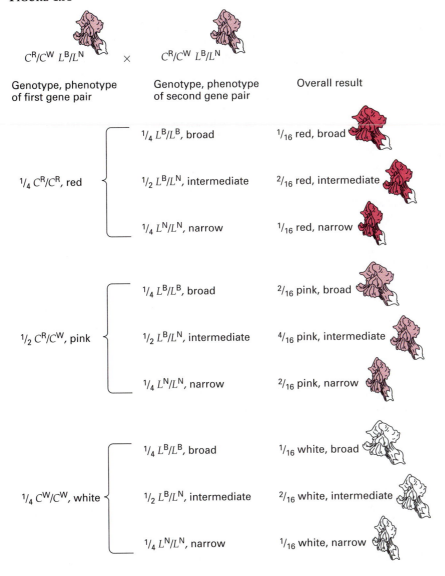

n is the number of heterozygous gene pairs involved in the $F_1 \times F_1$ cross.

Here we are asked specifically for the proportion of F_2 progeny that resemble the F_1 in appearance. We can calculate this proportion in a direct way, without needing to display all the possible genotypes and collect those classes with the appropriate phenotype. First, we calculate the frequency of pink-flowered plants in the F_2; then we determine the proportion of these plants that have the other two attributes. From a $C^R/C^W \times C^R/C^W$ cross we calculate that half of the progeny will be heterozygous C^R/C^W and therefore pink. Next, we determine the proportion of F_2 plants that are phenotypically like the F_1 with respect to height (i.e., tall). Either T/T or T/t plants will be tall, and so 3/4 of the F_2 will be tall. Similarly, 3/4 of the F_2 plants will be normal-flowered like the F_1s. To obtain the probability of occurrence of all three of these phenotypes together (i.e., pink, tall, normal), we must multiply the individual probabilities since the gene pairs segregate independently. The answer is $1/2 \times 3/4 \times 3/4$, or 9/32.

Q4.3a. An $F_1 \times F_1$ self gives a 9:7 phenotypic ratio in the F_2. What phenotypic ratio would you expect if you testcrossed the F_1?

b. Answer the same question for an $F_1 \times F_1$ cross that gives a 9:3:4 ratio.

c. Answer the same question for a 15:1 ratio.

A4.3 This question deals with epistatic effects. In answering the question, we must consider the interaction between the different genotypes in order to proceed with the testcross. Let us set up the general genotypes that we will deal with throughout. The simplest are allelic pairs a^+ and a, and b^+ and b, where the wild-type alleles are completely dominant to the other member of the pair.

a. A 9:7 ratio in the F_2 implies that both members of the F_1 are double heterozygotes and that epistasis is involved. Essentially, any genotype with a homozygous recessive condition has the same phenotype, so the 3, 3, and 1 parts of a 9:3:3:1 ratio are phenotypically combined into one class. In terms of genotype, 9/16 are $a^+/- b^+/-$ types, and the other 7/16 are $a^+/- b/b$, $a/a b^+/-$, and $a/a b/b$. (As always, the use of the "−"after a wild-type allele signifies that the same phenotype results, whether the missing allele is a wild type or a mutant.) Now the testcross asked for is $a^+/a b^+/b \times a/a b/b$. Following the same logic used in questions like this one, we can predict a 1:1:1:1 ratio of $a^+/a b^+/b : a^+/a b/b : a/a b^+/b : a/a b/b$. The first genotype will have the same phenotype as the 9/16 class of the F_2, but because of epistasis, the other three genotypes will have the same phenotype as the 7/16 class of the F_2. In sum, the answer is a phenotypic ratio of 1:3 in the progeny of a testcross of the F_1.

b. We are asked to answer the same question for a 9:3:4 ratio in the F_2. Again, this question involves a modified dihybrid ratio, where two classes of the 9:3:3:1 have the same phenotype. Complete dominance for each of the two gene pairs occurs here also, so the F_1 individuals are $a^+/a b^+/b$. Perhaps both the $a^+/- b^+/b$ and $a/a b/b$ classes in the F_2 will have the same phenotype, while the $a^+/- b^+/-$ and $a/a b^+/-$ classes will have phenotypes distinct from each other and from the interaction class. The genotypic ratio of a testcross of the F_1 is the same as in part (a) of the question. Considering them in the same order as we did in part (a), the second and fourth classes would have the same phenotype, owing to epistasis. So there are only three possible phenotypic classes instead of the four found in the testcross of a dihybrid F_1, where there is complete dominance and no interaction. The phenotypic ratio here is 1:1:2, where these phenotypes are listed in the same relative order as in the 9:3:4.

c. This question is yet another example of epistasis. Since $15 + 1 = 16$, this number gives the outcome of an F_1 self of a dihybrid where there is complete dominance for each gene pair and interaction between the dominant alleles. In this case the $a^+/- b^+/-$, $a^+/- b/b$, and $a/a b^+/-$ classes have one phenotype and include 15/16 of the F_2 progeny, and the $a/a b/b$ class has the other phenotype and 1/16 of the F_2. The genotypic results of a testcross of the F_1 are the same as in parts (a) and (b) of the question; that is, the F_2 progeny exhibit a 1:1:1:1 ratio of $a^+/a b^+/b : a^+/a b/b : a/a b^+/b : a/a b/b$. The first three classes have the same phenotype, which is the same as that of the 15/16 of the F_2s, and the last class has the other phenotype. The answer, then, is a 3:1 phenotypic ratio.

QUESTIONS AND PROBLEMS

4.1 In rabbits, C = agouti coat color, c^{ch} = chinchilla, c^h = Himalayan, and c = albino. The four alleles constitute a multiple allelic series. The agouti C is dominant to the three other alleles, c is recessive to all three other alleles, and chinchilla is dominant to Himalayan. Determine the phenotypes of progeny from the following crosses:

a. $C/C \times c/c$
b. $C/c^{ch} \times C/c$
c. $C/c \times C/c$
d. $C/c^h \times c^h/c$
e. $C/c^h \times c/c$
f. $c^{ch}/c^h \times c^h/c$
g. $c^h/c \times c/c$
h. $C/c^h \times C/c$
i. $C/c^h \times C/c^{ch}$

*4.2 If a given population of diploid organisms contains three, and only three, alleles of a particular gene (say w, $w1$, and $w2$), how many different diploid genotypes are possible in the populations? List all possible genotypes of diploids (consider *only* these three alleles).

4.3 The genetic basis of the ABO blood types seems most likely to be:

a. multiple alleles.

b. polyexpressive hemizygotes.

c. allelically excluded alternates.

d. three independently assorting genes.

4.4 In humans the three alleles I^A, I^B, and i constitute a multiple allelic series that determine the ABO blood group system, as we described in this chapter. For the following problems, state whether the child mentioned can actually be produced from the marriage. Explain your answer.

a. An O child from the marriage of two A individuals.

b. An O child from the marriage of an A to a B.

c. An AB child from the marriage of an A to an O.

d. An O child from the marriage of AB to an A.

e. An A child from the marriage of an AB to a B.

4.5 A man is blood type O,M. A woman is blood type A,M and her child is type A,MN. The aforesaid man cannot be the father of the child because:

a. O men cannot have type A children.

b. O men cannot have MN children.

c. An O man and an A woman cannot have an A child.

d. An M man and an M woman cannot have an MN child.

*4.6 A woman of blood group AB marries a man of blood group A whose father was group O. What is the probability that

a. their two children will both be group A?

b. one child will be group B and the other group O?

c. the first child will be a son of group AB and their second child a son of group B?

4.7 If a mother and her child belong to blood group O, what blood group could the father *not* belong to?

*4.8 A man of what blood group could not be a father to a child of blood type AB?

4.9 In snapdragons, red flower color (C^R) is incompletely dominant to white (C^W); the C^R/C^W heterozygotes are pink. A red-flowered snapdragon is crossed with a white-flowered one. Determine the flower color of (a) the F_1; (b) the F_2; (c) the progeny of a cross of the F_1 to the red parent; (d) the progeny of a cross of the F_1 to the white parent.

*4.10 In Shorthorn cattle the heterozygous condition of the alleles for red coat color (C^R) and white coat color (C^W) is roan coat color. If two roan cattle are mated, what proportion of the progeny will resemble their parents in coat color?

4.11 What progeny will a roan Shorthorn have if bred to (a) red; (b) roan; (c) white?

*4.12 In peaches, fuzzy skin (F) is completely dominant to smooth (nectarine) skin (f), and the heterozygous conditions of oval glands at the base of the leaves (G^O) and no glands (G^N) give round glands. A homozygous fuzzy, no-gland peach variety is bred to a smooth, oval-gland variety.

a. What will be the appearance of the F_1?

b. What will be the appearance of the F_2?

c. What will be the appearance of the offspring of a cross of the F_1 back to the smooth, oval-glanded parent?

4.13 In guinea pigs, short hair (L) is dominant to long hair (l), and the heterozygous conditions of yellow coat (C^Y) and white coat (C^W) gives cream coat. A short-haired, cream guinea pig is bred to a long-haired, white guinea pig, and a long-haired, cream baby guinea pig is produced. When the baby grows up, it is bred back to the short-haired, cream parent. What phenotypic classes and in what proportions are expected among the offspring?

4.14 The shape of radishes my be long (S^L/S^L), oval (S^L/S^S), or round (S^S/S^S), and the color of radishes may be red (C^R/C^R), purple (C^R/C^W), or white (C^W/C^W). If a long, red radish plant is crossed with a round, white plant, what will be the appearance of the F_1 and the F_2?

4.15 In poultry the dominant alleles for rose comb (R) and pea comb (P), if present together, give walnut comb. The recessive alleles of each gene, when present together in a homozygous state, give single comb. What will be the comb characters of the offspring of the following crosses?

a. $R/R\ P/p \times r/r\ P/p$

b. $r/r\ P/P \times R/r\ P/p$

c. $R/r\ p/p \times r/r\ P/p$

d. $R/r\ P/p \times R/r\ P/p$

e. $R/r\ p/p \times R/r\ p/p$

4.16 For the following crosses involving the comb character in poultry, determine the genotypes of the two parents:

a. A walnut crossed with a single produces offspring 1/4 of which are walnut, 1/4 rose, 1/4 pea, and 1/4 single.

b. A rose crossed with a walnut produces offspring 3/8 of which are walnut, 3/8 rose, 1/8 pea, and 1/8 single.

c. A rose crossed with a pea produces five walnut and six rose offspring.

d. A walnut crossed with a walnut produces one rose, two walnut, and one single offspring.

4.17 In poultry feathered shanks (F) are dominant to clean (f), and white plumage of white leghorns (I) is dominant to black (i).

a. A feathered-shanked, white, rose-combed bird crossed with a clean-shanked, white, walnut-combed bird produces these offspring: two feathered, white, rose; four clean, white, walnut; three feathered, black, pea; one clean, black, single; one feathered, white, single; two clean, white, rose. What are the genotypes of the parents?

b. A feathered-shanked, white, walnut-combed bird crossed with a clean-shanked, white, pea-combed bird produces a single offspring, which is clean-shanked, black, and single-combed. In further offspring from this cross, what proportion may be expected to resemble each parent, respectively?

*4.18 F_2 plants segregate 9/16 colored : 7/16 colorless. If a colored plant from the F_2 is chosen at random and selfed, what is the probability that there will be no segregation of the two phenotypes among its progeny?

*4.19 In peanuts, the plant may be either "bunch" or "runner." Two different strains of peanut, V4 and G2, in which "bunch" occurred were crossed with the following results:

V4 bunch × V4 bunch
↓
all bunch

G2 bunch × G2 bunch
↓
all bunch

The two true-breeding strains of bunch were crossed in the following way:

V4 bunch × G2 bunch
↓
F_1 runner

F_1 × F_1
↓
F_2 9 runner : 7 bunch

What is the genetic basis of the inheritance pattern of runner and bunch in the F_2?

*4.20 In rabbits, one enzyme (the product of a functional gene *A*) is needed to produce a substance needed for hearing. Another enzyme (the product of a functional gene *B*) is needed to produce another substance also required for normal hearing. The genes responsible for the two enzymes are not linked. Individuals homozygous for either one or both of the nonfunctional recessive alleles, *a* or *b*, are deaf.

a. If a large number of matings were made between two double heterozygotes, what phenotypic ratio would be expected in the progeny?

b. This phenotypic ratio is a result of what well-known phenomenon?

c. What phenotypic ratio would be expected if rabbits homozygous recessive for trait A and heterozygous for trait B were mated to rabbits heterozygous for both traits?

4.21 In Doodlewags, the dominant allele *S* causes solid coat color; the recessive allele *s* results in white spots on a colored background. The black coat color allele *B* is dominant to the brown allele *b*, but these genes are expressed only in the genotype *a/a*. Individuals that are *A/–* are yellow regardless of *B* alleles. Six pups are produced in a mating between a solid yellow male and a solid brown female. Their phenotypes are: 2 solid black, 1 spotted yellow, 1 spotted black, and 2 solid brown.

a. What are the genotypes of the male and female parents?

b. What is the probability that the next pup will be spotted brown?

4.22 The allele *l* in *Drosophila* is recessive and sex-linked, and lethal when homozygous or hemizygous (the condition in the male). If a female of genotype *L/l* is crossed with a normal male, what is the probability that the first two surviving progeny to be observed will be males?

*4.23 A locus in mice is involved with pigment production; when parents heterozygous at this locus are mated, 3/4 of the progeny are colored and 1/4 are albino. Another phenotype concerns the coat color produced in the mice; when two yellow mice are mated, 2/3 of the progeny are yellow and 1/3 are agouti. The albino mice cannot express whatever alleles they may have at the independently assorting agouti locus.

a. When yellow mice are crossed with albino, they produce an F_1 consisting of 1/2 albino, 1/3 yellow, and 1/6 agouti. What are the probable genotypes of the parents?

b. If yellow F_1 mice are crossed among themselves, what phenotypic ratio would you expect among the progeny? What proportion of the yellow progeny produced here would be expected to be true breeding?

4.24 In *Drosophila melanogaster*, a recessive autosomal allele, ebony (*e*), produces a black body color when homozygous, and an independently assorting autosomal allele, black (*b*), also produces a black body color when homozygous. Flies with genotypes $e/e\ b^+/–$, $e^+/–\ b/b$ and $e/e\ b/b$ are phenotypically identical with respect to body color. Flies with genotype $e^+/–\ b^+/–$ have a grey body color. If true-breeding $e/e\ b^+/b^+$ ebony flies are crossed with true-breeding $e^+/e^+\ b/b$ black flies,

a. what will be the phenotype of the F_1 flies?

b. what phenotypes and what proportions would occur in the F_2 generation?

c. what phenotypic ratios would you expect to find in the progeny of these backcrosses: (1) F_1 × true-breeding ebony and (2) F_1 × true-breeding black?

*4.25 In four-o'clock plants, two genes, *Y* and *R*, affect flower color. Neither is completely dominant, and the two interact on each other to produce seven different flower colors:

$Y/Y\ R/R$ = crimson	$Y/y\ R/R$ = magenta
$Y/Y\ R/r$ = orange-red	$Y/y\ R/r$ = magenta-rose
$Y/Y\ r/r$ = yellow	$Y/y\ r/r$ = pale yellow

$y/y\ R/R$, $y/y\ R/r$, and $y/y\ r/r$ = white

a. In a cross of a crimson-flowered plant with a white one ($y/y\ r/r$), what will be the appearances of the F_1 the F_2, and the offspring of the F_1 backcrossed to the crimson parent?

b. What will be the flower colors in the offspring of a cross of orange-red × pale yellow?

c. What will be the flower colors in the offspring of a cross of a yellow with a $y/y\ R/r$ white?

4.26 Two four-o'clock plants were crossed and gave the following offspring: 1/8 crimson, 1/8 orange-red, 1/4 magenta, 1/4 magenta-rose, and 1/4 white. Unfortunately, the person who made the crosses was color-blind and could not record the flower colors of the parents. From the results of the cross, deduce the genotypes and flower colors of the two parents.

*4.27 Genes A, B, and C are independently assorting and control production of a black pigment.

a. Assume that A, B, and C act in a pathway as follows:

$$\text{colorless} \xrightarrow{A} \xrightarrow{B} \xrightarrow{C} \text{black}$$

The alternative alleles that give abnormal functioning of these genes are designated a, b, and c, respectively. A black $A/A\ B/B\ C/C$ is crossed by a colorless $a/a\ b/b\ c/c$ to give a black F_1. The F_1 is selfed. What proportion of the F_2 is colorless? (Assume that the products of each step except the last are colorless, so only colorless and black phenotypes are observed.)

b. Assume that C produces an inhibitor that prevents the formation of black by destroying the ability of B to carry out its function, as follows:

$$\text{colorless} \xrightarrow{A} \xrightarrow{B} \text{black}$$
$$\uparrow$$
$$C\ \text{(inhibitor)}$$

A colorless $A/A\ B/B\ C/C$ individual is crossed with a colorless $a/a\ b/b\ c/c$, giving a colorless F_1. The F_1 is selfed to give an F_2. What is the ratio of colorless to black in the F_2? (Only colorless and black phenotypes are observed, as in part [a].)

4.28 In cats, two alleles (B, O) at an X-linked gene control whether black or orange pigment is deposited. A dominant allele at an autosomal gene I/i partially inhibits the pigment deposition, lightening the coat color either from black to grey or orange to pale orange. A dominant allele at the autosomal gene T/t determines whether a tabby, or vertically striped, pattern is present. The tabby pattern is dependent on a dominant agouti (A) allele for expression, with nonagouti (a) epistatic to tabby. The agouti allele also causes a speckled, rather than solid, color coat. Judy, a stray cat that has a speckled, tortoiseshell pattern with grey and pale orange spots and no trace of a tabby pattern, gives birth to four kittens. Of the three female offspring, two are solid grey while a third is speckled grey and light orange like her mother, but in addition, shows traces of a tabby pattern. The single male offspring is solid grey.

a. Explain how the tortoiseshell pattern arises in cats. That is, how can a cat have distinct patches of fur with different pigment deposits?

b. Cats with a tortoiseshell pattern are usually female. Explain why this is the case, and also why, when an unusual male tortoiseshell male cat is found, he is atypically large and typically not very swift.

c. What are possible genotype(s) of Judy and her kittens?

d. What phenotype(s) should be considered in assessing the neighborhood males for paternity?

4.29 In *Drosophila* a mutant strain has plum-colored eyes. A cross between a plum-eyed male and a plum-eyed female gives 2/3 plum-eyed and 1/3 red-eyed (wild-type) progeny flies. A second mutant strain of *Drosophila*, called stubble, has short bristles instead of the normal long bristles. A cross between a stubble female and a stubble male gives 2/3 stubble and 1/3 normal-bristled flies in the offspring. Assuming that the plum gene assorts independently from the stubble gene, what will be the phenotypes and their relative proportions in the progeny of a cross between two plum-eyed, stubble-bristled flies? (Both genes are autosomal.)

4.30 In *Drosophila*, a recessive, temperature-sensitive mutation at a gene on chromosome 2 called *transformer-2* (*tra-2*) causes XX individuals raised at 29°C to be transformed into phenotypic males. At 16°C, these individuals develop as normal females. The sex-type of XY individuals is unaffected by the *tra-2* mutation. Suppose you are given three true-breeding, unlabeled vials containing different strains of *Drosophila* all raised at 16°C. Two of the strains have white eyes, while one has red eyes. You are told that one of the white-eyed strains also carries the *tra-2* mutation. Devise two different methods to decide which white-eyed strain has the *tra-2* mutation. Is there a reason to prefer one method over the other?

4.31 Normal *Drosophila* have straight wings and smooth, well-ordered compound eyes. A strain with curly wings and rough eyes has the following properties. Interbreeding its males and females always gives progeny identical to the parents. An outcross of a male from the strain to a normal female gives 45 curly and 49 rough progeny. An outcross of a female from the strain to a normal male gives 53 curly and 47 rough progeny. Crossing a curly F_1 male and female from the first outcross gives 81 curly and 53 straight progeny. The same curly F_1 male mated to a normal female gives 57 curly and 61 normal progeny. Crossing a rough F_1 male and female from the first outcross gives 78 rough and 42 smooth progeny. When the same rough F_1 male is mated to a normal female, 46 rough and 48 normal progeny are recovered. Develop hypotheses to explain these data, and test them using chi-square tests.

*4.32 In sheep, white fleece (W) is dominant over black (w), and horned (H) is dominant over hornless (h) in males but recessive in females. If a homozygous horned white ram is bred to a homozygous hornless black ewe, what will be the appearance of the F_1 and the F_2?

4.33 A horned black ram bred to a hornless white ewe has the following offspring: Of the males 1/4 are horned, white; 1/4 are horned, black; 1/4 are hornless, white; and 1/4 are hornless, black. Of the females 1/2 are hornless and black, and 1/2 are hornless and white. What are the genotypes of the parents?

*4.34 A horned white ram is bred to the following four ewes and has one offspring by the first three and two by the fourth: Ewe A is hornless and black; the offspring is a horned white female. Ewe B is hornless and white; the offspring is a hornless black female. Ewe C is horned and black; the offspring is a horned white female. Ewe D is hornless and white; the offspring are one hornless black male and one horned white female. What are the genotypes of the five parents?

*4.35 Common pattern baldness is more frequent in males than in females. This appreciable difference in frequency is assumed to be due to

a. Y-linkage of this trait.

b. X-linked recessive mode of inheritance.

c. sex-influenced autosomal inheritance.

d. excessive beer-drinking in males, consumption of gin being approximately equal between the sexes.

4.36 King George III, who ruled England during the period of the revolutionary war in the United States, is an ancestor of Queen Elizabeth II (See Figure 3.37 on p. 85). Near the end of his life, he exhibited sporadic periods of "madness." In retrospect, it appears that he showed symptoms of porphyria, an autosomal dominant disorder of heme metabolism. In addition to "madness," the symptoms of porphyria, which include a variety of physical ailments that were shown by King George III, are sporadic and variable in severity, can be affected by diet, and, currently, can be treated with medication.

a. How would you describe this disease in terms of penetrance and expressivity?

b. If indeed King George III had porphyria, what is the chance that the current Prince of Wales (Charles) carries a disease allele? State all of your assumptions.

4.37 Jasper Rine and his colleagues at the University of California at Berkeley have initiated the Dog Genome Initiative to study canine genes and behavior. They mated Pepper, a vocal, highly affectionate, very social New-

foundland female that is not good at fetching tennis balls but loves water, to Gregor, a quiet, less affectionate, less social border collie that is exceptionally good at fetching tennis balls but avoids water. They obtained 7 F_1 and (thus far) 23 F_2 progeny. When the above behavioral traits were analyzed, all of the 7 F_1 were very similar, each showing a mixture of the parents' behavioral traits. When the behaviors of the F_2 were analyzed, differences were more evident. In particular, two of the F_2 dogs (Lucy and Saki) shared Pepper's love of water. (For more information, see "California Geneticists Are Going to the Dogs" by Donald McCaig in *Smithsonian*, Vol. 27, 1996, pp. 126–141.)

a. Develop hypotheses to explain these observations, and when appropriate, test them using a chi-square test.

b. What practical value might there be in studying the genes of canines?

4.38 Parkinson's disease is a progressive neurological disease that causes slowness of movement, stiffness, and shaking, and eventually leads to disability. It affects about two percent of the U.S. adult population over 50 years of age, and most often appears in individuals who are between their fifth and seventh decades. There has been considerable discussion among scientists as to whether the disease is caused by environmental factors, genetic factors, or both. Support for the environmental hypothesis stems from the observation that the disease appears not to have been reported until after the industrial revolution, as well as from the discovery that some chemicals can cause parkinsonian symptoms. Support for the genetic hypothesis stems from the analysis of pedigrees.

Consider the pedigree below (modified to protect patient confidentiality), which shows the incidence of parkinsonism in a family of European descent. The shaded portions of the pedigree indicate family members who reside in the United States. The remaining portions of the pedigree reside in Europe. Members of the U.S. branches of the family have not visited Europe for any extensive period of time since the initial emigration from Europe.

a. If the disease in this family has a genetic basis, what is its basis? Explain your answer.

b. Why might this pedigree be particularly helpful in distinguishing between an environmental and a genetic cause for Parkinson's disease?

c. What hesitations, if any, do you have about concluding that the disease has a genetic basis in some individuals?

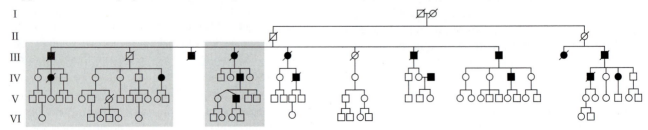

5 GENETIC MAPPING IN EUKARYOTES I

PRINCIPAL POINTS

~ The production of genetic recombinants results from physical exchanges between homologous chromosomes in meiosis. A chiasma is the site of crossing-over, and crossing-over is the reciprocal exchange of chromosome parts at corresponding positions along homologous chromosomes by symmetrical breakage and rejoining.

~ The proof that genetic recombination occurs when crossing-over takes place in meiosis came from experiments in which recombinants for genetic markers occurred only when exchanges of cytological markers occurred.

~ Crossing-over is a reciprocal event that in eukaryotes occurs at the four-chromatid stage in prophase I of meiosis.

~ The map distance between two genes is based on the frequency of recombination between the two genes. The recombination frequency is an approximation of the frequency of crossovers between the two genes. As distance between genes increases, the incidence of multiple crossovers causes the recombination frequency to be an underestimate of the crossover frequency and hence of the map distance. Mapping functions may be used to correct for this problem and hence to give a more accurate estimate of map distance.

~ The occurrence of a chiasma between two chromatids may physically impede the occurrence of a second chiasma nearby, a phenomenon called chiasma interference.

---

*I*n the examples discussed throughout Chapters 1–4, genes assorted independently during meiosis as a result of their location on nonhomologous chromosomes. In many instances, however, certain genes (and hence the phenotypic characters they control) are inherited together because they are located on the same chromosome. Genes on the same chromosome are said to exhibit **linkage** and are called **linked genes**. Linked genes are said to belong to a *linkage group* (Figure 5.1).

In classic genetic analysis, progeny from crosses between parents with different genetic characters are analyzed to determine the frequency with which the alleles in which the parents differed are associated in new combinations. Progeny in which the parental combinations of alleles are shown are called *parentals* and progeny in which the nonparental combinations of alleles are shown are called *recombinants*. For example, from *A B* and *a b* parents, the parental progeny are *A B* and *a b*, and the recombinant progeny are *A b* and *a B*. The process by which the recombinants are produced is called **genetic recombination**. When genetic analysis indicates that two genes show significantly less than 50 percent genetic recombination, the two genes are considered to be linked. Through testcrosses we can determine which genes are linked to each other and can then construct a *linkage map*, or *genetic map*, of each chromosome.

Classic genetic mapping has provided information that is useful in many aspects of genetic analysis. For

example, knowing the locations of genes on chromosomes has been useful in recombinant DNA research and in experiments directed toward understanding the DNA sequences in and around genes. These days the focus of mapping studies is on constructing genetic maps of genomes using not only gene markers but also *DNA markers* (regions of DNA in the genome that show allelic difference and that can be studied using recombinant DNA techniques; see Chapter 15). The goal of these modern studies is to generate high-resolution maps of the chromosomes that will be useful for investigating

~ **FIGURE 5.1**

Linked genes and linkage groups. Genes located on the same chromosome belong to the same linkage group.

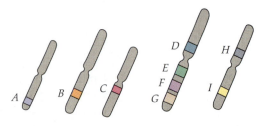

A, B, C:	Unlinked genes
D–E–F–G:	Genes on the same linkage group
H–I:	Genes on the same linkage group
[*A*], [*B*], [*C*], [*D–E–F–G*], [*H–I*]:	Five linkage groups

genes and their functions. The ultimate genetic maps will be of the base pair sequence of organisms' genomes (see Human Genome Project, Chapter 15, pp. 486–487). In this chapter, we discuss the discovery of genetic linkage, the evidence that genetic recombination involves a physical exchange of chromosome parts, and the basic methods of gene mapping analysis in eukaryotes.

DISCOVERY OF GENETIC LINKAGE

Linkage in the Sweet Pea

In 1905, William Bateson, E. R. Saunders, and R. C. Punnett, in their studies of heredity in the sweet pea, *Lathyrus odoratus*, discovered the first exception to the law of independent assortment of gene pairs. They crossed two true-breeding strains, one that produced purple flowers and long pollen and another that produced red flowers and round pollen. All F_1 plants had purple flowers and long pollen, indicating that purple flower color was dominant to red flower color and that long pollen was dominant to round pollen. The 381 plants of the F_2 consisted of 305 plants with purple petals and 76 plants with red petals. (This is *not* a good fit to the 3:1 ratio predicted for a cross of two heterozygotes [see chi-square analysis, Chapter 2], so presumably there was a difference in viability for the two phenotypic classes.) For the pollen-shape trait the F_2 plants segregated into 305 with long pollen and 76 with round pollen. Taken together, these results indicated that both the flower color trait and the pollen shape trait are controlled by single pairs of alleles at their respective genes.

From $F_1 \times F_1$ crosses done by Punnett in 1917 the F_2 progeny displayed the following phenotypes:

    4,831 (69.5 percent) purple flowers, long pollen
      390 (5.6 percent) purple flowers, round pollen
      393 (5.6 percent) red flowers, long pollen
    1,338 (19.3 percent) red flowers, round pollen

If the genes assorted independently, the expected proportions would be $9/16 : 3/16 : 3/16 : 1/16$ for the four phenotypes in the order given, or in percentages, 56.25 percent : 18.75 percent : 18.75 percent : 6.25 percent. However, the observed values differ significantly from those expected according to the independent assortment hypothesis. What constitutes a "significant" difference between observed results and the results expected on the basis of a particular hypothesis may be determined by a *chi-square test*, a statistical procedure described in Chapter 2 (p. 36).

Bateson, Saunders, and Punnett recognized that the dominant purple and long characters occurred together more often than predicted by Mendel's second law but could not explain why the discrepancy occurred. They tried to explain their exceptional results in terms of modified Mendelian ratios but without success. Today we know that the genes they were studying are located on the same chromosome. Thus the results can be explained as follows: If we symbolize the purple allele as P, the red allele as p, the long pollen allele as L, and the round pollen allele as l, the cross is $PP\ LL \times pp\ ll$. The F_1 double heterozygote is $Pp\ Ll$. If independent assortment were occurring (Figure 5.2), equal proportions of $P\ L$, $P\ l$, $p\ L$, and $p\ l$ gametes would be produced by the F_1 plants, and the resulting F_2 generation would exhibit the expected 9:3:3:1 ratio.

Instead, the observed ratio of F_2 phenotypes can occur if it is hypothesized that 44 percent of the gametes are $P\ L$, 44 percent are $p\ l$, and 6 percent each are $P\ l$ and $p\ L$. Figure 5.3 shows how the random combination of these gametes with these frequencies will give the frequencies of F_2 phenotypes with which we started. At the time of pollen and egg formation in the F_1 sweet peas, there was a tendency for the parental P and L alleles to stay together and for the parental p and l alleles to stay together. Since two *nonparental* combinations of phenotypes were produced (purple and round, and red and long), albeit at low frequency, $p\ L$ and $P\ l$ gametes must also have been

~ FIGURE 5.2

Expected results of $F_1 \times F_1$ cross if genes for flower color and pollen size were to assort independently. (As discussed, the two genes are actually linked.)

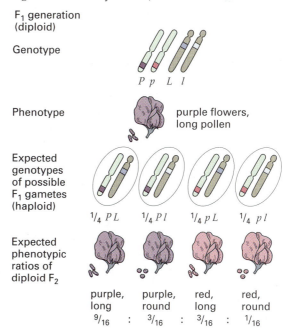

F1 generation (diploid)

Genotype

$P\ p$ $L\ l$

Phenotype — purple flowers, long pollen

Expected genotypes of possible F_1 gametes (haploid)

$1/4\ P\ L$ $1/4\ P\ l$ $1/4\ p\ L$ $1/4\ p\ l$

Expected phenotypic ratios of diploid F_2

purple, long	purple, round	red, long	red, round
$9/16$	$3/16$	$3/16$	$1/16$

~ FIGURE 5.3

Observed results indicating partial linkage of genes: (a) Under assumption that PL and pl gametes (pollen and eggs) each occur with a frequency of 44 percent and the Pl and pL gametes each occur with a frequency of 6 percent; (b) Production of gametes from F_1 double heterozygote; (c) Combination of gametes to produce the F_2 classes.

a)

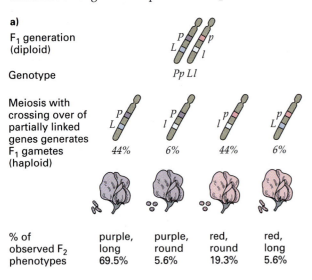

| % of observed F_2 phenotypes | purple, long 69.5% | purple, round 5.6% | red, round 19.3% | red, long 5.6% |

b) Random union of F_1 gametes to produce F_2

Gametes from F_1 heterozygote (%)	Pollen			
	PL 44	Pl 6	pl 44	pL 6
	Genotype frequencies in F_2 (%)			
Eggs PL 44	19.36	2.64	19.36	2.64
Pl 6	2.64	0.36	2.64	0.36
pl 44	19.36	2.64	19.36	2.64
pL 6	2.64	0.36	2.64	0.36

c) Phenotype frequencies in F_2 (%)

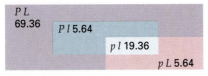

F_2 phenotype totals (%)	
PL 69.36	
Pl 5.64	
pl 19.36	
pL 5.64	

generated. Thus the linkage between the two alleles in these crosses was not complete. Such incomplete linkage occurs when homologous chromosomes exchange corresponding parts during meiosis through the process called *crossing-over* (see Chapter 3 and later discussion in this chapter). Figure 5.4 shows what the observed results would be if the linkage between the two alleles was complete. In the F_1 all are purple, long, and in the F_2 a 3:1 ratio is observed of purple, long : red, round.

~ FIGURE 5.4

Expected results of $F_1 \times F_1$ cross if genes are completely linked.

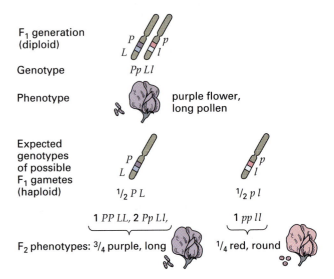

Morgan's Linkage Experiments with *Drosophila*

While we can show that the pea data were produced because the genes were linked on a single chromosome, we can also see that the F_2 results are not the best to use for calculating linkage relationships. Another approach had to be found. Our modern understanding of genetic linkage comes from the work of Thomas Hunt Morgan and his colleagues with linkage in *Drosophila melanogaster*, done around 1911. By 1911 Morgan had identified a number of X-linked mutants. From crosses of the kind we discussed in Chapter 4, Morgan showed that two recessive genes *w* (white eye) and *m* (miniature wing) were X-linked (both were on the X chromosome). Strains carrying these mutant genes were used in experiments to see whether or not genetic exchange occurred between the linked genes on the chromosome.

Morgan crossed a female white miniature (*w m / w m*) with a wild-type male (*w⁺ m⁺ //*) (Figure 5.5). For the former genotype, the slash signifies the pair of homologous chromosomes and indicates that the genes on either side of the slash are linked. For the latter genotype, since the genes are X-linked, a slash indicates the X chromosome and a bent slash indicates a Y chromosome. We will also use another special genetic symbolism for genes on the same chromosome: i.e., $\frac{a\,b}{a\,b}$ signify that genes *a* and *b* are on the same chromosome. With this system, X-linked genes in a female are indicated by appropriate allele symbols separated by one or two lines, for example,

$$\frac{w\,m}{w\,m} \quad \text{or} \quad \frac{w\,m}{w\,m}$$

and X-linked genes in a male are shown as, for example, $w\,m\,//$, where $/$ signifies the Y chromosome.

In the cross the F_1 males were white-eyed and had miniature wings (genotype $w\,m\,//$), while all females were wild type for both the eye color and wing size traits (genotype $w^+\,m^+/w\,m$). The F_1 flies were interbred and the 2,441 F_2 flies were analyzed. Note that in crosses of X-linked genes set up as in Figure 5.5, the $F_1 \times F_1$ is equivalent to doing a testcross since the F_1 males produce X-bearing gametes with recessive alleles of both genes, and Y-bearing gametes that have no alleles for the genes being studied. In the F_2, the most frequent phenotypic classes in both sexes were the grandparental phenotypes of mutant white eyes plus miniature wings or wild-type red eyes plus large wings. Conventionally, we refer to the original genotypes of the two chromosomes as **parental genotypes, parental classes,** or, more simply, **parentals.** The term is also used to describe phenotypes, so the original white miniature females and wild-type males in these particular crosses are defined as the parentals.

Morgan observed a significant number of flies with the nonparental phenotypic combinations of white eyes plus normal wings and red eyes plus miniature wings. Nonparental combinations of linked genes are called **recombinants;** the F_2 progeny of the sweet pea experiment in the previous section that had purple flowers and round pollen, and red flowers and long pollen, are recombinant. In all, 900 of 2,441 F_2 flies, or 36.9 percent, exhibited recombinant phenotypes. Fifty percent recombinant phenotypes was expected if independent assortment was the case;

~ FIGURE 5.5

Morgan's experimental crosses of white-eye and miniature-wing variants of *Drosophila melanogaster*, showing evidence of linkage and recombination in the X chromosome.

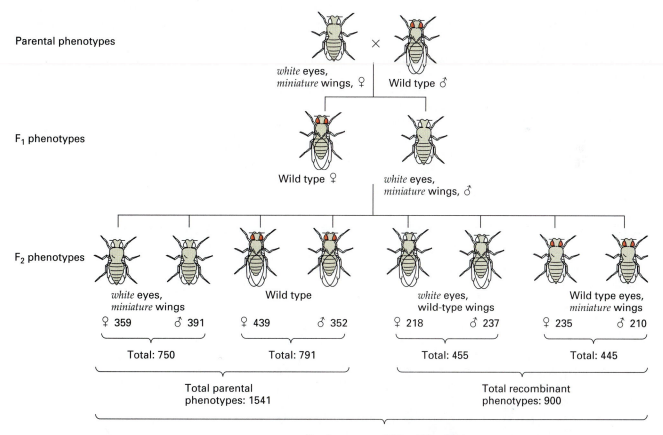

Parental phenotypes

white eyes, *miniature* wings, ♀   ×   Wild type ♂

F_1 phenotypes

Wild type ♀   *white* eyes, *miniature* wings, ♂

F_2 phenotypes

white eyes, *miniature* wings		Wild type		*white* eyes, wild-type wings		Wild type eyes, *miniature* wings	
♀ 359	♂ 391	♀ 439	♂ 352	♀ 218	♂ 237	♀ 235	♂ 210
Total: 750		Total: 791		Total: 455		Total: 445	

Total parental phenotypes: 1541   Total recombinant phenotypes: 900

Total progeny: 1541 + 900 = 2441

Percent recombinants: $^{900}/_{2441} \times 100 = 36.9$

thus the lower percentage observed is taken as evidence for linkage of the two genes. To explain the recombinants, Morgan proposed that, in meiosis, exchanges of genes had occurred between the two X chromosomes of the F_1 females. No such exchange needs to be postulated for sperm production since the males are hemizygous and no genetic exchange occurs between the nonhomologous X and Y chromosomes.

Morgan extended the work to include other sex-linked characters. He crossed a true-breeding white-eyed, yellow-bodied mutant female with a wild-type red-eyed, grey-bodied male. (Red eyes and a grey body are dominant traits.) As expected, the F_1 progeny consisted of wild-type females and white-eyed, yellow-bodied males; that is, the two parental phenotypic classes with the sexes reversed. Among the 2,205 F_2 flies obtained, the original grandparental combinations of white eyes plus yellow body and red eyes plus grey body were the most frequent phenotypic classes, while only 1.3 percent of the flies (29 of 2,205) showed recombinant phenotypes. Morgan concluded that the genes that controlled eye color and body color traits were more closely linked, and thus were separated by recombination less frequently, than were the genes for eye color and wing size.

Morgan's group conducted a large number of other crosses of this type, and the conclusions were always the same. *In each case the parental phenotypic classes were the most frequent while the recombinant classes occurred much less frequently.* Approximately equal numbers of each of the two parental classes were obtained, and similar results were obtained for the recombinant classes. Morgan's general conclusion was that *during meiosis, alleles of some genes assort*

together because they lie near each other on the same chromosome. To take this conclusion one step further, the closer two genes are on the chromosome, the more likely they are to remain together during meiosis. This is because the recombinants are produced as a result of crossing-over between homologous chromosomes during meiosis and the closer two genes are together the less likely there will be a recombination event between them.

Morgan also postulated a relationship between the production of the phenotypic classes in the F_2 progeny of the cross to chiasma formation during meiosis. Earlier, in 1909, F. Janssens had described *chiasmata* (singular = chiasma; cytologically observable reciprocal exchange between homologous chromosomes) during prophase I of meiosis in the salamander. Janssens thought (but could not prove) that the chiasmata might be sites of physical exchange between maternal and paternal homologues (see Chapter 3). Morgan hypothesized that the partial linkage occurred when two genes on the same chromosome were physically separated from one another by chiasma during meiosis. In 1912 Morgan and E. Cattell introduced the term *crossing-over* to describe the process of chromosomal interchange by which new combinations of linked genes (recombinants) arise.

The terminology relating to the physical exchange of homologous chromosome parts can be confusing. To clarify:

1. A chiasma (plural, *chiasmata*; see Figure 3.14, p. 60) is the place on a homologous pair of chromosomes at which a physical exchange is occurring; that is, it is the site of crossing-over.

~ **FIGURE 5.6**

Mechanism of crossing-over: a highly simplified diagram of a crossover between two nonsister chromatids during meiotic prophase, giving rise to recombinant (nonparental) combinations of linked genes.

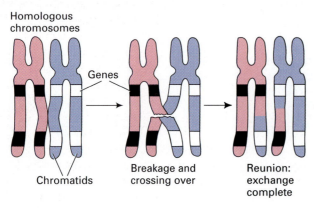

2. Crossing-over is the actual process of reciprocal exchange of chromosome segments at corresponding positions along homologous chromosomes; the process involves symmetrical breakage and rejoining.

3. Crossing-over is also defined as the events leading to genetic recombination between linked genes in both prokaryotes and eukaryotes.

Figure 5.6 (p. 138) presents a very simplified diagram of the process of crossing-over. The molecular basis of crossing over is discussed in Chapter 12.

$\mathcal{K}$EYNOTE

The production of genetic recombinants results from physical exchanges between homologous chromosomes during meiotic prophase I. A chiasma is the site of crossing-over. Crossing-over is the reciprocal exchange of chromosome parts at corresponding positions along homologous chromosomes by symmetrical breakage and rejoining. Crossing-over is also used to describe the events leading to genetic recombination between linked genes.

GENE RECOMBINATION AND THE ROLE OF CHROMOSOMAL EXCHANGE

Morgan's hypothesis that during meiosis physical exchange between chromosomes leads to genetic recombination is now universally accepted. At the time, however, it was only a hypothesis, and convincing evidence that the appearance of recombinants was associated with crossing-over was not obtained until the 1930s.

Corn Experiments

The first evidence for the association of gene recombination with chromosomal exchange came in 1931, from the work of Harriet B. Creighton and Barbara McClintock with corn (*Zea mays*). They performed an experiment in which the two chromosomes under study differed cytologically on both sides of the chromosomal segment in which the crossing-over event occurs.

Crosses were made with a strain of corn that was heterozygous for two genes on chromosome 9 (Figure 5.7a). One of the genes determines colored (*C*) or colorless (*c*) seeds. The other gene determines the forms of starch synthesized by the plants; standard-type plants (*Wx*) produced two forms of starch, amylose and amylopectin, while waxy plants (*wx*) produce only amylopectin. One of the chromosomes had a normal appearance and had the genotype *c Wx*. Its homolog had the genotype *C wx*, possessed a large, darkly staining knob at the end nearer *C*, and was longer than the *c Wx* chromosome because, in a previous generation of this particular strain, a piece of chromosome 8 had broken off and become attached to the end nearer *wx*. (The process of a chromosome segment breaking off from one chromosome and reattaching to another is called *translocation*—see Chapter 7.) Cytologically distinguishable features such as these are called **cytological markers**; the genes involved are called **genetic markers** or **gene markers**.

During meiosis, crossing-over occurs between the two loci (Figure 5.7b). When the two classes of recombinants, *c wx* and *C Wx*, that occurred in the progeny were examined, Creighton and McClintock found that whenever the genes had recombined, the cytological features (the knob and the extra piece) had also recombined (Figure 5.7c). No such physical exchange of cytological markers was evident in the parental (nonrecombinant) classes of progeny. These

~ FIGURE 5.7

Evidence for association of gene recombination with chromosomal exchange in corn. (a) Physical and genetic constitutions of the two chromosomes in the heterozygote; (b) Crossover occurs; and (c) Physical and genetic constitutions of the recombinant chromosomes.

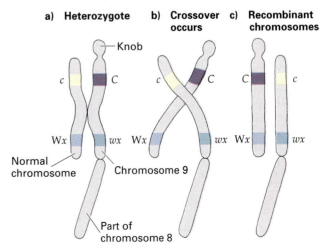

data provided strong evidence that genetic recombination is associated with physical exchange of parts between homologous chromosomes.

Drosophila Experiments

Within a few weeks of the publication of Creighton and McClintock's results, Curt Stern reported identi-

cal conclusions for experiments done with *Drosophila melanogaster*, indicating that the results from corn were not peculiar to that organism. In these fly experiments, the approach was similar: Strains containing appropriate genetic markers and cytological markers were crossed and the two types of markers were analyzed in the next generation (Figure 5.8).

In Stern's work, two linked gene loci were in-

~ FIGURE 5.8

Stern's experiment to demonstrate the relationship between genetic recombination and chromosomal exchange in *Drosophila melanogaster*.

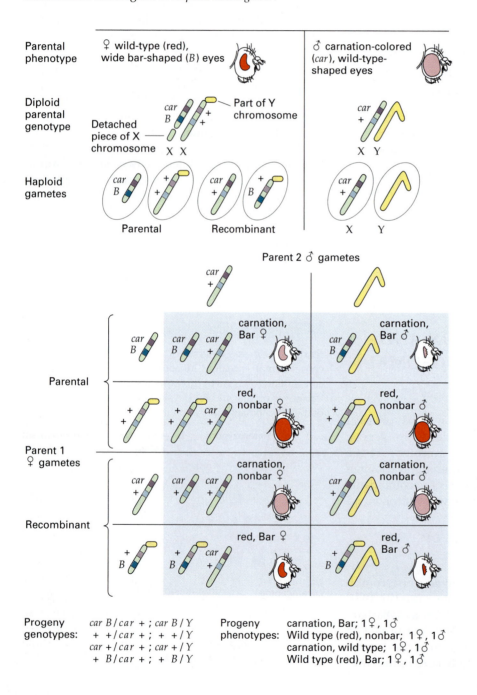

Progeny genotypes:	car B / car + ; car B / Y	Progeny phenotypes:	carnation, Bar; 1♀, 1♂
	+ + / car + ; + + / Y		Wild type (red), nonbar; 1♀, 1♂
	car + / car + ; car + / Y		carnation, wild type; 1♀, 1♂
	+ B / car + ; + B / Y		Wild type (red), Bar; 1♀, 1♂

volved; the *car* (carnation) gene and the *B* (bar-eye) gene. Mutants of the *car* gene are recessive and, when homozygous, result in a carnation-colored eye instead of the wild-type brick-red-colored eye. Mutants of the *B* gene are dominant, resulting in a bar-shaped eye instead of the round eye of the wild type. The *car* gene is located closer to the "left" end of the X chromosome than the *B* gene. In Stern's crosses, the male parent carried normal X and Y chromosomes with the recessive allele *car* (carnation-colored eye) and the wild-type allele of the *B* (bar-eye) gene. Since the male parent is hemizygous, these flies had nonbar (wild-type-shaped), carnation-colored eyes. The female parent had two abnormal and cytologically distinct X chromosomes. One X chromosome, on which were the wild-type alleles of both the *car* and the *B* genes, had a portion of the Y chromosome attached to it. The other X chromosome had the recessive *car* allele and the dominant *B* allele. In addition, the latter X chromosome was distinctly shorter than normal X since part of it had broken off and was attached to the small chromosome 4 (see the unattached piece in Figure 5.8). In females, the shape of the eye depends upon the number of copies of the *B* allele. In *B*/*B* homozygotes, the eye is very narrow, whereas in *B*/+ heterozygotes, the eye is kidney shaped, a condition known as wide-bar. Thus, phenotypically, the parental females had a wide-bar eye since they were *B*/+, and the eye was brick red (wild-type) in color since they were heterozygous +/*car*.

In gamete formation only two classes of sperm were produced: the Y-bearing and the X-bearing. The Y sperm had no genetic markers of relevance here, and the X sperm carried the *car* and *B*+ alleles. However, four classes of eggs were produced. Two—the parentals—resulted from meioses in which no crossing-over had occurred between the chromosomes, and the other two were recombinant gametes produced by crossing-over. As with the corn experiments, every case in which genetic recombination occurred was accompanied by an exchange of identifiable chromosome segments.

If no recombination occurred, the two phenotypic classes of progeny were: (1) brick red (wild-type) eye with normal (round) eye shape (i.e., wild-type for both genes); and (2) carnation-colored eye with a bar shape. No exchanges of chromosome parts were evident among these nonrecombinants. There were also two classes of recombinants: (1) carnation-colored, round eyes; and (2) brick-red (wild-type) colored, bar-shaped eyes. The carnation flies had a complete X chromosome, and the bar flies had a shorter than normal X chromosome to which a piece of the Y chromosome was attached, while the rest of the X chromsome was attached to chromosome 4. This chromosomal

makeup could only have resulted from physical exchanges of homologous chromosome parts.

There is no doubt, therefore, *that genetic recombination results from physical crossing-over between chromosomes.* (Note that, as with Bridges' work on nondisjunction, the application of a combined genetic and cytological approach generated an important and fundamental piece of information.)

*K*EYNOTE

The proof that genetic recombination occurs when crossing-over takes place during meiosis came from breeding experiments in which the parental chromosomes differed with respect to both genetic and cytological markers. Results of these experiments showed that whenever recombinant phenotypes occurred, the cytological markers indicated that crossing-over had also occurred.

Crossing-Over at the Four-Chromatid Stage of Meiosis

In Chapter 3 we learned that crossing-over occurs in prophase I of meiosis; further studies have shown that crossing-over occurs at the four-chromatid stage in prophase I (also called the *tetrad* stage). The organism *Neurospora crassa* (the orange bread mold) is used here to show how evidence was obtained to support this conclusion and, therefore, to disprove the alternative hypothesis that crossing-over occurred before meiosis in interphase.

LIFE CYCLE OF *NEUROSPORA CRASSA.* To understand the proof for crossing-over at the four-chromatid stage, we must understand the life cycle of *Neurospora crassa* (Figure 5.9). Then we can appreciate how suitable this organism is for the experiment.

Neurospora crassa is a mycelial-form fungus, meaning that it spreads over its growth medium in a weblike pattern. It produces a furry mycelial mat when it grows on bread, and its designation as an orange bread mold comes from the color of the asexual spores (called conidia) it produces when it grows. *Neurospora* has several properties that make it useful for genetic and biochemical studies: It is a haploid organism, so the effects of mutations may be seen directly, and its short life cycle facilitates study of the segregation of genetic defects.

~ FIGURE 5.9

Life cycle of the haploid, mycelial-form fungus *Neurospora crassa*. (Parts not to scale.)

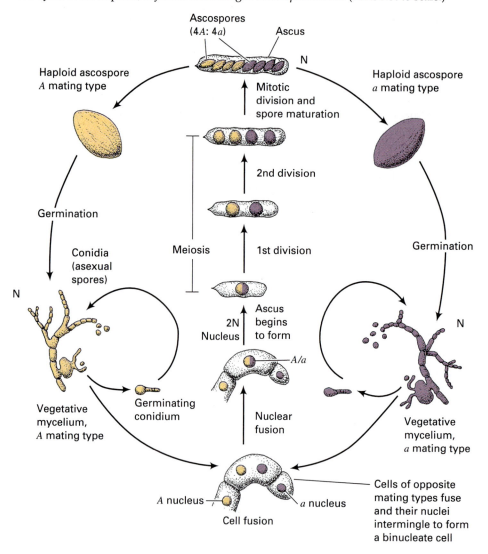

Neurospora can be propagated vegetatively (asexually) to produce unlimited populations of strains with particular genotypes for study. To propagate *Neurospora* asexually, pieces of the mycelial growth or the asexual spores (conidia) can be used to inoculate a suitable growth medium to give rise to a new mycelium. This growth involves only mitotic division of the nuclei and the production of new cell mass.

Neurospora crassa can also reproduce by sexual means. There are two mating types, called *A* and *a* (this is an example of a genic system of sex determination—see Chapter 3, p. 82). The two mating types look identical and can only be distinguished by the fact that strains of the *A* mating type do not mate with other *A* strains, and *a* strains do not mate with other *a* strains. Under normal conditions *Neurospora* reproduces asexually. However, if nutrients (particularly nitrogen sources) become seriously depleted in the growth medium and if both *A* and *a* mating strains

are present, the sexual cycle is initiated. Cells of the two mating types fuse, followed by fusion of two haploid nuclei to produce a transient *A/a* diploid nucleus, which is the only diploid stage of the life cycle. The diploid nucleus immediately undergoes meiosis and produces four haploid nuclei (two *A* and two *a*) within an elongating sac called an *ascus*. A subsequent mitotic division results in a linear arrangement of eight haploid nuclei around which spore walls form to produce eight sexual ascospores (four *A* and four *a*). Each ascus, then, contains all the products of the initial, single meiosis. When the ascus is ripe, the ascospores (sexual spores) are shot out of the ascus and out of the fruiting body that encloses the ascus to be dispersed by wind currents. A particularly important feature of sexual reproduction in *N. crassa* is that all the products of a single meiosis are found together in the ascus and can be isolated and cultured for analysis. Moreover, the order of the four spore

~ FIGURE 5.10

Experiment showing that crossing-over occurs at the four-chromatid stage of meiosis; two haploid strains with different genetic markers (*met* and *his*) are crossed. (a) If a single crossover occurred at the two-chromatid stage (i.e., in interphase, before chromosome duplication), the resulting ascus would contain only the two types of recombinant ascospores. This result differs from experimental findings that asci resulting from a single crossover contain both types of parental ascospores and both types of recombinant ascospores. (b) If a single crossover occurred at the four-chromatid stage, an ascus would be produced that contains both types of parental ascospores and both types of recombinant ascospores. This outcome matches observed results.

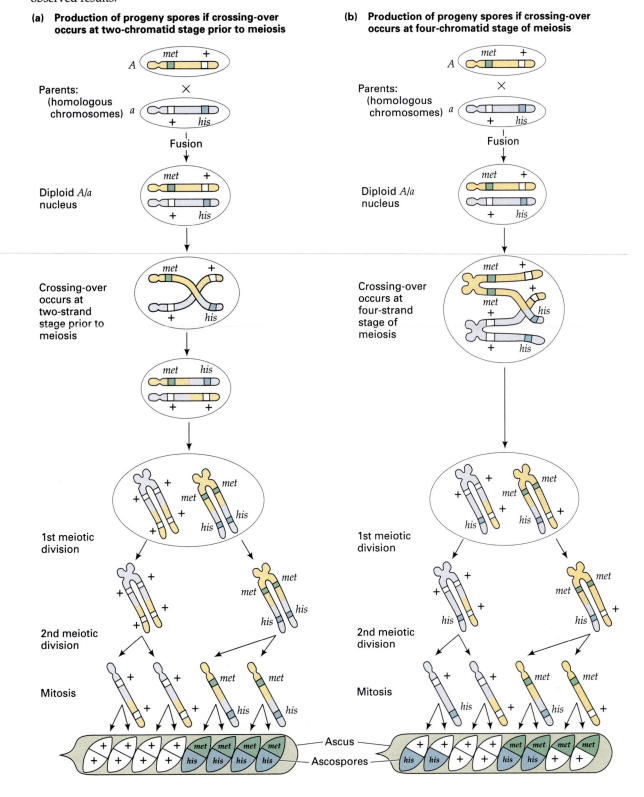

(a) Production of progeny spores if crossing-over occurs at two-chromatid stage prior to meiosis

(b) Production of progeny spores if crossing-over occurs at four-chromatid stage of meiosis

pairs within an ascus reflects exactly the orientation of the four chromatids of each tetrad at the metaphase plate in the first meiotic division: These are called **ordered tetrads**.

RECOMBINATION AT THE FOUR-CHROMATID STAGE IN *NEUROSPORA*.

To determine whether crossing-over occurs at the four-chromatid stage of prophase I, or prior to meiosis at the two-chromatid phase in interphase, crosses are made between haploid strains of *Neurospora* that differ genetically. A theoretical example is shown in Figure 5.10 (p. 143). The two strains have the following genetic constitution: One of them, of mating type *A*, has the genotype *met his*[+] (abbreviated *met* + in the figure)—it is unable to synthesize the amino acid methionine, so that compound must be provided in the growth medium in order for the strain to grow, but it is able to synthesize the amino acid histidine. This strain is wild type for all other genes. The other strain, of mating type *a*, has the genotype *met*[+] *his* (abbreviated + *his* in the figure)—it can synthesize methionine but is unable to synthesize histidine. Both genes are in the same linkage group; that is, on the same chromosome.

The diploid nucleus produced from the mating is heterozygous for the two genes. We can now make some theoretical predictions. In simplified terms, in meiosis, crossing-over will occur between the two genes—at a frequency that is a function of the distance between them on the chromosome. Recombinant progeny will be produced with the haploid genotypes of *met*[+] *his*[+] (abbreviated + + in the figure) and *met his*. The former is wild-type for both genes, and the latter is a double mutant strain that requires both methionine and histidine in the growth medium in order to grow. There is a relatively easy way to test the progeny strains for their phenotypes (which are equivalent to their genotypes since the organism is haploid). It is only necessary to determine whether they can grow on an unsupplemented medium, a medium with methionine, one with histidine, or one with both methionine and histidine.

If no crossing-over occurs, the resulting ascus will contain four spores of each parental type and no recombinant type spores. Figure 5.10a shows the consequences of crossovers occurring at the two-strand stage, that is, prior to meiosis in interphase: The resulting asci will contain only recombinant spores (four of each type). If, however, crossing-over occurs at the four-strand stage, the results would be as diagrammed in Figure 5.10b. Of the four chromatids, crossing-over occurs between just two, one from each parental strain. Those chromatids become recombinant for the gene markers and give rise to four recombinant spores in the ascus. The other two chromatids,

not involved in the crossover, give rise to four parental spores in the same ascus. The overall result is an ascus with four parental and four recombinant progeny spores, representing all four possible phenotypic classes. Asci like this are found routinely in such experiments.

These results provide unequivocal proof that crossing-over occurs at the tetrad stage of prophase I. This has to be the case because chromosomes enter meiosis already replicated into two chromatids; only later recombination events could produce such results. A given crossing-over event involves only two of the four strands, and those strands vary from crossover to crossover. Furthermore, the results showed that crossing over is a reciprocal process; that is, an equal exchange of portions of the chromosome.

KEYNOTE

Crossing-over occurs at the four-chromatid (tetrad) stage in prophase I in meiosis. The evidence for this principle came from the analysis of genetic experiments with the haploid, mycelial fungus *Neurospora crassa*, in which all the products of a single meiotic event are contained within a structure called the ascus. The crucial result was that a single ascus contained spores with both the parental genotypes and spores with both the recombinant genotypes.

CONSTRUCTION OF GENETIC MAPS

So far we have learned two significant concepts. The first is that genetic recombinants result from crossing-over between homologous chromosomes. More specifically, the number of genetic recombinants produced is characteristic of the two linked genes involved. The second concept is that crossing-over takes place at the tetrad stage in prophase I of meiosis and involves only two of the four chromatids for a given crossing-over event. We will now examine how genetic experiments can be used to determine the relative position of genes on chromosomes in eukaryotic organisms. This process is called **genetic mapping**.

Detecting Linkage Through Testcrosses

Before beginning any experiments to construct a **genetic map** of the relative positions of genes on a chromosome (also called a **linkage map**), geneticists must show that the genes under consideration are linked. If they are not linked, then the genes are assorting independently and may be on different chromosomes. Therefore, a way to test for linkage is to analyze the results of crosses to see whether the data deviate significantly from those expected by independent assortment.

The best cross to use to test for linkage is the testcross, a cross of one individual with an unknown genotype with another individual homozygous recessive for all genes involved. In this case, the distribution of phenotypes is the result of segregation events in only one of the two parents; the other parent contributes only recessive alleles to the progeny and those alleles do not contribute to the phenotype of the progeny. We saw in Chapter 2 that a testcross between $a^+/a\ b^+/b$ and $a/a\ b/b$, where genes a and b are unlinked, gives a 1:1:1:1 ratio of the four possible phenotypic classes $a^+\ b^+ : a^+\ b : a\ b^+ : a\ b$. A significant deviation from this ratio in the direction of too many parental types and too few recombinant types would suggest that the two genes are linked. It is important to know how large a deviation must be in order to be considered "significant." The *chi-square test* can be used to make such a decision. We described that test in Chapter 2, pages 36–37. Here we illustrate the use of the chi-square test for analyzing testcross data.

Consider data from a testcross involving fruit flies. In *Drosophila*, b is a recessive autosomal mutation which, when homozygous, results in black body color, and vg is a recessive autosomal mutation which, when homozygous, results in flies with vestigial (short, crumpled) wings. Wild-type flies have grey bodies and long, uncrumpled (normal) wings. True-breeding black, normal ($b/b\ vg^+/vg^+$) flies were crossed with true-breeding grey, vestigial ($b^+/b^+\ vg/vg$) flies. F$_1$ grey, normal ($b^+/b\ vg^+/vg$) female flies were testcrossed to black, vestigial ($b/b\ vg/vg$) male flies. (The female is the heterozygote in this testcross because, in *Drosophila*, no crossing-over occurs between *any* homologous pair of chromosomes in males.) The testcross progeny data were:

> 283 grey, normal
> 1,294 grey, vestigial
> 1,418 black, normal
> 241 black, vestigial

Total 3,236 flies

We hypothesize that the two genes are unlinked and use the chi-square test (see Chapter 2, pp. 36–37) to test the hypothesis, as shown in Table 5.1.

If the two genes are unlinked, then a testcross should result in a 1:1:1:1 ratio of the four phenotypic classes. Column 1 lists the four phenotypes expected in the progeny of the cross, column 2 lists the observed (o) numbers for each phenotype, and column 3 lists the expected (e) number for each phenotypic class, given the total number of progeny (3,236) and the hypothesis under evaluation (in this case 1:1:1:1). Column 4 lists the deviation value (d) calculated by subtracting the expected number (e) from the observed number (o) for each class to find differences. The sum of the d values is always zero.

Column 5 lists the deviation squared (d^2), and column 6 lists the deviation squared divided by the

---

~ TABLE 5.1

Chi-Square Test Used with Testcross Data to Test the Hypothesis That Two Genes Are Unlinked

(1) PHENOTYPES	(2) OBSERVED NUMBER (o)	(3) EXPECTED NUMBER (e)	(4) d	(5) d^2	(6) d^2/e
grey, normal	283	809	−526	276,676	342.00
grey, vestigial	1,294	809	485	235,225	290.76
black, normal	1,418	809	609	370,881	458.44
black, vestigial	241	809	−568	322,624	398.79
Total	3,236	3,236			1,489.99

(7) $\chi^2 = 1{,}489.99$ (8) df 3

expected number (d^2/e). The chi-square value, χ^2 (item 7 in the table), is given by the formula

$$\chi^2 = \Sigma \frac{d^2}{e}$$

In Table 5.1, χ^2 is the sum of the four values in column 6. In our example χ^2 is 1,490. The last value in the table, item 8, is the degrees of freedom (df) for the set of data; there are 3 degrees of freedom in this case.

The χ^2 value and the degrees of freedom are used with a table of chi-square probabilities (see Table 2.5, p. 37) to determine the probability (P) that the deviation of the observed values from the expected values is due to chance. For our example, $\chi^2 = 1,490$ with three degrees of freedom, the P value is much lower than 0.001; in fact it is not on the table. This is interpreted to mean that independent repetitions of this experiment would produce chance deviations from the expected as large as those observed much fewer than 1 out of 1,000 trials. As a reminder, if the probability of obtaining the observed χ^2 values is greater than 5 in 100 ($P > 0.05$), the deviation is considered not statistically significant and could have occurred by chance alone. If $P \leq 0.05$, we consider the deviation from the expected values to be statistically significant and not due to chance alone; the hypothesis may well be invalid. If $P \leq 0.01$, the deviation is highly statistically significant; the data are not consistent with the hypothesis. Thus, in this case we would reject the independent assortment hypothesis for our data and think of an alternative hypothesis, for example, that the genes are linked.

THE CONCEPT OF A GENETIC MAP. The data obtained by Morgan from *Drosophila* crosses indicated that the frequency of crossing-over (and hence of recombinants) for linked genes is characteristic of the gene pairs involved: For the X-linked genes white (w) and miniature (m), the frequency of crossing-over is 36.9 percent, and for the white and yellow (y) genes it is 1.3 percent. Moreover, the frequency of recombinants for two linked genes is the same, regardless of how the alleles of the two genes involved are arranged relative to each other on the homologous chromosomes. That is, in an individual doubly heterozygous for the w and m alleles, for example, the alleles can be arranged in two ways:

$$\frac{w^+ \; m^+}{w \; m} \quad \text{or} \quad \frac{w^+ \; m}{w \; m^+}$$

In the arrangement on the left, the two wild-type alleles are on one homolog and the two recessive mutant alleles are on the other homolog, an arrangement called **coupling** (or the *cis* configuration). Crossing-over between the two loci produces $w^+ m$ and $w m^+$

recombinants. (*Loci* is the plural of **locus**, the position of a gene on a genetic map.) In the arrangement on the right, each homolog carries the wild-type allele of one gene and the mutant allele of the other gene, an arrangement called **repulsion** (or the *trans* configuration). Crossing-over between the two genes produces $w^+ m^+$ and $w m$ recombinants. While the actual phenotypes of the recombinant classes are different for the two arrangements, the percentage of recombinants among the total progeny will be 36.9 percent in each case.

Morgan thought that the characteristic crossover frequencies for linked genes might be related to the physical distances separating the genes on the chromosome. In 1913 a student of Morgan's, Alfred Sturtevant, devised the testcross method to analyze the linkage relationships between genes. He suggested that the percentage of recombinants (produced by crossovers in gamete production) could be used as a quantitative measure of the genetic distance between two gene pairs on a genetic map (see Figure 5.11). This distance is measured in **map units** (mu). A crossover frequency of 1 percent between two genes was defined as 1 map unit. That is, one map unit is the distance between gene pairs for which 1 product out of 100 (1 percent) is recombinant. The map unit is sometimes called a **centimorgan** (cM) in honor of Morgan.

Sturtevant's concept of genetic map distance led him to the important discovery that on the map the genetic distances between a series of linked genes are additive. Thus the genes on a chromosome can be represented by a one-dimensional, genetic map that shows in linear order the genes belonging to the chromosome. Crossover and recombination values give the linear order of the genes on a chromosome and provide information about the genetic distance between any two genes. Genetically speaking, the farther apart two genes are, the greater will be the crossover frequency. Thus, in Figure 5.12, the probability of recombination occurring between genes A and B is much less than between genes B and C, because A and B are closer together than are B and C. The first genetic map ever constructed was based on recombination frequencies from *Drosophila* crosses involving the sex-linked genes w, m, and y discussed earlier, where w gives white eyes, m gives miniature wings, and y gives yellow body. From these mapping experiments, the recombination frequencies for the $w \times m$, $w \times y$, and $m \times y$ crosses were established as 32.6, 1.3, and 33.9 percent, respectively. The percentages are quantitative measures of the distance between the genes involved.

The logic used in constructing a genetic map of these three genes is as follows: The genes must be arranged at different points on a line that best accommodate the data. The recombination frequencies

~ FIGURE 5.11

Percentage of recombinants as a quantitative measure of the genetic distance (map unit) between two genes on a genetic map.

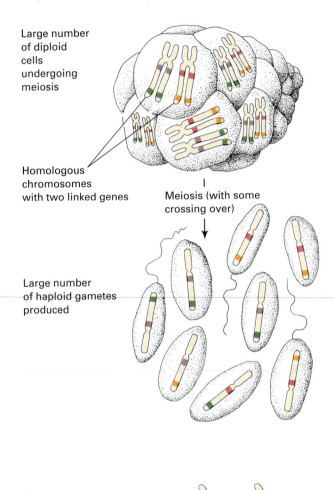

Large number of diploid cells undergoing meiosis

Homologous chromosomes with two linked genes

Meiosis (with some crossing over)

Large number of haploid gametes produced

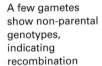

Most carry the parental genotype (no recombination)

A few gametes show non-parental genotypes, indicating recombination

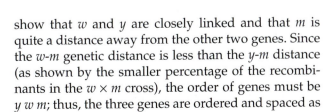

show that w and y are closely linked and that m is quite a distance away from the other two genes. Since the w-m genetic distance is less than the y-m distance (as shown by the smaller percentage of the recombinants in the $w \times m$ cross), the order of genes must be $y\ w\ m$; thus, the three genes are ordered and spaced as

~ FIGURE 5.12

Recombination between linked genes and map distance. The farther apart two genes are, the greater the number of possible sites for recombination. Thus, the probability of recombination occurring between genes A and B, for example, is much less than that between genes B and C. The percentage of recombinants can provide information about the relative genetic distance between two linked genes.

Recombination may occur at any point on chromosome arms

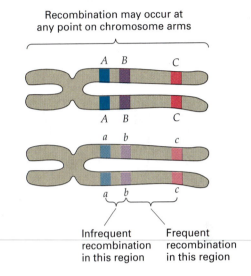

Infrequent recombination in this region

Frequent recombination in this region

follows, with 1.3 map units between y and w, and 32.6 map units between w and m:

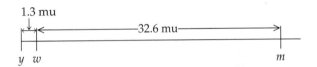

Recombination frequencies are used to construct genetic maps in all gene-mapping experiments, whether the organism is a eukaryote or a prokaryote.

Gene Mapping by Using Two-Point Testcrosses

From Sturtevant's work we have seen that the percentage of recombinants resulting from crossing-over is used as a measurement of the genetic distance between two linked genes. Quite simply, by carrying out two-point testcrosses such as those shown in Figure 5.13, we can determine the relative numbers of parental and recombinant classes in the progeny. For autosomal recessives (as in Figure 5.13), a double heterozygote is crossed with a doubly homozygous recessive mutant strain. When the double heterozygous $a^+\ b^+/a\ b$ F_1 progeny from a cross of $a^+\ b^+/a^+\ b^+$ with $a\ b/a\ b$ are testcrossed with $a\ b/a\ b$, four phenotypic classes are found among the F_2 progeny. Two of these classes have the parental phenotypes $a^+\ b^+$ and

~ FIGURE 5.13

Testcross to show that two genes are linked. Genes a and b are recessive mutant alleles linked on the same autosome. A homozygous $a^+ b^+/a^+ b^+$ individual is crossed with a homozygous recessive $a\,b/a\,b$ individual, and the doubly heterozygous F_1 progeny ($a^+ b^+/a\,b$) are testcrossed with homozygous $a\,b/a\,b$ individuals.

Recessive linked autosomal mutant alleles

phenotypes over recombinant phenotypes in the progeny of a testcross indicates linkage between the genes involved (as discussed previously).

Two-point testcrosses for the purposes of mapping are set up in similar ways for genes showing other mechanisms of inheritance. For autosomal dominants, the double heterozygous individuals are testcrossed with an individual homozygous for the recessive alleles which, in this case, are the wild-type alleles:

$$\frac{A\ \ B}{A^+ B^+} \times \frac{A^+\ B^+}{A^+ B^+}$$

For X-linked recessives a double heterozygous female is crossed with a hemizygous male carrying the recessive alleles:

$$\frac{a^+ b^+}{a\ \ b} \times \overrightarrow{a\ b}$$

And, for X-linked dominants a doubly heterozygous female is crossed with a male carrying wild-type alleles on the X chromosome:

$$\frac{A\ \ B}{A^+ B^+} \times \overrightarrow{A^+\ B^+}$$

For both X-linked cases, the females can be crossed with males of any genotype. If only male progeny are analyzed, the contribution of the male's X chromosome to the progeny is ignored.

In all cases a two-point testcross should yield a pair of parental types that occur with about equal frequencies, and a pair of recombinant types that also occur with equal frequencies. As indicated previously, the actual phenotypes will depend, of course, on the relative arrangement of the two allelic pairs in the homologous chromosomes; that is, whether they are in coupling (*cis*) or repulsion (*trans*). To get a count of the representatives of each progeny class (this will give the percentage of recombinant types), the following formula is used:

$$\frac{\text{number of recombinants}}{\text{total number of testcross progeny}} \times 100 = \frac{\text{percent}}{\text{recombinants}}$$

As was discussed previously, the value for the percentage of recombinants is usually directly converted into map units, so, for example, 11.3 percent recombination between two genes indicates that the two genes are 11.3 mu apart.

The two-point method of mapping is most accurate when the two genes examined are close together; when genes are far apart, there are inaccuracies, as we will see later. Large numbers of progeny must also be

$a\,b$ and derive from diploids in which no crossover had occurred. Since both classes result from the same no-crossover event, approximately equal numbers of these two types are expected (unless there are viability differences).

The other two F_2 phenotypic classes have recombinant phenotypes $a^+ b$ and $a\,b^+$, which derive from diploids in which a single crossover occurred between the chromosomes. Again, we expect approximately equal numbers of these two recombinant classes. Because a single crossover event occurs more rarely than no crossing over, an excess of parental

counted (scored) to ensure a high degree of accuracy. From Sturtevant's work we have already seen the logic behind constructing an actual linkage map from recombination frequencies involving all possible pairwise crosses of the genes under study. From mapping experiments carried out in all types of organisms, we know that genes are linearly arranged in linkage groups. There is a one-to-one correspondence of linkage groups and chromosomes, so the sequence of genes on the linkage group reflects the sequence of genes on the chromosome.

Generating a Genetic Map

A genetic map is generated from estimating the number of times a crossover event occurred in a particular segment of the chromosome out of all meioses examined. Since in many cases the probability of a crossing-over event is not uniform along a chromosome, we must be cautious about how far we extrapolate the genetic map (derived from data produced by genetic crosses) to the physical map of the chromosome (derived from determinations of the locations of genes along the chromosome itself; for example, from sequencing the DNA). Nonetheless, a simple working hypothesis is to consider crossovers as being randomly distributed along the chromosome.

The recombination frequencies observed between genes may also be used to predict the outcome of genetic crosses. For example, a recombination frequency of 20 percent between genes indicates that, for a doubly heterozygous genotype (such as $a^+ b^+/a\ b$), 20 percent of the gametes produced, on average, will be recombinants ($a^+ b$ and $a\ b^+$ for the example). Further, if we assume for simplicity that crossing over occurs randomly along a chromosome, more than one crossover may occur in a given region in a meiosis. The probabilities of these so-called **multiple crossovers** can readily be computed using the product rule (see Chapter 2, p. 28): the probability of two independent events occurring simultaneously is equal to the product of the individual probabilities of two single events. That is, the probability of two crossovers (called a **double crossover**) occurring between the genes in our example is

$$0.2 \times 0.2 = 0.04$$

The probability of three crossovers (a triple crossover) is

$$0.2 \times 0.2 \times 0.2 = 0.008$$

and so on.

For any testcross, the percentage of recombinants in the progeny cannot exceed 50 percent. That is, if the genes are assorting independently, an equal number of recombinants and parentals are *expected* in the progeny, so the frequency of recombinants is 50 percent. With such a value in hand, we state that the two genes are unlinked. Genes may be unlinked (that is, show 50 percent recombination) in two ways. First, the genes may be on different chromosomes, a case we discussed before. Second, *the genes may be on the same chromosome but lie so far apart that at least one crossover is almost certain to occur between them.*

The second case can be explained by referring to Figure 5.14, which shows the effects of single crossovers and double crossovers on the production of parental and recombinant chromatids. The loci are far apart on the same chromosome; as a result, in a given meiosis at least a single crossover (and usually more) always occurs between the two loci. For single crossovers occurring between any pair of non-sister chromatids, the result is two parental and two recombinant chromatids; that is, for two loci 50 percent of the products are recombinant (Figure 5.14a).

Double crossovers can involve two, three, or all four of the chromatids (Figure 5.14b). For double crossovers involving the same two non-sister chromatids (called a *two-strand double crossover*), all four resulting chromatids are parental for the two loci of interest. For *three-strand double crossovers* (double crossovers involving three of the four chromatids), two parental and two recombinant chromatids result. For a *four-strand double crossover*, all four resulting chromatids are recombinant. Considering all of the possible double crossover patterns together, 50 percent of the products are recombinant for the two loci. Similarly, for any multiple number of crossovers between loci that are far apart, examination of a large number of meioses will show 50 percent of the resulting chromatids are recombinant. This is the reason for the recombination frequency limit of 50 percent exhibited by unlinked genes on the same chromosome.

When genes are more closely linked together, however, there will be no crossovers between the two loci in some meioses, resulting in four parental chromatid products. In mapping linked genes, therefore, the map distance depends on the ratio of the meioses with no detectable crossovers to meioses with any number of crossovers between the loci.

The point is, if two genes show 50 percent recombination in a cross, it does not necessarily mean that they are on different chromosomes. More data would be needed to determine whether the genes are on the same chromosome or different chromosomes. One way to find out is to map a number of other genes in the linkage group. For example, if *a* and *m* show 50 percent recombination, perhaps we will find that *a* shows 27 percent recombination with *e*, and *e* shows

~ FIGURE 5.14

Demonstration that the recombination frequency between two genes located far apart on the same chromosome cannot exceed 50 percent. (a) Single crossovers produce one-half parental and one-half recombinant chromatids; (b) Double crossovers (two-strand, three-strand, and four-strand) collectively produce one-half parental and one-half recombinant chromatids.

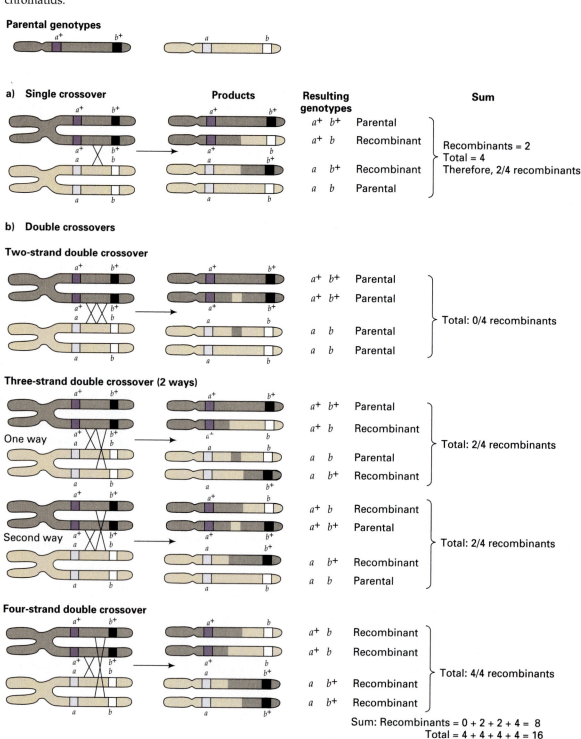

36 percent recombination with *m*. This result would indicate that *a* and *m* are in the same linkage group approximately 63 mu apart, as shown here:

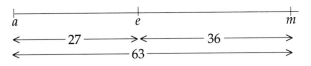

Double Crossovers

Consider a cross between a true-breeding yellow-bodied, white-eyed, miniature-winged female and a wild-type male *Drosophila*; i.e., *y w m/y w m* × *y⁺ w⁺ m⁺/⚲*. The F₁ flies are $y^+ w^+ m^+/y\ w\ m$ females (wild-type phenotype) and $y\ w\ m/⚲$ males (yellow, white, miniature phenotypes). Table 5.2 shows the data for the F₂ progeny resulting from an F₁ × F₁ cross. We have already shown that the gene order is *y-w-m*. The data show 29 exchanges between the genes for body color and eye color (*y* and *w* loci) and 719 exchanges between genes for eye color and wing size (*w* and *m* loci), giving a total of 748. However, only 746 exchanges are apparent between the body color and wing size genes. The discrepancy is due to the single normal-bodied, white-eyed, normal-winged fly. In the production of the egg from which this fly developed, two exchanges (a double crossover) must have occurred between the X chromosomes, with the result that the body color and wing size genes (normal body color and normal-length wings) appeared together, as in one grandparent, although the intervening gene (for white eyes) was derived from the other grandparent. This example shows that large crossover frequencies give an inaccurate map. Unless intervening genes are available, there is no way of detecting double crossovers (or even-numbered multiple crossovers).

To see the significance of the mechanics of double crossovers in genetic-mapping experiments, consider a hypothetical case of two allelic pairs (a^+/a and b^+/b) in coupling and separated by quite a distance on the same chromosome. Figure 5.15a shows that a single crossover results in recombination of the two allelic pairs. Figure 5.15b, top, shows that a two-strand crossover involving two of the four chromatids does not result in recombination of the allelic pairs, so only parental progeny result. Exactly the same progeny would result if there was no crossing over between the two chromatids in this region. However, the percentage of crossing-over between genes is a measure of the distance between them. Therefore, since the double crossover in Figure 5.15b, top, did not generate recombinants, two crossover events will be uncounted, and the estimate of map distance between genes *a* and *b* will be low. How low will depend on how many double crossovers occur per meiosis in the region being examined.

In sum, genetic map distance is derived from the average frequency of crossing over occurring between linked genes, whereas recombination frequency is a

---

~ TABLE 5.2

Data for the Progeny of a Cross Between a *y⁺ w⁺ m⁺/y w m* Female (Wild Type Phenotype) and a *y w m*/Y Male (Yellow, White, Miniature) *Drosophila*

PHENOTYPES				CROSSOVER		
BODY COLOR	EYE COLOR	WING SIZE	NUMBER	BODY COLOR AND EYE COLOR	EYE COLOR AND WING SIZE	BODY COLOR AND WING SIZE
normal	red (normal)	long (normal)	758	—	—	—
yellow	white	miniature	700	—	—	—
normal	red	miniature	401	—	401	401
yellow	white	long	317	—	317	317
normal	white	miniature	16	16	—	16
yellow	red	long	12	12	—	12
normal	white	long	1	1	1	—
yellow	red	miniature	0	0	0	—
		Total	2,205	29	719	746
		Percentage	100	1.3	32.6	33.8

~ FIGURE 5.15

Progeny of single and double crossovers. (a) A single crossover between linked genes generates recombinant gametes. (b) *Top*: A double crossover between linked genes gives parental gametes. Thus inaccurate map distances between genes result, since not all crossovers can be accounted for. *Bottom*: A possible solution to the double-crossover problem. The presence of a third allelic pair between the two allelic pairs of Figure 5.15a enables us to detect the double-crossover event. In a double crossover, the middle gene will change positions relative to the outside genes.

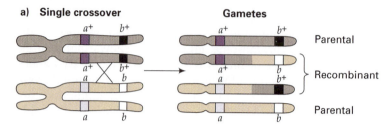

a)   Single crossover

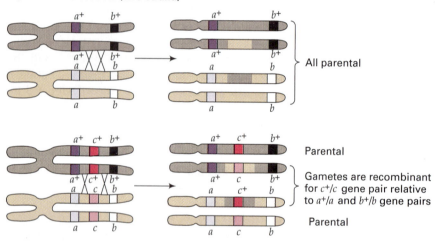

b)   Double crossover (two-strand)

measure of the crossovers that result in the detectable exchange of the gene markers. If no multiple crossovers occurred between genes, there would be a direct linear relationship between genetic map distance and recombination frequency. In practice, such a relationship is seen only when genetic map distances are small; that is, when genes are closely linked. As genes become farther apart, the chances of multiple crossovers between them increases, and there is no longer an exact linear relationship between map distance and recombination frequency. As a result, it is difficult to obtain an accurate measure of map distance when multiple crossovers are involved.

In many systems, it has been possible to derive mathematical formulas called **mapping functions**, which are used to correct the observed recombination values for the incidence of multiple crossovers. Mapping functions all require some basic assumptions about the frequency of crossovers compared

with distance between genes. Hence the usefulness of applying the mapping functions depends upon the validity of the assumptions. In operation, mapping functions convert experimentally observed recombination frequencies into genetic map distances that are additive by correcting for the effects of multiple crossovers, which can otherwise cause an underestimation of genetic distance.

Three-Point Cross

How can we avoid this pitfall? Double crossovers occur quite rarely within distances of 10 mu or less, and they occur more frequently in greater physical distances for certain chromosomal regions. Consequently, one way to get accurate map distances is to study closely linked genes. Another efficient way to obtain such data is to use a **three-point testcross** involving three genes within a relatively short section of a chromosome. (The evidence for the closely linked nature of three genes would derive from separate two-point testcrosses and the drawing of a preliminary genetic map.)

The potential advantage of the three-point test-cross can be seen in the theoretical case if we have a third allelic pair c^+/c between the a^+/a and b^+/b allelic pairs of Figure 5.15a, as diagrammed in Figure 5.15b, bottom. A two-strand double-crossover event between a and b might be detected by the recombination of the c^+/c allelic pair in relation to the other two allelic pairs. In fact, three-point testcrosses are routinely used as an efficient way of mapping genes both for their order in the chromosome and for the distances between them, as we will see in the next section.

$\mathcal{K}$EYNOTE

The map distance between two genes is based on the frequency of recombination between the two genes (an approximation of the frequency of crossovers between the two genes). By using appropriate genetic markers in crosses, geneticists are able to compute the recombination frequency between genes in chromosomes as the percentage of progeny showing the reciprocal recombinant phenotypes. The more closely the recombination frequency parallels the crossover frequency, the closer the genes are. But when the genetic distance increases, the incidence of multiple crossovers causes the recombination frequency to be an underestimate of the crossover frequency and hence of the map distance. Mapping functions may be used to correct for the effects of multiple crossovers and hence to give a more accurate estimate of map distance.

Mapping Chromosomes by Using Three-Point Testcrosses

In diploid organisms the three-point testcross is a cross of a triple heterozygote with a triply homozygous recessive. If the mutant genes in the cross are all recessive, a typical three-point testcross might be:

$$\frac{a^+ b^+ c^+}{a\ b\ c} \times \frac{a\ b\ c}{a\ b\ c}$$

If any of the mutant genes are dominant to the wild type, the *triply homozygous recessive parent in the test-cross will carry the wild-type allele of these genes*. For example, a testcross of a strain heterozygous for two recessive mutations (a and c) and one dominant mutation (B) might be

$$\frac{a^+ B^+ c^+}{a\ B\ c} \times \frac{a\ B^+ c}{a\ B^+ c}$$

In a testcross involving sex-linked genes, the female is the heterozygous strain (assuming that the female is the homogametic sex) and the male is hemizygous for the recessive markers.

Three-point mapping is also done in haploid eukaryotic organisms, but in this case a testcross is not needed. Thus in *Neurospora crassa* we might cross two haploid strains to generate the triply heterozygous diploid cell. A cross of an $a\ b\ c$ strain with an $a^+ b^+ c^+$ strain, for instance, would give an $a^+ b^+ c^+/a\ b\ c$ diploid cell. That cell goes through meiosis to generate haploid spores from which progeny *Neurospora* cultures grow. Since the cells are haploid, the phenotypes of the progeny are determined directly by the genotypes.

Let us suppose we have a plant in which there are three linked genes, all of which control fruit phenotypes. A recessive allele p of the first gene determines purple fruit color, in contrast to the yellow color of the wild type. A recessive allele r of the second gene results in a round fruit shape, as compared with an elongated fruit in the wild type. A recessive allele j of the third gene gives a juicy fruit instead of the wild-type dry fruit. The task before us is to determine the order of the genes on the chromosome and to determine the map distance between the genes. To do so, we must carry out the appropriate testcross of a triple heterozygote ($p^+ r^+ j^+/p\ r\ j$) with a triply homozygous recessive ($p\ r\ j/p\ r\ j$) and then count the different phenotypic classes in the progeny (Figure 5.16).

For each gene in the cross, two different phenotypes occur in the progeny; therefore, for the three genes a total of $(2)^3 = 8$ phenotypic classes will appear in the progeny, representing all possible combinations of phenotypes. In an actual experiment not all the phenotypic classes may be generated. The absence of a phenotypic class is also important information, and the experimenter should enter a 0 in the class for which no progeny are found.

ESTABLISHING THE ORDER OF GENES. The first step in finding the genetic map distances between the three genes is to determine the order in which the

~ FIGURE 5.16

Theoretical analysis of a three-point mapping, showing the testcross used and the resultant progeny.

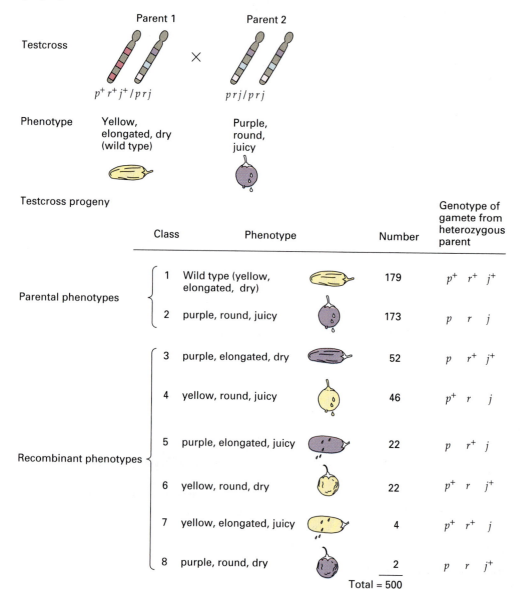

genes are located on the chromosome; that is, which gene is in the middle? Each progeny type is generated from a zygote formed by the fusion of the gametes from each parent. One parent carries the recessive alleles for all three genes; the other is heterozygous for all three genes. Therefore, the phenotype of the resulting individual is determined by the alleles in the gamete from the triply heterozygous parent; the gamete from the other parent will carry only recessive alleles. We know from the genotypes of the original parents that all three genes are in coupling (*cis*). Since the

heterozygous parent in the testcross was $p^+ r^+ j^+/p r j$, classes 1 and 2 in Figure 5.16 are parental progeny: Class 1 is produced by the fusion of a $p^+ r^+ j^+$ gamete with a $p r j$ gamete from the triply homozygous recessive parent. Class 2 is produced by the fusion of a $p r j$ gamete from the triply heterozygous parent and a $p r j$ gamete from the triply homozygous recessive parent. These classes are generated from meiosis in which no crossing-over occurs in the region of the chromosome in which the three genes are located.

The progeny classes deriving from a double

~ FIGURE 5.17

Rearrangement of the three genes of Figure 5.16 to $p\,j\,r$. The evidence is that a double crossover involving the same two chromatids (as it is shown in this figure) generates the least frequent pair of recombinant phenotypes (in this case class 7 with four progeny and class 8 with two progeny).

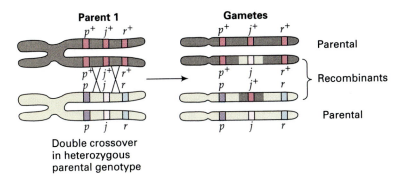

Double crossover
in heterozygous
parental genotype

crossover involving the same two chromatids between the two loci can often be found by inspecting the numbers of each phenotypic class. Since the frequency of a double crossover in a region is expected to be lower than the frequency of a single crossover, *double-crossover gametes are the least frequent class found.* Thus to identify the double-crossover progeny, we can examine the progeny to find the pair of classes that have the lowest number of representatives. In Figure 5.16 classes 7 and 8 are such a pair. The genotypes of the gametes from the heterozygous parent that give rise to these phenotypes are $p^+ r^+ j$ and $p\,r\,j^+$.

Referring now to Figure 5.15b, bottom, we see that a double crossover involving the same two chromatids changes the orientation of the gene in the center of the sequence (here, c^+/c) with respect to the two flanking allelic pairs. Therefore, genes p, r, and j must be arranged in such a way that the center gene switches to give classes 7 and 8. To determine the arrangement, we must check the relative organization of the genes in the parental heterozygote so that it is clear which genes are in coupling (*cis*) and which are in repulsion (*trans*). In this example, the parental (noncrossover) gametes are $p^+ r^+ j^+$ and $p\,r\,j$, so all are in coupling. The double-crossover gametes are $p^+ r^+ j$ and $p\,r\,j^+$, so the only possible gene order that is compatible with the data is $p\,j\,r$, with the genotype of the heterozygous parent being $p^+ j^+ r^+/p\,j\,r$. Figure 5.17 illustrates the generation of the double-crossover gametes from that parent.

CALCULATING THE MAP DISTANCES BETWEEN GENES.

Now the data can be rewritten as shown in Figure 5.18 to reflect the newly determined gene

order. For convenience in this analysis, the region between genes p and j is called region I and that between genes j and r is called region II.

Map distances can now be calculated as described previously. *The frequency of crossing-over*

~ FIGURE 5.18

Rewritten form of the testcross and testcross progeny of Figure 5.16, based on the actual gene order $p\,j\,r$.

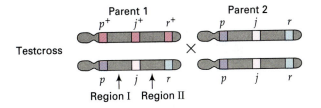

Region I   Region II

Testcross progeny

Class	Genotype of gamete from heterozygous parent			Number	Origin
1	p^+	j^+	r^+	179	Parentals, no crossover
2	p	j	r	173	
3	p	j^+	r^+	52	Recombinants, single crossover region I
4	p^+	j	r	46	
5	p	j	r^+	22	Recombinants, single crossover region II
6	p^+	j^+	r	22	
7	p^+	j	r^+	4	Recombinants, double crossover
8	p	j^+	r	2	

Total = 500

(genetic recombination) is computed between two genes at a time. For the *p-j* distance all the crossovers that occurred in region I must be added together. Thus we must consider the recombinant progeny resulting from a single crossover in that region (classes 3 and 4) and the recombinant progeny produced by a double crossover in which one crossover is between *p* and *j* and the other is between *j* and *r* (classes 7 and 8). The double crossovers must be included since each double crossover of that type includes a single crossover in region I and therefore is a recombination event between genes *p* and *j*. From Figure 5.18 there are 98 recombinant progeny in classes 3 and 4, and 6 in classes 7 and 8, giving a total of 104 progeny that result from a recombination event in region I. Since there are a total of 500 progeny, the percentage of progeny generated by crossing-over in region I is 20.8 percent, determined as follows (sco = single crossovers; dco = double crossovers):

$$\frac{\text{sco in region I } (p-j) + \text{dco}}{\text{total progeny}} \times 100$$

$$= \frac{(52 + 46) + (4 + 2)}{500} \times 100$$

$$= \frac{98 + 6}{500} \times 100$$

$$= \frac{104}{500} \times 100$$

$$= 20.8\%$$

In other words, the distance between genes *p* and *j* is 20.8 mu. This map distance, which is quite large, is chosen mainly for illustration. If this cross were an actual cross, we would have to be cautious about that map distance value, since it is in the range where double crossovers in which both crossovers occur between *p* and *j* might occur at a significant frequency and might go undetected. In an actual cross the value of 20.8 mu would probably underestimate the true distance.

The same method is used to obtain the map distance between genes *j* and *r*. That is, we calculate the frequency of crossovers in the cross that gave rise to progeny recombinant for genes *j* and *r* and directly relate that frequency to map distance. In this case all the crossovers that occurred in region II (see Figure 5.18) must be added (classes 5, 6, 7, and 8). The per-

centage of crossovers is calculated in the following manner:

$$\frac{\text{sco in region II } (j-r) + \text{dco}}{\text{total progeny}} \times 100$$

$$= \frac{(22 + 22) + (4 + 2)}{500} \times 100$$

$$= \frac{44 + 6}{500} \times 100$$

$$= \frac{50}{500} \times 100$$

$$= 10.0\%$$

Thus the map distance between genes *j* and *r* is 10.0 map units.

In summary, a genetic map of the three genes in the example has been generated (Figure 5.19). The example has illustrated that the three-point testcross is an effective way of establishing the order of genes and of calculating map distances. In calculating map distances from three-point testcross data, the double-crossover figure must be added to each of the single-crossover figures, since in each case a double crossover represents single-crossover events occurring simultaneously in regions I and II in the same meiosis.

To compute the map distance between the two outside genes, we simply add the two map distances. Thus in the example the *p-r* distance is 20.8 + 10.0 = 30.8 mu. This map distance can be computed directly from the data by combining the two formulas discussed previously. So the *p-r* distance is calculated in the following manner:

$$\text{distance} = \frac{(\text{sco in region I}) + (\text{dco}) + (\text{sco in region II}) + (\text{dco})}{\text{total progeny}} \times 100$$

$$= \frac{(\text{sco in region I}) + (\text{sco in region II}) + (2 \times \text{dco})}{\text{total progeny}} \times 100$$

$$= \frac{52 + 46 + 22 + 22 + 2(4 + 2)}{500} \times 100$$

$$= \frac{98 + 44 + 2(6)}{500} \times 100$$

$$= 30.8 \text{ map units}$$

~ FIGURE 5.19

Genetic linkage map of the *p j r* region of the chromosome is computed from the recombination data of Figure 5.18.

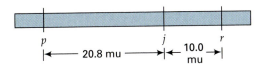

KEYNOTE

The map distance between genes is calculated from the results of testcrosses between strains carrying appropriate genetic markers. The most accurate testcross is the three-point testcross using a diploid organism in which a triple heterozygote is crossed with a homozygous recessive for all three genes. The most accurate map distances are obtained when the three genes are reasonably close to one another, with, say, 5–10 map units between each locus. The unit of genetic distance is the map unit (mu), which is the distance between gene pairs for which 1 product out of 100 is recombinant; in other words, a recombinant frequency of 1 percent is 1 mu. For illustration, part of the genetic map for *Drosophila melanogaster* is stylized in Figure 5.20 (p. 158).

INTERFERENCE AND COINCIDENCE. The map distances determined by three-point mapping are primarily useful in elaborating the overall organization of genes on a chromosome. In addition, the map distances obtained are useful in telling us a little about the recombination mechanisms themselves. For example, the map distance of 20.8 mu between gene *p* and *j* means that 20.8 percent of the gametes should result from crossing-over between the two gene loci. Similarly, the map distance of 10.0 mu between *j* and *r* indicates that 10.0 percent of the gametes should result from crossing-over between these two gene loci.

In the example of Figure 5.18, if crossing-over in region I is independent of crossing-over in region II, then by the product rule the probability of simultaneous crossing-over (i.e., a double crossover) in the two regions is equal to the product of the probabilities of the two events occurring separately. Thus if crossing-over in the two regions is independent, we have

$$\frac{\text{map distance, region I}}{100} \times \frac{\text{map distance, region II}}{100}$$

$$= 0.208 \times 0.100 = 0.0208$$

or 2.08 percent double crossovers are expected to occur. However, only 6/500 = 1.2 percent double crossovers occurred in this cross (in classes 7 and 8).

This discrepancy is not just a simple experimental error due to small sample size, erroneously scored progeny, and so on. It is characteristic of such mapping crosses that double-crossover progeny typically do not appear as often as the map distances between the genes lead us to expect. Thus, once a crossing-over event has occurred in one part of the meiotic tetrad, the probability of another crossing-over event occurring nearby is reduced, most probably by physical interference caused by the breaking and rejoining of the chromatids. This phenomenon is called **chiasma interference** (also **chromosomal interference**).

In most organisms, there is no way to predict the extent of interference, since it tends to vary from chromosome to chromosome and even between segments of the same chromosome. But for the gene maps to be useful in predicting the outcome of crosses, we should know the magnitude of the interference throughout the map. The usual way to express the extent of interference is as a **coefficient of coincidence**; that is

$$\text{coefficient of coincidence} = \frac{\text{observed double crossover frequency}}{\text{expected double crossover frequency}}$$

and

$$\text{interference} = 1 - \text{coefficient of coincidence}$$

For the portion of the map in our example the coefficient of coincidence is

$$0.012/0.0208 = 0.577$$

The coefficient of coincidence may be interpreted as follows: A coincidence of one means that in a given region all double crossovers occurred that were expected on the basis of two independent events; there is no

~ FIGURE 5.20

Part of the genetic map of *Drosophila melanogaster*, with illustrations of several commonly used genetic markers.

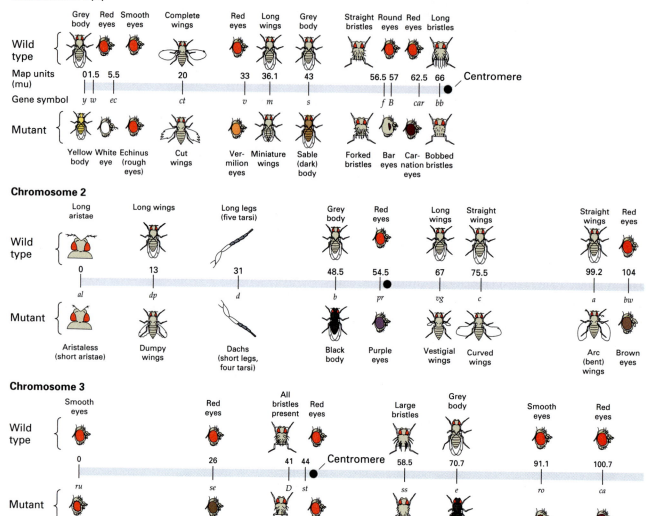

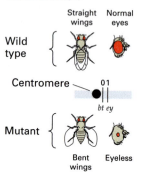

interference, so the interference value is zero. If the coefficient of coincidence is zero, none of the expected double crossovers occur. Here there is total interference, with one crossover completely preventing a second crossover in the region under examination; the interference value is one. These examples show that coincidence values and interference values are inversely related. In the example above the coefficient of coincidence of 0.577 means that the interference value is 0.423. Only 57.7 percent of the expected double crossovers took place in the cross. In some special cases, a crossover will actually facilitate the occurrence of a second crossover close by. In this situation the coefficient of coincidence can have a negative value because observed double crossovers exceed the expected double crossovers.

KEYNOTE

The occurrence of a chiasma between two chromatids may physically impede the occurrence of a second chiasma nearby, a phenomenon called chiasma interference. The extent of interference is expressed by the coefficient of coincidence, which is calculated by dividing the number of observed double crossovers by the number of expected double crossovers. The coefficient of coincidence ranges from zero to one, and the extent of interference is measured as 1 – coefficient of coincidence.

SUMMARY

In this chapter, linkage, crossing-over, and gene mapping in eukaryotes were discussed. The production of genetic recombinants results from physical exchanges between homologous chromosomes in meiosis. The site of crossing-over is called a chiasma, and the exchange of parts of chromatids is called crossing-over. Proof that genetic recombination occurs when crossing-over takes place in meiosis came from experiments in which recombinants for genetic markers occurred only when exchanges of cytological markers occurred. Crossing-over is a reciprocal event that, in eukaryotes, occurs at the four-strand stage in prophase I of meiosis.

Genetic mapping is the process of locating the position of genes in relation to one another on the chromosome. To map genes in this way, it is first necessary to show that genes are linked (located on the same chromosome), a characteristic indicated by the fact that they do not assort independently in crosses. The map distance between two linked genes is then calculated based on the frequency of recombination between the two genes. The recombination frequency is an approximation of the frequency of crossovers between the two genes. The most accurate map distances are determined for genes that are closely linked because, as map distance increases, the incidence of multiple crossovers causes the recombination frequency to be an underestimate of the crossover frequency, and hence of the map distance.

ANALYTICAL APPROACHES FOR SOLVING GENETICS PROBLEMS

Q5.1 In corn the gene for colored (C) seeds is completely dominant to the gene for colorless (c) seeds. Similarly, for the character of the endosperm (the part of the seed that contains the food stored for the embryo), a single gene pair controls whether the endosperm is full or shrunken. Full (S) is dominant to shrunken (s). A true-breeding colored, full-seeded plant was crossed with a colorless, shrunken-seeded one. The F_1 colored, full plants were test-crossed to the doubly recessive type, that is, colorless and shrunken. The result was as follows:

colored, full	4,032
colored, shrunken	149
colorless, full	152
colorless, shrunken	4,035
Total	8,368

Is there evidence that the gene for color and the gene for endosperm shape are linked? If so, what is the map distance between the two loci?

A5.1 The best approach is to diagram the cross using gene symbols:

P: colored and full × colorless and shrunken
       CC SS   *cc ss*
                 ↓

F_1:   colored and full
               Cc Ss

Testcross: colored and full × colorless and shrunken
         Cc Ss   *cc ss*

If the genes were unlinked, a 1:1:1:1 ratio of colored and full : colored and shrunken : colorless and full : colorless and shrunken would result in the progeny of this testcross. By inspection, we can see that the actual progeny deviate a great deal from this ratio, showing a 27:1:1:27 ratio. If we did a chi-square test (using the actual numbers, not the percentages or ratios), we would see immediately that the hypothesis that the genes are unlinked is invalid, and we must consider the two genes to be linked in coupling. More specifically, the parental combinations (colored plus full, and colorless plus shrunken) are more numerous than expected, while the recombinant types (colorless plus full, and colored plus shrunken) are correspondingly less numerous than expected. This result comes directly from the inequality of the four gamete types produced by meiosis in the colored and full F_1 parent.

Given that the two genes are linked, the crosses may be diagrammed to reflect their linkage as follows:

P: $\dfrac{C\,S}{C\,S}$ $\times$ $\dfrac{c\,s}{c\,s}$

$\downarrow$

F_1: $\dfrac{C\,S}{c\,s}$

Testcross: $\dfrac{C\,S}{c\,s}$ $\times$ $\dfrac{c\,s}{c\,s}$

To calculate the map distance between the two genes, we need to compute the frequency of crossovers in that region of the chromosome during meiosis. We cannot do that directly, but we can compute the percentage of recombinant progeny that must have resulted from such crossovers:

Parental types:	colored, full	4,032
	colorless, shrunken	4,035
		8,067
Recombinant types:	colored, shrunken	149
	colorless, full	152
		301

This calculation gives about 3.6 percent recombinant types (i.e. $301/8{,}368 \times 100$) and about 96.4 percent parental types (i.e. $8{,}067/8{,}368 \times 100$). Since the recombination frequency can be used directly as an indication of map distance, especially when the distance is small, we can conclude that the distance between the two genes is 3.6 mu (3.6 cM).

We would get approximately the same result if the two genes were in repulsion rather than in coupling. That is, the crossovers are occurring between homologous chromosomes, regardless of whether or not there are genetic differences in the two homologs that we, as

experimenters, use as markers in genetic crosses. This same cross in repulsion would be as follows:

P: colorless and full $\times$ colored and shrunken

$\dfrac{c\,S}{c\,S}$ $\qquad$ $\dfrac{C\,s}{C\,s}$

$\downarrow$

F_1: colored and full

$\dfrac{C\,s}{c\,S}$

Data from an actual testcross of the F_1 with colorless and shrunken (*cc ss*) gave 638 colored and full (recombinant) : 21,379 colored and shrunken (parental) : 21,906 colorless and full (parental) : 672 colorless and shrunken (recombinant) with a total of 44,595 progeny. Thus 2.94 percent were recombinants, for a map distance between the two genes of 2.94 mu, a figure reasonably close to the results of the cross made in coupling.

Q5.2 In the Chinese primrose, slate-colored flower (*s*) is recessive to blue flower (*S*); red stigma (*r*) is recessive to green stigma (*R*); and long style (*l*) is recessive to short style (*L*). All three genes involved are on the same chromosome. The F_1 of a cross between two true-breeding strains, when testcrossed, gave the following progeny:

PHENOTYPE	NUMBER OF PROGENY
slate flower, green stigma, short style	27
slate flower, red stigma, short style	85
blue flower, red stigma, short style	402
slate flower, red stigma, long style	977
slate flower, green stigma, long style	427
blue flower, green stigma, long style	95
blue flower, green stigma, short style	960
blue flower, red stigma, long style	27
Total	3,000

a. What were the genotypes of the parents in the cross of the two true-breeding strains?
b. Make a map of these genes, showing gene order, and distance between them.
c. Derive the coefficient of coincidence for interference between these genes.

A5.2a. With three gene pairs, eight phenotypic classes are expected, and eight are observed. The reciprocal pairs of classes with the most representatives are those resulting from no crossovers, and these pairs can tell us the genotypes of the original parents. The two classes are slate, red, long and blue, green, short. Thus the F_1 triply heterozygous parent of this generation must have been *S R L/s r l*, so the true-breeding parents were *S R L/S R L* (blue, green, short) and *s r l/s r l* (slate, red, long).

b. The order of the genes can be determined by inspecting the reciprocal pairs of phenotypic classes that represent the results of double crossing-over. These classes have the least numerous representatives, so the double-crossover classes are slate plus green plus short (*s R L*) and blue plus red plus long (*S r l*). The gene pair that has changed its position relative to the other two pairs of alleles is the central gene, *S/s* in this case. Therefore the order of genes is *R S L* (or *L S R*). We can diagram the F_1 testcross as follows:

$$\frac{R S L}{r s l} \times \frac{r s l}{r s l}$$

A single crossover between the *R* and *S* genes gives the green plus slate plus long (*R s l*) and red plus blue plus short (*r S L*) classes, which have 427 and 402 members, respectively, for a total of 829. The double-crossover classes have already been defined, and they yield 54 progeny. The map distance between *R* and *S* is given by the crossover frequency in that region, which is the sum of the single crossovers and double crossovers divided by the total number of progeny, then multiplied by 100 percent. Thus

$$\frac{829 + 54}{3{,}000} \times 100\% = \frac{883}{3{,}000} \times 100\%$$

$$= 29.43\% \text{ or } 29.43 \text{ map units}$$

With similar logic the distance between *S* and *L* is given by the crossover frequency in that region, which is the sum of the single-crossover and double-crossover progeny classes. The single-crossover progeny classes are green plus blue plus long (*R S l*) and red plus slate plus short (*r s L*), which have 95 and 85 members, respectively, for a total of 180. The map distance is given by

$$\frac{180 + 54}{3{,}000} \times 100\% = \frac{234}{3{,}000} \times 100\%$$

$$= 7.8\% \text{ or } 7.8 \text{ map units}$$

The data we have derived give us the following map:

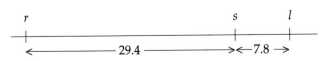

c. The coefficient of coincidence is given by

$$\frac{\text{frequency of observed double crossovers}}{\text{frequency of expected double crossovers}}$$

The observed frequency of observed double crossovers is $54/3{,}000 = 0.018$. The expected frequency of double crossovers is the product of the map distances between *r* and *s* and between *s* and *l*, that is, $0.294 \times 0.078 = 0.023$. The coefficient of coincidence, therefore, is $0.018/0.023 = 0.78$. In other words, 78 percent of the expected double crossovers did indeed take place; there was 22 percent interference.

QUESTIONS AND PROBLEMS

5.1 A cross $a^+a^+ \, b^+b^+ \times aa \, bb$ results in an F_1 of phenotype $a^+ \, b^+$; the following numbers are obtained in the F_2 (phenotypes):

$a^+ \, b^+$	110
$a^+ \, b$	16
$a \, b^+$	19
$a \, b$	15
Total	160

Are genes at the *a* and *b* loci linked or independent? What F_2 numbers would otherwise be expected?

*5.2 In corn a dihybrid for the recessives *a* and *b* is testcrossed. The distribution of the phenotypes was as follows:

A B	122
A b	118
a B	81
a b	79

Are the genes assorting independently? Test the hypothesis with a χ^2 test. Explain tentatively any deviation from expectation, and tell how you would test your explanation.

5.3 The F_1 from a cross of $A\,B/A\,B \times a\,b/a\,b$ is testcrossed, resulting in the following phenotypic ratios:

A B	308
A b	190
a b	292
a B	210

What is the frequency of recombination between genes *a* and *b*?

*5.4 In rabbits the English type of coat (white-spotted) is dominant over non-English (unspotted), and short hair is dominant over long hair (Angora). When homozygous English, short-haired rabbits were crossed with non-English Angoras and the F_1 crossed back to non-English Angoras, the following offspring were obtained: 72 English and short-haired, 69 non-English and Angora; 11 English and Angora; and 6 non-English and short-haired. What is the map distance between the genes for coat color and hair length?

5.5 In *Drosophila* the mutant black (*b*) has a black body, and the wild type has a grey body; the mutant vestigial (*vg*) has wings that are much shorter and crumpled compared to the long wings of the wild type. In the following cross, the true-breeding parents are given together with the counts of offspring of F_1 females × black and vestigial males:

P   black and normal × grey and vestigial
F_1 females × black and vestigial males

Progeny: grey, normal   283
         grey, vestigial   1,294
         black, normal   1,418
         black, vestigial   241

From these data, calculate the map distance between the black and vestigial genes.

*5.6 A gene controlling wing size is located on chromosome 2 in *Drosophila*. The recessive allele *vg* results in vestigial wings when homozygous; the vg^+ allele determines long wings. A new eye mutation, which we will call "maroon-like," is isolated. Homozygous maroon-like (*m/m*) results in maroon-colored eyes; the m^+ allele is bright red. The location of the *m* gene is unknown, and you are asked to design an experiment to determine whether *m* is located on chromosome 2.

You cross true-breeding virgin maroon females to true-breeding *vg/vg* males and obtain all wild-type F_1 progeny. Then you allow the F_1 to interbreed. As soon as the F_2 start to hatch, you begin to classify the flies, and among the first six newly hatched flies, you find four wild type; one vestigial-winged and red-eyed fly; and one vestigial-winged, maroon-eyed fly. You immediately draw the conclusions that (1) maroon-like is not X-linked and (2) maroon-like is not linked to vestigial. Based on this small sample, how could you tell? On what chromosome is *m* located? (Hint: There is no crossing-over in *Drosophila* males.)

*5.7 Use the following two-point recombination data to map the genes concerned. Show the order and the length of the shortest intervals.

GENE LOCI	% RECOMBINATION	GENE LOCI	% RECOMBINATION
a,b	50	*b,d*	13
a,c	15	*b,e*	50
a,d	38	*c,d*	50
a,e	8	*c,e*	7
b,c	50	*d,e*	45

5.8 Use the following two-point recombination data to map the genes concerned. Show the order and the length of the shortest intervals.

LOCI	% RECOMBINATION	LOCI	% RECOMBINATION
a,b	50	*c,d*	50
a,c	17	*c,e*	50
a,d	50	*c,f*	7
a,e	50	*c,g*	19
a,f	12	*d,e*	7
a,g	3	*d,f*	50
b,c	50	*d,g*	50
b,d	2	*e,f*	50
b,e	5	*e,g*	50
b,f	50	*f,g*	15
b,g	50		

5.9 The following data are from Bridges and Morgan's work on the recombination between the genes black, curved, purple, speck, and vestigial in chromosome 2 of *Drosophila*. On the basis of the data, map the chromosome for these five genes as accurately as possible. Remember that determinations for short distances are more accurate than those for long ones.

GENES IN CROSS	TOTAL PROGENY	NUMBER OF RECOMBINANTS
black, curved	62,679	14,237
black, purple	48,931	3,026
black, speck	685	326
black, vestigial	20,153	3,578
curved, purple	51,136	10,205
curved, speck	10,042	3,037
curved, vestigial	1,720	141
purple, speck	11,985	5,474
purple, vestigial	13,601	1,609
speck, vestigial	2,054	738

5.10 A corn plant known to be heterozygous at three loci is testcrossed. The progeny phenotypes and frequencies are as follows:

+	+	+	455
a	*b*	*c*	470
+	*b*	*c*	35
a	+	+	33
+	+	*c*	37
a	*b*	+	35
+	*b*	+	460
a	+	*c*	475
		Total	2,000

Give the gene arrangement, linkage relations, and map distances.

*5.11 Genes *a* and *b* are linked, with 10 percent recombination. What would be the phenotypes, and the probability of each, among progeny of the following cross?

$$\frac{a \; b^+}{a^+ b} \times \frac{a \; b}{a \; b}$$

*5.12 Genes *a* and *b* are sex-linked and are located 7 mu apart in the X chromosome of *Drosophila*. A female of genotype $a^+ b/a \; b^+$ is mated with a wild-type ($a^+ b^+$).
a. What is the probability that one of her sons will be either $a^+ b^+$ or $a \; b^+$ in phenotype?
b. What is the probability that one of her daughters will be $a^+ b^+$ in phenotype?

5.13 In *Drosophila*, *a* and *b* are linked autosomal genes whose recombination frequency in females is 5 percent; *c* and *d* are X-linked genes, located 10 map units apart. A homozygous dominant female is mated to a recessive male, and the daughters are testcrossed. Which of the following would you expect to observe in the testcross progeny?
a. Different ratios in males and females.
b. Nearly equal frequency of $a^+ b \; c^+ d^+$, $a^+ b \; c \; d$, $a \; b^+ c^+ d$, and $a \; b^+ c \; d$ classes.
c. Independent segregation of some genes with respect to others involved in the cross.
d. Double-crossover classes less frequent than expected because of interference between the two marked regions.

5.14 In maize the dominant genes *A* and *C* are both necessary for colored seeds. Homozygous recessive plants give colorless seed, regardless of the genes at the second locus. Genes *A* and *C* show independent segregation, while the recessive mutant gene waxy endosperm (*wx*) is linked with *C* (20 percent recombination). The dominant *Wx* allele results in starchy endosperm.
a. What phenotypic ratios would be expected when a plant of constitution *c Wx/C wx A/A* is testcrossed?
b. What phenotypic ratios would be expected when a plant of constitution *c Wx/C wx A/a* is testcrossed?

*5.15 Assume that genes *a* and *b* are linked and show 20 percent crossing-over.
a. If a homozygous *A B/A B* individual is crossed with an *a b/a b* individual, what will be the genotype of the F$_1$? What gametes will the F$_1$ produce and in what proportions? If the F$_1$ is testcrossed with a doubly homozygous recessive individual, what will be the proportions and genotypes of the offspring?
b. If, instead, the original cross is *A b/A b × a B/a B*, what will be the genotype of the F$_1$? What gametes will the F$_1$ produce and in what proportions? If the F$_1$ is test-crossed with a doubly homozygous recessive, what will be the proportions and genotypes of the offspring?

5.16 In tomatoes, tall vine is dominant over dwarf, and spherical fruit shape is dominant over pear shape. Vine height and fruit shape are linked, with a recombinant percentage of 20. A certain tall, spherical-fruited tomato plant is crossed with a dwarf, pear-fruited plant. The progeny are 81 tall, spherical; 79 dwarf, pear; 22 tall, pear; and 17 dwarf, spherical. Another tall and spherical plant crossed with a dwarf and pear plant produces 21 tall, pear; 18 dwarf, spherical; 5 tall, spherical; and 4 dwarf, pear. What are the genotypes of the two tall and spherical plants? If they were crossed, what would their offspring be?

*5.17 Genes *a* and *b* are in one chromosome, 20 mu apart; *c* and *d* are in another chromosome, 10 mu apart. Genes *e* and *f* are in yet another chromosome and are 30 mu apart. Cross a homozygous *A B C D E F* individual with an *a b c d e f* one, and cross the F$_1$ back to an *a b c d e f* individual. What are the chances of getting individuals of the following phenotypes in the progeny?
a. *A B C D E F*
b. *A B C d e f*
c. *A b c D E f*
d. *a B C d e f*
e. *a b c D e F*

*5.18 Genes *d* and *p* occupy loci 5 map units apart in the same autosomal linkage group. Gene *h* is a separate autosomal linkage group and therefore segregates independently of the other two. What types of offspring are expected, and what is the probability of each, when individuals of the following genotypes are testcrossed?

a. $\dfrac{D \; P}{d \; p} \dfrac{h}{h}$ **b.** $\dfrac{d \; P}{D \; p} \dfrac{H}{h}$

5.19 A hairy-winged (*h*) *Drosophila* female is mated with a yellow-bodied (*y*), white-eyed (*w*) male. The F$_1$ are all normal. The F$_1$ progeny are then crossed, and the F$_2$ that emerge are as follows:

Females:	wild type	757
	hairy	243
Males:	wild type	390
	hairy	130
	yellow	4
	white	3
	hairy, yellow	1
	hairy, white	2
	yellow, white	360
	hairy, yellow, white	110

Give genotypes of the parents and the F$_1$, and note the linkage relations and distances, where appropriate.

5.20   In the Maltese bippy, amiable (*A*) is dominant to nasty (*a*), benign (*B*) is dominant to active (*b*), and crazy (*C*) is dominant to sane (*c*). A true-breeding amiable, active, crazy bippy was mated, with some difficulty, to a true-breeding nasty, benign, sane bippy. An F_1 individual from this cross was then used in a testcross (to a nasty, active, sane bippy) and produced, in typical prolific bippy fashion, 4,000 offspring. From an ancient manuscript entitled *The Genetics of the Bippy, Maltese and Other*, you discover that all three genes are autosomal, *a* is linked to *b* but not to *c* and the map distance between *a* and *b* is 20 mu.

a.   Predict all the expected phenotypes and the numbers of each type from this cross.

b.   Which phenotypic classes would be missing had *a* and *b* shown complete linkage?

c.   Which phenotypic classes would be missing if *a* and *b* were unlinked?

d.   Again, assuming *a* and *b* to be unlinked, predict all the expected phenotypes of nasty bippies and the frequencies of each type resulting from a self-cross of the F_1.

5.21   Fill in the blanks. Continuous bars indicate linkage, and the order of linked genes is correct as shown. If all types of gametes are equally probable, write "none" in the right column headed "Least frequent classes." In the right column, show two gamete genotypes, unless all types are equally frequent, in which case write "none."

PARENT GENOTYPES	NUMBER OF DIFFERENT POSSIBLE GAMETES	LEAST FREQUENT CLASSES
$\dfrac{A\ b\ C}{a\ B\ c}$	_____	_____ _____
$\dfrac{A\ b\ C}{a\ B\ c}$	_____	_____ _____
$\dfrac{A\ b\ C\ D}{a\ B\ c\ d}$	_____	_____ _____
$\dfrac{A\ b\ C\ D\ E\ f}{a\ B\ C\ d\ e\ f}$	_____	_____ _____
$\dfrac{b\ D}{B\ d}$	_____	_____ _____

5.22   For each of the following tabulations of testcross progeny phenotypes and numbers, state which locus is in the middle, and reconstruct the genotype of the tested triple heterozygotes.

a.

A B C	191
a b c	180
A b c	5
a B C	5
A B c	21
a b C	31
A b C	104
a B c	109

b.

C D E	9
c d e	11
C d e	35
c D E	27
C D e	78
c d E	81
C d E	275
c D e	256

c.

F G H	110
f g h	114
F g h	37
f G H	33
F G h	202
f g H	185
F g H	4
f G h	0

*5.23   Genes at loci *f*, *m*, and *w* are linked, but their order is unknown. The F_1 heterozygotes from a cross of *FF MM WW* × *ff mm ww* are testcrossed. The most frequent phenotypes in testcross progeny will be *F M W* and *f m w* regardless of what the gene order turns out to be.

a.   What classes of testcross progeny (phenotypes) would be least frequent if locus *m* is in the middle?

b.   What classes would be least frequent if locus *f* is in the middle?

c.   What classes would be least frequent if locus *w* is in the middle?

5.24   The following numbers were obtained for testcross progeny in *Drosophila* (phenotypes):

+	*m*	+	218
w	+	*f*	236
+	+	*f*	168
w	*m*	+	178
+	*m*	*f*	95
w	+	+	101
+	+	+	3
w	*m*	*f*	1
		Total	1,000

Construct a genetic map.

*5.25   Three of the many recessive mutations in *Drosophila melanogaster* that affect body color, wing shape, or bristle morphology are black (*b*) body versus grey in the wild type, dumpy (*dp*), obliquely truncated wings versus long wings in the wild type, and hooked (*hk*) bristles at the tip versus not hooked in the wild type. From a cross of a dumpy female with a black and hooked male, all the F_1 were wild type for all three characters. The testcross of an F_1 female with a dumpy, black, hooked male gave the following results:

wild type	169
black	19
black, hooked	301
dumpy, hooked	21
hooked	8
hooked, dumpy, black	172
dumpy, black	6
dumpy	305
Total	1,000

a. Construct a genetic map of the linkage group (or groups) these genes occupy. If applicable, show the order and give the map distances between the genes.

b. (1) Determine the coefficient of coincidence for the portion of the chromosome involved in the cross. (2) How much interference is there?

5.26 In Chinese primroses long style (*l*) is recessive to short (*L*), red flower (*r*) is recessive to magenta (*R*), and red stigma (*rs*) is recessive to green (*Rs*). From a cross of homozygous short, magenta flower, and green stigma with long, red flower, and red stigma, the F₁ was crossed back to long, red flower, and red stigma. The following offspring were obtained:

STYLE	FLOWER	STIGMA	NUMBER
short	magenta	green	1,063
long	red	red	1,032
short	magenta	red	634
long	red	green	526
short	red	red	156
long	magenta	green	180
short	red	green	39
long	magenta	red	54

Map the genes involved.

5.27 The frequencies of gametes of different genotypes, determined by testcrossing a triple heterozygote, are as shown:

GAMETE GENOTYPE			%
+	+	+	12.9
a	b	c	13.5
+	+	c	6.9
a	b	+	6.5
+	b	c	26.4
a	+	+	27.2
a	+	c	3.1
+	b	+	3.5
		Total	100.0

a. Which gametes are known to have been involved in double crossovers?

b. Which gamete types have not been involved in any exchanges?

c. The order shown is not necessarily correct. Which gene locus is in the middle?

*5.28 Genes *a*, *b*, and *c* are recessive. Females heterozygous at these three loci are crossed to phenotypically wild-type males. The progeny are phenotypically as shown.

a. What is known of the genotype of the females' parents with respect to these three loci? Give gene order and the arrangement in the homologs.

b. What is known of the genotype of the male parents?

c. Map the three genes.

Daughters: all + + +

Sons:	+	+	+	23
	a	b	c	26
	+	+	c	45
	a	b	+	54
	+	b	c	427
	a	+	+	424
	a	+	c	1
	+	b	+	0
			Total	1,000

5.29 Two normal-looking *Drosophila* are crossed and yield the following phenotypes among the progeny:

Females:	+	+	+	2,000
Males:	+	+	+	3
	a	b	c	1
	+	b	c	839
	a	+	+	825
	a	b	+	86
	+	+	c	90
	a	+	c	81
	+	b	+	75
			Total	4,000

Give parental genotypes, gene arrangement in the female parent, map distances, and the coefficient of coincidence.

5.30 The questions below make use of this genetic map:

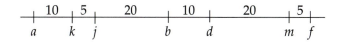

Calculate:

a. the frequency of *j b* gametes from a *J B/j b* genotype.

b. the frequency of *A M* gametes from an *a M/A m* genotype.

c. the frequency of *J B D* gametes from a *j B d/J b D* genotype.

d. the frequency of *J B d* gametes from a *j B d/J b D* genotype.

e. the frequency of *j b d/j b d* genotypes in a *j B d/J b D* × *j B d/J b D* mating.

f. the frequency of *A k F* gametes from an *A K F/a k f* genotype.

*5.31 A female *Drosophila* carries the recessive mutations *a* and *b* in repulsion on the X chromosome (she is heterozygous for both). She is also heterozygous for an X-linked recessive lethal allele, *l*. When she is mated to a true-breeding, normal male, she yields the following progeny:

Females: 1,000   + +

Males: 405   *a* +
           44   + *b*
           48   + +
            2   *a b*

Draw a chromosome map of the three genes in the proper order, and with map distances as nearly as you can calculate them.

*5.32   The following *Drosophila* cross is done:

$$\frac{a + b}{+ \ c \ +} \times \frac{a \ c \ b}{\longrightarrow}$$

Predict the numbers of phenotypes of male and female progeny which will emerge if the gene arrangement is as shown, the distance between *a* and *c* is 14 map units, the distance between *c* and *b* is 12 map units, the coefficient of coincidence is 0.3, and the number of progeny is 2,000.

5.33   A farmer who raises rabbits wants to break into the seasonal Easter market. He has stocks of two true-breeding lines. One is hollow, and long-eared, but not chocolate, while the second is solid, short-eared and chocolate. Hollow (*h*), long ears (*le*), and chocolate (*ch*) are all recessive, autosomal, and linked as in the following map:

The farmer can generate a trihybrid by crossing his two lines and at great expense he is able to obtain the services of a male homozygous recessive at all three loci to cross with his F₁ females.

The farmer has buyers for both solid and hollow bunnies; however, all must be chocolate and long eared. Assuming that interference is zero, if he needs 25 percent of the progeny of the desired phenotypes to be profitable, should he continue with his breeding? Calculate the percent of the total progeny that will be the desired phenotypes.

5.34   Three different semidominant mutations affect the tail of mice. They are linked genes, and all three are lethal in the embryo when homozygous. Fused-tail (*Fu*) and kinky-tail (*Ki*) mice have kinky-appearing tails, while brachyury (*T*) mice have short tails. A fourth gene, histocompatibility-2 (*H-2*), is linked to the three tail genes and is concerned with tissue transplantation. Mice that are *H-2/+* will accept tissue grafts, whereas *+/+* mice will not. In the following crosses the normal allele

is represented by a +. The phenotypes of the progeny are given for four crosses.

(1) $\dfrac{Fu \ +}{+ \ Ki} \times \dfrac{+ \ +}{+ \ +}$

Fused tail	106
Kinky tail	92
Normal tail	1
Fused–kinky tail	1

(2) $\dfrac{Fu \ H\text{-}2}{+ \ +} \times \dfrac{+ \ +}{+ \ +}$

Fused tail, accepts grafts	88
Normal tail, rejects graft	104
Normal tail, accepts graft	5
Fused tail, rejects graft	3

(3) $\dfrac{T \ H\text{-}2}{+ \ +} \times \dfrac{+ \ +}{+ \ +}$

Brachy tail, accepts graft	1,048
Normal tail, rejects graft	1,152
Brachy tail, rejects graft	138
Normal tail, accepts graft	162

(4) $\dfrac{Fu \ +}{+ \ T} \times \dfrac{+ \ +}{+ \ +}$

Fused tail	146
Brachy tail	130
Normal tail	14
Fused–brachy tail	10

Make a map of the four genes involved in these crosses, giving gene order and map distances between the genes.

*5.35   The cross in *Drosophila* of

$$\frac{a^+ \ b^+ \ c \ \ d \ \ e}{a \ \ b \ \ c^+ \ d^+ \ e^+} \ \times \ \frac{a \ \ b \ \ c \ \ d \ \ e}{a \ \ b \ \ c \ \ d \ \ e}$$

gave 1,000 progeny of the following 16 phenotypes:

	GENOTYPE	NUMBER
(1)	$a^+ \ b^+ \ c \ \ d \ \ e$	220
(2)	$a^+ \ b^+ \ c \ \ d \ \ e^+$	230
(3)	$a \ \ b \ \ c^+ \ d^+ \ e$	210
(4)	$a \ \ b \ \ c^+ \ d^+ \ e^+$	215
(5)	$a \ \ b^+ \ c^+ \ d^+ \ e$	12
(6)	$a \ \ b^+ \ c^+ \ d^+ \ e^+$	13
(7)	$a^+ \ b \ \ c \ \ d \ \ e^+$	16
(8)	$a^+ \ b \ \ c \ \ d \ \ e$	14
(9)	$a \ \ b^+ \ c^+ \ d \ \ e^+$	14
(10)	$a \ \ b^+ \ c^+ \ d \ \ e$	13
(11)	$a^+ \ b \ \ c \ \ d^+ \ e^+$	8
(12)	$a^+ \ b \ \ c \ \ d^+ \ e$	8
(13)	$a^+ \ b^+ \ c^+ \ d \ \ e^+$	7
(14)	$a^+ \ b^+ \ c^+ \ d \ \ e$	7
(15)	$a \ \ b \ \ c \ \ d^+ \ e^+$	6
(16)	$a \ \ b \ \ c \ \ d^+ \ e$	7

a. Draw a genetic map of the chromosome, indicating the linkage of the five genes and the number of map units separating each.

b. From the single-crossover frequencies, what would be the expected frequency of $a^+ b^+ c^+ d^+ e^+$ flies?

5.36 As shown in Figure 4.14, recessive alleles at either of the P/p or C/c genes can result in white-flowered sweet peas, but when at least one dominant allele at each gene is present, purple-flowered plants are produced. Two true-breeding white-flowered plants are crossed to give a purple F_1. If the F_1 is selfed, what progeny do you expect if P/p and C/c are (a) linked and 10 map units apart; (b) unlinked.

5.37 In *Drosophila*, many different mutations have been isolated that affect a normally deep-red eye color caused by the deposition of brown and bright-red pigments. Two X-linked recessive mutations are w (white eyes, map position 1.5) and *cho* (chocolate-brown eyes, map position 13.0), with w epistatic to *cho*.

a. A white-eyed female is crossed to a chocolate-eyed male, and the normal, red-eyed F_1 females are crossed to either wild-type or white-eyed males. Determine the frequency of the progeny types that are produced in each cross.

b. The recessive mutation st causes scarlet (bright-red) eyes, and maps to the third chromosome at position 44. Mutant flies with only st and *cho* alleles have white eyes, and w is epistatic to st. Suppose a true-breeding w male is crossed to a true-breeding *cho, st* female. Determine the frequency of the progeny types you would expect if their F_1 females are crossed to true-breeding scarlet-eyed males.

*5.38 Breeders of thoroughbred horses used for races keep extensive pedigree information. Such information can be useful to determine simple inheritance patterns (e.g., chestnut coat color has been determined to be recessive to bay coat color) and to speculate whether race horses that win very competitive races (so-called "classy" horses) share genetic traits. Sharpen Up was a chestnut stallion who was only somewhat successful as a race horse. He was retired from horse-racing at the age of four and put out to stud. His progeny were very successful: of 367 foals fathered in the United States, 43 were prize-winners in very competitive races, and of 260 foals fathered in England, 40 were prize-winners in very competitive races. A commentator who "analyzed" Sharpen Up's progeny (and that of other chestnut prize-winners) has suggested that "whatever gene combinations that produced class (winning horses) were tied to chestnut coat." Indeed, of the 83 progeny that have shown class (won very competitive races), about 45 were also chestnut in color. Use a chi-square test to assess whether there is any reason to believe that, if there is a gene (or genes) for "class," it is linked to the gene for chestnut coat color. Examine this issue using two different assumptions: (1) Sharpen Up was mated equally frequently to homozygous bay, heterozygous bay/chestnut and homozygous chestnut mares; (2) Sharpen Up was mated equally frequently to heterozygous bay/chestnut and homozygous chestnut mares. Carefully state any additional assumptions and your hypothesis.

6 GENETIC MAPPING IN EUKARYOTES II

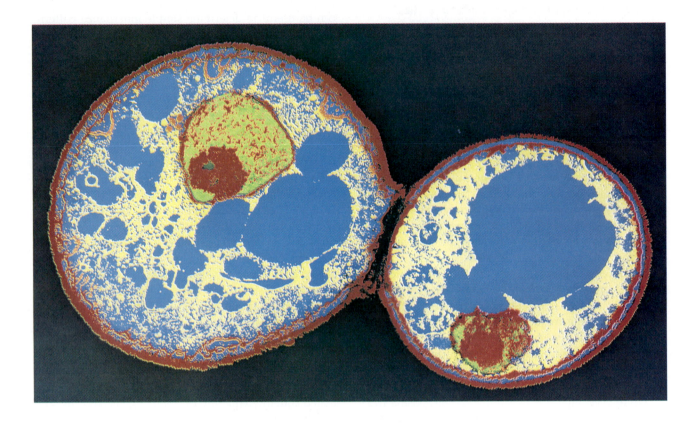

TETRAD ANALYSIS
Using Random-Spore Analysis to Map Genes in Haploid Eukaryotes
Using Tetrad Analysis to Map Two Linked Genes
Calculating Gene-Centromere Distance in Organisms with Linear Tetrads

MITOTIC RECOMBINATION
Discovery of Mitotic Recombination
Mechanism of Mitotic Crossing-Over
Mitotic Recombination in the Fungus *Aspergillus nidulans*

MAPPING THE HUMAN GENOME
Construction of Genetic Linkage Maps
Construction of Physical Maps
Integration of Genetic Linkage Maps and Physical Maps

PRINCIPAL POINTS

~ Tetrad analysis is a mapping technique that can only be used to map the genes of certain haploid eukaryotic organisms in which the products of a single meiosis, the meiotic tetrad, are contained within a single structure.

~ In organisms in which tetrads can be analyzed, the analysis of the relative proportion of tetrad types provides another way of computing the map distance between genes.

~ In organisms in which the meiotic tetrads are arranged in a linear fashion, it is easy to map the distance of a gene from its centromere.

~ Crossing-over can occur during mitosis (rarely) as well as during meiosis.

~ In certain organisms, parasexual genetic analysis is used to map genes. In these organisms, genetic recombination is achieved by means other than meiosis and fertilization, for example, by crossing-over during mitosis.

~ A major goal of human genetics research is to construct a highly detailed map of the human genome. Two types of maps are being constructed: genetic linkage maps and physical maps. Genetic linkage maps show the relative locations of gene and/or DNA markers as determined by genetic recombination analysis. Physical maps are maps of physically identifiable regions or markers along the DNA that are constructed without using genetic recombination analysis. The resolution of genetic linkage maps is increasing as more and more markers are found and analyzed. The resolution of physical maps depends upon the technique being used to construct the map. The markers on physical maps are being used as landmarks for specific explorations of the genome and as starting points for sequencing efforts.

---

*I*n the previous chapter we considered the classical principles for mapping genes in eukaryotes using recombination analysis. We saw that the outcome of crosses can be used to construct genetic maps with distances between genes given in map units (centimorgans) and that map distances are useful in predicting the outcome of other crosses. In this chapter we discuss the determination of map distance by tetrad analysis in certain appropriate haploid organisms, the rare incidence of crossing-over in mitosis, and an overview of the construction of genetic linkage maps and physical maps of the human genome.

TETRAD ANALYSIS

Tetrad analysis is a mapping technique that can be used to map the genes of those eukaryotic organisms in which the products of a single meiosis, the meiotic tetrad, are contained within a single structure. The eukaryotic organisms in which this phenomenon occurs are either fungi or single-celled algae, all of which are haploid. The orange bread mold *Neurospora crassa* and the yeast *Saccharomyces cerevisiae* (both fungi) and *Chlamydomonas reinhardtii* (a single-celled alga) are frequently used in tetrad analysis.

By analyzing the phenotypes of the meiotic tetrads, geneticists can directly infer the genotypes of each member of the tetrad. Haploid organisms exhibit no genetic dominance since there is only one copy of each gene; hence the genotype is expressed directly in the phenotype. Much valuable information about how and when genetic recombination (crossing-over) occurs and about the process of gene segregation has been discovered through tetrad analysis of various organisms, and we will outline this technique in the following sections.

Before we discuss the principles of tetrad analysis, it is important that we understand the life cycles of the organisms with which tetrad analysis can be done. The life cycle of *Neurospora crassa* was shown in Figure 5.9 (p. 142).

The life cycle of baker's yeast (also called the budding yeast), *Saccharomyces cerevisiae*, is diagrammed in Figure 6.1. Two mating types occur in yeast, **a** and α. The haploid vegetative cells of this organism reproduce mitotically, with the new cell arising from the parental cell by budding. Fusion of haploid **a** and α cells produces a diploid cell that is stable and that also reproduces by budding. Diploid **a**/α cells sporulate; that is, they go through meiosis. The four haploid meiotic products of a diploid cell, the ascospores, are contained within a roughly spherical ascus. Two of these ascospores are of mating type **a**, and two are of mating type α. When the ascus is ripe, the ascospores are released, and they germinate to produce haploid vegetative cells. On a solid medium each ascospore develops into a discrete colony. In yeast the four ascospores are arranged randomly within the ascus so that only unordered tetrads can be isolated from this organism (in contrast to *Neurospora crassa*, where the ascospores are linearly arranged

~ FIGURE 6.1

(a) Life cycle of the yeast *Saccharomyces cerevisiae*; (b) Colorized scanning electron micrograph of *Saccharomyces cerevisiae* cells, many of which have buds.

a)

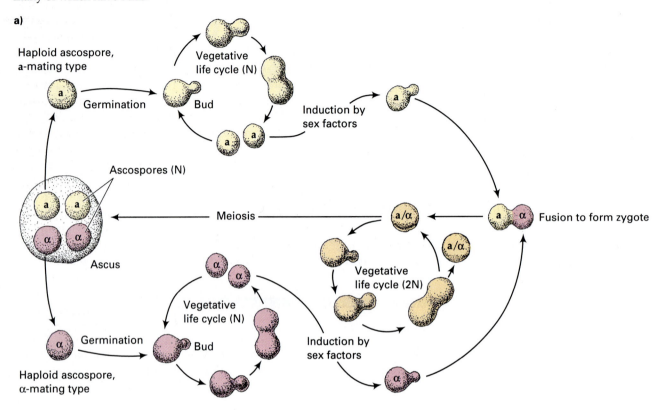

b)

in the ascus in a way that reflects the orientation of the four chromatids at metaphase I: see Figure 5.10, p. 143).

Like yeast, *Chlamydomonas reinhardtii* has haploid vegetative cells. Each individual is a single green algal cell that can swim freely as a result of the motion of its two flagella (Figure 21.10, p. 690). When nitrogen is limited, the cells change morphologically to become gametes so that mating is possible. There are two mating types, designated plus (mt^+) and minus (mt^-). Only gametes of opposite mating types (in this case, mt^+ and mt^-) can fuse to produce diploid zygotes. No fusion occurs between gametes of like mating types. After a maturation process, the zygote enters meiosis. The four haploid meiotic products are contained within a sac as an unordered tetrad, and there are two *mt*⁺ and two *mt*⁻ cells. When these cells are released, they are free-swimming, and by mitosis they give rise to clones of those meiotic products.

Using Random-Spore Analysis to Map Genes in Haploid Eukaryotes

In the three haploid eukaryotes we have discussed, the meiotic products can be collected after they have been released from the ascus (yeast, *Neurospora*) or sac (*Chlamydomonas*). In the case of the fungi (*Neurospora* and yeast), spores can be induced to germinate and the resulting cultures can be analyzed. The free-swimming meiotic products of *Chlamydomonas* can be analyzed directly. In fact, the haploid nature of the mature stages of all three organisms simplifies the analysis, since this stage is exactly equivalent to that of the gametes produced after meiosis in a diploid eukaryote. Thus we can make three-point crosses to map genes on a chromosome, using essentially the same approach as the one we used for a diploid eukaryote.

Figure 6.2 shows a three-point mapping cross in a haploid eukaryote to illustrate the similarities with three-point testcrosses in diploids. Here a wild-type (haploid) strain is crossed with a strain that carries three mutant genes in the same chromosome. The result is a diploid zygote that is triply heterozygous. When this zygote undergoes meiosis to produce the haploid progeny organisms, there is the potential to form eight genotypic (and hence phenotypic) classes in a fashion identical to that in three-point testcrosses involving diploid eukaryotes. The only difference is that the parents and the progeny are haploid, not diploid. Analysis for map distances is done exactly as was described in Chapter 5 for three-point testcrosses using diploids. Thus, random-spore analysis can be used with haploid eukaryotes to obtain data for drawing a genetic map of their chromosomes. This information is important for understanding the structural and functional organization of the genome in such organisms.

Using Tetrad Analysis to Map Two Linked Genes

Formally, random-spore analysis to map genes in *Neurospora*, yeast, and *Chlamydomonas* is little different from analysis of the progeny of testcrosses in diploid eukaryotes. Such analysis produces data for

~ FIGURE 6.2

Typical genetic cross for mapping three genes in a haploid organism such as yeast or *Neurospora*.

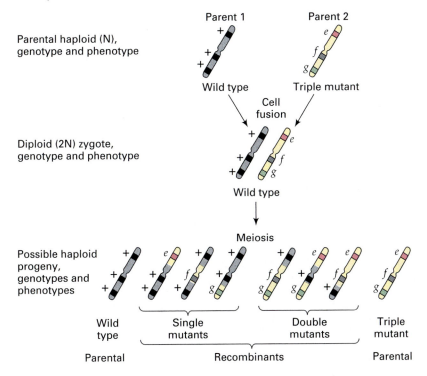

drawing genetic maps. Let us consider how we can determine the map distance between two linked genes by analyzing a number of complete meiotic tetrads. The analysis to be described requires only unordered tetrads, and hence is applicable to *Neurospora*, yeast, and *Chlamydomonas*.

By making an appropriate cross, a diploid zygote is constructed that is heterozygous for both genes, and after meiosis, the resulting tetrads are analyzed. In *Neurospora*, for example, the spores that represent a tetrad can be dissected out of an ascus by manual micromanipulation using a very fine glass needle. For yeast, the approach is similar, except that a mechanical micromanipulator with a carbon fiber needle is necessary since the spores are much smaller than those of *Neurospora*.

For a cross of $a^+ b^+ \times a\ b$ three possible tetrad types result, called **parental-ditype** (PD), **tetratype** (T), and **nonparental-ditype** (NPD) tetrads (Figure 6.3). The PD tetrads contain only two types of meiotic products,

both of which are of the parental type (hence the name parental ditype). The T tetrads contain all four possible types of meiotic products, that is, the two parental types $a^+ b^+$ and $a\ b$ and the two recombinant types $a^+ b$ and $a\ b^+$. The NPD tetrads contain two types of meiotic products, both of which are of the nonparental (recombinant) types $a^+ b$ and $a\ b^+$.

*K*EYNOTE

In organisms in which the products of a meiosis (the meiotic tetrad) are contained within a single structure, three types of tetrads are possible when two genes are segregating in a cross. The parental-ditype (PD) tetrad contains four nuclei, all of parental genotypes: two of one parent and two of the other parent. The nonparental-ditype (NPD) tetrad contains four nuclei, all of recombinant (non-

---

~ FIGURE 6.3

Three types of tetrads produced from a cross of $a^+ b^+$ (+ +) $\times a\ b$: parental-ditype (PD), tetratype (T), and nonparental-ditype (NPD) tetrads.

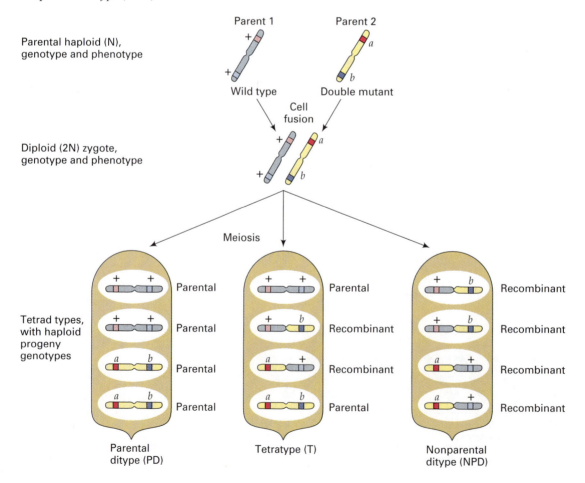

parental) genotypes, that is, two of each possible type. The tetratype (T) tetrad contains two parental and two recombinant nuclei (one of each parental type and one of each recombinant type).

Tetrad analysis helps to determine whether or not two genes are linked. This determination is based on the ways each tetrad type is produced when genes are unlinked and when genes are linked.

Figure 6.4 shows how PD, NPD, and T tetrads are produced when two genes are on different chromosomes. In this case the PD and NPD tetrads result from events in which no crossovers are involved; the relative metaphase plate orientation of the four chromatids for the two chromosomes determines whether a PD or an NPD tetrad results. Since the two sets of

~ FIGURE 6.4

Origin of tetrad types for a cross $a\ b \times +\ + (a^+\ b^+)$, in which the two genes are located on different, independently assorting chromosomes.

No crossover:
One metaphase alignment

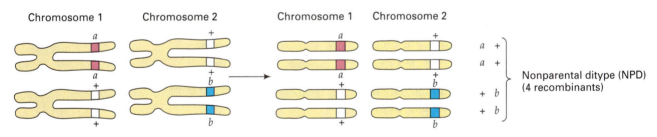

Alternative metaphase alignment

Single crossover:
Single crossover between *a* gene and its centromere

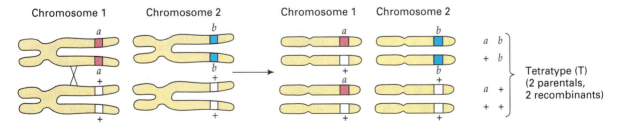

four chromatids align at the metaphase plate independently, the PD and NPD orientations shown occur with approximately equal frequency. Thus if two genes are unlinked, the frequency of PD tetrads will equal the frequency of NPD tetrads. T tetrads are produced when there is, for example, a single crossover between one of the genes and the centromere on that chromosome. The sum of all possible progeny types will equal 50 percent parental types and 50 percent recombinant types when the two genes are unlinked. That is, the overall frequency of PD tetrads = the overall frequency of NPD tetrads.

Figure 6.5 shows the origins of each tetrad type when two genes are linked on the same chromosome. If no crossing-over (Figure 6.5a) occurs between the genes, then a PD tetrad results. A single crossover (Figure 6.5b) produces two parental and two recombinant chromatids and hence a T tetrad. For double crossovers we must take into account the chromatid strands involved. In a two-strand double crossover (Figure 6.5c) the two crossover events between the two genes both involve the same two chromatids. This crossover results in a PD ascus, since no recombinant progeny are produced. Three-strand double crossovers (Figure 6.5d) involve three of the four chromatids; the two possible ways in which this event can happen are shown. In either case two recombinant and two parental progeny types are produced in each tetrad, which is a T tetrad. Lastly, in four-strand double crossovers (Figure 6.5e) each crossover event involves two distinct chromatids, so all four chromatids of the tetrad are involved. This crossover is the only way in which NPD tetrads are produced. Therefore since PD tetrads are produced either when there is no crossing-over or when there is a two-strand double crossover, and since NPD tetrads are only produced by four-strand double crossovers (which are 1/4 of all the possible double crossovers and, hence, are rare), two genes are considered to be linked if the frequency of PD tetrads is far greater than the frequency of NPD tetrads (i.e., PD ≫ NPD).

Once we know that two genes are linked, and once we have data on the relative numbers of each type of meiotic tetrad, the distance between the two genes can be computed by using a modification of the basic mapping formula:

$$\frac{\text{number of recombinants}}{\text{total number of progeny}} \times 100$$

In tetrad analysis we analyze types of tetrads, rather than individual progeny. To convert the basic mapping formula into tetrad terms, the recombination frequency between genes *a* and *b* becomes

$$\frac{1/2\,\text{T} + \text{NPD}}{\text{total tetrads}} \times 100$$

In this formula the 1/2 T and the NPD represent the only recombinants; the other 1/2 T and the PD represent the nonrecombinants (the parentals). Thus the formula does indeed compute the percentage of recombinants. For instance, if there are 200 tetrads with 140 PD, 48 T, and 12 NPD, the recombination frequency between the genes is

$$\frac{1/2(48) + 12}{200} \times 100 = 18\%$$

This conversion produces a formula for calculating the map distance between two linked genes by using the frequencies of the three possible types of tetrads rather than by analyzing individual progeny. If more than two genes are linked in a cross, the data may best be analyzed by considering two genes at a time and by classifying each tetrad into PD, NPD, and T for each pair.

KEYNOTE

In organisms in which all products of meiosis are contained within a single structure, the analysis of the relative proportion of tetrad types provides another way to compute the map distance between genes. If PD = NPD, the two genes are unlinked, whereas if PD > NPD, the two genes are linked. The general formula when two linked genes are being mapped is

$$\frac{1/2\,\text{T} + \text{NPD}}{\text{total tetrads}} \times 100$$

Calculating Gene-Centromere Distance in Organisms with Linear Tetrads

The tetrad analysis we have discussed so far is applicable to both linear tetrads or unordered tetrads (where the products are arranged randomly). *Neurospora* actually has eight spores in its linear ascus, since the four meiotic products undergo one more mitotic division before the ascospores are produced. Since mitotic division produces, in essence, clones, for the purposes of our genetic discussions the eight spores can be considered as four pairs and we will ignore the last mitotic division. The most interesting thing about ordered tetrads is that their genetic content directly reflects the orientation of the four chromatids of each chromosome pair in the diploid zygote nucleus at metaphase

~ FIGURE 6.5

Origin of tetrad types for a cross $a\ b \times + +\ (a^+\ b^+)$ in which both genes are located on the same chromosome. (a) No crossover; (b) Single crossover; (c–e) Three types of double crossovers.

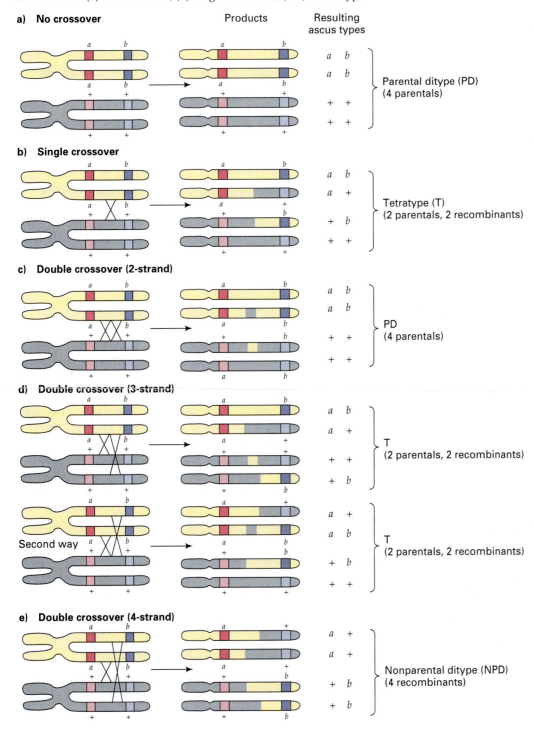

I. *This fact allows us to map the distance between genes and their centromeres*, which is not easy with unordered tetrads. Locating the centromeres on the genetic maps of chromosomes makes the maps more complete. In more complex organisms centromeres are located primarily through cytological studies (which are usually not possible in lower eukaryotes because their chromosomes are too small).

In this example we will map the position of the mating-type locus of *Neurospora* in relation to its centromere. Mating type is a function of which allele, *A* or *a*, is present at a locus in linkage group I. If a *Neurospora* strain of mating type *A* is crossed with one of mating type *a*, a diploid zygote of genotype *A/a* results. Figure 6.6 shows the various ways in which this zygote can give rise to the four meiotic products, the ascospores. For the purposes of illustration, ● will be used to indicate the centromere of the *A* parent and ○ will be used to indicate the centromere of the *a* parent. (In reality there is no difference between the two.)

If no crossing-over occurs between the mating-type locus and the centromere, the resulting ascospores have the genotypes shown in Figure 6.6a. An important point here is that the centromeres do not separate until just before the second meiotic division. Therefore the spores in the top half of the ascus always have the centromere from one parent (the ● centromere, in this case), and the spores in the other half of the ascus always have the centromere from the other parent (○, here).

Since the two types of centromeres segregate to different nuclear areas after the first meiotic division, we say that they show *first-division segregation*. And since no crossing-over occurs between the mating-type locus and the centromere in Figure 6.6a, each allelic pair also shows first-division segregation. That is, after one division, both copies of the *A* allele go to one pole, and both copies of the *a* allele go to the other pole—the *A* and *a* alleles have segregated into different nuclear areas after the first meiotic division. Note particularly that all four spores in an ascus showing first-division segregation of alleles are parental types: The *A* allele is on the chromosome with a ● centromere and the *a* allele is on the chromosome with the ○ centromere. Furthermore, since it is equally likely that the four chromatids in the diploid zygote will be rotated 180°, we expect equal numbers of first-division segregation asci in which the ● *A* spores are in the bottom half and the ○ *a* spores are in the top half.

To determine the map distance between a gene and its centromere, we must measure the crossover frequency between the two chromosomal sites. Thus we need to predict the consequences of a single crossover between the mating-type locus and the centromere. Figure 6.6b shows that four ascus types are produced. These four types are generated in equal frequencies, since they reflect the four possible orientations of the four chromatids in the diploid zygote at metaphase I. Each has first-division segregation of the centromere; that is, segregation occurs during meiosis I. By contrast, *A* and *a* in Figure 6.6b are both present in each of the two nuclear areas after the first division; they do not segregate into separate nuclei until the second division. This situation is called *second-*

~ Figure 6.6 (facing page)

Determination of gene-centromere distance of the mating-type locus in *Neurospora*. (a) Production of an ascus from a diploid zygote in which no crossing-over occurred between the centromere and the mating-type locus, and first-division segregation for the mating-type alleles. (In *Neurospora*, a mitotic division after the second meiotic division produces eight spores, thereby doubling each progeny type: i.e., the 2:2 ratio is actually 4:4. The mitotic division is not considered here for simplicity's sake.) (b) Production of asci after a single crossover occurs between the mating-type locus and its centromere. Chromosomes are shown after this crossover is complete. The asci show second-division segregation for the mating-type locus, and the four types of asci are produced in equal proportions.

division segregation for the gene; that is, segregation of *A* from *a* occurs during meiosis II. Here, the pattern of gene segregation depends upon which chromatids are involved in the crossover event. The 1:1:1:1 (*A a A a* and *a A a A*) and 1:2:1 (*A a a A* and *a A A a*) second-division segregation patterns are readily distinguishable from the usual 2:2 (*A A a a* and *a a A A*) first-division segregation pattern.

By analyzing ordered tetrads, we can count the number of asci that show second-division segregation for a particular gene marker. For the mating-type locus, about 14 percent of the asci show second-division segregation. This value must then be converted to a map distance using the formula

$$\frac{\text{gene-centromere}}{\text{distance}} = \frac{\text{percent of second-division tetrads}}{2}$$

That is, if we consider the centromere to be a gene marker (and in a sense it is), then the parental types are ● *A* and ○ *a*, and the recombinant types are ● *a* and ○ *A*. In a first-division segregation ascus (Figure 6.6a) all the spores are parentals, while in a second-division segregation ascus (Figure 6.6b) half are parentals and half are recombinants. Therefore to convert the tetrad data to crossover or recombination data, we divide the percentage of second-division segregation asci (14 percent) by 2. Thus the mating-type gene is 7 mu from the centromere of linkage group I.

In essence, the computation of gene-centromere distance is a special case of mapping the distance between two genes. If ordered tetrads are isolated, then not only can genes be mapped one to another, but each can be mapped to its centromere.

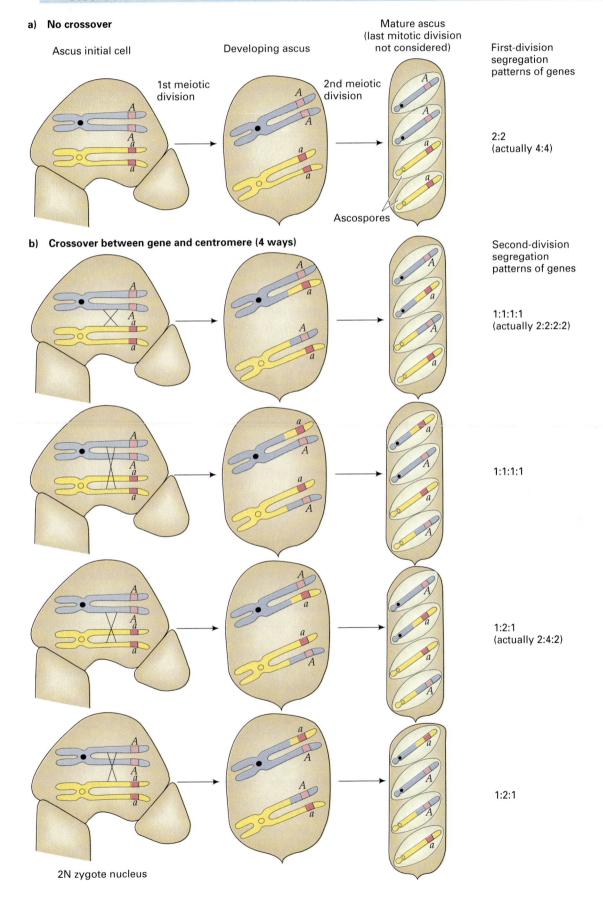

~ FIGURE 6.6

a) No crossover

Ascus initial cell

1st meiotic division

Developing ascus

2nd meiotic division

Mature ascus (last mitotic division not considered)

Ascospores

First-division segregation patterns of genes

2:2 (actually 4:4)

b) Crossover between gene and centromere (4 ways)

Second-division segregation patterns of genes

1:1:1:1 (actually 2:2:2:2)

1:1:1:1

1:2:1 (actually 2:4:2)

1:2:1

2N zygote nucleus

KEYNOTE

In some eukaryotic organisms the products of a meiosis are arranged in a specialized structure in a way that reflects the orientation of the four chromatids of each homologous pair of chromosomes at metaphase I. The ordered tetrads allow us to easily map the distance of a gene from its centromere. Where no crossover occurs between the gene and the centromere, the result is a first-division segregation tetrad, in which one parental type is found in half the ordered tetrad and the other parental type is found in the other half, giving a 2:2 segregation pattern. When a single crossover occurs between the gene and its centromere, several different tetrad segregation patterns are found, and each exemplifies second-division segregation. The gene-centromere map distance is computed as the percentage of second-division tetrads divided by 2.

MITOTIC RECOMBINATION

Discovery of Mitotic Recombination

Crossing-over can occur during mitosis as well as during meiosis. **Mitotic crossing-over** was first observed by Curt Stern in 1936, in crosses involving *Drosophila* strains carrying recessive sex-linked mutations that cause yellow body color (*y*) instead of the normal grey body color, and short, twisty bristles (singed, *sn*) instead of the normal long, curved bristles. In flies with a wild-type grey body color, all bristles are black; in yellow-bodied (mutant) flies the bristles are yellow.

From a cross of homozygous $y^+ sn/y^+ sn$ females (grey bodies, singed bristles) with $y\ sn^+//\wedge$ (yellow bodies, normal bristles), Stern found, as expected, that the female F$_1$ progeny were mostly wild type in appearance: they had grey bodies and normal bristles (genotype $y^+ sn/y\ sn^+$). However, some females had patches of yellow and/or singed bristles that were not explained by regular gene segregation. The origin of these flies could, however, have been explained by chromosome nondisjunction or by chromosomal loss. (These changes are somatic [body cells] rather than germline [gamete-producing cells] changes.)

Other females had **twin spots**, two adjacent regions of bristles, one showing the yellow phenotype and the other showing the singed phenotype; that is, a *mosaic* phenotype. Surrounding the twin spot, and constituting essentially the rest of the bristles, the phenotype was wild type. Figure 6.7 diagrams this twin-spot phenotype and contrasts it with the single-spot phenotypes. Stern reasoned that since the two parts of a twin spot were always adjacent, the twin spots must be the reciprocal products of the same genetic event. The best explanation was that they were generated by a *mitotic crossing-over* event, an event that occurs rarely.

The production of twin spots by mitotic crossing-over is shown in Figure 6.8. The starting point is a wild-type diploid cell that is heterozygous for both the *y* (yellow) and *sn* (singed) genes. The mutant alleles are on different homologs. The rare mitotic crossing-over event can occur either between the centromere and the *sn* locus (Figure 6.8, left), or between the *sn* and the *y* locus (Figure 6.8, right). (The mechanism of mitotic crossing-over will be described in the next section.) In the former case, if chromatids 1 and 3 segregate to one progeny nucleus, and chromatids 2 and 4 segregate to the other, one progeny cell is homozygous *y/y* and the other is homozygous *sn/sn*. When these divide and produce two tissue patches, one is yellow and the other is singed—a twin spot has been produced. The surrounding tissue, not involved in any mitotic crossing-over, will be wild type in phenotype. In the case of crossing-over between *sn* and *y*, and of chromatid segregation as above, the result is one progeny cell that is yellow in phenotype (*y/y*) and the other that is wild type in phenotype, producing a single yellow spot.

Mechanism of Mitotic Crossing-Over

Mitotic crossing-over (*mitotic recombination*) can only be studied in diploid cells. It is a process during mitosis that produces a progeny cell with a combination of genes that differs from that of the diploid parental cell that entered the mitotic cycle. *Mitotic crossing-over occurs at a stage similar to the four-strand stage of meiosis.* Recall that during mitosis each pair of homologous chromosomes replicates and that the two pairs of chromatids then align independently at the metaphase plate. When the chromatids separate, each progeny cell receives one copy of each parental homolog so that each cell has the same genetic constitution as the parental cell. Figure 6.9 (p. 180) shows the pattern of gene segregation for a theoretical cell heterozygous for all the genes on its one pair of homologous chromosomes. Both the parental and progeny cells are wild type in phenotype.

Rarely, after each chromosome has replicated and prior to metaphase, the two pairs of maternally

~ FIGURE 6.7

Body surface phenotype segregation in a *Drosophila* strain *y⁺ sn/y sn⁺*. The *sn* allele causes short, twisted (singed) bristles, and the *y* allele results in a yellow body coloration: (a) Single yellow spot in normal-body color background; (b) Twin spot of yellow color and singed bristles; (c) Single singed-bristle spot in normal-bristle phenotype background. (From *Principles of Genetics* by Robert H. Tamarin. Copyright © 1996. Reproduced with permission of The McGraw-Hill Companies.)

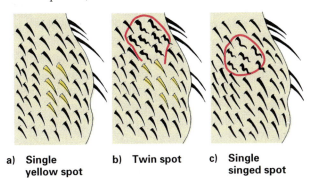

a) Single yellow spot **b) Twin spot** **c) Single singed spot**

~ FIGURE 6.8

Production of the twin spot and of the single yellow spot shown in Figure 6.7 by mitotic crossing-over.

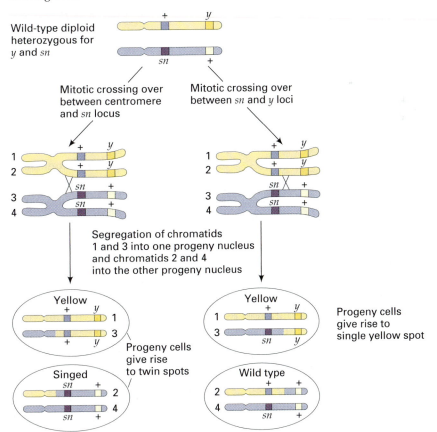

~ FIGURE 6.9

Normal mitotic segregation of genes in a theoretical diploid cell with one homologous pair of chromosomes. The cell is heterozygous for all five genes on the chromosome. During mitosis each homolog replicates, and the two pairs of chromatids align independently on the metaphase plate. Each progeny cell receives one chromatid of each pair and hence is heterozygous for all the five genes. The progeny are both genetically identical to the parental cell and phenotypically are wild type.

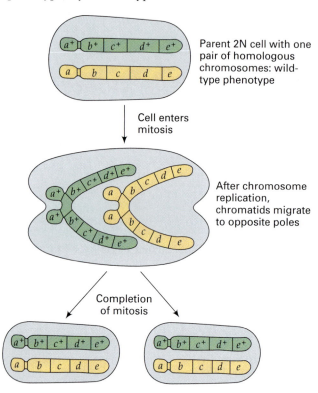

Parent 2N cell with one pair of homologous chromosomes: wild-type phenotype

Cell enters mitosis

After chromosome replication, chromatids migrate to opposite poles

Completion of mitosis

Progeny 2N cells, both with wild-type phenotype

derived and paternally derived chromatids come together to form a structure that is analogous to the four-strand stage in meiosis. It is during this stage that crossing-over can occur. Figure 6.10 diagrams the consequences of mitotic crossing-over between genes c and d in a cell of the same genotype as the cell in Figure 6.9. After the crossover has occurred, the two pairs of chromatids separate and migrate independently to the metaphase plate, resulting in two possible orientations; see parts (a) and (b) of Figure 6.10.

When type (a) completes mitosis, the two diploid cells 1 and 2 are produced. Progeny cell 1 is homozygous $d^+ e^+/d^+ e^+$ and therefore has the wild-type phenotypes associated with these two alleles. Progeny cell 2 is homozygous $d e/d e$ and hence expresses the mutant phenotypes associated with these two alleles. For both progeny cell types, all the other genes (a, b, and c) are heterozygous, and thus the phenotypes

expressed are those of the dominant, wild-type alleles of those genes: Progeny cell 1 is phenotypically like the parental cell. Progeny cell 2 has a new phenotype. In general, then, mitotic crossing-over can make all those genes distal to the crossover point (i.e., between the crossover and the end of the chromosome arm) homozygous if the chromatid pairs align appropriately at the metaphase plate. This phenomenon applies only to those genes on the *same* chromosome arm as the crossover; that is, to those genes from the centromere outward. Therefore the crossover between c and d has no effect on gene a, which is on the other arm of this chromosome.

When type (b) completes mitosis, the two diploid cells 3 and 4 are produced. In both progeny cells, as a result of the orientation of the chromatid pairs at metaphase, all genes are heterozygous, and the cells are phenotypically wild type. Nonetheless, inspection of Figure 6.10 shows that genetic recombination has occurred, since all the genes in cell 3 are not in coupling, as they were in the parental cell. Since genetic-mapping studies depend on the detection and counting of progeny with phenotypes that are recombined from those found in the parents, the progeny cells from type (a) metaphase are the ones to analyze in mitotic recombination studies.

KEYNOTE

Crossing-over can occur during mitosis as well as during meiosis, although it occurs much more rarely during mitosis. As in meiosis, mitotic crossing-over occurs at a four-strand stage. Single crossovers during mitosis can be detected in a heterozygote because loci distal to the crossover and on the same chromosome arm may become homozygous.

Mitotic Recombination in the Fungus *Aspergillus nidulans*

While there is always at least one crossover for each meiotic tetrad, mitotic crossing over occurs in less than 1 percent of mitotic divisions in those organisms in which it has been studied.

Mitotic recombination has been studied most extensively in fungi, particularly *Aspergillus nidulans*. *Aspergillus nidulans* is a mycelial-form fungus like *Neurospora*, and colonies of the wild type are greenish because the asexual spores (the conidia) are green. Unlike *Neurospora*, *Aspergillus* does not have two distinct mating types. Instead, any strain of *Aspergillus* is capable of self-mating. As a result, special procedures

~ **FIGURE 6.10**

Result of a mitosis of the same cell type as the cell in Figure 6.9 but in which a rare mitotic crossing-over event occurs.

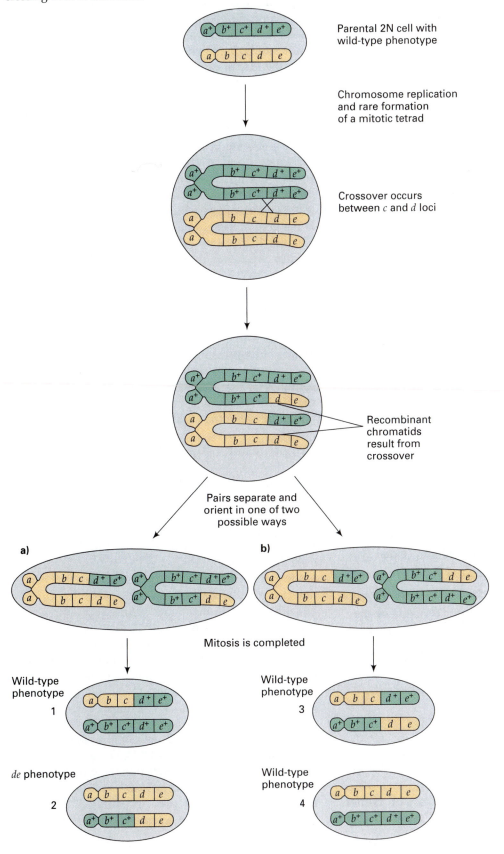

must be used to ensure that matings between identical nuclei do not obscure analysis of genetic recombinants. However, mitotic crossing-over does occur in *Aspergillus* at frequencies that make it possible to map genes by analyzing the numbers of recombinants. That is, the nuclei in the mycelium are typically haploid. When two haploid strains (e.g., with different genotypes) are mixed together, they will fuse. The result of fusion is a mycelium in which the two nuclear types from the two original haploid strains coexist and divide mitotically within the same cytoplasm. A cell—or a collection of cells, as in a mycelium—that possesses any number of genetically different nuclei in a common cytoplasm is called a **heterokaryon** (literally, "different nuclei"). The formation of heterokaryons in *Aspergillus* enables geneticists to construct heterozygous strains that can be used in mitotic recombination studies. Rarely, in a process called *diploidization*, two haploid nuclei in a heterokaryon will fuse to produce a diploid nucleus. Diploid nuclei will divide mitotically and, therefore, they can undergo mitotic crossing-over. In the way that we have already discussed, mitotic crossing-over produces mitotic genetic recombinants that can be analyzed to determine genetic distances between genes. In addition, the diploids often break down to haploids so new gene combinations can readily be detected by analysis of colonies that arise from uninucleate conidia produced by this fungus. A system such as this, in which genetic recombination (and, therefore, genetic variability) occurs by means other than meiotic recombination, is called a **parasexual system**.

ꓘEYNOTE

Genetic systems that achieve genetic recombination by means other than the regular alternation of meiosis and fertilization are called parasexual systems. For example, the parasexual system in fungi such as *Aspergillus* consists of (1) the formation of a heterokaryon by mycelial fusion; (2) fusion of the two haploid nuclei to give a diploid nucleus; and (3) mitotic crossing-over within the diploid nucleus to produce recombinants.

MAPPING THE HUMAN GENOME

Classical genetic mapping experiments with numerous organisms have localized many genes to their chromosome locations. Gene localization has been not an end point but a beginning point for further studies of the genes to determine their structures, functions, and regulatory aspects of their expression. Studies of the genes of organisms simpler than ourselves have been valuable in informing us in our studies of human genes, but an overriding desire has been the identification, localization, and characterization of the function of every gene in the human genome. The information we will glean from such investigations of the human genome is of obvious importance; for example, it would enable us to more easily understand the bases for genetic diseases that are known and any that may be discovered in the future. Localizing all the estimated 50,000–100,000 genes in the human genome and, in fact, obtaining the sequence of every base pair in the human genome is the goal of the Human Genome Project (HGP) (see Chapter 15, pp. 486–487). As part of the HGP, parallel studies are being done with selected model organisms, namely, *E. coli*, the yeast *Saccharomyces cerevisiae*, *Drosophila melanogaster*, the nematode *Caenorhabditis elegans*, and the mouse *Mus musculus*. Although much of the HGP involves DNA sequencing, the groundwork that must be done before sequencing commences involves mapping activities. In this section we provide an overview of some of the mapping activities of the human genome that are involved in the HGP. These mapping procedures, of course, are not exclusive to the HGP.

Construction of Genetic Linkage Maps

A *genetic linkage map* is exactly the kind of map we have discussed to this point in the text, that is, a map showing the relative locations of specific genetic markers determined by the analysis of recombinants from genetic crosses. A marker used in this mapping approach is any inherited characteristic that has different forms (alleles) among individuals and that can be detected readily. Markers may be genes of the kind we have discussed to this point in the text, but they also may be molecular markers, that is, DNA regions in the genome that may differ sufficiently between individuals so that they can be detected by molecular analysis of DNA isolated from family members (see Chapter 15 for some general discussions of detecting DNA differences between genomes). An example of a molecular marker is a restriction fragment length polymorphism (RFLP or "riff-lip"; see Chapter 15, p. 480). A RFLP results from changes in the DNA sequence so that a recognition site for a restriction enzyme that can cleave the DNA is added or lost. Digestion with the restriction enzyme results in different fragment lengths that are easily detectable in the laboratory.

Procedurally, human genetic linkage maps are constructed by analyzing the inheritance patterns of genetic markers and/or DNA markers in pedigrees for linkage. Due to the rarity of large, multigenerational pedigrees in which two genetic traits are segregating, it is hard to test for linkage and to calculate map distance between markers. So in most cases a statistical test known as the **lod** (logarithm of **od**ds) **score method,** invented in 1955 by mathematical geneticist Newton Morton, is used to test for possible linkage between two loci. The lod score method is usually done by computer programs that use pooled data from a number of pedigrees. A full discussion of this method is beyond the scope of this text, so only a brief presentation is given here.

The lod score method calculates and compares the probability that the pedigree results would have been obtained if two markers showed a certain degree of linkage (calculated from the pedigree data) and the probability that the results would have been obtained even if two markers were not linked. The results are expressed as the $\log_{10}$ of the ratio of the two probabilities; this is the lod score. By convention an hypothesis of linkage between two genes is accepted when the lod score is +3 or more because this means that the odds are 1,000:1 in favor of linkage between two genes or markers (because the $\log_{10}$ of 1,000 is +3). Similarly, an hypothesis of linkage between two genes is rejected when the lod score reaches −2.

Once linkage is established between two gene markers, a gene marker and a DNA marker, or two DNA markers, the map distance is computed from the recombination frequency giving the highest lod score. Thus, genetic linkage maps are built in map units (mu), or centimorgans. For the human genome, 1 mu corresponds, on average, to approximately 1 million base pairs (1 megabase, or 1 Mb). The most comprehensive genetic linkage map of the human genome produced to date is based on 5,264 molecular markers and has an average interval size between markers (average resolution) of 1.6 mu. A genetic linkage map of relatively closely linked DNA markers is valuable because it is a useful resource for mapping the locations of disease and other genes. A genetic linkage map of DNA markers is also valuable for the construction of the physical maps described in the next section.

Construction of Physical Maps

A **physical map** is a map of physically identifiable regions or markers on genomic DNA that is constructed without genetic recombination analysis. To place a marker on a physical map, the researcher first localizes it to an individual chromosome and then to the smallest possible subregion of the chromosome that can be resolved with present techniques. We discuss some examples of physical mapping in this section.

SOMATIC CELL HYBRIDIZATION. One way to localize a marker to an individual human chromosome is to use the **somatic cell hybridization** technique. In this technique cultured human cells are fused with cultured mouse cells to produce a *somatic cell hybrid.* The hybridization is set up using genetic markers and selection media so that the hybrid cells can grow and divide, but the parental cells cannot. Over the course of several divisions, the hybrid cells preferentially lose human chromosomes (the reason is not known) until eventually there is generated a *stable cell line* that contains all mouse chromosomes and a particular subset of human chromosomes. The number and particular set of human chromosomes vary with the cell line.

To map human genes by using hybrid somatic cell lines, we start with a human cell that carries one or more markers. Generally, we use markers that can readily be detected by analyzing tissue culture cells such as genes that control resistance to drugs, genes that code for enzymes or nonenzymatic proteins, or genes that determine nutritional requirements and DNA markers that can be detected by molecular analysis. We then analyze various stable hybrid cell lines for the presence or absence of the chosen markers. We correlate these data for a number of somatic cell lines with the presence or absence of particular chromosomes. Given enough cell lines, we can show that a particular marker is present only when one particular chromosome is present, and the marker is absent when that chromosome is absent. Markers shown to be linked to the same chromosome through this experimental approach are called **syntenic** ("together thread").

Many genes have been localized to individual chromosomes in the human genome using somatic cell hybridization. For example, through a combination of molecular biology techniques and classical human genetic linkage analysis, a particular DNA marker called G8 was found to be closely linked to the Huntington disease (*HD*) gene. Hybrid human-mouse cells were made and allowed to produce stable lines. Those lines that contained the G8 DNA marker were analyzed to determine their human chromosome composition with the following results:

Hybrid Cell Line	Human Chromosomes Present
WIL-5	4 17 18 21 X
WIL-6	4 5 6 7 8 10 11 14 17 19 20 21 X
NSL-15	2 4 5 7 8 12 13 14 15 17 18 19 21 22 X
ATR-13	1 2 3 4 5 6 7 8 10 12 13 14 15 16 17 18 19
XTR-22	2 4 5 6 8 10 11 18 19 20 21 22

If we look at the cell line with the fewest number of human chromosomes (WIL-5) and compare the chromosomes present with those in the other lines, we see that chromosome 4 is present in all five lines. We must be sure, though, that no other chromosome is also present in the five lines: chromosome 17 is not in XTR-22, chromosome 18 is not in WIL-6, chromosome 21 is not in ATR-13, and the X chromosome is not present in ATR-13 or XTR-22. Therefore, we are safe in concluding that the G8 DNA marker—and therefore the Huntington gene—is on chromosome 4.

In some instances further localization of genes to particular regions of chromosomes has been possible using cells with chromosomal mutations. For example, in some cells a part of a chromosome may spontaneously break off and be lost (*deleted*) or become attached to another chromosome (*translocated*). If either of these events occurs, we can form hybrid somatic cell lines involving the human cell with the chromosomal abnormality. The loss or addition of various pieces of a particular chromosome can then be correlated with the presence or absence of the genetic markers.

LOW-RESOLUTION PHYSICAL MAPPING: CHROMOSOMAL BANDING PATTERNS.

At the subchromosomal level *low-resolution physical mapping* has produced a map of human chromosomes with each chromosome characterized by a banding pattern visualized under the microscope after staining. (See Chapter 11, pp. 324–326 for a discussion of chromosome banding techniques.) The average size of a band is about 10 Mb. A standard nomenclature has been established for the chromosomes based on the banding patterns so that scientists can talk about gene and marker locations with reference to specific regions and subregions. Each chromosome has two arms separated by the centromere. The smaller arm is designated p and the larger arm is designated q. Numbered regions and numbered subregions are then assigned from the centromere outward; that is, region 1 is closest to the centromere. For example, the breast cancer susceptibility gene *BRCA1* is at location 17q21, meaning it is on the long arm of chromosome 17 in region 21.

MEDIUM- TO HIGH-RESOLUTION PHYSICAL MAPPING USING FISH.

A relatively new method allows individual eukaryotic chromosomes to be colored fluorescently at the location of specific gene or DNA sequences. In this method human metaphase chromosomes adhering to a microscope slide are treated to cause the two DNA strands to separate but stay in the same physical location. Specific DNA sequences are cloned and tagged with fluorescent chemicals. The tagged DNA sequences—called DNA probes—are added to the chromosomes where they will bind (hybridize) to the region from which they were cloned. This procedure is called *fluorescence in situ hybridization* and given the acronym FISH. (*In situ* hybridization is not a new technique. What is new is the use of fluorescent tags; the probes were traditionally labeled with tritium, a radioactive isotope of hydrogen.) In this way the chromosome sites corresponding to the probe can be identified by the fluorescent emissions of the tag. By using chemicals that fluoresce at different wavelengths, we can use a number of different probes in the same experiment. We can visualize where the probes have bound to the chromosomes by using fluorescence and digital imaging microscopy—essentially computer imaging analysis of the sample examined under a fluorescence microscope.

Figure 6.11 shows the results of FISH with six different probes. The probe colors are not the true colors from the fluorescence, but are pseudocolors generated by the computerized imaging device. The complete chromosomes are visualized by staining them with a chemical that colors all the DNA blue (in this case), making it possible to identify particular chromosomes on the basis of their size and morphology. Each chromosome to which a probe has hybridized has two dots because the DNA in metaphase chromosomes is already duplicated in preparation for cell division. For example, the yellow dots identify an uncharacterized DNA sequence on chromosome 5, the green dots identify the Down syndrome region on chromosome 21, and the red dots identify the Duchenne muscular dystrophy gene on the X chromosome.

With metaphase chromosomes, FISH methods have a resolution of 2–5 Mb for DNA markers. Thus, it

~ FIGURE 6.11

An example of the results of fluorescent *in situ* hybridization (FISH) in which fluorescently tagged DNA probes were hybridized to human metaphase chromosomes.

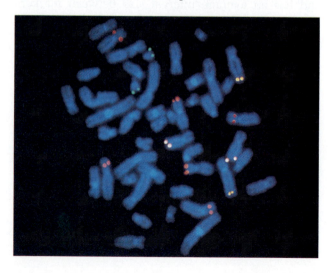

is possible to localize a marker to a subregion of an average chromosome band using FISH. A new improvement of the FISH method has made it possible to use chromosomes in interphase, where they are much more extended than they are in metaphase. With interphase FISH, map resolution is improved to approximately 50–500 kb. Even higher resolution physical mapping using FISH can be done with chromosomes that have been extended artificially; with this method markers can be resolved in a range of 5–700 kb.

HIGH-RESOLUTION PHYSICAL MAPS OF SE-QUENCE-TAGGED SITES. A **sequence-tagged site (STS)** is a short segment of DNA, about 60–1,000 bp long, that defines a unique position in the human genome. A particular STS is detected by using a procedure called the polymerase chain reaction (PCR), which is described in detail in Chapter 15, pp. 476–477. In brief, PCR permits the amplification of any segment of DNA that is flanked by a pair of DNA primers. Thus, what makes each STS unique is the sequences of the pair of primers used in the PCR. The amplified DNA is detected by gel electrophoresis (see Chapter 15, pp. 469–470).

One of the stated primary goals of the Human Genome Project is to construct a physical map of the human genome with 30,000 STSs by 1998. As of the end of 1996, 22,582 STSs have been mapped, providing sufficient information to begin an orderly effort to sequence the genome. Moreover, the STS map provides landmarks in the form of the STSs themselves (analogous to having a large number of individually identifiable pins on a printed map) that tell us exactly where we are in the genome and so enables us to "move around" the genome with reference to those landmarks. This is very useful to do, for example, in efforts to map genes with reference to STSs. Information about the STSs and the STS map itself has been made available via the World Wide Web to genome researchers worldwide, making it possible for any research group to explore regions of interest in the genome by starting from the nearest STSs.

Integration of Genetic Linkage Maps and Physical Maps

As the resolution of the physical maps increases, it will become easier to locate genes with respect to DNA markers and thereby integrate the genetic linkage map with the physical maps. Newly discovered genes will then be more easily mapped to their locations on the chromosomes. Several disease genes have already been cloned by homing in on their chromosomal location through the use of DNA markers, a process called **positional cloning.** An example—that

of cloning the gene responsible for cystic fibrosis—is described in Chapter 15, pp. 481–486.

As a summary, Figure 6.12 shows a map of human chromosomes, indicating some of the details of the banding patterns and presenting a selection of the known locations of mutations causing disease.

KEYNOTE

Extensive efforts are being made to map the human genome. Two types of maps are being constructed: genetic linkage maps and physical maps. Physical maps have different resolutions, depending on the technique being used. A high-resolution physical map of thousands of sequence-tagged sites that define unique positions in the genome has been constructed. The STSs are landmarks used as jumping-off points for further genome exploration and for sequencing efforts. Future integration of high-resolution physical maps with genetic linkage maps will make it easier to map newly discovered genes to their chromosome locations.

SUMMARY

In this chapter we learned how to map genes in certain microorganisms by using tetrad analysis. In a subset of those microorganisms, the meiotic tetrads are ordered in a way that reflects the orientation of the four chromatids of each homologous pair of chromosomes at metaphase I. Ordered tetrads make it possible to map a gene's location relative to its centromere. In this chapter we also learned about the rare incidence of crossing-over in mitosis and how mitotic recombination may be used in certain organisms to map genes.

A major research effort called the Human Genome Project is currently underway to map and sequence the human genome. Two types of maps are being constructed: genetic linkage maps and physical maps. Genetic linkage maps, constructed using genetic recombination analysis, show the relative positions of gene and/or DNA markers. Physical maps, constructed using non-recombination analysis, show the locations of DNA markers. The ultimate physical map will be the complete sequence of the human genome. Physical maps currently being constructed have different resolutions depending upon the method being used in their construction. In the future, high resolution physical maps will be integrated with detailed genetic linkage maps, making it easier to map newly discovered genes.

~ FIGURE 6.12

Genetic map of human chromosomes showing some of the details of the banding patterns. The map presents a selection of the known locations of mutations causing disease. Note: The chromosomes are not drawn to scale.

● DNA test currently available; ■ Gene mapped but not yet isolated; ○ DNA test being developed;
◆ Diagnosis available through family linkage studies of DNA markers

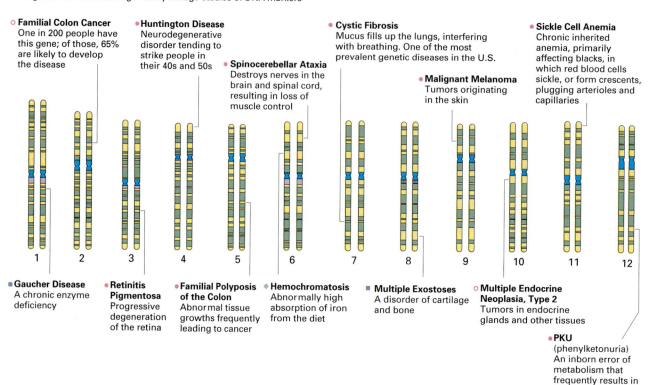

○ **Familial Colon Cancer**
One in 200 people have this gene; of those, 65% are likely to develop the disease

● **Huntington Disease**
Neurodegenerative disorder tending to strike people in their 40s and 50s

● **Spinocerebellar Ataxia**
Destroys nerves in the brain and spinal cord, resulting in loss of muscle control

● **Cystic Fibrosis**
Mucus fills up the lungs, interfering with breathing. One of the most prevalent genetic diseases in the U.S.

● **Malignant Melanoma**
Tumors originating in the skin

● **Sickle Cell Anemia**
Chronic inherited anemia, primarily affecting blacks, in which red blood cells sickle, or form crescents, plugging arterioles and capillaries

■ **Gaucher Disease**
A chronic enzyme deficiency

● **Retinitis Pigmentosa**
Progressive degeneration of the retina

● **Familial Polyposis of the Colon**
Abnormal tissue growths frequently leading to cancer

◆ **Hemochromatosis**
Abnormally high absorption of iron from the diet

■ **Multiple Exostoses**
A disorder of cartilage and bone

○ **Multiple Endocrine Neoplasia, Type 2**
Tumors in endocrine glands and other tissues

● **PKU**
(phenylketonuria)
An inborn error of metabolism that frequently results in mental retardation

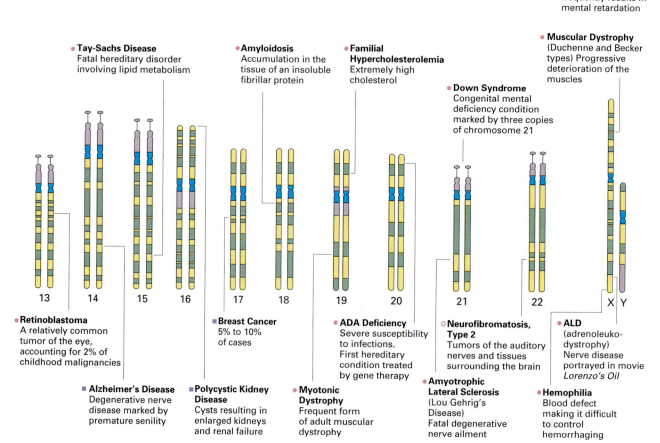

● **Tay-Sachs Disease**
Fatal hereditary disorder involving lipid metabolism

● **Amyloidosis**
Accumulation in the tissue of an insoluble fibrillar protein

● **Familial Hypercholesterolemia**
Extremely high cholesterol

● **Down Syndrome**
Congenital mental deficiency condition marked by three copies of chromosome 21

● **Muscular Dystrophy**
(Duchenne and Becker types) Progressive deterioration of the muscles

● **Retinoblastoma**
A relatively common tumor of the eye, accounting for 2% of childhood malignancies

■ **Breast Cancer**
5% to 10% of cases

● **ADA Deficiency**
Severe susceptibility to infections. First hereditary condition treated by gene therapy

○ **Neurofibromatosis, Type 2**
Tumors of the auditory nerves and tissues surrounding the brain

● **ALD**
(adrenoleukodystrophy)
Nerve disease portrayed in movie *Lorenzo's Oil*

■ **Alzheimer's Disease**
Degenerative nerve disease marked by premature senility

■ **Polycystic Kidney Disease**
Cysts resulting in enlarged kidneys and renal failure

● **Myotonic Dystrophy**
Frequent form of adult muscular dystrophy

● **Amyotrophic Lateral Sclerosis**
(Lou Gehrig's Disease)
Fatal degenerative nerve ailment

● **Hemophilia**
Blood defect making it difficult to control hemorrhaging

ANALYTICAL APPROACHES FOR SOLVING GENETICS PROBLEMS

Q6.1 A *Neurospora* strain that required both adenine (*ad*) and tryptophan (*trp*) for growth was mated to a wild-type strain (*ad⁺ trp⁺*), and this cross produced seven types of ordered tetrads in the frequencies shown:

Spore pair 1:	*ad trp*	*ad +*	*ad trp*	*ad trp*
Spore pair 2:	*ad trp*	*ad +*	*ad +*	*+ trp*
Spore pair 3:	*+ +*	*+ trp*	*+ trp*	*ad +*
Spore pair 4:	*+ +*	*+ trp*	*+ +*	*+ +*
TYPE	(1) 63	(2) 3	(3) 15	(4) 9

Spore pair 1:	*ad trp*	*ad +*	*+ trp*
Spore pair 2:	*+ +*	*+ trp*	*+ +*
Spore pair 3:	*ad trp*	*ad +*	*ad trp*
Spore pair 4:	*+ +*	*+ trp*	*ad +*
TYPE	(5) 3	(6) 1	(7) 6

a. Determine the gene-centromere distance for the two genes.

b. From the data given, calculate the map distance between the two genes.

A6.1a. The gene-centromere distance is given by the formula

$$\frac{\text{percent second-division tetrads}}{2} = x \text{ mu from centromere}$$

For the *ad* gene tetrads 4, 5, and 6 show second-division segregation; the total number of such tetrads is 13. There are 100 tetrads, so the *ad* gene is (13/2)% = 6.5 mu from its centromere. For the *trp* gene tetrads, 3, 5, 6, and 7 show second-division segregation, and the total number of such tetrads is 25, indicating the *trp* gene is 12.5 mu from its centromere.

b. The linkage relationship between the genes can be determined by analyzing the relative number of parental-ditype (PD), nonparental-ditype (NPD), and tetratype (T) tetrads. Tetrads 1 and 5 are PD, 2 and 6 are NPD, and 3, 4, and 7 are T. If two genes are unlinked, the frequency of PD tetrads will approximately equal the frequency of NPD tetrads. If two genes are linked, the frequency of PD tetrads will greatly exceed that of the NPD tetrads. Here the latter case prevails, so the two genes must be linked. The map distance between two genes is given by the general formula

$$\frac{1/2\,T + NPD}{\text{total}} \times 100$$

For this example the number of T tetrads is 30, and the number of NPD tetrads is 4. Thus the map distance between *ad* and *trp* is

$$\frac{(1/2 \times 30) + 4}{100} \times 100 = 19 \text{ mu}$$

Thus we have the following map, with ● indicating the centromere:

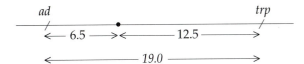

Q6.2 In *Aspergillus*, forced diploids were constructed between a wild-type strain and a strain containing the mutant genes *y* (yellow), *w* (white), *pro* (proline requirement), *met* (methionine requirement), and *ad* (adenine requirement). All these genes are known to be in a single chromosome.

Homozygous yellow and homozygous diploid white segregants were isolated and analyzed for the presence of the other gene markers. The following phenotypic results were obtained:

y/y segregants:	*w⁺ pro⁺ met⁺ ad⁺*	15
	w⁺ pro met⁺ ad⁺	28
w/w segregants:	*y⁺ pro⁺ met ad*	6
	y⁺ pro⁺ met⁺ ad	12

Draw a map of the chromosome, giving the order of the genes and the position of the centromere.

A6.2 In mitotic recombination a single crossover renders all gene loci distal to that point homozygous. In this regard the crossover events in one chromosome arm are independent of those in the other chromosome arm. Therefore we must inspect the data with these concepts in mind.

There are two classes of *y/y* segregants: wild-type segregants and proline-requiring segregants, which are homozygous for the *pro* gene. Thus of the four loci other than yellow, only *pro* is in the same chromosome arm as *y*. Furthermore, since not all the *y/y* segregants are *pro* in phenotype, the *pro* locus must be closer to the centromere than the *y* locus, as shown in the following map:

$$\underbrace{}_{y}\underbrace{}_{pro}●$$

A single crossover between *pro* and *y* will give *y/y* segregants that are wild-type for all other genes, whereas a single crossover between the centromere and *pro* will give homozygosity for both *pro* and *y*—hence, the *pro* requirement.

Similar logic can be applied to the w/w segregants. Again, there are two classes. Both classes are also phenotypically *ad*, indicating that the *ad* locus is further from the centromere than the *w* locus. Hence every time *w* becomes homozygous, so does *ad*. The remaining gene to be located is *met*. Some of the w/w segregants are *met+* and some are *met*, so the *met* gene is closer to the centromere than the *w* gene. The reasoning here is analogous to that for the placement of the *pro* gene in the other arm. Taking all the conclusions together, we have the following gene order:

QUESTIONS AND PROBLEMS

6.1 In *Saccharomyces*, *Neurospora*, and *Chlamydomonas*, what meiotic events give rise to PD, NPD, and T tetrads?

6.2 What important item of information regarding crossing-over can be obtained from tetrad analysis (as in *Neurospora*) but not from single-strand analysis (as in *Drosophila*)?

6.3 A cross was made between a pantothenate-requiring (*pan*) strain and a lysine-requiring (*lys*) strain of *Neurospora crassa*, and 750 random ascospores were analyzed for their ability to grow on a minimal medium (a medium lacking pantothenate and lysine). Thirty colonies subsequently grew. Map the *pan* and *lys* loci.

*6.4 In *Neurospora* the following crosses yielded the progeny as shown:

$$a^+ b \times a\, b^+ \rightarrow$$

981	$a^+ b$
1,000	$a\ b^+$
10	$a^+ b^+$
9	$a\ b$
2,000	

$$a^+ c \times a\, c^+ \rightarrow$$

850	$a^+ c$
833	$a\ c^+$
169	$a^+ c^+$
148	$a\ c$
2,000	

$$b^+ c \times b\, c^+ \rightarrow$$

850	$b^+ c$
850	$b\ c^+$
140	$b^+ c^+$
160	$b\ c$
2,000	

What is the probable gene order and what are the approximate map distances between adjacent genes?

6.5 Four different albino strains of *Neurospora* were each crossed to the wild type. All crosses resulted in half wild-type and half albino progeny. Crosses were made between the first strain and the other three with the following results:

1×2:	975 albino, 25 wild type
1×3:	1000 albino
1×4:	750 albino, 250 wild type

Which mutations represent different genes, and which genes are linked? How did you arrive at your conclusions?

*6.6 Genes *met* and *thi* are linked in *Neurospora crassa*; we wish to locate *arg* with respect to *met* and *thi*. From the cross *arg ++ × + thi met*, the following random ascospore isolates were obtained. Map these three genes.

arg thi met	26		*arg* + +	51
arg thi +	17		+ *thi* +	4
arg + *met*	3		+ + *met*	14
+ *thi met*	56		+ + +	29

6.7 Given a *Neurospora* zygote of the constitution shown in the figure below, diagram the significant events producing an ascus where the *A* alleles segregate at the first division and the *B* alleles segregate at the second division.

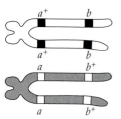

6.8 Double exchanges between two loci can be of several types, called two-strand, three-strand, and four-strand doubles.

a. Four recombination gametes would be produced from a tetrad in which the first of two exchanges is depicted in the figure below. Draw in the second exchange.

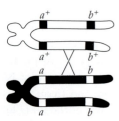

b. In the following figure, draw in the second exchange so that four nonrecombination gametes would result.

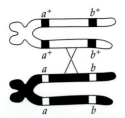

c. Other possible double-crossover types would result in two recombination and two parental gametes. If all possible multiple-crossover types occur at random, the frequency of recombination between genes at two loci will never exceed a certain percentage, regardless of how far apart they are on the chromosome. What is that percentage? Explain the reason why the percentage cannot theoretically exceed that value.

*6.9 A cross between a pink (p^-) yeast strain of mating type **a** and a grey strain (p^+) of mating type α produced the following tetrads:

18	p^+ **a**	p^+ **a**	p^- α	p^- α
8	p^+ **a**	p^- **a**	p^+ α	p^- α
20	p^+ α	p^+ α	p^- **a**	p^- **a**

On the basis of these results, are the p and mating type genes on separate chromosomes?

6.10 In *Neurospora* the peach gene (*pe*) is on one chromosome and the colonial gene (*col*) is on another. Disregarding the occurrence of chiasmata, what kinds of tetrads (asci) would you expect and in what proportions if these two strains are crossed?

6.11 The following unordered asci were obtained from the cross *leu* + × + *rib* in yeast. Draw the linkage map and determine the map distance.

110	45	6	39
leu +	*leu rib*	+ +	*leu* +
+ *rib*	*leu* +	*leu rib*	+ *rib*
leu +	+ +	*leu rib*	+ +
+ *rib*	+ *rib*	+ +	*leu rib*

*6.12 The genes *a*, *b*, and *c* are on the same chromosome arm in *Neurospora crassa*. The following ordered asci were obtained from the cross *a b* + × + + *c*

45	5	146	1
a b c	*a b* +	*a b* +	*a b* +
+ *b c*	*a* + +	*a b* +	+ + +
a + +	+ *b c*	+ + *c*	*a b c*
+ + *c*	+ + *c*	+ + *c*	+ + *c*

10	20	15	58
a b +	*a b* +	*a b* +	*a b* +
a + *c*	+ + *c*	*a b c*	+ *b* +
+ *b* +	*a b* +	+ + +	*a* + *c*
+ + *c*	+ + *c*	+ + *c*	+ + *c*

a. Determine the correct gene order.
b. Determine all gene-gene and gene-centromere distances.

*6.13 If 10 percent of the asci analyzed in a particular two-point cross of *Neurospora crassa* show that an exchange has occurred between two loci (i.e., exhibit second division segregation), what is the map distance between the two loci?

6.14 An *Aspergillus* diploid was obtained from a heterokaryon made by fusing together the haploid strains $a^+ b c i$ and $a b^+ c^+ i^+$, where the recessive allele *a* determines a requirement for growth substance A, the recessive allele *b* determines a requirement for growth substance B, the recessive allele *c* determines a requirement for growth substance C, and the recessive allele *i* determines *resistance* to an inhibitor I. When diploid asexual spores were plated on a medium lacking substance A, but containing the inhibitor I, 43 diploid colonies resistant to the inhibitor grew. Analysis of the colonies indicated that 17 required neither B nor C, 14 required C but not B, and 11 required both B and C. What is the relationship between the *b*, *c*, and *i* genes and the centromere?

*6.15 The frequency of mitotic recombination in experimental organisms can be increased by exposing them to low levels of ionizing radiation (such as X rays) during development. Hans Becker used this method to examine the patterns of clones produced by mitotic recombination in the *Drosophila* retina. What type of spots would be produced in the *Drosophila* retina if you were to irradiate a developing *Drosophila* female obtained from crossing a white-eyed male with a cherry-eyed female? (See Table 4.2 [p. 102] for a description of the *w* and wch alleles.)

6.16 The accompanying table shows the only human chromosomes present in stable human-mouse cell hybrid lines.

		HUMAN CHROMOSOMES			
		2	4	10	19
	A	−	+	+	−
Hybrid lines	*B*	+	−	+	+
	C	−	+	+	+
	D	+	+	−	−

The presence of four enzymes, I, II, III, and IV, was investigated: I was present in *A*, *B*, and *C* but absent in *D*; II was in *B* and *D* but absent in *A* and *C*; III was in *A*, *C*, and *D* but not in *B*; and IV was in *B* and *C* but not in *A* and *D*. On what chromosomes are the genes for the four enzymes?

7 CHROMOSOMAL MUTATIONS

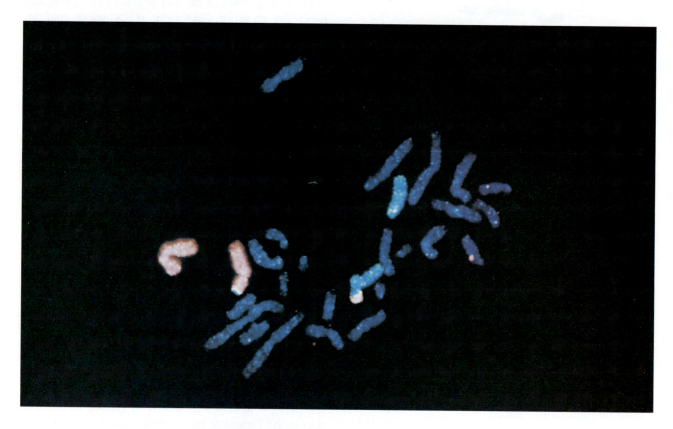

PRINCIPAL POINTS

~ Chromosomal mutations are variations from the wild-type condition in either chromosome number or chromosome structure. Chromosomal mutations can occur spontaneously, or they can be induced by chemical or radiation treatment.

~ Chromosomal mutations may involve parts of chromosomes rather than whole chromosomes. The four major types of such mutations are deletions (loss of a DNA segment), duplications (duplication of a DNA segment), inversions (change in orientation of a DNA segment within a chromosome without loss of DNA material), and translocations (movement of a DNA segment to another location in the genome without loss of DNA material).

~ Variations in the chromosome number of a cell or an organism give rise to aneuploidy, monoploidy, and polyploidy. Aneuploidy is the state in which the cell has one, two, or more whole chromosomes more or fewer than the basic number of the species involved. Monoploidy is the state in which a normally diploid cell has only a haploid set of chromosomes, and polyploidy is the state in which there are more than the normal number of sets of chromosomes present.

~ A change in chromosome number or chromosome structure may have serious, even lethal, consequences to the organism. In eukaryotes, abnormal phenotypes typically result from problems with chromosome segregation during meiosis, from gene disruptions at chromosome breakage sites, or from altered gene expression levels when gene dosage is altered. Some human tumors, for example, are associated with chromosomal mutations, either a change in the number of chromosomes, or a change in chromosome structure.

~ Some changes in chromosome organization occur naturally as mechanisms of altering gene expression, often as part of a developmental program. Examples of such changes are amplification of genes, deletion of a chromosome or part of a chromosome, inversion of a chromosome segment that results in a switch in transcription from one gene or set of genes to an alternative set, and transposition of a gene from a silent location in the genome to an active location where transcription then occurs.

In the previous six chapters we have learned many of the fundamental principles of transmission genetics, as applied to eukaryotes. With our understanding of the relationship between genes and chromosomes, we now consider chromosomal mutations, which are changes in normal chromosome structure or chromosome number. Chromosomal mutations affect both prokaryotes and eukaryotes, as well as viruses. In this chapter we will focus on chromosomal mutations in eukaryotes and discuss some of the human disease syndromes that result from chromosomal mutations. The association of genetic defects with changes in chromosome structure or chromosome number indicates that not all genetic defects result from simple mutations of single genes. In this chapter we also briefly discuss some examples of changes in chromosome organization that affect gene expression. (The molecular basis of gene expression will be covered in later chapters.) The study of chromosome behavior in normal cells and of how this behavior is altered in cells with chromosomal mutations involves both genetic and cytological approaches and hence falls into the field of *cytogenetics*.

TYPES OF CHROMOSOMAL MUTATIONS

With the exception of gametes, most cells of the same eukaryotic organism characteristically have the same number of chromosomes. (There are many exceptions, for example, gametes, endosperm cells in plants and so on. However, for the purposes of discussion these exceptions need not concern us.) Further, the organization and the number of genes on the chromosomes of an organism are the same from cell to cell. These characteristics of chromosome number and gene organization are the same for all members of the same species. Deviations are known as **chromosomal muta-**

tions (or **chromosomal aberrations**). Chromosomal mutations are *variations from the wild-type condition in either chromosome structure or chromosome number.* In both prokaryotes and eukaryotes, chromosomal mutations arise spontaneously or can be induced experimentally by chemical or radiation mutagens; they can be detected in both types of organisms by genetic analysis, that is, by changes in the linkage arrangements of genes. In eukaryotes, chromosomal mutations can also often be detected cytologically during mitosis and meiosis. The ability to detect different kinds of chromosomal mutations depends on the size, structure, and number of chromosomes involved. For example, under the microscope the chromosomes in the human karyotype show a range of sizes and morphologies (the latter related to centromere position) and a variety of distinctive banding patterns after different types of staining. (Karyotype analysis and chromosome banding are discussed more fully in Chapter 11.) This makes it possible to identify specific changes in chromosome structure (providing the changes are large enough to be seen under the microscope) or chromosome number.

We often have an impression that reproduction in humans occurs without significant problems affecting chromosome structure or number. After all, the vast majority of babies appear normal, as does the majority of the adult population. However, chromosomal mutations are more common than we once thought, and they contribute very significantly to spontaneously aborted pregnancies and stillbirths. For example, major chromosomal mutations are present in about half of spontaneous abortions, and a visible chromosomal mutation is present in about 6 out of 1,000 live births. Moreover, about 1 in 7 fertilizations ends in spontaneous abortion, so about 15 percent of all conceptions contain chromosomal mutations. Other studies have shown that about 11 percent of men with serious fertility problems and about 6 percent of people institutionalized with mental deficiencies have chromosomal mutations. Indeed, chromosomal mutations are significant causes of developmental disorders.

₭EYNOTE

Chromosomal mutations are variations from the wild-type condition in either chromosome number or chromosome structure. Chromosomal mutations

can occur spontaneously, or they can be induced by chemical or radiation treatment.

VARIATIONS IN CHROMOSOME STRUCTURE

Variations in chromosome structure involve changes in parts of chromosomes rather than changes in the number of chromosomes or sets of chromosomes in a genome. There are four common types of such mutations: *deletions* and *duplications* (both of which involve a change in the amount of DNA on a chromosome), *inversions* (which involve a change in the arrangement of a chromosomal segment), and *translocations* (which involve a change in the location of a chromosomal segment). Duplication, inversion, and translocation mutations can change back ("revert") to the wild-type state by a reversal of the process by which they were formed. However, deletion mutations cannot revert because a whole segment of chromosome is missing.

All four classes of chromosomal structure mutations are initiated by one or more breaks in the chromosome. If a break occurs within a gene, then a gene mutation has been produced, the consequence of which depends on the function of the gene and the time of its expression. Wherever the break occurs, the breakage process leaves broken ends without the specialized sequences found at the ends of chromosomes (the telomeres) that prevent degradation by exonucleases and "stickiness." As a result, the broken end of a chromosome is "sticky," meaning that it may adhere to other broken chromosome ends or even to the normal ends of other chromosomes. This stickiness property can help us understand the formation of the types of chromosomal structure mutations discussed in the following sections.

We have learned a lot about changes in chromosome structure from the study of **polytene chromosomes**, a special type of chromosome found in certain insects—for example, *Drosophila* (see Figure 17.22 and Chapter 17, pp. 562–564). Polytene chromosomes consist of chromatid bundles resulting from repeated cycles of chromosome duplication without nuclear or cell division. Polytene chromosomes may be a thousand times the size of corresponding chromosomes at meiosis or in the nuclei of ordinary somatic cells, and are easily detectable through microscopic examination. In each polytene chromosome, the two homologous chromosomes are tightly paired; therefore the number of polytene chromosomes per cell is half the diploid number of chromosomes. The number of duplicated copies of homologous chromosome pairs per polytene chromosome is species-specific.

As a result of the intimate pairing of the multiple copies of chromatids, characteristic banding patterns are easily seen, enabling cytogeneticists to identify unambiguously any segment of a chromosome. In *Drosophila melanogaster*, for example, more than 5,000 bands and interbands can be counted in the four polytene chromosomes. Each band was originally thought to represent a single protein-coding gene and the region between bands was thought to represent intergenic DNA. It is now known that each band contains an average of 30,000 base pairs (30 kb) of DNA, enough to encode several average-sized proteins. DNA cloning and sequencing studies have shown that many bands contain a number of genes (up to seven) that are transcribed independently. Genes are also found in the interbands. Polytene chromosomes will appear often in this chapter, as it is relatively easy to see the different chromosomal mutations in salivary gland chromosomes. The changes occur in most chromosomes and most organisms, but are most easily seen in *Drosophila* salivary gland chromosomes.

Deletion

A **deletion** is a chromosomal mutation involving the loss of a segment of a chromosome (Figure 7.1). The deleted segment may be located anywhere along the chromosome. A deletion starts where breaks occur in chromosomes. These breaks may be induced, for example, by agents such as heat, radiation (especially ionizing radiation: see Chapter 19), viruses, chemi-

cals, transposable elements (see Chapter 20), or by errors in recombination. Because a segment of chromosome is missing, deletion mutations cannot revert to the wild-type state.

The consequences of the deletion depend on the genes or parts of genes that have been removed. In diploid organisms the effects may be lessened by the presence of a copy of the lost set of genes in the homologous chromosome. However, if the homolog contains recessive genes with deleterious effects, the consequences can be severe. We are most knowledgeable about deletions that cause a detectable phenotypic change. If the deletion involves the loss of the centromere, for example, the result is an acentric chromosome, which is usually lost during meiosis. This leads to the deletion of an entire chromosome from the genome. Depending on the organism, this chromosome loss may have very serious or lethal consequences. For example, there are no known cases of living humans who have one whole chromosome of a homologous pair of autosomes deleted from the genome. (Recall from Chapter 3 that XO human females are viable.)

In organisms in which karyotype analysis (analysis of the chromosome complement: see Chapter 3, p. 77) is practical, deletions can be detected by that procedure, if the deficiencies are large enough. In that case, a mismatched pair of homologous chromosomes is seen with one shorter than the other. In heterozygous individuals, deletions result in unpaired loops when the two homologous chromosomes pair at meiosis. Cytological analysis can show such unpaired loops; the loops may often be readily seen in polytene chromosomes (Figure 7.2).

Deletions can be used to determine the physical location of a gene on a chromosome. In *Drosophila*, for example, the banding patterns of polytene chromosomes are very useful visible landmarks for such *deletion mapping* of genes (see also Chapter 8, pp. 250–252). The principle of the method is that the deletion of the dominant allele of a heterozygote results in the manifestation of the phenotype of the recessive allele. This unexpected expression of a recessive trait, caused by the absence of a dominant allele, is called **pseudodominance**. Figure 7.3 shows how Demerec and Hoover in 1936 used deletion mapping to localize genes to specific physical sites on *Drosophila* polytene chromosomes. The fly strain studied was heterozygous for the X-linked recessive mutations *y*, *ac*, and *sc*. Genetic analysis had shown that the three loci were linked at the left end of the X chromosome. The banding pattern of the left end of the X chromosome is shown in Figure 7.3a. The regions labeled A, B, and C

~ **FIGURE 7.1**

A deletion of a chromosome segment (here, *D*).

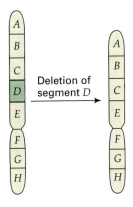

~ FIGURE 7.2

Cytological effects at meiosis of heterozygosity for a deletion. Paired *Drosophila* salivary gland polytene chromosomes in a strain with a heterozygous deletion, showing the unpaired region; numbers refer to bands, which are known to include genes.

Paired polytene chromosomes

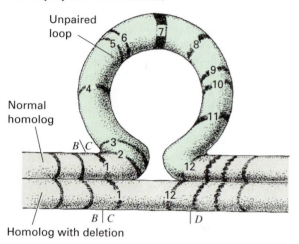

are major cytological subdivisions of the X chromosome and the numbers within each region refer to the chromosome bands. Recall from the discussion above that a single polytene chromosome is actually a tightly fused pair of homologous chromosomes. Deletions of this region of the chromosome were used to localize the gene loci. In strain 260-1, bands A1-7 and B1-4 are deleted, so that pseudodominance is observed for *y*, *ac*, and *sc*. The extent of the deletion is shown in Figure 7.3a, and the appearance of the polytene chromosomes in the deletion heterozygote is shown in Figure 7.3b. In strain 260-2, bands A1-7 and B1 are deleted from the chromosome bearing the wild-type alleles so that pseudodominance is observed for *y* and *ac*. The extent of this deletion is shown in Figure 7.3a also, and the appearance of the polytene chromosomes in the deletion heterozygote is shown in Figure 7.3c. Since the wild-type *sc* locus was lost in deletion strain 260-1, but it was not lost in deletion strain 260-2, *sc* must be located in the region of the X chromosome that distinguishes the two deletions; namely, bands B2-B4 (see Figure 7.3a). This method of analysis was used to construct the detailed physical map of *Drosophila* polytene chromosomes that has been so valuable for geneticists in the decades that have followed.

~ FIGURE 7.3

Use of deletions to determine the physical locations of genes on *Drosophila* polytene chromosomes. (a) Cytological appearance of the left end of the X chromosome heterozygous for the recessive mutations *y*, *ac*, and *sc* showing major regions A, B, and C, and the chromosome bands they contained. The region 260-1 shows the extent of a deletion which produced pseudodominance for *y*, *ac*, and *sc*, and region 260-2 shows the extent of a deletion which produced pseudodominance for *y* and *ac*. (b) Cytological appearance of the polytene X chromosome in flies heterozygous for the 260-1 deletion. These flies show pseudodominance for *y*, *ac*, and *sc*. (c) Cytological appearance of the polytene X chromosome in flies heterozygous for the 260-2 deletion. These flies show pseudodominance for *y* and *ac*.

a) Wild type

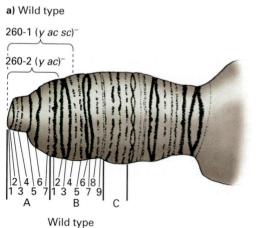

b) Polytene X chromosome in flies heterozygous for the 260-1 deletion

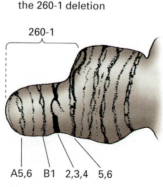

c) Polytene X chromosome in flies heterozygous for the 260-2 deletion

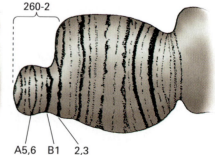

~ FIGURE 7.4

Cri-du-chat syndrome results from the deletion of part of one of the copies of human chromosome 5. (a) Karyotype of individual with cri-du-chat syndrome; (b) Photograph of a child with cri-du-chat syndrome.

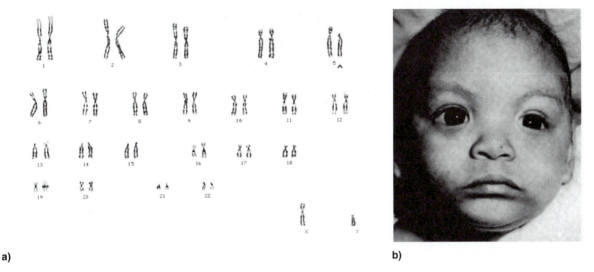

a)

b)

A number of human disorders are caused by deletions of chromosome segments. In many cases the abnormalities are found in heterozygous individuals; homozygotes for deletions are often lethal if the deletion is large. This points out that, in humans at least, the number of copies of genes is important for normal development and function. One human disorder caused by a heterozygous deletion is *cri-du-chat syndrome*, which results from an observable deletion of part of the short arm of chromosome 5, one of the larger human chromosomes (Figure 7.4). Children with cri-du-chat syndrome are severely mentally retarded, have a number of physical abnormalities, and cry with a sound like the mew of a cat (hence the name, which is French for "cry of the cat").

Another example is *Prader-Willi syndrome*, which results from heterozygosity for a deletion of part of the long arm of chromosome 15. Infants with this syndrome are weak because their sucking reflex is poor, making feeding difficult. As a result, growth is poor. By age 5 to 6, for reasons not yet understood, Prader-Willi children become compulsive eaters and this results in obesity and related health problems. If untreated, afflicted individuals can feed themselves to death. Other phenotypes associated with the syndrome include poor sexual development in males, behavioral problems, and mental retardation. Many individuals with the syndrome go undiagnosed, so its frequency of occurrence is not known.

Duplication

A **duplication** is a chromosomal mutation that results in the doubling of a segment of a chromosome (Figure 7.5). Duplications have played an important role in the evolution of gene families (see Chapter 22, p. 762). The size of the duplicated segment may vary

~ FIGURE 7.5

Duplication, with a chromosome segment (here, *BC*) repeated.

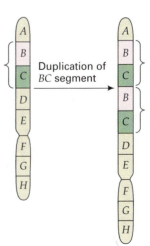

~ FIGURE 7.6

Forms of chromosome duplications are tandem, reverse tandem, and terminal tandem duplications.

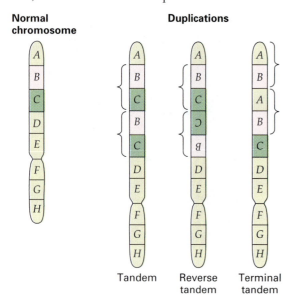

considerably, and duplicated segments may occur at different locations in the genome or in a *tandem* configuration (that is, adjacent to each other). When the order of genes in the duplicated segment is the opposite of the order of the original, it is a *reverse tandem*

duplication; when the duplicated segments are tandemly arranged at the end of a chromosome, it is a *terminal tandem duplication* (Figure 7.6). Heterozygous duplications result in unpaired loops similar to those described for chromosome deletions and, hence, may be detected cytologically.

Duplications of particular genetic regions can have unique phenotypic effects, as in the *Bar* mutant on the X chromosome of *Drosophila melanogaster*, first studied by Alfred Sturtevant and Thomas H. Morgan in the 1920s. In strains homozygous for the *Bar* mutation (not to be confused with the Barr body), the number of facets of the compound eye is fewer than for the normal eye (shown in Figure 7.7a), giving the eye a bar-shaped (slitlike) rather than an oval appearance (shown in Figure 7.7b). *Bar* resembles an incompletely dominant mutation since females heterozygous for *Bar* have more facets and hence a somewhat larger bar-shaped eye (sometimes referred to as "wide-*Bar*" eye) than do females homozygous for *Bar*. Males hemizygous for *Bar* have very small eyes like those of homozygous *Bar* females. Cytological examination of polytene chromosomes by Calvin Bridges and Herman J. Müller in the 1930s showed that the *Bar* trait is the result of a duplication of a small segment (16A) of the X chromosome (Figure 7.7b).

A duplication such as *Bar* probably arose as a result of a process called *unequal crossing-over* (see

~ FIGURE 7.7

Chromosome constitutions of *Drosophila* strains, showing the relationship between duplications of region 16A of the X chromosome and the production of reduced-eye size phenotypes: (a) Wild type; (b) Homozygous *Bar* mutant; (c) Homozygous *double-Bar* mutant.

Figure 7.8). Unequal crossing-over may result when homologous chromosomes pair inaccurately, perhaps because similar DNA sequences occur in neighboring regions of chromosomes. Crossing-over in the mispaired region results in gametes with a duplication or a deletion. Figure 7.8a shows how the duplicated 16A segment of the *Drosophila* X chromosome that is associated with *Bar* (Figure 7.7b) could have arisen.

Homozygous *Bar* flies are not perfectly true-breeding, since rarely (about 1 in 1,600 progeny) a reversion to wild type occurs. In the same frequency, new mutant flies, called *double-Bar*, are produced; their eyes are even more reduced than those of *Bar*. Cytologically, *double-Bar* mutants are homozygous for chromosomes with *three* copies of the 16A segment in contrast to the one copy in the wild type, and two copies in the *Bar* mutant (Figure 7.7c). As shown in Figure 7.8b, unequal crossing-over in a homozygous *Bar* strain can produce a gamete with a chromosome containing three copies of the 16A segment.

Duplications have played an important role in the evolution of multigene families. For example, hemoglobin molecules contain two copies each of two different subunits, the α-globin polypeptide and the β-globin polypeptide. At different developmental stages from the embryo to the adult, a human individual has different hemoglobin molecules assembled from different types of α-globin and β-globin polypeptides. The genes for each of the α-globin type of polypeptides are clustered together on one chromosome, while the genes for each of the β-globin type of polypeptides are clustered together on another chromosome. (See Chapter 17 for further discussion.) The sequences of the genes in the α-globin family are all very similar, as are the sequences of the β-globin family of genes. It is thought that each family evolved from a different ancestral gene by duplication and subsequent sequence divergence.

Inversion

An **inversion** is a chromosomal mutation that results when a segment of a chromosome is excised and then reintegrated in an orientation 180 degrees from the original orientation (Figure 7.9). When the inverted segment includes the centromere, the inversion is called a **pericentric inversion** (Figure 7.9a). When the inverted segment occurs on one chromosome arm and does not include the centromere, the inversion is called a **paracentric inversion** (Figure 7.9b). Whether the inversion is pericentric or paracentric has important consequences for the appearance of the chromosome and the behavior of inversions during crossing-over, as we will see.

In general, genetic material is not lost when an inversion takes place, although there can be

~ **FIGURE 7.8**

Unequal crossing-over and the *Bar* mutant of *Drosophila*. (a) Unequal crossing-over involving the 16A segment of the *Drosophila* X chromosome generates a chromosome with two copies of 16A (a duplication) and a chromosome with no copies of 16A (a deletion). A fly homozygous for two copies of 16A are *Bar* mutants. (b) Unequal crossing-over in a homozygous *Bar* mutant can produce a chromosome with three copies of 16A and a chromosome with one copy of 16A. Homozygosity for the chromosome with three copies of 16A is a *double-Bar* mutant, and homozygosity for the chromosome with one copy of 16A is wild type.

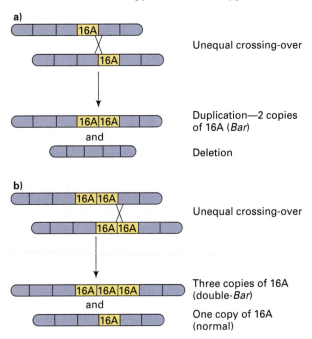

phenotypic consequences when the break points (inversion points) occur within genes or within regions that control gene expression. Also, since gene order may affect the regulation of gene expression, inversions can disrupt such regulation. Homozygous inversions can be seen because of the non-wild-type linkage relationships that result for the genes within the inverted segment and the genes that flank the inverted segment. For example, if the order of genes on the normal chromosome is *ABCDEFGH* and the *BCD* segment is inverted, the gene order will now be *ADCBEFGH* with *D* now more closely linked to *A* than to *E*, and *B* now more closely linked to *E* than to *A* (see Figure 7.9b).

The meiotic consequences of a chromosome inversion depend on whether the inversion occurs in a homozygote or a heterozygote. If the inversion is homozygous (e.g., *ADCBEFGH/ADCBEFGH*, where the *BCD* segment is the inverted segment), then meiosis will take place normally and there are no problems

~ **FIGURE 7.9**

Inversions. (a) Pericentric inversion; (b) Paracentric inversion.

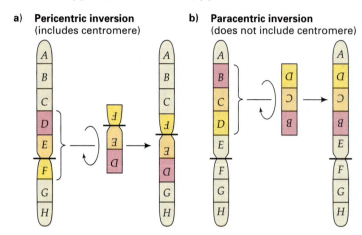

related to gene duplications or deletions. However, crossing-over within inversion heterozygotes (e.g., *ABCDEFGH/ACDBEFGH*, where one chromosome has an inverted *BCD* segment) has serious genetic consequences. Moreover, the recombinant chromosomes are different for crossing-over in paracentric as opposed to pericentric inversion heterozygotes. In such inversion heterozygotes, the homologous chromosomes attempt to pair so that the best possible base-pairing occurs. Because of the inverted segment on one homolog, pairing of homologous chromosomes requires the formation of loops containing the inverted segments, called inversion loops (Figure 7.10a; also see Figure 7.11). Inversion heterozygotes, then, may be identified by looking for those loops.

Heterozygous inversions can also be identified in genetic experiments, since recombination is reduced significantly or suppressed. Actually, the frequency of crossing-over is not diminished in inversion heterozygotes in comparison with normal cells, but gametes derived from recombined chromatids are inviable. The inviability results from an *unbalanced* set of genes in the gamete—one or more genes are missing and/or one or more genes are present in two copies instead of one (see Figure 7.10b). If no crossovers occur in the inversion loop of a heterozygous inversion, then all resulting gametes receive a complete set of genes (two gametes with a normal gene order, *ABCDEFGH*, and two gametes with the inverted segment, *ACDBEFGH*) and they are all viable. Figure 7.10b shows the effects of a single crossover in the inversion loop of an individual heterozygous for a paracentric inversion. During the first meiotic anaphase, the two centromeres migrate to opposite poles of the cell. Because of the crossover between genes *B* and *C* in the inversion loop, one

recombinant chromatid becomes stretched across the cell as the two centromeres begin to migrate in anaphase, forming a **dicentric bridge**, that is, a chromosome with two centromeres (a **dicentric chromosome**). With continued migration, the dicentric bridge breaks because of the tension. The other recombinant product of the crossover event is a chromosome without a centromere (an acentric fragment). This acentric fragment is unable to continue through meiosis and is usually lost (i.e., it is not found in the gametes).

In the second meiotic division, the chromosomes are segregated to the four gametes. Two of the gametes have complete sets of genes and are viable: the gamete with the normal order of genes (*ABCDEFGH*) and the gamete with the inverted segment of genes (*ACDBEFGH*). The other two gametes are inviable because they are unbalanced: many genes are deleted. Thus, *the only gametes that can give rise to viable progeny are those containing the chromosomes that were not involved in the crossover.* However, in many cases in female animals, the dicentric chromosomes or acentric fragments arising as a result of inversion are shunted to the polar bodies, so the reduction in fertility may not be so great.

The consequences of a single crossover in the inversion loop of an individual heterozygous for a pericentric inversion are shown in Figure 7.11. The crossover event and the ensuing meiotic divisions result in two viable gametes, with the nonrecombinant chromosomes *ABCDEFGH* (normal) and *ACDBEFGH* (inversion), and in two recombinant gametes that are inviable, each as a result of the deletion of some genes and the duplication of other genes.

Some crossover events within an inversion loop do not affect gamete viability. For example, a double crossover close together and involving the same two

~ FIGURE 7.10

Consequences of a paracentric inversion: (a) Photomicrograph of an inversion loop in polytene chromosomes of a strain of *Drosophila melanogaster* that is heterozygous for a paracentric inversion; (b) Meiotic products resulting from a single crossover within a heterozygous, paracentric inversion loop. Crossing-over occurs at the four-strand stage involving two nonsister homologous chromatids.

a)

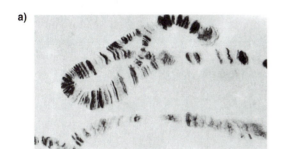

b) Products of meiotic crossover

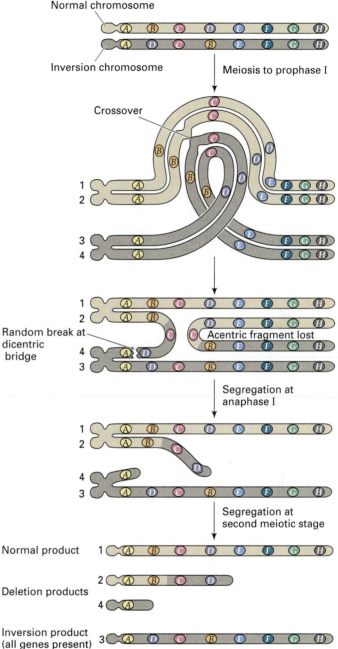

chromatids (a two-strand double crossover; see Chapter 6) produces four viable gametes. A second exception occurs when the duplicated and deleted segments of the recombinant chromatids do not affect gene expression and hence viability to a significant degree, as when the chromosome segments involved are very small. Also, recent studies with mammals show that inverted segments may remain unpaired. Since crossing-over cannot occur between unpaired segments, the generation of inviable gametes is avoided.

Translocation

A **translocation** is a chromosomal mutation in which there is a change in position of chromosome segments and the gene sequences they contain (Figure 7.12). There is no gain or loss of genetic material involved in a translocation. Two simple kinds of translocations occur. One kind involves a change in position of a chromosome segment within the same chromosome: this is called an *intrachromosomal* (within a chromo-

~ **FIGURE 7.11**

Meiotic products resulting from a single crossover within a heterozygous, pericentric inversion loop. Crossing-over occurs at the four-strand stage, involving two nonsister homologous chromatids.

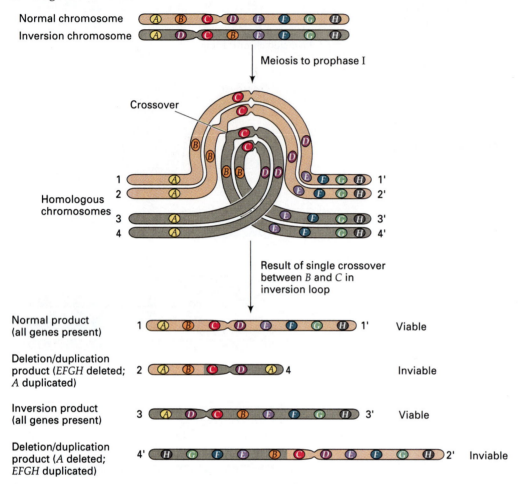

some) *translocation* (Figure 7.12a). The other kind involves the transfer of a chromosome segment from one chromosome into a nonhomologous chromosome: this is called an *interchromosomal* (between chromosomes) *translocation* (Figure 7.12b and c). If this latter translocation involves the transfer of a segment in one direction from one chromosome to another, it is a *nonreciprocal translocation* (Figure 7.12b); if it involves the exchange of segments between the two chromosomes it is a *reciprocal translocation* (Figure 7.12c).

In organisms homozygous for the translocations, the genetic consequence is an alteration in the linkage relationships of genes. For example, in the nonreciprocal intrachromosomal translocation shown in Figure 7.12a, the *BC* segment has moved to the other chromosome arm and has become inserted between the *F* and *G* segments. As a result, genes in the *F* and *G* segments are now farther apart than they are in the normal strain, and genes in the *A* and *D* segments are now more closely linked. Similarly, in reciprocal

translocations new linkage relationships are produced.

Translocations typically affect the products of meiosis. In many cases, some of the gametes produced are unbalanced in that they have duplications and/or deletions and, are, in many cases inviable. There are other cases in which gametes are viable, such as familial Down syndrome, resulting from a duplication stemming from a translocation (see later in the chapter). We focus here on reciprocal translocations, since they are the most frequent and the most important in genetic studies.

In strains *homozygous* for a reciprocal translocation, meiosis takes place normally, since all chromosome pairs can pair properly and crossing-over does not produce any abnormal chromatids. In strains *heterozygous* for a reciprocal translocation, however, all homologous chromosome parts pair as best they can. Because there is one set of normal chromosomes (N) and one set of translocated chromosomes (T)

~ FIGURE 7.12

Translocations: (a) Nonreciprocal intrachromosomal; (b) Nonreciprocal interchromosomal; (c) Reciprocal interchromosomal.

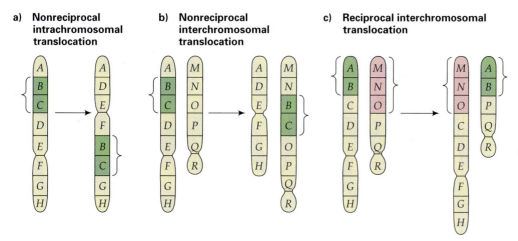

involved, the result is a crosslike configuration in meiotic prophase I (Figure 7.13). These crosslike figures consist of four associated chromosomes, each chromosome being partially homologous to two other chromosomes in the group.

Segregation at anaphase I may occur in three different ways. In one way, termed *alternate segregation*, alternate centromeres segregate to the same pole (Figure 7.13, left: N_1 and N_2 to one pole, T_1 and T_2 to the other pole). This produces two gametes, each of which is viable because it contains a complete set of genes, no more, no less. One of these gametes has two normal chromosomes and the other has two translocated chromosomes. In the second way, termed *adjacent-1 segregation*, adjacent *nonhomologous* centromeres migrate to the same pole (Figure 7.13, middle: N_1 and T_2 to one pole, N_2 and T_1 to the other pole). Both gametes produced contain gene duplications and deletions and are often inviable. Adjacent-1 segregation occurs about as frequently as alternate segregation. In the third way, termed *adjacent-2 segregation*, different pairs of adjacent *homologous* centromeres migrate to the same pole (Figure 7.13, right: N_1 and T_1 to one pole, N_2 and T_2 to the other pole). Both products have gene duplications and deletions and are always inviable. Adjacent-2 segregation seldom occurs. In sum, of the six theoretically possible gametes, the two from alternate segregation are functional, the two from adjacent-1 segregation are inviable (because of gene duplications and deficiencies), and the two from adjacent-2 seldom occur and are inviable if they do. And, since alternate segregation and adjacent-1 segregation occur with about equal frequency, the term *semisterility* is applied to this condition. (This term is also used for inversion heterozygotes.)

In practice, animal gametes that have duplicated or deleted chromosome segments may function, but the zygotes formed by such gametes typically die. The gametes may function normally and viable offspring may result if the duplicated and deleted chromosome segments are small. In plants, pollen grains with duplicated or deleted chromosome segments typically do not develop completely and hence are nonfunctional.

TRANSLOCATIONS AND HUMAN TUMORS. A number of human tumors have been shown to be associated with chromosomal mutations, either through a change in the number of chromosomes through nondisjunction, or through a change in chromosome structure(s) involving deletions, duplications, inversions, and translocations. It is not always clear whether the tumor is caused by chromosomal mutations, or whether the chromosomal mutation results from the growth activities of the tumor cell. There is, however, support for the first hypothesis. In some cases, for example, a tumor cell that exhibits a chromosomal abnormality early in its inception will develop other chromosomal mutations in time, and this often correlates with a progression to an uncontrolled growth state. Examples of tumors associated with consistent chromosome translocations are chronic myelogenous leukemia (reciprocal translocation involving chromosomes 9 and 22), and Burkitt's lymphoma (reciprocal translocation involving chromosomes 8 and 14).

Chronic myelogenous leukemia (CML) is an invariably fatal cancer involving uncontrolled replication of myeloblasts (stem cells for white blood cells). Ninety percent of chronic myelogenous leukemia patients have a chromosomal mutation in the leukemic cells called the *Philadelphia chromosome* (Ph^1), so named

~ FIGURE 7.13

Meiosis in a translocation heterozygote in which no crossover occurs.

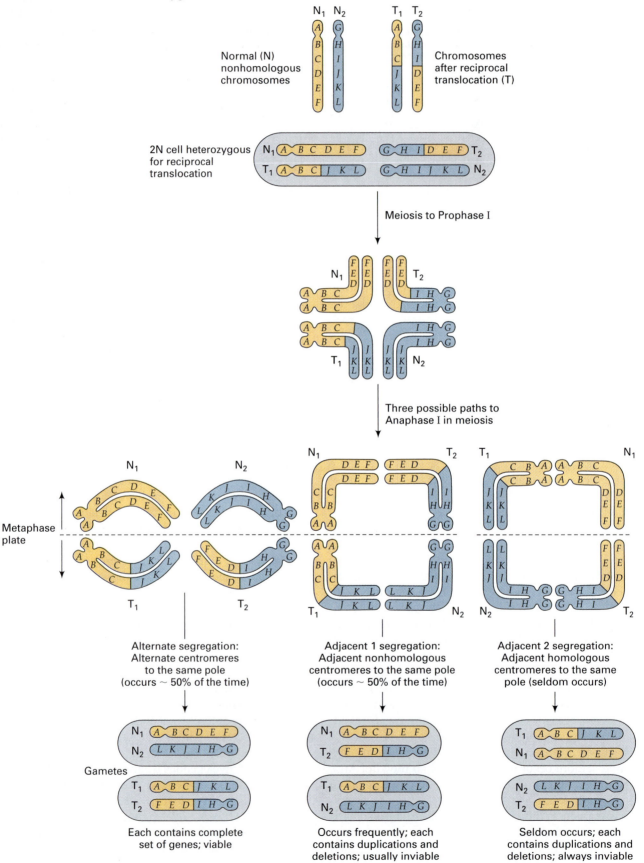

because the discovery was made in the city of Philadelphia. The Philadelphia chromosome results from a reciprocal translocation event in which a part of the long arm of chromosome 22 (the second smallest human chromosome) is translocated to chromosome 9, and a small part from the tip of chromosome 9 is translocated to chromosome 22 (Figure 7.14). Thus, chronic myelogenous leukemia actually results from two chromosomal mutations, one involving chromosome 22 and the other involving chromosome 9. This reciprocal translocation event apparently activates genes (called *oncogenes*; see Chapter 18), which initiate the transition from a differentiated cell to a tumor cell with an uncontrolled pattern of growth. Specifically, the *c-abl* ("c-able"—cellular *Abelson*) oncogene, normally located on chromosome 9, is translocated to chromosome 22 in CML patients (see

Figure 7.14). The translocation event positions *c-abl* within the *bcr* (breakpoint *cluster* region) gene. This hybrid oncogene arrangement somehow causes the uncontrolled expression of both genes, and this produces the leukemia.

Burkitt's lymphoma, a particularly common disease in Africa, is a virus-induced tumor that affects cells of the immune system called B cells. Characteristically, the malignant B cells secrete antibodies. Ninety percent of tumors in Burkitt's lymphoma patients are associated with a reciprocal translocation involving chromosomes 8 and 14 (Figure 7.15). As in the case of chronic myelogenous leukemia and the Philadelphia chromosome, there is evidence that a cancer-causing oncogene (in this case *c-myc* ["c-mick"]) becomes activated as a result of the reciprocal translocation event. The c-*myc* oncogene is

~ **FIGURE 7.14**

Origin of the Philadelphia chromosome in chronic myelogenous leukemia by a reciprocal translocation involving chromosomes 9 and 22. The arrows show the sites of the breakage points.

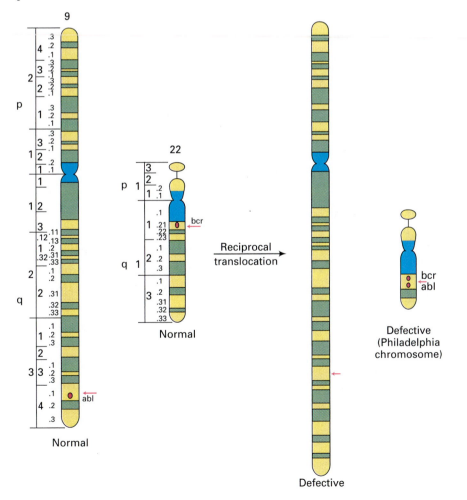

~ FIGURE 7.15

The common reciprocal translocation involving chromosomes 8 and 14 seen in Burkitt's lymphoma. The breakage points are indicated by arrows. The approximate locations of the *c-myc* oncogene and the *Ig-VCμ* immunoglobulin genes are indicated. (Adapted with permission from J. J. Yunis, "The Chromosomal Basis of Human Neoplasi" in *Science* 221 (1983). Copyright © 1983 American Association for the Advancement of Science.)

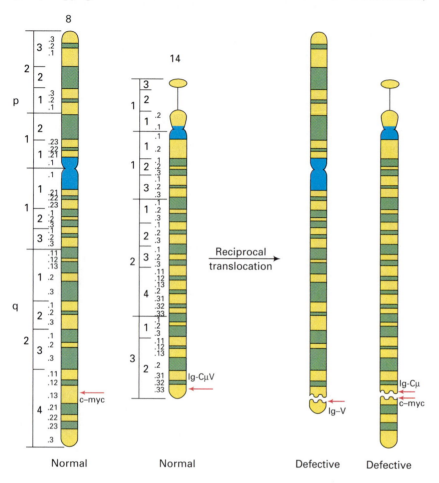

normally located on chromosome 8. In translocations involving chromosome 14, the oncogene plus a distal part of the chromosome exchanges with the end of the long arm of chromosome 14. The breakpoint on chromosome 14 occurs within the immunoglobulin genes *Ig-VCμ* so that *Ig-Cμ* now is near *c-myc*. This translocation event results in altered transcription of the *c-myc* gene and is associated with the transition of the cell in which the translocation has occurred to the tumor state. In addition, the translocation results in the expression of the immunoglobulin gene (*Ig-Cμ*) and this is responsible for the observed secretion of antibodies (immunoglobulin molecules) associated with the disease.

𝒦EYNOTE

Chromosomal mutations may involve parts of individual chromosomes rather than whole chromosomes or sets of chromosomes. The four major types of structural alterations are deletions and duplications (both of which involve a change in the amount of DNA on a chromosome), inversions (which involve no change in the amount of DNA on a chromosome but rather a change in the arrangement of a chromosomal segment), and translocations (which also involve no change in the amount

of DNA but involve a change in location of one or more DNA segments).

Position Effect

Unless inversions or translocations involve breaks within a gene, these chromosomal mutations do not produce mutant phenotypes. Rather, as we have seen, these mutations have significant consequences in meiosis when they are heterozygous with normal sequences. In some cases, however, phenotypic effects of inversions or translocations occur because of a different phenomenon called **position effect**; that is, a change in the phenotypic expression of one or more genes as a result of a change in position in the genome.

For example, position effect may be exhibited if a gene, normally located in euchromatin (chromosomal regions, representing most of the genome, that are condensed during division but that become uncoiled during interphase), is brought near heterochromatin (chromosomal regions that remain condensed throughout the cell cycle and are genetically inactive) by a chromosomal rearrangement. (Transcription of genes typically occurs in euchromatin but not in heterochromatin.) An example of this kind of position effect involves the X-linked white-eye (w) locus in *Drosophila*. One inversion moves the w^+ gene from a euchromatic region near the end of the X chromosome to a position next to the heterochromatin at the centromere of the X. In a w^+ male, or in a w^+/w female where the w^+ is involved in the inversion, the eye exhibits a mottled pattern of red and white rather than being completely red as expected. The explanation is that, in flies with the inversion, some eye-cell clones have the w^+ allele inactivated because of the position effect of w^+ near heterochromatin. Those clones produce white spots in the eyes. Clones of cells in which the w^+ allele is not inactivated produce red spots in the eye. Since the inactivation event is variable, the eye exhibits a mottled pattern of red and white spots.

Position effect of a different kind is seen for *Bar* eye of *Drosophila*. Recall from earlier in the chapter that a wild-type X chromosome has one copy of segment 16A, a *Bar* chromosome has two copies of 16A, and a *double-Bar* chromosome has three copies of 16A (see Figure 7.7). This allows experiments to be done to investigate the possibly different effects of segment 16A in various numerical and positional combina-tions. For example, through standard genetics, flies can be bred that are *Bar* homozygotes (each chromosome with two 16A segments) or are *double-Bar*/+ heterozygotes (one chromosome with three 16A segments and one chromosome with one 16A segment). In each case there is a total of four 16A segments. A study by Alfred Sturtevant in 1925 showed that these two chromosomal organizations do not produce the same physiological results. As Figure 7.16 shows, the average number of facets in the homozygous *Bar* strain is 68, while that in the *double-Bar*/+ heterozy-gote is 45. In some fashion three 16A segments side by side exert a greater effect on eye development than do the same number of segments when located on different chromosomes. This position effect phenome-non provided one of the earliest indications that rearrangements of chromosome segments can affect gene expression.

Fragile Sites and Fragile X Syndrome

When human cells are grown in culture, some of the chromosomes develop narrowings or gaps called *fragile sites*. Over forty fragile sites have been identi-

~ FIGURE 7.16

Position effect and *Bar* eye in *Drosophila*. In homozygous *Bar* females (left), there are four copies of X chromosome segment 16A, two on each homolog; these flies have an average of 68 facets in their reduced eyes. In females heterozygous for *double-Bar* and normal (right), there are also four copies of segment 16A, three on the *double-Bar* chromosome and one on the normal chromosome; even though the number of segments is the same, these flies have smaller eyes, with an average number of 45 facets.

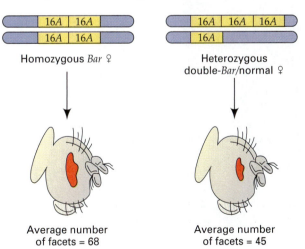

Homozygous *Bar* ♀   Heterozygous double-*Bar*/normal ♀

Average number of facets = 68   Average number of facets = 45

fied since the first one was discovered in 1965. One human condition related to a fragile site is *fragile X syndrome,* in which the X chromosome is prone to breakage. Discovered by Herbert Lubs in 1969, this particular fragile site is in the long arm of the X chromosome at position Xq27.3 as shown in Figure 7.17. After Down syndrome, fragile X syndrome is the leading cause of mental retardation in the United States, with an incidence of about 1 in 1,250 males and 1 in 2,500 females (heterozygotes). Since the X chromosome is involved, it is males who predominantly exhibit this type of mental retardation.

The fragile X chromosome is inherited as a typical Mendelian gene. Male offspring of carrier females have a 50 percent chance of receiving a fragile X chromosome. However, only 80 percent of those males receiving a fragile X chromosome are mentally retarded; the rest are normal. These phenotypically normal males are called *normal transmitting males* (NTMs) because they can pass on the fragile X chromosome to their daughters. The sons of those daughters frequently show symptoms of mental retardation. Female offspring of carrier (heterozygous) females also have a 50 percent chance of inheriting a fragile X

chromosome. Some of the carrier females (up to 33 percent) show mild mental retardation.

The localization of the fragile site responsible for fragile X syndrome to a particular region of the X chromosome was achieved using cytogenetics, linkage studies, and somatic cell analysis (see Chapter 6, p. 183). Modern molecular techniques brought to bear on this disease have resulted in an understanding of the disease at the level of DNA. There is a repeated three-base-pair sequence $\frac{CGG}{GCC}$ which we will simply call a CGG repeat, an example of a triplet repeat, in a gene called *FMR-1* (fragile X mental retardation-1) located at the fragile X site. Normal individuals have an average of 29 CGG repeats (range from 6 to 54) in the coding region of the *FMR-1* gene. Normal transmitting males and their daughters, as well as some carrier females, have a significantly larger number of CGG repeats, ranging from 55 to 200 copies. Since these individuals do not show symptoms of fragile X syndrome, the increased number of repeats they have is called premutations. Males and females with fragile X syndrome have even larger numbers of the CGG repeats, ranging from 200 to 1,300 copies; these are considered to be the full mutations. We may conclude, therefore, that there is a mechanism which results in an unusual amplification of the triplet repeat, CGG, in the *FMR-1* gene, and below a certain threshold number of copies (about 200 or fewer) no clinical symptoms result, while above that threshold number of copies (greater than 200) clinical symptoms are produced. Interestingly, amplification of the CGG repeats does not occur in males, only in females. Therefore, a normal transmitting male (who has the premutation) transmits his X chromosome to his daughter. During his daughter's meiosis the triplets amplify, and she can transmit the amplified X to her son. Thus, affected males inherit the mutation from their grandfather. As yet, the function of the *FMR-1* gene, in which the triplet repeat amplification occurs, is unknown so we do not yet understand how such amplification produces mental retardation.

Triplet repeat amplification as a cause of disease is not unique to fragile X syndrome. In the past two years triplet repeat amplification has been shown to cause myotonic dystrophy, spinobulbar muscular atrophy (also called Kennedy disease), and Huntington disease (see Chapter 4). Each of these cases differs from fragile X syndrome in that the amplification can occur in both sexes at each generation. Again, there is a threshold number of triplet repeat copies above which symptoms of the disease are produced.

~ FIGURE 7.17

(a) Scanning electron micrograph and (b) diagram of a human X chromosome showing the location of the fragile site responsible for fragile X syndrome. (From Gerald Stine, *The New Human Genetics.* Copyright © 1989. Reproduced by permission of The McGraw-Hill Companies.)

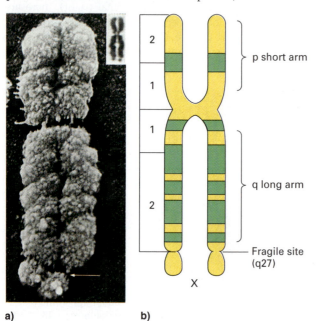

VARIATIONS IN CHROMOSOME NUMBER

When an organism or cell has one complete set of chromosomes, or an exact multiple of complete sets, that organism or cell is said to be **euploid**. Thus, eukaryotic organisms that are normally diploid (e.g., humans, fruit flies), and eukaryotic organisms that are normally haploid (e.g., yeast) are all euploids. Chromosome mutations that result in variations in the number of chromosome sets occur in nature and the resulting organism or cells are also euploid. Chromosome mutations may also occur which result in variations in the number of individual chromosomes or in variations in parts of chromosomes; these conditions are examples of **aneuploidy**. An aneuploid organism or cell has a chromosome number that is not an exact multiple of the haploid set of chromosome, or chromosomes with a part or parts duplicated or deleted (the latter were considered earlier in this chapter). Both euploid and aneuploid variations affecting whole chromosomes will be discussed in this section.

Changes in One or a Few Chromosomes

While we will focus our discussions on eukaryotes and, in particular, on diploid eukaryotes, changes in chromosome number can also occur in haploid organisms. Nondisjunction of one or more chromosomes during meiosis I or meiosis II is typically responsible for generating gametes with abnormal numbers of chromosomes. Nondisjunction was discussed in Chapter 3 in the context of unusual complements of X chromosomes, with Figure 3.27 (p. 73) illustrating the consequences of nondisjunction at the first and second meiotic divisions. Referring to Figure 3.27 and considering just one particular chromosome, it can be seen that nondisjunction at meiosis I produces four abnormal gametes; that is, two gametes with a chromosome duplicated and two gametes with the corresponding chromosome missing. Nondisjunction at meiosis I can produce in a male a gamete with both the X and the Y chromosome and in a female a gamete with both sets of homologs (and thus possible heterozygotes in a gamete). Fusion of the former gamete type with a normal gamete will produce a zygote with three copies of the particular chromosome instead of the normal two and, unless nondisjunction has involved other chromosomes, two copies of all other chromosomes. Similarly, fusion of the latter gamete type with a normal gamete will produce a zygote with only one copy of the particular chromosome instead of the normal two, and two copies of all other chromosomes. Nondisjunction in meiosis II (see Figure 3.27) results in two normal gametes and two abnormal gametes; that is, a single gamete with two sister chromosomes and one gamete with that same chromosome missing. Fusion of these with normal gametes will give the zygote types just discussed. Unlike the case with nondisjunction at meiosis I, some normal gametes are produced by nondisjunction at meiosis II; specifically, two of the four gametes are normal. More complicated gametic chromosome compositions result when more than one chromosome is involved in nondisjunction, and/or when nondisjunction occurs in both meiotic divisions. Further, nondisjunction can occur in mitosis, giving rise to somatic cells with unusual chromosome complements.

ANEUPLOIDY. In aneuploidy, one or several chromosomes are lost from or added to the normal set of chromosomes (Figure 7.18). In most cases, aneuploidy is lethal in animals, so in mammals it is detected mainly in aborted fetuses. As we will see below, plants are more often aneuploid. The condition can occur, for example, from the loss of individual chromosomes in mitosis or meiosis. In this case nuclei are formed with fewer than the normal number of chromosomes. Aneuploidy may also result from nondisjunction, the irregular distribution of sister chromatids during mitosis or of homologous chromosomes during meiosis (see Chapter 3). In nondisjunction, one progeny nucleus with more and one with less than the normal number of chromosomes are produced.

In diploid organisms, aneuploid variations take four main forms (see Figure 7.18):

1. **Nullisomy** (a nullisomic cell) involves a loss of one homologous chromosome pair; that is, the cell is 2N − 2. (This could arise, for example, if nondisjunction occurs for the same chromosome in meiosis in both parents producing gamete each with no copies of that chromosome, and one copy of all other chromosomes in the set.)
2. **Monosomy** (a monosomic cell) involves a loss of a single chromosome; that is, the cell is 2N − 1. (This can arise, for example, if nondisjunction in meiosis in parent produces a gamete with no copies of a particular chromosome, and one copy of all other chromosomes in the set; see Figure 3.27, p. 73.)
3. **Trisomy** (a trisomic cell) involves a single extra chromosome; that is, the cell has three copies of one chromosome type and two copies of every other chromosome type. A trisomic cell is 2N + 1.

~ FIGURE 7.18

Normal (theoretical) set of metaphase chromosomes in a diploid (2N) organism (top) and examples of aneuploidy (bottom).

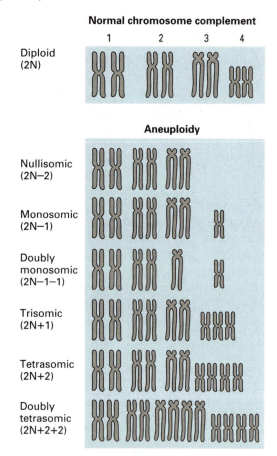

Normal chromosome complement

Diploid (2N)

Aneuploidy

Nullisomic (2N−2)

Monosomic (2N−1)

Doubly monosomic (2N−1−1)

Trisomic (2N+1)

Tetrasomic (2N+2)

Doubly tetrasomic (2N+2+2)

~ FIGURE 7.19

Meiotic segregation possibilities in a trisomic individual of genotype $+/+/a$, when two chromosomes migrate to one pole and one goes to the other pole, and assuming no crossing-over between the a locus and its centromere.

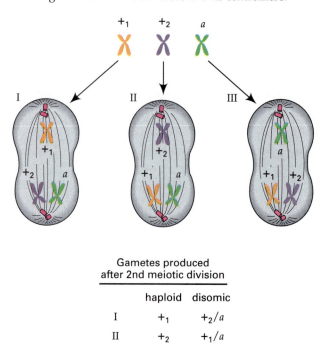

Gametes produced after 2nd meiotic division

	haploid	disomic
I	$+_1$	$+_2/a$
II	$+_2$	$+_1/a$
III	a	$+_1/+_2$

In sum: $2 +/a : 2 + : 1 +/+ : 1 a$

(This can arise, for example, if nondisjunction in meiosis in parent produces a gamete with two copies of a particular chromosome, and one copy of all other chromosomes in the set; see Figure 3.27.)

4. **Tetrasomy** (a tetrasomic cell) involves an extra chromosome pair, resulting in the presence of four copies of one chromosome type and two copies of every other chromosome type. A tetrasomic cell is 2N + 2. (This could arise, for example, if nondisjunction occurs for the same chromosome in meiosis in both parents producing gamete each with two copies of that chromosome, and one copy of all other chromosomes in the set.)

Aneuploidy may involve the loss or the addition of more than one specific chromosome or chromosome pair. For example, a *double monosomic* has two

chromosomes present in only one copy each; that is, it is 2N − 1 − 1, and a *double tetrasomic* has two chromosomes present in four copies each; that is, it is 2N + 2 + 2. In both of these cases, one explanation is that meiotic nondisjunction involved two different chromosomes in one parent's gamete production.

All forms of aneuploidy have serious consequences in meiosis. Monosomics, for example, will produce two kinds of haploid gametes, N and N − 1. Alternatively, the odd, unpaired chromosome in the 2N − 1 cell may be lost during meiotic anaphase and not be included in either daughter nucleus, thereby producing two N − 1 gametes. There are more segregation possibilities for trisomics in meiosis. Consider a trisomic of genotype $+/+/a$ in an organism that can tolerate trisomy, and assume no crossing-over between the a locus and its centromere. As shown in Figure 7.19, random segregation of the three types of chromosomes produces four genotypic classes of gametes as follows: 2 (+ a) : 2 (+) : 1 (+ +) : 1 (a). In a cross of a $+/+/a$ trisomic to an a/a individual, the predicted phenotypic ratio among the progeny is 5 wild type : 1 mutant (*a*). This ratio is seen in many actual crosses of this kind. However, such crosses in tomato, corn, and some other plants do not produce results conforming to the predictions. The reasons for these exceptions are not completely known, but they could involve crossing-over between the gene locus and its centromere, or decreased ability of *n* + 1 pollen to function in fertilization.

In the following sections we examine some examples of aneuploidy as they are found in the human population. Table 7.1 presents a summary of various aneuploid abnormalities for autosomes and for sex chromosomes in the human population. Examples of aneuploidy of the X and Y chromosomes were discussed in Chapter 3. Recall that, in mammals, aneuploidy of the sex chromosomes is more often found than aneuploidy of the autosomes because of a dosage compensation mechanism (lyonization) by which excess X chromosomes are inactivated.

In humans, autosomal monosomy is only found rarely. Presumably, monosomic embryos do not develop significantly and are lost early in pregnancy. This early lethality, in some cases, may be related to unmasking of recessive lethals (see Chapter 4, p. 116) In contrast, autosomal trisomy accounts for about one-half of chromosomal abnormalities producing fetal deaths. In fact, only a few autosomal trisomies are seen in live births. Most of these (trisomy 8, 13, and 18) result in early death. Only in trisomy 21 (Down syndrome) does survival to adulthood occur.

~ TABLE 7.1

Aneuploid Abnormalities in the Human Population

CHROMOSOMES	SYNDROME	FREQUENCY AT BIRTH
Autosomes		
Trisomic 21	Down	14.3/10,000
Trisomic 13	Patau	2/10,000
Trisomic 18	Edwards	2.5/10,000
Sex Chromosomes, Females		
XO, monosomic	Turner	4/10,000 females
XXX, trisomic	Viable; most are fertile	
XXXX, tetrasomic		14.3/10,000 females
XXXXX, pentasomic		
Sex Chromosomes, Males		
XYY, trisomic	Normal	25/10,000 males
XXY, trisomic		
XXYY, tetrasomic	Klinefelter	40/10,000
XXXY, tetrasomic		

Trisomy-21.   **Trisomy-21** occurs when there are three copies of chromosome 21, as shown in the karyotype in Figure 7.20a. Trisomy-21 occurs with a frequency of about 3,510 per 1 million conceptions and about 1,430 per 1 million live births. Individuals with trisomy-21

~ FIGURE 7.20

Trisomy-21 (Down syndrome): (a) Karyotype; (b) Individual.

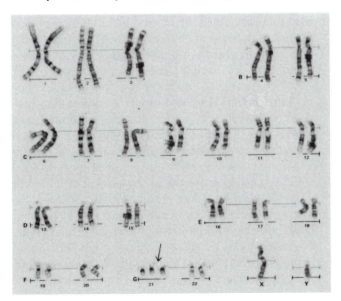

a)

b)

have Down syndrome and exhibit such abnormalities as low IQ, epicanthal folds over the eyes, short and broad hands, and below-average height. Figure 7.20b is a photograph of a Down syndrome individual.

Trisomy-21 is probably more common than other trisomies because chromosome 21 is a very small chromosome with fewer genes than most of the other chromosomes. The disruptions caused by the extra dose of chromosome 21 genes are less severe than would be the case with larger autosomes (nonsex chromosomes). Trisomies for all autosomes have been detected among fetuses that are aborted spontaneously, and it is from such fetuses that we have obtained knowledge about this sort of chromosomal mutation. Such studies have also revealed that all autosomal monosomies in humans are lethal.

The relationship between the age of the mother and the probability of her having a trisomy-21 individual is indicated in Table 7.2. (The correlation with age of the father is much more tenuous.) During the development of a female child before birth, the eggs in the ovary go through meiosis and arrest in prophase I. In a fertile female, each month at ovulation the nucleus of a secondary oocyte (see Chapter 3) begins the second meiotic division but progresses only to metaphase, when division again stops. If a sperm penetrates the secondary oocyte, the second meiotic division is completed. The probability of nondisjunction increases with the length of time the egg is in the ovary. It is important, then, that older mothers-to-be consider testing to determine whether the fetus has a normal complement of chromosomes, for example, by conducting amniocentesis or chorionic villus sampling (see Chapter 9).

Down syndrome individuals can also result from a different sort of chromosomal mutation called **Robertsonian translocation,** which produces three copies of the long arm of chromosome 21. This form of Down syndrome is called *familial Down syndrome.*

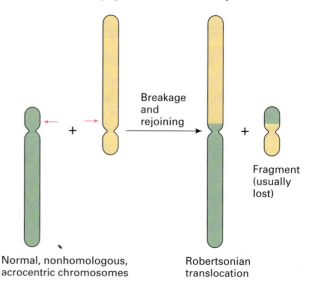

~ FIGURE 7.21

Production of a Robertsonian translocation by chromosome breakage and rejoining involving two acrocentric chromosomes. The breakage points are indicated by arrows.

Breakage and rejoining

Fragment (usually lost)

Normal, nonhomologous, acrocentric chromosomes

Robertsonian translocation

A Robertsonian translocation is a type of nonreciprocal translocation in which the long arms of two nonhomologous acrocentric chromosomes (chromosomes with centromeres near their ends) become attached to a single centromere (Figure 7.21). The short arms of the two acrocentric chromosomes become attached to form the reciprocal product, which typically contains nonessential genes and is usually lost within a few cell divisions. In humans, when a Robertsonian translocation joins the long arm of chromosome 21 with the long arm of chromosome 14 (or 15), the heterozygous carrier is phenotypically normal, since there are two copies of all major chromosome arms and hence two copies of all essential genes involved.

However, there is a high risk of Down syndrome among the offspring of these carriers. Figure 7.22 shows the gametes produced by the heterozygous carrier parent and by the normal parent and the chromosome constitutions of the offspring generated. The normal parent produces gametes with one copy each of the relevant chromosomes 14 and 21. The heterozygous carrier parent produces three reciprocal pairs of gametes, each pair produced as a result of different segregation of the three chromosomes involved. As Figure 7.22 shows, the zygotes produced by pairing these gametes with gametes of normal chromosomal constitution are as follows: one-sixth have normal chromosomes 14 and 21; one-sixth are heterozygous carriers like the parent and are phenotypically normal; one-sixth are inviable because of monosomy for

~ TABLE 7.2

Relationship Between Age of Mother and Risk of a Trisomy-21

AGE OF MOTHER	RISK OF TRISOMY-21 IN CHILD
16–26	7.7/10,000
27–34	4/10,000
35–39	29/10,000
40–44	100/10,000
45–47	333/10,000
All mothers combined	15/10,000

chromosome 14; one-sixth are inviable because of monosomy for chromosome 21; one-sixth are inviable because of trisomy for chromosome 14; and one-sixth are trisomy-21 and therefore produce a Down syndrome individual. (Note that these latter individuals actually have the normal diploid number of 46 chromosomes but because of the Robertsonian translocation, they have three copies of the long arm of chromosome 21, which is sufficient to give the Down syndrome symptoms. Similarly, the trisomy-14 zygotes shown in Figure 7.22 have 46 chromosomes, but they have three copies of the long arm of chromosome 14. Apparently the dosage of the genes involved on this larger chromosome is more critical, and consequently, trisomy-14 individuals are inviable.) In sum, one-half of the zygotes produced are inviable and theoretically one-third of the viable zygotes will give rise to a familial Down syndrome individual—a much higher risk than that for nonfamilial Down syndrome associated with the mother's age. In practice, the observed risk is lower.

Trisomy-13. Trisomy-13 produces Patau syndrome (Figure 7.23). About 2 in 10,000 live births produce individuals with trisomy-13. Characteristics of individuals with trisomy-13 include cleft lip and palate, small eyes, polydactyly (extra fingers and toes), mental and developmental retardation, and cardiac anomalies, among many other abnormalities. Most die before the age of three months.

Trisomy-18. Trisomy-18 produces Edwards syndrome (Figure 7.24). It occurs in about 2.5 in 10,000 live births. For reasons that are not known, about 80 percent of Edwards syndrome infants are female. Individuals with trisomy-18 are small at birth and have multiple congenital malformations affecting almost every organ system in the body. Clenched fists, elongated skull, low-set malformed ears, mental and developmental retardation, and many other abnormalities are associated with the syndrome. Ninety percent of infants with trisomy-18 die within six months, often from cardiac problems.

Changes in Complete Sets of Chromosomes

Monoploidy and **polyploidy** involve variations from the normal state in the number of complete sets of chromosomes. Since the number of complete sets of chromosomes is involved in each case, monoploids and polyploids are both euploids, as described earlier in the chapter. Monoploidy and polyploidy are lethal

~ **FIGURE 7.22**

The three segregation patterns of a heterozygous Robertsonian translocation involving the human chromosomes 14 and 21. Fusion of the resulting gametes with gametes from a normal parent produces zygotes with various combinations of normal and translocated chromosomes.

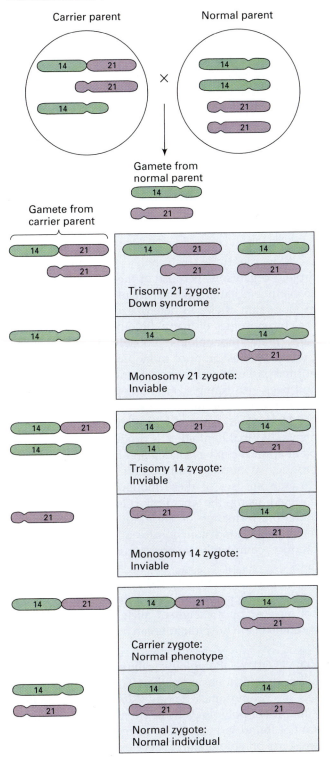

~ FIGURE 7.23

Trisomy-13 (Patau syndrome): (a) Karyotype; (b) Individual.

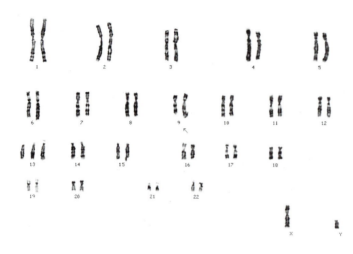

a)

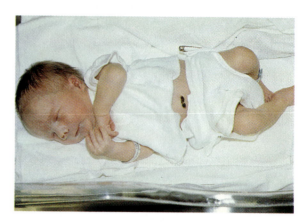

b)

for most animal species, but are tolerated more readily by plants. Both have played significant roles in plant speciation.

Changes in complete sets of chromosomes can result, for example, when meiotic nondisjunction occurs such that all chromosomes are involved. If such nondisjunction occurs at meiosis I, one-half of the gametes will have no chromosome sets, and the other half will have two chromosome sets (refer to Figure 3.27, p. 73). If such nondisjunction occurs at meiosis II, one-half of the gametes will have the normal one set of chromosomes, one-quarter will have two sets of chromosomes, and one-quarter will have no chromosome sets (see Figure 3.27). Fusion of a gamete with two chromosome sets with a normal gamete will produce a polyploid zygote, in this case one with three sets of chromosomes, which is a *triploid* (3N). Similarly, fusion of two gametes, each with two chromosome sets, will produce a *tetraploid* (4N) zygote. Polyploidy of somatic cells can also occur following mitotic nondisjunction of complete chromosome sets. Monoploid individuals, by contrast, typically develop from unfertilized eggs.

MONOPLOIDY. A monoploid individual has only one set of chromosomes instead of the usual two sets (Figure 7.25a). Monoploidy is sometimes called haploidy, although haploidy is typically used to describe the chromosome complement of gametes. Monoploidy is typical of gametes, some fungi, and males of haplo/diplo species (ants, bees, wasps).

Monoploidy is seen only rarely in normally diploid organisms. Because of the presence of recessive lethal mutations (which are usually counteracted by dominant alleles in heterozygous individuals) in the chromosomes of many diploid eukaryotic organisms, many monoploids probably do not survive. Certain species produce monoploid organisms as a normal part of their life cycle. Some male wasps, ants, and bees, for example, are monoploid because they develop from unfertilized eggs.

Monoploids are used in plant-breeding experiments. Normal haploid cells produced by meiosis in plant anthers can be isolated and induced to grow to produce monoploid cultures for study. The single chromosome set of these monoploids then can be doubled using the chemical colchicine (which inhibits the formation of the mitotic spindle, thereby resulting in nondisjunction for all chromosomes) to produce completely homozygous diploid breeding lines.

Cells of a monoploid individual are very useful for the production of mutants because mutations are not masked by a dominant allele on a homologous chromosome. Thus, mutants can be isolated directly.

POLYPLOIDY. Polyploidy is the situation in which a cell or organism has more than its normal number of sets of chromosomes (Figure 7.25b). Again the best examples of polyploids come from plants. Polyploids are not found in most living animals, although there are some polyploid animal species such as the North

~ FIGURE 7.24

Trisomy-18 (Edwards syndrome): (a) Karyotype; (b) Individual.

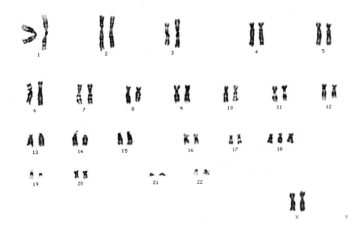

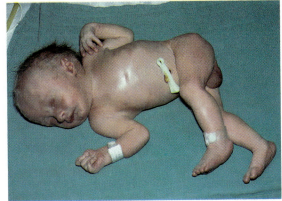

a)

b)

~ FIGURE 7.25

Variations in number of complete chromosome sets: (a) Monoploidy (one set of chromosomes instead of two); (b) Polyploidy (more than the normal number of chromosomes).

Normal chromosome complement

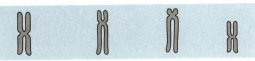

Diploid (2N)

a) Monoploidy
(only one set of chromosomes)

Monoploid (N)

b) Polyploidy
(more than the normal number of sets of chromosomes)

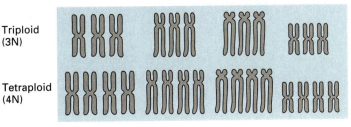

Triploid (3N)

Tetraploid (4N)

American sucker (a freshwater fish), salmon, and some salamanders.

Polyploids may arise spontaneously or be induced experimentally. They often occur as a result of a breakdown of the spindle apparatus in one or more meiotic divisions or in mitotic divisions. For example, the drug colchicine inhibits the formation of the mitotic spindle. When this drug is administered to somatic cells, the chromosomes duplicate but mitotic nondisjunction occurs so that the chromosomes cannot be distributed to progeny nuclei. As a result, a cell is produced with twice the normal number of sets of chromosomes.

There are two general classes of polyploids: those that have an *even* number of chromosome sets and those that have an *odd* number of sets. Polyploids with an even number of chromosome sets have a better chance of being at least partially fertile, since there is the potential for homologs to pair during meiosis. Polyploids with an odd number of chromosome sets always have an unpaired chromosome for each chromosome type, so the probability of a balanced gamete is extremely low; such organisms are usually sterile.

Let us consider the consequences of polyploidy on gene segregation to the offspring. In triploids, the nucleus has a complete set of trisomics. As a result, triploids are very unstable in meiosis because, like in trisomics, two of the three homologous chromosomes go to one pole, and the other goes to the other pole. What happens for each triple set of chromosomes in the genome is independent, so the probability of producing balanced gametes with either a haploid or diploid set of chromosomes is small; many of the gametes will be unbalanced with one copy of one chromosome, two copies of another, and so on. In general, the probability of a triploid producing a haploid gamete is $(1/2)^n$, where n is the number of chromosomes. Genotypically, a triploid organism with genotype $+/+/a$ for a locus near the centromere, would produce haploid and diploid gametes in the ratio 2 (+) : 1 (+ +) : 2 (+ a) : 1 (a) (Figure 7.26). (This is like the segregation pattern for trisomics described earlier.) Selfing such an organism would produce diploid, triploid, and tetraploid offspring with a phenotypic ratio of 35 + : 1 a (see Figure 7.26). Of course, much more complicated segregation patterns will be seen if there is heterozygosity for genes on the other chromosomes.

In tetraploids, the nucleus has four copies of each chromosome. Tetraploids are relatively stable in meiosis because all chromosomes have a chance to pair, unlike the situation in triploidy. A tetraploid organism with the heterozygous genotype $+/+/a/a$,

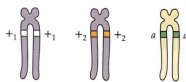

~ **FIGURE 7.26**

Production of gametes in a triploid that has the genotype $+/+/a$, and the production of offspring in a 35 + : 1 a phenotypic ratio upon selfing.

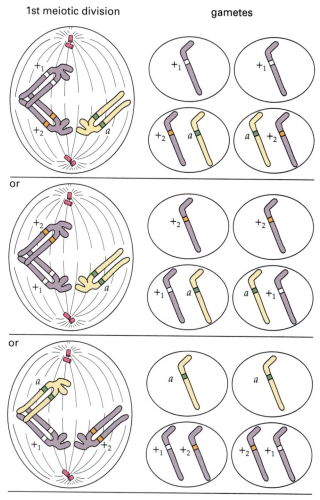

3 homologous chromosomes in triploid

1st meiotic division   gametes

or

or

In sum: gametic ratio = 4+ : 2+/+ : 4+/a : 2a
                       = 2+ : 1+/+ : 2+/a : 1a

	2+	1+/+	2+/a	1a
2+	4+/+	2+/+/+	4+/+/a	2+/a
1+/+	2+/+/+	1+/+/+/+	2+/+/+/a	1+/+/a
2+/a	4+/+/a	2+/+/+/a	4+/+/a/a	2+/a/a
1a	2+/a	1+/+/a	2+/a/a	1a/a

selfing of triploid

phenotypes = 35+ : 1a

for example, would produce diploid gametes in the ratio 1 (+ +) : 4 (+ a) : 1 (a a) (Figure 7.27). Selfing such a tetraploid will produce offspring with a phenotypic ratio of 35 + : 1 a (see Figure 7.27).

In humans, the most common type of polyploidy is triploidy. Triploidy is always lethal. Triploidy is seen in 15 to 20 percent of spontaneous abortions and in about 1 in 10,000 live births, but most die within one month. Triploid infants have many abnormalities, including a characteristic enlarged head. Tetraploidy in humans is always lethal, usually before birth. It is seen in about 5 percent of spontaneous abortions. Very rarely a tetraploid human will be born, but such an individual does not survive long.

Plants are more "tolerant" of polyploidy for two reasons. First, sex determination is less sensitive to polyploidy in plants than in animals. Second, many plants undergo self-fertilization so if a plant is produced with an even polyploid number of chromosome sets (e.g., 4N) it can still produce fertile gametes and reproduce.

Two types of polyploidy are encountered in plants. In **autopolyploidy** all the sets of chromosomes originate in the same species. The condition probably results from a defect in meiosis, which leads to diploid or triploid gametes. If a diploid gamete fuses with a normal haploid gamete, the zygote and the organism that develops from it will have three sets of chromosomes: it will be triploid. The cultivated banana is an example of a triploid autopolyploid plant. Because it has an odd number of chromosome sets, the gametes have a variable number of chromosomes and few fertile seeds are set, thereby making most bananas seedless and highly palatable. Because of the triploid state, cultivated bananas are propagated vegetatively (i.e., by cuttings). In general, the development of "seedless" fruits relies on "odd-number" polyploidy. Triploidy has also been found in grasses, garden flowers, crop plants, and forest trees.

In **allopolyploidy** the sets of chromosomes involved come from different, though usually related, species. This situation can arise if two different species interbreed to produce an organism with one haploid set of each parent's chromosomes and then both chromosome sets double. For example, fusion of haploid gametes of two diploid plants that can cross may produce an $N_1 + N_2$ hybrid plant which will have a haploid set of chromosomes from plant 1 and a haploid set from plant 2. However, because of the differences between the two chromosome sets, pairing of chromosomes does not occur at meiosis and no viable gametes are produced. As a result, the hybrid plants

~ **FIGURE 7.27**

Production of gametes in a tetraploid that has the genotype +/+/a/a, and the production of offspring in a 35 + : 1 a phenotypic ratio upon selfing.

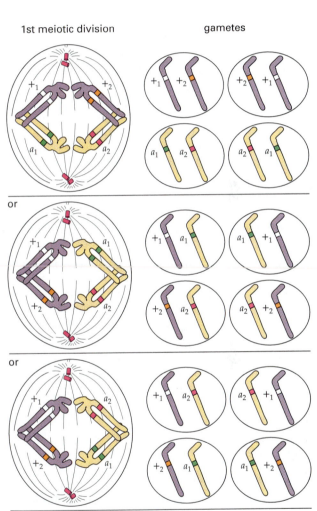

In sum: gametic ratio = 2+/+ : 8+/a : 2a/a
= 1+/+ : 4+/a : 1a/a

	1+/+	4+/a	1a/a
1+/+	1+/+/+/+	4+/+/+/a	1+/+/a/a
4+/a	4+/+/+/a	16+/+/a/a	4+/a/a/a
1a/a	1+/+/a/a	4+/a/a/a	1a/a/a/a

selfing of autotetraploid

phenotypes = 35 + : 1a

are sterile. Rarely, through a division error, the two sets of chromosomes double, producing tissues of $2N_1 + 2N_2$ genotype. (That is, the cells in the tissue have a diploid set of chromosomes from plant 1 and a diploid set from plant 2.) Each diploid set can function normally in meiosis, so that gametes produced from the $(2N_1 + 2N_2)$ plant are $(N_1 + N_2)$. Fusion of two gametes like this can produce fully fertile, allotetraploid, $2N_1 + 2N_2$ plants.

A classic example of allopolyploidy resulted from crosses made between cabbages (*Brassica oleracea*) and radishes (*Raphanus sativus*) by Karpechenko in 1928. Both parents have a chromosome number of 18, and the F_1 hybrids also have 18 chromosomes, 9 from each parent. These hybrids are morphologically intermediate between cabbages and radishes. The F_1 plants are mostly sterile because of the failure of chromosomes to pair at meiosis. However, a few seeds are produced through meiotic errors and some of those seeds are fertile. The somatic cells of the plants produced from the seeds have 36 chromosomes; that is, full diploid sets of chromosomes from both the cabbage and the radish. These plants are completely fertile; this breeding species has been named *Raphnobrassica*, a fusion of the two parental genus names.

Lastly, all commercial grains, most crops, and many common commercial flowers are polyploid. In fact, polyploidy is the rule rather than the exception in agriculture. For example, the cultivated bread wheat, *Triticum aestivum*, is an allohexaploid with 42 chromosomes. This plant species is descended from three distinct diploid species, each with a diploid set of 14 chromosomes. Meiosis is normal because only homologous chromosomes pair, so the plant is fertile.

*K*EYNOTE

Variations in the chromosome number of a cell or an organism give rise to aneuploidy, monoploidy, and polyploidy. In aneuploidy a cell or organism has one, two, or a few whole chromosomes more or less than the basic number of the species under study. In monoploidy an organism that is usually diploid has only one set of chromosomes. And in polyploidy an organism has more than the normal number of complete sets of chromosomes. Any or all of these abnormal conditions may have serious consequences to the organism.

CHROMOSOME REARRANGEMENTS THAT ALTER GENE EXPRESSION

Up to this point we have discussed chromosomal mutations; that is, abnormal alterations in the organization of genes in the chromosomes. Now we will describe some chromosomal rearrangements that occur in nature as mechanisms of altering gene expression. Many of these chromosomal rearrangements are part of the normal developmental program of an organism, and thus belong more in the area of developmental biology than genetics.

Amplification or Deletion of Genes

The programmed amplification or deletion of genes is found widely in nature. Amplification may involve all or a large part of the genome, or only a small set of genes. The former is exemplified by polytenization of chromosomes in insect salivary glands and the latter by the amplification of the 18S, 5.8S, and 28S rRNA transcription units in frog eggs. Programmed deletions may involve a systematic loss of whole chromosomes (e.g., the loss of an X chromosome in some organisms as a way to achieve the same dosage compensation that other organisms acquire through X-chromosome inactivation), or the loss of one or a few genes.

Gene amplification generally leads to the synthesis of more gene product, not to a change in expression of the genes involved. The effects of deletions on gene expression vary, however. If a gene is completely deleted, then the expression of that gene is abolished, but deletions can also include the regions adjacent to genes that are part of the system for regulating the expression of that gene and, hence, lead to either an increase or decrease in gene expression.

Inversions That Alter Gene Expression

Inversion as a mechanism of regulation of gene expression is rare, only involving a few genes in bacteria and their viruses. One way that an inversion can function as a regulator is as follows: A single promoter flanks a piece of DNA with genes *A* and *B* in opposite orientations. If the *A* gene is near the promoter, then the *A* gene is expressed. If the segment of DNA inverts, the *B* gene is brought next to the promoter, and the *B* gene is expressed instead of the *A* gene. Thus, the inversion event acts as a genetic switch. This type of inversion control of gene expression is seen in phage Mu, in which two types of tail fibers can be made, enabling the phage to infect more than one type of bacterial host.

Transpositions That Alter Gene Expression

Transpositions are movements of DNA segments from one location to another in the genome. Programmed transpositions that alter gene expression have been described for a number of systems, two of which will be mentioned here.

1. *Antigenic variation in Trypanosomes.* Trypanosomes are parasitic protozoa that can multiply for long periods in the bloodstream of mammalian hosts. The parasite evades the host immune defense system by repeatedly altering its surface coat so that it avoids attack by antibodies. The surface coat consists of a single glycoprotein species, and changes result from the transposition of a silent copy of a new surface protein gene to the expression site, at which surface protein genes can be transcribed. With at least 103 silent surface protein genes in the genome, an extremely large repertoire of possible surface coats is possible.

2. *Mating-type switching in yeast.* Yeast has two mating types, **a** and α (see Chapter 6). Mating type is controlled by the DNA sequence present at the mating types locus, *MAT*. Under genetic control, haploid cells may frequently switch mating type during growth. Figure 7.28a shows the organization of mating-type genes on chromosome III of an **a** cell. The α phenotype is specified by a 2.5-kb (1 kb = 1,000 bp) segment of the *MAT*α locus. An identical copy of the 2.5-kb, α-specific sequence is present on chromosome III at a locus called *HML*α. This DNA is not expressed because associated with it is a regulatory sequence *HMLE* that represses transcription; that is, *HML*α is a "silent," unexpressed copy of the *MAT*α sequence. On the opposite side of the active *MAT*α locus is a locus called *HMR***a**, which contains all of the DNA necessary for the mating-type **a** phenotypes. *HMR***a** transcription is repressed by an associated silencing sequence, *HMRE*.

   Mating-type switching in yeast involves replacing the mating-type sequence at the active *MAT* locus with a *copy* of a sequence with opposite mating-type information. Thus, in an α cell, a copy of the *HMR***a** sequence replaces the active sequence at *MAT*. Since there is no repression of genes at the *MAT* locus, the *HMR***a** sequence at that locus (now called *MAT***a**) is expressed and this leads to the cell expressing the mating-type **a** phenotype (Figure 7.28b). The next switch of the cell replaces the active **a** sequence at *MAT* with a copy of the *HML*α to produce an active *MAT*α mating-type locus again; the cell now exhibits the mating-type α phenotype (Figure 7.28c).

~ FIGURE 7.28

(a) Organization of mating-type genes on chromosome III of a mating-type α yeast cell. The *HMLE* sequence represses transcription of *HML*α, and the *HMRE* sequence represses transcription *HMR***a**. (b) Mating-type switching from α to **a**; (c) Mating-type switching from **a** to α.

a)

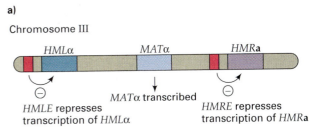

Chromosome III

HMLα   MATα   HMRa

⊖ HMLE represses transcription of HMLα

MATα transcribed

⊖ HMRE represses transcription of HMRa

b)

Mating-type **a** sequence is copied from *HMR***a**

HMLα   MATa   HMRa

⊖

MATa transcribed

⊖

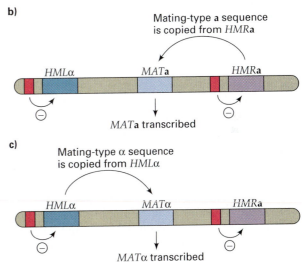

c)

Mating-type α sequence is copied from *HML*α

HMLα   MATα   HMRa

⊖

MATα transcribed

⊖

KEYNOTE

Some chromosome organizational changes occur in nature as mechanisms for altering gene expression. Often these changes function as genetic switches during development. Examples of such changes include amplification of the entire genome, amplification of one or a few genes, deletion of a chromosome or part of a chromosome, inversion of a chromosome segment causing a switch of transcription from one gene to another, and transposition of a gene from a silent location to an active location, where transcription can occur.

SUMMARY

In this chapter we have considered several kinds of chromosomal mutations. Chromosomal mutations are defined as variations from the wild-type conditions in either chromosome structure or chromosome number. They may occur spontaneously or their frequency can be increased by exposure to radiation or to chemical mutagens. Variations in chromosome structure are exemplified by those mutations in which changes from the normal state occur in parts of individual chromosomes rather than whole chromosomes or sets of chromosomes in the genome. The four major types of structural mutations are (1) deletion, in which a chromosome segment is lost; (2) duplication, in which more copies of a chromosome segment are present than in the normal state; (3) inversion, in which the orientation of a chromosome segment is opposite that of the wild type; and (4) translocation, in which a chromosome segment has moved to a new location in the genome. The consequences of these kinds of structural mutations depend on the specific mutation involved. Firstly, each kind of mutation involves one or more breaks in a chromosome. If a break occurs within a gene, then a gene mutation has been produced, the consequence of which depends on the function of the gene and the time of its expression. In the case of deletions, genes may be lost and multiple mutant phenotypes may result. In some cases deletions and duplications result in lethality or severe defects as a consequence of a deviation from the normal gene dosage.

Secondly, chromosomal mutations in the heterozygous condition can result in production of some gametes that are inviable because of duplications and/or deficiencies. Commonly this is seen following meiotic crossovers that produce inversions and translocations in heterozygotes.

Variations in chromosome number involve departure from the normal diploid (or haploid) state of the organism. For diploids, three classes of such mutations are observed: (1) aneuploidy, in which one to a few whole chromosomes are lost from or added to the normal chromosome set; (2) monoploidy, in which only one set of chromosomes is present in a usually diploid organism; and (3) polyploidy, in which a cell or organism has more than its normal number of sets of chromosomes. The consequences of these chromosomal mutations depend on the organism. In general, plants are more tolerant than animals of variations in the number of chromosome sets; for example, wheat is hexaploid. While some animal species are naturally polyploid, in the majority of instances monoploidy and polyploidy are lethal, probably because gene expression problems occur when abnormal numbers of gene copies are present. Even in viable individuals, viable gametes may not result because of segregation problems during meiosis.

Lastly, we mentioned a few examples of natural systems in which changes in chromosome organization are associated with altered gene expression. These changes include amplification of chromosomes or chromosome segments, deletions, inversions, and transpositions. A number of these changes function as genetic switches during development.

ANALYTICAL APPROACHES FOR SOLVING GENETICS PROBLEMS

Q7.1 Diagram the meiotic pairing behavior of the four chromatids in an inversion heterozygote $a\,b\,c\,d\,e\,f\,g$/$a'\,b'\,f'\,e'\,d'\,c'\,g'$. Assume that the centromere is to the left of gene a. Next, diagram the early anaphase configuration if a crossover occurred between genes d and e.

A7.1 This question requires a knowledge of meiosis and the ability to draw and manipulate an appropriate inversion loop. Part (a) of the figure to the right shows the diagram for the meiotic pairing.

Note that the lower pair of chromatids (a', b', etc.) must loop over in order for all the genes to align; this looping is characteristic of the pairing behavior expected for an inversion heterozygote.

Once the first diagram has been constructed, answering the second part of the question is straightforward. We diagram the crossover, then trace each chromatid from the centromere end until the other end is

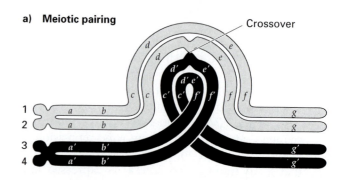

a) Meiotic pairing

Crossover

b) Early anaphase

Dicentric bridge

Acentric fragment lost

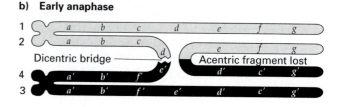

reached. It is convenient to distinguish maternal and paternal genes, perhaps by a' versus a, and so on, as we did in part (a) in the figure. The result of the crossover between d and e is shown in part (b).

In anaphase I of meiosis the two centromeres, each with two chromatids attached, migrate toward the opposite poles of the cell. At anaphase the noncrossover chromatids (top and bottom chromatids in the figure) segregate to the poles normally. As a result of the single crossover between the other two chromatids, however, unusual chromatid configurations are produced, and these configurations are found by tracing the chromatids from left to right. If we begin by tracing the second chromatid from the top, we get $_\circ a\ b\ c\ d\ e'\ f'\ b'\ a'_\circ$, a dicentric chromatid (where $\circ$ is a centromere); in other words, we have a single chromatid attached to two centromeres. This chromatid also has duplications and deletions for some of the genes. Thus during anaphase this so-called dicentric chromosome becomes stretched between the two poles of the cells as the centromeres separate, and the chromosome will eventually break at a random location. The other product of the single crossover event is an acentric fragment (without a centromere) that can be traced starting from the right with the second chromatid from the top. This chromatid $(g\ f\ e\ d'\ c'\ g')$ contains neither a complete set of genes nor a centromere—it is an acentric fragment that will be lost as meiosis continues.

Thus the consequence of a crossover event within the inversion in an inversion heterozygote is the production of gametes with duplicated or deleted genes. These gametes often will be inviable. Viable gametes are produced, however, from the noncrossover chromatids: One of these chromatids (1 in part [b] of the figure) has the normal gene sequence, and the other (3 in part [b] of the figure) has the inverted gene sequence.

Q7.2 Eyeless is a recessive gene (ey) on chromosome 4 of *Drosophila melanogaster*. Flies homozygous for ey have tiny eyes or no eyes at all. A male fly trisomic for chromosome 4 with the genotype $+/+/ey$ is crossed with a normal, eyeless female of genotype ey/ey. What are the expected genotypic and phenotypic ratios that would result from random assortment of the chromosomes to the gametes?

A7.2 To answer this question, we must apply our understanding of meiosis to the unusual situation of a trisomic cell. Regarding the ey/ey female, only one gamete class can be produced, namely, eggs of genotype ey. Gamete production with respect to the trisomy for chromosome 4 occurs by a random segregation pattern in which two chromosomes migrate to one pole and the other chromosome migrates to the other pole during meiosis I. (This pattern is similar to the meiotic segregation pattern shown in secondary nondisjunction of XXY cells; see Chapter 3.) Three types of segregation are possible in the formation of gametes in the trisomy, as

shown in part (a) of Figure 7.A. The union of these sperm at random with eggs of genotype ey occurs as shown in part (b).

The resulting genotypic and phenotypic ratios are illustrated in part (c).

FIGURE 7.A

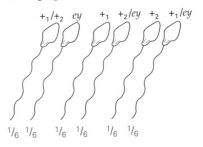

a) Segregation

$+_1/+_2\ ey\qquad +_1\qquad +_2/ey\qquad +_2\qquad +_1/ey$

$\frac{1}{6}\quad \frac{1}{6}\qquad \frac{1}{6}\quad \frac{1}{6}\qquad \frac{1}{6}\quad \frac{1}{6}$

b) Union

		Eggs ey	Phenotype
	$+/+$	$+/+/ey$	$+$
	ey	ey/ey	ey
Sperm	$+$	$+/ey$	$+$
	$+/ey$	$+/ey/ey$	$+$
	$+$	$+/ey$	$+$
	$+/ey$	$+/ey/ey$	$+$

c) Summary of genotypes and phenotypes

Ratios:	Genotypes	Phenotype
	$\frac{1}{6}\ +/+/ey$	$\frac{5}{6}$ wild type
	$\frac{1}{3}\ +/ey/ey$	$\frac{1}{6}$ eyeless
	$\frac{1}{3}\ +/ey$	
	$\frac{1}{6}\ ey/ey$	

QUESTIONS AND PROBLEMS

*7.1 A normal chromosome has the following gene sequence:

$$A\ B\ C\ D\ _\circ\ E\ F\ G\ H$$

Determine the chromosomal mutation illustrated by each of the following chromosomes:

a. $A\ B\ C\ F\ E\ _\circ\ D\ G\ H$

b. $A\ D\ _\circ\ E\ F\ B\ C\ G\ H$

c. $A\ B\ C\ D\ _\circ\ E\ F\ E\ F\ G\ H$

d. $A\ B\ C\ D\ _\circ\ E\ F\ F\ E\ G\ H$

e. $A\ B\ D\ _\circ\ E\ F\ G\ H$

*7.2 Distinguish between pericentric and paracentric inversions.

7.3 Very small deletions behave in some instances like recessive mutations. Why are some recessive mutations known not to be deletions?

7.4 What would be the effect on protein structure, if a small inversion were to occur within the coding region of a structural gene?

*7.5 Inversions are known to affect crossing-over. The following homologs with the indicated gene order are given:

●$A B C D E$

○$A D C B E$

a. Diagram the alignment of these chromosomes during meiosis.
b. Diagram the results of a single crossover between homologous genes B and C in the inversion.
c. Considering the position of the centromere, what is this sort of inversion called?

7.6 Single crossovers within the inversion loop of inversion heterozygotes give rise to chromatids with duplications and deletions. What happens when, within the inversion loop, there is a two-strand double crossover in such an inversion heterozygote when the centromere is outside the inversion loop?

7.7 An inversion heterozygote possesses one chromosome with genes in the normal order:

○$a b c d e f g h$

It also contains one chromosome with genes in the inverted order:

○$a b f e d c g h$

A four-strand double crossover occurs in the areas $e - f$ and $c - d$. Diagram and label the four strands at synapsis (showing the crossovers) and at the first meiotic anaphase.

*7.8 The following gene arrangements in a particular chromosome are found in *Drosophila* populations in different geographical regions. Assuming the arrangement in part a is the original arrangement, in what sequence did the various inversion types most likely arise?

a. $A B C D E F G H I$
b. $H E F B A G C D I$
c. $A B F E D C G H I$
d. $A B F C G H E D I$
e. $A B F E H G C D I$

*7.9 Human abnormalities associated with chromosomal mutations often exhibit a range of symptoms, of which only some subsets are shown in a particular individual. Recombinant 8 [Rec(8)] syndrome is an inherited chromosomal abnormality found primarily in individuals of Hispanic origin. Phenotypic characteristics associated with the syndrome can include congenital heart disease, urinary system abnormalities, eye abnormalities, hearing loss, and abnormal muscle tone. Most reported cases of Rec(8) have been found in the offspring of phenotypically normal parents who are heterozygous for an inversion of chromosome 8 with breakpoints at p23.1 and q22.1. Individuals with Rec(8) syndrome typically have a duplication of part of 8q (from q22.1 to the terminus of the q arm) and a deletion of 8p (from p23.1 to the terminus of the p arm).

a. Using diagrams, explain why individuals with Rec(8) syndrome typically have a duplication and a deletion for part of chromosome 8.
b. An individual is heterozygous for an inversion on chromosome 8 with breakpoints at p23.1 and q22.1. What is the chance, if a crossover occurs within the inverted region during a particular meiosis, that the resulting offspring will have Rec(8) syndrome?
c. Why might the phenotypes of Rec(8) individuals vary?
d. A child with some, but not all, of the symptoms of Rec(8) syndrome is referred to a human geneticist. The karyotype of the child reveals heterozygosity for a large pericentric inversion in chromosome 8 with breakpoints at p23.1 and q22.1. Cytogenetic analysis of her phenotypically normal mother and maternal grandmother reveals a similar karyotype. According to the child's mother, the father has a normal phenotype, but he is unavailable for examination. Propose at least two explanations for why the child, but not her mother or maternal grandmother, is affected with some of the symptoms of Rec(8) syndrome. (Hint: consider the limitations of karyotype analysis utilizing G-banding methods and what is unknown about the father.)

*7.10 A particular species of plant that had been subjected to radiation for a long period of time in order to produce chromosome mutations was then inbred for many generations until it is homozygous for all of these mutations. It was then crossed to the original unirradiated plant and the meiotic process of the F_1 hybrids was examined. It was noticed that the following structures occurred, at low frequency, in anaphase I of the hybrid: a cell (cell A) with a dicentric chromosome (bridge) and a fragment, and another cell (cell B) with a dicentric chromosome with two bridges and two fragments.

a. What kind of chromosome mutation occurred in the irradiated plant? In your answer, indicate where the centromeres are.
b. Explain in words and with a clear diagram where crossover(s) occurred, and how the bridge chromosome of cell A arose.
c. Explain in words and with a clear diagram where crossover(s) occurred, and how the double bridge chromosome of cell B arose.

7.11 On a normal-ordered chromosome, two loci a and b lie 15 map units apart on the left arm of a metacentric chromosome. A third locus c lies 10 map units to the right of b on the right arm of the chromosome. What frequency of progeny phenotypes do you expect to see in a testcross of an $a\ b\ c/a^+\ b^+\ c^+$ individual if the $a^+\ b^+\ c^+$ chromosome:

a. has a normal order.

b. has an inversion with breakpoints just proximal (towards the centromere) to a and just distal (away from the centromere) to b.

c. has an inversion with breakpoints just proximal to a and just proximal to c.

d. has an inversion with breakpoints just distal to a and just distal to c.

*7.12 Mr. and Mrs. Lambert have not yet been able to produce a viable child. They have had two miscarriages and one severely defective child who died soon after birth. Studies of banded chromosomes of father, mother, and child showed all chromosomes were normal except for pair number 6. The number 6 chromosomes of mother, father, and child are shown in the following figure:

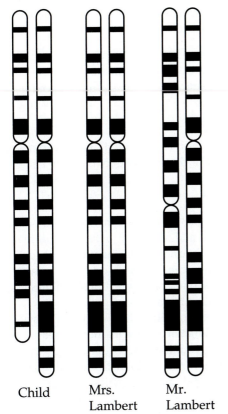

Child   Mrs.
Lambert   Mr.
Lambert

a. Does either parent have an abnormal chromosome? If so, what is the abnormality?

b. How did the chromosomes of the child arise? Be specific as to what events in the parents gave rise to these chromosomes.

c. Why is the child not phenotypically normal?

d. What can be predicted about future conceptions by this couple?

7.13 Mr. and Mrs. Simpson have been trying for years to have a child but have been unable to conceive. They consulted a physician, and tests revealed that Mr. Simpson had a markedly reduced sperm count. His chromosomes were studied, and a testicular biopsy was done as well. His chromosomes proved to be normal, except for pair 12. The following figure shows Mrs. Simpson's normal pair of number 12 chromosomes and Mr. Simpson's number 12 chromosomes.

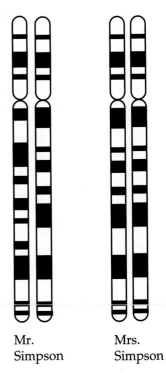

Mr.
Simpson   Mrs.
Simpson

a. What is the nature of the abnormality in Mr. Simpson's chromosomes of pair number 12?

b. What abnormal feature would you expect to see in the testicular biopsy? (cells in various stages of meiosis can be seen)

c. Why is Mr. Simpson's sperm count low?

d. What can be done about it?

7.14 In the early 1930s, Sinitskaya recovered a highly rearranged X chromosome in the progeny of a *Drosophila* male exposed to a large dose of ionizing radiation. The rearranged X chromosome resulted from two simultaneously occurring (superimposed) paracentric inversions. Diagram each of the inversions individually, and then diagram a chromosome with both inversions superimposed, using Figure 5.20 (p. 158), which illustrates where some of the relevant loci are on a normal-ordered chromosome, and the following information. The first inversion had a breakpoint in the *sc* locus, which lies very near *y* at map position 0.0 (not illustrated in the figure), and another breakpoint between the *bb* locus and the centromere. The second inversion had one breakpoint just distal to *ct* and another breakpoint just proximal to *m*.

*7.15   Chromosome I in maize has the gene sequence *ABCDEF*, whereas chromosome II has the sequence *MNOPQR*. A reciprocal translocation resulted in *ABCPQR* and *MNODEF*. Diagram the expected pachytene configuration of the F$_1$ of a cross of homozygotes of these two arrangements (see p. 59).

7.16   Diagram the pairing behavior at prophase of meiosis I of a translocation heterozygote that has normal chromosomes of gene order *abcdefg* and *tuvwxyz* and has the translocated chromosomes *abcdvwxyz* and *tuefg*. Assume that the centromere is at the left end of all chromosomes.

*7.17   Mr. and Mrs. Denton have been trying for several years to have a child. They have experienced a series of miscarriages, and last year they had a child with multiple congenital defects. The child died within days of birth. The birth of this child prompted the Dentons' physician to order a chromosome study of parents and child. The results of the study are shown in the figure below. Chromosome banding was done, and all chromosomes were normal in these individuals except some copies of number 6 and number 12. The number 6 and number 12 chromosomes of mother, father, and child are shown in the figure (the number 6 chromosomes are the larger pair).

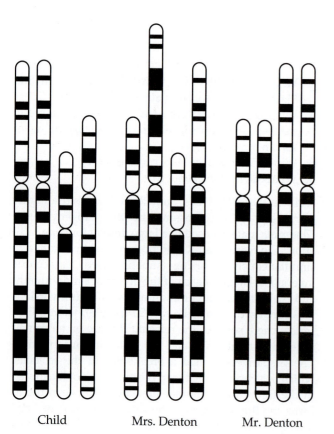

Child   Mrs. Denton   Mr. Denton

a. Does either parent have an abnormal karyotype? If so, which parent has it, and what is the nature of the abnormality?
b. How did the child's karyotype arise (what pairing and segregation events took place in the parents)?
c. Why is the child phenotypically defective?
d. What can this couple expect to occur in subsequent conceptions?
e. What medical help, if any, can be offered to them?

7.18   Irradiation of *Drosophila* sperm produces translocations between the X chromosome and autosomes, and between the Y chromosome and autosomes, and between different autosomes. Translocations between the X and Y chromosomes are not produced. Explain the absence of X-Y translocations.

7.19   Define the terms *aneuploidy, monoploidy,* and *polyploidy*.

7.20   If a normal diploid cell is 2N, what is the chromosome content of the following:
a. a nullisomic
b. a monosomic
c. a double monosomic
d. a tetrasomic
e. a double trisomic
f. a tetraploid
g. a hexaploid

*7.21   In humans, how many chromosomes would be typical of nuclei of cells that are
a. monosomic
b. trisomic
c. monoploid
d. triploid
e. tetrasomic

*7.22   An individual with 47 chromosomes, including an additional chromosome 15, is said to be
a. triplet
b. trisomic
c. triploid
d. tricycle

*7.23   A color-blind man marries a homozygous normal woman and after four joyful years of marriage they have two children. Unfortunately, both children have Turner syndrome, although one has normal vision and one is color-blind. The type of color blindness involved is a sex-linked recessive trait.
a. For the color-blind child with Turner syndrome, did nondisjunction occur in the mother or the father? Explain your answer.
b. For the Turner child with normal vision, in which parent did nondisjunction occur? Explain your answer.

*7.24   The frequency of chromosome loss in *Drosophila* can be increased by a recessive chromosome 2 mutation called *pal*. The mutation causes the preferential loss of chromosomes contributed to a zygote by *pal/pal* fathers. The paternally contributed chromosomes are lost during the first few mitotic divisions after fertilization. What phenotypic consequences do you expect in offspring of the following crosses? (Keep in mind how sex is determined in *Drosophila*, that loss of an entire chromosome 2 or chromosome 3 is lethal, and that loss of one copy of the small chromosome 4 is tolerated.)

a. X chromosome loss at the first mitotic division in a cross between a true-breeding *yellow* (recessive, X-linked mutation causing yellow body color) female and a *pal/pal* father.

b. Chromosome 4 loss at the first mitotic division in a cross between a true-breeding *eyeless* (recessive mutation on chromosome 4 causing reduced eye size) female and a *pal/pal* father.

c. Chromosome 3 loss at the first mitotic division in a cross between a true-breeding *ebony* (recessive mutation on chromosome 3 causing black body color) female and a *pal/pal* father.

7.25 Assume that x is a new mutant gene in corn. A female x/x plant is crossed with a triplo-10 individual (trisomic for chromosome 10) carrying only dominant alleles at the x locus. Trisomic progeny are recovered and crossed back to the x/x female plant.

a. What ratio of dominant to recessive phenotypes is expected if the x locus is *not* on chromosome 10?

b. What ratio of dominant to recessive phenotypes is expected if the x locus *is* on chromosome 10?

7.26 Why are polyploids with even multiples of the chromosome set generally more fertile than polyploids with odd multiples of the chromosome set?

7.27 Select the correct answer from the key below for the following statement: One plant species (N = 11) and another (N = 19) produced an allotetraploid.

I. The chromosome number of this allotetraploid is 30.

II. The number of nuclear linkage groups of this allotetraploid is 30.

KEY:

A. Statement I is true and Statement II is true.

B. Statement I is true, but Statement II is false.

C. Statement I is false, but Statement II is true.

D. Statement I is false and Statement II is false.

*7.28 According to Mendel's first law, genes A and a segregate from each other and appear in equal numbers among the gametes. But Mendel did not know that his plants were diploid. In fact, since plants are frequently tetraploid, he could have been unlucky enough to have started with peas that were 4N rather than 2N. Let us assume that Mendel's peas were tetraploid, that every gamete contains two alleles, and that the distribution of alleles to the gamete is random. Suppose we have a cross of *AA AA* × *aa aa*, where A is dominant, regardless of the number of a alleles present in an individual.

a. What will be the genotype of the F_1?

b. If the F_1 is selfed, what will be the phenotypic ratios in the F_2?

7.29 What phenotypic ratio of A to a is expected if *AA aa* plants are testcrossed against *aa aa* individuals? (Assume that the dominant phenotype is expressed whenever at least one A is present, no crossing-over occurs, and each gamete receives two chromosomes.)

7.30 The root tip cells of an autotetraploid plant contain 48 chromosomes. How many chromosomes were contained by the gametes of the diploid from which this plant was derived?

7.31 A number of species of the birch genus have a somatic chromosome number of 28. The paper birch is reported as occurring with several different chromosome numbers; *fertile* individuals with the somatic numbers 56, 70, and 84 are known. How should the 28-chromosome individuals be designated with regard to chromosome number?

*7.32 How many chromosomes would be found in somatic cells of an allotetraploid derived from two plants, one with N = 7 and the other with N = 10?

7.33 Plant species A has a haploid complement of 4 chromosomes. A related species B has 5. In a geographical region where A and B are both present, C plants are found that have some characters of both species and somatic cells with 18 chromosomes. What is the chromosome constitution of the C plants likely to be? With what plants would they have to be crossed in order to produce fertile seed?

8 GENETIC MAPPING IN BACTERIA AND BACTERIOPHAGES

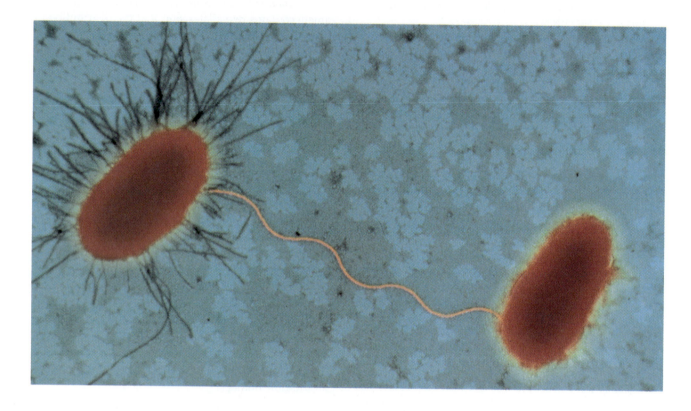

PRINCIPAL POINTS

~ Transformation is the transfer of genetic material between organisms by means of small extracellular pieces of DNA. By means of genetic recombination, part of the transforming DNA molecule can exchange with a portion of the recipient's chromosomal DNA. Transformation can be used experimentally to determine gene order and map distance between genes.

~ Conjugation is a process in which there is a unidirectional transfer of genetic information through direct cellular contact between a donor and a recipient bacterial cell. The donor state is conferred by the presence of a plasmid called an *F* factor. Conjugation results in the transfer of the *F* factor from donor to recipient.

~ The *F* factor can integrate into the bacterial chromosome. Strains in which this has occurred—*Hfr* strains—can conjugate with recipient strains and transfer of the bacterial chromosome ensues. The sequential order of bacterial genes can be determined by the relative times at which donor genes enter the recipient during conjugation.

~ Transduction is a process whereby bacteriophages (phages) mediate the transfer of bacterial DNA from one bacterium (the donor) to another (the recipient). Transduction can be used experimentally to map bacterial genes.

~ The same principles used to map eukaryotic genes are used to map phage genes. That is, genetic material is exchanged between strains differing in genetic markers and recombinants are detected and counted.

~ The same general principles of recombinational mapping can be applied to mapping the distance between mutational sites in different genes (intergenic mapping) and to mapping mutational sites within the same gene (intragenic mapping).

~ As a result of fine-structure analysis of the *rII* region of bacteriophage T4 and from other experiments, it was determined that the unit of mutation and of recombination is the DNA base pair.

~ The number of units of function (genes) that cause a particular mutant phenotype is determined by the complementation, or *cis-trans*, test. If two viral mutants, each carrying a mutation in a different gene, are combined in a single host cell, the mutations will complement and a wild-type function will result. If two mutants, each carrying a mutation in the same gene, are combined, the mutations will not complement and the mutant phenotype will still be expressed.

---

*I*n Chapters 5 and 6 we considered the principles of genetic mapping in eukaryotic organisms. To map genes classically in bacteria and bacteriophages, geneticists use essentially the same experimental strategies. Crosses are made between strains that differ in genetic markers, and recombinants, the products of the exchange of genetic material, are detected and counted. The analysis of data obtained from such crosses is the same as in eukaryotes: the frequency with which exchanges occur between two sets of genes relates to the map distance between the two loci. In this chapter we describe the classic genetic analysis of bacteria and bacteriophages. Recently, the focus has changed from genetic mapping of genes to determining the DNA sequences of genes and genomes. At this writing, complete sequences have been determined for the *E. coli* genome, for the genomes of two other bacteria, and for an archaeon microbe (see Chapter 15, pp. 486–487). With genome sequences, scientists can look for genes directly, thereby providing the ultimate in genetic maps—that is, information at the nucleotide level about the organization of genes in the genome.

We also describe a series of classic genetic experiments which investigated the fine structure of the gene: that is, the detailed molecular organization of the gene as it relates to the mutational, recombinational, and functional events in which the gene is involved. A bacteriophage gene was the subject of these experiments.

GENETIC ANALYSIS OF BACTERIA

Genetic material can be transferred between bacteria by three main processes: transformation, conjugation, and transduction. In each case: (1) transfer is unidirectional; (2) unlike eukaryotes, no true diploid zygote is formed; and (3) only genes included in the circular chromosome will be inherited stably. It is possible to map bacterial genes by using any one of these methods. However, not all methods can be used for all species of bacterium, and the size of the region that can be mapped varies according to method.

Among bacteria, *E. coli* is used extensively for genetic and molecular analysis. It is found most commonly in the large intestines of most animals (including humans). This bacterium is a good subject for study since it can be grown on a simple, defined medium and can be handled with simple microbiological techniques.

E. coli is a cylindrical organism about 1–3 μm long and 0.5 μm in diameter (Figure 8.1); that is, like most bacteria, it is small compared with eukaryotic cells.[1] Its cytoplasm is full of ribosomes (organelles on which proteins are made), and a single circular DNA chromosome is in a central region called the **nucleoid.** As in all prokaryotes, there is no membrane between the bacterium's nucleoid region and the rest of the cell.

Like other bacteria, *E. coli* can be grown both in a liquid culture medium and on the surface of growth medium solidified with agar. Genetic analysis of bacteria typically is done by spreading ("plating") cells on agar-solidified media. Wherever a single bacterium lands on the agar surface it will grow and divide repeatedly, resulting in the formation of a visible cluster of genetically identical cells called a *colony* (Figure 8.2). Each colony consists of a clone of the original cell; that is, the colony contains genetically identical copies of the cell that initiated the colony. The concentration of bacterial cells in a culture (the *titer*) can be determined by spreading known volumes of the culture or of a known dilution of the culture on the agar surface, incubating the plates, and then counting the number of resulting colonies.

~ **FIGURE 8.1**

Colorized scanning electron micrograph of several *Escherichia coli* bacteria.

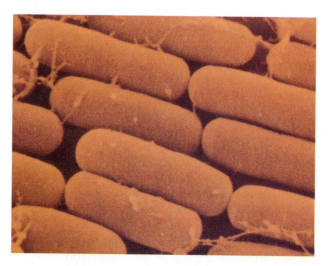

[1]The largest bacterium is *Epulopiscium fishelsoni,* the surgeonfish symbiont. With a cell size as large as 800 μm long and 60 μm wide, it is a million times larger than *E. coli* and larger than some eukaryotic cells.

~ **FIGURE 8.2**

Bacterial colonies growing on a nutrient medium in a Petri dish.

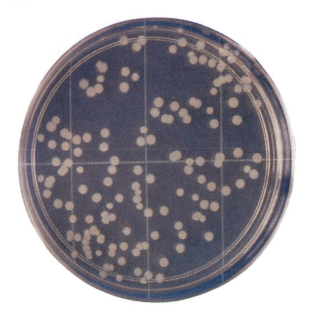

The composition of the culture medium used depends upon the experiment and the genotypes of the strains being used. Each bacterial species (or any other microorganism, such as yeast or *Neurospora*) has a characteristic *minimal medium* on which it will grow. A minimal medium contains the simplest set of ingredients that the microorganism can use to synthesize all of the molecules (e.g., amino acids, vitamins, and DNA and RNA precursors) needed for growth and reproduction. The minimal medium for wild-type *E. coli*, for example, consists of a sugar (a carbon source), and some salts and trace elements. By contrast, the *complete medium* for a microorganism supplies all the building blocks and vitamins needed to synthesize macromolecules required for growth and reproduction.

Genetic analysis of bacteria (and other microorganisms) typically involves studies of mutants that are defective in their abilities to make one or more molecules essential for growth. Strains that are unable to synthesize essential nutrients are called **auxotrophs** (also called *auxotrophic mutants, nutritional mutants,* or *biochemical mutants*). A strain that is wild type and thus can synthesize all essential nutrients is called a **prototroph**; prototrophs need no nutritional supplements in the growth medium. By definition, the wild type or prototroph grows on the minimal medium for that organism, whereas an auxotroph grows on complete medium or on minimal medium plus the appropriate nutritional supplement or supplements. For example, consider the *E. coli* strain with the genotype *trp ade thi+*. In this way of noting geno-type, the gene symbol indicates a particular nutrition-al requirement. If the genotype has a superscript +, then the gene is wild type, indicating that the bacteria are able to synthesize the nutritional substance. If no superscript is present, the gene is mutant and the cul-ture medium must be supplemented with this partic-ular substance. Thus, this strain can grow on the *E. coli* complete medium but not on the minimal medium. It will also grow on minimal medium supplemented with the amino acid tryptophan (*trp*) and the purine (base needed for RNA and DNA precursor synthesis) adenine (*ade*), but without the vitamin thiamine (*thi*). By varying the nutrient composition of the medium it is possible to detect and quantify the various geno-typic classes in a genetic analysis.

In genetic experiments with microorganisms such as *E. coli*, crosses are made between strains differing in genotype (and, therefore, phenotype), and progeny are analyzed for parental and recombinant pheno-types. When auxotrophic mutations are involved, the determination of parental and progeny phenotypes (and, therefore, genotypes, since bacteria are haploid) involves testing colonies for their growth require-ments. One convenient procedure that can be used for such tests is *replica plating* (see Figure 19.27, pp. 644–646). In the replica plating technique, samples from a culture (e.g., after a genetic cross) are plated onto a complete medium. Upon incubation, colonies grow wherever a cell has landed on the medium. The pattern of the colonies is transferred onto sterile velve-teen cloth. Replicas of the original colony pattern on the cloth are then made by gently pressing new plates onto the velveteen. If the new plate contains minimal medium, only prototrophic colonies can grow. So by comparing the patterns on the original master plate with those on the minimal medium replica plate, researchers can readily identify auxotrophic colonies, since they will be on the master plate but not on the minimal medium plate. Using other plates containing minimal medium plus combinations of nutritional supplements appropriate for the strain or strains in-volved, the phenotypes/genotypes of all the auxo-trophic colonies can be determined.

BACTERIAL TRANSFORMATION

Transformation is the transfer of genetic material between organisms by means of extracellular pieces of DNA. In bacterial transformation, fragments of free DNA are taken up by a living bacterium, bringing about a stable genetic change in the recipient cell. Bacterial transformation was first observed by Frederick Griffith in 1928, and in 1944 Oswald Avery and his colleagues showed that DNA was responsible

for the genetic change that was observed (see Chapter 10). Bacterial transformation is generally used to map genes in bacterial species in which mapping by other methods (i.e., conjugation or transduction) is not pos-sible. In mapping experiments using transformation, DNA from a donor bacterial strain is extracted and purified. The extraction process breaks the DNA into small linear double-stranded fragments, and this genetic material is then added to a suspension of recipient bacteria with a different genotype. The donor DNA that is taken up by the recipient cell may recombine with the homologous parts of the recipi-ent's chromosome to produce a recombinant chromo-some. Those recipients whose phenotypes are changed by a recombination event are called **transfor-mants.** In principle, any bacterial strain can serve as a donor strain, and any bacterial strain can serve as a recipient. Only if there are genetic differences between donors and recipients will transformants be detected.

Most bacterial species can probably undergo transformational recombination. However, there is a great deal of variability in the ability to take up DNA. Over the years, scientists have found ways to increase the efficiency of uptake for a number of bacterial species. Typically, cells are treated chemically, or are exposed to a strong electric field in a process called *electroporation*, making the cell membrane more per-meable to DNA. Cells prepared for taking up DNA by transformation are called *competent cells*. As a result of such experimentation, it has become routine to transform bacterial species like *E. coli* in the test tube with DNA of prokaryotic or eukaryotic origin, and that ability has been extremely important in advancing our knowledge of gene function in both prokaryotes and eukaryotes.

There are two types of bacterial transformation: *natural transformation*, in which bacteria are naturally able to take up DNA and be genetically transformed by it; and *engineered transformation*, in which bacteria have been altered to enable them to take up and be genetically transformed by added DNA. *Bacillus sub-tilis* exemplifies bacteria in which natural transforma-tion occurs; *E. coli* exemplifies bacteria in which engi-neered transformation occurs. *Bacillus subtilis* is a cylindrical, spore-forming bacterium, about 3–8 μm long, and 1–1.5 μm wide. It can usually be isolated from water and sewage systems.

Only a small proportion of the cells involved in transformation will actually take up DNA. Let us con-sider an example of natural transformation of *Bacillus subtilis* (Figure 8.3). (Other systems may differ in the details of the process.) The donor double-stranded DNA fragment is wild type (*a+*) for a mutant allele *a* in the recipient cell (Figure 8.3a). During DNA uptake, one of the two DNA strands is degraded so that only one intact linear DNA strand ultimately is

~ FIGURE 8.3

Natural transformation in *Bacillus subtilis*: (a) Linear donor double-stranded bacterial DNA fragment, which carries the a^+ allele and the recipient bacterium with the a allele; (b) One of the donor DNA strands enters the recipient; (c) The single, linear DNA strand pairs with the homologous region of the recipient's chromosome, forming a triple-stranded structure; (d) A double crossover produces a recombinant a^+/a recipient chromosome and a linear a DNA fragment. The linear fragment is degraded and, by replication, one-half of the progeny are a^+ transformants and one-half are a nontransformants.

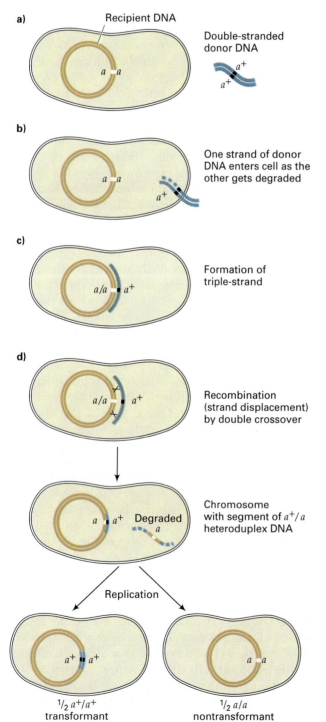

a)   Recipient DNA
      Double-stranded donor DNA
      a — a
      a^+
      a^+

b) a — a
      One strand of donor DNA enters cell as the other gets degraded
      a^+

c) a/a — a^+
      Formation of triple-strand

d) a/a a^+
      Recombination (strand displacement) by double crossover

      a a^+ Degraded a
      Chromosome with segment of a^+/a heteroduplex DNA

      Replication

      a^+ a^+
      $\frac{1}{2}\ a^+/a^+$
      transformant

      a — a
      $\frac{1}{2}\ a/a$
      nontransformant

found within the cell (Figure 8.3b). This single, linear strand pairs with the homologous DNA of the recipient cell's circular chromosome, forming base pairs with the recipient's DNA strand with which it is complementary (Figure 8.3c). Recombination can then occur by a double crossover event involving the single-stranded DNA strand of the donor and the double-stranded DNA of the recipient (Figure 8.3d). The result is a recombinant recipient chromosome: in the region between the two crossovers, one DNA strand has the donor a^+ DNA segment, and the other strand has the recipient a DNA segment. In other words, in that region, *the two DNA strands are part donor, part recipient for the genetic information.* A region of DNA with different sequence information on the two strands is called **heteroduplex DNA**. (The other product of the double crossover event is a single-stranded piece of DNA carrying an a DNA segment; that DNA fragment is degraded.) After one round of replication of the recipient chromosome, one progeny chromosome has donor genetic information on both DNA strands and is an a^+ transformant. The other progeny chromosome has recipient genetic information on both DNA strands and is an a nontransformant. Equal numbers of a^+ transformants and a nontransformants are expected. Given highly competent recipient cells and an excess of DNA fragments, transformation of most genes will occur at a frequency of about 1 cell in every 10^3 cells.

It is possible to determine gene linkage, gene order, and map distance through transformation. The principles are as follows. If two genes, x^+ and y^+, are far apart on the donor chromosome, they will always be found on different DNA fragments. Thus, given an $x^+ y^+$ donor and an $x\ y$ recipient, the probability of simultaneous transformation (*cotransformation*) of the recipient to $x^+ y^+$ (from the product rule) is the product of the probability of transformation with each gene alone. If transformation occurred at a frequency of 1 in 10^3 cells per gene (see above), $x^+ y^+$ transformants would be expected to appear at a frequency of 1 in 10^6 recipient cells ($10^{-3} \times 10^{-3}$). Therefore, if two genes are close enough that they often are carried on the same DNA fragment, the cotransformation frequency would be close to the frequency of transformation of a single gene. As determined experimentally, if the frequency of cotransformation of two genes is substantially higher than the products of the two individual transformation frequencies, the two genes must be close together.

Gene order can also be determined from cotransformation data (Figure 8.4). For example, if genes p and q are often transmitted to the recipient together, then these two genes must be relatively closely linked. Similarly, if genes q and o are often transmit-

~ FIGURE 8.4

Demonstration of determining gene order by cotransformation.

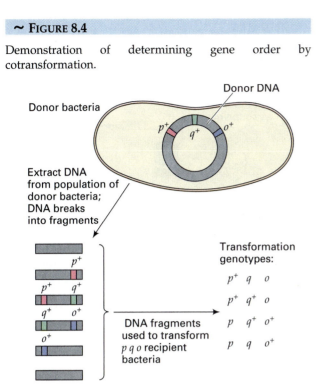

ted together, those two genes must be close to one another. To determine gene order, we now need information about genes p and o. Theoretically, there are two possible orders: p-o-q, and p-q-o. If the order is p-o-q, then p and o should be cotransformed because they are more closely linked than p and q, whereas if the order is p-q-o then p and o should be cotransformed rarely or not at all because they are relatively far apart. The data show no cotransformants for p and o, indicating that the gene order must be p-q-o.

Since geneticists can control the size of DNA fragments used in transformation experiments, the probability of cotransformation can be related to the average molecular size of the transforming DNA. By relating cotransformation frequency to average size of the transforming DNA, one can derive a physical map of the genes (i.e., a map of the relative physical locations of genes along the DNA). This specialized process is used for only a relatively few organisms of genetic organisms so it will not be discussed in this text.

𝒦EYNOTE

Transformation is the transfer of small extracellular pieces of DNA between organisms. In transforma-

tion, DNA is extracted from a donor strain and added to recipient cells. A DNA fragment taken up by the recipient cell may associate with the homologous region of the recipient's chromosome. Part of the transforming DNA molecule can exchange with part of the recipient's chromosomal DNA. Frequent cotransformation of donor genes indicates close physical linkage of those genes. Analysis of cotransformants can be used to determine gene order and map distance between genes. Transformation has been used to construct genetic maps for bacterial species for which conjugation or transduction analyses are not possible.

CONJUGATION IN BACTERIA

Discovery of Conjugation in *E. coli*

Conjugation is a process in which there is a unidirectional transfer of genetic information through direct cellular contact between a donor bacterial cell and a recipient bacterial cell. The contact is followed by the formation of a bridge physically connecting both cells. Then a segment (rarely all) of the donor's chromosome may be transferred into the recipient and may undergo genetic recombination with a homologous chromosome segment of the recipient cell. The recipients receiving donor DNA are called **transconjugants**.

Conjugation was discovered in 1946 by Joshua Lederberg and Edward Tatum. They studied two *E. coli* strains that differed in their nutritional requirements. Strain A had the genotype *met bio thr⁺ leu⁺ thi⁺*, and strain B had the genotype *met⁺ bio⁺ thr leu thi*. Recall from our earlier discussion that the gene symbol indicates a particular nutritional requirement. Here, strain A can only grow on a medium supplemented with the amino acid methionine (*met*) and the vitamin biotin (*bio*) but does not need the amino acids threonine (*thr*) or leucine (*leu*), or the vitamin thiamine (*thi*). In contrast, strain B can only grow on a medium supplemented with threonine, leucine, and thiamine, but does not require methionine or biotin.

In the experiment shown in Figure 8.5, Lederberg and Tatum wanted to determine whether two strains of *E. coli* could exchange genetic material. They mixed

~ FIGURE 8.5

Lederberg and Tatum experiment showing that sexual recombination occurs between cells of *E. coli*. After the cells from group A and group B have been mixed and the mixture plated, a few colonies grow on the minimal medium, indicating that they can now make the essential constituents. These colonies are recombinants produced by an exchange of genetic material between the strains.

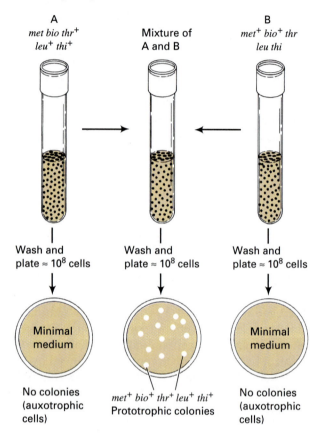

A
met bio thr⁺ leu⁺ thi⁺

Mixture of A and B

B
met⁺ bio⁺ thr leu thi

Wash and plate ≈ 10^8 cells

Wash and plate ≈ 10^8 cells

Wash and plate ≈ 10^8 cells

Minimal medium

Minimal medium

No colonies (auxotrophic cells)

met⁺ bio⁺ thr⁺ leu⁺ thi⁺
Prototrophic colonies

No colonies (auxotrophic cells)

strains A and B together and plated them onto minimal medium. When the two strains were plated separately in control experiments, neither strain could grow without the appropriate nutritional supplement. The mixed culture, however, gave rise to some prototrophic colonies. Genotypically, these prototrophic strains were *met⁺ bio⁺ thr⁺ leu⁺ thi⁺*, and they arose at a frequency of about 1 in 10 million cells (1×10^{-7}). Since no colonies appeared on the control plates, mutation was ruled out as the cause of the prototrophic colonies. The mixing, then, is a genetic cross from which recombinants can result.

What mechanism was involved in the genetic exchange? To determine whether cell-to-cell contact was needed for the genetic exchange to occur, Bernard Davis performed an experiment using a U tube apparatus (Figure 8.6). Strains A and B were placed in a liquid medium on either side of the U sep-

~ FIGURE 8.6

Davis's U tube experiment showing that physical contact between the two bacterial strains of the earlier experiment was needed in order for genetic exchange to occur.

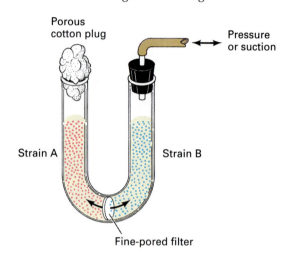

Porous cotton plug

Pressure or suction

Strain A

Strain B

Fine-pored filter

arated by a fine filter whose pores were too small to allow bacteria to move through. After several hours of incubation, during which the medium was moved between compartments by alternating suction and pressure, Davis plated the cells on minimal medium to check for the appearance of prototrophic colonies. No prototrophic colonies appeared. The interpretation was that cell-to-cell contact was required in order for the previously observed genetic exchange to occur. The genetic exchange could not have resulted from something secreted by the cells. Such experiments led to the proposal that *E. coli* has a type of mating system, called **conjugation**, in which genetic material can be transferred between bacteria that are transiently connected.

The Sex Factor *F*

In 1953 William Hayes obtained evidence that the genetic exchange in *E. coli* occurs in only one direction, with one cell acting as a donor and the other cell acting as a recipient. Hayes proposed that the transfer of genetic material between the strains is mediated by a *sex factor* (also called a fertility factor) named *F* that the donor cell possesses (F^+) and the recipient cell lacks (F^-). The *F* factor found in *E. coli* is an example of a **plasmid**, a self-replicating, circular, double-stranded piece of DNA found distinct from the main bacterial chromosome. A wide range of different plasmids occur among bacterial species.

In crosses between F^+ and F^- cells of *E. coli*, recombinants for chromosomal (nonplasmid) gene

~ FIGURE 8.7

Electron micrograph of bacterial conjugation between an F^+ donor and F^- recipient *E. coli* bacterium. (The spherical structure on the upper *F* pilus bridging the two bacteria is a spherical donor-specific RNA phage MS-2.) Magnification 30,000×.

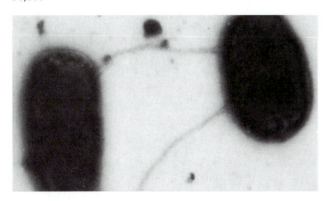

markers are rare, but the *F* factor is transferred from the F^+ to the F^- cell during the conjugation process. About a fortieth of the size of the host chromosome, the *F* factor contains a region of DNA, called the **origin** (or O), that is required for its replication within *E. coli* as well as a number of genes, including those that specify hairlike cell surface components called **F-pili** (singular, *F*-pilus), or sex-pili, which allow the physical union of F^+ and F^- cells to take place.

When F^+ and F^- cells are mixed, they may conjugate ("mate"). Typically only a small proportion of the bacteria in the mixture will conjugate. Figure 8.7 shows an electron micrograph of conjugation between an F^+ donor and an F^- recipient and Figure 8.8a1 diagrams the event. No conjugation can occur between two cells of the same mating type (i.e., two F^+ bacteria or two F^- bacteria). Genetic material is transferred from donor to recipient during conjugation. One strand of the *F* factor is nicked at the origin, and DNA replication proceeds from that point (Figure 8.8a2). Beginning at the origin, a single strand of DNA is transferred to the *F* cell as replication takes place (Figure 8.8a3). Think of the process like a roll of paper towels unraveling. The origin is the first stretch of DNA unwound; as unwinding continues, replication maintains the remaining circular *F* factor in a double-stranded form. Once the *F* factor DNA enters the F^- recipient, the complementary strand is synthesized (Figure 8.8a4). In the transfer process the origin is always transferred first, followed by the rest of the *F* factor. When the complete *F* factor has been transferred, the F^- cell becomes an F^+ cell as a result of the *F* factor genes (Figure 8.8a5). Continued conjugation between F^+ and F^- cells leads to an increase in the proportion of F^+ cells in the population. Replication

of the *F* factor can be inhibited by exposure to some chemicals, causing F^+ cells to revert to F^-. This is called *curing*. In $F^+ \times F^-$ crosses, none of the bacterial chromosome is transferred; only the *F* factor is transferred.

𝒦EYNOTE

Some *E. coli* bacteria possess a plasmid, called the *F* factor, that is required for mating. *E. coli* cells containing the *F* factor are designated F^+ and those without it are F^-. The F^+ cells (donors) can mate with F^- cells (recipients) in a process called conjugation, which leads to the one-way transfer of a copy of the *F* factor from donor to recipient during replication of the *F* factor. As a result, both donor and recipient are F^+. None of the bacterial chromosome is transferred during $F^+ \times F^-$ conjugation.

High-Frequency Recombination Strains

After geneticists established that F^- cells can be converted to the F^+ state by the transfer of the *F* factor, the next question was how conjugation resulted in the formation of recombinants for *chromosomal* genes. This puzzle was solved when William Hayes and Luca Cavalli-Sforza, working separately, isolated unusual donor strains from F^+ stock. In a cross of these donor strains with an F^- strain having different chromosomal gene markers, the frequency of recombinants for the chromosomal genes was about a thousand times greater than normally seen in $F^+ \times F^-$ crosses. This new, highly recombinant type of F^+ donor strain was called an **Hfr**, or **high-frequency recombination**, strain. The two particular strains characterized by Hayes and Cavalli-Sforza differed in the order in which bacterial genes were transferred; they were later called *HfrH* and *HfrC*, after their discoverers.

The interpretation of the high frequency transfer of chromosomal genes in *Hfr* strains is that *F* factor has integrated into the bacterial chromosome by a rare single crossover event (Figure 8.8b1–2). Plasmids such as *F* that are also capable of integrating into the bacterial chromosomes are called **episomes**. When the *F* factor is integrated, it no longer replicates independently, but is replicated as part of the host chromosome. Since the *F* factor genes are still functional, *Hfr* cells will conjugate with F^- cells (Figure 8.8b3) but not with F^+ cells. When mating happens, events simi-

~ FIGURE 8.8

Transfer of genetic material during conjugation in *E. coli*. (a) Transfer of the *F* factor from donor to recipient cell during $F^+ \times F^-$ matings. (b) Production of *Hfr* strain by integration of *F* factor and transfer of bacterial genes from donor to recipient cell during $Hfr \times F^-$ matings.

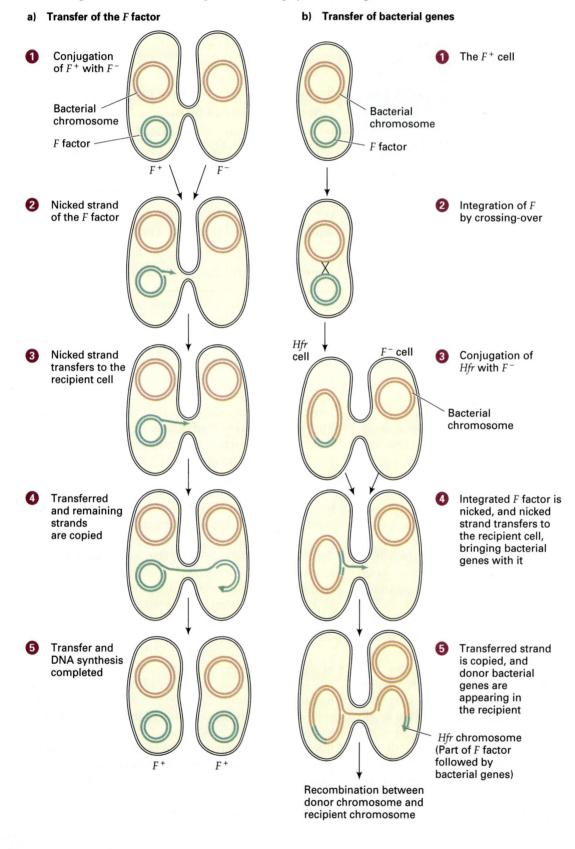

a) **Transfer of the *F* factor**

1. Conjugation of F^+ with F^-

   Bacterial chromosome

   F factor

   F^+ F^-

2. Nicked strand of the *F* factor

3. Nicked strand transfers to the recipient cell

4. Transferred and remaining strands are copied

5. Transfer and DNA synthesis completed

   F^+ F^+

b) **Transfer of bacterial genes**

1. The F^+ cell

   Bacterial chromosome

   F factor

2. Integration of *F* by crossing-over

   Hfr cell F^- cell

3. Conjugation of *Hfr* with F^-

   Bacterial chromosome

4. Integrated *F* factor is nicked, and nicked strand transfers to the recipient cell, bringing bacterial genes with it

5. Transferred strand is copied, and donor bacterial genes are appearing in the recipient

   Hfr chromosome (Part of *F* factor followed by bacterial genes)

   Recombination between donor chromosome and recipient chromosome

lar to those in the $F^+ \times F^-$ matings occur. The integrated F factor becomes nicked at the origin and replication begins (Figure 8.8b4). During replication *part of the F factor first moves into the recipient cell* where the transferred strand is copied. With time, the attached donor bacterial chromosome begins to be transferred into the recipient. If there are differences between the genes on the donor chromosome and those of the recipient chromosome, recombinants can be isolated (Figure 8.8b5). This recombination process occurs by means of double crossovers between the donor DNA that has entered the recipient cell and the recipient chromosome, similar to that described for transformation. In a double crossover, a segment of donor DNA is exchanged for the homologous segment of recipient DNA. If instead only a single crossover occurs between the donor linear fragment and the circular recipient chromosome, the result would be a linear molecule. Since linear bacterial chromosomes cannot replicate, this type of crossing-over event does not lead to productive recombinants.

In $Hfr \times F^-$ matings the F^- cell almost never acquires the Hfr phenotype. In order for the recipient cell to become Hfr, it must receive a complete copy of the F factor. However, only part of the F factor is transferred at the beginning of conjugation; the rest of the F factor is at the end of the donor chromosome. *All* of the donor chromosome would have to be transferred in order for a complete functional F factor to be found in the recipient, and that would require about 100 minutes at 37°C. This is an extremely rare event because all the while the bacteria are conjugating, they are "jiggling" around by Brownian motion, so there is a very high probability that the mating pair will break apart long before the second part of the F factor is transferred.

The low-frequency recombination of chromosomal gene markers in $F^+ \times F^-$ crosses can be understood when we consider that only about 1 in 10,000 F^+ cells in a population become Hfr cells by F factor integration. The reverse process, excision of the F factor, also occurs spontaneously and at low frequency, producing an F^+ cell from an Hfr cell. In excision, the F factor loops out of the Hfr chromosome, and by a single crossing-over event (just like the integration event), a circular host chromosome and a circular extrachromosomal F factor are generated.

F' Factors

Occasionally, the excision of the F factor from the host chromosome is not precise. As a result, the excised F factor may contain a relatively small section of the host chromosome that was adjacent to the F factor where it integrated. Since the F factor integrates at

one of many sites on the chromosomes, many different host chromosome segments may be picked up if an F factor excises incorrectly. Consider an *E. coli* strain in which the F factor has integrated next to the lac^+ region, a set of genes required for the breakdown of lactose (Figure 8.9a). If the looping out is not precise, then the adjacent lac^+ host chromosomal genes may be included in the loop (Figure 8.9b). Then by a single crossover the looped-out DNA will be separat-

~ FIGURE 8.9

Production of an F' factor: (a) Region of bacterial chromosome into which the F factor has integrated; (b) The F factor looping out incorrectly, so it includes a piece of bacterial chromosome, the lac^+ genes; (c) Excision, in which a single crossover between the looped-out DNA segment and the rest of the bacterial chromosome (i.e., the reverse of integration) results in an F' factor, called F' (*lac*).

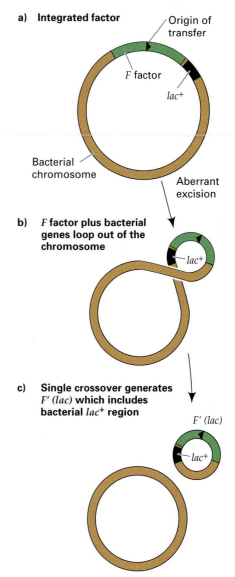

a) **Integrated factor**

Origin of transfer

F factor

lac^+

Bacterial chromosome

Aberrant excision

b) *F* **factor plus bacterial genes loop out of the chromosome**

lac^+

c) **Single crossover generates** F' (*lac*) **which includes bacterial** lac^+ **region**

F' (*lac*)

lac^+

ed from the host chromosome (Figure 8.9c) to produce an *F* factor that also contains the *lac*⁺ genes of the host chromosome. *F* factors containing bacterial genes are called *F'* (F prime) factors, and they are named for the genes they have picked up. For example, an *F'* with the *lac* genes is called *F' (lac)*.

Cells with *F'* factors can conjugate with *F⁻* cells since all the F factor functions are present. As in ordinary conjugation, a copy of the *F'* factor is transferred to the *F⁻* cell, which then becomes *F'*. The recipient also receives a copy of the bacterial gene(s) on the F factor (*lac* in our example). Since the recipient has its own copy of that DNA, the resulting cell line will be partially diploid (*merodiploid*), having two copies of one or a few genes and only one copy of all the others. This particular type of conjugation is called **F-duction**, or *sexduction*.

Using Conjugation and Interrupted Mating to Map Bacterial Genes

As was the case for genetic analysis using transformation, in genetic analysis using conjugation, one strain acts as the donor and another strain acts as the recipient. In genetic analysis using conjugation, the determining factor for whether a strain is a donor is the presence of an integrated *F* factor in the chromosome, making the strain an *Hfr* strain. Thus, in planning conjugation experiments to map genes, experimenters will construct an appropriate donor strain by introducing an F factor through *F⁺ × F⁻* mating, and then selecting an *Hfr* derivative.

In the late 1950s François Jacob and Elie Wollman studied the transfer of chromosomal genes from *Hfr* strains to *F⁻* cells. Their experimental design involved making an *Hfr × F⁻* mating and, at various times after conjugation began, breaking apart the conjugating pairs using a kitchen blender. This is called *interrupted mating*. The separated cells were then analyzed to determine the times at which donor genes entered the recipients and produced genetic recombinants by double crossovers between the donor fragment and the recipient chromosome (Figure 8.10a). That is, if gene *a* enters the recipient after 9 minutes, and gene *b* after 15 minutes, genes *a* and *b* are considered to be 6 minutes apart on the genetic map. Thus, by measuring the times at which donor genes enter recipients, the order of genes on the chromosome and the map distances between genes (with map units measured in minutes) can be determined. The genetic map of large chromosomal segments of *E. coli* was constructed in just this way.

The use of interrupted mating to map bacterial genes may be illustrated by considering one of Jacob and Wollman's experiments, which involved the following cross:

Donor:
HfrH thr⁺ leu⁺ azi^r ton^r lac⁺ gal⁺ str^s

Recipient:
F⁻ thr leu azi^s ton^s lac gal str^r

(The superscript ^s means "sensitive" and the superscript ^r means "resistant.")

The *HfrH* strain is prototrophic and is sensitive to the antibiotic streptomycin. The *F⁻* cell carries a streptomycin resistance gene and also a number of mutant genes. These genes cause the *F⁻* to be auxotrophic for threonine (*thr*) and leucine (*leu*), to be sensitive to sodium azide (*azi^s*) and to infection by bacteriophage T1 (*ton^s*), and to be unable to ferment lactose (*lac*) or galactose (*gal*).

In such a conjugation experiment the two cell types are mixed together in a nutrient medium at 37°C, the normal growth temperature for *E. coli*. After a few minutes have passed to allow the *HfrH* and *F⁻* cells to pair, the culture is diluted to prevent the formation of new mating pairs. This procedure ensures some synchrony in the timing of chromosome transfer between cells. Samples are removed from the mating mixture at various times and are then agitated to break the pairs apart. These transconjugants are plated on a selective medium designed to allow recombinant types to grow and divide while selecting against both the *HfrH* and *F⁻* parental types.

For this particular cross the medium contains streptomycin to kill the *HfrH* and lacks threonine and leucine so that the parental *F⁻* cannot grow. In this cross the threonine (*thr⁺*) and leucine (*leu⁺*) genes are the first donor genes to be transferred to the *F⁻* to produce a merodiploid, so recombinants formed by the exchange of those genes with the *thr leu* genes of the *F⁻* recipient grow on the selective medium. Appropriate media can be used to test for the appearance of other donor genes (*azi^r, ton^r, lac⁺,* and *gal⁺*) among the selected *thr⁺ leu⁺ str^r* transconjugants. For example, medium with sodium azide added can test for the presence of *azi^r* from the donor, and so on.

Figure 8.10b shows an example of the results. The transconjugants here are *thr⁺, leu⁺,* and *str^r*. The first unselected marker gene to be transferred is *azi^r*, and recombinants for this gene are seen at about 8 minutes. The second gene transferred is *ton^r* at 10 minutes, followed by *lac⁺* at about 17 minutes and *gal⁺* at about 25 minutes. Figure 8.10b also shows the rates of appearance and the maximum frequencies obtained for each recombinant type. The following conclusions can be made:

~ FIGURE 8.10

Interrupted-mating experiment involving the cross *HfrH thr⁺ leu⁺ azi*ᵗ *ton*ᵗ *lac⁺ gal⁺ str*ˢ × *F⁻ thr leu azi*ˢ *ton*ˢ *lac gal str*ᵗ. The progressive transfer of donor genes with time is illustrated. Recombinants are generated by an exchange of a donor fragment with the homologous recipient fragment resulting from a double crossover event. (a) At various times after mating commences, the conjugating pairs are broken apart and the transconjugant cells are plated on selective media to determine which genes have been transferred from the *Hfr* to the *F⁻*. (b) The graph shows the appearance of donor genetic markers in the *F⁻* cells as a function of time; that is, after the selected *thr⁺* and *leu⁺* genes have entered.

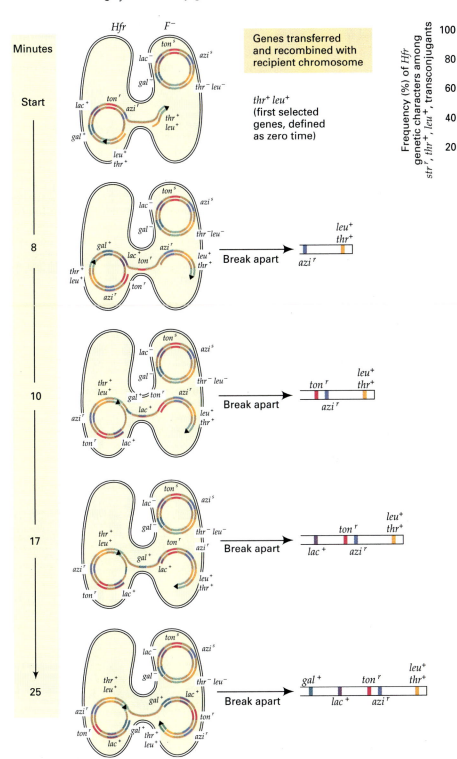

a) **Progressive transfer of donor genes to recipient during *Hfr* × *F⁻* conjugation**

b) **Appearance of donor genetic markers in recipient as a function of time**

1. As the time for conjugation increases, both the rates of appearance and the maximum frequencies of recombinants decrease.
2. The rate of transfer from mating couple to mating couple is not constant.
3. The maximum frequency of recombinants becomes smaller the later the gene enters the recipient because with time there is an increasing chance that mating pairs will break apart.

Note that there is a fundamental difference between the recombination process just described for merodiploids and that observed in meiosis: in merodiploid recombination one product is lost, whereas in meiotic recombination both reciprocal products are recovered.

In this experiment, each gene from an *Hfr* bacterium appears in recombinants at a different but reproducible time after mating begins, and the time intervals between gene appearances are used as a measure of genetic distance. From such data we can conclude that an *Hfr* chromosome is transferred into the *F⁻* cell in a linear way. As with *F* factor transfer, the transfer starts at the origin within the integrated *F* factor. Genes located far from the origin tend not to be transferred because of the high probability that the conjugating pairs will break apart before their transfer can occur. Thus from the experiment described (*HfrH* × *F⁻*) and from the time intervals, the map in Figure 8.11 may be constructed. Again, the map units are in "minutes"; the entire *E. coli* chromosome requires about 100 minutes for transfer.

Circularity of the *E. coli* Map

Only one *F* factor is integrated in each *Hfr* strain. Different *Hfr* strains have the *F* factor integrated at different locations and in different orientations in the chromosome. For a particular bacterium, the integration of the *F* factor to produce an *Hfr* cell is somewhat random. Since it is the *F* factor that is responsible for the transfer of donor genes into the recipient strain, the *Hfr* strains differ with respect to (1) where the transfer of donor genes begins, and (2) the direction, and therefore order of transfer of donor genes. Figure 8.12a shows the order of chromosomal gene transfer for four different *Hfr* strains *H*, *1*, *2*, and *3*. In each case, only one *Hfr* strain was used to cross with the recipient, and the order of gene transfer and the time between the appearance of each gene in the recipient was determined. The genetic distance in time units

~ FIGURE 8.12

Interrupted-mating experiments with a variety of *Hfr* strains, showing that the *E. coli* linkage map is circular: (a) Orders of gene transfer for the *Hfr* strains *H*, *1*, *2*, and *3*; (b) Alignment of gene transfer for the *Hfr* strains; (c) Circular *E. coli* chromosome map derived from the *Hfr* gene transfer data. The map is a composite of various locations of integrated *F* factors. A given *Hfr* strain has only one integrated *F* factor. The "arrowhead" is transferred first.

a)   Orders of gene transfer

Hfr strains:

H	origin – *thr* – *pro* – *lac* – *pur* – *gal* – *his* – *gly* – *thi*
1	origin – *thr* – *thi* – *gly* – *his* – *gal* – *pur* – *lac* – *pro*
2	origin – *pro* – *thr* – *thi* – *gly* – *his* – *gal* – *pur* – *lac*
3	origin – *pur* – *lac* – *pro* – *thr* – *thi* – *gly* – *his* – *gal*

b)   Alignment of gene transfer for the *Hfr* strains

H	*thr* – *pro* – *lac* – *pur* – *gal* – *his* – *gly* – *thi*
1	*pro* – *lac* – *pur* – *gal* – *his* – *gly* – *thi* – *thr*
2	*lac* – *pur* – *gal* – *his* – *gly* – *thi* – *thr* – *pro*
3	*gal* – *his* – *gly* – *thi* – *thr* – *pro* – *lac* – *pur*

c)   Circular *E. coli* chromosome map derived from *Hfr* gene transfer data

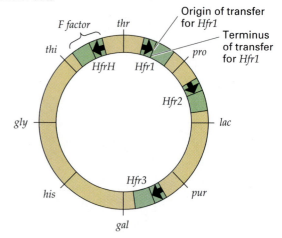

~ FIGURE 8.11

Chromosome map of the genes in the interrupted-mating experiment of Figure 8.10. The marker positions represent the time of entry of the genes into the recipient during the experiment. Thus the map distance is given in minutes. The *azi* marker, for example, appeared about 8 minutes after conjugation began, whereas the *gal* marker appeared after about 25 minutes.

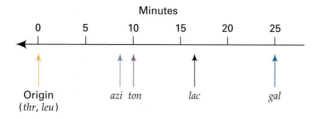

between a particular pair of genes is constant, within experimental error, no matter which *Hfr* strain is used as donor; for example, the genetic distance between *thr* and *pro* is the same in *H, 1, 2,* and *3*. This validates the use of time units as a measure of genetic distance in *E. coli*.

From this sort of data, a genetic map of the chromosome is constructed by aligning the genes transferred by each *Hfr* as shown in Figure 8.12b. In view of the overlap of the genes, the simplest map that can be drawn from these data is a circular one, as shown in Figure 8.12c. The map, then, is a composite of the results of the individual matings. The circularity of the map was itself a truly significant finding, since all previous genetic maps of eukaryotic chromosomes were linear.

By using conjugation experiments, a complete genetic map of the *E. coli* chromosome was constructed; it is 100 minutes long (Figure 8.30). As with genetic maps of other organisms, this map provides information about the relative locations of *E. coli* genes on the circular chromosome. Using recombinant DNA techniques (introduced in Chapter 1 and discussed fully in Chapter 15), *physical maps* of chromosomes can also be constructed. Physical maps provide information about the relative locations of cloned DNA sequences on the chromosome. Such a physical map has been constructed for the *E. coli* chromosome. In brief, the entire chromosome was cut up into overlapping fragments using restriction enzymes (enzymes that cut DNA at specific sites), and the fragments were then cloned in a cloning vector. Each cloned fragment was analyzed to determine the sites for cleavage by a number of other restriction enzymes, and computer analysis sorted the cloned fragments based on the restriction site data to establish an order of the fragments around the chromosome. These ordered clones can be exploited for the isolation of any desired *E. coli* genes if their genetic map location is known. As we have just seen, such a genetic map has been constructed by conjugation analysis.

ᛕEYNOTE

The circular *F* factor can integrate into the circular bacterial chromosome by a single crossover event. Strains in which this integration has happened can conjugate with *F⁻* strains, and transfer of the bacterial chromosome occurs. The strains containing the integrated *F* factor are called *Hfr* (high-frequency recombination) strains. In *Hfr* × *F⁻* matings the chromosome is transferred in a one-way fashion from

the *Hfr* cell to the *F⁻* cell, beginning at a specific site called the origin (O). The farther a gene is from O, the later it is transferred to the *F⁻*, and this is the basis for mapping genes by their times of entry into the *F⁻* cell. Conjugation and interrupted mating allow mapping of large chromosome segments.

TRANSDUCTION IN BACTERIA

Transduction (literally "leading across") is a process by which bacteriophages (bacterial viruses: phages, for short) function as intermediaries in the transfer of bacterial genetic information from one bacterium (the donor) to another (the recipient); such phages are called **phage vectors**. Since the amount of DNA a phage can carry is limited, the amount of genetic material that can be transferred is usually less than 1 percent of that in the bacterial chromosome. Once the donor genetic material has been introduced into the recipient, it may undergo genetic recombination with a homologous region of the recipient chromosome. The recipients inheriting donor DNA in this way are called **transductants**.

Bacteriophages: An Introduction

Most bacterial strains are susceptible to infection by specific phages. For example, *E. coli* is susceptible to infection by DNA-containing phages such as T2, T4, T5, T6, T7 and λ (lambda), among others. As a result of extensive study, most of the genes located on the chromosomes of the commonly used phages have been identified and mapped.

The structure of all phages is simple. A phage contains its genetic material (either DNA or RNA) in a single chromosome surrounded by a coat of protein molecules. Variation in the number and organization of the proteins gives the phages their characteristic appearances. Figure 8.13 presents electron micrographs and diagrams of two phages of great genetic significance, T4 (Figure 8.13a) and λ (Figure 8.13b). Phage T4 is one of a series of arbitrarily numbered phages with similar properties named T2, T4, and T6 (collectively called T-even phages). It has a number of distinct structural components: a head (which contains DNA), a core, a sheath, a base plate, and tail fibers (the latter two structures enable the phage to attach itself to a bacterium). The head and all other structural components consist of proteins.

~ FIGURE 8.13

Electron micrographs and diagrams of two bacteriophages (1 nm = 10^{-9} m): (a) T4 phage, which is representative of T-even phages; (b) λ phage.

a) T4 phage

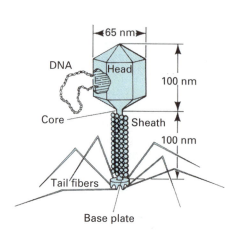

b) λ phage

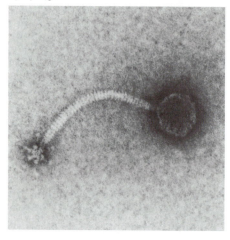

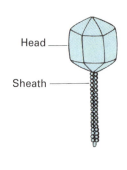

The life cycles of the T4 and the λ phages are not identical. Figure 8.14 presents the stages of the T4 cycle. First (1), the phage particle attaches to the surface of the bacterial cell, and the phage chromosome is injected into the bacterium in a series of events that include a springlike contraction of the phage sheath. Then (2–5), by the action of phage genes, the phage takes over the bacterium, breaking down the bacterial chromosome and directing its growth and reproductive mechanisms to produce progeny phages. Finally (6), the progeny phages are eventually released from the bacterium as the cell is broken open (lysed). The suspension of released progeny phages is called a **phage lysate.** This type of phage life cycle is called the **lytic cycle,** and phages that always follow that cycle when they infect bacteria are called **virulent phages.**

The λ life cycle, shown in Figure 8.15 (p. 240), is more complex than that of a T4 phage. Phage λ has a structure similar to that of T4. When its DNA is injected into *E. coli*, there are two paths that the phage can follow. One is a lytic cycle, exactly like that of the T phages. The other is the **lysogenic pathway** (or *lysogenic cycle*). In the lysogenic pathway the λ chromosome does not replicate; instead, it inserts (integrates) itself physically into a specific region of the host cell's chromosome, much like *F* factor integration. In this integrated state the phage chromosome is called a **prophage.** Every time the host cell chromosome replicates, the integrated λ chromosome is also replicated. The bacterium that contains a phage in the prophage state is said to be **lysogenic** for that phage; the phenomenon of the insertion of a phage chromosome into a bacterial chromosome is called **lysogeny.**

~ FIGURE 8.14

Lytic life cycle of a virulent phage, such as T2 or T4.

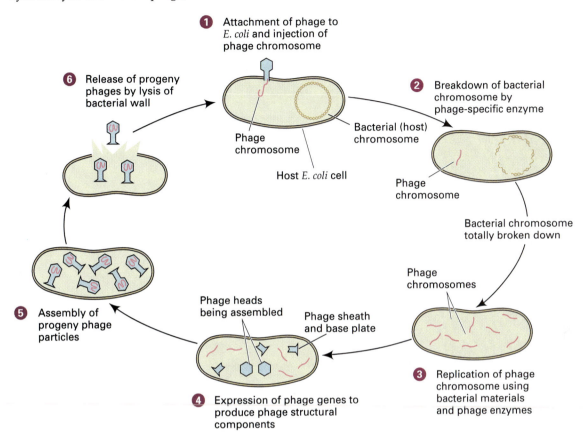

1. Attachment of phage to *E. coli* and injection of phage chromosome

2. Breakdown of bacterial chromosome by phage-specific enzyme

Bacterial (host) chromosome

Phage chromosome

Host *E. coli* cell

Phage chromosome

6. Release of progeny phages by lysis of bacterial wall

Bacterial chromosome totally broken down

Phage chromosomes

3. Replication of phage chromosome using bacterial materials and phage enzymes

5. Assembly of progeny phage particles

Phage heads being assembled

Phage sheath and base plate

4. Expression of phage genes to produce phage structural components

Phages that have a choice between lytic and lysogenic pathways are called *temperate phages*. The prophage state is maintained by the action of a specific gene product (a repressor protein) that prevents expression of λ genes essential for the lytic cycle (see Chapter 16). When the repressor that maintains the prophage state is destroyed, for example by environmental factors such as ultraviolet light irradiation, then the lytic cycle is induced, and progeny λ phages are released.

Transduction Mapping of Bacterial Chromosomes

Transduction is a useful mechanism for mapping bacterial genes. Often, the useful phages are temperate phages, because they do not kill the bacteria they infect. Examples are phages P1 and λ for *E. coli* and phage P22 for *Salmonella typhimurium*. Two types of transduction occur: In **generalized transduction** any gene may be transferred between bacteria; in **specialized transduction** only specific genes are transferred.

GENERALIZED TRANSDUCTION. The discovery of generalized transduction is credited to Joshua and Esther Lederberg and Norton Zinder. In 1952 these researchers were testing to see whether conjugation occurred in the bacterial species *Salmonella typhimurium*, which causes typhoid fever in animals. Their experiment was similar to the one that showed conjugation existed in *E. coli*. They mixed together two multiple auxotrophic strains and looked for the appearance of prototrophs. One strain was *phe⁺ trp⁺ tyr⁺ met his*, which required methionine and histidine, and the other strain was *phe trp tyr met⁺ his⁺*, which required phenylalanine, tryptophan, and tyrosine. When they crossed these two auxotrophic strains, they found prototrophic recombinants *phe⁺ trp⁺ tyr⁺ met⁺ his⁺* at a low frequency.

When Zinder and the Lederbergs used the U tube apparatus (which helped establish conjugation in *E. coli*), they still found prototrophs. This result indicated that recombinants were being produced by a mechanism that did not require cell-to-cell contact.

~ FIGURE 8.15

Life cycle of a temperate phage, such as λ. When a temperate phage infects a cell, the phage may go through either the lytic or lysogenic cycle.

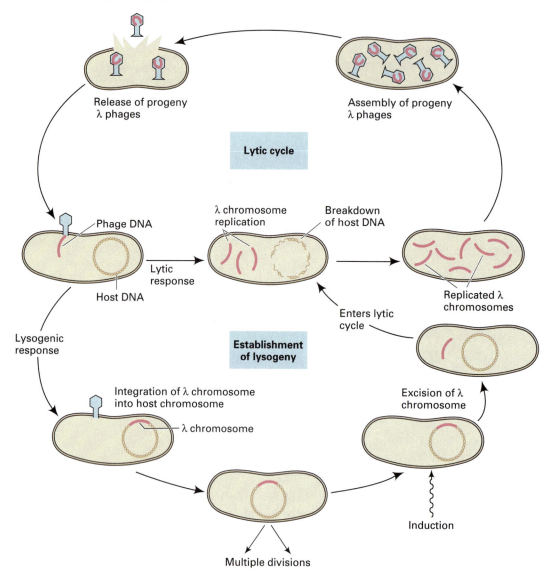

The interpretation was that the agent responsible for the formation of recombinants was a *filterable agent*, since it could pass through a filter with pores small enough to block bacteria. In this particular case the filterable agent was identified as the temperate phage P22; that is, all attempts at separating the transducing activity from phage P22 failed.

The mechanism for generalized transduction by P1 is shown in Figure 8.16. Normally, the P1 phage enters the lysogenic state when it infects *E. coli*, existing in the prophage state in these cells (i.e., the phage genes required for taking the phage through the lytic cycle are suppressed). As we have said, the mecha-

nism that maintains this state breaks down occasionally, and the phage genome goes through the lytic cycle and produces progeny phages. During the lytic cycle, the bacterial DNA is degraded, and on rare occasions, instead of the phage DNA being packaged into the phage head, a piece of bacterial DNA is so packaged (see Figure 8.16). Since this event is rare, only a very small proportion of the progeny phages carry bacterial genetic material. These phages are called **transducing phages** since they are the vehicles by which genetic material is carried between bacteria. The population of phages in the phage lysate, consisting mostly of normal phages but with about 1 in 10^5

~ **FIGURE 8.16**

Generalized transduction between strains of *E. coli*: (1) Wild-type donor cell of *E. coli* infected with the temperate bacteriophage P1; (2) The host cell DNA is broken up during the lytic cycle; (3) During assembly of progeny phages, some pieces of the bacterial chromosome are incorporated into some of the progeny phages to produce transducing phages; (4) Following cell lysis, a low frequency of transducing phages is found in the phage lysate; (5) The transducing phage infecting an auxotrophic recipient bacterium; (6) A double-crossover event results in the exchange of the donor *a*⁺ gene with the recipient mutant *a* gene; (7) The result is a stable *a*⁺ transductant, with all descendants of that cell having the same genotype.

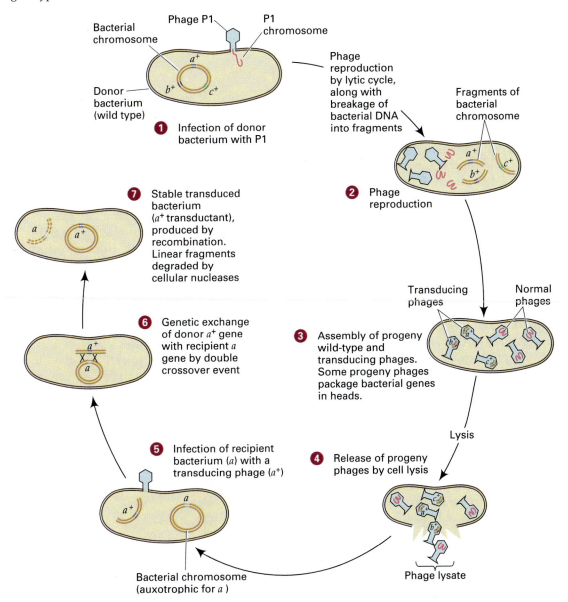

transducing phages present, can now be used to infect a new population of bacteria. Recipient bacteria that show genetic recombination of the transferred donor genes are called *transductants*. As with transformation and conjugation, the recombinants are produced by a double crossover, resulting in an exchange of a donor DNA segment with a homologous segment of the recipient chromosome.

Typically, a transduction experiment is designed so that the donor cell type and the recipient cell type have different genetic markers; then the transduction events can be followed. The strategy behind a transduction experiment is similar to the strategy behind a transformation experiment—and, in fact, behind all genetic mapping experiments. For example, if the donor cell is *thr*⁺ and the recipient cell is *thr*, prototrophic transductants can be detected, since the cell no longer requires threonine in order to grow. In this way, researchers are able to pick out the extremely low number of transductants from among all the cells present by selecting for those cells that are able to do something the nontransduced cells cannot, namely, grow on a minimal medium. In this case, *thr*⁺ is called a *selected marker*. Other markers in the experiment are called *unselected markers*.

The process just described is called *generalized transduction* because the piece of bacterial DNA that the phage erroneously picks up is a *random* piece of the fragmented bacterial chromosome. Thus any genes can be transduced: only an appropriate phage and bacterial strains carrying different genetic markers are needed. Just as in transformation, generalized transduction may be used to determine gene order and to map the distance between genes. The logic is identical for the two processes. In a typical transduction experiment, transductants for one of two or more donor markers are selected and these transductants are then analyzed for the presence or absence of other donor markers.

Consider a donor bacterium with genotype *a*⁺ *b*⁺ and a recipient bacterium with genotype *a b*. If we do a transduction experiment and select for *a*⁺ transductants, we can then test whether they are also recombinant for *b*⁺. Since the production of a transducing phage is a rare event, and since the amount of bacterial DNA the phage can carry is limited by the capacity of the phage head, only if the genes are close enough to be included in the transducing phage, can an *a*⁺ *b*⁺ transductant—called a *cotransductant*—be produced. Thus, **cotransduction** of two or more genes is a good indication that the genes are closely linked. With phage P1, cotransduction occurs for gene markers no more than about 1.5 minutes apart on the *E. coli* genetic map.

Gene order and map distances between cotransduced genes may be determined by using generalized transduction, and it is by this procedure that fine structure (i.e., detailed) linkage maps of bacterial chromosomes have been constructed. For example, consider the mapping of some *E. coli* genes by using transduction with the temperate phage P1. The donor *E. coli* strain is *leu*⁺ *thr*⁺ *azi*ʳ (able to grow on a minimal medium and resistant to the metabolic poison sodium azide). The recipient cell is *leu thr azi*ˢ (needs leucine and threonine as supplements in the medium and sensitive to azide). The P1 phages are grown on the bacterial donor cells, and the phage lysate is used to infect the recipient bacterial cells. Transductants are selected for any one of the donor markers and then analyzed for the presence of the other unselected markers (in this case, two). Typical data from such an experiment are shown in Table 8.1.

Consider the *leu*⁺ selected transductants. Of these transductants 50 percent exhibit *azi*ʳ and 2 percent have *thr*⁺. For the *thr*⁺ transductants 3 percent are *leu*⁺ and 0 percent are *azi*ʳ. The simplest interpretation is that the *leu* gene is closer to the *thr* gene than is the *azi* gene. The order of genes then is

$$\text{thr} \qquad\qquad \text{leu} \quad \text{azi}$$

The transductants are produced by crossing-over between the piece of donor bacterial chromosome brought in by the infecting phage and the homologous region on the recipient bacterial chromosome. The infected donor DNA "finds" the region of the recipient chromosome to which it is homologous, and the exchange of parts is accomplished by double (or other even-numbered) crossovers (see Figure 8.16).

Map distance can be obtained from transduction experiments involving two or more genes. As before, transductants for one of two or more donor markers are selected and these transductants are then analyzed for the presence or absence of other donor markers. For example, transduction from an *a*⁺ *b*⁺ donor to an *a b* recipient produces various transductants for *a*⁺ and *b*⁺, namely *a*⁺ *b*, *a b*⁺, and *a*⁺ *b*⁺. If we select for one or other donor markers, we can determine linkage information for the two genes. If we select for *a*⁺ transductants, map distance (expressed as *cotransduction frequency*) between genes *a* and *b* is given by:

$$\frac{\text{number of single-gene transductants}}{\text{number of total transductants}} \times 100\%$$

$$= \frac{(a^+\, b)}{(a^+\, b) + (a^+\, b^+)} \times 100\%$$

~ **TABLE 8.1**

Transduction Data for Deducing Gene Order

SELECTED MARKER	UNSELECTED MARKERS
leu⁺	50% = *azi*⁺
	2% = *thr*⁺
thr⁺	3% = *leu*⁺
	0% = *azi*ʳ

If we select for b^+ transductants, map distance between a and b is given by

$$\frac{(a\ b^+)}{(a\ b^+) + (a^+\ b^+)} \times 100\%$$

This method of gene mapping produces map distances only if the genes involved are close enough on the chromosome so that they can be cotransduced.

SPECIALIZED TRANSDUCTION. Some temperate bacteriophages can transduce only certain sections of the bacterial chromosome, in contrast to generalized transducing phages, which can carry any part of the bacterial chromosome. An example of such a **specialized transducing phage** is λ, which infects *E. coli*.

The life cycle of λ was described earlier (see Figure 8.15). In the bacterial cell, the λ genome integrates into the bacterial chromosome at a specific site between the *gal* region and the *bio* region, producing a *lysogen* (Figure 8.17a). That site on the *E. coli* chromosome is called *att* λ (attachment site for lambda), and it is homologous with a site called *att* in the λ DNA. By a single-crossover event the λ chromosome integrates. In the integrated state the phage, now called a *prophage*, is maintained by the action of a phage-encoded repressor protein.

The particular *E. coli* strain that λ lysogenizes is *E. coli K12*, and when it contains the λ prophage, it is designated *E. coli K12(λ)*. Let us focus just on the *gal* gene and assume that the particular K12 strain that λ lysogenized is *gal⁺*; that is, it can ferment galactose as a carbon source. This phenotype is readily detectable by plating the cells on a solid medium containing galactose as a carbon source as well as a dye that changes color in response to the products of galactose fermentation. On this medium the *gal⁺* colonies are pink, while *gal* colonies are white. If we induce the prophage (see p. 238)—that is, reverse the inhibition of phage functions—the lytic cycle is initiated.

When the lytic cycle is initiated, the phage chromosome loops out, generating a separate circular λ chromosome by a single-crossover event at the *att* λ/*att* sites (Figure 8.17b). In most cases the excision of the phage chromosome is precise, so that the complete λ chromosome is produced (Figure 8.17b1). In rare cases the excision events are not precise: crossing-over between the phage chromosome and the bacterial chromosome at sites other than the homologous recognition sites results in an abnormal circular DNA product (Figure 8.17b2). In the case diagrammed, a piece of λ chromosome has been left in the bacterial chromosome, while a piece of bacterial chromosome, including the *gal⁺* gene, has been added

to the rest of the λ chromosome. Since a bacterial gene (or genes) is included in a progeny phage, we have a transducing phage, called, in this case, λ*d gal⁺*. The *d* stands for "defective" since not all phage genes are present, and the *gal* indicates that the bacterial host cell *gal* gene has been acquired. This event is similar to *F'* production by defective excision of the *F* factor. The λ*d gal⁺* can replicate and lyse the host cell in which it is produced, however, since all λ genes are still present: some are on the phage chromosome while the others are in the bacterial chromosome.

Since the abnormal looping-out phenomenon is a rare event, the phage lysate produced from the initial infection diagrammed contains mostly normal phages and a few transducing phages ($1/10^5$). The phage lysate can be used to infect bacteria that are *gal*. The phage lysate contains mostly wild-type phages and relatively few *gal⁺* transducing phages. Because of the small proportion of transducing phages, the lysate is called a *low-frequency transducing* (LFT) lysate.

Infection of the *gal* bacterial cells with the LFT lysate produces two types of transductants (Figure 8.17c). In one type the wild-type λ integrates at its normal *att* λ site, and then the λ*d gal⁺* phage integrates by a crossing-over event within the common λ sequences to produce a double lysogen (Figure 8.17c1). In this case both types of phages are integrated in the bacterial chromosome, and the bacterium is heterozygous *gal⁺/gal* and hence can ferment galactose.

This type of transductant is unstable because phage growth can be initiated by induction. The wild-type λ has a complete set of genes for virus replication so it controls the outlooping and replication of itself and the λ*d gal⁺*. In this capacity the wild-type λ phage acts as a *helper phage*. Potentially one-half of the progeny phages will be λ*d gal⁺* and so this new lysate is called a *high-frequency transducing* (HFT) lysate.

The second type of transductant produced by the initial lysate is stable: These transductants are produced when only a λ*d gal⁺* phage infects a cell (Figure 8.17c2). The *gal⁺* gene carried by the phage may be exchanged for the bacterial *gal* gene by a double-crossover event. Such a transductant is stable because the bacterial chromosome contains only one type of gal gene and no phage genes are integrated.

Because of the mechanisms involved, specialized transduction can only transduce small segments of the bacterial chromosome that are on either side of the prophage. Specialized transduction is used for moving *specific* genes between bacteria; for example, for constructing strains with particular genotypes.

~ FIGURE 8.17

Specialized transduction by bacteriophage λ (for detail see text): (a) Production of a lysogenic bacterial strain by crossing-over in the region of homology between the circular bacterial chromosome (att λ) and the circularized phage chromosome (att). (b) Production of initial low frequency transducing (LFT) lysate: induction of the lysogenic bacterium causes out-looping. Normal outlooping (1) produces normal λ and rare abnormal outlooping (2) produces a transducing λd gal⁺ phages. (c) Transduction of gal bacteria by the initial lysate, produces either (1) unstable transductants by integration of both λ and λd gal⁺ (resulting in a double lysogen) or (2) stable transductants (single lysogens) by crossing-over around the gal region. Induction of the unstable double lysogen produces about equal numbers of λd gal⁺ and wild-type λ phages; the result is a high-frequency transducing (HFT) lysate.

a) Production of lysogen

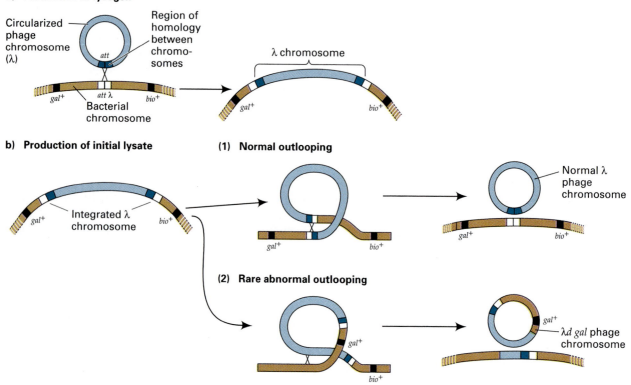

b) Production of initial lysate

c) Transduction of gal bacteria by initial lysate, consisting of λ and λd gal phage

1) Lysogenic transductant

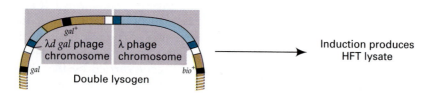

2) Transductant produced by recombination

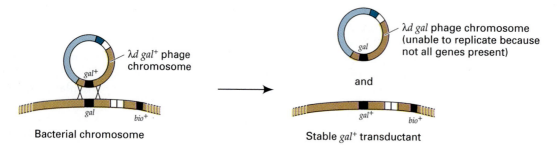

~ FIGURE 8.18

Plaques of the *E. coli* bacteriophages (a) T4 and (b) λ.

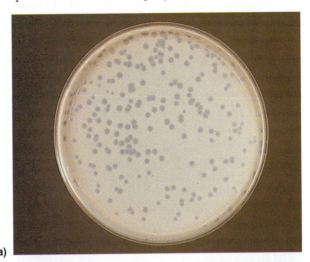

a)

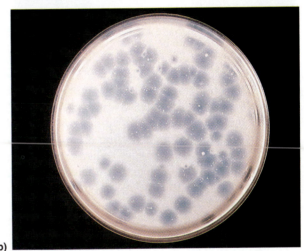

b)

KEYNOTE

Transduction is the process by which bacterio-phages mediate the transfer of genetic information from one bacterium (the donor) to another (the recipient). The capacity of the phage particle is limited, so the amount of DNA transferred is usually less than 1 percent of that in the bacterial chromosome. In generalized transduction any bacterial gene can be accidentally incorporated into the transducing phage during the phage life cycle and subsequently transferred to a recipient bacterium. Specialized transduction is mediated by temperate phages (such as λ) in which prophages associate with only one site of the bacterial chromosome. In this case the transducing phage is generated by abnormal excision of the prophage from the host chromosome so that the prophage includes both bacterial and phage genes. Transduction allows the fine-structure mapping of small chromosome segments.

MAPPING GENES IN BACTERIOPHAGES

The same principles used to map eukaryotic genes are used to map phage genes. Crosses are made between phage strains that differ in genetic markers, and the proportion of recombinants among the total progeny is determined. The basic procedure of mapping phage genes in two-, three-, or four-gene crosses involves the mixed infection of bacteria with phages of different genotypes.

In phages this basic experimental design must be adapted to the phage life cycle. First, we must be able to count phage types. We do so by plating a mixture of phages and bacteria on a solid medium. The concentration of bacteria is chosen so that an entire "lawn" of bacteria grows. Phages are present in much lower concentrations. Each phage infects a bacterium and goes through the lytic cycle. The released progeny phages infect neighboring bacteria, and the lytic cycle is repeated. The result is a cleared patch in the lawn of bacteria. The clearing is called a **plaque**, and a single plaque derives from one of the original bacteriophages that was plated. Figure 8.18 shows plaques of the *E. coli* phages T4 and λ; note the T4 plaque is clear while the λ plaque has a turbid center.

Next, we must be able to distinguish phage phenotypes easily. Since individual phages are visible only under the electron microscope, mutations that affect phage morphology cannot be used effectively. However, several mutations affect the phage life cycle, giving rise to differences in the appearances of plaques on a bacterial lawn. For example, one of the earliest studies of phage recombination was done in the late 1940s by Alfred Hershey and R. Rotman. They used strains of T2 that differed in two ways: plaque morphology (that is, the appearance of the *plaques*, not the phages) and host range (that is, which bacterial strain the phage can lyse). One phage strain, genotype $h^+ r$, was wild-type for the host range gene (h^+, able to lyse the *B* strain but not the *B/2* strain of *E. coli*; that is, strain *B* is the *permissive host* and strain

$B/2$ is the *nonpermissive host* for h^+ phages) and mutant for the plaque morphology gene (r, producing large plaques with distinct borders). The other phage strain, $h r^+$, was mutant for the host range (able to lyse both the B and the $B/2$ strains of *E. coli*) and wild-type for the plaque morphology gene (r^+, producing small plaques with fuzzy borders). In addition, when plated on a lawn containing both the B and $B/2$ strains, any phage carrying the mutant host range allele h (infects both B and $B/2$) resulted in clear plaques, while phages carrying the wild-type h^+ allele resulted in cloudy plaques. The latter characteristic arises because phages bearing the h^+ allele can only infect the B bacteria, leaving a background cloudiness of uninfected $B/2$ bacteria.

To map these two genes, we must make a genetic cross. The cross is achieved by adding both types of phages ($h^+ r$ and $h r^+$) to a culture of *E. coli* strain B (which both parents can infect), making sure that there is a high enough concentration of each phage type to ensure that a high proportion of bacteria are infected simultaneously with both phages (= high *multiplicity* of infection, or moi; see Figure 8.19a). Once the two genomes are within the bacterial cell, each will replicate (Figure 8.19b). If an $h^+ r$ and an $h r^+$ chromosome come together, a crossover can occur between the two gene loci to produce $h^+ r^+$ and $h r$ recombinant chromosomes (Figure 8.19c), which are assembled into progeny phages. When the bacterium lyses, the recombinant progeny are released into the medium, along with non-recombinant (parental) phages (Figure 8.19d).

After the life cycle is completed, the progeny phages are plated onto a bacterial lawn containing a mixture of *E. coli* strains B and $B/2$. Four plaque phenotypes are found from this experiment (Figure 8.20), two parental types and two recombinant types. The parental type $h r^+$ gives a small plaque with a fuzzy border; the other parental $h^+ r$ gives a large plaque with a distinct border. The reciprocal recombinant types give recombined phenotypes: The $h^+ r^+$ plaques are cloudy and small, and the $h r$ plaques are clear and large.

Once the progeny plaques are counted, we can calculate the recombination frequency for recombination events occurring between h and r. In this case the frequency is given by

$$\frac{(h^+ r^+) + (h\, r) \text{ plaques}}{\text{total plaques}} \times 100$$

Making the same assumption used in mapping eukaryotic genes, namely, that crossovers occur randomly along the chromosome, the recombination frequency then represents the map distance (in map units) between the phage genes. In addition, other

~ FIGURE 8.19

Schematic of the principles of performing a genetic cross with bacteriophages. (a) Bacteria of *E. coli* strain B are coinfected with the two parental bacteriophages, $h^+ r$ and $h r^+$. (b) Replication of both parental chromosomes. (c) There is pairing of some chromosomes of each parental type, and crossing-over takes place between the two gene loci to produce $h^+ r^+$ and $h r$ recombinants. (d) Progeny phages are assembled and are released into the medium when the bacteria lyse; both parental and recombinant phages are found among the progeny.

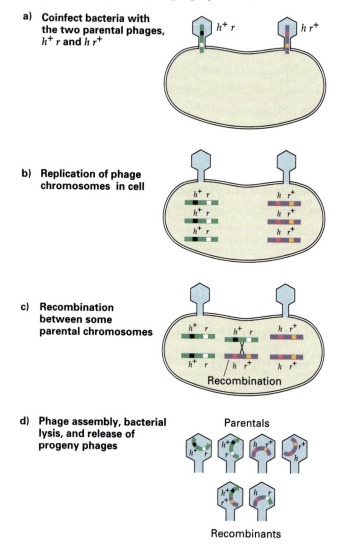

a) **Coinfect bacteria with the two parental phages, $h^+ r$ and $h r^+$**

b) **Replication of phage chromosomes in cell**

c) **Recombination between some parental chromosomes**

Recombination

d) **Phage assembly, bacterial lysis, and release of progeny phages**

Parentals

Recombinants

experiments have shown that the recombination frequency is the same when the parental strains are $h^+ r^+$ and $h r$, supporting this conclusion. Soon after these first experiments were performed, a number of other T2 genes were mapped in much the same way. It should be noted, though, that some infected *E. coli* cells will contain only one parental type or the other, so the actual frequency of recombination will be underestimated.

~ FIGURE 8.20

Plaques produced by progeny of a cross of T2 strains $h\,r^+ \times h^+\,r$. Four plaque phenotypes, representing both parental types and the two recombinants, may be discerned. The parental $h\,r^+$ type produces clear, small plaques, and the parental $h^+\,r$ type produces cloudy, large plaques. The recombinant $h\,r$ type produces a clear, large plaque, and the recombinant $h^+\,r^+$ type produces a cloudy, small plaque.

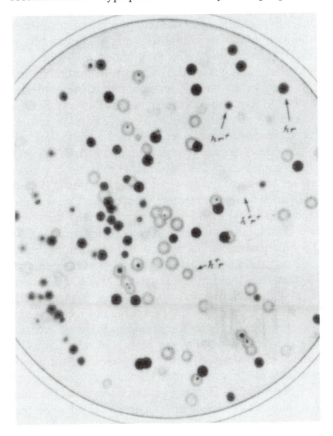

KEYNOTE

The same principles used to map eukaryotic genes are used to map phage genes. That is, genetic material is exchanged between strains differing in genetic markers and recombinants are detected and counted.

FINE-STRUCTURE ANALYSIS OF A BACTERIOPHAGE GENE

All the genetic-mapping experiments described in Chapters 5 and 6, and in this chapter for both bacteria and bacteriophages, involved the use of mutant alleles of different genes. As we have seen, the recombi-

national mapping of the distance between genes, called *intergenic mapping* (inter = between), can be used to construct chromosome maps for organisms. Indeed, the early picture of a gene was that it was like a bead on a string with mutation changing a bead from wild type to mutant or vice versa, and with recombination occurring between the beads. We now know, of course, that the gene is subdivisible by mutation and recombination and that the same general principles of recombinational mapping can be applied to mapping the distance between mutational sites within the same gene, a process called *intragenic mapping* (intra = within).

The first evidence that the gene is subdivisible by mutation and recombination came from the work of C. P. Oliver in 1940. Oliver worked with two mutations that were considered to be alleles of the X-linked *lozenge* (*lz*) locus of *Drosophila*; that is, females heterozygous for the two mutations showed the mutant lozenge-shaped eye phenotype. When female flies heterozygous for these two alleles were crossed with male flies hemizygous for either allele, progeny flies with wild-type eyes were seen with a frequency of about 0.2 percent. Oliver showed that these wild-type offspring had resulted from recombination between the alleles. In other words, he had shown that the gene was divisible by recombination, rather than an indivisible "bead on a string." Using genetic symbols, the last cross was $\dfrac{lz^A\ +}{+\ \ lz^B} \times \dfrac{lz^A\ +}{\longrightarrow}$ where lz^A and lz^B are the two lozenge alleles. Recombination in the female between the two alleles will produce + + gamete and, thence, wild-type progeny.

Oliver's discovery spawned investigations of the detailed organization of alleles within a gene. As we now know, such intragenic mapping is possible because each gene consists of many nucleotide pairs of DNA linearly arranged along the chromosome. Much of the impetus to analyze the fine details of gene structure came from the elegantly detailed work by Seymour Benzer in the 1950s and 1960s on the *rII* region of bacteriophage T4. His genetic experiments revealed much about the relationship between mapping and gene structure. In his experiments Benzer used phage T4, mainly because bacteriophages produce large numbers of progeny, which facilitates the potential determination of very low recombination frequencies. Benzer's initial experiments involved **fine-structure mapping**, the detailed genetic mapping of sites within a gene.

When cells of *E. coli* strain B or strain K12(λ), growing on solid medium in a Petri dish, are infected with wild-type T4, small turbid plaques with fuzzy edges are produced as a result of the phage life cycle (Figure 8.21). (Strain *E. coli* K12(λ) contains the λ phage chromosome incorporated into the bacterial

The r^+ and mutant *rII* plaques on a lawn of *E. coli* B. The r^+ plaque is turbid with a fuzzy edge; the *rII* plaque is larger, clear, and has a distinct boundary.

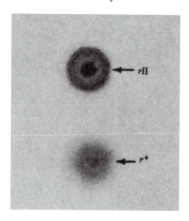

chromosome.) On the other hand, when cells of *E. coli* strain *B* are infected with *r* (rapid-lysis) mutants of phage T4, large, clear plaques with distinct edges are produced (see Figure 8.21). Using *intergenic*-mapping experiments, the *r* mutations were found to map at several locations in the phage's genome, defining several *r* genes. (See Figure 8.31, p. 258, for a map of phage T4 including the locations of the *r* genes.)

In his intragenic mapping experiments Benzer used phage strains carrying mutations of the *rII* region. Of special importance to the experiments that we are about to describe was Benzer's 1953 finding that such *rII* mutants, in addition to their *plaque morphology* phenotype, also have distinct *host range properties*. Specifically, whereas wild-type T4 is able to grow in and lyse cells of either *E. coli* B or $K12(\lambda)$, *rII* mutants can grow in and lyse cells of *B*, producing large, clear plaques on solid medium (see Figure 8.21) but are unable to grow in cells of $K12(\lambda)$. Strain B is the permissive host for *rII* mutants, while strain $K12(\lambda)$ is the nonpermissive host. Thus *rII* mutants are conditional mutants, because they can grow under one set of conditions but not others. The reasons for the inability of *rII* phages to multiply in the $K12(\lambda)$ strain are not known.

Recombination Analysis of *rII* Mutants

Benzer realized that the growth defect of *rII* mutants on *E. coli* $K12(\lambda)$ could serve as a powerful selective tool for detecting the presence of a very small proportion of r^+ phages within a large population of *rII* mutants. Between 1953 and 1963 Benzer made a collection of thousands of independently isolated *rII* mutants, some of which had arisen spontaneously and some of which had been induced by treating r^+ phages with mutagens.

Initially, Benzer set out to construct a fine-structure genetic map of the *rII* region. To do so, he crossed 60 independently isolated *rII* mutants in all possible combinations in using *E. coli* B as the permissive host, then collected the progeny phages once the cells had lysed (Figure 8.22). (For a phage such as T4, about 10^{10}–10^{11} phages per milliliter of phage lysate would be typical.) For each cross he plated a sample of the phage progeny on *E. coli* B, the permissive host, in order to count the total number of progeny phage per milliliter. He plated another sample of the phage progeny on *E. coli* $K12(\lambda)$, the nonpermissive host, to find the frequency of the very rare r^+ recombinants resulting from genetic recombination between the two *rII* mutants used in the cross. In this way Benzer was able to calculate the percentage of very rare r^+ recombinants between closely linked genetic sites.

For each cross of two *rII* mutants, such as *rII*1 and *rII*2 (Figure 8.23, p. 250), four genetic classes of progeny are possible: the two parental *rII* types (*rII*1 and *rII*2) and the two recombinant types. One of the recombinant types is the wild type (r^+), and the other is a double mutant carrying both *rII* mutations (*rII*1, 2), which has an *r* phenotype and is phenotypically indistinguishable from one or the other of the parental *rII* mutants. Recall from Chapters 5 and 6 and earlier in this chapter that the map distance between two mutations is given by the percentage of recombinant progeny among all the progeny produced by crosses of the two mutants. For the crosses of *rII* mutants, a single crossover event between the two mutant genes produces the r^+ and double *rII* mutant recombinants (see Figure 8.23). Therefore the frequencies of the two recombinant classes of phages are expected to be the same. Consequently, the total number of recombinants from each *rII* × *rII* cross is approximated by twice the number of r^+ plaques counted on plates of strain $K12(\lambda)$. The general formula for the map distance between two *rII* mutations is

$$\frac{2 \times \text{number of } r^+ \text{ recombinants}}{\text{total number of progeny}} \times 100\%$$

= map distance (in map units)

An important control was done for each cross. Each *rII* parent alone was used to infect the permissive *E. coli* B host and the progeny tested on plates of B and of $K12(\lambda)$. Just as a mutation can occur to produce an *rII* mutant from the r^+, so a mutation can occur for an *rII* mutant to change back (revert) to the r^+. Thus, it is extremely important to calculate the reversion frequencies for the two *rII* mutations in a cross and subtract the combined value from the computed recombination frequency. Fortunately, for enabling the experimental results to be analyzed to

~ FIGURE 8.22

~ FIGURE 8.22

Benzer's general procedure for determining the number of r^+ recombinants from a cross involving two rII mutants of T4.

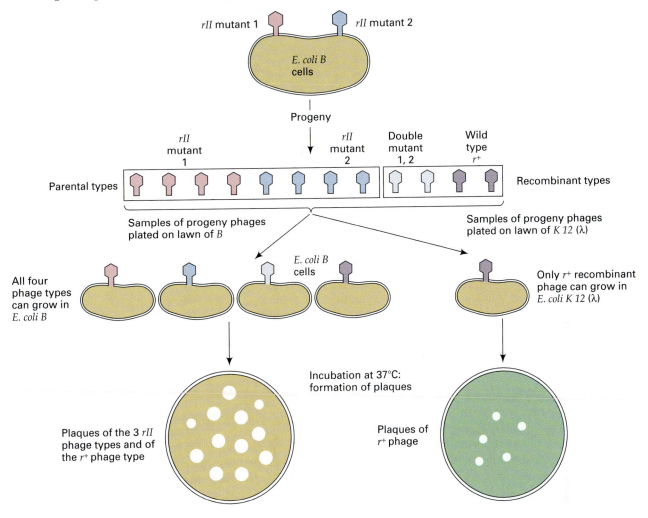

give accurate recombination frequencies, the reversion frequency for an rII mutation is at least an order of magnitude lower than the smallest recombination frequency that was found.

From the recombination data obtained from all possible pairwise crosses of the initial set of 60 rII mutants, Benzer was able to construct a linear genetic map (Figure 8.24, p. 251). Some pairs produced no r^+ recombinants when they were crossed. This result was interpreted to mean that those pairs carried mutations at exactly the same site; that is, the same nucleotide pair in the DNA had been changed; there was no possibility of recombination between the mutations. Mutations that change the same nucleotide pair within a gene are termed *homoallelic*. However, most pairs of rII mutants did produce r^+ recombinants when crossed, indicating that they carried different altered nucleotide pairs in the DNA.

Mutations that change different nucleotide pairs within a gene are termed *heteroallelic*. The map showed that the lowest frequency with which r^+ recombinants were formed in any pairwise crosses of rII mutants carrying heteroallelic mutations was 0.01 percent.

The minimum map distance of 0.01 percent can be used to make a rough calculation of the molecular distance—the distance in base pairs—between mutant markers. The circular genetic map of phage T4 is known to be about 1,500 map units. If two rII mutants produce 0.01 percent r^+ recombinants, this means that the mutations are separated by 0.02 map units, or by about $0.02/1500 = 1.3 \times 10^{-5}$ of the total T4 genome. Since the total T4 genome contains about 2×10^5 base pairs, the smallest recombination distance that was observed was $(1.3 \times 10^{-5}) \times (2 \times 10^5)$, or about 3 base pairs. That means Benzer's data had shown that

~ **FIGURE 8.23**

Production of parental and recombinant progeny from a cross of two *rII* mutants with mutations in different sites within the *rII* region. Progeny phages of parental genotype are produced if no crossing-over occurs, whereas progeny phages with recombinant genotypes are produced if a single crossover occurs between the sites of the two mutations.

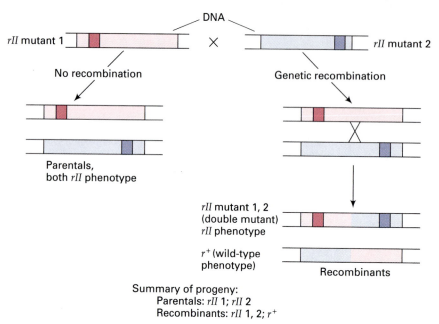

Summary of progeny:
Parentals: *rII* 1; *rII* 2
Recombinants: *rII* 1, 2; *r*⁺

genetic recombination can occur within distances on the order of 3 base pairs. Later experiments by others showed conclusively that recombination can occur between mutations that affect adjacent base pairs in the DNA. That is, genetic experiments have shown that the base pair is both the *unit of mutation* and the *unit of recombination.*

𝒦EYNOTE

The same general principles of recombinational mapping can be applied to mapping the distance between mutational sites in different genes (intergenic mapping) and to mapping mutational sites within the same gene (intragenic mapping).

Deletion Mapping

Following his initial series of crossing experiments, Benzer continued to map over 3,000 *rII* mutants in order to complete his fine-structure map. To map these 3,000 mutants would have required approxi-

mately 5 million crosses—an overwhelming task even in phages, with which up to 50 crosses can be done per day. Therefore Benzer developed some genetic tricks to simplify his mapping studies. These tricks involved the use of *deletion mapping* to localize unknown mutations.

Most of the *rII* mutants isolated by Benzer were **point mutants**; their phenotype resulted from an alteration of a single nucleotide pair. A point mutant can revert to the wild-type state spontaneously or after treatment with an appropriate mutagen. However, some of Benzer's *rII* mutants did not revert, nor did they produce *r*⁺ recombinants in crosses with a number of *rII* point mutants that were known to be located at different places on the *rII* map. These mutants were *deletion mutants*, which had lost a segment of DNA. Benzer found a wide range in the extent and location of deleted genetic material among the *rII* deletion mutants he studied. Some deletion mutants are shown in Figure 8.25, p. 252.

In actual practice an unknown *rII* point mutant was first crossed with each of the seven standard deletion mutants that defined seven main segments of the *rII* region (segments *A1-A6* and *B* in Figure 8.25). For example, if an *rII* point mutant produced *r*⁺ recombinants when crossed with deletion mutants *rA105* (deficient in *A6* and *B*) and *r638* (deficient in *B*),

~ FIGURE 8.24

Preliminary fine-structure genetic map of the *rII* region of phage T4 derived by Benzer from crosses of an initial set of 60 *rII* mutants. Lower levels in the figure show finer detail of the map. In the lowest level, the numbered vertical lines indicate individual *rII* point mutants, the *(colored)* rectangles indicate the individual *rII* deletion mutants 47, 312, 295, 164, 196, 187 and 102, and the decimals indicate the percentage of *r⁺* recombinants found in crosses between the two *rII* mutants connected by an arrow.

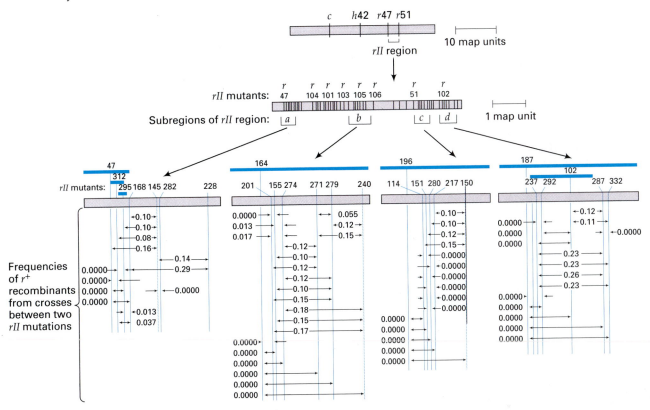

but did not produce *r⁺* recombinants when crossed with deletion mutants *r1272* (deficient in all segments), *r1241* (deficient in *A2-A6* and *B*), *rJ3* (deficient in *A3-A6* and *B*), *rPT1* (deficient in *A4-A6* and *B*), and *rPB242* (deficient in *A5-A6* and *B*), the point mutation must be in the segment of DNA that the five nonrecombinant deletion mutants lack. *r⁺* recombinants cannot be produced in crosses with deletion mutants if the deleted segment overlaps the region of DNA containing the point mutation. All five nonrecombinant deletion mutants lack the segment *A5*, and both recombinant deletion mutants contain this segment, so the point mutation must be in the *A5* region.

Once the main segment in which the mutation occurred was known, the point mutant was crossed with each of the relevant secondary set of reference deletions, *r1605*, *r1589*, and *rPB230* (see Figure 8.25). With segment *A5*, for example, three deletions divide

A5 into the four subsegments *A5a* through *A5d*. The presence or absence of *r⁺* recombinants in the progeny of the crosses of the *A5 rII* mutant with the secondary set of deletions enabled Benzer to localize the mutation more precisely to a smaller region of the DNA. For example, if the mutation was in segment *A5c*, then *r⁺* recombinants were produced with deletion *rPB230* but not with either of the other two deletions. Other deletion mutants defined even smaller regions of each of the four subsegments *A5a* through *A5d*; for example, *A5c* was divided into *A5c1*, *A5c2a1*, *A5C2a2*, and *A5c2b* by deletions *r1993*, *r1695*, and *r1168*.

In three sequential sets of crosses of point mutants with deletion mutants, it was possible to localize any given *rII* point mutant to one of 47 regions defined by the deletions, as shown in Figure 8.26, p. 253. Then all those point mutants within a given region could be

~ FIGURE 8.25

Segmental subdivision of the *rII* region of phage T4 by means of deletion. Level I shows the whole T4 genetic map. In Level II seven deletions define seven segments of the *rII* region. In Level III three deletions define four subsegments of the *A5* segments. In Level IV three deletions define four subsegments of the *A5c* subsegment. Level V shows the order and spacing of the sites of the *rII* mutations in the *A5c2a2* subsegment, as established by pairwise crosses of seven point mutants. Level VI is a model of the DNA double helix, indicating the approximate scale of the level V map.

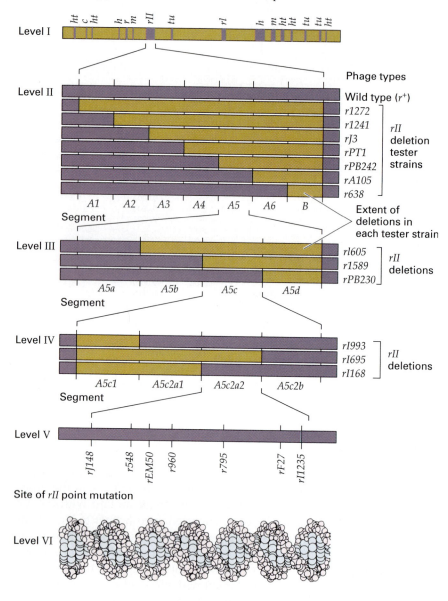

crossed in all possible pairwise crosses to construct a detailed genetic map. In this way Benzer used the more than 3,000 *rII* mutants to prove that the *rII* region is subdivisible into more than 300 mutable sites that were separable by recombination (Figure 8.27). The distribution of mutants is not random; certain sites (called *hot spots*) are represented by a large number of independently isolated point mutants.

KEYNOTE

As a result of fine-structure analysis of the *rII* region of bacteriophage T4 as well as other experiments, it was determined that the unit of mutation and of recombination is the base pair in DNA.

~ **FIGURE 8.26**

Map of deletions used to divide the *rII* region into 47 small segments. Here, those segments are shown as small boxes at the bottom of the figure.

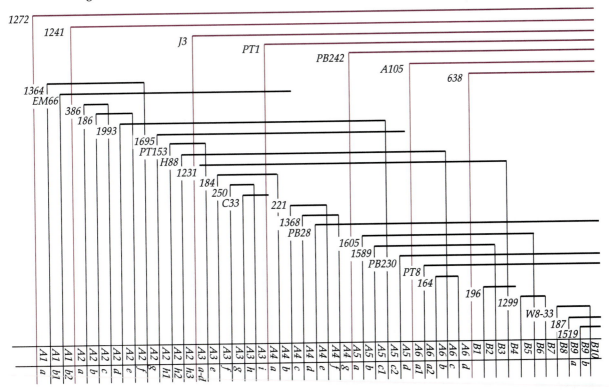

~ **FIGURE 8.27**

Fine-structure map of the *rII* region derived from Benzer's extensive experiments. The number of independently isolated mutations that mapped to a given site is indicated by the number of blocks at the site. Hot spots are represented by a large number of blocks.

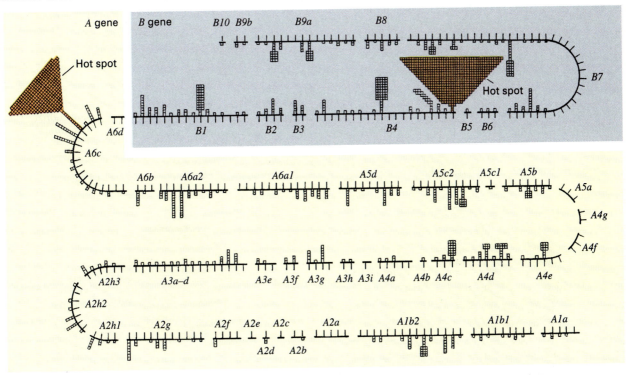

Defining Genes by Complementation Tests

From the classical point of view the gene is a *unit of function*; that is, each gene specifies one function. Benzer designed genetic experiments to determine whether this classical view was indeed true of the *rII* region of the bacteriophage. To find out whether two different *rII* mutants belonged to the same gene (unit of function), Benzer adapted the **cis-trans** or **complementation test** developed by Edward Lewis to study the nature of the functional unit of the gene in *Drosophila*. For clarification of the following discussion, it will help to know that the complementation tests indicated that the *rII* region actually consists of two genes (units of function), *rIIA* and *rIIB*. A mutation at any point in either gene will produce the *rII* phenotypes [the production of large, clear plaques on *E. coli* B and the inability to grow in *E. coli* K12(λ)]. In other words, there are two distinct genes in the *rII* region, *rIIA* and *rIIB*, each of which specifies a functional product needed for growth in *E. coli* K12(λ).

The complementation test is used to establish how many units of function (genes) define a given set of mutations that express the same mutant phenotypes. In Benzer's work with the *rII* mutants, the nonpermissive strain *K12(λ)* was infected with a pair of *rII* mutant phages to see whether the two mutants, each unable by itself to grow in strain *K12(λ)*, are able to "work together" to produce progeny phages. If the phages do produce progeny, the two mutants are said to complement each other, meaning that the two mutations must be in different genes (units of function) that specify different functional products. That is, those two products work together to allow progeny to be produced. If no progeny phages are produced, the mutants are not complementary, indicating that the mutations are in the same functional unit. Here, both mutants produce the same defective product, so the phage life cycle cannot proceed and no progeny phages result. (Note that genetic recombination is not necessary for complementation to occur; if genetic recombination does take place, a few plaques may occur on the lawn, but if complementation occurs, the entire lawn of bacteria will be lysed.)

These two situations are diagrammed in Figure

~ **FIGURE 8.28**

Complementation tests for determining the units of function in the *rII* region of phage T4; the nonpermissive host *E. coli* K12(λ) is infected with two different *rII* mutants: (a) Complementation occurs; (b) Complementation does not occur.

a) **Complementation**

Phage with mutation in *rIIA*   *E. coli* K12(λ)   Phage with mutation in *rIIB*

rIIA   rIIB
rIIA   rIIB

Defective A product (nonfunctional)   Defective B product (nonfunctional)

Functional B product   Functional A product

Progeny phage produced (lysis of host)

Plaques formed on lawn of *E. coli* B

b) **No complementation**

Phage with mutation in *rIIA*   *E. coli* K12(λ)   Phage with mutation in *rIIA*

rIIA   rIIB
rIIA   rIIB

Defective A product (nonfunctional)

Functional B product   Functional B product

No progeny phage produced (no lysis of host)

No plaques formed on lawn of *E. coli* B

8.28. In the first case the bacterium is infected with two phage genomes, one with a mutation in the *rIIA* gene and the other with a mutation in the *rIIB* gene (Figure 8.28a). The *rIIA* mutant makes a nonfunctional *A* product and a functional *B* product, while the *rIIB* mutant makes a functional *A* product and a nonfunctional *B* product. So, complementation occurs because the *rIIA* mutant still makes a functional *B* product and the *rIIB* mutant makes a functional *A* product; that is, between them functional *A* and functional *B* products are made. In other words, the two mutants work together to make both products necessary for phage propagation in *E. coli* K12(λ), so progeny phages are assembled and released. In the second case the bacterium is infected with two phage genomes, each with a different mutation in the same gene, *rIIA* (Figure 8.28b). Here, no complementation occurs because, while both produce a functional *rIIB* product, neither of the mutants makes functional *A* product, so the *A* function cannot take place. As a result, phage reproduction in *E. coli* K12(λ) does not occur.

On the basis of the results of such complementation tests, Benzer found that *rII* mutants fall into two units of function, *rIIA* and *rIIB* (also called complementation groups, which directly correspond to genes). That is, all *rIIA* mutants complement all *rIIB* mutants, but *rIIA* mutants fail to complement other *rIIA* mutants, and *rIIB* mutants fail to complement other *rIIB* mutants. The dividing line between the *rIIA* and *rIIB* units of function is indicated in the fine-structure map of Figure 8.27. Point mutants and deletion mutants in the *rII* region obey the same rules in the complementation tests. The only exceptions are deletions that span parts of both the *A* and the *B* functional units. Such deletion mutants do not complement either *A* or *B* mutants.

For the complementation test examples shown in Figure 8.28, each of the two phages that co-infect the nonpermissive *E. coli* strain K12(λ) carries one *rII* mutation, a configuration of mutations called the *trans* configuration. In this configuration, the two mutations are carried by different phages. Thus if the mutations do not complement in *trans*, they must be in the same functional unit. As a control, it is usual to coinfect *E. coli* K12(λ) with an *r*+ (wild-type) phage and an *rII* mutant phage carrying both mutations to see whether the expected wild-type function results. When both mutations under investigation are carried on the same chromosome, the configuration is called the *cis* configuration of mutations. (Because of the *cis* and *trans* configurations of mutations used, the complementation test is also called the *cis-trans* test.) In the *cis* test, the *r*+ is expected to be dominant over the two mutations carried by the *rII* mutant phage, so

progeny phages will be produced. Therefore, failure to produce progeny would not prove the mutations are in different functional genes.

Benzer referred to the genetic unit of function revealed by the *cis-trans* test as the *cistron*. A cistron may be defined as the smallest segment of DNA that encodes a piece of RNA (see Chapter 13 on transcription). At the present time *gene* is commonly used and *cistron* is being used less. Nonetheless, *gene* and *cistron* are equivalent in referring to the genetic unit of function. It is appropriate to refer to the *A* and *B* functional units as the *rIIA* and *rIIB* cistrons or genes. Presumably, their two products act in common processes necessary for T4 propagation in strain K12(λ). Genetically, the *rIIA* cistron is about 6 map units and 800 base pairs long, and the *rIIB* cistron is about 4 map units and 500 base pairs long.

The complementation test is commonly used to define the functional units (complementation groups or genes) for mutants with the same phenotype. The principles for performing a complementation test are always the same; only the practical details of performing the test are organism-specific. For example, in yeast one could select two haploid cells that are of different mating types (**a** and α) and that carry different mutations conferring the same mutant phenotype. Mating these two types would produce a diploid, which would then be analyzed for complementation of the two mutations. In animal cells, two cells, each exhibiting the same mutant phenotype, can be fused together and analyzed; a wild-type phenotype indicates that complementation has occurred. Again, in neither of these cases is recombination necessary for complementation to occur.

Let us consider an example of complementation in a diploid organism. (The cross is diagrammed in Figure 8.29, and the example comes from Question 4.24, p. 130.) Two true-breeding mutant strains of *Drosophila melanogaster* have black body color instead of the wild-type grey-yellow. When the two strains are crossed, all of the F_1 flies have wild-type body color. How can these data be interpreted? The simplest explanation is that complementation has occurred between mutations in two genes, each of which is involved in the body color phenotype. That is, a recessive autosomal gene, *ebony* (*e*), when homozygous, produces a black body color. On another autosome a different recessive gene, *black* (*b*), which also produces a black body color when homozygous. Since the two parents are homozygotes, they are genotypically $e/e\ b^+/b^+$ and $e^+/e^+\ b/b$, and each is phenotypically black. The F_1 genotype is $e^+/e\ b^+/b$, which is equivalent to the *trans* configuration of the *rII* cistron experiments. The F_1 have wild-type body color because complementation has

~ **FIGURE 8.29**

Complementation between two black body mutations of *Drosophila melanogaster*.

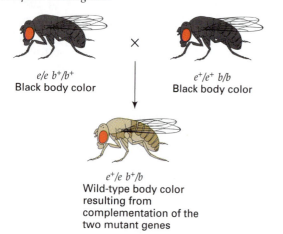

e/e b⁺/b⁺
Black body color

×

e⁺/e⁺ b/b
Black body color

e⁺/e b⁺/b
Wild-type body color
resulting from
complementation of the
two mutant genes

occurred. As in the phage experiments, no recombination is involved in the complementation of body-color genes; the double heterozygote was produced simply by the fusion of gametes produced by the two true-breeding parents.

What if the F₁ from the cross between two independently isolated, true-breeding recessive *black* mutant strains were all phenotypically black? The interpretation would be that, since the mutant phenotype was exhibited, the two mutations involved did not show complementation, and therefore the mutations are in the same complementation group; that is, they are different mutant alleles of the same gene.

𝒦EYNOTE

The complementation, or *cis-trans*, test is used to determine how many units of function (genes) define a given set of mutations expressing the same mutant phenotypes. If two mutants, each carrying a mutation in a different gene, are combined, the mutations will complement and a wild-type function will result. If two mutants, each carrying a mutation in the same gene, are combined, the mutations will not complement and the mutant phenotype will be exhibited.

SUMMARY

In this chapter, we have seen how genetic mapping can be accomplished in prokaryotes such as bacteria and bacteriophages. The same experimental strategy is used for all gene mapping; that is, genetic material is exchanged between strains differing in genetic markers and recombinants are detected and counted. In bacteria, the mechanism of gene transfer may be transformation, conjugation, or transduction. In each process, there is a donor strain and a recipient strain.

Transformation is the transfer of genetic material as small extracellular pieces of DNA between organisms. Conjugation is a plasmid-mediated process in which there is a unidirectional transfer of genetic information through direct cellular contact between a donor and a recipient. Transduction is a process whereby bacteriophages mediate the transfer of bacterial DNA from the donor bacterium to the recipient. Bacteriophage DNA may be mapped by infecting bacteria simultaneously with two phage strains and analyzing the resulting phage progeny for parental and recombinant phenotypes. Formally, this method of mapping is the same as that used for mapping genes in haploid eukaryotes. As a summary, two linkage maps of prokaryotes of genetic significance are shown: *E. coli* (Figure 8.30) and bacteriophage T4 (Figure 8.31, p. 258).

Insights into the relationships between mapping and gene structure were obtained from a fine structure analysis of the bacteriophage T4 *rII* region. The mutational sites within a gene were mapped through intragenic mapping. The resulting map indicated that the unit of mutation and the unit of recombination are the same—that is, the base pair in DNA. These definitions replaced the classical definition that genes were indivisible by mutation and recombination.

The number of units of function (genes) is determined by complementation tests. Given a set of mutations expressing the same mutant phenotype, two mutants are combined and the phenotype is determined. If the phenotype is wild type, the two mutations have complemented and must be in different units of function. If the phenotype is mutant, the two mutations have not complemented and must be in the same unit of function.

~ FIGURE 8.30

Circular genetic map of *E. coli K12* mutations based on conjugation experiments. Units are in minutes timed from an arbitrary origin at 12 o'clock. The total map size is 100 minutes. The inner circle shows the origins and directions of transfer of the chromosome in a number of *Hfr* strains; strains *HfrH* and *HfrC* mentioned in the text (p. 231) are at 97 and 12 minutes, respectively. Asterisks indicate genes whose positions are only approximate. (From *Genetics*, 3rd ed. by Monroe W. Strickberger. Copyright © 1985. Reprinted by permission of Prentice-Hall, Inc., Upper Saddle River, NJ.)

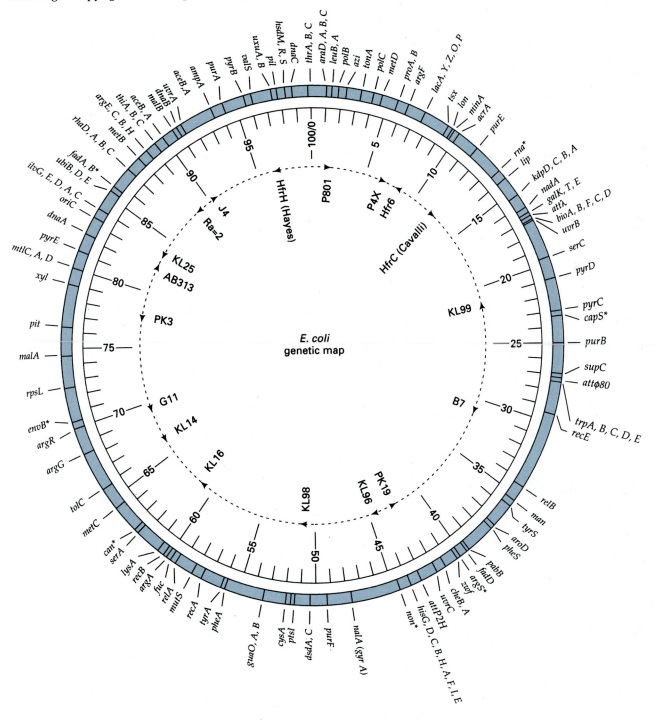

~ FIGURE 8.31

Genetic map of the bacteriophage T4. Note the clustering of genes of related function around the perimeter of the map. Units are in kilobase pairs (kb). (1 kb = 1,000 base pairs.)

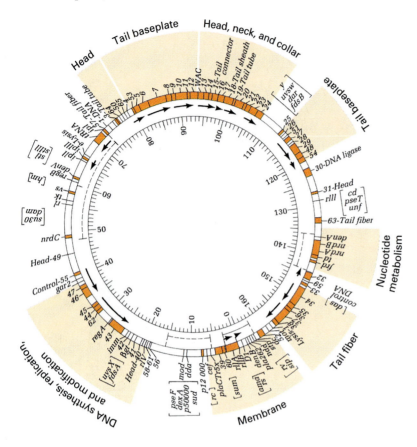

ANALYTICAL APPROACHES FOR SOLVING GENETICS PROBLEMS

Q8.1 In a transformation experiment, donor DNA from an $a^+ b^+$ strain was used to transform a recipient strain of genotype $a\ b$. The transformed classes were isolated and their frequencies determined to be:

$a^+ b^+$	307
$a^+ b$	215
$a\ b^+$	278

The total number of transformants was 800. What is the frequency with which the b locus is cotransformed with the a locus?

A8.1 The frequency with which b^+ is cotransformed with the a^+ gene is calculated using values for the total number of a^+ transformants, and the number of transformants for both a^+ and b^+. The formula is:

$$\frac{\text{number of } a^+ b^+ \text{ cotransformants}}{\text{total number of } a^+ \text{ transformants}} \times 100\%$$

The $a^+ b^+$ cotransformants number 307. The a^+ transformants are represented by two classes: $a^+ b^+$ (307) and $a^+ b$ (215), for a total of 522. The $a\ b^+$ class is irrelevant to the question because they are not transformants for a^+. Thus, the cotransformation frequency for a^+ and b^+ is $307/522 \times 100 = 58.8\%$.

Q8.2 In *E. coli* the following *Hfr* strains donate the markers shown in the order given:

Hfr Strain	Order of Gene Transfer
1	G E B D N A
2	P Y L G E B
3	X T J F P Y
4	B E G L Y P

All the *Hfr* strains were derived from the same F^+ strain. What is the order of genes in the original F^+ chromosome?

A8.2 This question is an exercise in piecing together various segments of the circumference of a circle. The best approach is to draw a circle and label it with the genes transferred from one *Hfr* and then to see which of the other *Hfr*'s transfers an overlapping set. For example, *Hfr* 1 transfers *E*, then *B*, then *D*, and so on; and *Hfr* 4 transfers *B* then *E*, and so forth. Now we can juxtapose the two sets of genes transferred by the two *Hfr*'s and deduce that the polarities of transfer are opposite:

Hfr 1	G E B D N A
Hfr 4	P Y L G E B

Extending this reasoning to the other *Hfr*'s, we can draw an unambiguous map (see the figure below), with the arrowheads indicating the order of transfer.

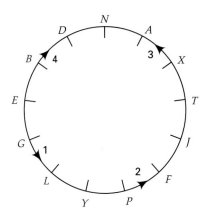

The same logic would be used if the question gave the relative time units of entry of each of the genes. In that case we would expect that the time "distance" between any two genes would be approximately the same regardless of the order of transfer or how far the genes were from the origin.

Q8.3 In a transduction experiment, the donor was $c^+ d^+ e^+$ and the recipient was $c\ d\ e$. Selection was for c^+. The four classes of transductants from this experiment were:

Class	Genetic Composition	Number of Individuals
1	$c^+\ d^+\ e^+$	57
2	$c^+\ d^+\ e$	76
3	$c^+\ d\ e$	365
4	$c^+\ d\ e^+$	2
	Total	500

a. Determine the cotransduction frequency for c^+ and d^+.

b. Determine the cotransduction frequency for c^+ and e^+.

c. Which of the cotransduction frequencies calculated in (a) and (b) represents the greater actual distance between genes? Why?

A8.3a. The analysis is similar to the cotransformation frequency analysis described in Q8.1. The formula for the cotransduction frequency for c^+ and d^+ is:

$$\frac{\text{number of } c^+ d^+ \text{ cotransductants}}{\text{total number of } c^+ \text{ transductants}} \times 100\%$$

From the data presented, Classes 1 and 2 are the $c^+ d^+$ cotransductants, and the total number of c^+ transductants is the sum of Classes 1 through 4. Thus the number of c^+ and d^+ transductants is $57 + 76 = 133$, and the cotransductant frequency is $133/500 \times 100 = 26.6\%$.

b. The analysis is identical in approach to (a). The formula for the cotransduction frequency for c^+ and e^+ is:

$$\frac{\text{number of } c^+ e^+ \text{ cotransductants}}{\text{total number of } c^+ \text{ transductants}} \times 100\%$$

From the data presented, Classes 1 and 4 are the $c^+ e^+$ cotransductants, and the total number of c^+ transductants is the sum of Classes 1 through 4. Thus the number of c^+ and e^+ transductants is $57 + 2 = 59$, and the cotransductant frequency is $59/500 \times 100 = 11.8\%$.

c. The greater actual distance is for the c^+ and e^+ genes. The principle involved is that the closer two genes are on the chromosome, the greater the chance that they will be cotransduced. Thus, as the distance between genes increases, concomitantly the cotransduction frequency decreases. Since the $c^+ e^+$ cotransduction frequency is 11.8 percent and the $c^+ d^+$ cotransduction frequency is 26.6 percent, genes c^+ and e^+ are farther apart than genes c^+ and d^+.

Q8.4 Five different *rII* deletion strains of phage T4 were tested for recombination by pairwise crossing in *E. coli* B. The following results were obtained, where $+ = r^+$ recombinants produced, and $0 =$ no r^+ recombinants produced:

	A	B	C	D	E
E	0	+	0	+	0
D	0	0	0	0	
C	0	0	0		
B	+	0			
A	0				

Draw a deletion map compatible with these data.

A8.4 The principle here is that if two deletion mutations overlap, then no r^+ recombinants can be produced. Conversely, if two deletion mutations do not overlap, then r^+ recombinants can be produced. To approach a question of this kind, we must draw overlapping and nonoverlapping lines from the given data.

Starting with A and B, these two deletions do not overlap since r^+ recombinants are produced. Therefore these two mutations can be represented as follows:

_____A_____   _____B_____

The next deletion, C, does not produce r^+ recombinants with any of the other four deletions. We must conclude, therefore, that C is an extensive deletion that overlaps all of the other four, with endpoints that cannot be determined from the data given. One possibility is as follows:

_____C_____

_____A_____   _____B_____

Deletion D does not produce r^+ recombinants with A, B, or C, but it does with E. In turn, E produces r^+ recombinants with B and D but not with A or C. Thus D must overlap both A and B but not E, and E must overlap A and C but not B. A compatible map for this situation follows. Other maps can be drawn in terms of the endpoints of the deletions.

_____C_____

_____A_____   _____B_____

_____E_____   _____D_____

Q8.5 Seven different *rII* point mutants (*1* to *7*) of phage T4 were tested for recombination crosses in *E. coli* B with the five deletion strains described in Question 4. The following results were obtained, where + = r^+ recombinants produced and 0 = no r^+ recombinants produced:

	A	B	C	D	E
1	0	+	0	+	+
2	+	0	0	+	+
3	0	+	0	+	0
4	+	+	0	+	0
5	+	0	0	0	+
6	0	+	0	0	+
7	+	+	0	0	+

In which regions of the map can you place the seven point mutations?

A8.5 If an r^+ recombinant is produced, the *rII* point mutation cannot overlap the region missing in the deletion mutation with which it was crossed. Thus the table of results enables us to localize the point mutations to the regions defined by the deletion mutants. Potentially, the results define the relative extent of deletion overlap. For example, point mutation 7 produces r^+ recombinants with A, B, and E but not with D. Logically, then, 7 is located in the region defined by the part of deletion D that is not involved in the overlap with A and B. Similarly, point mutation 4 gives r^+ recombinants with A, D, and B but not with E. Thus 4 must be in a region defined by a segment of deletion E that does not overlap deletion A. Furthermore, since 4 does not produce r^+ recombinants with C either, deletion C must overlap the site defined by point mutation 4. This result, then, refines the deletion map with regard to the E, C, and A endpoints. The map we can draw from the matrix of results is as follows:

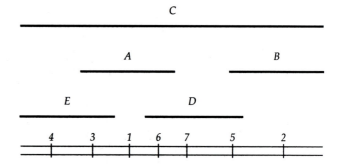

QUESTIONS AND PROBLEMS

*8.1 If an *E. coli* auxotroph A could grow only on a medium containing thymine, and an auxotroph B could grow only on a medium containing leucine, how would you test whether DNA from A could transform B?

*8.2 In $F^+ \times F^-$ crosses the F^- recipient is converted to a donor with very high frequency. However, it is rare for a recipient to become a donor in $Hfr \times F^-$ crosses. Explain why.

*8.3 With the technique of interrupted mating four *Hfr* strains were tested for the sequence in which they transmitted a number of different genes to an *F⁻* strain. Each *Hfr* strain was found to transmit its genes in a unique sequence, as shown in the accompanying table (only the first six genes transmitted were scored for each strain).

ORDER OF TRANSMISSION	*Hfr* STRAIN 1	2	3	4
First	*O*	*R*	*E*	*O*
	F	*H*	*M*	*G*
	B	*M*	*H*	*X*
	A	*E*	*R*	*C*
	E	*A*	*C*	*R*
Last	*M*	*B*	*X*	*H*

What is the gene sequence in the original strain from which these *Hfr* strains derive? Indicate on your diagram the origin and polarity of each of the four *Hfr*'s.

8.4 At time zero an *Hfr* strain (strain 1) was mixed with an *F⁻* strain, and at various times after mixing, samples were removed and agitated to separate conjugating cells. The cross may be written as:

$$Hfr\ 1:\quad a^+\ b^+\ c^+\ d^+\ e^+\ f^+\ g^+\ h^+\ str^s$$
$$F^-:\quad\ a\ \ \ b\ \ \ c\ \ \ d\ \ \ e\ \ \ f\ \ \ g\ \ \ h\ \ \ str^r$$

(No order is implied in listing the markers.)

The samples were then plated onto selective media to measure the frequency of *h⁺ str^r* recombinants that had received certain genes from the *Hfr* cell. The graph of the number of recombinants against time is shown in the accompanying figure.

a. Indicate whether each of the following statements is true or false.

i. All *F⁻* cells which received *a⁺* from the *Hfr* in the chromosome transfer process must also have received *b⁺*.

ii. The order of gene transfer from *Hfr* to *F⁻* was *a⁺* (first), then *g⁺*, then *b⁺*, then *e⁺*, then *h⁺*.

iii. Most *e⁺ str^r* recombinants are likely to be *Hfr* cells.

iv. None of the *b⁺ str^r* recombinants plated at 15 minutes are also *a⁺*.

b. Draw a linear map of the *Hfr* chromosome indicating the

i. point of insertion, or origin;

ii. order of the genes *a⁺, b⁺, e⁺, g⁺,* and *h⁺*;

iii. shortest distance between consecutive genes on the chromosomes.

*8.5 Three different prototrophic strains (#1, #2 and #3) that are all sensitive to the antibiotic streptomycin are isolated. Each is individually mixed with an auxotrophic *F⁻* strain that is *a b c d e f g h* (and therefore requires compounds A, B, C, D, E, F, G, and H to grow) and resistant to the antibiotic streptomycin. At one-minute intervals after the initial mixing, a sample of the mixture is removed, shaken violently, and plated on media to select for *c⁺ str^R* recombinants. Recombinants are then tested for the presence of other genes. The following results are obtained:

Strain #1 × *F⁻*: No *c⁺* recombinants are ever obtained, even after 25 minutes.

Strain #2 × *F⁻*: *c⁺* recombinants are obtained at 6 min., *g⁺* at 8 min., *h⁺* at 11 min., *a⁺* at 14 min., *b⁺* at 16 min. No *d⁺, e⁺,* and *f⁺* recombinants are obtained.

Strain #3 × *F⁻*: *c⁺* recombinants are obtained at the 1-minute time point and *c⁺ g⁺* recombinants are obtained on or after the 3-minute time point. No *a⁺, b⁺, d⁺, e⁺, f⁺* or *h⁺* recombinants are obtained.

If *c⁺* recombinants obtained at the 16-minute time point from the cross involving strain #2 are mixed with an *amp^R* (ampicillin-resistant) *F⁻* strain, no *c⁺ amp^R* recombinants are ever recovered. However, if *c⁺* recombinants obtained at the 16-minute time point from the cross involving strain #3 are mixed with an *amp^R F⁻* strain, *amp^R c⁺* recombinants can be recovered after 1 minute of mating.

a. How was the initial selection for *c⁺ str^R* recombinants done? How were the subsequent selections done?

b. Use these data to ascertain, as best you can, whether each strain is *F⁻, Hfr, F⁺,* or *F'*. If these data do not allow you to make an unambiguous determination, indicate the possibilities.

c. To the extent you can, draw a map of the possible chromosome(s) present in each of strains #1, #2, and #3. Indicate the location and distance between genes *a⁺, b⁺, d⁺, e⁺, f⁺* and *h⁺* as best you can.

8.6 You are given a prototrophic *str^R* (streptomycin-resistant) *Hfr* strain and an *amp^R* (ampicillin-resistant) *F⁻* auxotrophic strain that requires leucine (*leu*), arginine (*arg*), lysine (*lys*), purine (*pur*), and biotin (*bio*).

a. Devise a strategy to determine quickly which gene (*leu⁺, arg⁺, lys⁺, pur⁺,* or *bio⁺*) lies closest to the F-factor origin of replication.

b. Even when the prototrophic *Hfr* strain is mixed with the *F⁻* strain for very long periods of time, streptomycin resistance is not transferred. State two hypotheses that explain this finding.

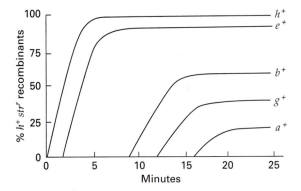

8.7 Distinguish between generalized and specialized transduction.

*8.8 Indicate whether each of the following occurs or is a characteristic of

Generalized Transduction	(GT)
Specialized Transduction	(ST)
Occurs in Both	(B)
Occurs in Neither	(N)

a. Phage carries DNA of bacterial or viral DNA origin, never both.

b. Phage carries viral DNA covalently linked to bacterial DNA.

c. Phage integrates into a specific site on the host chromosome.

d. Phage integrates at a random site on the host chromosome.

e. "Headful" of bacterial DNA encoated.

f. Host lysogenized.

g. Prophage state exists.

h. Temperate phage involved.

i. Virulent phage involved.

8.9 Consider the following transduction data:

DONOR	RECIPIENT	SELECTED MARKER	UNSELECTED MARKER	%
aceF+ dhl	aceF dhl+	aceF+	dhl	88
aceF+ leu	aceF leu+	aceF+	leu	34

Is *dhl* or *leu* closer to *aceF*?

*8.10 Consider the following data with P1 transduction:

DONOR	RECIPIENT	SELECTED MARKER	UNSELECTED MARKER	%
aroA pyrD+	aroA+ pyrD	pyrD+	aroA	5
aroA+ cmlB	aroA cmlB+	aroA+	cmlB	26
cmlB pyrD+	cmlB+ pyrD	pyrD+	cmlB	54

Choose the correct order:

a. *aroA - cmlB - pyrD*

b. *aroA - pyrD - cmlB*

c. *cmlB - aroA - pyrD*

8.11 Order the mutants *trp*, *pyrF* and *qts* on the basis of the following three factor transduction cross:

Donor	trp+ pyr+ qts
Recipient	trp pyr qts+
Selected Marker	trp+

UNSELECTED MARKERS	NUMBER
pyr+ qts+	22
pyr+ qts	10
pyr qts+	68
pyr qts	0

*8.12 Order *cheA*, *cheB*, *eda* and *supD* from the following data:

MARKERS	% COTRANSDUCTION
cheA-eda	15
cheA-supD	5
cheB-eda	28
cheB-supD	2.7
eda-supD	0

*8.13 Wild-type phage T4 grows on both *E. coli B* and *E. coli K12(λ)*, producing turbid plaques. The *rII* mutants of T4 grow on *E. coli B*, producing clear plaques, but do not grow on *E. coli K12(λ)*. This host range property permits the detection of a very low number of *r+* phages among a large number of *rII* phages. With this sensitive system it is possible to determine the genetic distance between two mutations within the same gene, in this case the *rII* locus. Suppose *E. coli B* is mixedly infected with *rIIx* and *rIIy*, two separate mutants in the *rII* locus. Suitable dilutions of progeny phages are plated on *E. coli B* and *E. coli K12(λ)*. A 0.1-mL sample of a thousandfold dilution plated on *E. coli B* produced 672 plaques. A 0.2-mL sample of undiluted phage plated on *E. coli K12(λ)* produced 470 turbid plaques. What is the genetic distance between the two *rII* mutations?

8.14 Construct a map from the following two factor phage cross data (show map distance):

CROSS	% RECOMBINATION
r1 × r2	0.10
r1 × r3	0.05
r1 × r4	0.19
r2 × r3	0.15
r2 × r4	0.10
r3 × r4	0.23

*8.15 The following two-factor crosses were made to analyze the genetic linkage between four genes in phage λ: *c,mi,s,* and *co*.

PARENTS	PROGENY
c + × + mi	1,213 c +, 1,205 + mi, 84 + + , 75 c mi
c + × + s	566 c +, 808 + s, 19 + +, 20 c s
co + × + mi	5,162 co +, 6,510 + mi, 311 + +, 341 co mi
mi + × + s	502 mi +, 647 + s, 65 + +, 56 mi s

Construct a genetic map of the four genes.

8.16 Three gene loci in T4 that affect plaque morphology in easily distinguishable ways are *r* (rapid lysis), *m* (minute), and *tu* (turbid). A culture of *E. coli* is mixedly infected with two types of phage *r m tu* and *r+ m+ tu+*. Progeny phage are collected and the following genotype classes are found:

CLASS			NUMBER
r^+	m^+	tu^+	3,729
r^+	m^+	tu	965
r^+	m	tu^+	520
r^+	m^+	tu^+	172
r^+	m	tu	162
r	m^+	tu	474
r	m	tu^+	853
r	m	tu	3,467
			10,342

Construct a map of the three genes. What is the coefficient of coincidence, and what does the value suggest?

*8.17 The *rII* mutants of bacteriophage T4 grow in *E. coli* B but not in *E. coli* K12(λ). The *E. coli* strain *B* is doubly infected with two *rII* mutants. A 6×10^7 dilution of the lysate is plated on *E. coli* B. A 2×10^5 dilution is plated on *E. coli* K12(λ). Twelve plaques appeared on strain K12(λ), and 16 on strain *B*. Calculate the amount of recombination between these two mutants.

8.18 Wild-type (r^+) strains of T4 produce turbid plaques, whereas *rII* mutant strains produce larger, clearer plaques. Five *rII* mutations (*a-e*) in the *A* cistron of the *rII* region of T4 give the following percentages of wild-type recombinants in two-point crosses:

CROSS	% OF WILD-TYPE RECOMBINANTS	CROSS	% OF WILD-TYPE RECOMBINANTS
$a \times b$	0.2	$e \times d$	0.7
$a \times c$	0.9	$e \times c$	1.2
$a \times d$	0.4	$e \times b$	0.5
$b \times c$	0.7	$b \times d$	0.2
$e \times a$	0.3	$d \times c$	0.5

What is the order of the mutational sites and what are the map distances between the sites?

*8.19 Given the following map with point mutants, and given the data in the table below, draw a topological representation of deletion mutants *r21, r22, r23, r24* and *r25*. (Be sure to show the endpoints of the deletions.)

(+ = r^+ recombinants are obtained. 0 = r^+ recombinants are not obtained.)

Map:

r12 r16 r11 r15 r13 r14 r17

DELETION MUTANTS	POINT MUTANTS						
	r11	*r12*	*r13*	*r14*	*r15*	*r16*	*r17*
r21	0	+	0	+	0	+	+
r22	+	+	0	0	+	+	0
r23	0	0	0	+	0	0	+
r24	+	+	0	0	+	+	+
r25	+	+	0	0	0	+	+

8.20 A set of seven different *rII* deletion mutants of bacteriophage T4, *1* through *7*, were mapped, with the following result:

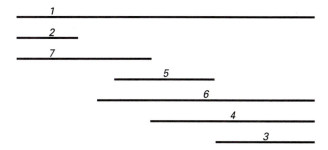

Five *rII* point mutants were crossed with each of the deletions, with the following results, where + = r^+ recombinants, were obtained, 0 = no r^+ recombinants were obtained:

POINT MUTANTS	DELETION MUTANTS						
	1	2	3	4	5	6	7
a	0	+	+	+	0	0	0
b	0	0	+	+	+	+	0
c	0	+	+	0	0	0	+
d	0	+	0	0	+	0	+
e	0	+	+	+	+	0	0

Map the locations of the point mutants.

*8.21 Given the following deletion map with deletions *r31, r32, r33, r34, r35* and *r36*, place the point mutants *r41, r42* etc., on the map. Be sure you show where they lie with respect to end points of the deletions.

r31

r32

r33

r34

r35   r36

POINT MUTANTS	DELETION MUTANTS (+ = RECOMBINANTS PRODUCED; 0= NO r^+ RECOMBINANTS PRODUCED)					
	r31	*r32*	*r33*	*r34*	*r35*	*r36*
r41	0	0	0	0	+	0
r42	0	0	0	+	0	+
r43	0	0	+	+	+	0
r44	0	0	0	0	+	+
r45	0	+	0	+	+	+
r46	0	0	+	0	+	0

Show the dividing line between the *A* cistron and the *B* cistron on your map above from the following data [+ = growth on strain *K12(λ)*, 0 = no growth on strain *K12(λ)*]:

MUTANT	COMPLEMENTATION WITH	
	rIIA	*rIIB*
*r*31	0	0
*r*32	0	0
*r*33	0	+
*r*34	0	0
*r*35	0	+
*r*36	0	0
*r*41	0	+
*r*42	0	+
*r*43	+	0
*r*44	0	+
*r*45	0	+
*r*46	0	+

8.22 Mutants in the *ade2* gene of yeast require adenine and are pink because of the intracellular accumulation of a red pigment. Diploid strains were made by mating haploid mutant strains. The diploids exhibited the following phenotypes:

CROSS	DIPLOID PHENOTYPES
1 × 2	pink, adenine-requiring
1 × 3	white, prototrophic
1 × 4	white, prototrophic
3 × 4	pink, adenine-requiring

How many genes are defined by the four different mutants? Explain.

*8.23 In *Drosophila* mutants *A, B, C, D, E, F,* and *G* all have the same phenotype: the absence of red pigment in the eyes. In pairwise combinations in complementation tests the following results were produced, where + = complementation and − = no complementation.

	A	*B*	*C*	*D*	*E*	*F*	*G*
G	+	−	+	+	+	+	−
F	−	+	+	−	+	−	
E	+	+	−	+	−		
D	−	+	+	−			
C	+	+	−				
B	+	−					
A	−						

a. How many genes are present?
b. Which mutants have defects in the same gene?

8.24 A homozygous white-eyed Martian fly (w_1/w_1) is crossed to a homozygous white-eyed fly from a different stock (w_2/w_2). It is well known that wild-type Martian flies have red eyes. This cross produces all white-eyed progeny. State whether the following is true or false. Explain your answer.

a. w_1 and w_2 are allelic genes.
b. w_1 and w_2 are non-allelic.
c. w_1 and w_2 affect the same function.
d. The cross was a complementation test.
e. The cross was a cis-trans test.
f. w_1 and w_2, are allelic by terms of the functional test.

The F$_1$ white-eyed flies were allowed to interbreed, and when you classified the F$_1$ you found 20,000 white-eyed flies and ten red-eyed progeny. Concerned about contamination, you repeat the experiment and get exactly the same results. How can you best account for the presence of the red-eyed progeny? As part of your explanation give the genotypes of the F$_1$ and F$_2$ generation flies.

8.25 A breeder obtains three strains (A, B, and C) of true-breeding, white-flowered sweet peas. When A and C or B and C are crossed, the F$_1$ are all purple-flowered. When either of these purple F$_1$ are selfed, about 44 percent of the progeny are white and the rest are purple. When strains A and B are crossed, the F$_1$ progeny are all white. When these white F$_1$ are selfed, only 0.015 percent of the progeny are purple. Explain these data in terms of complementation and intragenic recombination (Hint: see Figure 4.14, p. 113).

8.26 Propose a genetic explanation for the ugly duckling phenomenon: Two proud white parents have a rare black offspring amidst a prolific number of white offspring.

8.27 Two independently isolated *Drosophila* mutations, *yellow*-1 and *yellow*-2, lighten the body color. *yellow*-1 causes the normally gray cuticle and the normally black bristles to become yellow. *yellow*-2 causes the cuticle to become yellow but does not affect the color of the bristles. Devise two different ways to determine if *yellow*-1 and *yellow*-2 are alleles at one gene, or alleles at different genes. (Hint: see Figure 7.3, p. 194.)

8.28 A large number of mutations in *Drosophila* alter the normal deep-red eye color. As was discussed in Chapter 4 (pp. 100–102), even alleles at one gene (*white*, *w*) can display a variety of phenotypes.

You are given a wild-type, deep-red strain and six independently isolated, true-breeding mutant strains that have varying shades of brown eyes, with the assurance that each mutant strain has only a single mutation. How would you proceed to determine:

a. whether the mutation in each strain is dominant or recessive?
b. how many different genes are affected in the six mutant strains?
c. which mutants, if any, are allelic?
d. whether any of these mutants are alleles at genes already known to affect eye color? (See Figure 5.20, p. 158).

9 THE BEGINNINGS OF MOLECULAR GENETICS: GENE FUNCTION

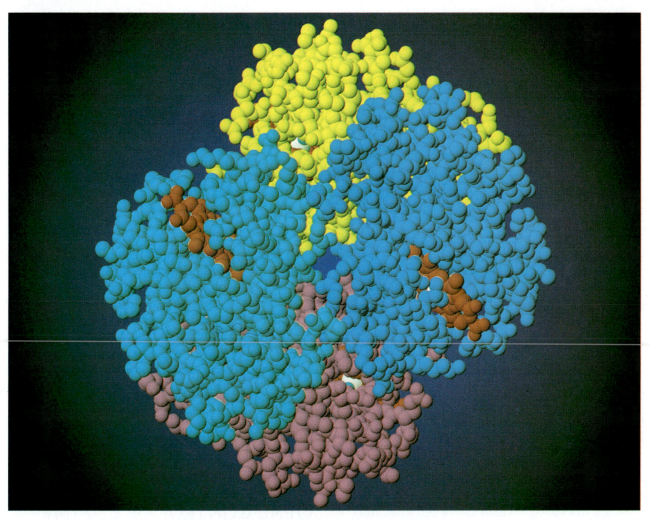

PRINCIPAL POINTS

~ There is a specific relationship between genes and enzymes, historically embodied in Beadle and Tatum's "one gene–one enzyme" hypothesis, which states that each gene controls the synthesis or activity of a single enzyme. Since we now know that enzymes may consist of more than one polypeptide and that genes code for individual polypeptide chains, a more modern title for this hypothesis is "one gene–one polypeptide."

~ Many human genetic diseases are caused by deficiencies in enzyme activities. Most of these diseases are inherited as recessive traits.

~ From the study of alterations in proteins other than enzymes, convincing evidence was obtained that genes control the structures of all proteins.

~ Genetic counseling is the analysis of the risk that prospective parents may produce a child with a genetic defect, and the presentation to appropriate family members of the available options for avoiding or minimizing those possible risks. Early detection of a genetic disease is done by carrier detection, and by fetal analysis.

*I*n the preceding chapters we have explored many aspects of genes. We have learned that genes are defined by mutations, that genes segregate in genetic crosses in the same way chromosomes segregate during meiosis, and that genes can be located—mapped—relative to one another on chromosomes in both prokaryotes and eukaryotes by analyzing the results of appropriate genetic crosses. We have also learned that alleles can have differing relationships: some are recessive to the wild-type allele, some are dominant, and others are codominant or incompletely dominant. Throughout all of these discussions we have considered the gene as an abstract entity, a factor that is located on a chromosome and that can give rise to a phenotype. We know now, of course, that a gene is a stretch of DNA in a chromosome and that that section of DNA contains information for a specific product: this concept will be developed more fully in the following chapters. The entire set of genes in the genome specifies a vast array of products that are responsible for all of a cell's and organism's structures and functions. In the popular press, the genetic information contained within an organism's genome is called the "blueprint for life." We now need to describe how a gene functions; that is, how the information in the DNA is used to specify a product and how that product is involved in the determination of a phenotype.

We saw in Chapter 8 that complementation tests can be used to assign mutations that result in the same phenotype to units of function—the genes. In this chapter we will examine gene function: what do genes code for, and how do we know what they code for? We will present some of the classical evidence that genes code for enzymes and for other proteins. In particular, we examine the involvement of certain sets of genes in directing and controlling a particular biochemical pathway; that is, the series of enzyme-catalyzed steps required to break down or synthesize a particular chemical compound. Instead of viewing the gene in isolation, as in the past few chapters, we will see that the gene must often work in cooperation with other genes in order for cells to function properly. The experiments we will discuss represent the beginnings of molecular genetics, in that their goal was to understand better a gene at the molecular level. In the following chapters our modern understanding of gene structure and function will be developed further.

GENE CONTROL OF ENZYME STRUCTURE

Our studies so far have shown how genetic analysis can provide insights about the relationships between mutations and phenotypic change. In this section we analyze some of the classical genetic data that showed that genes code for enzymes and for nonenzymatic proteins.

Garrod's Hypothesis of Inborn Errors of Metabolism

In 1902, Archibald Garrod, an English physician, provided the first evidence of a specific relationship between genes and enzymes. Garrod studied *alkap-*

tonuria, a human disease characterized by urine that turns black upon exposure to the air and by a tendency to develop arthritis later in life.

In studying the occurrence of alkaptonuria in families of individuals with the disease, Garrod and his colleague William Bateson discovered two interesting facts that indicated to them that alkaptonuria is a genetically controlled trait: (1) Several members of the same families frequently had alkaptonuria, and (2) the disease was much more common among children of marriages involving first cousins than among children of marriages between unrelated partners. This finding was significant because first cousins have 1/8 of their genes in common and, therefore, the chances are greater for recessive alleles to be homozygous in children of first-cousin marriages.

Garrod found that people with alkaptonuria excrete in their urine all the homogentisic acid (HA) they produce, whereas normal people excrete none. Moreover, he showed that it is HA in the urine that turns black in the air. This result indicated to Garrod that normal people are able to metabolize HA but that people with alkaptonuria cannot. That means that alkaptonurics lack the enzyme that metabolizes HA. Figure 9.1 shows part of the biochemical pathway in which HA is involved and the step that is blocked in people with alkaptonuria. From his data and the genetic evidence, Garrod concluded that alkaptonuria is a genetic disease caused by the absence of a particular enzyme necessary for the metabolism of HA. In Garrod's terms, this disease is an example of an *inborn error of metabolism*. The mutation responsible for alkaptonuria is recessive, so only people homozygous for the mutant gene express the defect. The gene is on an autosome, but which autosome is unknown.

Garrod also studied three other human genetic diseases that affected biochemical processes, and in each case he was able to conclude correctly that a metabolic pathway was blocked. An important aspect of Garrod's analysis of these human diseases was his understanding that the position of a block in a metabolic pathway can be determined by the accumulation of the chemical compound (HA in the case of alkaptonuria) that precedes the blocked step. However, the significance of Garrod's work was not appreciated by his contemporaries.

One Gene–One Enzyme Hypothesis

George Beadle and Edward Tatum in 1942 heralded the beginnings of biochemical genetics, a branch of genetics that combines genetics and biochemistry to elucidate the nature of metabolic pathways. Results of their studies involving the fungus *Neurospora crassa*

~ FIGURE 9.1

Part of the phenylalanine-tyrosine metabolic pathways. The numbers indicate steps in the pathways catalyzed by enzymes. *(Left)* The biochemical steps that operate in normal individuals. *(Right)* The metabolic block found in individuals with alkaptonuria.

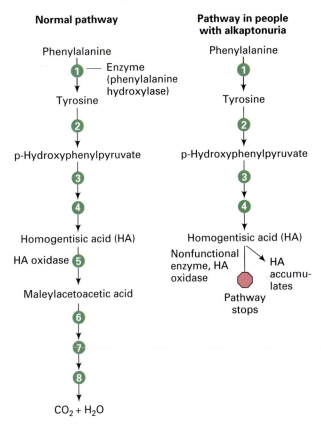

showed a direct relationship between genes and enzymes and led to the *one gene–one enzyme hypothesis*, a very important landmark in the history of genetics. Beadle and Tatum later received the Nobel Prize in recognition of this work.

ISOLATION OF NUTRITIONAL MUTANTS OF *NEUROSPORA*.   In Chapter 5 we discussed the life cycle of *Neurospora crassa* (see Figure 5.9, p. 142). This fungus has simple growth requirements. Wild-type *Neurospora*, by definition, is prototrophic; it can grow on a minimal medium that contains only inorganic salts, a carbon source (such as glucose or sucrose), and the vitamin biotin. Beadle and Tatum reasoned that *Neurospora* synthesized the materials it needed for growth (amino acids, nucleotides, vitamins, nucleic acids, proteins, etc.) from the various chemicals present in the minimal medium. They also realized that it should be possible to isolate nutritional mutants (i.e., auxotrophs) of *Neurospora* that required

~ **FIGURE 9.2**

Method devised by Beadle and Tatum to isolate auxotrophic mutations in *Neurospora.* Here, the mutant strain isolated is a tryptophan auxotroph.

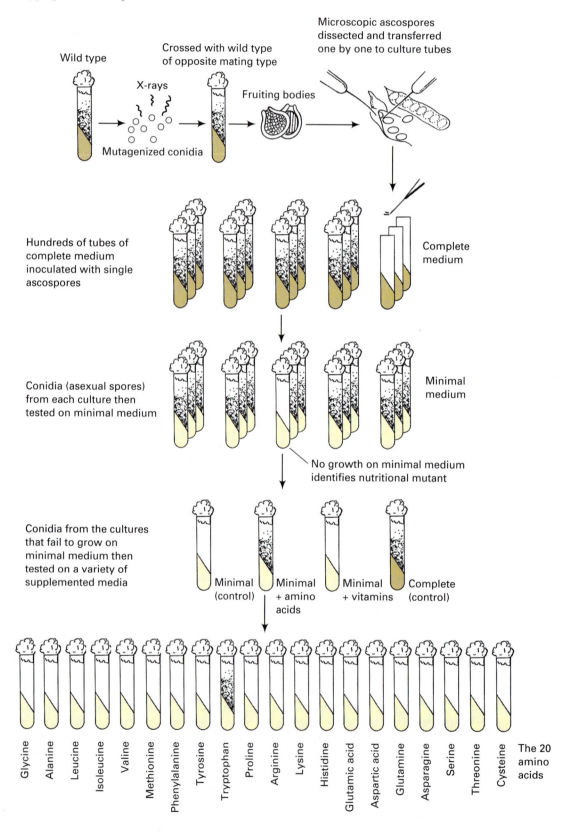

nutritional supplements in order to grow. These aux-otrophic mutants could be isolated because they would not grow on the minimal medium.

Figure 9.2 shows how Beadle and Tatum isolated and characterized nutritional mutants. They treated asexual spores (conidia) with X rays to induce genetic mutants. Then they crossed the organisms derived from the surviving, mutagen-treated spores with a wild-type (prototrophic) strain of the opposite mating type. This cross was done because they wanted to study the genetics of biochemical events, and they had to identify those nutritional mutations that were heritable. By crossing the mutagenized spores with the wild type, they ensured that any nutritional mutant they isolated had segregated in a cross and therefore had a genetic basis rather than a nongenetic reason for requiring the nutrient.

One progeny spore per ascus from the crosses was allowed to germinate in a nutritionally complete medium that contained all necessary amino acids, purines, pyrimidines, and vitamins, in addition to the sucrose, salts, and biotin found in minimal medium. Thus, any strain that could not make one or more nec-essary compounds from the basic ingredients found in minimal medium could still grow by using the compounds supplied in the growth medium. Each culture grown from a progeny spore was then tested for growth on minimal medium. Those strains that did not grow were assumed to be auxotrophic mutants. These mutants were, in turn, individually tested for their ability to grow on various supple-mented minimal media. In this screening two media were used: minimal medium plus amino acids, and minimal medium plus vitamins. Theoretically, an amino acid auxotroph—a mutant strain that has lost the ability to synthesize a particular amino acid—would grow on minimal medium plus amino acids but not on minimal medium plus vitamins or on min-imal medium. Similarly, vitamin auxotrophs would only grow on minimal medium plus vitamins.

Having categorized the strains as non-aux-otrophs, amino acid auxotrophs, and vitamin aux-otrophs, Beadle and Tatum next conducted a second round of screening to determine which specific sub-stance each auxotrophic strain needed for growth. Let us suppose an amino acid auxotroph was identified. To determine which of the 20 amino acids was required, the strain was inoculated into 20 tubes, each containing minimal medium plus one of the 20 amino acids. In the example shown in Figure 9.2, a trypto-phan auxotroph was identified because it grew only in the tube containing minimal medium plus trypto-phan.

Figure 9.3 shows the method used to confirm that an auxotrophic mutation identified by the above pro-cedures has a genetic rather than a nongenetic basis. In this case, a tryptophan auxotroph is crossed with a wild-type strain. The spores from a single meiotic division (a meiotic tetrad) can be isolated and ana-lyzed as described in Chapter 6. If the auxotrophy is caused by a single gene mutation, then half the spores should be wild type and half should be auxotrophic. This hypothesis is tested by first germinating the spores individually on minimal medium plus trypto-phan (so all will grow) and then checking for the abil-ity of each to grow on minimal medium. Only the wild-type strains will grow on minimal medium; the auxotrophic strains will not. If, however, the trypto-phan auxotroph is not the result of a mutation, then all the spores should grow on minimal medium—or at least there will not be a 4:4 segregation of wild type : auxotroph when tetrads are tested for growth on minimal medium.

~ **FIGURE 9.3**

Procedure used to confirm the genetic basis of a nutritional defect in *Neurospora*.

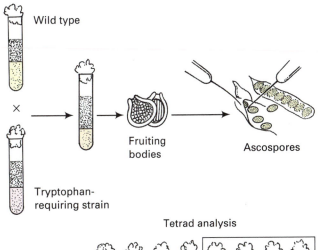

Wild type

×

Tryptophan-requiring strain

Fruiting bodies

Ascospores

Tetrad analysis

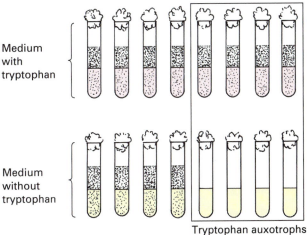

Medium with tryptophan

Medium without tryptophan

Tryptophan auxotrophs

~ TABLE 9.1

Growth Responses of Arginine Auxotrophs

MUTANT STRAINS	GROWTH RESPONSE ON MINIMAL MEDIUM AND . . .				
	NOTHING	ORNITHINE	CITRULLINE	ARGININO-SUCCINATE	ARGININE
Wild-type	+	+	+	+	+
argE	−	+	+	+	+
argF	−	−	+	+	+
argG	−	−	−	+	+
argH	−	−	−	−	+

GENETIC DISSECTION OF A BIOCHEMICAL PATHWAY. Once Beadle and Tatum had isolated and identified nutritional mutants, they investigated the biochemical pathways affected by the mutations. They assumed that *Neurospora* cells, like all cells, function through the interaction of the products of a very large number of genes. Further, they proposed that wild-type *Neurospora* converted the constituents of minimal medium into amino acids and other required compounds by a series of reactions that were organized into pathways. In this way the synthesis of cellular components occurred by a series of small steps, each catalyzed by an enzyme. An example follows of Beadle and Tatum's analysis that led to an understanding of the relationship between genes and enzymes.

We consider the genetic dissection of the pathway for the biosynthesis of the amino acid arginine in *Neurospora crassa*. Starting with a set of arginine auxotrophs, genetic crosses and complementation tests (see Chapter 8) show that four genes are involved; a mutation in any one of them gives rise to auxotrophy for arginine. These four genes in a wild-type cell are designated *argE+*, *argF+*, *argG+*, and *argH+*. Next, the growth pattern of the four mutant strains on media supplemented with presumed arginine precursors was determined, with the results shown in Table 9.1.

(The presumed precursors must be able to be taken up by the cells in order for this analysis to be completed.) These results are the primary data used for analysis. By definition all four mutant strains can grow on arginine, and none can grow on unsupplemented minimal medium.

The analysis is based on some fundamental principles. Consider the hypothetical pathway shown in Figure 9.4 in which a precursor compound is converted to an end product C via two intermediates A and B. Each step in the pathway is catalyzed by a different enzyme. Beadle and Tatum proposed that a specific gene encodes each enzyme. They called this proposed relationship between an organism's genes and the enzymes that carry out the steps in a biochemical pathway the **one gene–one enzyme hypothesis.** Gene mutations that result in the loss of enzyme activity lead to the accumulation of precursors in the pathway (and to possible side reactions), as well as to the absence of the end product of the pathway.

The one gene–one enzyme hypothesis allows us to make some predictions about the consequences of mutations affecting enzymes in the pathway. For example, if a mutation in gene C results in a nonfunctional enzyme C, then intermediate product B in the pathway cannot be converted to the end product C and growth does not occur unless C is added to the

~ FIGURE 9.4

Hypothetical biochemical pathway for the conversion of a precursor substrate to an end product C in three enzyme-catalyzed steps. Each enzyme is coded for by one gene.

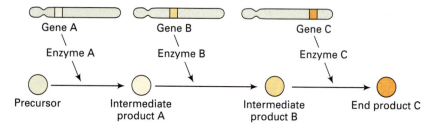

culture medium. Intermediate product B also accumulates since it cannot be converted to C. Mutations in gene *A* or gene *B* also will result in auxotrophy for end product C.

The three auxotrophs *A, B,* and *C* are distinguishable on other grounds, namely, their ability or inability to grow on various intermediates in the pathway. Gene *C* mutants can grow only if supplemented with end product C; gene *B* mutants can grow if supplemented with either B or C; and gene *A* mutants can grow on either A, B, or C. A pattern of pathway precursor accumulation occurs in these strains. Gene *C* mutants accumulate product B; gene *B* mutants accumulate product A; and gene *A* mutants accumulate the precursor for the pathway. In actual experiments in which the pathway is unknown, the sequence of steps in a pathway can be deduced from the pattern of growth supplementation and precursor accumulation. In general, the farther along in a pathway a mutant strain is blocked genetically, the fewer intermediate compounds permit the strain to grow. If a mutant strain is blocked at early steps, a larger number of intermediates enable the strain to grow.

Returning to the auxotroph data in Table 9.1, the *argH* mutant strain grows when supplemented with arginine but not when supplemented with any of the intermediates. This means that the *argH* gene codes for the enzyme that controls the last step in the pathway, which leads to the formation of arginine. The *argG* mutant strain grows on media supplemented with arginine or argininosuccinate; the *argF* mutant strain grows on media supplemented with arginine, argininosuccinate, or citrulline; and the *argE* strain grows on media supplemented with arginine, argininosuccinate, citrulline, or ornithine. Using the principles just described, we conclude that argininosuccinate must be the immediate precursor of arginine, since argininosuccinate permits all but the *argH* mutant to grow.

By continuing this reasoning, the pathway shown in Figure 9.5 can be proposed. Gene *argF*+ codes for the enzyme that converts ornithine to citrulline. Therefore, an *argF* mutant strain can grow on minimal medium plus citrulline, argininosuccinate, or arginine. Since there is an accumulation of the intermediate compound produced prior to the step blocked by the genetic mutation, that information can be used to confirm the conclusions about the sequence of steps in a pathway. The *argF* mutant strains, for example, accumulate ornithine, and hence ornithine is before citrulline, argininosuccinate, and arginine in the pathway.

With this approach we have genetically dissected a biochemical pathway—that is, we have determined the sequence of steps in the pathway and related each step to a specific gene or genes. However, more than one gene may control each step in a pathway. An enzyme may have two or more different polypeptide chains, each of which is coded for by a specific gene. In this case, more than one gene specifies that enzyme and thus that step in the pathway. Therefore Beadle and Tatum's original hypothesis has been changed to the **one gene–one polypeptide hypothesis.** We should also be aware that some biochemical pathways are branched and, hence, their analysis is more complicated.

ℛEYNOTE

A number of classical studies indicated the specific relationship between genes and enzymes, eventually embodied in Beadle and Tatum's one gene–one enzyme hypothesis, which states that each gene controls the synthesis or activity of a single enzyme. Since we now know that enzymes may consist of more than one polypeptide and that genes code for individual polypeptide chains, a

~ FIGURE 9.5

Arginine biosynthetic pathway, showing the four genes in *Neurospora crassa* that code for the enzymes that catalyze each reaction. (The genes are not on the same chromosome.)

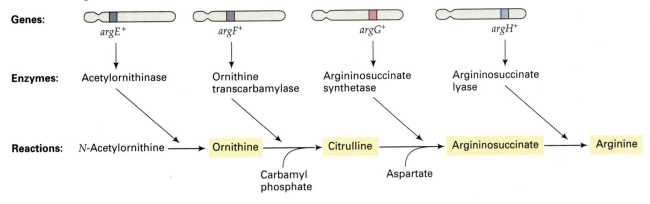

more modern description for this hypothesis is the one gene–one polypeptide hypothesis.

GENETICALLY BASED ENZYME DEFICIENCIES IN HUMANS

Many human genetic diseases are caused by a single gene mutation that alters the function of an enzyme. In general, an enzyme deficiency caused by a mutation may have either simple or **pleiotropic** (wide-reaching) consequences. Table 9.2 presents several of these diseases (which Garrod would have called inborn errors of metabolism). Studies of these diseases have offered further evidence that many genes code for enzymes. Some of these genetic diseases will be discussed in the following sections.

Phenylketonuria

Like alkaptonuria, *phenylketonuria* (*PKU*) is an inborn error of metabolism. An individual with PKU is called a *phenylketonuric*. PKU occurs in about 1 in 12,000 Caucasian births. PKU is most commonly caused by a recessive mutation of a gene on chromosome 1 (an autosome), and individuals must therefore be homozygous for the mutation in order to exhibit the condition. The mutation is in the gene for phenylalanine hydroxylase. The absence of that enzyme activity prevents the conversion of the amino acid phenylalanine to the amino acid tyrosine (Figure 9.6a and b). Phenylalanine is one of the *essential amino acids*; that is, it is an amino acid that must be included in the diet since humans are unable to synthesize it. The phenylalanine is required to make our own proteins, but excess amounts are harmful and are converted to tyrosine for further metabolism. Thus a serious problem for children born with PKU is that the absence of phenylalanine hydroxylase results in the accumulation of the phenylalanine they ingest. (PKU children are unaffected prior to birth or at birth because excess phenylalanine that accumulates is metabolized by maternal enzymes.) However, unlike the accumulated precursor HA for alkaptonuria, which is excreted in the urine, the accumulated phenylalanine in phenylketonurics is converted by a secondary pathway to phenylpyruvic acid. This substance drastically affects the cells of the central nervous system and produces serious symptoms: severe mental retardation, a slow growth rate, and early death.

A phenylketonuric also cannot make tyrosine, an amino acid required for protein synthesis and for the production of the hormones thyroxine and adrenaline and the skin pigment melanin. This aspect of the phenotype is not very serious because tyrosine can be

~ **FIGURE 9.6**

Phenylalanine-tyrosine metabolic pathways: (a) Biochemical steps that operate in normal individuals; (b) Metabolic block in the pathway exhibited by individuals with phenylketonuria—phenylalanine cannot be metabolized to tyrosine and unusual metabolites of phenylalanine accumulate; (c) Metabolic block in the pathway exhibited by individuals with albinism—no or very little melanin pigment is produced.

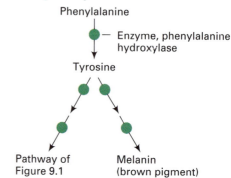

a) **Normal pathway**

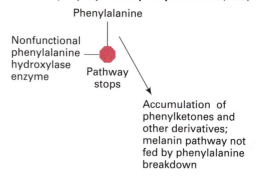

b) **Pathway in people with phenylketonuria (PKU)**

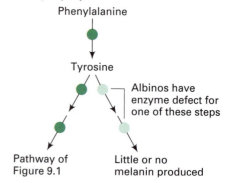

c) **Pathway in people with albinism**

obtained from food. Yet food does not normally have a lot of tyrosine. As a result, people with PKU make relatively little melanin, since their bodies can use only tyrosine (rather than phenylalanine and tyrosine) for melanin synthesis. Hence people with PKU tend to have very fair skin and blue eyes (even if they have brown-eye genes). In addition, PKU individuals have relatively low adrenaline levels.

∼ TABLE 9.2

Selected Human Genetic Disorders with Demonstrated Enzyme Deficiencies

GENETIC DEFECT	ENZYME DEFICIENCY
Acatalasia I and II[a]	Catalase
Acid phosphatase deficiency	Acid phosphatase
Alkaptonuria	Homogentisic acid oxidase
Ataxia, intermittent[a]	Pyruvate decarboxylase
Cystic fibrosis	Cystic fibrosis transmembrane conductance regulator (CFTR)
Cataract	Galactokinase
Citrullinemia[a]	Argininosuccinate synthetase
Disaccharide intolerance	Invertase
Fructose intolerance	Fructose-1-phosphate aldolase
Fructosuria	Liver fructokinase
Galactosemia[a]	Galactose-1-phosphate uridyl transferase
Gaucher disease[a]	Glucocerebroside oxidase
G6PD deficiency (favism)[a]	Glucose-6-phosphate dehydrogenase
Glycogen storage disease I	Glucose-6-phosphatase
Glycogen storage disease II[a]	α-1,4-Glucosidase
Glycogen storage disease III[a]	6-Phosphofructokinase
Glycogen storage disease IV[a]	Amylo-1:4, 1:6-transglucosidase
Gout, primary	Hypoxanthine-guanine phosphoribosyltransferase
Hemolytic anemia[a]	Glutathione peroxidase or glutathione reductase or glutathione synthetase or hexokinase or pyruvate kinase
Hypervalinemia[a]	Valine transaminase
Hypoglycemia and acidosis	Fructose-1,6-diphosphatase
Immunodeficiency	Uridine monophosphate kinase
Intestinal lactase deficiency (adult)	Lactase
Ketoacidosis[a]	Succinyl CoA:3-ketoacid CoA-transferase
Kidney tubular acidosis with deafness	Carbonic anhydrase B
Leigh's necrotizing encephalomelopathy[a]	Pyruvate carboxylase
Lesch-Nyhan syndrome[a]	Hypoxanthine phosphoribosyltransferase
Lysine intolerance	Lysine: NAD-oxidoreductase
Male pseudohermaphroditism	Testicular 17,20-desmolase
Maple sugar urine disease[a]	Keto acid decarboxylase
Muscular dystrophy	Acetylcholinesterase or acetylcholine transferase or creatinine phosphokinase
Niemann-Pick disease[a]	Sphingomyelin hydrolase
Orotic aciduria[a]	Orotidylic decarboxylase
Phenylketonuria[a]	Phenylalanine hydroxylase
Porphyria, acute[a]	Uroporphyrinogen III synthetase
Porphyria, congenital erythropoietic[a]	Uroporphyrinogen III cosynthetase
Pulmonary emphysema	α-l-Antitrypsin
Pyridoxine-dependent infantile convulsions	Glutamic acid decarboxylase
Pyridoxine-responsive anemia	λ-Aminolevulinic synthetase
Ricketts, vitamin D-dependent	25-Hydroxycholecalciferol 1-hydroxylase
Tay-Sachs disease[a]	Hexosaminidase A
Thyroid hormone synthesis, defect in	Iodide peroxidase or deiodinase
Tyrosinemia	p-Hydroxyphenylpyruvate oxidase
Xeroderma pigmentosum[a]	DNA-specific endonuclease (repair enzyme)

[a]Prenatal diagnosis possible.

The symptoms of PKU depend upon the amount of phenylalanine which accumulates, so this genetic disease can be managed by controlling the dietary intake of that amino acid. A mixture of individual amino acids with a very controlled amount of phenylalanine is used as a protein substitute in the PKU diet. However, since it is impossible to devise a diet completely devoid of protein—and most proteins contain phenylalanine—the PKU diet will contain some phenylalanine. The PKU diet is expensive, costing more than $5,000 per year. The diet needs to be continued at least until adolescence to ensure full intellectual development. Additionally, female phenylketonurics are usually advised to maintain the restricted diet strictly until after completion of reproduction because of the potential complication of high maternal phenylalanine on the developing fetus. Given the very serious consequences of allowing PKU to go untreated in this way, all states require that newborns be screened for PKU. The screen—the Guthrie test—is conducted by placing a drop of blood on a filter containing a phenylalanine analog and bacteria that can only grow if excess phenylalanine is present.

You may have noticed that foods and drinks containing the artificial sweetener NutraSweet carry a warning that people with PKU should not use them. NutraSweet is aspartame, a chemical consisting of the amino acid aspartic acid attached to the amino acid phenylalanine. This combination signals to your taste receptors that the substance is sweet, yet it is not sugar and does not have the calories of sugar. Once ingested, aspartame is broken down to aspartic acid and phenylalanine, so given enough aspartame there can be very serious effects for a phenylketonuric.

Albinism

Albinism (see Figure 2.21, p. 39) is caused by an autosomal recessive mutation, and individuals must be homozygous for the mutation to exhibit the condition. About 1 in 33,000 Caucasians and 1 in 28,000 African-Americans in the United States has albinism. The mutation affects a gene for an enzyme used in the pathway from tyrosine to the brown pigment melanin (see Figure 9.6c). Melanin absorbs light in the UV range and is important in protecting the skin against harmful UV radiation from the sun. Persons with albinism produce no melanin so they have white skin, white hair, red eyes (because of a lack of pigment in the iris), and are very light-sensitive. No apparent problems result from the accumulation of precursors in the pathway prior to the block.

There are at least two kinds of albinism, since at least two biochemical steps can be blocked to prevent melanin formation. Thus, two parents with albinism that has resulted from different enzyme deficiencies for two different steps of the pathway can produce normal children as a result of complementation of the two nonallelic mutations (see Chapter 8, p. 254).

Lesch-Nyhan Syndrome

Lesch-Nyhan Syndrome is an ultimately fatal human trait caused by a recessive mutation in an X chromosome gene located at Xq26–27. The gene spans 44 kb (44,000 base pairs) and encodes a polypeptide of 218 amino acids. An estimated 1 in 10,000 males exhibit Lesch-Nyhan syndrome. Because this disease is often lethal before reproductive age, virtually no Lesch-Nyhan females (who would be homozygous mutants) are seen. Carrier females (who usually receive the mutant allele from their mothers) may not be symptomless because of random X inactivation (lyonization: see Chapter 3, p. 79).

Lesch-Nyhan syndrome is the result of a deficiency in the enzyme hypoxanthine-guanine phosphoribosyl transferase (HGPRT), an enzyme essential to the utilization of purines. When the biosynthetic pathway is highly impaired, as in this case, excess purines accumulate, and are converted to uric acid. At birth, Lesch-Nyhan individuals are healthy and develop normally for several months. The uric acid excreted in the urine leads to the deposition of orange uric acid crystals in the diapers, an indication of this genetic disease. Between three and eight months, delays in motor development lead to weak muscles. Later, the muscle tone changes radically, producing uncontrollable movements and involuntary spasms, seriously affecting feeding activities.

After two or three years, as a result of severe neurological complications, Lesch-Nyhan children begin to show extremely bizarre activity, such as compulsive biting of fingers, lips, and the inside of the mouth. This self-mutilation is difficult to control and is not without severe pain to those afflicted. Typically, behavior toward others becomes aggressive. There are no available therapies for these neurological complications. In intelligence tests these patients score in the severely retarded region, although the difficulties they show in communicating with others may be a major contributing factor to the low scores. Most Lesch-Nyhan individuals die before they reach their twenties, usually from pneumonia, kidney failure, or uremia (uric acid in the blood).

Understandably, the elevated uric acid levels that result from an HGPRT deficiency might give rise to uremia, kidney failure, and mental deficiency. No clear explanation has been proposed, however, of how such a deficiency can lead to self-mutilating behavior.

Tay-Sachs Disease

Lysosomes are membrane-bound organelles that contain 40 or more different digestive enzymes. These enzymes catalyze the breakdown of nucleic acids, proteins, polysaccharides, and lipids. A number of human diseases are caused by mutations in genes that code for lysosomal enzymes. These diseases, collectively called *lysosomal-storage diseases*, are generally caused by recessive mutations.

The best-known genetic disease of this type is *Tay-Sachs disease* (also called infantile amaurotic idiocy), which is caused by homozygosity for a rare recessive mutation that maps to chromosome 15. While Tay-Sachs disease is rare in the population as a whole, it has a relatively high incidence in Ashkenazi Jews of Central European origin with about 1 in 3,600 of their children having the disease.

The gene in question (*hex A*) codes for the enzyme *N*-acetylhexosaminidase A, which catalyzes the reaction shown in Figure 9.7. This enzyme cleaves a terminal *N*-acetylgalactosamine group from a brain ganglioside. In Tay-Sachs infants the enzyme is nonfunctional, so the unprocessed ganglioside accumulates in the brain cells. This accumulation causes cerebral degeneration and death, usually by age three.

Individuals with Tay-Sachs disease exhibit a number of different clinical symptoms, illustrating the pleiotropic effects of the gene mutation. Typically, the symptom first recognized is an unusually enhanced reaction to sharp sounds. Early diagnosis is also made possible by the presence on the retina of a cherry-colored spot surrounded by a white halo. About a year after birth, there is rapid neurological degeneration as the uncleaved ganglioside accumulates and the brain begins to lose control over normal function and activities. This degeneration involves generalized paralysis, blindness, a progressive loss of hearing, and serious feeding problems. By two years of age the infants are essentially immobile, and death occurs at about three to four years of age, often from respiratory infections. There is no known cure for Tay-Sachs disease.

～ FIGURE 9.7

Diagram of the biochemical step for the conversion of the brain ganglioside G_{M2} to the ganglioside G_{M3}, catalyzed by the enzyme *N*-acetylhexosaminidase A (hex A): (a) Normal pathway; (b) Pathway in individuals with Tay-Sachs disease, where the activity of hex A is deficient; the result is an abnormal buildup of ganglioside G_{M2} in the brain cells.

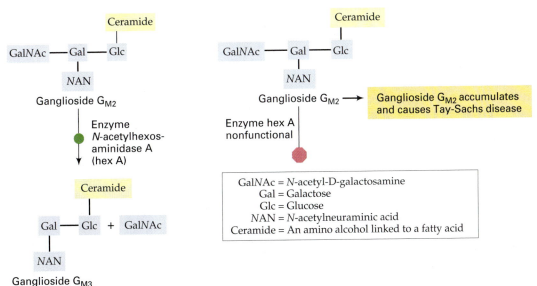

$\mathcal{K}$EYNOTE

Many human genetic diseases are caused by deficiencies in enzyme activities. Most of these diseases are inherited as recessive traits.

GENE CONTROL OF PROTEIN STRUCTURE

In the first section of this chapter, evidence was presented that genes code for enzymes; this evidence emphasized the historical significance of those studies in advancing our understanding of gene function. While all enzymes are proteins, however, not all proteins are enzymes. To understand completely how genes function, we will next look at the experimental evidence that genes are responsible for the structure of nonenzymatic proteins such as hemoglobin. Nonenzymatic proteins are often easier to study than enzymes. Enzymes are usually present in small amounts, while nonenzymatic proteins occur in relatively large quantities in the cell, which makes them easier to isolate and purify.

Sickle-Cell Anemia: Symptoms and Causes

Sickle-cell anemia is a genetically based human disease that results from an amino acid substitution in the hemoglobin protein that transports oxygen in the blood. Sickle-cell anemia was first described in 1910 by J. Herrick. He found that red blood cells from individuals with the disease lose their characteristic disc shape in conditions of low oxygen tension and assume the shape of a sickle (Figure 9.8). The sickled red blood cells are more fragile than normal red blood cells and tend to break easily. In addition, sickled cells are not as flexible as normal cells and therefore tend to clog the capillaries rather than squeeze through them. As a consequence, blood circulation is impaired, and tissues served by the capillaries become deprived of oxygen. Although oxygen deprivation occurs particularly at the extremities, the heart, lungs, brain, kidneys, gastrointestinal tract, muscles, and joints can also suffer from oxygen deprivation and its subsequent damage. The afflicted individual may therefore suffer from a variety of health problems, including heart failure, pneumonia, paralysis, kidney failure, abdominal pain, and rheumatism.

Sickle-cell anemia is caused by changes in hemoglobin, the oxygen-transporting protein in red blood cells. As shown in Figure 9.9, hemoglobin consists of four polypeptide chains—two α polypeptides and two β polypeptides—each of which is associated with a heme group (involved in the binding of oxygen). Sickle-cell anemia is caused by homozygosity for a mutation in the β-globin gene which encodes the β polypeptide of hemoglobin. At the molecular level, the sickle-cell mutation β^S is codominant with the wild-type allele β^A. Thus people who are heterozygous $\beta^A\beta^S$ make two types of hemoglobins. One, called Hb-A (hemoglobin A), is completely normal, with two normal α chains and two normal β chains specified by two wild-type α-globin genes and one wild-type β-globin gene (β^A). The other, called Hb-S, is the defective hemoglobin, with two normal α chains specified by wild-type α-globin genes and two abnormal β chains specified by the mutant β-globin gene β^S. The $\beta^A\beta^S$ heterozygotes are said to have *sickle-cell trait*, though under normal conditions they usually show few symptoms of the disease. However, after a sharp drop in oxygen tension (e.g., in an unpressurized aircraft climbing into the atmosphere), sickling of red blood cells may occur, giving rise to symptoms similar to as those found in people with severe anemia.

The evidence that an abnormal hemoglobin molecule was present in individuals with sickle-cell anemia was obtained by Linus Pauling and his co-workers in the 1950s. They used the procedure of electrophoresis to separate the electrically charged protein molecules. In this technique, proteins of the same molecular weight but different charges migrate at different rates. Pauling isolated hemoglobin from the red blood cells of three groups: normal individuals, individuals with sickle-cell trait, and individuals with sickle-cell anemia. He then subjected each sample to electrophoresis (Figure 9.10). Under the electrophoretic conditions Pauling used, the hemoglobin from normal people (Hb-A) migrated more slowly than the hemoglobin from people with sickle-cell anemia (Hb-S). The hemoglobin from sickle-cell trait individuals behaved electrophoretically like a 1:1

~ **FIGURE 9.8**

Colorized scanning electron micrograph of (left) normal and (right) sickled red blood cells.

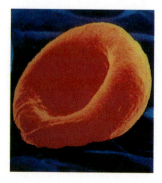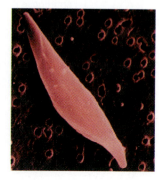

~ FIGURE 9.9

The hemoglobin molecule, showing the two α polypeptides and two β polypeptides, each of which is associated with a heme group.

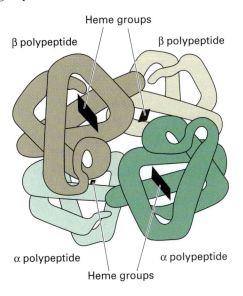

~ FIGURE 9.10

Electrophoresis of hemoglobin found *(left)* in normal $\beta^A\beta^A$ individuals, *(center)* in $\beta^A\beta^S$ individuals with sickle-cell trait, and *(right)* in $\beta^S\beta^S$ individuals who exhibit sickle-cell anemia. The two hemoglobins migrate to different positions in an electric field and hence must differ in electric charge.

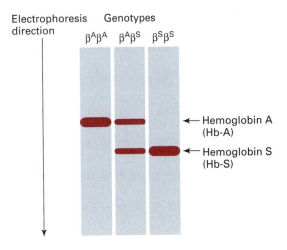

mixture of Hb-A and Hb-S consistent with the finding (above) that heterozygous individuals make both types of hemoglobin. (Note: Only one type of β chain is found in any one hemoglobin molecule, so only two types of hemoglobin molecules are possible here; one has two normal β chains, and the other has two mutant β chains.) Pauling concluded that sickle-cell anemia results from a mutation that alters the chemical structure of the hemoglobin molecule. This experiment was one of the first rigorous proofs that protein structure is controlled by genes.

The precise molecular change in the hemoglobin molecule of sickle-cell anemia individuals was determined by J. Ingram in 1957. He determined the amino acid sequences of the α and β polypeptide chains of the two types of hemoglobin, Hb-A and Hb-S, and found that the α chain was identical in the two. In the β chain, however, Ingram found a single amino acid substitution at the sixth position from the N terminal end. Figure 9.11 shows this substitution, which is the replacement of the acidic amino acid glutamic acid with the neutral amino acid valine. This particular substitution of amino acids causes the β polypeptide to fold up in a different way. In turn, this leads to sickling of the red blood cells in individuals with sickle-cell anemia. The explanation for the amino acid change is that the β^S mutation involves a single base pair change in the β-globin gene so that valine is encoded instead of glutamic acid. This finding provided strong evidence that genes specify the amino acid sequence of polypeptide chains. Furthermore,

since the hemoglobin protein consists of two types of polypeptides, α and β, it could be proposed that two genes are needed to encode hemoglobin. The α-globin and β-globin genes are located on different autosomes.

Other Hemoglobin Mutants

Many other mutant hemoglobins have been detected in general screening programs in which hemoglobin is isolated from red blood cells and subjected to electrophoresis. Over two hundred hemoglobin mutants have been detected through electrophoresis; Figure 9.12 lists some of these mutants along with the amino acid substitutions that have been identified. Some mutations affect the α chain and others the β chain, and there is a wide variety in the types of amino acid substitutions that occur. From the changes in DNA that are assumed to be responsible for the substitutions, it is apparent that a single base change is involved in each case.

Not all the identified hemoglobin mutants have as drastic an effect as the sickle-cell anemia mutant because the consequences of the substitution depend both on the amino acids involved and on the location of the mutant amino acid in the chain. For example, in the Hb-C hemoglobin molecule there is a change in the β chain from a glutamic acid to a lysine. Like the Hb-S change, this substitution involves the sixth amino acid from the N terminal end of the polypeptide. But unlike the Hb-S change, this change is not a

~ FIGURE 9.11

The first seven N terminal amino acids in normal and sickled hemoglobin β polypeptides, showing the single amino acid change from glutamic acid to valine at the sixth position.

		1	2	3	4	5	6	7
Normal β polypeptide, Hb-A	H_3N^+—	Val —	His —	Leu —	Thr —	Pro —	Glu —	Glu ···

Changes to ↓

		1	2	3	4	5	6	7
Sickle-cell β polypeptide, Hb-S	H_3N^+—	Val —	His —	Leu —	Thr —	Pro —	Val —	Glu ···

serious defect because both amino acids are hydrophilic so the conformation of the hemoglobin molecule is not as drastically altered. Therefore people homozygous for the β^C mutation experience only a mild form of anemia.

~ FIGURE 9.12

Examples of amino acid substitutions found in (a) α and (b) β polypeptides of various human hemoglobin variants.

a) α-chain

Amino acid position

	1	2	16	30	57	58	68	141
Normal	Val	Leu	Lys	Glu	Gly	His	Asn	Arg
Hb variants:								
HbI	Val	Leu	Asp	Glu	Gly	His	Asn	Arg
Hb-G Honolulu	Val	Leu	Lys	Gln	Gly	His	Asn	Arg
Hb Norfolk	Val	Leu	Lys	Glu	Asp	His	Asn	Arg
Hb-M Boston	Val	Leu	Lys	Glu	Gly	Tyr	Asn	Arg
Hb-G Philadelphia	Val	Leu	Lys	Glu	Gly	His	Lys	Arg

b) β-chain

Amino acid position

	1	2	3	6	7	26	63	67	121	146
Normal	Val	His	Leu	Glu	Glu	Glu	His	Val	Glu	His
Hb variants:										
Hb-S	Val	His	Leu	Val	Glu	Glu	His	Val	Glu	His
Hb-C	Val	His	Leu	Lys	Glu	Glu	His	Val	Glu	His
Hb-G San Jose	Val	His	Leu	Glu	Gly	Glu	His	Val	Glu	His
Hb-E	Val	His	Leu	Glu	Glu	Lys	His	Val	Glu	His
Hb-M Saskatoon	Val	His	Leu	Glu	Glu	Glu	Tyr	Val	Glu	His
Hb Zurich	Val	His	Leu	Glu	Glu	Glu	Arg	Val	Glu	His
Hb-M Milwaukee-1	Val	His	Leu	Glu	Glu	Glu	His	Glu	Glu	His
Hb-D β Punjab	Val	His	Leu	Glu	Glu	Glu	His	Val	Gln	His

Biochemical Genetics of the Human ABO Blood Groups

The mechanism of inheritance of the human ABO blood groups was mentioned in Chapter 4. To summarize, the blood group depends on the presence of chemical substances called antigens on the red blood cell surface. When antigens are injected into a host organism, they may be recognized as foreign and removed from the circulation by the host's antibodies. The molecular basis of the ABO blood groups further supports the direct relationship between genes and amino acid sequences in proteins.

An individual generally has antibodies circulating in blood serum that are directed toward those antigens that the individual is not carrying on his or her red blood cells. People of blood group A (genotypes I^A/I^A or I^A/i) carry the A antigen on their blood cells, and in their serum they have antibodies against the B antigen (the β antibody). Similarly, people of blood group B (genotypes I^B/I^B or I^B/i) carry the B antigen and have antibodies against the A antigen (the α antibody). An individual with AB blood type (genotype I^A/I^B), however, has both the A and B antigens but neither α nor β circulating antibodies. Finally, an individual of blood group O (genotype i/i) has neither the A nor the B antigens on their blood cells and so has both the α and β antibodies in circulation. As we learned in Chapter 4, I^A, I^B, and i are all alleles of one gene (the ABO locus) and hence constitute a multiple allelic series.

The ABO locus is involved in the production of enzymes that are used in the biosynthesis of polysaccharides. The enzymes, called *glycosyltransferases*, act to add sugar groups to a preexisting polysaccharide. The polysaccharides that are important here are those that attach to lipids to produce compounds called glycolipids. The glycolipids then associate with red blood cell membranes to form the blood group antigens. Figure 9.13 shows the terminal sugar structures of the polysaccharide components of the glycolipids involved in the ABO blood type system. Most individuals produce a glycolipid called the H antigen, which has the terminal sugar sequence shown in the figure. The I^A allele produces a glycosyltransferase enzyme called α-N-acetylgalactosamyl transferase, which recognizes the H antigen and adds the sugar α-N-acetylgalactosamine to the end of the polysaccharide to produce the A antigen. The I^B allele, on the

other hand, produces an α-D-galactosyltransferase, which also recognizes the H antigen but adds galactose to its polysaccharide to produce the B antigen. (Note that this small difference in the structure of the A and B antigens is sufficient to induce an antibody response.) In both cases some H antigen remains unconverted.

For the I^A/I^B heterozygote, both enzymes are produced, so some H antigen is converted to the A antigen and some to the B antigen. The blood cell has

~ **FIGURE 9.13**

Terminal sugar sequences in the polysaccharide chains of glycolipids involved in the human ABO blood types (a) H antigen; (b) A antigen; (c) B antigen.

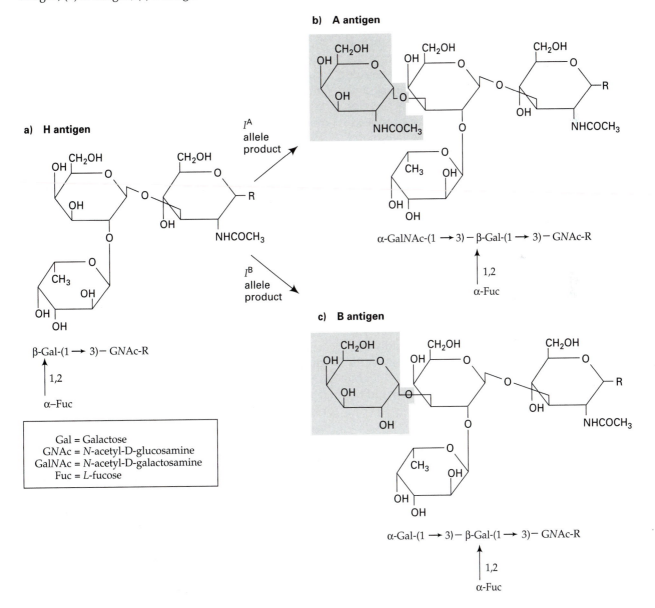

both antigens on the surface, so the individual is of blood type AB. It is the presence of both surface antigens that provides the molecular basis of the codominance of the I^A and I^B alleles (see Chapter 4, p. 104).

People who are homozygous for the i allele produce no enzymes to convert the polysaccharide component of the H antigen glycolipid. As a consequence, their red blood cells carry neither the A nor the B antigen, although they do carry the H antigen. The H antigen does not elicit an antibody response in people of other blood groups because its polysaccharide component is the basic component of the A and the B antigens as well, so it is not perceived as a foreign substance. People who are heterozygous for the i allele will have the blood type of the other allele. For example, in I^B/i individuals the I^B allele will result in the conversion of some of the H antigen to the B antigen, determining the blood type of the individual.

The H antigen is produced by the action of the dominant H allele at a locus that is distinct from the ABO locus. Individuals homozygous for the recessive mutant allele, h, do not make the H antigen, and therefore, regardless of the presence of I^A or I^B alleles at the ABO locus, no A or B antigens can be produced. Thus, these very rare h/h individuals are like blood type O individuals, in the sense that they lack A and B antigens; they are said to have the Bombay blood type. Note, however, that the blood of people who have the Bombay blood type is not identical to that of people with blood group O. This is because individuals with the Bombay blood groups produce anti-O antibodies, while individuals with blood group O do not. That is, the anti-O antibodies are made because of the H antigen found in blood group O individuals; obviously, blood group O individuals do not make antibodies against their own H antigens. As a consequence of the Bombay blood group phenomenon, it is theoretically possible for two blood type O individuals to produce a child who has blood type A or B. That is, if one parent is h/h $I^B/-$ (i.e., h/h I^B/I^B or h/h I^B/i), and the other is H/H i/i, the child could be H/h I^B/i, and, therefore, be blood type B.

Cystic Fibrosis

Cystic fibrosis (CF) is a human disease that causes pancreatic, pulmonary, and digestive dysfunction in children and young adults. Typical of the disease is an abnormally high viscosity of secreted mucus. Male adults with CF are usually infertile, while female adults are often infertile. CF is managed by hitting the chest to help clear infected secretions that accumulate in the lungs, and by giving antibiotics to treat the infections. CF is a lethal disease; with present management procedures, life expectancy is about forty years.

CF is caused by homozygosity for an autosomal recessive mutation located on the long arm of chromosome seven. CF is the most common lethal autosomal recessive disease among Caucasians, with about 1 in 1,800 newborns having the disease. About 1 in 25 Caucasians is estimated to be a heterozygous carrier. In the African-American population, about 1 in 17,000 newborns have CF, and in Asians, the CF frequency is 1 in 90,000 newborns.

The defective gene product in CF patients was not identified by biochemical analysis as was the case for PKU and many other diseases, but by a combination of genetic and modern molecular biology techniques. Since CF is a genetic disease, it can be followed in pedigrees of families which express the disease. Researchers were able to correlate the segregation of particular DNA fragments (generated by cutting nuclear DNA with a restriction enzyme) with the inheritance of the CF mutation. (This method is called restriction fragment length polymorphism [RFLP] mapping, and is described in more detail in Chapter 15.) One fragment that showed linkage with the CF mutation enabled investigators to locate the CF gene to chromosome 7. Two other DNA fragments were subsequently found that were more closely linked to the CF gene. They gave the scientists two DNA landmarks in the genome to search for and clone the gene. By looking carefully at cloned DNA fragments spanning the over 1,500 kb (1,500,000 base pairs) between the landmarks on chromosome 7, L.-C. Tsui and John Riordan, as well as Francis Collins's group, found the CF gene and cloned it from both a normal individual and from a CF patient. (The cloning of the CF gene is described in more detail in Chapter 15.)

Analysis of the DNA sequences of the cloned genes showed that, in patients with a serious form of CF, the most common mutation—ΔF508—is the deletion of three consecutive base pairs in the ATP–binding, nucleotide binding fold (NBF) region toward the 5' end of the gene. Since each amino acid in a protein is specified by three base pairs in the DNA, this finding means that one amino acid is missing near middle of the CF protein of the patients. But what does the CF protein do? Given the DNA sequence of the gene, it was possible to determine the amino acid sequence of the protein it encoded and then to make some predictions about the type and three-dimensional structure of that protein. The analysis indicated that the 1480-amino acid CF protein is associated with membranes. The proposed structure for the CF protein—called cystic fibrosis transmembrane conductance regulator

(CFTR)—is shown in Figure 9.14. By comparing the amino acid sequence of the CF protein with the amino acid sequences of other proteins in a computer database, CFTR protein was found to be homologous to a large family of proteins involved in active transport of materials across cell membranes. Functional analysis of the protein has shown that it is a chloride channel in certain cell membranes. In people with CF, the mutated CF gene results in an abnormal CFTR protein, and this results in abnormal salt transport across membranes. Because of this the symptoms of CF occur, starting with abnormal mucus secretion.

𝒦EYNOTE

From the study of alterations in proteins other than enzymes, such as those in hemoglobin responsible for sickle-cell anemia, convincing evidence was obtained that genes control the structures of all proteins.

GENETIC COUNSELING

We have learned that many human genetic diseases are caused by enzyme or protein defects; those defects ultimately are the result of mutations at the DNA level. Many other genetic diseases arise from chromosome defects. It is now possible to assay for many enzyme or protein deficiencies, and for many of the DNA changes associated with genetic diseases and thereby to determine whether or not an individual has a genetic disease or is a carrier for that disease. Further, it is possible to determine whether individuals have any chromosomal defects. **Genetic counseling** is the analysis of the probability that individuals have a genetic defect, or of the risk that prospective parents may produce a child with a genetic defect. In the latter case, genetic counseling involves presenting the available options for avoiding or minimizing those possible risks. Thus, genetic counseling provides people with an understanding of the genetic problems that are or may be inherent in their families or prospective families. Counseling uti-

～ FIGURE 9.14

Proposed structure for cystic fibrosis transmembrane conductance regulator (CFTR). The protein has two hydrophobic segments that span the plasma membrane and after each segment is a nucleotide-binding fold (NBF) region that binds ATP. The site of the amino acid deletion resulting from the three-nucleotide-pair deletion in the CF gene most commonly seen in patients with severe cystic fibrosis is in the first (toward the amino end) NBF; this is the ΔF508 mutation. The central portion of the molecule contains sites that can be phosphorylated by the enzymes protein kinase A and protein kinase C.

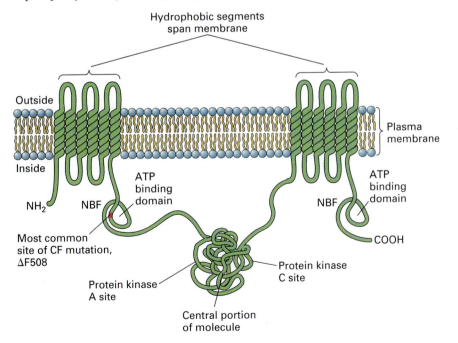

lizes a wide range of information on human heredity. In many instances, the risk of having a child with a genetic defect may be stated in terms of rather precise probabilities; in others, where the role of heredity is not sufficiently clear, the risk may be estimated only in general terms. It is the responsibility of the genetic counselors to supply their clients with clear, unemotional, and nonprescriptive statements based on the family history and on their knowledge of all relevant scientific information and of the probable risks of giving birth to a child with a genetic defect. In sum, genetic counseling is a lot more than the simple presentation of risk facts and figures to patients; it is the prevention of disease, the relief of pain, and the maintenance of health, all of which are goals of the medical profession in general.

Genetic counseling generally starts with pedigree analysis of both families to determine the likelihood that a genetic disease is present in either group. (Pedigree analysis is described in Chapters 2 and 3.) Once evidence is found of a genetic disease in a family or families, prospective parents need to be informed of the probability that they will produce a child with that disease. Early detection of a genetic disease occurs at one or both of two levels. One is the detection of heterozygotes (carriers) of recessive mutations, and the other is the determination of whether or not the developing fetus shows the biochemical defect. Tests that measure enzyme activities or protein amounts are limited to those genetic diseases in which the biochemical defect is expressed in the parents and/or the developing fetus. Tests that measure actual changes in the DNA do not depend upon expression of the gene in the parents or fetus. Detection of enzyme and protein changes is discussed in this section, while detection of DNA changes is discussed in Chapter 15.

Although we can identify carriers of many genetic diseases and can determine if fetuses have a genetic disease, in most cases there is no cure for the diseases. Carrier detection and fetal analysis serve mainly to inform parents of the risks and probabilities of having a child with the defect.

Carrier Detection

Carrier detection is the detection of individuals who are heterozygous for a recessive gene mutation. The heterozygous carrier of a mutant gene usually is normal in phenotype. Nonetheless, the individual carries a recessive mutation that can be passed on to the progeny. In the case of a recessive mutation that has serious deleterious consequences to an individual homozygous for that mutation, there is great value in

determining whether two people who are contemplating having a child are both carriers, because in that situation one-quarter of the children would be born with the trait. Carrier detection can be used in those cases in which a gene product (protein or enzyme) can be assayed. In those cases, the heterozygote (carrier) is expected to have approximately half of the enzyme activity or protein amount as homozygous normal individuals. In Chapter 15, we will see how carriers can be detected by DNA tests.

Fetal Analysis

A second important aspect of genetic counseling is finding out whether a fetus is normal. **Amniocentesis** is one way in which this can be done (Figure 9.15). As a fetus develops in the amniotic sac, it is surrounded by amniotic fluid, which serves to cushion it against shock. In amniocentesis a sample of the amniotic fluid is taken by carefully inserting a syringe needle through the mother's uterine wall and into the amniotic sac. The amniotic fluid contains cells that have sloughed off the fetus's skin; these cells can be cultured in the laboratory. Once cultured, they may be examined for protein or enzyme alterations or deficiencies, DNA changes, and chromosomal abnormalities. Amniocentesis is possible at any stage of pregnancy, but the quantity of amniotic fluid available and the increased risk to the fetus makes it impractical to perform the procedure before the twelfth week of pregnancy. Because amniocentesis is complicated and costly, it is primarily used in high-risk cases.

Another method for fetal analysis is **chorionic villus sampling** (Figure 9.16). The procedure can be done between the eighth and twelfth week of pregnancy, that is, earlier than amniocentesis. The chorion is a membrane layer surrounding the fetus, and it consists entirely of embryonic tissue. Hence, the cells obtained reflect the genetic makeup of the fetus. A chorionic villus tissue sample may be taken from the developing placenta through the abdomen (as in amniocentesis) or via the vagina using biopsy forceps or a flexible catheter, aided by ultrasound. The latter is the preferred method. Once the tissue sample is obtained, the analysis is similar to that used in amniocentesis. Advantages of the technique are that the time of testing permits the parents to learn if the fetus has a genetic defect earlier in the pregnancy than is the case with amniocentesis, and that it is not necessary to culture cells to obtain enough to do the biochemical assays. Fetal loss and inaccurate diagnoses due to the presence of maternal cells are more common in chorionic villus sampling than in amniocentesis, however.

KEYNOTE

Genetic counseling is the analysis of the risk that prospective parents may produce a child with a genetic defect, and the presentation to appropriate family members of the available options for avoiding or minimizing those possible risks. Early detection of a genetic disease is done by carrier detection, and by fetal analysis.

SUMMARY

In this chapter we began to think of the gene in molecular rather than abstract terms. Specifically, we discussed gene function—what genes code for, and the evidence for that coding. We considered a number of classical experiments that showed that genes code for enzymes and for nonenzymatic proteins. A number of these experiments were done prior to the unequivocal proof that DNA is the genetic material and before the elucidation of DNA structure.

As early as 1902 there was evidence for a specific relationship between genes and enzymes. This evidence was obtained by Archibald Garrod in his investigations of "inborn errors of metabolism," human genetic diseases that resulted in enzyme deficiencies. The specific relationship between genes and enzymes is historically embodied in the one gene–one enzyme hypothesis proposed by Beadle and Tatum to explain their data on auxotrophic mutants of *Neurospora*. The one gene–one enzyme hypothesis states that each gene controls the synthesis

~ FIGURE 9.15

Steps in amniocentesis, a procedure used for prenatal diagnosis of genetic defects.

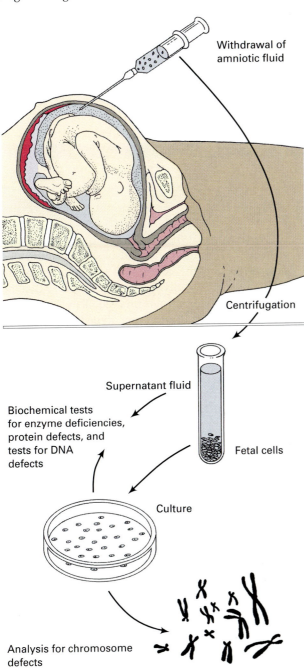

Withdrawal of amniotic fluid

Centrifugation

Supernatant fluid

Biochemical tests for enzyme deficiencies, protein defects, and tests for DNA defects

Fetal cells

Culture

Analysis for chromosome defects

~ FIGURE 9.16

Chorionic villus sampling, a procedure used for early prenatal diagnosis of genetic defects.

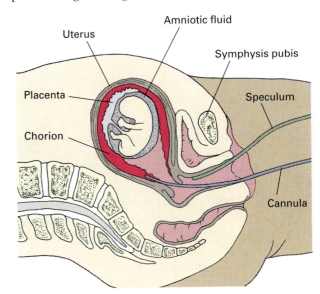

Uterus

Amniotic fluid

Symphysis pubis

Placenta

Speculum

Chorion

Cannula

or activity of a single enzyme. Since some enzymes and nonenzymatic proteins consist of more than one polypeptide (e.g. hemoglobin), a more modern maxim is one gene–one polypeptide. Beadle and Tatum's methods for dissecting a biochemical pathway genetically—that is, to determine the sequence of steps in the pathway and relating each step to a specific gene or genes—are classical. Their experiments are considered to be landmarks in the development of molecular genetics.

In this chapter we also discussed a number of examples of genetically based enzyme deficiencies that give rise to genetic diseases in humans. Many of these are the result of homozygosity for recessive mutant genes. The severity of the disease depends on the effects of the loss of function of the particular enzyme involved. Thus, albinism is a relatively mild genetic disease, while Tay-Sachs disease is lethal. To make our understanding of gene function more complete, we examined experimental evidence that genes control the structure of nonenzymatic proteins as well as enzymes. That is, while all enzymes are proteins, not all proteins are enzymes.

Given the knowledge we currently have about a large number of genetically based enzyme and protein deficiencies in humans, it is possible to make some predictions about the existence of certain genetic diseases in individuals or in families. Genetic counseling is the analysis of the probability that individuals have a genetic defect, or of the risk that prospective parents may produce a child with a genetic defect, and the presentation to the individuals involved of any available options for avoiding or minimizing those possible risks. Techniques available to genetic counselors include amniocentesis and chorionic villus sampling.

ANALYTICAL APPROACHES FOR SOLVING GENETICS PROBLEMS

Q9.1 k^+, l^+, and m^+ are independently assorting genes that control the production of a red pigment. These three genes act in a biochemical pathway as follows:

$$\text{colorless 1} \xrightarrow{k^+} \text{colorless 2} \xrightarrow{l^+} \text{orange} \xrightarrow{m^+} \text{red}$$

The mutant alleles that produce abnormal functioning of these genes are k, l, and m; each is recessive to its wild-type counterpart. A red individual homozygous for all three wild-type alleles is crossed with a colorless individual that is homozygous for all three recessive mutant alleles. The F_1 is red. The F_1 is then selfed to produce the F_2 generation.

a. What proportion of the F_2 are colorless?
b. What proportion of the F_2 are orange?
c. What proportion of the F_2 are red?

A9.1a. There are two ways to answer this question. One is to determine all of the genotypes that can specify the colorless phenotypes, and the other is to use subtractive logic, in which the proportions of orange and red are first calculated and these proportions are subtracted from one to give the proportion of colorless progeny in the F_2. We will first consider the second method.

To produce an orange phenotype three things are needed: (1) at least one wild-type k^+ allele must be present so that the colorless 1 to colorless 2 step can occur; (2) at least one wild-type l^+ allele must be present so that the colorless 2 to orange step can occur; and (3) the individual must be m/m so that the orange to red step cannot proceed. Thus, an orange phenotype results from the genotype $k^+/- \, l^+/- \, m/m$. From an $F_1 \times F_1$ cross, the probability of getting an individual with that genotype is $3/4 \times 3/4 \times 1/4 = 9/64$.

To produce a red phenotype three things are needed: (1) at least one wild-type k^+ allele must be present so that the colorless 1 to colorless 2 step can occur; (2) at least one wild-type l^+ allele must be present so that the colorless 2 to orange step can occur; and (3) at least one $m+$ allele must be present so that the orange to red step can proceed. Thus, a red phenotype results from the genotype $k^+/- \, l^+/- \, m^+/-$. From an $F_1 \times F_1$ self, the probability of getting an individual with that genotype is $3/4 \times 3/4 \times 3/4 = 27/64$.

By default, all other F_2 individuals are colorless. The proportion of F_2 individuals that are colorless is, therefore, $1 - 9/64 - 27/64 = 1 - 36/64 = 28/64$.

Using the first approach to calculate the proportion of F_2 colorless, we must determine all genotypes that will give a colorless phenotype and then add up the probabilities of each occurring. The genotypes and their probabilities are as follows:

GENOTYPES	PROBABILITIES	
$k/k\ l/l\ m/m$	1/64	(cannot convert colorless 1)
$k/k\ l/l\ m^+/-$	3/64	(cannot convert colorless 1)
$k/k\ l^+/-\ m/m$	3/64	(cannot convert colorless 1)
$k/k\ l^+/-\ m^+/-$	9/64	(cannot convert colorless 1)
$k^+/-\ l/l\ m/m$	3/64	(cannot convert colorless 2)
$k^+/-\ l/l\ m^+/-$	9/64	(cannot convert colorless 2)
Total	28/64	

b. The proportion of the F_2 was calculated in (a), above, i.e., 9/64.

c. The proportion of the F_2 was calculated in (a), above, i.e., 27/64.

Q9.2 A number of auxotrophic mutant strains were isolated from wild-type, haploid yeast. These strains responded to the addition of certain nutritional supplements to minimal culture medium either by growth (+) or no growth (0). The following table gives the growth patterns for single gene mutant strains:

MUTANT STRAINS	SUPPLEMENTS ADDED TO MINIMAL CULTURE MEDIUM				
	B	A	R	T	S
1	+	0	+	0	0
2	+	+	+	+	0
3	+	0	+	+	0
4	0	0	+	0	0

Diagram a biochemical pathway which is consistent with the data, indicating where in the pathway each mutant strain is blocked.

A9.2 The data to be analyzed are very similar to those discussed in the text for Beadle and Tatum's analysis of *Neurospora* auxotrophic mutants, from which they proposed the one gene–one enzyme hypothesis. Recall that an important principle is that the later in the pathway a mutant is blocked, the fewer nutritional supplements can be added to allow growth. In the data given, we must assume that the nutritional supplements are not listed necessarily in the order in which they appear in the pathway.

Analysis of the data indicates that all four strains will grow if given R, and that none will grow if given S. From this we can conclude that R is likely to be the end product of the pathway (all mutants should grow if given the end product) and that S is likely to be the first compound in the pathway (none of the mutants should grow given the first compound in the pathway). Thus, the pathway as deduced so far is:

$$S \longrightarrow [B,A,T] \longrightarrow R$$

where the order of B, A, and T are as yet undetermined.

Now let us consider each of the mutant strains, and see how their growth phenotypes can help define the biochemical pathway:

Strain 1 will grow only if given B or R. Therefore, the defective enzyme in strain 1 must act somewhere prior to the formation of B and R, and after the substances A, T and S. Since we have deduced that R is the end product of the pathway, we can propose that B is the immediate precursor to R, and that strain I cannot make B. The pathway so far is:

$$S \longrightarrow [A,T] \xrightarrow{1} B \longrightarrow R$$

Strain 2 will grow on all compounds except S, the first compound in the pathway. Thus, the defective enzyme in strain 2 must act to convert S to the next compound in the pathway, which is either A or T. We do not know yet whether A or T follows S in the pathway, but the growth data at least allow us to conclude where strain 2 is blocked in the pathway, that is:

$$S \xrightarrow{2} [A,T] \xrightarrow{1} B \longrightarrow R$$

Strain 3 will grow on B, R and T, but not on A or S. We know that R is the end product and S is the first compound in the pathway. This mutant strain allows us to determine the order of A and T in the pathway. That is, since strain 3 grows on T but not A, T must be later in the pathway than A, and the defective enzyme in 3 must be blocked in the yeast's ability to convert A to T. The pathway now is:

$$S \xrightarrow{2} A \xrightarrow{3} T \xrightarrow{1} B \longrightarrow R$$

Strain 4 will grow only if given the deduced end product R. Therefore, the defective enzyme produced by the mutant gene in strain 4 must act before the formation of R, and after the formation of A, T, and B from the first compound S. The mutation in 4 must be blocked in the last step of the biochemical pathway in the conversion of B to R. The final deduced pathway, and the positions of the mutant blocks are:

$$S \xrightarrow{2} A \xrightarrow{3} T \xrightarrow{1} B \xrightarrow{4} R$$

Q9.3 The ABO blood group locus determines the production of blood group antigens. People known as secretors have the same AB blood group antigens in their saliva as are found on the surfaces of the red blood cells. Nonsecretors do not have these antigens in their saliva. The secretion phenotype is controlled by a single pair of alleles: *Se* (secretor) and *se* (nonsecretor). The *Se* allele is completely dominant to the *se* allele. The secretor locus is unlinked to the ABO locus.

A child has the blood group O, M and is a secretor. Five matings are indicated in the following list. Explain for each case whether or not the child could have resulted from the mating.

a. O, M secretor × O, MN, secretor
b. A, MN, nonsecretor × B, MN, secretor
c. A, M, secretor × A, M, secretor
d. AB, MN, secretor × O, M, nonsecretor
e. A, M, nonsecretor × A, M, nonsecretor

A9.3 The child must be i/i L^M/L^M $Se/-$, where the secretor locus can be either homozygous or heterozygous for the dominant *Se* allele.

a. The genotypes of the parents are i/i L^M/L^M $Se/-$ and i/i L^M/L^N $Se/-$, so it is possible to produce the child described. There are no major assumptions that we have to make, since it does not matter whether the parents are homozygous or heterozygous for the secretor locus

b. The genotypes of the parents are $I^A/-$ L^M/L^N se/se and $I^B/-$ L^M/L^N $Se/-$, so it is possible to produce the child. The assumptions here are that both parents carry the i allele and that the secretor parent is heterozygous at that locus.

c. The parents are $I^A/-$ L^M/L^M $Se/-$ and $I^A/-$ L^M/L^M $Se/-$. On the assumption that both parents are heterozygous I^A/i, it is possible to produce the child. Since both parents are secretors, the genotype is no problem with respect to producing a secretor child.

d. The parents are I^A/I^B L^M/L^N $Se/-$ and i/i L^M/L^M se/se. The child could not be produced from this mating since one parent is AB in blood group, meaning that the child must be either A or B blood group. There is no problem with either of the other two loci.

e. The parents are $I^A/-L^M/L^M$ se/se and $I^A/-$ L^M/L^M se/se. The child could not be produced from this mating since both parents are nonsecretors and hence are homozygous se/se. Therefore, all children must be se/se and be nonsecretors. There is no problem with either of the other two loci.

QUESTIONS AND PROBLEMS

9.1 Phenylketonuria (PKU) is an inheritable metabolic disease of humans; its symptoms include mental deficiency. This phenotypic effect is due to:
a. accumulation of phenylketones in the blood
b. accumulation of sugar in the blood
c. deficiency of phenylketones in the blood
d. deficiency of phenylketones in the diet

*9.2 If a person were homozygous for both PKU (phenylketonuria) and AKU (alkaptonuria), would you expect him or her to exhibit the symptoms of PKU or AKU or both? Refer to the pathway below.

Phenylalanine
↓ (blocked in PKU)
tyrosine → DOPA → melanin
↓
ρ-Hydroxyphenylpyruvic acid
↓
Homogentisic acid
↓ (blocked in AKU)
Maleylacetoacetic acid

9.3 Refer to the pathway shown in Question 9.2. What effect, if any, would you expect PKU (phenylketonuria) and AKU (alkaptonuria) to have on pigment formation?

*9.4 a^+, b^+, c^+, and d^+ are independently assorting Mendelian genes controlling the production of a black pigment. The alternate alleles that give abnormal functioning of these genes are a, b, c, and d. A black individual of genotype a^+/a^+ b^+/b^+ c^+/c^+ d^+/d^+ is crossed with a colorless individual of genotype a/a b/b c/c d/d to produce a black F_1. $F_1 \times F_1$ crosses are then done. Assume that a^+, b^+, c^+, and d^+ act in a pathway as follows:

$$a^+ \qquad b^+ \qquad c^+ \qquad d^+$$
colorless ⟶ colorless ⟶ colorless ⟶ brown ⟶ black

a. What proportion of the F_2 are colorless?
b. What proportion of the F_2 are brown?

9.5 Using the genetic information given in Problem 9.4, now assume that a^+, b^+, and c^+ act in a pathway as follows:

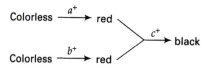

Black can be produced only if both red pigments are present; that is, C converts the two red pigments together into a black pigment.

a. What proportion of the F_2 are colorless?
b. What proportion of the F_2 are red?
c. What proportion of the F_2 are black?

9.6a. Three genes on different chromosomes are responsible for three enzymes that catalyze the same reaction in corn:

$$\text{colorless compound} \xrightarrow{a^+, b^+, c^+} \text{red compound}$$

The normal functioning of any one of these genes is sufficient to convert the colorless compound to the red compound. The abnormal functioning of these genes is designated by *a*, *b*, and *c*, respectively. A red $a^+/a^+\ b^+/b^+\ c^+/c^+$ is crossed by a colorless $a/a\ b/b\ c/c$ to give a red F_1, $a^+/a\ b^+/b\ c^+/c$. The F_1 is selfed. What proportion of the F_2 are colorless?

b. It turns out that another step is involved in the pathway. It is controlled by gene d^+, which assorts independently of a^+, b^+, and c^+:

$$\underset{\substack{\text{colorless}\\ \text{compound 1}}}{} \xrightarrow{d^+} \underset{\substack{\text{colorless}\\ \text{compound 2}}}{} \xrightarrow{a^+,\ b^+,\ c^+} \underset{\substack{\text{red}\\ \text{compound}}}{}$$

The inability to convert colorless 1 to colorless 2 is designated *d*. A red $a^+/a^+\ b^+/b^+\ c^+/c^+\ d^+/d^+$ is crossed by a colorless $a/a\ b/b\ c/c\ d/d$. The F_1 are all red. The red F_1's are now selfed. What proportion of the F_2 are colorless?

9.7 In *Drosophila*, the recessive allele *bw* causes a brown eye, and the (unlinked) recessive allele *st* causes a scarlet eye. Flies homozygous for both recessives have white eyes. The genotypes and corresponding phenotypes, then, are as follows:

$bw^+/-$	$st^+/-$	red eye
bw/bw	$st^+/-$	brown
$bw^+/-$	st/st	scarlet
bw/bw	st/st	white

Outline a hypothetical biochemical pathway that would produce this type of gene interaction. Demonstrate why each genotype shows its specific phenotype.

*9.8 In J. R. R. Tolkien's *The Lord of the Rings*, the Black Riders of Mordor ride steeds with eyes of fire. As a geneticist, you are very interested in the inheritance of the fire-red eye color. You discover that the eyes contain two types of pigments, brown and red, that are usually bound to core granules in the eye. In wild-type steeds precursors are converted by these granules to the above pigments, but in steeds homozygous for the recessive X-linked gene *w* (white eye), the granules remain unconverted and a white eye results. The metabolic pathways for the synthesis of the two pigments are

shown in Figure 9.A. Each step of the pathway is controlled by a gene: A mutation *v* results in vermilion eyes; *cn* results in cinnabar eyes; *st* results in scarlet eyes; *bw* results in brown eyes; and *se* results in black eyes. All these mutations are recessive to their wild-type alleles and all are unlinked. For the following genotypes, show the phenotypes and proportions of steeds that would be obtained in the F_1 of the given matings.

a. $w/w\ bw^+/bw^+\ st/st \times w^+/Y\ bw/bw\ st^+/st^+$

b. $w^+/w^+\ se/se\ bw/bw \times w/Y\ se^+/se^+\ bw^+/bw^+$

c. $w^+/w^+\ v^+/v^+\ bw/bw \times w/Y\ v/v\ bw/bw$

d. $w^+/w^+\ bw^+/bw\ st^+/st \times w/Y\ bw/bw\ st/st$

*9.9 Upon infection of *E. coli* with bacteriophage T4, a series of biochemical pathways result in the formation of mature progeny phages. The phages are released following lysis of the bacterial host cells. Let us suppose that the following pathway exists:

$$A \xrightarrow{\text{enzyme}} B \xrightarrow{\text{enzyme}} \text{mature phage}$$

Let us also suppose that we have two temperature-sensitive mutants that involve the two enzymes catalyzing these sequential steps. One of the mutations is cold-sensitive (*cs*) in that no mature phages are produced at 17°C. The other is heat-sensitive (*hs*) in that no mature phages are produced at 42°C. Normal progeny phages are produced when phages carrying either of the mutations infect bacteria at 30°C. However, let us assume that we do not know the sequence of the two mutations. Two models are therefore possible:

$$(1)\ A \xrightarrow{hs} B \xrightarrow{cs} \text{phage}$$
$$(2)\ A \xrightarrow{cs} B \xrightarrow{hs} \text{phage}$$

Outline how you would experimentally determine which model is the correct model without artificially lysing phage-infected bacteria.

FIGURE 9.A

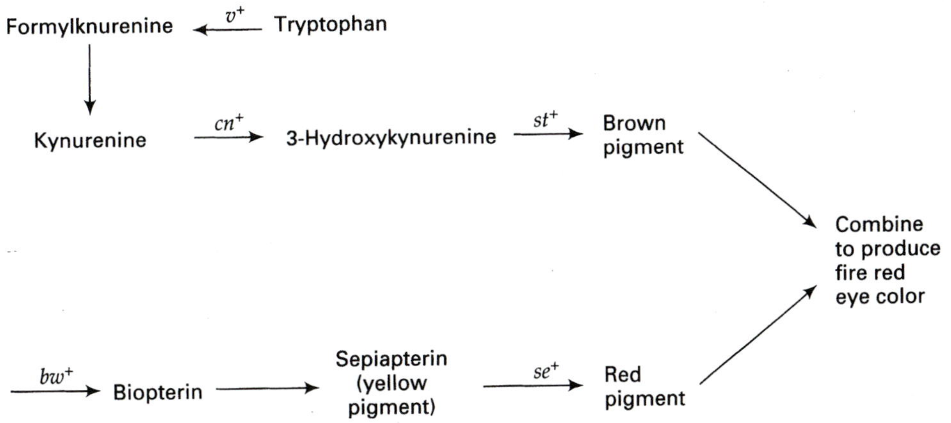

$$\text{Formylknurenine} \xleftarrow{v^+} \text{Tryptophan}$$

Kynurenine $\xrightarrow{cn^+}$ 3-Hydroxykynurenine $\xrightarrow{st^+}$ Brown pigment

$\xrightarrow{bw^+}$ Biopterin $\longrightarrow$ Sepiapterin (yellow pigment) $\xrightarrow{se^+}$ Red pigment

Combine to produce fire red eye color

9.10 Four mutant strains of *E. coli (a, b, c,* and *d)* all require substance X in order to grow. Four plates were prepared, as shown in Figure 9.B. In each case the medium was minimal, with just a trace amount of substance X, to allow a small amount of growth of the mutant cells. On plate *a,* cells of mutant strain *a* were spread over the agar, and grew to form a thin lawn. On plate *b* the lawn is composed of mutant *b* cells, and so on. On each plate, cells of the four mutant types were inoculated over the lawn, as indicated by the circles. Dark circles indicates luxuriant growth occurred. That is, this experiment tests whether the bacterial strain spread on the plate can "feed" the four strains inoculated on the plate, allowing them to grow. What do these results show about the relationship of the four mutants to the metabolic pathway leading to substance X?

*9.11 The following table indicates what enzyme is deficient in six different complementing mutants of *E. coli,* none of which can grown on minimal medium. All of them will grow if tryptophan (Trp) is added to the medium.

MUTANT	ENZYME MISSING
trpE	anthranilate synthetase
trpA	tryptophan synthetase
trpF	IGP synthetase
trpB	tryptophan synthetase
trpD	PRA transferase
trpC	PRA isomerase

Each of the plates in Figure 9.C shows the results of streaking three of the mutants on minimal medium with just a trace of added tryptophan. Heavy shading indicates regions of heavy growth, indicating that in order to permit a few cycles of replication a strain can be fed by the strain streaked next to it on the plate. In what order do the enzymes listed above act in the tryptophan synthetic pathway?

9.12 Refer to the list of mutants and enzymes given in Problem 9.11, and explain how it can be that two different complementing mutants (*trpA* and *trpB*) can affect the activity of the same enzyme. How will two such mutants be related in terms of their position within the metabolic pathway?

*9.13 Two mutant strains of *Neurospora* lack the ability to make compound Z. When crossed, the strains usually yield asci of two types: (1) those with spores that are all mutant and (2) those with four wild-type and four mutant spores. The two types occur in a 1:1 ratio.

a. Let *c* represent one mutant, and let *d* represent the other. What are the genotypes of the two mutant strains?

b. Are *c* and *d* linked?

c. Wild-type strains can make compound Z from the constituents of the minimal medium. Mutant *c* can make Z if supplied with X but not if supplied with Y, while mutant *d* can make Z from either X or Y. Construct the simplest linear pathway of the synthesis of Z from the precursors X and Y, and show where the pathway is blocked by mutations *c* and *d*.

9.14 The following growth responses (where + = growth and 0 = no growth) of mutants *1-4* were seen on the related biosynthetic intermediates A, B, C, D, and E. Assume all intermediates are able to enter the cell, that each mutant carries only one mutation, and that all mutants affect steps after B in the pathway.

FIGURE 9.B

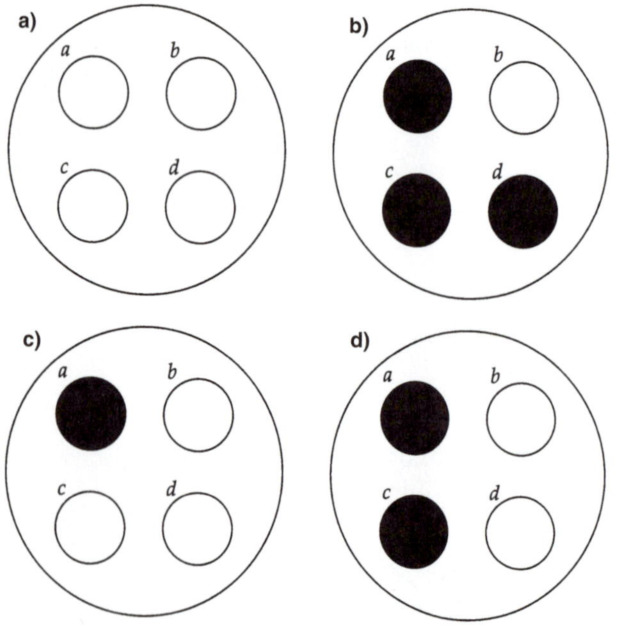

FIGURE 9.C

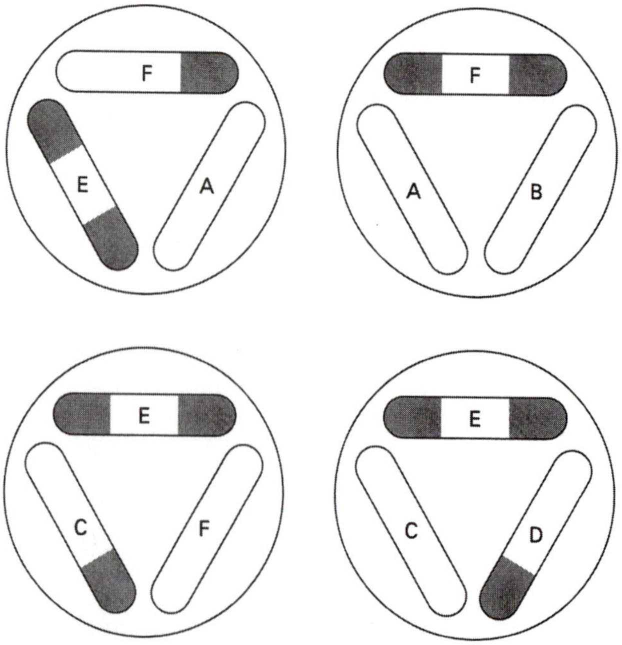

	GROWTH ON				
MUTANT	A	B	C	D	E
1	+	0	0	0	0
2	0	0	0	+	0
3	0	0	+	0	0
4	0	0	0	+	+

Which of the schemes in the figure fits best with the data with regard to the biosynthetic pathway?

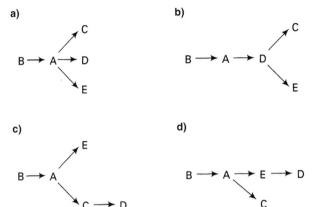

a)
b)
c)
d)

9.15 Four strains of *Neurospora*, all of which require arginine but have unknown genetic constitution, have the following characteristics. The nutrition and accumulation characteristics are as follows:

			GROWTH ON		
STRAIN	MINIMAL MEDIUM	ORNITHINE	CITRULLINE	ARGININE	ACCUMULATES
1	–	–	+	+	Ornithine
2	–	–	–	+	Citrulline
3	–	–	–	+	Citrulline
4	–	–	–	+	Ornithine

The pairwise complementation tests of the four strains gave the following results (+ = growth on minimal medium and 0 = no growth on minimal medium):

	4	3	2	1
1	0	+	+	0
2	0	0	0	
3	0	0		
4	0			

Crosses among mutants yielded prototrophs in the following percentages:

1 × 2: 25 percent
1 × 3: 25 percent
1 × 4: none detected among 1 million ascospores
2 × 3: 0.002 percent
2 × 4: 0.001 percent
3 × 4: none detected among 1 million ascospores

Analyze the data and answer the following questions.

a. How many distinct mutational sites are represented among these four strains?

b. In this collection of strains, how many types of polypeptide chains (normally found in the wild type) are affected by mutations?

c. Write the genotypes of the four strains, using a consistent and informative set of symbols.

d. Determine the map distances between all pairs of linked mutations.

e. Determine the percentage of prototrophs that would be expected among ascospores of the following types: (1) strain 1 × wild type; (2) strain 2 × wild type; (3) strain 3 × wild type; (4) strain 4 × wild type

*9.16 Alleles at a wild-type gene can be thought of as giving a normal phenotype because they confer a particular, normal amount of gene function. Mutant alleles can alter the level of function in a variety of ways. So-called loss-of-function alleles can be thought of as eliminating or decreasing gene function. These include amorphic alleles (that eliminate gene function) and hypomorphic alleles (that decrease gene function). A hypermorphic allele is an overproducer that makes more of the wild-type product. So-called gain-of-function alleles have novel functions. These include antimorphic alleles (that are antagonistic to the function of a wild-type allele) and neomorphic alleles (that provide a new, substantially altered function to the gene).

Consider the following hypothetical situation. In the production of purple pigment in sweet peas in Figure 4.14 (p. 113), the wild-type (+) product of the *P* gene is an enzyme P that converts white pigment to purple pigment. This enzyme activity can be measured in tissue extracts and mixtures of tissue extracts. The table below describes the results obtained when enzyme activity is measured in extracts and mixtures of extracts from strains with different *P* genotypes (all are *CC*). Complete the columns of the table, filling in: (a) the phenotype expected in a homozygote; (b) the phenotype expected in a heterozygote (allele/+); (c) the phenotype expected in a hemizygote (an individual that is heterozygous for the allele and a deletion for the locus); (d) the classification of the allele using the above definitions.

GENOTYPE	PERCENT OF +/+ ACTIVITY	PERCENT OF +/+ ACTIVITY WHEN MIXED 50:50 WITH +/+ EXTRACT	(a) HOMO-ZYGOTE PHENO-TYPE	(b) HETERO-ZYGOTE PHENO-TYPE	(c) HEMI-ZYGOTE PHENO-TYPE	(d) ALLELE CLASSI-FICATION
p^+/p^+	100	100	___	___	___	___
p^1/p^1	20	60	___	___	___	___
p^2/p^2	0	50	___	___	___	___
p^3/p^3	300	200	___	___	___	___
p^4/p^4	0	5	___	___	___	___
p^5/p^5	0^a	50^b	___	___	___	___

[a] Produces red, not purple pigment
[b] Produces red and purple pigments

9.17 A breeder of Irish setters has a particularly valuable show dog that he knows is descended from the famous bitch Rheona Didona, who carried a recessive gene for atrophy of the retina. Before he puts the dog to stud, he must ensure that it is not a carrier for this allele. How should he proceed?

*9.18 Suppose you were on a jury to decide the following case:

The Jones family claims that Baby Jane, given to them at the hospital, does not belong to them but to the Smith family, and that the Smith's baby Joan really belongs to the Jones family. It is alleged that the two babies were accidentally exchanged soon after birth. The Smiths deny that such an exchange has been made. Blood group determinations show the following results:

    Mrs. Jones, AB
    Mr. Jones, O
    Mrs. Smith, A
    Mr. Smith, O
    Baby Jane, A
    Baby Joan, O

Which baby belongs to which family?

*9.19 Glutathione (GSH) is important for a number of biological functions including prevention of oxidative damage in red blood cells, synthesis of deoxyribonucleotides from ribonucleotides, transport of some amino acids into cells, and maintenance of protein conformation. Mutations that have lowered levels of glutathione synthetase (GSS), a key enzyme in the synthesis of glutathione, result in one of two clinically distinguishable disorders. The severe form is characterized by massive urinary excretion of 5-oxoproline (a chemical derived from a synthetic precursor to glutathione), metabolic acidosis (an inability to regulate physiological pH appropriately), anemia, and central nervous system damage. The mild form is characterized solely by anemia. The characterization of GSS activity and the GSS protein in two affected patients, each with normal parents, is shown below.

Patient	Disease Form	GSS Activity in Fibroblasts (percent of normal)	Effect of Mutation on GSS Protein
1	severe	9%	Arginine at position 267 replaced by tryptophan
2	mild	50%	Aspartate at position 219 replaced by glycine

a. What pattern of inheritance do you expect these disorders to exhibit?

b. Explain the relationship of the form of the disease to the level of GSS activity.

c. How can two different amino acid substitutions lead to dramatically different phenotypes?

d. Why is 5-oxoproline only produced in significant amounts in the severe form of the disorder?

e. Is there evidence that the mutations causing the severe and mild forms of the disease are allelic?

f. How might you design a test to aid in prenatal diagnosis of this disease?

9.20 Some methods used to gather fetal material for prenatal diagnosis are invasive and therefore pose a small but very real risk to the fetus.

a. What specific risks and problems are associated with chorionic villus sampling and amniocentesis?

b. How are these risks balanced with the benefits of each procedure?

c. Fetal cells are reportedly present in maternal circulation after about eight weeks of pregnancy. However, the number of cells is very low, perhaps no more than one in several million maternal cells. To date, it has not been possible to isolate fetal cells in sufficient quantities for routine genetic analysis. If the problems associated with isolating fetal cells from maternal blood were overcome, and sufficiently sensitive methods were developed to perform genetic tests on a small number of cells, what would be the benefits of performing genetic tests on fetal cells isolated from maternal blood?

9.21 Refer back to Figure 9.12. What would you expect the phenotype to be in individuals heterozygous for the following two hemoglobin mutations?

a. Hb Norfolk and Hb-S

b. Hb-C and Hb-S

9.22 In evaluating my teacher, my sincere opinion is that:

a. He/she is a swell person whom I would be glad to have as a brother-in-law/sister-in-law.

b. He/she is an excellent example of how tough it is when you do not have either genetics or environment going for you.

c. He/she may have okay DNA to start with, but somehow all the important genes got turned off.

d. He/she ought to be preserved in tissue culture for the benefit of other generations.

10 THE STRUCTURE OF GENETIC MATERIAL

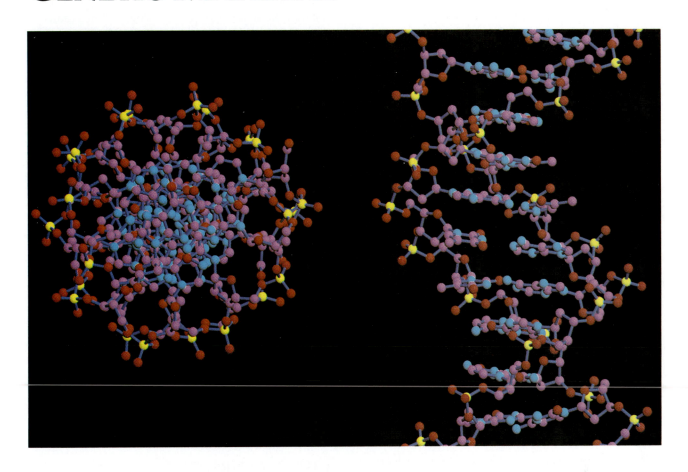

THE NATURE OF GENETIC MATERIAL: DNA AND RNA
   The Discovery of DNA as Genetic Material
   The Discovery of RNA as Genetic Material
THE CHEMICAL COMPOSITION OF DNA AND RNA
   The Physical Structure of DNA: The Double Helix
   Other DNA Structures
   DNA in the Cell
   Bends in DNA

PRINCIPAL POINTS

~ Genetic material must contain all the information for the cell structure and function of an organism. Genetic material must also replicate accurately so that progeny cells have the same genetic information as the parental cell. In addition, genetic material must be capable of variation, a basis for evolutionary change. A series of experiments proved that the genetic material of organisms consists of one of two types of nucleic acids, DNA or RNA. Of the two, DNA is most common; only certain viruses have RNA as their genetic material.

~ DNA and RNA are macromolecules composed of smaller building blocks called nucleotides. Each nucleotide consists of a 5-carbon sugar (deoxyribose in DNA, ribose in RNA) to which is attached one of four nitrogenous bases and a phosphate group.

~ In DNA the four nitrogenous bases are adenine, guanine, cytosine, and thymine, while in RNA the four bases are adenine, guanine, cytosine, and uracil.

~ According to Watson and Crick's model, the DNA molecule consists of two polynucleotide chains (polymers of nucleotides) joined by hydrogen bonds between pairs of bases (A and T, G and C) in a double helix. The diameter of the helix is 2 nm, and there are 10 base pairs in each complete turn (3.4 nm).

~ Three major types of double-helical DNA have been identified by X-ray diffraction analysis, the right-handed A- and B-DNAs, and the left-handed Z-DNA. The common form found in cells is B-DNA. A-DNA probably does not exist in cells. Z-DNA may exist in cells in stretches of DNA that are particularly rich in guanine and cytosine. The functional significance of Z-DNA is unknown.

*I*n the previous nine chapters we have learned much about how traits are inherited from one generation to another, how variations in allele organization along the chromosome arise (through recombination), and how the location of chromosomal genes governing certain traits has been determined (by gene mapping). In all our discussions, however, we have taken as given the existence and the function of the genetic material. In the next several chapters we will explore the molecular structure and function of genetic material—both **DNA (deoxyribonucleic acid)** and **RNA (ribonucleic acid)**—and we will examine in detail the mechanisms by which genetic information is transmitted from generation to generation. We will see exactly what the genetic message is, and we will learn how DNA directs the manufacture of proteins, including enzymes.

In this chapter we begin our study of the molecular aspects of genetics by discussing the properties of the genetic material, the evidence that DNA and RNA are genetic material, and the structure of DNA and RNA molecules.

THE NATURE OF GENETIC MATERIAL: DNA AND RNA

Long before DNA and RNA were proved to carry genetic informatioxn, geneticists recognized that specific molecules must fulfill that function. They postulated that the material responsible for heritable traits would have to have three principal characteristics:

1. It must contain, in a stable form, *the information* for an organism's cell structure, function, development, and reproduction.
2. It must *replicate accurately* so that progeny cells have the same genetic information as the parental cell.
3. It must be capable of *variation*. Without variation (such as through mutation and recombination), organisms would be incapable of change and adaptation, and evolution could not occur.

From experiments in the mid- to late 1800s and the early 1900s, many scientists suspected that genetic material was composed of protein. Chromosomes were first seen under the microscope in the latter part of the nineteenth century. Chemical analysis over the first 40 years of this century revealed that the nucleus contained a unique molecular constituent, deoxyribonucleic acid (DNA). It was not understood, however, that DNA was the chemical constituent of genes. Rather, DNA was believed to be a molecular framework for a theoretical class of special proteins that carried genetic information. In the 1940s, George Beadle and Edward Tatum obtained evidence from experiments with the orange bread mold, *Neurospora crassa*, that genes function by controlling the synthesis of specific enzymes (the "one gene–one enzyme" hypothesis), thereby proving a hypothesis proposed by Archibald Garrod in 1902 (see Chapter 9 for a detailed discussion). Evidence gathered from experiments in the middle of this century showed unequivocally that the genetic material of living organisms and many viruses consisted, not of protein, but of double-stranded DNA; in other viruses the genetic material may be single-stranded DNA, double-stranded RNA, or single-stranded RNA. The discovery that DNA was the genetic material lead to the explosion of knowledge about molecular aspects of biology.

The Discovery of DNA as Genetic Material

One of the first studies (done in 1928) that ultimately led to the identification of DNA as the genetic material involved the bacterium *Streptococcus pneumoniae* (also called pneumococcus) (Figure 10.1). Two strains of pneumococcus were used. The smooth (*S*) strain is infectious (virulent) and results in the death of the infected animal. Although this was not known in 1928, each bacterial cell of the *S* type is surrounded by a polysaccharide coat or capsule that gives the strain its infectious properties and results in the smooth, shiny appearance of *S* colonies. The rough (*R*) strain is noninfectious (avirulent) because *R* cells lack the polysaccharide coat; colonies of this strain are nonshiny.

The *S* strains can be classified further into *IIS* and *IIIS*, which have clearly defined differences in the chemical composition of the polysaccharide coat. The differences between *IIS* and *IIIS* are genetically determined, as are the *S* and *R* states. In about one *S* cell out of 10^7 there will be a mutational change from *S* to *R*. Rarely will the *R* colony that results give rise to an *S* cell by a second mutation. The *S* colony that devel-

~ FIGURE 10.1

Transmission electron micrograph of the bacterium *Streptococcus pneumoniae*. (Many of the cells are in the process of dividing.)

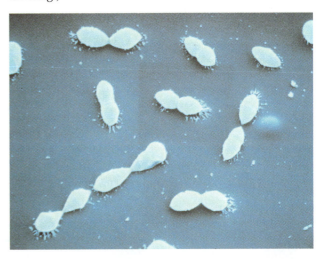

ops from the *R* mutant has the same capsule type as the *S* strain from which the *R* mutant was originally derived. That is, if the original was *IIS*, then the reversion of the *R* strain that derives from it will also be *IIS*, not *IIIS*.

In 1928 Frederick Griffith injected mice with *R* bacteria, produced by mutation of type *IIS* bacteria (Figure 10.2). Genetically these bacteria lacked the ability to make the polysaccharide coat and, if a coat could be made at all, had the genes for making a type II polysaccharide coat. The *R* bacteria did not affect the mice, and after a while the bacteria disappeared from the animals' bloodstreams. Griffith also injected mice with living type *IIIS* bacteria. Those animals died, and living type *IIIS* bacteria could be isolated from their blood. If the type *IIIS* bacteria were heat-killed before injecting them, however, the mice survived. These two experiments showed that the bacteria had to be alive and had to possess the polysaccharide coat in order for them to be infectious.

Lastly, Griffith injected mice with a mixture of living *R* bacteria (derived from *IIS*) and heat-killed type *IIIS* bacteria. In this case the mice died, and living *S* bacteria were present in the blood. These bacteria were all of type *IIIS* and therefore could not have arisen by mutation of the *R* bacteria, since mutation would have produced *IIS* colonies. Rather, Griffith concluded that some *R* bacteria had somehow been transformed into smooth, infectious type *IIIS* cells by interaction with the dead type *IIIS* cells. The transformed *IIIS* cells retained their infectious properties and capsule type in successive generations, indicating that the transformation was stable. Griffith believed

~ **FIGURE 10.2**

Griffith's transformation experiment. Mice injected with type *IIIS* pneumococcus died, whereas mice injected with either type *IIR* or heat-killed type *IIIS* bacteria survived. When injected with a mixture of living type *IIR* and heat-killed type *IIIS* bacteria, however, the mice died.

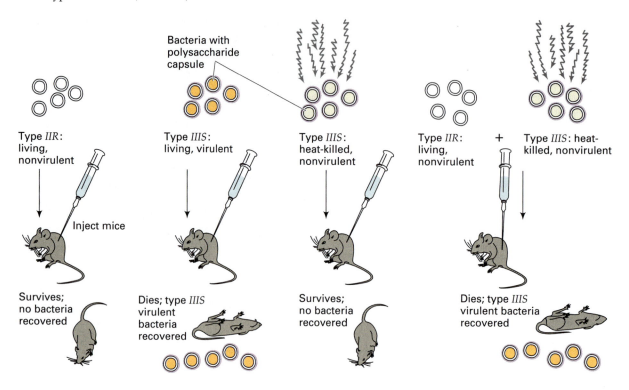

that the unknown agent responsible for the change in genetic material was a protein. He referred to the agent as the *transforming principle.* (See Chapter 8 for a discussion of bacterial transformation.)

From a modern molecular perspective, the data indicate that the genetic material of the virulent bacteria was not destroyed by the heat treatment and that it was released from the heat-killed cells and entered the living avirulent cells where recombination occurred to generate a virulent transformant (Figure 10.3). Subsequent research has verified this genetic hypothesis. That is, the *S* gene codes for a key enzyme needed for the synthesis of the carbohydrate-containing capsule. The *R* allele of the gene produces no active enzyme, so the capsule is not made.

THE NATURE OF THE TRANSFORMING PRINCIPLE.
In 1944, Oswald T. Avery, Colin M. MacLeod, and Maclyn McCarty showed that the transforming principle was not protein but was instead DNA. They used a system in which the transformation of bacteria from *R* to *S* type was done in the test tube. In their

experiments they lysed type *IIIS* cells and separated the lysate into the various cellular macromolecular components such as lipids, polysaccharides, proteins, and the nucleic acids RNA and DNA by the method outlined in Figure 10.4. They tested each component to see whether it contained the transforming principle by checking whether or not it could transform living *R* bacteria derived from *IIS*. The nucleic acids (at this point not separated into DNA and RNA) were the only components of the type *IIIS* cells that could transform the *R* cells into *IIIS*.

Avery and his colleagues then used specific nucleases—enzymes that degrade nucleic acids—to determine whether DNA or RNA was the transforming principle (Figure 10.5). When they treated the nucleic acids with **ribonuclease (RNase)**, which degrades RNA but not DNA, the transforming activity was still present. However, when they used **deoxyribonuclease (DNase)**, which degrades only DNA, no transformation resulted. These results strongly suggested that DNA was the genetic material. Although Avery's work was important, it was crit-

~ **FIGURE 10.3**

Transformation of a genetic characteristic of a bacterial cell (*Streptococcus pneumoniae*) by addition of heat-killed cells of a genetically different strain. In the example, an *R* cell receives a chromosome fragment containing the *S* gene. Since most *R* cells receive other chromosomal fragments, the efficiency of transformation for a given gene is usually less than 1 percent.

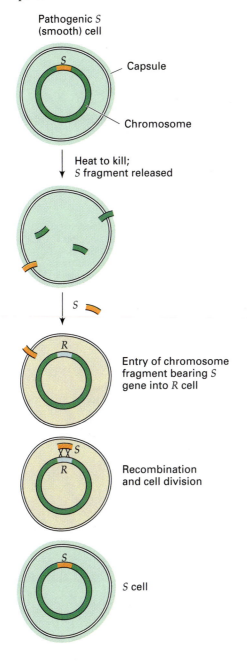

Pathogenic *S* (smooth) cell

Capsule

Chromosome

Heat to kill; *S* fragment released

S

Entry of chromosome fragment bearing *S* gene into *R* cell

Recombination and cell division

S cell

icized because the nucleic acids isolated from the bacteria were not completely pure and, in fact, contained some protein contamination.

CONSTANCY OF CHROMOSOMAL DNA AMOUNT. Support for the hypothesis that DNA was the genetic material came from the results of experiments that showed that DNA was located almost exclusively in the nucleus of eukaryotic cells but only in cell locations where chromosomes, the carriers of genetic information, were present. In addition, Alfred Mirsky and Hans Ris in 1949 demonstrated by cytochemical analysis that the amount of DNA per diploid set of chromosomes was constant for a given organism. Further, the diploid DNA amount was shown to be equal to twice the amount of DNA found in the haploid sperm. All of these observations relate directly to the inheritance patterns of genes.

Comparative studies indicated that different organisms have different amounts of DNA. We now know that the DNA content of certain cells of an organism can vary depending on the tissue of origin. In general, though, with the exception of spontaneous chromosome loss or breakage, the DNA content is usually a multiple of the DNA content of a zygote cell. For example, the DNA content of some human cells may be from two to four times that of a diploid cell, and the DNA content of the root nodule cells of leguminous plants such as the pea is characteristically double that of the cells of the rest of the plant.

Another aspect of DNA strongly supported its role as the genetic material. That is, DNA is metabolically stable, meaning that it is not rapidly degraded like many other cellular molecules.

THE HERSHEY AND CHASE BACTERIOPHAGE EXPERIMENTS. In 1953, Alfred D. Hershey and Martha Chase reported experimental results showing DNA to be genetic material. They were studying the replication of bacteriophage T2 to see which phage components were needed to complete the life cycle. (See Figure 8.14, p. 239, for the life cycle of this bacteriophage.) At the time it was known that T2 infected *Escherichia coli* by first attaching itself with its tail fibers, then injecting into the bacterium some genetic material that somehow controlled the production of progeny T2 particles. Ultimately, the bacterium lysed, releasing 100 to 200 phages that could infect other bacteria. However, the nature of T2's genetic material was not known.

Hershey and Chase grew cells of *E. coli* in media containing either a radioactive isotope of phosphorus (^{32}P) or a radioactive isotope of sulfur (^{35}S) (Figure 10.6a). They used these isotopes because DNA contains phosphorus but no sulfur, and protein contains sulfur but no phosphorus. They infected the bacteria with T2 and collected the progeny phages that were produced. At this point, Hershey and Chase had two

~ FIGURE 10.4

Chemical method used in the original isolation of a chemically pure transforming agent.

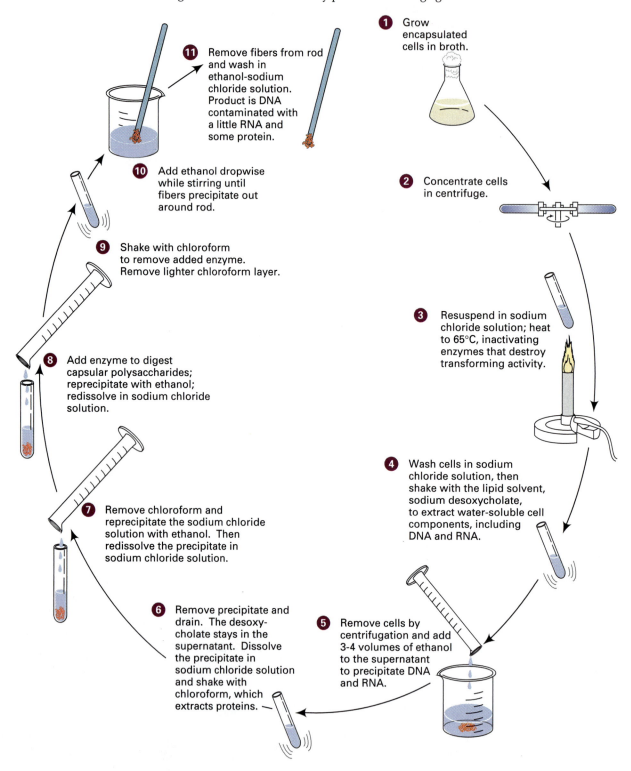

1 Grow encapsulated cells in broth.

2 Concentrate cells in centrifuge.

3 Resuspend in sodium chloride solution; heat to 65°C, inactivating enzymes that destroy transforming activity.

4 Wash cells in sodium chloride solution, then shake with the lipid solvent, sodium desoxycholate, to extract water-soluble cell components, including DNA and RNA.

5 Remove cells by centrifugation and add 3-4 volumes of ethanol to the supernatant to precipitate DNA and RNA.

6 Remove precipitate and drain. The desoxycholate stays in the supernatant. Dissolve the precipitate in sodium chloride solution and shake with chloroform, which extracts proteins.

7 Remove chloroform and reprecipitate the sodium chloride solution with ethanol. Then redissolve the precipitate in sodium chloride solution.

8 Add enzyme to digest capsular polysaccharides; reprecipitate with ethanol; redissolve in sodium chloride solution.

9 Shake with chloroform to remove added enzyme. Remove lighter chloroform layer.

10 Add ethanol dropwise while stirring until fibers precipitate out around rod.

11 Remove fibers from rod and wash in ethanol-sodium chloride solution. Product is DNA contaminated with a little RNA and some protein.

batches of T2; one had the proteins radioactively labeled with ^{35}S, and the other had the DNA labeled with ^{32}P.

Since they knew that each T2 phage consisted of only DNA and protein, they knew that one of these two classes of molecules must be the genetic material.

~ FIGURE 10.5

~ FIGURE 10.5

Experiment that showed DNA, and not RNA, was the transforming principle. When the nucleic acid mixture of DNA and RNA was treated with ribonuclease (RNase), S transformants still resulted. However, when the DNA and RNA mixture was treated with deoxyribonuclease (DNase), no S transformants resulted. (Note: R colonies resulting from untransformed cells are present on both plates; they have been omitted from the drawings.)

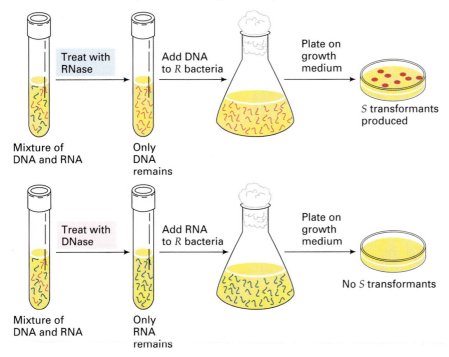

To determine which class it was, they infected unlabeled E. coli with the two types of radioactively labeled T2 with the results shown in Figure 10.6b. When the infecting phage was ^{32}P-labeled, most of the radioactivity could be found within the bacteria soon after infection. Very little could be found in protein parts of the phage (the phage ghosts) released from the cell surface after mixing the cells in a kitchen blender. After lysis, some of the ^{32}P was found in the progeny phages. In contrast, after infecting E. coli with ^{35}S-labeled T2, virtually none of the radioactivity appeared within the cell, and none was found in the progeny phage particles. Most of the radioactivity could be found in the phage ghosts released after treating the cultures in the kitchen blender.

Since genes serve as the blueprint for making the progeny virus particles, it was also presumed that the blueprint must get into the bacterial cell in order for new phage particles to be built. Therefore, since *it was DNA and not protein that entered the cell*, as evidenced by the presence of ^{32}P and the absence of ^{35}S, Hershey and Chase reasoned that DNA *must be the material responsible for the function and reproduction of phage T2*. The protein, they hypothesized, provided a structural framework to contain the DNA and the specialized structures required to inject the DNA into the bacterial cell.

The Discovery of RNA as Genetic Material

Most of the organisms and viruses discussed in this book (such as humans, *Drosophila*, yeast, E. coli, and phage T2) have DNA as their genetic material. However, some bacterial viruses (e.g., Qβ), some animal viruses (e.g., poliovirus), and some plant viruses (e.g., tobacco mosaic virus) have RNA as their genetic material. No known prokaryotic or eukaryotic organism has RNA as its genetic material. The following classical experiment showed that RNA is the genetic material of the *tobacco mosaic virus* (TMV), a virus that causes lesions on the leaves of the tobacco plant.

Like phage T2, TMV contains two chemical components, in this case RNA and protein in a spiral (helical) configuration (Figure 10.7). The protein surrounds the RNA core, protects it from attack by nucleases, and along with the RNA core, functions in the infection of the plant cells. Many varieties of TMV are known; they differ in the plants they infect and in the extent to which they affect these plants.

Hershey-Chase experiment: (a) The production of T2 phages either with (1) ^{32}P-labeled DNA or with (2) ^{35}S-labeled protein; (b) The experimental evidence showing that DNA is the genetic material in T2: (1) The ^{32}P is found within the bacteria and appears in progeny phages, (2) while the ^{35}S is not found within the bacteria and is released with the phage ghosts.

a)   Preparation of radioactively labeled T2 bacteriophage

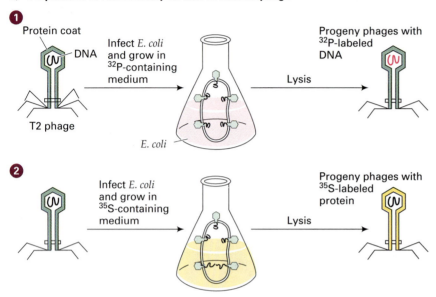

b)   Experiment that showed DNA to be the genetic material of T2

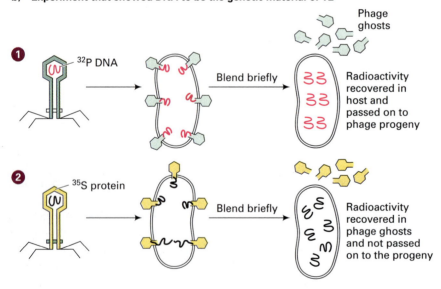

In 1956, A. Gierer and G. Schramm showed that when tobacco plants were inoculated with the purified RNA of TMV (i.e., RNA without the protein coat), they developed typical virus-induced lesions. This result indicated strongly that RNA was the genetic material of TMV. This conclusion was supported by the observation that no lesions were produced when the RNA had been degraded by treat-

ment with ribonuclease and then injected into the plant.

In an experiment in 1957, Heinz Fraenkel-Conrat and B. Singer confirmed Gierer and Schramm's conclusions. They isolated the RNA and protein components of two distinct TMV strains and reconstituted the RNA of one type with the protein of the other type, and vice versa (Figure 10.8). They then infected

~ FIGURE 10.7

Typical tobacco mosaic virus (TMV) particle. The helical RNA core is surrounded by a helical arrangement of protein subunits.

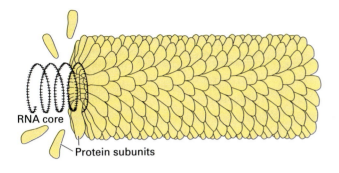

RNA core

Protein subunits

the tobacco leaves with the two hybrid viruses. The progeny viruses isolated from the resulting lesions were of the type specified by the RNA and not by the protein. These results conclusively showed that RNA is the genetic material of TMV.

KEYNOTE

Genetic material must contain all the information for the cell structure and function of an organism and must replicate accurately so that progeny cells have the same genetic information as the parental cell. In addition, genetic material must be capable of variation, one of the bases for evolutionary change. A series of experiments proved that the genetic material of organisms consists of one of two types of nucleic acids, DNA or RNA. Of the two, DNA is most common; only certain viruses have RNA as their genetic material.

~ FIGURE 10.8

Demonstration that RNA is the genetic material in tobacco mosaic virus (TMV). Hybrid particles were made from the protein subunits of one TMV strain and the RNA of a different TMV strain. Tobacco leaves were infected with the reconstituted hybrid viruses, and the progeny viruses isolated from the resulting leaf lesions were analyzed. The progeny viruses always had protein subunits specified by the RNA component; that is, the character of the protein coat cannot be transmitted from the hybrid particles to their progeny.

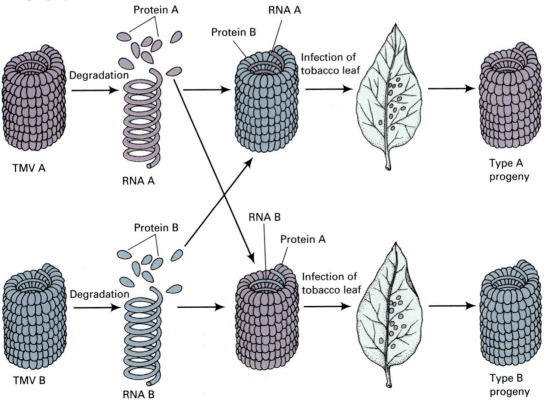

THE CHEMICAL COMPOSITION OF DNA AND RNA

Both DNA and RNA are **macromolecules**, which means that they have a molecular weight of at least a few thousand daltons (1 dalton is equivalent to a twelfth of the mass of the carbon 12 atom, or 1.67×10^{-24} g). Both DNA and RNA are polymeric molecules made up of four different monomeric units called **nucleotides**. Each nucleotide consists of three distinct parts: (1) a **pentose** (5-carbon) sugar, (2) a **nitrogenous** (nitrogen-containing) **base**, and (3) a **phosphate group**. Because they can be isolated from nuclei, and because they are acidic, these macromolecules are called nucleic acids.

For RNA the pentose sugar is **ribose**, and for DNA the sugar is **deoxyribose** (Figure 10.9). The two sugars differ by the chemical groups attached to the 2' carbon: a hydroxyl group (OH) in ribose, a hydrogen atom (H) in deoxyribose. (The carbon atoms in the pentose sugar are numbered 1' to 5' to distinguish them from the carbon and nitrogen atoms in the rings of the bases.)

The nitrogenous bases fall into two classes, the **purines** and the **pyrimidines**. In DNA the purines are **adenine** (A) and **guanine** (G), and the pyrimidines are **thymine** (T) and **cytosine** (C). The RNA molecule also contains adenine, guanine, and cytosine, but thymine is replaced by **uracil** (U). The chemical struc-

~ FIGURE 10.9

Structures of ribose and deoxyribose, the pentose sugars of RNA and DNA, respectively. The differences between the two sugars are highlighted.

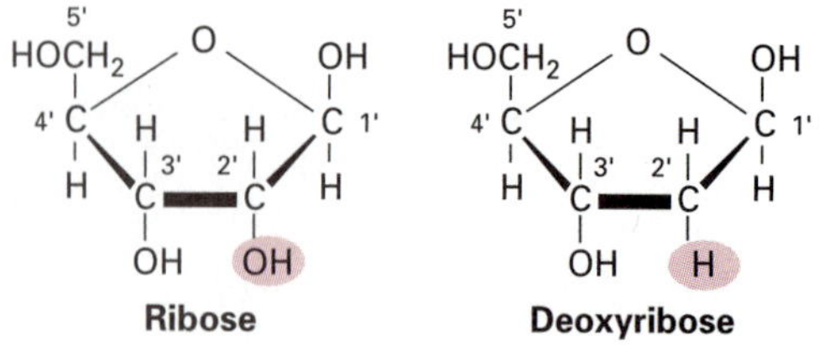

Ribose   Deoxyribose

tures of these five bases are given in Figure 10.10. (The carbons and nitrogens of the purine rings are numbered 1 to 9, and those of the pyrimidines are numbered 1 to 6.) Note that thymine contains a methyl group (CH_3) not found in uracil.

In DNA and RNA, bases are always attached to the 1' carbon of the pentose sugar by a covalent bond. The purine bases are bonded at the 9 nitrogen, while the pyrimidines bond at the 1 nitrogen. The phosphate group (PO_4^{2-}) is attached to the 5' carbon of the sugar in both DNA and RNA. Examples of a DNA nucleotide (a **deoxyribonucleotide**) and an RNA nucleotide (a **ribonucleotide**) are shown in Figure 10.11a. For future discussions, remember that a *nucleotide*, the basic building block of the DNA and

~ FIGURE 10.10

Structures of the nitrogenous bases in DNA and RNA. The parent compounds are purine (top), and pyrimidine (bottom). Differences between the bases are highlighted.

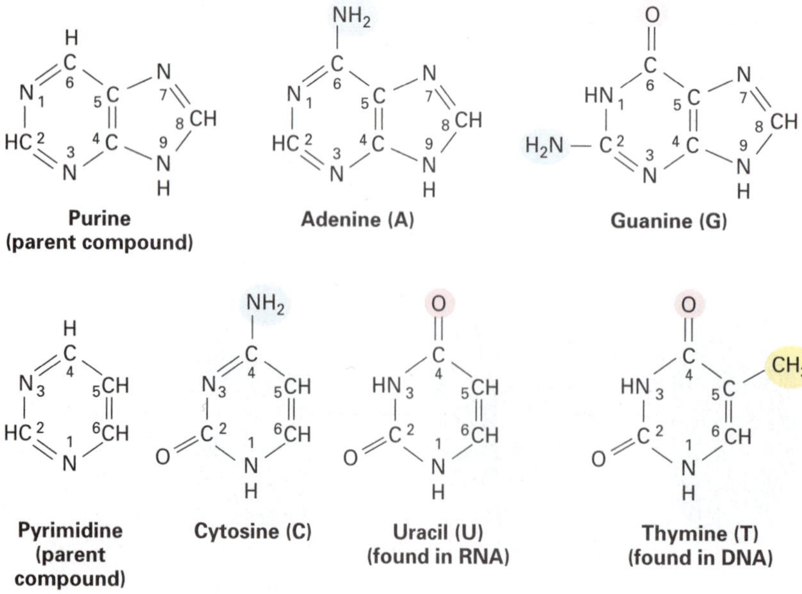

Purine (parent compound)   Adenine (A)   Guanine (G)

Pyrimidine (parent compound)   Cytosine (C)   Uracil (U) (found in RNA)   Thymine (T) (found in DNA)

~ FIGURE 10.11

Chemical structures of DNA and RNA. (a) Basic structures of DNA and RNA nucleosides (sugar plus base) and nucleotides (sugar plus base plus phosphate group), the basic building blocks of DNA and RNA molecules. Here the phosphate groups are orange, the sugars are red, and the bases are brown. (b) A segment of a polynucleotide chain, in this case a single strand of DNA. The deoxyribose sugars are linked by phosphodiester bonds (shaded) between the 3' carbon of one sugar and the 5' carbon of the next sugar.

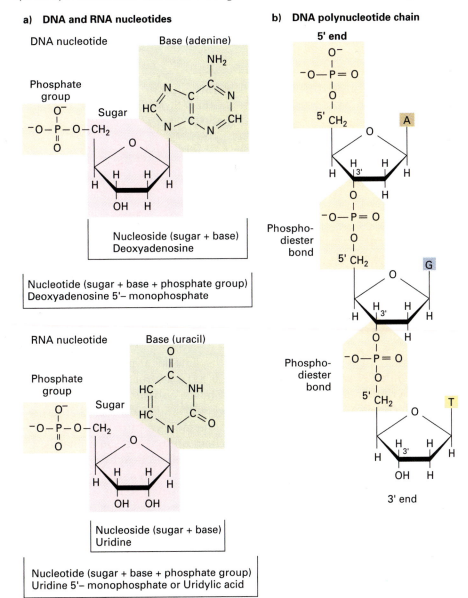

RNA molecules, consists of the sugar, a base, and a phosphate group. The sugar plus base only is called a *nucleoside*, so a nucleotide is also called a **nucleoside phosphate**. A complete listing of the names for the bases, nucleosides, and nucleotides is presented in Table 10.1. The four deoxyribonucleotide subunits of DNA are deoxyadenosine 5'-monophosphate (dAMP), deoxyguanosine 5'-monophosphate (dGMP), deoxycytidine 5'-monophosphate (dCMP), and deoxythymidine 5'-monophosphate (dTMP). The four ribonucleotide subunits of RNA are adenosine 5'-monophosphate (AMP), guanosine 5'-monophosphate (GMP), cytidine 5'-monophosphate (CMP), and uridine 5'-monophosphate (UMP).

~ TABLE 10.1

Names of the Base, Nucleoside, and Nucleotide Components Found in DNA and RNA

| | | BASE: PURINES (PU) | | BASE: PYRIMIDINES (PY) | | |
		ADENINE (A)	GUANINE (G)	CYTOSINE (C)	THYMINE (T) (DEOXYRIBOSE ONLY)	URACIL (U) (RIBOSE ONLY)
DNA	Nucleoside: deoxyribose + base	Deoxy-adenosine (dA)	Deoxy-guanosine (dG)	Deoxycytidine (dC)	Thymidine (dT)	
	Nucleotide: deoxyribose + base + phosphate group	Deoxyadenylic acid or deoxy-adenosine monophosphate (dAMP)	Deoxyguanylic acid or deoxy-guanosine monophosphate (dGMP)	Deoxycytidylic acid or deoxy-cytidine monophosphate (dCMP)	Thymidylic acid or thymidine mono-phosphate (TMP)	
RNA	Nucleoside: ribose + base	Adenosine (A)	Guanosine (G)	Cytidine (C)		Uridine (U)
	Nucleotide: ribose + base + phosphate group	Adenylic acid or adenosine monophos-phate (AMP)	Guanylic acid or guanosine monophos-phate (GMP)	Cytidylic acid or cytidine monophos-phate (CMP)		Uridylic acid or uridine monophos-phate (UMP)

Because DNA contains the pentose sugar deoxyribose and the pyrimidine thymine, while RNA contains the pentose sugar ribose and the pyrimidine uracil, these two nucleic acids have different chemical and biological properties. The presence of the 2' OH, for example, enables the RNA to be degraded with alkali, while DNA is resistant to that treatment. Also, the cellular enzymes that catalyze nucleic acid synthesis (polymerases) and nucleic acid degradation (nucleases) are usually DNA-specific or RNA-specific. The differences between the two molecules permit them to be separated and purified relatively easily for study in the laboratory as we saw earlier in the discussion of the Avery, MacLeod, and McCarty experiment.

To form polynucleotides of either DNA or RNA, nucleotides are linked together by a covalent bond between the phosphate group (which is attached to the 5' carbon of the sugar ring) of one nucleotide and the 3' carbon of the pentose sugar of another nucleotide. These 5'-3' phosphate linkages are called **phosphodiester bonds**. A short polynucleotide chain is diagrammed in Figure 10.11b. The phosphodiester bonds are relatively strong, and as a consequence, the repeated sugar-phosphate-sugar-phosphate backbone of DNA and RNA is a stable structure.

To understand how a polynucleotide chain is synthesized (which we will study in a later chapter), we must be aware of one more feature of the chain: The two ends of the chain are not the same; that is, the chain has a 5' carbon (with a phosphate group on it) at one end and a 3' carbon (with a hydroxyl group on it) at the other end, as shown in 10.11b. This asymmetry is referred to as *polarity* of the chain.

KEYNOTE

DNA and RNA occur in nature as macromolecules composed of smaller building blocks called nucleotides. Each nucleotide consists of a 5-carbon sugar (deoxyribose in DNA, ribose in RNA) to which is attached a phosphate group and one of four nitrogenous bases—adenine, guanine, cytosine, and thymine (in DNA) or adenine, guanine, cytosine and uracil (in RNA).

~ **FIGURE 10.12**

James Watson (*left*) and Francis Crick (*right*) with the 1953 model of DNA structure.

The Physical Structure of DNA: The Double Helix

In 1953 James D. Watson and Francis H. C. Crick (Figure 10.12) published a paper in which they proposed a model for the physical and chemical structure of the DNA molecule. According to their model, most DNA consists of two polynucleotide chains wound around each other in a right-handed (clockwise) helix. In generating their model, Watson and Crick used three main pieces of evidence:

1. The DNA molecule was known to be composed of bases, sugars, and phosphate groups linked together as a **polynucleotide** (deoxyribonucleotide) chain.

2. By chemical treatment Erwin Chargaff had hydrolyzed the DNA of a number of organisms and had quantified the purines and pyrimidines released. His studies showed that in all the (double-stranded) DNAs the amount of the purines was equal to the amount of the pyrimidines. More important, the amount of adenine (A) was equal to that of thymine (T), and the amount of guanine (G) was equal to that of cytosine (C). These equivalencies have become known as Chargaff's rules. In comparisons of DNAs from different organisms, the A/T and G/C ratios are always the same, although the (A + T)/(G + C) ratio (typically presented as %GC) varies (see Table 10.2).

3. Rosalind Franklin, working with Maurice H. F. Wilkins, studied isolated fibers of DNA by using the X-ray diffraction technique, a procedure in which a beam of parallel X rays is directed on a regular, repeating array of atoms. The beam is diffracted ("broken up") by the atoms in a pattern that is characteristic of the atomic weight and the spatial arrangement of the molecules. The diffracted X rays are recorded on a photographic plate. By analyzing the photograph, Franklin could obtain information about the molecule's atomic structure. The analysis of X-ray diffraction patterns is extremely complicated. As a result, given diffraction patterns can usually be interpreted in more than one way, and models built of the analyzed molecules may not be accurate. Moreover, since the experiments usually use molecules in a crystalline or fiber formation, the structures deduced may not precisely reflect the form of the molecules in the cell.

The diffraction patterns obtained by directing X rays along the length of drawn-out fibers of DNA indicated that the molecule is organized in a highly

~ **TABLE 10.2**

Base Compositions of DNAs from Various Organisms

DNA Origin	PERCENTAGE OF BASE IN DNA				RATIOS		
	A	T	G	C	A/T	G/C	(A+T)/(G+C)
Human (sperm)	31.0	31.5	19.1	18.4	0.98	1.03	1.67
Corn (*Zea mays*)	25.6	25.3	24.5	24.6	1.01	1.00	1.04
Drosophila	27.3	27.6	22.5	22.5	0.99	1.00	1.22
Euglena nucleus	22.6	24.4	27.7	25.8	0.93	1.07	0.88
Escherichia coli	26.1	23.9	24.9	25.1	1.09	0.99	1.00

~ FIGURE 10.13

X-ray diffraction analysis of DNA: The X-ray diffraction pattern of DNA that Watson and Crick used in developing their double-helix model. The dark areas that form an X shape in the center of the photograph indicate the helical nature of DNA. The dark crescents at the top and bottom of the photograph indicate the 0.34-nm distance between the base pairs.

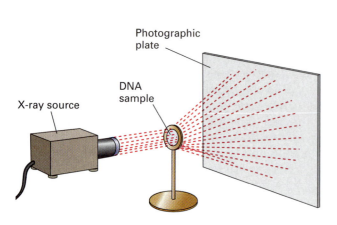

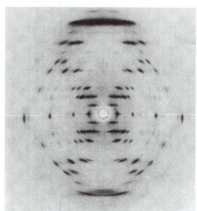

X-ray diffraction pattern

ordered, helical structure. An example of DNA's X-ray diffraction pattern and the method by which it was obtained are illustrated in Figure 10.13. Franklin interpreted these kinds of data to mean that DNA was a helical structure which had two distinctive regularities of 0.34 nm and 3.4 nm along the axis of the molecule (1 nanometer (nm) = $1/10^9$ meter = 10 Ångstrom units (Å).

Watson and Crick considered all the evidence just described and began to build three-dimensional models for the structure of DNA. The model they devised, which fit all the known data on the composition of the DNA molecule, is the now-famous double-helix model for DNA. Unquestionably, the determination of the structure of DNA was a momentous occasion in biology, leading directly to many Nobel prize–winning discoveries in molecular biology. Figure 10.14a shows a three-dimensional model of the DNA molecule, and Figure 10.14b is a diagram of the DNA molecule, showing the arrangement of the sugar-phosphate backbone and base pairs in a stylized way.

The double-helical model of DNA proposed by Watson and Crick has the following main features:

1. The DNA molecule consists of two polynucleotide chains wound around each other in a right-handed double helix; that is, viewed on end, the two strands wind around each other in a clockwise (right-handed) fashion.
2. The diameter of the helix is 2 nm.
3. The two chains are *antiparallel* (show *opposite*

polarity); that is, the two strands are oriented in opposite directions with one strand oriented in the 5' to 3' way, while the other strand is oriented 3' to 5'. To put in simpler terms, if the 5' end is the "head" of the chain and the 3' end is the "tail" of the chain, antiparallel means that the head of one chain is against the tail of the other chain and *vice versa*.

4. The sugar-phosphate backbones are on the outsides of the double helix, while the bases are oriented toward the central axis (see Figure 10.14). The bases of both chains are flat structures oriented perpendicularly to the long axis of the DNA; that is, the bases are stacked like pennies on top of one another (except for the "twist" of the helix).

5. The bases of the opposite strands are bonded together by relatively weak hydrogen bonds. The only specific pairings observed are A with T (two hydrogen bonds: Figure 10.15a) and G with C (three hydrogen bonds: Figure 10.15b). The weak hydrogen bonds make it relatively easy to separate the two strands of the DNA, for example, by heating. Breaking the A-T base pair by heating is easier than breaking the G-C base pair because A-T has two hydrogen bonds and G-C has three hydrogen bonds. The A-T and G-C base pairs are the only ones that can fit the physical dimensions of the helical model, and they are totally in accord with Chargaff's rules. The specific A-T and G-C pairs are called **complementary base pairs,**

∼ FIGURE 10.14

Molecular structure of DNA: (a) Three-dimensional molecular model of DNA as prepared by Watson and Crick; (b) Stylized representation of the DNA double helix; (c) Schematic showing the major and minor grooves in double-helical DNA.

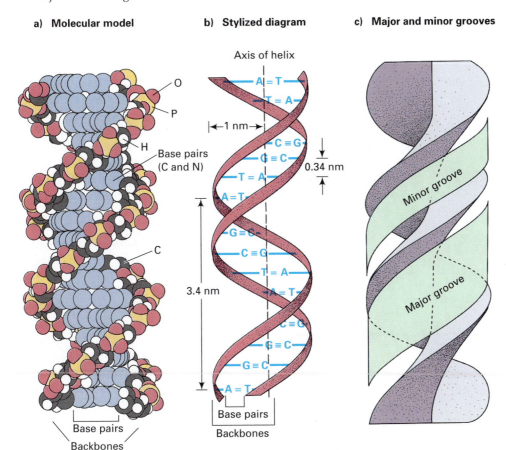

a) **Molecular model**   b) **Stylized diagram**   c) **Major and minor grooves**

so the nucleotide sequence in one strand dictates the nucleotide sequence of the other. For instance, if one chain has the sequence 5'-TATTCCGA-3', the opposite, antiparallel chain must bear the sequence 3'-ATAAGGCT-5'.

6. The base pairs are 0.34 nm apart in the DNA helix. A complete (360°) turn of the helix takes 3.4 nm; therefore, there are 10 base pairs per turn. Each base pair, then, is twisted 36° clockwise with respect to the previous base pair.

7. Because of the way the bases bond with each other, the two sugar-phosphate backbones of the double helix are not equally spaced along the helical axis. This results in grooves of unequal size between the backbones called the *major groove* (the wider groove of the two) and the *minor groove* (the narrower groove of the two)(Figure 10.14c). Both of these grooves are large enough to allow protein molecules to make contact with the bases. The phenomenon of proteins "reading" specific base-pair sequences is common to many molecular processes; we will encounter many examples in the following chapters.

ᛕEYNOTE

According to Watson and Crick's model, the DNA molecule consists of two polynucleotide chains joined by hydrogen bonds between pairs of bases (A and T, G and C) in a double helix. The diameter of the helix is 2 nm, and there are 10 base pairs in each complete turn (3.4 nm). The double-helix model was proposed by Watson and Crick from chemical and physical analyses of DNA.

~ FIGURE 10.15

Structures of the complementary base pairs found in DNA. In both cases a purine pairs with a pyrimidine: (a) The adenine-thymine bases, which pair through two hydrogen bonds; (b) The guanine-cytosine bases, which pair through three hydrogen bonds.

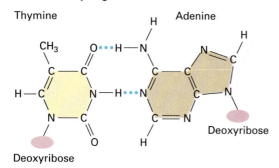

a) Adenine-thymine base (Double hydrogen bond)

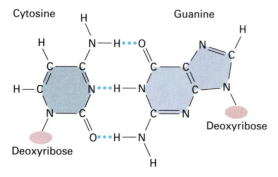

b) Guanine-cytosine base (Triple hydrogen bond)

Other DNA Structures

A more thorough analysis of the early X-ray diffraction patterns of DNA fibers revealed that DNA can exist in different forms, depending on the conditions. When humidity is relatively high, DNA is in what is called the B form (the double-helical form deduced by Watson and Crick). When the humidity is relatively low, DNA is found in the A form. Both of these DNA forms involve right-handed helices; they are compared in this section.

An inherent problem in analyzing DNA fibers by X-ray diffraction is that the DNA strands are generally not regularly ordered around and along the fiber axis. Thus the DNA models generated from the resulting data show averaged helical structures. Moreover, the data cannot reveal any localized structural variations that might result from particular base sequences. This limitation has been overcome in recent years because methods have been developed to

synthesize short DNA molecules (called **oligomers**, *oligo* meaning "few") of defined sequences. The pure DNA oligomers can be crystallized and analyzed by single-crystal X-ray diffraction. These more modern approaches have made possible a more detailed analysis of A-DNA and B-DNA. In addition, other forms of DNA have been identified in the laboratory, the most intriguing of which is Z-DNA because of its totally unexpected structure. Z-DNA has a left-handed helix and a zigzag sugar-phosphate backbone. (The latter property gave this DNA form its "Z" designation.) Figure 10.16 shows space-filling models, and Figure 10.17 shows perspective drawings of A-, B-, and Z-DNA, based on single-crystal studies. Some of the key structural features of each DNA type as deduced from the single-crystal studies are given in Table 10.3.

FEATURES OF A-DNA AND B-DNA. A-DNA and B-DNA are right-handed double helices with 10.9 and 10.0 base pairs per 360° turn of the helix, respectively. These values correspond closely with the data derived from DNA fiber studies. The bases are inclined away from the perpendicular from the helix axis (which would be 0°) by 13° in A-DNA and 2° in B-DNA.

The single crystal studies confirmed the fiber studies' conclusion that the A-DNA double helix is short and wide with a narrow, very deep major groove and a wide, shallow minor groove. (Think of these descriptions in terms of canyons—narrow versus wide describes the distance from rim to rim, and shallow vs. deep describes the distance from the rim down to the bottom of the canyon.) The B-DNA double helix is thinner and taller for the same number of base pairs than A-DNA, with a wide major groove and a narrow minor groove; both grooves are of similar depths.

FEATURES OF Z-DNA. Z-DNA was discovered when single crystals of a short length of DNA with alternating cytosine and guanine nucleotides (i.e., CGCGCG... / GCGCGC...) were subjected to X-ray diffraction analysis. Many other arrangements than alternating Gs and Cs can assume the Z-DNA conformation; all have an alternating purine-pyrimidine nucleotide sequence. However, it appears that stretches of DNA with alternating As and Ts do not form Z-DNA. Z-DNA is a left-handed helix.

In Z-DNA there are 12.0 base pairs per complete helical turn. The bases are inclined away from the perpendicular with the helix axis by 8.8 degrees. The Z-DNA helix is thin and elongated with a deep, minor groove. The major groove is pushed to the surface of the helix so that it is not really evident as a dis-

~ FIGURE 10.16

Space-filling models of (a) A-DNA, (b) B-DNA, and (c) Z-DNA.

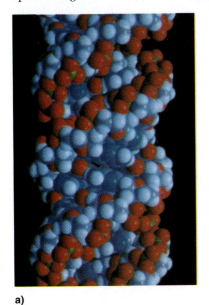
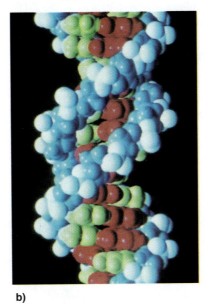
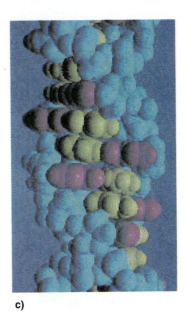

a)   b)   c)

tinct groove. In Z-DNA the negatively charged phosphate groups are brought closer together than in B-DNA, and this causes the phosphate groups to repel each other and this makes Z-DNA less stable than B-DNA.

DNA in the Cell

In solution, DNA usually is found in the B form and hence B-DNA is the form typically found in cells. Since A-DNA is found only when the DNA is in dehydrated states, it is unlikely that any lengthy sections of A-DNA exist within cells.

While the formation of Z-DNA was shown unequivocally *in vitro* by a variety of physical techniques, the existence of Z-DNA in living cells has long been a topic of debate among scientists. If so, what is the function of that Z-DNA? Some studies have confirmed the existence of Z-DNA in cells. For example, indirect evidence for the presence of Z-DNA has come from the identification of Z-DNA—binding proteins in cells, e.g., in nuclei of *Drosophila*, human, wheat and bacterial cells. It is hypothesized that these proteins stabilize the DNA in the Z-DNA form.

If Z-DNA is truly present in cells, what is its function? As we will discuss in the next three chapters,

~ TABLE 10.3

Properties of A-DNA, B-DNA, and Z-DNA

PROPERTY	A-DNA	B-DNA	Z-DNA
Helix direction	Right-handed	Right-handed	Left-handed
Base pairs per helix turn	10.9	10.0	12.0
Overall morphology	Short and wide	Longer and thinner	Elongated and thin
Major groove	Extremely narrow and very deep	Wide and of intermediate depth	Flattened out on helix surface
Minor groove	Very wide and shallow	Narrow and of intermediate depth	Extremely narrow and very deep
Helix axis location	Major groove	Through base pairs	Minor groove
Base inclination from helix axis (degrees)	13.0	2.0	8.8

~ FIGURE 10.17

Comparison of (a) A-DNA, (b) B-DNA, and (c) Z-DNA. Short segments of each DNA type are shown. The bases are red and the backbones are blue. A-DNA and B-DNA are right-handed helices, while Z-DNA is a left-handed helix. Note the zig-zag course of the sugar-phosphate backbone chain in the Z-DNA.

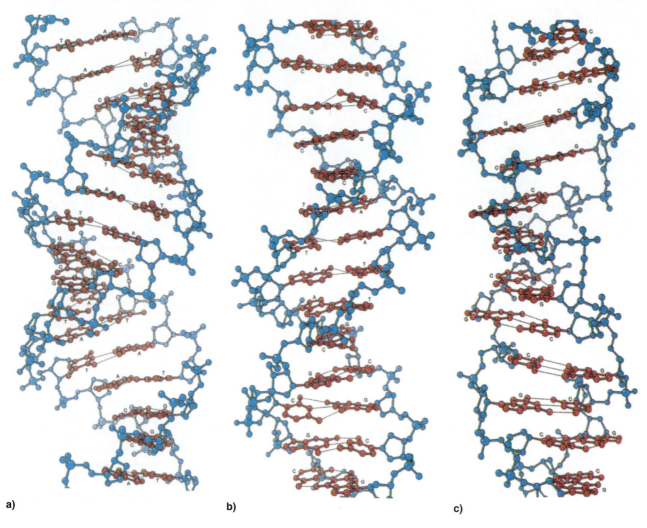

a)   b)   c)

DNA must become unwound as part of the processes of packaging into chromosomes, replication, and transcription. It has been proposed that Z-DNA provides a stretch of left-handed turns that can be used to help unwind right-handed turns in B-DNA. In a similar vein, it has also been proposed that Z-DNA may be involved in local DNA unwinding required for genetic recombination events. There is evidence to support the role of Z-DNA in at least some replication and transcription events. For example, a control sequence in the genome of SV40 (simian virus 40; a double-stranded DNA virus that infects monkeys) becomes significantly unwound during replication and transcription. Within the control sequence there are three alternating purine-pyrimidine stretches that could assume the Z-DNA form—one copy of

$\frac{\text{ACACACAT}}{\text{TGTGTGTA}}$ and two copies of $\frac{\text{ATGCATGC}}{\text{TACGTACG}}$. Since the control sequence is essential for both replication and transcription, transition from B-DNA to Z-DNA in these particular DNA stretches may have important biological effects. Finally, it is hypothesized that Z-DNA is the more stable form under extreme environmental conditions.

KEYNOTE

X-ray diffraction analysis of single crystals of DNA oligomers of known sequence has produced detailed molecular models of three DNA types:

A-DNA, B-DNA, and Z-DNA. A-DNA and B-DNA are both right-handed double helices while Z-DNA is a left-handed double helix. The number of base pairs per helical turn are 10.9, 10.0, and 12.0 for A-, B-, and Z-DNA, respectively. A-DNA is a short and broad molecule, B-DNA is a longer and thinner molecule, and Z-DNA is elongated and thin. The three DNA forms differ in a number of molecular attributes, including dimensions of the major and minor grooves and location of the helix axis. B-DNA is the form found predominantly in cells. Z-DNA has also been found in the chromosomes of some organisms but its role in cellular function, if any, is not clear. Since A-DNA forms only under relatively dehydrated conditions, it appears unlikely that this DNA type exists in a stable form in cells.

Bends in DNA

Even though the DNA molecule is drawn in textbooks as a very straight, helical molecule, implying that it is a rigid rodlike molecule, this is actually not so. In solution, B-DNA is quite a flexible molecule, undergoing shape changes as it responds to the local environmental ionic and temperature conditions. The ability of B-DNA to bend slightly is enhanced by the presence of specific base-pair sequences. The sequence CAAAAAT, for example, particularly if it is repeated three or four times separated by 10 or so base pairs, has been shown to cause DNA to bend. More often, DNA bending has been shown to result from binding of specific proteins to specific DNA sequences, as the following examples illustrate: (1) DNA bending plays an important role in certain enzyme-mediated DNA recombination events, such as the integration of the lambda phage genome into the *E. coli* chromosome; (2) certain restriction endonucleases—enzymes that cleave DNA at specific sequences (see Chapters 1 and 15)—bend DNA when they bind to their recognition sequences; and (3) many prokaryotic and eukaryotic proteins that bind to specific regulatory DNA sequences involved in the control of expression of genes cause the DNA to bend when they bind. For example, when the hormone estrogen interacts with its target cells, it binds to a receptor in the cell. The hormone-receptor complex then binds to specific regulatory sequences associated with the genes that estrogen controls. The DNA in the regulatory sequence bends as a result of the binding and it is speculated that the bending somehow turns on the transcription of the adjacent gene.

SUMMARY

In this chapter we have learned that DNA or RNA can be the genetic material. All prokaryotic and eukaryotic organisms, and most viruses have DNA as their genetic material, while some viruses have RNA as their genetic material. The form of the genetic material varies among organisms and their viruses. In prokaryotic and eukaryotic organisms, the DNA is always double-stranded, while in viruses the genetic material may be double- or single-stranded DNA or RNA, depending on the virus.

Chemical analysis revealed that DNA and RNA are macromolecules composed of smaller building blocks called nucleotides. Each nucleotide consists of a 5-carbon sugar (deoxyribose in DNA, ribose in RNA) to which is attached one of four nitrogenous bases and a phosphate group. In DNA the four nitrogenous bases are adenine, guanine, cytosine, and thymine, while in RNA the four bases are adenine, guanine, cytosine, and uracil. Thymine and uracil differ only in a single methyl group which is present in thymine and absent in uracil. From chemical and physical analysis, it was determined that the DNA molecule consists of two polynucleotide chains joined by hydrogen bonds between pairs of bases (A and T, G and C) in a double helix. (The double helix model was first proposed by Watson and Crick.) The diameter of the helix is 2 nm, and there are 10 base pairs in each complete turn of the helix (3.4 nm); each base pair is thus twisted 36° relative to the preceding base pair in the molecule.

A number of different types of double-helical DNA have been identified by X-ray diffraction analysis. The three major types are the right-handed A- and B-DNAs, and the left-handed Z-DNA. The common form found in cells is B-DNA (the form analyzed by Watson and Crick). A-DNA probably does not exist in cells. Z-DNA may exist in cells in stretches of DNA that are particularly rich in guanine and cytosine. The functional significance of Z-DNA is unknown.

Lastly, B-DNA in solution is a flexible molecule, capable of undergoing shape changes. Certain sequences have been shown to enhance the ability of B-DNA to bend. DNA bending may also result from the binding of specific proteins to specific DNA sequences.

ANALYTICAL APPROACHES FOR SOLVING GENETICS PROBLEMS

Q10.1 The linear chromosome of phage T2 is 52 μm long. The chromosome consists of double-stranded DNA, with 0.34 nm between each base pair. The average weight of a base pair is 660 daltons. What is the molecular weight of the T2 molecule?

A10.1 This question involves the careful conversion of different units of measurement. The first step is to put the lengths in the same units: 52 μm is 52 millionths of a meter, or $52,000 \times 10^{-9}$ m, or 52,000 nm. One base occupies 0.34 nm in the double helix, so the number of base pairs in this chromosome is 52,000 divided by 0.34, or 152,941 base pairs. Each base pair, on the average, weighs 660 daltons, and therefore the molecular weight of the chromosome is $152,941 \times 660 = 1.01 \times 10^8$ daltons, or 101 million daltons.

The human genome contains 3×10^9 bp of DNA for a total length of about 1 meter distributed among 23 chromosomes. The average length of the double helix in a human chromosome is 3.8 cm, which is 3.8 hundredths of a meter or 38 million nm—substantially longer than the T2 chromosome! There are over 111.7 million base pairs in the average human chromosome.

Q10.2 The accompanying table lists the relative percentages of bases of nucleic acids isolated from different species. For each one, what type of nucleic acid is involved? Is it double- or single-stranded? Explain your answer.

SPECIES	ADENINE	GUANINE	THYMINE	CYTOSINE	URACIL
(i)	21	29	21	29	0
(ii)	29	21	29	21	0
(iii)	21	21	29	29	0
(iv)	21	29	0	29	21
(v)	21	29	0	21	29

A10.2 This question focuses on the base-pairing rules and the difference between DNA and RNA. In analyzing the data, we should determine first whether the nucleic acid is RNA or DNA, and then whether it is double- or single-stranded. If the nucleic acid has thymine, it is DNA; if it has uracil, it is RNA. Thus species (i), (ii), and (iii) must have DNA as their genetic material, and species (iv) and (v) must have RNA as their genetic material.

Next, the data must be analyzed for strandedness. Double-stranded DNA must have equal percentages of A and T and of G and C. Similarly, double-stranded RNA must have equal percentages of A and U and of G

and C. Hence species (i) and (ii) have double-stranded DNA, while species (iii) must have single-stranded DNA since the base-pairing rules are violated, with A = G and T = C but A ≠ T and G ≠ C. As for the RNA-containing species, (iv) contains double-stranded RNA since A = U and G = C, and (v) must contain single-stranded RNA.

Q10.3 Here are four characteristics of one strand (the "original" strand) of a particular long double-stranded DNA molecule:

a. 35 percent of the adenine-containing nucleotides (As) have guanine-containing nucleotides (Gs) on their 3' sides;

b. 30 percent of the As have Ts as their 3' neighbors;

c. 25 percent of the As have Cs; and

d. 10 percent of the As have As as their 3' neighbors.
   Use this information to answer the following questions as completely as possible, explaining your reasoning in each case.

   i. In the complementary DNA strand, what will be the frequencies of the various bases on the 3' side of A?
   ii. In the complementary strand, what will be the frequencies of the various bases on the 3' side of T?
   iii. In the complementary strand, what will be the frequency of each kind of base on the 5' side of T?
   iv. Why is the percentage of A not equal to the percentage of T (and the percentage of C not equal to the percentage of G) among the 3' neighbors of A in the original DNA strand described?

A10.3a. This cannot be answered without more information. Although we know that the As neighbored by Ts in the original strand will correspond to As neighbored by Ts in the complementary strand, there will be additional As in the complementary strand about whose neighbors we know nothing.

b. This also cannot be answered. All the As in the original strand correspond to Ts in the complementary strand, but we know only about the 5' neighbors of these Ts, not the 3' neighbors.

c. On the original strand, 35 percent were 5'-AG-3', so on the complementary strand, 35 percent of the sequences will be 3'-TC-5'. So, 35 percent of the bases on the 5' side of T will be C. Similarly, on the original strand, 30 percent were 5'-AT-3', 25 percent were 5'-AC-3', and 10 percent were 5'-AA-3', meaning that on the complementary strand, 30 percent of the sequences were 3'-TA-5', 25 percent were 3'-TG-5', and 10 percent were 3'-TT-5'. So, 30 percent of the bases on the 5' side of T will be A, 25 percent will be G and 10 percent will be T.

d. The A = T and G = C rule only applies when considering both strands of a double stranded DNA. Here we are considering only the original single strand of DNA.

QUESTIONS AND PROBLEMS

10.1 The experiment by Griffith by which a mixture of dead and live bacteria were injected into mice demonstrated (choose the right answer):
a. DNA is double stranded.
b. mRNA of eukaryotes differs from mRNA of prokaryotes.
c. bacterial transformation.
d. bacteria can recover from heat treatment if live helper cells are present.

10.2 In the 1920s while working with *Streptococcus pneumoniae*, the agent that causes pneumonia, Griffith discovered an interesting phenomenon. In the experiments mice were injected with different types of bacteria. For each of the following bacteria type(s) injected, indicate whether the mice lived or died:
a. type *IIR*
b. type *IIIS*
c. heat-killed *IIIS*
d. type *IIR* + heat-killed *IIIS*

10.3 Several years after Griffith described the transforming principle, Avery, MacLeod, and McCarty investigated the same phenomenon.
a. Describe their experiments.
b. What did their experiments demonstrate beyond Griffith's?
c. How were enzymes used as a control in their experiments?

*10.4 The Hershey-Chase experiment in which T2 phages grown in bacteria on media containing either radioactively labeled sulfur or radioactively labeled phosphorus demonstrated that (choose the correct answer):
a. The coat material of the phage controls the kind of DNA replicated in the host cell.
b. The nucleic acid of the phage contains the genetic information.
c. The prophage state is necessary for generalized transduction.
d. Conjugation is dependent upon some of the cells being *Hfr*.
e. A metaphase chromosome is composed of two chromatids, each containing a single DNA molecule.

*10.5 Hershey and Chase showed that when phages were labeled with ^{32}P and ^{35}S, the ^{35}S remained outside the cell and could be removed without affecting the course of infection, whereas the ^{32}P entered the cell and could be recovered in progeny phages. What distribution of isotope would you expect to see if parental phages were labeled with isotopes of
a. C?
b. N?
c. H?
Explain your answer.

10.6 What is the evidence that the genetic material of TMV (tobacco mosaic virus) is RNA?

10.7 The X-ray diffraction data obtained by Rosalind Franklin suggested (choose the correct answer):
a. DNA is a helix with a pattern which repeats every 3.4 nanometers.
b. purines are hydrogen bonded to pyrimidines.
c. replication of DNA is semiconservative.
d. mRNA of eukaryotes differs from mRNA of prokaryotes.

*10.8 In DNA and RNA, which carbon atoms of the sugar molecule are connected by a phosphodiester bond?

10.9 Which base is unique to DNA, and which base is unique to RNA?

10.10 How do nucleosides and nucleotides differ?

10.11 What chemical group is found at the 5' end of a DNA chain? At the 3' end of a DNA chain?

10.12 What evidence do we have that in the helical form of the DNA molecule the base pairs are composed of one purine and one pyrimidine?

*10.13 What evidence is there to substantiate the statement: "There are only two base pair combinations in DNA, A-T and C-G"?

10.14 How many different kinds of nucleotides are there in DNA molecules?

*10.15 What is the base sequence of the DNA strand that would be complementary to the following single-stranded DNA molecules?
a. 5' A G T T A C C T G A T C G T A 3'
b. 5' T T C T C A A G A A T T C C A 3'

*10.16 Is an adenine-thymine or a guanine-cytosine base pair harder to break apart? Explain your answer.

10.17 The double-helix model of DNA, as suggested by Watson and Crick, was based on a variety of lines of evidence gathered on DNA by other researchers. The facts fell into the following two general categories; give three examples of each:

a. chemical composition
b. physical structure

*10.18   For double-stranded DNA, which of the following base ratios always equals 1?
a. $(A + T)/(G + C)$
b. $(A + G)/(C + T)$
c. C/G
d. $(G + T)/(A + C)$
e. A/G

10.19   If the ratio of $(A + T)$ to $(G + C)$ in a particular DNA is 1.00, does this result indicate that the DNA is most likely constituted of two complementary strands of DNA or a single strand of DNA, or is more information necessary?

10.20   Explain whether the $(A + T)/(G + C)$ ratio in double-stranded DNA is expected to be the same as the $(A + C)/(G + T)$ ratio.

*10.21   The percent cytosine in a double-stranded DNA is 17. What is the percent of adenine in that DNA?

10.22   Upon analysis, a double-stranded DNA molecule was found to contain 32 percent thymine. What percent of this same molecule would be made up of cytosine?

10.23   A sample of double-stranded DNA has 62 percent GC. What is the percentage of A in the DNA?

10.24   A double-stranded DNA polynucleotide contains 80 thymidylic acid and 110 deoxyguanylic acid residues. What is the total nucleotide number in this DNA fragment?

10.25   Analysis of DNA from a bacterial virus indicates that it contains 33 percent A, 26 percent T, 18 percent G, and 23 percent C. Interpret these data.

10.26   The following are melting temperatures for different double-stranded DNA molecules. Arrange these molecules from lower to higher content of GC pairs.
a. 73°C
b. 69°C
c. 84°C
d. 78°C
e. 82°C

10.27   What is a DNA oligomer?

10.28   The genetic material of bacteriophage ΦX174 is single-stranded DNA. What base equalities or inequalities might we expect for single-stranded DNA?

10.29   Through analysis of single crystal X-ray diffraction patterns of synthesized DNA oligomers, different forms of DNA have been identified. These forms include A-DNA, B-DNA, and Z-DNA, and each has unique molecular attributes.
a. How do the major and minor grooves of B-DNA and Z-DNA differ?
b. How do the helical properties of these forms of DNA differ?
c. What functions might Z-DNA have?

*10.30   What evidence is there that DNA in cells is not a rigid, inflexible molecule? Why is DNA flexibility important?

10.31   If a virus particle contains double-stranded DNA with 200,000 base pairs, how many complete 360° turns occur in this molecule?

*10.32   A double-stranded DNA molecule is 100,000 base pairs (100 kilobases) long.
a. How many nucleotides does it contain?
b. How many complete turns are there in the molecule?
c. How long is the DNA molecule?

10.33   Organisms have vastly different amounts of genetic material. *E. coli* has about 4.6 million base pairs of DNA in one circular chromosome, the haploid budding yeast (*S. cerevisiae*) has 12,057,500 base pairs of DNA in 17 chromosomes, and the gametes of humans have about 2.75 billion base pairs of DNA in 23 chromosomes.
a. If all of the DNA were B-DNA, what would be the average length of a chromosome in these cells?
b. On average, how many complete turns would be in each chromosome?
c. Would your answers to (a) and (b) be significantly different if the DNA were composed of, say, 20% Z-DNA and 80% B-DNA?
d. What implications do your answers to these questions have for the packaging of DNA in cells?

*10.34   If nucleotides were arranged at random in a single-stranded RNA 10^6 nucleotides long, and if the base composition of this RNA was 20 percent A, 25 percent C, 25 percent U and 30 percent G, how many times would you expect the specific sequence 5'-GUUA-3' to occur?

11 THE ORGANIZATION OF DNA IN CHROMOSOMES

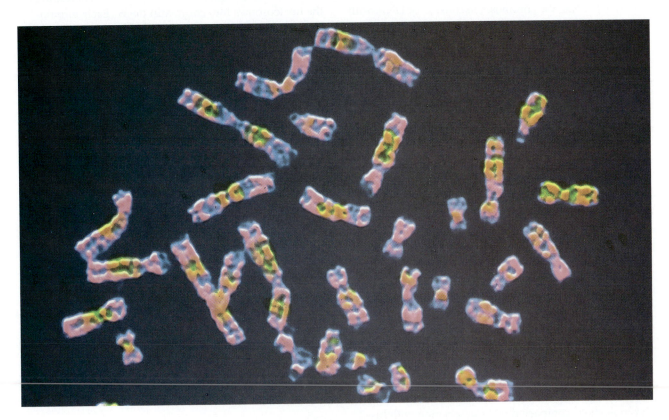

Principal Points

~ In prokaryotes and viruses in which the genetic material is DNA, the chromosome consists of DNA with some proteins associated with it.

~ In bacteria the chromosome is a circular, double-stranded DNA molecule that is compacted by supercoiling of the DNA helix. In the *E. coli* chromosome about 100 independent looped domains of supercoiled DNA have been identified.

~ The complete set of metaphase chromosomes in a eukaryotic cell is called its karyotype. The karyotype is species-specific.

~ The amount of DNA in a prokaryotic chromosome or the haploid amount of DNA in a eukaryotic cell is called the C value. There is not a direct relationship between the C value and the structural or organizational complexity of an organism.

~ The nuclear chromosomes of eukaryotes are complexes of DNA and histone and nonhistone chromosomal proteins. Each chromosome consists of one linear, unbroken, double-stranded DNA molecule running throughout its length; the DNA is variously coiled and folded. The histones are constant from cell to cell within an organism, while the nonhistones vary significantly among cell types.

~ The large amount of DNA present in the eukaryotic chromosome is compacted by its association with his-

tones in nucleosomes, and by higher levels of folding of the nucleosomes into chromatin fibers. Each chromosome contains a large number of looped domains of 30-nm chromatin fibers attached to a protein scaffold. The functional state of the chromosome is related to the extent of coiling: The more condensed a part of a chromosome is, the less likely it is that the genes in that region will be active.

~ The centromere region of each eukaryotic chromosome is responsible for the accurate segregation of the replicated chromosome to the daughter cells during mitosis and meiosis. The DNA sequences of centromeres are species-specific.

~ The ends of chromosomes, the telomeres, often are associated with the nuclear envelope. Telomeres in a species share a common sequence. Characteristically, sequences at or very close to the extreme ends of the chromosomal DNA consist of simple, relatively short, tandemly repeated sequences.

~ Prokaryotic genomes consist mostly of unique-sequence DNA, with only a few repeated sequences and genes. Eukaryotes have both unique and repetitive sequences in the genome. The spectrum of complexity of repetitive DNA sequences among eukaryotes is extensive.

---

*I*n Chapter 10 we discussed the structures of DNA. In this chapter we examine how the DNA is organized into chromosomes. The chromosomes of eukaryotes are much more complex than the chromosomes of prokaryotes. Eukaryotic chromosomes consist of a highly ordered complex of DNA and proteins, with special regions—centromeres and telomeres—that are of particular importance for chromosome function. Ultimately, geneticists will need an even greater understanding of chromosome structure than they have now before the regulation of gene expression can be completely understood.

Structural Characteristics of Bacterial and Viral Chromosomes

The Watson and Crick double-helix model of DNA structure does not by itself describe the structural characteristics of chromosomes. In the following sec-

tion we will examine the organization of DNA molecules in chromosomes of bacteria and viruses. An organism's genes are discrete regions of longer molecules of DNA (perhaps a thousand to over a million base pairs). It is important to understand how DNA is organized in chromosomes in order to understand how gene expression is controlled.

Bacterial Chromosomes

In Chapter 8 we learned that the DNA of the bacterium *Escherichia coli* is located in a central region called the nucleoid. If an *E. coli* cell is lysed (broken open) gently, the DNA is released in a highly folded state. The double-stranded DNA is present as a single chromosome,[1] approximately 1,100 μm long (4×10^3 kb

---

[1]The number of chromosomes per cell in *E. coli* actually depends on the growth rate. Fast-growing cells contain two to four copies of the chromosome, whereas slow-growing cells contain one to two copies of the chromosome. Since all of the chromosomes are identical, each cell behaves as a haploid.

[1 kb = 1 kilobase pairs = 1,000 base pairs]), which contains all the genes necessary for the bacterium to grow and survive in a variety of natural and laboratory environments. What are the properties of this chromosome?

THE *E. COLI* SUPERCOILED CHROMOSOME.

The amount of DNA in the *E. coli* chromosome is approximately 1,000 times the length of the *E. coli* cell. How is all that DNA packaged into the nucleoid region of the cell? To answer that question we must learn about the properties of linear and circular forms of DNA.

An early prediction was that when DNA isolated from a bacterium or virus is centrifuged, all of the DNA would sediment at the same rate, forming a band in the centrifuge tube at the end of the experiment. However, in 1963, Jerome Vinograd obtained an unexpected result. When he centrifuged circular DNA from polyoma virus, two bands, not one, were observed. His investigations of this finding led to an understanding that circular DNA can exist in a *relaxed* or a *supercoiled form*. These forms are called *topoisomers* and are explained in the following paragraphs.

Let us consider a 208-base-pair linear piece of DNA in the B-form (Figure 11.1a). Such a molecule has two free ends and, since there are 10.4 base pairs per helical turn of B-form DNA, there are 20 helical turns in this molecule. If we now simply join the two ends, we have produced a circular DNA molecule (Figure 11.1b). This circular DNA molecule is said to be *relaxed*. Alternatively, if we first untwist one end of the linear DNA molecule by two turns (Figure 11.1c), and then join the two ends, the circular DNA molecule produced will have 18 helical turns and a small unwound region (Figure 11.1d). Such a structure is not energetically favored, and will switch to a structure with 20 helical turns and two superhelical turns (Figure 11.1e). This structure is said to be *supercoiled*;

~ FIGURE 11.1

Illustration of DNA supercoiling. (a) A linear B-form DNA with 20 helical turns (208 base pairs); (b) Relaxed circular DNA produced by joining the two ends of the linear molecule of (a); (c) The linear DNA molecule of (a) unwound from one end by two helical turns; (d) A possible circular DNA molecule produced by joining the two ends of the linear molecule of (c) showing 18 helical turns and a short unwound region; (e) The more energetically favored form of (d), a supercoiled DNA with 20 helical turns and two superhelical turns.

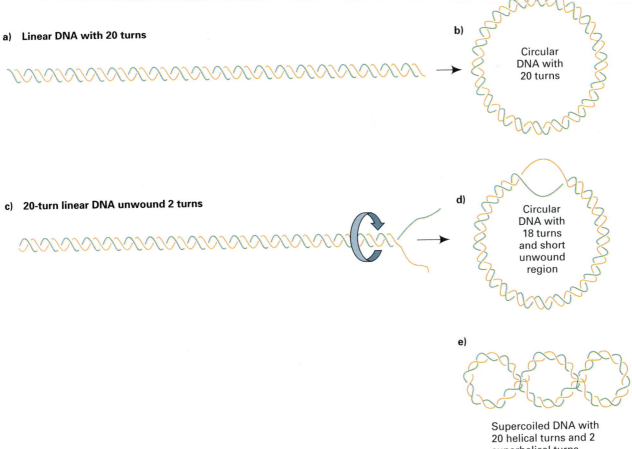

a) **Linear DNA with 20 turns**

b) Circular DNA with 20 turns

c) **20-turn linear DNA unwound 2 turns**

d) Circular DNA with 18 turns and short unwound region

e) Supercoiled DNA with 20 helical turns and 2 superhelical turns

that is, the DNA has supercoils introduced into it by having the double helix twisted in space about its own axis. The very act of supercoiling produces tension in the DNA molecule. Therefore, if a break—called a *nick*—is introduced into the sugar-phosphate backbone of a supercoiled circular DNA molecule, spontaneous untwisting of the molecule produces a relaxed DNA circle. Figure 11.2 pictures relaxed and supercoiled circular DNA of a plasmid to show how much more compact a supercoiled molecule is compared to a relaxed molecule. Supercoiling can also occur in a linear DNA molecule. This is not intuitively obvious. That is, if we twist a length of rope on one end without holding the other end, the rope just spins in the air and remains linear (i.e., relaxed). However, for a large, linear DNA molecule, supercoiling occurs in localized regions and the ends behave as if they are fixed due to the drag of the rest of the molecule.

There are two types of supercoiling, *negative supercoiling* and *positive supercoiling*. In negative supercoiling, the DNA is untwisted; that is, in the *opposite* direction from the clockwise turns of the right-handed double helix. To visualize this, think of the DNA double helix as a spiral staircase that turns in a clockwise direction. If you untwist the spiral staircase by one complete turn, you have *the same number of stairs to climb but you have one less 360° turn to make*; this is a negative supercoil. In positive super-

coiling, the DNA is twisted more tightly, that is, in the same direction as the clockwise turns of the helix. In the spiral staircase analogy, you would have the same number of stairs to climb, but now there is one more 360° turn to make. Either type of supercoiling causes the DNA to become more compact.

Given this information, we can explain the packaging of the *E. coli* chromosome into the nucleoid region of the cell as follows: the *E. coli* chromosome is circular and is extensively supercoiled—negatively supercoiled, in this case—resulting in a highly compact molecule.

The amount and type of DNA supercoiling is brought about by enzymes called **topoisomerases**. These enzymes convert one topological form of DNA into another. In *E. coli*, topoisomerase II (also called *DNA gyrase*) untwists relaxed DNA to produce negatively supercoiled DNA. Topoisomerase I does the opposite; that is, it converts negatively supercoiled DNA to the relaxed state. Topoisomerase I cannot take a relaxed DNA and make it positively supercoiled. With both topoisomerases in the cell, then, DNA can be interchanged between the negatively supercoiled and relaxed states. By making highly negatively supercoiled DNA, topoisomerase II enables the *E. coli* chromosome to pack into a very compact state. Equivalent topoisomerases that carry out the two functions just described are found in all organisms.

PROTEINS COMPLEXED TO BACTERIAL CHROMO-SOMES. Eukaryotic DNA is complexed with a number of discrete structural proteins called histones, which serve to compact the DNA into the chromosome structures characteristic of eukaryotic nuclei (see p. 328–330).

Are there structural proteins associated with bacterial chromosomes? When *E. coli* cells are lysed, the DNA of the nucleoid region is released into the medium (Figure 11.3). That DNA is organized into loops. The DNA in the loops is folded into a more compact shape than would be the case for extended DNA molecules. The explanation for this structure is that the DNA, which is an acidic (negatively charged) molecule, is associated with basic (positively charged) structural proteins. Several of these DNA-binding, structural proteins have been isolated from *E. coli* and characterized. Two of these proteins—HU and H—resemble two of the histone structural proteins found associated with eukaryotic DNA. Based on analyses of the *E. coli* proteins and of the properties of nucleoid DNA, a model for the structure of the bacterial chromosome has been proposed (Figure 11.4). In the model, the bacterial chromosome has about 100 independent *domains*. Each domain consists of a loop of about 40,000 base pairs (40-kilobase pairs [kb] =

∼ FIGURE 11.2

Electron micrographs of bacteriophage plasmid DNA: (a) Relaxed (nonsupercoiled) DNA; (b) Supercoiled DNA. Both molecules are shown at the same magnification.

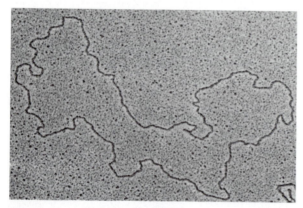

a)

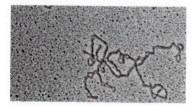

b)

the middle and single-stranded on each end. The single-stranded ends are complementary to each other and will reanneal to form a circle of double-stranded DNA with a circumference that is the same as the length of the linear T2 chromosomes. The circles are visualized by electron microscopy.

There is also evidence that each linear T2 or T4 DNA molecule has the same sequence of nucleotides at the two ends of the molecule. This *terminal redundancy*, as it is called, is illustrated in Figure 11.6. (Because of circular permutation, the two ends of each DNA molecule are different in sequence from molecule to molecule.)

How are the circularly permuted and terminally redundant T2 and T4 chromosomes found in populations of those phages generated from a single parental phage? The answer is found in the mechanism used to package DNA into the phage heads (Figure 11.7, p. 320). After the phage chromosome is injected into the host bacterium, it replicates several times. This replication produces a number of chromosomes, all of which have the same *terminally redundant* sequence as

~ FIGURE 11.6

Examples of terminally redundant double-stranded DNA molecules. (Because of circular permutation, the two ends of each DNA molecule are different in sequence from molecule to molecule.)

Terminal redundancy and circular permutation

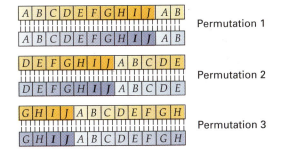

~ TABLE 11.1

Characteristics of the Chromosomes of Selected Viruses

Virus	Host	Structure and Type of Genetic Material	Chromosome Description	Length of Chromosome (µm)	%GC
T-even phages	E. coli	Double-stranded DNA	Linear; circularly permuted and terminally redundant	60	35
T7	E. coli	Double-stranded DNA	Linear; unique sequence	12	48
λ	E. coli	Double-stranded DNA	Linear; single-stranded "sticky ends"	16	49
P22	Salmonella typhimurium	Double-stranded DNA	Linear; unique sequence	14	48
ΦX174	E. coli	Single-stranded DNA	Circular	1.8	A25 G24 T33 C18
Qβ	E. coli	Single-stranded RNA	Linear	1.4	A22 G24 T29 C25
Reovirus	Mammals	Double-stranded RNA	Several pieces	8.3	A38 G17 U38 C17
SV40	Human	Double-stranded DNA	Supercoiled ring	1.7	41
Murine leukemia virus (Moloney)	Mouse	Single-stranded RNA	Linear	2.8	A25 G25 U23 C27
Tobacco mosaic virus (TMV)	Tobacco	Single-stranded RNA	Linear	2.2	A30 G25 U26 C19

~ FIGURE 11.7

Circularly permuted and terminally redundant T4 progeny chromosomes produced experimentally by making concatameric molecules and cleaving them into pieces large enough to fill the phage heads.

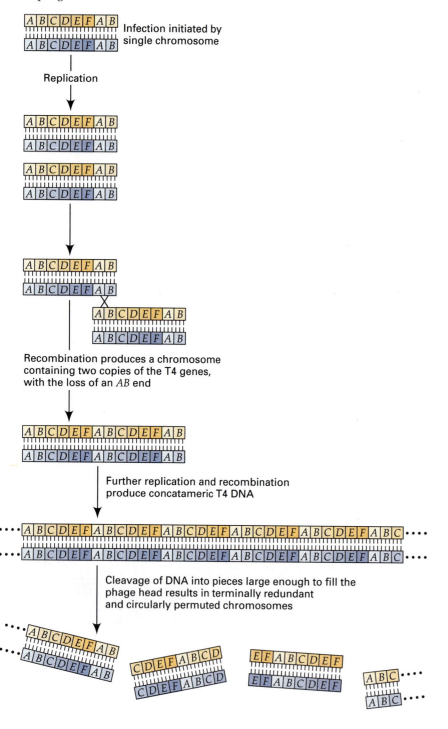

the parental DNA. Molecular recombination occurs between the DNA molecules at the terminally redundant ends, splicing the chromosomes together into very long molecules, called *concatamers*. These structures repeat the base sequence of the original unit phage chromosome in a tandem fashion. The concatamers undergo several rounds of replication, and late in infection the DNA is packaged into the phage heads that have been produced concurrently. The packaging is done by the *headful*; the length of DNA that is put into each head is determined by the volume of the head. Since the head can hold a little more than a genome's worth of DNA, we see how each phage chromosome contains a terminally redundant region. In addition, the successive clipping of the concatamer molecules into headful lengths leads to the circular permutation of the population of progeny phage chromosomes. In other words, each phage contains the same amount of DNA, with the same sequences represented, and with terminal redundancy. In a population of phage chromosomes, individual chromosomes will have different terminal sequences as a result of different permutations of the same sequence.

Bacteriophage ΦX174 Chromosome

The virulent phage ΦX174 is a DNA phage that infects *E. coli*. It has attracted the attention of geneticists because it is smaller than other phages. An electron micrograph of ΦX174 and diagrams of its structure are shown in Figure 11.8. The ΦX174 phage is an icosahedron consisting of protein subunits surrounding the genetic material. At each vertex (angular point) of the protein coat, there is a spike, which is involved in the infection process. Unlike the T-even phages and the λ phage, ΦX174 does not have a contractile sheath, baseplate, or tail fibers. The study of this phage has provided valuable information about the molecular biology of prokaryotic DNA replication.

In 1959 Robert Sinsheimer found that the DNA of ΦXl74 has a base composition that does not fit the complementary base-pairing rules. That is, instead of having A = T and G = C as is the case for double-stranded DNA, the ratio of bases is 25A:33T:24G:18C. The interpretation was that ΦX174 chromosome is not double-stranded but single-stranded. In addition, if the chromosome is treated with an exonuclease (an enzyme which digests DNA from free ends), the DNA is not affected. The simplest explanation is that the ΦXl74 chromosome is circular, a fact that has been corroborated by electron microscopy. Further, the sequence of nucleotides in the chromosome has been determined in its entirety, and it is known to contain 5,386 nucleotides.

ΦX174 makes eleven proteins, with a total molecular weight of 262,000 daltons. If the entire genome encodes proteins, then the 5,386-nucleotide genome would code for 1,795 amino acids (i.e., 5,386 ÷ 3, where 3 nucleotides codes for one amino acid). The average molecular weight of an amino acid is approximately 110 daltons, so 1,795 amino acids would have a molec-

---

~ **FIGURE 11.8**

Bacteriophage ΦXl74: (a) Electron micrograph of ΦX174 phage particles; (b) Models of the ΦX174 phage particle: top—the view along an axis of 2-fold symmetry; bottom—view along axis of 5-fold symmetry; (c) Single-stranded circular DNA chromosome of ΦXl74.

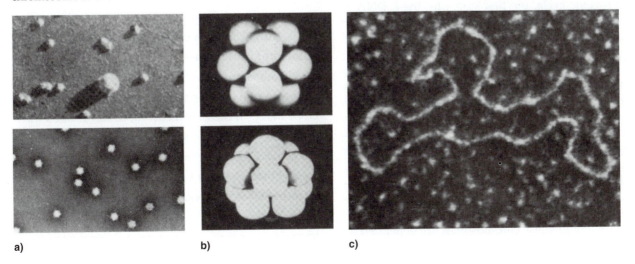

a)   b)   c)

ular weight of approximately 197,000 daltons. This discrepancy is explained by the fact that the ΦX174 genome has a special, and spatially very efficient, organization of genes. In some cases entire genes are located within the sequences of other genes, and in other cases the end of one gene overlaps the start of the next gene on the chromosome. Only four small sequences of the genome—spacers—do not specify amino acids. The arrangement of the ΦX174 genes on the chromosome is shown in Figure 11.9. Five different proteins are specified from shared base sequences, a state known as *overlapping genes*. For example, protein A is encoded by the full-length *A* gene, while protein A* is made by starting translation of the *A* gene part way along its length in the same reading frame. (A reading frame is one of three ways a base sequence may be grouped as a series of codons.) Protein B is encoded by a sequence that is entirely within the sequence for protein A, but it is in a different reading frame. In the same way, gene *E* is entirely within the sequence of gene *D*. The sequence for the yet more complicated gene *K* begins near the end of gene *A*, including all of gene *B*, and terminating in gene *C*. The reading frame of gene *K* is different from either *A* or *C*.

Bacteriophage λ Chromosome

The structure and life cycle of the temperate bacteriophage λ was described in Chapter 8. This phage has been studied extensively for a number of years, so many of its gene functions are well understood.

The phage λ chromosome is double-stranded DNA, is linear, and has no structural proteins associated with it. The two ends of the λ DNA molecule are single-stranded, and complementary. The single-stranded terminus is 12 nucleotides long; its sequence is shown in Figure 11.10a.

Regardless of whether λ goes through the lytic or the lysogenic cycle, the first step after the λ DNA is injected into the host cell is the conversion of the linear molecule into a circular molecule (Figure 11.10a). The naturally complementary ends ("sticky ends") pair and the single-stranded gaps are bonded in a reaction catalyzed by the enzyme DNA ligase. The paired ends are called the *cos* sequence. In the lysogenic cycle the circular DNA finds a particular site in the *E. coli* chromosome, and by a crossing-over event the DNA is integrated into the main chromosome, and by a crossing-over event the DNA is integrated into the main chromosome (see Chapter 8).

In the lytic cycle the DNA replicates and produces a long concatameric molecule similar to the one for T2 and T4. It is from this concatameric structure that progeny phage λ chromosomes are generated as

~ FIGURE 11.9

Gene organization of the bacteriophage ΦXl74 chromosome showing the overlapping genes. The small spacers between some genes are shown in taupe. For details, see text.

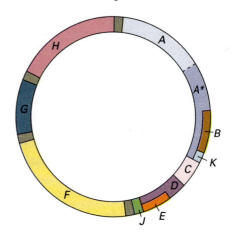

follows: The phage λ chromosome has a gene called *ter* (for "terminus-generating activity": Figure 11.10b), the product of which is a DNA endonuclease (an enzyme that digests a nucleic acid chain by cutting somewhere along its length rather than at the termini). The endonuclease recognizes the *cos* sequence. Once *ter* is aligned on the DNA at the *cos* site, the endonuclease makes a staggered cut such that linear λ chromosomes with the correct complementary, 12-base-long, single-stranded ends are produced. The chromosomes are then packaged in the assembled phage heads, and progeny λ phages are assembled, then released from the cell when it lyses.

KEYNOTE

In prokaryotes in which the genetic material is DNA, the chromosome consists of DNA with a number of proteins associated with it. In bacteria the chromosome is a circular, double-stranded DNA molecule that is compacted by supercoiling of the DNA helix. In the *E. coli* chromosome, about a hundred independent looped domains of supercoiled DNA have been identified. In bacteriophages the genetic material may be double-stranded or single-stranded DNA or RNA. Phage chromosomes can be linear or circular.

~ FIGURE 11.10

λ chromosome structure varies at stages of lytic infection of *E. coli*: (a) Parts of the λ chromosome showing the nucleotide sequence of the two single-stranded, complementary ("sticky") ends, and the chromosome circularizing after infection by pairing of the ends, with the single-stranded gaps filled in to produce a covalently closed circle; (b) Generation of the "sticky" ends of the λ DNA during the lytic cycle. During replication of the λ chromosome, a giant concatameric DNA molecule is produced; it contains tandem repeats of the λ genome. The diagram shows the "join" between two adjacent λ chromosomes and the extent of the *cos* sequence. The *cos* sequence is recognized by the *ter* gene product, an endonuclease that makes two cuts at the sites shown by the arrows. These cuts produce a complete λ chromosome from the concatamer.

a) Linear λ chromosome (~48,000 base pairs) forms circular λ chromosome

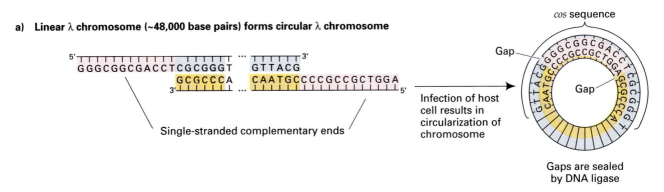

b) Production of progeny, linear λ chromosomes from concatamers (multiple copies linked end-to-end at complementary ends)

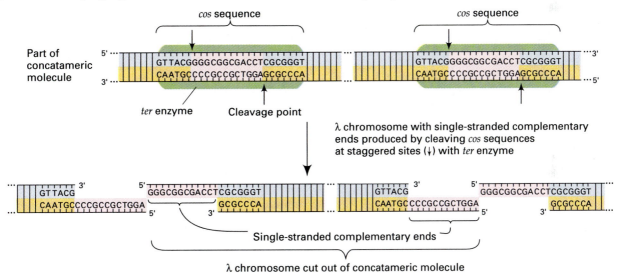

STRUCTURAL CHARACTERISTICS OF EUKARYOTIC CHROMOSOMES

A fundamental difference between prokaryotes and eukaryotes is that prokaryotes have a single type of chromosome (sometimes present in more than one copy in the cell), while most eukaryotes have a diploid number of chromosomes in almost all somatic cells. A photograph of a stained set of metaphase chromosomes from a human male is shown in Figure 11.11. (Metaphase chromosomes are used because they are the most compact form of eukaryotic chromosomes and, therefore, are easily seen under the microscope after staining.) The number of human chromosomes is 46. Recall that this is because humans are diploid (2N) organisms, possessing one haploid (N) set of chromosomes (23 chromosomes) from the egg, and another haploid set from the sperm.

~ FIGURE 11.11

Mitotic metaphase chromosomes of a human male.

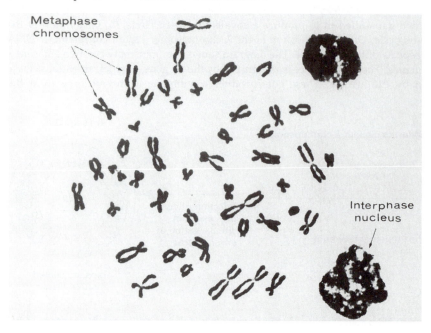

The Karyotype

A complete set of all the metaphase chromosomes in a cell is called its **karyotype** (literally, "nucleus type"). For most organisms, all cells have the same karyotype. However, the karyotype is species-specific, so a wide range of number, size, and shape of metaphase chromosomes is seen among eukaryotic organisms. Even closely related organisms may have quite different karyotypes. Table 11.2 gives the chromosome numbers of selected eukaryotes. The diploid number of chromosomes in humans is 46, in chimpanzees it is 48, in cats it is 38, and in garden peas it is 14. In the fruit fly, *Drosophila melanogaster*, it is 8. Examples of chromosome numbers for haploid organisms are: 16 in the green alga *Chlamydomonas reinhardtii*, 17 in baker's yeast *Saccharomyces cerevisiae*, and 7 in the orange bread mold *Neurospora crassa*.

Figure 11.12 shows the karyotype for the cell of a normal human male. It is customary, particularly with human chromosomes, to arrange chromosomes in order according to size and position of the centromere. This karyotype shows 46 chromosomes: 2 pairs of each of the 22 autosomes and 1 of each of the X and Y sex chromosomes. In a human karyotype, the chromosomes are numbered for easy identification. Conventionally, the largest pair of homologous chromosomes is designated 1, the next largest 2, and so on. Although chromosome 21 is actually smaller than chromosome 22, it is called 21 for historical reasons.

In humans chromosomes 1 through 22 are called autosomes to distinguish them from the pair of sex chromosomes. Formally, the sex chromosomes constitute pair 23 even though, in humans at least, they do not fit properly in the size scale. As shown in Figure 11.12, the X chromosome is a large metacentric chromosome, and the Y chromosome is a much smaller chromosome.

A knowledge of the size, overall morphology, and banding patterns (discussed below) of chromosomes permits geneticists to identify certain chromosome mutations that correlate with congenital abnormalities or dysfunctions. For example, on rare occasions chromosomes undergo changes in morphology, such as by gaining or losing a portion of a chromosome or by exchanging pieces with a nonhomologous chromosome. In addition, changes in the number of chromosomes in the karyotype may occur because of an error in cell division, for example, as a result of chromosome nondisjunction (see Chapter 3, p. 73). Chromosome mutations are discussed more fully in Chapter 7.

Chromosomal Banding Patterns

The human metaphase chromosomes shown in Figure 11.12 were seen after staining the chromosomes with orcein and Giemsa stains. This material stains the chromosomes uniformly, making it quite

~ TABLE 11.2

Chromosome Numbers in Eukaryotic Organisms

SCIENTIFIC NAME	COMMON NAME	NUMBER OF CHROMOSOMES[a]
DIPLOID ORGANISMS		
Animals		
Homo sapiens	Human	46
Pan troglodytes	Chimpanzee	48
Equus caballus	Horse	64
Bos taurus	Cattle	60
Canis familiaris	Dog	78
Felis domesticus	Cat	38
Oryctolagus cuniculus	Rabbit	44
Rattus norvegicus	Rat	42
Mus musculus	House mouse	40
Meleagris gallopavo	Turkey	82
Xenopus laevis	Toad	36
Caenorhabditis elegans	Nematode	11 ♂/12 hermaphrodite
Musca domestica	Housefly	12
Drosophila melanogaster	Fruit fly	8
Plants		
Pinus ponderosa	Ponderosa pine	24
Pisum sativum	Garden pea	14
Solanum tuberosum	Potato	48
Nicotiana tabacum	Tobacco	48
Triticum aestivum	Bread wheat	42
Zea mays	Maize	20
Vicia faba	Broad bean	12
HAPLOID ORGANISMS		
Chlamydomonas reinhardtii	Unicellular alga	16
Neurospora crassa	Orange bread mold	7
Aspergillus nidulans	Green bread mold	8
Saccharomyces cerevisiae	Brewer's yeast	17

[a]For the diploid organisms, the diploid number of chromosomes is given.

difficult to distinguish chromosomes that are similar in size and general morphology. In Figure 11.12 groups of chromosomes with similar morphologies are given letter designations (A through G).

Fortunately, a number of procedures have been developed that stain certain regions or *bands* of the chromosomes more intensely than other regions.

Banding patterns are specific for each chromosome, enabling each chromosome in the karyotype to be distinguished clearly. This makes it easier to distinguish chromosomes of similar sizes and shapes.

One of these staining techniques is called **G banding**. In this procedure, metaphase chromosomes are first treated with mild heat or proteolytic enzymes

~ FIGURE 11.12

Human male metaphase chromosomes arranged as a karyotype. Note that groups of chromosomes with similar morphologies are arranged under letter designations (A through G). This arrangement is based on the size of the condensed chromosomes. In the male all chromosomes except the X and Y sex chromosomes are present in pairs.

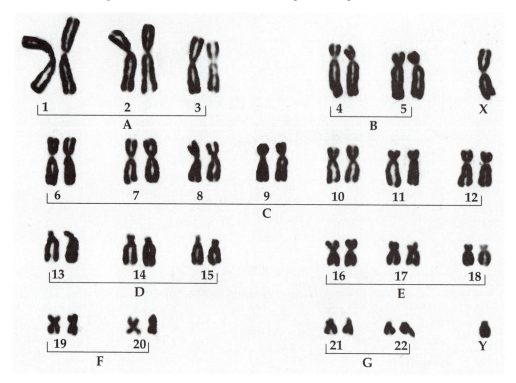

(enzymes that digest proteins) to partially digest the chromosomal proteins, and then stained with Giemsa stain to produce dark bands called G bands (Figure 11.13). G bands reflect regions of DNA that are rich in adenine and thymine. In humans, approximately 300 G bands can be distinguished in metaphase chromosomes, while approximately 2,000 G bands can be distinguished in chromosomes from prophase. The G bands are stable and are even visible in scanning electron micrographs of chromosomes as constrictions at the banding regions.

Another staining technique is called **Q banding**. In this procedure, chromosomes are stained with a quinacrine dye in a complex process in which the dye binds preferentially to AT-rich regions of DNA. This procedure produces what are known as Q bands that can be visualized by using a special fluorescence microscope. G and Q bands have the same locations on the chromosomes.

The purpose of staining the chromosomes to produce banding patterns is to distinguish the individual chromosomes for cytogenetic analysis, and to provide landmarks along the chromosomes for locating genes. Conventionally, drawings of human chromosomes show the G banding pattern.

Cellular DNA Content and the Structural or Organizational Complexity of the Organism

The total amount of DNA in the haploid genome is characteristic of each living species, prokaryote or eukaryote, and is known as its **C value**. Table 11.3 gives the C values for a selection of prokaryotes and eukaryotes as well as the length of the DNA as calculated from the number of base pairs. Figure 11.14 (p. 329) presents a summary of the range of C values found in a variety of animals. The "tree" on the left of the figure represents the evolutionary relationships between the animals. Within any given taxon, there is a general trend toward increasing amounts of DNA as organisms are further removed from the original ancestor.

The C value data show that the amount of DNA found among organisms varies considerably. It can be seen that there may or may not be significant variation in DNA amount among related organisms. For example, mammals, birds, and reptiles each show little variation, while amphibians, insects, and plants each vary over a wide range, often tenfold or so. Note also that there is not a direct relationship between the

~ FIGURE 11.13

G banding in a karyotype of human male metaphase chromosomes.

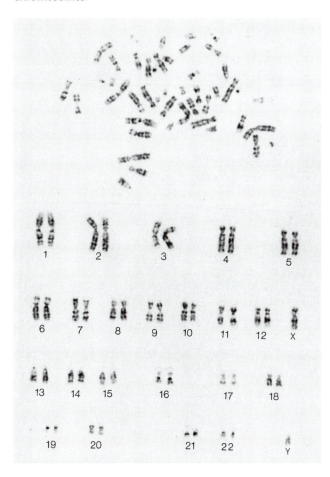

C value and the structural or organizational complexity of the organism. At least one reason for this is variation in the amount of repetitive-sequence DNA in the genome (pp. 336–338).

KEYNOTE

In eukaryotes the complete set of metaphase chromosomes in a cell is called its karyotype. The karyotype is species-specific. The amount of DNA in a prokaryotic chromosome or the haploid (N) amount of DNA in a eukaryotic cell is called the C value. The amount of genetic material varies greatly among prokaryotes and eukaryotes. There is not a direct relationship between the C value and the structural or organizational complexity of the organism.

The Molecular Structure of the Eukaryotic Chromosome

In comparison with a prokaryotic cell, a eukaryotic cell contains a large amount of DNA in its nucleus. A human cell, for example, has more than a thousand times as much DNA as does *E. coli*. We saw earlier that the *E. coli* chromosome (about 1 mm long in its fully extended state) is supercoiled so that its 4×10^6 base pairs of DNA can be packaged into the nucleoid region. By contrast, the human cell has 5.5×10^9 base pairs of DNA in its diploid nucleus. Without the compacting of this DNA, accomplished by the specific association between DNA and proteins in the chromosomes, the DNA of the chromosomes of a single human cell would be over 6 feet long (about 200 cm) if the molecules were placed end to end!

Each eukaryotic chromosome consists of one linear, unbroken, double-stranded DNA molecule running throughout its length and contains about twice as much protein by weight as DNA. The chromosome consists of a complex of DNA, chromosomal proteins, and RNA; this is called **chromatin**. The fundamental structure of chromatin is essentially identical in all eukaryotes.

There are two types of chromatin: **euchromatin** and **heterochromatin**. Euchromatin is chromatin that stains lightly. It is uncoiled during interphase, but becomes condensed during mitosis. Most of the genome consists of euchromatin. Heterochromatin, on the other hand, is chromatin that stains darkly. This occurs because the chromatin region involved is more highly condensed than is the case with euchromatin. Functionally, euchromatin is genetically active (i.e., it contains genes that are being expressed), whereas heterochromatin is genetically inactive, either because it contains no genes or because the genes it does contain cannot be expressed. Characteristically, heterochromatin replicates later in the S phase of the cell cycle than euchromatin, a result of the higher degree of chromosome condensation in heterochromatin.

Heterochromatin is found in all eukaryotic species near centromeres, at *telomeres* (the ends of the chromosomes), and elsewhere in a species-specific manner. (Telomeres are discussed later in this chapter.) Two classes of heterochromatin can be distinguished. **Constitutive heterochromatin** is always genetically inactive, and is found at homologous sites on chromosome pairs. Centromeric and telomeric heterochromatin are examples of constitutive heterochromatin. **Facultative heterochromatin** is chromatin that has the potential to become condensed to the heterochromatin state. It may contain genes that are made inactivate when the chromatin becomes

~ TABLE 11.3

Haploid DNA Content (the C Value) of Selected Prokaryotes and Eukaryotes

| | C VALUE | |
ORGANISM	BASE PAIRS (bp)	LENGTH[a]
PROKARYOTES		
Bacteria:		
Escherichia coli	4.1×10^6	1.4 mm
Salmonella typhimurium	1.1×10^7	3.8 mm
Viruses and phages:		
ΦX174 (double-stranded form)	5,386	1.8 μm
λ	4.65×10^4	16 μm
T2	1.75×10^5	60 μm
SV40	5,226	1.7 μm
EUKARYOTES		
Vertebrates:		
Human	2.75×10^9	94 cm
Mouse	2.2×10^9	75 cm
Frog	2.25×10^{10}	7.7 m
Puffer fish	4×10^8	13.6 cm
Invertebrates:		
Sea urchin	8×10^8	27.2 cm
Drosophila melanogaster	1.75×10^8	6.0 cm
Caenorhabditis elegans	8×10^7	2.7 cm
Plants:		
Lilium longiflorum (lily)	3×10^{11}	100 m
Zea mays (maize)	2.7×10^9	93 cm
Arabidopsis thaliana	1×10^8	3.4 cm
Fungi:		
Neurospora crassa	2.7×10^7	9.2 mm
Saccharomyces cerevisiae	1.75×10^7	6.0 mm

[a]Calculated from the formula 10 bp = 3.4 nm.

condensed. Barr bodies (inactivated X chromosomes in female mammals—see Chapter 3) are examples of facultative heterochromatin.

There are two major types of proteins associated with DNA in chromatin: **histones** and **nonhistones**. Histones and nonhistones play an important role in determining the physical structure of the chromosome. The DNA is wrapped around a core of histone molecules, and the nonhistones are somehow associated with that complex. Various studies have shown that some of the nonhistones have a structural role in the chromosomes. If the histones are removed from the chromosome, for example, the DNA unravels and is displaced from the complex, but a skeleton of nonhistone proteins in the shape of the chromosome remains.

THE HISTONES. The histones are the most abundant proteins associated with chromosomes. We will see in Chapter 17 that histones play an important role in the regulation of gene expression.

The histones are relatively small basic proteins; that is, at the normal pH of a cell, the histones have a net positive charge, thus facilitating their binding to the negatively charged DNA. This positive charge is found mainly on the $-NH_3^+$ groups of side chains of the basic amino acids lysine and arginine (see Figure 14.2, p. 421). In fact, histone molecules consist of approximately 25 percent arginine and lysine. This is a higher percentage than is found in most other proteins.

Five main types of histones are associated with eukaryotic DNA: H1, H2A, H2B, H3, and H4. Weight for weight there is about an equal amount of histone

~ FIGURE 11.14

The amount of DNA per haploid chromosome set in a variety of animals. Variation within certain phyla is quite wide (as in insects and amphibians). The "tree" on the left of the figure indicates the evolutionary relationship of the animals.

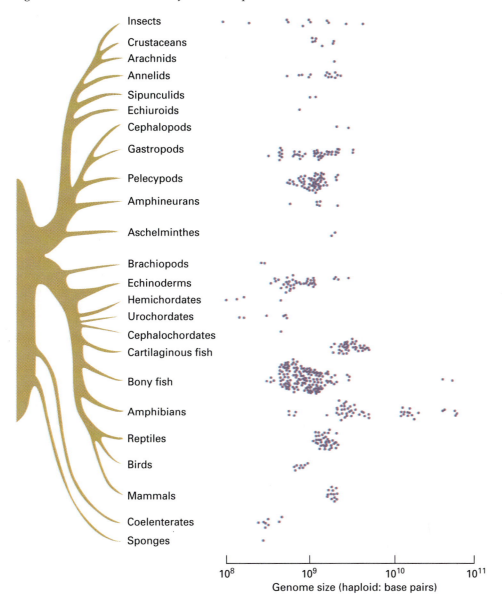

and of DNA in chromatin. The amount and proportions of these histones relative to DNA amount are relatively constant from cell to cell in all eukaryotic organisms. The properties of the histones from calf thymus DNA are given in Table 11.4. The amino acid sequences of the four histones H2A, H2B, H3, and H4 have been determined for a wide variety of organisms. Comparison of the sequences has indicated, in some cases, a high degree of conservation in the sequences among distantly related species. For example, only two amino acid differences exist in the H4 proteins of cows and peas, and there is only one

~ TABLE 11.4

Characteristics of Histones from Calf Thymus DNA

HISTONE TYPE	BASIC AMINO ACIDS		NUMBER OF AMINO ACIDS	MOLECULAR WEIGHT (DALTONS)
	LYS	ARG		
H1	29%	1%	215	23,000
H2A	11%	9%	129	13,960
H2B	16%	6%	125	13,775
H3	10%	13%	135	15,340
H4	11%	14%	102	11,280

amino acid difference between sea urchin and calf thymus histone H3. (However, there are 11 amino acid differences in the histone H3 of yeast and *Tetrahymena*, a protozoan.) In fact these four histones are among the most highly conserved of all known proteins. The remarkable degree of amino acid sequence similarity for the H2A, H2B, H3, and H4 histones among organisms is a strong indicator that histones perform the same basic role in organizing the DNA in the chromosomes of all eukaryotes.

There are several different but closely related H1 histones in each cell. This class of histones has a central core region of amino acids that has been conserved through evolution, but the rest of the molecule has been much less conserved. Moreover, in some tissues H1 is not present, being replaced by yet another histone protein. For example, in avian red blood cells (which, unlike mammalian red blood cells, have nuclei), a histone designated H5 is present instead of H1.

NONHISTONE CHROMOSOMAL PROTEINS.

Nonhistones are all the proteins associated with DNA apart from the histones. There are a very large number of different types of nonhistone chromosomal proteins. Some nonhistones play a structural role, and others are transiently associated with chromatin. The enzymes and proteins involved with replication, transcription, and the regulation of gene expression are examples of the latter type.

As a class, the nonhistones are very different from the histones. Nonhistones are usually acidic proteins—that is, proteins with a net negative charge—and are likely to bind to positively charged histones in the chromatin. Each eukaryotic cell has many different nonhistones in the nucleus, and weight for weight there may be as much nonhistone protein present as DNA and histone proteins combined. In contrast to the histones, the nonhistone proteins differ markedly in number and type from cell type to cell type within an organism, at different times in the same cell type, and from organism to organism. Relative to the mass of DNA in chromosomes, the nonhistone proteins also range widely, from more than 100 percent of the DNA mass to less than 50 percent. Well-studied among the nonhistones are the high-mobility group (HMG) proteins, so named because of their rapid electrophoretic mobility in polyacrylamide gel electrophoresis. The HMG proteins are abundant and heterogeneous in chromatin, commonly binding to the minor groove of DNA and bending the DNA. DNA bending is an important step in the formation of higher-order structures of chromatin (discussed later), suggesting that HMG proteins have a role in that process.

KEYNOTE

The nuclear chromosomes of eukaryotes are complexes of DNA, histone proteins, and nonhistone chromosomal proteins. Each chromosome consists of one linear, unbroken, double-stranded DNA molecule running throughout the length of the chromosome. These are five main types of histones—H1, H2A, H2B, H3, and H4—which are constant from cell to cell within an organism. Nonhistones, of which there are a large number, vary significantly among cell types, both within and among organisms as well as with time in the same cell type.

NUCLEOSOMES.

The DNA helix of each chromosome is coiled within the nucleus in a nonrandom way. There are several levels of packing that enable chromosomes several millimeters or even centimeters long to fit into a nucleus that is a few micrometers in diameter. The simplest level of packing involves the winding of DNA around histones in a structure called a **nucleosome**, and the most highly packed state involves higher-order coiling and looping of the DNA-histone complexes, as exemplified by the metaphase chromosomes. In general, the expression of a gene is facilitated by a change in the chromosome from a high level of coiling to a lower level of coiling. That is, the more condensed a part of a chromosome is, the less likely it is that the genes in that region will be active. We need to understand how chromosome regions undergo transitions between levels of coiling, therefore, in order to understand the regulation of gene expression completely.

In one recent model for the nucleosome, there are two molecules of each of the four core histones, H2A, H2B, H3, and H4 (called the *histone octamer*), a single molecule of the linker histone H1, and about 180 bp of DNA. In the test tube the core histones are capable of self-assembling into a histone octamer which is a flat-ended, cylindrical particle about 11 nm in diameter and 5.7 mm thick. The DNA appears to wind around the outside of the nucleosome core about one and three-quarters times, thereby compacting the DNA by a factor of about seven. The exact path of the DNA around the histone octamer, the position of the H1 linker histone in the nucleosome, and the position of the DNA that connects adjacent nucleosomes—called *linker DNA*—remain to be elucidated. A dia-

~ FIGURE 11.15

A possible nucleosome structure.

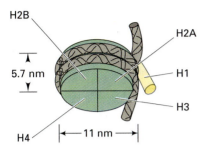

~ FIGURE 11.16

Electron micrograph of unraveled chromatin showing the nucleosomes in a "beads on a string" morphology.

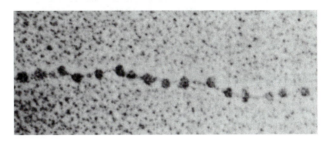

gram of a possible nucleosome structure is shown in Figure 11.15.

Electron microscopy reveals the DNA-protein complex as 10-nm chromatin fibers (10-nm **nucleofilaments**), that is, fibers with a diameter of about 10 nm. In its most unraveled state, the chromatin fiber of a nucleofilament has the appearance of "beads on a string," where the beads are the nucleosomes (Figure 11.16). The thinner "thread" connecting the "beads" is naked linker DNA.

HIGHER-ORDER STRUCTURES IN CHROMATIN.
In the living cell, chromatin typically does not exist in a "beads-on-a-string" nucleofilament; rather, it is kept in a more highly compacted state. The details of these higher levels of folding, which are found predominantly in the chromosomes, are still incompletely understood. Moreover, the details of the changes in chromatin folding that take place as a cell goes from interphase to mitosis (or meiosis) and back to interphase are not well understood. There is general agreement, derived from microscopic observations, that the next level of packing above the nucleosome is the *30-nm chromatin fiber* (Figure 11.17). One major

problem that has hindered investigations is that as chromatin compaction proceeds, the individual nucleosomes can no longer be distinguished, and hence their arrangement in the 30-nm fiber becomes much more difficult to define. Nonetheless, a number of models for the 30-nm fiber have been proposed based on biophysical and biochemical data derived from studies of both compact and relaxed (unraveled) chromatin, one of which is shown in Figure 11.18. Histone H1 must play an important role in the formation of the 30-nm fiber, since H1-depleted chromatin can form 10-nm fibers, but not 30-nm fibers. However, the precise role of histone H1 in forming the 30-nm fiber is currently unknown.

Coiling of the 10-nm nucleofilament to produce the 30-nm fiber creates a structure with 6 nucleosomes per turn, and compacts the DNA another sixfold. However, this does not provide sufficient packing of the chromosomes to explain the degree to which chromosomes are condensed in the cell nucleus. That is, an average human chromosome would extend approximately 1 mm as a 30-nm fiber, and, as such, would be 200 times longer than the diameter of the nucleus (about 5 μm). Thus, further folding of the

~ FIGURE 11.17

Electron micrograph of a 30-nm chromatin fiber.

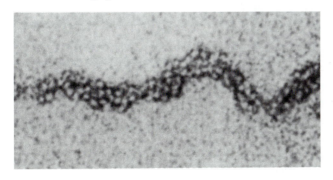

~ FIGURE 11.18

A model for the packaging of nucleosomes into the 30-nm chromatin fiber.

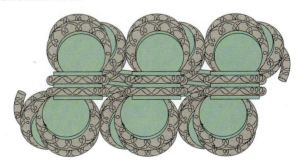

chromatin fiber must occur. However, the next levels of packing beyond the 30-nm chromatin fiber are not clearly understood. An interphase chromosome may have a diameter of 300 nm while a metaphase chromosome may have a diameter of 700 nm.

A major step in modeling the structure of metaphase chromosomes came when data were obtained showing that metaphase chromosomes from which the histones have been removed still retain a residual folded structure with *looped domains* of 30-nm fiber DNA extending from a condensed protein lattice. (Recall that bacterial chromosomal DNA is also organized into looped domains.) The central structure, called the *chromosome scaffold,* actually has the shape of the metaphase chromosome (Figure 11.19), a shape that remains even when the DNA is digested away by nucleases.

Figure 11.20 presents a schematic of a series of looped domains of DNA in a chromosome. The looped domains extend at an angle from the main chromosome axis and are anchored to a filamentous structural framework inside the nuclear envelope called the *nuclear matrix.* The amount of DNA in each loop ranges from tens to hundreds of kilobase pairs. An average human chromosome could contain approximately 2,000 looped domains.

~ FIGURE 11.20

Schematic model for the organization of 30-nm chromatin fiber into looped domains that are anchored to a nonhistone protein scaffold.

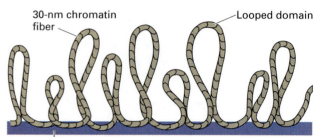

By treating nuclei to remove lipids, proteins, DNA, and RNA, the nuclear matrix can be isolated. The matrix preparation consists mostly of protein, with a small proportion of nucleic acids. The DNA sequences attached to the proteins in the matrix are called *MARs* (*m*atrix *a*ttachment *r*egions). Analysis of the sequences of the MARs has shown that a large proportion of them flank transcriptionally active genes and actively replicating regions. The organization of chromatin into loops appears to be important,

~ FIGURE 11.19

Electron micrograph of a metaphase chromosome depleted of histones. The chromosome maintains its general shape by a nonhistone protein scaffolding from which loops of DNA protrude.

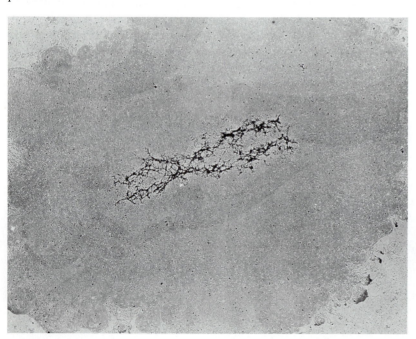

then, for compacting the chromatin fiber in the chromosome and for the regulation of gene expression. In this vein, it has been hypothesized that each loop is an independent unit of transcription and replication that functions independently of neighboring loops.

Figure 11.21 shows the different orders of DNA packing that could give rise to the highly condensed metaphase chromosome. Interphase chromosomes would be packed less tightly than metaphase chromosomes. Very little is known about the mechanisms involved in the transitions between interphase and metaphase chromosome morphologies, or about the more subtle transitions in localized chromosome regions when genes are turned on or off.

$\mathcal{K}$EYNOTE

The large amount of DNA present in the eukaryotic chromosome is compacted by its association with histones in nucleosomes and by higher levels of folding of the nucleosomes into chromatin fibers. Each chromosome contains a large number of looped domains of 30-nm chromatin fibers attached to a protein scaffold. The functional state of the chromosome is related to the extent of coiling: The more condensed a part of a chromosome is, the less likely it is that the genes in that region will be active.

Centromeres and Telomeres

So far in this chapter we have described eukaryotic chromosomes as linear structures each containing a single linear DNA molecule wrapped around histones and associated with nonhistone proteins. In terms of size, number, and morphology the chromosome complement of an organism is species-specific. Nonetheless, as described in Chapter 3, all chromosomes behave similarly at the time of cell division. In mitosis, for example, the centromeres of all the replicated chromosomes become aligned at the metaphase plate, the sister chromatids separate at the centromeres, and one chromatid (now daughter chromosome) of each pair is distributed to each daughter cell. The behavior of chromosomes in mitosis and meiosis is a function of the centromeres, the sites at which chromosomes attach to the mitotic and meiotic spindles, and the *telomeres*, the ends of the chromosomes. Centromeres and telomeres are discussed in the following two sections.

~ **FIGURE 11.21**

Schematic drawing of the many different orders of chromatin packing that are thought to give rise to the highly condensed metaphase chromosome.

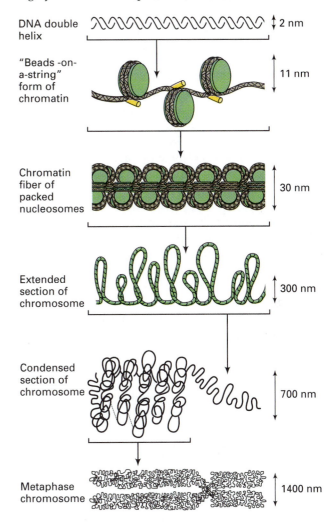

CENTROMERES. The centromere region of each chromosome is responsible for the accurate segregation of the replicated chromosomes to the daughter cells during mitosis and meiosis. The error frequency for this process is low but significant, and varies from organism to organism. In yeast, for example, segregation errors (i.e., nondisjunction—see Chapter 3) occur at a frequency of 1 in 10^5 or less.

In humans nondisjunction occurs much more frequently. However, most zygotes derived from nondisjunctional gametes die early *in utero* (trisomy 21 and some X chromosome abnormalities are exceptions). Nondisjunction of chromosomes occurs when the centromere fails to function properly. If the centromere is absent, the chromosome loses its ability to attach to the spindle and thus will migrate through the cell randomly during the cell division process. Such acentric chromosomes often are degraded so that they are "lost" to the segregation process. That is, the chromosome will replicate but the two chromosome copies will not always segregate properly to the daughter cells.

In many higher eukaryotes the centromere is seen as a constricted region at one point along the chromosome. The region contains the kinetochore (see Chapter 3) to which the spindle fibers attach during cell division, and gives the chromosome its characteristic appearance at metaphase. In most eukaryotes several spindle fibers attach to each centromere, whereas in the yeast *Saccharomyces cerevisiae*, the organism in which centromeres have been best characterized, only one spindle fiber attaches to each centromere.

The DNA sequences (called *CEN* sequences, after the *cen*tromere) of the yeast centromeres have been determined. Though each centromere has the same function, the *CEN* regions are very similar, but not identical, to one another in nucleotide sequence and organization. Figure 11.22a presents the key sequence information for four yeast centromeres, and Figure 11.22b gives the consensus sequence (i.e., the sequence giving the base most commonly found at each position) for thirteen yeast centromeres. The common core centromere region consists of three sequence domains (centromere DNA elements or CDEs). CDEII, a 76-to-86 bp region of high A + T content (greater than 90 percent A + T bases) is the largest domain. Flanking to one side is CDEI, which is a conserved RTCACRTG sequence (where R is a purine,

i.e., either A or G), and to the other side is CDEIII, a 25 bp conserved sequence domain that is A-T rich. Centromere sequences have been determined for a number of other organisms; they are different both from those of yeast and from each other. The centromeres of the fission yeast, *Schizosaccharomyces pombe,* for example, are 40 to 80 kb long with complex arrangements of several repeated sequences. Thus, while centromeres carry out the same function in all eukaryotes, there is no common sequence that is responsible for that function.

How are centromere sequences organized in chromatin? In yeast, mutational studies have led to the identification of various proteins that interact with the centromere and with the spindle microtubule in the kinetochore structure. Indeed, since spindle microtubules must attach to centromeres, it was expected that specific proteins would be involved. A hypothetical model for the yeast kinetochore structure based on the mutational data is shown in Figure 11.23. In this model, CBF1 (centromere binding factor 1) binds to CDEI and a complex of proteins comprising CBF3 binds to CDEIII. The longer CDEII sequence may be wrapped once around a histone octamer. Perhaps through a linker protein, CBF3 binds to the end of a microtubule; other proteins may also be bound around the end of the microtubule. Confirmation of this model must await detailed molecular analysis.

TELOMERES. A telomere is the region of DNA at each end of a linear chromosome. It is required for replication and stability of that chromosome. That the

---

~ FIGURE 11.22

(a) Regions of sequence homology between *CEN3, CEN4, CEN6,* and *CEN11* in the centromere DNA of the yeast, *Saccharomyces cerevisiae*. The centromere DNA elements I-III (CDEI-CDEIII) are set up in a spatial arrangement that is nearly identical in the centromeres of four different chromosomes; (b) Consensus sequence for thirteen yeast centromeres. (R = a purine; N = any base.)

a) Specific centromere sequences

CDE regions: I   II   III

CEN3   GTCACATG ← 84 bp (93% AT) → TGTATTTGATTTCCGAAAGTTAAAA

CEN4   GTCACATG ← 78 bp (93% AT) → TGTTTATGATTACCGAAACATAAAA

CEN6   ATCACGTG ← 84 bp (94% AT) → AGTTTTTGTTTTCCGAAGATGTAAA

CEN11 GTCACATG ← 84 bp (94% AT) → TGTTCATGATTTCCGAACGTATAAA

← 8 bp →   ←——— 25 bp ———→

b) Consensus centromere sequence for 13 centromeres analyzed

RTCACRTG ← 78-86 bp(>90% AT) → TGT^T_AT^T_ATGNNTTCCGAANNNNNAAA

← 8 bp →   ←——— 25 bp ———→

~ FIGURE 11.23

Hypothetical model for the kinetochore of yeast, showing the relationship of the spindle fiber microtubule to the centromere DNA elements. (Adapted with permission from Pluta et al., *Science* 270 (1995):1591–94. Copyright © 1995 American Association for the Advancement of Science.)

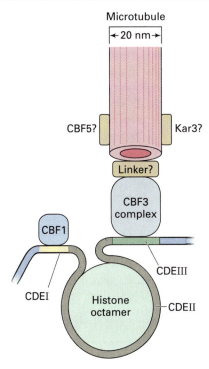

~ TABLE 11.5

Tandemly Repeated Telomeric DNA Sequences of Eukaryotes

ORGANISM	5'→3' SEQUENCE[a]
Holotrichous Ciliates	
Tetrahymena	
Glaucoma	. . . TTGGGG
Paramecium	
Hypotrichous Ciliates	
Stylonichia	. . . TTTTGGG
Oxytricha	
Flagellates	
Trypanosoma	
Leptomonas	. . . TTAGGG
Leishmania	
Crithidia	
Slime Molds	
Physarum	. . . T_nAGGG
Dictyostelium	. . . AG_{1-8}
Fungus	
Saccharomyces cerevisiae	. . . TG_{1-3}
Higher Eukaryotes	
Humans	. . . TTAGGG
Arabidopsis	. . . TTTAGGG

[a]The 5' to 3' strand sequence is directed from the interior of the linear chromosome to its end.

ends of chromosomes were special regions was indicated by genetic analysis of chromosomal mutations (see Chapter 7). That is, there are many examples of chromosomal mutations resulting from "stickiness" of the ends of broken chromosomes. However, the telomeres of chromosomes are not "sticky" so rearrangements in which the end of a chromosome is moved to an internal position are not seen. Thus, telomeres must function to stabilize unbroken chromosomes against chromosomal mutations.

Telomeres are characteristically, but not necessarily, heterochromatic. In most organisms that have been examined, the telomeres are positioned just inside the nuclear envelope, and are often found associated with each other as well as with the nuclear envelope.

All telomeres in a given species share a common sequence. Most telomeric sequences may be divided into two types:

1. **Simple telomeric sequences** are at the extreme ends of the chromosomal DNA molecules. These sequences are species-specific and consist of simple, tandemly repeated DNA sequences. Simple telomeric sequences are the essential functional components of telomeric regions in that they are sufficient to supply a chromosomal end with stability. Table 11.5 gives the DNA sequences for the simple telomeric sequences found in a number of eukaryotes. In the ciliate *Tetrahymena*, for example, reading from the interior of the chromosome to its end, the repeated sequence consists of elements that are 5'-TTGGGG-3' elements, and in the flagellate *Trypanosoma*, the repeated sequence is 5'-TTAGGG-3'. Figure 11.24 shows the telomere sequence at the end of a *Tetrahymena* chromosome. Much remains to be learned about these sequences in general and the properties of chromosomal ends. We will see in the next chapter the role these telomere sequences have in the replication of eukaryotic chromosomes.

~ FIGURE 11.24

Telomere sequence at the ends of *Tetrahymena* chromosomes.

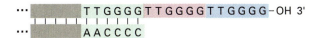

2. **Telomere-associated sequences** are regions near but not at the ends of chromosomes. These sequences often contain repeated but still complex DNA sequences extending many thousands of base pairs in from the chromosome end.

While the telomeres of most eukaryotes contain short, simple, repeated sequences, the telomeres of *Drosophila* are quite different. That is, *Drosophila* telomeres consist of *transposons,* DNA sequences that can move to other locations in the genome. These telomere sequences belong to the LINE family of repeated sequences found in eukaryotic genomes (see p. 338). Centromere-proximal to these unusual telomeres are tandemly repeated sequences that are structurally and functionally analogous to the telomere-associated sequences found in other eukaryotes.

Keynote

The centromere region of each eukaryotic chromosome is responsible for the accurate segregation of the replicated chromosome to the daughter cells during both mitosis and meiosis. The DNA sequences of yeast centromeres (*CEN* sequences) are very similar in length and sequence.

The ends of chromosomes, the telomeres, often are associated with the nuclear envelope. Telomeres in a species share a common sequence.

Characteristically, sequences at or very close to the extreme ends of the chromosomal DNA consist of simple, relatively short, tandemly repeated sequences. Repeated, often complex, DNA sequences, called telomere-associated sequences, are found farther in from the chromosome ends.

Unique-Sequence and Repetitive-Sequence DNA in Eukaryotic Chromosomes

Now that we know about the basic structure of DNA and its organization in chromosomes, we can think about the distribution of certain sequences in the genomes of prokaryotes and eukaryotes. From molecular analyses, geneticists have determined the relative abundances of different types of sequences in DNA from a variety of organisms. Such analyses have

provided relatively sketchy information about the distribution of the various classes of sequences in genomes. As the complete DNA sequences of more and more eukaryotic genomes are determined (that of yeast was reported in 1996), we will develop a precise understanding of the organization of unique-sequence and repetitive-sequence DNA in genomes.

For convenience, three different types of sequences have been named: **unique** (or single-copy, present in one to a few copies per genome), **moderately repetitive** (present in a few to as many as 10^3 to 10^5 times in the genome), and **highly repetitive** (present as many as 10^5 to 10^7 times), although there are really no discrete boundaries between them. In prokaryotes, with the exception of the ribosomal RNA genes, tRNA genes, and a few other sequences, all of the genome is present as unique-sequence DNA. Eukaryotic genomes, on the other hand, consist of both unique-sequence and repetitive-sequence DNA, with the latter typically being quite complex in number of types, number of copies, and distribution. Using the three sequence types just defined, the human genome can be described as consisting of about 64 percent unique-sequence DNA, 25 percent moderately repetitive DNA, and 10 percent highly repetitive DNA. By contrast, the green frog has 22 percent unique sequence DNA, 67 percent moderately repetitive DNA, and 9 percent highly repetitive DNA. However, it is more instructive to discuss the classes of DNA with respect to their arrangement in the chromosome and to their functions, rather than with respect to their relative abundance in the genome. Thus, in the following section we will first discuss unique-sequence DNA, and then the various arrangements of repetitive-sequence DNA.

Unique-Sequence DNA

Unique sequences (sometimes called single-copy sequences) are defined as sequences present as single copies in the genome. (Thus there are two copies per diploid cell.) Actually, in current usage the term usually applies to sequences that have one to a few copies per genome. Most of the genes that we know about—those that code for proteins in the cell—are in the unique-sequence class of DNA. Conversely, not all unique-sequence material contains protein-coding sequences.

Repetitive-Sequence DNA

In this section we will divide the subject into two and discuss tandemly repeated DNA sequences and dispersed repeated DNA sequences.

TANDEMLY REPEATED DNA. *Tandemly repeated DNA* refers to reiterated sequences that are tandemly arranged (arranged one after the other) in the genome. Tandemly repeated DNA is quite common in eukaryotic genomes, and a number of different examples are known. The genes for ribosomal RNA (rRNA: see Chapter 13), and transfer RNA (tRNA: see Chapter 13), for example, are tandemly repeated in one or more clusters in most eukaryotes. For instance, there are about 450 copies of the major class of rRNA genes in the clawed toad (*Xenopus laevis*), 160–200 copies in humans, 260 copies in the sea urchin, and 3,900 copies in the garden pea. The histone genes are repeated in eukaryotes and, except for birds and mammals where they are dispersed in the genome, the repeated copies tend to be linked. As for copy number, yeast has two copies of the histone genes, *Drosophila* has 100 copies, *Xenopus* has 25 copies, and humans have 5 to 20 copies.

Examples are also known of tandemly repeated genes that are related, but not identical. In vertebrates, hemoglobin consists of two copies each of an α-globin and a β-globin polypeptide. These polypeptides are encoded by multiple genes—a *multigene family*—with each gene encoding a related but different polypeptide. Thus, there is an α-globin family of genes and a β-globin family of genes. In humans, the three genes of the α-globin family are located in a 25-kb region of DNA on chromosome 16, and the five genes of the β-globin family are located in a 65-kb region of DNA on chromosome 11. These two families of genes are expressed in a developmentally timed order so that hemoglobin molecules are assembled with oxygen-carrying abilities appropriate for the developmental stage from embryo to adult (see Chapter 17, p. 561).

Lastly, some tandemly repeated sequences are not associated with genes. In fact, the greatest amounts of tandemly repeated DNA are associated with centromeres and telomeres (see pp. 333–336). At each centromere, for example, hundreds to thousands of copies of simple, relatively short, tandemly repeated sequences (i.e., highly repetitive sequences) may be found. Indeed, a significant proportion of the eukaryotic genome may be comprised of the highly repeated sequences found at centromeres: 16 percent in *Drosophila melanogaster*, 8 percent in the mouse, about 50 percent in the kangaroo rat, and about 5 percent in humans. Since these sequences often have a %GC content (base ratio) different from that of the majority of the DNA in the genome, they will usually form a distinct band (or bands) when the chromosomes are fragmented and centrifuged in a cesium chloride (CsCl) solution that separates DNA on the basis of density (and therefore %GC content—see Box 12.1, p. 347);

such bands are called *satellite bands*, and the DNA contained therein is known as **satellite DNA**. Figure 11.25 shows a satellite DNA band from a mouse. Note that the number of satellite DNA bands is species-specific with some bands involving repeated DNA sequences other than centromeric repeated DNA. Subclasses of satellite DNA, minisatellite sequences, and microsatellite sequences have proved useful in analyzing genomes and for DNA typing studies (see Chapter 15).

DISPERSED REPEATED SEQUENCES. *Dispersed repeated sequences* refers to those reiterated sequences that are dispersed in the genome, rather than being tandemly arranged. Examples of both gene sequences and non-gene sequences are found in this class of repeated sequences. For example, members of some multigene families may be dispersed through the genome. As was mentioned above, the repeated histone genes of mammals and birds are dispersed in the genome. Typically, dispersed repeated genes are present in relatively low copy number, often less than 50. Transposons—DNA elements that can move to other locations in the genome—are examples of another type of dispersed repeated sequence. Transposons are found in all eukaryotic organisms—they are discussed in Chapter 20. Also, many eukaryotic DNAs contain dispersed repeated sequences that are most likely not genes; these

~ **FIGURE 11.25**

Fractionation of mouse genomic DNA by CsCl centrifugation showing the separation into a main band and a peak of satellite DNA of lighter density comprising about 8 percent of the total DNA.

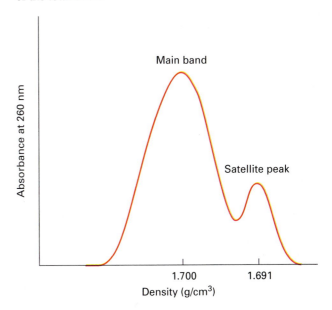

sequences may be reiterated as many as one million times. We will focus our attention on this latter class in the following discussion.

Molecular analysis of genomes using restriction enzymes, cloning, and sequencing has provided information about the distribution of dispersed repeated sequences in a variety of organisms. The picture that has emerged is one of families of repeated sequences interspersed through the genome with unique sequence DNA. Often small numbers of families have very high copy numbers and make up most of the dispersed repeated sequences in the genome. Two general interspersion patterns are encountered. In one, the families have sequences 100 to 500 bp long—these sequences are called **SINEs** for **short** *inter***spersed repeats**. In the other, the families have sequences about 5,000 bp or more long—these sequences are called LINEs for **long** *inter***spersed repeats**. All eukaryotic organisms have SINEs and LINEs, although the relative proportions vary considerably. *Drosophila* and birds, for example, have mostly LINEs, while humans and frogs have mostly SINEs.

The SINEs pattern of repeated sequences is found in a diverse array of eukaryotic species, including mammals, amphibians, and sea urchins. Each species with SINEs has its own characteristic array of SINE families. A well-studied SINE family is the Alu family of certain primates. This family is named for the restriction site for *Alu*I ("Al-you-one") typically found in the repeated sequence. In humans, the Alu family is the most abundant SINE family in the genome, consisting of 300-bp sequences repeated as many as 9×10^5 times, and comprising about 9 percent of the total haploid DNA. One Alu repeat is located about every 5,000 bp in the genome! As you might imagine, cloned segments of human DNA have a good chance of containing at least one Alu sequence.

The locations of a number of Alu repeats in the human genome have been determined. Examples are found in a variety of places, including in introns, in the DNA flanking genes, and in satellite DNA. Alu sequences (and other SINE sequences) that are located in introns or in the regions flanking protein-coding genes may be transcribed. As a result, the population of transcribed RNA found in the nucleus contains many transcripts of Alu (and other SINE) sequences interspersed among other unique sequences. The Alu/SINE parts of the transcripts are removed as the initial transcripts are processed to produce mature mRNA molecules, so relatively few Alu/SINE sequences are found in cytoplasmic mRNA.

Mammalian genomes have many copies of a particular LINE family of repeated sequence, the LINE-1 family, in addition to many SINE families. Other LINE families may be present also, but they are much less abundant than LINE-1. LINE-1 family members are up to 7,000 bp long, although a wide range of sizes is seen. Like Alu repeats/SINEs, LINE-1 sequences are found in introns, in the sequences flanking genes, and in satellite DNA, and they may also be transcribed if they are in those locations. There is evidence that some of the longer LINE-1 family sequences are transposons; that is, sequences that are capable of transposing—moving from one location to another in the genome.

While we have a lot of information about the sequence organization and distribution of SINEs and LINEs, we have very little knowledge about the function(s) of those sequences. One hypothesis, in fact, is that most of these sequences have no function at all. Another hypothesis is that some of the repeated sequences, or their transcripts, or both, are somehow involved in mechanisms for regulating gene expression. At this time, it is certainly clear that some repeated sequences have no known function, and there is inconclusive evidence for roles of repeated sequences in gene regulation.

$\mathcal{K}$EYNOTE

Prokaryotic genomes consist mostly of unique-sequence DNA, with only a few sequences and genes repeated; eukaryotes have both unique and repetitive sequences in the genome. Particularly in higher eukaryotes, the spectrum of complexity of the repetitive DNA sequences is extensive. The highly repetitive sequences tend to be localized to heterochromatic regions, for example around centromeres, whereas the unique and moderately repetitive sequences tend to be interspersed.

SUMMARY

In this chapter we have learned some aspects of the molecular organization of prokaryotic, viral, and eukaryotic chromosomes. Such structural information is important to know in order to understand gene regulation mechanisms more completely.

The chromosomes of prokaryotic organisms consist of circular, double-stranded DNA molecules that are complexed with a number of proteins. The chromosome is compacted within the cell by supercoiling of the DNA helix and the formation of looped domains of supercoiled DNA. Examples of chromosome organization in bacteriophages were discussed, including circular permutation and terminal redundancy of double-stranded DNA in T-even phages, sticky ends in λ double-stranded DNA and their involvement in the phage life cycle, and circular, single-stranded DNA in ΦX174. Each chromosome form is replicated in a particular way to generate progeny chromosomes.

A distinguishing feature of eukaryotic chromosomes is that DNA is distributed among a number of chromosomes. The set of metaphase chromosomes in a cell is called its karyotype. We saw in an earlier chapter how karyotype analysis is useful for the diagnosis of certain human chromosome aberrations. This analysis is aided by the ability to stain human chromosomes so that each chromosome is distinguishable by its banding pattern.

The DNA in the genome specifies an organism's structure, function, and reproduction. The amount of DNA in a genome (made up of a prokaryotic chromosome or of the haploid set of chromosomes in eukaryotes) is called the C value. There is no direct relationship between the C value and the structural or organizational complexity of an organism.

The organization of DNA in eukaryotic chromosomes was described in this chapter. The nuclear chromosomes of eukaryotes are complexes of DNA with histone and nonhistone chromosomal proteins. Each chromosome consists of one linear, double-stranded DNA molecule that extends along its length. As a class, the histones are constant from cell to cell within an organism and have been evolutionarily conserved. Nonhistone chromosomal proteins, on the other hand, vary significantly among cell types, and among organisms. The DNA in a eukaryotic chromosome is highly compacted by its association with histones to form nucleosomes and by the several higher levels of folding of nucleosomes into chromatin fibers. The most highly compacted chromosome structure is seen in metaphase chromosomes, while the least compacted structure is seen in interphase chromosomes. The factors controlling the transitions between different levels of chromosome folding as cells go through the cell division cycle are not known. Like prokaryotic chromosomes, eukaryotic chromosomes are organized into a large number of looped domains. These loops consist of 30-nm chromatin fibers and are attached to a protein scaffold.

Important elements for the chromosome segregation in mitosis and meiosis are centromeres and telomeres. Significant progress has been made in defining the sequences of centromeres and telomeres; for example, the DNA sequences of centromeres are relatively complex and are species-specific, and the DNA sequences of telomeres are simple, relatively short, tandemly repeated sequences that are also species-specific. Thus, common chromosome activities in different eukaryotes are controlled by quite different DNA sequences.

Molecular analysis has provided information about the distribution of DNA sequences in genomes. Such analysis has revealed that prokaryotic and viral genomes consist mostly of unique-sequence DNA, with only a few repeated sequences. In contrast, the genomes of eukaryotes contain both unique-sequence DNA and repeated-sequence DNA and there is a wide spectrum of complexity of the repeated DNA sequences among eukaryotes. Many genes are found in unique-sequence DNA, but not all unique-sequence DNA contains genes.

Repeated DNA sequences may be tandemly arranged or interspersed with unique sequence DNA in the eukaryotic genome. Examples of gene families are known for both of these classes of repeated sequences. For example, ribosomal RNA genes are tandemly repeated in most eukaryotes, while histone genes are tandemly repeated in some species and dispersed in other species. The greatest amounts of tandemly repeated DNA are not associated with genes but with centromeres and telomeres. At each centromere, for example, hundreds to thousands of copies of simple, relatively short, tandemly repeated sequences may be found. There is no known function for centromeric repetitive DNA. Because they are highly repetitive, the centromeric sequences often have a %GC content (base ratio) different from that of the majority of the DNA in the genome, so they will usually form a satellite DNA band (or bands) when centrifuged in a cesium chloride (CsCl) gradient.

Dispersed repeated sequences as a class contains both gene sequences and non-gene sequences. Dispersed repeated gene sequences typically are present in relatively low copy numbers (less than 50) while dispersed non-gene sequences may be present in hundreds of thousands of copies. Eukaryotic species have characteristic families of these repeated sequences interspersed through the genome with unique sequence DNA. The two general interspersion families are SINEs (short *in*terspersed repeats) in which the repeated sequences are 100–500 bp long, and LINEs (long *in*terspersed repeats) in which the repeated sequences about 5,000 bp or more long. SINEs and LINEs are found in all eukaryotes, although the relative proportions vary considerably. While much knowledge has been accumulated about SINEs and LINEs in a number of eukaryotic genomes, little is known about the functions of these sequences.

ANALYTICAL APPROACHES FOR SOLVING GENETICS PROBLEMS

Q11.1 When double-stranded DNA is heated to 100°C, the two strands separate because the hydrogen bonds between the strands break. When cooled again, the two strands typically do not come back together again because they do not collide with one another in the solution in such a way that complementary base sequences can form. In other words, heated DNA tends to remain as single strands. Consider the DNA double helix:

```
┌┬┬┬┬┬┬┬┬┬┬┬┬┬┬┐
 G C G C G C G C G C G C G C
 C G C G C G C G C G C G C G
└┴┴┴┴┴┴┴┴┴┴┴┴┴┴┘
```

If this DNA is heated to 100°C and then cooled, what might be the structure of the single strands? Assume in formulating your answer that the concentration of DNA is so low that the two strands never find one another.

A11.1 This question serves two purposes. First, it reinforces certain information about double-stranded DNA, and, second, it poses a problem that can be solved by simple logic.

We can analyze the base sequences themselves to see whether there is anything special about them in order to avoid an answer of "nothing significant happens." The DNA is a 14-base-pair segment of alternating GC and CG base pairs. By examining just one of the strands, we can see that there is an axis of symmetry at the midpoint such that it is possible for the single strand to form a double-stranded DNA molecule by intrastrand ("within strand") base pairing. The result is a double-stranded hairpin structure, as shown in the following diagram (from the top strand; the other strand will also form a hairpin structure):

Q11.2 An organism has a haploid genome of 10^{10} nucleotide pairs, of which 70 percent is unique-sequence DNA with a copy number of one, 20 percent is moderately repetitive DNA with an average copy number of 1,000, and 10 percent is highly repetitive DNA with an average copy number of 10^6. Assuming that an average DNA sequence is 10^3 nucleotide pairs, how many different sequences are in each of the three DNA classes?

A11.2a. *Unique-sequence DNA*: the total DNA in this class is 70 percent of 10^{10} nucleotide pairs = 0.7×10^{10} = 7×10^9. Only one of each sequence exists so the number of different sequences = $(7 \times 10^9)/10^3$ = $\underline{7 \times 10^6}$.

b. *Moderately repetitive DNA*: DNA in this class is 20 percent of 10^{10} bp = 0.2×10^{10} = 2×10^9. Given an average DNA sequence length of 10^3 bp, there are $(2 \times 10^9)/10^3$ = 2×10^6 sequences in this class. Since the average copy number is 1,000, the number of *different* sequences = $(2 \times 10^6)/10^3$ = $\underline{2 \times 10^3 = 2,000}$.

c. *Highly repetitive DNA*: DNA in this class is 10 percent of 10^{10} bp = 0.1×10^{10} = 1×10^9. Given an average DNA sequence length of 10^3 bp, there are $(1 \times 10^9)/10^3$ = 1×10^6 sequences in this class. Since the average copy number is 10^6 the number of different sequences = $(1 \times 10^6)/10^6$ = $\underline{1}$.

QUESTIONS AND PROBLEMS

*11.1 Two double-stranded DNA molecules from a population of T2 phages were heat-denatured to produce the following four single-stranded DNAs:

$$1 \text{ T A G C T C C} \longrightarrow \qquad 3 \text{ G C T C C T A} \longrightarrow$$

and

$$2 \text{ A T C G A G G} \longleftarrow \qquad 4 \text{ C G A G G A T} \longleftarrow$$

These separated strands were allowed to renature. Diagram the structures of the renatured molecules most likely to appear when (a) strand 2 renatures with strand 3 and (b) strand 3 renatures with strand 4. Mark the strands and indicate sequences and polarity.

11.2 Capital letters represent regions in the chromosome of phage T4. A particular *E. coli* cell was infected by a single T4 chromosome with the sequence ABCDEFAB, but before this chromosome could replicate it suffered a deletion of the E region. This did not interfere with phage replication. What would you expect to be the chromosome sequence(s) of the progeny phage produced upon lysis of this cell? Explain your reasoning.

*11.3a. If you were to denature and renature a population of normal T4 chromosomes, what kinds of structures would form?

b. How would the results differ if the chromosomes were from T7?

c. From λ?

11.4 What are topoisomerases?

11.5 In typical human fibroblasts in culture, the G_1 period of the cell cycle lasts about 10 h, S lasts about 9 h, G_2 takes 4 h, and M takes 1 h. Imagine you were to do an experiment in which you added radioactive (^{3}H) thymidine to the medium and left it there for 5 min (pulse), and then washed it out and put in ordinary medium (chase).

a. What percentage of cells would you expect to become labeled by incorporating the ^{3}H-thymidine into their DNA?

b. How long would you have to wait after removing the ^{3}H medium before you would see labeled metaphase chromosomes?

c. Would one or both chromatids be labeled?

d. How long would you have to wait if you wanted to see metaphase chromosomes containing ^{3}H in the regions of the chromosomes that replicated at the beginning of the S period?

11.6 Assume you did the experiment in Question 11.5, but left the radioactive medium on the cells for 16 h instead of 5 min. How would your answers to the above questions change?

11.7 Karyotype analysis performed on cells cultured from an amniotic fluid sample reveals that the cells contain 47 chromosomes. The stained chromosomes are classified into groups and the arrangement shows 6 chromosomes in A, 4 in B, 16 in C, 6 in D, 6 in E, 4 in F, and 5 in the G group. Based on the above information:

a. What could be the genotype of the fetus? If more than one possibility exists, give all.

b. How would you proceed to distinguish between the possibilities?

11.8 What is the relationship between cellular DNA content and the structural or organizational complexity of the organism?

11.9 Match the DNA type with the chromatin type in the following table. (More than one DNA type may match a given chromatin type.)

DNA TYPE	CHROMATIN TYPE
Barr body (inactivated DNA)	Euchromatin
Centromere	Facultative heterochromatin
Telomere	Constitutive heterochromatin
Most expressed genes	

11.10 Eukaryotic chromosomes contain (choose the best answer):

a. protein.

b. DNA and protein.

c. DNA, RNA, histone, and nonhistone protein.

d. DNA, RNA, and histone.

e. DNA and histone.

11.11 In a particular eukaryotic chromosome (choose the best answer):

a. Heterochromatin and euchromatin are regions where genes make functional gene products (i.e., where they are active).

b. Heterochromatin is active but euchromatin is inactive.

c. Heterochromatin is inactive but euchromatin is active.

d. Both regions are inactive.

11.12 List four major features that distinguish eukaryotic chromosomes from prokaryotic chromosomes.

11.13 From the list below, identify

a. three features that both eukaryotic and bacterial chromosomes have in common.

b. four features that eukaryotic chromosomes have but which are not found in bacterial chromosomes.

c. one feature that bacterial chromosomes have but which are not found in eukaryotic chromosomes.

Words for Matching

A. centromeres	F. nonhistone protein scaffolds
B. hexose sugars	G. DNA
C. amino acids	H. nucleosomes
D. supercoiling	I. circular chromosome
E. telomeres	J. looping

11.14 A nucleosome core particle is composed of

a. Two molecules each of H2A, H2B, H3, and H4.

b. One molecule each of H1, H2A, H2B, H3, and H4.

c. Two molecules each of H1, H2, H3, and H4.

11.15 Discuss the structure and role of nucleosomes.

11.16 Answer the questions below after setting up the following "rope trick." Start with a belt (representing a DNA molecule; imagine the phosphodiester backbones lying along the top and bottom edges of the belt) and a soda can. Holding the belt buckle at the bottom of the can, wrap the belt flat against the side of the can, counterclockwise three times around the can. Now remove the "core" soda can and, holding the ends of the belt, pull the ends of the belt taut. After some reflection, answer the following questions.

a. Did you make a left- or a right-handed helix?

b. How many helical turns were present in the coiled belt before it was pulled taut?

c. How many helical turns were present in the coiled belt after it was pulled taut?

d. Why does the belt appear more twisted when pulled taut?

e. About what percentage of the length of the belt was decreased by this packaging?

f. Is the DNA of a linear chromsome that is coiled around histones supercoiled?

g. Why are topoisomerases necessary to package linear chromosomes?

11.17 Arrange the following figures in increasing order of eukaryotic chromosome condensation. Designate the simplest chromosome organization as number 1 and the most complex chromosome organization as number 6.

A. Chromatin fiber of packed nucleosomes

B. Metaphase chromosome

C. DNA double helix

D. Extended section of chromosome

E. "Beads-on-a-string" form of chromatin

F. Condensed section of chromosome

***11.18** What are the main molecular features of yeast centromeres?

11.19 What are telomeres?

11.20 Would you expect to find most protein coding genes in unique-sequence DNA, in moderately repetitive DNA, or in highly repetitive DNA?

11.21 Would you expect to find ribosomal RNA genes in unique-sequence DNA, in moderately repetitive DNA, or in highly repetitive DNA?

11.22 Both histone and non-histone proteins are essential for DNA packaging in eukaryotic cells. However, these classes of proteins are fundamentally dissimilar in a number of ways. Describe how these classes of proteins differ in terms of

a. their protein characteristics.

b. their presence and abundance in cells.

c. their interactions with DNA.

d. their role in DNA packaging.

***11.23** Rearrangements at the end of 16p (the short arm of chromosome 16) underlie a variety of common human genetic disorders, including α-thalassemia (a defect in hemoglobin metabolism caused by mutations in the α-globin gene), mental retardation, and the adult form of polycystic kidney disease. Recently, the determination of approximately 285 kilobase pairs (kb) of DNA sequence at the end of human chromosome 16p has allowed for very detailed analysis of the structure of this chromosome region. The first functional gene lies about 44 kb from the region of simple telomeric sequences, and about 8 kb from the telomere-associated sequences. Analysis of sequences proximal (moving towards the centromere) to the first gene reveals a sinusoidal variation in GC content, with GC-rich regions associated with gene-rich areas and AT-rich regions associated with Alu-dense areas. The α-globin gene lies about 130 kb from the telomere-associated sequences.

a. Discuss these findings in light of the current view of telomere structure and function as presented in the text.

b. In what different ways might a rearrangement with a breakpoint at the end of 16p affect the expression of the α-globin gene?

c. What new information have these data revealed about the distribution of SINEs in the terminus of 16p?

11.24 Many different types of middle repetitive DNA sequences exist. Some of these have been associated with specific functions; for others either the function is not known or they simply have no function.

a. List three specific types of middle repetitive DNA sequences and give their characteristics and function(s), if known.

b. What functional significance might there be to the fact that some genes are repeated?

12 DNA REPLICATION AND RECOMBINATION

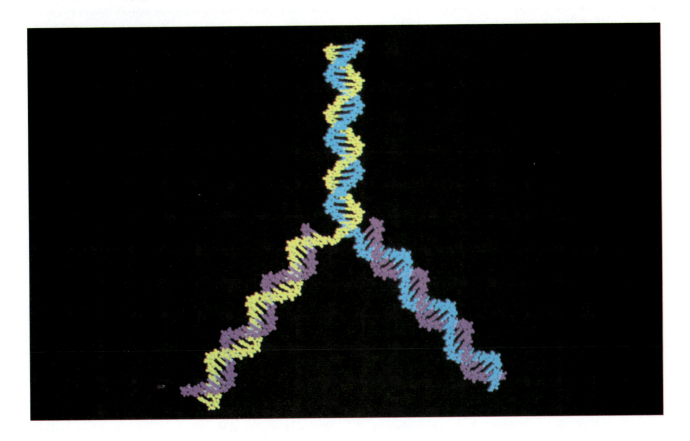

Principal Points

~ DNA replication in prokaryotes and eukaryotes occurs by a semiconservative mechanism in which the two strands of a DNA double helix separate and a new complementary strand of DNA is synthesized on each of the two parental template strands. This mechanism ensures the faithful copying of the genetic information at each cell division.

~ The enzymes that catalyze the synthesis of DNA are called DNA polymerases. All DNA polymerases perform the same DNA synthesis reaction; new strands are made in the 5'-to-3' direction using deoxyribonucleoside 5'-triphosphate (dNTP) precursors.

~ DNA polymerases are incapable of initiating the synthesis of a new DNA strand. All new DNA strands need a short primer of RNA, the synthesis of which is catalyzed by the enzyme DNA primase.

~ DNA replication in *E. coli* requires two DNA polymerases and several other enzymes and proteins. In both prokaryotes and eukaryotes, synthesis of DNA on one template strand is continuous while it is discontinuous on the other template strand, a process called semidiscontinuous replication.

~ In eukaryotes, DNA replication occurs in the S phase of the cell cycle and is biochemically and molecularly very similar to the replication process in prokaryotes. To enable long chromosomes to replicate efficiently, replication of DNA is initiated at a large number of sites along the chromosomes.

~ Special enzymes—telomerases—are used to replicate the ends of chromosomes in eukaryotes. A telomerase is a complex of proteins and RNA. The RNA acts as a template for synthesis of the complementary telomere repeat of the chromosome. In mammals, that telomerase activity is limited to immortal cells (such as tumor cells). The absence of telomerase activity in other cells results in progressive shortening of chromosome ends as the cell divides, thereby limiting the number of cell divisions before the cell dies.

~ When the DNA replicates, new nucleosomes are assembled so the whole chromosome duplicates.

~ In both prokaryotes and eukaryotes, genetic recombination involves the breakage and rejoining of homologous DNA double helices.

~ Numerous models have been proposed to describe the molecular events involved with genetic recombination. The Holliday model for genetic recombination involves a precise alignment of homologous DNA sequences of two parental double helices, followed by a series of enzyme-catalyzed reactions, which can lead to the generation of DNA molecules that are recombinant for loci flanking the recombination site. Some of the enzymes involved in genetic recombination are also involved in DNA replication.

~ Mismatch repair of heteroduplex DNA that is an intermediate in genetic recombination can result in a non-Mendelian segregation of alleles, typically a 3:1 or 1:3 ratio rather than the expected 2:2 ratio. This is called gene conversion.

*I*n Chapter 10 we learned that one of the essential properties of genetic material is that it must replicate accurately so that progeny cells have the same genetic information as the parental cell. In this chapter we will learn about the mechanics of DNA replication and chromosome duplication in prokaryotes and eukaryotes and about some of the enzymes and other proteins required for replication. Some of the replication enzymes are also involved in the repair of damage to DNA, a topic we discuss in Chapter 19. Also, we will learn some of the basic molecular details of DNA recombination.

DNA Replication in Prokaryotes

Early Models for DNA Replication

In Chapter 10 we present Watson and Crick's double-helix model for DNA. Watson and Crick reasoned that replication of the DNA would be straightforward if their model was correct. That is, by unwinding the DNA molecule and separating the two strands, each strand could be a template for the synthesis of a new, complementary strand of DNA. As the DNA double helix is progressively unwound from one end, the base sequence of the new strand would be determined by the base sequence of the template strand, following complementary base-pairing rules. When replication is completed, there would be two progeny DNA double helices, each consisting of one parental DNA strand and one new DNA strand. This model for DNA replication is known as the **semiconservative model** since each progeny molecule retains one of the parental strands (Figure 12.1a).

At the time, two other models for DNA replication were proposed, the **conservative model** (Figure 12.1b) and the **dispersive model** (Figure 12.1c). In the conservative model, the two parental strands of DNA remain together or reanneal after replication and as a whole serve as a template for the synthesis of new progeny DNA double helices. Thus, one of the two progeny DNA molecules is actually the parental double-stranded DNA molecule, and the other consists of totally new material. In the dispersive model, the parental double helix is cleaved into double-stranded DNA segments which act as templates for the synthesis of new double-stranded DNA segments. Somehow, the segments reassemble into complete DNA double helices, with parental and progeny DNA segments interspersed. Thus, while the two progeny DNAs are identical with respect to base pair sequence, the parental DNA has actually become dispersed throughout both progeny molecules.

The Meselson-Stahl Experiment

Five years after Watson and Crick developed their model, Matthew Meselson and Frank Stahl obtained experimental evidence that the semiconservative replication model was correct. To examine the process of DNA replication, Meselson and Stahl used the bacterium *Escherichia coli* because it can be grown easily and quickly in a minimal medium. In Meselson and Stahl's experiment (Figure 12.2), *E. coli* was grown for several generations in a minimal medium in which the only nitrogen source was $^{15}NH_4Cl$ (ammonium chloride). In this compound the normal isotope of nitrogen, ^{14}N, is replaced with ^{15}N, the heavy isotope. (Note: Density is weight/volume so ^{15}N, with one extra neutron in its nucleus, is 1/14 more dense than ^{14}N.) As a result, all the bacteria's cellular nitrogen-containing compounds, including

~ FIGURE 12.1

Three models for the replication of DNA: (a) The semiconservative model (the correct model); (b) The conservative model; (c) The dispersive model. The parental strands are shown in taupe and the newly synthesized strands are shown in red.

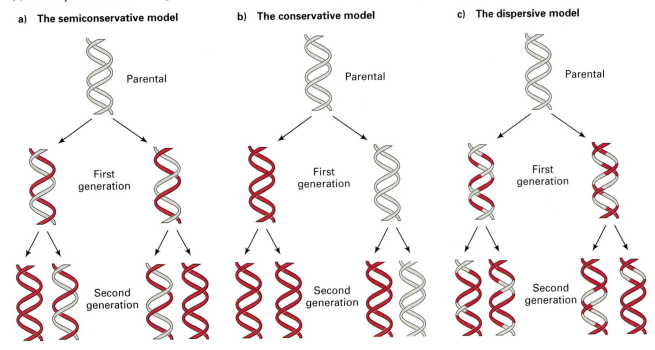

a) **The semiconservative model**

b) **The conservative model**

c) **The dispersive model**

~ FIGURE 12.2

The Meselson-Stahl experiment: The demonstration of semiconservative replication in *E. coli*. Cells were grown in ^{15}N-containing medium for several generations, and then transferred to ^{14}N-containing medium. At various times over several generations, samples were taken; the DNA was extracted and analyzed by CsCl equilibrium density gradient centrifugation. Shown in the figure are a schematic interpretation of the DNA composition at various generations, photographs of the DNA bands, and a densitometric scan of the bands.

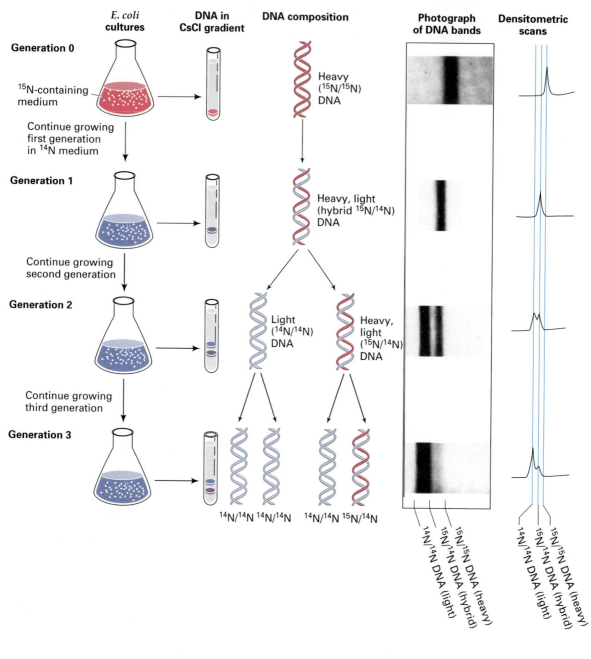

every purine and pyrimidine base in its DNA, contained ^{15}N instead of ^{14}N. ^{15}N DNA can be separated from ^{14}N DNA by using equilibrium density gradient centrifugation (described in Box 12.1). Briefly, in this technique a solution of cesium chloride (CsCl) is cen-trifuged at high speed, causing the cesium chloride to form a density gradient. If DNA is present in the solution, it will band to a position where its buoyant density is the same as that of the surrounding cesium chloride.

Box 12.1

Equilibrium Density Gradient Centrifugation

In experiments involving equilibrium density gradient centrifugation, a concentrated solution of cesium chloride (CsCl) is centrifuged at high speed. The opposing forces of sedimentation and diffusion produce a stable, linear concentration gradient of the CsCl. The actual densities of CsCl at the extremes of the gradient are related to the CsCl concentration that is centrifuged.

For example, to examine DNA of density 1.70 g/cm^3 (a typical density for DNA), one makes a

gradient that spans that density, for example from 1.60 to 1.80 g/cm^3. If the DNA is mixed with the CsCl when the density gradient is generated in the centrifuge, the DNA will come to equilibrium at the point in the gradient where its buoyant density equals the density of the surrounding CsCl (see Box Figure 12.1). The DNA is said to have banded in the gradient. If DNAs are present that have different densities, as is the case with ^{15}N-DNA and ^{14}N-DNA, then they will band (i.e., come to equilibrium) in different positions. The DNA is detected as UV-absorbing material.

The method can also be used experimentally to purify DNAs on the basis of their different densities. It has been used, for example, to fractionate nuclear and mitochondrial DNA in extracts of eukaryotic cells and to separate bacterial plasmid DNA from the host bacterium's DNA.

~ BOX FIGURE 12.1

Schematic diagram for separating DNAs of different buoyant densities by equilibrium centrifugation in a cesium chloride density gradient. The separation of satellite DNA (containing highly repetitive sequences) and main band DNA (the remainder of the DNA) from a eukaryote is illustrated.

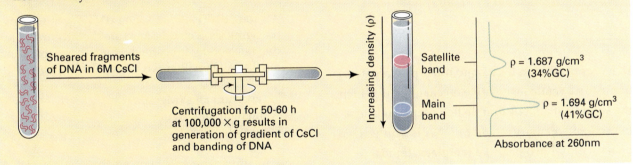

As the next step in Meselson and Stahl's experiment, the ^{15}N-labeled bacteria were transferred to a medium containing nitrogen in the normal ^{14}N form. The bacteria were allowed to replicate in the new conditions for several generations. Throughout the period of growth in the ^{14}N medium, samples of E. coli were taken, lysed to release the cellular contents, and the DNA was analyzed in cesium chloride density gradients (see Figure 12.2). After one generation in ^{14}N medium, all the DNA had a density that was exactly intermediate between that of totally ^{15}N DNA and totally ^{14}N DNA. After two generations, half the DNA was of the intermediate density and half was of the density of DNA containing entirely ^{14}N. These observations, presented in Figure 12.2, and those for subsequent generations were exactly what the semiconservative model predicted.

If the conservative model of DNA replication had been correct, after one generation two bands of DNA would be seen (see Figure 12.1b). One band would be in the heavy-density position of the gradient, containing parental DNA molecules, and both strands would consist of ^{15}N-labeled DNA only. The other band would be in the light-density position, containing progeny DNA molecules with both strands totally ^{14}N-labeled. In subsequent generations, the heavy parental DNA band would be seen at each generation, and in the amount found at the start of the experiment. All new DNA molecules would have both strands totally labeled with ^{14}N. Hence, the amount of DNA in the light-density position would increase with each generation. In the conservative model of DNA replication, then, the most significant prediction was that *at no time would any DNA of intermediate density have been found.*

The fact that intermediate-density DNA *was* found ruled out the conservative model.

In the dispersive model for DNA replication, the parental DNA is scattered in double-stranded segments throughout the progeny DNA molecules (see Figure 12.1c). According to this model, all DNA present after one generation in ^{14}N-containing medium would be of intermediate density, and this was seen in the Meselson-Stahl experiment. Thus, the dispersive model could not be ruled out after just one generation of replication. After a second generation in ^{14}N-containing medium, the dispersive model predicted that DNA segments from the first generation would be dispersed through the progeny DNA double helices produced. Thus, the ^{15}N-^{15}N DNA segments dispersed among new ^{14}N-^{14}N DNA after one generation would then be distributed among twice as many DNA molecules after two generations. As a result, the DNA molecules would be found at one band located halfway between the intermediate-density and light-density band positions. With subsequent generations, there would continue to be one band and it would become lighter in density with each generation. Such a slow shift in DNA density was not seen in the results of the Meselson-Stahl experiment and therefore the dispersive model was ruled out.

KEYNOTE

DNA replication in *E. coli* and other prokaryotes occurs by a semiconservative mechanism in which the strands of a DNA double helix separate and a new complementary strand of DNA is synthesized on each of the two parental template strands. Semiconservative replication results in two double-stranded DNA molecules, each of which has one strand from the parent molecule and one newly synthesized strand. This mechanism ensures the faithful copying of the genetic information at each cell division.

Enzymes Involved in DNA Synthesis

In 1955, Arthur Kornberg and his colleagues set out to find the enzymes that were necessary for the DNA replication so that they could dissect the reactions involved in detail. Kornberg's approach was to identify all the necessary ingredients required for the synthesis of *E. coli* DNA in vitro. The first successful synthesis of DNA was accomplished in a reaction mixture containing DNA fragments, a mixture of four deoxyribonucleoside 5'-triphosphate precursors (dATP, dGTP, dTTP, and dCTP collectively abbreviated dNTP, for *deoxyribo-nucleoside triphosphate*), and a lysate prepared from *E. coli* cells. So that he could measure the very small amount of DNA expected to be synthesized in the reaction, Kornberg used radioactively labeled dNTPs in the reaction mixture.

Realizing that a crucial component (or components) for DNA synthesis must be present in the *E. coli* lysate, Kornberg next analyzed the lysate in order to find that component. In doing so, he isolated an enzyme that was capable of DNA synthesis. This enzyme was originally called the *Kornberg enzyme*, but it is now most commonly called **DNA polymerase I**. (By definition, the enzymes that catalyze the synthesis of DNA are called **DNA polymerases**.)

Once DNA polymerase I was purified, more detailed information could be obtained about DNA synthesis in vitro. The first things studied were the various ingredients necessary for the synthesis of DNA in vitro. Researchers found that four components were needed for the in vitro synthesis of DNA. If any one of the following four components were omitted, DNA synthesis would not occur:

1. All four dNTPs (If any one dNTP was missing, no synthesis occurs.) These are the precursors for the nucleotide (phosphate-sugar-base) building blocks of DNA that we discussed in Chapter 10 (pp. 300–302);
2. Magnesium ions (Mg^{2+});
3. A fragment of DNA;
4. DNA polymerase I.

Subsequent experiments showed that the new DNA that was made in vitro was a faithful base pair–for–base pair copy of the original DNA. That is, the original DNA acted as a template for synthesis of the new DNA.

ROLE OF DNA POLYMERASES. All DNA polymerases catalyze the polymerization of nucleotide precursors (deoxyribonucleotides) into a DNA chain. The reaction that they catalyze (in the presence of magnesium ions) is

$$\{(dNMP)_n\} + dNTP \xrightleftharpoons[\text{DNA polymerase}]{} \{(dNMP)_{n+1}\} + PP_i$$

Here the DNA is represented as a string of a number (n) of deoxyribonucleoside 5'-monophosphates (dNMP). The next nucleotide to be added is the precursor, dNTP. The DNA polymerase catalyzes a reaction in which one nucleotide is added to the existing DNA chain with the release of inorganic pyrophosphate

(PP$_i$). The enzyme then repeats its action until the new DNA chain is complete.

The action of DNA polymerase in synthesizing a DNA chain is shown at the molecular level in Figure 12.3a. The same reaction in shorthand notation is shown in Figure 12.3b. The reaction has three main features:

1. At the growing end of the DNA chain, DNA polymerase catalyzes the formation of a phosphodiester bond between the 3'-OH group of the deoxyribose on the last nucleotide and the 5'-phosphate of the deoxyribonucleoside 5'-triphosphate (dNTP) precursor. The formation of the phosphodiester bond results in the release of two of three phosphates from the dNTP. The important concept here is that *the lengthening DNA chain acts as a primer in the reaction.* (A primer is a preexisting polynucleotide chain in DNA replication to which new nucleotides can be added.)

2. The addition of nucleotides to the chain is not random; each deoxyribonucleotide is selected by the DNA polymerase, which is always bound to the DNA and which moves along the template strand as the polynucleotide chain is lengthened. The polymerase finds the precursor nucleotide (dNTP) that can form a complementary base pair with the nucleotide on the template strand of DNA. Since the DNA polymerase is bound to the template DNA, it ensures that the correct precursor has been chosen. This does not occur with 100 percent accuracy, but the error frequency is extremely low.

3. The direction of synthesis of the new DNA chain is only from 5' to 3' because of the properties of DNA polymerase.

All known DNA polymerases from both prokaryotes and eukaryotes carry out the same reaction and will make new DNA copies from any double-stranded DNA added to the reaction mixture, provided all the necessary ingredients are present. Thus, *E. coli* DNA polymerase can replicate human DNA faithfully, and *vice versa.*

One of the best understood systems of DNA replication is that of *E. coli.* For several years after the discovery of DNA polymerase I (also called the *Kornberg enzyme*), scientists believed that this enzyme was the only DNA replication enzyme in *E. coli.* However, genetic studies proved that this was not so. In general, one way to study the action of a particular enzyme *in vivo* is to induce a mutation in the gene that codes for that particular enzyme. In this way the phenotypic consequences of the mutation can be compared with the wild-type phenotype. A mutation in the gene coding for an enzyme that is as essential to cell function as DNA polymerase, for instance, would be expected to be lethal. The first DNA polymerase I mutant, *polA1*, was isolated in 1969 by Peter DeLucia and John Cairns. These mutants show less than 1 percent of normal polymerizing activity, and near-normal 5'→3' exonuclease activity. Unexpectedly, *E. coli* cells carrying the *polA1* mutation grew and divided normally. However, *polA1* mutants have a high mutation rate when exposed to UV light and chemical mutagens. This was interpreted to mean that DNA polymerase I has an important function in repairing damaged (chemically changed) DNA.

To study the consequences of mutations in genes coding for essential proteins and enzymes, geneticists find it easiest to work with **conditional mutants**, that is, mutant organisms that are normal under one set of growth conditions but that become seriously impaired or die under other growth conditions. The most common types of conditional mutants are those that are temperature-sensitive—mutant organisms that function normally until the temperature is raised past some threshold level, at which time some temperature-sensitive defect is manifested. It was expected that the temperature-sensitive DNA polymerase I mutants would die at elevated temperatures. At *E. coli*'s normal growth temperature of 37°C, temperature-sensitive *polAex1* mutant strains produce DNA polymerase I with normal catalytic activity. At 42°C, however, the temperature-sensitive DNA polymerase I has near-normal polymerizing activity but is defective in 5'→3' exonuclease activity. At 42°C, temperature-sensitive *polAex1* mutants are lethal, showing that 5'→3' exonuclease activity is essential for DNA replication. Taken together, the results of studies of the *polA1* and *polAex1* mutants indicated that there must be other DNA-polymerizing enzymes in the cell. With improvements in preparing cell extracts and enzyme assay techniques, two new *E. coli* DNA polymerases were identified. Martin Gefter, Rolf Knippers, and C. C. Richardson, all working independently, discovered DNA polymerase II in 1970, and Tom Kornberg and Gefter, working together, discovered DNA polymerase III in 1971.

DNA polymerase I is encoded by the *polA* gene and consists of one polypeptide. DNA polymerase II is encoded by the *polB* gene and also consists of one polypeptide. The complete DNA polymerase III enzyme consists of ten polypeptide subunits coded for by many genes. Three of these subunits constitute the catalytic core of the enzyme: (1) α [alpha] (130,000 daltons [Da] encoded by the *dnaE* gene); (2) ε [epsilon] (27,500 Da, encoded by the *dnaQ* gene); and (3) θ [theta] (10,000 Da; encoded by the *holE* gene). In every *E. coli* cell there are about 400 molecules of DNA polymerase I and 10–20 molecules of DNA polymerase III. The number of molecules of DNA polymerase II is not exactly known, but it is believed

~ Figure 12.3

DNA chain elongation catalyzed by DNA polymerase. (a) Mechanism at molecular level; (b) The same mechanism, using a shorthand method to represent DNA.

a) Mechanism of DNA elongation

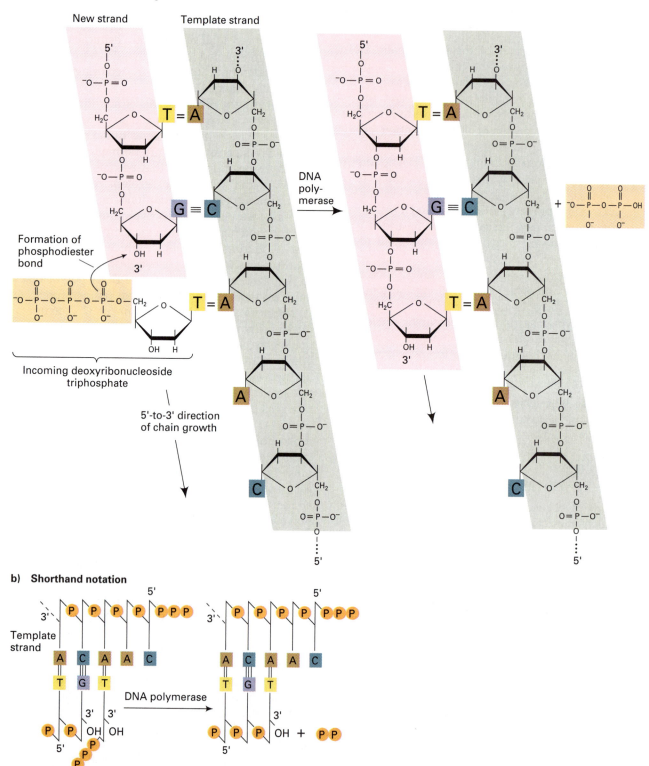

b) Shorthand notation

to be on the same order of magnitude as DNA polymerase III.

A comparison of the important properties of the three *E. coli* DNA polymerases is given in Table 12.1. Both DNA polymerase I and III are known to be involved in DNA replication, but the role of DNA polymerase II in the cell is unknown at this time. All three *E. coli* DNA polymerases have 3'→5' exonuclease activity; that is, they can catalyze the removal of nucleotides from the 3' end of a DNA chain. This means that the DNA polymerases check the accuracy of the most recently assembled base pair. If the wrong base pair has been put together (occurs at a frequency of about 10^{-6} for both DNA polymerase I and DNA polymerase III), DNA synthesis may stop. In such cases, the 3'→5' exonuclease activity catalyzes the excision of the erroneous nucleotide on the primer strand. The polymerase activity then catalyzes the formation of the correct base pair. Thus in DNA replication, 3'→5' exonuclease activity is a **proofreading mechanism** that helps keep the frequency of DNA replication errors very low. With proofreading, replication errors by DNA polymerase I or III are reduced to lower than 10^{-9}.

In addition, DNA polymerase I has 5'→3' exonuclease activity and can remove nucleotides from the 5' end of a DNA strand or of an RNA primer strand. This activity, also important in DNA replication, will be examined later in this chapter.

𝒦EYNOTE

The enzymes that catalyze the synthesis of DNA are called DNA polymerases. Three DNA polymerases, I, II, and III, have been identified in *E. coli*: I and III are known to be involved in DNA replication. The three enzymes differ in a number of properties, including size, number of molecules per cell, and proofreading ability.

Molecular Details of DNA Replication in Prokaryotes

Genetic and biochemical analyses have been used to identify many of the proteins—such as DNA polymerases—involved in the replication of DNA in *E. coli* and its phages and plasmids. Temperature-sensitive mutants have been particularly useful in enabling researchers to study biological functions that are vital to the cell since the functions stop or slow down significantly at elevated temperatures. Once temperature-sensitive mutants have been obtained, they are screened for the defective function involved, such as DNA replication. The genes involved are then mapped on the chromosomes and biochemical comparisons of wild-type and mutants are done to define and characterize the replication proteins encoded. Table 12.2 presents the functions of some of the *E. coli* replication genes isolated in this way, as well as some key DNA sequences involved in the replication process. Some of the encoded proteins function also in other cellular processes such as the repair of damaged DNA and genetic recombination. In this section, we discuss some of the molecular aspects of DNA replication in prokaryotes, focusing largely on *E. coli*.

INITIATION OF DNA REPLICATION. The initiation of DNA replication in prokaryotes requires the

~ TABLE 12.1

Comparison of the Structural and Functional Characteristics of the *E. coli* DNA Polymerases I, II, and III

DNA POLYMERASE	PROPERTIES					
	POLYMERIZATION: 5'→3'	EXONUCLEASE: 3'→5'	EXONUCLEASE: 5'→3'	MOLECULAR WEIGHT (DALTONS)	MOLECULES PER CELL (APPROXIMATELY)	STRUCTURAL GENES
I	Yes	Yes	Yes	103,000	400	*polA*
II	Yes	Yes	No	90,000	?	*polB*
III	Yes	Yes	No	Core of 130,000, 27,500, and 10,000 subunits; 7 other subunits[a]	10–20	*dnaE*, *dnaQ*, and *holE* for core

[a]Polymerase III consists of a catalytic core of α (alpha: 130,000 Da; *dnaE*), ∈ (epsilon: 27,500 Da; *dnaQ*; responsible for 3'→5' exonuclease activity), and θ (theta: 10,000 Da; *holE*), and 7 other subunits: τ (tau: 71,000 Da; *dnaX*), γ (gamma: 47,500 Da; *dnaX*), δ (delta: 35,000 Da; *holA*), δ' (delta prime: 33,000 Da; *holB*), χ (chi: 15,000 Da; *holC*), ψ (psi: 12,000 Da; *holD*), and β (beta: 40,600 Da; *dnaN*).

~ TABLE 12.2

Functions of Some of the Genes and DNA Sequences Involved in DNA Replication in *E. coli*

GENE PRODUCT AND/OR FUNCTION	GENE
DNA polymerase I	*polA*
DNA polymerase II	*polB*
DNA polymerase III	*dnaE, dnaQ, dnaX, dnaN, dnaD*
	holA→E
Initiator protein; binds to *oriC*	*dnaA*
IHF protein—DNA binding protein; binds to *oriC*	*himA*
FIS protein—DNA binding protein; binds to *oriC*	*fis*
Helicase and activator of primase	*dnaB*
Complexes with *dnaB* protein and delivers it to DNA	*dnaC*
Primase—makes RNA primer for extension by DNA polymerase III	*dnaG*
Single-stranded binding (SSB) proteins—bind to unwound single-stranded arms of replication forks	*ssb*
DNA ligase—seal single-stranded gaps	*lig*
Gyrase (type II topoisomerase)—replication swivel to avoid tangling of DNA as replication fork advances	*gyrA, gyrB*
Origin of chromosomal replication	*oriC*
Terminus of chromosomal replication	*ter*
TBP (*ter* binding protein)—stalls replication forks	*tus*

local separation of the two DNA strands (a process called *denaturation*) at a specific DNA sequence called an **origin of replication**.

Origin of Replication. Replication of prokaryotic and viral DNA usually starts at a specific site on the chromosome, the **origin**. At that site the double helix denatures into single strands, exposing the bases for the synthesis of new strands. The local denaturing of the DNA produces what is called a **replication bubble**. The segments of untwisted single strands upon which the new strands are made (following complementary base pairing rules) are called the **template strands**. In the circular *E. coli* chromosome there is a single origin of replication, called *oriC*, from which replication proceeds bidirectionally. The *oriC* region is located at about 84 minutes on the genetic map (see Figure 8.30, p. 257). The region has been cloned and sequenced. A minimal sequence of about 245 nucleotide pairs is required for the initiation of replication in *E. coli*.

INITIATION OF DNA SYNTHESIS. Figure 12.4 shows a model for the formation of a replication bubble at a replication origin in *E. coli*, and the initiation of the new DNA strand. This model is based on in vitro studies of DNA replication.

First, supercoiled DNA is relaxed by topoisomerase. Then, initiator proteins bind to the parental DNA molecule at the origin of replication sequence. The initiator proteins then wrap the DNA around them to form a large complex (Figure 12.4, part 1). Physical untwisting of the two DNA strands causes the double helix to denature in that region. The untwisting is catalyzed by the enzyme **DNA helicase** (product of the *dnaB* gene), which first becomes bound to the initiator proteins (Figure 12.4, part 2), then is loaded onto the DNA itself (Figure 12.4, part 3). The untwisting reaction requires energy derived from the hydrolysis of ATP. Next, the enzyme **DNA primase** (product of *dnaG* gene) binds to the helicase and the denatured DNA (Figure 12.4, part 4).

Primase is important in DNA replication because no known DNA polymerases can initiate the synthesis of a DNA strand; they can only catalyze the addition of deoxyribonucleotides to a preexisting strand. Primase is activated by DNA helicase, then synthesizes a short **RNA primer** (11 ± 1 nucleotides) that is required for the initiation of DNA synthesis (Figure 12.4, part 5). The primers start with two purine nucleotides, most frequently AG. The complex of the primase and the helicase with the DNA is called the **primosome**. The primers function as a preexisting polynucleotide chain to which new deoxyribonucleotides can be added by reactions catalyzed by DNA polymerase (Figure 12.4, part 5). The RNA primers themselves do not remain as part of the new DNA chain; they are

~ FIGURE 12.4

Schematic model for the formation of a replication bubble at a replication origin in *E. coli* and the initiation of the new DNA strand.

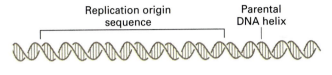

Replication origin sequence   **Parental DNA helix**

❶ Initiator proteins bind to replication origin

Initiator proteins

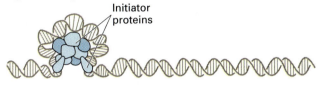

❷ DNA helicase binds to initiator proteins

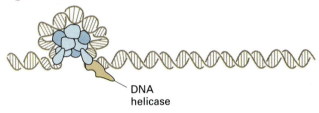

DNA helicase

❸ Helicase loads onto DNA

❹ Helicase denatures helix and binds with DNA primase to form primosome

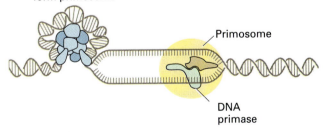

Primosome

DNA primase

❺ Primase synthesizes RNA primer, which is extended as DNA chain by DNA polymerase

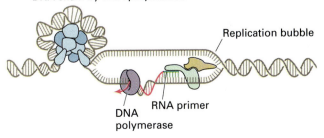

Replication bubble

DNA polymerase   RNA primer

---

removed and replaced with DNA by the action of DNA polymerase, as we shall discuss later.

We must be clear about the difference between *template* and *primer* with respect to DNA replication. A template strand is the one on which the new strand is based, following complementary base pairing rules; thus, the base sequence of the template strand directs the synthesis of a new strand with a complementary base sequence. A primer is a short segment of nucleotides bound to the template strand. The primer nucleotides serve as a substrate for the action of DNA polymerase, which extends the primer as a new DNA strand, the sequence of which is complementary to the template strand.

𝒦EYNOTE

No DNA polymerase can initiate the synthesis of a new DNA chain. Instead, the initiation of DNA synthesis first involves the denaturation of double-stranded DNA at an origin of replication, catalyzed by DNA helicase. Next, DNA primase binds to the helicase and the denatured DNA and synthesizes a short RNA primer. The RNA primer is extended by DNA polymerase as new DNA is made. The RNA primer is later removed.

SEMIDISCONTINUOUS DNA REPLICATION. When a double-stranded DNA molecule unwinds to expose the two single-stranded template strands for DNA replication, a Y-shaped structure called a **replication fork** is formed. A replication fork moves in one direction. This unidirectional movement poses a problem, since DNA polymerases can only make new DNA in the 5'-to-3' direction, yet the two DNA strands are of opposite polarity. As the DNA helix unwinds to provide templates for new synthesis, the new DNA cannot be polymerized continuously on the 3'-to-5' strand.

The work of Reiji and Tuneko Okazaki and their colleagues suggested a model to explain this result. They added a radioactive DNA precursor (^{3}H-thymidine) to cultures of *E. coli* for 0.5 percent of a generation time. Next, they added a large amount of nonradioactive thymidine to prevent the incorporation of any more radioactivity into the DNA. They followed what happened to the radioactively labeled thymidine. At various times they extracted the DNA and determined the size of the newly labeled molecules. At intervals soon after the labeling period, most of the radioactive ^{3}H-thymidine was present in relatively low-molecular-weight DNA about 100–1,000 nucleotides long. As time increased, a greater and greater proportion of the labeled molecules was found in DNA of a high molecular weight. These results indicated that DNA replication normally involves the synthesis of short DNA segments, called **Okazaki fragments**, which are subsequently linked together by the action of DNA polymerase to remove the RNA primers. Removal is followed by the action of an enzyme called **DNA ligase**, which catalyzes formation of the final phosphodiester bond between Okazaki fragments to form a long polynucleotide chain. In other words, Okazaki's group had shown that DNA replication is **discontinuous**.

We can relate the discontinuous nature of DNA synthesis to the replication of the circular DNA chromosome in *E. coli*. At *oriC* (see p. 352) the DNA denatures to expose the two template strands. Denaturation produces a replication bubble that is actually *two* Y-shaped replication forks linked head to head. Thus, DNA synthesis takes place in both directions (bidirectionally) away from the origin point. (Note that for some other circular chromosomes—certain plasmids, for example—replication is unidirectional.)

The early stages of **bidirectional replication** are shown in Figure 12.5. This figure also shows how discontinuous replication occurs. As the replication fork migrates, synthesis of one new strand (the **leading strand**), is continuous since the 3'-to-5' template strand (the leading strand template) is being copied. Synthesis of the other new strand (the **lagging strand**) must be discontinuous because the helix must untwist to expose a new segment of 3'-to-5' template (the lagging strand template) so that the new strand can be made in the 5'-to-3' direction. Since one new DNA strand is synthesized continuously and the other discontinuously, DNA replication as a whole is considered to occur in a **semidiscontinuous** fashion.

DNA Replication Model.   We will now consider a model for DNA replication which incorporates all of the facts we have just discussed (Figure 12.6). The model involves a single replication fork; however, the events described for a single replication fork also apply to the two replication forks formed during bidirectional replication of circular bacterial chromosomes.

The key steps in DNA replication are as follows:
1. *Relaxation of supercoiled DNA.* This event is catalyzed by topoisomerase and is important for allowing the replication fork to move.
2. *Denaturation and untwisting of the double helix* (Figures 12.4 and 12.6a). These events are catalyzed by DNA helicases. When bound to single-stranded DNA, helicase hydrolyzes ATP. This causes a change in shape of the enzyme molecule,

~ **FIGURE 12.5**

Bidirectional DNA replication. Synthesis of DNA is initiated at the origin and proceeds in the 5'-to-3' direction while the two replication forks are migrating in opposite directions. Replication is continuous on the leading strand and semidiscontinuous on the lagging strand. These events occur in the replication of all DNA in prokaryotes and eukaryotes.

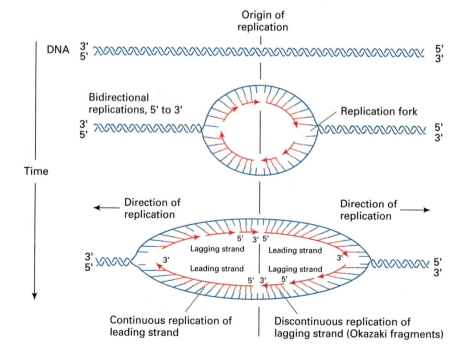

~ FIGURE 12.6

Model for the events occurring around the replication fork of the *E. coli* chromosome. (a) Untwisting; (b) Initiation; (c) Further untwisting and elongation of the new DNA strands; (d) Further untwisting and continued DNA synthesis; (e) Removal of the primer by DNA polymerase I; (f) Joining of adjacent DNA fragments by the action of DNA ligase. Green = RNA; red = new DNA.

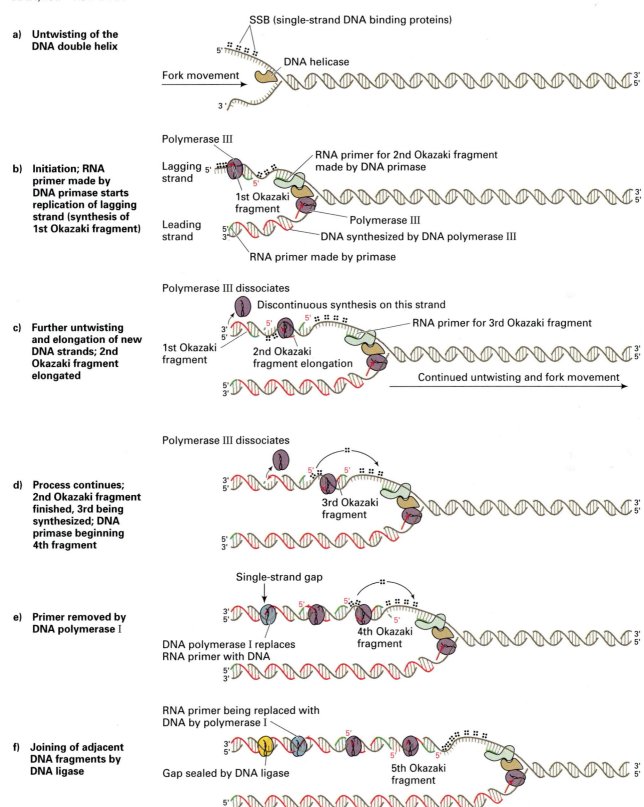

a) **Untwisting of the DNA double helix**

b) **Initiation; RNA primer made by DNA primase starts replication of lagging strand (synthesis of 1st Okazaki fragment)**

c) **Further untwisting and elongation of new DNA strands; 2nd Okazaki fragment elongated**

d) **Process continues; 2nd Okazaki fragment finished, 3rd being synthesized; DNA primase beginning 4th fragment**

e) **Primer removed by DNA polymerase I**

f) **Joining of adjacent DNA fragments by DNA ligase**

enabling the enzyme to move along a DNA single strand. By repeated ATP hydrolysis, the helicases can move along the single strand and untwist any double-stranded DNA they encounter. Two helicases are bound at the replication fork, one to the lagging-strand template strand and the other to the leading-strand template strand. Since the replication fork moves in one direction, the two helicases must move in opposite directions along their respective DNA strands, one in the 5'-to-3' direction, and the other in the 3'-to-5' direction.

3. *Stabilization of the single-stranded DNA in the replication fork.* In the single-stranded form, DNA is a flexible molecule. Therefore, as the helicases untwist the double helix, the resulting single-stranded DNA could potentially reform a double-helical molecule or fold on itself and form hydrogen bonds between bases at different parts of the same strand (called intramolecular base pairing). In either case, the double-stranded regions would impede the path of DNA polymerase as it synthesizes a new strand on the exposed DNA template. This is avoided by the action of **single-strand DNA-binding (SSB) proteins** (also called **helix-destabilizing proteins**) that bind to single-stranded DNA without covering the bases so they are still readable by DNA polymerase (Figure 12.6). The SSB proteins also help the DNA unwinding process by stabilizing the single-stranded DNA to which they bind. In *E. coli* the SSB protein (encoded by the *ssb* gene) is a tetramer of four identical subunits that binds to a 32-nucleotide segment of DNA. More than 200 of the proteins bind to each replication fork.

4. *Initiation of synthesis of new DNA strands.* The primase binds to helicase and the DNA to form the primosome and synthesizes the RNA primer (Figures 12.4 and 12.6b). The RNA primers are lengthened by the action of DNA polymerase III, which synthesizes new DNA chains complementary to the template strands (Figure 12.6b and 12.6c). To maintain the 5'-to-3' polarity of DNA synthesis, and one overall direction of replication fork movement, the direction of DNA synthesis is different on the two template strands. That is, the leading strand is synthesized in the same direction as the direction of fork movement while the lagging strand is synthesized in the opposite direction. Each new piece of DNA synthesized on the lagging strand template is an Okazaki fragment.

5. *Elongation of the new DNA strands.* The DNA is untwisted further by the helicases (Figure 12.6c). On the leading strand template (bottom strand of Figure 12.6), the new leading strand is synthe-

sized continuously as indicated above. Since DNA synthesis can only proceed in the 5'-to-3' direction, however, the DNA polymerase synthesis reaction on the lagging-strand template (top strand of Figure 12.6) has gone as far as it can. For DNA replication to continue on that strand, a new initiation of DNA synthesis must occur on the single-stranded template that has been produced by the unwinding of the double helix. As before, an RNA primer is made, and this occurs close to the replication fork, catalyzed by the primase which is still bound to the helicase. The primer is lengthened by the action of DNA polymerase III, which displaces SSB proteins as it synthesizes the new Okazaki fragment.

In Figure 12.6d the process repeats itself: The DNA untwists, continuous DNA synthesis occurs on the leading-strand template, and discontinuous DNA synthesis occurs on the lagging-strand template.

6. *Joining of Okazaki fragments on the lagging strand to make a continuous strand.* Eventually, the unconnected Okazaki fragments on the lagging-strand template are joined into a continuous DNA strand. This requires the activities of two enzymes, DNA polymerase I and **DNA ligase**. Consider two adjacent Okazaki fragments. The 3' end of the newer DNA fragment is adjacent to, but not joined to, the primer at the 5' end of the previously made fragment. DNA polymerase III dissociates from the DNA, and DNA polymerase I continues the 5'-to-3' synthesis of the newer DNA fragment made by DNA polymerase III, simultaneously removing the primer section of the older fragment (Figure 12.6e). The removal of the RNA primer takes place nucleotide by nucleotide by the 5'→3' exonuclease activity of DNA polymerase I.

When DNA polymerase I has completed replacement of RNA primer nucleotides with DNA nucleotides, a single-stranded gap exists between adjacent nucleotides on the DNA strand between the two fragments. The two fragments are joined into one continuous DNA strand by the enzyme DNA ligase. The result is a longer DNA strand (Figure 12.6f). The catalytic reaction of DNA ligase is diagrammed in Figure 12.7. The whole process is repeated until all the DNA is replicated.

In sum, DNA replication in *E. coli* is a complicated process. It has been shown that the key replication proteins are closely associated to form a **replication machine** or **replisome** (Figure 12.8). Figure 12.8 shows the lagging-strand DNA folded so that the DNA polymerase III of the lagging strand is com-

~ FIGURE 12.7

Action of DNA ligase in sealing the gap between adjacent DNA, fragments (e.g., Okazaki fragments) to form a longer, covalently continuous chain. The DNA ligase catalyzes the formation of a phosphodiester bond between the 3'-OH and the 5'-phosphate groups on either side of a gap, sealing the gap.

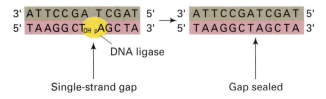

plexed with the DNA polymerase III on the leading strand. The folding also brings the 3' end of each completed Okazaki fragment near the site where the next Okazaki fragment will start. The primase-helicase complex (the primosome) moves with the fork, synthesizing new RNA primers as it proceeds. Similarly, because the lagging strand polymerase is complexed with the other replication proteins at the fork, that polymerase can be continually reused at the same replication fork, synthesizing a string of Okazaki fragments as it moves with the rest of the replication machine. That is, the complex of replication proteins that forms at the replication fork moves as a unit along the DNA, and enables new DNA to be synthesized efficiently on both the leading-strand template and lagging-strand template.

PROOFREADING: CORRECTING ERRORS IN DNA REPLICATION.

Occasionally an error is made in DNA replication and an incorrect nucleotide is incor-

porated into the DNA chain being synthesized. The mismatched base has a very high probability of being excised by the 3'→5' exonuclease activity of the DNA polymerase before the next base in the chain is added. This process, or proofreading ability, results in an extremely low error frequency in inserting the wrong base during DNA replication (less than 10^{-9} for DNA polymerases I and III). Somehow the mismatched base pair at the growing 3' end of the new DNA strand triggers the 3'→5' exonuclease activity, perhaps because it bulges like the frayed end of a rope rather than lying tightly beside the template strand. The DNA polymerase then backs up one base on the template strand, removing the incorrect base on the new strand. (This process resembles that of the backspace delete key on a computer keyboard which is used to erase the incorrect character.) The DNA polymerase then resumes the forward direction and inserts the correct character.

REPLICATION OF CIRCULAR DNA AND THE TWISTING PROBLEM.

In *E. coli* the parental DNA strands remain in a circular form throughout the replication cycle. This is true of many but not all circular DNA molecules. During replication, these circular DNA molecules exhibit a theta-like (θ) shape because of a replicating bubble's initiation at the replication origin (Figure 12.9).

As the two DNA strands in a circular chromosome untwist for replication, positive supercoils will form elsewhere in the molecule. By analogy, if you take a circular piece of double-stranded rope and try to separate the two strands at one point, the rope will become tightly supercoiled at the opposite side of the circle. For the replication fork to move, then, the

~ FIGURE 12.8

Model for the "replication machine," or replisome, the complex of key replication proteins, with the DNA at the replication fork. The DNA polymerase III on the lagging strand template (top of figure) is just finishing the synthesis of an Okazaki fragment.

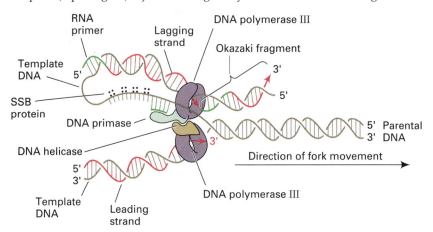

~ **FIGURE 12.9**

Bidirectional replication of circular DNA molecules.

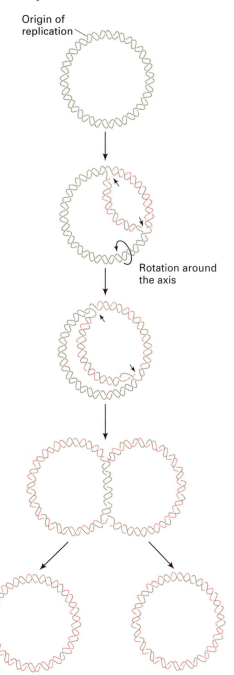

Origin of replication

Rotation around the axis

tively supercoiled DNA into relaxed DNA. Topoisomerases play an important role in the replication process by preventing excessively supercoiled DNA from forming and thereby allowing both parental strands to remain intact during the replication cycle as the replication fork migrates. That is, the unreplicated part of the theta structure ahead of the replication fork repeatedly has negatively supercoils introduced into it by the action of DNA gyrase (a type II topoisomerase) relieving the positive supercoiling that occurs as DNA is untwisted during replication. Figure 12.11 shows twisted replicating molecules of the circular DNA of the animal virus SV40: Topoisomerases function in this system to prevent excessive twisting.

ROLLING CIRCLE REPLICATION OF DNA. The rolling circle model of DNA replication applies to the replication of several viral DNAs such as ΦX174 and λ (see Figure 8.15, p. 240) and to the replication of the *E. coli* F factor during conjugation and transfer of donor DNA to a recipient (see Figures 8.8, p. 232, and 8.10, p. 235).

The rolling circle model for DNA replication is shown in Figure 12.12. The first step is the generation of a specific cut (nick) in one of the two strands at the origin of replication. The 5' end of the cut strand is then displaced from the circular molecule. This creates a replication fork structure and leaves a single-stranded stretch of DNA that serves as a template for the addition of deoxyribonucleotides to the free 3' end by DNA polymerase III using the intact circular DNA as a template. This new DNA synthesis occurs continuously as the 5' cut end continues to be displaced from the circular molecule; thus, the intact circular DNA is acting as the leading strand template.

The 5' end of the cut DNA strand is rolled out as a free "tongue" of increasing length as replication proceeds. (This is analogous to pulling out the end of a roll of paper towels.) This single-stranded DNA tongue becomes covered with SSB proteins. New DNA synthesis on the displaced DNA occurs in the 5'-to-3' direction; this means new synthesis occurs from the circle out toward the end of the displaced DNA. With further displacement, new DNA synthesis again must begin at the circle and move outward along the displaced DNA strand. Thus, synthesis on this strand occurs discontinuously; i.e., the displaced strand is the lagging strand template (see Figure 12.6). That is, primase synthesizes short RNA primers that are extended as DNA (Okazaki fragments) by DNA polymerase III. The RNA primers are ultimately removed and adjacent Okazaki fragments are joined through the action of DNA ligase. As the single-stranded DNA tongue rolls out, DNA synthesis continues on the circular DNA template.

chromosome ahead of the fork must rotate. Given a rate of movement of the replication fork of 500 nucleotides per second, at 10 base pairs per turn, the helix ahead of the fork has to rotate at 50 revolutions per second, or 3,000 rpm (Figure 12.10)!

The twisting problem is solved by the action of topoisomerases (see Chapter 11), enzymes that introduce negative supercoils into DNA or convert nega-

~ Figure 12.10

Illustration of the twisting problem in DNA replication. If the replication fork moves at 500 nucleotides per second, the DNA helix ahead of the fork rotates at 50 revolutions per second, or 3,000 rpm.

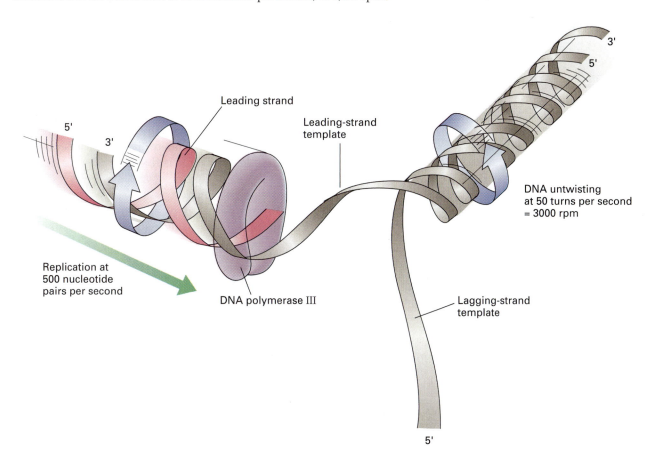

~ Figure 12.11

Twisted replicating molecules of circular SV40 DNA. (a) Electron micrograph of an early phase of replication; (b) Diagram showing the unreplicated, supercoiled parent strands and the portions already replicated.

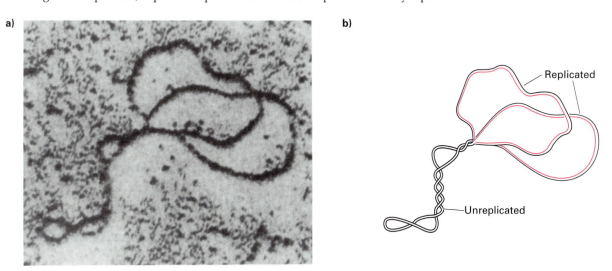

~ FIGURE 12.12

The replication process of double-stranded circular DNA molecules through the rolling circle mechanism. The active force that unwinds the 5' tail is the movement of the replisome propelled by its helicase components.

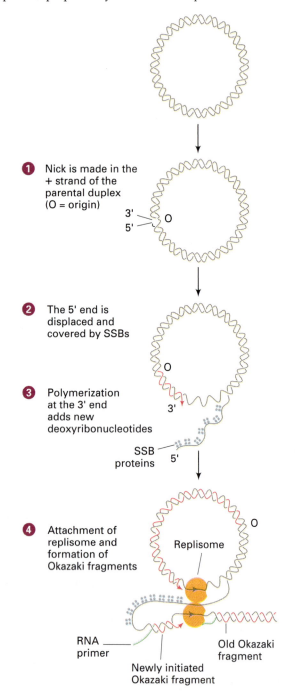

1 Nick is made in the + strand of the parental duplex (O = origin)

3'
5'
O

2 The 5' end is displaced and covered by SSBs

O

3 Polymerization at the 3' end adds new deoxyribonucleotides

3'

SSB proteins   5'

4 Attachment of replisome and formation of Okazaki fragments

Replisome   O

RNA primer

Old Okazaki fragment

Newly initiated Okazaki fragment

Since the parental DNA circle can continue to roll, it is possible to generate a linear double-stranded DNA molecule that is longer than the circumference of the circle. For example, in the later stages of phage lambda DNA replication, linear "tongues" are produced by rolling circle replication that are many times the cir-

cumference of the original circle. These molecules are cut into individual linear λ chromosomes by the *ter* enzyme (see Figure 11.10, p. 326) and those unit-length molecules are then packaged into phage heads.

Rolling circle DNA replication is also used in the transfer of DNA from donor to recipient cell during conjugation of *E. coli* cells (Figure 12.13). In this case, leading strand DNA synthesis serves to maintain a complete double-stranded copy of the *F* factor or *Hfr* donor chromosome in the donor cell. The displaced 5' tongue passes through the conjugation tube as a single-stranded molecule, entering the recipient cell in that form. It is not known, however, whether the single-stranded molecule is first converted to double-stranded DNA by lagging-strand synthesis and then undergoes recombination with the recipient's chromosome, or whether it recombines with the recipient's chromosome as a single-stranded molecule.

𝒦EYNOTE

Replication of DNA in *E. coli* requires at least two of the DNA polymerases and several other enzymes and proteins. The DNA helix is untwisted by DNA helicase to provide templates for the synthesis of new DNA. Since new DNA is made in the 5'-to-3' direction, chain growth is continuous on one strand and discontinuous (i.e., in segments that are later joined) on the other strand. This semidiscontinuous model is applicable to many other prokaryotic replication systems, each of which differs in the number and properties of the enzymes and proteins required. In the replication of circular DNA molecules, topoisomerases act to prevent the DNA from tangling ahead of the replication fork as the DNA untwists.

~ FIGURE 12.13

Transfer of single-stranded DNA from a donor bacterium into a recipient, via the rolling circle mechanism.

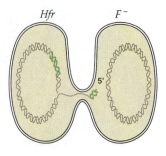

Hfr   *F⁻*

5'

DNA REPLICATION IN EUKARYOTES

The biochemistry and molecular biology of DNA replication are very similar in prokaryotes and eukaryotes. However, the added complication in eukaryotes is that DNA is not found in only one chromosome but is distributed among many chromosomes, each of which is a complex aggregate of DNA and proteins. In each cell division cycle, each of these chromosomes must be faithfully duplicated and a copy of each distributed to each of the two progeny cells. This means that both the DNA and the histones must be doubled with each division cycle.

The cell cycle is qualitatively the same from eukaryote to eukaryote and tissue to tissue within one individual, although there are significant differences both in the relative amount of time spent in each phase of the cycle and in the total time spent in a particular cycle. Among higher eukaryotes, for example, some cells divide once every 3 hours and some cells divide once every 200 hours; human cells in culture divide approximately once every 24 hours.

Recall from Chapter 3 that the cell cycle in most somatic cells of higher eukaryotes is divided into four stages: gap 1 (G_1), synthesis (S), gap 2 (G_2), and mitosis (M) (see Figure 3.5, p. 52). The DNA replicates and the chromosomes duplicate during the S phase, and the progeny chromosomes segregate into daughter cells during the M phase. Although DNA is neither replicating nor being segregated during G_1 and G_2, the cell is metabolically active and is growing. Events involved in the subsequent synthesis and mitotic activity also occur during G_1 and G_2.

Collectively, the G_1, S, and G_2 phases of the eukaryotic cell cycle constitute interphase, the time between divisions, while division (mitosis or meiosis) itself occurs within the M phase.

Genetic and molecular studies have elucidated a number of steps involved in controlling the progression of the cell through the cell cycle. A full discussion of this belongs more appropriately in a cell biology text, so only a simplified overview is given here. The focus will be on yeasts, for which most details are known; mammalian systems work similarly, although they seem to involve more complex components.

Progression through the cell cycle is tightly controlled through the functions of products of many genes in an elaborate system of checks and balances (Figure 12.14). As a cell proceeds through G_1, it is gearing up for the process of DNA replication and chromosome duplication in the S phase. A major checkpoint in G_1 determines whether the cell is able to or should continue into S. In yeasts, this checkpoint is known as

START, and in mammalian cells it is known as the G_1 checkpoint. Unless a cell grows to a large enough size and the environment is favorable, it will stay in G_1. Another major checkpoint, called the G_2 checkpoint both in yeasts and mammalian cells, occurs at the junction between G_2 and M. Unless all the DNA has replicated, the cell is big enough, and the environment is favorable, the cell is unable to enter the mitotic phase of the cell cycle. A third checkpoint occurs during M. That is, the chromosomes must be attached properly to the mitotic spindle in order to trigger the separation of chromatids and the completion of mitosis.

Key components involved in the regulatory events that occur at the checkpoints are proteins known as *cyclins* (named because their concentration increases and decreases in a regular pattern through the cell cycle; changes in the rates of synthesis and proteolysis of the proteins produce the pattern) and enzymes known as *cyclin-dependent kinases* (Cdks). In yeasts, a single Cdk functions at both the G_1 and the G_2 checkpoints, while in mammalian cells at least two Cdks are involved at each checkpoint. In budding yeast (*Saccharomyces cerevisiae*) the Cdk is called CDC28 after the gene that encodes it, while in fission yeast (*Schizosaccharomyces pombe*) it is called cdc2. Specificity of the activity at each checkpoint is determined by the type of cyclins involved.

At START, one or more G_1 *cyclins* bind to the CDC28/cdc2 kinase and activate it. The Cdk then phosphorylates key proteins that are needed for progression into S. Once the cyclin has activated the Cdk, the level of the cyclins decreases as a result of increased proteolysis. A similar process occurs at the G_2 checkpoint, when one or more *mitotic cyclins* bind to the Cdk to form the *M-phase promoting factor* (MPF). Then, when other enzymes phosphorylate and dephosphorylate it, MPF is activated and stimulates the events needed to move the cell into M. During mitosis, just after metaphase, the mitotic cyclin is degraded, which leads to the inactivation of MPF and allows the cell to complete mitosis.

As indicated at the beginning of this discussion, the regulatory events controlling the cell cycle are complex. Many more steps and pathways are involved than have been described. We will return to this subject when we discuss the genetics of cancer in Chapter 18, for some genes that normally control steps in the cell cycle are often mutated in cancer.

Molecular Details of DNA Synthesis in Eukaryotes

As we saw earlier, many of the enzymes and proteins involved in prokaryote DNA replication have been identified. Less is known about the enzymes and pro-

~ **FIGURE 12.14**

Some of the molecular events that control progression through the cell cycle in yeasts.

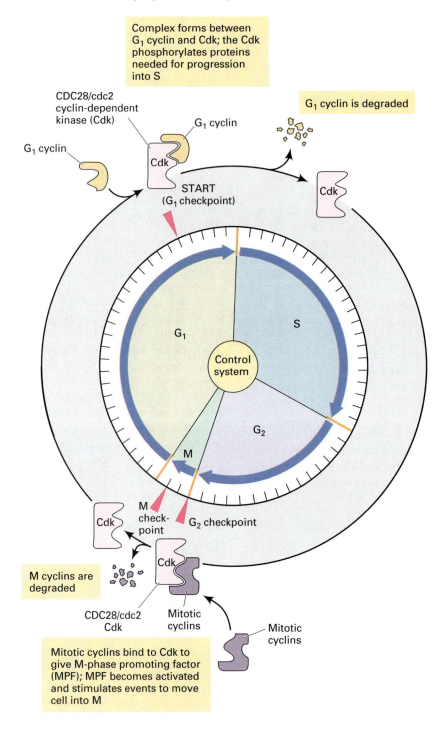

teins involved in eukaryotic DNA replication. It is clear, however, that the sequential steps described for DNA synthesis in prokaryotes also occur for DNA synthesis in eukaryotes, namely, denaturation of the DNA double helix and the semiconservative, semi-

discontinuous replication of the DNA. Next we discuss the features of DNA replication that are largely unique to eukaryotes; we will not repeat the molecular details of DNA replication that are common between prokaryotes and eukaryotes.

SEMICONSERVATIVE DNA REPLICATION IN EU-KARYOTES. Semiconservative replication of DNA in eukaryotic chromosomes can be visualized using a special staining procedure. The experimental system is Chinese hamster ovary (CHO) cells growing in tissue culture. At the beginning of the experiment, the chemical 5-bromodeoxyuridine (BUdR) is added. BUdR is a *base analog*, with a structure very similar to a normal base in DNA, which is thymine in this case. During replication, wherever a T is called for in the new strand, BUdR can be incorporated instead. After two rounds of replication in the presence of BUdR, the mitotic chromosomes are stained with a fluorescent dye and Giemsa stain with the result shown in Figure 12.15. Sister chromatids are visible, one darkly stained and one lightly stained. The dark-light staining pattern brings to mind the costumes of harlequins, hence the chromosomes are called **harlequin chromosomes.** We can interpret the photograph with the knowledge that a chromatid with a DNA double helix consisting of two BUdR-labeled strands stains less intensely than a chromatid with a DNA double helix consisting of one BUdR-labeled strand and one T-labeled strand. That is, if semiconservative replication occurs, after one round of replication two progeny DNAs will be produced, each with one T-labeled strand and one BUdR-labeled strand. Then, after a second round of replication in the presence of BUdR, the result will be one sister chromatid consisting of a DNA with one T-labeled and one BUdR-labeled strand (darkly staining), and the other sister chromatid consisting of a DNA with two BUdR-labeled strands (lightly staining). The staining pattern observed is consistent with DNA replication occurring in a semiconservative manner.

INITIATION OF DNA REPLICATION. Replicons. Each eukaryotic chromosome consists of one linear DNA double helix. If there was only one origin of replication per chromosome, the replication of each chromosome would take many, many hours. For example, there are 2.75×10^9 base pairs of DNA in the haploid human genome (23 chromosomes), and the average chromosome is roughly 10^8 base pairs long. With a replication rate of 2 kilobases (2,000 bases) per minute in human cells, it would take approximately 830 hours to replicate one chromosome. If each cell cycle were at least that long for a developing human embryo, the gestation period would be many years instead of nine months.

Actual measurements show that the chromosomes in eukaryotes replicate much faster than would be the case with only one origin of replication per chromosome. The diploid set of chromosomes in *Drosophila* embryos, for example, replicates in 3 minutes. This is 6 times faster than the replication of the *E. coli* chromosome, even though there is about 100 times more DNA in *Drosophila* than there is in *E. coli*.

The rapidity of eukaryotic chromosome duplication is possible because DNA replication is initiated at many origins of replication throughout the genome. At each origin of replication, the DNA denatures (as in *E. coli*). Replication proceeds bidirectionally, and the DNA double helix opens to expose single strands that act as templates for new DNA synthesis. Eventually, each replication fork will run into an adjacent replication fork, initiated at an adjacent origin of replication. In eukaryotes the stretch of DNA from the origin of replication to the two termini of replication (where adjacent replication forks fuse) on each side of the origin is called a **replicon** or a **replication unit.** Figure 12.16a presents an electron micrograph showing a large number of replicons on a piece of *Drosophila* DNA; Figure 12.16b is an interpretive drawing of that micrograph. The piece of DNA shown is about 500 kb long (about a tenth the size of the *E. coli* chromosome), and many replicons are present. Table 12.3 presents the number of replicons, their average size, and the rate of replication fork movement for a number of organisms. Note that the replicon size is much smaller, and the rate of fork movement is much slower in eukaryotic organisms

~ **FIGURE 12.15**

Visualization of semiconservative DNA replication in eukaryotes. Shown are harlequin chromosomes in Chinese hamster ovary cells that have been allowed to go through two rounds of DNA replication in the presence of the base analog 5-bromodeoxyuridine, followed by staining.

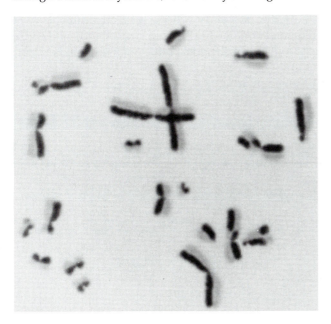

~ FIGURE 12.16

Replicating DNA of *Drosophila melanogaster*: (a) Electron micrograph showing replication units (replicons); (b) An interpretation of the electron micrograph shown in (a).

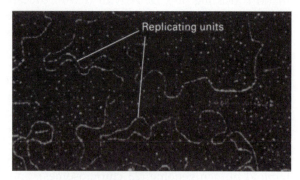

a)

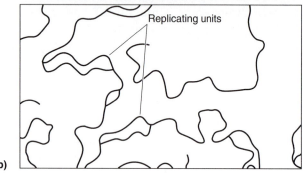

b)

than in bacteria. In general, each replicating eukaryotic chromosome has many replicons. Also, the number of origins and the average size of the replicon can vary with the nature of the cell, for example, its growth rate and tissue of origin.

Replication of DNA does not occur simultaneously in all the replicons in an organism's genome. Instead, a temporal ordering of initiation events occurs during the replication phase. For many cell types a particular temporal ordering of initiation is characteristic of the cell type; that is, the same pattern

occurs, cell generation after cell generation. The diagram in Figure 12.17 represents the temporal ordering of initiation phenomena from autoradiographic studies. The figure shows one segment of one chromosome in which there are three replicons that always begin replicating at distinct times. When the replication forks fuse at the margins of adjacent replicons, the chromosome has replicated into two sister chromatids. In general, replication of a segment of chromosomal DNA occurs following the synchronous activation of a cluster of origins.

Replication Origins.   Are there specific sequences that function as origins of replication in eukaryotes? In the yeast *Saccharomyces cerevisiae*, specific sequences have been identified that, when they are included as part of an extrachromosomal, circular DNA molecule, confer upon that molecule the ability to replicate autonomously within the yeast cell. These sequences are called **autonomously replicating sequences**, or **ARSs**. A variety of sequences from other eukaryotic organisms are also able to function as ARSs in yeast, although there is no evidence that these DNA fragments are origins of replication in the organisms from which they were isolated.

It is believed that at least some ARSs do correspond to origins of replication that function within the cell. On the other side of the coin, it appears that not all ARSs that are known to be located along a well-mapped chromosomal segment are used for replication initiation. Molecular analysis has shown that ARSs consist of multiple elements. Four elements have been identified: A, B1, B2, and B3. The A element is found in all ARSs along with two or more of the other elements. The A element plays an essential role in the function of the ARS, while the other elements affect the frequency with which an origin is used. Organizationally, the A along with the B1 and/or B2 elements form the core of the ARS to which a multi-

~ TABLE 12.3

Comparison of Bacterial and Eukaryote Replicons

ORGANISM	NO. OF REPLICONS	AVERAGE LENGTH	FORK MOVEMENT
Bacterium (*E. coli*)	1	4,200 kb	50,000 bp/min
Yeast (*Saccharomyces cerevisiae*)	500	40 kb	3,600 bp/min
Fruit fly (*Drosophila melanogaster*)	3,500	40 kb	2,600 bp/min
Toad (*Xenopus laevis*)	15,000	2,000 kb	500 bp/min
Mouse (*Mus musculus*)	25,000	150 kb	2,200 bp/min
Plant (bean: *Vicia faba*)	35,000	300 kb	n.a.[a]

[a] n.a.—value not available.

~ FIGURE 12.17

Temporal ordering of DNA replication initiation events in replication units of eukaryotic chromosomes.

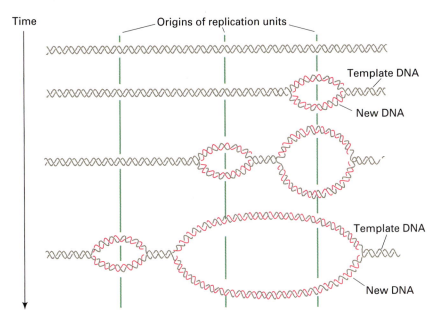

protein origin recognition complex (ORC) binds; the B3 element enhances replication through the binding of Abf1 (ARS-binding factor 1).

In mammals and other complex eukaryotes, origins of replication sequences are not as well defined, although progress is being made in characterizing some common elements in replicons of a number of eukaryotes.

Denaturation of DNA. As in prokaryotes, topoisomerases are used to facilitate DNA untwisting during the replication process. As we saw earlier, in *E. coli*, enzymes called helicases catalyze the untwisting of the DNA double helix to expose single strands of DNA that act as templates in DNA replication. Untwisting enzymes appear to be involved in eukaryotic DNA replication as well. For example, helicases have been identified in a variety of eukaryotic systems including those of humans, mice, rats, calves, and higher plants. The activity of these enzymes increases about twenty-fold when a cell begins DNA synthesis, which indicates that they might have a role in DNA replication; however, they have not yet been shown directly to be involved.

Once the helix unwinds and single-stranded regions are exposed, we expect that proteins similar to the SSB (single-strand binding) proteins in prokaryotes might attach and prevent the DNA from renaturing. Since a large number of proteins have the ability to stick to DNA (such as histones and DNA

polymerases), it is difficult to identify exactly which proteins are SSBs.

Single-strand binding proteins *have* been isolated from a number of eukaryotic cells such as yeast, fungi, amphibians, rodents, calf, and human cells. In mammalian cells the SSB proteins have been shown to bind to the single-stranded DNA and leave the bases exposed while simultaneously keeping the DNA in an extended configuration and increasing the affinity of the DNA polymerase for the template. Thus these SSB proteins stimulate polymerase activity (at least in mammals) as measured by the amount of DNA synthesized.

REPLICATION ENZYMES AND PROTEINS. Five different DNA polymerases have been identified in mammalian cells, designated α (alpha), β (beta), δ (delta), γ (gamma) and ∈ (epsilon). Their properties are summarized in Table 12.4. The α, β, δ, and ∈ polymerases are located in the nucleus, while the γ polymerase is found in the mitochondrion. DNA polymerases α and δ are the two enzymes responsible for nuclear DNA replication. They function essentially like the *E. coli* DNA polymerase III. DNA polymerase β serves a DNA repair function, and DNA polymerase γ catalyzes mitochondrial DNA replication. Both nuclear replication polymerases α and δ can use RNA primers for the initiation of DNA synthesis; δ can also use a DNA primer. The mitochondrial polymerase γ uses an RNA primer. Uniquely among the

~ TABLE 12.4

Properties of DNA Polymerases in Mammalian Cells

Property	DNA POLYMERASE				
	α	β	δ	ε	γ
Location in Cell:					
Nucleus	+	+	+	+	−
Mitochondrion	−	−	−	−	+
Polymerization activity:					
5'-to-3'	+	+	+	+	+
Primase associated with enzyme:	+	−	−	−	−
Exonuclease activity:					
3'-to-5' proofreading	−	−	+	+	+

DNA polymerases, the α polymerase also has primase activity; a separate enzyme is used for synthesizing a primer in DNA synthesis catalyzed by the other DNA polymerases. It is not known whether or not these enzymes have 5'-to-3' exonuclease activity. Only δ, ε, and γ have been shown to have proofreading (3'-to-5' exonuclease) activity.

Three DNA polymerases—α, δ, and ε—have been identified and studied in yeast. These polymerases are encoded by the *POL1*, *POL2*, and *POL3* genes, respectively. Mutation studies have shown that all three genes are essential for viability, and that all three DNA polymerases are required for chromosomal replication. For example, when temperature-sensitive *pol1* mutant yeast strains are shifted to the nonpermissive temperature, growth stops because DNA replication is blocked, indicating that DNA polymerase α is indispensable for that process.

₭EYNOTE

In eukaryotes DNA replication occurs in the S phase of the cell cycle and is similar to the replication process in prokaryotic cells. Synthesis of DNA is initiated by RNA primers, occurs in the 5'-to-3' direction, is catalyzed by DNA polymerases, requires a large number of other enzymes and proteins, and is a semiconservative and semidiscontinuous process. Replication of DNA is initiated at a large number of sites throughout the chromosomes.

Replicating the Ends of Chromosomes

The nuclear chromosomes of eukaryotes are linear with one double-stranded DNA running from one end

to the other. We learned earlier in this chapter that DNA polymerases can only synthesize new DNA by extending a primer that is made by primase. This means there are special problems in replicating the ends of eukaryotic chromosomes, as illustrated schematically in Figure 12.18. A parental chromosome (Figure 12.18a) is replicated semiconservatively. After replication (Figure 12.18b), segments of new DNA still with their RNA primers are hydrogen-bonded to the parental strands. Most of the RNA primers are removed, the resulting gaps are filled in by DNA polymerase, and the new DNA fragments are joined by DNA ligase. However, at the very 5' end of each new chromosomal DNA strand (at the chromosome telomeres), the gaps left by removal of the RNA primers remain because DNA polymerase cannot initiate new DNA synthesis. If nothing was done about these gaps, then in successive rounds of DNA replication, the chromosomes would get shorter and shorter.

In fact, there is a special mechanism for replicating the ends of the chromosomes. Recall that most eukaryotic chromosomes have tandemly repeated, simple sequences at their telomeres (see Chapter 11, p. 337). The work of Elizabeth Blackburn and Carol W. Greider has shown that an enzyme called *telomerase* maintains chromosome lengths by adding telomere repeats to the chromosome ends (Figure 12.19). This mechanism does not involve the regular replication machinery. Let us see how this occurs. In the protozoan *Tetrahymena* the repeated sequence is 5'-TTGGGG-3' reading toward the end of the DNA on the top strand in Figure 12.19. We start the special events at the stage shown in Figure 12.18c; that is, where a chromosome end has been produced with a gap at the end of the chromosome at the 5' end of the new DNA (Figure 12.19a). Telomerase is an enzyme made up of both protein and RNA. The RNA component of telomerase includes a base sequence that is complementary to the telomere repeat unit. Because of this, the telomerase binds specifically to the overhanging telomere repeat at the end of the chromosome (Figure 12.19b). Next, the telomerase catalyzes the synthesis of three nucleotides of new DNA—TTG—using the telomerase RNA as a template (Figure 12.19c). The telomerase then slides toward the end of the chromosome so its AAC at the 3' end of the RNA template now pairs with the newly synthesized TTG on the DNA (Figure 12.19d). Telomerase then makes the rest of the TTGGGG telomere repeat (Figure 12.19e). The process is repeated to add more telomere repeats. In this way, the chromosome is lengthened by the addition of a number of telomere repeats. Then, by primer synthesis and DNA synthesis catalyzed by DNA polymerase in the conventional way, the former gap is filled in, and the new chromo-

~ FIGURE 12.18

The problem of replicating completely a linear chromosome in eukaryotes. (a) Schematic diagram of a parent double-stranded DNA molecule representing the full length of a chromosome; (b) After semiconservative replication, new DNA segments hydrogen-bonded to the template strands have RNA primers at their 5' ends; (c) The RNA primers are removed, DNA polymerase fills the resulting gaps, and DNA ligase joins the adjacent fragments. However, at the two telomeres there are still gaps at the 5' ends of the new DNA resulting from RNA primer removal because no new DNA synthesis could fill them in.

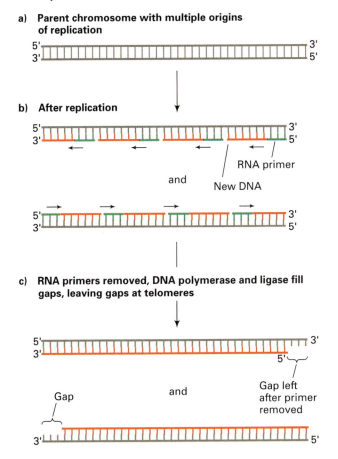

a) **Parent chromosome with multiple origins of replication**

b) **After replication**

RNA primer

and

New DNA

c) **RNA primers removed, DNA polymerase and ligase fill gaps, leaving gaps at telomeres**

Gap

and

Gap left after primer removed

somal DNA is lengthened (Figure 12.19f). After removal of the RNA primer a new 5' gap is left (Figure 12.19g), but any net shortening of the chromosome has been averted.

Introducing mutant telomerase RNA genes with certain template bases changed into cells showed that telomerase RNA is used as a template for the synthesis of new chromosomal telomere repeats. The new repeats made by the mutated telomerase had sequences complementary to the altered RNA, rather than the normal sequence.

Telomere length, while not identical from chromosome end to chromosome end, is nonetheless regulated to an average length for the organism and cell type. In wild-type yeast, for example, the simple telomere sequences (TG_{1-3}) occupy an average of about 300 bp. Mutants that affect telomere length have been identified. For example, deletion mutations of the *TLC1* gene (that encodes the telomerase RNA) or mutations of the *EST1* (*Ever Shorter Telomeres*) gene cause telomeres to shorten continuously until the cells die. This phenotype provides evidence that telomerase activity is required for long-term cell viability. The product of the *EST1* gene, the protein Est1, is either a component of the telomere RNA-protein complex or a separate factor essential for telomerase function. Mutations of the *TEL1* and *TEL2* genes cause cells to maintain their telomeres at a new, shorter-than-wild-type length. Thus, it is clear that telomere length is regulated genetically.

Current evidence suggests many levels of regulation of telomere activity and telomere length. Significant attention is being given, for example, to the observation that telomerase activity in mammals is limited to immortal cells (such as tumor cells). The absence of telomerase activity in other cells results in progressive shortening of chromosome ends during successive divisions because of the failure to replicate those ends, and in a limited number of cell divisions before the cell will die.

KEYNOTE

Special enzymes—telomerases—are used to replicate the ends of chromosomes in eukaryotes. A telomerase is a complex of proteins and RNA. The RNA acts as a template for synthesis of the complementary telomere repeat of the chromosome.

Assembly of New DNA into Nucleosomes

Eukaryotic DNA is complexed with histones in nucleosome structures, which are the basic units of chromosomes (see Chapter 11). Therefore, when the DNA is replicated, the histone complement must be doubled so that all nucleosomes are duplicated. This involves two processes: the synthesis of new histone proteins and the assembly of new nucleosomes.

HISTONE SYNTHESIS. Most histone synthesis is coordinated with DNA replication. The transcription of the genes for the five histones is initiated near the end of the G_1 phase, just prior to S. Translation of the histone mRNAs occurs throughout S, producing the histones to be assembled into nucleosomes as the chromosomes are duplicated. As DNA replication

~ FIGURE 12.19

Synthesis of telomeric DNA by telomerase. The example is for *Tetrahymena* telomeres. The process is described in the text. (a) Starting point: chromosome end with 5' gap left after primer removal; (b) Binding of telomerase to the overhanging telomere repeat at the end of the chromosome; (c) Synthesis of three-nucleotide DNA segment at chromosome end using the RNA template of telomerase; (d) The telomerase moves so that the RNA template can bind to the newly synthesized TTG in a different way. (e) Telomerase catalyzes the synthesis of a new telomere repeat using the RNA template. The process is repeated to add more telomere repeats. (f) After telomerase has left, new DNA is made on the template starting with an RNA primer. After the primer is removed (g), the result is a longer chromosome than at the start, with a new 5' gap.

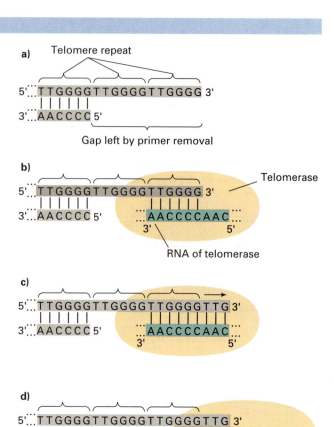

approaches completion, transcription of histone mRNAs is turned off, and the already synthesized mRNAs are soon broken down by normal cellular processes. As a result, histone synthesis ceases.

NUCLEOSOMES AND NEWLY SYNTHESIZED DNA.

Newly replicated DNA from *Drosophila* has been examined by electron microscopy to determine the distribution of nucleosomes. The results of these studies show that the diameter of the nucleosomes and their spacing along the DNA fibers produces the identical, beaded chromatin appearance in both unreplicated DNA and newly replicated DNA. In other words, new DNA is assembled into nucleosomes virtually immediately.

Nonetheless, for replication to proceed, nucleosomes must disassemble during the short time when a replication fork passes. Measurements indicate that there is a nucleosome-free zone about 200 to 300 base pairs around replication forks. But, are new nucleosomes made from an all-new set of histones with the old nucleosomes conserved (conserved nucleosome segregation), or do old and new histones become mixed in both the old and new nucleosomes (semiconserved nucleosome segregation)? To distinguish between these two models, experiments were done in which cells were grown in a medium containing heavy (density labeled) amino acids so that histones and, therefore, nucleosomes were heavy. Then the cells were transferred to a medium containing light (nondensity labeled) amino acids so that new histones would have normal density. Nucleosomes were then analyzed for their density. The prediction was

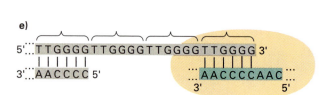

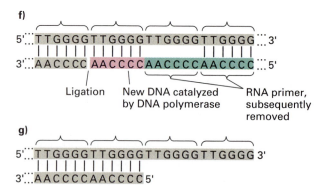

that, if nucleosome segregation was conserved, there would be two populations of nucleosomes; one would have heavy density (the old nucleosomes before the replication forks) and the other would have normal density (the new nucleosomes assembled after the replication forks have passed). If nucleosome segregation was semiconserved, the prediction was that there would be one population of nucleosomes of intermediate density between heavy and normal nucleosomes because of the reassembly of nucleosomes from old and newly synthesized histones. The experimental data indicated nucleosomes of intermediate density, suggesting that the semiconserved nucleosome segregation model is correct. However, this model is not completely accepted. Data from different kinds of experiments have been interpreted to mean that the conserved model is correct, with complete nucleosome cores going randomly to the two DNA strands after the replication fork.

KEYNOTE

Eukaryotic nuclear DNA is complexed with histones and organized into nucleosomes, the basic units of chromosomes. When the DNA replicates, new histones are synthesized and are assembled into nucleosomes. It is not known precisely how nucleosomes segregate from cell generation to cell generation.

DNA RECOMBINATION

In chapters 5, 6, and 8, we discuss genetic recombination and the mapping of genes in prokaryotes and eukaryotes. We learned that crossing-over involves the physical exchange of homologous chromosome parts as a result of breakage and rejoining of homologous DNA double helices. Now that we have an understanding of DNA structure (Chapter 10), the organization of DNA into chromosomes (Chapter 11), and DNA replication (this chapter), we return to a consideration of genetic recombination, this time discussing some molecular aspects.

Crossing-over: Breakage and Rejoining of DNA

Crossing-over in eukaryotes occurs in prophase I of meiosis (see Chapter 3, p. 57). Crossing-over involves the breaking and rejoining of DNA at the same position in two homologous DNA molecules. No base pairs are added to or deleted from the DNA as a result of the breakage and rejoining events. Thus a very early step in recombination is the generation of a break in each double helix involved. One possible sequence of events is that cellular enzymes introduce breaks along a pair of synapsed chromosomes in meiosis, thereby stimulating the recombination process. Some evidence for this is that the artificial introduction of breaks in DNA, such as by UV- or X-irradiation, results in a significant stimulation of crossing-over. However, exact molecular details of the initiation of recombination are not known at this time.

KEYNOTE

In both prokaryotes and eukaryotes, genetic recombination involves the breakage and rejoining of homologous DNA double helices.

The Holliday Model for Recombination

In the mid-1960s Robin Holliday proposed a model for reciprocal recombination. Since then, the *Holliday model* has been refined by other geneticists, notably Matthew Meselson and Charles Radding, and T. Orr-Weaver and Jack Szostak. To give a flavor for the recombination process at the molecular level, we present the Holliday model here.

The Holliday model is diagrammed in Figure 12.20 for genetically distinguishable homologous chromosomes, one with alleles a^+ and b^+ at opposite ends, and the other with alleles a and b. The two DNA double helices in the figure participate in the recombination event. The first stage of the recombination process is *recognition and alignment* (Figure 12.20, part 1), in which two homologous DNA double helices become precisely aligned so that no genetic material is added or deleted to either DNA in the subsequent exchange events. (In eukaryotes, this synapsis is accompanied by the formation of a synaptonemal complex, which mediates the meiotic exchange of genetic information. See Chapter 3.) In the second stage, one strand of each double helix breaks; each broken strand then invades the opposite double helix and base pairs with the complementary nucleotides of the invaded helix (Figure 12.20, part 2). Each of these steps is catalyzed by specific enzymes. The process leaves gaps that are sealed by DNA polymerase and DNA ligase, producing what is called a *Holliday intermediate*, with an internal branch point

~ FIGURE 12.20

Holliday model for reciprocal genetic recombination. Shown are two homologous DNA double-helices that participate in the recombination process.

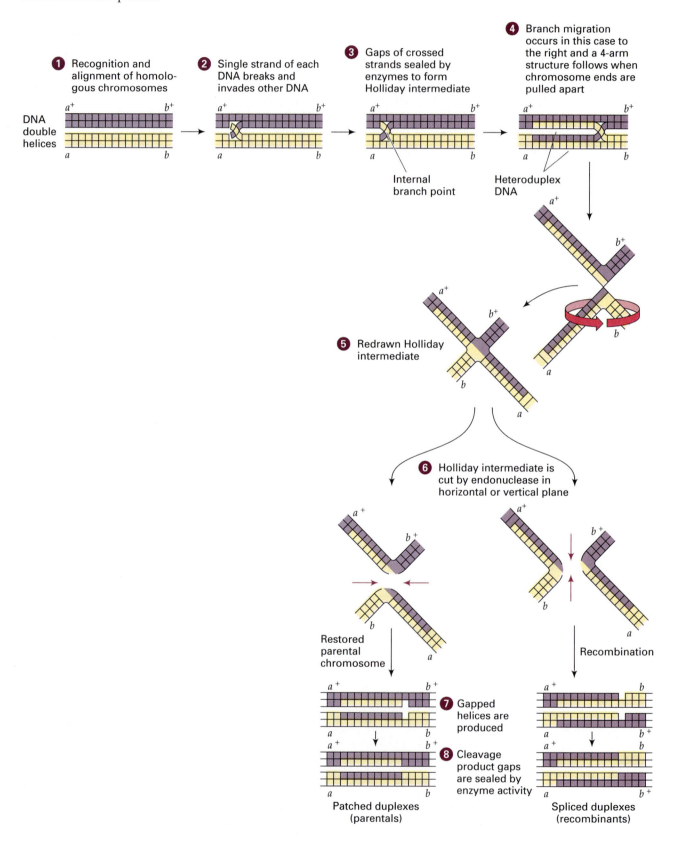

1 Recognition and alignment of homologous chromosomes

2 Single strand of each DNA breaks and invades other DNA

3 Gaps of crossed strands sealed by enzymes to form Holliday intermediate

4 Branch migration occurs in this case to the right and a 4-arm structure follows when chromosome ends are pulled apart

Internal branch point

Heteroduplex DNA

5 Redrawn Holliday intermediate

6 Holliday intermediate is cut by endonuclease in horizontal or vertical plane

Restored parental chromosome

Recombination

7 Gapped helices are produced

8 Cleavage product gaps are sealed by enzyme activity

Patched duplexes (parentals)

Spliced duplexes (recombinants)

(Figure 12.20, part 3). The hybrid DNA molecules evident at this stage are called *heteroduplexes*; that is, the two strands of the double-stranded DNA molecules do not have completely complementary sequences. The two DNA double helices in the Holliday intermediate can rotate, causing the branch point to move to the right or to the left. Figure 12.20, part 4, shows a branch migration event that has occurred to the right. The four-armed structure for the DNA strands is produced simply by pulling the four chromosome ends apart. Holliday intermediate structures have been demonstrated to be very stable, with no or very few unpaired bases, even in the region of the branch. Figure 12.21 shows an electron micrograph of a Holliday intermediate.

The result of branch migration is the generation of complementary regions of hybrid DNA in both double helices (diagrammed as stretches of DNA helices with two different colors in Figure 12.20, parts 5 through 8). Careful measurements have indicated that branch migrations can occur very rapidly, perhaps moving as much as 50 base pairs per second in eukaryotes. Thus, long regions of hybrid DNA could be formed in the few seconds that might elapse between the formation of a Holliday intermediate and the cleavage and ligation phase of the model that occurs next. These heteroduplexes may well contain mismatched base pairs; the mismatches are subject to repair.

The *cleavage and ligation* phase of recombination is best visualized if the Holliday intermediate is redrawn so that no DNA strand passes over or under another DNA strand. Thus, if the four-armed Holliday intermediate following Figure 12.20, part 4 is taken as a starting point and the lower two arms are rotated 180° relative to the upper arms, the structure shown in Figure 12.20, part 5 is produced.

Next, the Holliday intermediate is cut by enzymes at two points in the single-stranded DNA region of the branch point (Figure 12.20, part 5). The cuts can be in either the horizontal or vertical planes; both kinds of cuts occur with equal probability. Endonuclease cleavage in the horizontal plane (Figure 12.20, part 6, left) produces the two double helices shown in Figure 12.20, part 7, left. In each helix is a single-stranded gap. DNA ligase seals the gaps to produce the double helices shown in Figure 12.20, part 8, left. Since each of the resulting helices contains a segment of single-stranded DNA from the other helix, flanked by nonrecombinant DNA, these double helices are called *patched duplexes*.

If the endonuclease cleavage in Figure 12.20, part 5 is in the vertical plane (Figure 12.20, part 6, right), the gapped double helices of Figure 12.20, part 7, right are produced. In this case, there are segments of hybrid DNA in each duplex, but they are formed by what looks like a splicing together of two helices. The result is the double helices in Figure 12.20, part 8, right, which are called *spliced duplexes*.

In the example diagrammed in Figure 12.20, the parental duplexes contained different genetic markers at the ends of the molecules. One parent was $a^+ b^+$ and the other was $a\ b$, as would be the case for a doubly heterozygous parent. However, the reciprocal recombination events shown in Figure 12.20 result in different products. In the patched duplexes (Figure 12.20, part 8, left) the markers are $a^+ b^+$ and $a\ b$, which is the parental configuration. However, in the spliced duplexes (Figure 12.20, part 8, right), the markers are recombinant, that is, $a^+ b$ and $a\ b^+$. Since the enzymes that cut in the branch region during the final cleavage and ligation phase (Figure 12.20, part 6, left and right) cut randomly with respect to the plane of the cut (i.e., horizontal or vertical), the Holliday model predicts that a physical exchange between two gene loci on homologous chromosomes should result in the genetic exchange of the outside chromosome markers about half of the time.

It is important to note that the Holliday model is only one model to explain the events of recombination. While its basic features are generally accepted, a number of other models have been proposed that attempt to explain recombination either at a more detailed level or recombination in special systems. Discussion of these other models is beyond the scope of this text.

KEYNOTE

The Holliday model for genetic recombination involves a precise alignment of homologous DNA sequences of two parental double helices, endonucleolytic cleavage, invasion of the other helix by each broken end, ligation to produce the Holliday intermediate, branch migration, and finally, cleavage and ligation to resolve the Holliday intermediate into the recombinant double helices. Depending on the orientation of the cleavage events, the resulting double helices will be patched duplexes or spliced duplexes. If the recombination events occur between heterozygous loci, the patched duplexes produced are parental, whereas the spliced duplexes are recombinant for the two loci.

~ FIGURE 12.21

Electron micrograph of a Holliday intermediate with some single-stranded DNA in the branch point region.

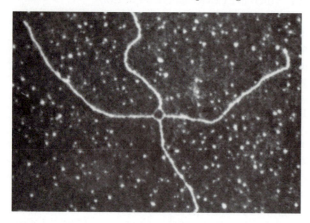

Gene Conversion and Mismatch Repair

In Chapter 6 we discuss tetrad analysis, in which all products of each meiotic event can be isolated and analyzed. Tetrad analysis is limited to relatively few organisms; among them are fungi such as *Saccharomyces cerevisiae* (yeast) and *Neurospora crassa*. Recall that from a cross of $a\ b \times a^+\ b^+$ in which a and b are linked on the same chromosome, but still reasonably distant, three types of tetrads can result: parental ditype (PD), $a\ b$, $a\ b$, $a^+\ b^+$, $a^+\ b^+$; tetratype (T), $a\ b^+$, $a\ b$, $a^+\ b^+$, $a^+\ b$; and the rare nonparental ditype (NPD), $a\ b^+$, $a\ b^+$, $a^+\ b$, $a^+\ b$. For each tetrad type, there is a 2:2 segregation of the alleles.

Occasionally, the Mendelian law of 2:2 segregation of alleles is not obeyed in that 3:1 and 1:3 ratios are seen. Tetrads showing these unusual ratios are proposed to derive from a process called *gene conversion*. Any pair of alleles is subject to gene conversion. Consider, for example, the cross $a^+\ m\ b^+ \times a\ m^+\ b$, where gene m is located between genes a and b. The generation by *mismatch repair* of a tetrad showing gene conversion is diagrammed in Figure 12.22. The starting point is parental homologous chromosomes synapsed in meiotic prophase (Figure 12.22, part 1). If a recombination event occurs between the two inner chromatids (Figure 12.22, part 2), a patched duplex can be produced with two mismatches. That is, in these mismatches (heteroduplexes), one of the two DNA strands has the m sequence from one parent and other DNA strand in the helix has the m^+ sequence of the other parent. Both mismatches can be excised and repaired as shown in Figure 12.22, part 3. That is, an exonuclease removes a segment of one DNA strand of the molecule, DNA polymerase catalyzes new

DNA synthesis, and DNA ligase seals the gap between the new DNA and the preexisting DNA. Following the subsequent two meiotic divisions, a tetrad is produced, showing a 3:1 gene conversion for the + allele of m, while the outside genetic markers a^+/a and b^+/b retain their parental configuration Figure 12.22, part 4).

Lastly, if two markers are very close together, they undergo *co-conversion* so that both will exhibit a

~ FIGURE 12.22

Gene conversion by mismatch repair at two sites. (1) Parent homologs; (2) Recombination between inner two chromatids produces a patched duplex with two mismatches; (3) Both mismatches are repaired by excision and DNA synthesis; (4) Two meiotic divisions produce a tetrad with a 3:1 conversion for the m^+ allele.

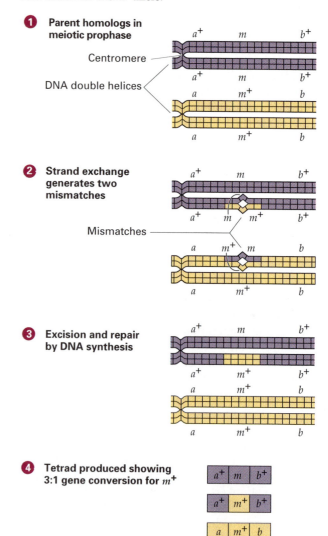

❶ **Parent homologs in meiotic prophase**

Centromere

DNA double helices

❷ **Strand exchange generates two mismatches**

Mismatches

❸ **Excision and repair by DNA synthesis**

❹ **Tetrad produced showing 3:1 gene conversion for m^+**

1:3 or 3:1 segregation in a tetrad, while outside markers will segregate in a normal 2:2 ratio.

KEYNOTE

In organisms in which all four products of meiosis can be analyzed, mismatch repair of heteroduplex DNA can result in a non-Mendelian segregation of alleles, typically a 3:1 or 1:3 ratio rather than a 2:2 segregation. This phenomenon is known as gene conversion. Co-conversion of alleles can occur if the two alleles are very close together.

SUMMARY

In this chapter we discussed DNA replication and DNA recombination. Many aspects of DNA replication are similar in prokaryotes and eukaryotes, for example, a semiconservative and semidiscontinuous mechanism, synthesis of new DNA in the 5'-to-3' direction, and the use of RNA primers to initiate DNA chains. The enzymes that catalyze the synthesis of DNA are the DNA polymerases. In *E. coli* there are three DNA polymerases, two of which are known to be involved in DNA replication along with several other enzymes and proteins. The DNA polymerases have 3'-to-5' exonuclease activity, which permits proofreading to take place during DNA synthesis if an incorrect nucleotide is inserted opposite the template strand. In eukaryotes, two DNA polymerases are involved in nuclear DNA replication. Neither has associated proofreading activity, that function presumably being the property of a separate protein.

In prokaryotes, DNA replication begins at specific chromosomal sites. Such sites are known also for yeast. A prokaryotic chromosome has one initiation site for DNA replication, while a eukaryotic chromosome has several initiation sites dividing the chromosome into replication units, or replicons. It is not clear whether the initiation sites used are the same sequences from cell generation to cell generation. The existence of replicons means that DNA replication of the entire set of chromosomes in a eukaryotic organism can proceed relatively quickly, in some cases faster than the single *E. coli* chromosome, despite the presence of orders of magnitude more DNA.

Since eukaryotic chromosomes are linear, there is a special problem of maintaining the lengths of chromosomes because removal of RNA primers results in a shorter new DNA strand on the lagging-strand template. This problem is overcome by using special enzymes to maintain the length of chromosomes. These enzymes are called telomerases. Telomerases are a combination of proteins and RNA. The RNA component acts as a template to guide the synthesis of new telomere repeat units at the chromosome ends.

Replication of DNA in eukaryotes occurs in the S phase of the cell division cycle. Eukaryotic chromosomes are complexes of DNA with histones and non-histone chromosome proteins, so not only must DNA be replicated but the chromosome structure must also be duplicated. In particular, the nucleosome organization of chromosomes must be duplicated as the replication forks migrate. It is known that nucleosomes are "mature" soon after a replication fork passes, although it is not known precisely how nucleosomes segregate from cell generation to cell generation.

A number of models have been proposed to describe the molecular events involved in the breakage and rejoining of DNA during crossing-over. One model, developed by Holliday, is described in this chapter. All molecular models for genetic recombination involve new DNA synthesis in small regions participating in crossing-over. In a later chapter (Chapter 19) we will see that DNA synthesis may also be involved in the repair of certain genetic damage. Thus, three major cellular processes involve DNA synthesis: DNA replication, genetic recombination, and DNA repair.

ANALYTICAL APPROACHES FOR SOLVING GENETICS PROBLEMS

Q12.1a. Meselson and Stahl used ^{15}N-labeled DNA to prove that DNA replicates semiconservatively. The method of analysis was cesium chloride equilibrium density gradient centrifugation, in which bacterial DNA labeled in both strands with ^{15}N (the heavy isotope of nitrogen) bands to a different position in the gradient than DNA labeled in both strands with ^{14}N (the normal isotope of nitrogen). Starting with a mixture of ^{15}N-containing and ^{14}N-containing DNA, then, two bands result after CsCl density gradient centrifugation. If both DNAs are first heated to 100°C and cooled slowly before centrifuging, the result is different. Two bands are seen in exactly the same positions as before, and a new third band is seen at a position halfway between the other two bands. From its position relative to the other two bands, the new band is interpreted to be intermediate in density between the other two bands. Explain the existence of the three bands in the gradient.

b. DNA from *E. coli* containing ^{15}N in both strands is mixed with DNA from another bacterial species, *Bacillus subtilis*, containing ^{14}N in both strands. Two bands are seen after CsCl density gradient centrifugation. If the two DNAs are mixed, heated to 100°C, slowly cooled, and then centrifuged, two bands again result. These bands are in the same positions as in the unheated DNA experiment. Explain these results.

A12.1a. When DNA is heated to 100°C, the DNA is denatured to single strands (see Chapter 11). If denatured DNA is allowed to cool slowly, renaturation of complementary strands occurs to produce double-stranded DNA again. Thus, when mixed, denatured ^{15}N-^{15}N DNA and ^{14}N-^{14}N DNA from the same species is cooled slowly, the single strands pair randomly during renaturation so that ^{15}N-^{15}N, ^{14}N-^{14}N, and ^{15}N-^{14}N double-stranded DNA is produced. The latter DNA will have a density intermediate between those of the two other DNA types, accounting for the third band. Theoretically, if all DNA strands pair randomly, there should be a distribution of 1:2:1 of ^{15}N-^{15}N, ^{15}N-^{14}N, and ^{14}N-^{14}N DNAs, and this should be reflected in the relative intensities of the bands.

b. DNAs from different bacterial species have different sequences. In other words, DNA from one species typically is not complementary to DNA from another species. Therefore only two bands are seen because only the two *E. coli* DNA strands can renature to form ^{15}N-^{15}N DNA and only the two *B. subtilis* DNA strands can renature to form ^{14}N-^{14}N DNA. No ^{15}N-^{14}N hybrid DNA can form, so in this case there is no third band of intermediate density.

Q12.2 What would be the effect upon chromosome replication in *E. coli* strains carrying deletions of the following genes?
a. *dnaE*
b. *polA*
c. *dnaG*
d. *lig*
e. *ssb*
f. *oriC*

A12.2 When genes are deleted, the function encoded by those genes is lost. All of the genes listed in the question are involved in DNA replication in *E. coli*, and their functions are briefly described in Table 12.2 (p. 352), and discussed further in the text.

a. *dnaE* encodes the largest subunit (α) of the catalytic core of DNA polymerase III, the principal DNA polymerase in *E. coli* that is responsible for elongation of DNA chains. A deletion of the *dnaE* gene would undoubtedly lead to a nonfunctional DNA polymerase III. In the absence of DNA polymerase III activity, DNA strands could not be synthesized from RNA primers; hence, synthesis of new DNA strands could not occur, and there would be no chromosome replication.

b. *polA* encodes DNA polymerase I, which is used in DNA synthesis to extend DNA chains made by DNA polymerase III while simultaneously excising the RNA primer by 5'-to-3' exonuclease activity. As discussed in the text, in mutant strains lacking the originally studied DNA polymerase, DNA polymerase I, chromosome replication still occurred. Thus, chromosome replication would occur normally in an *E. coli* strain carrying a deletion of *polA*.

c. *dnaG* encodes DNA primase, the enzyme that synthesized the RNA primer on the DNA template. Without the synthesis of the short RNA primer, DNA polymerase III cannot initiate DNA synthesis and, therefore, chromosome replication will not take place.

d. *lig* encodes DNA ligase, the enzyme that catalyzes the ligation of Okazaki fragments together. In a strain carrying a deletion of *lig*, DNA synthesis would occur, but stable progeny chromosomes would not result because the Okazaki fragments could not be ligated together, so the lagging strand synthesized discontinuously on the lagging-strand template would be in fragments.

e. *ssb* encodes the single-strand binding proteins which bind to and stabilize the single-stranded DNA regions produced as the DNA is unwound at the replication fork. In the absence of single-strand binding proteins, impeded or absent DNA replication would result because the replication bubble could not be kept open.

f. *oriC* is the origin of replication region in *E. coli*, that is, the location at which chromosome replication initiates. Without the origin, the initiator protein cannot bind, no replication bubble can form, and therefore chromosome replication cannot take place.

QUESTIONS AND PROBLEMS

12.1 Compare and contrast the conservative and semiconservative models for DNA replication.

12.2 Describe the Meselson and Stahl experiment, and explain how it showed that DNA replication is semiconservative.

*12.3 In the Meselson and Stahl experiment, ^{15}N-labeled cells were shifted to ^{14}N medium, at what we can designate as generation 0.
a. For the semiconservative model of replication, what proportion of ^{15}N-^{15}N, ^{15}N-^{14}N, and ^{14}N-^{14}N would you expect to find at generations 1, 2, 3, 4, 6, and 8?
b. Answer the above question in terms of the conservative model of DNA replication.

12.4 Suppose *E. coli* cells are grown on an ^{15}N medium for many generations. Then they are quickly shifted to an ^{14}N medium, and DNA is extracted from the samples taken after one, two, and three generations. The extracted DNA is subjected to equilibrium density gradient centrifugation in CsCl. In figures (a) and (b), using the reference positions of pure ^{15}N and pure ^{14}N DNA as guides, indicate where the bands of DNA would equilibrate if replication were semiconservative or conservative.

a) **Semiconservative model**

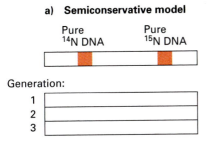

b) **Conservative model**

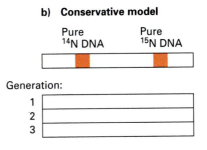

12.5 A spaceship lands on earth and with it a sample of extraterrestrial bacteria. You are assigned the task of determining the mechanism of DNA replication in this organism.

You grow the bacteria in unlabeled medium for several generations, then grow it in presence of ^{15}N for exactly one generation. You extract the DNA and subject it to CsCl centrifugation. The banding pattern you find is shown in the figure below.

It appears to you that this is evidence that DNA replicates in the semiconservative manner, but this result does not prove that this is so. Why? What other experiment could you perform (using the same sample and technique of CsCl centrifugation) that would further distinguish between semiconservative and dispersive modes of replication?

*12.6 The elegant Meselson-Stahl experiment was among the first experiments to contribute to what is now a highly detailed understanding of DNA replication. Reconsider this experiment in light of current molecular models by answering the following questions.
a. Does the fact that DNA replication is semiconservative *a priori* require it to be semidiscontinuous?
b. Does the fact that DNA replication is semidiscontinuous ensure that it is also semiconservative?
c. Do any properties of known DNA polymerases ensure that DNA is synthesized semiconservatively?

*12.7 List the components necessary to make DNA in vitro by using the enzyme system isolated by Kornberg.

*12.8 Give two lines of evidence that the Kornberg enzyme is not the enzyme involved in the replication of DNA for the duplication of chromosomes in growth of *E. coli*.

12.9 Kornberg isolated DNA polymerase I from *E. coli*. DNA polymerase I has an essential function in DNA replication. Which of the following is that function?
a. filling gaps left by the removal of RNA primer
b. filling in gaps where introns are removed
c. the formation of stem loops in tRNA
d. recognition of rho factor for the initiation of transcription
e. production of poly(A) tails on eukaryotic mRNAs

12.10 Assume you have a DNA molecule with the base sequence T-A-T-C-A going from the 5' to the 3' end of one of the polynucleotide chains. The building blocks of the DNA are drawn as in the following figure. Use this shorthand system to diagram the completed double-stranded DNA molecule, as proposed by Watson and Crick.

12.11 Base analogs are compounds that resemble the natural bases found in DNA and RNA, but are not normally found in those macromolecules. Base analogs can replace their normal counterparts in DNA during in vitro DNA synthesis. Four base analogs were studied for their effects on in vitro DNA synthesis using the *E. coli* DNA polymerase. The results were as follows, with the amounts of DNA synthesized expressed as percentages of that synthesized from normal bases only.

	NORMAL BASES SUBSTITUTED BY THE ANALOG			
ANALOG	A	T	C	G
A	0	0	0	25
B	0	54	0	0
C	0	0	100	0
D	0	97	0	0

Which bases are analogs of adenine? of thymine? of cytosine? of guanine?

12.12 Describe the semidiscontinuous model for DNA replication. What is the evidence showing that DNA synthesis is discontinuous on at least one template strand?

*12.13 Distinguish between a primer strand and a template strand.

*12.14 The length of the *E. coli* chromosome is about 1,100 μm.

a. How many base pairs does the *E. coli* chromosome have?
b. How many complete turns of the helix does this chromosome have?
c. If this chromosome replicated unidirectionally and if it completed one round of replication in 60 minutes, how many revolutions per minute would the chromosome be turning during the replication process?
d. The *E. coli* chromosome, like many others, replicates bidirectionally. Draw a simple diagram of a replicating *E. coli* chromosome that is halfway through the round of replication. Be sure to distinguish new and old DNA strands.

12.15 In *E. coli* the replication fork moves forward at 500 nucleotide pairs per second. How fast is the DNA ahead of the replication fork rotating?

12.16 A diploid organism has 4.5×10^8 base pairs in its DNA. This DNA is replicated in 3 minutes. Assuming all replication forks move at a rate of 10^4 base pairs per minute, how many replicons (replication units) are present in this organism's genome?

12.17 The following events, steps or reactions occur during *E. coli* DNA replication. For each entry in Column A, select the appropriate entry in Column B. Each entry in A may have more than one answer, and each entry in B can be used more than once.

Column A	Column B
___ a. Unwinds the double helix	A Polymerase I
___ b. Prevents reassociation of complementary bases	B Polymerase III
___ c. Is an RNA polymerase	C Helicase
___ d. Is a DNA polymerase	D Primase
___ e. Is the "repair" enzyme	E Ligase
___ f. Is the major elongation enzyme	F SSB protein
___ g. A 5'-to-3' polymerase	G Gyrase
___ h. A 3'-to-5' polymerase	H None of these
___ i. Has 5'-to-3' exonuclease function	
___ j. Has 3'-to-5' exonuclease function	
___ k. Bonds free 3'-OH end of a polynucleotide to a free 5' monophosphate end of polynucleotide	
___ l. Bonds 3'-OH end of a polynucleotide to a free 5' nucleotide triphosphate	
___ m. Separates daughter molecules and causes supercoiling	

12.18 Compare and contrast the three *E. coli* DNA polymerases with respect to their enzymatic activities.

*12.19 In *E. coli*, distinguish between the activities of primase; single-stranded, binding protein; helicase; DNA ligase; DNA polymerase I; and DNA polymerase III in DNA replication.

12.20 Describe the molecular action of the enzyme DNA ligase. What properties would you expect an *E. coli* cell to have if it had a temperature-sensitive mutation in the gene for DNA ligase?

12.21 Chromosome replication in *E. coli* commences from a constant point, called the origin of replication. It is known that DNA replication is bidirectional. Devise a biochemical experiment to prove that the *E. coli* chromosome replicates bidirectionally. (*Hint*: Assume that the amount of gene product is directly proportional to the number of genes.)

12.22 What property of DNA replication was indicated by the presence of Okazaki fragments?

12.23 A space probe returns from Jupiter and brings with it a new microorganism for study. It has double-stranded DNA as its genetic material. However, studies of replication of the alien DNA reveal that, while the process is semiconservative, DNA synthesis is continuous on both the leading-strand and the lagging-strand templates. What conclusion(s) can you draw from that result?

*12.24 Compare and contrast eukaryotic and prokaryotic DNA polymerases.

12.25 Draw a eukaryotic chromosome as it would appear at each of the following cell cycle stages. Show both DNA strands, and use different line styles for old and newly synthesized DNA.
a. G_1
b. anaphase of mitosis
c. G_2
d. anaphase of meiosis I
e. anaphase of meiosis II

12.26 In eukaryotes, cellular proliferation is tightly controlled. Explain why this is important and outline how key regulatory proteins control the progression of the cell cycle.

*12.27 Autoradiography is a technique which allows radioactive areas of chromosomes to be observed under the microscope. The slide is covered with a photographic emulsion, which is exposed by radioactive decay. In regions of exposure the emulsion forms silver grains upon being developed. The tiny silver grains can be seen on top of the (much larger) chromosomes. Devise a method for finding out which regions in the human karyotype replicate during the last 30 min of the S period. (Assume a cell cycle in which the cell spends 10 h in G_1, 9 h in S, 4 h in G_2 and 1 h in M.)

12.28 In Figure 12.15, semiconservative DNA replication is visualized in eukaryotic cells using the harlequin chromosome-staining technique.
a. Explain what the harlequin chromosome-staining technique is, and how it provides evidence for semiconservative DNA replication in eukaryotes.
b. Propose an hypothesis to explain why, in Figure 12.15, some chromatids appear to contain segments of both T-labeled DNA and BUdR-labeled DNA, while others appear to be entirely T-labeled or BUdR-labeled.

12.29 When the eukaryotic chromosome duplicates, the nucleosome structures must duplicate. Discuss the synthesis of histones in the cell cycle, and discuss the model for the assembly of new nucleosomes at the replication forks.

12.30 A mutant *Tetrahymena* has an altered repeated sequence in its telomeric DNA. What change in the telomerase enzyme would have this phenotype?

12.31 What evidence is there that telomere length is under genetic control? Why might such control be important?

*12.32 How is mismatch repair related to recombination and repair of DNA?

12.33 What is gene conversion? How does the Holliday model for genetic recombination allow for gene conversion?

*12.34 Crosses were made between strains, each of which carried one of three different alleles of the same gene, *a*, in yeast. For each cross, some unusual tetrads resulted at low frequencies. Explain the origin of each of these tetrads:

Cross:	a1 a2+	a1 a3+	a2 a3+
	×	×	×
	a1+ a2	a1+ a3	a2+ a3
Tetrads:	a1+ a2	a1+ a3	a2+ a3
	a1+ a2+	a1+ a3	a2+ a3+
	a1 a2	a1+ a3+	a2 a3+
	a1 a2+	a1 a3+	a2 a3+

12.35 From a cross of *y1 y2+* × *y1+ y2*, where *y1* and *y2* are both alleles of the same gene in yeast, the following tetrad type occurs at very low frequencies:

$$y1^+ \quad y2$$
$$y1 \quad y2$$
$$y1 \quad y2$$
$$y1 \quad y2^+$$

Explain the origin of this tetrad at the molecular level.

12.36   In *Neurospora* the *a*, *b*, and *c* loci are situated in the same arm of a particular chromosome. The location of *a* is near the centromere; *b* is near the middle, and *c* is near the telomere of the arm. Among the asci resulting from a cross of *ABC* x *abc*, the following ascus was found (the 8 spores are indicated in the order in which they were arranged in the ascus): *ABC, ABC, ABc, ABc, aBC, aBC, abc, abc*. How might this ascus have arisen?

12.37   In the population of asci produced in question 12.36 an ascus was found containing, in this order, the spores *ABC, ABC, ABc, Abc, aBC, aBC, abc, abc*. How could this ascus have arisen?

13 TRANSCRIPTION, RNA MOLECULES, AND RNA PROCESSING

PRINCIPAL POINTS

~ Transcription, the process of transcribing DNA base pair sequences into RNA base sequences, is similar in prokaryotes and eukaryotes. The DNA unwinds in a short region next to a gene, and an RNA polymerase catalyzes the synthesis of an RNA molecule in the 5'-to-3' direction along the 3'-to-5' template strand of DNA. Generally, only one strand of the double-stranded DNA is transcribed into an RNA molecule.

~ In *E. coli*, a single RNA polymerase synthesizes mRNA, tRNA, and rRNA. Eukaryotes have three distinct nuclear RNA polymerases, each of which transcribes different gene types: RNA polymerase I transcribes the genes for the 28S, 18S, and 5.8S ribosomal RNAs; RNA polymerase II transcribes mRNA genes and some snRNA genes; and RNA polymerase III transcribes genes for the 5S rRNAs, the tRNAs, and the remaining snRNAs.

~ In *E. coli*, the initiation of transcription of protein-coding genes requires the holoenzyme form of RNA polymerase (core enzyme + sigma factor) binding to the promoter. Once transcription has begun, the sigma factor dissociates and RNA synthesis is completed by the RNA polymerase core enzyme. Termination of transcription is signaled by specific sequences in the DNA. Two types of termination sequences are found and a particular gene will have one or the other. One type of terminator is recognized by the RNA polymerase alone, and the other type is recognized by the enzyme in association with the *rho* factor.

~ The transcripts of protein-coding genes are linear precursor mRNAs (pre-mRNAs). Prokaryotic mRNAs are modified little once they are transcribed, while eukaryotic mRNAs are modified by the addition of a 5' cap and a 3' poly(A) tail. Most eukaryotic pre-mRNAs contain sequences called introns, which do not code for amino acids. The introns are removed as the primary transcript is processed to produce the mature, functional mRNA molecule. The amino acid–coding segments separated by introns are called exons.

~ Introns are removed from pre-mRNAs in a series of well-defined steps. Introns typically begin with a 5' GU and end with a 3' AG. Intron removal begins with the cleavage of the pre-mRNA at the 5' splice junction. The free 5' end of the intron loops back and bonds to an A

nucleotide in the branch-point consensus sequence, which is located upstream of the 3' splice junction. Cleavage at the 3' splice junction releases the intron, which is shaped like a lariat. The exons that flanked the intron are spliced together once the intron is excised. The removal of introns from eukaryotic pre-mRNA occurs in the nucleus in complexes called spliceosomes. The spliceosome consists of several small nuclear ribonucleoprotein particles (snRNPs) bound specifically to each intron.

~ Ribosomes are the cellular organelles on which protein synthesis takes place. In both prokaryotes and eukaryotes, ribosomes consist of two unequally sized subunits. Each subunit consists of a complex between one or more ribosomal RNA (rRNA) molecules and many ribosomal proteins. Eukaryotic ribosomes are larger and more complex than prokaryotic ribosomes.

~ In eukaryotes, three of the four rRNAs are encoded by tandem arrays of transcription units. Each transcription unit produces a single pre-rRNA molecule with the three rRNAs separated by spacer sequences. The individual rRNAs are generated by processing of the pre-rRNA to remove the spacers. The fourth rRNA is encoded by separate genes.

~ In the precursor rRNAs of some organisms there are introns, the RNA sequences of which fold into a secondary structure that excises itself, a process called self-splicing. This process does not involve protein enzymes.

~ RNA polymerase III transcribes 5S rRNA genes (5S rDNA), tRNA genes (tDNA), and some snRNA genes. For 5S rDNA and tDNA, the promoter for RNA polymerase III is located within the gene itself. The internal promoter is called the internal control region and consists of different combinations of functional domains depending on the class of gene involved. The domains are binding sites for transcription factors required for transcription by RNA polymerase III.

~ Transfer RNA (tRNA) molecules bring amino acids to the ribosomes where the amino acids are polymerized into a polypeptide chain. All tRNAs are about the same length, contain a number of modified bases, and have similar three-dimensional shapes. tRNAs are made as pre-tRNA molecules containing 5'-leader and 3'-trailer sequences, which are removed by enzymes.

*T*he structure, function, development, and reproduction of an organism depends on the properties of the proteins present in each cell and tissue. A protein consists of one or more chains of amino

acids. Each chain of amino acids is called a polypeptide. The sequence of amino acids in a polypeptide chain is coded for by a gene; that is, a specific base-pair sequence in DNA. (Recall the one gene–one polypeptide hypothesis discussed in Chapter 9, pp. 267–271.) When a protein is needed in the cell, the genetic code for that protein's amino acid sequence must be read from the DNA and processed into the finished protein. Two major steps occur in the process of protein synthesis: transcription and translation.

Transcription is the transfer of information from a double-stranded, template DNA molecule to a single-stranded RNA molecule. **Translation** (protein synthesis; see Chapter 14) is the conversion in the cell of the messenger RNA (mRNA) base sequence information into the amino acid sequence of a polypeptide. Unlike DNA replication, which typically occurs during only part of the cell cycle (at least in eukaryotes), transcription and translation generally occur throughout the cell cycle (although they are much reduced during the M phase of the cell cycle). In this chapter we examine the transcription process itself, and the structures and properties of the different RNA classes; that is, messenger RNA, transfer RNA, small nuclear RNA, and ribosomal RNA. We will see that the initial RNA transcripts (the **primary transcripts** or **precursor RNA molecules [pre-RNAs]**) of most genes must be modified and/or processed to produce the biologically active (mature) RNA molecules.

THE TRANSCRIPTION PROCESS

In 1956, three years after Watson and Crick proposed their double-helix model for DNA, Crick gave the name *Central Dogma* to the two-step process of DNA → RNA → protein (transcription followed by translation) for the synthesis of proteins encoded by DNA.

In this section we will discuss how an RNA chain is synthesized, and introduce the different classes of RNA and the genes that code for them.

RNA Synthesis

Only some of the DNA is transcribed at any one time. That is, the genome of an organism consists of specific sequences of base pairs distributed among a number of chromosomes. Genes include sequences of base pairs that are transcribed and thus the transcription process is also referred to as *gene expression*. Associated with each gene are base-pair sequences called **gene regulatory elements**, which are involved in the regulation of gene expression (see Chapters 16 and 17).

In both prokaryotes and eukaryotes, transcription is catalyzed by an enzyme called **RNA polymerase** (Figure 13.1). The DNA double helix must unwind for a short region next to the gene before transcription can begin. In prokaryotes this occurs as part of the function of RNA polymerase, while in eukaryotes it is brought about by other proteins that bind to the DNA at the start point for transcription.

Generally, only one of the two DNA strands is transcribed into an RNA. If each of the two DNA strands of a gene was transcribed to produce an RNA molecule, each gene would produce two RNA products that were complementary in sequence. Translation of the two RNAs would produce two very different proteins.

During transcription, RNA is synthesized in the 5'-to-3' direction. The 3'-to-5' DNA strand that is read to make the RNA strand is called the *template strand*. The 5'-to-3' DNA strand complementary to the template strand that has the *same* polarity as the resulting RNA strand is called the *nontemplate strand*.

~ FIGURE 13.1

Transcription process. The DNA double helix is denatured by RNA polymerase in prokaryotes, or by other proteins in eukaryotes. RNA polymerase then catalyzes the synthesis of a single-stranded RNA chain, beginning at the "start of transcription" point. The RNA chain is made in the 5'-to-3' direction, using only one strand of the DNA as a template to determine the base sequence.

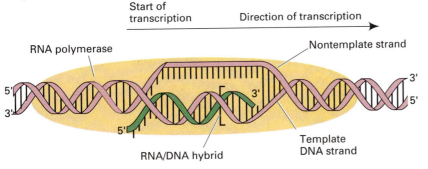

In transcription, the RNA precursors are the ribonucleoside triphosphates ATP, GTP, CTP, and UTP, collectively called NTPs (*nucleoside triphosphates*). RNA synthesis occurs by polymerization reactions that are very similar to the polymerization reactions involved in DNA synthesis (Figure 13.2; DNA polymerization is shown in Figure 12.3, p. 350). The next nucleotide to be added to the chain is selected by the RNA polymerase for its ability to pair with the exposed base on the DNA template strand. Unlike DNA polymerases, RNA polymerases can initiate new polynucleotide chains (i.e., no primer is needed), and they have no proofreading abilities.

Recall that RNA chains contain nucleotides with the base uracil instead of thymine. Therefore where an A nucleotide occurs on the DNA template chain, a U nucleotide is placed in the RNA chain instead of a T.

As an example, if the template DNA strand reads

3'-A T A C T G G A C-5',

then the RNA chain will be synthesized in the 5'-to-3' direction and will have the sequence

5'-U A U G A C C U G-3',

and the DNA nontemplate strand is

5'-T A T G A C C T G-3',

~ **FIGURE 13.2**

Chemical reaction involved in the RNA polymerase-catalyzed synthesis of RNA on a DNA template strand.

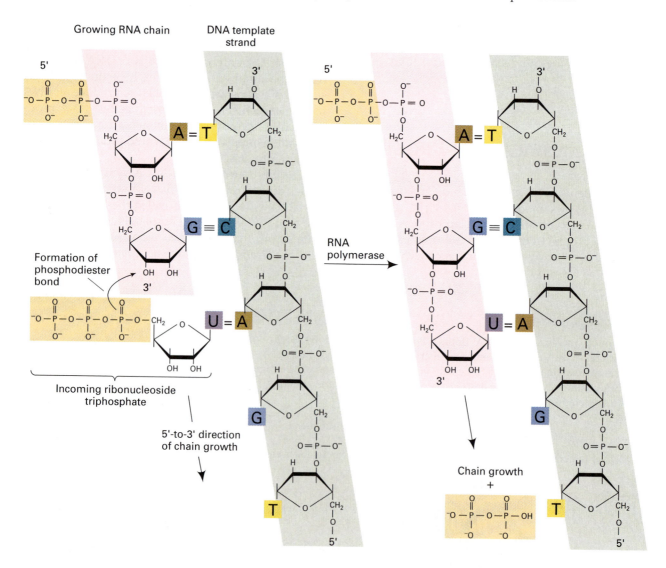

KEYNOTE

Transcription, the process of transcribing DNA base-pair sequences into RNA base sequences, exhibits basic similarities in prokaryotes and eukaryotes. The DNA unwinds in a short region next to a gene, and an RNA polymerase catalyzes the synthesis of an RNA molecule in the 5'-to-3' direction along the 3'-to-5' template strand of the DNA. Generally, only one strand of the double-stranded DNA is transcribed into an RNA molecule.

Classes of RNA and the Genes That Code for Them

Four different major classes of RNA molecules or transcripts are produced by transcription: **messenger RNA (mRNA)**, **transfer RNA (tRNA)**, **ribosomal RNA (rRNA)**, and **small nuclear RNA (snRNA)**. The mRNAs, tRNAs, and rRNAs are found in both prokaryotes and eukaryotes, while snRNAs are found only in eukaryotes. The classes of RNAs differ in their stability (i.e., lifetime in the cell before they are degraded). Where found, tRNAs, rRNAs, and snRNAs are stable RNAs, relatively speaking, since they have long lifetimes. In bacteria, mRNAs typically have a short lifetime, while in eukaryotes mRNAs with a range of lifetimes from very short to very long are known. The role of mRNA stability in the regulation of gene expression is discussed in chapters 16 and 17.

Only the mRNA molecule is translated to produce a polypeptide. Translation of mRNAs occurs on cellular organelles called **ribosomes**. (Translation is described in Chapter 14.) Ribosomes are composed of proteins and three (prokaryotes) or four (eukaryotes) rRNA molecules. The tRNAs bring amino acids to the ribosome, where they are matched to the mRNA and the amino acids they carry are assembled into a polypeptide chain. The snRNAs are involved in processing of mRNA precursor molecules in eukaryotes (discussed later in this chapter).

The primary transcripts of the genes for the RNAs generally are precursor-RNA (pre-RNA) molecules. After transcription, pre-RNA molecules are modified to produce the mature, functional RNAs. Two general types of modifications occur: (1) chemical modifications, in which the bases are altered; and (2) RNA processing, in which sequences of the precursor are removed in a precise, orderly way. In bacterial cells, the production of mature RNAs primarily involves chemical modification events, while in eukaryotic cells both chemical modification and RNA processing events occur.

A gene that codes for an mRNA molecule, and hence for a protein, is called a **structural gene**, or a *protein-coding gene*. In this and later chapters we discuss primarily structural genes. The genes that code for tRNA, rRNA, and snRNA molecules are different from structural genes because their RNA transcripts (the products of transcription) are the final products of gene expression. That is, these RNAs function as RNA molecules—they are not translated into proteins.

In bacteria, only one type of RNA polymerase is used to transcribe the structural genes and the genes for tRNA and rRNA, whereas in eukaryotes three different RNA polymerases transcribe the genes for the four types of RNAs (Table 13.1). **RNA polymerase I**, located exclusively in the nucleolus, catalyzes the synthesis of three of the RNAs found in ribosomes, the organelles responsible for protein synthesis: the 28S, 18S, and 5.8S ribosomal RNA (rRNA) molecules. (In brief, the S value of a protein, RNA, or DNA molecule derives from the rate at which the molecules sediment in a gradient of sucrose in a centrifuge—see Box 13.1.) The rate of sedimentation of a molecule is related both to the molecular weight of the molecule and to its three-dimensional configuration. Thus, S values give a very rough comparison of molecular sizes. **RNA polymerase II**, found only in the nucleoplasm of the nucleus, synthesizes messenger RNAs (mRNAs) and some snRNAs, some of which are involved in RNA processing events. **RNA polymerase III** (also found only in the nucleoplasm) synthesizes the following: (1) the transfer RNAs (tRNAs), which bring amino acids to the ribosome; (2) 5S rRNA, a small rRNA molecule found in each ribosome and (3) the remaining small nuclear RNAs (snRNA) not made by RNA polymerase II, some of which are involved in RNA processing events.

～ TABLE 13.1

Properties of Eukaryotic RNA Polymerases

RNA POLYMERASE	LOCATION	PRODUCTS	α-AMANITIN SENSITIVITY
I	Nucleolus	28S, 18S, 5.8S rRNAs	Insensitive
II	Nucleus	mRNA Some snRNAs	Highly sensitive
III	Nucleus	tRNA 5S rRNA Some snRNAs	Intermediate sensitivity

Box 13.1
Sucrose Gradient Centrifugation

Sucrose gradient centrifugation is usually used to separate the components present in a mixture based on their relative rates of sedimentation. With this procedure we can measure the rates at which the components sediment in the gradient under centrifugal forces. In sucrose gradient centrifugation the sedimentation rates are converted to **Svedberg units**, using a formula the derivation of which is beyond the scope of this book. Svedberg units, or simply **S values**, are then used as a rough indication of relative sizes of the components being analyzed.

The rate of sedimentation of a component in a sucrose gradient is related both to the molecular weight of the component and to its three-dimensional configuration. Two components of the same molecular weight, for example, will have different S values if one is highly compact and thus sediments rapidly, while the other has a more extended shape and thus sediments slowly. Sucrose gradient centrifugation is typically used to separate, or estimate, the sizes of RNA molecules, ribosomes, ribosomal subunits, proteins, various cellular organelles, and so on.

The sucrose gradient centrifugation method involves a tubeful of sucrose in which the concentration of sucrose increases toward the bottom of the tube. For the separation of different-sized RNA molecules, for example, a continuous gradient of sucrose concentration, ranging from 10 percent at the top to 30 percent at the bottom, is prepared in a centrifuge tube (Box Figure 13.1a). Next, a small amount of the sample, in this case a solution containing the RNAs, is very carefully layered on top of the gradient (Box Figure 13.1b). The sucrose gradient is then centrifuged at high speeds for several hours, during which time the RNA molecules move through the gradient at different rates depending on their molecular weight and configuration (Box Figure 13.1c). At the completion of centrifugation, the RNAs of similar S

~ BOX FIGURE 13.1

Sucrose density gradient centrifugation technique for separating and isolating RNA molecules in a mixture. (a) 10 to 30 percent sucrose density gradient; (b) RNA sample added to top of gradient; (c) Three RNA types separate into bands after centrifugation; (d) Fractionation of gradient to collect the RNA types.

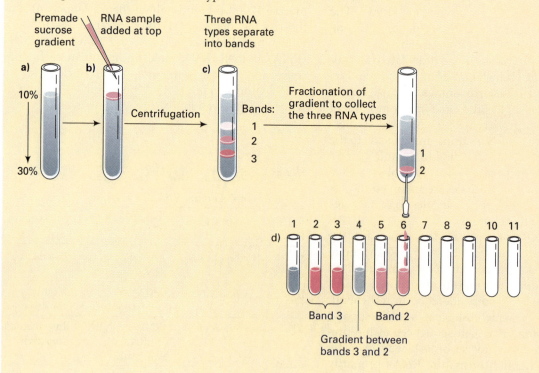

values are located in discrete zones or bands in the gradient. These bands of RNA are not visible to the eye, but are determined after the analysis of the gradient as will now be described.

To collect the different RNA types, the bottom of the tube is punctured with a needle, and the drops that run out are collected by using a fraction collector (Box Figure 13.1d). Fractions containing the RNAs are identified by measuring the degree to which each fraction absorbs ultraviolet light. Fractions with RNA will absorb ultraviolet light, while those without RNA will not. The

relative positions the RNAs occupy in the gradient indicate the S values of the RNAs: Those with higher S values (larger and/or more compact) are found closer to the bottom of the gradient than those with lower S values.

Note that sucrose gradient centrifugation separates molecules on the basis of their relative rates of sedimentation. Hence it differs from cesium chloride equilibrium density gradient centrifugation (see Chapter 12, p. 347), in which molecules are separated on the basis of buoyant density and not size.

Another characteristic helps distinguish between the three RNA polymerases: their different sensitivities to inhibition by α-amanitin, a product of the poisonous mushroom *Amanita phalloides*. RNA polymerase I is insensitive to inhibition by α-amanitin, RNA polymerase II exhibits the greatest sensitivity to inhibition, and RNA polymerase III shows intermediate sensitivity to α-amanitin inhibition. These properties have made it possible to isolate and purify the three different RNA polymerase classes.

Compared with what we know about the structure and function of the RNA polymerases in *E. coli* and a number of other prokaryotes, less is known about the structure and function of eukaryotic RNA polymerases. One reason for this is that few RNA polymerase mutants are known in eukaryotes. Another reason is that the amount of RNA polymerase in a eukaryotic cell is relatively low, thus making purification of the polymerase very difficult. In calf thymus, for example, RNA polymerases constitute only 0.05 percent of the total cellular protein, whereas in *E. coli* RNA polymerase accounts for 1 percent of the total cellular protein. What we do know is that all eukaryotic RNA polymerases consist of several subunits (two large and at least four small). The subunits of any given RNA polymerase type (I, II, or III) have apparently been conserved through evolution, since they are very similar in size and activity (where that is known) in eukaryotes ranging from yeast to humans.

KEYNOTE

In *E. coli*, a single RNA polymerase synthesizes mRNA, tRNA, and rRNA. Eukaryotes have three distinct nuclear RNA polymerases, each of which transcribes different gene types: RNA polymerase I transcribes the genes for the 28S, 18S, and 5.8S ribosomal RNAs; RNA polymerase II transcribes mRNA genes and some snRNA genes; and RNA polymerase III transcribes genes for the 5S rRNAs, the tRNAs, and the remaining snRNAs.

TRANSCRIPTION OF PROTEIN-CODING GENES

In this section we discuss the transcription of protein-coding genes in prokaryotes (focusing on *E. coli*) and in eukaryotes.

Prokaryotes

A prokaryotic protein-coding gene may be divided into three sequences with respect to its transcription (Figure 13.3):

1. A sequence upstream of the start of the RNA coding sequence called the **promoter** with which RNA polymerase interacts. This interaction determines the start point for transcription.
2. The RNA coding sequence; that is, the sequence of DNA base pairs transcribed by RNA polymerase into the single-stranded mRNA transcript.
3. A sequence downstream of the end of the RNA coding sequence specifies where transcription will stop. This sequence is called a transcription **terminator sequence**, or more simply, a **terminator**.

INITIATION OF TRANSCRIPTION.   From comparisons of sequences upstream of coding sequences and from studies of the effects of specific base pair changes, two DNA sequences in most promoters of *E. coli* genes have been shown to be critical for specifying the initiation of transcription. These sequences

~ FIGURE 13.3

Organization of a gene in terms of transcription into three regions: promoter, RNA coding sequence, and terminator. The promoter is considered to be "upstream" of the coding sequence, and the terminator is considered to be "downstream" of the coding sequence.

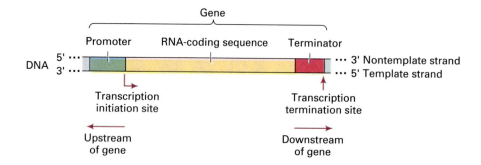

generally are found at −35 and −10; that is, centered at 35 and 10 base pairs upstream from the base pair at which transcription starts. (Conventionally, this initial base pair is designated as +1; base pairs upstream from the initial base pair, such as promoters, are given negative numbers, while those downstream from the initial base pair are given positive numbers.) From examination of the promoters of a large number of genes, the **consensus sequence** (i.e., the sequence indicating which nucleotides are found most frequently at each position) for the −35 region is

5'-TTGACA-3'.

The consensus sequence for the −10 region (also called the **Pribnow box** after the researcher who first discovered it) is

5'-TATAAT-3'.

For transcription to begin, a form of RNA polymerase called the *holoenzyme* (or *complete enzyme*) must bind to the promoter. This holoenzyme consists of the **core enzyme** form of RNA polymerase (which has four polypeptides—two α, β, and β') bound to another polypeptide called the *sigma factor* (σ). The sigma factor is essential for recognition of a promoter sequence; if the sigma factor is not present, the core enzyme binds to DNA in various places but does not initiate transcription efficiently from any of them. An *E. coli* cell contains about three thousand copies of the holoenzyme.

The RNA polymerase holoenzyme binds to a promoter in two distinct steps. First, it binds loosely to the −35 region of the promoter while the DNA is still in double-helical form (Figure 13.4a). The second step involves a shift to a tighter binding between RNA polymerase and DNA. This shift accompanies a local untwisting of about 17 base pairs of the DNA cen-

tered around the −10 region (Figure 13.4b). This untwisting represents almost two complete helical turns of DNA. Note that the −10 region is all AT base pairs; having only two hydrogen bonds such base pairs are easier to break apart than GC base pairs, which have three. Once the RNA polymerase is bound at the −10 region, it is correctly oriented to begin transcription at the right nucleotide (Figure 13.4b).

Since promoters differ slightly in their actual sequence, the efficiency of RNA polymerase binding varies. As a result, the rate at which transcription is initiated varies from gene to gene, a fact that helps to explain why different genes have different rates of expression at the RNA level. For example, a −10 region sequence of 5'-GATACT-3' will have a lower rate of transcription initiation than the 5'-TATAAT-3' because the ability of RNA polymerase to recognize and bind to the former sequence is lower.

The situation just described is what happens at most *E. coli* promoters. In fact there are several different sigma factors in *E. coli*. These sigma factors play an important role in regulating gene expression. Each sigma factor binds to the core RNA polymerase and permits the resulting holoenzyme to recognize different promoters. Most promoters have the recognition sequences we have just discussed; these are recognized by a sigma factor with a molecular weight of 70,000 Da, called σ^{70}. Under conditions of high heat (a heat shock) and other forms of stress, a different sigma factor called σ^{32} (molecular weight 32,000 Da) increases in amount. This directs some RNA polymerase molecules to bind to the promoters of genes that encode proteins required to cope with the stress. These promoters, like those recognized by the main sigma factor, have two recognition sequences, in this case CCCCC at −39, and TATAAATA at −16. Under conditions of limiting nitrogen, a third sigma factor, σ^{54} (molecular weight 54,000), is produced. This

~ FIGURE 13.4

Action of *E. coli* RNA polymerase in the initiation and elongation stages of transcription. (a) In initiation, the RNA polymerase holoenzyme first binds loosely to the promoter at the −35 region. (b) As initiation continues, RNA polymerase binds more tightly to the promoter at the −10 region, accompanied by a local untwisting of about 17 bp centered around the −10 region. At this point, the RNA polymerase is correctly oriented to begin transcription at +1. (c) After 8 to 9 polymerizations have occurred, the sigma factor dissociates from the core enzyme. (d) As the RNA polymerase elongates the new RNA chain, the enzyme untwists the DNA ahead of it; as the double helix reforms behind the enzyme, the RNA is displaced away from the DNA.

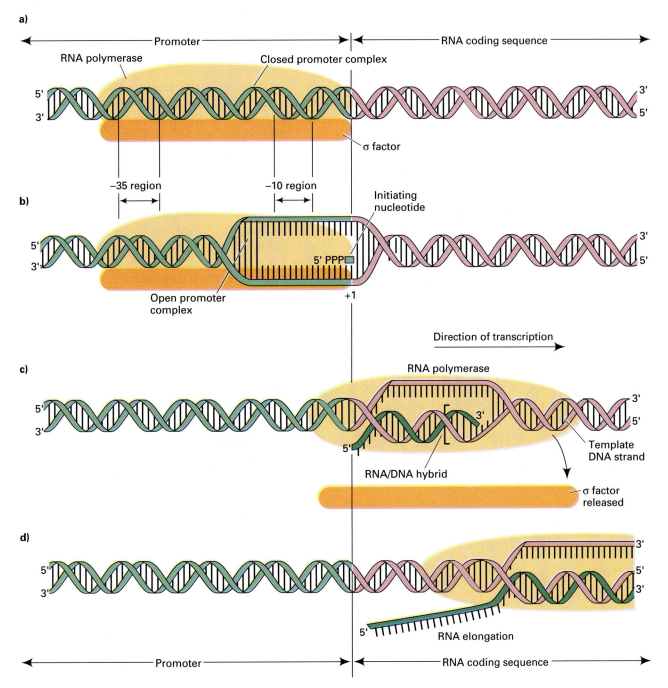

sigma factor recognizes promoters with the consensus sequences GTGGC at −26 and TTGCA at −14. A fourth sigma factor, σ²³ (molecular weight 23,000) is made when cells are infected with phage T4. This sigma factor recognizes promoters with the consensus sequence TATAATA at −15. Other sigma factors are produced under various other conditions but they have not yet been well characterized.

ELONGATION OF THE RNA CHAIN. RNA synthesis takes place in a region of the DNA that has denatured to form a transcription bubble. (This bubble is similar to a DNA replication bubble.) Once 8 or 9 RNA nucleotides have been linked together, the sigma factor dissociates from the RNA polymerase core enzyme (Figure 13.4c) and can be used again in other transcription initiation reactions. The core enzyme completes the transcription of the gene.

As the core RNA polymerase moves along, it untwists the DNA double helix ahead of it. This produces torsional stress in the DNA much like untwisting the middle part of a stretch of rope does, so that the DNA helix reforms behind the enzyme as the enzyme moves along the template (Figure 13.4d). About 17 base pairs of the DNA are kept unpaired as transcription continues, about the same as are initially untwisted. Within the untwisted region, some bases of RNA are bonded to the DNA in a temporary RNA-DNA hybrid while the rest of the RNA is displaced away from the DNA (Figure 13.4d). Transcription proceeds at a rate averaging 30 to 50 nucleotides per second.

TERMINATION OF TRANSCRIPTION. Termination of the transcription of a prokaryotic gene is signaled by controlling elements called *terminators*. One important protein involved in the termination of transcription of some genes in *E. coli* is *rho* (ρ). The terminators of such genes are called *rho-dependent terminators* (also called type II terminators). At many other terminators the core RNA polymerase itself can carry out the termination events. Those terminators are called *rho-independent terminators* (also called type I terminators).

Rho-independent terminators consist of sequences with twofold symmetry that are about 16 to 20 base pairs upstream of the transcription termination point. A sequence with twofold symmetry is one that is approximately self-complementary about its center, that is, one half of the sequence is complementary to the other half. Thus, the transcript of the region with twofold symmetry can form a hairpin loop as shown in Figure 13.5. This sequence is followed by a string of about four to eight AT base pairs giving rise to a series of Us in the RNA transcript which is just upstream of the transcription termination point. The combination of the hairpin loop followed by the string of Us (transcribed from a string of AT base pairs) in the RNA leads to transcription termination. While it is not clear how the AT string and the hairpin loop bring about *rho*-independent termination, it is possible that the rapid formation of the hairpin loop destabilizes the RNA-DNA hybrid in the terminator region, which in turn causes the release of the RNA and transcription termination.

Rho-dependent terminators lack the AT string found in *rho*-independent terminators, and many cannot form hairpin structures. The model for this type of termination is as follows. The *rho* factor is a protein

~ **FIGURE 13.5**

Sequence of a ρ-independent terminator and structure of the terminated RNA. The light brown-shaded mutations partially or completely prevent termination.

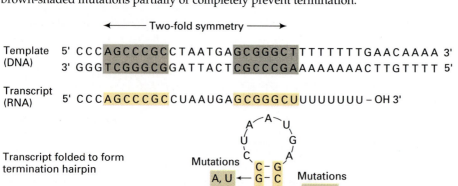

with two domains: one domain binds to RNA, and the other domain binds to ATP. For transcription termination, *rho* first binds to ATP, and is activated by it. The activated *rho* then binds to its recognition sequences in the termination region when that region of the RNA transcript is synthesized. When *rho* binds, ATP is hydrolyzed, and somehow the RNA transcript is unwound from the DNA template. Transcription then stops and both the transcript and the RNA polymerase are released from the DNA template.

In sum, three key events occur at both *rho*-dependent and *rho*-independent terminators: (1) RNA synthesis stops, (2) the RNA chain is released from the DNA, and (3) RNA polymerase is released from the DNA.

KEYNOTE

In *E. coli*, the initiation of transcription of protein-coding genes requires the holoenzyme form of RNA polymerase (core enzyme + sigma factor) binding to the promoter. Once transcription has begun, the sigma factor dissociates and RNA synthesis is completed by the RNA polymerase core enzyme. Termination of transcription is signaled by specific sequences in the DNA. Two types of termination sequences are found and a particular gene will have one or the other. One type of terminator is recognized by the RNA polymerase alone, and the other type is recognized by the enzyme in association with the *rho* factor.

Eukaryotes

TRANSCRIPTION OF PROTEIN-CODING GENES BY RNA POLYMERASE II.

In eukaryotes, RNA polymerase II transcribes protein-coding (structural) genes. The product of transcription is a **precursor-mRNA (pre-mRNA) molecule**; that is, a transcript that must be modified and/or processed to produce the mature, functional mRNA molecule. As stated earlier, RNA polymerase II also transcribes some snRNA genes.

Regulatory Elements. A protein-coding gene may have a large assortment of DNA sequences involved in the regulation of that gene's transcription. These DNA sequences are called *regulatory elements*; they are located both upstream and downstream of the RNA initiation site for the gene. Each gene in an animal cell, for example, has a particular combination of positive regulatory elements (for activating transcription) and negative regulatory elements (for repressing transcription) adjacent to it. These regulatory elements can bind specific **transcription factors** (specific proteins required for the initiation of transcription by a eukaryotic RNA polymerase) and **regulatory factors** (proteins involved in the activation or repression of transcription of the gene). Generally, the regulatory elements are located within several hundred base pairs from the site of initiation of transcription, usually upstream from that point. However, some regulatory elements are much farther away. By definition the regulatory element adjacent to the transcription start site is the *promoter*.

The promoters of protein-coding genes have been analyzed in two principal ways. One way was to isolate single-base-pair mutations at every base pair position for a 100 or so base pairs upstream from the start point of transcription and to examine those mutants in terms of their effects on transcription. It was presumed that mutations that caused significant effects on transcription would define important promoter elements. The second way was to compare the DNA sequences upstream of a number of protein-coding genes to see if there were any regions with similar sequences. The results of these experiments indicated that the promoter of a protein-coding gene is arranged as a series of **promoter elements** (also called *promoter modules*). Starting closest to the transcription initiation site, the promoter elements are the **TATA box** or **TATA element** (also called the **Goldberg-Hogness box** after its discoverers), the **CAAT box**, and the **GC box** (Figure 13.6). The elements are named for the general DNA base sequences they contain. Promoters can contain various combinations of promoter elements, but no one element is found in all promoters. For example, not all promoters contain a TATA element, and some promoters contain more than one copy of a CAAT element or a GC element. All this argues that the elements provide promoter function but that no element is essential for all promoters.

The TATA box has the consensus sequence 5'-TATAAA-3'. (By convention, this sequence is given as seen on the nontemplate strand.) In higher eukaryotes, the TATA box is almost always located at position −30. Mutations in the TATA element result in relatively little decrease in transcription, although the RNA initiation point often is changed. In promoters that lack a TATA element, typically there is no unique initiation point for RNA synthesis. Thus, it seems that the TATA element functions to specify a particular start point for RNA transcription.

The CAAT box has the consensus sequence 5'-GGCCAATCT-3'. The CAAT box is found at approximately −75 in many genes, but it can also function at a

~ FIGURE 13.6

Promoter elements (modules) for a eukaryotic protein-coding gene transcribed by RNA polymerase II. Each promoter element has a different function in transcription. The DNA sequences between the elements are not important for the transcription process. Transcription factors bind to the elements to promote or repress transcription.

number of other locations. Mutations in the CAAT element cause a very marked reduction in the rate of transcription, indicating that the CAAT element plays a very important role in the initiation of transcription.

The GC box has the consensus sequence 5'-GGGCGG-3' and is located at −90. Often there is more than one copy of a GC box in a promoter, and the elements may function in either orientation; i.e., facing toward the gene $\binom{5'\text{-GGGCGG-3'}}{3'\ \text{CCCGCC-5'}}$ ...gene), or away from the gene $\binom{5'\text{-CCGCCC-3'}}{3'\ \text{GGCGGG-5'}}$ ...gene). The GC boxes appear to help bind the RNA polymerase near the transcription start point.

Although the promoter elements are crucial for determining whether transcription can occur, **enhancers**, or **enhancer elements**, are required for maximal transcription of the gene to occur. A gene's enhancer helps control transcription from the gene's promoter. Enhancers function in either orientation (i.e., facing toward or away from the gene) and at a large distance from the gene, often more than 1,000 base pairs from the promoter. In animal cells, enhancers can activate genes when the enhancers are either upstream or downstream from the RNA initiation site. In most cases, though, the enhancers are found upstream of the gene. Similar elements that have essentially the same properties as enhancer elements, except that they repress rather than activate gene transcription, are called **silencer elements**. Silencers are much less common than enhancers.

There is no consensus sequence for a eukaryotic enhancer element, and the mechanism of enhancer action is not well understood. It is clear that regulatory proteins bind to the enhancer elements, and which regulatory proteins bind depends on the DNA sequence of the enhancer element. A model for how the enhancers affect transcription from a distance is that specific regulatory proteins bind to the enhancer element and the DNA then forms a loop so that the

enhancer-bound regulatory proteins interact with the regulatory proteins and transcription factors bound to the promoter elements. Through the interactions of all of the proteins, transcription is either activated (enhancers) or repressed (silencers).

In yeast, there are elements functionally similar to enhancers that are called **upstream activator sequences** (**UASs**). Like enhancers, UASs can function in either orientation and at variable distances upstream of the promoter. However, unlike enhancers, UASs cannot function when located downstream of the promoter.

Transcription Events.   RNA polymerase II is unable to recognize the promoter on its own. Instead, **specific transcription factors (TFs)** are needed to bring about the initiation of transcription by RNA polymerase II. In general, all eukaryotic RNA polymerases require transcription factors for transcription initiation. The transcription factors are named for the RNA polymerase with which they work: TFI for RNA polymerase I, TFII for RNA polymerase II, and TFIII for RNA polymerase III. Because a number of different transcription factors are involved with transcription by each polymerase, the TFs are also lettered A, B, C, and so on.

For protein-coding genes, some transcription factors bind to specific DNA sequences of the promoter, while others appear to bind to the RNA polymerase II when it initiates transcription. Figure 13.7 shows the events that may occur during the initiation of transcription by RNA polymerase II. The first step is the binding of TFIID (= the TATA-binding protein [TBP] plus TBP-associated factors [TAFs]) to the TATA element to form the *initial committed complex*. This complex acts as a binding site for TFIIB which then recruits RNA polymerase II and TFIIF to produce the *minimal transcription initiation complex*. Next TFIIE and TFIIH bind to produce the *complete transcription initia-*

tion complex, and transcription begins. The rate of this basal transcription is modulated by the effects of other proteins binding to other promoter elements and to the enhancer or silencer.

The termination of transcription of eukaryotic protein-coding genes occurs in a different way from that of bacterial protein-coding genes. We will discuss that process in our discussion of eukaryotic mRNA molecules.

*K*EYNOTE

Protein-coding genes in eukaryotes are transcribed by RNA polymerase II. The promoter for these genes consists of different combinations of promoter elements or modules, depending on the gene. These promoter elements are crucial for determining whether transcription can occur. They are the sites for interaction of transcription factors and reg-ulatory factors. Another important element associated with genes is the enhancer, which, through interaction with regulatory factors, functions to facilitate maximum transcription of the gene with which it is associated.

mRNA Molecules

Figure 13.8 shows the general structure of the mature, biologically active mRNA as it exists in both prokaryotic and eukaryotic cells. The mRNA molecule has three main parts. At the 5' end is a **leader sequence,** or 5' untranslated region (5' UTR), which is constant in length for a given mRNA type but varies in length between different mRNA types. Within this leader sequence is the coded information that the ribosome reads to orient it correctly for beginning protein synthesis; none of the bases of the leader sequence are

~ **FIGURE 13.7**

Schematic of the events that may occur during the initiation of transcription catalyzed by RNA polymerase II. See text for detailed description of events.

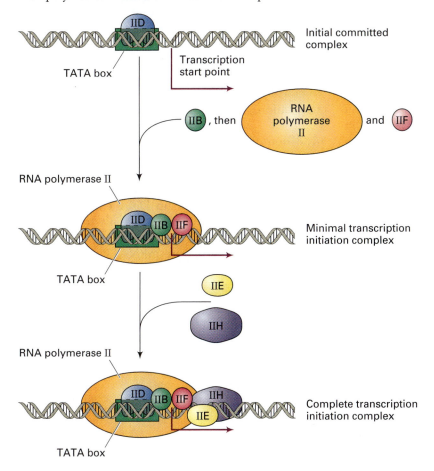

translated into amino acids. Following the 5' leader sequence is the actual **coding sequence** of the mRNA; this sequence determines the amino acid sequence of a protein during translation. The coding sequence varies in length, depending on the length of the protein for which it codes. Following the amino acid–coding sequence, and constituting the rest of the mRNA at the 3' end of the molecule, is an untranslated **trailer sequence,** or 3' untranslated region (3' UTR). The trailer sequence also varies in length from mRNA to mRNA.

The production of functioning mRNA is fundamentally different in prokaryotes and eukaryotes. In prokaryotes (Figure 13.9a) the RNA transcript functions directly as the mRNA molecule for translation but, in eukaryotes (Figure 13.9b), the RNA transcript must be modified in the nucleus by a series of events known as *RNA processing* in order to produce the mature mRNA. In addition, since prokaryotes lack a nucleus, an mRNA begins to be translated on ribosomes before it has been completely transcribed; this process is called *coupled transcription and translation* (Figure 13.10). In eukaryotes, however, the mRNA must migrate from the nucleus to the cytoplasm (where the ribosomes are located) before it can be translated. Thus, a eukaryotic mRNA is always completely transcribed and processed before it is translated.

PRODUCTION OF MATURE mRNA IN PROKARY-OTES. As indicated above, in prokaryotes the initial transcript of a protein-coding gene *is* the mature mRNA molecule. That is, an exact point-by-point relationship exists between the order of base pairs in the gene and the order of the corresponding bases in the mature mRNA. This is referred to as *colinearity* between a gene and its primary transcript.

PRODUCTION OF MATURE mRNA IN EUKARY-OTES. Unlike prokaryotic mRNAs, eukaryotic mRNAs are usually modified at both the 5' and 3' ends. These *posttranscriptional modifications* are catalyzed by specific enzymes. In addition, many pro-

tein-coding genes have insertions of non-amino acid–coding sequences called **introns** between amino acid–coding sequences, or **exons** (also called **coding sequences**). The term *intron* is derived from *interven*ing sequence that is *not* translated into an amino acid sequence, and the term *exon* is the term derived from *ex*pressed sequence that *is* translated into an amino acid sequence. Both exons and introns are copied into the primary mRNA transcript—the pre-mRNA—from which they are removed in the processing of pre-mRNA to the mature mRNA molecule. The synthesis and processing of pre-mRNA is discussed in this section.

5' Capping. The 5' end of the eukaryotic pre-mRNA is modified by the addition of a *cap* in a process called **5' capping**. 5' capping involves the addition of a guanine nucleotide (most commonly 7-methyl guanosine [m^7G] to the terminal 5' nucleotide by an unusual 5'-to-5' linkage as opposed to the usual 5'-to-3' linkage) and the addition of two methyl groups (CH_3) to the first two nucleotides of the RNA chain (Figure 13.11). 5' capping takes place as follows: RNA polymerase II starts transcription at the base to which the cap is added. That DNA site is called the *cap site.* There is no DNA template for the 5' cap. When the growing RNA transcript is about 20 to 30 nucleotides long, a methylated cap structure is added at the 5' end of the transcript by a *capping enzyme.* The 5' cap remains as the pre-mRNA is processed to the mature mRNA. The cap is essential for the ribosome to bind to the 5' end of the mRNA, an initial step of translation.

Addition of the 3' Poly(A) Tail. The posttranscriptional modification of the 3' ends of eukaryotic pre-mRNAs is usually the addition of a sequence of about 50 to 250 adenine nucleotides. This sequence is called a **poly(A) tail**. There is no DNA template for the poly(A) tails, and they are added only to pre-mRNA molecules. The poly(A) tail remains as the pre-mRNA is processed to the mature mRNA. Poly(A) tails are

~ **FIGURE 13.8**

General structure of mature, biologically active mRNA as found in both prokaryotic and eukaryotic cells.

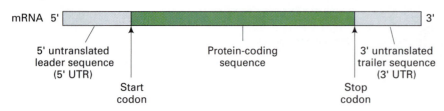

~ FIGURE 13.9

Processes for synthesis of functional mRNA in prokaryotes and eukaryotes. (a) In prokaryotes the mRNA synthesized by RNA polymerase does not have to be processed before it can be translated by ribosomes. Also, since there is no nuclear membrane, translation of the mRNA can begin while transcription continues, resulting in a coupling of the transcriptional and translational processes. (b) In eukaryotes, the primary RNA transcript is a precursor-mRNA (pre-mRNA) molecule, which is processed in the nucleus (addition of 5' cap and 3' poly(A) tail, and removal of introns to produce the mature, functioning mRNA molecule). Only when that mRNA is transported to the cytoplasm can translation occur.

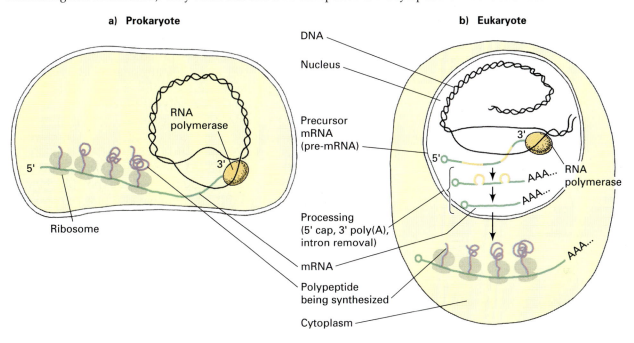

~ FIGURE 13.10

Electron micrograph of coupled mRNA transcription and translation in *E. coli.* From the faint DNA strand, a number of mRNA molecules are emerging. Since the length of the mRNA molecules increases from left to right, we assume that this strand is a single gene with a promoter to the left of the photograph. The globular structures on the mRNAs are ribosomes, which are translating the message into a protein before the synthesis of the mRNA is finished.

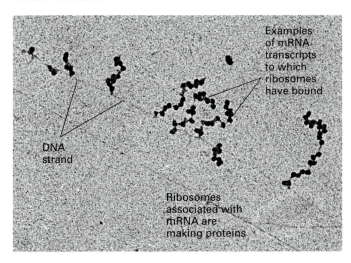

~ FIGURE 13.11

Cap structure at the 5' end of a eukaryotic mRNA. The cap results from the addition of a guanine nucleotide and two methyl groups.

found on most, but not all, of the mRNAs of all eukaryotic species. For instance, histone mRNAs in mammalian cells have no poly(A) tails.

The addition of the poly(A) tail is part of the mechanism for signaling the 3' end of the mRNA. In prokaryotes, specific transcription termination sequences specify the end of an mRNA molecule, but in eukaryotes no such termination sequences are found in the DNA corresponding to the end of an mRNA molecule. Instead, mRNA transcription continues, in some cases for hundreds or thousands of nucleotides, past a site called the **poly(A) site,** where cleavage generates a 3' OH end to which the poly(A) tail is added. The location of the poly(A) site is signaled in the transcript by an AAUAAA sequence 10 to 30 nucleotides upstream. In mammalian cells,

poly(A) addition occurs as follows (Figure 13.12): A number of proteins including the three- or four-polypeptide CPSF (cleavage and polyadenylation specificity factor) protein, the three-polypeptide CstF (cleavage stimulation factor) protein, and two multi-subunit cleavage factor proteins (CFI and CFII) are required for binding the poly(A) signal and cleaving the RNA. CPSF binds to the AAUAAA signal, and CstF binds to a GU-rich or U-rich sequence (GU/U in Figure 13.12) downstream of the poly(A) site. CPSF and CstF also bind to each other, producing a loop in the RNA. CFI and CFII are bound near the actual cleavage site. Once the RNA is cleaved, the single-polypeptide enzyme **poly(A) polymerase** (PAP) uses ATP as a substrate and catalyzes the addition of A nucleotides to the 3' end of the RNA to produce the poly(A) tail. During this process, PAP is bound to CPSF. As the poly(A) tail is synthesized, it becomes bound by the single-polypeptide poly(A) binding protein II (PABII); this protein also plays a regulatory role in controlling poly(A) tail length. In other organisms, the 3' end-processing reactions for pre-mRNA are similar to those in mammalian cells although details may differ. Clearly, polyadenylation of mRNA is a complex process and requires the products of quite a large number of genes.

The poly(A) tail is important for determining the stability of the mRNA. For example, mRNAs can be injected into frog oocytes where they will be translated. Globin mRNA, with its normal poly(A) tail, remains active in frog oocytes for a much longer time than globin mRNA from which the poly(A) tail has been removed because the non-poly(A) mRNA is more rapidly degraded. Thus, the poly(A) tail likely protects mRNAs from degradation by ribonucleases that are present in the cytoplasm.

Introns.   Pre-mRNAs often contain long insertions of non-amino-acid-coding sequences. These noncoding sequences are transcribed from the introns of the gene. Introns must be excised from each pre-mRNA in order to convert the transcript into a mature mRNA molecule that can be translated into a complete polypeptide. The mature mRNA, then, contains in a contiguous form the exon sequences that in the gene were separated by intron sequences.

One of the early experiments that elegantly showed the existence of introns was done by Philip Leder's group. They studied the β-globin genes in cultured mouse cells. The β-globin gene encodes the β-globin polypeptide that is part of a hemoglobin protein molecule. Leder's group analyzed the β-globin mRNA by sucrose gradient centrifugation (see Box 13.1). The data indicated that the β-globin mRNA has a size of about 10S, which is about the size expected if

~ **FIGURE 13.12**

Schematic diagram of the 3' end formation of mRNA and the addition of the poly(A) tail to that end in mammals. In eukaryotes, the formation of the 3' end of an mRNA is produced by cleavage of the lengthening RNA chain. This process is signaled by specific nucleotide sequences in the RNA. See text for a discussion of the proteins involved.

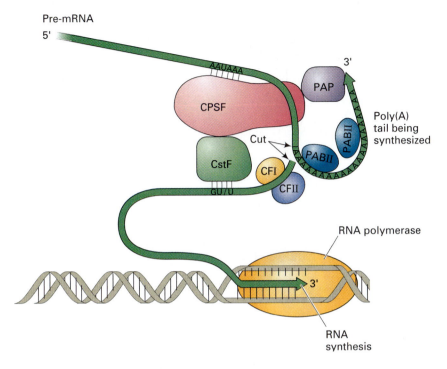

virtually all of the mRNA codes for the 147–amino acid β-globin protein. Sequence analysis indicated the mature mRNA is 0.7-kb (700 nucleotides) long.

Next, they studied the organization of the gene encoding the mRNA using **R-looping** experiments. In R-looping, RNA is hybridized with the double-stranded DNA which encoded the RNA. Under the conditions used, RNA-DNA hybrids are more stable than DNA-DNA hybrids. Thus, the RNA binds to its complementary sequence in the template DNA strand, displacing a loop of single-stranded DNA called an *R loop*. The DNA outside the region of the paired RNA and DNA remains double-stranded. The displaced R loops are visualized by electron microscopy.

Figure 13.13a shows the result of hybridizing the 0.7-kb β-globin mRNA with DNA containing the β-globin gene. Two R loops were seen to flank a loop of double-stranded DNA. This result indicated that there are two sequences in the β-globin gene that are complementary to the two ends of the mRNA. Between these two sequences, though, is a sequence that remains as double-stranded DNA because it is *not* complementary to any part of the mRNA. The conclusion was that the mouse β-globin gene and the β-globin mRNA were not colinear.

At the time introns were discovered, it was known that the nucleus contains a large population of RNA molecules of various sizes. These RNA molecules are called **heterogeneous nuclear RNA**, or **hnRNA**, and it

was thought that hnRNAs included pre-mRNA molecules. We now know that to be the case. A 15S (1.5-kb) RNA molecule was isolated from nuclear hnRNA that was the β-globin pre-mRNA, the primary transcript of the β-globin gene. Like the mature mRNA, the pre-mRNA has a 5' cap and a 3' poly(A) tail. When the 1.5-kb pre-mRNA was hybridized to DNA containing the β-globin gene in an R-looping experiment, one continuous R loop was seen (Figure 13.13b). This result indicated that the pre-mRNA was colinear with the gene that encoded it. The interpretation of this series of experiments was that the β-globin gene contains an intron of about 800 nucleotide pairs. Transcription of the gene results in a 1.5-kb pre-mRNA containing both exon and intron sequences. This RNA is found only in the nucleus. The intron sequence is excised by processing events, and the flanking exon sequences are spliced together to produce a mature mRNA. (Note: Subsequent research showed that the β-globin gene contains two introns; the second, smaller intron was not detected in the early research.)

The scientific community was shocked by the discovery of introns. That is, it was generally accepted that the gene sequence was completely colinear with the amino acid sequence of the encoded protein. Thus, the finding that genes could be "in pieces" was one of those highly significant discoveries that changed our thinking about genes markedly. From the years of research done since the initial discovery

~ FIGURE 13.13

Demonstration that the mouse β-globin gene contains an intron: (a) R loops formed between 10S mature β-globin mRNA and the β-globin gene. (b) R loop formed between 15S β-globin pre-mRNA and the β-globin gene. Interpretative diagrams appear alongside the micrographs.

a)   10S β-globin mRNA

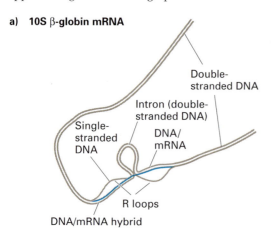

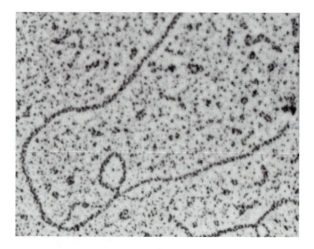

b)   15S β-globin pre-mRNA

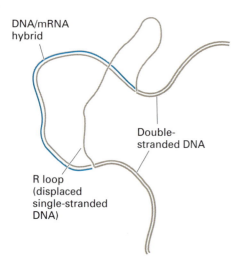

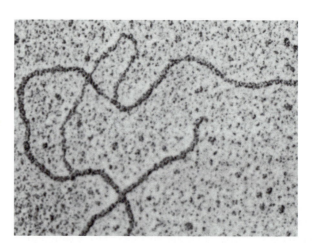

of introns, we have learned that many eukaryotic protein-coding genes contain introns. Typically, higher eukaryotes tend to have more genes with introns and longer introns than do lower eukaryotes. Interestingly, some bacteriophage genes have been shown to contain introns. For example, there is a single intron in the bacteriophage T4 thymidylate synthetase gene.

As an aside, we raise the question: What is a gene? Up until the discovery of introns, geneticists assumed that a gene was a contiguous stretch of DNA base pairs that was transcribed into a mature mRNA. That mRNA, in turn, was translated into an amino acid sequence. Prokaryotic genes fit this definition closely. However, the primary transcripts of many eukaryotic genes are not the molecules that are translated. Before it can be translated, introns must be excised, and the adjacent segments—the exons—must be spliced together to produce the mature mRNA. If we define a gene as only the amino acid–coding regions of the DNA, then the presence of introns clearly means that eukaryotic genes are in pieces. On the other hand, if we define a gene as that region of DNA corresponding to the pre-mRNA, then the whole stretch of coding sequences (exons) plus introns constitute a gene. Both definitions may be encountered in your studies, and it is important to keep the distinctions between the two in mind.

KEYNOTE

The transcripts of protein-coding genes are messenger RNAs or their precursors. These molecules are linear and vary widely in length in correspondence to the variation in the size of the polypeptides they specify and whether they contain introns. Prokaryotic mRNAs are not modified once they are transcribed, whereas most eukaryotic mRNAs are modified by the addition of a cap at the 5' end and a poly(A) tail at the 3' end. Many eukaryotic pre-mRNAs contain non–amino acid–coding sequences called introns, which must be excised from the mRNA transcript to make a mature, functional mRNA molecule. The amino acid–coding segments separated by introns are called exons.

Production of mRNA from Pre-mRNA. A model for mRNA production from genes with introns is dia-

grammed in Figure 13.14. The sequence of steps is the same for genes without introns, except for the step involving intron removal. In brief, the steps are the transcription of the gene by RNA polymerase II, the addition of the methylated 5'-cap and the poly(A) tail to produce the pre-mRNA molecule, and finally the processing of the pre-mRNA in the nucleus to remove the introns and splice the exons together to produce the mature mRNA. 5' capping and 3' poly(A) addition were described earlier. Here we focus on intron removal.

Introns in pre-mRNAs are looped out with the aid of snRNAs (see p. 399); the loop is then removed by nuclease cleavage. The adjacent exons are ligated together to generate a contiguous molecule. These events are called **mRNA splicing**. For mRNA splicing to take place, there must be some way for the machinery involved to determine what is an intron and what is an exon. In fact, introns typically begin with 5'-GU and end with AG-3'. More than just those nucleotides

(see p. 399)

~ **FIGURE 13.14**

General sequence of steps in the formation of eukaryotic mRNA. Not all steps are necessary for all mRNAs.

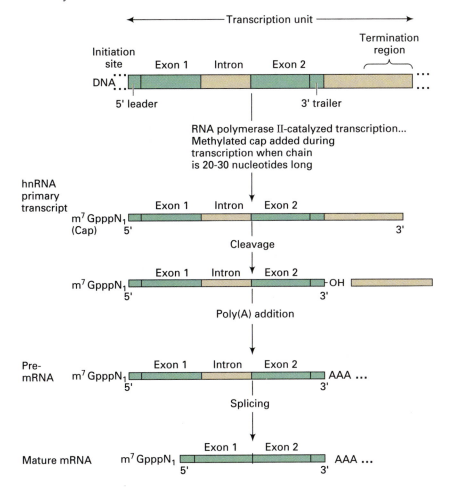

are needed to specify a splicing junction between an intron and an exon; the 5' splice junction probably involves at least seven nucleotides and the 3' splice junction involves at least ten nucleotides of intron sequence.

Figure 13.15 diagrams the sequence of events involved in splicing two exons (1 and 2) together with the elimination of an intron. The organization of the pre-mRNA is shown in Figure 13.15a. The first step in splicing is a cleavage at the 5' splice junction that results in the separation of exon 1 from an RNA molecule that contains the intron and exon 2. The free 5' end of the intron loops and becomes joined to an A nucleotide that is part of a sequence called a

branch-point sequence, which is located upstream of the 3' splice junction (Figure 13.15b). Because of its resemblance to the rope cowboys use, the looped back structure is called an *RNA lariat structure*. For those unfamiliar with the word *lariat*, think of the structure as a ρ. In mammalian cells, the branch-point consensus sequence is YNCURAY (where Y is a pyrimidine, R is a purine, and N is any base). The key A nucleotide in the sequence is located 18 to 38 nucleotides upstream of the 3' splice junction. In yeast, the branch-point sequence is UACUAAC (the italic A is where the 5' end of the intron bonds), although the position of the branch-point sequence position is more variable.

~ FIGURE 13.15

Details of intron removal from a pre-mRNA molecule. At the 5' end of an intron is the sequence GU and at the 3' end is the sequence AG. Eighteen to 38 nucleotides upstream from the 3' end of the intron is an A nucleotide located within the branch-point sequence, which, in mammals is YNCURAY, where Y = pyrimidine, N = any base, R = purine, and A = adenine. (a) Intron removal begins by a cleavage event at the first exon-intron junction. The G at the released 5' of the intron folds back and forms an unusual 2'→5' bond with the A of the branch-point sequence. (b) This reaction produces a lariat-shaped intermediate. (c) Cleavage at the 3' intron-exon junction and ligation of the two exons completes the removal of the intron. Inset: Branch-point junction structure involving an unusual 2'→5' phosphodiester bond.

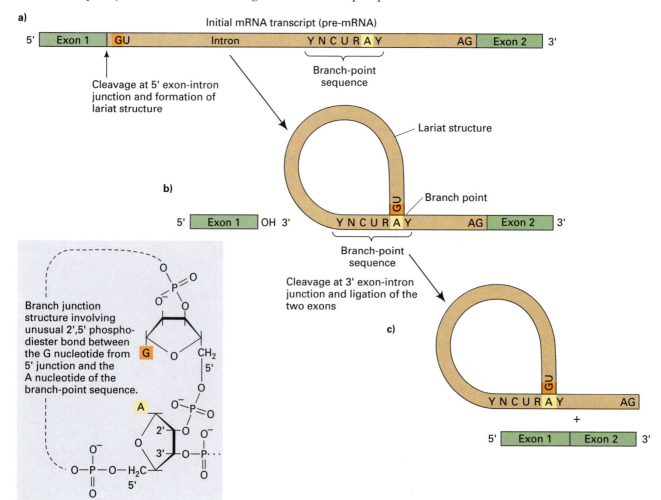

~ FIGURE 13.16

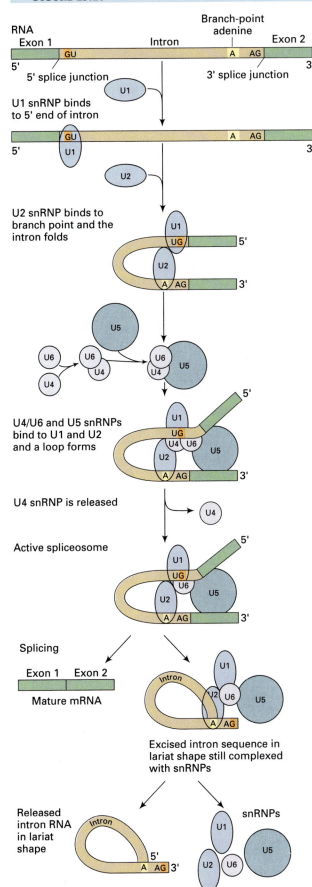

RNA
Exon 1   Intron   Branch-point adenine   Exon 2

5' splice junction   3' splice junction

U1 snRNP binds to 5' end of intron

U2 snRNP binds to branch point and the intron folds

U4/U6 and U5 snRNPs bind to U1 and U2 and a loop forms

U4 snRNP is released

Active spliceosome

Splicing

Exon 1   Exon 2

Mature mRNA

Excised intron sequence in lariat shape still complexed with snRNPs

Released intron RNA in lariat shape

snRNPs

Model for spliceosome assembly and intron removal.

The branch point in the RNA that produces the lariat structure involves an unusual 2'-5' phosphodiester bond formed between the 2' OH of the adenine nucleotide in the branch-point sequence and the 5' phosphate of the guanine nucleotide at the end of the intron (Figure 13.15 inset). The A itself remains in normal 3'-5' linkage with its adjacent nucleotides of the intron.

Next, mRNA and a precisely excised intron (still in lariat shape), appear simultaneously as a result of cleavage at the 3' splice junction and ligation of the two coding sequences (Figure 13.15c). The lariat RNA is subsequently converted to a linear molecule by the action of a de-branching enzyme and is then degraded.

The splicing of pre-mRNA molecules occurs exclusively in the nucleus. The splicing events occur in splicing complexes called **spliceosomes**, which consist of the pre-mRNA bound to **small nuclear ribonucleoprotein particles** (**snRNPs**; sometimes called *snurps* by researchers). These are small nuclear RNAs (snRNAs) associated with proteins. There are six principal snRNAs (named U1–U6) in the nucleus, and they are associated with six to ten proteins each to form the snRNPs. Some of the proteins are specific to particular snRNPs, while others are common to all snRNPs. U4 and U6 snRNAs are found within the same snRNP (U4/U6 snRNP), while the others are found within their own special snRNPs. Each snRNP type is abundant in the nucleus, with at least 10^5 copies per cell.

A model for the assembly of a spliceosome and its role in removal of an intron is shown in Figure 13.16. The steps are as follows:

1. U1 snRNP binds to the 5' splice site. This binding is primarily the result of base pairing of the U1 snRNPs to the 5' splice site sequence. Evidence for this conclusion (and similar conclusions made for other snRNA/pre-mRNA interactions) has come from mutational studies. That is, if a mutation is made in the 5' splice junction sequence so that base pairing with U1 snRNA is impaired, no splicing will occur. The mutation is suppressed if a complementary mutation is made in the gene for U1 snRNA; in this case pairing is restored, albeit with both sequences mutated from the wild-type, and splicing occurs.
2. U2 snRNP binds to the branch-point region.
3. A preassembled U4/U6/U5 particle joins the complex. This occurs through association of the particle with the bound U1 and U2 snRNPs.
4. U4 snRNP dissociates from the complex, and this results in the formation of the active spliceosome.

KEYNOTE

Introns are removed from pre-mRNAs in a series of well-defined steps. Introns typically begin with a 5' GU and end with a 3' AG. Intron removal begins with the cleavage of the pre-mRNA at the 5' splice junction. The free 5' end of the intron loops back and bonds to an A nucleotide in the branch-point consensus sequence, which is located upstream of the 3' splice junction. Cleavage at the 3' splice junction releases the intron, which is shaped like a lariat. The exons that flanked the intron are spliced together once the intron is excised. The removal of introns from eukaryotic pre-mRNA occurs in the nucleus in complexes called spliceosomes. The spliceosome consists of several small nuclear ribonucleoprotein particles (snRNPs) bound specifically to each intron.

TRANSCRIPTION OF OTHER GENES

In this section we discuss the transcription of non–protein-coding genes, focusing on genes for tRNA and for rRNA, and the production of mature tRNA and rRNA molecules from their precursors.

Ribosomal RNA and Ribosomes

Ribosomes are the organelles within the cell on which protein synthesis takes place. Each cell contains thousands of ribosomes. Ribosomes bind to mRNA and facilitate the binding of the tRNA to the mRNA so that the polypeptide chain can be synthesized. In this section we discuss what ribosomes are and how the rRNA molecules they contain are produced.

RIBOSOME STRUCTURE. Ribosomes are complex structures, and we have yet to understand fully how they function. In both prokaryotes and eukaryotes the ribosomes consist of two unequally sized subunits, (the large and small ribosomal subunits), each of which consists of a complex between RNA molecules and proteins. Each subunit contains at least one **ribosomal RNA (rRNA)** molecule and a large number of **ribosomal proteins**.

The Bacterial Ribosome. The *E. coli* ribosome is used as a model of a prokaryotic ribosome. The *E. coli* ribosome has a size of 70S, and the sizes of the two subunits are 50S (large subunit) and 30S (small subunit). (Recall from Box 13.1 that the S value is a measure of the rate of sedimentation of a component in a centrifuge, and that this is related both to the molecular weight and the three-dimensional shape of the component. Thus, the reason that the 50S and 30S subunits together give a 70S ribosome is that, when the two subunits are together, they actually have a three-dimensional shape that makes the ribosome sediment more slowly in the centrifuge than the sum of the two parts would predict.) Figure 13.17 presents two views of the complete (70S) ribosome of *E. coli*.

~ FIGURE 13.17

Two views of a model of the complete (70S) ribosome of *E. coli*. The small (30S) ribosomal subunit is yellow and the large (50S) ribosomal subunit is red.

Clearly the 30S and 50S ribosomal subunits have distinct and recognizable three-dimensional shapes.

Approximately two-thirds of the *E. coli* ribosome consists of ribosomal RNA (rRNA), the rest consisting of ribosomal proteins. The large 50S subunit has 34 different proteins and a 23S rRNA (2,904 nucleotides) and a 5S rRNA (120 nucleotides). The small 30S ribosomal subunit has 20 different proteins and a 16S rRNA (1,542 nucleotides). Transcription of the ribosomal protein genes occurs by the mechanism we have already discussed for protein-coding genes.

The Eukaryotic Ribosome. In general, eukaryotic ribosomes are larger and more complex than their prokaryotic counterparts. The size of the ribosome and the molecular weights of the rRNA molecules differ from organism to organism. Simple eukaryotes have the smallest ribosomes (although they are larger than ribosomes found in *E. coli*), while mammals have the largest ribosomes. Nonetheless, all eukaryotic ribosomes have many common structural and chemical features. We will use the mammalian ribosome as a model for discussion.

Mammalian ribosomes have a size of 80S (versus 70S for the bacterial ribosomes) consisting of a large 60S subunit and a small 40S subunit (Figure 13.18). The 80S mammalian ribosome consists of about equal weights of rRNA and ribosomal proteins. There are four rRNA types: 18S (~1,900 nucleotides), 28S (~4,700 nucleotides), 5.8S (156 nucleotides), and 5S (120 nucleotides) (see Figure 13.18). The 40S subunit contains the 18S rRNA, and the large 60S subunit contains the 28S, 5.8S, and 5S rRNAs. The 5.8S rRNA is hydrogen-bonded to the 28S rRNA in the functional ribosome. The eukaryotic 5.8S rRNA has been shown to be homologous, both in sequence and structure, to the 5' end of bacterial 23S rRNA, and the eukaryotic 5S rRNA has been shown to be evolutionarily related to bacterial 5S rRNA. There are about 85 different ribosomal proteins in the mammalian ribosome with about 35 in the small subunit and about 50 in the large subunit. Transcription of the ribosomal protein genes occurs by the mechanism we have already discussed for protein-coding genes.

KEYNOTE

Ribosomes, the organelles within the cell in which protein synthesis takes place, consist of two unequally sized subunits in both prokaryotes and eukaryotes. Each subunit contains ribosomal RNA and ribosomal proteins. Prokaryotic ribosomes contain three distinct rRNA molecules, whereas eukaryotic cytoplasmic ribosomes (larger and more complex) contain four.

TRANSCRIPTION OF PROKARYOTIC rRNA GENES. In prokaryotes and eukaryotes the regions of DNA that contain the genes for rRNA are called **ribosomal DNA (rDNA).** In *E. coli*, production of equal amounts of the three rRNAs is ensured by the transcription of the three adjacent genes for 16S, 23S, and 5S in rDNA into a *single* **precursor rRNA (pre-rRNA)** molecule. One rRNA transcription unit consists of one gene each for the three rRNAs. *E. coli* has seven such transcription units (*rrn* regions) scattered on the chromosome.

Figure 13.19a shows the general organization of a transcription unit. In each transcription unit, the three rRNA genes are arranged in the order 16S-23S-5S. In

~ **FIGURE 13.18**

Composition of whole ribosomes and of ribosomal subunits in mammalian cells.

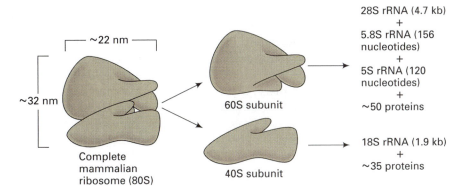

28S rRNA (4.7 kb)
+
5.8S rRNA (156 nucleotides)
+
5S rRNA (120 nucleotides)
+
~50 proteins

18S rRNA (1.9 kb)
+
~35 proteins

~ FIGURE 13.19

(a) General organization of an *E. coli* rRNA transcription unit (an *rrn* region) and (b) the scheme for the synthesis and processing of a precursor rRNA (p30S) to the mature 16S, 23S, and 5S rRNAs of *E. coli*. (The tRNAs have been omitted from this depiction of the processing scheme.)

a) *E. coli* **rRNA transcription unit**

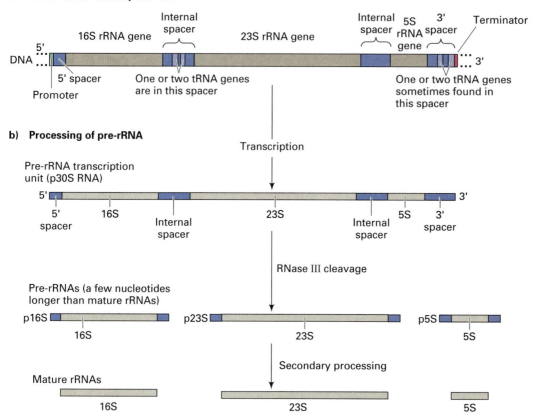

b) **Processing of pre-rRNA**

all seven transcription units, one or two tRNA genes are also found in the internal spacer between the 16S and 23S rRNA coding sequences. Another one or two tRNA genes are found in the 3' spacer between the end of the 5S rRNA gene and the 3' end in three of the seven transcription units. During RNA synthesis from the transcription units, the tRNA genes are transcribed as part of the pre-rRNA molecule. The tRNA sequences are then removed from the precursor by the action of specific pre-tRNA processing enzymes (see the discussion of tRNA synthesis, pp. 410–412).

Since there is only one RNA polymerase in *E. coli*, transcription of an rRNA transcription unit occurs in the same way as we described for protein-coding genes. The product of transcription is a 30S pre-rRNA (p30S), which contains a 5' leader sequence, the 16S, 23S, and 5S rRNA sequences (each separated by spacer sequences), and a 3' trailer sequence (Figure 13.19b). The p30S molecule is cleaved by RNase III to produce the p16S, p23S, and p5S precursor molecules that are longer than the mature RNAs. In normal

cells, RNase III cleaves the transcript of the rRNA transcription unit to produce the three rRNA precursors while transcription is still occurring. As a result, the p30S is not seen in normal cells. However, in mutants that have a temperature-sensitive RNase III activity, the p30S molecule accumulates at the high temperature. Through studies of these mutants, the pathway of p30S synthesis and processing has been discovered. Thus we know that the mature 16S, 23S, and 5S rRNAs are produced from their precursors by the action of other specific processing enzymes (Figure 13.19b).

As the rRNA genes are being transcribed, the pre-rRNA transcript rapidly becomes associated with ribosomal proteins. The cleavage of the transcript takes place, then, within a complex formed between the rRNA transcript (as it is being transcribed) and ribosomal proteins. It is most probably directed by the conformational state of the rRNA-ribosomal protein particle at the time of the cleavage. In this way, the functional ribosomal subunits are assembled.

TRANSCRIPTION OF EUKARYOTIC rRNA GENES. rRNA Gene Organization. Most eukaryotes that have been examined have a large number of copies of the genes for each of the four rRNA species 18S, 5.8S, 28S, and 5S. The genes for 18S, 5.8S, and 28S rRNAs are usually found adjacent to one another in the order 18S, 5.8S, 28S, with each set of three genes repeated many times to form tandem arrays called **rDNA repeat units** (Figure 13.20a). There are one or more clusters of rDNA repeat units in the genome, and around each cluster a nucleolus is formed. Within each nucleolus, the rRNAs are synthesized and they associate with ribosomal proteins to produce the ribosomal subunits. The ribosomal subunits are transported to the cytoplasm where they function in protein synthesis; they are not functional in the nucleus.

There are typically 100 to 1,000 copies of the rDNA repeat unit, the number varying from eukaryote to eukaryote. Thus the rRNA genes are representative of the moderately repetitive DNA in the genome (see Chapter 11). For example, yeast has 140 rRNA gene sets and human cells have 1250.

~ FIGURE 13.20

(a) Generalized diagram of a eukaryotic, ribosomal DNA repeat unit. The coding sequences for 18S, 5.8S, and 28S rRNAs are indicated in light brown. NTS = nontranscribed spacer; ETS = external transcribed spacer; ITS = internal transcribed spacer. (b) The transcription and processing of 45S pre-rRNA in HeLa cells to produce the mature 18S, 5.8S, and 28S rRNAs.

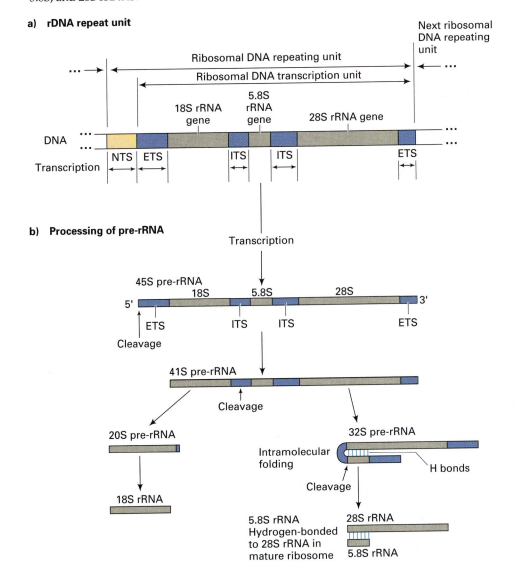

In most organisms the 5S rRNA genes are located in the genome at a site or sites distinct from the sites for the other rRNA genes. In humans, for example, the 5S genes are clustered in one location, whereas in *Xenopus* they are in clusters scattered throughout the genome. The number of 5S genes relative to the other genes also sets no pattern. The organism may have more of them, the same number, or less of them than the other rRNA genes. In some organisms, such as yeast (*Saccharomyces cerevisiae*) and the cellular slime mold (*Dictyostelium discoideum*), the 5S rRNA genes are interspersed with the other rRNA gene sets.

Introns have been found in the rRNA genes of only a few organisms (e.g., *Drosophila*, the slime mold *Physarum*, and the protozoan *Tetrahymena*). Thus, they are certainly not as widespread as the introns in mRNA genes. The introns are excised from the pre-rRNA transcript as it is processed to produce the mature rRNAs. The splicing reactions involved are different from those involved in removing introns from pre-mRNA and pre-tRNA. The removal of introns from *Tetrahymena* pre-rRNAs occurs in a very special way and will be discussed later.

Transcription of rDNA Repeat Units by RNA Polymerase I.   Each rDNA repeat unit is transcribed by RNA polymerase I to produce a large precursor-rRNA (pre-rRNA) molecule (45S pre-rRNA in human HeLa cells)(Figure 13.20b). This occurs in the nucleolus where ribosomes are assembled. An electron micrograph of pre-rRNA being transcribed from an rDNA repeat unit is shown in Figure 13.21. The pre-rRNA contains the 18S, 5.8S, and 28S rRNA

sequences and sequences between, and flanking, those three sequences. The latter are called **spacer sequences**. The external transcribed spacers—ETSs—are transcribed sequences that are located immediately upstream of the 5' end of the 18S sequence and downstream of the 3' end of the 28S sequence. The internal transcribed spacers—ITSs—are transcribed sequences that are located on either side of the 5.8S sequence; that is, between the 18S sequence and the 5.8S sequence, and between the 5.8S sequence and the 28S rRNA sequence. Between each rDNA repeat unit is the **nontranscribed spacer (NTS)** sequence which is not transcribed (see Figure 13.20a).

While RNA polymerases II and III each transcribe more than one type of gene, RNA polymerase I is unique in that it only transcribes the rDNA repeat units. The promoter for RNA polymerase I is upstream of the transcription initiation site in the NTS. In the human rDNA promoter, there are two domains: (1) a core promoter element that overlaps the start of the rRNA transcript extending from +7 to −45; and (2) an upstream control element (UCE from −107 to −186). Like other RNA polymerases, RNA polymerase I itself does not bind to the promoter; instead, specific transcription factors bind and form a complex to which RNA polymerase I binds and transcription begins. In humans, two transcription factors have been identified that are necessary for transcription to occur (Figure 13.22). Human upstream binding factor (hUBF) is a sequence-specific DNA-binding protein that binds to both elements of the human rDNA promoter to activate transcription. A second protein, SL1, is needed for promoter recognition and initiation of transcription by RNA polymerase I. SL1 consists of TBP (TATA binding protein), a component of TFIID in RNA polymerase II transcription (see pp. 390–391) and three TBP-associated factors (TAFs) that are different from those in TFIID. The latter presumably make it possible for SL1 to bind to the promoter, given that there is no TATA box. Once hUBF and SL1 are bound to the promoter, RNA polymerase I binds and transcription can begin.

Termination of transcription of the pre-rRNA involves termination sites located downstream of the rDNA transcription unit in the next NTS.

~ **FIGURE 13.21**

Electron micrograph of transcription of pre-rRNA molecules from rRNA genes in an oocyte from the spotted newt, *Triturus viridescens*. Transcription of the rRNA gene is from left to right.

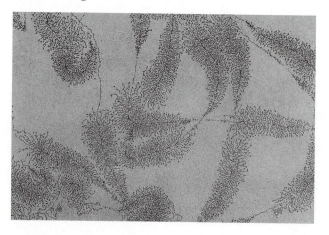

KEYNOTE

RNA polymerase I transcribes the 18S, 5.8S, and 28S rRNA sequences into a single precursor molecule. A tandem array of 18S + 5.8S + 28S rRNA transcription units is found in most eukaryotic organ-

~ FIGURE 13.22

Involvement of transcription factors in the initiation of human rDNA transcription by RNA polymerase I. (For explanation, see text.)

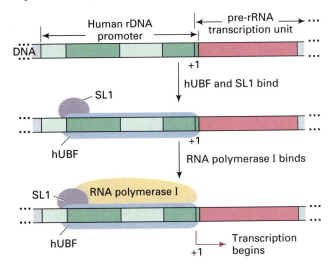

isms. Each transcription unit is separated from the next by a nontranscribed spacer (NTS) sequence. The promoter and terminator for each transcription unit are located within the NTS. As for the other eukaryotic RNA polymerases, specific transcription factors are required for the initiation of transcription by RNA polymerase I.

Processing of Pre-rRNA. To produce the mature 18S, 5.8S, and 28S rRNAs, the pre-rRNA is processed at specific sites by special ribonucleases to remove ITS and ETS sequences. As an example, Figure 13.20b shows the pre-rRNA-processing pathway in human HeLa cells. The first cleavage removes the 5' ETS sequence and produces a precursor molecule containing all three rRNA sequences. A second cleavage produces the 20S precursor to 18S rRNA and the 32S precursor to 28S and 5.8S rRNAs. The 18S rRNA is produced from the 20S precursor by the removal of the ITS sequence. In the processing of the 32S precursor, the molecule folds so that the 5.8S sequence hydrogen-bonds to the 28S sequence. Then the ITS sequences are removed from the 5' end and from between the 5.8S and 28S sequences, and the ETS is removed from the 3' end.

All the pre-rRNA-processing events take place in complexes formed between the pre-rRNA, 5S rRNA, and the ribosomal proteins. The 5S rRNA is produced by transcription of the 5S rRNA genes by RNA poly-

merase III (see p. 407), and the ribosomal proteins are produced by transcription of the ribosomal protein genes by RNA polymerase II and the subsequent translation of the mRNAs. As pre-rRNA processing proceeds, the complexes undergo shape changes resulting in the formation of the 60S and 40S ribosomal subunits. Once assembled, the two ribosomal subunits migrate out of the nucleolus, through the nucleus, and into the cytoplasm, where they associate with mRNAs and tRNAs and begin the process of protein synthesis.

𝒦EYNOTE

Three of the four rRNAs in eukaryotic ribosomes, the 18S, 5.8S, and 28S rRNAs, are transcribed from the rDNA onto a single pre-rRNA molecule. In addition to the rRNA coding sequences, the pre-rRNA contains spacer sequences located at the ends of the molecule (external transcribed spacers) and internally between the rRNA sequences (internal transcribed spacers). The spacers are removed as the pre-rRNA is processed in the nucleolus to produce the mature rRNAs. The fourth rRNA, 5S rRNA, is transcribed separately from the other three rRNAs and is imported into the nucleolus where it is assembled with the mature rRNAs and the ribosomal proteins to produce the functional ribosomal subunits.

Self-Splicing of Introns in Tetrahymena Pre-rRNA. In some species of *Tetrahymena*, for example, *T. thermophila*, the genes for the 28S rRNA are all interrupted by a 413-bp intron. The intron sequence is removed during processing of the pre-rRNA in the nucleolus. The excision of the intron unexpectedly was shown to occur by a *protein-independent reaction* in which the RNA intron becomes folded into a secondary structure that promotes its own excision. This process is called **self-splicing** and was discovered by Tom Cech and his research group. The self-splicing of the *Tetrahymena* pre-rRNA intron was the first example of what is now called **group I-intron self-splicing**. Other self-splicing group I introns have been found in rRNA genes and some mRNA genes in the mitochondria of yeast, *Neurospora*, and other fungi, in rRNA genes of all insect species that have been examined, in rRNA genes, and some mRNA and tRNA genes in bacteriophages.

Figure 13.23 diagrams the self-splicing reaction for the group I intron in *Tetrahymena* pre-rRNA. The steps are as follows:

1. The pre-rRNA is cleaved at the 5' splice junction as guanosine is added to the 5' end of the intron. (It is the addition of a G that characterizes group I introns. Another class of self-splicing introns, the group II introns, involve a reactive A in the intron removal process.)
2. The intron is cleaved at the 3' splice junction.
3. The two exons are spliced together.
4. The excised intron circularizes to produce a lariat molecule which is cleaved to produce a circular RNA and a short linear piece of RNA.

In the processing of *Tetrahymena* pre-rRNA, two separate events occur: (1) The precursor is cleaved to remove spacer sequences from the mature rRNAs, and (2) the intron in the 28S rRNA sequence is removed and the two 28S rRNA parts spliced together. The two processes are not identical. *Removal of the spacer sequences releases rRNAs that remain separate. Intron removal, by contrast, results in the splicing together of the RNA sequences that flanked the intron.*

The self-splicing activity of the intron RNA sequence can not be considered an enzyme activity. That is, while it catalyzes the reaction, it is not regenerated in its original form at the end of the reaction, as is the case with protein enzymes. Recently, however, the *Tetrahymena* intron RNA has been modified in the laboratory so that it can function as an enzyme in the true sense, that is, catalytically. This **RNA enzyme** is called a **ribozyme**. Ribozymes can be used experimentally to cleave RNA molecules at specific sequences.

The discovery that RNA can act like a protein was an extremely important and exciting landmark in biology. In particular, the discovery has revolutionized theories about the origin of life on this planet. That is, previous theories proposed that proteins were required for replication of the first nucleic acid molecules. The new theories propose that the first nucleic acid was self-replicating.

KEYNOTE

In some precursor rRNAs there are introns, the RNA sequences of which fold into a secondary

---

~ FIGURE 13.23

Self-splicing reaction for the group I intron in *Tetrahymena* pre-rRNA.

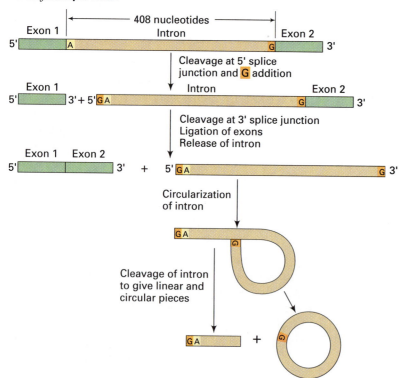

structure that excises itself, a process called self-splicing. The self-splicing reaction requires guanine but does not involve any proteins.

Transcription of 5S rRNA Genes.

As we have just learned, the 5S rRNA genes (5S rDNA) are repeated genes transcribed separately from the rDNA repeat units. The 5S rRNA genes encode the 120-nucleotide 5S rRNA; one 5S rRNA is found in each large ribosomal subunit. The 5S rRNA genes are transcribed by RNA polymerase III; this enzyme also transcribes tRNA genes and some snRNA genes.

The 5S rRNA genes were the first eukaryotic genes for which the promoter structure and transcription factor requirements were determined. Based on the knowledge of upstream promoters for bacterial genes, the expectation was that the 5S rRNA genes would also have an upstream promoter. The promoter location was investigated by making a series of overlapping deletions that removed either various segments of the upstream sequences and of the gene itself, or of the downstream sequences (Figure 13.24). The effect of the deletions on the ability of the gene to be transcribed was then determined. The unexpected result was that the promoter for RNA polymerase III in these genes is *within the gene itself*. More specifically, the experiments showed that removal of any part of a 40-to-50 nucleotide pair sequence from position 50 to 97 of the gene reduced transcription markedly. Unlike the arrangements we have discussed previously, the promoter for RNA polymerase III in 5S rRNA genes is *within the gene itself*. This internal promoter is called the **internal control region (ICR).** The tRNA genes (tDNA), which encode the 75–90

nucleotide tRNAs, also have internal promoters, although snRNA genes typically have promoters upstream of the genes.

The 5S rDNA and tDNA ICRs each have two functional domains, *boxA* and *boxC* for 5S rDNA and *boxA* and *boxB* for tDNA. The ICRs function through interaction with the transcription factors TFIIIA, TFIIIB, and TFIIIC. Figure 13.25 shows a model for formation of a transcription initiation complex on a 5S rDNA ICR. TFIIIA first binds to *boxC* of the ICR (Figure 13.25, part 1) and this allows TFIIIC to bind to *boxA* region (Figure 13.25, part 2). TFIIIB then binds to the other TFs and not to DNA (Figure 13.25, part 3). TFIIIB positions RNA polymerase III correctly on the gene; that is, TFIIIB functions as a *transcription initiation factor* (Figure 13.25, part 4). RNA polymerase III

Use of overlapping deletions for determining the location of the promoter for 5S rRNA genes. (For explanation, see text.)

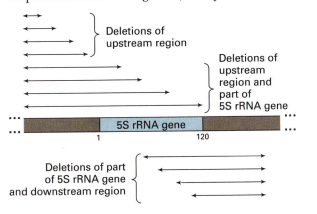

Model for the formation of a transcription initiation complex on a 5S rDNA ICR. See text for detailed description of events.

❶ TFIIIA binds to *box C*

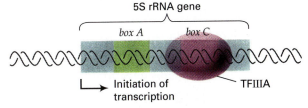

❷ This facilitates binding of TFIIIC to *box A*

❸ TFIIIB binds to other TFs, but not to DNA

❹ TFIIIB positions RNA polymerase III on gene

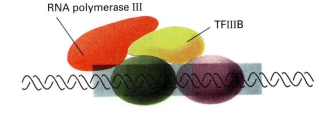

then initiates transcription 50 base pairs *upstream* from the beginning of *boxA*, that is, at the beginning of the gene. While the transcription factors are stably positioned on the gene, they interact with one RNA polymerase III molecule after another to facilitate repeated transcription of the whole gene.

Termination of transcription for both types of genes involves simple sequences at the 3'-ends of the genes that signal the release of RNA polymerase III. For 5S rDNA, on the coding strand there is a cluster of four or more T nucleotides surrounded by GC-rich sequences. For tDNA, a cluster of T nucleotides is used to signal transcription termination.

Transcription of 5S rDNA directly produces the mature 5S rRNA; that is, there are no extra sequences that must be removed. Transcription of tDNA produces a **pre-tRNA molecule** which has extra sequences that must be removed to produce the mature tRNA. This is discussed later.

a particular tRNA molecule to bind a particular amino acid. All tRNA molecules are also extensively modified chemically after transcription (see p. 411).

The nucleotide sequences of all tRNAs can be arranged into what is called a *cloverleaf model of tRNA*. This model is a secondary structure in that only two dimensions are used to display the sequence. Figure 13.26 shows the general features of the cloverleaf model of yeast alanine tRNA. The cloverleaf results from complementary-base pairing between different sections of the molecule which results in four base-paired "stems" separated by four loops, I, II, III, and IV. (Some tRNAs do not have loop III.) Loop II contains within it the three-nucleotide sequence called the **anticodon**, which pairs with a codon (three-nucleotide sequence) in mRNA by complementary-base pairing during translation. This codon-anticodon pairing is crucial for adding the correct amino

𝒦EYNOTE

RNA polymerase III transcribes 5S rRNA genes (5S rDNA), tRNA genes (tDNA), and some snRNA genes. For 5S rDNA and tDNA, the promoter for RNA polymerase III is located within the gene itself. The internal promoter is called the internal control region and consists of different combinations of functional domains depending on the class of gene involved. The domains are binding sites for transcription factors required for transcription by RNA polymerase III.

Transfer RNA

Transfer RNAs (tRNAs) function in both prokaryotes and eukaryotes to bring amino acids to the ribosome-mRNA complex, where they are polymerized into protein chains in the translation (protein synthesis) process. In this section we discuss the structure and biosynthesis of tRNAs.

STRUCTURE OF tRNA. Transfer RNA molecules constitute between 10 and 15 percent of the total cellular RNA in both prokaryotes and eukaryotes. They have a size of about 4S, and they consist of a single chain of 75 to 90 nucleotides. Each type of tRNA molecule has a different sequence, although all tRNAs have the sequence 5'-CCA-3' at their 3' ends. The differences in nucleotide sequences explain the ability of

~ **FIGURE 13.26**

Cloverleaf structure of yeast alanine tRNA. Py = pyrimidine. Modified bases are: I = inosine; T = ribothymidine; ψ = pseudouridine; D = dihydrouridine; GMe = methylguanosine; GMe2 = dimethylguanosine; IMe = methylinosine.

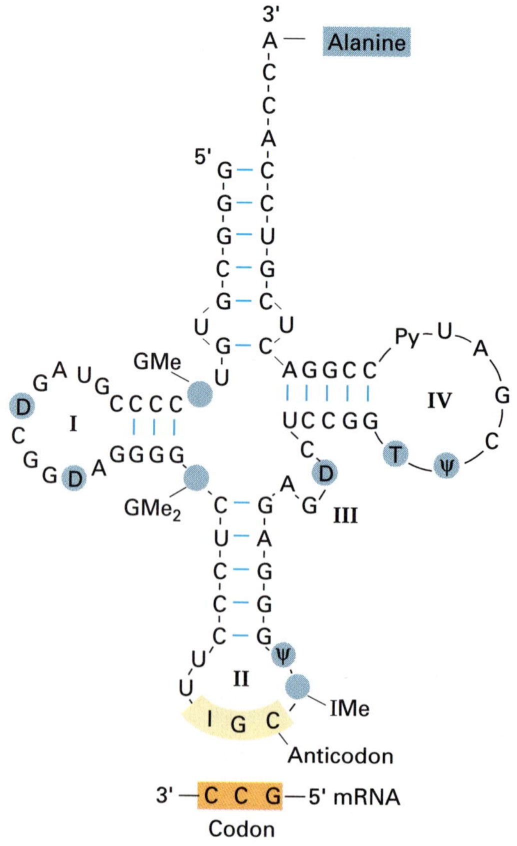

acid (as specified by the mRNA) to the growing polypeptide chain. Loop IV contains the sequence 5' T-ψ-C 3', which is universal among tRNAs.

Because tRNA molecules are so small, they can be crystallized, and X-ray crystallography can be used to develop a three-dimensional model. Figure 13.27 shows the tertiary-structure model for yeast tRNA.Phe (this terminology indicates the amino acid specified by the anticodon of the tRNA, in this case phenylalanine). All other tRNAs that have been examined show similar three-dimensional structures.

The crystallography data showed that all hydrogen-bonded stem structures proposed in the cloverleaf model do exist in the tRNA molecule. In addition, the results showed that there is other hydrogen bonding that folds the cloverleaf into a more compact shape, like an upside-down "L." In this *L*-shaped structure the 3' end of the tRNA (the end to which the amino acid attaches) is at the opposite end of the "L" from the anticodon loop.

tRNA GENES.   A three-base sequence (codon) in an mRNA specifies each amino acid to be added to a polypeptide chain. While only 20 different amino

acids can be used to make a protein, 61 different codons are used in an mRNA to specify the 20 amino acids. Three additional codons do not specify amino acids but instead are used as termination signals for protein synthesis. Since each amino acid–specifying codon must be matched by an appropriate anticodon on a tRNA molecule, any cell could contain 61 different types of tRNA. Thus, for many amino acids, there is more than one tRNA with the appropriate anticodon; this will be described in more detail in Chapter 14. So that the instructions for the manufacture of such large numbers of tRNAs can be coded, the genome has many copies of tRNA genes. tRNA genes are found scattered around the genome in single copies and, in some cases, grouped together in what are called *gene clusters* similar to the rDNA repeat units discussed earlier.

On the *E. coli* chromosome, specific tRNA genes are present in one copy, while others are present in two or more copies, a situation called **gene redundancy**. Recall also that some tRNA genes are found in the rRNA transcription units (see p. 402). In general, many more tRNA genes occur in eukaryotes than in prokaryotes. In yeast, for example, about 400

~ **FIGURE 13.27**

(a) Schematic of the three-dimensional structure of yeast phenylalanine tRNA as determined by X-ray diffraction of tRNA crystals. Note the characteristic upside-down L-shaped structure. (b) Photograph of a space-filling molecular model of yeast phenylalanine tRNA. The CCA end of the molecule is at the upper right, and the anticodon loop is at the bottom.

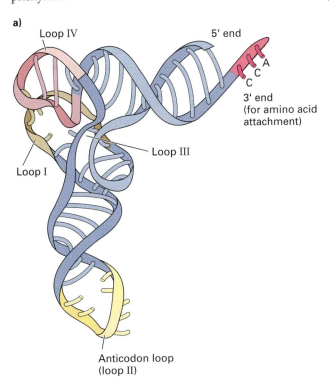
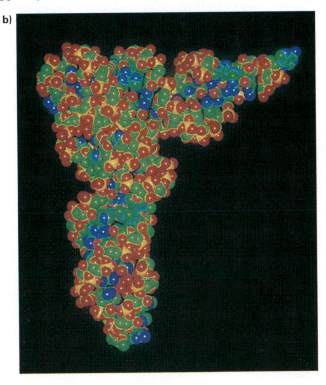

tRNA genes occur in the genome and *Xenopus laevis* has more than 200 copies of *each* tRNA gene per genome. As a class, then, eukaryotic tRNA genes are found in the moderately repetitive class of DNA (see Chapter 11).

Some tRNA genes in certain eukaryotic organisms contain introns. For example, about 10 percent of the 400 tRNA genes in yeast have introns. Depending on the tRNA, the intron is 14 to 60 base pairs long and is almost always located between the first and second nucleotides 3' to the anticodon. The anticodon itself often, but not always, pairs with intron sequences in the pre-tRNA. For reasons that are not understood, introns are found only in tRNAs for the amino acids tyrosine, phenylalanine, tryptophan, lysine, proline, serine, leucine, and isoleucine.

The first tRNA gene in which an intron was found was the yeast gene for tRNA.Tyr, the gene for

the tRNA that carries the amino acid tyrosine. Figure 13.28 shows the nucleotide sequence for the initial transcript of the tRNA gene, the pre-tRNA (Figure 13.28a), and the mature tRNA (Figure 13.28b), as well as the organization of the two sequences into cloverleaf models. The 14-base intron is located in the pre-tRNA just to the 3' side of the anticodon. The presence of the transcript of the intron sequence in the pre-tRNA results in a significant change in the anticodon loop, as Figure 13.28a shows. Removal of the intron involves cleavage with a specific endonuclease. The RNA pieces generated by intron removal are then spliced together by an enzyme called **RNA ligase**.

TRANSCRIPTION OF tRNA GENES AND THE PRODUCTION OF tRNA. In *E. coli* the tRNA genes are transcribed by the same RNA polymerase that transcribes all the genes; hence the promoter and ter-

~ **FIGURE 13.28**

Cloverleaf models for yeast precursor tRNA.Tyr and mature tRNA.Tyr. (a) The 14-base intron in the precursor, located adjacent to the anticodon-coding region. (b) The mature tRNA.Tyr molecule produced after the intron has been removed.

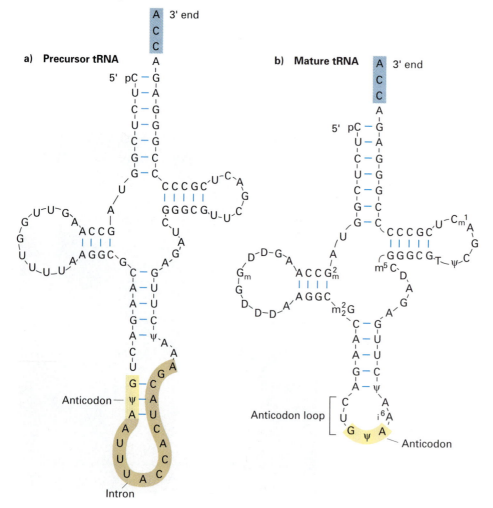

minator sequences are similar to other gene classes. In eukaryotes the tRNA genes are transcribed by RNA polymerase III. Recall from our discussion of 5S rDNA transcription by RNA polymerase III that eukaryotic tRNA genes have internal promoters called internal control regions.

The primary transcripts of tRNA genes in both prokaryotes and eukaryotes are the precursor tRNAs (pre-tRNAs). The pre-tRNAs are extensively modified and processed to produce the mature tRNAs. Two types of modification occur: (1) addition of a 5'-CCA-3' sequence to the 3' end; and (2) extensive chemical modification of a number of nucleotides at many places within the chain. Neither of these modifications occur in pre-mRNAs or pre-rRNAs. The type and extent of the modifications vary from tRNA to tRNA, but typically include the addition of methyl groups to specific bases, the reduction of certain uridines—say, to dihydrouridine—or the rearrangement of some uridines to produce pseudouridine (ψ). Figure 13.29 gives examples of the modified bases found in tRNAs. These modifications are brought about by specific enzyme action. The specific modifications result in the particular two- and three-dimensional configurations of tRNAs, which in turn determine the functions of those molecules; specifically, their ability to pick up a specific amino acid and bind to ribosomes.

As is the case with mRNAs, pre-tRNAs are longer than mature tRNAs and have a 5' leader sequence and a 3' trailer sequence. Unlike mRNA, in which the leader and trailer sequences remain, these sequences are removed by specific enzymes during the processing of pre-tRNA into mature tRNA. In eukaryotes the pre-tRNA processing occurs in the nucleus, so only mature tRNAs are found in the cytoplasm. Figure 13.30 shows the nucleotide sequence of a prokaryotic pre-tRNA and the leader and trailer sequences that are removed during the processing steps. A similar arrangement exists in eukaryotic systems.

In prokaryotes, but not in eukaryotes, a cluster of tRNA genes may be transcribed to produce a single RNA transcript containing a number of tRNA sequences. The general organization of the multi-tRNA pre-tRNA molecules is

$$5'\text{-leader–}(\text{tRNA–spacer})_n\text{–tRNA–trailer-}3'$$

where n is a number of (tRNA–spacer)s characteristic

~ **FIGURE 13.29**

Some modified bases found in tRNA molecules.

Inosine (I)

1-Methylinosine (IMe)

N₂-Dimethylguanosine (GMe₂)

1-Methylguanosine (GMe)

Ribothymidine (T)   Dihydrouridine (D)   Pseudouridine (ψ)

~ **FIGURE 13.30**

Nucleotide sequence of a prokaryotic precursor tRNA (*E. coli* pre-tRNA.Tyr), showing the nucleotides removed during the processing steps.

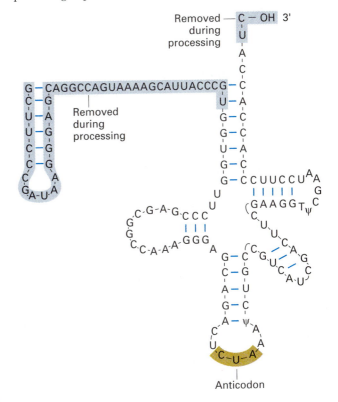

Anticodon

of a cluster. In such cases the leader, the trailer, and the spacer sequences are removed by specific enzymes.

Figure 13.31 gives an example of an *E. coli* pre-tRNA molecule that contains two tRNA sequences. In such cases the leader, the trailer, and the spacer sequences must be removed during the processing of the pre-tRNA to produce two discrete, mature tRNAs. The non-tRNA sequences are removed from the pre-tRNA molecules by the action of specific enzymes. In *E. coli* at least two enzymes are needed. The enzyme RNase P catalyzes removal of the 5' leader sequences, and RNase Q catalyzes the removal of the 3' trailer sequence. RNase P has been shown to be a ribonucleoprotein enzyme, that is, a complex between an RNA and a protein. Another enzyme (or enzymes) is involved in removing the spacer region between clustered tRNAs. Little is known about the enzymes needed for the removal of the terminal noncoding sequences of eukaryotic pre-tRNAs, except that they must be located in the nucleus, since pre-tRNAs are not found in the cytoplasm.

KEYNOTE

Molecules of transfer RNA bring amino acids to the ribosomes, where the amino acids are polymerized into a protein chain. All the tRNA molecules are between 75 and 90 nucleotides long, contain a number of modified bases, and have similar three-dimensional shapes. A 5'-CCA-3' sequence is found at the 3' end of all tRNAs. tRNAs are made as pre-tRNA molecules containing 5'-leader and 3'-trailer sequences, both of which are removed by enzyme activity. Some eukaryotic pre-tRNAs contain introns, which are removed in processing steps that are different from pre-mRNA processing steps.

SUMMARY

Transcription

When a gene is expressed, the DNA base pair sequence is transcribed into the base sequence of an RNA molecule. Four major classes of RNA transcripts are produced by transcription of four classes of genes: messenger RNA (mRNA), transfer RNA (tRNA),

ribosomal RNA (rRNA) and small nuclear RNA (snRNA). snRNA is found only in eukaryotes, while the other three classes are found in both prokaryotes and eukaryotes. Only mRNA is subsequently translated to produce a protein molecule.

When a gene is transcribed, only one of the two DNA strands is copied. The direction of RNA synthesis is 5' to 3' and the reaction is catalyzed by RNA polymerase. In addition to the coding sequences, genes contain other sequences important for the regulation of transcription, including promoter sequences and terminator sequences. Promoter sequences specify where transcription of the gene is to begin, and terminator sequences specify where transcription is to stop.

In bacteria there is only one type of RNA polymerase, hence all classes of RNA are synthesized by the same enzyme. Consequently, the promoters for all three classes of genes are very similar. The promoter is recognized by a complex between the RNA polymerase core enzyme and a protein factor called sigma. Once transcription is initiated correctly, the sigma factor dissociates from the enzyme and is reused in other transcription initiation events. Termination occurs in one of two ways.

In eukaryotes there are three different RNA polymerases located in the nucleus. RNA polymerase I, located exclusively in the nucleolus, transcribes the 18S, 5.8S, and 28S rRNA sequences. These rRNAs are part of ribosomes. RNA polymerase II, located in the nucleoplasm, transcribes protein-coding genes (into mRNAs) and some snRNA genes. RNA polymerase III, located in the nucleoplasm, transcribes tRNA genes, 5S rRNA genes, and the remaining snRNA genes. None of the three eukaryotic RNA polymerases directly binds to promoters. Rather, the promoter sequences for the genes they transcribe are first recognized by specific transcription factors. The transcription factors bind to the DNA and facilitate binding of the polymerase to the transcription factor–DNA complex for correct initiation of transcription.

The promoters for the three types of RNA polymerase differ. For genes transcribed by RNA polymerase II, the promoter consists of a number of sequence modules located within a relatively short distance upstream of the transcription initiation point. These modules serve as binding sites for transcription factors and for regulatory proteins that function to regulate transcription. The 18S, 5.8S, and 28S rRNA sequences transcribed by RNA polymerase I are organized into a single transcription unit so that transcription produces a precursor RNA molecule containing all three rRNAs plus extra RNA material. Uniquely for 5S rRNA genes and tRNA genes transcribed by RNA polymerase III, the promoter is locat-

~ FIGURE 13.31

A tRNA precursor for *E. coli* that contains two tRNA sequences, one for tRNA.Ser and the other for tRNA.Thr. The nucleotides removed during processing are shown in blue.

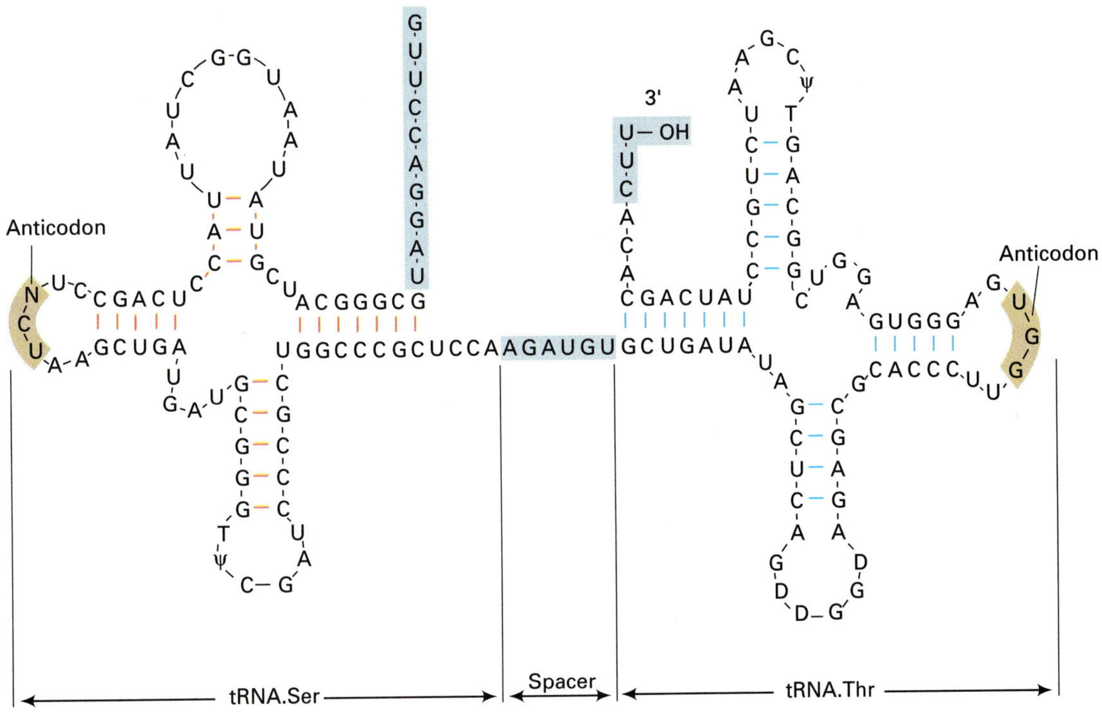

ed within the gene itself. For all three RNA polymerases, transcription factors first recognize the promoter; then the polymerase binds to initiate transcription.

In sum, while the transcription process is very similar in prokaryotes and eukaryotes, the molecular components of the transcription process itself differ considerably. Even within eukaryotes, three different promoters have evolved, along with three distinct RNA polymerases. Much remains to be learned about the associated transcription factors and regulatory proteins before we have a complete understanding of transcription processes and their regulation.

RNA Molecules and RNA Processing

We discussed the structure, synthesis, and function of mRNA, tRNA, and rRNA. Each mRNA encodes the amino acid sequence of a polypeptide chain. The nucleotide sequence of the mRNA is translated into the amino acid sequence of the polypeptide chain on ribosomes. Ribosomes consist of two unequal-sized subunits, each of which contains both rRNA and protein molecules. The amino acids that are assembled into proteins are brought to the ribosome attached to tRNA molecules.

There are three main parts to mRNAs: a 5' leader sequence, the amino acid coding sequence, and the 3' trailer sequence. Since all three sequences vary from mRNA to mRNA, as a group mRNAs show extensive variability in length. In prokaryotes the primary gene transcript functions directly as the mRNA molecule, while in eukaryotes the primary RNA transcript must be modified in the nucleus by RNA modifying and processing events to produce the mature mRNA. These events are the addition of a 5' methylated cap, the addition of a 3' poly(A) tail, and the removal of any introns (internal sequences that do not code for amino acids) that are present. The latter is known as RNA splicing and involves specific interactions with snRNPs in molecular structures called spliceosomes. Only when all processing events have been completed is the mRNA functional; at that point, it leaves the nucleus and can be translated in the cytoplasm.

Ribosomal RNAs are important structural components of ribosomes, the cellular organelles on which protein synthesis takes place. In both prokaryotes and eukaryotes, ribosomes consist of two unequal-sized subunits. The subunits consist of both rRNA and ribosomal proteins. In each case the smaller subunit contains one rRNA molecule, 16S in prokaryotes, and 18S in eukaryotes. The prokaryotic larger subunit contains 23S and 5S rRNAs, while the eukaryotic larger subunit contains 28S, 5.8S, and 5S rRNAs.

In prokaryotes the genes for the 16S, 23S, and 5S rRNAs are transcribed into a single pre-rRNA molecule. The tRNA genes are found in the spacer regions of each transcription unit. The pre-rRNA molecules are processed in a number of enzymatically catalyzed steps to remove the noncoding sequences found at the ends of the molecules and between the rRNA sequences. At the same time the tRNAs are released. All of the processing events occur while the pre-rRNA is associating with the ribosomal proteins so that, when processing is complete, the functional 50S and 30S subunits have been assembled.

In eukaryotes there are many copies of the genes for each of the four rRNAs. The 18S, 5.8S, and 28S rRNA genes comprise a transcription unit, and many such transcription units are organized in a tandem array. The 5S rRNA genes are also present in many copies but they are usually located elsewhere in the genome. The 18S, 5.8S, and 28S sequences are transcribed into pre-rRNA molecules which, in addition to the rRNA sequences, contain spacer sequences at the ends and between the rRNA sequences. The spacers are removed by specific processing events that take place while the rRNA sequences are associated with ribosomal proteins. The 5S rRNA, which is transcribed elsewhere in the nucleus, becomes associated with the assembling 60S subunit so that, at the end of the processing steps, functional 60S and 40S ribosomal subunits have been produced. All of these events occur in the nucleolus. The mature ribosomal subunits then exit the nucleus and participate in protein synthesis in the cytoplasm.

In the pre-rRNA of *Tetrahymena* the 28S rRNA sequence is interrupted by an intron. The intron is removed during processing of the pre-rRNA in the nucleolus. The excision of this particular intron (and a few other examples in other systems) occurs by a protein-independent reaction in which the RNA sequence of the intron folds into a secondary structure that promotes its own excision. This process is called self-splicing.

All tRNAs carry out the same function; that is, bringing amino acids to the ribosomes. Thus, all tRNAs are very similar in length (75 to 90 nucleotides) and in secondary and tertiary structure. tRNAs are usually synthesized as precursor molecules that have extra nucleotides at both the 5' and 3' ends. The extra nucleotides are removed by specific processing events. In addition a number of the bases are modified during the maturation process and a 5'-CCA-3' sequence is added to the 3' end of all tRNAs. Some tRNA genes in eukaryotes contain introns; these are removed from the transcripts of those genes by splicing events that are different from those used for intron removal from precursor mRNAs.

In sum, the functional RNA molecules of the cell are typically transcribed as precursor molecules that include extra nucleotides that are removed by specific processing events. In prokaryotes the processing events are confined to the tRNAs and rRNAs. In eukaryotes, RNA processing is much more complex, particularly with mRNAs.

ANALYTICAL APPROACHES FOR SOLVING GENETICS PROBLEMS

Q13.1 If two RNA molecules have complementary base sequences, they can hybridize to form a double-stranded structure just as DNA can. Imagine that in a particular region of the genome of a certain bacterium one DNA strand is transcribed to give rise to the mRNA for protein A, while the other DNA strand is transcribed to give rise to the mRNA for protein B.
a. Would there be any problem in expressing these genes?
b. What would you see in protein B if a mutation occurred which affected the structure of protein A?

A13.1a. mRNA A and mRNA B would have complementary sequences, so they might hybridize with each other and not be available for translation.

b. Every mutation in gene A would also be a mutation in gene B, so protein B might also be abnormal.

Q13.2 Compare and contrast the following two events in terms of what their consequences would be. Event (1): an incorrect nucleotide is inserted into the new DNA strand during replication, and not corrected by the proofreading or repair systems before the next replication. Event (2): an incorrect nucleotide is inserted into an mRNA during transcription.

A13.2 Event (1) would result in a mutation, assuming it occurred within a gene. The mistake would be inherited by future generations, and would affect the structure of all mRNA molecules transcribed from the region and therefore all molecules of the corresponding protein could be affected.

Event (2) would produce a single aberrant mRNA. This could produce a few aberrant protein molecules. Additional normal protein molecules would exist because other, normal mRNAs would have been transcribed. The abnormal mRNA would soon be degraded. The mRNA mutation would not be hereditary.

Q13.3 You are given four different RNA samples. Sample I has a short lifetime; sample II has a homogeneous molecular weight; sample III is produced by processing of a larger precursor RNA; and sample IV has an additional sequence added onto the original transcript. For each sample, state whether the RNA could be rRNA, mRNA, or tRNA. If it is not one of those, state what it might be. Note that for each sample more than one RNA could apply. Give reasons for your choices.

A13.3 Sample I: A short lifetime is characteristic of mRNAs in prokaryotes, and many mRNAs in eukaryotes. Heterogeneous nuclear RNA of eukaryotes also has a short lifetime. Sample II: rRNA and tRNA species have homogeneous molecular weights since they carry out specific functions within the cell for which their length and three dimensional configuration are important. Messenger RNA and heterogeneous nuclear RNA are heterogeneous in length. Sample III: rRNA, tRNA, and some mRNAs are all produced by the processing of a larger precursor RNA molecule. Sample IV: Both tRNA and eukaryotic mRNA may have additional sequences added after they are transcribed. For tRNA this sequence is the CCA at the 3' end, and for mRNA this sequence is the poly(A) tail at the 3' end.

QUESTIONS AND PROBLEMS

*13.1 Describe the differences between DNA and RNA.

13.2 Compare and contrast DNA polymerases and RNA polymerases.

13.3 All base pairs in the genome are replicated during the DNA synthesis phase of the cell cycle, but only *some* of the base pairs are transcribed into RNA. How is it determined *which* base pairs of the genome are transcribed into RNA?

13.4 Discuss the structure and function of the *E. coli* RNA polymerase. In your answer, be sure to distinguish between RNA core polymerase and RNA core polymerase-sigma factor complex.

*13.5 Discuss the similarities and differences between the *E. coli* RNA polymerase and eukaryotic RNA polymerases.

13.6 Discuss the molecular events involved in the termination of RNA transcription in prokaryotes. In what ways is this process fundamentally different in eukaryotes?

*13.7 Which classes of RNA do each of the three eukaryotic RNA polymerases synthesize? What are the functions of the different RNA types in the cell?

13.8 *E. coli* RNA polymerase is able to transcribe all the genes of *E. coli*, but the three eukaryotic RNA polymerases transcribe only specific, non-overlapping subsets of eukaryotic genes. What mechanisms are used to restrict the transcription of each of the three eukaryotic polymerases to a particular subset of eukaryotic genes?

13.9 What is the Pribnow box? The Goldberg-Hogness box (TATA element)?

*13.10 What is an enhancer element?

13.11 A piece of mouse DNA was sequenced as follows (a space is inserted after every 10th base for ease in counting; "..." means a lot of unspecified bases):

AGAGGGCGGT CCGTATCGGC   CAATCTGCTC   ACAGGGCGGA
TTCACACGTT GTTATA TAAA   TGACTGGGCG   T ACCCCAGGG
TTCGAGTATT CTATCGTATG   GTGCACCTGA   CT(...)
GCTCACAAGT ACCACTAAGC(...).

What can you see in this sequence to indicate it might be all or part of a transcription unit?

13.12 Compare and contrast the structures of prokaryotic and eukaryotic mRNAs.

*13.13 Compare the structures of the three classes of RNA found in the cell.

13.14 Many eukaryotic mRNAs, but not prokaryotic mRNAs, contain introns. What is the evidence for the presence of introns in genes? Describe how these sequences are removed during the production of mature mRNA.

13.15 How is the mechanism of group I intron removal different from the mechanism used to remove the introns in most eukaryotic mRNAs? Speculate as to why these different mechanisms for intron removal might have evolved and how each might be advantageous to a eukaryotic cell.

13.16 Distinguish between leader sequence, trailer sequence, coding sequence, intron, spacer sequence, nontranscribed spacer sequence, external transcribed spacer sequence, and internal transcribed sequence. Give examples of actual molecules in your answer.

FIGURE 13.A

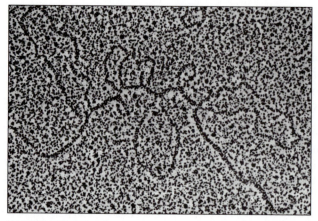

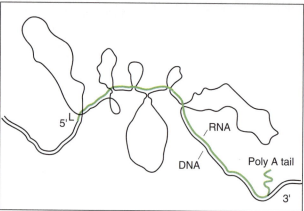

13.17 Figure 13A shows the electron micrograph and interpretative diagram resulting from an R-looping experiment in which the mature chicken ovalbumin mRNA was hybridized with the ovalbumin gene. What would be the minimum number of phosphodiester bonds that would have to be cut to produce the mRNA from the pre-mRNA?

*13.18 Discuss the posttranscriptional modifications that take place on the primary transcripts of tRNA, rRNA, and protein-coding genes.

13.19 Describe the organization of the ribosomal DNA repeating unit of a higher eukaryotic cell.

*13.20 Which of the following kinds of mutations would be likely to be recessive lethals in humans? Explain your reasoning.
a. Deletion of the U1 genes
b. Deletion within intron 2 of β-globin
c. Deletion of 4 bases at the end of intron 2 and 3 bases at the beginning of exon 3 in β-globin.

13.21 The diagram in Figure 13.B shows the transcribed region of a typical eukaryotic protein-coding gene:

What is the size (in bases) of the fully processed, mature mRNA? Assume in your calculations a poly(A) tail of 200 As.

13.22 Most human obesity does not follow Mendelian inheritance patterns, as body fat content is determined by a number of interacting genes and environmental variables. Insights into how specific genes function to regulate body fat content have come from studies of mutant, obese mice. In one mutant strain, *tubby (tub)*, obesity is inherited as a recessive trait. Comparison of the DNA sequence of the *tub*⁺ and *tub* alleles has revealed a single base-pair change: within the transcribed region, a 5' GC base pair has been mutated to a TA base pair. The mutation causes an alteration of the initial 5' base of the first intron. Therefore, in the *tub/tub* mutant, a longer transcript is found. Propose a molecularly based explanation for how a single base change

FIGURE 13.B

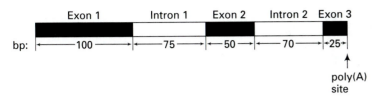

causes a non-functional gene product to be produced, why a longer transcript is found in *tub/tub* mutants, and why the *tub* mutant is recessive.

13.23 Which of the following could occur in a single mutational event in a human? Explain.
a. Deletion of 10 copies of the 5S ribosomal RNA genes only
b. Deletion of 10 copies of the 18S rRNA genes only
c. Simultaneous deletion of 10 copies of the 18S, 5.8S, and 28S rRNA genes only
d. Simultaneous deletion of 10 copies each of the 18S, 5.8S, 28S, and 5S rRNA genes.

13.24 During DNA replication in a mammalian cell a mistake occurs: 10 wrong nucleotides are inserted into a 28S rRNA gene, and this mistake is not corrected. What will likely be the effect on the cell?

13.25 Give the correct answers, noting that each blank may have more than one correct answer, and that each answer (1 through 4) could be used more than once.

Answers:
1. Eukaryotic mRNAs
2. Prokaryotic mRNAs
3. Transfer RNAs
4. Ribosomal RNAs

a. _____ Have a cloverleaf structure
b. _____ Are synthesized by RNA polymerases
c. _____ Display an anticodon each
d. _____ Are the template of genetic information during protein synthesis
e. _____ Contain exons and introns
f. _____ There are four types of these in eukaryotes and only three types in *E. coli*
g. _____ They get charged with an amino acid by aminoacyl-tRNA synthetase
h. _____ Contain unusual or modified nitrogenous bases
i. _____ Are capped on their 5' end and polyadenylated on their 3' end.

14 THE GENETIC CODE AND THE TRANSLATION OF THE GENETIC MESSAGE

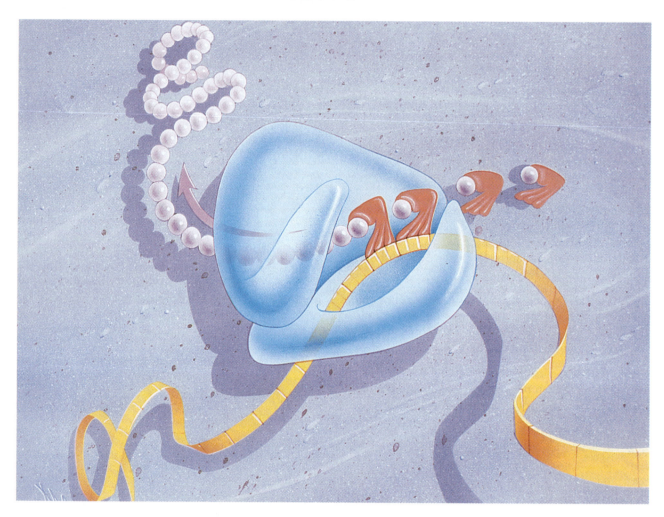

PROTEIN STRUCTURE
Chemical Structure of Proteins
Molecular Structure of Proteins

THE NATURE OF THE GENETIC CODE
The Genetic Code Is a Triplet Code
Deciphering the Genetic Code
Nature and Characteristics of the Genetic Code

TRANSLATION OF THE GENETIC MESSAGE
Aminoacyl-tRNA Molecules
Initiation of Translation
Elongation of the Polypeptide Chain
Termination of Translation

PROTEIN SORTING IN THE CELL
Proteins Distributed by the Endoplasmic Reticulum
Proteins Transported into Mitochondria and Chloroplasts
Proteins Transported into the Nucleus

PRINCIPAL POINTS

~ A protein consists of one or more macromolecular subunits called polypeptides, which are themselves composed of smaller building blocks called amino acids. The amino acids are linked together in the polypeptide by peptide bonds.

~ The primary amino acid sequence of a protein determines its secondary, tertiary, and quaternary structure, and hence its functional state.

~ The genetic code is a triplet code in which each three-nucleotide codon in an mRNA specifies one amino acid. Some amino acids are represented by more than one codon. The code is almost universal, and it is read without gaps in successive, nonoverlapping codons.

~ Translation of the mRNA into a protein chain occurs on ribosomes. Amino acids are brought to the ribosome on tRNA molecules. The correct amino acid sequence is achieved by the specific binding of each amino acid to its specific tRNA, and by the specific binding between the codon of the mRNA and the complementary anticodon of the tRNA.

~ In prokaryotes and eukaryotes, AUG (methionine) is the initiator codon for the start of translation. Elongation of the protein chain involves peptide bond formation between the amino acid attached to the tRNA

in a particular site of the ribosome called the A site and the growing polypeptide attached to the tRNA in an adjacent site called the P site. Once the peptide bond has formed, the ribosome translocates one codon along the mRNA in preparation for the next tRNA with its bound amino acid to bind to the next available codon now occupying the A site.

~ Translation continues until a chain-terminating codon (UAG, UAA, or UGA) is reached in the mRNA. These codons are read by one or more release factor proteins and then the polypeptide is released from the ribosome and the other components of the protein synthesis machinery dissociate.

~ In eukaryotes, proteins are found free in the cytoplasm, as well as in the various cell compartments such as the nucleus, mitochondria, chloroplasts, and secretory vesicles. Special mechanisms exist to sort proteins to their appropriate cell compartments. For example, proteins to be secreted have N-terminal signal sequences that facilitate their entry into the endoplasmic reticulum for later sorting in the Golgi complex and beyond. Proteins destined for the nucleus, mitochondria, or chloroplasts each have specific sequences that program their localization to those compartments.

*T*he information for the proteins found in a cell is encoded in the structural genes of the cell's genome. Expression of a protein-coding gene occurs by transcription of the gene to produce an mRNA (discussed in Chapter 13), followed by **translation** of the mRNA; that is, the conversion of the mRNA base sequence information into an amino acid sequence of a polypeptide. The DNA base pair information that specifies the amino acid sequence of a polypeptide is called the **genetic code**.

In this chapter we will study the translation of the genetic message and how the information for the amino acid sequence of proteins is encoded in the nucleotide sequence of messenger RNA. We will see that three classes of RNA—messenger RNA, transfer RNA, and ribosomal RNA—are involved in the translation process. We start with a discussion of the structure of proteins, the end products of the translation process.

PROTEIN STRUCTURE

Chemical Structure of Proteins

A **protein** is one of a group of high-molecular-weight, nitrogen-containing organic compounds of complex shape and composition. Each cell type has a characteristic set of proteins that give that cell type its functional properties. A protein consists of one or more macromolecular subunits called **polypeptides**, which are themselves composed of smaller building blocks, the **amino acids,** linked together to form long chains. Each molecule of a given protein consists of the same number and kind of polypeptide chains and each of these in turn is composed of the same number, kind, and sequence of amino acids. It is the sequence of amino acids in a polypeptide that gives the polypeptide its three-dimensional shape and its properties in the cell.

The basic chemical structural units of proteins are amino acids. With the exception of proline, the amino acids have a common structure, which is shown in Figure 14.1. The structure consists of a central carbon atom (α-carbon, α meaning alpha) to which is bonded an amino group (NH_2), a carboxyl group (COOH), and a hydrogen atom. At the pH commonly found within cells, the NH_2 and COOH groups of the free amino acids are in a charged state; that is, $-NH_3^+$ and $-COO^-$, respectively.

Bound to the α-carbon, each amino acid has an additional chemical group, called the *R group*. It is the R group that varies from one amino acid to another and gives each amino acid its distinctive properties. Since different polypeptides have different sequences and proportions of amino acids, the organization of the R groups gives a polypeptide its structural and functional properties.

There are 20 amino acids used to make proteins; their names, three-letter abbreviations, and chemical structures are shown in Figure 14.2. The 20 amino acids are divided into subgroups, based on whether the R group is acidic (e.g., aspartic acid), basic (e.g., lysine), neutral-polar (e.g., serine), or neutral-nonpolar (e.g., leucine).

The amino acids of a polypeptide are held together by a **peptide bond**, a covalent bond that joins the carboxyl group of one amino acid to the amino group of another amino acid. The formation of the peptide bond is the result of the reaction depicted in Figure 14.3. A polypeptide, then, is a linear, unbranched molecule that consists of many amino acids (usually 100 or more) joined by peptide bonds. Every polypeptide has a free α-amino group at one end (called the N terminus, or the N terminal end) and a free α-carboxyl group at the other end (called the C terminus, or the C terminal end). Polypeptides have polarity: because the polypeptide is constructed that way, the N terminal end is defined as the beginning of a polypeptide chain.

Molecular Structure of Proteins

The molecular structure of a protein is relatively complex; there are four levels of structural organization, as shown in Figure 14.4.

1. The *primary structure* of the polypeptide chain that constitutes a protein is the amino acid sequence (Figure 14.4a). The amino acid sequence is directly determined by the base pair sequence of the gene that encodes it.

2. The *secondary structure* of a protein refers to the folding and twisting of a single polypeptide chain into a variety of shapes. A polypeptide's secondary structure is the result of weak bonds (e.g., electrostatic or hydrogen) between NH and CO groups of amino acids that are near each other on the chain. One type of secondary structure found in regions of many polypeptides is the α-*helix*, a structure discovered by Linus Pauling and Robert Corey in 1951. Figure 14.4b shows the α-helix and diagrams the hydrogen bonding between the NH group of one amino acid (i.e., an NH group that is part of a peptide bond) and the CO group (also part of a peptide bond) of an amino acid that is four amino acids away in the chain. The repeated formation of this bonding results in the helical coiling of the chain. The α-helix content of proteins varies.

   Another type of secondary structure is the β-pleated sheet (not illustrated). Also discovered by Pauling and Corey in 1951, the β-pleated sheet involves a polypeptide chain or chains folded in a zigzag way with parallel regions or chains linked by hydrogen bonds. A number of proteins contain a mixture of α-helical and β-pleated sheet regions.

3. A protein's *tertiary structure* (Figure 14.4c) is the three-dimensional structure into which the secondary structure is folded. The three-dimensional shape of a polypeptide is often called its *conformation*. Tertiary folding is a direct property of the amino acid sequence of the chain and hence is related to the distribution of the R groups along the chain. Figure 14.4c shows the tertiary structure of the β polypeptide of hemoglobin (see below).

4. *Quaternary structure* of a protein is shown in Figure 14.4d. Note that primary, secondary, and tertiary structures all refer to single polypeptide chains. As we said earlier, however, proteins may consist of more than one polypeptide chain; such proteins are called multimeric ("many subunits") proteins. Quaternary structure is found only in proteins having more than one polypep-

~ **FIGURE 14.1**

General structural formula for an amino acid.

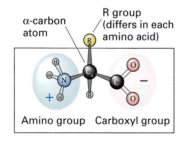

Structures common to all amino acids

~ FIGURE 14.2

Structures of the 20 naturally occurring amino acids. Below each amino acid name are its three-letter and one-letter abbreviations.

Acidic

Neutral, nonpolar

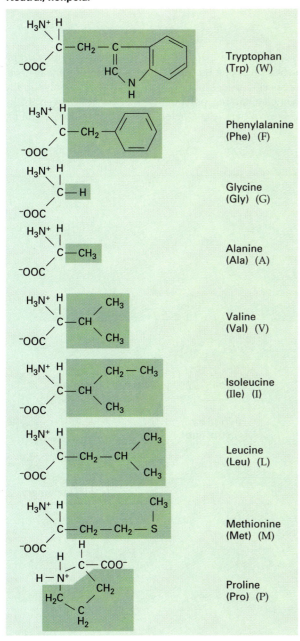

Basic

Neutral, polar

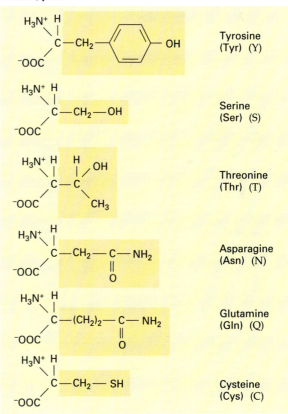

~ FIGURE 14.3

Mechanism for peptide bond formation between the carboxyl group of one amino acid and the amino group of another amino acid.

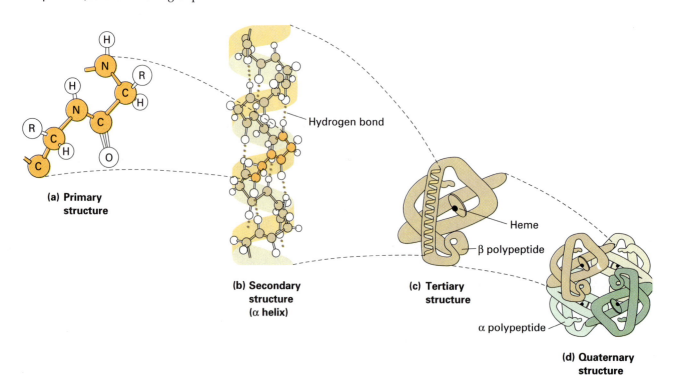

~ FIGURE 14.4

Four levels of protein structure: (a) Primary, the sequence of amino acids in a polypeptide chain. (b) Secondary, the folding and twisting of a single polypeptide chain into a variety of shapes. Shown is one type of secondary structure: the α-helix. Both structures are stabilized by hydrogen bonds. (c) Tertiary, the specific three-dimensional folding of the polypeptide chain. Shown here is the β polypeptide chain of hemoglobin, a heme-containing polypeptide that carries oxygen in the blood. (d) Quaternary, the specific aggregate of polypeptide chains. Shown here is hemoglobin; it consists of two α chains, two β chains, and four heme groups.

tide chain. The term refers to how polypeptides are packaged into the whole protein molecule. Shown in Figure 14.4d is the quaternary structure of a well-known example of a multimeric protein, the oxygen-carrying protein hemoglobin, which consists of four polypeptide chains (two α polypeptides and two β polypeptides), each of which is associated with a heme group (involved in the binding of oxygen). The α and β polypeptides have different amino acid sequences. The α polypeptide contains 141 amino acids, and the β polypeptide contains 146 amino acids. In the quaternary structure of hemoglobin, each α chain is in contact with each β chain; however, little interaction occurs between the two α chains or between the two β chains.

KEYNOTE

A protein consists of one or more molecular subunits called polypeptides, which are themselves composed of smaller building blocks, the amino acids, linked together by peptide bonds to form long chains. The primary amino acid sequence of a protein determines its secondary, tertiary, and quaternary structure and hence its functional state.

THE NATURE OF THE GENETIC CODE

How do nucleotides in the mRNA molecule specify the amino acid sequence in proteins? With four different nucleotides (A, C, G, U), a three-letter code generates 64 possible codons, yet there are only 20 different amino acids. If it were a one-letter code, only four amino acids could be encoded. If it were a two-letter code, then only 16 (4 × 4) amino acids could be encoded. A three-letter code, however, generates 64 (4 × 4 × 4) possible codes, more than enough to code for the 20 amino acids. The assumption of a three-letter code also suggests that some amino acids may be specified by more than one codon, which is in fact the case.

The Genetic Code Is a Triplet Code

The evidence that the genetic code is a triplet code; that is, that a set of three nucleotides (a **codon**) in mRNA code for one amino acid in a polypeptide chain came from genetic experiments done by Francis Crick, Leslie Barnett, Sidney Brenner, and R. Watts-Tobin in

the early 1960s. The experiments used the bacteriophage T4. Recall from Chapter 8 that *rII* mutants of T4 produce clear plaques on *E. coli B*, whereas the wild type *r*[+] strain produces turbid plaques. Further, in contrast to the *r*[+] strain, *rII* mutants are unable to reproduce in *E. coli K12(λ)*.

Crick and his colleagues began with an *rII* mutant strain that had been produced by treating the *r*[+] strain with a chemical that causes mutations (a *mutagen*—discussed in more detail in Chapter 18). The mutagen used was the chemical proflavin, which induces mutations by causing the addition or deletion of a base pair in the DNA. Such mutations are called *frameshift mutations* (see Chapter 19, pp. 620 and 621). They reasoned that, if the *rII* mutant phenotypes resulted from either an addition or a deletion, treatment of the *rII* mutant with proflavin could reverse the mutation to the wild-type—*r*[+]—state. The process of changing a mutant back to the wild-type state is called **reversion**, and the wild type produced in this way is called a *revertant*. So, if the original mutation was an addition, it could be corrected by a deletion. Thus, they were able to isolate a number of *r*[+] revertant strains by plating a population of *rII* phages that had been treated with proflavin onto a lawn of *E. coli K12(λ)* in which only *r*[+] phages can grow. This made it very easy to isolate the low number of *r*[+] revertants produced by the proflavin treatment.

Some of the revertants resulted from an exact correction of the original mutation; that is, an addition corrected the deletion, or a deletion corrected the addition. A second type of revertant was much more useful for determining the nature of the genetic code. This revertant type resulted from a second mutation within the *rII* gene very close to, but distinct from, the original mutation site. If, for example, the first mutation was a deletion of a single base pair, the reversion of this mutation in this way would involve an addition of a base pair nearby. Figure 14.5a shows a hypothetical segment of DNA. For the purposes of discussion, we have assumed that the code is a triplet code. Thus, the mRNA transcript of the DNA would be read ACG ACG ACG, and so on, giving a polypeptide with a string of identical amino acids—threonine—each specified by ACG. If proflavin treatment causes a deletion of the second AT base pair, the mRNA will now read ACG CGA CGA, etc., giving a polypeptide starting with the amino acid specified by ACG (threonine) following by a string of amino acids that are specified by CGA (arginine) (Figure 14.5b). In essence, the reading frame of the message has been changed; this mutation is a frameshift mutation—the message is out of frame by one. Reversion of this mutation can occur by adding a base pair nearby. For example, the insertion of a GC base pair after the GC in the third triplet results in a mRNA that is read as ACG CGA

~ **FIGURE 14.5**

Reversion of a deletion frameshift mutation by a nearby addition mutation. (a) Hypothetical segment of normal DNA, mRNA transcript and polypeptide in the wild type. (b) Effect of a deletion mutation on the amino acid sequence of a polypeptide. The reading frame is disrupted. (c) Reversion of the deletion mutation by an addition mutation. The reading frame is restored, leaving a short segment of incorrect amino acids.

a)   Wild type

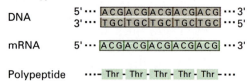

b)   Frameshift mutation by deletion

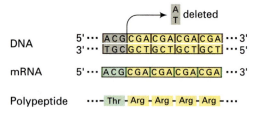

c)   Reversion of deletion mutation by addition

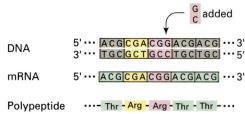

CGG ACG ACG, and so on. This gives a polypeptide consisting mostly of the amino acid specified by ACG (threonine) but with two wrong amino acids; those specified by the CGA and CGG (both arginine). Thus, the second mutation has restored the reading frame and a nearly wild-type polypeptide is produced. As long as the incorrect amino acids in the short segment between the mutations do not significantly affect the function of the polypeptide, the double mutant will have a normal or a near-normal phenotype.

In short, an addition mutation can be reverted by a nearby deletion mutation and a deletion mutation can be reverted by a nearby addition mutation. We symbolize addition mutations as + mutations and deletion mutations as − mutations. The next step Crick and his colleagues took was to combine genetically distinct *rII* mutations of the same type (either all + or all − mutations[1]) in various numbers to see whether any combinations reverted the *rII* phenotypes. Figure 14.6 gives a hypothetical presentation of the type of results they obtained, showing the effects of the mutations just on the mRNA. The figure shows a 30-nucleotide segment of mRNA that codes for 10 different amino acids in the polypeptide. If we add three base pairs at nearby locations in the DNA coding for this mRNA segment, the result will be a 33-nucleotide segment that codes for 11 amino acids; that is, one more than the original. Note, though, that the amino acids between the first and third insertions are not the same as the wild-type mRNA. In essence, the reading frame is correct before the first insertion and

---

[1]Crick and his colleagues did not know if an *rII* mutant resulted from a + or a − mutation. But they did know which of their single-mutant *rII* strains were of one sign and which were of the other sign. That is, all mutants of one sign (e.g., +) could be reverted by nearby mutants of the other sign (i.e., −), and vice versa.

~ **FIGURE 14.6**

Hypothetical example showing how three nearby + (addition) mutations restore the reading frame, giving normal, or near-normal function. The mutations are shown here at the level of the mRNA.

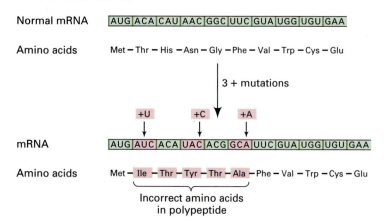

again after the third insertion. The incorrect amino acids between those points may result in a not quite wild-type phenotype for the revertant.

Crick and his colleagues found that the combination of either three nearby + mutations or three nearby − mutations gave r^+ revertants. No other multiple combinations worked, except multiples of three. They concluded, therefore, that *the genetic code is a triplet code.*

Deciphering the Genetic Code

With the evidence in hand that the code was a triplet code, the next step was to determine how each three-letter codon corresponded to an amino acid. The exact relationship of the 64 codons to the 20 amino acids was determined by experiments done mostly in the laboratories of Marshall Nirenberg and Ghobind Khorana. Essential to these experiments was the use of **cell-free, protein-synthesizing systems**, which contained ribosomes, tRNAs with amino acids attached, and all the necessary protein factors for polypeptide synthesis. These cell-free, synthesizing systems were assembled from components isolated and purified from *Escherichia coli* (Figure 14.7). Protein synthesis in these systems is inefficient, in that very little protein is made. Therefore, radioactively labeled amino acids have to be used to measure the incorporation of amino acid into new proteins.

Figure 14.8 shows the typical time course for the incorporation of radioactive amino acids into proteins in an in vitro system. Protein synthesis takes place rapidly for about 15 minutes, then stops gradually because the mRNA that was in the *E. coli* extract (the endogenous mRNA) degrades. If fresh mRNA is added, protein synthesis resumes and proceeds until the new mRNA (the exogenous mRNA) is degraded. Thus, once the *E. coli* extract has been made and been incubated to allow the system's original endogenous

mRNA to degrade, it becomes an excellent system for testing the coding capacity of synthetic mRNAs. The system's usefulness lies in the fact that a single type of mRNA molecule can be added to the system, enabling the investigators to analyze the result of this addition on the composition of the polypeptide produced. As we shall see, the mRNAs used were synthetically made in the laboratory; their relative simplicity generated valuable information about the relationships between codons and amino acids that would not have been possible with the very complex populations of natural mRNAs.

In one approach to establish which codons specify which amino acids, synthetic mRNAs containing one, two, or three different types of bases were made and added to cell-free, protein-synthesizing systems. The polypeptides made in these systems were then analyzed. When the synthetic mRNA contained only one type of base, the results were unambiguous. Synthetic poly(U) mRNA, for example, directed the synthesis of a polypeptide consisting of phenylalanines (a polyphenylalanine chain). Since the genetic code is a triplet code, this result indicated that UUU is a codon for phenylalanine. Similarly, a synthetic poly(A) mRNA directed the synthesis of a polylysine, and poly(C) directed the synthesis of polyproline, indicating that AAA is a codon for lysine and CCC is a codon for proline. The results from poly(G) were inconclusive since the poly(G) folds up upon itself, so it cannot be translated in vitro.

Synthetic mRNAs made by the random incorporation of two different bases (called *random copolymers*) were also analyzed in the cell-free, protein-synthesizing systems. When mixed copolymers are made, the bases are incorporated into the synthetic molecule in a random way. Thus poly(AC) molecules can contain eight different codons (CCC, CCA, CAC, ACC, CAA, ACA, AAC, and AAA), and the propor-

---

~ **FIGURE 14.7**

Experimental details of a cell-free, protein-synthesizing system.

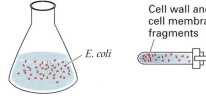 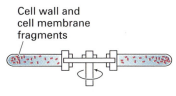  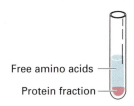

1 *E. coli* cells are collected by centrifugation and lysed. DNase is added to degrade DNA.

2 Cell lysate is centrifuged to pellet cell wall and cell membrane fragments. Supernatant, containing polysomes, ribosomes, mRNA, tRNA, and enzymes, is transferred to test tubes. ATP, GTP, and radioactive amino acids are added to the tubes, which are incubated at 37°C for various times.

tRNA
ATP and GTP
Enzymes
Free ribosome subunits and polyribosomes containing mRNA
Radioactive amino acids

3 Proteins are precipitated by acid, leaving amino acids in solution. Radioactivity in the precipitate indicates the amount of amino acids incorporated into newly synthesized protein.

~ FIGURE 14.8

Typical time course for the incorporation of radioactive amino acids into proteins in an in vitro system. Initially radioactivity is incorporated into proteins until the endogenous mRNA breaks down. Incorporation of radioactive amino acids into proteins resumes when exogenous mRNA is added and continues until that mRNA breaks down.

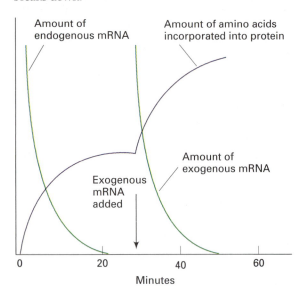

tions of each depend on the A-to-C ratio used to make the polymer. In the cell-free, protein-synthesizing system, poly(AC) synthetic mRNAs caused the incorporation of asparagine, glutamine, histidine, and threonine into polypeptides, in addition to the lysine expected from AAA codons and the proline expected from CCC codons. The proportions of asparagine, glutamine, histidine, and threonine incorporated into the polypeptides produced depended on the A/C ratio used to make the mRNA, and these observations were used to deduce information about the codons that specify the amino acids. For example, since an AC random copolymer containing much more A than C resulted in the incorporation of many more asparagines than histidines, researchers concluded that asparagine is coded by two As and one C and histidine by two Cs and one A. With experiments of this kind, the base composition (*but not the base sequence*) of the codons for a number of amino acids was determined.

Another experimental approach also used copolymers, but these copolymers had been synthesized so that they had a known sequence, not a random one. For example, a repeating copolymer of U and C gives a synthetic mRNA of UCUCUCUCUC. When this copolymer is tested in a cell-free, protein-synthesizing system, the resulting polypeptide had a repeating amino acid pattern of leucine-serine-leucine-serine. From this result, researchers conclude that UCU and CUC specify leucine and serine, although they could not determine from this information alone which coded for which.

Yet another approach used a *ribosome-binding assay*, developed in 1964 by Nirenberg and Philip Leder. This assay depended on the fact that, in the absence of protein synthesis, specific tRNA molecules will bind to complexes formed between ribosomes and mRNAs. For example, when synthetic mRNA poly(U) is mixed with ribosomes, it forms a poly(U)-ribosome complex, and only tRNA.Phe (i.e., the tRNA that will bring phenylalanine to an mRNA and that has the appropriate anticodon, AAA, for the UUU codon) will bind to the UUU codon. Importantly, the specific binding of the appropriate tRNA to the mRNA-ribosome complex does not require the presence of long mRNA molecules; the binding of a trinucleotide (i.e., a codon-length piece of RNA) will suffice. The discovery of the trinucleotide-binding property made it possible to determine relatively easily the specific relationships between many codons and the amino acids for which they code. Note that in this particular approach, the *specific nucleotide sequence of the codon is determined*, not merely the nucleotide composition. Using the ribosome-binding assay, many of the ambiguities that had arisen from other approaches were resolved. For example, UCU was found to promote the binding of a tRNA.Ser, that is, a tRNA that brings the amino acid serine to the message. Thus UCU must code for serine. Similarly, CUC causes tRNA.Leu binding, so CUC codes for leucine. All in all, about 50 codons were clearly identified by using this approach.

In sum, no single approach produced an unambiguous set of codon assignments, but information obtained through all the approaches enabled all codon assignments to be deduced with high degrees of certainty.

Nature and Characteristics of the Genetic Code

From the types of experiments just described, all 64 codons were assigned with the result shown in Figure 14.9. Each codon is written as it appears in mRNA and reads in a 5'-to-3' direction. The characteristics of the genetic code are as follows:

1. *The code is a triplet code.* Each mRNA codon that specifies an amino acid in a polypeptide chain consists of three nucleotides.

~ FIGURE 14.9

The genetic code. Of the 64 codons, 61 each specify one of 20 amino acids. One of those 61 codons, AUG, is used in the initiation of protein synthesis. The other 3 codons are chain-terminating codons and do not specify any amino acid. (*UGA encodes selenocysteine in certain cases [see page 429].)

Second letter

	U	C	A	G	
U	UUU Phe (F) UUC UUA Leu UUG (L)	UCU UCC Ser UCA (S) UCG	UAU Tyr UAC (Y) UAA **Stop** UAG **Stop**	UGU Cys UGC (C) UGA **Stop*** UGG Trp (W)	U C A G
C	CUU CUC Leu CUA (L) CUG	CCU CCC Pro CCA (P) CCG	CAU His CAC (H) CAA Gln CAG (Q)	CGU CGC Arg CGA (R) CGG	U C A G
A	AUU AUC Ile AUA (I) AUG Met (M)	ACU ACC Thr ACA (T) ACG	AAU Asn AAC (N) AAA Lys AAG (K)	AGU Ser AGC (S) AGA Arg AGG (R)	U C A G
G	GUU GUC Val GUA (V) GUG	GCU GCC Ala GCA (A) GCG	GAU Asp GAC (A) GAA Glu GAG (E)	GGU GGC Gly GGA (G) GGG	U C A G

(First letter on left axis, Third letter on right axis)

■ = Chain termination codon (stop)

■ = Initiation codon

2. *The code is comma-free; that is, it is continuous.* The mRNA is read continuously, three nucleotides (one codon) at a time, without skipping any nucleotides of the message.

3. *The code is nonoverlapping.* The mRNA is read in successive groups of three nucleotides. A message of AAGAAGAAG . . . in the cell would be read as lysine-lysine-lysine. . . , which is what the AAG specifies. Theoretically, three readings are possible from this message, depending on where the reading is begun (i.e., the *reading frame* in which message is read); namely, the repeating AAG, the repeating AGA, and the repeating GAA. Later in this chapter we will examine the mechanisms in the cell that ensure that the translation of the genetic code in an mRNA begins at the correct point.

4. *The code is almost universal.* All organisms share the same genetic language. Thus, for example, lysine is coded for by AAA or AAG in the mRNA of all organisms, arginine by CGU, CGC, CGA, CGG, AGA, and AGG, and so on. Hence

we can isolate an mRNA from one organism, translate it by using the machinery isolated from another organism, and produce the protein as if it had been translated in the original organism. The code, however, is not completely universal. For example, the mitochondria of some organisms, such as mammals, have minor changes in the code, as does the nuclear genome of the protozoan *Tetrahymena* (discussed in Chapter 21).

5. *The code is degenerate.* With two exceptions (AUG, which codes for methionine, and UGG, which codes for tryptophan), more than one codon occurs for each amino acid. This multiple coding is called the **degeneracy** of the code. A close examination of Figure 14.9 reveals particular patterns in this degeneracy. Thus, when the first two nucleotides in a codon are identical and the third letter is U or C, the codon always codes for the same amino acid. For example, UUU and UUC specify phenylalanine; similarly, CAU and CAC specify histidine. Also, when the first two nucleotides in a codon are identical and the third letter is A or G, the same amino acid is often specified. For example, UUA and UUG specify leucine, and AAA and AAG specify lysine. In a few cases, when the first two nucleotides in a codon are identical and the base in the third position is U, C, A, or G, the same amino acid is often specified. For example, CUU, CUC, CUA, and CUG all code for leucine.

Even though there is degeneracy of the code, that does not mean that all codons are used equally. Studies have shown that codon usage is not random. Rather, some codons are used repeatedly, while others are almost never used. Figure 14.10 illustrates the results of a study of a number of genes in animals. This study showed, for example, a significant bias toward the use of certain codons for particular amino acids. For example, UUC is used more frequently to code for phenylalanine than UUU.

6. *The code has start and stop signals.* Specific start and stop signals for protein synthesis are contained in the code. In both eukaryotes and prokaryotes, AUG (which codes for methionine) is usually the start codon for protein synthesis, although in some cases GUG (which normally codes for valine) is used. Thus, in examining the sequence of a particular mRNA for the location of the amino acid–coding sequence, we look near the 5' end for the occurrence of the AUG start codon and begin reading the amino acid sequence from there.

As Figure 14.9 shows, only 61 of the 64 codons specify amino acids; these codons are called the **sense codons**. The other three

~ FIGURE 14.10

Codon usage in a number of genes of animals. The numbers next to the codons are the instances of the codons in 1,001 codons tabulated.

First letter	Second letter: U	Second letter: C	Second letter: A	Second letter: G	Third letter
U	13 UUU Phe	16 UCU	10 UAU Tyr	10 UGU Cys	U
	28 UUC	18 UCC Ser	23 UAC	13 UGC	C
	2 UUA Leu	9 UCA	UAA Stop	UGA Stop	A
	9 UUG	2 UCG	UAG Stop	12 UGG Trp	G
C	9 CUU	14 CCU	10 CAU His	8 CGU	U
	27 CUC Leu	17 CCC Pro	21 CAC	11 CGC Arg	C
	7 CUA	10 CCA	10 CAA Gln	4 CGA	A
	47 CUG	5 CCG	28 CAG	5 CGG	G
A	11 AUU	15 ACU	8 AAU Asn	12 AGU Ser	U
	24 AUC Ile	28 ACC Thr	28 AAC	21 AGC	C
	4 AUA	11 ACA	19 AAA Lys	8 AGA Arg	A
	16 AUG Met	6 ACG	49 AAG	10 AGG	G
G	9 GUU	28 GCU	16 GAU Asp	22 GGU	U
	21 GUC Val	38 GCC Ala	24 GAC	32 GGC Gly	C
	5 GUA	14 GCA	21 GAA Glu	16 GGA	A
	33 GUG	6 GCG	34 GAG	11 GGG	G

codons—UAG (amber), UAA (ochre), and UGA (opal)—do not specify an amino acid, and no tRNAs in normal cells carry the appropriate anticodons. These three codons are the **stop codons**, also called **nonsense codons**, or **chain-terminating codons**. They are used singly or in tandem groups (UAG UAA, for example) to specify the end of the translation process of a polypeptide chain. Thus, when we read a particular mRNA sequence, we look for the presence of a stop codon *in the same reading frame as* the AUG start codon to determine where the amino acid-coding sequence for the polypeptide ends. We will examine starting and stopping protein synthesis when we study the details of the translation process.

7. *Wobble occurs in the anticodon.* Since 61 sense codons specify amino acids in mRNA, a total of 61 tRNA molecules could have the appropriate anticodons. Theoretically, though, the complete set of 61 sense codons can be read by fewer than 61 distinct tRNAs because of the wobble in the anticodon. The **wobble hypothesis**, proposed by Francis Crick, is shown in Table 14.1. Sequence analysis had shown that the base at the 5' end of the anticodon (complementary to the base at the 3' end of the codon; i.e., the third letter) is not as constrained as the other two bases. This feature allows for less exact base pairing so that the base at the 5' end of the anticodon can potentially pair with one of three different bases at the 3' end of the codon; it can wobble. As Table 14.1 shows, no single tRNA molecule can recognize four different codons. But if the tRNA molecule contains the modified nucleoside inosine (see Figure 13.29, p. 411) at the 5' end of the anticodon, then three different codons can be read by that one tRNA. Figure 14.11a gives an example of how a single leucine tRNA can read two different leucine codons by base-pairing wobble, and Figure

~ TABLE 14.1

Wobble in the Genetic Code

NUCLEOTIDE AT 5' END OF ANTICODON		NUCLEOTIDE AT 3' END OF CODON
G	can pair with	U or C
C	can pair with	G
A	can pair with	U
U	can pair with	A or G
I (inosine)	can pair with	A, U, or C

~ **FIGURE 14.11**

Example of base-pairing wobble. (a) Two different leucine codons (CUC, CUU) can be read by the same leucine tRNA molecule, contrary to regular base-pairing rules. (b) Three different glycine codons (GGU, GGC, GGA) can be read by the same glycine tRNA molecule by base-pairing wobble involving inosine in the anticodon.

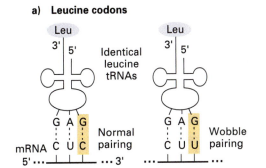

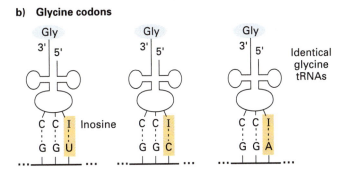

14.11b shows how a single glycine tRNA can read three different glycine codons by wobble pairing.

8. The genetic code is essentially unambiguous. It is very rare that one codon specifies more than one amino acid. An example of this was given in point 6, where we saw that GUG can sometimes function as a start codon in protein synthesis. When it is a start codon, the amino acid it specifies is methionine, whereas when GUG is anywhere else in coding sequence of the mRNA, it specifies valine. This is an example of *site-specific variation in codon translation*, also called *context effect in codon translation*. In other words, the translation of a particular codon may be affected by the nucleotides surrounding it in the mRNA. Another example of context effect involves the incorporation of the modified amino acid selenocysteine at certain internal UGA codons in mRNAs encoding selenoproteins. These mRNAs are found in prokaryotes and eukaryotes ranging from *E. coli* to humans. Normally, UGA is a stop codon, but a UGA located internally in the coding sequence in the appropriate nucleotide sequence context can be read as selenocysteine. Selenocysteine is a modified form of the amino acid cysteine (see Figure 14.2) in which the sulfur atom of the R group is replaced by a selenium atom. Selenocysteine is an essential component of selenoproteins, a small number of proteins many of which function as enzymes in oxidation-reduction reactions and have the selenocysteine in the active site. The incorporation of selenocysteine into protein has been studied most extensively in *E. coli*, where four genes (*selA, B, C,* and *D*) required for the process have been identified by genetic studies. Mutations in the *sel* gene leads to a deficiency in selenoprotein synthesis. The mechanism for incorporat-

ing selenocysteine at an internal UGA codon will be discussed later in the chapter.

KEYNOTE

The genetic code is a triplet code in which each codon (a set of three contiguous bases) in an mRNA specifies one amino acid. Since 64 code words are possible and since 20 amino acids exist, some amino acids are specified by more than one codon. The genetic code in mRNA is read without gaps and in successive, nonoverlapping codons. The code is universal: The same codons specify the same amino acids in most systems. Protein synthesis is typically initiated by codon AUG (methionine), and it is terminated by three codons singly or in combination: UAG, UAA, UGA (these codons do not code for any amino acid).

TRANSLATION OF THE GENETIC MESSAGE

In brief, protein synthesis takes place on ribosomes, where the genetic message encoded in mRNA is translated. The mRNA molecule is translated in the 5'-to-3' direction (the same direction in which it is made), and the polypeptide is made in the N-terminal to C-terminal direction. Amino acids are brought to the ribosome bound to tRNA molecules. The correct amino acid sequence is achieved as a result of (1) the specific binding of each amino acid to its own specific tRNA, and (2) the amino acid specific binding between the codon of the mRNA and the complementary anticodon in the tRNA.

That the mRNA codon recognizes the tRNA anticodon and not the amino acid carried by the tRNA was proved by G. von Ehrenstein, B. Weisblum, and S. Benzer. They attached cysteine to tRNA.Cys in vitro; then they chemically converted the attached cysteine to alanine. The resulting Ala-tRNA.Cys was used in the in vitro synthesis of hemoglobin. In vivo, the α and β chains of hemoglobin each contain one cysteine. When the hemoglobin made in vitro was examined, however, the amino acid alanine was found in both chains at the positions normally occupied by cysteine. This result could only mean that the Ala-tRNA.Cys had read the cysteine codon and had inserted the amino acid it carried, in this case alanine. Therefore, the researchers concluded that the *specificity of codon recognition lies in the tRNA molecule and not in the amino acid it carries.*

Aminoacyl-tRNA Molecules

The important function of tRNA molecules in protein synthesis is that they bring specific amino acids to the mRNA-ribosomal complex so that the correct polypeptide chain can be assembled. In this section we see how an amino acid becomes attached to its appropriate tRNA molecule to produce an aminoacyl-tRNA molecule.

ATTACHMENT OF AMINO ACID TO tRNA. The correct amino acid is attached to the tRNA by an enzyme called an **aminoacyl-tRNA synthetase**. Since 20 different amino acids exist, 20 types of aminoacyl-tRNA synthetases also exist. Further, since degeneracy is common in the genetic code (that is, a single amino acid may be specified by more than one codon), all the tRNAs that are specific for a particular amino acid must have a common recognition site or identity site so that the amino acid can be added by the same aminoacyl-tRNA synthetase. The aminoacyl-tRNA synthetases themselves are highly specific in their function, so few errors are made in the bonding of an amino acid to its tRNA.

In general, the sequence differences among the tRNAs specify which enzyme will recognize the tRNA. However, no single rule exists for recognition by the aminoacyl tRNA synthetases. For *E. coli* tRNA.Val and tRNA.Met, the recognition site is the anticodon itself, but this obvious location is not the recognition site for all tRNAs. In other *E. coli* tRNAs, the recognition sites have been localized by mutational studies to various positions, for example to a single base pair in the acceptor stem (the stem of the tRNA to which the amino acid attaches) or to a combination of nucleotides in different parts of a tRNA.

Figure 14.12 shows how an amino acid is attached to a tRNA molecule to produce an **aminoacyl-tRNA**, in this case Ser-tRNA, and presents the generalized structure of an aminoacyl-tRNA molecule. The amino acid attaches at the 3' end of the tRNA by a linkage between the carboxyl group of the amino acid and the 3'-OH or 2'-OH group of the ribose of the adenine nucleotide found at the end of every tRNA (the amino acid is attached to the 3'-OH in the figure). The act of adding the amino acid to the tRNA is called **charging**, and the product is commonly referred to as a **charged tRNA** or *aminoacylated-tRNA*.

𝒦EYNOTE

Protein synthesis occurs on ribosomes, where the genetic message encoded in mRNA is translated. Amino acids are brought to the ribosome on charged tRNA molecules. The correct amino acid sequence is achieved as a result of (1) the specific binding of each amino acid to its own specific tRNA, and (2) the specific binding between the codon of the mRNA and the complementary anticodon in the tRNA.

The three basic stages of protein synthesis—*initiation*, *elongation*, and *termination*—are similar in prokaryotes and eukaryotes. In the following sections we discuss each of these stages in turn, concentrating, as before, on the processes in *E. coli*, with which most work has been done. In the discussions we note where significant differences in translation occur in prokaryotes and eukaryotes.

Initiation of Translation

INITIATOR CODON AND INITIATOR tRNA. In both prokaryotes and eukaryotes, translation usually begins at the AUG initiator codon in the mRNA, which specifies methionine. (In a few mRNAs, translation begins at a GUG codon.) As a result, newly made proteins in both types of organisms begin with methionine; in some cases this methionine is subsequently removed.

~ **FIGURE 14.12**

Molecular details of the attachment of an amino acid to a tRNA molecule. (a) In the first step the amino acid (serine here) reacts with ATP to produce an aminoacyl-AMP complex. This reaction is catalyzed by an aminoacyl-tRNA synthetase (seryl-tRNA synthetase). (b) In the second step (also catalyzed by aminoacyl-tRNA synthetase) the aminoacyl-AMP complex then reacts with the appropriate tRNA molecule (tRNA.Ser) to produce an aminoacyl-tRNA (Ser-tRNA.Ser). (c) In the general molecular structure of an aminoacyl-tRNA molecule (charged tRNA), the carboxyl group of the amino acid is attached to the 3'-OH or 2'-OH group of the 3' terminal adenine nucleotide of the tRNA.

a) First step

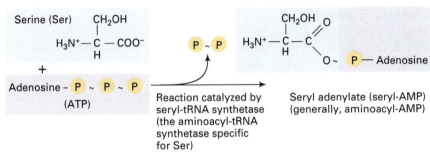

Reaction catalyzed by seryl-tRNA synthetase (the aminoacyl-tRNA synthetase specific for Ser)

Seryl adenylate (seryl-AMP) (generally, aminoacyl-AMP)

b) Second step

Ser — AMP + [tRNA diagram 3' 5'] → [tRNA diagram Ser 5'] + AMP

Seryl-tRNA synthetase

Seryl-AMP   tRNA.Ser   Ser-tRNA.Ser

c) General structure

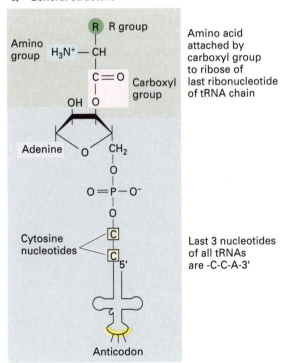

Amino acid attached by carboxyl group to ribose of last ribonucleotide of tRNA chain

Last 3 nucleotides of all tRNAs are -C-C-A-3'

Prokaryotes. In prokaryotes, the initiator methionine is a modified form of methionine, called **formylmethionine** (abbreviated fMet), in which a formyl group has been added to the methionine's amino group. The fMet is brought to the ribosome attached to a special tRNA, called tRNA.fMet, which has the anticodon 5'-CAU-3' to bind to the AUG start codon. This tRNA is special, since it is involved specifically with the initiation process of protein synthesis. The aminoacylation of tRNA.fMet occurs as follows: First, methionyl-tRNA synthetase catalyzes the addition of methionine to the tRNA. Then an enzyme called *transformylase* catalyzes the addition of the formyl group to the methionine. The resulting molecule is designated fMet-tRNA.fMet (this nomenclature indicates that the tRNA is specific for the attachment of fMet and that, in fact, fMet is attached to it).

When an AUG codon in an mRNA molecule is encountered at a position other than at the start of the amino acid–coding sequence, another species of tRNA is used to insert methionine at that point in the polypeptide chain. This tRNA is called tRNA.Met, and it is charged by the same aminoacyl-tRNA synthetase as is tRNA.fMet to produce Met-tRNA.Met. However, tRNA.Met and tRNA.fMet molecules are coded for by different genes and have different sequences. We will see later in this chapter how the two tRNAs are used differently.

Eukaryotes. In eukaryotes, protein synthesis initiation occurs in much the same way. That is, AUG is the usual initiation codon, and hence the N-terminal amino acid in polypeptides is methionine. The N-terminal methionine is not formylated. Nonetheless, a special methionine tRNA is used for initiation of translation in eukaryotes, and a different species of tRNA.Met is used to read AUG codons elsewhere in an mRNA molecule.

FORMATION OF THE INITIATION COMPLEX. The initiation of translation is very similar in prokaryotes and eukaryotes. That is, in both types of organisms, the small ribosomal subunit first binds to the mRNA with the aid of special proteins, and then the large ribosomal subunit binds to produce an *initiation complex* that has all the components for beginning protein synthesis.

Prokaryotes. In addition to the AUG initiation codon, the initiation of protein synthesis in prokaryotes requires other information coded in the base sequence of an mRNA molecule upstream (to the 5' side) of the initiation codon. These sequences serve to align the ribosome on the message in the proper reading frame so that polypeptide synthesis can proceed correctly. To identify these aligning sequences, researchers sequenced a number of mRNAs in the region upstream from the AUG initiation codon. To do so, they allowed ribosomal subunits to bind to the mRNA under conditions in which protein synthesis could not commence. Under these conditions the 30S ribosomal subunits freeze on the mRNA message at the **ribosome-binding site**, the site at which the ribosome becomes oriented in the correct reading frame for the initiation of protein synthesis. When the subunits are bound to the mRNAs, the message is protected from RNAse (ribonuclease) attack, and the protected regions can be isolated and studied.

Figure 14.13 shows the sequences of some prokaryotic ribosome-binding sites. The AUG codon

---

~ FIGURE 14.13

Sequences of some prokaryotic ribosome-binding sites. The initiation codon, AUG, is yellow. (Note that GUG can be used as an initiator codon in some systems.) The purple regions indicate the regions of contiguous complementarity between mRNA and the 3' end of 16S rRNA.

		Binding site sequences					Initiation codon			
Phage R17 A protein	UCC	UAG	GAG	GUU	UGA	CCU	AUG	CGA	GCU	UUU
Phage Qβ replicase	UAA	CUA	AGG	AUG	AAA	UGC	AUG	UCU	AAG	ACA
Phage λ Cro	AUG	UAC	UAA	GGA	GGU	UGU	AUG	GAA	CAA	CGC
Phage φX174 A	AAU	CUU	GGA	GGC	UUU	UUU	AUG	GUU	CGU	UCU
E. coli trpB	AUA	UUA	AGG	AAA	GGA	ACA	AUG	ACA	ACA	UUA
E. coli lacZ	UUC	ACA	CAG	GAA	ACA	GCU	AUG	ACC	AUG	AUU
E. coli RNA polymerase β	AGC	GAG	CUG	AGG	AAC	CCU	AUG	GUU	UAC	UCC

can be clearly identified. Most of the binding sites have a purine-rich sequence about 8 to 12 nucleotides upstream from the initiation codon. Evidence from the work of John Shine and Lynn Dalgarno indicates that this purine-rich sequence, and other nucleotides in this region, are complementary to a pyrimidine-rich region (which always contains the sequence CCUCC) at the 3' end of 16S rRNA (Figure 14.14). The mRNA region that binds in this way has become known as the **Shine-Dalgarno sequence**. The model is that the formation of complementary base pairs between the mRNA and 16S rRNA allows the ribosome to locate the true sequence in the mRNA for the initiation of protein synthesis. There is genetic evidence that the model is correct. That is, if the Shine-Dalgarno sequence is mutated so that its possible pairing with the 16S rRNA sequence is significantly diminished or abolished, the particular mRNA involved cannot be translated. Likewise, if the rRNA sequence complementary to the Shine-Dalgarno sequence is mutated, mRNA translation does not occur. Since it can be argued that the loss of translatability as a result of mutations in one or other RNA partner could occur because of effects unrelated to the loss of pairing of the two RNA segments, a more elegant experiment was done. That is, mutations were made in the Shine-Dalgarno sequence to abolish pairing with the wild-type rRNA sequence and compensating mutations were made in the rRNA sequence so that the two mutated sequences could pair. In this case, mRNA translation occurred normally, indicating the importance of the pairing of the two RNA segments. (This type of experiment, in which compensating mutations are made in two sequences that are

hypothesized to interact, has been used in a number of other systems to explore the roles of specific interactions in biological functions.)

As Figure 14.15 shows, translation in *E. coli* commences with the formation of a 30S initiation complex which involves recognition of the AUG initiation codon by the small ribosome subunit as just discussed. At the beginning of this process, three protein **initiation factors**, IF1, IF2, and IF3, are bound to the 30S ribosomal subunit along with a molecule of GTP (guanosine triphosphate). The fMet–tRNA.fMet and the mRNA then attach to the 30S–IF–GTP complex to form the *30S initiation complex*. IF3 is released as a result of this process. Next, the 50S subunit binds, leading to GTP hydrolysis and the release of IF1 and IF2. The final complex is called the *70S initiation complex*. All three IF molecules are recycled for use in other initiation reactions. The 70S ribosome has two binding sites for aminoacyl-tRNA, the peptidyl (P), and aminoacyl (A) sites; the fMet-tRNA.fMet is bound to the mRNA in the P site.

Eukaryotes. Shine-Dalgarno sequences are not found in eukaryotic mRNAs. Instead, the eukaryotic ribosome uses another way to begin translating an mRNA. First, a **eukaryotic initiator factor** eIF4A, a multimer of several proteins including the *cap binding protein* (CBP), binds to the cap at the 5' end of the mRNA (see Chapter 13). Then, a complex of the 40S ribosomal subunit with the initiator Met-tRNA.Met, several eIFs, and GTP binds, along with other eIFs, and moves along the mRNA, scanning for the initiator AUG codon. The AUG codon is embedded in a short sequence—called the Kozak sequence—that indicates it is the initiator codon. This is called the *scanning model* for initiation. The AUG codon is usually, but not always, the first AUG codon from the 5' end of the mRNA. Once it finds this AUG, the 40S subunit binds to it and then the 60S ribosomal subunit binds, displacing the eIFs, producing the *80S initiation complex*. Protein synthesis then begins at the AUG codon. Like its prokaryotic counterpart, the eukaryotic 80S ribosome has a P site and an A site and the initiator Met-tRNA.Met is bound to the mRNA in the P site.

Elongation of the Polypeptide Chain

After initiation, the elongation phase of translation begins. Figure 14.16 diagrams these events as they take place in prokaryotes. This phase has three steps:

1. the binding of aminoacyl-tRNA (charged tRNA) to the ribosome
2. the formation of a peptide bond

~ **FIGURE 14.14**

Sequences involved in the binding of ribosomes to the mRNA in the initiation of protein synthesis in prokaryotes. (a) Nucleotide sequence at the 3' end of *E. coli* 16S rRNA. (b) Example of how the 3' end of 16S rRNA can base-pair with the nucleotide sequence 5' upstream from the AUG initiation codon.

a) Sequence at 3' end of 16S rRNA

3'... AUUCCUCCAUAG ...5'

b) Example of mRNA leader and 16S rRNA pairing

Shine-Dalgarno sequence   Initiation codon

5'...UGUACUAAGGAGGUUGUAUGGAACAACGC ... 3'
        |||||||||
        3'... AUUCCUCCAUAG...5'
        16S rRNA 3' end

~ FIGURE 14.15

Initiation of protein synthesis in prokaryotes. A 30S ribosomal subunit, complexed with initiation factors and GTP, binds to mRNA and fMet-tRNA.fMet to form a 30S initiation complex. Next, the 50S ribosomal subunit binds, forming a 70S initiation complex. During this event, the initiation factors are released and GTP is hydrolyzed.

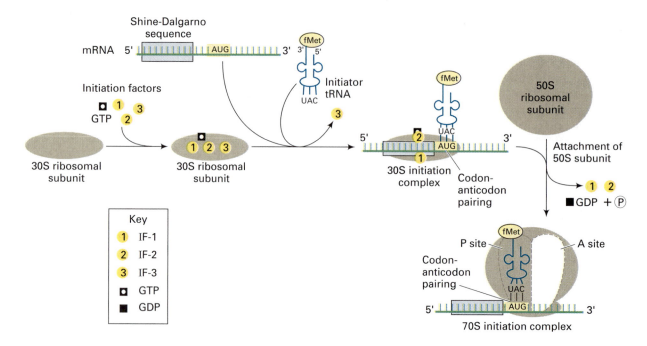

3. the movement (translocation) of the ribosome along the mRNA, one codon at a time

BINDING OF AMINOACYL-tRNA. At the start of the elongation phase in prokaryotes, the fMet-tRNA.fMet is hydrogen-bonded to the AUG initiation codon in the peptidyl (P) site of the ribosome in the 70S initiation complex. fMet-tRNA.fMet is the only tRNA to bind before the complete ribosome is formed. The orientation of this tRNA-codon complex exposes the next codon in the mRNA in the aminoacyl (A) site (Figure 14.16, part 1). In Figure 14.16 this codon (UCC) specifies serine (Ser).

Next, the appropriate aminoacyl-tRNA (in this case Ser-tRNA.Ser) binds to the exposed mRNA codon in the A site (Figure 14.16, part 2). This aminoacyl-tRNA is brought to the ribosome complexed with the protein elongation factor EF-Tu and a molecule of GTP. When the aminoacyl-tRNA binds to the codon in the A site, the GTP is hydrolyzed and EF-Tu is released bound to the GDP produced.

As shown in Figure 14.16, part 2, EF-Tu is recycled. First, a second elongation factor, EF-Ts, binds to EF-Tu and displaces the GDP. Next, GTP binds to the EF-Tu–EF-Ts complex to produce an EF-Tu–GTP complex simultaneously with the release of EF-Ts. The aminoacyl-tRNA binds to the EF-Tu–GTP, and that complex can then bind to the A site in the ribosome when the appropriate codon is exposed.

PEPTIDE BOND FORMATION. In Figure 14.16, part 3 fMet-tRNA.fMet is bound in the P site of the bacterial ribosomes, and the second specified aminoacyl-tRNA (in this case Ser-tRNA.Ser) is bound in the A site. The ribosome maintains the two aminoacyl-tRNAs in the correct positions so that a peptide bond can form between the two amino acids (Figure 14.16, part 3).

The two steps involved in the formation of a peptide bond are shown in Figure 14.17. The first step is the breakage of the bond between the carboxyl group of the amino acid and the tRNA in the P site. In this case the breakage is between the fMet and its tRNA. The second step is the formation of the peptide bond between the now-freed fMet and the Ser attached to the tRNA in the A site. This reaction is catalyzed by **peptidyl transferase.** For many years this enzyme activity was thought to be a result of the interaction of a few ribosomal proteins of the 50S ribosomal subunit. However, Harry Noller and his colleagues in 1992 found that when most of the proteins of the 50S ribosomal subunit were removed, leaving only the

~ **FIGURE 14.16**

Elongation stage of polypeptide synthesis in prokaryotes.

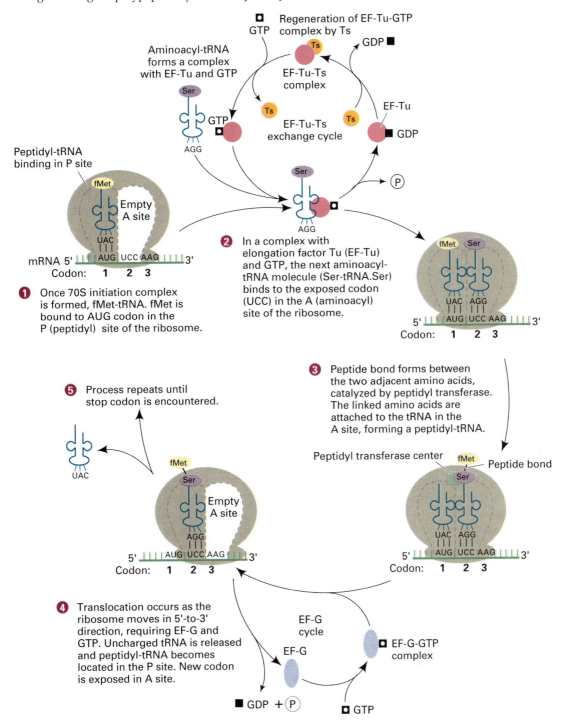

ribosomal RNA, peptidyl transferase activity could still be measured. In addition, this activity was inhibited by the antibiotics chloramphenicol and carbomycin, both of which are known specifically to inhibit peptidyl transferase activity. Further, when the rRNA was treated with ribonuclease T1 which degrades RNA but not protein, the peptidyl transferase activity was lost. These results suggest that the 23S rRNA molecule of the large ribosomal subunit is intimately involved with the peptidyl transferase

~ FIGURE 14.17

The formation of a peptide bond between the first two amino acids (fMet and Ser) of a polypeptide chain is catalyzed on the ribosome by peptidyl transferase. (a) Adjacent aminoacyl-tRNAs bound to the mRNA at the ribosome. (b) Following peptide bond formation, an uncharged tRNA is in the P site and a dipeptidyl-tRNA is in the A site.

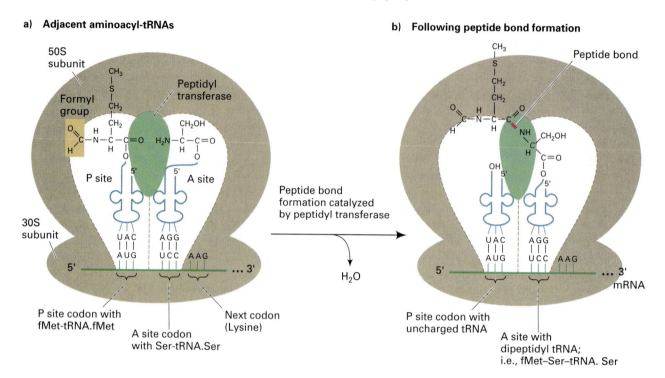

a) **Adjacent aminoacyl-tRNAs**

b) **Following peptide bond formation**

activity, and may in fact be that enzyme. This exciting finding has expanded our understanding of the role particular RNA molecules can have in catalytic activities previously thought to be the function only of proteins. Recall, for instance, the catalytic role of RNA in the self-splicing of a *Tetrahymena* rRNA intron discussed in Chapter 13 (pp. 405–406).

Once the peptide bond has formed (Figure 14.16, part 3), a tRNA without an attached amino acid (an uncharged tRNA) is left in the P site. The tRNA in the A site, now called peptidyl-tRNA, has the first two amino acids of the polypeptide chain attached to it, in this case fMet-Ser.

TRANSLOCATION.   The last step in the elongation cycle is **translocation** (Figure 14.16, part 4). Once the peptide bond is formed and the growing polypeptide chain is on the tRNA in the A site, the ribosome moves one codon along the mRNA toward the 3' end. In prokaryotes, translocation requires the activity of another protein elongation factor, EF-G. An EF-G–GTP complex binds to the ribosome, and translocation then takes place along with movement of the uncharged tRNA away from the P site. Based on studies of the binding of tRNAs to ribosomes, a 50S subunit site

called E (for *Exit*) was recently proposed to be involved in the release of the uncharged tRNA from the *E. coli* ribosome. The model is that the uncharged tRNA moves from the P site and then binds to the E site, effectively blocking the next aminoacyl-tRNA from binding to the A site until translocation is complete and the peptidyl-tRNA is bound correctly in the P site. (For simplicity, the E site is not shown in the figures.) After translocation, the EF-G is then released in a reaction requiring GTP hydrolysis; EF-G can then be reused, as shown in Figure 14.16, part 4. During the translocation step the peptidyl-tRNA remains attached to its codon on the mRNA. And since the ribosome has moved, the peptidyl-tRNA is now located in the P site (hence the name, the *peptidyl* site). The exact mechanism for the physical translocation of the ribosome is not known.

After translocation is completed, the A site is vacant. An aminoacyl-tRNA with the correct anticodon binds to the newly exposed codon in the A site, using the process already described (Figure 14.16, part 5). The whole process is repeated until translation terminates at a stop codon.

In eukaryotes, the elongation and translocation steps are similar to those in prokaryotes, although

~ **FIGURE 14.18**

Electron micrograph and diagram of a polysome, a number of ribosomes each translating the same mRNA sequentially.

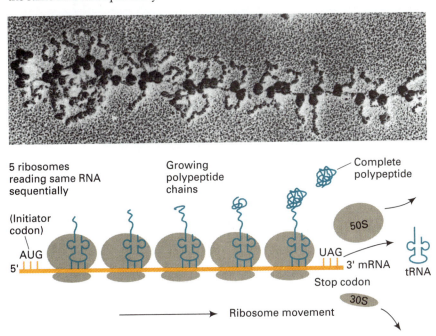

there are differences in the number and properties of elongation factors and in the exact sequences of events.

In both prokaryotes and eukaryotes, once the ribosome moves away from the initiation site on the mRNA, the initiation site is open for another initiation event to occur. Thus many ribosomes may simultaneously be translating each mRNA. The complex between an mRNA molecule and all the ribosomes that are translating it simultaneously is called a **polyribosome** or **polysome** (Figure 14.18). An average mRNA may have eight to ten ribosomes synthesizing protein from it. Simultaneous translation enables a large amount of protein to be produced from each mRNA molecule.

KEYNOTE

The AUG (methionine) initiator codon signals the start of translation in prokaryotes and eukaryotes. Elongation proceeds when a peptide bond forms between the amino acid attached to the tRNA in the A site of the ribosome and the growing polypeptide attached to the tRNA in the P site. This reaction is catalyzed by peptidyl transferase which is an activity of the 23S rRNA molecule of the large ribosomal subunit. Translocation occurs when the now uncharged tRNA in the P site is released from the ribosome and the ribosome moves one codon down the mRNA.

Termination of Translation

Elongation continues until the polypeptide coded for in the mRNA is completed. The termination of translation is diagrammed in Figure 14.19. The end of a polypeptide chain is signaled by one of three stop codons, UAG, UAA, and UGA, which are the same in prokaryotes and eukaryotes (Figure 14.19, part 1). The stop codons do not code for any amino acid, and so no tRNAs in the cell have anticodons for them. The ribosome recognizes a chain termination codon only with the help of proteins called **termination factors**, or **release factors** (RF), which read the chain termination codons (Figure 14.19, part 2), and then initiate a series of specific termination events.

~ FIGURE 14.19

Termination of translation. The ribosome recognizes a chain termination codon (UAG) with the aid of release factors. The release factor reads the stop codon, and this initiates a series of specific termination events leading to the release of the completed polypeptide.

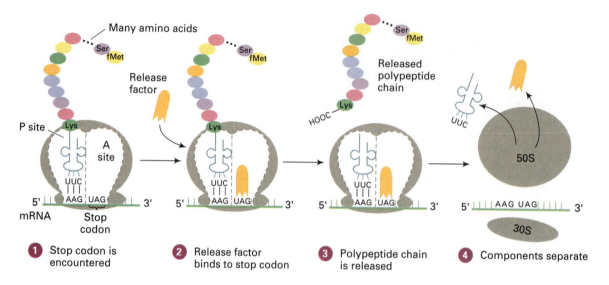

1 Stop codon is encountered

2 Release factor binds to stop codon

3 Polypeptide chain is released

4 Components separate

E. coli has three RFs—RF1, RF2, and RF3—and each is a single polypeptide. Factor RF1 recognizes UAA and UAG, while RF2 recognizes UAA and UGA. Thus these two release factors have overlapping specificity in codon recognition. Factor RF3, which does not recognize any of the stop codons, stimulates the termination events. In eukaryotes only one release factor, called eukaryotic release factor (eRF), is needed. eRF recognizes all three stop codons. No stimulatory factor analogous to RF3 has been found in eukaryotes.

The specific termination events triggered by the release factors are: (1) release of the polypeptide from the tRNA in the P site of the ribosome in a reaction catalyzed by peptidyl transferase (Figure 14.19, part 3), and then (2) release of the tRNA from the ribosome and dissociation of the two ribosomal subunits from the mRNA (Figure 14.19, part 4). In both prokaryotes and eukaryotes, the initiating amino acid—fMet and Met, respectively—are usually cleaved from the completed polypeptide.

As we discussed earlier in the chapter (p. 429), the modified amino acid selenocysteine may be incorporated into proteins if there is a UGA codon in an appropriate nucleotide context located internally in the coding sequence of the mRNA. How is the internal UGA codon read as selenocysteine and not as a chain termination codon? In *E. coli*, the products of the *selA*, *B*, *C*, and *D* genes are required for this event. The *selC* gene encodes a special tRNA which has the anticodon 5'-UCA-3'. This tRNA is a modified serine tRNA and is charged with serine by the seryl-tRNA synthetase. The products of the *selA* and *selD* genes then convert the Ser-tRNA.Ser to selenocysteine-tRNA.Ser. For this charged tRNA to be used at an internal UGA codon, the protein product of the *selB* gene is needed. This protein functions as a specialized EF-Tu factor, replacing that factor to bring about the binding of the selenocysteine-tRNA.Ser to the ribosome and the UGA codon. Presumably, the special components required for this reaction compete successfully with the release factors so that the polypeptide chain is not terminated at the internal UGA codon. At normal termination UGA codons, the selenocysteine incorporation system does not work because the nucleotide context in which the stop codon is found is not appropriate.

KEYNOTE

Protein synthesis continues until a chain-terminating codon is located in the A site of the ribosome. These codons are read by one or more release proteins. Then the polypeptide and its tRNA are released from the ribosome, and the ribosome disengages from the reading frame of the mRNA.

PROTEIN SORTING IN THE CELL

In eukaryotes, some proteins may be secreted, depending on cell type, while other proteins need to be located in different cell compartments in order to function. That is, eukaryotic cells are compartmentalized, and the examination of the contents of these compartments (such as the nucleus, mitochondria, chloroplasts, and lysosomes) shows that each contains a specific set of proteins. Therefore proteins must end up in the compartments in which they are to function. The sorting of proteins to their appropriate compartments is under genetic control. Similarly, in prokaryotes, certain proteins become localized in the membrane and others are secreted. Here, we will focus on eukaryotes, and briefly describe protein transport into the ER (endoplasmic reticulum), the mitochondria, the chloroplasts, and the nucleus.

Proteins Distributed by the Endoplasmic Reticulum

Electron microscopic studies showed that secretory cells, such as those found in the liver and pancreas, have an extensive membrane structure called the endoplasmic reticulum (ER). Some of the ER appears rough, due to the presence of ribosomes on the membrane. This observation has resulted in the terms *membrane-bound* and *free* ribosomes. Proteins synthesized on the rough ER are extruded into the space between the two membranes of the ER (the cisternal space) and are then transferred to the Golgi complex. Proteins to be secreted are packaged into secretory vesicles. From these vesicles the proteins are secreted to the outside of the cell by the fusion of the vesicles with the cell membrane. Figure 14.20 is a simple diagram of all these events. For our purposes, it is necessary to understand that initially a common mechanism translocates proteins destined to be secreted from the cell, proteins that will become embedded in the plasma membrane, and proteins that are packaged into lysosomes. That is, initially the proteins are moved into the ER cisternal space. Thereafter, sorting to their final destinations occurs primarily at the Golgi complex.

In 1975, Gunther Blobel and his colleagues found that all the major proteins secreted by the pancreas initially contain extra amino acids at the amino terminal end. As a result of their studies, Blobel and B. Dobberstein proposed the **signal hypothesis**, which states that the secretion of proteins out of a cell occurs through the binding of a hydrophobic, amino terminal extension (the **signal sequence**) to the membrane and the subsequent removal and degradation of the extension in the cisternal space of the ER.

Our current understanding of the signal hypothesis is as follows (Figure 14.21): The growing proteins (that is, those being synthesized on ribosomes) that are to be secreted from the cell or inserted into membranes all have an N-terminal extension of about 15 to 30 amino acids, called the *signal sequence*. Only proteins that have signal sequences can be transferred across or inserted into the membrane of the rough ER while translation is taking place. Proteins without signal sequences remain in the cytoplasm.

When a protein destined for the ER exposes its signal sequence, a cytoplasmic receptor particle called the **signal recognition particle** (SRP: a complex of the small RNA molecule 7S RNA, with six proteins) recognizes the signal sequence, binds to it, and blocks further translation of the mRNA. The temporary halt in protein synthesis brought about by SRP occurs when the polypeptide chain is long enough so that the signal sequence has completely emerged from the ribosome and can be recognized by SRP. Translation stops until the nascent polypeptide-SRP-ribosome-mRNA complex reaches and binds to the ER. The SRP recognizes an integral membrane protein of the ER

~ **FIGURE 14.20**

Movement of secretory proteins. These proteins move from their site of synthesis on the rough endoplasmic reticulum (RER), through the Golgi apparatus, into vesicles, and then into the extracellular space. The arrow indicates the general path of the proteins to be secreted.

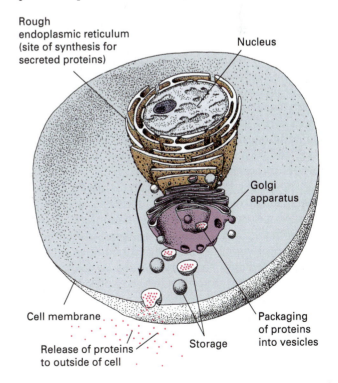

~ **FIGURE 14.21**

Model for the translocation of proteins into the endoplasmic reticulum in eukaryotes.

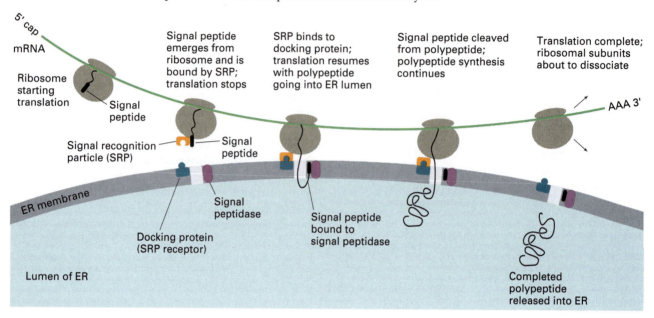

called the **docking protein** (also called the *signal recognition protein* receptor). The association of the SRP and docking protein facilitates the binding of polypeptide's signal sequence and associated ribosomes to the ER. Translation resumes, and the SRP is released. The growing polypeptide (with its signal sequence) now is translocated through the membrane into the cisternal space of the ER. The transport of the protein molecule into the ER simultaneously with its synthesis is an example of **cotranslational transport**.

Once the signal sequence is fully into the cisternal space of the ER, it is removed from the polypeptide by the action of an enzyme called **signal peptidase**. When the complete polypeptide is entirely within the ER cisternal space, it is typically modified further by the addition of specific carbohydrate groups to produce *glycoproteins*. The glycoproteins must now be sorted for their final destinations.

The Golgi complex is the main distribution center where most of the sorting decisions are made. Proteins targeted for the plasma membrane are carried from the Golgi complex to the plasma membrane in transport vesicles. Proteins destined to be secreted are packaged into secretory storage vesicles, which form by budding from the Golgi complex. The secretory vesicles migrate to the cell surface where they fuse with the plasma membrane and release their packaged proteins to the outside of the cell.

Lysosomal proteins are collected into a specific class of Golgi export vesicles, which eventually become functional lysosomes.

Currently, significant progress is being made in defining in yeast the precise steps involved in protein translocation across the ER membrane and the subsequent sorting events through the study of defined genetic mutants called *sec* (secretion) mutants, which affect different stages of those processes. This kind of combined genetic-cell biology approach will contribute significant new knowledge to this area in the future.

Proteins Transported into Mitochondria and Chloroplasts

A large number of proteins find their way into mitochondria and chloroplasts. These proteins include the ribosomal proteins of the organellar ribosomes, some enzymes of the electron transport chain in mitochondria, and some photosynthesis enzymes in chloroplasts.

In contrast to proteins that are translated simultaneously with entry into the ER, proteins destined for the mitochondria and chloroplasts are synthesized completely before the import process begins. That is, ribosomes that make these proteins do not become associated with either organelle's outer membranes.

Transport in which synthesis of the protein is completed before import into the organelle takes place is called **posttranslational transport**.

Proteins imported into mitochondria and chloroplasts are made as precursor proteins, which contain extra sequences at their N-terminal ends called **transit sequences** that are both necessary and sufficient for transport. The precursors associate with specific receptors on the outer membrane of the appropriate organelle. They are then transported into the organelle. Once inside the organelle, the transit sequence is removed from the protein by an enzyme, **transit peptidase**.

Proteins Transported into the Nucleus

Many proteins find their way into the nucleus, including the ribosomal proteins that are needed for ribosome assembly, histone and nonhistone chromosomal proteins, gene regulatory proteins, and the many enzymes needed for DNA replication, DNA repair, recombination, transcription, and RNA processing. Except for small proteins (<60 kDa), the nucleus is extremely selective about which proteins it allows in. Like the other proteins we have discussed, nuclear proteins have simple sequences—*nuclear localization sequences*—that specify their translocation into the nucleus.

Like mitochondrial and chloroplast proteins, proteins destined for the nucleus are transported by a posttranslational transport mechanism. The recognition of the nuclear localization sequence occurs at the nuclear pores in the nuclear envelope and then the protein is transported through the pore into the nucleus.

Unlike the other compartmentalized proteins we have discussed, the nuclear localization sequences of many nuclear proteins are not removed once they enter the nucleus. The reason for this is that each time the cell divides, the nuclear envelope is degraded, then reforms prior to cytokinesis. Since, during cell division, the nuclear proteins are free in the cytoplasm, they must retain their ability to reenter the nucleus selectively once the nuclear envelope reforms.

𝒦EYNOTE

Eukaryotic proteins that enter the endoplasmic reticulum, the mitochondrion, the chloroplast, or the nucleus are distinguished from proteins that remain free in the cytoplasm by the presence of specific targeting sequences. Proteins that enter the ER, for instance, have signal sequences at their N-terminal ends. Characteristically, the signal sequences contain a significant number of hydrophobic amino acids. Proteins destined for the ER are translocated into the cisternal space of the ER while they are being synthesized on the ribosomes (i.e., cotranslationally). Once in the ER, the signal sequence is removed by signal peptidase. They are then sorted to their final destinations by the Golgi complex.

Proteins destined for the mitochondrion, the chloroplast, and the nucleus also have specific sequences to direct them into their respective organelles. Such proteins enter their target organelles only after they have been synthesized completely; that is, posttranslationally. Once mitochondrial or chloroplast proteins are inside the respective organelle, the special sequence called a *transit sequence*, is removed. For nuclear proteins, however, the nuclear localization sequence remains, to allow those proteins to become localized in the nucleus again after cell division.

SUMMARY

In this chapter we discussed the features of the genetic code and the process of translation of an mRNA to produce a protein. At the molecular level, we learned that a protein consists of one or more macromolecular subunits called polypeptides, which are themselves composed of smaller building blocks called amino acids. The amino acids are linked together in the polypeptide by peptide bonds. The amino acid sequence of a protein determines its secondary, tertiary, and quaternary structure, and hence its functional state.

RNA consists of four different subunits (nucleotides), while protein consists of 20 different subunits (amino acids). It was clearly very important to determine how amino acid information was coded in the nucleotide sequence of mRNA. From a number of experiments using synthetic mRNAs, the genetic code was found to have the following features: it is a triplet code (meaning that a sequence of three nucleotides specifies one amino acid), it is comma-free, it is nonoverlapping, it is almost universal, and, with two exceptions, it is degenerate. Of the 64 codons, 61 specify amino acids, while the other three codons are chain-terminating codons. One codon, AUG (for methionine) is typically used to specify the first amino acid in a protein chain.

Protein synthesis occurs on ribosomes, where the genetic message encoded in mRNA is translated. Amino acids are brought to the ribosome on charged tRNA molecules. Each tRNA has an anticodon, which binds specifically to a codon in the mRNA. As a result, the correct amino acid sequence is achieved by (1) the specific binding of each amino acid to its own specific tRNA, and (2) the specific binding between the codon of the mRNA and the complementary anticodon in the tRNA.

In both prokaryotes and eukaryotes an AUG codon is the initiator codon for the start of protein synthesis. In prokaryotes, the initiation of protein synthesis requires a sequence upstream of the AUG codon, to which the small ribosomal subunit binds. This sequence is the Shine-Dalgarno sequence, and it binds specifically to the 3' end of the 16S rRNA of the small ribosomal subunit, thereby associating the small subunit with the mRNA. No functionally equivalent sequence occurs in eukaryotic mRNAs; instead, the ribosomes load onto the mRNA at the 5' end of the mRNA and scan towards the 3' end, initiating translation at the first AUG codon that is embedded in the correct sequence context they encounter.

In both prokaryotes and eukaryotes, the initiation of protein synthesis requires protein factors called initiation factors. Initiation factors are bound to the ribosome-mRNA complex during the initiation phase and dissociate once the polypeptide chain has been initiated. During the elongation phase, the polypeptide chain is elongated one amino acid at a time. This occurs simultaneously with the movement of the ribosome towards the 3' end of the mRNA one codon at a time.

Protein factors called elongation factors play important catalytic roles. The signal for polypeptide chain growth to stop is the presence of a chain-terminating codon (UAG, UAA, or UGA) in the mRNA. No naturally occurring tRNA has an anticodon that can read a chain-terminating codon. Instead, specific protein factors called release factors read the stop codon and initiate the events characteristic of protein synthesis termination; namely, the

release of the completed polypeptide from the ribosome, the release of the tRNA from the ribosome, and the dissociation of the two ribosomal subunits from the mRNA.

Eukaryotic proteins that enter the endoplasmic reticulum, the mitochondrion, the chloroplast, or the nucleus are different from proteins that remain free in the cytoplasm in that they possess specific targeting sequences, that is, sequences that are required for localizing them into the appropriate cellular compartment. Proteins destined for the ER, for example, have specific hydrophobic signal sequences at their amino ends. When a protein being synthesized exposes its signal sequence, a signal recognition particle (SRP) binds to it and blocks further translation of the mRNA. Translation is blocked until the nascent polypeptide–SRP–ribosome–mRNA complex reaches and binds to the ER by interaction between the SRP and an integral membrane protein of the ER called the docking protein. Resumption of translation then results in the growing polypeptide being translocated into the cisternal space of the ER where the signal sequence is removed. In other words, transport of a protein into the ER occurs in a cotranslational manner. Once inside the ER, the protein is then sorted for its final location by the Golgi complex.

Proteins destined for the mitochondrion, chloroplast, or nucleus also have specific sequences to direct them into their respective organelles. These sequences are called transit sequences. In these cases only completed proteins are transported into the organelles; that is, transport is posttranslational. Once mitochondrial or chloroplast proteins are inside the respective organelle, this transit sequence is removed. For nuclear proteins, however, the nuclear localization sequence remains attached so that those proteins can become transported again to the nucleus after cell division.

In sum, protein synthesis is a complex process involving the interaction between three major classes of RNA (mRNA, tRNA, and rRNA) and a large number of accessory protein factors that act catalytically in the process. With very few exceptions the genetic code that specifies the amino acids for each mRNA codon is the same in every organism, prokaryotic and eukaryotic. By repeated translation of an mRNA molecule by a string of ribosomes (producing a polysome), a large number of identical protein molecules can be produced. Thus, from a single gene, large quantities of a protein can be produced by two amplification steps: (1) the production of multiple mRNAs from the gene; and (2) the production of many protein molecules by repeated translation of each mRNA.

ANALYTICAL APPROACHES FOR SOLVING GENETICS PROBLEMS

Q14.1a. How many of the 64 codon permutations can be made from the three nucleotides A, U, and G?

b. How many of the 64 codon permutations can be made from the four nucleotides A, U, G, and C, with one or more Cs in each codon?

A14.1a. This question involves probability. There are four bases, so the probability of a cytosine at the first position in a codon is 1/4. Conversely, the probability of a base other than cytosine in the first position is $(1 - 1/4) = 3/4$. These same probabilities apply to the other two positions in the codon. Therefore the probability of a codon without a cytosine is $(3/4)^3 = 27/64$.

b. This question involves the relative frequency of codons that have one or more cytosines. We have already calculated the probability of a codon not having a cytosine, so all the remaining codons have one or more cytosines. The answer to this question, therefore, is $(1 - 27/64) = 37/64$.

Q14.2 Random copolymers were used in some of the experiments directed toward deciphering the genetic code. For each of the following ribonucleotide mixtures, give the expected codons and their frequencies, and give the expected proportions of the amino acids that would be found in a polypeptide directed by the copolymer in a cell-free, protein-synthesizing system.

a. 2U:1C
b. 1U:1C:2G

A14.2a. The probability of a U at any position in a codon is 2/3, and the probability of a C at any position in a codon is 1/3. Thus, the codons, and their relative frequencies, and the amino acids for which they code are:

$$UUU = (2/3)(2/3)(2/3) = 8/27 = 0.296 = 29.6\% \text{ Phe}$$
$$UUC = (2/3)(2/3)(1/3) = 4/27 = 0.148 = 14.8\% \text{ Phe}$$
$$UCC = (2/3)(1/3)(1/3) = 2/27 = 0.0743 = 7.43\% \text{ Ser}$$
$$UCU = (2/3)(1/3)(2/3) = 4/27 = 0.148 = 14.8\% \text{ Ser}$$
$$CUU = (1/3)(2/3)(2/3) = 4/27 = 0.14 = 14.8\% \text{ Leu}$$
$$CUC = (1/3)(2/3)(1/3) = 2/27 = 0.0743 = 7.43\% \text{ Leu}$$
$$CCU = (1/3)(1/3)(2/3) = 2/27 = 0.0743 = 7.43\% \text{ Pro}$$
$$CCC = (1/3)(1/3)(1/3) = 1/27 = 0.037 = 3.7\% \text{ Pro}$$

In sum, 44.4% Phe, 22.23% Ser, 22.23% Leu, 11.13%

Pro. (Does not quite add up to 100% because of rounding-off errors.)

b. The probability of a U at any position in a codon is 1/4, the probability of a C at any position in a codon is 1/4, and the probability of a G at any position in a codon is 1/2. Thus, the codons and their relative frequencies, and the amino acids for which they code are:

$$UUU = (1/4)(1/4)(1/4) = 1/64 = 1.56\% \text{ Phe}$$
$$UUC = (1/4)(1/4)(1/4) = 1/64 = 1.56\% \text{ Phe}$$
$$UCU = (1/4)(1/4)(1/4) = 1/64 = 1.56\% \text{ Ser}$$
$$UCC = (1/4)(1/4)(1/4) = 1/64 = 1.56\% \text{ Ser}$$
$$CUU = (1/4)(1/4)(1/4) = 1/64 = 1.56\% \text{ Leu}$$
$$CUC = (1/4)(1/4)(1/4) = 1/64 = 1.56\% \text{ Leu}$$
$$CCU = (1/4)(1/4)(1/4) = 1/64 = 1.56\% \text{ Pro}$$
$$CCC = (1/4)(1/4)(1/4) = 1/64 = 1.56\% \text{ Pro}$$
$$UUG = (1/4)(1/4)(1/2) = 2/64 = 3.13\% \text{ Leu}$$
$$UGU = (1/4)(1/2)(1/4) = 2/64 = 3.13\% \text{ Cys}$$
$$UGG = (1/4)(1/2)(1/2) = 4/64 = 6.25\% \text{ Trp}$$
$$GUU = (1/2)(1/4)(1/4) = 2/64 = 3.13\% \text{ Val}$$
$$GUG = (1/2)(1/4)(1/2) = 4/64 = 6.25\% \text{ Val}$$
$$GGU = (1/2)(1/2)(1/4) = 4/64 = 6.25\% \text{ Gly}$$
$$GGG = (1/2)(1/2)(1/2) = 8/64 = 12.5\% \text{ Gly}$$
$$CCG = (1/4)(1/4)(1/2) = 2/64 = 3.13\% \text{ Pro}$$
$$CGC = (1/4)(1/2)(1/4) = 2/64 = 3.13\% \text{ Arg}$$
$$CGG = (1/4)(1/2)(1/2) = 4/64 = 6.25\% \text{ Arg}$$
$$GCC = (1/2)(1/4)(1/4) = 2/64 = 3.13\% \text{ Ala}$$
$$GCG = (1/2)(1/4)(1/2) = 4/64 = 6.25\% \text{ Ala}$$
$$GGC = (1/2)(1/2)(1/4) = 4/64 = 6.25\% \text{ Gly}$$
$$UCG = (1/4)(1/4)(1/2) = 2/64 = 3.13\% \text{ Ser}$$
$$UGC = (1/4)(1/2)(1/4) = 2/64 = 3.13\% \text{ Cys}$$
$$CUG = (1/4)(1/4)(1/2) = 2/64 = 3.13\% \text{ Leu}$$
$$CGU = (1/4)(1/2)(1/4) = 2/64 = 3.13\% \text{ Arg}$$
$$GUC = (1/2)(1/4)(1/4) = 2/64 = 3.13\% \text{ Val}$$
$$GCU = (1/2)(1/4)(1/4) = 2/64 = 3.13\% \text{ Ala}$$

In sum, 3.12% Phe, 6.25% Ser, 9.38% Leu, 6.25% Pro, 6.26% Cys, 6.25% Trp, 12.5% Val, 12.5% Gly, 12.5% Arg, 12.5% Ala.

QUESTIONS AND PROBLEMS

*14.1   The form of genetic information used directly in protein synthesis is (choose the correct answer):
a.   DNA
b.   mRNA
c.   rRNA
d.   Ribosomes

*14.2   Proteins are (choose the correct answer):
a.   Branched chains of nucleotides
b.   Linear, folded chains of nucleotides
c.   Linear, folded chains of amino acids
d.   Invariable enzymes

14.3   Various proteins are treated as specified. For each case indicate what level(s) of protein structure would change as the result of the treatment.
a.   Hemoglobin is stored in a hot incubator at 80°C.
b.   Egg white (albumin) is boiled.
c.   RNase (a single polypeptide enzyme) is heated to 100°C.
d.   Meat in your stomach is digested (gastric juices contain proteolytic enzymes).
e.   In the β polypeptide chain of hemoglobin, the amino acid valine replaces glutamic acid at the number six position.

14.4   The process in which ribosomes engage is (choose the correct answer):
a.   Replication
b.   Transcription
c.   Translation
d.   Disjunction
e.   Cell division

14.5   What are the characteristics of the genetic code?

14.6   Base-pairing wobble occurs in the interaction between the anticodon of the tRNAs and the codons. On the theoretical level, determine the minimum number of tRNAs needed to read the 61 sense codons.

14.7   Antibiotics have been very useful in elucidating the steps of protein synthesis. If you have an artificial messenger of the sequence of AUGUUUUUUUUUUUU..., it will produce the following polypeptide in a cell-free, protein-synthesizing system: fMet-Phe-Phe-Phe. ...In your search for new antibiotics you find one called putyermycin, which blocks protein synthesis. When you try it with your artificial mRNA in a cell-free system, the product is fMet-Phe. What step in protein synthesis does putyermycin affect? Why?

14.8   Describe the reactions involved in the aminoacylation (charging) of a tRNA molecule.

14.9   Compare and contrast the following in prokaryotes and eukaryotes:

a.   protein synthesis initiation
b.   protein synthesis termination

14.10   Discuss the two species of methionine tRNA, and describe how they differ in structure and function. In your answer, include a discussion of how each of these tRNAs binds to the ribosome.

*14.11   Random copolymers were used in some of the experiments that revealed the characteristics of the genetic code. For each of the following ribonucleotide mixtures, give the expected codons and their frequencies, and give the expected proportions of the amino acids that would be found in a polypeptide directed by the copolymer in a cell-free, protein-synthesizing system:
a.   4 A : 6 C
b.   4 G : 1 C
c.   1 A : 3 U : 1 C
d.   1 A : 1 U : 1 G : 1 C

*14.12   Other features of the reading of mRNA into proteins being the same as they are now (i.e., codons must exist for 20 different amino acids), what would the minimum WORD (CODON) SIZE be if the number of different bases in the mRNA were, instead of four:
a.   two
b.   three
c.   five

14.13   Suppose that at stage A in the evolution of the genetic code only the first two nucleotides in the coding triplets led to unique differences and that any nucleotide could occupy the third position. Then, suppose there was a stage B in which differences in meaning arose depending upon whether a purine (A or G) or pyrimidine (C or T) was present at the third position. Without reference to the number of amino acids or multiplicity of tRNA molecules, how many triplets of different meaning can be constructed out of the code at stage A? at stage B?

*14.14   A gene makes a polypeptide 30 amino acids long containing an alternating sequence of phenylalanine and tyrosine. What are the sequences of nucleotides corresponding to this sequence in the following:
a.   The DNA strand which is read to produce the mRNA, assuming Phe = UUU and Tyr = UAU in mRNA
b.   The DNA strand which is not read
c.   tRNA

14.15   A segment of a polypeptide chain is Arg-Gly-Ser-Phe-Val-Asp-Arg. It is encoded by the following segment of DNA:

Which strand is the template strand? Label each strand with its correct polarity (5' and 3').

*14.16 Two populations of RNAs are made by the random combination of nucleotides. In population A the RNAs contain only A and G nucleotides (3A:1G), while in population B the RNAs contain only A and U nucleotides (3A:1U). In what ways *other than amino acid content* will the proteins produced by translating the population A RNAs differ from those produced by translating the population B RNAs?

14.17 In *E. coli* a particular tRNA normally has the anticodon 5'-GGG-3', but because of a mutation in the tRNA gene, the mutant tRNA has the anticodon 5'-GGA-3'.
a. What amino acid would this tRNA carry?
b. What codon would the normal tRNA recognize?
c. What codon would the mutant tRNA recognize?
d. What would be the effect of the mutation on the proteins in the cell?

*14.18 A particular protein found in *E. coli* normally has the N-terminal sequence Met-Val-Ser-Ser-Pro-Met-Gly-Ala-Ala-Met-Ser. . . . In a particular cell a mutation alters the anticodon of a particular tRNA from 5'-GAU-3' to 5'-CAU-3'. What would be the N-terminal amino acid sequence of this protein in the mutant cell? Explain your reasoning.

14.19 The gene encoding an *E. coli* tRNA containing the anticodon 5'-GUA-3' mutates so that the anticodon now is 5'-UUA-3'. What will be the effect of this mutation? Explain your reasoning.

*14.20 The normal sequence of the coding region of a particular mRNA is shown below, along with several mutant versions of the same mRNA. Indicate what protein would be formed in each case. (. . . = many [a multiple of 3] unspecified bases.)

Normal: AUGUUCUCUAAUUAC(. . .)AUGGGGUGGGUGUAG
Mutant *a* AUGUUCUCUAAUUAG(. . .)AUGGGGUGGGUGUAG
Mutant *b* AGGUUCUCUAAUUAC(. . .)AUGGGGUGGGUGUAG
Mutant *c* AUGUUCUCGAAUUAC(. . .)AUGGGGUGGGUGUAG
Mutant *d* AUGUUCUCUAAAUAC(. . .)AUGGGGUGGGUGUAG
Mutant *e* AUGUUCUCUAAUUC(. . .)AUGGGGUGGGUGUAG
Mutant *f* AUGUUCUCUAAUUAC(. . .)AUGGGGUGGGUGUGG

14.21 The normal sequence of a particular protein is given below, along with several mutant versions of it. For each mutant, explain what mutation occurred in the coding sequence of the gene.

Normal: Met-Gly-Glu-Thr-Lys-Val-Val-. . . -Pro
Mutant 1: Met-Gly
Mutant 2: Met-Gly-Glu-Asp
Mutant 3: Met-Gly-Arg-Leu-Lys
Mutant 4: Met-Arg-Glu-Thr-Lys-Val-Val-. . . -Pro

14.22 The N-terminus of a protein has the sequence Met-His-Arg-Arg-Lys-Val-His-Gly-Gly. A molecular biologist wishes to synthesize a DNA chain that could encode this portion of the protein. How many possible DNA sequences could encode this polypeptide?

14.23 In the recessive condition in humans known as sickle cell anemia, the β-globin polypeptide of hemoglobin is found to be abnormal. The only difference between it and the normal β-globin is that the 6th amino acid from the N-terminal is valine, whereas the normal β-globin has glutamic acid at this position. Explain how this occurred.

*14.24 As discussed in Chapter 9 (pages 279–281 and Figure 9.15), cystic fibrosis is an autosomal recessive disease in which the cystic fibrosis transmembrane conductance regulator (CFTR) protein is abnormal. The RNA coding sequence of this gene spans about 250,000 base pairs of DNA. The CFTR protein, with 1,480 amino acids, is translated from an mRNA of about 6,500 bases. The most common mutation in this gene results in a protein that is missing a phenylalanine at position 508 (ΔF508).
a. Why is the RNA coding sequence of this gene so much larger than the mRNA from which the CFTR protein is translated?
b. About what percent of the mRNA together comprises 5' untranslated leader and 3' untranslated trailer sequences?
c. At the DNA level, what alteration would you expect to find in the ΔF508 mutation?
d. What consequences might you expect if the DNA alteration that you describe in (c) were to occur at random in the protein coding region of the CFTR gene?

14.25 Antibiotics have been useful in determining whether cellular events depend on transcription or translation. For example, actinomycin D is used to block transcription, and cycloheximide (in eukaryotes) is used to block translation. In some cases, though, surprising results are obtained after antibiotics are administered. The addition of actinomycin D, for example, may result in an increase, not a decrease, in the activity of a particular enzyme. Discuss how this result might come about.

14.26 In the last several years, a set of diseases have been shown to be caused by triplet repeat amplification (for review, see Chapter 7, p. 206). Amplification of the following triplet repeats have been associated with disease phenotypes: 5'-CAG-3', 5'-CGG-3', 5'-GAA-3', and 5'-CTG-3'. CGG expansions have been found 5' to coding sequence of a gene, CAG expansions have been found within coding regions, GAA expansions have been found in intronic regions, and CTG and CGG expansions have been found 3' to the gene's coding sequence. Speculate how each of these triplet repeat expansions might result in abnormal gene function.

15 RECOMBINANT DNA TECHNOLOGY AND THE MANIPULATION OF DNA

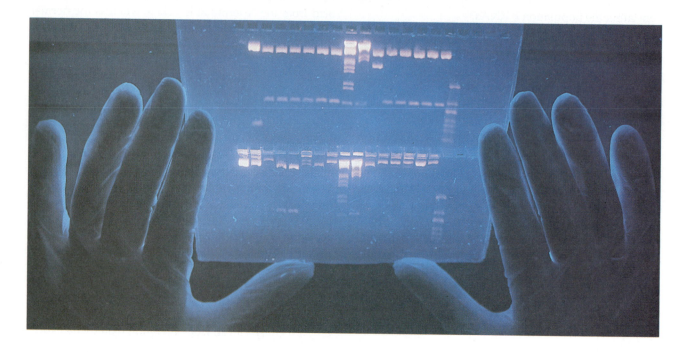

APPLICATIONS OF RECOMBINANT DNA TECHNOLOGY
Analysis of Biological Processes
Diagnosis of Human Genetic Diseases by DNA Analysis
Isolation of Human Genes
Human Genome Project
DNA Typing
Gene Therapy
Commercial Products
Genetic Engineering of Plants

PRINCIPAL POINTS

~ Genes are cloned by splicing DNA from an organism into a cloning vector (a DNA molecule capable of replication in a host organism) to make a recombinant DNA molecule, and then introducing that molecule into a host cell in which it will replicate. Essential to cloning are restriction enzymes (restriction endonucleases). These enzymes are useful for cloning because they recognize specific nucleotide pair sequences in DNA (restriction sites) and cleave at a specific point within the sequence.

~ Different kinds of cloning vectors have been developed to construct and clone recombinant DNA molecules. All cloning vectors must: (1) replicate within at least one host organism; (2) have one or more restriction sites into which foreign DNA can be inserted; and (3) have one or more dominant selectable markers to detect those cells which contain the vectors. The main classes of vectors are plasmids, viruses, and cosmids. Shuttle vectors have the above properties but in addition are able to replicate in more than one type of host. Yeast artificial chromosomes (YACs) enable DNA fragments several hundred kilobase pairs long to be cloned in yeast.

~ Genomic libraries are collections of clones that contain one copy of every DNA sequence in an organism's genome. Genomic libraries are useful, for example, for isolating specific genes and for studying the organization of the genome.

~ Individual chromosomes can be isolated and chromosome-specific libraries made from them. If a gene has been localized to a specific chromosome by genetic means, the existence of chromosome-specific libraries makes it easier to isolate a clone of the gene.

~ DNA copies, called complementary DNAs (cDNAs), can be made from mRNA molecules isolated from the cell. The cDNAs can be cloned using appropriate cloning vectors.

~ Specific sequences in genomic libraries and cDNA libraries can be identified using a number of approaches, including the use of specific DNA or cDNA probes,

heterologous probes, specific antibodies, complementation of mutations, or oligonucleotide probes.

~ Genes and cloned DNA sequences can be analyzed to determine the arrangement and specific locations of restriction sites, a process called restriction mapping. Gene transcripts can be analyzed to determine tissue specificity and the level of gene expression using recombinant DNA procedures.

~ Rapid methods have been developed for determining the sequence of a cloned piece of DNA. A commonly used method, the dideoxy or Sanger procedure, uses enzymatic synthesis of a new DNA chain on a template DNA strand. With this procedure, synthesis of new strands is stopped by the incorporation of a dideoxy analog of the normal deoxyribonucleotide. Using four different dideoxy analogs, the new strands stop at all possible nucleotide positions, thereby allowing the complete DNA sequence to be determined.

~ The polymerase chain reaction (PCR) uses synthetic oligonucleotides to amplify a specific segment of DNA many thousand fold in an automated procedure. A major benefit of PCR is that small amounts of DNA are needed. Thus, DNA from a single cell can be amplified using PCR. PCR is finding increasing applications both in research and in the commercial arena, including generating specific DNA segments for cloning or for sequencing, and for amplifying DNA to detect the presence of specific genetic defects.

~ There are many applications for recombinant DNA technology and related procedures. For example, with appropriate probes, detection of specific genetic diseases is possible, and many products in the clinical, veterinary, and agricultural areas have been developed. A project is underway to determine the complete sequence of the human genome. The knowledge obtained will expand greatly our understanding of human genetics. With continued advances in genetic engineering of plants, it is expected that many types of improved crops (increased yields, disease resistance) will be forthcoming.

*E*xperimental procedures have been developed that enable researchers to construct **recombinant DNA molecules** in test tubes. In this process, genetic material from two different sources is combined into a single DNA molecule. The technology has opened the way for new and exciting research possibilities and affirms the plausibility of **genetic engineering**, that is, the alteration of the genetic constitution of cells or individuals by directed and selective modification, insertion, or deletion of an individual gene or genes. In this chapter we describe gene cloning and discuss the manipulation of DNA using recombinant DNA techniques. We also present some examples of how **recombinant DNA technology** is furthering our knowledge of the structure and function of the prokaryotic and eukaryotic genomes, how cloned DNA sequences are being used for genetic diagnosis, and how the ability to manipulate DNA sequences has opened new ways for producing commercial products.

Gene Cloning

In brief, genes are cloned by taking a piece of DNA from an organism and splicing it into a **cloning vector** to make a recombinant DNA molecule. A cloning vector is an artificially constructed DNA molecule capable of replication in a host organism, such as a bacterium, and into which a piece of DNA to be studied can be specifically inserted at known positions. The recombinant DNA molecule is introduced into a host such as *E. coli*, yeast, animal cell, or plant cell. Replication of the recombinant DNA molecule (**molecular cloning**) occurs in the host cell, thereby producing many identical copies. Why is this useful? There are many reasons. For example, suppose we want to study the gene for a particular human protein in order to determine its DNA sequence, and how its expression is regulated. Each human cell contains only two copies of that gene, making it an extremely difficult, almost impossible task to isolate enough copies of the gene for analysis. By contrast, an essentially unlimited number of copies of the gene can be produced by cloning. Similarly, if we can clone a gene selectively, we can design experiments for manipulating the gene (for example, changing its DNA sequence), or for the synthesis of large amounts of the gene's products. For instance, human insulin (called *humulin*), produced from a recombinant DNA molecule, is now substituted for insulin isolated from pig pancreases in the treatment of diabetes. In this section we describe how DNA sequences can be cloned.

Restriction Enzymes

Recombinant DNA molecules could not be constructed without the use of **restriction enzymes** (or **restriction endonucleases**). The important feature of a restriction enzyme is that it is able to cleave double-stranded DNA molecules at a specific nucleotide pair sequence called a *restriction site*, or restriction enzyme recognition sequence. Restriction enzymes are used to produce a pool of discrete DNA molecules to be cloned. Restriction enzymes are also used to analyze the positioning of restriction sites in a piece of cloned DNA or in a segment of DNA in the genome (see pp. 469–473).

Although one restriction enzyme has been found in the green alga *Chlorella,* most restriction enzymes are found in bacteria. Their natural function in bacteria is to protect the organism against invading viruses. That is, the bacterium modifies its own restriction sites (by methylation) so that the restriction enzyme it makes cannot cut its DNA. Then, when a virus injects its DNA, the restriction enzyme is able to cleave the viral DNA without affecting the bacterial DNA.

Over 400 different restriction enzymes have been isolated. Restriction enzymes are named for the organism from which they are isolated. Conventionally, a three-letter system is used, italicized or underlined, followed by a roman numeral. Additional letters are sometimes added to signify a particular strain of the bacterial species from which the enzymes were obtained. For example, *Bgl*II is from *Bacillus globigi*; *Eco*RI is from *E. coli* strain RY13, and *Hind*III is from *Haemophilus influenzae* strain Rd. The names are pronounced in particular ways but those pronunciations follow no set pattern. For example, *Bam*HI is "bam-H-one," *Bgl*II is "bagel-two," *Eco*RI is "echo-R-one," *Hind*III is "hin-D-three," *Hha*I is "ha-ha-one," and *Hpa*II is "hepa-two." The pronunciation of other enzymes will be given when they are introduced.

Restriction enzymes recognize a specific nucleotide pair sequence in DNA and cleave the DNA within that sequence. In some cases, restriction enzymes isolated from different bacteria recognize and cleave the same DNA sequences; these enzymes are called *isoschizomers.*

The recognition sequences for restriction enzymes generally have an axis of symmetry through the midpoint of the recognition sequence: the base sequence from 5' to 3' on one DNA strand is the same as the base sequence from 5' to 3' on the complementary DNA strand (Figure 15.1). Thus, the sequences are said to have *twofold rotational symmetry* or to be *palindromic.* Because of this property we can determine the complete restriction enzyme sequence if we know only the first half of the nucleotides, starting from the 5' end. So, for an enzyme that recognizes a six-nucleotide-pair sequence, if we are given the sequence 5'-CTG-3', we can deduce that there must be a symmetrical sequence on the opposite DNA strand because of two-fold rotation symmetry, giving

5' C T G ? ? ? 3'
3' ? ? ? G T C 5'

Filling in the rest by the base pairing rules, we get

5' C T G C A G 3'
3' G A C G T C 5'

Since each restriction enzyme cuts DNA at an enzyme-specific sequence, the number of cuts the enzyme makes in a particular DNA molecule or molecules depends on the number of times the particular recognition/cleavage sequence is present in the DNA. To help us understand the consequences of cutting DNA with a restriction enzyme, let us consider what happens when we cut a number of copies of the same genome with one type of restriction enzyme. The restriction enzyme will cut the DNA at its specific restriction sites which are distributed through the genome. Although this will produce millions of fragments of different sizes, all of the identical chromosomal DNAs in the multiple genome copies will be cut at identical recognition sequences.

The cleavage sites for a number of restriction enzymes are shown in Table 15.1. The most commonly used restriction enzymes recognize four nucleotide pairs (e.g., *Hae*III ["hay-three"], *Hha*I) or six nucleotide pairs (e.g., *Bam*HI, *Eco*RI). Several enzymes have been found that recognize eight-nucleotide-pair sequences (e.g., *Not*I ["not-one"]). Other classes of enzymes do not quite fit our model in that the recognition site is not symmetrical about the center. *Hin*fI ["hin-f-one"], for example, recognizes a five-nucleotide-pair sequence in which there is symmetry in the two base pairs on either side of the central base pair, but the central base pair is obviously asymmetrical within the sequence. *Bst*XI ["b-s-t-x-one"] is representative of a number of restriction enzymes with a nonspecific spacer region between symmetrical sequences (see Table 15.1).

For DNA with random distribution of nucleotide pairs, there is a clear relationship between the number of nucleotide pairs in the recognition sequence and the frequency of cutting the DNA. That is, an enzyme that recognizes a specific four-nucleotide-pair sequence will cut more frequently than one that recognizes a five-nucleotide-pair sequence, and so on. Let us consider this more mathematically.

Consider DNA with 50 percent GC and a random distribution of nucleotide pairs. For that DNA, there is an equal chance of finding one of the four possible nucleotide pairs $\frac{G}{C}$, $\frac{C}{G}$, $\frac{A}{T}$, or $\frac{T}{A}$, at any one position. Consider the restriction enzyme *Hae*III which recognizes the sequence $\frac{5'\text{-GGCC-}3'}{3'\text{-CCGG-}5'}$. The probability of this sequence occurring in DNA is computed as follows:

1st nucleotide pair: $\frac{G}{C}$, probability $= 1/4$

2nd nucleotide pair: $\frac{G}{C}$, probability $= 1/4$

3rd nucleotide pair: $\frac{C}{G}$, probability $= 1/4$

4th nucleotide pair: $\frac{C}{G}$, probability $= 1/4$

The probability of finding any one of the nucleotide pairs is independent of the probability of finding one of the other nucleotide pairs. Therefore, according to the product rule, the probability of finding the *Hae*III restriction site in DNA with a random distribution of nucleotide pairs is $1/4 \times 1/4 \times 1/4 \times 1/4 = 1/256$. In short, the recognition sequence for *Hae*III occurs on the average once every 256 base pairs in such a piece of DNA.

In general, the probability of occurrence of a restriction site is given by the formula $(1/4)^n$ where n is the number of nucleotide pairs in the recognition sequence. These values are given in Table 15.2.

~ **FIGURE 15.1**

Restriction enzyme recognition sequence in DNA, showing symmetry of the sequence about the center point. The sequence is a palindrome, reading the same from left to right (5'-to-3') on the top strand (GAATTC, here), as it does from right to left (5'-to-3') on the bottom strand. Shown here is the recognition sequence for the restriction enzyme *Eco*RI.

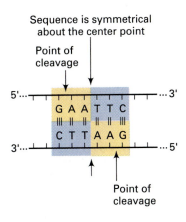

~ **TABLE 15.1**

Characteristics of Some Restriction Endonucleases

	ENZYME NAME	PRONUN-CIATION	ORGANISM IN WHICH ENZYME IS FOUND	RECOGNITION SEQUENCE AND POSITION OF CUT[a]
Enzymes with 6-bp Recognition Sequences	*Bam*HI	"bam-H-one"	*Bacillus amyloliquefaciens H*	5' G↓G A T C C 3' 3' C C T A G↑G 5'
	*Bgl*II	"bagel-two"	*Bacillus globigi*	A↓G A T C T T C T A G↑A
	*Eco*RI	"echo-R-one"	*E. coli RY13*	G↓A A T T C C T T A A↑G
	*Hae*II	"hay-two"	*Haemophilus aegyptius*	R G C G C↓Y Y↑C G C G R
	*Hind*III	"hin-D-three"	*Haemophilus influenzae* R_d	A↓A G C T T T T C G A↑A
	*Pst*I	"P-S-T-one"	*Providencia stuartii*	C T G C A↓G G↑A C G T C
	*Sal*I	"sal-one"	*Streptomyces albus G*	G↓T C G A C C A G C T↑G
	*Sma*I	"sma-one"	*Serratia marcescens*	C C C↓G G G G G G↑C C C
Enzymes with 4-bp Recognition Sequences	*Hae*III	"hay-three"	*Haemophilus egyptius*	G G↓C C C C↑G G
	*Hha*I	"ha-ha-one"	*Haemophilus hemolyticus*	G C G↓C C↑G C G
	*Hpa*II	"hepa-two"	*Haemophilus parainfluenzae*	C↓C G G G G C↑C
	*Sau*3A	"sow-three-A"	*Staphylococcus aureus 3A*	↓G A T C C T A G↑
Enzymes with 8-bp Recognition Sequences	*Not*I	"not-one"	*Nocardia otitidis-caviarum*	G C↓G G C C G C C G C C G G↑C G
Enzymes with Recognition Sequences That Are Not Symmetrical	*Bst*XI	"b-s-t-x-one"	*Bacillus stearothermophilus*	C C A N N N N N↓N T G G G G T N N N N N↑N A C C

[a]In this column the two strands of DNA are shown with the sites of cleavage indicated by arrows. Since there is an axis of twofold rotational symmetry in each recognition sequence, the DNA molecules resulting from the cleavage are symmetrical.   Key: R = purine; Y = pyrimidine; N = any base

Note, though, that DNA from living organisms does not have a random distribution of base pairs, and most such DNAs are not composed of 50 percent GC. Rather, some DNA is GC-rich and other DNA is AT-rich. The latter, for example, would have few sites for *Hae*III, so this enzyme would cut AT-rich DNA much less frequently than the table indicates. Thus, the predicted frequencies of cutting in Table 15.2 should be used only as a rough guide in planning experiments. Note that enzymes with eight-nucleotide-pair restriction sites cut, on the average, only once every 65,476 bp. Because they cut relatively infrequently, they are useful for cutting large genomes—human, for example—into relatively large pieces.

As Table 15.1 indicates, some enzymes, such as *Hae*III or *Sma*I ("sma-one"), cut both strands of DNA between the same two nucleotide pairs to produce *blunt ends* (Figure 15.2a), while others, such as *Eco*RI, *Bam*HI, and *Hin*dIII, make staggered cuts in the symmetrical nucleotide pair sequence to produce *sticky* or *staggered ends*. The staggered ends may either have an overhanging 5' end, as in the case of cleavage with *Bam*HI or *Eco*RI (Figure 15.2b), or an overhanging 3' end, as in the case of cleavage with *Hae*II ("hay-two") or *Pst*I ("P-S-T-one") (Figure 15.2c). In each case, the restriction enzyme makes its cut by cleaving the DNA backbone between the 3' carbon of the deoxyribose sugar and the phosphate on the subsequent deoxyribose in the chain (Figure 15.3). Thus, all restriction enzyme-cut DNA fragments have a phosphate on their 5' ends and a hydroxyl group on their 3' ends.

Restriction enzymes that produce staggered ends are of particular value in cloning DNA fragments because every DNA fragment generated by cutting a piece of DNA with the same restriction enzyme has the same base sequence at the two staggered ends. Therefore, if the ends of two pieces of DNA produced by the action of the same restriction enzyme (such as *Eco*RI)—for example, cloning vector, and a chromoso-

~ FIGURE 15.2

Examples of how restriction enzymes cleave DNA. (a) *Sma*I results in blunt ends; (b) *Bam*HI results in overhanging ("sticky") 5' ends; (c) *Pst*I results in overhanging ("sticky") 3' ends.

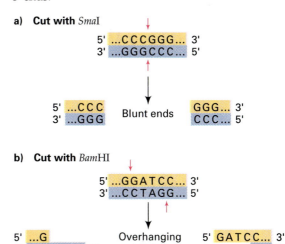

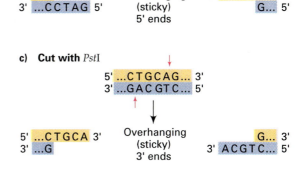

~ TABLE 15.2

Occurrence of Restriction Sites for Restriction Enzymes in DNA with Randomly Distributed Nucleotide Pairs

NUCLEOTIDE PAIRS IN RESTRICTION SITE	PROBABILITY OF OCCURRENCE
4	$(1/4)^4$ = 1 in 256 bp
5	$(1/4)^5$ = 1 in 1,024 bp
6	$(1/4)^6$ = 1 in 4,096 bp
8	$(1/4)^8$ = 1 in 65,476 bp
n	$(1/4)^n$

mal DNA fragment—come together in solution, perfect base pairing occurs; that is, the two single-stranded DNA are said to *anneal* (Figure 15.4). In the presence of DNA ligase, the sugar-phosphate backbones can then be sealed to produce a whole DNA molecule with a reconstituted restriction enzyme site. Even DNA fragments with blunt ends can be ligated together with DNA ligase at high concentrations of the enzyme. This ligation of two DNA fragments with identical sticky ends or of two DNA fragments with blunt ends by DNA ligase is the principle behind the formation of recombinant DNA molecules.

~ FIGURE 15.3

Cleavage of DNA with a restriction enzyme results in DNA fragments with a phosphate on their 5' ends and a 3' hydroxyl on their 3' ends.

~ FIGURE 15.4

Cleavage of DNA by the restriction enzyme *Eco*RI. *Eco*RI makes staggered, symmetrical cuts in DNA, leaving "sticky" ends. A DNA fragment with a sticky end produced by *Eco*RI digestion can bind by complementary base pairing (anneal) to any other DNA fragment with a sticky end produced by *Eco*RI cleavage. The gaps may then be sealed by DNA ligase.

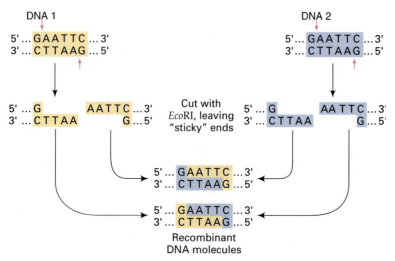

KEYNOTE

Genes are cloned by splicing DNA from an organism into a cloning vector to make a recombinant DNA molecule, and then introducing that molecule into a host cell in which it will replicate. Essential to cloning are restriction enzymes, or restriction endonucleases. These enzymes are useful for cloning because they recognize specific nucleotide pair sequences in DNA (restriction sites) and cleave at a specific point within the sequence. Cleavage of the DNA with a restriction enzyme can be staggered (producing DNA fragments with single-stranded, "sticky," ends) or blunt.

Cloning Vectors and the Cloning of DNA

Three major types of vectors are commonly used for cloning DNA sequences in bacterial cells: plasmids, bacteriophages, and cosmids. Each differs in the way it must be manipulated to clone pieces of DNA and in the maximum amount of DNA that can be cloned using it. Each type of vector has been specially constructed in the laboratory to possess the features necessary to make it an efficient cloning vector.

PLASMID CLONING VECTORS. Plasmids (see Chapter 8 and Chapter 20) are extrachromosomal genetic elements that replicate autonomously within host cells. Their DNA is circular and double-stranded and carries the sequences (origins or *ori*) required for the replication of the plasmid, as well as for the plasmid's other functions. The particular plasmids used for cloning experiments are all derivatives of naturally occurring plasmids that have been "engineered" to have features that facilitate gene cloning. Because they are most commonly used, here we focus on features of *E. coli* plasmid cloning vectors.

An *E. coli* plasmid cloning vector must have three features:

1. An *ori* sequence, which allows the plasmid to replicate in *E. coli* because it provides a DNA sequence that is recognized by the replication enzymes in the cell.
2. A dominant *selectable marker*, which enables *E. coli* cells that carry the plasmid to be easily distinguished from cells that lack the plasmid. The usual selectable marker is a gene that confers an antibiotic resistance phenotype upon the *E. coli*

host, for example, the amp^R gene for ampicillin resistance, the tet^R gene for tetracycline resistance, or the cam^R gene for chloramphenicol resistance. When plasmids carrying antibiotic resistance genes such as these are introduced by transformation (see Chapter 8) into a plasmid-free and therefore antibiotic-sensitive *E. coli* host cell, the uptake of an antibiotic-resistance plasmid can be detected easily since the host cell will grow in medium to which the appropriate antibiotic has been added.

3. *Unique restriction enzyme cleavage sites*—sites present just once in the vector—for the insertion of the DNA sequences that are to be cloned. Cloning most commonly involves cutting the plasmid at one of the unique sites with the appropriate restriction enzyme, and splicing into that site a piece of DNA that has been cut with the same enzyme.

As an example, Figure 15.5 diagrams the plasmid vector pUC19 ("puck-19"). This 2,686 bp vector has the following features that make it useful for cloning DNA:

1. It has a high copy number, approaching 100 copies per cell.
2. It has the amp^R selectable marker.
3. It contains a number of unique restriction sites clustered in one region. Such a region of clustered unique restriction sites is called a **polylinker** or **multiple cloning site**.
4. The polylinker is inserted near the 5' end of a fragment of the *E. coli* β-galactosidase gene *lacZ*+ gene. The insertion was engineered so that functional β-galactosidase is produced when pUC19 is introduced into a *lacZ*− mutant cell that makes nonfunctional β-galactosidase. Then, when a piece of DNA is cloned into the polylinker, the β-galactosidase reading frame is disrupted so functional β-galactosidase cannot be produced in *E. coli*. A simple color test is used to distinguish colonies of *E. coli* containing pUC19 with no inserted DNA (blue) from colonies containing pUC19 with inserted DNA (white). This test provides a rapid, visual way to identify cells containing recombinant DNA molecules.

Similar vectors are available with variants on these features, for example with different arrays of unique restriction sites in the polylinker, and with phage promoters flanking the polylinker-*lacZ*+ region. The phage promoters, such as those for T7, T3, or SP6 DNA-dependent RNA polymerases, are used to make RNA copies of the cloned DNA sequences *in vitro* in

~ FIGURE 15.5

Restriction map of the plasmid cloning vector pUC19. This plasmid has an origin of replication (*ori*), an *amp*^R selectable marker, and a polylinker located within the β-galactosidase gene, *lacZ*^+. In a bacterial cell, pUC19 without a DNA insert can make β-galactosidase, while pUC19 with a DNA insert cannot make β-galactosidase. This makes it possible to use a color selection system for distinguishing vectors with and without DNA inserts (see text).

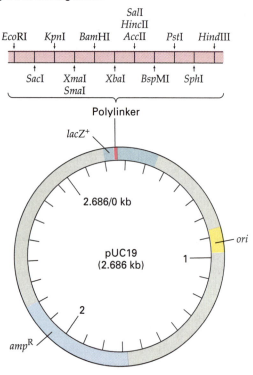

pUC19 cloning vector

ori = Origin of replication sequence
amp^R = Ampicillin resistance gene
lacZ^+ = β-galactosidase gene

the presence of the appropriate polymerase. If the RNA is made radioactive by the incorporation of ^32P-NTPs in the reaction mixture, it can be used as a probe in further analytical techniques. DNA fragments of up to a few kilobase pairs are efficiently cloned in plasmid vectors, although they may accept fragments of 5 to 10 kb. Plasmids carrying larger DNA fragments are often unstable and tend to lose most of the inserted DNA.

Figure 15.6 illustrates how a piece of DNA can be inserted into a plasmid cloning vector such as pUC19. In the first step, pUC19 is cut with a restriction enzyme that has a site in the polylinker. Next, the piece of DNA to be cloned is generated by cutting high-molecular-weight DNA with the same restriction

enzyme. Since restriction sites are *nonrandomly arranged* in DNA, fragments of various sizes are produced. The DNA fragments are mixed with the cut vector and randomly allowed to join; in some cases, the DNA fragment becomes inserted between the two cut ends of the plasmid. The resulting recombinant DNA plasmid is introduced into an *E. coli* host by transformation. Resulting ampicillin-resistant colonies indicate cells that were transformed by plasmids, and white colonies (versus blue) on color selection plates indicate the recombinant DNA clones (versus plasmids that were resealed without including a DNA fragment).

PHAGE CLONING VECTORS. Commonly used phage cloning vectors are derivatives of bacteriophage λ (lambda: see Chapter 8) which have been engineered so that the lytic cycle is possible, but lysogeny is not possible. The λ cloning vectors possess restriction enzyme sites so that DNA fragments can be cloned using them. One such vector, the lambda replacement vector (Figure 15.7), has a chromosome in which there is a "left" arm and a "right" arm that collectively contain all the essential genes for the lytic cycle. Between the two arms is a disposable segment of DNA ("disposable" since it does not contain any genes needed for phage propagation). The junctions between the disposable central segment and the two arms each have a cleavage site for one restriction enzyme, here *Eco*RI; no other *Eco*RI sites are present in the λ replacement vector's DNA.

Cloning a DNA fragment using a λ replacement vector is achieved as follows: First the λ vector is cut with the enzyme *Eco*RI to separate the two arms from the replaceable segment. Next, the DNA fragments to be cloned are generated by cutting high-molecular-weight DNA from a given organism with *Eco*RI. The foreign DNA fragments are then mixed with the λ DNA fragments and the pieces ligated together by DNA ligase.

The only splicing events that produce viable, functional λ chromosomes are those in which the foreign DNA is inserted between the left arm and right arm and in which the total length of the recombined DNA fragments is approximately 37 to 52 kb. Since only DNA fragments of about 37 to 52 kb can be packaged into lambda particles, DNA molecules that are either too long or too short are unable to be packaged. Given the number of essential genes that must be present on the left and right arms for phage propagation, only about 15 kb of DNA can be inserted into a λ replacement vector. Lambda cloning vectors are commonly used for initial cloning of genes of interest since it is easier to screen for plaques with the desired sequence than it is to screen bacterial colonies.

~ **FIGURE 15.6**

Insertion of a piece of DNA into the plasmid cloning vector pUC19 to produce a recombinant DNA molecule. pUC19 contains several unique restriction enzyme sites localized in a polylinker. Insertion of a DNA fragment into the polylinker disrupts the β-galactosidase gene, making it possible to distinguish colonies containing pUC19 without an inserted piece of DNA (blue) from colonies containing pUC19 with an inserted piece of DNA (white).

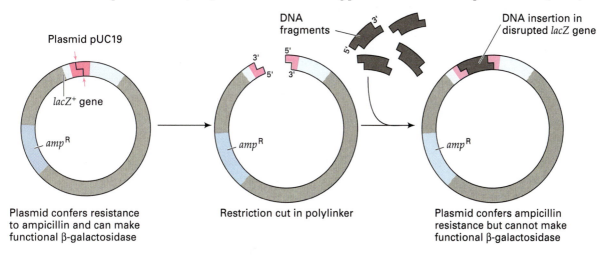

Plasmid confers resistance to ampicillin and can make functional β-galactosidase

Restriction cut in polylinker

Plasmid confers ampicillin resistance but cannot make functional β-galactosidase

COSMID CLONING VECTORS. DNA fragments of approximately 45 kb—for example, some of the large genes found in eukaryotes—can be cloned in vectors called **cosmids** (Figure 15.8). Unlike plasmid and phage vectors, which are derived from naturally occurring genetic elements, cosmids do not occur naturally—they have been constructed by combining features of both plasmid cloning vectors and phage λ. That is, a cosmid has an *ori* sequence to permit its replication in *E. coli*, a dominant selectable marker such as *amp*^{R}, and unique restriction sites for the insertion and cloning of DNA fragments. In addition, cosmids have a *cos* site that is derived from phage λ. Recall from Chapter 11 that the *cos* site is the site at which multiple copies of the λ genome, attached in one long piece called a concatamer, are cleaved into 48-kb pieces to be packaged into phage heads. Recall also that the cut at the *cos* site is staggered and results in a linear DNA molecule with sticky (single-stranded), complementary ends. In fact, all that is needed for packaging DNA into a phage head are *cos* sites that are a specified distance apart on a linear molecule, or a *cos* site in a circular DNA of the appropriate size. Thus the *cos* sites on a cosmid permit packaging of DNA into a λ phage particle that facilitates the introduction of large DNA molecules into a bacterial cell. This property is absent from plasmid cloning vectors.

Lambda *cos* sites have been cloned into plasmid cloning vectors to produce cosmids as small as 5 kb.

When a DNA fragment of about 32–47 kb is cloned into such a cosmid, the recombinant DNA molecule is then the right size to be packaged into a phage head. The phage is then used to introduce the recombinant cosmid into the *E. coli* host cell, where it replicates as a plasmid does. Recombinant cosmids that are either too small (<37 kb) or too large (> ~52 kb) cannot be packaged.

SHUTTLE VECTORS. So far all the cloning vectors we've discussed (plasmids, λ phage, and cosmids) must be used to clone DNA within *E. coli* cells. Other vectors have been developed to introduce recombinant DNA molecules into a variety of prokaryotic and eukaryotic organisms. There are vectors that can be used to transform mammalian cells in culture, as well as vectors to transform other animal cells, and vectors to transform plant cells. The most developed vectors are those used to transform yeast cells. A **shuttle vector** is a cloning vector that allows it to replicate in two or more host organisms. Shuttle vectors are used for experiments in which recombinant DNA is to be introduced into organisms other than *E. coli*. Figure 15.9 diagrams the yeast–*E. coli* shuttle vector YEp24, which can be introduced into yeast or *E. coli* cells.

Like *E. coli* cloning vectors, YEp24 has an *ori* sequence that allows it to replicate in *E. coli* and has dominant selectable markers that confer ampicillin

~ FIGURE 15.7

Scheme for using phage λ DNA as a cloning vector. Foreign DNA fragments of approximately 15 kb in length can be cloned in λ vectors.

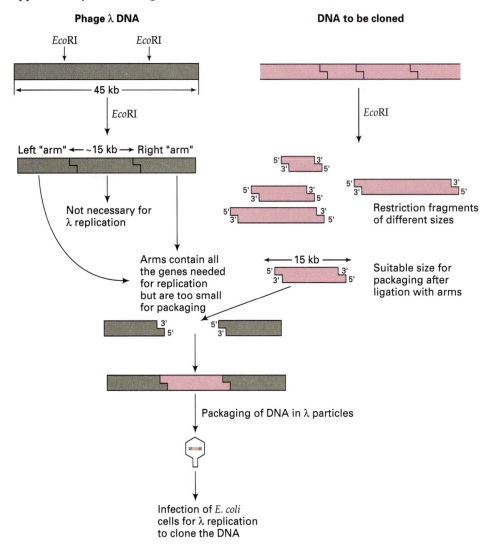

~ FIGURE 15.8

Cosmid cloning vehicle.

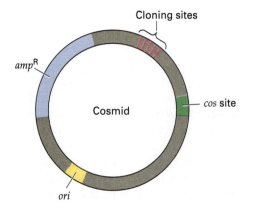

resistance and tetracycline resistance upon *E. coli* cells that contain this vector. YEp24 also contains the selectable marker *URA3* (a wild-type yeast gene for an enzyme required for uracil biosynthesis—in yeast, wild-type dominant genes are symbolized by uppercase letters and the corresponding recessive mutant genes by lowercase letters). This marker enables yeast *ura3* mutant host cells containing YEp24 to be identified. That is, if YEp24 transforms a yeast cell carrying a *ura3* mutation, the yeast cell's phenotype would be changed from uracil-requiring to uracil-independent by the presence of the *URA3* gene in the YEp24 vector. YEp24 also carries a yeast-specific sequence, the two-micron circle (2μ), that allows it to replicate autonomously in a yeast cell. Thus, YEp24 is able to

~ FIGURE 15.9

The yeast *E. coli* shuttle vector YEp24. Unique restriction enzyme cleavage sites are shown with names highlighted in green.

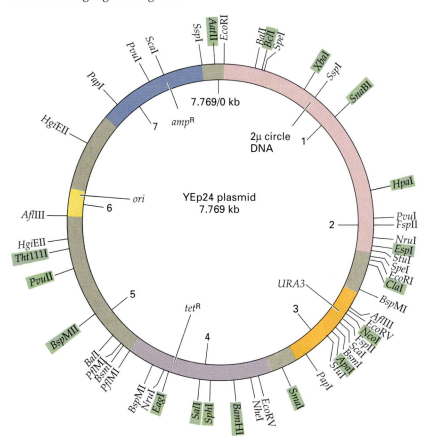

replicate in both yeast and *E. coli*. Not all shuttle vectors have the ability to replicate in the nonbacterial host. Those that do not will typically integrate into a host cell's chromosome and be replicated as that chromosome replicates.

YEAST ARTIFICIAL CHROMOSOMES (YACs). Eukaryotic chromosomes typically contain hundreds of kilobase pairs of DNA. Each eukaryotic chromosome replicates accurately once each cell division cycle because of the combined properties of the two telomeres, the centromere, and the origins of replication along the chromosome. **Yeast artificial chromosomes (YACs)** (Figure 15.10) are cloning vectors that enable—as their name suggests—artificial chromosomes to be made and cloned in yeast cells. YACs are linear vectors that have the following features:

1. A yeast telomere at each end;
2. A yeast centromere sequence (*CEN*);

3. A selectable marker on each arm for detecting the plasmid in yeast (e.g., *TRP1* and *URA3* for tryptophan and uracil independence, respectively);
4. An origin of replication sequence—*ARS* (*autonomously replicating sequence*)—that allows it to replicate in a yeast cell;
5. Restriction sites unique to the YAC that can be used for inserting foreign DNA.

YAC vectors can accommodate DNA fragments that are several hundred kilobase pairs long; that is, much longer than the fragments that can be cloned in the other vectors we have discussed. Thus, YACs are not used for routine cloning experiments. Rather, they are used to clone very large DNA fragments, for example, in creating physical maps of large genomes such as the human genome and for positional cloning (described on p. 481). YAC clones are made by ligating very high molecular weight DNA pieces to YAC arms generated by cutting with a restriction enzyme

~ FIGURE 15.10

Example of a yeast artificial chromosome (YAC) cloning vector. It contains a yeast telomere (*TEL*) at each end, a yeast centromere sequence (*CEN*), a yeast selectable marker for each arm (here *TRP1* and *URA3*), a sequence that allows autonomous replication in yeast (*ARS*), and restriction sites for cloning.

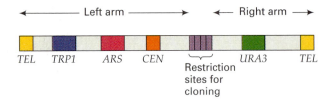

Left arm ⟶   ⟵ Right arm ⟶

TEL   *TRP1*   *ARS*   *CEN*   Restriction sites for cloning   *URA3*   *TEL*

in the cloning site. Clones are introduced into yeast by transformation; by selecting for both *TRP1* and *URA3*, it can be ensured that the clone has both the left and right arms of the YAC vector.

₭EYNOTE

Different kinds of vectors have been developed to construct and clone recombinant DNA molecules. To be useful, all vectors must be able to replicate within their host organism, they must have one or more restriction enzyme cleavage sites at which foreign DNA fragments can be inserted, and they must have one or more dominant selectable markers. Three major types of cloning vectors are used to clone recombinant DNA in *E. coli*: plasmids, which are introduced by transformation into the *E. coli* host, where they replicate; and bacteriophage λ and cosmids, both of which are packaged into λ phage particles, which, in turn, inject the DNA into the *E. coli* host. In λ clones, the injected λ DNA replicates and progeny phages are produced, while in cosmid clones, the cosmids replicate like plasmids.

Shuttle vectors are cloning vectors that can also be used to clone recombinant DNA in more than one host. Yeast artificial chromosomes enable DNA fragments several hundred kilobase pairs long to be cloned in yeast.

CONSTRUCTION OF GENOMIC LIBRARIES, CHROMOSOME LIBRARIES, AND cDNA LIBRARIES

Researchers are interested in cloning individual genes so that those genes can be studied in detail. It is easy to isolate DNA from cells of an organism and to cleave that DNA into fragments with restriction enzymes. The restriction fragments collectively represent the entire genome of the organism, and they may be cloned in a cloning vector. The collection of clones that hopefully contains at least one copy of every DNA sequence in the genome is called a **genomic library**. A genomic library can be searched to identify and single out for study a recombinant DNA molecule that contains a particular gene or DNA sequence of interest. This will be described a little later in the chapter. Genomic libraries have been made for many organisms of genetic interest, including humans (see discussion of Human Genome Project, pp. 486–487). Related to genomic libraries are *chromosome libraries,* which are collections of clones of fragments of individual chromosomes. Also, mRNA molecules can be isolated from cells, and DNA copies called **complementary DNA (cDNA)** can be made and cloned. The following sections describe the construction of these different types of libraries.

Genomic Libraries

A genomic library is a collection of clones that, it is hoped, contains one copy of every DNA sequence in the genome. Researchers typically want to study a specific gene rather than a whole chromosome. One approach to obtaining a clone of a gene is to isolate it from a genomic library through the use of a specific probe. In this section we focus on the construction of genomic libraries of eukaryotic DNA.

PRODUCTION OF GENOMIC LIBRARIES. There are three ways to produce genomic libraries:

1. Genomic DNA is completely digested by a restriction enzyme, and the resulting DNA fragments are then cloned in a cloning vector. This technique does have a drawback. If the specific gene the researcher wants to study contains restriction sites for the enzyme, the gene will be split into two or more fragments when the DNA is digested by the restriction enzyme. In this case, the gene would then be cloned in two or more

pieces. Another drawback is that the average size of the fragment produced by digestion of eukaryotic DNA with restriction enzymes is relatively small (about 4 kb for restriction enzymes that have six-base-pair recognition sequences: see Table 15.2). Thus, an entire library would need to contain a very large number of recombinant DNA molecules, and screening for the specific gene would be very laborious.

2. The problems of genes split into fragments and the large number of recombinant DNA molecules can be minimized by cloning longer DNA fragments. Longer DNA fragments can be generated by mechanically shearing high-molecular-weight (usually 100 to 150 kb) DNA. For example, passage of the DNA through a syringe needle will produce a population of overlapping DNA fragments. However, since the ends of the resulting DNA fragments have not been generated by cutting with restriction enzymes, additional enzymatic manipulations are necessary to add appropriate ends to the molecules for insertion into vector cloning sites.

3. Another approach for producing DNA fragments of appropriate size for constructing a genomic library is to perform a *partial digestion* of the DNA with restriction enzymes that recognize frequently occurring four-base-pair recognition sequences (Figure 15.11a). Partial digestion means that only a portion of the available restriction sites is actually cut with the enzyme. This is achieved by limiting the amount of the enzyme used and/or the time of incubation with the DNA. The ideal result of partial digestion is a population of overlapping fragments representing the entire genome. Sucrose gradient centrifugation or agarose gel electrophoresis (see pp. 469–470) is then used to collect fragments of the desired size for cloning. Those fragments can be cloned directly since the ends of the fragments were produced by restriction enzyme digestion. For example, if the DNA is digested with the enzyme *Sau*3A, which has the recognition sequence $\begin{array}{l}5'\text{-GATC-}5'\\3'\text{-CTAG-}5'\end{array}$, the ends are complementary to the ends produced by digestion of a cloning vector with *Bam*HI, which has the recognition sequence 5'-GGATCC-3' (Figure 15.11b). That is, in

$$\begin{array}{c}\downarrow\\5'\text{-GATC-}3'\\3'\text{-CTAG-}5'\\\uparrow\end{array}$$

*Sau*3A cuts to the left of the upper G and to the right of the lower G to give a 5' overhang with the sequence 5' GATC . . . 3', as follows:

$$\begin{array}{llll}5'\text{-} & \text{and} & 5'\text{ GATC-}3'\\3'\text{-CTAG }5' & & \text{-}5'\end{array}$$

In the sequence:

$$\begin{array}{c}\downarrow\\5'\text{-GGATCC-}3'\\3'\text{-CCTAGG-}5'\\\uparrow\end{array}$$

*Bam*HI cuts between the two G nucleotides also to give a 5' overhang with the sequence 5' GATC . . . 3', as follows:

$$\begin{array}{llll}5'\text{-G} & \text{and} & 5'\text{ GATCC-}3'\\3'\text{-CCTAG }3' & & \text{G-}5'\end{array}$$

~ **FIGURE 15.11**

Use of a restriction enzyme to produce DNA fragments of appropriate size for constructing a genomic library.

a) Partial digestion of DNA by a restriction enzyme, e.g., *Sau*3A, generates a series of overlapping fragments, each with identical 5' GATC sticky ends

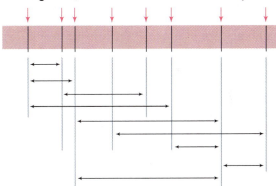

b) Resulting fragments may be inserted into *Bam*HI site of plasmid cloning vector

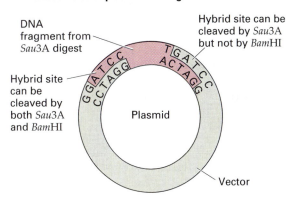

The *Sau*3A and *Bam*HI "sticky" ends can pair to produce a hybrid recognition site.[1]

The recombinant DNA molecules produced by ligating the *Sau*3A-cut fragments and the *Bam*HI-cut vectors together are used to transform *E. coli*. In the case of plasmid and cosmid libraries, the transformants are plated on selective medium to clone the sequences. Each colony that is produced almost always represents a different cloned DNA sequence since each bacterium that gave rise to a colony most likely contained a different recombinant DNA molecule.

For λ genomic libraries, the recombinant DNA is packaged into λ particles. Those particles are used to infect a culture of *E. coli*, and each DNA fragment is cloned by the repeated rounds of infection and lysis that each original phage goes through in the culture. Eventually, the culture becomes transparent as all the bacteria have been lysed and a population of progeny λ phages, with a concentration of 10^{10} to 10^{11} phages/ml, is produced, with many, many representatives of each of the recombinant DNA molecules that were originally constructed in the test tube.

The aim of the methods just described is to produce a library of recombinant molecules that is as complete as possible. However, not all sequences of the eukaryotic genome are equally represented in such a library. For example, if the restriction sites in a particular region are very far apart or extremely close together, the chances of obtaining a fragment of clonable size are small. Additionally, some regions of eukaryotic chromosomes may contain sequences that affect the ability of vectors containing them to replicate in *E. coli*; these sequences would then be lost from the library.

Lastly, the probability of having at least one copy of any DNA sequence represented in the genomic library can be calculated from the formula

$$N = \frac{\ln(1 - P)}{\ln(1 - f)}$$

where N is the necessary number of recombinant DNA molecules, P is the probability desired, and f is the fractional proportion of the genome in a single recombinant DNA molecule (i.e., f is the average size, in kilobase pairs, of the fragments used to make the library divided by the size of the genome, in kilobase pairs). For example, for a 99 percent chance that a particular yeast DNA fragment is represented in a library of 40-kb fragments in a cosmid genomic library, where the yeast genome size is about 12,000 kb,

$$N = \frac{\ln(1 - 0.99)}{\ln[1 - (40/12{,}000)]}$$

$$= 1{,}409$$

That is, 1,409 recombinant DNA molecules of 40-kb size would be needed to be 99 percent sure that a given DNA sequence was represented in the library.

𝒦EYNOTE

A genomic library is a collection of clones that contains at least one copy of every DNA sequence in an organism's genome. Like regular book libraries, genomic libraries are great resources of information—in this case, the information is about the genome. They are used for isolating specific genes and for studying the organization of the genome, among many other things.

Chromosome Libraries

The genome size of many organisms, including humans, is so large that many thousands of clones are needed to represent the entire genome. This makes searching the library for a gene of interest very time-consuming. One approach for reducing the screening time of large genomes is to make libraries of the individual chromosomes in the genome. In humans, this gives 24 different libraries, one each for the 22 autosomes, the X and the Y. Then, if a gene has been localized to a chromosome by genetic means, researchers can restrict their attention to the library of that chromosome when they search for its DNA sequence.

---

[1]Since the hybrid site contains a 5' GATC 3' sequence, it can be cleaved by *Sau*3A. However, whether it can be cleaved by *Bam*HI depends on the base pair "inside" the cloned *Sau*3A-digested fragment. If it *is* a C/G nucleotide pair, then the hybrid site will be as follows:

5'- G G A T C C -3'
3'- C C T A G G -5'

the recognition site for *Bam*HI. This is the case with the left-hand hybrid site in Figure 15.11b. If any other nucleotide pair is next along the *Sau*3A fragment, the hybrid site will *not* be a *Bam*HI cleavage site (e.g., the right-hand hybrid site in Figure 15.11b).

Individual chromosomes of an organism can be separated if their morphologies and sizes are distinct enough, as is the case for human chromosomes. One procedure currently used to isolate large chromosomes individually is *flow cytometry*. In this procedure, chromosomes from cells in the mitotic phase of the cell cycle are stained with a fluorescent dye. Chromosomes released from cells are passed through a laser beam connected to a light detector. This system sorts and fractionates the chromosomes based on their differences in dye binding and resulting light scattering. Approximately 100 chromosomes can be sorted and isolated per second. Once the chromosomes have been fractionated, a library of each chromosome type can be made by cutting the chromosomal DNA with restriction enzymes and inserting the fragments into a cloning vector. As a result of the application of these procedures, libraries of DNA prepared from all human chromosomes are now available to researchers.

𝒦EYNOTE

In certain organisms, individual chromosomes can be isolated and chromosome-specific libraries made from them. If a gene has been localized to a specific chromosome by genetic means, the existence of chromosome-specific libraries make it easier to isolate a clone of the gene.

cDNA Libraries

DNA copies, called **complementary DNA (cDNA)**, can be made from mRNA molecules isolated from cells (as described below). These cDNA molecules can then be cloned. Thus, if a specific mRNA molecule can be isolated, the corresponding cDNA can be made and cloned. The analysis of that cloned cDNA molecule can then provide information about the gene that encoded the mRNA. More typically, the entire mRNA population of a cell is isolated and a corresponding set of cDNA molecules is made and inserted into a cloning vector to produce a **cDNA library**. Since a cDNA library reflects the gene activity of the cell type at the time the mRNAs are isolated, the construction and analysis of cDNA libraries is useful for comparing gene activities in different cell

types of the same organism, because there would be similarities and differences in the clones represented in the cDNA libraries of each cell type.

Recognize that the clones in the cDNA library represent the *mature mRNAs* found in the cell. In eukaryotes, mature mRNAs are processed molecules, so the sequences obtained are *not* equivalent to gene clones. In particular, intron sequences are typically present in gene clones but *not* in cDNA clones. For any mRNA, cDNA clones can be useful for subsequently isolating the gene that codes for that mRNA. The gene clone can provide more information than can the cDNA clone, for example, on the presence and arrangement of introns, and on the regulatory sequences associated with the gene.

SYNTHESIS OF cDNA MOLECULES. cDNA libraries are mostly made from eukaryotic mRNAs. This can be achieved relatively easily because, uniquely among the RNAs found in eukaryotes, only mRNAs contain a poly(A) tail (see Chapter 13, p. 394). These poly(A)+ mRNAs can be purified from the mixture by passing the RNA molecules over a column to which short chains of deoxythymidylic acid, called *oligo(dT) chains*, have been attached. As the RNA molecules pass through an oligo(dT) column, the poly(A) tails on the mRNA molecules form complementary base pairs with the oligo(dT) chains. As a result the mRNAs are captured on the column while the other RNAs pass through. The captured mRNAs are subsequently released and collected, for example, by decreasing the ionic strength of the buffer passing over the column so that the hydrogen bonds will be disrupted. This method results in significant enrichment of poly(A)+ mRNAs in the mixed RNA population to about 50 percent versus about 3 percent in the cell.

Once an enriched population of mRNA molecules has been isolated, double-stranded complementary DNA (cDNA) copies are made *in vitro* (Figure 15.12). First, a short oligo(dT) chain is hybridized to the poly(A) tail at the 3' end of each mRNA strand. The oligo(dT) acts as a primer for **reverse transcriptase** (RNA-dependent DNA polymerase), which makes a complementary DNA copy of the mRNA strand. (The enzyme's name comes from the fact that it catalyzes a reaction that is the reverse of transcription: i.e., RNA → DNA.) Next, RNase H, DNA polymerase I, and DNA ligase are used to synthesize the second DNA strand. RNase H degrades the RNA strand in the hybrid DNA-mRNA, DNA polymerase I makes new DNA fragments using the partially degraded RNA fragments as primers, and finally DNA ligase ligates the new DNA fragments together

~ FIGURE 15.12

~ FIGURE 15.12

The synthesis of double-stranded complementary DNA (cDNA) from a polyadenylated mRNA, using reverse transcriptase RNase H, DNA polymerase I, and DNA ligase.

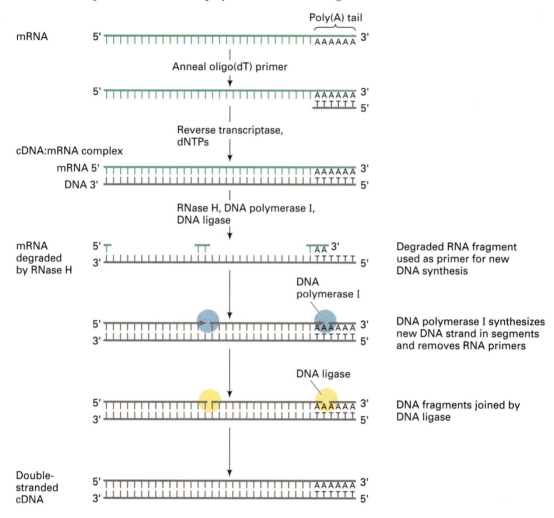

to make a complete chain. The result is a double-stranded cDNA molecule, the sequence of which is derived from the original poly(A)+ mRNA molecule.

PRODUCTION OF cDNA LIBRARIES. Once cDNA molecules have been synthesized, they must be cloned. Figure 15.13 illustrates the cloning of cDNA using a **restriction site linker**, or **linker**, which is a relatively short, double-stranded piece of DNA (oligodeoxyribonucleotide) about 8 to 12 nucleotide pairs long. The linker contains a restriction site; for example, the linker shown in Figure 15.13 contains the *Bam*HI restriction site. Both the cDNA molecules and the linkers have blunt ends, and they can be ligated together at high concentrations of T4 DNA ligase. Sticky ends are produced in the cDNA molecule by cleaving the cDNA (with linkers now at each end) with *Bam*HI. The resulting DNA is inserted into a

cloning vector that has also been cleaved with *Bam*HI and the recombinant DNA molecule produced is transformed into an *E. coli* host cell for cloning.

KEYNOTE

Given a population of mRNAs purified from a cell, it is possible to make DNA copies of those mRNA molecules. First, the enzyme reverse transcriptase makes a single-stranded DNA copy of the mRNA; then RNase H, DNA polymerase I, and DNA ligase are used to make a double-stranded DNA copy called complementary DNA (cDNA). This cDNA can be spliced into cloning vectors using restriction site linkers.

~ FIGURE 15.13

The cloning of cDNA by using *Bam*HI linkers.

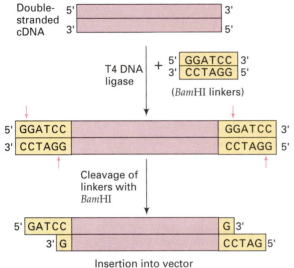

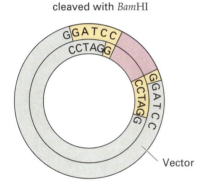

IDENTIFYING SPECIFIC CLONED SEQUENCES IN cDNA LIBRARIES AND GENOMIC LIBRARIES

Identifying Specific Cloned Sequences in a cDNA Library

A cDNA clone can be identified in a cDNA library in a number of ways. One way is to select the cDNA clone that codes for a specific protein (Figure 15.14). This approach requires that a protein can be purified in sufficient quantities so that antibodies can be made against the protein. It also requires that the cDNAs be cloned in a special kind of vector called an *expression*

vector (Figure 15.14, part 1). In such a vector, the cDNA is placed next to a promoter sequence and to translation start signals so that in the host an mRNA will be transcribed corresponding to the cDNA and the encoded protein produced by translation of the mRNA. The expression vector also contains a transcription termination signal downstream of the cDNA sequence.

Theoretically, each individual *E. coli* cell in a population transformed with a cDNA library in an expression vector contains a different cDNA clone (Figure 15.14, part 2). When the cells are plated, each bacterium will give rise to a colony (Figure 15.14, part 3). These clones are preserved, for example, by picking each colony off the plate and placing it into the medium in a well of a microtiter dish (Figure 15.14, part 4). Replicas of the set of clones are placed (printed) onto a membrane filter that has been placed on a Petri plate of selective medium appropriate for the recombinant molecules (Figure 15.14, part 5). Incubation of the plate produces colonies growing on the filter in the same pattern produced by the clones in the microtiter dish. The filter is peeled from the dish and treated to lyse the cells *in situ* (Figure 15.14, part 6). The proteins within the cell, including those expressed from the cDNA, become stuck to the filter. The filter is then incubated with an antibody to the protein of interest (Figure 15.14, part 7).

If the antibody is radioactively labeled, any clones that expressed the protein of interest can be identified by placing the dried filter against X-ray film, leaving it in the dark for a period of time (from one hour to overnight) to produce what is called an *autoradiogram* (Figure 15.14, part 8). The process is called *autoradiography*. When the film is developed, dark spots are seen wherever the radioactive probe is bound to the filter in the antibody reaction (see Figure 15.14, part 8). (The dark spots are the result of the decay of the radioactive atoms changing the silver grains of the film.) Once a cDNA clone for a protein of interest has been identified, it can be used as a hybridization probe, for example, to analyze the genome of the same or other organisms for homologous sequences, to isolate the nuclear gene for the mRNA from a genomic library, or to quantify mRNA production.

Identifying Specific Cloned Sequences in a Genomic Library

Given the existence of a probe, such as a cloned cDNA probe, it is possible to identify in a genomic library the cloned gene that codes for the mRNA molecule from which the cDNA was made, and then to isolate it for characterization.

~ Figure 15.14

Identifying specific cDNA plasmids in a cDNA library.

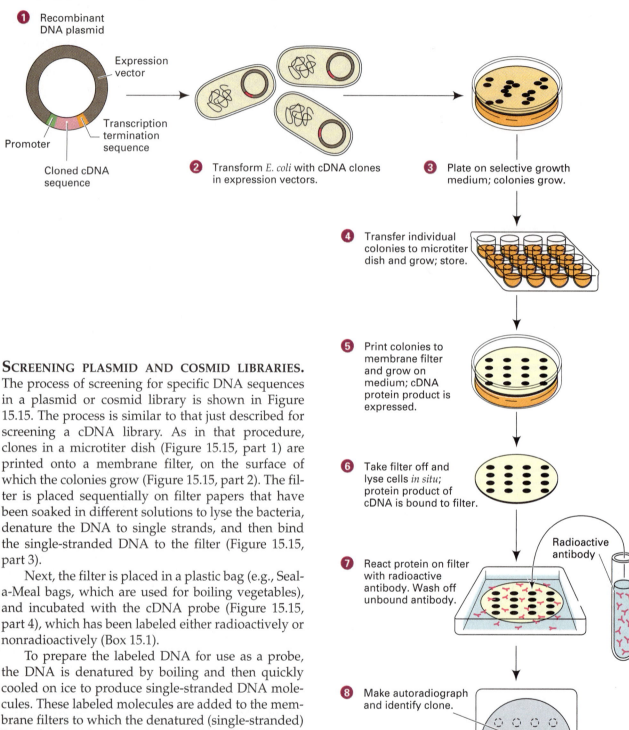

1 Recombinant DNA plasmid

Expression vector

Transcription termination sequence

Promoter

Cloned cDNA sequence

2 Transform *E. coli* with cDNA clones in expression vectors.

3 Plate on selective growth medium; colonies grow.

4 Transfer individual colonies to microtiter dish and grow; store.

5 Print colonies to membrane filter and grow on medium; cDNA protein product is expressed.

6 Take filter off and lyse cells *in situ*; protein product of cDNA is bound to filter.

7 React protein on filter with radioactive antibody. Wash off unbound antibody.

Radioactive antibody

8 Make autoradiograph and identify clone.

Screening plasmid and cosmid libraries.

The process of screening for specific DNA sequences in a plasmid or cosmid library is shown in Figure 15.15. The process is similar to that just described for screening a cDNA library. As in that procedure, clones in a microtiter dish (Figure 15.15, part 1) are printed onto a membrane filter, on the surface of which the colonies grow (Figure 15.15, part 2). The filter is placed sequentially on filter papers that have been soaked in different solutions to lyse the bacteria, denature the DNA to single strands, and then bind the single-stranded DNA to the filter (Figure 15.15, part 3).

Next, the filter is placed in a plastic bag (e.g., Seal-a-Meal bags, which are used for boiling vegetables), and incubated with the cDNA probe (Figure 15.15, part 4), which has been labeled either radioactively or nonradioactively (Box 15.1).

To prepare the labeled DNA for use as a probe, the DNA is denatured by boiling and then quickly cooled on ice to produce single-stranded DNA molecules. These labeled molecules are added to the membrane filters to which the denatured (single-stranded) DNA from each colony has been bound. Wherever the two sequences are complementary, DNA-DNA hybrids will form between the probe and the colony DNA. If the cDNA probe is derived from the mRNA

~ FIGURE 15.15

Using DNA probes to screen plasmid and cosmid libraries for specific DNA sequences.

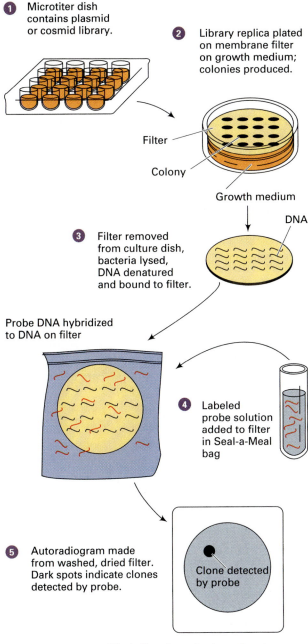

1 Microtiter dish contains plasmid or cosmid library.

2 Library replica plated on membrane filter on growth medium; colonies produced.

Filter

Colony

Growth medium

DNA

3 Filter removed from culture dish, bacteria lysed, DNA denatured and bound to filter.

Probe DNA hybridized to DNA on filter

4 Labeled probe solution added to filter in Seal-a-Meal bag

5 Autoradiogram made from washed, dried filter. Dark spots indicate clones detected by probe.

Clone detected by probe

Wash filter free of unbound probe. Detect hybridization by autoradiography for radioactively labeled probes or by chemiluminescent detection for nonradioactively labeled probe.

for β-globin, for example, that probe will hybridize with DNA bound to the filter that encodes the β-globin mRNA, that is, the β-globin gene. After sufficient time for hybridization has elapsed (approximately 24 hours), the filters are washed to remove unbound probe, and subjected to the detection procedure appropriate for whether the probe was radioactive or nonradioactive (see Box 15.1); that is, autoradiography for a radioactive probe, or chemiluminescent or colorimetric detection for a nonradioactive probe (Figure 15.15, part 5). From the position(s) of the spot(s) on the film or filter, the locations of the original bacterial culture(s) in the microtiter dish(es) can be determined and the clone(s) of interest isolated for further characterization.

SCREENING λ GENOMIC LIBRARIES. Since plasmids and cosmids are cloned in bacteria, which grow on plates to produce colonies, while phages are cloned in bacteria, killing them to produce plaques, a slightly different screening method is needed for λ genomic libraries. That is, to screen for a clone of interest in a genomic library made in λ, samples of the phage suspension are plated on a bacterial lawn to produce plaques (Figure 15.16, p. 468). Each plaque consists of a pool of progeny phages produced by successive rounds of infection and lysis of bacteria initiated originally by a single phage infection. Each time a bacterium is lysed by a phage infection, not only are progeny phages released, but many copies of replicated λ DNA that were not packaged into particles are also released. A piece of membrane filter is laid carefully on top of the plate and left there for a few minutes, permitting the λ DNAs to bind to the filter; this is called making a *plaque lift*. The filter is removed and placed on blotting paper soaked in an alkaline solution, which denatures the DNA into single strands. After neutralization, the filters are incubated with a labeled probe, as described for the screening of the plasmid and cosmid libraries.

A number of plaque lifts can be made from each plate of λ plaques, allowing the investigator to screen the same library simultaneously with a number of different probes. Evidence of hybridization of the probe with DNA from a plaque is seen as a spot on the film after autoradiography or chemiluminescent detection. The position of the spot on the film is used to find the plaque on the plate from which the plaque lift was made. Phages with the cloned DNA fragment of interest can be isolated from the plaque and propagated in a bacterial culture to produce large quantities of phage DNA for analysis.

BOX 15.1
Labeling of DNA

DNA can be labeled either radioactively or non-radioactively. Typically it has been more common to label DNA radioactively, but with increasing regulations concerning the disposal of radioactive materials and the health concerns of exposure to radioactive compounds, great strides have been made in developing nonradioactive methods of labeling DNA that produce probes that are as sensitive as radioactive probes in seeking out the target DNA. Thus, it is now possible to detect as little as 0.1 picogram (0.1×10^{-12} gm) of DNA with either radioactive or nonradioactive probes. We will now discuss briefly some methods for preparing radioactively labeled and nonradioactively labeled DNA probes.

Radioactive Labeling of DNA

A DNA probe may be labeled radioactively by the *random primer method* (Box Figure 15.1). The DNA is denatured to single strands by boiling and quick cooling on ice. DNA primers, six nucleotides long (hexanucleotides), synthetically made by the random incorporation of nucleotides, are annealed to the DNA. The *hexanucleotide random primers* will pair with complementary sequences in the DNA, and this will occur at many locations because all possible hexanucleotide sequences are present. The primers are elongated by the Klenow fragment of DNA polymerase I that uses dNTP precursors, one or more of which is radioactively labeled. Typically, the label is ^{32}P located in the phosphate group that is attached to the 5' carbon of the deoxyribose sugar. This phosphate group is called the α-phosphate because it is the first in the chain of three; the α-phosphate is used in forming the phosphodiester bonds of the sugar-phosphate backbone. (The Klenow fragment, named for the person who discovered it, lacks 5'-to-3' exonuclease activity that would otherwise remove the short primers, but still has the 3'-to-5' proofreading activity.)

    After using the radioactive DNA probe in an experiment, detection depends upon the properties of the radioactive isotope. For example, if a ^{32}P-labeled probe has hybridized with a target DNA sequence on a membrane filter, the filter is

~ BOX FIGURE 15.1
Random primer method of radioactively labeling DNA.

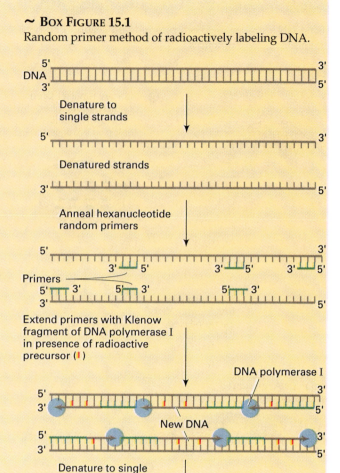

placed against a piece of X-ray film and the sandwich is placed in the dark. Every location on the filter where there is [32]P (a spot, band, etc.) is then detected as a black region on the X-ray film after it is developed. This process is called *autoradiography* and the resulting "picture" of radioactive signal is called an *autoradiogram*.

Nonradioactive Labeling of DNA

Random primer labeling may also be used to prepare nonradioactively labeled DNA probes. The difference from preparing radioactively labeled DNA is that a special DNA precursor molecule is used rather than a ^{32}P-labeled precursor. For example, in one of many labeling systems, digoxigenin-dUTP (DIG-dUTP) is added to the dATP, dCTP, dGTP, and dTTP precursor mixture. Digoxigenin is a steroid, and it is linked to dUTP—deoxyuridine 5'-triphosphate. During DNA synthesis, DIG-dUTP may be incorporated opposite A nucleotides on the template DNA strand.

The nonradioactively labeled DNA may be used in experiments in the same way as is radioactively labeled DNA. Detection is different, however. Once the DIG-dUTP-labeled probe has bound to target DNA on a filter, for example, an anti-DIG-AP conjugate is added. The anti-DIG part of the conjugate is an antibody that will react specifically with DIG, and the AP part of the conjugate is the enzyme alkaline phosphatase. Wherever the DIG-labeled DNA is hybridized to target DNA on the filter, the anti-DIG-AP conju-

gate will bind to form a DNA-DIG–anti-DIG-AP complex. The location(s) of the probe-target hybrid is then visualized by substrates that react with the alkaline phosphatase. For sensitivity matching that of radioactively labeled probes, a chemiluminescent substrate is used. Such a substrate produces light in a reaction catalyzed by alkaline phosphatase, and detection involves exposing X-ray film much like making an autoradiogram. If great sensitivity is not necessary, colorimetric substrates for the enzyme are used. In this case, spots or bands develop directly on the filter as purple/blue regions as the enzyme reaction proceeds.

Identifying Specific DNA Sequences in Libraries Using Heterologous Probes

cDNA probes may be used to identify and isolate specific genes, and a very large number of genes have been cloned from both prokaryotes and eukaryotes in this way. It is also possible to identify specific genes in a genomic library by using clones of equivalent genes from other organisms as probes. Such probes are called *heterologous probes* and their effectiveness depends upon a good degree of homology between the probes and the genes. For that reason, the greatest success with this approach has come either with highly conserved genes or with probes from a species closely related to the organism from which a particular gene is to be isolated.

Identifying Genes in Libraries by Complementation of Mutations

For those organisms in which genetic systems of analysis have been well developed and for which there are well defined mutations, it is possible to clone genes by complementation of those mutations. For example, the yeast *Saccharomyces cerevisiae* has been extremely well exploited genetically, a large number of mutations have been generated and characterized, and integrative and replicative transformation systems using *E. coli*–yeast shuttle vectors (see p. 455–457) have been developed.

To clone a yeast gene by complementation, first a genomic library is made of DNA fragments from the wild-type yeast strain in a replicative shuttle vector such as YEp24 (see Figure 15.9). The library is used to

transform a host yeast strain carrying a mutation to enable transformants to be selected—*ura3* in the case of YEp24—and a mutation in the gene for which the wild-type gene is to be cloned. Let us consider the cloning of the *ARG1* gene, the wild-type gene for an enzyme needed for arginine biosynthesis (Figure 15.17) by complementation of an *arg1* mutation. A yeast strain carrying the *arg1* mutation would have an inactive enzyme for arginine biosynthesis and, hence, a growth requirement for arginine. A genomic library is made using DNA isolated from a wild-type (*ARG1*) yeast strain, (Figure 15.17, parts 1 and 2). When a population of *ura3 arg1* yeast cells is transformed with the YEp24 genomic library (Figure 15.17, part 3), some cells will receive plasmids containing the normal (*ARG1*) gene for the arginine biosynthesis enzyme. The plasmid's *ARG1* gene will be transcribed, and the resultant mRNA will be translated to produce a normal, functional enzyme for arginine biosynthesis. The cell will be able to grow on minimal medium—that is, in the absence of arginine—despite the presence of a defective *arg1* gene on the cell's chromosomes (Figure 15.17, part 4). The *ARG1* gene is said to overcome the functional defect of the *arg1* mutation by *complementation* of that mutation (Figure 15.17, parts 5 and 6). The plasmid is then isolated from the cells and the cloned gene is characterized.

Identifying Genes or cDNAs in Libraries Using Oligonucleotide Probes

A number of genes have been isolated from libraries by using synthetically made oligonucleotide probes. (Oligonucleotides—literally, few nucleotides—are rel-

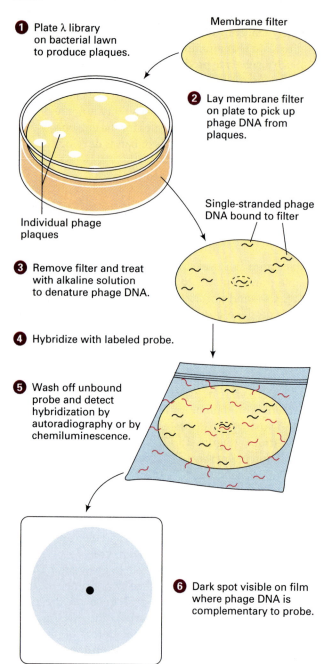

~ FIGURE 15.16

Screening a bacteriophage λ library for a specific gene clone.

1 Plate λ library on bacterial lawn to produce plaques.

Membrane filter

2 Lay membrane filter on plate to pick up phage DNA from plaques.

Individual phage plaques

Single-stranded phage DNA bound to filter

3 Remove filter and treat with alkaline solution to denature phage DNA.

4 Hybridize with labeled probe.

5 Wash off unbound probe and detect hybridization by autoradiography or by chemiluminescence.

6 Dark spot visible on film where phage DNA is complementary to probe.

versions of the gene that are available in *GenBank*, a computer database where sequences are deposited and made available to researchers worldwide. Then, since the genetic code is universal, oligonucleotides about 20 nucleotides long can be designed which, if translated, would give the known amino acid sequence. Because of the degeneracy of the genetic code—up to six different codons can specify a given amino acid—a number of different oligonucleotides are made, all of which could encode the targeted amino acid sequence. These mixed oligonucleotides are labeled and used as probes to search the libraries with the hope that at least one of the oligonucleotides will detect the gene or cDNA of interest. If the probe is labeled radioactively, detection is by autoradiography, whereas if the probe is labeled nonradioactively, detection is either colorimetric or by chemiluminescence (see Box 15.1). While not successful all of the time, oligonucleotide-based library screening has been extremely fruitful and has allowed many genes to be cloned for which previous genetic information was lacking.

KEYNOTE

Specific sequences in cDNA libraries and genomic libraries can be identified using a number of approaches, including the use of specific antibodies, cDNA probes, heterologous probes, complementation of mutations, and mixed oligonucleotide probes.

ANALYSIS OF GENES AND GENE TRANSCRIPTS

Cloned DNA sequences are resources for experiments designed to answer many kinds of biological questions. The following experimental techniques will be described in this section:

1. *The cloned DNA may be mapped with respect to the number and arrangement of restriction sites.* The resulting map, analogous to the arrangement of genes on a linkage map, is called a **restriction map**. This is commonly done for both cloned chromosomal genes (contain introns) and cloned cDNAs (correspond to mRNAs, so no introns) to produce a restriction map of the cloned DNA.

atively short, single-stranded pieces of DNA.) This method requires that at least some of the amino acid sequence is known for the protein encoded by the gene and, ideally, that the amino acid substitutions associated with specific mutations have been identified. In that case, it may be possible that a consensus sequence can be determined from previously cloned

~ FIGURE 15.17

Example of cloning a gene by complementation of mutations: the cloning of the yeast *ARG1* gene.

1 High molecular-weight DNA from wild-type (*ARG1*) yeast strain.

2 Make genomic library of fragments in a yeast-*E. coli* vector such as YEp24.

URA3 selectable marker

Yeast DNA   *ARG1* gene

3 Transform *ura3 arg1* yeast strain.

4 Plate on minimal medium. Only cells with plasmid containing *ARG1* gene can grow.

Yeast colonies containing recombinant DNA molecule with *ARG1* gene.

5 Yeast chromosomal *arg1* makes defective enzyme.

6 Complementation occurs because *ARG1* in vector produces functional enzyme.

Such information might be useful for making clones of subsections of the gene or the cDNA, or for comparing the gene and the cDNA.

2. *A cloned cDNA or a cloned gene may be used to analyze the transcription of the corresponding gene in the cell.* For example, we can study the size of the initial transcript of the gene, the processing steps it goes through (if any) to produce a mature RNA, the amount of the gene transcript, the time of expression in the cell cycle, and the time and amount of expression in different tissues and/or during development.

3. *The complete sequence of the cloned DNA may be determined.* In the case of a cloned gene, the sequence information can be useful in studies of how the expression of the gene is regulated, for example. DNA sequences may also be compared with other DNA sequences in a computer database to determine the extent of similarity between related genes. If the cloned DNA sequence is not a known gene, a computer search might provide insights into what the sequence might be. Further, the DNA sequence of a protein-coding gene can be "translated" by computer to provide information about the properties of the protein for which it codes. Such information can be helpful for an investigator who wishes to isolate and study an unknown protein product of a gene for which a clone is available.

Restriction Enzyme Analysis of Cloned DNA Sequences

Cloned DNA sequences are often analyzed to determine the arrangement and specific locations of restriction sites. Because cloned DNA sequences represent a homogeneous population of DNA molecules, restriction enzyme cleavage of cloned DNA sequences produces a relatively small number of discretely sized DNA fragments. These DNA fragments can easily be seen following agarose gel electrophoresis and ethidium bromide staining of the DNA fragments (described in the next paragraph), permitting restriction maps to be constructed without the need for hybridization with a labeled probe and subsequent detection.

EXAMPLE OF RESTRICTION MAPPING. Let us consider the construction of a relatively simple restriction map (Figure 15.18). We have cloned a 5.0-kb piece of DNA and wish to construct a restriction map of it (Figure 15.18, part 1). One sample of the DNA is digested with *Eco*RI, a second sample is digested with *Bam*HI, and a third sample is digested with both *Eco*RI and *Bam*HI. The DNA restriction fragments of each reaction are separated according to their molecular size by agarose gel electrophoresis; controls are a sample of the same DNA uncut with any enzyme, and DNA fragments of known size ("DNA fragment size markers," or simply "size markers") so that the sizes of the unknown DNA fragments can be computed (Figure 15.18, part 2). The gel is a rectangular, horizontal slab of agarose (a firm, gelatinous material), which has a matrix of pores through which DNA passes in an electric field. Each gel is cast by boiling a buffered agarose solution and allowing it to cool in a mold. A toothed comb is used to divide the gel into a number of lanes so that a number of different samples can be analyzed simultaneously. Since DNA is negatively charged due to its phosphates, the DNA migrates toward the positive pole. Because small DNA fragments can "squirm" more readily through the small pores in the gel, the small fragments move through the gel more rapidly than large DNA fragments. Thus, the smallest fragments migrate the farthest distance while the largest fragments migrate the least. After electrophoresis, the gel is stained with ethidium bromide and photographed under ultraviolet light. From the photograph the distance each DNA band migrated can be measured. For the marker, the molecular size of each DNA band is known, so a calibration curve can be drawn of DNA size (in log kb) versus migration distance (in mm) (Figure 15.18, part 3). The migration distances for the DNA bands from our uncut and cut DNA are then used with the calibration curve to determine the molecular sizes of the DNA fragments in the bands (Figure 15.18, part 4). For our theoretical example the results are shown in Figure 15.18, part 5.

The results are analyzed as follows (Figure 15.18, parts 6–8):

1. When the 5.0-kb DNA is cut with *Eco*RI, 4.5-kb and 0.5-kb DNA fragments result. This indicates that there is only one restriction site for *Eco*RI in the DNA and that this site is 0.5 kb from one end of the molecule.

2. Using similar logic, there is one restriction site for *Bam*HI located 2 kb from one end of the molecule.

3. At this point, we know there is one restriction site for each enzyme, but we do not know the relationship between the two. We can, however, make two models (see Figure 15.18, part 7). In model A, the *Eco*RI site is 0.5 kb from one end and the *Bam*HI site is 2.0 kb from that same end. In model B, the *Eco*RI site is 0.5 kb from one end and the *Bam*HI site is 3.0 kb from that end (i.e., 2.0 kb from the other end). Model A predicts that cutting with *Eco*RI and *Bam*HI will produce three fragments of 0.5, 1.5, and 3.0 kb (going from left-to-right along the DNA). In model B, cutting with both enzymes will produce three fragments of 0.5, 2.5, and 2.0 kb. The actual data show three

fragments with sizes 2.5, 2.0 and 0.5 kb, thereby validating model B.

In real situations, restriction mapping typically involves data that are much more complicated—for example, involving more restriction enzymes and a number of sites for each enzyme. Analysis may be done with a complete plasmid (which is circular) or with the cloned sequence or part of it cut out of the plasmid and purified by *preparative agarose gel electrophoresis*. In this technique, large quantities of DNA are cut with appropriate restriction enzymes, and the fragments are separated by agarose gel electrophoresis. After staining with ethidium bromide, the bands can be visualized under an ultraviolet light. At that time the bands can be physically cut out of the gel and the DNA extracted for analysis.

Restriction Enzyme Analysis of Genes in the Genome

As part of the analysis of genes, it is often useful to determine the arrangement and specific locations of restriction sites. This information is useful, for example, *for comparing homologous genes in different species, for analyzing intron organization, or for planning experiments to clone parts of a gene, such as its promoter or controlling sequences, into a vector.* The arrangement of restriction sites in a gene can be analyzed without actually cloning the gene by using a cDNA probe or a closely related heterologous gene probe. The process of analysis proceeds as follows:

1. Samples of high-molecular-weight DNA are cut with different restriction enzymes (Figure 15.19, parts 1 and 2), each of which will produce DNA fragments of different lengths (depending on the locations of the restriction sites on the DNA molecules).

2. The DNA restriction fragments are separated according to their molecular size by agarose gel electrophoresis (Figure 15.19, part 3).

   After electrophoresis, the DNA is stained with ethidium bromide so that it can be seen under ultraviolet light illumination. When total cellular DNA is digested with a restriction enzyme, the result is usually a continuous smear of fluorescence down the length of the gel lane due to the fact that the enzyme produces fragments of all sizes.

3. The DNA fragments are transferred to a membrane filter (Figure 15.19, part 4). The transfer to the membrane filter is done by the **Southern blot technique** (named after its inventor, Edward Southern). In brief, the gel is soaked in an alka-

~ FIGURE 15.18

Construction of a restriction map for *Eco*RI and *Bam*HI in a DNA fragment.

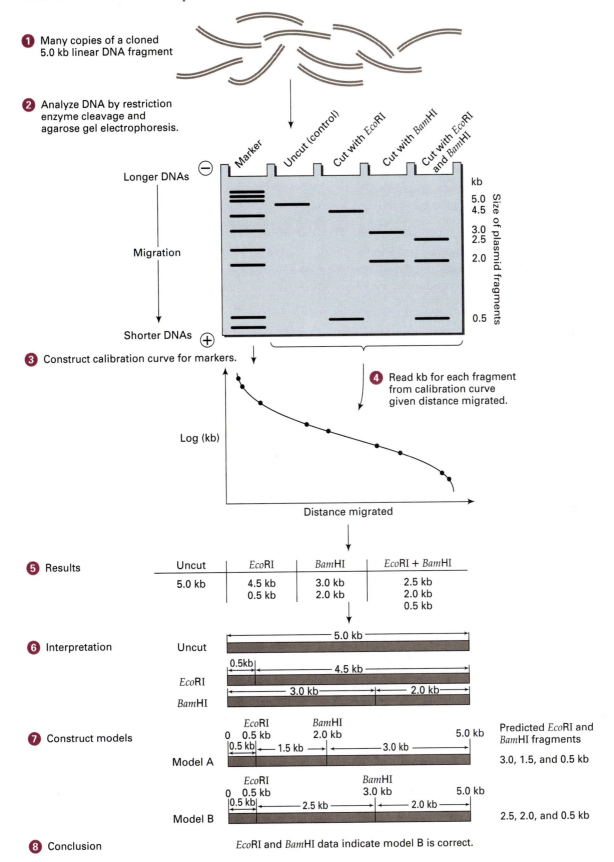

1 Many copies of a cloned 5.0 kb linear DNA fragment

2 Analyze DNA by restriction enzyme cleavage and agarose gel electrophoresis.

Longer DNAs

Migration

Shorter DNAs

Marker

Uncut (control)

Cut with *Eco*RI

Cut with *Bam*HI

Cut with *Eco*RI and *Bam*HI

kb

5.0
4.5

3.0
2.5

2.0

0.5

Size of plasmid fragments

3 Construct calibration curve for markers.

4 Read kb for each fragment from calibration curve given distance migrated.

Log (kb)

Distance migrated

5 Results

Uncut	*Eco*RI	*Bam*HI	*Eco*RI + *Bam*HI
5.0 kb	4.5 kb	3.0 kb	2.5 kb
	0.5 kb	2.0 kb	2.0 kb
			0.5 kb

6 Interpretation

Uncut

5.0 kb

*Eco*RI

0.5kb

4.5 kb

*Bam*HI

3.0 kb ———— 2.0 kb

7 Construct models

Model A

*Eco*RI *Bam*HI
0   0.5 kb   2.0 kb   5.0 kb
0.5 kb ← 1.5 kb → ← 3.0 kb →

Predicted *Eco*RI and *Bam*HI fragments

3.0, 1.5, and 0.5 kb

Model B

*Eco*RI   *Bam*HI
0   0.5 kb   3.0 kb   5.0 kb
0.5 kb ← 2.5 kb → ← 2.0 kb →

2.5, 2.0, and 0.5 kb

8 Conclusion

*Eco*RI and *Bam*HI data indicate model B is correct.

~ FIGURE 15.19

Southern blot procedure for the analysis of cellular DNA for the presence of sequences complementary to a radioactive probe, such as a cDNA molecule made from an isolated mRNA molecule. The hybrids, shown as three bands in this theoretical example, are visualized by autoradiography.

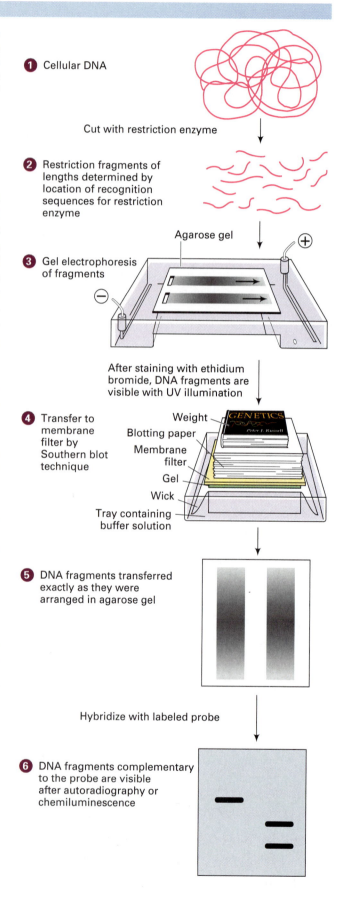

line solution to denature the double-stranded DNA into single strands. The gel is neutralized and placed on a piece of blotting paper that spans a glass plate. The ends of the paper are in a container of buffer and act as wicks. A piece of membrane filter is laid down so that it covers the gel. Additional sheets of blotting paper (or paper towels) and a weight are stacked on top of the membrane filter. The buffer solution in the bottom tray is wicked up by the blotting paper, passed through the gel, through the membrane filter, and finally into the stack of blotting paper. During this process, the DNA fragments are picked up by the buffer and transferred from the gel to the membrane filter, to which they adhere because of membrane filter's chemical properties. The fragments on the filter are arranged in exactly the same way as they were in the gel (Figure 15.19, part 5).

4. Once the Southern blot technique is completed, the single-stranded DNA fragments are fixed to the membrane filter by UV crosslinking or baking it in an 80°C oven for two hours. The filter is then placed in a hybridization buffer, and any radioactively labeled or nonradioactively labeled probe added to the membrane filter at this point will bind to any complementary DNA fragment (Figure 15.19, part 6).

5. After the hybridization of the probe and the DNA fragments, the filter is washed to remove probe molecules that have not hybridized, and detection of the probe is carried out in a way appropriate for whether the probe is radioactive or nonradioactive to determine the position(s) of the hybrids (Figure 15.19, part 6). If DNA fragment size markers (e.g., λ DNA cut with *Hin*dIII or with *Hin*dIII and *Eco*RI) are separated in a different lane in the agarose gel electrophoresis process, the sizes of the genomic restriction fragments that hybridized with the probe can be calculated. From the fragment sizes obtained, a **restriction map** can be generated to show the relative positions of the restriction sites. Suppose,

for example, that using only *Bam*HI produces a DNA fragment of 3 kb that hybridizes with the radioactive probe. If a combination of *Bam*HI and *Pst*I is then used and produces two DNA fragments, one of 1 kb and the other of 2 kb, we would deduce that the 3-kb *Bam*HI fragment contains a *Pst*I restriction site 1 kb from one end and 2 kb from the other end. Further analysis with other enzymes, individually and combined, enables the researcher to construct a map of all the enzyme sites relative to all other sites.

Analysis of Gene Transcripts

A related blotting technique to the Southern blot technique (called **northern blot analysis**) has been developed to analyze RNA rather than DNA. (In this case, the name is derived, not from a person, but to indicate that it is a technique related to the Southern blot technique.) In northern blot analysis, RNA extracted from a cell is separated by size using gel electrophoresis, and the RNA molecules are transferred and bound to a filter in a procedure that is essentially identical to Southern blotting in which DNA is transferred. After hybridization with a labeled probe, and the appropriate detection system used, bands show the locations of RNA species that were complementary to the probe. Given appropriate RNA size markers, the sizes of the RNA species identified with the probe can be determined.

Northern blot analysis is useful in several kinds of experiments. For example, northern blot analysis can reveal the size or sizes of the specific mRNA encoded by a gene. In some cases, a number of different mRNA species encoded by the same gene have been identified in this way, suggesting that either different promoter sites can be used, different terminator sites can be used, or alternative mRNA processing can occur. Northern blot analysis can also be used to investigate whether or not an mRNA species is present in a cell type or tissue, and how much of it is present. This type of experiment is useful for determining levels of gene activity, for instance, during development, in different cell types of an organism, or in cells before and after they are subjected to various physiological stimuli.

*K*EYNOTE

Genes and cloned DNA sequences are often analyzed to determine the arrangement and specific locations of restriction sites. The analytical process involves cleavage of the DNA with restriction enzymes, followed by separation of the resulting DNA fragments by agarose gel electrophoresis, and staining of the DNA fragments with ethidium bromide so that they may be visualized with ultraviolet light. The DNA fragments produced by cleavage of cloned DNA sequences can be seen as discrete bands, enabling restriction maps to be constructed based on the calculated molecular lengths of the DNA in the bands. DNA fragments produced by cleavage of genomic DNA show a wide range of sizes, resulting in a continuous smear of DNA fragments in the gel. In this case, specific gene fragments can only be visualized by transferring the DNA fragments to a nitrocellulose filter in a procedure called the Southern blot technique, hybridizing a specific radioactive probe with the DNA fragments, and detecting the hybrids by autoradiography. At that point, a restriction map can be made.

DNA SEQUENCE ANALYSIS

Cloned DNA fragments may be analyzed to determine the nucleotide pair sequence of the DNA. This information is useful, for example, for identifying gene sequences and regulatory sequences within the fragment, and for comparing the sequences of homologous genes from different organisms.

Two techniques for the *rapid sequencing of DNA molecules* were developed in the 1970s. These techniques revolutionized molecular biology by making it possible to obtain the sequences of any DNA segment of interest. One method, developed by Allan Maxam and Walter Gilbert, uses chemicals to cleave DNA at different sites and is called **Maxam-Gilbert sequencing**, after its developers. The other method, called **dideoxy sequencing** or **Sanger sequencing**, was developed by Fred Sanger and involves enzymatic extension of a short primer. The dideoxy method is more commonly used today and will be detailed here.

Dideoxy (Sanger) DNA Sequencing

Both linear DNA and circular DNA can be sequenced using the dideoxy DNA sequencing method. Linear DNA fragments can be generated, for example, by cutting plasmid DNA with a restriction enzyme or enzymes. Circular DNA to be sequenced is found most commonly in the form of supercoiled plasmids and that DNA is the most efficient substrate for dideoxy sequencing.

The principle of dideoxy DNA sequencing is straightforward. The DNA is first denatured to single strands by treatment with alkali or by heat treatment. Next, a short oligonucleotide primer is annealed to one of the two DNA strands (Figure 15.20). The oligonucleotide is designed so that its 3' end will be next to the DNA sequence of interest. The oligonucleotide acts as a *primer* for DNA synthesis, and the 5'-to-3' orientation chosen ensures that the DNA made is a *complementary copy of the DNA sequence of interest* (see Figure 15.20). Consider, for example, DNA fragments cloned into the plasmid cloning vector, pUC19 (see Figure 15.6). By having a pair of oligonucleotide primers complementary to the DNA flanking the polylinker, any DNA insert can be sequenced from each end. Any other DNA fragment can be sequenced provided some sequence information is available from which a primer can be made.

For each sequencing experiment, four separate reactions are set up with single-stranded DNA to which the primer has been annealed. Each reaction contains the four normal precursors of DNA, that is, dATP, dTTP, dCTP, and dGTP, and DNA polymerase (T7 DNA polymerase is commonly used). The precursors are labeled either radioactively with ^{32}P, ^{33}P, or ^{35}S, or nonradioactively so that newly synthesized DNA can be easily detected. The reactions differ in the presence of a different modified nucleotide called a **dideoxy nucleotide** (Figure 15.21). The only difference between a dideoxy nucleotide and the deoxynucleotides normally used in DNA synthesis is that a dideoxy nucleotide has a 3'-H rather than a 3'-OH on the deoxyribose sugar. If a dideoxy nucleoside triphosphate (ddNTP) is used in a DNA synthesis reaction, the dideoxy nucleotide can be incorporated into the growing DNA chain. However, once that happens, no further DNA synthesis can then occur because the *absence of a 3'-OH prevents the formation of a phosphodiester bond with an incoming DNA precursor.*

The four sequencing reactions, then, use ddA, ddG, ddC, or ddT; that is, they have *one of* ddATP, ddGTP, ddCTP, and ddTTP, in addition to the four normal precursors dATP, dGTP, dCTP, and dTTP. So that some DNA synthesis occurs in the dideoxy sequencing reactions, only a small proportion of the precursors are dideoxy precursors. Generally, the dideoxy precursor is present in about one one-hundredth the amount of the normal precursor. The primer is extended by DNA polymerase, and when a particular nucleotide is specified by the template

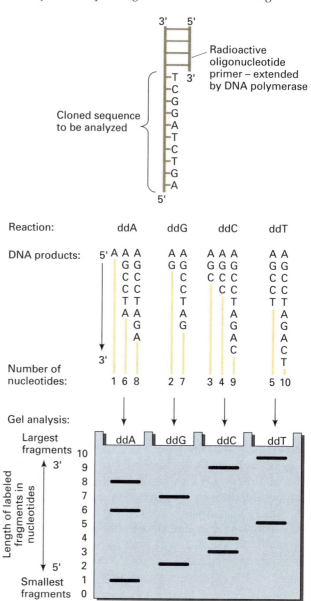

~ **FIGURE 15.20**

Dideoxy DNA sequencing of a theoretical DNA fragment.

Sequence deduced from banding pattern of autoradiogram made from gel:

5' A-G-C-C-T-A-G-A-C-T 3'

strand, there is a small chance that the dideoxy nucleotide will be incorporated instead of the normal nucleotide in the appropriate reaction mixture. For example, if an A is specified by the template strand a ddA could be incorporated rather than dA in the reaction mixture. Once a dideoxy nucleotide is incorporated, elongation of the chain stops. In a population of molecules in the same DNA synthesis reaction,

~ FIGURE 15.21

A dideoxy nucleotide DNA precursor.

Dideoxynucleoside
triphosphate

(Normal DNA precursor
has OH at 3' position)

~ FIGURE 15.22

Autoradiogram of a dideoxy sequencing gel. The letters over the lanes (A, C, G, and T) correspond to the particular dideoxy nucleotide used in the sequencing reaction analyzed in the lane.

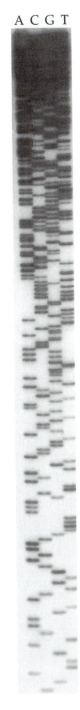

then, new DNA chains will stop at all possible positions where the nucleotide is required because of the incorporation of the dideoxy nucleotide at different positions. In the ddA reaction, the many different chains that are produced all end with ddA, all chains end with ddG in the ddG reaction, and so on (see Figure 15.20). It is important to realize that each DNA chain synthesized started from the same fixed point and ended at a particular base, the latter determined by the dideoxy nucleotide incorporated. The DNA chains in each reaction mixture are separated by polyacrylamide gel electrophoresis and the locations of the DNA bands are revealed by autoradiography (in the case of radioactive DNA sequencing experiments). The DNA sequence of the newly synthesized strand is determined from the autoradiogram by reading from the bottom to the top to give the sequence in 5'-to-3' orientation. In the example, the band that moved the farthest ended with ddA, the band that moved the second farthest ended with ddG, etc. The complete sequence determined is 5'-AGCCTAGACT-3'; this is complementary to the sequence on the template strand (see Figure 15.20).

An example of a dideoxy sequencing gel result is shown in Figure 15.22. There the various DNA chains that are synthesized in each of the dideoxy reactions are given to illustrate the principle involved in dideoxy sequencing. The 5'-to-3' DNA sequence is read from the bottom to the top of the gel since the smallest fragment migrates the fastest.

KEYNOTE

Rapid methods have been developed for determining the sequence of a cloned piece of DNA. One method, the Maxam-Gilbert procedure, uses specific chemicals to modify and cleave the DNA chain at specific nucleotides. The other method, the dideoxy or Sanger procedure, uses enzymatic synthesis of a new DNA chain on a cloned template DNA strand. With this procedure, synthesis of new strands is stopped by the incorporation of a dideoxy analog of the normal deoxyribonucleotide. Using four different dideoxy analogs, the new strands stop at all possible nucleotide positions, thereby allowing the complete DNA sequence to be determined.

Recently, automated procedures have been developed that enable DNA sequencing to proceed much more rapidly than the manual dideoxy method. In automated procedures, four separate dideoxy reactions are done, in which each of the four dideoxy nucleotides is labeled with a different fluorescent molecule. The four products then are combined into one tube, and the DNA fragments are subjected to electrophoresis in a *single* lane on a polyacrylamide gel. Since all four dideoxy nucleotides are used, every length of DNA chain synthesized is present in that same lane. Then, by passing the gel lane through a laser device that excites the fluorescent labels and determines which particular fluorescent label is present at each position, the sequence can be read automatically. Such procedures are of great utility as research teams proceed to determine the complete sequences of various genomes, including that of humans.

Analysis of DNA Sequences

Sequences determined by any sequencing method are usually entered into computer databases. The databases are analyzed, for example, for restriction site locations, and to compare a variety of sequences, homologous regions, controlling site similarities, and so on. Computer programs can search DNA sequences for possible protein-coding regions by looking for an initiator codon in frame with a chain terminating codon (called an open reading frame, or ORF). (Note that the discovery of a possible ORF does not mean that that DNA sequence encodes a protein in the cell. Many experiments would have to be done to see if that is the case. And the failure to detect an ORF does not necessarily mean that the DNA sequence has no function; it may represent a regulatory region.) Other programs can be used to translate a cloned DNA sequence theoretically into an amino acid sequence. Analysis of the amino acid sequences can also be instructive in making predictions about the structure and function of the protein. This is possible because the sequences of all sequenced proteins are in computer databases, enabling researchers to make detailed and rapid comparisons by computer. For example, many DNA binding proteins have particular secondary structure motifs and the detection of such a motif in a gene sequence would suggest that the gene encodes for a DNA binding protein. Since many regulatory proteins are DNA binding proteins, this conclusion would be of significant interest. Such a finding would then suggest the direction of future experimentation to define the actual function of the gene product.

POLYMERASE CHAIN REACTION (PCR)

The generation of large numbers of identical copies of DNA by the construction and cloning of recombinant DNA molecules was made possible in the 1970s. Indeed, recombinant DNA techniques revolutionized molecular genetics by making it possible to analyze genes and their functions in new ways. However, the cloning of DNA is time consuming, involving the insertion of DNA into cloning vectors and typically the screening of libraries to detect specific DNA sequences. In the mid-1980s, the **polymerase chain reaction (PCR)** was developed and this has resulted in yet a new revolution in the way genes may be analyzed. PCR is a method for producing an extremely large number of copies of a *specific* DNA sequence from a DNA mixture without having to clone it, a process called *amplification*. That is, PCR permits the selective amplification of DNA sequences. Kary Mullis, who developed PCR, was awarded the Nobel Prize in 1993.

The starting point for PCR is the DNA mixture containing the DNA sequence to be amplified, and a pair of oligonucleotide primers that flank that DNA sequence of interest (Figure 15.23). The primers are made synthetically, so some information must be available about the specific DNA sequence of interest so that it can be amplified. In brief, the PCR procedure is as follows:

1. Denature the DNA to single strands by incubating at 94°C. Cool (to 37–55°C, depending on how well the base sequence of the primers matches the

base sequence of the DNA) and anneal the specific pair of primers (primer A and primer B) that flank the targeted DNA sequence (Figure 15.23, part 1).

2. Extend the primers with DNA polymerase (Figure 15.23, part 2). For this, a special heat-resistant DNA polymerase called *Taq* ("tack") *polymerase* is used.

3. Repeat the heating cycle to denature the DNA to single strands and cooling to anneal new primers (Figure 15.23, part 3). (The further amplification of the original strands is omitted in the remainder of the figure.)

4. Repeat the primer extension with *Taq* DNA polymerase (Figure 15.23, part 4). In each of the two double-stranded molecules produced in the figure, one strand is of unit length; that is, it is the length of DNA between the 5' end of primer A and the 5' end of primer B—the length of the target DNA. The other strand in both molecules is longer than unit length.

5. Repeat the denaturation of DNA and anneal new primers (Figure 15.23, part 5). (For simplification, the further amplification of the longer-than-unit-length strands continues with linear increase only and is omitted in the rest of the figure.)

6. Repeat the primer extension with *Taq* DNA polymerase (Figure 15.23, part 6). This produces unit-length, double-stranded DNA. Note that it took three cycles to produce the two molecules of target-length DNA. Repeated denaturation, annealing, extension cycles results in the geometric increase of the unit-length DNA. Amplification of the longer-than-unit-length DNA occurs simultaneously, but only in a linear fashion.

Using PCR, the amount of new DNA generated increases geometrically. Starting with one molecule of DNA, one cycle of PCR produces two molecules, two cycles produces four molecules, and three cycles produces eight molecules, two of which are the target DNA. A further ten cycles produces 1,024 copies (2^{10}) of the target DNA and in 20 cycles there will be 1,048,576 copies (2^{20}) of the target DNA! The procedure is rapid, each cycle taking only a few minutes using a *thermal cycler*, a machine that automatically cycles through the temperature changes in a programmed way.

One disadvantage of the PCR procedure is that *Taq* polymerase does not have proofreading properties, so that errors are introduced into the DNA copies at low frequencies. Of course, should an error be introduced in an early cycle of PCR, then all subsequent copies made from that DNA will also have that error. PCR is also susceptible to contamination. That is, if the reaction mixture becomes contaminated with other DNA to which the primers can bind, then that DNA can also be amplified as well as the targeted DNA.

There are many applications for PCR, including amplifying DNA for cloning, amplifying DNA from genomic DNA preparations for sequencing without cloning, mapping DNA segments, disease diagnosis, sex determination of embryos, forensics, studies of molecular evolution, etc. In each case, some DNA sequence information must be available so that appropriate pairs of primers can be synthesized. In disease diagnosis, for example, PCR can be used to detect bacterial pathogens or viral pathogens such as HIV (human immunodeficiency virus, the causative agent of AIDS), human cytomegalovirus (a common virus infecting immunocompromised patients), and hepatitis B virus. PCR can also be used in genetic disease diagnosis, which is discussed in the next section. This makes it easier to detect single-copy sequences in total DNA isolated from cells.

In forensics, for example, PCR can be used to amplify trace amounts of DNA in samples such as hair, blood, or semen collected from a crime scene. The amplified DNA can then be analyzed and compared with DNA from a victim and a suspect, and the results can be used to implicate or exonerate the suspect in the crime. This analysis of DNA—called *DNA typing*—is discussed in more detail on pp. 487–489. Another interesting application of PCR is amplifying "ancient" DNA for analysis, using samples of tissues preserved many hundreds or thousands of years ago—for example, 40-million-year-old insects in amber and 40,000-year-old mammoths. PCR makes it possible to amplify selected DNA sequences and then analyze the sequences of those DNA molecules for comparison with contemporary DNA samples. These analyses enable us to make evolutionary comparisons between ancient forebears and present-day descendants.

KEYNOTE

The polymerase chain reaction (PCR) uses specific oligonucleotides to amplify a specific segment of DNA many thousand fold in an automated procedure. PCR is finding increasing applications both in research and in the commercial arena, including generating specific DNA segments for cloning or for sequencing, and for amplifying DNA to detect the presence of specific genetic defects.

~ FIGURE 15.23

The polymerase chain reaction (PCR) for selective amplification of DNA sequences.

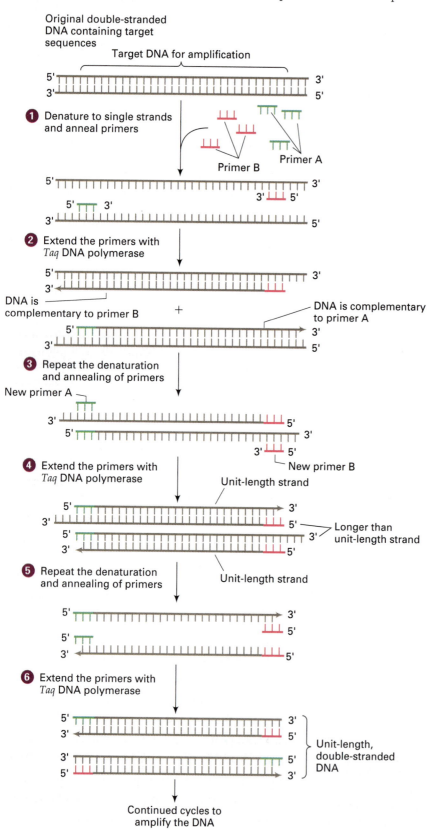

APPLICATIONS OF RECOMBINANT DNA TECHNOLOGY

Recombinant DNA and PCR technologies have many applications, including the diagnosis of human and animal genetic diseases; the synthesis of commercially important products such as human insulin, and human growth hormone; *in vitro* modifications of genes; and genetic engineering of plants. Some of the important applications will be briefly outlined in this section.

Analysis of Biological Processes

One of the fundamental and widespread applications of recombinant DNA and PCR methodologies is in basic research to explore biological functions. Questions in most areas of biology are being addressed with these techniques. In genetics, researchers are investigating such things as the functional organization of genes and the regulation of gene expression. In developmental biology, key regulatory genes and target genes responsible for developmental events are being discovered and analyzed, and gene changes associated with aging and cancer are being investigated. In evolutionary biology, DNA sequence analysis is adding new information about the evolutionary relationships between organisms. It is rare today to find a research scientist who is not aware of advances being made in his or her field through the use of recombinant DNA and PCR techniques.

Diagnosis of Human Genetic Diseases by DNA Analysis

We have learned in this text that many human genetic diseases are caused by enzyme or protein defects; those defects ultimately are the result of mutations at the DNA level. For an increasingly greater number of genetic diseases, we can screen individuals for the actual DNA mutation, rather than for the resulting biochemical change. Such information is useful in *genetic counseling*, the procedure whereby the risks of prospective parents having a child who expresses a genetic disease are evaluated and explained to them. Recall from our discussion that, once evidence is found of a genetic disease in a family or families, prospective parents need to be informed of the probability that they will produce a child with that disease.

Early detection of a genetic disease typically occurs at two levels: (1) the detection of heterozygotes (carriers) of recessive mutations; and (2) the determination of whether or not the developing fetus shows the defect. The principles of genetic counseling, and these two levels of detection of a genetic disease, were described in Chapter 9 (pp. 281–283).

For example, we can use DNA probes and other molecular techniques to detect the presence of a mutant allele or a DNA marker associated with a particular genetic disease such as Duchenne muscular dystrophy, Huntington disease, hemophilia, cystic fibrosis, Tay-Sachs disease and sickle-cell anemia. Unlike analysis for enzyme or protein defects, these techniques are limited not by whether a gene product is expressed in the developing fetus, but by whether or not there is a known DNA difference that can be used to distinguish the wild type from the mutant allele. It is important to understand that the probes only detect the presence of particular mutant alleles or markers linked to those alleles that are correlated with the development of a disease. Detecting the presence of an allele or marker associated with a disease is not a diagnosis of the disease itself; it only indicates that there is a probability that disease may develop. This is especially relevant given the fact that many disease-causing alleles exhibit much less than 100 percent penetrance (see Chapter 4, pp. 117–119).

Recombinant DNA and/or PCR approaches for detecting genetic disease require cellular DNA as the starting point. Such DNA can be isolated from fetal cells obtained by amniocentesis or chorionic villus sampling, and from blood samples of children and adults. The DNA may be digested with a restriction enzyme, producing restriction fragments of lengths determined by the locations of the restriction sites along the DNA molecules. The restriction fragments are analyzed as described earlier: the fragments are separated according to size by agarose gel electrophoresis, then transferred to a membrane filter by the Southern blot technique for hybridization with a specific labeled DNA probe. Alternatively, specific regions of the isolated DNA may be amplified by PCR (provided sequence information is available for primer design), and then analyzed, for example, by restriction enzyme analysis. In this case no blotting or probing is required.

These analytical procedures are most useful when the genetic mutation that causes a disease is associated with a change in the number or distribution of restriction sites, either within the gene or in a flanking region. That is, in the human genome, and the genomes of other eukaryotes, different restriction maps may be found among individuals for the same region of a chromosome (i.e., detected by the same probe). The different restriction maps result from different patterns of distribution of restriction sites and are called **restriction fragment length polymorphisms**, or **RFLPs** ("riff-lips"), because they are detected by the presence of restriction fragments of different lengths on gels. (*Polymorphism* means the existence of many different forms; here a region of DNA that can have different lengths on different homologs and/or in different individuals.) RFLPs arise, for example, by the addition or deletion of DNA between **restriction sites**, or by base pair changes that create or abolish a restriction site sequence. A restriction map is independent of gene function, so a RFLP is detected whether the DNA sequence change responsible affects a detectable phenotype or not. A RFLP can be used as a genetic marker in the same way as the "conventional" genetic markers we have discussed previously. In this case we assay the DNA—that is, the genotype—directly in the form of a restriction map. Moreover, because we are looking directly at DNA, both parental types are seen in heterozygotes, so carriers can easily be identified.

RFLPs are useful both for mapping chromosome regions and in human disease diagnosis. For the latter there are many cases of a RFLP being associated with a gene known to cause a disease, as the following example illustrates. The genetic disease *sickle-cell anemia* (discussed in more detail in Chapter 9) results from a single base pair change in the gene for the hemoglobin's β-globin polypeptide, resulting in an abnormal form of hemoglobin, Hb-S, instead of the normal Hb-A form. This change from $\frac{A}{T}$ to $\frac{T}{A}$ changes the codon from GAG to GTG; this results in the substitution of a valine for a glutamic acid in the sixth amino acid of the polypeptide, which, in turn, produces abnormal associations of hemoglobin molecules, sickling of the red blood cells, tissue damage, and sometimes death.

Using a cDNA probe for human β-globin, a RFLP has been shown for the restriction enzyme *Dde*I ("d-d-e-one"). That is, the base-pair change for the sickle-cell mutation results in the loss of a *Dde*I restriction site that is present in DNA of normal individuals (Figure 15.24a). In normal individuals, there are three *Dde*I sites: one is upstream of the start of the β-globin gene, and the other two are within the coding sequence itself. When DNA from normal individuals is cut with *Dde*I, the fragments separated by gel electrophoresis and transferred to a membrane filter by the Southern blot technique, and then probed with the 5' end of a cloned β-globin gene, two fragments of 175 bp and 201 bp are seen (Figure 15.24b). DNA from individuals with sickle-cell anemia analyzed in the same way

gives one fragment of 376 bp because of the loss of the *Dde*I site. Heterozygotes can be detected by the presence of three bands of 376 bp, 201 bp, and 175 bp.

Not all RFLPs result from changes in restriction sites directly related to the gene mutations. Many result from changes to the DNA flanking the gene, sometimes a fair distance away. This is the case for a RFLP that is related to the genetic disease PKU (see Chapter 9). Recall that PKU results from a deficiency in the activity of the enzyme phenylalanine hydroxylase. Following digestion of genomic DNA with *Hpa*I, Southern blotting, and probing with a cDNA probe derived from phenylalanine hydroxylase mRNA, different-sized restriction fragments are produced from DNA isolated from PKU individuals and from DNA isolated from homozygous normal individuals. This RFLP results from a difference outside the coding region of the gene, in this case to the 3' side of the gene. The RFLP can be used for diagnosing PKU in fetuses following amniocentesis or chorionic villus sampling. In these cases, detection of the genetic disease relies on the flanking RFLP segregating most of the time with the gene mutation. Recombination occurring between the RFLP and the gene of interest can occur, of course, and this can cause some difficulty in interpreting the results.

Other examples of human genetic diseases for which recombinant DNA technology can or will soon provide early detection include four types of thalassemia (hemoglobin diseases resulting in anemia): α-antitrypsin deficiency (a deficiency of a serum protein), hemophilia A, hemophilia B, Huntington disease, cystic fibrosis, and Duchenne muscular dystrophy (a progressive disease resulting in muscle atrophy and muscle dysfunction).

The beauty of the recombinant DNA approach is that it directly assays for a DNA genotype (i.e., RFLP), so detection does not depend on the expression of the gene (phenotype). To illustrate the present power of the approach, it is now possible to isolate DNA from one cell of an 8-cell stage blastula, diagnose whether or not a mutant gene for a genetic disease is present using PCR and probing, and then to reimplant the blastula if it is normal.

In general, the recombinant DNA approach is not limited to detecting genetic diseases in which the gene involved is active in the parent or the fetus at the time of analysis. For example, phenylalanine hydroxylase, the enzyme defective in individuals with PKU, is found in the liver but is not found either in blood serum or in fibroblast cells, the cells usually cultured following amniocentesis. The DNA of fibroblast cells can be analyzed, however, for RFLPs, and both individuals with the mutant allele or marker associated with the disease *and carriers* can be detected.

~ **FIGURE 15.24**

Detection of sickle-cell gene by the *Dde*I restriction fragment length polymorphism. (a) Diagrams of DNA segments showing the *Dde*I restriction sites; (b) Schematic drawing of the results of analysis of DNA cut with *Dde*I, subjected to gel electrophoresis, blotted, and probed with a β-globin probe.

a) *Dde* **I restriction sites**

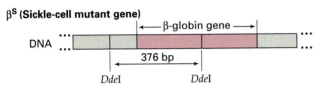

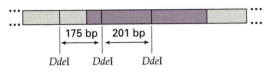

b) *Dde* **I fragments detected on a Southern blot by probing with beginning of β-globin gene**

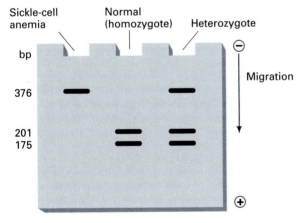

Isolation of Human Genes

The diagnosis of human genetic diseases is very valuable for modern medicine. However, more important is providing a cure for genetic diseases. In order to move in that direction, it is necessary to have in hand information about the normal function of the gene involved in the disease and about the mechanism by which disease symptoms occur when the gene is

mutated. In classical biochemical genetics studies applied to many organisms, researchers were able to define biochemical changes resulting from a gene mutation and, hence, to get a good understanding of how the mutant phenotype resulted (e.g., see Chapter 9). With a defined gene product, it is relatively easy to clone the gene, for example, by using antibodies made against the gene product to screen a cDNA library made in an expression vector (see p. 463 and Figure 15.14). However, for most human genetic diseases, the gene product that is altered is unknown, so this poses a significant problem for cloning the gene involved. Fortunately, there are now a number of techniques available to solve this problem. Some examples follow:

1. Probes from the homologous animal gene have been used to clone the HGPRT gene involved in Lesch-Nyhan syndrome (see Chapter 9, p. 274) and the phenylalanine hydroxylase gene involved in phenylketonuria (PKU) (see Chapter 9, p. 272).
2. Linkage analysis provided a starting point for cloning the cystic fibrosis (CF) gene, and linkage analysis and chromosome mutation analysis provided a starting point for cloning the dystrophin gene involved in Duchenne muscular dystrophy (DMD).

In general, the linkage analysis approach involves identifying a RFLP marker (see p. 480) that is linked to the disease phenotype, and then homing in on the gene starting from the marker location on the chromosome. The isolation of a gene associated with a genetic disease on the basis of its approximate chromosomal position is called **positional cloning**. An earlier term was *reverse genetics*, because it involved isolating the gene without knowledge of the gene product. We will describe some of the techniques that are often used in this approach in the context of the story of the cloning of the gene involved in cystic fibrosis (CF).

CLONING THE CF GENE. Cystic fibrosis is a disease caused by an autosomal recessive mutation. Disease symptoms and genetic properties of the disease were introduced in Chapter 9. The CF gene was the first human disease gene to be cloned solely by positional cloning. The effort took four years and the involvement of many laboratories.

Identifying RFLP Markers Linked to the CF Gene. Initially, a large number of individuals in CF pedigrees were screened with a large number of RFLPs to determine if any were linked genetically to the CF gene. This was done by tracking the inheritance of the CF gene (both homozygotes and heterozygotes) in the families, and simultaneously analyzing their DNA by Southern blot analysis and hybridizing with probes to identify any RFLP marker that showed genetic linkage to the CF locus. One such RFLP was found that showed weak linkage to the CF locus.

Identifying the Chromosome on Which the CF Gene Is Located. The RFLP marker was used to identify the chromosome on which the CF gene is located. This was done by *in situ hybridization*, a technique in which chromosomes are spread on a microscope slide and hybridized with a labeled probe. Using human metaphase chromosomes and probes labeled with ^{3}H, hybridization to a chromosome is shown by a positive signal after autoradiography. By using a ^{3}H-labeled RFLP probe, it was shown that chromosome 7 was the location of the CF gene.

More recently, fluorescent *in situ* hybridization (FISH) techniques have been developed so that regions as close as 50 kb may be distinguished on interphase chromosomes with fluorescently labeled probes (see Chapter 6, pp. 184–185 and Figure 6.11). This new technique makes it easier to home in on the chromosome location of a probe sequence.

Identifying the Chromosome Region Where the CF Gene Is Located. Once the chromosome on which the CF gene is located was known, the next step was to use known chromosome 7 RFLPs to find ones that were closely linked to the CF gene. This linkage analysis led to the identification of two closely linked flanking markers (i.e., one marker on each side of the CF gene) called *met* and D7S8. The *met* marker is a known proto-oncogene; that is, a gene involved in cell differentiation control (see Chapter 18). The D7S8 marker is a randomly cloned DNA fragment (also called an *anonymous probe*). This marker illustrates the nomenclature used for such cloned fragments: D is for "DNA segment," the number is the number of the chromosome (7, here), S means that it is a single-copy (unique) sequence as opposed to a repetitive sequence (which would be designated NF for "numerous fragments"), and the final number is a sequential number assigned at the Human Gene Mapping Workshops. The two flanking markers were known to be located at region 7q31-q32 (7 = chromosome 7; q = the long arm [p is the short arm]; 31-32 = subregions 31 and 32) so this localized the CF gene to that stretch of chromosome 7.

Cloning the CF Gene Between the Flanking Markers. With flanking markers in hand, it is theo-

retically a straightforward task to clone a gene. The flanking markers are used as starting points for cloning the DNA in between, and in that DNA should be the gene of interest, in this case the CF gene. However, this is not always as simple as it seems on paper. In this particular case, the two flanking markers are about 1.5 map units (cM) apart according to genetic linkage analysis. In the human genome, 1 map unit is equivalent to approximately 1 million base pairs of DNA. This is an average number derived from mapping in many areas of the genome. However, it is well established that genetic recombination is *not* constant throughout the genome, so the estimate of 1.5 million base pairs (megabase pairs = Mb) between the markers is very rough. This amount of DNA is obviously a large amount to search through to find a gene of interest. Fortunately, in subsequent research two other cloned DNA probes detected polymorphisms linked to CF and reduced the targeted span of DNA to about 500 kb.

The approach often used to find a gene between flanking markers is **chromosome walking**. Chromosome walking is a process used to identify adjacent clones in a genomic library (Figure 15.25). A chromosome walk is done as follows: An initial cloned DNA fragment—for example, one of the flanking markers—is used to begin the walk. A labeled end piece of the clone (right end in the figure) is used to screen a lambda or cosmid genomic library for clones that hybridize with it. The labeled probe should find all clones that overlap the original clone. These clones can be analyzed by restriction mapping to determine the extent of overlap. Then, a new labeled probe can be made from the right end of a clone with minimal overlap and the library is screened again. By repeating this process over and over, we can walk along the chromosome clone by clone.

There are some limitations to chromosome walking. Firstly, there is the potential problem of repetitive DNA in the cloned fragments. For example, if the end piece of a clone that is used as a probe is repetitive DNA where the sequence involved is scattered throughout the genome, then a number of clones will be identified in a genomic library and most will not overlap the original clone. Thus, it is very important that the probe used to find an overlapping clone be unique-sequence DNA, and this can be checked fairly easily.

Secondly, the length of each step of the walk is limited. Consider cosmid libraries with 50-kb DNA inserts, for example. If we assume an overlap between adjacent clones in a walk of about 15 kb, then each step in the walk is about 35 kb. We have just learned that quite small genetic distances are actually quite large physical distances. In the case of the CF cloning project, for example, 15 steps in the walk would be necessary to span the 500 kb between the flanking markers. While the procedures for each step are not complicated, they are time consuming.

A procedure related to chromosome walking is also used to move along chromosomes. Called *chromosome jumping* (Figure 15.26), it is a technique to cross large amounts of DNA, particularly those that are otherwise unclonable (e.g., long stretches of repet-

~ **FIGURE 15.25**

Chromosome walking. (From *Biochemistry* by Donald Voet and Judith G. Voet. Copyright © 1990 Donald Voet and Judith G. Voet. Reprinted by permission of John Wiley & Sons, Inc.)

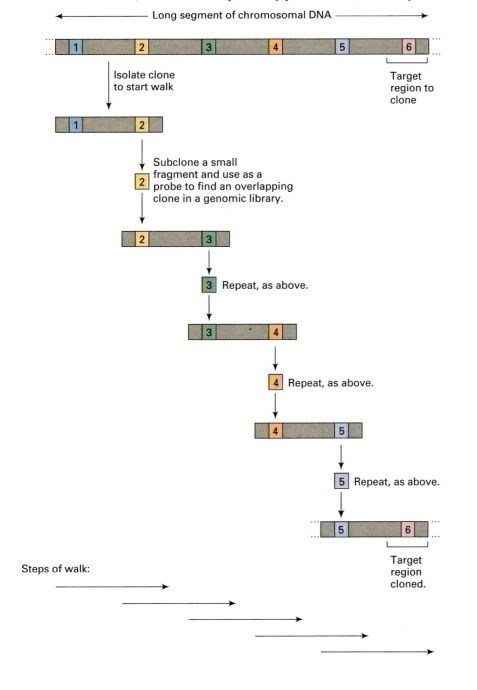

~ FIGURE 15.26

Chromosome jumping.

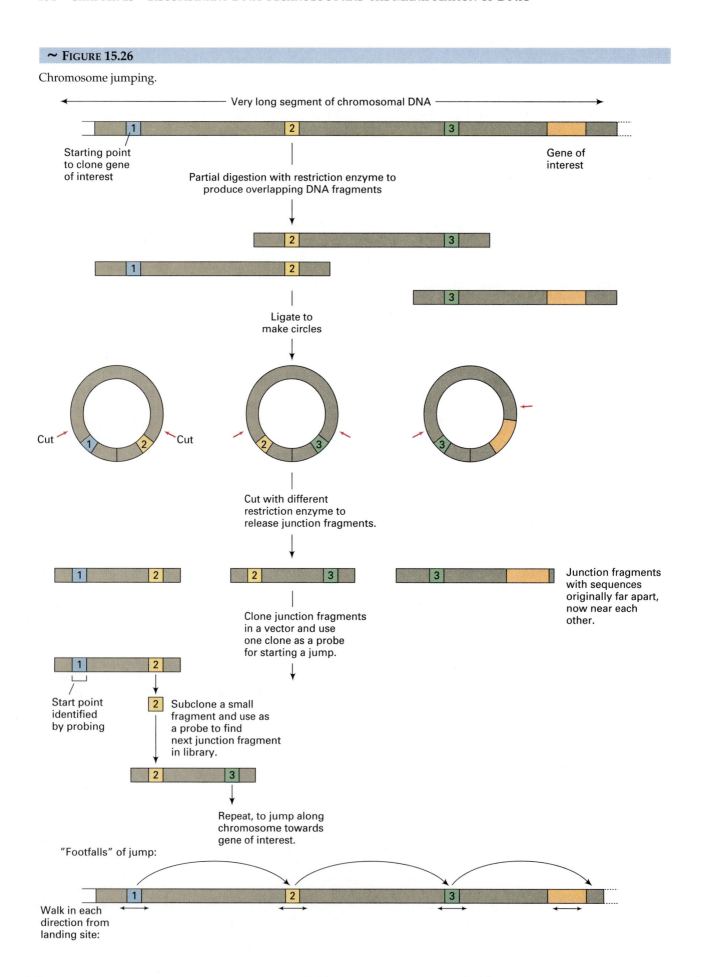

itive DNA). So, while in chromosome walking, each "step" is an overlapping DNA clone, in chromosome jumping, each "jump" is from one chromosome location to another without touching down on the intervening DNA.

Chromosome jumping is done as follows (see Figure 15.26). High-molecular-weight genomic DNA is isolated and partially digested with a restriction enzyme to produce overlapping DNA fragments. One fragment has a DNA sequence that is the starting point for a jump near one end; for example, it contains a marker known to be linked to a gene of interest. The fragments are then circularized by DNA ligase. This brings close together DNA sequences that were originally near the two ends of the linear fragment. Next, the circular molecules are cut to release the junction segments which are separately cloned in lambda or cosmid vectors to form what is called a *jumping library*. Use of such a library is straightforward. First, a probe made from a marker that is the starting point for a jump is used to find a hybridizing clone in the jumping library. That clone will contain the marker fragment joined to a DNA fragment that originally was quite a distance away along the chromosome (see Figure 15.26). The distance between the two DNA fragments equals the length of the DNA that connected the fragments during the circularization step. The DNA fragment identified by the jump can in turn be used to make another jump along the chromosome in the same, ultimately generating markers spaced along the chromosome segment being analyzed. Of course, each marker defined by the jump can be used as the starting point for chromosome walks in each direction, so the combination of chromosome jumping and chromosome walking enables an investigator to move almost at will along the chromosome and home in on regions of particular interest.

For cloning the CF gene, seven chromosome jumps were made towards the locus and chromosome walks were made from each jump site to identify overlapping clones. In the end, a large number of lambda and cosmid clones were isolated that spanned over 500 kb of DNA of the CF region.

Identifying the CF Gene in the Cloned DNA. The clones spanning the CF region were known to include the CF gene itself, but how does one identify the particular gene of interest in the set of clones? This is a general problem that gene cloners face. First, one has to find genes, and then one has to prove that a gene is the gene being targeted.

There are some approaches that can be used to home in on genes in the clones. One is to use cloned DNA as probes to see if they can hybridize with sequences in other species. The reasoning is that genes are conserved in sequence among related species, whereas non–gene sequences are likely not to be as conserved. The experiment is to isolate DNA from other organisms (e.g., mouse, hamster, chicken), digest it with restriction enzymes, separate the fragments by agarose gel electrophoresis, transfer the fragments to a membrane filter by Southern blotting, and hybridize with a labeled probe. Because the blot contains DNA from a variety of organisms, it is often called a *zoo blot*. In the CF project, five probes from the CF region cross-hybridized with DNA sequences from other organisms, identifying them as possible candidates for the CF gene. Two of the probes were ruled out based on linkage analysis, and a third was ruled out because it proved to be a pseudogene, that is, a sequence resembling a functional gene but lacking appropriate expression signals (see p. 560).

Another obvious property of protein-coding genes is that they produce mRNAs when they are expressed. Thus, it is expected that a DNA probe made from a protein-coding gene or part of such a gene should hybridize with mRNAs on a northern blot (see p. 473). Using this test, a fourth probe was ruled out, leaving only one left. This fifth probe was sequenced and was found to contain a cluster of C and G nucleotides, called *CpG islands*. Since the promoters of many protein-coding genes are known to contain CpG islands, this discovery was an encouraging sign that the CF gene was nearby.

Next, a probe made from this fifth sequence was used to screen a cDNA library made from mRNA molecules isolated from cultured normal sweat gland cells. This tissue type was used as the source of mRNA because the symptoms of CF were believed to result from a defect in sodium and chloride transport and such transport is an important function of sweat gland cells. The probe identified a single, positive clone on northern blots and the subsequent study of that clone and overlapping clones established a candidate cDNA clone of about 6,500 bp indicating, of course, a mRNA size of the same size. The cDNA clone was used to analyze the genomic clones in more detail with the result that the candidate CF gene was shown to span approximately 250 kb of DNA and involve 24 exons.

It was still necessary to prove definitively that the cDNA was derived from the CF gene. Ideal would be a functional assay involving the complementation test (see p. 467) to show that the cloned DNA could complement known CF mutations, but this cannot be done with human patients. Theoretically it is possible to do complementation tests with tissue cultures, providing there is a functional assay to show that complementation has occurred. Since the product was not known for the CF gene, a complementation test was not possible.

It *is* feasible to use cDNA as a probe to compare normal individuals and CF patients in the hope of detecting mutations such as deletions or other significant abnormalities. However, no major differences were seen in Southern blot analysis of genomic DNA or northern blot analysis of mRNA. Confirmation that the cDNA was derived from the CF gene had to come from comparing the DNA sequences of the candidate CF gene in a normal individual with that of a CF patient. The expectation was that mutational changes would be obvious if the correct gene had been identified. This proved to be the case—a 3 base-pair deletion was detected in the CF patient.

Defect in Cystic Fibrosis. With the CF gene identified, researchers then investigated more fully the nature of the mutations responsible for the disease itself. It was found that 68 percent of patients had the 3-bp deletion mentioned above, which results in the loss of the amino acid phenylalanine in the protein encoded by the gene. The remaining CF patients show over 60 different mutations, so the prospect of the development of a simple DNA-based diagnostic test for CF appears not to be simple.

We started the discussion of positional cloning with the premise that this type of cloning is done when the protein product is not known. Using the CF gene as an example, it is instructive to see what can be done with DNA sequence information to make predictions about protein products. By computer analysis of the protein encoded by the CF mRNA, it was determined that the protein consists of two similar motifs, each with a domain having properties consistent with membrane association, and a domain possibly involved with binding ATP. The phenylalanine deletion just mentioned is in the middle of the first predicted domain. The proposed structure for the CF protein—called cystic fibrosis transmembrane conductance regulator (CFTR)—derived from the computer-derived properties of the protein domain was described in Chapter 9 (pp. 280–281 and Figure 9.14).

CLONING OTHER HUMAN GENES. Generally similar approaches have been used to clone other human genes, with a particular focus on those causing human diseases. For example, the Huntington disease (HD) gene on chromosome 4 has been cloned and analyzed. Analysis of the normal HD gene and the HD genes from a number of Huntington patients revealed that the molecular basis of the disease is a triplet mutation like that for fragile X syndrome (see Chapter 7, pp. 205–206). That is, normal individuals have 11 to 34 repeats of the sequence CAG within the protein-coding sequence, whereas Huntington patients have 42 to 66 copies of the CAG repeat. It is not yet known how the extra CAG copies and the resultant extra amino acids in the encoded protein cause Huntington disease. Unfortunately, computer analysis of the protein encoded by the HD showed that the protein is unrelated to other proteins in the database, making it difficult at this point to understand the cellular cause of the disease.

ҠEYNOTE

The isolation of human genes, particularly those associated with disease, is possible using an array of molecular techniques. Where the gene product is not known, the starting point for cloning is knowledge of the genetic linkage between the disease locus and a DNA marker or markers. The isolation of a gene associated with a genetic disease on the basis of its approximate chromosomal position is called positional cloning.

Human Genome Project

The **Human Genome Project** is an extensive, collaborative effort to map all of the estimated 50,000 to 100,000 human genes, and to obtain the sequence of the complete 3 billion (3×10^9) nucleotide pairs of the genome. In the United States, the Human Genome Project is being overseen primarily by the National Center of Human Genome Research (a part of the National Institutes of Health) and the Department of Energy. Officially, the Human Genome Project began on October 1, 1990. The current director of the project is Francis Collins, after leadership during the first two years of the project by James D. Watson.

Associated with the Human Genome Project are parallel efforts to obtain gene maps and complete sequences of the genomes of a number of other model organisms, including *E. coli*, yeast, *Drosophila melanogaster*, the plant *Arabidopsis thaliani*, the nematode *Caenorhabditis elegans*, and mouse. Sequencing of the complete yeast genome was completed in early 1996 by an international consortium of scientists. The genome is 12,057 kb (excluding repetitive DNA) distributed among the organism's 16 chromosomes; it contains an estimated 6,000 genes. The yeast genome was the first eukaryotic genome to be sequenced completely. The complete sequences of two prokaryotic genomes were reported in 1995: the genome of the bacterium *Haemophilus influenzae* is 1,830 kb and the genome of the bacterium *Mycoplasma genitalium* is 580 kb. In early 1997, the complete 4,600-kb genome of *E.*

coli was reported. Late in 1996, the complete 1,660-kb genome sequence of the microbe *Methanococcus jannaschii* was reported. This organism belongs to a group of organisms called the Archaea (archaeons), many of which live in extreme conditions such as hot springs and deep sea vents. *Methanococcus jannaschii,* for example, was isolated from a deep sea vent in the Pacific Ocean at a site where the temperature is close to the boiling point of water and where the pressure is 245 times greater than at sea level. A surprising 56 percent of the 1,738 genes the organism contains are entirely new to science, adding support to the growing evidence that the Archaea represent a third kingdom of life (along with the Bacteria, prokaryotic cells that have no nucleus; and the Eukarya, organisms with nucleated cells).

A major goal of the Human Genome Project is to generate a highly detailed map of the human genome. The progressive generation of this genetic map has come from studies of polymorphic DNA markers; that is, loci in the genome where the different "alleles" are detectable differences in DNA length. We have already discussed DNA markers resulting from restriction fragment length polymorphism (pp. 479–480). The first human genetic map for the Human Genome Project—published in 1987—was based on such RFLP markers.

Other DNA markers called *microsatellite markers* have been very useful recently in the construction of the most comprehensive genetic map of the human genome. Microsatellite markers are essentially *short tandem repeat polymorphisms of simple sequences,* 2–5 base pairs long. They are called microsatellites because they are a subset of the satellite sequences in highly repetitive sequence DNA (see Chapter 11, p. 336). The map—completed in 1996—consists of 5,264 short tandem repeat polymorphisms involving the common microsatellite sequence AC/TG. In the DNA, AC is on one strand and TG is on the complementary strand, thus:

$$5'\text{-A C A C A C A C-}3'$$
$$3'\text{-T G T G T G T G-}5'$$

In other words, the polymorphisms result from different numbers of this 2 base-pair repeat on the two homologs at each microsatellite locus. By mapping the different microsatellite loci throughout the genome with respect to each other, the genetic map was constructed. The average genetic interval between markers was calculated to be 1.6 map units. Interestingly, the genetic length of the genome is different between the sexes: 4,396.9 mu in females and 2,729.7 mu in males, for a sex-averaged genetic length of 3,699.2 mu. Given that the total genome length is calculated to be 3,154 million base pairs, there are approximately 1 million base pairs of DNA per map unit, showing that even closely linked genes are physically quite a distance apart.

A second major goal of the project is to determine the entire nucleotide-by-nucleotide sequence of the human genome. The target date for completion of this phase of the project is the year 2005.

DNA Typing

Everyone is familiar with the use of fingerprints in forensic science. The principle is that no two individuals have the same fingerprints so that fingerprints left at the scene of a crime are important evidence in a criminal investigation. Similarly, no two human individuals (except identical twins) have exactly the same genome, base pair for base pair, and this has led to the development of DNA techniques for use in forensic science, in paternity and maternity testing, and elsewhere. Such human uses might seem unfounded, at first, since it is clear that the similarities in the DNA of different individuals greatly exceed the differences, as we would expect of a genome which specifies the human species. In fact, estimates indicate that about 1 in 1,000 base pairs is the site of polymorphisms among humans. For **DNA typing** (also called **DNA fingerprinting**, or *DNA profiling*)—the use of DNA analysis to identify an individual—scientists may use *highly polymorphic markers* scattered throughout the genome. Each marker consists of a restriction fragment within which are short, identical segments of DNA tandemly arranged head to tail. Differences between individuals result from a great variation in the number of tandem repeats, called *variable number of tandem repeats,* or VNTRs (sometimes called *minisatellite sequences*). These are similar to the microsatellite sequences just described; in this case the repeating unit is 15–100 base pairs long. The use of VNTRs as markers is illustrated in Figure 15.27. Individual A is heterozygous for two alleles at a VNTR locus flanked by restriction sites. One allele has three copies of the repeat sequence and the other allele has ten copies. Individual B is homozygous for a 6-repeat allele. When genomic DNA is digested with the enzyme, restriction fragments of different sizes are produced because of the VNTRs. The results of the digests can be visualized by use of a probe for the particular repeat sequence at the marker locus. (A probe that is specific for the VNTR sequences at one locus in the genome is called a *monomorphic probe.* Probes that detect VNTR sequences at a number of loci in the genome are known as *polymorphic probes.*) In this way two DNA bands would be detected for individual A and one DNA band would be detected for individual B.

~ FIGURE 15.27

Illustration of the concept involved in using VNTRs (variable number of tandem repeats) as DNA markers.

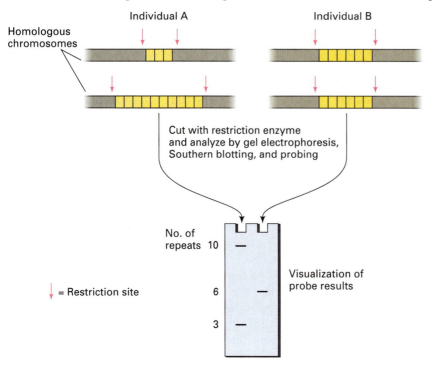

Let us now consider an example of using DNA typing in a paternity case. In this fictional scenario, a mother of a new baby has accused a particular man of being the father of her child, and the court will decide the case based on evidence from DNA typing. The DNA typing proceeds as follows (Figure 15.28): DNA samples are obtained from each individual involved in the case (Figure 15.28, part 1). In a paternity case, the simplest source of DNA is from a blood sample. The DNA is cut with the appropriate restriction enzyme, the resulting fragments are separated by electrophoresis (part 2), transferred to a membrane filter by the Southern blot technique (part 3), and probed with a labeled VNTR probe (parts 4 and 5). The DNA banding pattern after autoradiography or chemiluminescence detection is then analyzed to compare the samples (part 6).

The data may be interpreted as follows: Two DNA fragments are detected for the mother, so she is heterozygous for one particular pair of alleles at the VNTR locus under investigation. Likewise, two DNA fragments are detected for the baby so the baby is also heterozygous. One of the fragments for the baby matches the larger of the fragments for the mother, while the other fragment for the baby is much larger, indicating many more repeats in that allele. This pattern makes sense because the baby should receive one allele from its mother and one from its father. The important question is whether the paternal allele of the baby matches an allele from the alleged father. Inspection of that lane in the autoradiogram leads us to conclude that the answer is "yes," for there are two fragments, one in common with the larger fragment of the baby, and the other distinct from other fragments already discussed.

Do these data prove that the accused is the father of the child? The data do indicate that the accused shares an allele with the baby, but they do not prove that he contributed that allele to the genome of the baby, though he obviously could have. If the man had no alleles in common with the baby, the DNA typing data would have proved that he is not the father. To establish positive identity through DNA typing is much more difficult. In fact, it boils down to calculating the relative odds that the allele came from the accused or from another person. This calculation depends upon knowing the frequencies of VNTR alleles identified by the probe in the ethnic population from which the accused comes. Most of the legal arguments stem from here. That is, good estimates of VNTR allele frequencies are known for only a limited array of ethnic groups, so calculations of probability of guilt give numbers of questionable accuracy in many cases. To try to minimize possible inaccuracy, investigators use a number of different probes (often five or more) so that the combined probabilities calculated for the set can be high enough to convince the courts that the accused is guilty, even allowing for problems with knowing true VNTR allele frequencies for the population in question. In our paternity case, we would likely be more persuaded that the accused was the father of the child if the data for five different monomorphic probes all indicated that he could have contributed a particular allele to the child.

~ FIGURE 15.28

Procedure for DNA typing (DNA fingerprinting) as used for a paternity case.

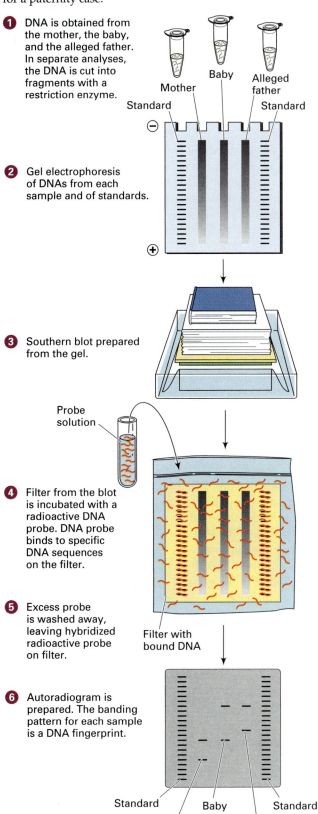

1 DNA is obtained from the mother, the baby, and the alleged father. In separate analyses, the DNA is cut into fragments with a restriction enzyme.

Standard Mother Baby Alleged father Standard

2 Gel electrophoresis of DNAs from each sample and of standards.

3 Southern blot prepared from the gel.

Probe solution

4 Filter from the blot is incubated with a radioactive DNA probe. DNA probe binds to specific DNA sequences on the filter.

5 Excess probe is washed away, leaving hybridized radioactive probe on filter.

Filter with bound DNA

6 Autoradiogram is prepared. The banding pattern for each sample is a DNA fingerprint.

Standard Mother Baby Alleged father Standard

DNA typing is more and more commonly used in forensic analysis. In murder, rape, and other violent cases, DNA is taken from crime scenes and compared with that of victims and suspects. For example, for a murder, DNA can be isolated from blood or hair at the scene, and for a rape DNA can be taken from a semen sample. If only miniscule amounts of DNA can be collected, sometimes PCR can be used to amplify the DNA for the typing experiments.

DNA typing is not yet a generally accepted method for proving guilt in the U.S. courts. It is clear that the scientific basis for the method is not in question; rather, DNA evidence is most commonly rejected for other reasons, such as possible errors in collecting or processing the evidence or weak population statistics. There are, however, many examples where DNA typing has excluded an accused individual because of mismatches, and such conclusions are now widely accepted by U.S. courts. Another application of DNA typing is in the U.S. military, where DNA fingerprints are being substituted for the traditional "dog tag." DNA typing is also used extensively outside of human applications—for example, in basic population genetics experiments, animal poaching cases, proving pedigree status in certain breeds of horses, and in conservation biology studies of endangered species.

Gene Therapy

Is it possible to treat genetic diseases? These diseases result from the phenotypes caused by the mutant gene in somatic cells. Theoretically, two types of gene therapy are possible: (1) Somatic cell therapy, in which somatic cells are modified genetically to correct a genetic defect; and (2) germ-line cell therapy, in which the germ line cells are modified to correct a genetic defect. Somatic cell therapy results in a treatment for the genetic disease in the individual, but not a cure, since progeny could still inherit the mutant gene. Germ-line cell therapy, however, would result in a cure since the mutant gene(s) would be replaced by the normal gene. While both somatic cell therapy and germ-line cell therapy have been demonstrated in non-human organisms, only somatic cell therapy has been attempted in humans because of the ethical issues raised by germ-line cell therapy.

The most promising candidates for somatic cell therapy are genetic disorders that result from a simple defect of a single gene, and for which the cloned normal gene is available. Gene therapy involving somatic cells proceeds as follows: (1) A sample of the individual's mutant cells is taken; (2) normal, wild-type copies of the mutant gene are introduced into the cells and the cells reintroduced into the individual, where, it is hoped, the cells will function to produce normal gene product and the symptoms of the genetic disease will be treated successfully.

The source of the mutant cells is likely to vary with the genetic disease. For example, blood disorders, such as thalassemia or sickle-cell disease, would require modification of blood-line cells isolated from the bone marrow. For genetic diseases affecting circulating proteins, a promising approach is the gene therapy of skin fibroblasts, constituents of the dermis (the lower layer of the skin). The modified fibroblasts can easily be implanted back into the dermis, where the tissue becomes vascularized allowing gene products to be distributed.

A cell that has had a gene introduced into it by artificial means is called a *transgenic* cell, and the gene involved is called a *transgene*. The introduction of normal genes into a mutant cell poses several problems. First, procedures to introduce DNA into cells (transformation) typically are inefficient; perhaps only one in 1,000 or 100,000 cells will receive the gene of interest. Thus, a large population of cells must be obtained from the individual in order for gene therapy to be attempted. Present procedures use special viral-related vectors to facilitate the gene uptake. Second, for those cells which take up the cloned gene, the fate of the "foreign" DNA cannot be predicted. In some cases, the mutant gene will be replaced by the normal gene, while in others, the normal gene will integrate into the chromosome at a site distinct from that of the gene locus of the mutant gene. In the first case, the gene therapy will be successful provided that the gene is expressed. In the second case, successful treatment of the disease will only result if (1) the introduced gene is expressed; and (2) the resident mutant gene produces no or very little protein, or a lot of inactive protein—that is, the mutant phenotype results from no or low gene activity, as is common for recessive mutants. In terms of the mutant gene, if the disease results not from low gene activity, but the production of an altered protein which is an inhibitor or interferes with other regulatory proteins, then the presence of a normal gene in addition to the mutant gene in the genome may not help treat the disease. The mutants here would likely be known dominant mutants.

Somatic cell therapy has been repeatedly demonstrated in experimental animals such as mouse, rat, and rabbit. In recent years, a few gene therapy trials have been done with humans. For example, in 1990, a 4-year-old girl suffering from severe combined immunodeficiency (SCID), which results from a deficiency in adenosine deaminase (ADA), an enzyme required for normal function of the immune system, was subjected to gene therapy involving somatic cells. Here, T cells (cells involved in the immune system) were isolated from the girl, grown in the laboratory, and the normal ADA gene was introduced using a viral vector. The "engineered" cells were then reintroduced into the patient. Since T cells have a finite life in the body, continued infusions of engineered cells have been necessary. It is clear that the introduced ADA gene is expressed, probably throughout the life of the T cell. As a result, the patient's immune system is functioning more normally and she now gets no more than the average number of infections, compared with many more than the average number before the therapy. Essentially, the gene therapy treatment has enabled her to live a more normal life.

It is clear that, with time, many other genetic diseases will become targets of somatic cell therapy. Human genetic diseases that could likely be early candidates for gene therapy are, for example, sickle-cell anemia, thalassemias, phenylketonuria, Lesch-Nyhan syndrome, cancer, and cystic fibrosis. For example, after successful experiments with rats, human clinical trials are in process for transferring the normal CF gene to cystic fibrosis patients. As the "tricks" are learned to target genes to replace their mutant counterparts and to regulate the expression of the introduced genes, increasing success in treating genetic diseases will undoubtedly result. However, many scientific, ethical, and legal questions will have to be addressed before the routine implementation of gene therapy.

Commercial Products

Many new biotechnology companies have emerged since the mid-1970s, when basic recombinant DNA techniques were developed. The first was Genentech. With older, established pharmaceutical and chemical companies, these companies are focusing on using recombinant DNA technology to make a wide array of commercial products that are useful in businesses in the human health industry, and agriculture sector, as well as in the general marketplace. Some examples are as follows:

1. Tissue plasminogen activator (TPA)—used to prevent or reverse blood clots, therefore preventing strokes, heart attacks, or pulmonary embolisms.
2. Human growth hormone—used to treat pituitary dwarfism.
3. Tissue growth factor-beta (TGF-β)—promotes new blood vessel and epidermal growth; hence, potentially useful for wound healing and burns.
4. Human blood clotting factor VIII—to treat hemophiliacs.
5. Human insulin ("humulin")—to treat insulin-dependent diabetes.
6. DNase—to treat cystic fibrosis.

7. Platelet derived growth factor (PDGF) for treatment of chronic skin ulcers in human diabetic patients.
8. Bovine growth hormone—to increase cattle and dairy yields.
9. Recombinant vaccines—for treatment of human and animal viral diseases (e.g., hepatitis B in humans).
10. Genetically engineered bacteria and other microorganisms for improved production of, for example, industrial enzymes (e.g., amylases to break down starch to glucose), citric acid (flavoring), and ethanol.
11. Genetically engineered bacteria that can accelerate the degradation of oil pollutants or of certain chemicals in toxic wastes (e.g., dioxin).

Genetic Engineering of Plants

For many centuries the traditional genetic engineering of plants involved selective breeding experiments in which plants with desirable traits were used as parents for the next generation in order to reproduce offspring with those traits. As a result, humans have produced hardy varieties of plants (e.g., corn, wheat, oats) and have been successful in breeding varieties with increased yields, all by using standard plant breeding techniques. (Similar experiments have also been done with animals, e.g., dogs and horses, to produce desired breeds.) In this section, we will discuss briefly and selectively the application of recombinant DNA technology to plant breeding.

TRANSFORMATION OF PLANT CELLS. Historically, introducing genes into plant cells has been a problem that has placed plant genetic engineering's rate of progress behind bacterial, fungal, and animal genetic engineering. Work is being done, however, to resolve this problem. Currently, one solution that has been found is to exploit features of a soil bacterium, *Agrobacterium tumefaciens*, which infects many kinds of plants. Techniques take advantage of a natural mechanism in the bacterium for transferring a defined segment of DNA into the chromosome of the plant.

Agrobacterium tumefaciens causes crown gall disease, characterized by tumors (the gall) at wounding sites. Most dicotyledonous plants (called *dicots*) are susceptible to crown gall disease, but monocotyledonous plants are not. (A plant is said to be a dicot if the embryo of the seed has two cotyledons; examples are potatoes, petunias, and apples. In monocots, the embryo of the seed has a single cotyledon; examples are corn, wheat, and grass.) *Agrobacterium tumefaciens* transforms plant cells at the wound site, causing the cells to grow and divide autonomously and, therefore, to produce the tumor.

The transformation of plant cells is mediated by a plasmid in the *Agrobacterium* called the *Ti plasmid* (the Ti stands for tumor-inducing) (Figure 15.29). Ti plasmids are circular DNA plasmids analogous to pUC19, but, in comparison, Ti plasmids are enormous in size (about 200 kb versus 2.69 kb for pUC19).

The interaction between the infecting bacterium and the plant cell of the host stimulates the bacterium to excise a 30-kb region of the Ti plasmid called T-

~ FIGURE 15.29

Formation of tumors (crown galls) in plants by infection with certain species of *Agrobacterium*. Tumors are induced by the Ti plasmid, which is carried by the bacterium and which integrates some of its DNA (the T, or transforming DNA) into the plant cell's chromosome.

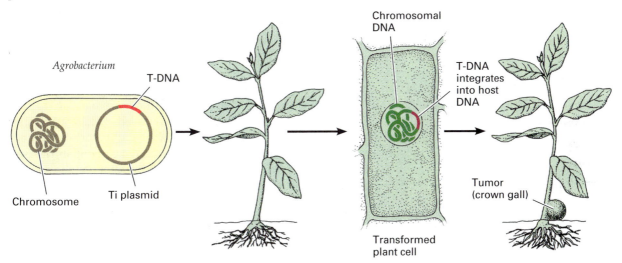

DNA (so-called because it is transforming DNA). T-DNA is flanked by two repeated 25-bp sequences called *borders* that are involved in T-DNA excision. Excision is initiated by a nick in one strand of the right-hand border sequence. A second nick in the left-hand border sequence releases a single-stranded T-DNA molecule, which is then transferred from the bacterium to the nucleus of the plant cell by a process analogous to bacterial conjugation. Once in the plant cell nucleus, the T-DNA integrates into the nuclear genome. As a result, the plant cell acquires the genes found on the T-DNA, including the genes for plant cell transformation. However, the genes needed for the excision, transfer, and integration of the T-DNA into the host plant cell are not part of the T-DNA. Instead, they are found elsewhere on the Ti plasmid, in a region called the *vir* (for *virulence*) region.

Using recombinant DNA approaches, researchers have found that excision, transfer, and integration of the T-DNA require only the 25-bp terminal repeat sequences, because the alteration or removal of the rest of the T-DNA does not affect those processes. As a result, the Ti plasmid and the T-DNA it contains is a useful vector for introducing new DNA sequences into the nuclear genome of somatic cells from susceptible plant species. Since any genes placed between the 25-bp borders will integrate into the host genome, a variety of transformation vectors have been derived from the Ti plasmid and T-DNA for the efficient introduction of genes into plants.

While the T-DNA based transformation system is very effective for dicotyledonous plants, it is not effective for monocotyledonous plants because they are not part of the normal host range of *Agrobacterium tumefaciens*. This is a serious limitation because most crop plants are monocotyledonous. Fortunately, alternative transformation procedures have been developed in which the DNA is delivered into the cell physically rather than by a plasmid vector. In the *electroporation* method, DNA is added to plant cell protoplasts and the mixture is "shocked" with high voltage to introduce the DNA into the cell. After the cells are grown in tissue culture to allow them to regenerate their cell walls and begin growing again, appropriate procedures can be applied to select for the cells that were successfully transformed. Another method involves the *gene gun* (made by Bio-Listics, an obvious pun on shooting). In this method, DNA is coated onto the surface of tiny tungsten beads which are placed on the end of a plastic bullet. The bullet is fired by a special particle gun. The bullet hits a plate at the end of the barrel of the gun, and the tungsten beads are propelled through a small hole in the plate into a chamber in which target cells have been placed. The force of the "shot" is sufficient to introduce the DNA-carrying beads into the cells. Selection techniques can then be applied to isolate successfully transformed cells, and these can be used for regeneration of whole plants.

APPLICATIONS FOR PLANT GENETIC ENGINEERING. In the near future we can expect a range of plants to be produced that have been engineered with recombinant DNA techniques. Of particular value will be crop plants that have increased yield, insect pest resistance, and herbicide tolerance (to enable fields to be sprayed to kill weeds but not the crop). Already, rice has been genetically engineered to produce strains resistant to the damaging rice stripe virus, and wheat has been engineered to be resistant to a particular herbicide.

Let us briefly consider, as an example, generating transgenic plants that are tolerant to the broad-spectrum herbicide Roundup. This herbicide contains the active ingredient glyphosate which kills plants by inhibiting EPSPS, a chloroplast enzyme required for the biosynthesis of essential aromatic amino acids. Roundup is used widely because it is active in relatively low doses and is degraded rapidly in the environment by microbes in the soil. Approaches for making transgenic, Roundup-tolerant plants include: (1) introducing a mutated bacterial form of EPSPS that is resistant to the herbicide so that the aromatic amino acids can still be synthesized even when the chloroplast enzyme is inhibited; and (2) introducing genes that encode enzymes which convert the herbicide to an inactive form. Monsanto brought Roundup Ready soybeans to market in 1996.

All genetically engineered plants must be tested extensively before their future release for commercial use. With more sophisticated approaches, some of which are already available, it will be possible to make transgenic plants that control the expression of genes in different tissues. An example would be controlling the rate at which cut flowers die, or the time at which fruit ripens. Already approved for market is the "Flavr Savr" tomato (see Figure 1.3, p. 4), genetically engineered by Calgene Inc. in collaboration with the Campbell Soup Company. Commercially produced tomatoes are picked while in an unripe state so they can be shipped without bruising. Prior to shipping, they are exposed to ethylene gas which initiates the ripening process so that they arrive in the ripened state at the store. Such prematurely picked, artificially ripened tomatoes do not have the flavor of tomatoes picked when they are ripe. Calgene scientists devised a way to block the tomato from making the normal amount of polygalacturonase (PG), a fruit-softening enzyme. Their approach was to introduce into the plant a copy of the PG gene that was backward in its orientation with respect to the promoter and the terminator. When this gene is transcribed, the mRNA is complementary to the mRNA produced by the normal gene: it is called an *antisense* mRNA.[2] In the cell, the antisense mRNA binds to the normal, "sense" mRNA, preventing much of it from being translated. Thus, much less PG enzyme is produced and the tomato can remain longer on the vine without getting too soft for handling. Once picked, the Flavr Savr tomato is also less susceptible to bruising on the way to the store or to overripening in the store. The Flavr Savr tomato tastes better than store-ripened tomatoes and much like home-grown tomatoes.

KEYNOTE

Recombinant DNA technology is finding ever-increasing applications throughout the world. In the basic research laboratory, processes across all areas of biology are being studied. With appropriate probes, a number of genetic diseases can now be diagnosed using recombinant DNA methods, genes can be isolated even if only linkage information is known, and DNA fingerprinting is being used in forensic science. In addition, many products in the clinical, veterinary, and agricultural areas can be synthesized in commercial quantities using recombinant DNA procedures. Genetic engineering of plants is also readily possible using recombinant DNA technology. It is expected that many types of improved crops will result from applications of this new technology.

SUMMARY

In this chapter we have discussed some of the procedures involved in recombinant DNA technology and the manipulation of DNA. Collectively, these procedures are also referred to as genetic engineering, at least in the popular press. We have seen how it is possible to cut DNA at specific sites using restriction enzymes, how DNA can be cloned into specially constructed vectors, and how the cloned DNA can be analyzed in various ways. Through the construction of genomic libraries and cDNA libraries, and the application of screening procedures to those libraries, a large number of genes have been cloned and identified from a wide variety of organisms.

Restriction mapping analysis has provided detailed molecular maps of genes and chromosomes analogous to the genetic maps constructed on the basis of recombination analysis. Rapid DNA sequencing methods have given us an enormous amount of information about the DNA organization of genes, both their coding sequences and regulatory sequences. The amount of DNA sequence information available is growing at an extremely rapid rate and computer databases of such sequences are available for researchers to analyze. For example, when a new gene is sequenced, it is of interest to determine if it has any DNA sequences in common with other sequenced genes that are included in the database. In this way, potential families of proteins or parts of proteins have been identified. Such groups may have related functions in the organisms from which they are derived.

In the mid-1980s, a new technique called polymerase chain reaction (PCR) was developed. Given some sequence information about a DNA fragment, synthetic oligonucleotides can be made which can be used to amplify large amounts of the DNA fragment from the genome in a repeated cycle of DNA denaturation (strand separation), annealing of oligonucleotides to act as primers, and extension of the primers with a special DNA polymerase. Numerous applications have been rapidly found for PCR, including cloning rare pieces of DNA, preparing DNA for sequencing without cloning, and genetic disease diagnosis.

Recombinant DNA technology and PCR are being widely applied in both basic research and commercial areas. There is probably not an area of basic biology in which these revolutionary molecular techniques have not been applied to the questions being asked. The human genome project has the mandate to generate a complete map of all of the genes in the human genome, and to obtain the complete sequence of the human genome. The knowledge obtained from this ambitious, long-range project will contribute markedly to our understanding of human genetics.

Recombinant DNA and PCR techniques are also being used, for example, to develop new pharmaceuticals (including drugs, other therapeutics, and vaccines), to develop new tools for diagnosis of infectious and genetic diseases, for human gene therapy, in forensic analysis (e.g., analyzing DNA from a crime scene to match with a suspect), and in agricultural areas (e.g., improving disease resistance and yields of livestock and crops). While products generated by genetic engineering have been slower coming to the market than originally expected, there is an undiminished enthusiasm for the genesis and commercialization of a wide array of useful products in the present decade.

---

[2]The use of antisense mRNA to prevent or inhibit the translation of a natural mRNA is called antisense technology . This technology is being tested in a number of systems, for example, as a means of controlling genetic diseases.

ANALYTICAL APPROACHES FOR SOLVING GENETICS PROBLEMS

Although this is a rather descriptive area, it is often necessary to interpret data derived from restriction enzyme analysis of DNA fragments in order to generate a restriction map, that is, a map of the locations of restriction enzymes. The logic used for this type of analysis is very similar to that used in generating a genetic map of loci from two-point mapping crosses.

Q15.1 A piece of DNA 900 bp long is cloned and then cut out of the vector for analysis. Digestion of this linear piece of DNA with three different restriction enzymes singly and in all possible combinations of pairs gave the following restriction fragment size data:

ENZYME(S)	RESTRICTION FRAGMENT SIZES
*Eco*RI	200 bp, 700 bp
*Hind*III	300 bp, 600 bp
*Bam*HI	50 bp, 350 bp, 500 bp
*Eco*RI + *Hind*III	100 bp, 200 bp, 600 bp
*Eco*RI + *Bam*HI	50 bp, 150 bp, 200 bp, 500 bp
*Hind*III + *Bam*HI	50 bp, 100 bp, 250 bp, 500 bp

Construct a restriction map from these data.

A15.1 The approach to this kind of problem is to consider a pair of enzymes and to analyze the data from the single and double digestions. First let us consider the *Eco*RI and *Hind*III data. Cutting with *Eco*RI produces two fragments, one of 200 bp and the other of 700 bp, while cutting with *Hind*III also produces two fragments, one of 300 bp and the other of 600 bp. Thus, we know that both restriction sites are asymmetrically located along the linear DNA fragment with the *Eco*RI site 200 bp from an end and the *Hind*III site 300 bp from an end. When we consider the *Eco*RI + *Hind*III data we can determine the positions of these two restriction sites relative to one another. If, for example, the *Eco*RI site is 200 bp from the fragment end, and the *Hind*III site is 300 bp from that same end, then we would predict that cutting with both enzymes would produce three fragments of sizes 200 bp (end to *Eco*RI site), 100 bp (*Eco*RI site to *Hind*III site), and 600 bp (*Hind*III site to other end). On the other end, if the *Eco*RI site is 200 bp from one fragment end and the *Hind*III site is 300 bp from the other fragment end, cutting with both enzymes would produce three fragments of sizes 200 bp (end to *Eco*RI site), 400 bp (*Eco*RI site to *Hind*III site), and 300 bp (*Hind*III site to end). The actual data support the first model.

Now we pick another pair of enzymes, for example, *Hind*III and *Bam*HI. Cutting with *Hind*III produces fragments of 300 bp and 600 bp as we have seen, and cutting with *Bam*HI produces three fragments of sizes 50 bp, 350

bp, and 500 bp, indicating that there are two *Bam*HI sites in the DNA fragment. Again the double digestion products are useful in locating the sites. Double digestion with *Hind*III and *Bam*HI produces four fragments of 50 bp, 100 bp, 250 bp, and 500 bp. The simplest interpretation of the data is that the 300-bp *Hind*III fragment is cut into the 50-bp and 250-bp fragments by *Bam*HI, and that the 600-bp *Hind*III fragment is cut into the 100-bp and 500-bp fragments by *Bam*HI. Thus, the restriction map shown in the accompanying figure can be drawn:

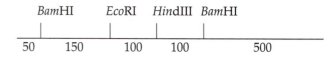

The *Bam*HI + *Eco*RI data are compatible with this model.

Q15.2 The recessive allele, *bw*, when homozygous, results in brown eyes in *Drosophila*, in contrast to the wild-type bright red eye color. A restriction fragment length polymorphism (RFLP) for a particular DNA region results in either two restriction fragments (type I) or one restriction fragment (type II) when *Drosophila* DNA is cut with restriction enzyme C, the fragments separated by DNA electrophoresis, blotted to a membrane filter, and probed with a particular DNA probe.

A true-breeding brown-eyed fly with type I DNA was crossed with a true-breeding, wild-type fly with type II DNA. The F₁ flies had wild-type eye color and exhibited both type I and type II DNA patterns. The F₁ flies were crossed with true-breeding brown-eyed flies with type I DNA, and the progeny were scored for eye color and RFLP type. The results were as follows:

CLASS	PHENOTYPES	NUMBER
1	Red eyes, type I and II DNA	184
2	Red eyes, type I DNA	21
3	Brown eyes, type I DNA	168
4	Brown eyes, type I and II DNA	27
	Total progeny	400

Analyze these data.

A15.2 The eye color mutation is a familiar genetic marker. The restriction fragment length polymorphisms (RFLPs) are also genetic markers and can be analyzed just like any gene marker. If we symbolize the type I DNA as I and the type II DNA as II, the F₁ cross is:

$$\frac{bw^+ \; \text{II}}{bw \;\; \text{I}} \times \frac{bw \;\; \text{I}}{bw^+ \; \text{II}}$$

(The cross is drawn as if the markers were linked; of course, we have yet to show this.) This is a testcross with the exception that DNA markers do not exhibit dominance or recessiveness. If the eye color and RFLP markers are unlinked, the result would be equal numbers of the four progeny classes. However, the data show a great excess of these two classes: (a) red eyes, type I and II DNA; and (b) brown eyes, type I DNA. Their origin was the pairing of F_1 bw^+ II and bw I gametes with bw I gametes to give bw^+ II/bw I and bw I/bw I progeny genotypes, respectively. Similarly, the progeny (a) red eyes, type I DNA and (b) brown eyes, type I and II DNA derive from pairing bw^+ I and bw II gametes with bw I gametes to give bw^+ I/bw I and bw II/bw I progeny, respectively. These two latter classes occur with about equal frequency, that is, a frequency much lower than those for the other two classes. The simplest explanation is that the brown eye color gene and the RFLP marker are linked, so the F_1 cross was a mapping cross, much like those we analyzed in Chapter 5. Classes 1 and 3 are the parentals, and classes 2 and 4 are the recombinants. The map distribution between bw and the RFLP is, therefore:

$$\frac{21 + 27}{\text{total}} \times 100\%$$

$$= \frac{48}{400} \times 100\%$$

$$= 12 \text{ map units}$$

QUESTIONS AND PROBLEMS

*15.1 A new restriction endonuclease is isolated from a bacterium. This enzyme cuts DNA into fragments that average 4,096 base pairs long. Like all other known restriction enzymes, the new one recognizes a sequence in DNA that has twofold rotational symmetry. From the information given, how many base pairs of DNA constitute the recognition sequence for the new enzyme?

15.2 An endonuclease called AvrII ("a-v-r-two") cuts DNA whenever it finds the sequence $\begin{smallmatrix} 5'\text{-CCTAGG-3'} \\ 3'\text{-GGATCC-5'} \end{smallmatrix}$.

About how many cuts would AvrII make in the human genome, which is about 3×10^9 base pairs long and about 40% GC?

15.3 About 40 percent of the base pairs in human DNA are GC. On the average, how far apart (in terms of base pairs) will the following sequences be?
a. two BamHI sites
b. two EcoRI sites
c. two NotI sites
d. two HaeIII sites

15.4 What are the features of plasmid cloning vectors that make them useful for constructing and cloning recombinant DNA molecules?

15.5 *E. coli* is a commonly used host for the propagation of cloned sequences in plasmid vectors. Wild-type *E. coli* turns out to be quite an unsuitable host, however—not only are the plasmid vectors "engineered," so is the host bacterium. Based on your understanding of how DNA is cloned in plasmid vectors, speculate how each of the following bacterial mutations might enable a better host to be engineered.

MUTATION	DESCRIPTION
recA	Gene central to recombination and DNA repair. Mutation eliminates general recombination and renders bacteria sensitive to UV light.
traD36	Mutation inactivates conjugal transfer of *F'* episome.
rpsL	Mutation affects ribosomal protein S12, and makes cells resistant to streptomycin.
endA	Normal gene encodes the DNA-specific endonuclease I. Mutation eliminates endonuclease function.
mcrA	Normal gene encodes the McrA restriction endonuclease that restricts DNA when it is methylated at the internal cytosine in the site CCGG.
mrr	Normal gene encodes the Mrr restriction endonuclease that restricts DNA when it is methylated at the methylated adenine of GAC or CAG sites.
hsdR	Normal gene encodes EcoK restriction endonuclease, which restricts DNA that is *not* methylated at certain A residues.

15.6 Considerable effort has been spent on developing cloning vectors that replicate in organisms other than *E. coli.*
a. Describe several different reasons one might want to clone DNA in an organism other than *E. coli.*
b. What is a shuttle vector and why is it used?
c. Describe the salient features of a vector that could be used for cloning DNA in yeast or plants.

*15.7 Genomic libraries are important resources for isolating genes of interest and for studying the functional organization of chromosomes. List the steps you would use to make a genomic library of yeast in a lambda vector.

15.8 The human genome contains about 3×10^9 bp of DNA. How many 40 kb pieces would you have to clone into a library if you wanted to be 90 percent certain of including a particular sequence?

*15.9 Suppose a researcher wishes to clone the genomic sequences that include a human gene for which a cDNA has already been obtained. The researcher has available a variety of genomic libraries that can be screened with a probe made from the cDNA, as described in Figures 15.15 and 15.16.

a. Assuming that each library has an equally good representation of the 3×10^9 base pairs in a haploid human genome, about how many clones should be screened if the researcher wants to be 95 percent sure of obtaining at least one hybridizing clone and
   i. the library is a plasmid library with inserts that are, on average, 7 kb?
   ii. the library is a lambda library with inserts that are, on average, 15 kb?
   iii. the library is a cosmid library with inserts that are, on average, 40 kb?
   iv. the library is a YAC library with inserts that are, on average, 350 kb?
b. What advantages and disadvantages are there to screening these different libraries?
c. What kinds of information might be gathered from the analysis of genomic DNA clones that could not be gathered from the analysis of cDNA clones?

15.10 What is a cDNA library and from what cellular genetic material is it derived? How is a cDNA library used in cloning particular genes?

15.11 A colleague has sent you a 4.5-kb DNA fragment excised from a plasmid cloning vector with the enzymes *Pst*I and *Bgl*II (see Table 15.1 for a description of these enzymes and the sites they recognize). Your colleague tells you that within the fragment there is an *Eco*RI site that lies 0.49 kb from the *Pst*I site. Suppose you wanted to clone this DNA fragment into the plasmid vector pUC19 (described in Figure 15.5).
a. How do you proceed?
b. How would you verify that you have cloned the correct fragment, and determine the orientation of the fragment within the pUC19 cloning vector?

*15.12 Suppose you wanted to produce human insulin (a peptide hormone) by cloning. Assume that this could be done by inserting the human insulin gene into a bacterial host, where, given the appropriate conditions, the human gene would be transcribed then translated into human insulin. Which do you think it would be best to use as your source of the gene, human genomic insulin DNA or a cDNA copy of this gene? Explain your choice.

15.13 You are given a genomic library of yeast prepared in a bacterial plasmid vector. You are also given a cloned cDNA for human actin, a protein which is conserved in protein sequences among eukaryotes. Outline how you would use these resources to attempt to identify the yeast actin gene.

*15.14 A cDNA library is made with mRNA isolated from liver tissue. When a cloned cDNA from that library is digested with the enzymes *Eco*RI (E), *Hind*III (H), and *Bam*HI (B), the restriction map shown in Figure 15.A, part a) is obtained. When this cDNA is used to screen a cDNA library made with mRNA from brain tissue, three identical cDNAs with the restriction map shown in Figure 15.A, part b) are obtained. When either cDNA is used to synthesize a uniformly labeled ^{32}P-labeled probe and the probe is allowed to hybridize to a Southern blot prepared from genomic DNA digested singly with the enzymes *Eco*RI, *Hind*III and *Bam*HI, an autoradiograph shows the pattern of bands in Figure 15.A, part c). When either cDNA is used to synthesize a uniformly labeled ^{32}P-labeled probe and used to probe a northern blot prepared with poly-A$^+$ RNA isolated from liver and brain tissues, the pattern of bands in Figure 15.A, part d) is seen. Fully analyze these data and then answer the following questions.

FIGURE 15.A

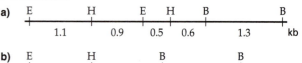

a)

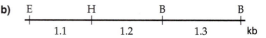

b)

c)

d)

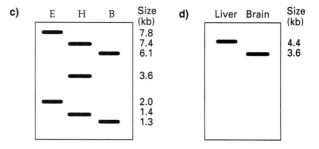

a. Do cDNAs derive from the same gene?
b. Why are different sized bands seen on the northern blot?
c. Why do the cDNAs have different restriction maps?
d. Why are some of the bands seen on the whole genome Southern blot different sizes than some of the restriction fragments in the cDNAs?

15.15 A researcher digests genomic DNA with the restriction enzyme *Eco*RI, separates it by size on an agarose gel, and transfers the DNA fragments in the gel to a membrane filter using the Southern blot procedure. What result would she expect to see if the source of the DNA and the probe for the blot is as described below?
a. The genomic DNA is from a normal human. The probe is a 2.0-kb DNA fragment obtained by excision with the enzyme *Eco*RI from a cosmid containing single-copy genomic DNA.
b. The genomic DNA is from a normal human. The probe is a 5.0-kb DNA fragment that is a copy of a LINE sequence that has an internal *Eco*RI site.
c. The genomic DNA is from a normal human. The probe is a 5.0-kb DNA fragment that is a copy of a LINE sequence that lacks an internal *Eco*RI site.

d. The genomic DNA is from a human heterozygous for a translocation between chromosomes 14 and 21. The probe is a 3.0-kb DNA fragment that is obtained by excision with the enzyme *Eco*RI from a cosmid containing single-copy genomic DNA from a normal chromosome 14. The translocation breakpoint on chromosome 14 lies within the 3.0-kb genomic DNA fragment.

e. The genomic DNA is from a normal female. The probe is a 5.0-kb DNA fragment containing part of the *TDF* gene (see p. 80, Chapter 3).

15.16 Restriction endonucleases are used to construct restriction maps of linear or circular pieces of DNA. The DNA is usually produced in large amounts by recombinant DNA techniques. The generation of restriction maps is similar to the process of putting the pieces of a jigsaw puzzle together. Suppose we have a circular piece of double-stranded DNA that is 5,000 base pairs long. If this DNA is digested completely with restriction enzyme I, four DNA fragments are generated: fragment *a* is 2,000 base pairs long; fragment *b* is 1,400 base pairs long; *c* is 900 base pairs long; and *d* is 700 base pairs long. If, instead, the DNA is incubated with the enzyme for a short time, the result is incomplete digestion of the DNA, not every restriction enzyme site in every DNA molecule will be cut by the enzyme, and all possible combinations of adjacent fragments can be produced. From an incomplete digestion experiment of this type, fragments of DNA were produced from the circular piece of DNA, which contained the following combinations of the above fragments: *a-d-b*, *d-a-c*, *c-b-d*, *a-c*, *d-a*, *d-b*, and *b-c*. Lastly, after digesting the original circular DNA to completion with restriction enzyme I, the DNA fragments were treated with restriction enzyme II under conditions conducive to complete digestion. The resulting fragments were: 1,400, 1,200, 900, 800, 400, and 300. Analyze all the data to locate the restriction enzyme sites as accurately as possible.

*15.17 A piece of DNA 5,000 bp long is digested with restriction enzymes A and B, singly and together. The DNA fragments produced were separated by DNA electrophoresis and their sizes were calculated, with the following results:

DIGESTION WITH		
A	B	A + B
2,100 bp	2,500 bp	1,900 bp
1,400 bp	1,300 bp	1,000 bp
1,000 bp	1,200 bp	800 bp
500 bp		600 bp
		500 bp
		200 bp

Each A fragment was extracted from the gel and digested with enzyme B, and each B fragment was extracted from the gel and digested with enzyme A. The sizes of the resulting DNA fragments were determined by gel electrophoresis, with the following results.

A FRAGMENT	FRAGMENTS PRODUCED BY DIGESTION WITH B	B FRAGMENT	FRAGMENT PRODUCED BY DIGESTION WITH A
2,100 bp →	1,900, 200 bp	2,500 bp →	1,900, 600 bp
1,400 bp →	800, 600 bp	1,300 bp →	800, 500 bp
1,000 bp →	1,000 bp	1,200 bp →	1,000, 200 bp
500 bp →	500 bp		

Construct a restriction map of the 5,000 bp DNA fragment.

*15.18 Draw the banding pattern you would expect to see on a DNA-sequencing gel if you annealed the primer 5'-C-T-A-G-G-3' to the following single-stranded DNA fragment and carried out a dideoxy sequencing experiment. Assume the dNTP precursors were all labeled.

3'-G-A-T-C-C-A-A-G-T-C-T-A-C-G-T-A-T-A-G-G-C-C-5'.

15.19 DNA was prepared from small samples of white blood cells from a large number of people. Ten different patterns were seen when these DNAs were all digested with *Eco*RI, then subjected to electrophoresis and Southern blotting. Finally, the blot was probed with a radioactively labeled cloned human sequence. The figure below shows the ten DNA patterns taken from ten people.

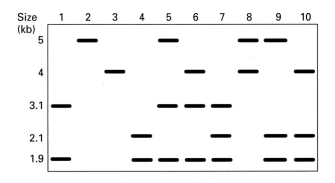

a. Explain the hybridization patterns seen in the ten people in terms of variation in *Eco*RI sites.

b. If the individuals whose DNA samples are in lanes 1 and 6 on the blot were to produce offspring together, what bands would you expect to see in DNA samples from these offspring?

15.20 Imagine that you have been able to clone the structural gene for an enzyme in a catecholamine biosynthetic pathway from the adrenal gland of rats. How could you use this cloned DNA as a probe to determine whether this same gene functions in the brain?

FIGURE 15.B

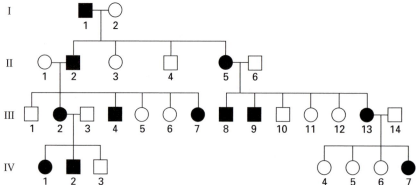

FIGURE 15.C

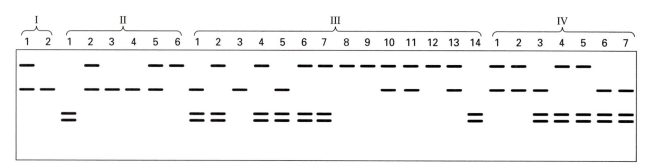

*15.21 Filled symbols in the pedigree (Figure 15.B) indicate people with a rare autosomal dominant genetic disease. DNA samples were prepared from each of the individuals in the pedigree. The samples were restricted, electrophoresed, blotted, and probed with a cloned human sequence called DS12-88, with the results shown in Figure 15.C. Do the data in these two figures support the hypothesis that the locus for the disease that is segregating in this family is linked to the region homologous to DS12-88? Make your answer quantitative, and explain your reasoning.

15.22 Imagine that you find a RFLP in the rat genomic region homologous to your cloned catecholamine synthetic gene from question 15.21, and that in a population of rats displaying this polymorphism there is also a behavioral variation. You find that some of the rats are normally calm and placid, but others are hyperactive, nervous, and easily startled. Your hypothesis is that the behavioral difference seen is caused by variations in your gene. How could you use your cloned sequence to test this hypothesis?

*15.23 The maps of the sites for restriction enzyme R in the wild type and the mutated cystic fibrosis genes are shown schematically in the following figure:

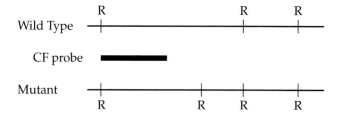

Samples of DNA obtained from a fetus (F) and her parents (M and P) were analyzed by gel electrophoresis followed by the Southern blot technique and hybridization with the radioactively labeled probe designated "CF probe" in the above figure. The autoradiographic results are shown in the following figure:

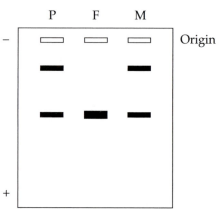

Given that cystic fibrosis is a recessive mutation, will the fetus be affected? Explain.

*15.24 The PCR technique was used to amplify two genomic regions (homologous to probes A and B) in DNA from *Neurospora*. Two strains of opposite mating type (strains J and K) were found to differ from each other in both amplified regions, as shown in Figure 15.D, when the amplified DNA was cut with *Eco*RI. Strain J was crossed to strain K, 100 asci resulting from the cross were dissected, and the PCR reaction was done on DNA from the individual spores in each ascus. Six different patterns of distribution of DNA types within asci were seen as shown in Figure 15.E. Only four patterns are shown for each ascus. Remember that spores 1 and 2 in the ascus are ordinarily identical, as are spores 3 and 4, 5 and 6, and 7 and 8. In Figure 15.E the band pattern displayed by spores 1 and 2 is indicated in the lane designated "1." The pattern shown by spores 3 and 4 is in lane

2, the pattern shown by spores 5 and 6 is in lane 3, and the pattern shown by spores 7 and 8 is in lane 4. Draw a map showing the relationships among the region detected by probe A, the region homologous to probe B, and any relevant centromeres.

15.25 The polymerase chain reaction (PCR) is widely applied in molecular genetics. It has often provided a faster, more sensitive means to answer a question than more traditional methods such as Southern blotting or DNA cloning.

a. Describe the polymerase chain reaction, including the components and products involved.

b. Describe a specific application of PCR in the cloning or analysis of a prokaryotic or eukaryotic gene. How might the use of PCR in this application replace, augment, speed, improve the sensitivity of, or simplify an alternative method of analysis?

FIGURE 15.D

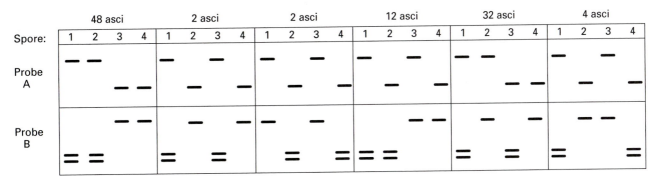

FIGURE 15.E

15.26 One application of DNA fingerprinting technology has been to identify stolen children and return them to their parents. Bobby Larson was taken from a supermarket parking lot in New Jersey in 1978, when he was 4 years old. In 1990, a 16-year-old boy called Ronald Scott was found in California, living with a couple named Susan and James Scott, who claimed to be his parents. Authorities suspected that Susan and James might be the kidnappers, and that Ronald Scott might be Bobby Larson. DNA samples were obtained from Mr. and Mrs. Larson, and from Ronald, Susan, and James Scott. Then DNA fingerprinting was done, using a probe for a particular VNTR family, with the results shown in Figure 15.F. From the information in the figure, what can you say about the parentage of Ronald Scott? Explain.

FIGURE 15.F

16 REGULATION OF GENE EXPRESSION IN BACTERIA AND BACTERIOPHAGES

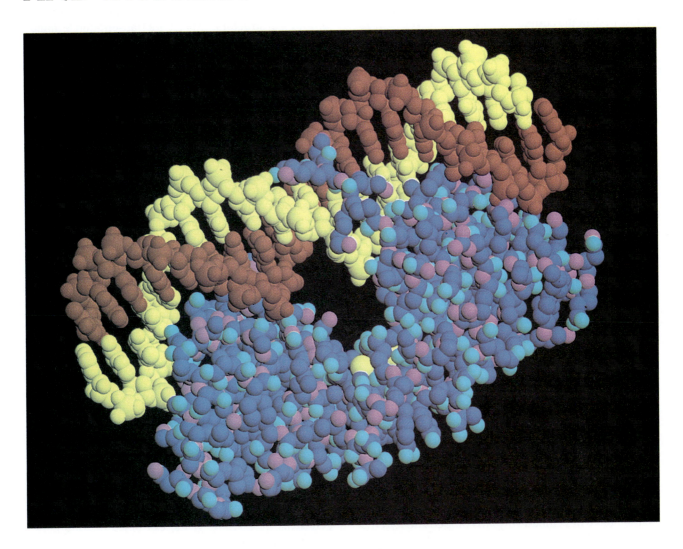

PRINCIPAL POINTS

~ From studies of the synthesis of the lactose-utilizing enzymes of *E. coli* a model was generated that is the basis for the regulation of gene expression in a large number of bacterial and bacteriophage systems. In the lactose system the addition of lactose to the cells brings about a rapid synthesis of three enzymes. In the absence of lactose the synthesis of the three enzymes is turned off. The genes for the enzymes are contiguous on the *E. coli* chromosome and are adjacent to a controlling site (an operator) and a single promoter. The genes, the operator, and the promoter constitute an operon. Transcription of the genes results in a single polygenic mRNA. A regulatory gene is associated with an operon. To turn on gene expression in the lactose system, a lactose metabolite binds with a repressor protein (the product of the regulatory gene), inactivating it and preventing it from binding to the operator. As a result, RNA polymerase can bind to the promoter and transcribe the three genes as a single polygenic mRNA.

~ Expression of a number of bacterial amino acid synthesis operons is controlled by a repressor-operator system and through attenuation at a second controlling site called an attenuator. The repressor-operator system

functions essentially like that for the *lac* operon, except that the addition of amino acid to the cell activates the repressor, thereby turning the operon off. An attenuator is located between the operator region and the first tryptophan structural gene; it is a transcription termination site that allows only a fraction of RNA polymerases to transcribe the rest of the operon. Attenuation involves a coupling between transcription and translation, and the formation of particular RNA secondary structures that signal whether or not transcription can continue.

~ Bacteriophages such as lambda are especially adapted for undergoing reproduction within a bacterial host. Many of the genes related to the production of progeny phages, or to the establishment or reversal of lysogeny of temperate phages, are organized into operons. These operons, like bacterial operons, are controlled through the interaction of regulatory proteins with operators that are adjacent to clusters of structural genes. Phage lambda has been an excellent model for studying the genetic switch that controls the choice between lytic and lysogenic pathways in a lysogenic phage.

O rganisms live and reproduce in changing environments. Through evolutionary processes, they have developed ways to compensate for environmental changes and hence to function in a variety of environments. One way that an organism can adjust to a new environment is to alter its gene activity so that gene products appropriate to the new conditions are synthesized. As a result, the organism is optimally adjusted to grow and reproduce in that environment. This is particularly evident in free-living bacteria.

A change in the set of gene products synthesized in a prokaryotic cell involves regulatory mechanisms that control gene expression. Genes whose activity is controlled in response to the needs of a cell or organism are called *regulated genes*. An organism also possesses a large number of genes whose products are essential to the normal functioning of a growing and dividing cell, no matter what the life-supporting envi-

ronmental conditions are. These genes are always active in growing cells and are known as **constitutive genes** or *housekeeping genes*; examples include genes that code for the enzymes needed for protein synthesis and glucose metabolism. Recognize, though, that all genes are regulated on some level, and that there is a continuum of regulation. If the environment changes suddenly so that it is no longer conducive to normal cell function, the expression of all genes, including constitutive genes, will be reduced by regulatory mechanisms. Thus, the distinction between regulated and constitutive genes is somewhat arbitrary.

In this chapter we examine some of the mechanisms by which gene expression is regulated in bacteria and bacteriophages. In the wild a bacterium lives in a potentially unstable environment that does not ensure a constant food supply or access to the same types of nutrients at all times. To adapt to alterations in their environments, bacteria have evolved several regulatory mechanisms for turning off genes whose products are not needed and for turning on the genes whose products are needed in each environment. In this chapter we learn in detail about some of the basic gene regulation mechanisms in bacteria and how this regulation relates to the organization of genes in the bacterial genome. Studies of bacterial gene regulation have provided important insights into how genes are regulated in higher organisms, including humans.

Since bacteriophages rely on the activities of a host bacterium to reproduce, they must direct their own reproductive cycle in a carefully programmed way; this cycle is directed through the regulation of gene expression. In this chapter we examine specifically the gene regulation events required for controlling the choice between the lytic and lysogenic pathways.

As we discuss the specific examples of gene regulation in bacteria and their viruses, we will learn that turning genes on and off involves specific interactions between regulatory proteins and DNA sequences. We have already seen many examples of the importance of specific protein–nucleic acid interactions to cell function. The long list includes the interaction of DNA polymerase with DNA origins during replication, as well as the binding of RNA polymerase with promoters, restriction enzymes with DNA, and release factors with stop codons on mRNA. And, in Chapter 13, we discussed how the transcription of different sets of *E. coli* genes is regulated by different sigma factors interacting with the RNA polymerase. Thus, the interactions involved in gene regulation are simply further examples of a common theme in cell function. In the next chapter, we see that this theme extends also to gene regulation in eukaryotes.

GENE REGULATION OF LACTOSE UTILIZATION IN *E. COLI*

When gene expression is "turned on" in a bacterium by the addition of a substance (such as lactose) to the medium, the genes are said to be *inducible.* The regulatory substance that brings about this gene induction is called an **inducer.** The inducer is an example of a class of small molecules, called **effectors** or **effector molecules,** that are involved in the control of expression of many regulated genes.

Figure 16.1 diagrams an inducible gene. Transcription of an inducible gene occurs only in response to a particular regulatory event occurring at a specific DNA sequence near the protein-coding sequence. This DNA sequence is called a **controlling site.** The

~ **FIGURE 16.1**

General organization of an inducible gene.

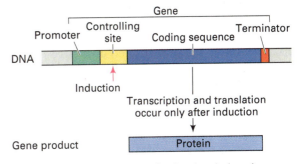

Inducible genes are expressed only when induced

regulatory event typically involves an inducer—a molecule that causes the cell to turn on the synthesis of specific enzymes or nonenzymatic proteins—and a regulatory protein. When the regulatory event occurs, RNA polymerase binds to the promoter (usually adjacent to the controlling site and between the controlling site and the coding sequence) and initiates transcription. The gene is "turned on," mRNA is made, and the protein coded for by the gene is produced. The controlling site itself does not code for any product. The phenomenon of producing a gene product only in response to an inducer is called **induction.** We will now discuss the induction of expression of genes required for lactose utilization in *E. coli.*

Lactose as a Carbon Source for *E. coli*

E. coli is able to grow in a simple medium containing salts (including a nitrogen source) and a carbon source such as glucose. These chemicals provide molecules that can be manipulated by the enzymatic machinery of the cell to produce everything the cell needs to grow and reproduce, such as nucleic acids, proteins, and lipids. The energy for these biochemical reactions in the cell comes from the metabolism of glucose, a process that is of central importance to the functioning of a bacterial cell and of cells of all organisms. The enzymes required for glucose metabolism are coded for by constitutive genes.

If lactose, or one of several other sugars, is provided to *E. coli* as a carbon source instead of glucose, a number of enzymes are rapidly synthesized; these enzymes are needed for the metabolism of this particular sugar. (The same series of events, each involving a "sugar-specific" set of enzymes, would be triggered by other sugars as well.) The enzymes are synthesized because the genes that code for them become actively transcribed in the presence of the sugar; these same genes are inactive if the sugar is absent. In other words, the genes are regulated genes whose products are needed only at certain times.

Biochemical analysis had shown that when lactose (a disaccharide consisting of the two component monosaccharides, glucose and galactose) is the sole carbon source in the growth medium, three proteins are synthesized:

1. β-*galactosidase*. This enzyme has two functions (Figure 16.2): (a) It catalyzes the breakdown of lactose into its two component monosaccharides, glucose and galactose; and (b) it catalyzes the isomerization ("conversion to a different form") of lactose to *allolactose*, a compound important in the regulation of expression of the lactose utilization genes. (In the cell the galactose is converted to glucose through the action of enzymes encoded by a gene system specific for galactose catabolism. The glucose is then utilized by constitutively produced enzymes.)

2. *Lactose permease* (also called *M protein*). This protein is found in the *E. coli* cytoplasmic membrane and is needed for the active transport of lactose from the growth medium into the cell.

3. *Transacetylase*. The function of this enzyme is poorly understood.

In a wild-type *E. coli* that is growing in a medium containing glucose (or another carbon source) but no lactose, only a few molecules of each of the above three enzymes are produced, indicating a low level of expression of the three genes that code for the proteins. For example, only an average of three molecules of β-galactosidase is present in the cell under these conditions. If lactose but no glucose is present in the growth medium, the number of molecules of each of the three enzymes increases coordinately about a thousandfold (e.g., to about 3,000 molecules of β-galactosidase per cell). This occurs because the three essentially inactive genes are now being actively transcribed and the resulting mRNA translated. This

---

~ FIGURE 16.2

Reactions catalyzed by the enzyme β-galactosidase. Lactose brought into the cell by the permease is either converted to glucose and galactose (*top*) or to allolactose (*bottom*), the true inducer for the lactose operon of *E. coli*.

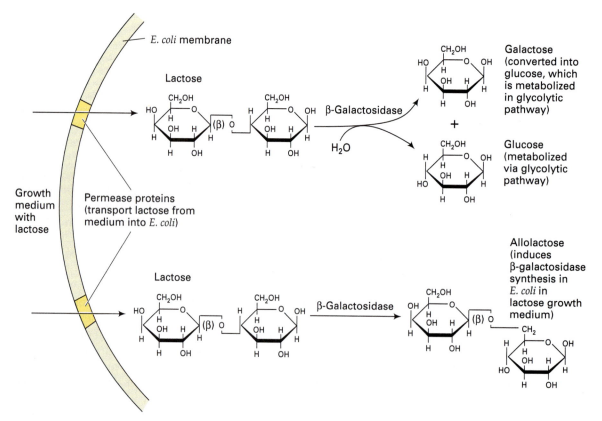

process is called **coordinate induction**. The inducer molecule directly responsible for the increased production of the three enzymes is actually allolactose, not lactose. Allolactose is produced from lactose by β-galactosidase as we have already discussed. Further, the mRNAs for the enzymes have a relatively short half-life, so the transcripts must be made continually in order for the enzymes to be produced. So, when lactose is no longer present, transcription of the three genes is stopped and any mRNAs already present are broken down thereby resulting in a drastic reduction in the amounts of the three proteins in the cell. In the following section, we will describe the key features of this gene regulation system.

Experimental Evidence for the Regulation of the *lac* Genes

Our basic understanding of the organization of the genes, the controlling sites involved in lactose utilization, and the control of expression of the *lac* genes came largely from the genetic experiments of François Jacob and Jacques Monod, for which they received the Nobel Prize.

MUTATIONS IN THE PROTEIN-CODING GENES. Mutations in the protein-coding genes were obtained following treatment with mutagens (chemicals that induce mutations). As a result of these genetic studies, the β-galactosidase gene was named *lacZ*, the permease gene *lacY*, and the transacetylase gene *lacA*. The *lacZ⁻*, *lacY⁻*, and *lacA⁻* mutations obtained were used to map the locations of the three genes using standard genetic procedures. These experiments established that the order of the structural genes in the genome was *lacZ-lacY-lacA* and that the three genes were tightly linked in a cluster.

In Chapter 14 some classes of mutations that affect the reading of the genetic message are discussed. For *lac* genes, missense mutations, which result in the substitution of one amino acid for another in a polypeptide, only affect the function of the product of the gene to which the mutations map. That is, *lacZ* missense mutations that result in a nonfunctional or partially functional β-galactosidase have no effect on permease and transacetylase function. Similarly, a *lacY* missense mutation that results in a nonfunctional or partially functional permease has no effect on β-galactosidase and transacetylase functions, and so on.

It was expected that nonsense (chain-terminating) mutations would act similarly for the *lac* genes. That was not the case. Nonsense mutations mapping in the *lacZ* gene not only resulted in loss of function of β-galactosidase, but also in the loss of permease and transacetylase activities. The *lacY* nonsense mutations resulted in nonfunctional permease and no transacetylase activity, but β-galactosidase activity was unaffected. Lastly, *lacA* nonsense mutants exhibited no transacetylase activity, but rather the normal β-galactosidase and permease activities. That is, nonsense mutations in the cluster of three genes involved in lactose utilization have different effects, depending on where they are located within the cluster: the nonsense mutations are said to exhibit *polar effects*, and the phenomenon is called **polarity**. (Nonsense mutations that show polar effects are often called *polar mutations*.)

The interpretation of the polar effects of nonsense mutations in the *lac* structural genes is that all three genes are clustered on the genome and are transcribed onto a single mRNA molecule—called a **polygenic mRNA** or **polycistronic mRNA**—rather than onto three separate mRNAs. That is, RNA polymerase initiates transcription at a single promoter and a polygenic mRNA is synthesized with the gene transcripts in the order 5'-*lacZ⁺-lacY⁺-lacA⁺*-3'. In translation, a ribosome can only load onto the polygenic mRNA at the 5' end and not at the internally located start codons. The ribosome first synthesizes β-galactosidase, then slides a short distance along the mRNA until it recognizes the initiation sequence for permease, synthesizes permease, slides along the mRNA until it recognizes the initiation sequence for transacetylase, synthesizes transacetylase, and finally dissociates from the mRNA (Figure 16.3a).

A nonsense mutation in the *lacZ* gene exerts its effect on translation of a polygenic mRNA as shown in Figure 16.3b. The ribosome begins to translate the *lacZ* sequence and stops at the premature nonsense codon; a partially completed and nonfunctional β-galactosidase is released. The ribosome continues to slide along the polygenic mRNA but typically dissociates before the start codon for permease because of the greater distance. Thus no ribosomes (or at least, very few ribosomes) translate the downstream permease and transacetylase sequences, so no functional enzymes can be produced. One can see by extension how a *lacY* (permease) nonsense mutation will affect downstream transacetylase production but not affect upstream β-galactosidase translation.

MUTATIONS AFFECTING THE REGULATION OF GENE EXPRESSION. Of special interest to Jacob and Monod were mutations that affected the regulation of all three lactose utilization genes. In wild-type *E. coli* the three gene products are induced coordinately only when lactose is present. Jacob and Monod isolated a number of mutants in which all gene products of the lactose operon were synthesized *constitutively*;

~ **FIGURE 16.3**

Translation of the polygenic mRNA encoded by *lac* utilization genes in (a) Wild-type *E. coli* and (b) A mutant strain with a nonsense mutation in the β-galactosidase (*lacZ*) gene.

a) Wild type

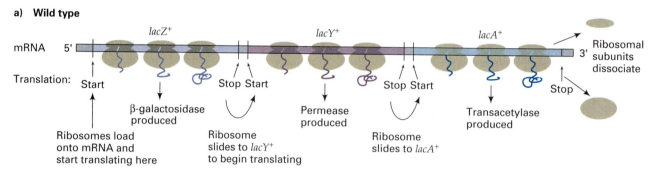

b) *lacZ⁻* nonsense mutant

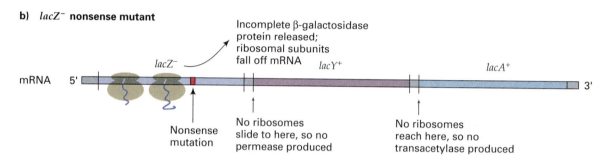

that is, all three enzymes were synthesized whether or not the inducer was present. Jacob and Monod hypothesized that the mutations in the mutant strains were regulatory mutations that affected the normal mechanisms responsible for controlling the expression of the structural genes for the enzymes. Jacob and Monod identified two classes of constitutive mutations: One class mapped to a small DNA region adjacent to the *lacZ* gene that they called the **operator** (*lacO*). The other class mapped to a gene-sized DNA region a short distance away that they called the *lacI* gene or *lac* **repressor** gene. Figure 16.4 diagrams the **organization** of the *lac* structural gene cluster and the associated regulatory elements. We will discuss this complex, called the *lac* operon, a little later.

Operator Mutations.   The mutations of the operator were called operator-constitutive, or *lacO*ᶜ, mutations. All *lacO*ᶜ mutants synthesize the lactose utilization enzymes in the presence or absence of lactose. Through the use of partial diploid strains (i.e., F' strains; see Figure 8.9, p. 233), Jacob and Monod were able to define better the role of the operator in regulating the expression of the *lac* genes. One such partial diploid was $\frac{F' \; lacO^+ \; lacZ^- \; lacY^+}{lacO^c \; lacZ^+ \; lacY^-}$ (both gene sets have a normal promoter, and the lacA gene is omitted because it is not important to our discussions).

One *lac* region in the partial diploid has a normal operator (*lacO⁺*), a mutated β-galactosidase gene (*lacZ⁻*), and a normal permease gene (*lacY⁺*). The other *lac* region has a constitutive operator mutation (*lacOᶜ*), a normal β-galactosidase gene (*lacZ⁺*), and a mutated permease gene (*lacY⁻*). This partial diploid was tested for the production of β-galactosidase (from the *lacZ⁺* gene) and of permease (from the *lacY⁺* gene), both in the presence and the absence of the inducer allolactose.

For this partial diploid, β-galactosidase is synthesized in the absence of inducer, but permease is not. Only when lactose is added to the culture and the allolactose inducer is produced does permease synthesis occur. That is, the *lacZ⁺* gene (which is on the same DNA as *lacOᶜ*) is *constitutively expressed* (i.e., the gene is active in the presence or absence of inducer) whereas the *lacY⁺* gene is under normal inducible control (i.e., *the gene is inactive in the absence of inducer and active in the presence of inducer*). In other words, a *lacOᶜ mutation affects only those genes downstream from it on the same DNA strand.* Similarly, the *lacO⁺* region only controls those *lac* structural genes adjacent to it and has no effect on the genes on the other chromosome strand. This phenomenon of a gene or DNA sequence controlling only genes that are on the same, contiguous piece of DNA is called **cis-dominance**. Thus the *lacOᶜ* mutation is cis-dominant since the defect affects the adjacent genes only and cannot be overcome by a normal *lacO⁺* region elsewhere in the genome.

Jacob and Monod also concluded from these results that the operator region does not produce a diffusible product that functions in the cell. If it did, then in the $lacO^+/lacO^c$ diploid state, one or the other of the regions would have controlled all the lactose utilization genes regardless of their location.

lacI Gene Regulatory Mutations.

The second class of *lac* constitutive mutants defined the *lacI* gene which, we now know, encodes the lactose repressor protein. Again, the use of partial diploid strains illuminated the normal function of the *lacI* gene; namely, encoding a repressor protein that can diffuse through the cell and bind to operator sites.

The partial diploid here is $\dfrac{lacI^+ \ lacO^+ \ lacZ^- \ lacY^+}{lacI^- \ lacO^+ \ lacZ^+ \ lacY^-}$; both gene sets have normal operators and normal promoters. As before, this partial diploid was tested for structural gene expression in the absence and the presence of the inducer. In the absence of inducer, no β-galactosidase or permease was produced; both were synthesized in the presence of inducer. In other words, the expression of both genes was inducible. This means that the $lacI^+$ gene in the cell can overcome the defect of the $lacI^-$ mutation. Since the two lacI genes are located on different chromosomes (that is, they are in a trans configuration) the $lacI^+$ is said to be **trans-dominant** to $lacI^-$.

Since the $lacI^+$ gene controlled the genes on the other chromosome strand, the *lacI* gene must produce a diffusible product; that is, a product that could move through the cell. Jacob and Monod proposed that the $lacI^+$ gene produces a functional **repressor molecule** (hence, the *lacI* gene is also called the *lac* repressor gene) and that no functional repressor molecules are produced in $lacI^-$ mutants. Thus in a haploid bacterial strain that has a $lacI^-$ mutation, the *lac* operon is constitutive. In a partial diploid with both a $lacI^+$ and a $lacI^-$, however, the functional repressor molecules produced by the $lacI^+$ gene control the

expression of both *lac* operons present in the cell, making both operons inducible. As we will soon see, the repressor encoded by the lac^+ gene exerts its regulatory action by binding to the operator.

Promoter Mutations.

The promoter for the structural genes (located at the *lacZ* gene end of the cluster of *lac* genes; see Figure 16.4) is also susceptible to mutation. Most of the known promoter mutants (P_{lac}^-) affect all three structural genes. Even in the presence of inducer, the lactose utilization enzymes are not made–or are made only at very low rates. Since the promoter is the recognition sequence for RNA polymerase and does not code for any product, the effect of a P_{lac} mutation is confined to the genes that it controls on the same DNA strand. The P_{lac}^- mutations are another example of cis-dominant mutations.

Jacob and Monod's Operon Model for the Regulation of the *lac* Genes

Based on the results of their studies of genetic mutations affecting the regulation of the synthesis of the lactose utilization enzymes, Jacob and Monod proposed their now-classical **operon model**. By definition, an **operon** is a *cluster of genes, the expressions of which are regulated together by operator-regulator protein interactions, plus the operator region itself and the promoter.* (Note that the promoter was not part of Jacob and Monod's original model; its existence was demonstrated by later studies.) The order of the controlling elements and genes in the *lac* operon is promoter-operator-lacZ-lacY-lacA, and the regulatory gene *lacI* is located close to the structural genes, just upstream of the promoter (see Figure 16.4). The *lacI* gene has its own promoter and terminator. The promoter for *lacI* is a weak promoter, so relatively few repressor molecules are present in the cell. The *lac* repressor protein encoded by *lacI* is made constitutively but its ability to bind to the operator is affected by the presence of inducer.

~ **FIGURE 16.4**

Organization of the *lac* genes of *E. coli* and the associated regulatory elements, the operator, promoter, and regulatory gene. The complex is called the *lac* operon.

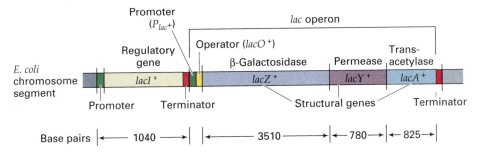

~ **FIGURE 16.5**

Functional state of the *lac* operon in wild-type *E. coli* growing in the absence of lactose.

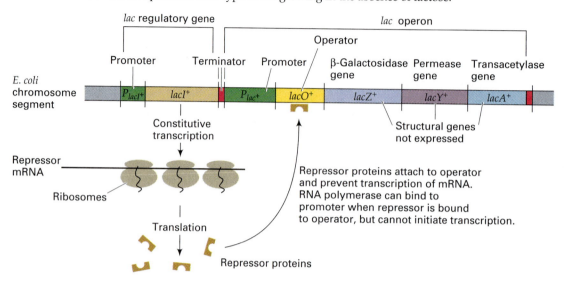

The following description of the Jacob-Monod model for the regulation of the *lac* operon has been embellished with more up-to-date molecular information. Figure 16.5 diagrams the state of the *lac* operon in wild-type *E. coli* growing in the absence of lactose. In this case RNA polymerase molecules bind to the repressor gene (*lacI⁺*) promoter and transcribe the *lacI* gene. Translation produces a polypeptide consisting of 360 amino acids. Four of these polypeptides associate together to form the functional repressor protein.

The functional repressor binds to the operator (*lacO⁺*). When the repressor is bound to the operator, RNA polymerase can bind to the operon's promoter but is blocked from initiating transcription of the protein-coding genes. The *lac* operon is said to be under *negative control* since the binding of the repressor at the operator site blocks transcription of the structural genes. (The low level of transcription of the genes that results in the presence of a few molecules of each protein, even in the absence of inducer, occurs because repressors do not just bind and stay; they bind and unbind. In the split second when one repressor unbinds and before another binds, an RNA polymerase can initiate transcription of the operon, even in the absence of inducer.)

When wild-type *E. coli* grows in the presence of lactose as the sole carbon source (Figure 16.6), some of the lactose transported into the cell is converted by existing molecules of β-galactosidase into a metabolite of lactose called allolactose (see Figure 16.2). (Recall that allolactose, not lactose, is the true inducer

of the *lac* operon protein-coding genes.) Besides having a recognition site for the *lac* operator, the *lac* repressor protein also has a recognition site for allolactose. When allolactose binds to the repressor, it changes the shape of the repressor—this is called an *allosteric shift*. As a result, the repressor loses its affinity for the *lac* operator, and it dissociates from the site. Free repressor proteins are also altered so that they cannot bind to the operator. Thus, the allolactose "induces" the production of the *lac* operon enzymes.

In the absence of repressor, RNA polymerase is now able to bind to the operon's promoter and initiate the synthesis of a single polygenic mRNA molecule that contains the transcripts for the *lacZ⁺*, *lacY⁺*, and *lacA⁺* genes. The polygenic mRNA for the *lac* operon is translated by a string of ribosomes to produce the three proteins specified by the operon. This efficient mechanism ensures the coordinate (simultaneous) production of proteins of related function.

EFFECT OF *lacO^c* MUTATIONS. The *lacO^c* mutations lead to constitutive expression of the *lac* operon genes and are cis-dominant to *lacO⁺*. The explanation of the phenotype of *lacO^c* mutations is that base pair alterations of the operator DNA sequence make it unrecognizable by the repressor protein. Since the repressor cannot bind, the structural genes physically linked to the *lacO^c* mutation become constitutively expressed. This is illustrated in Figure 16.7.

EFFECTS OF *lacI* GENE MUTATIONS. The effect of *lacI⁻* mutations on *lac* operon expression in a haploid

~ FIGURE 16.6

Functional state of the *lac* operon in wild-type *E. coli* growing in the presence of lactose as the sole carbon source.

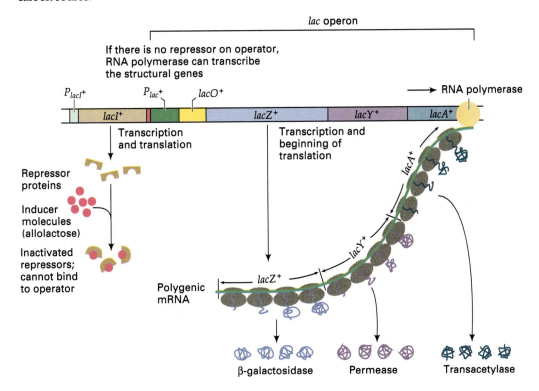

strain is shown in Figure 16.8a. The *lacI* mutations map within the repressor structural gene and result in amino acid changes in the repressor. The repressor's shape is changed, and it cannot now recognize and bind to the operator. As a consequence, transcription cannot be prevented, even in the absence of lactose, and there is constitutive expression of the *lac* operon.

The dominance of the *lacI⁺* (wild-type) gene over *lacI⁻* mutants is illustrated for the partial diploid described earlier, $\dfrac{lacI^+\ lacO^+\ lacZ^-\ lacY^+}{lacI^-\ lacO^+\ lacZ^+\ lacY^-}$ in Figure 16.8b and c. In the absence of the inducer (Figure 16.8b), the defective *lacI* repressor is unable to bind to either normal operator (*lacO⁺*) in the cell. But sufficient normal repressors, produced from the *lacI⁺* gene, are present which bind to the two operators and block transcription of both operons. When inducer is present (Figure 16.8c), the wild-type repressors are inactivated, so both operons are transcribed. One produces a defective β-galactosidase and a normal permease, while the other produces a normal β-galactosidase and a defective permease; between them, active β-galactosidase and permease are produced. Thus, in *lacI⁺/ lacI⁻* partial diploids, both operons present in the cell are under inducible control.

Other classes of *lacI* gene mutants have subsequently been identified since the time Jacob and Monod studied the *lacI* class of mutants. One of these classes, the *lacIˢ* (*superrepressor*) mutants, shows no production of *lac* enzymes in the presence or absence of lactose. In partial diploids with a *lacI⁺/lacIˢ* genotype, the *lacIˢ* allele is trans-dominant, affecting both operon copies. Figure 16.9 (p. 513) diagrams the effect of *lacIˢ* mutations in the partial diploid. The explanation here is that the mutant repressor gene produces a superrepressor protein that can bind to the operator, but cannot recognize the inducer allolactose. Therefore the mutant superrepressors bind to the operators regardless of whether the inducer molecule is present or absent, and transcription of the operons can never occur even in the presence of inducer. The presence of normal repressors in the cell has no effect on this situation, since once a *lacIˢ* repressor is on the operator, the repressor cannot be induced to fall off. Basal levels of transcription will occur since the superrepressor is not permanently (covalently) bound to the operator. Cells with a *lacIˢ* mutation cannot use lactose as a carbon source.

A third type of repressor gene mutation is the *lacI⁻ᵈ* (*dominance*) class. These missense mutations are

~ FIGURE 16.7

Cis-dominant effect of *lacO^c* mutation in a partial-diploid strain of *E. coli*. (a) In the absence of the inducer, the *lacO^+* operon is turned off, while the *lacO^c* operon produces functional β-galactosidase from the *lacZ^+* gene and nonfunctional permease molecules from the *lacY^-* gene. (b) In the presence of the inducer, the functional β-galactosidase and defective permease are produced from the *lacO^c* operon, while the *lacO^+* operon produces nonfunctional β-galactosidase from the *lacZ^-* gene and functional permease from the *lacY^+* gene. Between the two operons in the cell, functional β-galactosidase and permease are produced.

a)   Partial diploid in the absence of inducer

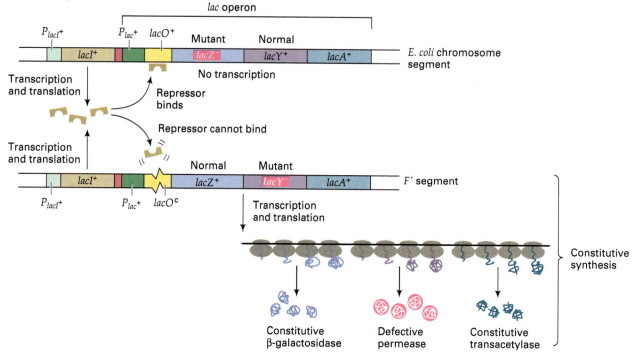

b)   Partial diploid in the presence of inducer

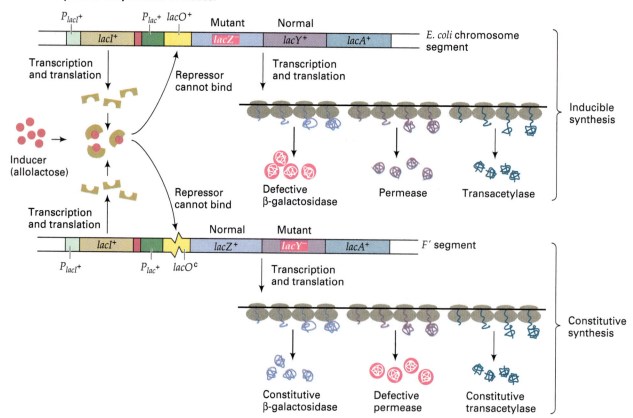

clustered toward the 5' end of the *lacI* gene. In haploid cells the *lacI*$^{-d}$ mutants have a constitutive phenotype like the other *lacI*$^-$ mutants; the *lac* enzymes are made in the presence or absence of lactose. Unlike the *lacI*$^-$ mutations, the *lacI*$^{-d}$ mutations are trans-dominant to *lacI*$^+$ in *lacI*$^{-d}$/*lacI*$^+$ partial diploids, so *lac* enzymes are produced constitutively even in the presence of the *lacI*$^+$ gene.

The dominance of *lacI*$^{-d}$ mutants relates to the structure of the *lac* repressor. The repressor protein is a tetramer consisting of four identical polypeptides. There are only about a dozen repressor molecules in the cell. In the *lacI*$^{-d}$ mutants, the repressor subunits do not combine normally, so no complete repressor is formed, and no operator-specific binding is possible. The *lacI*$^{-d}$/*lacI*$^+$ diploids have a mixture of normal and mutant polypeptides, which combine randomly to form repressor tetramers. The presence of one or more defective polypeptides in the repressor is enough to block normal binding to the operator. So, there is a good chance that no normal repressor proteins will be produced, since there are so few molecules per cell. As a result of the absence (or virtual absence) of complete, functional repressors, a constitutive enzyme phenotype results.

Lastly, some mutations in the repressor gene promoter affect the expression of the repressor gene itself. We mentioned earlier that the extent of transcription of a gene is a function of the affinity of that gene's promoter for RNA polymerase molecules. Clearly, since relatively few repressor molecules are synthesized in the wild-type *E. coli* cells, the repressor gene promoter must be of low affinity. As in any other region of the DNA, the repressor gene promoter is subject to base pair changes by mutation. Base pair mutations have been found that decrease and that increase transcription rates. (Remember that the repressor gene is a constitutively expressed gene under promoter control.)

~ FIGURE 16.8

Effect of a *lacI*$^-$ mutation: (a) On *lac* operon expression in a haploid cell, where mutant, inactive repressor molecules that cannot bind to the operator *lacO*$^+$ are produced, so the structural genes are transcribed constitutively. In a partial diploid strain *lacI*$^+$ *lacO*$^+$ *lacZ*$^-$ *lacY*$^+$/ *lacI*$^-$ *lacO*$^+$ *lacZ*$^+$ *lacY*$^-$ in (b) the absence or (c) the presence of inducer (see p. 512).

a) Haploid strain (in presence or absence of inducer)

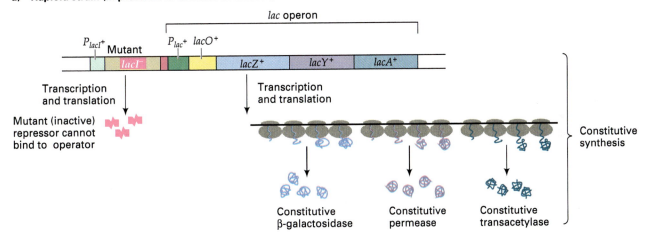

b) Partial diploid in the absence of inducer

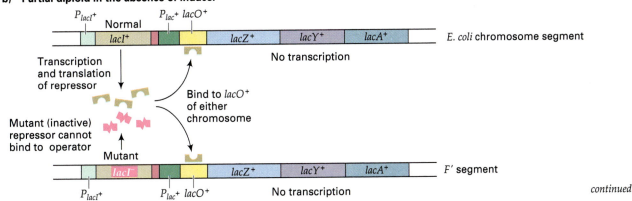

continued

~ FIGURE 16.8 continued

c) Partial diploid in the presence of inducer

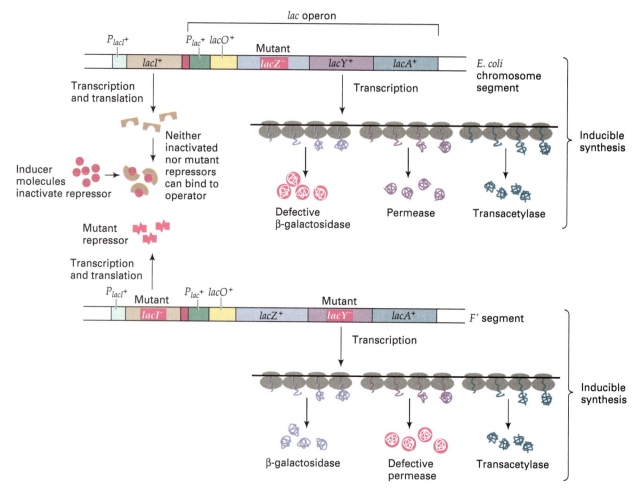

The most useful mutants in this regard are *lacI*^{Q} and *lacI*^{SQ} mutants (where Q stands for "quantity" and SQ for "superquantity"). Both mutations result in an increase in the rate of transcription of the repressor gene, with the *lacI*^{SQ} mutants giving the greater increase. These mutants were useful historically because they produce large numbers of repressor molecules, which facilitated their isolation and purification and, consequently, the determination of the amino acid sequence of the constituent polypeptide subunit. Since *lacI*^{Q} and *lacI*^{SQ} mutants produce more repressor molecules than the wild type, these mutants reduce the efficiency of induction of the *lac* operon. These mutants can be induced, however, at very high lactose concentrations.

The mutants of the *lacI* gene point out the functions of the product of that gene, the repressor. Specifically, the repressor is involved in three recognition interactions, any one of which can be affected by mutation: (1) binding of the repressor to the operator region; (2) binding of the inducer to the repressor; and (3) binding of individual repressor polypeptides

to each other to form the active repressor tetramer. The repressor tetramer is shown in Figure 16.10.

Positive Control of the *lac* Operon

In the *lac* operon the repressor protein exerts a negative effect on the expression of the *lac* operon by blocking RNA polymerase's action if the inducer is absent. Thus, the *lac* operon is said to be under *negative control*. Several years after Jacob and Monod proposed their operon model, researchers also found a positive control system that regulates the *lac* operon, a system that functions to turn on the expression of the operon. This system ensures that the *lac* operon will be expressed only if lactose is the sole carbon source *but not if glucose is present as well*.

If both glucose and lactose are present in the medium, the glucose is used preferentially and the *lac* operon is not expressed. The *lac* operon is repressed under these conditions because the concentration of a *positive regulator* that binds to the *lac* operon to make transcription possible is reduced in the presence of

~ FIGURE 16.9

Dominant effect of *lacI*ˢ mutation over wild-type *lacI⁺* in a *lacI⁺ lacO⁺ lacZ⁺ lacY⁺ lacA⁺/lacIˢ lacO⁺ lacZ⁺ lacY⁺ lacA⁺* partial-diploid cell growing in the presence of lactose.

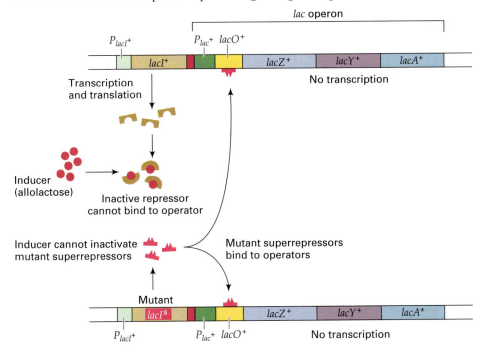

glucose. Figure 16.11 shows the mechanism involved. This process is called **catabolite repression (glucose effect)**, and functions as follows: First, a protein called *CAP* (catabolite gene activator protein) binds with *cAMP* (cyclic AMP, or cyclic adenosine 3',5'-monophosphate; see Figure 16.12), to form a CAP-cAMP complex. This complex is the positive-regulator molecule. The CAP protein itself is a dimer that consists of two identical polypeptides. Next, the CAP-cAMP complex binds to a specific site in the DNA, called the *CAP site*, which is located upstream of the promoter site. This binding causes the DNA to bend (see Chapter 10, p. 309). DNA bending facilitates protein-protein interactions between the activating domain of CAP and target sites on RNA polymerase, which lead to the activation of transcription.

When glucose is present in the medium, catabolite repression occurs. In catabolite repression, the expression of the lactose operon is turned off even if lactose is present in the medium. This occurs because the presence of glucose causes the amount of cAMP in the cell to be greatly reduced. As a result, insufficient CAP-cAMP complex is available to facilitate RNA polymerase binding to the *lac* operon promoter,

~ FIGURE 16.10

Molecular model of the *lac* repressor tetramer. The four individual monomers are colored green, violet, red, and yellow.

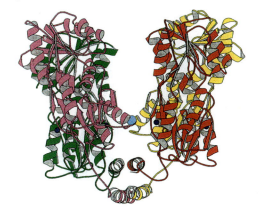

and transcription is lowered significantly, even though repressors are removed from the operator by the presence of allolactose. In other words, RNA polymerase cannot bind to the promoter without the aid of the CAP-cAMP complex. That cAMP plays a crucial role in catabolite repression was shown by a number of experiments, including one in which transcription of the *lac* operon was restored by the addition of cAMP to the cell, even though glucose was still present.

The model is that catabolite repression acts on adenylate cyclase, the enzyme that makes cAMP in the cell (see Figure 16.12). In *E. coli*, adenylate cyclase is activated by the phosphorylated form of an enzyme called III^{Glc}. When glucose is transported across the cell membrane into the cell, it triggers a series of events that includes the dephosphorylation of III^{Glc}. As a result, adenylate cyclase is inactivated and no new cAMP is produced. This, along with the break-

down of cAMP by phosphodiesterase, reduces the level of cAMP in the cell.

Catabolite repression occurs in the same way in a number of other bacterial operons related to catabolism of sugars other than glucose. These operons all have in common a CAP site in their promoters to which a specific CAP-cAMP complex binds to facilitate RNA polymerase binding.

Molecular Details of *lac* Operon Regulation

From DNA- and RNA-sequencing experiments much information is now available concerning the nucleotide sequences of the significant *lac* operon regulatory sequences. One general approach to obtain this information has been to purify the protein known to bind to a regulatory site and to let it bind to isolated *lac* operon DNA *in vitro*. For example, if the repressor

~ **FIGURE 16.11**

Role of cyclic AMP (cAMP) in the functioning of glucose-sensitive operons such as the lactose operon of *E. coli*.

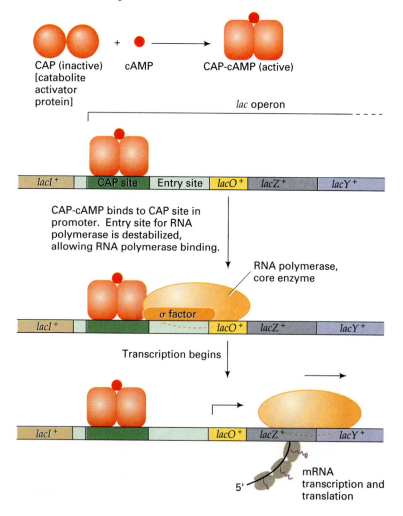

~ FIGURE 16.12

Structure of cyclic AMP (cAMP, or cyclic adenosine 3',5' monophosphate). cAMP is synthesized from ATP in a reaction catalyzed by adenylate cyclase, and it is broken down in a reaction catalyzed by phosphodiesterase.

is bound to the *lac* operator, it will protect that region of the operon from deoxyribonuclease digestion. If DNase is allowed to digest the rest of the DNA, the operator sequence can be isolated, cloned by using recombinant DNA technology, and sequenced.

PROMOTER REGION OF THE *lac* REPRESSOR GENE (*lacI*). Figure 16.13 shows the nucleotide pair sequence for the *lacI* gene promoter region, the sequence for the 5' end of the repressor mRNA, and the first few amino acids of the repressor protein itself. The nucleotide sequence of the repressor mRNA can be aligned with this promoter sequence, with its start approximately in the middle. As usual with all gene transcripts, translation does not start right at the end of the mRNA molecule: The translation start sequence (ribosome binding site) is a Shine-Dalgarno sequence (AGGG) at 12 to 9 bases upstream from the start codon (Chapter 14). In this unusual case, the start codon is GUG rather than AUG at nucleotides 27 to 29 from the 5' end of the messenger. Figure 16.13 also shows the single base pair change found for a particular *lacI*Q mutant; this change, from CG to TA, brings about a tenfold increase in repressor production.

***lac* OPERON CONTROLLING SITES.** Figure 16.14 shows the nucleotide pair sequence of the *lac* operon controlling sites. The orientation of this sequence was put together from several different pieces of information. First, the amino acid sequences of the repressor protein and of β-galactosidase are completely known, and this information enables us to identify the coding regions of the *lacI* gene and of the *lacZ*⁺ gene. Then the other regions were identified on the basis of "protection" experiments of the kind described previously. Here CAP-cAMP complex, RNA polymerase, and repressor protein were used separately to bind to the DNA, and DNase-resistant regions were then sequenced. (Although studies indicate the results that follow, we should perhaps be cautious before we defi-

~ FIGURE 16.13

Base-pair sequences of the *lac* operon *lacI*⁺ gene promoter (P_{lac+}) and of the 5' end of the repressor mRNA. Also shown is the amino acid sequence of the first part of the repressor protein itself. Note that GUG is the initiation codon for methionine in this case.

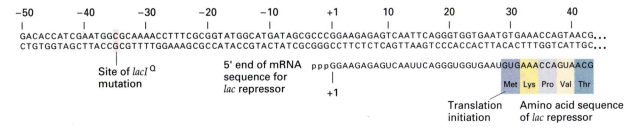

nitely assume that the discrete boundaries between the various parts of the controlling region are real.)

The beginning of the promoter region is defined as position –84 in the figure (i.e., 84 base pairs upstream from the mRNA initiation site), immediately next to the stop codon for the *lacI* gene. The consensus sequence matches for the CAP-cAMP binding site are nucleotide pairs –54 to –58, and –65 to –69, and the polymerase binding site spans nucleotide pairs –47 to –8, including –10 and –35 consensus sequence matches (see Figure 16.11). Together, the region from –8 to –84 that includes the CAP protein and the RNA polymerase interaction sites (including a Pribnow box) essentially define the *lac* operon promoter region.

Immediately next to the promoter region is the operator. The region protected by the repressor protein is the area containing nucleotide pairs –3 to +21. When the repressor is bound to the operator, RNA polymerase can bind to the DNA, but it cannot transcribe the genes.

The β-galactosidase mRNA has a *leader region* before the start codon is encountered. The actual start of the mRNA here is nucleotide pair +1 in Figure 16.14, which is very close to the beginning of the repressor binding site. Transcription of the *lac* operon includes a large proportion of the operator region in addition to the protein-coding genes themselves. The AUG start codon for β-galactosidase, which defines the beginning of the *lacZ* coding sequence, is at

nucleotide pairs +39 to +41. Thus the first 38 bases of the *lac* mRNA are not translated.

Figure 16.14 also shows the sites of base pair substitutions that have been identified for some of the *lacO*c mutations that have been studied. In each case a single base pair change is responsible for the altered regulatory control of the *lac* operon.

In conclusion, the lactose operon has proved to be a model system for understanding gene regulation in prokaryotic organisms. Jacob and Monod's original work on this system had a great impact on further studies. As the first molecular model for the regulation of gene expression in any organism, it sparked numerous studies in both prokaryotes and eukaryotes to see whether operons were generally the case.

𝒦EYNOTE

Studies of the synthesis of the lactose-utilizing enzymes of *E. coli* generated a model that is the basis for the regulation of gene expression in a large number of bacterial and bacteriophage systems. In the lactose system the addition of lactose to the cells brings about a rapid synthesis of three enzymes. The genes for these enzymes are contiguous on the *E. coli* chromosome and are adjacent to a

~ **FIGURE 16.14**

Base pair sequence of the controlling sites, promoter and operator, for the lactose operon of *E. coli*. Also shown are locations of some known *lacO*c mutations (indicated by arrows).

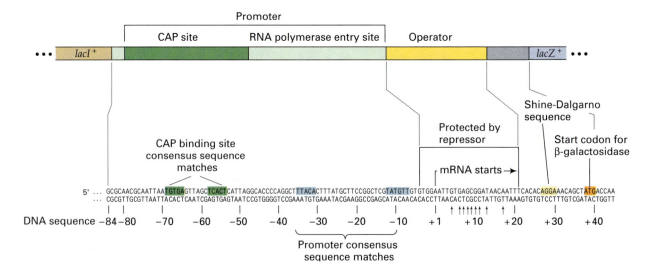

controlling site (an operator) and a single promoter. The genes, the operator, and the promoter constitute an operon, which is transcribed as a single unit. In the absence of lactose the operon is turned off by a repressor.

A positive control system also regulates the *lac* operon. That is, CAP-cAMP binds to the promoter and this facilitates binding of RNA polymerase to the promoter. If glucose is present, however, no CAP-cAMP is produced so RNA polymerase cannot bind and the *lac* genes are not transcribed.

TRYPTOPHAN OPERON OF *E. COLI*

Just as glucose may not always be available in a bacterial growth medium for use as a carbon source, all necessary amino acids may not be present in a growth medium to enable bacteria to assemble proteins. If an amino acid is missing, a bacterium has certain operons and other gene systems that enable the bacterium to manufacture that amino acid so that it may grow and reproduce. As we learned in Chapter 9, each step in the biosynthetic pathway through which amino acids are assembled is catalyzed by a specific enzyme coded by a specific gene or genes.

When an amino acid is present in the medium, the genes encoding the enzymes for that amino acid's biosynthetic pathway are turned off. If an amino acid is not present in the medium, however, those genes must be turned on in order for the biosynthetic enzymes to be made. Unlike the *lac* operon, where gene activity is induced when a chemical (lactose) is added to the medium, in this case there is a repression of gene activity when a chemical (an amino acid) is added. We refer to amino acid biosynthesis operons controlled in this way as *repressible operons*. In general, repressible operons function to turn off anabolic (biosynthetic) pathways when the end product is readily available. One repressible operon in *E. coli* that has been extensively studied is the operon for the biosynthesis of the amino acid tryptophan (Trp). Although the regulation of the *trp* operon shows some basic similarities to the regulation of the classical *lac* operon, some intriguing differences also appear to be common among similar, repressible, amino acid biosynthesis operons in bacteria.

Gene Organization of the Tryptophan Biosynthesis Genes

Figure 16.15 shows the organization of the controlling sites and of the genes that code for the tryptophan biosynthetic enzymes and how they relate to the biosynthetic steps. Much of the work that we will discuss is that of Charles Yanofsky and his collaborators.

Five structural genes (A–E) occur in the tryptophan operon. The promoter and operator regions are closely integrated in the DNA and are upstream from the *trpE* gene. Between the promoter-operator region and *trpE* is a 162–base pair region called *trpL*, or the leader region. Within *trpL*, relatively close to *trpE*, is an *attenuator site* (*att*) that plays an important role in the regulation of the tryptophan operon, as we will see later.

The entire tryptophan operon is approximately 7,000 base pairs long. Transcription of the operon results in the production of a polygenic mRNA for the five structural genes. Each of these transcripts is translated to an equal extent.

Regulation of the *trp* Operon

Two regulatory mechanisms are involved in controlling the expression of the *trp* operon. One mechanism uses a repressor/operator interaction, and the other determines whether initiated transcripts include the structural genes or are terminated before those genes.

EXPRESSION OF THE *trp* OPERON IN THE PRESENCE OF TRYPTOPHAN. The regulatory gene for the *trp* operon is *trpR*; it is located at some distance from the operon (and therefore does not appear in Figure 16.15). The product of *trpR* is an *aporepressor protein*, which alone cannot bind to the operator. When tryptophan is abundant in the growth medium, it binds to the aporepressor and converts it to an active repressor. (Tryptophan is an example of an *effector* molecule, just as allolactose is the effector molecule for the *lac* operon.) The active repressor binds to the operator and prevents the initiation of transcription of the *trp* operon protein-coding genes by RNA polymerase. As a result, the tryptophan biosynthesis enzymes are not produced. By repression, transcription of the *trp* operon can be reduced about seventyfold.

EXPRESSION OF THE *trp* OPERON IN THE ABSENCE OF TRYPTOPHAN OR IN THE PRESENCE OF LOW AMOUNTS OF TRYPTOPHAN. The second regulatory mechanism is involved in the expression of the *trp* operon under conditions of tryptophan starvation or tryptophan limitation. Under severe tryptophan star-

~ FIGURE 16.15

Organization of controlling sites and the structural genes of the *E. coli* tryptophan operon. Also shown are the steps catalyzed by the products of the structural genes *trpA, trpB, trpC, trpD,* and *trpE.*

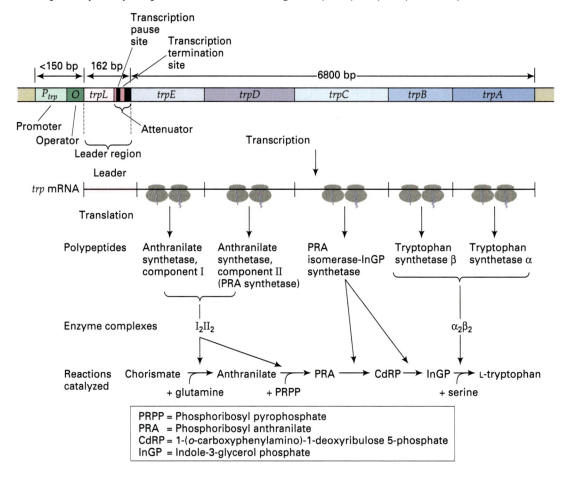

vation, the *trp* genes are expressed maximally, while under less severe starvation conditions, the *trp* genes are expressed at less than maximal levels. This is accomplished by a mechanism that controls the ratio of the transcripts that include the five *trp* structural genes to those that are terminated before the structural genes. That is, some RNA transcripts are of the complete operon—the leader region plus the structural genes—while others are short, 140-bp transcripts that have terminated at the attenuator site within the *trpL* region (see Figure 16.15). The short transcripts have terminated by a process called **attenuation**. The proportion of the transcripts that include the structural genes is inversely related to the amount of tryptophan in the cell—the more tryptophan there is, the greater the proportion of short transcripts. Attenuation can reduce transcription of the *trp* operon by eightfold to tenfold. Thus repression and attenuation together

can regulate the transcription of the *trp* operon over a range of about 560-fold to 700-fold.

MOLECULAR MODEL FOR ATTENUATION. The entire DNA sequence of the *trp* operon is known. Analysis of the sequence of the leader region indicated that a transcript of that region could be translated to produce a short polypeptide. Just before the stop codon of the transcript are two adjacent codons for tryptophan. As we will see, these Trp codons play an important role in attenuation.

There are four regions of the leader peptide mRNA that can form secondary structures by complementary base pairing (Figure 16.16). Pairing of regions 1 and 2 results in a *pause signal*, pairing of 3 and 4 is a *termination of transcription signal* (similar to the *rho*-independent signal for termination of transcription; see Chapter 13, pp. 388–389), and pairing of

~ FIGURE 16.16

Four regions of the tryptophan operon leader mRNA and the alternate secondary structures they can form by complementary base pairing.

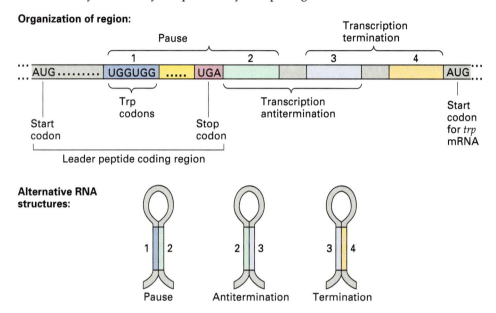

2 and 3 is an *antitermination signal* for transcription to continue. The role of these signals will now be explained.

Crucial to the attenuation model is the fact that transcription and translation are tightly coupled in prokaryotes. (This is possible because of the absence of a nuclear envelope and mRNA molecules in bacteria do not undergo the extensive processing characteristic of eukaryotic mRNAs.) In the *trp* regulatory system, translation of the leader mRNA transcript is occurring just behind the point at which RNA polymerase is at work extending the transcript. This coupling of transcription and translation is brought about by a pause of the RNA polymerase caused by the pairing of RNA regions 1 and 2 just after they have been synthesized (see Figure 16.16). The pause lasts long enough for the ribosome to load onto the mRNA and to begin translating the leader peptide, and this results in the tight coupling of transcription and translation.

As coupled transcription/translation continues, the position of the ribosome on the leader transcript plays an important role in the regulation of transcription termination at the attenuator. If the cells are starved for tryptophan, then the amount of Trp-tRNA.Trp molecules (i.e., charged tryptophanyl-tRNA) drops dramatically, because very few tryptophan molecules are available for the aminoacylation of the tRNA.Trp molecules. As a result, a ribosome

translating the leader transcript stalls at the tandem Trp codons in region 1 because the next specified amino acid in the peptide cannot be added to the growing chain; the leader peptide cannot be completed (Figure 16.17a). Since the ribosome now "covers" region 1 of the attenuator region, the 1-2 stem-loop cannot form because region 1 is no longer available. However, RNA region 2 will pair with RNA region 3 once region 3 is synthesized. Because region 3 is paired with region 2, region 3 cannot then pair with region 4 when it is synthesized. The 2:3 pairing is an antitermination signal because the termination signal of 3 paired with 4 does not form, and this allows RNA polymerase to continue past the attenuator and transcribe the structural genes.

If, instead, sufficient tryptophan is present so that the ribosome can translate the Trp codons (Figure 16.17b), the ribosome stalls before the stop codon for the leader peptide. Since the ribosome is then covering part of RNA region 2, region 2 is unable to pair with region 3, and region 3 is then able to pair with region 4 when it is synthesized. The bonding of region 3 with region 4 is a termination signal for RNA polymerase to stop transcription. The 3:4 structure is referred to as the attenuator. The key signal for attenuation is the level of Trp-tRNA.Trp in the cell, since that determines where the ribosome will stall on the leader transcript; i.e., either at the two Trp codons or at the stop codon.

~ Figure 16.17

Models for attenuation in the tryptophan operon of *E. coli;* the taupe structures are ribosomes that are translating the leader transcript: (a) Tryptophan-starved cells; (b) Non–tryptophan-starved cells.

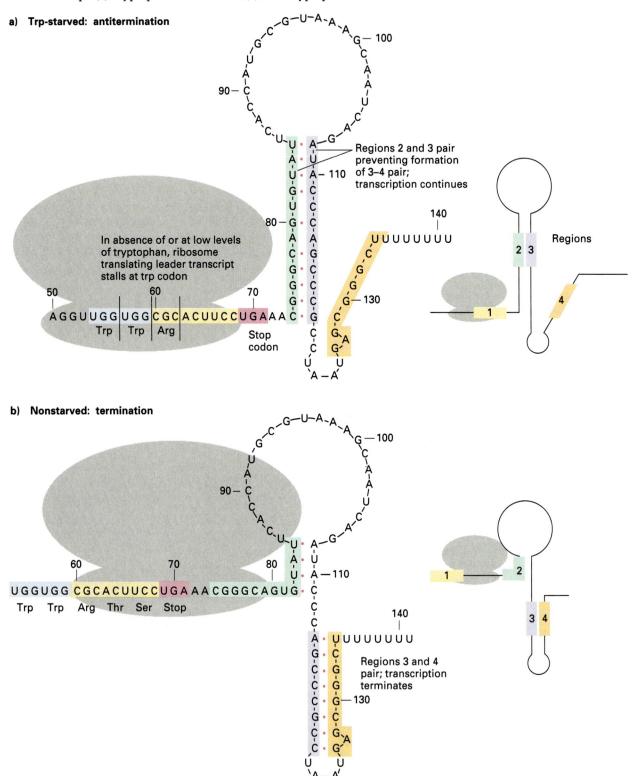

Genetic evidence for the attenuation model has been obtained through the study of mutants. One type of mutant shows less efficient transcription termination at the attenuator, increasing structural gene expression. The mutations involved are single base-pair changes leading to the base changes in the leader transcript shown in Figure 16.18. In each case the change is in the regions of 3:4 pairing; each causes a disruption of the pairing so that the structure is less stable. In the less stable state, the structure is less able to prevent transcription from proceeding into the structural genes.

Further direct evidence for the attenuation model came from DNA manipulations in which the DNA sequences for the two Trp codons were changed to encode another amino acid. In those mutant strains, attenuation was *not* seen in response to changing levels of tryptophan, but it *was* seen in response to changing levels of the amino acid now specified by the codons.

FINE-SCALE REGULATION OF THE *trp* OPERON. Everything we have said so far relates to regulation at the gene expression level in the *trp* operon. One other mechanism operates in addition to this regulation to give an overall regulatory system for gene expression. This mechanism is the degradation of the polygenic mRNA transcript of the protein-coding genes. As is the case with the lactose operon and with most other prokaryotic genes, the transcripts have a relatively short half-life, so mRNA must continually be made in order for the enzymes to be produced. The short half-life allows the cell to change its physiological state when the environment changes. When tryptophan is added to the medium, transcription termination at the attenuator and repression at the operator block further transcription of the structural genes, and the structural gene mRNAs already made are rapidly degraded ("turned over"). In this way the cell does not make unnecessary materials.

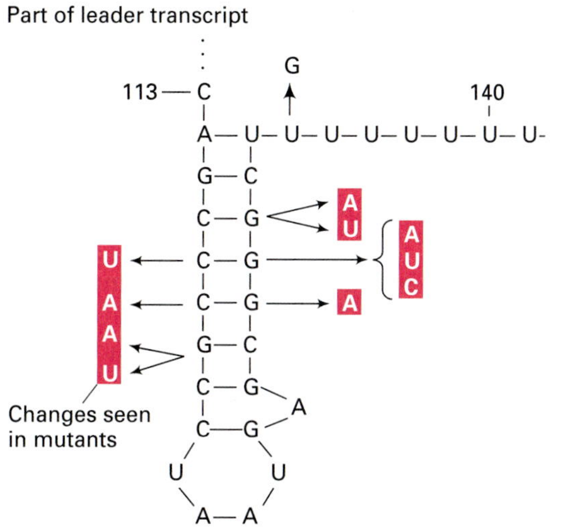

Another part of the overall regulatory system is a fine-tuning system. Gene regulation (turning genes on and off) involves control processes that function in response to long-term changes in the environment. Cells must also respond to short-term changes that occur. If, for example, tryptophan being made by the tryptophan biosynthetic pathway accumulates at a rate faster than it is being used, a mechanism called *feedback inhibition* (also called end-product inhibition) is triggered.

In feedback inhibition the end product of a biosynthetic pathway can often be recognized by the first enzyme in the biosynthetic pathway. When too much tryptophan is produced, for example, the tryptophan binds to the first enzyme in the tryptophan biosynthetic pathway, thereby altering the enzyme's three-dimensional conformation. (The technical term for this configurational change is *allosteric shift*, and the phenomenon is called *allostery*. The change in properties of the *lac* repressor in the presence of lactose is also the result of an allosteric shift.) When the shape of the enzyme is changed, the function of the enzyme is impaired, because the substrate for the enzyme can no longer bind to the enzyme. As a result, the tryptophan pathway is temporarily turned off at the first step in the biosynthetic pathway. The enzyme remains nonfunctional until the tryptophan level in the cell drops and the tryptophan dissociates from the enzyme. This step reverses the feedback inhibition of the enzyme's function, and the pathway is restarted. This feedback mechanism is a common attribute of many biosynthetic pathways in cells of all kinds (both prokaryotic and eukaryotic), serving to halt the flow of intermediates in a pathway.

Regulation of Other Amino Acid Biosynthesis Operons

Attenuation has been shown to be involved in the genetic regulation of a number of other amino acid biosynthetic operons of *E. coli* and of *Salmonella typhimurium*. The molecular information that has been gathered for those operons indicates that the attenua-

~ **FIGURE 16.18**

In the *trpL* region, mutation sites that show less efficient transcription at the attenuator site. The mutations map to DNA regions that correspond to regions 3 and 4 in the RNA.

Part of leader transcript

~ **FIGURE 16.19**

Predicted amino acid sequences of the leader peptides of a number of attenuator-controlled bacterial operons. Shown are the peptides for the *phe*A, *his*, *leu*, *thr*, and *ilv* (isoleucine and valine) operons of *E. coli* or *Salmonella typhimurium*. The amino acids that regulate the respective operons are highlighted in yellow.

pheA: Met – Lys – His – Ile – Pro – Phe – Phe – Phe – Ala – Phe – Phe – Phe – Thr – Phe – Pro – –

his: Met – Thr – Arg – Val – Gln – Phe – Lys – His – His – His – His – His – His – His – Pro – Asp – –

leu: Met – Ser – His – Ile – Val – Arg – Phe – Thr – Gly – Leu – Leu – Leu – Leu – Asn – Ala – Phe –
   Ile – Val – Arg – Gly – Arg – Pro – Val – Gly – Ile – Gln – His – –

thr: Met – Lys – Arg – Ile – Ser – Thr – Thr – Ile – Thr – Thr – Thr – Ile – Thr – Ile – Thr – Thr –
   Gly – Asn – Gly – Ala – Gly – –

ilv: Met – Thr – Ala – Leu – Leu – Arg – Val – Ile – Ser – Leu – Val – Val – Ile – Ser – Val – Val –
   Val – Ile – Ile – Ile – Pro – Pro – Cys – Gly – Ala – Ala – Leu – Gly – Arg – Gly – Lys – Ala – –

tion mechanism is very similar in each case. That is, in every case a leader sequence exists in which there are two or more codons for the particular amino acid, the synthesis of which is controlled by the enzymes encoded by the operon (Figure 16.19). For example, the histidine operon of *E. coli* has a string of 7 histidines in the leader peptide, and 7 of the 15 amino acids in the leader for the phenylalanine A operon of *E. coli* are phenylalanine. Interestingly, some of the operons use only attenuation and not repression. For example, no histidine repressor protein, or any mutation that gives a phenotype that we would expect of an altered repressor protein, has been found for the histidine operon of *S. typhimurium*.

Attenuation has also been shown to regulate a number of genes not involved with amino acid biosynthesis—for example, the ribosomal protein gene S10, rRNA operons (*rrn*), and the *ampC* gene of *E. coli* (for ampicillin resistance), as well as some animal virus and animal cell genes. Among all these examples, a number of attenuation mechanisms have been found to be involved.

located in the leader region between the operator region and the first *trp* structural gene. The attenuator acts to terminate transcription dependent upon the concentration of tryptophan. In the presence of large amounts of tryptophan, attenuation is very effective; that is, enough Trp-tRNA.Trp is present so that the ribosome can move past the attenuator and allow the leader transcript to form a secondary structure that causes transcription to be blocked. In the absence of or at low amounts of tryptophan the ribosomes stall at the attenuator, and the leader transcript forms a secondary structure that permits continued transcription activity.

SUMMARY OF OPERON FUNCTION

There are numerous examples of operons in bacteria and their bacteriophages. The following generalizations can be made about the regulation of operons:

1. For most, but not all operons, a regulator protein (e.g., the *lac* repressor) plays a key role in the regulation process since it is able to bind to a controlling site in the operon, the operator.
2. The transcription of a set of clustered structural genes is controlled by an adjacent operator through interaction with a regulator protein

𝒦EYNOTE

Regulation of the tryptophan (*trp*) operon of *E. coli* is at the level of initiating or completing a transcript of the operon. This is accomplished through a repressor-operator system, which responds to free tryptophan levels, and through attenuation at a second controlling site called an attenuator, which responds to Trp-tRNA.Trp levels. The attenuator is

which can exert positive or negative control depending upon the operon.

3. The trigger for changing the state of an operon from off to on and vice versa is an effector molecule (e.g., the inducer allolactose in the case of the *lac* operon), which controls the conditions under which the regulator protein will or will not bind to the operator.

It should be pointed out, however, that not in all cases are genes for a related function clustered in the prokaryotic genome.

GENE REGULATION IN BACTERIOPHAGES

Because bacteriophages exist by parasitizing bacteria, they themselves do not need to code for the required enzymes for all the biosynthetic pathways needed to make progeny bacteriophages. Instead, many or all the essential components for phage reproduction are provided by the host cell, and the use of those components is controlled by the products of phage genes. Most genes of a phage, then, code for products that control the life cycle and the production of progeny phage particles. Much is known about gene regulation in a number of bacteriophages. In this section we will discuss the regulation of gene expression as it relates to the lytic cycle and lysogeny in bacteriophage lambda (λ).

Regulation of Gene Expression in Phage Lambda

As we learned in Chapter 8, phage lambda is a temperate phage. In this section we examine the regulatory mechanisms that determine whether a λ phage enters the lytic or lysogenic cycle. This system is an excellent model for a developmental switch and as such has contributed to our thinking about how developmental switches operate in eukaryotic systems (see Chapter 17).

FUNCTIONAL ORGANIZATION OF THE λ GENOME. Figure 16.20 shows the genetic map of the λ phage. The mature λ chromosome is linear, and has complementary "sticky" ends. Once free in the host cell, the λ chromosome circularizes. Thus it is conventional to show the genetic map in a circular form.

In λ, genes with related function are clustered in the genome. The genes for DNA replication that are active during the lytic cycle are clustered in the early right operon at about 1 o'clock, whereas the genes for lysogeny (those controlling integration and excision of the λ chromosome) are clustered in the early left operon at 10 to 12 o'clock. The genes for the various structural components of the phage particles (active only in the lytic cycle)—namely, the heads and the tails—are clustered together in the late operon (from 2 to 7 o'clock).

EARLY TRANSCRIPTION EVENTS. Soon after λ infects *E. coli*, a choice is made between the lytic and lysogenic pathways. This decision depends upon a sophisticated *genetic switch* which involves competition between the products of the *cI* gene (the repressor) and the *cro* gene (the Cro protein). (The *cI*, *cII*, and *cIII* genes were named because mutants of these genes produce a clear plaque phenotype.) If the repressor dominates, the lysogenic pathway is followed; if the Cro protein dominates, the lytic pathway is followed. When the lysogenic pathway is followed, the λ chromosome integrates into the *E. coli* chromosome at a specific site and no progeny phages are produced (see Chapter 8). In this integrated, prophage state, the lytic pathway genes are repressed and the λ genome replicates only when the *E. coli* chromosome replicates. When the lytic pathway is followed, progeny phages are assembled, the bacterial cells are lysed, and the phages are released.

Different genes are expressed when λ follows the lytic or the lysogenic pathway. When the λ chromosome first infects a cell, however, some stages of phage growth are the same, regardless of whether the lytic or lysogenic pathway is followed. First, the linear chromosome becomes circular through the pairing of the 12-nucleotide-long complementary ends and the action of DNA ligase. Since no repressor molecules have been made to prevent expression of the genes for the lytic pathway, transcription begins at promoters P_L, and P_R (Figure 16.21, part 1). (P_L is the promoter for leftward transcription of the left early operon, and P_R is the promoter for rightward transcription of the right early operon.) Promoter P_L is on a different DNA strand from the P_R promoter, and hence P_L is oriented in an opposite direction from P_R. As a result, transcription occurs in opposite directions, counterclockwise for P_L and clockwise for P_R.

From P_R, the first gene to be transcribed is *cro* (control of *r*epressor and *o*ther), the product of which is the Cro protein. This protein plays an important role in setting the genetic switch to the lytic pathway. From P_L, the first gene to be transcribed is *N*. The resulting N protein is a transcription *antiterminator* that allows RNA synthesis to proceed past certain transcription terminators. This process is called

~ **FIGURE 16.20**

A map of phage λ, showing the major genes. (Promoters discussed in text: P_R = promoter for rightward transcription of the right early operon; P_L = promoter for leftward transcription of the left early operon; P_{RM} = promoter for repressor mainterance; and P_{RE} = promoter for repressor establishment.)

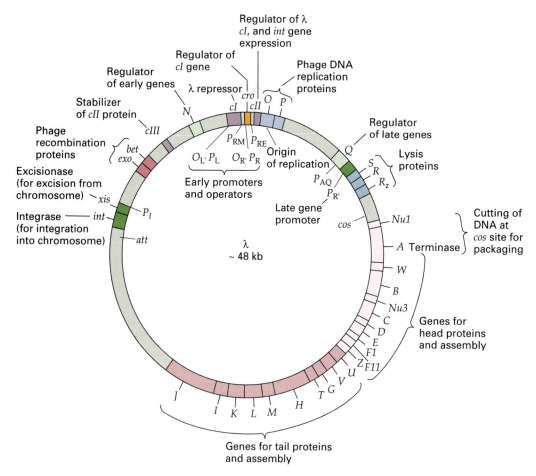

antitermination or readthrough transcription. N protein acts to allow RNA polymerase to proceed leftward of gene N, and rightward of gene cro, thereby including all of the early genes (Figure 16.21, part 2). One of the genes transcribed as a result of the action of N protein is cII. The cII protein turns on the following genes: cI (encodes the λ repressor), O and P (encodes two DNA replication proteins), and Q (encodes a protein needed to turn on late genes for lysis and phage particle proteins). The Q protein functions as another antiterminator protein, permitting transcription to continue into the late genes, which are involved in the lytic pathway. However, only when the switch is set to the lytic pathway and transcription continues from PR for a sufficient time does enough Q protein accumulate to function effectively.

THE LYSOGENIC PATHWAY. After the early transcription events, either the lysogenic or lytic pathway is followed (Figure 16.21, part 3). Setting the switch for the lysogenic pathway is made in the following way.

The establishment of lysogeny requires the protein products of the cII (right early operon) and cIII (left early operon; see Figure 16.20) genes. The cII protein (stabilized by cIII protein) activates transcription of the cI gene (located between the P_L and P_R promoters—see Figure 16.21, part 4a) from a promoter called P_{RE} (promoter for repressor establishment), located to the right of the cI–cro region. This transcription takes place in a counterclockwise direction as described below. The product of the cI gene, the λ repressor protein, blocks the expression of the genes necessary for

~ **FIGURE 16.21**

Expression of λ genes after infecting *E. coli* and the transcriptional events that occur when either the lysogenic or lytic pathways are followed.

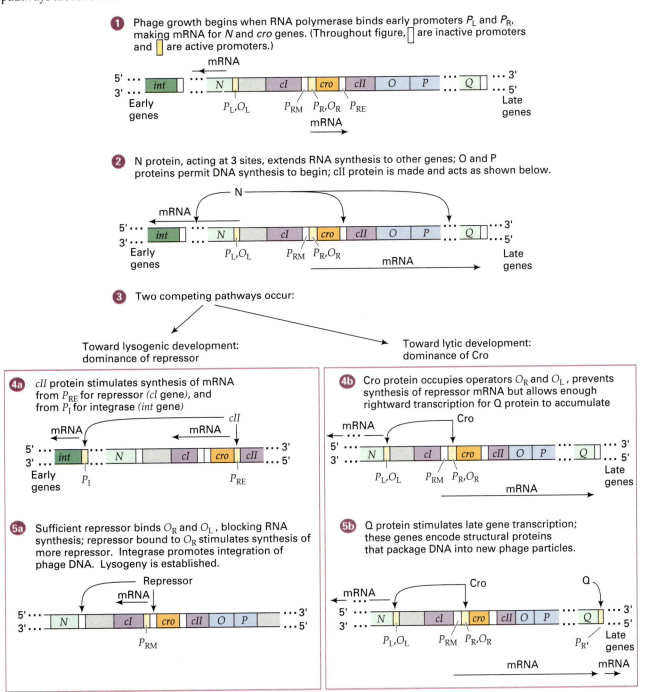

chromosome replication, progeny phage assembly, and cell lysis, thereby maintaining the lysogenic state.

From the work of Mark Ptashne, the λ repressor is known to consist of two nearly equal-sized domains, the amino domain and the carboxyl domain (Figure 16.22a). The functional λ repressor is a dimer of the polypeptide, formed largely by contacts between the carboxyl domains (Figure 16.22b). As soon as enough λ repressor is produced within the cell, it binds to two operator regions, O_L and O_R (see Figure 16.20), whose

~ **FIGURE 16.22**

(a) The λ repressor showing the amino (N) and carboxyl (C) domains and the short connecting segment; (b) Repressor monomers form dimers, which can dissociate to monomers.

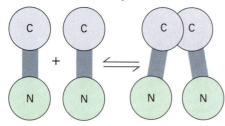

a) λ **repressor**

Carboxyl domain

Amino domain

236

132

92

1

Amino acids

b) **Dimerization of** λ **repressor**

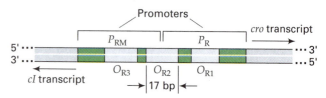

sequences overlap the P_L and P_R promoters, respectively. The binding of the λ repressor prevents the further transcription by RNA polymerase of the early operons controlled by P_L and P_R. As a result, transcription of the N and cro genes is blocked, and since the two proteins specified by these two genes are unstable, the levels of these two proteins in the cell drop dramatically. The binding of the λ repressor also blocks the N protein–controlled antitermination transcription of the O, P, and Q genes. Furthermore, a repressor bound to O stimulates the synthesis of more repressor mRNA from a different promoter, P_{RM}, thereby maintaining repressor concentrations in the cell (Figure 16.21, part 5a). Thus if enough λ repressors are present, lysogeny is established by the binding of the repressor to the O_L and O_R operator regions, followed by integration of λ DNA catalyzed by integrase.

The O_L and O_R operators each have three binding sites for the repressor, designated O_{L1}, O_{L2}, and O_{L3}, and O_{R1}, O_{R2}, and O_{R3}, respectively. Interestingly, the repressor does not bind with equal strength to the binding sites. For O_L, the repressor binds most strongly to O_{L1}, less strongly to O_{L2}, and least strongly to O_{L3}. In other words, the relative binding strength (binding affinity) of the repressor for the binding sites in O_L, is $O_{L1} > O_{L2} > O_{L3}$. For the O_R operator, the relative binding strength of the repressor for the binding sites is $O_{R1} > O_{R2} > O_{R3}$. Let us consider the events that occur at the right operator (O_R) sites, which are major sites where the genetic switch controlling the choice

between the lysogenic and lytic pathways operates; O_L sites are not part of the switch.

Figure 16.23 shows the arrangement of the O_R repressor binding sites relative to the cI (λ repressor) and cro (control of repressor and other) genes. Note that the O_{R1} site overlaps the P_R promoter and that the O_{R3} site overlaps the P_{RM} (RM = repressor maintenance) promoter. Because of the different binding affinities of the three binding sites for repressor, if we started with a repressor-free operator, a repressor would bind first to O_{R1} (Figure 16.24a). With a repressor bound to O_{R1}, the affinity of O_{R2} for the second repressor increases because the second repressor dimer not only binds to O_{R2}, but it also touches the repressor that is bound at O_{R1} (Figure 16.24b). At higher repressor concentrations, repressor will also bind to O_{R3} (Figure 16.24c). This last repressor to bind does not touch either of the two repressors already bound.

For setting the switch to the lysogenic state, then, repressor is typically present at concentrations so that repressor is bound to both O_{R1} and O_{R2}. This prevents RNA polymerase from binding to P_R (see Figure 16.20 and Figure 16.21, part 5a) and further transcribing the cro gene. However, since no repressor is bound to O_{R3}, RNA polymerase can bind to P_{RM}, and the repressor gene cI can be transcribed (Figure 16.25). In other words, λ repressor bound to O_{R1} and O_{R2} acts as a positive regulator for its own transcription. This occurs as a result of direct interaction between λ repressor bound to O_{R2} and RNA polymerase at P_{RM}. As a result, the amount of repressor in the cell increases, so a repressor molecule will bind also to the O_{R3} site. This blocks repressor synthesis by preventing RNA polymerase from binding and transcribing the cI gene. This causes the amount of repressor to drop in the cell and the repressor bound to O_{R3} falls off. RNA polymerase can then bind again to P_{RM}, and cI transcription can resume. In other words, the λ repressor plays a negative regulatory role by blocking transcription of the cro gene, and a positive regulatory role by maintaining optimum repressor levels in the

~ **FIGURE 16.23**

Part of the chromosome, showing the two promoters P_{RM} and P_R and the right operator (O_R), which overlaps them. Transcription from P_{RM} occurs leftward to transcribe the repressor gene (cI). Transcription from P_R proceeds rightward to transcribe the cro gene.

Promoters

P_{RM} P_R cro transcript

5' ···
3' ···
··· 3'
··· 5'

O_{R3} O_{R2} O_{R1}

cI transcript   17 bp

~ FIGURE 16.24

A repressor binds to three sites in the right operator (O_R). (a) The affinity of repressor for O_{R1} is about ten times higher than for O_{R2} or O_{R3}. (b) Once repressor is bound to O_{R1}, a second repressor rapidly binds to O_{R2}. (c) Repressor binds to O_{R3} only at high repressor concentrations.

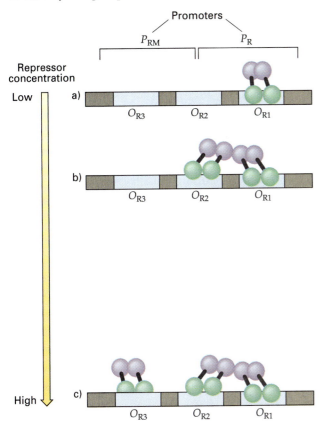

~ FIGURE 16.25

In the lysogenic state, repressors are bound to O_{R1} and O_{R2}, but not to O_{R3}, which allows RNA polymerase to bind to P_{RM} and transcribe the *cI* (repressor) gene. Polymerase cannot bind to P_R since that promoter is covered by repressors; hence the *cro* gene is not transcribed.

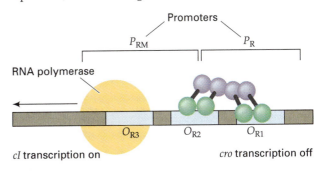

selves, separating the amino and carboxyl domains. (The role of RecA in the repair of DNA damage is described in Chapter 19, pp. 616–617.) This event causes the chain reaction shown in Figure 16.26: the cleaved monomers are unable to form dimers and, as a result, when repressors fall off the operator through the process of normal dissociation, they are not replaced.

The absence of repressor at O_{R1} that results allows RNA polymerase to bind at P_R and the *cro* gene is then further transcribed. The resulting Cro protein then acts

cell by modulating RNA polymerase binding at P_{RM}. As long as enough repressor is present to bind to O_{R1} and O_{R2}, the lysogenic state is maintained.

THE LYTIC PATHWAY. The lysogenic pathway is favored when enough λ repressor is made so that early promoters are turned off, thereby repressing all the genes needed for the lytic pathway. One important lytic pathway gene that is repressed is *Q*; the Q protein is a positive regulatory protein required for the production of lysis proteins and phage coat proteins.

Let us consider the induction of the lytic pathway caused by ultraviolet light irradiation. Inducers such as ultraviolet light typically damage the DNA, and this somehow causes a change in the function of the bacterial protein RecA (product of the *recA* gene). Normally RecA functions in DNA recombination, but when DNA is damaged, RecA is activated and stimulates the λ repressor monomers to cleave them-

~ FIGURE 16.26

Activated RecA protein causes the λ repressor to cleave itself. As a result, when repressors dissociate from operators they are not replaced, *cro* is transcribed, and induction of lytic growth follows.

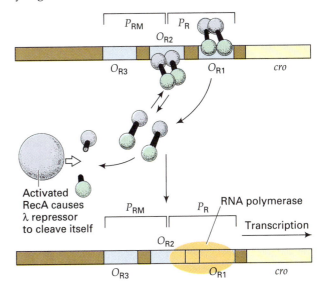

~ **FIGURE 16.27**

Cro protein binds to the same three sites in the right operator (O_R) as λ repressor, but with different affinities for the three sites. (a) The affinity of Cro protein for O_{R3} is about ten times higher than for O_{R2} or O_{R1}. (b) Once a Cro protein is bound to O_{R3}, the second dimer binds with equal chance to O_{R2} or O_{R1}. (c) At high Cro dimer concentration, all three sites are occupied.

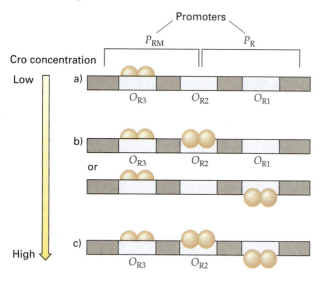

to decrease RNA synthesis from P_L and P_R, and this reduces synthesis of cII protein, the regulator of λ repressor synthesis, and blocks synthesis of λ repressor mRNA from P_{RM} (Figure 16.21, part 4b). At the same time, transcription of the right early operon genes from P_R is decreased, but enough Q proteins are accumulated to set the switch for transcription of the late genes for starting the lytic pathway (Figure 16.21, part 5b).

How does Cro protein work? Cro is a 66–amino acid protein that is folded into one globular domain. The functional form of Cro is a dimer, and each dimer can bind to the same 17-nucleotide-pair operator binding sites to which the repressor can bind. Cro dimers bind independently to the three O_R sites, although the order of affinity for the three sites is *exactly opposite* that of the λ repressor dimer; that is, the strongest affinity of Cro protein is for O_{R3} (Figure 16.27a), and after that the affinity is approximately equal for O_{R2} and O_{R1} (Figure 16.27b). Unlike the λ repressor, which has both a positive and negative regulatory action, Cro acts only as a negative regulator. The first Cro synthesized binds to O_{R3} and this blocks the ability of RNA polymerase to bind to P_{RM}, thereby stopping further repressor synthesis. At this point, the switch has been set to the lytic pathway. At higher

Cro concentrations, O_{R2} and/or O_{R1} are filled, and this turns down transcription of *N*, *cII*, and *cro* itself.

As with the repressor protein, the Cro protein controls its own transcription. At high Cro protein concentrations, all three O_R binding sites become occupied by Cro protein (Figure 16.27c), blocking transcription initiation at the P_R promoter. This results in decreased transcription of the *cro* gene, and of the other early lytic pathway genes.

In summary, lambda uses complex regulatory systems to direct either the lytic or the lysogenic pathway. The choice is made between the two pathways, using an elaborate genetic switch, most of the details of which are now in hand. This switch involves two regulatory proteins with opposite affinities for three binding sites within each of the two major operators that control the λ life cycle. Which pathway is followed depends upon which regulatory protein "wins" in terms of binding to the those two operators.

$\mathcal{K}$EYNOTE

Bacteriophages are especially adapted for undergoing their reproduction within a bacterial host. Many of the genes related to the production of progeny phages, or to the establishment or reversal of lysogeny in temperate phages, are organized into operons. These operons (like bacterial operons) are controlled through the interaction of regulatory proteins with operators that are adjacent to clusters of structural genes.

SUMMARY

Bacteria typically live in environments that can change rapidly. To function efficiently in the changing environment, bacteria have evolved special systems for regulating gene expression in response to the dictates of conditions. Generally, when a particular function is needed, the genes for that function (e.g., the lactose utilization genes when lactose is present as a carbon source, or the tryptophan biosynthesis genes when tryptophan is absent from the medium) are coordinately transcribed so that the relevant enzymes are coordinately produced. Conversely, when the conditions change so that function is not needed, gene transcription is turned off.

From studies of the regulation of expression of the three lactose-utilizing genes in *E. coli* bacteria, Jacob and Monod developed a model that is the basis for the regulation of gene expression in a large number of bacterial and bacteriophage systems. The genes for the enzymes are contiguous in the chromosome and are adjacent to a controlling site (an operator) and a single promoter. This complex constitutes a transcriptional regulatory unit called an operon. A regulator gene, which may or may not be nearby, is associated with the operon. The addition of an appropriate substrate to the cell results in the coordinate induction of the operon's structural genes. With respect to the *lac* operon, induction occurs as follows: lactose binds with a repressor protein that is encoded by the regulator gene, inactivating it and preventing it from binding to the operator. As a result, RNA polymerase can transcribe the three genes onto a single polygenic mRNA. As long as lactose is present, mRNA continues to be produced and the enzymes are made. When lactose is no longer present, the repressor protein is no longer inactivated, and it binds to the operator, thereby preventing RNA polymerase from transcribing the *lac* genes.

If both glucose and lactose are present in the medium, the lactose operon is not induced because glucose (which requires less energy to metabolize than does lactose) is the preferred energy source. This phenomenon is called catabolite repression and involves cellular levels of cyclic AMP. That is, in the presence of lactose and the absence of glucose, cAMP complexes with CAP to form a positive regulator needed for RNA polymerase to bind to the promoter. Addition of glucose results in a lowering of cAMP level so no CAP-cAMP complex is produced and, therefore, RNA polymerase cannot bind to the promoter and transcribe the *lac* genes.

The genes for a number of bacterial amino acid biosynthesis pathways are also arranged in operons. Expression of these operons is accomplished by a repressor-operator system or, in some cases, through attenuation at a second controlling site called an attenuator, or both. The tryptophan operon is an example of an operon with both types of transcription regulation systems. The repressor-operator system functions essentially like that of the *lac* operon, except that the addition of tryptophan to the cell activates the repressor, thereby turning the operon off. This makes sense since the role of the gene products is to make the amino acid. The attenuator is located downstream from the operator in a leader region that is translated. The attenuator is a transcription termina-

tion site that allows only a fraction of RNA polymerases to transcribe the rest of the operon. Attenuation involves a coupling between transcription and translation, and the formation of particular RNA secondary structures that signal whether or not transcription can continue.

Key to the attenuation phenomenon is the presence of multiple copies of codons for the amino acid synthesized by the enzymes encoded by the operon. When enough of the amino acid is present in the cell, enough charged tRNAs are produced so that the ribosome can translate the key codons in the leader region, and this causes the RNA that is being made by the RNA polymerase ahead of the ribosome to assume a secondary structure which signals transcription to stop. However, when the cell is starved for that amino acid, there are insufficient charged tRNAs to be used at the key codons so that the ribosome stalls at that point. As a result the RNA ahead of that point assumes a secondary structure that permits continued transcription into the structural genes. The combination of repressor-operator regulation and attenuation control permits a fine degree of control of transcription of the operon.

Operons are also extensively used by bacteriophages to coordinate the synthesis of proteins with related functions. Many of the genes related to the production of progeny phages, or to the establishment or reversal of lysogeny of temperate phages, are organized into operons. These operons, like bacterial operons, are controlled through the interaction of regulatory proteins with operators that are adjacent to clusters of structural genes. Bacteriophage lambda has been an excellent model for studying the elaborate genetic switch that controls the choice between lytic and lysogenic pathways in a lysogenic phage. The switch involves two regulatory proteins, the lambda repressor and the Cro protein, which have opposite affinities for three binding sites within each of the two major operators that control the lambda life cycle.

In sum, operons are commonly encountered in bacteria and their viruses. They provide a simple way to coordinate the transcription of genes with related function at the transcriptional level. Generally, there is an interaction between an effector molecule (such as lactose or tryptophan) and a repressor molecule. That interaction may inactivate (lactose) or activate (tryptophan) the repressor with respect to its ability to bind to the operator, respectively. If no repressor is bound to the operator, RNA polymerase can bind to the adjacent promoter and the genes can be transcribed. Operons have not been found in eukaryotic systems.

ANALYTICAL APPROACHES FOR SOLVING GENETICS PROBLEMS

Q16.1 In the laboratory you are given ten strains of *E. coli* with the following lactose operon genotypes, where I = *lacI* (the repressor gene), $P = P_{lac}$ (the promoter), O = *lacO* (the operator), and Z = *lacZ* (the β-galactosidase gene).

For each strain, predict whether β-galactosidase will be produced (a) if lactose is absent from the growth medium, and (b) if lactose is present in the growth medium.

1. $I^+ P^+ O^+ Z^+$

2. $I^- P^+ O^+ Z^+$

3. $I^+ P^+ O^c Z^+$

4. $I^- P^+ O^c Z^+$

5. $I^+ P^+ O^c Z^-$

6. $\dfrac{F'\ I^+ P^+ O^c Z^-}{I^+ P^+ O^+ Z^+}$

7. $\dfrac{F'\ I^+ P^+ O^+ Z^-}{I^+ P^+ O^c Z^+}$

8. $\dfrac{F'\ I^+ P^+ O^+ Z^+}{I^- P^+ O^+ Z^-}$

9. $\dfrac{F'\ I^+ P^+ O^c Z^-}{I^- P^+ O^+ Z^+}$

10. $\dfrac{F'\ I^- P^+ O^+ Z^-}{I^- P^+ O^c Z^+}$

(Note: In the partial-diploid strains (6–10) one copy of the *lac* operon is in the host chromosome and the other copy is in the *F* episome.)

A16.1 The answers are as follows, where + = β-galactosidase produced and − = it is not produced:

GENOTYPE	NONINDUCED: LACTOSE ABSENT	INDUCED: LACTOSE PRESENT
(1)	−	+
(2)	+	+
(3)	+	+
(4)	+	+
(5)	−	−
(6)	−	+
(7)	+	+
(8)	−	+
(9)	−	+
(10)	+	+

To answer this question completely, we need a good understanding of how the lactose operon is regulated in the wild type and of the consequences of particular mutations on the regulation of the operon.

Strain (1) is the standard wild-type operon. No enzyme is produced in the absence of lactose, since the repressor produced by the I^+ gene binds to the operator (O^+) and blocks the initiation of transcription. When lactose is added, it binds to the repressor, changing its conformation so that it no longer can bind to the O^+ region, thereby facilitating transcription of the structural genes by RNA polymerase.

Strain (2) is a haploid strain with a mutation in the *lacI* gene (I^-). The consequence is that the repressor protein cannot bind to the (normal) operator region, so there is no inhibition of transcription, even in the absence of lactose. This strain, then, is constitutive, meaning that the functional enzyme is produced by the *lacZ*+ gene in the presence or absence of lactose.

Strain (3) is the other possible constitutive mutant. In this case the repressor gene is a wild type and the β-galactosidase gene *lacZ*+ is a wild type, but there is a mutation in the operator region (O^c). Therefore repressor protein cannot bind to the operator, and the transcription occurs in the presence or absence of lactose.

Strain (4) carries both of the regulatory mutations of the previous two strains. Functional repressor is not produced, but even if it were, the operator is changed so that it cannot bind. The consequence is the same: constitutive enzyme production.

Strain (5) produces functional repressor, but the operator (O^c) is mutated. Therefore transcription cannot be blocked, and lactose polygenic mRNA is produced in the presence or absence of lactose. However, because there is also a mutation in the β-galactosidase gene (Z^-), no functional enzyme is generated.

In the partial-diploid strain (6) one lactose operon is completely wild type, and the other carries a constitutive operator mutation and a mutant β-galactosidase gene. In the absence of lactose no functional enzyme is produced. For the wild-type operon the repressor binds to the operator and blocks transcription. For the operon with the two mutations, the operator region is mutated and cannot bind repressor, so the mRNA for the mutated operon is produced; however, the *lacZ* gene is also mutated so that functional enzyme cannot be produced. In the presence of lactose, functional enzyme is produced, since repression of the wild-type operon is relieved so that the Z^+ gene can be transcribed. This type of strain provided one of the pieces of evidence that the operator region does not produce a diffusible substance.

In diploid (7) functional enzyme will be produced in the presence or absence of lactose, because one of the operons has an Oc mutation that does not respond to a repressor and that is linked to a wild-type Z^+ gene. That operon is transcribed constitutively. The other operon is inducible, but because there is a Z^- mutation, no functional enzyme is produced.

Diploid (8) has a wild-type operon and an operon with an I^- regulatory mutation and a Z^- mutation. The I^+ gene product is diffusible so that it can bind to the O^+ region of both operons, thereby putting both operons under inducer control. This strain demonstrates that the I^+ gene is trans-dominant to a I^- mutation. In this case the particular location of the one Z^- mutation is irrelevant: The same result would have been obtained had the Z^+ and Z^- been switched between the two operons. In this diploid, β-galactosidase is not produced unless lactose is present.

In strain (9) β-galactosidase is produced only when the lactose is present, because the O^c region controls only those genes that are adjacent to it on the same chromosome (cis dominance), and in this case one of the adjacent genes is Z^-, which codes for a nonfunctional enzyme. The diploid is heterozygous I^+/I^-, but I^+ is trans-dominant, as discussed for strain (8). Thus the only normal Z^+ gene is under inducer control.

Diploid (10) has defective repressor protein as well as an O^c mutation adjacent to a Z^+ gene. On the latter ground alone this diploid is constitutive. The other operon is also constitutively transcribed, but because there is a Z^- mutation, no functional enzyme is generated from it.

QUESTIONS AND PROBLEMS

16.1 How does lactose bring about the induction of synthesis of β-galactosidase, permease, and transacetylase? Why does this event not occur when glucose is also in the medium?

16.2 Operons produce polygenic mRNA when they are active. What is a polygenic mRNA? What advantages, if any, do they confer on a cell in terms of its function?

*16.3 If an *E. coli* mutant strain synthesizes β-galactosidase whether or not the inducer is present, what genetic defect(s) might be responsible for this phenotype?

*16.4 Distinguish the effects you would expect from (a) a missense mutation and (b) a nonsense mutation in the *lacZ* (β-galactosidase) gene of the *lac* operon.

16.5 The elucidation of the regulatory mechanisms associated with the enzymes of lactose utilization in *E. coli* was a landmark in our understanding of regulatory processes in microorganisms. In formulating the operon hypothesis as applied to the lactose system, Jacob and Monod found that results from particular partial-diploid strains were invaluable. Specifically, in terms of the operon hypothesis, what information did the partial diploids provide that the haploids could not?

*16.6 For the *E. coli lac* operon, write the partial-diploid genotype for a strain that will produce β-galactosidase constitutively and permease by induction.

16.7 Mutants were instrumental in the elaboration of the model for the regulation of the lactose operon.
a. Discuss why *lacO^c* mutants are cis-dominant but not trans-dominant.
b. Explain why *lacI^s* mutants are trans-dominant to the wild-type *lacI^+* allele but *lacI^-* mutants are recessive.
c. Discuss the consequences of mutations in the repressor gene promoter as compared with mutations in the structural gene promoter.

*16.8 This question involves the lactose operon of *E. coli* where $I = lacI$ (the repressor gene), $P = P_{lac}$ (the promoter), $O = lacO$ (the operator), $Z = lacZ$ (the β-galactosidase gene), and $Y = lacY$ (the permease gene). Complete Table 16.A (p. 532), using + to indicate if the enzyme in question will be synthesized and – to indicate if the enzyme will not be synthesized.

16.9 A new sugar, sugarose, induces the synthesis of two enzymes from the *sug* operon of *E. coli*. Some properties of deletion mutations affecting the appearance of these enzymes are as follows (here, + = enzyme induced normally, i.e., synthesized only in the presence of the inducer; C = enzyme synthesized constitutively; 0 = enzyme cannot be detected):

MUTATION OF	ENZYME 1	ENZYME 2
Gene *A*	+	0
Gene *B*	0	+
Gene *C*	0	0
Gene *D*	C	C

a. The genes are adjacent in the order *ABCD*. Which gene is most likely to be the structural gene for enzyme 1?
b. Complementation studies using partial-diploid (*F'*) strains were made. The episome (*F'*) and chromo-

Table 16.A

Genotype	Inducer Absent:		Inducer Present:	
	β-galactosidase	Permease	β-galactosidase	Permease
a. $I^+ P^+ O^+ Z^+ Y^+$				
b. $I^+ P^+ O^+ Z^- Y^+$				
c. $I^+ P^+ O^+ Z^+ Y^-$				
d. $I^- P^+ O^+ Z^+ Y^+$				
e. $I^s P^+ O^+ Z^+ Y^+$				
f. $I^+ P^+ O^c Z^+ Y^+$				
g. $I^s P^+ O^c Z^+ Y^+$				
h. $I^+ P^+ O^c Z^+ Y^-$				
i. $I^{-d} P^+ O^+ Z^+ Y^+$				
j. $\dfrac{I^- P^+ O^+ Z^+ Y^+}{I^+ P^+ O^+ Z^- Y^-}$				
k. $\dfrac{I^- P^+ O^+ Z^+ Y^-}{I^+ P^+ O^+ Z^- Y^+}$				
l. $\dfrac{I^s P^+ O^+ Z^+ Y^-}{I^+ P^+ O^+ Z^- Y^+}$				
m. $\dfrac{I^+ P^+ O^c Z^- Y^+}{I^+ P^+ O^+ Z^+ Y^-}$				
n. $\dfrac{I^- P^+ O^c Z^+ Y^-}{I^+ P^+ O^+ Z^- Y^+}$				
o. $\dfrac{I^s P^+ O^+ Z^+ Y^+}{I^+ P^+ O^c Z^+ Y^+}$				
p. $\dfrac{I^{-d} P^+ O^+ Z^+ Y^-}{I^+ P^+ O^+ Z^- Y^+}$				
q. $\dfrac{I^+ P^- O^c Z^+ Y^-}{I^+ O^+ O^+ Z^- Y^+}$				
r. $\dfrac{I^+ P^- O^+ Z^+ Y^-}{I^+ P^+ O^c Z^- Z^+}$				
s. $\dfrac{I^- P^- O^+ Z^+ Y^+}{I^+ P^+ O^+ Z^- Y^-}$				
t. $\dfrac{I^- P^+ O^+ Z^+ Y^-}{I^+ P^- O^+ Z^- Y^+}$				

some each carried one set of *sug* genes. The results were as follows (symbols are the same as in previous table):

GENOTYPE OF F'	CHROMOSOME	ENZYME 1	ENZYME 2
$A^+ B^- C^+ D^+$	$A^- B^+ C^+ D^+$	+	+
$A^+ B^- C^- D^+$	$A^- B^+ C^+ D^+$	+	0
$A^- B^+ C^- D^+$	$A^+ B^- C^+ D^+$	0	+
$A^- B^+ C^+ D^+$	$A^+ B^- C^+ D^-$	+	+

From all the evidence given, determine whether the following statements are true or false:

1. It is possible that gene *D* is a structural gene for one of the two enzymes.
2. It is possible that gene *D* produces a repressor.
3. It is possible that gene *D* produces a cytoplasmic product required to induce genes *A* and *B*.
4. It is possible that gene *D* is an operator locus for the *sug* operon.

5. The evidence is also consistent with the possibility that gene *C* could be a gene that produces a cytoplasmic product required to induce genes *A* and *B*.

6. The evidence is also consistent with the possibility that gene *C* could be the controlling end of the *sug* operon (end from which mRNA synthesis presumably commences).

16.10 Four different polar mutations, *1*, *2*, *3*, and *4*, in the *lacZ* gene of the lactose operon were isolated following mutagenesis of *E. coli*. Each caused total loss of β-galactosidase activity. Two revertant mutants, due to suppressor mutations in genes unlinked to the *lac* operon, were isolated from each of the four strains: suppressor mutations of polar mutation 1 are *1A* and *1B*; those of polar mutation 2 are *2A* and *2B*; and so on. Each of the eight suppressor mutations was then tested, by appropriate crosses, for its ability to suppress each of the four polar mutations; the test involved examining the ability of a strain carrying the polar mutation and the suppressor mutation to grow with lactose as the sole carbon source. The results follow (+ = growth on lactose and – = no growth):

POLAR MUTATION	SUPPRESSOR MUTATION							
	1A	1B	2A	2B	3A	3B	4A	4B
1	+	+	+	+	+	+	+	+
2	+	–	+	+	+	+	–	–
3	+	–	+	–	+	+	–	–
4	+	+	+	+	+	+	+	+

A mutation to a UAG codon is called an amber nonsense mutation, and a mutation to a UAA codon is called an ochre nonsense mutation. Suppressor mutations allowing reading of UAG and UAA are called amber and ochre suppressors, respectively.

a. Which of the polar mutations are probably amber? Which are probably ochre?

b. Which of the suppressor mutations are probably amber suppressors? Which are probably ochre suppressors?

c. How would you explain the anomalous failure of suppressor *2B* to permit growth with polar mutation *3*? How could you test your explanation most easily?

d. Explain precisely why ochre suppressors suppress amber mutants but amber suppressors do not suppress ochre mutants.

*16.11 What consequences would a mutation in the catabolite activator protein (CAP) gene of *E. coli* have for the expression of a wild-type *lac* operon?

16.12 The lactose operon is an inducible operon, whereas the tryptophan operon is a repressible operon. Discuss the differences between these two types of operons.

*16.13 In the presence of high intracellular concentrations of tryptophan, only short transcripts of the *trp* operon are synthesized because of attenuation of transcription 5' to the structural genes. This is mediated by the recognition of two Trp codons in the leader sequence. If these codons were mutated to be amber (UAG) nonsense mutations, what effect would this have on the regulation of the operon in the presence or absence of tryptophan? Explain.

16.14 In the bacterium *Salmonella typhimurium* seven of the genes coding for histidine biosynthetic enzymes are located adjacent to one another in the chromosome. If excess histidine is present in the medium, the synthesis in all seven enzymes is coordinately repressed, whereas in the absence of histidine all seven genes are coordinately expressed. Most mutations in this region of the chromosome result in the loss of activity of only one of the enzymes. However, mutations mapping to one end of the gene cluster result in the loss of all seven enzymes, even though none of the structural genes have been lost. What is the counterpart of these mutations in the *lac* operon system?

16.15 Upon infecting an *E. coli* cell, bacteriophage λ has a choice between the lytic and lysogenic pathways. Discuss the molecular events that determine which pathway is taken.

16.16 How do the lambda repressor protein and the Cro protein regulate their own synthesis?

*16.17 If a mutation in the phage lambda *cI* gene results in a nonfunctional *cI* gene product, what phenotype would you expect the phage to exhibit?

16.18 Bacteriophage λ can form a stable association with the bacterial chromosome because the virus manufactures a repressor. This repressor prevents the virus from replicating its DNA, making lysozyme and all the other tools used to destroy the bacterium. When you induce the virus with UV light, you destroy the repressor, and the virus goes through its normal lytic cycle. This repressor is the product of a gene called the *cI* gene and is a part of the wild-type viral genome. A bacterium that is lysogenic for λ⁺ is full of repressor substance, which confers immunity against any λ virus added to these bacteria. These added viruses can inject their DNA, but the repressor from the resident virus prevents replication, presumably by binding to an operator on the incoming virus. Thus this system has many analogous elements to the lactose operon. We could diagram a virus as shown in the figure. Several mutations of the *cI* gene are known. The c_i mutation results in an inactive repressor.

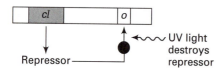

a. If you infect *E. coli* with λ containing a c_i mutation, can it lysogenize (form a stable association with the bacterial chromosome)? Why?

b. If you infect a bacterium simultaneously with a wild-type c^+ and a c_i mutant of λ, can you obtain stable lysogeny? Why?

c. Another class of mutants called c^{IN} makes a repressor that is insensitive to UV destruction. Will you be able to induce a bacterium lysogenic for c^{IN} with UV light? Why?

16.19 To regulate the growth of lambda, the lambda repressor must form a dimer. As more repressor monomers are synthesized, more dimers can form, and as the concentration of monomers is decreased, dimers dissociate to maintain an equilibrium between repressor dimers and repressor monomers in the cell.

a. How will binding of repressor dimers to O_{R1}, O_{R2}, and O_{R3} be affected by a linear increase in the production of repressor monomers?

b. How will transcription from the promoters P_R and P_{RM} be affected by a linear increase in the production of repressor monomers?

c. After a bacterium is exposed to UV light, a lytic pathway is efficiently induced. How does each of the following properties of the lambda repressor contribute to this efficient induction?

   i. It binds operator sites only as a dimer.

   ii. It has different affinities for operator sites.

16.20 In the early 1970s, Allen Campbell showed that the regions of the lambda genome between the *J* and *N* genes could be replaced experimentally by a variety of segments of *E. coli* DNA. Consider what you have learned about the genetic pathways involved in lysogeny and lytic growth, and the fact that substitution of these regions effectively results in deletion mutations for the genes in them.

a. What effects on lambda growth would you expect to see in such substitution mutations? What phenotypes might these mutations have?

b. How might you use this information to design a cloning vector in lambda?

c. Suppose that a gene needed for lytic growth was deleted in an early step in the production of a lambda cloning vector. How might this deficit be remedied to insure that the engineered lambda vector would be capable of lytic growth?

*16.21 The Cro and lambda repressor proteins have different affinities for specific sites within the operators O_L and O_R. In contrast, other proteins, such as those within nucleosomes, bind DNA without site specificity.

a. How might some proteins bind DNA without site specificity?

b. How might some proteins bind DNA with site specificity?

c. How might some proteins bind DNA at multiple sites and with different affinities for each?

16.22 Lambda mutations that eliminate the normal functions of the *cI*, *cII*, and *cIII* genes all show the same phenotype (see question 16.17 to review the phenotype of *cI* mutants). How could you identify whether a new mutation showing this phenotype was a mutation at the *cI*, *cII* or *cIII* gene?

*16.23 While much of the regulation of prokaryotic gene expression occurs at the level of transcription initiation, there is also significant control at the level of transcript elongation via antitermination or attenuation.

a. What is the utility to a prokaryotic organism of having these two different control mechanisms employed at the same operon?

b. How is transcription antitermination at the tryptophan operon in *E. coli* different from the transcription antitermination in lambda that is mediated by the N protein?

c. What phenotype would you expect for the following mutations that affect antitermination mechanisms?

   i. A base-substitution mutation in the tryptophan operon that destabilized the 2:3 stem loop pairing depicted in Figure 16.16.

   ii. A mutation in lambda that resulted in a nonfunctional N protein.

17 REGULATION OF GENE EXPRESSION AND DEVELOPMENT IN EUKARYOTES

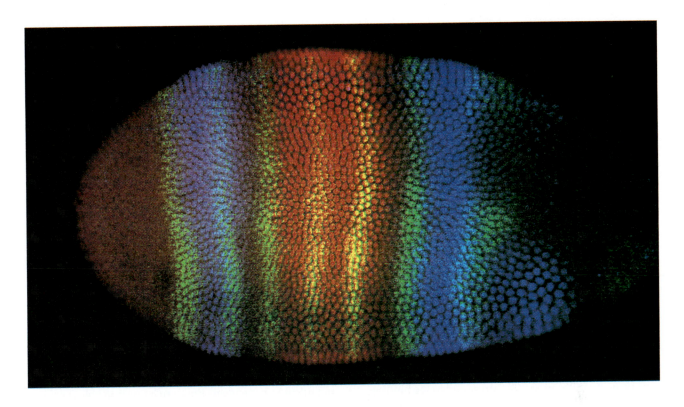

LEVELS OF CONTROL OF GENE EXPRESSION IN EUKARYOTES
    Transcriptional Control
    RNA Processing Control
    Transport Control
    mRNA Translation Control
    mRNA Degradation Control
    Protein Degradation Control

GENE REGULATION IN DEVELOPMENT AND DIFFERENTIATION
    Genomic Activity in Higher Eukaryotes
    Constancy of DNA in the Genome During Development
    Differential Gene Activity Among Tissues and During Development
    Immunogenetics and Chromosome Rearrangements During Development

GENETIC REGULATION OF DEVELOPMENT IN *DROSOPHILA*
    Drosophila Developmental Stages
    Embryonic Development
    Imaginal Discs
    Homeotic Genes

PRINCIPAL POINTS

~ There are no operons in eukaryotes. However, genes for related functions are often regulated coordinately.

~ In eukaryotes, gene expression is regulated at a number of distinct levels. That is, there are regulatory systems for the control of transcription, of precursor-RNA processing, transport of the mature RNA out of the nucleus, translation of the mRNAs, degradation of the mature RNAs, and degradation of the protein products.

~ Eukaryotic protein-coding genes contain both promoter elements and enhancer elements. Some promoter elements are required for transcription to begin. Other promoter elements have a regulatory function; these are specialized with respect to the gene they control, binding specific regulatory proteins that modulate expression of the gene. Specific regulatory proteins bind also to the enhancer elements and activate transcription through their interaction with proteins bound to the promoter elements. Enhancer elements and promoter elements appear to bind many of the same proteins. This implies that both types of regulatory elements affect transcription by a similar mechanism, probably involving interactions of the regulatory proteins.

~ Transcriptionally active chromatin is more sensitive to digestion with DNase I than transcriptionally inactive chromatin. This sensitivity is a result of a loosened DNA-protein structure in the areas of transcriptional activity. For active genes, certain sites—called hypersensitive sites—are highly sensitive to DNase I digestion. These hypersensitive sites probably correspond to the binding sites for RNA polymerase and regulatory proteins.

~ Histones repress transcription by assembling nucleosomes on TATA boxes associated with genes. Gene activation occurs when proteins become bound to enhancers and disrupt the nucleosomes on the TATA boxes, thereby allowing the appropriate proteins to bind to the promoter elements to initiate transcription.

~ The DNA of many eukaryotes has been shown to be methylated at a certain proportion of the bases. Methyl groups are added mostly to cytosines. In those eukaryotes, transcriptionally active genes generally exhibit lower levels of DNA methylation than transcriptionally inactive genes.

~ In higher eukaryotes, steroid hormones regulate the expression of particular sets of genes. To function in this short-term regulatory system, a steroid hormone enters a cell. If that cell contains a receptor molecule specific for the hormone, the hormone binds to the receptor and activates it, and the hormone-receptor complex binds to hormone regulatory elements next to genes in the nucle-us, thereby regulating the expression of those genes. Other hormones (polypeptide hormones) act at the cell surface, activating a system to produce cyclic AMP, a substance that acts as a second messenger to control gene activity. The specificity of hormone action results from the presence of hormone receptors in only certain cell types and by interactions of steroid-receptor complexes with cell type–specific regulatory proteins.

~ Regulation at the level of RNA processing operates in the production of mature mRNA molecules from precursor-mRNA molecules. Two regulatory events that exemplify this level of control are choice of poly(A) site and choice of splice site for intron removal. In both cases different types of mRNAs are produced, depending upon the choices made.

~ mRNA transport from the nucleus to the cytoplasm is another important control point in eukaryotic gene regulation. A spliceosome retention model has been proposed for this regulation. In this model, spliceosome assembly on a precursor-mRNA competes with nuclear export of the molecule. Thus, during intron removal, the spliceosome serves to keep the immature RNA in the nucleus but, when all introns have been removed and the spliceosome has dissociated, the free mature mRNA can interact with the nuclear pore and exit.

~ Gene expression is regulated also by mRNA translation control and by mRNA degradation control. The latter is believed to be a major control point in the regulation of gene expression. Structural features of individual mRNAs have been shown to be responsible for the range of mRNA degradation rates observed, although the precise roles of cellular factors and enzymes have yet to be determined.

~ Long-term regulation is involved in controlling events that activate and repress genes during development and differentiation. Those two processes result from differential gene activity of a genome that contains a constant amount of DNA, rather than from a programmed loss of genetic information.

~ Antibodies are specialized proteins called immunoglobulins, which bind specifically to antigens (*anti*body *gen*erators—chemicals recognized as foreign by an organism which induce an immune response). Immunity against a particular antigen results from clonal selection, in which cells already making the required antibody are stimulated to proliferate. Antibody molecules consist of two light chains and two heavy chains. The amino acid sequence of one domain of each type chain is variable; this variation is responsible for the different antigen-binding site on different antibody molecules. In germ-line DNA the coding regions for immunoglobulin chains are scattered in tandem arrays of gene segments. During development, somatic recombination occurs to bring particular gene segments together to form functional antibody chain genes. A large number of different antibody chain genes result from the many possible ways in which the gene segments can recombine.

~ *Drosophila* has become a relatively tractable model system in which to study the genetic control of development. *Drosophila* body structures result from specific gradients in the egg, and the subsequent determination of embryo segments that directly correspond to adult body segments. Both processes are under genetic control as defined by mutations that disrupt the development events. Studies of the mutants indicate that

Drosophila development is directed by a temporal regulatory cascade.

~ A number of *Drosophila* adult structures develop from imaginal discs, which are determined early in larval development. Rarely, the determined state of an imaginal disc changes so that a different adult structure is produced; this is called transdetermination.

~ Once the basic segmentation pattern has been laid down in *Drosophila*, homeotic genes determine the developmental identity of the segments. Homeotic genes share common DNA sequences called homeoboxes. Homeoboxes have been found in developmental genes in other organisms, and the homeodomains—the regions of the proteins the homeoboxes encode—probably play a role in regulating transcription by binding to specific DNA sequences.

As we learned in Chapter 16, gene expression in prokaryotes is commonly regulated in a unit of protein-coding sequences and adjacent controlling sites (promoter and operator), collectively called an operon. The molecular mechanisms in operon function provide a relatively simple means of controlling the coordinate synthesis of proteins with related functions. In eukaryotes, gene expression is also regulated in a unit of protein-coding sequences and adjacent controlling sites. However, in eukaryotic genes there is nothing formally like the prokaryotic operator region, so eukaryotes are not considered to have operons. In this chapter we look at some of the well-studied regulatory systems in eukaryotes.

Eukaryotic gene regulation falls into two categories. Short-term regulation involves regulatory events in which gene sets are quickly turned on or off in response to changes in environmental or physiological conditions in the cell's or organism's environment. Long-term gene regulation involves regulatory events other than those required for rapid adjustment to local environmental or physiological changes; that is, those events that are required for an organism to develop and differentiate. Both of these categories of gene regulation are addressed in this chapter.

LEVELS OF CONTROL OF GENE EXPRESSION IN EUKARYOTES

Prokaryotic organisms are unicellular and free-living. As we discussed in Chapter 16, the control of gene expression in prokaryotes occurs primarily at the

transcriptional level, with some control at the translational level. Transcriptional control is accomplished through the interaction of regulatory proteins with upstream regulatory DNA sequences. A rapid transition between protein synthesis and no protein synthesis is achieved by both switching off gene transcription and rapid degradation of the mRNA molecules involved.

There are both unicellular and multicellular eukaryotes. In both eukaryote types the control of gene expression is more complicated than is the case with prokaryotes. Figure 17.1 diagrams some of the levels at which the expression of protein-coding genes can be regulated in eukaryotes: *transcriptional control, RNA processing control, transport control, mRNA translation control, RNA degradation control*, and *protein degradation control*. We will consider each of these in turn. Keep in mind that these control processes help coordinate the generation of new proteins in different cells at different times.

Transcriptional Control

GENERAL ASPECTS OF TRANSCRIPTIONAL CONTROL. Transcriptional control regulates whether or not a gene is to be transcribed and the rate at which transcripts are produced.

Protein-coding eukaryotic genes contain both promoter elements and enhancers (see Chapter 13). Recall that particular proteins bind to promoters and enhancers to facilitate the initiation of transcription by RNA polymerase II. The promoter elements are located just upstream of the site at which transcrip-

~ FIGURE 17.1

Levels at which gene expression can be controlled in eukaryotes.

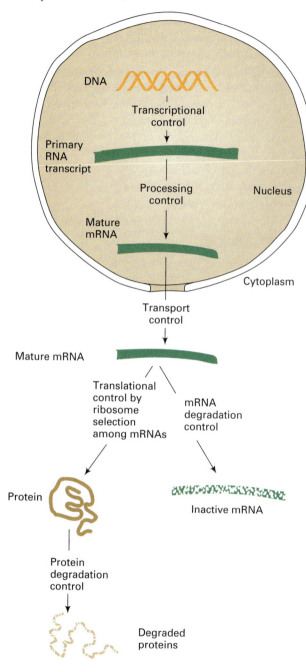

begin. Other promoter elements control whether or not transcription of the gene occurs: specific regulatory proteins (transcription factors) bind to these elements (Figure 17.2). If transcription is activated, the promoter element is a *positive regulatory element* and the protein is a *positive regulatory protein.* If transcription is turned off, the promoter element is a *negative regulatory element* and the protein is a *negative regulatory protein.*

A regulatory promoter element is specialized with respect to the gene (or genes) it controls because it binds a signaling molecule that is involved in the regulation of that gene's expression. Depending on the particular gene, there can be one, a few, or many regulatory promoter elements, since under various conditions there may be one, a few, or many regulatory proteins that control the gene's expression. The remarkable specificity of regulatory proteins in binding to their specific regulatory element in the DNA and to no others ensures careful control of which genes are turned on and which are turned off.

While promoter elements are crucial for determining whether transcription can occur, enhancer elements determine that maximal transcription of the gene occurs. Recall that enhancer elements were originally identified because they activate transcription from the promoter linked to the element, even though the enhancer may be some distance away (see Chapter 13). Regulatory proteins bind to the enhancer elements, and which regulatory proteins bind depends on the DNA sequence of the enhancer element. A model for how the enhancers affect transcription from a distance sets out the following pattern. Specific transcription factors bind to the enhancer element; the DNA then forms a loop so that the enhancer-bound regulatory proteins interact with the regulatory proteins and transcription factors bound to the promoter elements. Through the interactions of the proteins, transcription is either activated or repressed.

Both promoters and enhancers are important in regulating transcription of a gene. Each regulatory promoter element and enhancer element binds a special regulatory protein. Some regulatory proteins are found in most or all cell types, while others are found in only a limited number of cell types. Because some of the regulatory proteins activate transcription when they bind to the enhancer or promoter element, while others repress transcription, the net effect of a regulatory element on transcription depends on the combination of different proteins bound. If positive regulatory proteins are bound at both the enhancer and promoter elements, the result is activation of tran-

tion begins. The enhancers are usually some distance away, either upstream or downstream. We can think of the different promoter elements as modules that function in the regulation of expression of the gene. Certain promoter elements, such as the TATA element, are required to specify where transcription is to

~ **FIGURE 17.2**

Schematics of (a) positive regulation of gene transcription and (b) negative regulation of gene transcription.

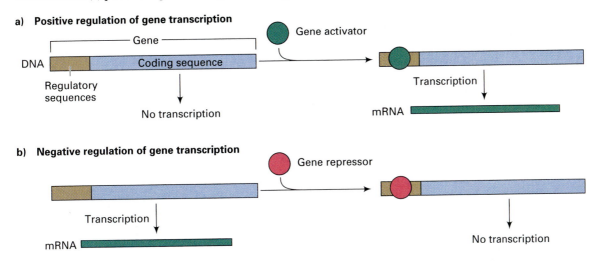

scription. However, if a negative regulatory protein binds to the enhancer and a positive regulatory protein binds to the promoter element, the result will depend on the interaction between the two regulatory proteins. If the negative regulatory protein has a strong effect, the gene will be repressed. In this case, the enhancer is acting as a silencer element.

Enhancer elements and promoter elements appear to bind many of the same proteins. This implies that both types of regulatory elements affect transcription by a similar mechanism, probably involving interactions of the regulatory proteins as described before. Interestingly, there appear to be a relatively small number of regulatory proteins that control transcription. Therefore, by combining relatively few regulatory proteins in particular ways, the transcription of different arrays of genes is regulated, and a large number of cell types is specified. This is called **combinatorial gene regulation**.

*K*EYNOTE

Eukaryotic protein-coding genes contain both promoter elements and enhancer elements. Some promoter elements are required for transcription to begin. Other promoter elements have a regulatory function; these are specialized for the gene they control, binding specific regulatory proteins that control expression of the gene. Specific regulatory proteins bind also to the enhancer elements and activate transcription through their interaction with proteins bound to the promoter elements. Enhancer elements and promoter elements appear to bind many of the same proteins. This implies that both types of regulatory elements affect transcription by a similar mechanism, probably involving interactions of the regulatory proteins.

DNA BINDING PROTEINS AND THE REGULATION OF TRANSCRIPTION. Many of the transcription factors and regulatory proteins involved in the process of transcription initiation in eukaryotes (and prokaryotes) are DNA-binding proteins. They interact with specific DNA sequences and exert an effect on transcription initiation. For a number of these proteins, but by no means all, certain well-defined structural motifs of the protein are responsible for binding of the protein to the DNA. Examples of some of the structural motifs (called *DNA-binding domains*), the helix-turn-helix, zinc finger, and leucine zipper, are shown in Figure 17.3.

CHROMOSOME CHANGES AND TRANSCRIPTIONAL CONTROL. As we learned in Chapter 11, the eukaryotic chromosome consists of DNA complexed with histones (to form nucleosomes) and nonhistone chromosomal proteins. In this section we will learn

~ FIGURE 17.3

Examples of the structural motifs (DNA-binding domains) found in DNA-binding proteins such as transcription factors and transcription regulator proteins: a) Zinc finger motif: shown is a computer-generated ribbon diagram. Zinc fingers constitute the DNA recognition domains of many DNA regulatory proteins and are so-named for their resemblance to fingers projecting from the protein. Characteristically two cysteine amino acids and two histidine amino acids are positioned to bind a zinc molecule. The region containing the histidines is in the form of an α-helix; it is this coiled coil that binds in the major groove of DNA; b) Leucine zipper motif: shown is a computer-generated model. Leucine zipper proteins are dimers with each leucine zipper domain consisting of two helical regions. The name derives from the presence of leucines (L) at every seventh position in the region at the carboxy end of the protein. This positioning puts the leucines all on the same face of the amino acid helix and facilitates the binding together of the two proteins to form a coiled coil because of the hydrophobicity of the leucines. The amino-terminal recognition helices of the proteins have a high level of positively charged amino acids (+)—this end of the dimer binds to the DNA; c) Helix-turn-helix motif: shown is a computer-generated model of the helix-turn-helix-containing bacteriophage lambda repressor protein (two polypeptides colored dark blue and yellow) bound to DNA.

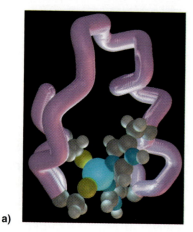
a)

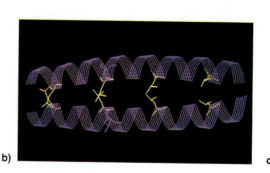
b)

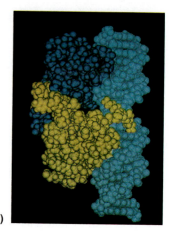
c)

about the relationship between chromosome structure and transcriptional control.

DNase Sensitivity and Gene Expression. It was hypothesized that the chromosome structure in the area of a gene being transcribed was looser than that in the area of a transcriptionally inactive gene. This hypothesis was examined by testing the susceptibility of transcriptionally active and transcriptionally inactive chromosome regions to digestion by DNase I (an endonuclease). DNase I can only digest DNA to which it is accessible, so if the hypothesis is correct, transcriptionally active DNA should be more readily digested than transcriptionally inactive DNA.

A study of globin and ovalbumin synthesis in chickens exemplifies a test of the hypothesis. In the nuclei of **erythroblasts** (red blood cell precursors) from chick embryos, the globin genes are active and hemoglobin is made, while the ovalbumin genes are inactive. Conversely, in hen oviduct nuclei, the ovalbumin gene is actively transcribed but the globin genes are not.

The experimental approach used to demonstrate changes in DNase I sensitivity of DNA sequences involves the use of specific cloned genes as probes and the Southern blotting technique (see Chapter 15). Figure 17.4 shows how this approach was used for the work with the chicken globin gene and ovalbumin gene just described. Chromatin was extracted from erythroblast nuclei and, as a control, from cells growing in culture that were not making globin. Samples were treated with different concentrations of DNase I; DNA was then extracted from each chromatin sample and digested with the restriction enzyme *Bam*HI (see Chapter 15), which cuts the DNA at sites on either side of the globin gene sequence. The two *Bam*HI sites are 4.5 kb apart on the chromosomal DNA. The resulting DNA fragments were separated by agarose gel electrophoresis and transferred to nitrocellulose filters using the Southern blotting technique. The filters were hybridized with a radioactive globin DNA probe to determine the sizes of the globin gene DNA fragments that were present following DNase digestion.

~ FIGURE 17.4

Relationship of gene expression and sensitivity to DNase. (a) In erythroblasts, which make globin, globin DNA is sensitive to DNase I. (b) In erythroblasts, which do not make ovalbumin, ovalbumin DNA is resistant to DNase.

a) DNA from 14-day erythroblasts making globin and probed with the β-globin gene

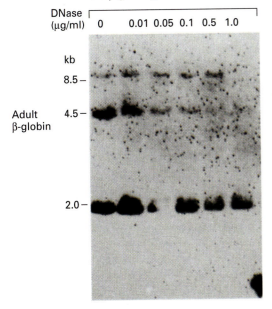

b) DNA from 14-day erythroblasts making globin and probed with the ovalbumin gene

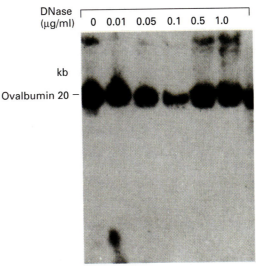

The results were as follows: For chromatin extracted from erythroblasts, a gradual disappearance of the 4.5-kb *Bam*HI-*Bam*HI, globin gene-containing DNA fragments was seen with increasing DNase concentration (Figure 17.4a), indicating that the chromatin segment containing the globin gene was sensitive to DNase digestion. That is, the chromatin organization was loosened sufficiently in the globin gene region such that the globin DNA sequence was accessible to DNase I. By contrast, in chromatin isolated from control cells not making globin, the 4.5-kb *Bam*HI DNA fragment was not digested at any DNase concentration. This result indicated that, in cells not making globin, the chromatin containing the globin gene was still highly condensed; as a result, the globin gene sequence was inaccessible to DNase so no degradation occurred. Moreover, analysis of the ovalbumin gene (which is also flanked on either side by *Bam*HI sites so that cutting with *Bam*HI produces a 20-kb DNA fragment) in chromatin from erythroblasts using a cloned ovalbumin gene probe showed that the ovalbumin gene was not sensitive to digestion by DNase I, as evidenced by the presence of the 20-kb *Bam*HI-*Bam*HI DNA fragment at all DNase I concentrations (Figure 17.4b). The latter result is explained by the fact that ovalbumin is not made in erythroblasts, so that region of the genome has not become loosened.

From similar experiments with other systems it has been found that almost all transcriptionally active genes have an increased DNase I sensitivity, indicating a looser chromosome structure. The extent of the DNase I-sensitive region varies with the particular gene, ranging from a few kilobase pairs around the gene to as much as 20 kb of flanking DNA. Note, however, that the increased DNase I sensitivity does not mean that the DNA is not organized into nucleosomes, only that the chromosome is less highly coiled in those regions. In fact, recent experiments show that the histone octamer can "step around" the transcribing RNA polymerase, indicating that there is no reason for the nucleosome to disassemble for transcription to occur.

DNase Hypersensitive Sites. More detailed studies of the regions of DNA around transcriptionally active genes have shown that certain sites, called **hypersensitive sites** or **hypersensitive regions**, are even more highly sensitive to digestion by DNase I. These sites or regions are typically the first to be cut with DNase I. Most, but not all, DNase-hypersensitive sites are in the regions upstream from the start of transcription, probably corresponding to the DNA sequences where RNA polymerase and other gene regulatory proteins bind.

KEYNOTE

Chromosome regions that are transcriptionally active have looser DNA-protein structures than chromosome regions that are transcriptionally inactive, resulting in sensitivity of the DNA to digestion by DNase. The promoter regions of active genes typically have an even looser DNA-protein structure, resulting in hypersensitivity to DNase.

Histones and Gene Regulation.   In the chromosome, DNA is wrapped around histones to form nucleosomes. In the evolutionary sense, the histones are extremely conserved; their amino acid sequence, structure, and function are all very similar among all eukaryotic organisms. Furthermore, from cell to cell in an organism, the five different types of histones are arranged in an essentially constant way along the DNA. Clearly, this uniform distribution of histones along the DNA gives a system nowhere near the specificity needed for controlling the expression of thousands of genes. Nonetheless, histones do become modified, primarily through acetylation and phosphorylation, thereby altering their ability to bind to the DNA. If the gene that they cover is to be expressed, the histones must be modified to loosen their grip on the DNA or be displaced from the DNA so that the DNA strands can interact with transcription factors and/or regulatory proteins. In essence, the histones act as *general repressors of transcription*. That is, when promoter elements (particularly the TATA box, which is important for transcription initiation: see Chapter 13) are associated with the histones of a nucleosome core, they cannot be found by the regulatory proteins and transcription factors that must bind to them to initiate transcription. (These regulatory proteins and transcription factors are in the nonhistone class of chromosomal proteins.)

Some test-tube experiments gave important insights into how histones influence gene expression. These experiments investigated what happened to transcription when histones, promoter-binding proteins, and enhancer-binding proteins interact with DNA. In brief, the results were as follows:

1. If DNA is mixed simultaneously with histones and promoter-binding proteins, the histones compete more strongly for promoters and form nucleosomes at TATA boxes. As a result, the promoter-binding proteins cannot bind and transcription cannot occur.
2. If DNA is mixed first with promoter-binding proteins, the proteins assemble on TATA boxes and other promoter elements and block nucleosome assembly on those sites when histones are subsequently added. As a result, transcription occurs.
3. If DNA is mixed *simultaneously* with histones, promoter-binding proteins, and enhancer-binding proteins, the enhancer-binding proteins bind to enhancers and help promoter-binding proteins to bind to TATA boxes by blocking access by the histones. As a result, transcription occurs.

These results indicate that histones are effective repressors of transcription, but other proteins can overcome that repression.

How does gene activation occur in the cell? All of the details are not known at this time. One model is as follows: Histones block transcription by forming nucleosomes on TATA boxes. Promoter-binding proteins are unable to disrupt the nucleosomes on the TATA boxes. However, enhancer-binding proteins bind to the enhancers (displacing any histones that are bound there) and interact with the histones in the TATA-bound nucleosomes. This causes the nucleosome core particle to break up, and promoter-binding proteins can now bind to the freed DNA.

Other research has established a correlation between the level of histone acetylation and transcriptional activity. Specifically, histones in transcriptionally active chromatin are hyperacetylated, while histones in transcriptionally inactive chromatin are hypoacetylated. How might increased histone acetylation loosen chromatin organization and help facilitate transcription? Acetylation occurs on lysines located in regions of the histones on the outside of the nucleosome. When the histones become acetylated, nucleosome conformation is altered and higher-order chromatin structure is destabilized. As a result, the DNA associated with the nucleosomes becomes more accessible to transcription factors and the generally repressive action of histones on transcription is overcome.

KEYNOTE

Histones repress transcription by assembling nucleosomes on TATA boxes associated with genes. Gene activation occurs when proteins become bound to enhancers and disrupt the nucleosomes on the

TATA boxes, thereby allowing the appropriate proteins to bind to the promoter elements to initiate transcription.

DNA METHYLATION AND TRANSCRIPTIONAL CONTROL. Once DNA has been replicated within the eukaryotic cell, a small proportion of bases becomes chemically modified through the action of enzymes. Potentially, modified DNA bases could act as signals in the processes in which DNA is involved, such as replication, repair, and transcription, although in most cases we have no information about this activity. A particular modified base has received significant attention since it has been correlated with gene activity in a number of cases. This base, 5-methylcytosine (5^mC), is generated by the modification of DNA shortly after replication by the action of the enzyme DNA methylase (Figure 17.5).

The percentage of 5^mC in eukaryotic DNA varies over a wide range. In mammalian DNA, for example, about 3 percent of cytosines are present as 5^mC. The DNAs of less complex eukaryotes, such as *Drosophila*, *Tetrahymena*, and yeast, appear to contain very little, if any, 5^mC.

The distribution of 5^mC is nonrandom, with most of the 5^mC (greater than 90 percent in mammalian DNA) being found in the sequence CG. Since this sequence has twofold rotational symmetry in DNA, the cytosines are symmetrically located on opposite strands of the DNA molecule. These CG sequences form part of some restriction sites, which allow the use of restriction enzymes to study methylation of a segment of DNA, because many enzymes with cytosine in their recognition sequence fail to cleave the DNA when cytosine is methylated. The enzyme *Hpa*II ("hepa-two"), for example, will cleave DNA at the sequence 5'-CCGG-3' but not if the internal cytosine of the two is methylated, that is, if it is 5'-C^{m}CGG-3'. The enzyme *Msp*I ("M-S-P-one") also will cleave the same CCGG sequence, but unlike *Hpa*II, it *will* cleave the methylated sequence C^{m}CGG.

The use of *Hpa*II and *Msp*I in analyzing the extent of methylation of a segment of DNA is illustrated by the following theoretical example. Consider a fragment of genomic DNA that contains three CCGG sequences (Figure 17.6). If the sequences are not methylated (Figure 17.6a), then both *Hpa*II and *Msp*I will cleave the DNA at these sequences to produce four DNA fragments of discrete sizes. These DNA fragments can be identified following (1) electrophoresis of the digested DNA on agarose gels, (2)

Southern blotting to a membrane filter, and (3) probing the fragments with a labeled DNA probe that will hybridize with any and all of the fragments. In this case DNA fragments of identical sizes will be produced from both digests. If one of the CCGG sequences is methylated—for example, the first one from the left (Figure 17.6b)—then all three sequences will be cleaved by *Msp*I, whereas only the two unmethylated sequences will be cleaved by *Hpa*II. As a result, the array of DNA fragments produced by digestion of the same DNA piece with the two enzymes will be different, as indicated in the stylized blot result. In actual practice, a piece of DNA is digested separately by the two enzymes *Hpa*II and *Msp*I, and if the results show that the *Msp*I has digested the DNA into more pieces than *Hpa*II, the interpretation is that one of the CCGG sequences is methylated; that is, C^{m}CGG.

With the *Hpa*II/*Msp*I approach a large number of systems have been examined in an attempt to determine whether there is a relationship between gene methylation and transcriptional activity of the gene. For at least thirty genes examined a negative correlation exists between DNA methylation and transcription; that is, there is a lower level of methylated DNA (hypomethylated DNA) in transcriptionally active genes compared with transcriptionally inactive genes. We must exercise caution in taking this broad generalization too far, because not all methylated C nucleotides in a gene region become demethylated when the gene is expressed. And, as was stated earlier, some organisms do not have significant amounts of methylated C in their DNA. Further, the fact that a correlation exists between the level of DNA methylation and the transcriptional state of a gene provides no information about cause and effect. In other words, we do not know whether methylation changes are necessary for the onset of transcriptional activity (in those organisms in which there is signifi-

~ **FIGURE 17.5**

Production of 5-methylcytosine from cytosine in DNA by the action of the enzyme DNA methylase.

Cytosine (in DNA)　　　　**5-Methylcytosine (5mC)**

~ FIGURE 17.6

Effect of 5-methylcytosine on cleavage of DNA with *Hpa*II and *Msp*I: (a) Unmethylated (CCGG) sequence; (b) Methylated (C^mCGG) sequence (the methyl group is on the second C).

a) Unmethylated

DNA fragment, no methylated CCGG sequences; all sites can be cleaved with *Hpa*II and *Msp*I

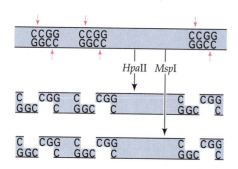

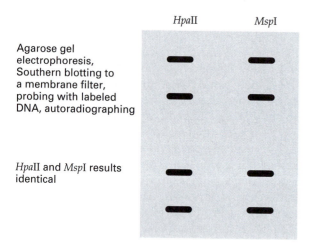

Agarose gel electrophoresis, Southern blotting to a membrane filter, probing with labeled DNA, autoradiographing

*Hpa*II and *Msp*I results identical

b) Methylated

DNA fragment with one methylated C^mCGG site that can be cleaved by *Msp*I but not by *Hpa*II

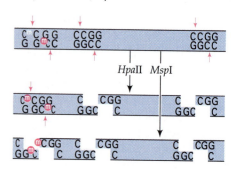

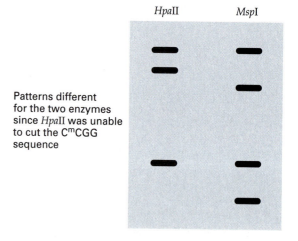

Patterns different for the two enzymes since *Hpa*II was unable to cut the C^mCGG sequence

cant DNA methylation) or whether the changes are the result of transcriptional activity initiated by other means.

Nonetheless, there are now some observations indicating the importance of methylation in gene expression. For example, a gene encoding an enzyme that adds methyl groups to DNA has been shown to be essential for development in mice; that is, mice homozygous for mutations in that gene die at an early stage. Also, methylation is involved in the development of fragile X syndrome, which we discussed in Chapter 7, pp. 205–206. Recall that this syndrome is the leading cause of inherited mental retardation. The syndrome develops following expansion of a triplet repeat (a repeated 3-base-pair sequence) in the *FMR-1* gene and abnormal methylation of the gene to the point that transcription of the *FMR-1* gene is silenced.

𝕂EYNOTE

The DNA of many eukaryotes has been shown to be methylated at a certain proportion of the bases, with most methylations occurring on cytosine residues. Transcriptionally active genes exhibit lower levels of DNA methylation than transcriptionally inactive genes.

TRANSCRIPTIONAL CONTROL OF GENE EXPRESSION IN EUKARYOTES. In this section we look at some examples of the short-term regulation of gene expression in eukaryotes. This regulation involves rapid responses to changes in environmental or physiological conditions. The responses occur at the transcriptional level and/or at the level of post-translational modulation of protein activity. We will concentrate on the former here. As we examine these systems, keep in mind how one form of short-term regulation of gene expression in bacteria involves operons so that you can compare the molecular events involved.

Galactose-Utilizing Genes of Yeast. In the galactose-utilizing system of yeast, three genes, *GAL1*, *GAL7*, and *GAL10*, encode enzymes that function in a pathway to metabolize the sugar galactose (Figure 17.7a). The enzymes are galactokinase (*GAL1* gene), galactose transferase (*GAL7* gene), and galactose epimerase (*GAL10* gene). (Note that genes are in italics while the proteins they encode are not.) Once D-glucose 6-phosphate is made, it enters the glycolytic pathway, where it is acted upon by enzymes that are always present in the cell as a result of constitutive gene expression. In the absence of galactose, the *GAL* genes are not transcribed. When galactose is added, there is a rapid, coordinate induction of transcription of the *GAL* genes, and therefore a rapid production of the three galactose-utilizing enzymes.

Genetic studies have shown that the *GAL1*, *GAL7*, and *GAL10* genes are located near each other, but do not constitute an operon (Figure 17.7b). Genetic studies have also identified an unlinked regulatory gene, *GAL4*, which encodes the GAL4 protein. The GAL4 protein is similar to the lambda repressor in that there are two domains. One is a DNA-binding domain and the other is for activating transcription. The DNA binding domain of GAL4 is a zinc finger. GAL4 binds to a specific sequence in the DNA called *upstream activator sequence–galactose* (UAS$_G$); this sequence is a promoter element. A UAS$_G$ is located between the divergently transcribed *GAL1* and *GAL10* genes. The UAS$_G$ consists of four similar, 17-bp sequences, each of which shows twofold symmetry, and each of which binds GAL4. GAL4 binds to each 17-bp sequence as a dimer (much like the lambda repressor and Cro protein do to their target sequence; see Chapter 16). Transcription occurs in both directions from the UAS$_G$—the *GAL10* gene is transcribed to the left, and the *GAL1* gene is transcribed to the right (see Figure 17.7b).

GAL4 is similar in functional organization to lambda repressor; that is, there is a DNA-binding domain and a domain required for activating transcription. This was shown in a "domain-swapping" experiment. Using recombinant DNA techniques, different DNA fragments from the *GAL4* genes were fused to DNA fragments from bacterial repressors,

The galactose metabolizing pathway of yeast. (a) The steps catalyzed by the enzymes encoded by the *GAL* genes; (b) The organization of the *GAL* protein-coding genes. The arrows indicate the direction of transcription. (UDP = uridine diphosphate)

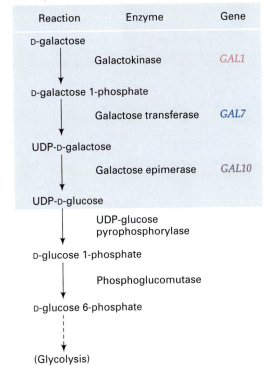

a) **Pathway**

Reaction	Enzyme	Gene
D-galactose		
	Galactokinase	*GAL1*
D-galactose 1-phosphate		
	Galactose transferase	*GAL7*
UDP-D-galactose		
	Galactose epimerase	*GAL10*
UDP-D-glucose		
	UDP-glucose pyrophosphorylase	
D-glucose 1-phosphate		
	Phosphoglucomutase	
D-glucose 6-phosphate		
(Glycolysis)		

b) **Gene organization**

UAS$_G$

enabling protein hybrids to be made. One protein hybrid had the DNA binding domain of a bacterial repressor and the GAL4 activator domain. This hybrid could not function in yeast because there are no sequences recognized by the bacterial repressor DNA binding domain. However, if a DNA sequence to which the bacterial repressor DNA binding domain can bind is placed next to a gene and introduced into yeast, the hybrid protein can bind to it, and if galactose is added, transcription of the gene adjacent to the bacterial repressor sequence is induced.

A model for the regulation of transcription of the GAL genes is presented in Figure 17.8. In the absence of galactose, a GAL4 dimer is bound to a UAS$_G$. Another protein, GAL80 (encoded by the *GAL80*

~ **Figure 17.8**

Model for the activation of the *GAL* genes of yeast.

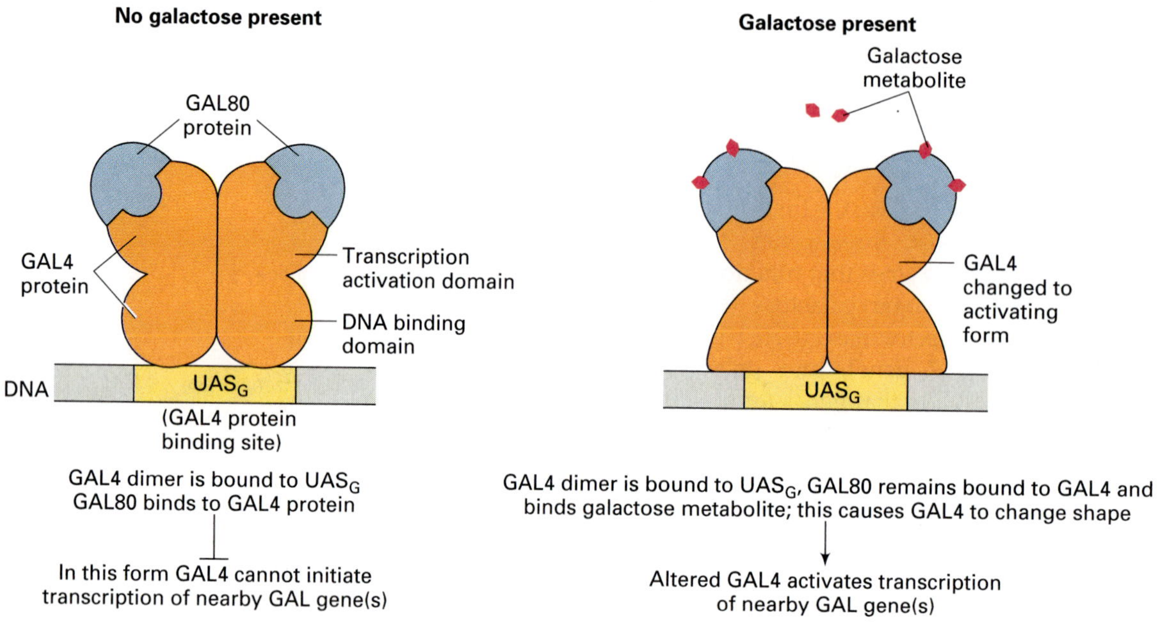

No galactose present

GAL80 protein

GAL4 protein

Transcription activation domain

DNA binding domain

DNA

UAS_G

(GAL4 protein binding site)

GAL4 dimer is bound to UAS_G
GAL80 binds to GAL4 protein

In this form GAL4 cannot initiate transcription of nearby GAL gene(s)

Galactose present

Galactose metabolite

GAL4 changed to activating form

UAS_G

GAL4 dimer is bound to UAS_G, GAL80 remains bound to GAL4 and binds galactose metabolite; this causes GAL4 to change shape

Altered GAL4 activates transcription of nearby GAL gene(s)

gene) is bound to the domain of GAL4 that is needed to turn on transcription. No transcription can occur under these circumstances. When galactose is added, a metabolite of galactose binds to GAL80 and this causes GAL4 to become phosphorylated (i.e., phosphate groups are added to certain amino acids). This changes the shape of the GAL4-GAL80 complex so that GAL4 is changed to the activating form needed to turn on transcription of the *GAL* enzyme-coding genes. In this system, then, the GAL4 protein acts as a positive regulator (activator), and galactose is an effector molecule.

Steroid Hormone Regulation of Gene Expression in Animals. Short-term regulation enables an organism or cell to adapt rapidly to changes in its physiological environment so that it can function as optimally as possible. While lower eukaryotes exhibit some cell specialization, higher eukaryotes are differentiated into a number of cell types, each of which carries out a specialized function or functions. The cells of higher eukaryotes are not exposed to rapid changes in environment as are cells of bacteria and of microbial eukaryotes. The environment to which most cells of higher eukaryotes are exposed, the intercellular fluid, is relatively constant in the nutrients, ions, and other important molecules it supplies. The intercellular fluid, in a sense, is analogous to a bacterium's growth medium. The constancy of the cell's environment is, in part, maintained through the action of chemicals called *hormones*, which are secreted by vari-

ous cells in response to signals and which circulate in the blood until they stimulate their target cells. Elaborate feedback loops control the amount of hormone secreted, controlling the response so that appropriate levels of chemicals in the blood and tissues are maintained.

A hormone, then, is an effector molecule that is produced in low concentrations by one cell and that causes a physiological response in another cell. As shown in Figure 17.9, some hormones—for example, steroid hormones—act by binding to a specific cytoplasmic receptor called a steroid hormone receptor (SHR), and then the complex binds directly to the cell's genome to regulate gene expression. Other hormones—for example, polypeptide hormones—may act at the cell surface to activate a membrane-bound enzyme, adenyl cyclase, which produces cyclic AMP (cAMP) from ATP (see Chapter 16). The cAMP acts as an intracellular signaling compound (called a *second messenger*) to activate the cellular events associated with the hormone. The exact molecular mechanisms of these reactions are not known.

A key to hormone action is that hormones act on specific target cells that have receptors capable of recognizing and binding that particular hormone. For most of the polypeptide hormones (e.g., insulin, ACTH, vasopressin), the receptors are on the cell surface, whereas the receptors for steroid hormones are inside the cell.

Since the action of steroid hormones on gene expression has been well studied, we shall concen-

~ **FIGURE 17.9**

Mechanisms of action of polypeptide hormones and steroid hormones.

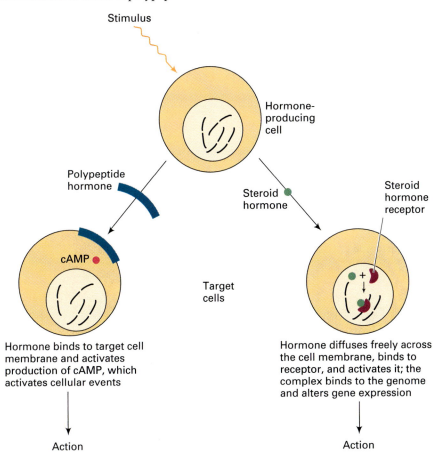

Stimulus

Hormone-producing cell

Polypeptide hormone

Steroid hormone

Steroid hormone receptor

cAMP

Target cells

Hormone binds to target cell membrane and activates production of cAMP, which activates cellular events

Hormone diffuses freely across the cell membrane, binds to receptor, and activates it; the complex binds to the genome and alters gene expression

Action

Action

trate on steroid hormones here. Steroid hormones have been shown to be important to the development and physiological regulation of organisms ranging from fungi to humans. Figure 17.10 gives the structures of four of the most common mammalian steroid hormones. All have a common four-ring structure; the differences in the side groups are responsible for their different physiological effects.

Table 17.1 presents examples of the induction of specific proteins by selected steroid hormones. As can be seen, there are tissue-specific effects. Estrogen, for example, induces the synthesis of the protein prolactin in the rat pituitary gland, of the protein vitellogenin in frog liver, and of the proteins conalbumin, lysozyme, ovalbumin, and ovomucoid in the hen oviduct. Glucocorticoids, for example, induce the synthesis of growth hormone in the rat pituitary gland and of the enzyme phosphoenolpyruvate carboxykinase in the rat kidney. The specificity of the response to steroid hormones is controlled by the hormone receptors. That is, only target tissues that contain receptors for a particular steroid hormone can

respond to that hormone, and this directly controls the sites of the hormone's action. With the exception of receptors for the steroid hormone glucocorticoid, which are widely distributed among tissue types, steroid receptors are found in a limited number of target tissues. While steroid hormones have well-substantiated effects on transcription, steroids also can affect the stability of mRNAs, and possibly the processing of mRNA precursors.

Mammalian cells contain between 10,000 and 100,000 steroid hormone receptor molecules. The SHRs are proteins with a very high affinity for their respective steroid hormones. All steroid hormones work in the same general way. For example, in the absence of a particular steroid hormone, the appropriate SHR is found in the cell associated with a large multiprotein complex of proteins called *chaperones*, one of which is Hsp90. In this association, the SHR is inactive. Data from studies of mutant yeast strains suggest that chaperone proteins have an active role in keeping the SHRs functional. When a steroid hormone such as glucocorticoid enters a cell, it binds to its specific SHR molecule,

~ FIGURE 17.10

Structures of some mammalian steroid hormones. (a) Hydrocortisone, which helps regulate carbohydrate and protein metabolism; (b) Aldosterone, which regulates salt and water balance; (c) Testosterone, which is used for the production and maintenance of male sexual characteristics; (d) Progesterone, which, with estrogen, prepares and maintains the uterine lining for embryo implantation.

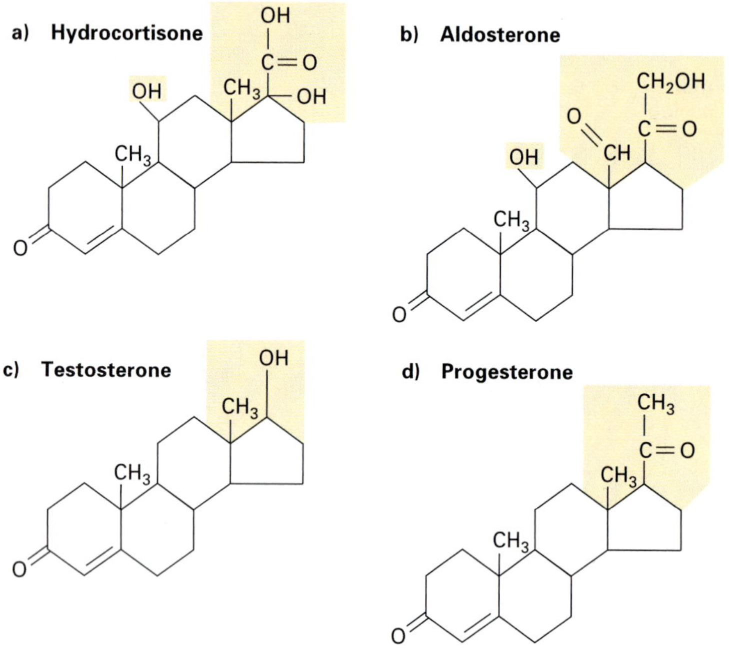

~ TABLE 17.1

Examples of Proteins Induced by Steroid Hormones

HORMONE	TISSUE	INDUCED PROTEIN
Estrogen	Oviduct (hen)	Conalbumin[a] Lysozyme[a] Ovalbumin[a] Ovomucoid[a]
	Liver (rooster)	Vitellogenin Apo-very low density Lipoprotein (apoVLDL)
	Liver (frog)	Vitellogenin
	Pituitary gland (rat)	Prolactin
Glucocorticoids	Liver (rat)	Tyrosine aminotransferase Tryptophan oxygenase
	Kidney (rat)	Phosphoenolypyruvate carboxykinase
	Pituitary gland (rat)	Growth hormone
	Oviduct (hen)	Avidin[a] Conalbumin[a] Lysozyme[a] Ovomucoid[a]
	Uterus (rat)	Uteroglobin
Testosterone	Liver (rat)	α-2u Globulin
	Prostate gland (rat)	Aldolase

[a]Egg-white proteins.

~ FIGURE 17.11

Model for the action of the steroid hormone glucocorticoid in mammalian cells.

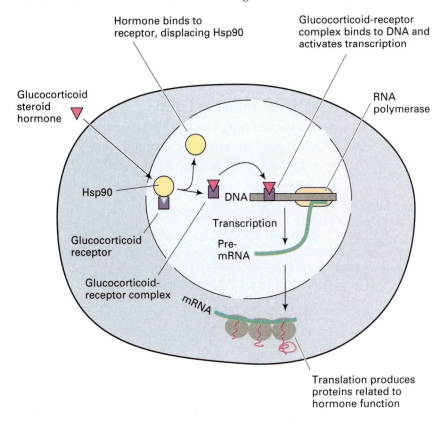

Hormone binds to receptor, displacing Hsp90

Glucocorticoid-receptor complex binds to DNA and activates transcription

Glucocorticoid steroid hormone

RNA polymerase

Hsp90

DNA

Transcription

Glucocorticoid receptor

Pre-mRNA

Glucocorticoid-receptor complex

mRNA

Translation produces proteins related to hormone function

displacing Hsp90, and forming a glucocorticoid-SHR complex (Figure 17.11). In this diagram, the SHR is located in the nucleus, although there is a controversy concerning the exact location of SHRs in the cell. That is, SHRs are found in both the cytoplasm and the nucleus, and it is not clear whether both are active in binding the steroid hormone, although the current thinking is that the nuclear SHRs are the active ones. When the steroid hormone is bound to the SHR, the receptor protein becomes activated, and the complex is now found only in the nucleus. The steroid-activated receptor complex now binds to specific DNA regulatory sequences, activating and/or repressing the transcription of the specific genes controlled by the hormone. For genes turned on by the hormone, the new mRNAs appear within minutes after a steroid hormone encounters its target cell, enabling new proteins to be produced rapidly. The DNA binding domains of many steroid hormone receptor proteins are zinc fingers.

All the genes regulated by a specific steroid hormone have in common a DNA sequence to which the steroid-receptor complex binds. The binding regions are called **steroid hormone response elements** or **HREs**. The H in the acronym is replaced with another letter to indicate the specific steroid involved. Thus, GRE is the glucocorticoid response element and ERE

is the estrogen response element. The HREs are located, often in multiple copies, in the enhancer regions of genes. The GRE, for example, is located about 250 bp upstream from the transcription starting point. The consensus sequence for GRE is AGAACANNNT-GTTCT, where N is any nucleotide. The ERE consensus sequence is AGGTCANNNTGACCT. Note that, for both of these HREs, the sequences on each side of the Ns are complementary; that is, the sequence shows twofold symmetry.

How the hormone-receptor complexes, once bound to the correct HREs, regulate transcriptional levels is not completely known. Potentially, functional interactions arise among the hormone-receptor complexes and with transcription factors in the transcription initiation complex. To this end, recall that multiple HREs are present for many genes and therefore multiple hormone-receptor complexes can bind to each gene. These interactions may facilitate the initiation of transcription by RNA polymerase II. It is presumed that each steroid hormone regulates its specific transcriptional activation by the same general mechanism. The unique action of each type of steroid results from the different receptor proteins and HREs involved.

Lastly, it is of particular interest that in different types of cells the same steroid hormone may activate different sets of genes, even though the various cells have the same SHR. This is because many regulatory proteins bind to both promoter elements and enhancers to regulate gene expression (see the discussions earlier in this chapter and Chapter 13). Thus a steroid-receptor complex can activate a gene only if the correct array of other regulatory proteins is present. Since the other regulatory proteins are specific for the cell type, different patterns of gene expression can result.

In sum, steroid hormones act as positive effector molecules, and SHRs act as regulatory molecules. When the two combine, the resulting complex binds to DNA and regulates gene transcription, resulting in a large (and specific) increase or decrease in cellular mRNA levels. The specific responses characteristic of each steroid hormone result from the fact that receptors are found in only certain cell types, and each of those cell types contains different arrays of other cell type–specific regulatory proteins that interact with the steroid-receptor complex to activate specific genes.

Similar systems operate to control gene expression in response to other signals. For example, an unusual increase in temperature brings about what is called a *heat shock response* in which the transcription of some genes is turned off and that of some other genes (about twenty or so) is turned on. The latter are called *heat shock genes*. The heat shock response is seen in both prokaryotes and eukaryotes. In eukaryotes, the heat shock genes have in common a specific response element called the *heat shock response element* or HSE. The consensus sequence for the HSE is CNGAANTTCNG, where N is any nucleotide. Unlike the steroid response elements, the HSEs are located in the promoter region of the gene, usually about 15 bp upstream from the initiation of transcription. There may be multiple copies of the HSE associated with some heat shock genes. How does the system work? In heat-shocked cells, a specific transcription factor, heat shock transcription factor (HSTF), normally present in an inactive form, becomes activated and binds to DNA at sites including the HSEs and this begins the process of transcribing the adjacent genes.

KEYNOTE

In higher eukaryotes one of the well-studied systems of short-term gene regulation is the control of enzyme synthesis by hormones. Polypeptide hormones act at the cell surface, activating a system to produce cAMP, which acts as a second messenger to control gene activity. Steroid hormones exert their action by forming a complex with a specific receptor protein, thereby activating the receptor; the complex then binds directly to the cell's genome to regulate the expression of specific genes. The specificity of hormone action is caused by the presence of hormone receptors in only certain cell types and by interactions of steroid-receptor complexes with cell type–specific regulatory proteins.

Hormone Control of Gene Expression in Plants. Many chemicals have been identified that play important roles in controlling growth and development in plants; these chemicals are called *plant hormones*. Plant hormones fall into five main classes, namely ethylene, abscisic acid, auxins, cytokinins, and gibberellins (Figure 17.12). These extracellular hormones are responsible for many activities. Ethylene, for example, induces fruit ripening, flower maturation, and plant senescence, among other things. The molecular actions of plant hormones are generally ill-defined.

Among the best-characterized plant hormones are *gibberellins*. Like many steroid hormones in mammals, gibberellins stimulate transcription and thereby result in an increase in the production of specific proteins. Those proteins are responsible for profound changes in plant cell form and cell differentiation.

Gibberellins have been shown to have many effects when applied to plants, such as making certain mutant dwarf plants grow tall or making normal plants grow taller. Gibberellins are also important because they stimulate germination of some seeds; for example, those of barley (Figure 17.13). In this process the embryo of the seed produces gibberellins that diffuse to the aleurone layer, the outermost layer of the endosperm. As a result of the effect of the gibberellins on gene expression, particularly on the gene for the enzyme α-amylase, the aleurone layer synthesizes and secretes α-amylase, which breaks down the endosperm, releasing nutrients that enable the embryo to grow. By the time the endosperm has been exhausted, the young, developing plant can rely on photosynthesis to obtain nutrients for continued growth.

Even though the action of some gibberellins has been well described, no gibberellin receptor has been detected to date, and the molecular aspects of gibberellins' action are incompletely understood.

~ FIGURE 17.12

Chemical structures of five plant hormones.

Ethylene

Abscisic acid

Indoleacetic acid (an auxin)

Zeatin (a cytokinin)

Gibberellic acid (a gibberellin)

~ FIGURE 17.13

Effect of gibberellins on the germination of barley seeds.

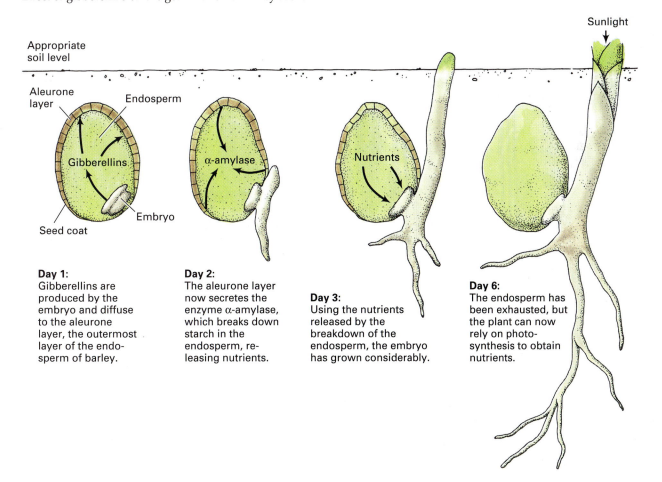

Day 1: Gibberellins are produced by the embryo and diffuse to the aleurone layer, the outermost layer of the endosperm of barley.

Day 2: The aleurone layer now secretes the enzyme α-amylase, which breaks down starch in the endosperm, releasing nutrients.

Day 3: Using the nutrients released by the breakdown of the endosperm, the embryo has grown considerably.

Day 6: The endosperm has been exhausted, but the plant can now rely on photosynthesis to obtain nutrients.

RNA Processing Control

RNA processing control regulates the production of mature RNA molecules from precursor-RNA molecules. As we discussed in Chapter 13, all three major classes of RNA molecules, which include mRNA, tRNA, and rRNA, can be synthesized as precursor molecules. Perhaps the most significant range of control is exhibited by the mRNA class and involves control of processing of mRNA precursor molecules. We will discuss two types of RNA processing control: choice of alternative poly(A) sites (Figure 17.14a) and choice of alternative splice sites (Figure 17.14b). Small nuclear ribonucleoprotein particles (snRNAs) may well play a role in differential processing.

CHOICE OF A POLY(A) SITE. Recall that the poly(A) site is a sequence in a pre-mRNA molecule that specifies the position at which a poly(A) tail is

added (Chapter 13, p. 550). A classic example of regulation by choice of poly(A) site concerns the production of immunoglobulin M (IgM), one of the immunoglobulin molecules involved in the immune response system during B cell differentiation. Five copies of an IgM monomer aggregate to form the functional pentameric IgM molecule. The IgM molecules, each of which consists of two copies of a small protein (the light chain) and two copies of a large protein (the heavy chain), are manufactured throughout the life of a cell (Figure 17.15). The heavy chain is encoded by the μ gene. Transcription of this gene produces a premRNA that can be processed at two different poly(A) addition sites to produce heavy chains of different lengths. If the cell is in an early developmental stage (B cells), the longer heavy chain is produced and the resulting IgM associates with the cell membrane. If the cell is in a later developmental stage (plasma cells), the shorter heavy chain is produced and the

~ FIGURE 17.14

Models for control of the processing of mRNA precursors in eukaryotes by choice of (a) poly(A) sites or (b) splice sites. (p = transcription initiation site [promoter]; t = transcription termination site.)

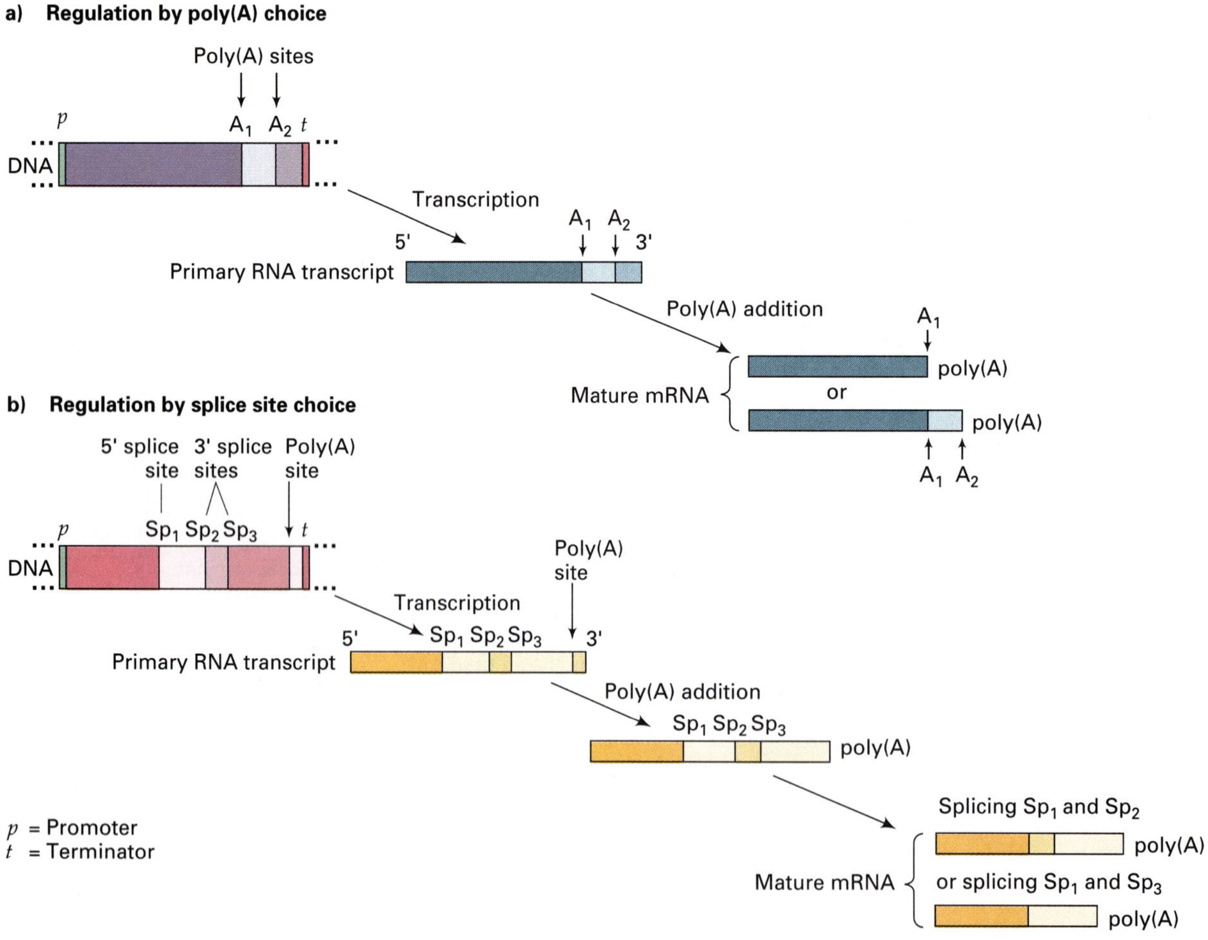

~ FIGURE 17.15

Structure of an IgM monomer. Two light (L) chains and two heavy (H) chains are linked by disulfide bridges. Each type of chain has a variable domain (V_L and V_H) characterized by variations in amino acid sequences among IgM molecules. The light chains have one constant domain (C_L), while the heavy chains have four constant domains (C_H). In all IgM molecules, constant domains have the same amino acid sequence. The variable domains collectively constitute the antigen-binding sites. Functional IgM is a pentamer of the structure shown here.

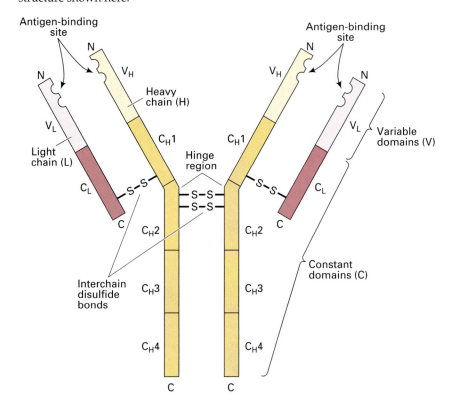

resulting IgM is secreted from the cell.

Recent evidence has shown that alternative processing of IgM pre-mRNA at different poly(A) sites is regulated by one of the factors involved in polyadenylation. Recall from our discussion in Chapter 13 (p. 551) that poly(A) addition involves the function of a number of proteins, one of which is the three-polypeptide CstF (cleavage stimulation factor) protein. CstF binds to the GU/U sequence downstream of the poly(A) site and is required for cleavage of the pre-mRNA molecule. The three polypeptides of CstF are 77, 64, and 50 kDa in size. Experiments have shown that the accumulation of the 64-kDa subunit (CstF-64) is repressed in B cells, so that only inactive complexes of 77- and 50-kDa subunits can form. That this situation is responsible for the processing pattern seen in B cells was demonstrated by experiments in which overexpression of a CstF-64 gene introduced into B cells caused processing to switch to the poly(A) site typically used in plasma cells. The model derived

from the data is that CstF has a higher affinity for the poly(A) site used to make the membrane-associated form of IgM (μm site) than the secreted form of IgM (μs site). When CstF-64 synthesis is repressed and functional CstF is limiting, as in B cells, the weaker μs site is not recognized efficiently because of its lower affinity for CstF. As a result, polyadenylation occurs at the μm site to produce the longer IgM molecule. When synthesis of CstF-64 is normal and levels of functional CstF increase, recognition of the μs site is much more efficient and processing occurs there frequently to produce the shorter IgM molecule.

CHOICE OF SPLICE SITE. A well-studied example of gene regulation by differential splicing of pre-mRNA involves sex determination in *Drosophila* (see Chapter 3). Recall that sex is determined by the X-chromosome:autosome (X:A) ratio. A number of mutations disrupt normal sex determination. Their study has led to a *regulation cascade model* for sex

~ FIGURE 17.16

Regulatory cascade for sex determination in *Drosophila*. For details, see text.

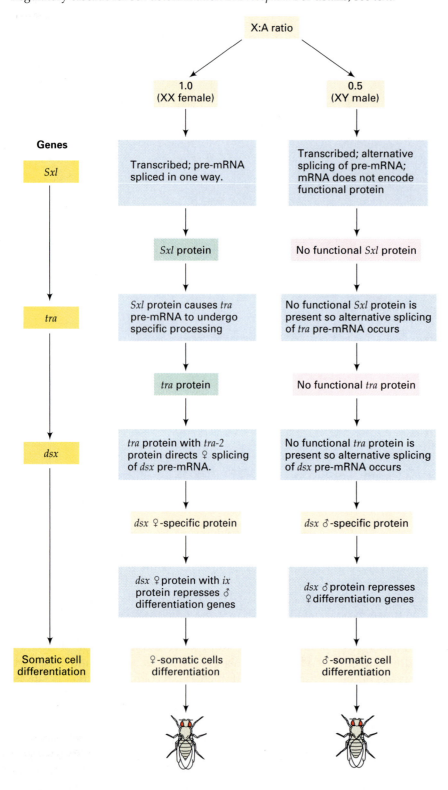

determination in *Drosophila* (Figure 17.16). First, the X:A ratio is read during development, in an as yet unknown way. For wild-type *Drosophila*, the ratio for females (XX) is 2X:2 autosome sets, and the ratio for males (XY) is 1X:2 autosome sets. This information is transmitted to the sex determination genes, which make the choice between the alternative female and male developmental pathways, starting with the master regulatory gene *Sxl* (*sex-lethal*).

In both males and females *Sxl* is transcribed to give a pre-mRNA with 8 exons. If X:A = 1 (a female), RNA processing results in the removal of all introns and, in addition, the removal of exon number 3. If X:A = 0.5 (a male), *alternative splicing* occurs in which all introns are removed but exon number 3 is left in the mature mRNA. Since exon number 3 contains a stop codon in the reading frame, the *Sxl* protein encoded by the mRNA is functional in females, but nonfunctional in males. The *Sxl* protein causes the *tra* (*transformer*) gene pre-mRNA to undergo female-specific splicing. Translation of the resulting mRNA produces a protein which, along with the *tra-2* protein, causes the *dsx* (*doublesex*) pre-mRNA to be spliced to produce a female-specific mRNA. The resulting *dsx* protein interacts with the *ix* (*intersex*) protein to turn off the genes for male differentiation and set the switch so that female somatic cells are produced. If no Sxl protein is produced, alternative splicing of *tra* pre-mRNA produces a mRNA from which no functional protein is produced. Since *tra-2* cannot act alone, alternative splicing of *dsx* pre-mRNA occurs and a male-specific protein is made from the resulting mRNA. This protein is a repressor of female differentiation genes, so the switch is set to produce male somatic cells.

KEYNOTE

Gene expression in eukaryotes can be regulated at the level of RNA processing. This type of regulation operates to determine the production of mature RNA molecules from precursor-RNA molecules. Two regulatory events that exemplify this level of control are choice of poly(A) site and choice of splice site. In both cases different types of mRNAs are produced, depending upon the choices made. There are known examples of both types of regulation in developmental systems.

Transport Control

The next level of control is **transport control**, which is the regulation of the number of transcripts that exit the nucleus to the cytoplasm. A profound difference between prokaryotes and eukaryotes is the presence of a nucleus with a surrounding membrane in the latter. This nuclear membrane potentially is a control point in gene expression.

We know that primary transcripts are processed extensively in the nucleus. Experiments have also shown that perhaps one half of the primary transcripts of protein-coding genes (the hnRNA population) never leave the nucleus, but are degraded there. How is transport of mature mRNA from the nucleus regulated? It seems that the mRNA exits through the nuclear pores, but very little is known about the export process itself or of the signals required for nuclear retention and export. Recall that processing of pre-mRNA involves snRNPs (small nuclear ribonucleoprotein particles: see Chapter 13, p. 399). There is some evidence that snRNPs are important for retaining mRNAs in the nucleus. For example, in mutants of yeast that prevent the assembly of spliceosomes (the splicing complexes formed by the association of several snRNPs bound to the pre-mRNA: see Chapter 13, p. 399), mRNA export from the nucleus is facilitated. This has led to a *spliceosome retention* model. In this model, spliceosome assembly competes with nuclear export. Thus, when pre-mRNAs are in spliceosomes undergoing processing, the RNA is retained in the nucleus, unable to interact with the nuclear pore. However, when processing is complete, the mRNA dissociates from the spliceosome, and the spliceosome remains associated with the intron. The free mRNA is able to interact with the nuclear pore, but the intron cannot. It is still not clear whether mRNAs need a specific export signal, or whether they are exported by default. It does seem, however, that the mRNAs must have the correctly methylated 5' cap in order to be exported to the cytoplasm.

KEYNOTE

The transport of mRNAs from the nucleus to the cytoplasm is another important control point in eukaryotic gene regulation. A spliceosome retention model has been proposed for this regulation. In this model, spliceosome assembly on a precursor-mRNA competes with nuclear export of the molecule. Thus, during intron removal, the spliceosome serves to keep the immature RNA in the nucleus but, when all introns have been removed and the spliceosome has dissociated, the free mature mRNA can interact with the nuclear pore and exit.

mRNA Translation Control

Messenger RNA molecules are subject to **translational control** by ribosome selection among mRNAs. Thus, differential translation can greatly affect gene expression. For example, mRNAs are stored in many vertebrate and invertebrate unfertilized eggs. In the unfertilized state, the rate of protein synthesis is very slow; however, protein synthesis increases significantly after fertilization. Since this increase occurs without new mRNA synthesis, it can be concluded that translational control is responsible. The mechanisms involved are not completely understood, but typically, stored mRNAs are associated with proteins that serve both to protect the mRNAs and to inhibit their translation. Further, the poly(A) tail is known to promote the initiation of translation. It has been found that, in general, stored, inactive mRNAs have shorter poly(A) tails (15–90 As) than active mRNAs. Thus, mRNAs synthesized in growing oocytes that are destined for storage and later translation have short poly(A) tails.

In principle, an mRNA molecule can have a short poly(A) tail either because only a short string of A nucleotides was added at the time of polyadenylation or because a normal-length poly(A) tail was added that was subsequently trimmed. At least for some messages stored in growing oocytes of mouse and frog, the latter mechanism is involved. In one example, examination of one particular mRNA in this class has shown that the pre-mRNA still in the process of intron removal has a long poly(A) tail (300–400 As) while the mature, stored message has a short poly(A) tail (40–60 As). It has been shown that the decrease in length of the poly(A) tail for this message class occurs rapidly in the cytoplasm by a deadenylation enzyme. What pinpoints a particular mRNA for rapid deadenylation, rather than a default, slow decrease in poly(A) length, is a sequence in the 3' untranslated region (3' UTR) of the mRNA upstream of the AAUAAA polyadenylation sequence. This signal for deadenylation is called *the adenylate/uridylate (AU)-rich element* (ARE) and has the consensus sequence UUUUUAU. Interestingly, to activate a stored mRNA in this class, a cytoplasmic polyadenylation enzyme recognizes the ARE and adds 150 A nucleotides or so. Thus, the same sequence element is used to control poly(A) tail length, and therefore mRNA translatability, at different times and in opposite ways.

mRNA Degradation Control

Once in the cytoplasm, all RNA species are subjected to **degradation control**, in which the rate of RNA breakdown (also called RNA turnover) is regulated. Usually, both rRNA (in ribosomes) and tRNA are very stable species. By contrast, mRNA molecules exhibit a diverse range of stability, with some mRNA types known to be stable for many months while others degrade within minutes. The stability of particular mRNA molecules may change in response to regulatory signals. For example, the addition of a regulatory molecule to a cell type can lead to an increase in synthesis of a particular protein or proteins. This is accomplished by an increase in the rate of transcription of the gene(s) involved and/or an increase in the stability of the mRNAs produced. Table 17.2 presents examples of systems in which changes in mRNA stability for a number of cell types occur in the presence and absence of specific effector molecules.

mRNA degradation is believed to be a major control point in the regulation of gene expression in eukaryotes. Various sequences or structures have

~ **TABLE 17.2**

Examples of Tissues or Cells in Which Regulation of mRNA Stability Occurs in Response to Specific Effector Molecules[a]

mRNA	Tissue or Cell	Regulatory Y Signal (=Effector Molecule)	Half-life of mRNA	
			With Effector	Without Effector
Vitellogenin	Liver (frog)	Estrogen	500 h	16 h
Vitellogenin	Liver (hen)	Estrogen	~24 h	<3 h
Apo-very low density lipoprotein (apoVLDL)	Liver (hen)	Estrogen	~20–24 h	<3 h
Ovalbumin, conalbumin	Oviduct (hen)	Estrogen, progesterone	>24 h	2–5 h
Casein	Mammary gland (rat)	Prolactin	92 h	5 h
Prostatic steroid-binding protein	Prostate (rat)	Androgen	Increases 30X	

[a]Note that the effector molecule in each case results in an increase in transcription as well as stabilization of the mRNA.

been shown to affect the half-lives of mRNAs, including the AU-rich elements (ARE) discussed earlier and various secondary structures. Two major mRNA decay pathways are the *deadenylation-dependent decay* and *deadenylation-independent decay pathways*. In the deadenylation-dependent decay pathway, the poly(A) tails are deadenylated until the tails are too short (5–15 As) to bind PAB (poly(A) binding protein). In yeast, the product of the *PAN1* gene, PAB-dependent poly(A) nuclease, may catalyze the deadenylation. Once the tail is almost removed, the 5' cap structure is removed in a step called *decapping*. Decapping is an enzyme-catalyzed process; in yeast, the decapping enzyme, or at least an essential part of it, is encoded by the *DCP1* gene. After an mRNA molecule is decapped, it is degraded from the 5' end by a 5'-to-3' exonuclease. In yeast, this enzyme—encoded by the *XRN1* gene—is very aggressive, attesting to the importance of the 5' cap in protecting active mRNAs in the cell.

Yeast strains with a mutant *dcp1* gene are viable, and mRNA degradation still occurs. This provides evidence for the existence of mRNA degradation pathways other than the pathway just described. In these deadenylation-independent decay pathways, mRNAs may be decapped without being deadenylated, thereby exposing them to rapid degradation by 5'-to-3' exonucleases, or they may be cleaved internally by endonucleases without being deadenylated, and then broken down further.

Note that, while our understanding of mRNA degradation in yeast is becoming clearer, the details of mRNA degradation in mammalian cells are not as well known. Both deadenylation-dependent and deadenylation-independent decay pathways are known to exist in mammals, and decapping is an important step in at least the former pathway. Searches are currently underway to identify possible homologs of the yeast genes already shown to have key roles in mRNA degradation.

Protein Degradation Control

Regulatory mechanisms also exist at the posttranslational level to determine the lifetime of a protein. This topic is rather peripheral to our discussion of gene expression, so it will only be discussed very briefly here.

A wide variety of possibilities exist to regulate the amount of a particular protein in a cell. For example, a constitutively produced mRNA may be continuously translated with the level of protein product controlled by the degradation rate of that protein. At another extreme, a short-lived mRNA may encode a protein that is very stable so that it persists for very long periods in the cell. Proteins in the lens of higher vertebrate eyes, for example, fall into the latter situation. Their mRNAs have long since degraded, while the protein itself persists usually for the lifetime of the individual. On the other hand, steroid receptors and heat-shock proteins—the products themselves of relatively unstable mRNAs—have relatively short half-lives.

Degradation of proteins (*proteolysis*) in eukaryotes has been shown to require the protein cofactor, *ubiquitin* (a protein found apparently ubiquitously). The binding of ubiquitin to a protein identifies it for degradation by proteolytic enzymes. Ubiquitin is released intact during the degradation process enabling it to be used to tag other proteins for degradation.

But how is a protein initially targeted for ubiquitin binding? The amino acid at the N-terminus of a protein is the key. In what has become known as the *N-end rule*, the particular N-terminal amino acid relates directly to the half-life of the protein. In a yeast test system in which the lifetime of the same protein was measured with different N-terminal amino acids, arginine, lysine, phenylalanine, leucine, and tryptophan each specified a half-life of 3 minutes or less, while cysteine, alanine, serine, threonine, glycine, valine, proline, and methionine all specified a half-life of more than 20 hours. The same general hierarchy is seen in an *E. coli* system. The N-terminal amino acid directs the rate at which ubiquitin molecules can bind to the protein which, in turn, determines the time at which the protein will be degraded.

𝒦EYNOTE

Gene expression is regulated also by mRNA translation control and by mRNA degradation control. The latter is believed to be a major control point in the regulation of gene expression. Structural features of individual mRNAs have been shown to be responsible for the range of mRNA degradation rates, although the precise roles of cellular factors and enzymes have yet to be determined.

In sum, in prokaryotes, control of gene expression occurs mainly at the transcriptional level, in association with rapid turnover of mRNA molecules. In eukaryotes, gene expression is regulated at transcriptional, post-transcriptional and post-translational levels. Regulatory systems appear to exist for the control of transcription, a precursor-RNA processing, transport out of the nucleus, degradation of mature RNA species, translation of the mRNA, and degradation of the protein product. The intertwining of the regulatory events at these different levels leads to the fine tuning of the amount of the controlled protein in the cell.

GENE REGULATION IN DEVELOPMENT AND DIFFERENTIATION

Most higher eukaryotes have many different cells, tissues, and organs with specialized functions, and yet all cells of the same organism have the same genotype. During an organism's growth throughout its life cycle, cells which were genetically identical become physiologically and phenotypically differentiated in terms of their structure and function.

Two terms are used in the description of long-term gene regulation. **Development** refers to the process of regulated growth that results from the genome's interaction with cytoplasm and the external environment and that involves a programmed sequence of phenotypic events that are typically irreversible. The total of the phenotypic changes constitutes the life cycle of an organism. **Differentiation**, the most spectacular aspect of development, involves the formation of different types of cells, tissues, and organs through the processes of specific regulation of gene expression; differentiated cells have characteristic structural and functional properties.

At a very general level the processes in differentiation and development are the result of a highly programmed pattern of gene activation and gene repression. However, we are a long way from understanding the molecular bases for the differentiation and development events. We can describe these events well at the morphological level, and we can describe them to some extent at the biochemical level. We are only in the relatively early stages of knowing the details of how the complex activation-repression patterns are programmed, because of the greater complexity of the processes in comparison with the processes of bacterial operons.

In the following sections, we will discuss only selected aspects of gene regulation in development and differentiation, since much of this area belongs more appropriately in a developmental biology course.

Genomic Activity in Higher Eukaryotes

As discussed in Chapter 11, the genomes of higher eukaryotes are much larger than those of prokaryotes, and they have a large amount of highly repetitive and moderately repetitive DNA. About 20 to 40 percent of the genomic DNA of higher eukaryotes is highly repetitive DNA. This DNA does not appear to encode proteins. The remaining 60 to 80 percent of the DNA is distributed between the moderately repetitive and unique-sequence DNA.

What is the function of the moderately repetitive and unique-sequence DNA? In his studies of the sea urchin, Eric Davidson isolated the species' total mRNA, and by determining the total percentage of unique-sequence DNA to which this RNA hybridized, he was able to quantify the amount of DNA that encodes structural information (proteins). Davidson used his method to study mRNAs obtained from populations of sea urchins in all stages of development and from mature sea urchin tissues. The results indicated the percentage of unique-sequence DNA to which RNA hybridized at various stages of development. Maximal transcription was found to occur in the oocyte, with about 6 percent of the unique-sequence DNA hybridized. Transcriptional activity remains high during the blastula, gastrula, and larval stages of development (>2 percent unique-sequence DNA hybridized) and then drops to low levels (<0.8 percent unique sequence DNA hybridized) in mature tissue such as the intestine and tubefoot. These results mean that at any one time a maximum of about 6 percent of the unique sequence DNA is transcriptionally active and producing proteins.

Because higher eukaryotes are more complex than prokaryotes, it was expected that a larger proportion of their DNA would encode structural information. Thus, the fact that the majority of the genome in sea urchins (and in other eukaryotes) is not transcribed was a surprise. By studying the locations of transcribed and nontranscribed regions of the genome, we have discovered that the structural genes (which are transcribed) are located within long stretches of DNA that are not transcribed. The function of the large amount of nontranscribed DNA in higher eukaryotes (less is found in lower eukaryotes) has long been controversial. It could be "junk" DNA left over from evolutionary changes, for example, or perhaps it could have important regulatory roles as yet undetermined.

Constancy of DNA in the Genome During Development

In early studies of the genetic processes associated with differentiation and development, an important area of research focused on whether differentiation and development involve a *loss* of genetic information (i.e., of DNA), or whether these processes involve a programmed sequence of gene activation and repression involving a constant genome; that is, a genome which is the same in the adult as it is in the zygote. The most direct way to decide is to determine whether the genome of a differentiated cell contains the same genetic information found in the zygote. Two studies that were done to answer this question, one involving plants and the other involving animals, will now be described.

REGENERATION OF CARROT PLANTS FROM MATURE SINGLE CELLS.

In the 1950s, Frederick Steward dissociated the tissue of a carrot to separate the cells, and then attempted to culture new carrot plants from those cells by using plant tissue culture techniques (Figure 17.17). Mature plants with edible carrots were successfully produced (cloned) from phloem cells. That the mature cells had the potential to act as zygotes and develop into complete plants indicated that mature cells had all the DNA found in zygotes. Steward's findings supported the notion that the DNA content of a cell remains constant during development and differentiation and provided evidence *against* the notion that development and differentiation involve losses of genetic information.

NUCLEAR TRANSPLANTATION EXPERIMENTS WITH SHEEP: CLONING A MAMMAL.

An objection to the general applicability of the results of the carrot experiments is that plants are much more able to propagate themselves vegetatively than are animals—horticulturists have long been able to regenerate plants from cuttings, for example. Researchers also had to determine whether the DNA content of animal cells remained constant during development or whether this condition was particular to plants. Some success came in 1975 when John Gurdon and his colleagues showed that a nucleus from a skin cell of an adult frog injected into an enucleated egg could direct

development to the tadpole stage. No adults were produced in that experiment, however, leaving unresolved the question of whether a nucleus from adult differentiated tissue is genetically capable of directing development from the egg cell stage. That question was answered in the affirmative in 1997, when Ian Wilmut and his colleagues reported the cloning of a mammal (sheep) starting with an adult cell. This means that an adult nucleus has all the genetic information necessary to direct development again from the start—the nucleus is said to be totipotent, or to demonstrate **totipotency.**

Wilmut's group tested the ability of nuclei from embryonic cells, fetal cells, and adult cells to direct the development of sheep. Their experimental approach, illustrated in Figure 17.18, was as follows:

1. Embryonic cells, fetal fibroblast (muscle-forming) cells, and mammary epithelial cells from donor ewes (Poll Dorset, Black Welsh, and Finn Dorset breeds, respectively) were grown in tissue culture, and then induced to enter the quiescent state (the G_0 phase of the cell cycle) by reducing the concentration of the growth serum.

2. The cells were fused with enucleated oocytes (egg cells) and the fusion cells were allowed to grow and divide for six days to produce embryos.

3. The embryos were implanted into recipient ewes and the establishment of pregnancy and its progression was monitored.

The results were as follows: Four of 385 embryo-derived cells, two of 172 fetal fibroblast-derived cells, and one of 277 adult mammary epithelium-derived cells gave rise to live lambs. The most significant of these results is the last, because it demonstrates the totipotency of the adult nucleus. That lamb, designated 6LL3 and named Dolly, is perfectly normal at 18 months

~ FIGURE 17.17

Cloning of a mature carrot plant from a cell of a mature carrot.

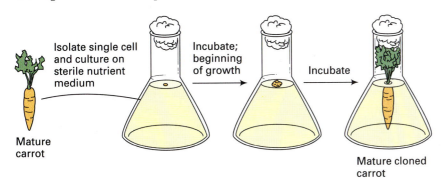

~ **FIGURE 17.18**

Representation of Wilmut's sheep cloning experiment, which showed the totipotency of the nucleus of a differentiated, adult cell.

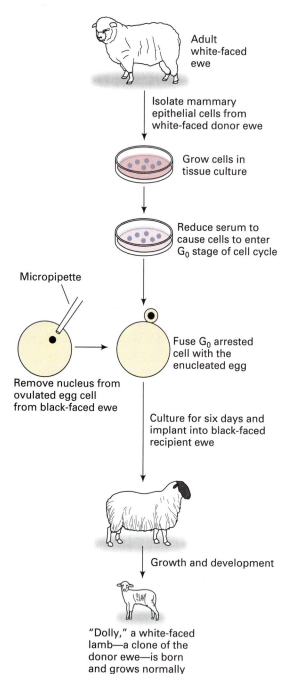

Adult white-faced ewe

Isolate mammary epithelial cells from white-faced donor ewe

Grow cells in tissue culture

Reduce serum to cause cells to enter G$_0$ stage of cell cycle

Micropipette

Fuse G$_0$ arrested cell with the enucleated egg

Remove nucleus from ovulated egg cell from black-faced ewe

Culture for six days and implant into black-faced recipient ewe

Growth and development

"Dolly," a white-faced lamb—a clone of the donor ewe—is born and grows normally

contained the nucleus from a (whiteface) Finn Dorset ewe and was implanted into a Scottish Blackface recipient ewe; Dolly is morphologically Finn Dorset. Second, and more definitive, analysis of polymorphic microsatellite DNA markers (see Chapter 15, p. 558) at four loci showed that the DNA of Dolly perfectly matched that of the donor mammary epithelial cells and not that of the recipient ewe.

In sum, while the success rate for the experiment was not high (probably for technical reasons), the highly significant accomplishment here was the development of a live lamb directed by an adult nucleus. When the result of the experiment was published, concerns were raised internationally about the possible cloning of humans. In their stories, the media picked various professional athletes, politicians, or other celebrities as theoretical subjects for cloning. We learned earlier in this text (Chapter 4, pp. 125–126), however, that genes provide only a potential blueprint for the adult individual, with many influences coming from the environment. Thus, it is extremely unlikely that cloning could produce an exact replica of any given person. Nevertheless, cloning technology is undoubtedly applicable to humans, and the ethical issues it raises will continue to be debated.

𝒦EYNOTE

Long-term regulation is involved in controlling the events that activate and repress genes during development and differentiation. Development and differentiation result from differential gene activity of a genome that contains a constant amount of DNA, from the zygote stage to the mature organism stage. Thus these events do not result from a loss of genetic information.

Differential Gene Activity Among Tissues and During Development

Specialized cell types have different cell morphologies; for example, nerve cells are clearly different from intestinal epithelial cells. Cells from different organs and tissues can also be distinguished easily. Since the amount of DNA typically remains constant during development, the simplest hypothesis is that the phenotypic differences between different cell types reflect differential gene activity. The following examples discuss some of the evidence supporting this hypothesis.

old (the time of this writing). Dolly progressed normally to sexual maturity and she is pregnant with delivery expected in early summer 1998.

Evidence that Dolly was truly the result of the cell fusion experiment is of two kinds: First, the fusion cell

~ **FIGURE 17.19**

Molecular organization of the human α-globin and β-globin genes.

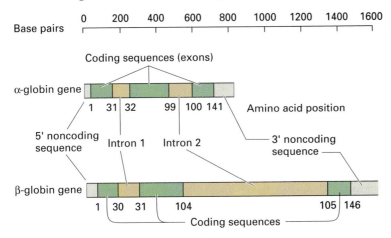

HEMOGLOBIN TYPES AND HUMAN DEVELOPMENT.

Human adult hemoglobin Hb-A has been examined in this book in many contexts. Hb-A is a tetrameric protein made up of two α and two β polypeptides, where each type of polypeptide is coded by a separate gene, α and β. The two genes appear to have arisen during evolution by duplication of a single ancestral gene, followed by alteration of the base sequences in each gene. The organization of the genes (Figure 17.19) shows that each contains two introns (intron 1 and intron 2), which are transcribed with the coding sequences but are removed to produce a mature mRNA transcript. While the introns are of different sizes in the two genes, they are placed in very similar positions in the two genes.

Hemoglobin Hb-A is only one type of hemoglobin found in humans. Genetic studies have shown that several distinct genes code for α- and β-like globin polypeptides which form different types of hemoglobin at different times during human development. Figure 17.20 shows the globin chains synthesized at different stages of human development. In the human embryo the hemoglobin initially made in the yolk sac consists of two ζ (zeta) polypeptides and two ε (epsilon) polypeptides. From comparisons of the amino acid sequences, ζ is an α-like polypeptide, and ε is a β-like polypeptide. After about three months of development, synthesis of embryonic hemoglobin ceases (i.e., the ζ and ε genes are no longer transcribed), and the site of hemoglobin synthesis shifts to the fetal liver and spleen. Here, *fetal hemoglobin* (Hb-F) is made. Hb-F contains two α polypeptides, and two β-like γ (gamma) polypeptides, either two γA or two γG. γA and γG differ from each other by only one amino acid (out of 146 amino acids) and are each coded for by distinct genes.

Fetal hemoglobin is made until just before birth, when synthesis of the two types of γ chains stops, and the site of hemoglobin synthesis switches to the bone marrow. In that tissue, α and β polypeptides are made, along with some β-like δ (delta) polypeptides. In the newborn through adult human, most of the hemoglobin is our familiar $\alpha_2\beta_2$ tetramer (Hb-A), with about one in 40 molecules having the constitution $\alpha_2\delta_2$ (Hb-A2). Thus, globin gene expression switches during human development, and this switching must involve a sophisticated gene regulatory system that turns appropriate globin genes on and off over a long time period.

In the genome, the α-like genes (two α genes and one ζ gene) are all on chromosome 16, and the β-like

~ **FIGURE 17.20**

Comparison of synthesis of different globin chains at given stages of embryonic, fetal, and postnatal development.

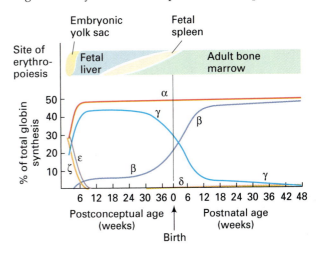

genes (ε, γA, γG, δ, and β) are all on chromosome 11 (Figure 17.21). Note that between the ζ and α genes on chromosome 16 are three sequences, one of which has a nucleotide sequence that is very similar to the ζ sequence, and the other two of which closely resemble the α sequence. However, these sequences do not have all the features required for producing a functional polypeptide product. Sequences which are very similar to known functional genes but which themselves are defective (they cannot produce a functional product), are called *pseudogenes*. Pseudogenes, such as these, which have recognizable intron sequences in them are thought to be nonfunctional relics of gene duplication and gene modification that occurred some time ago in evolutionary time. Pseudogenes lacking introns are thought to have arisen through reverse transcription. We learned in an earlier chapter that a DNA copy of RNA can be made with the enzyme reverse transcriptase. Conceivably a DNA copy of an mRNA molecule could be made by reverse transcriptase activity and that cDNA could then integrate into the genome. That integrated DNA would be similar in sequence to the gene from which the mRNA was transcribed except it would lack introns and would have no associated promoter elements to enable it to be transcribed. As a nonfunctional gene sequence it would be subject to mutational events and thus its sequence could change significantly over evolutionary time. The β-like gene cluster has a pseudogene located between the γ genes and the δ gene.

Significantly, the α-like genes and the β-like genes are each arranged in the chromosome in an order that exactly parallels the timing in which the genes are transcribed during human development. Recall that embryonic hemoglobin consists of ζ and ε polypeptides; these genes are the first functional genes at the left of the clusters. Next, the α and γ genes are transcribed to produce fetal hemoglobin (Hb-F), and if

one moves from left to right, these genes are the next functional genes that can be transcribed from the clusters. Lastly the δ and β polypeptides are produced, and these genes are last in line in the β-like globin gene cluster. Although such an arrangement surely must occur by more than coincidence, there is no insight as yet about how the gene order relates to the regulation of expression of these genes during development.

POLYTENE CHROMOSOME PUFFS DURING DIPTERAN (TWO-WINGED FLY) DEVELOPMENT. Polytene **chromosomes** are a special type of chromosome that consists of a bundle of chromatids. These chromatid bundles result from repeated cycles of chromosome duplication without nuclear division or chromosome segregation (a process called *endoreduplication*), and they are readily visible after staining under the light microscope. Polytene chromosomes are characteristic of certain tissues of insects in the order Diptera, for example, the salivary glands in the larval stages, and they may be 1,000 times the thickness of corresponding chromosomes at meiosis or in the nuclei of ordinary somatic cells. After staining, distinct and characteristic bands, called *chromomeres*, can be seen along each chromosome (Figure 17.22). Genes have been shown to be located in both bands and interbands.

Specific chromomeres have been shown to exhibit *puffs* (localized unravelings of the chromosome to give a "puffy" appearance) at characteristic and reproducible times during development (Figure 17.23). In other words, the puffs appear and disappear in specific patterns at certain chromosomal loci as development proceeds, so it is fair to say that they are developmentally controlled. (Note that puffing occurs as a result of very high levels of gene expression. Most genes that are expressed do so at low levels and do not puff.)

~ **FIGURE 17.21**

Linkage maps of human globin gene clusters. The function, if any, of the α-globin φ1 gene is not known. (Map is not to scale.)

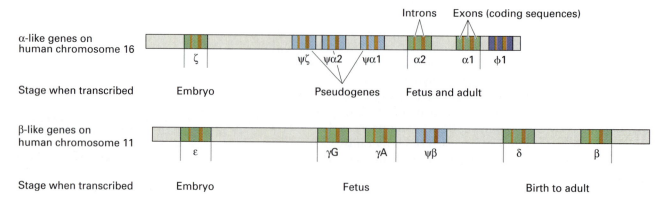

~ FIGURE 17.22

Diagram of the complete set of *Drosophila* polytene chromosomes in a single salivary gland cell. There are four chromosome pairs, but each pair is tightly synapsed so that only a single chromosome is seen for each pair. The four chromosome pairs are linked together by regions near their centromeres to produce a large *chromocenter*.

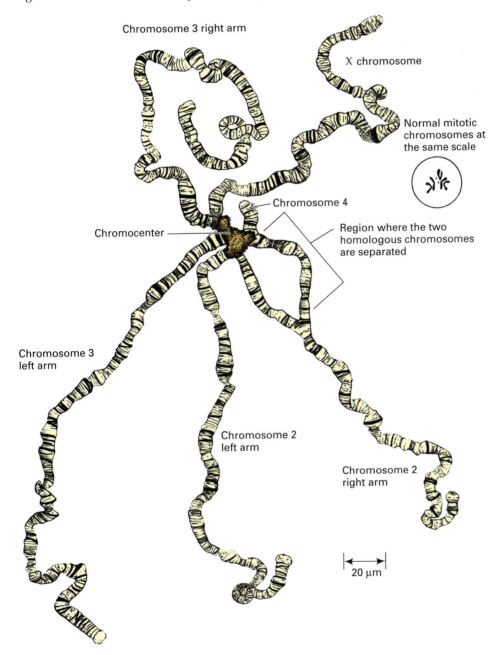

While we do not know what molecular event is responsible for the induction of a puff, hormonal control is involved in many cases. Dipteran larvae go through different stages, each separated by a molt. The molting process is controlled by the steroid molting hormone *ecdysone*, which is made in the larvae's prothoracic gland. The prothoracic gland is, in turn, stimulated to produce ecdysone by the action of a brain hormone synthesized by neurosecretory cells. The levels of ecdysone increase and decrease periodically during larval development.

Electron microscopic studies have shown that puffing involves a loosening of the tightly coiled DNA into long, looped structures more accessible to

~ FIGURE 17.23

Light micrograph of a polytene chromosome from *Chironomus*, showing two puffs that result from localized uncoiling of the chromosome structure and indicate transcription of those regions. DNA is shown in blue, RNA in red/violet.

RNA polymerase. At the molecular level this event must involve an uncoiling of the supercoiled DNA-protein structure of the chromosome.

Evidence that the puffs are the visual manifestation of transcriptionally active genes has come from radioactive tracer experiments. If radioactive uridine (an RNA precursor) is added to developing flies, it is incorporated into the RNA being synthesized. If salivary glands are dissected from larvae and their polytene chromosomes are prepared and autoradiographed, radioactivity is localized at the puff sites, indicating that RNA, presumably transcripts, is associated with the puffs. That is, when a polytene chromosome gene is to be transcribed to produce a protein required for salivary gland activities during development, the chromosome structure loosens in order to permit efficient transcription of that region of the DNA. When transcription is completed, the chromosome reassumes its compact configuration and the puff disappears.

A model for the control of sequential gene activation by ecdysone is as follows. Ecdysone binds to a receptor protein and this complex binds to both early (the early-puffing genes) and late genes (the expression of which is seen later in development). The complex turns on the early genes and represses the late genes. One (or more) early gene encodes a protein which accumulates during development. When the level of this protein reaches a certain threshold, it displaces the ecdysone-receptor complex from both early and late genes. This turns off the early genes and removes the repression (i.e., turns on) the late genes. In support of the model, some of the early genes have been shown to encode DNA-binding proteins, prod-

ucts expected of regulatory genes. Further, an ecdysone receptor gene has been cloned and shown to encode a steroidlike receptor protein.

In sum, polytene puffs, the sites of RNA synthesis in Dipteran insects, are directly related to the developmental processes in the organism, and at least some of the puffs are regulated by hormone action. We can conclude, therefore, that at least some gene regulation of developmental processes in Dipteran flies is done by hormones, and that this regulation occurs at the transcriptional level.

Immunogenetics and Chromosome Rearrangements During Development

In this section we discuss how antibodies, the proteins which mediate the humoral immune response, are encoded in the DNA. In this system, DNA rearrangements occur in cells which produce the genes from which antibody polypeptides are transcribed. These rearrangements are examples of *loss* of genetic information during development.

SYNOPSIS OF THE IMMUNE SYSTEM. The following is a brief list of the important properties of the immune system.

1. All vertebrates have an *immune system*. The immune system provides protection against infectious agents such as viruses, bacteria, fungi, and protozoans.
2. The immune system distinguishes between molecules that are *foreign* from molecules that are *self.*
3. Any substance that elicits an immune response is called an **antigen** (*anti*body *gen*erator).
4. The cells that are responsible for immune specificity are *lymphocytes*, specifically *T cells* and *B cells*. We focus our discussion on B cells. B lymphocytes develop in the adult bone marrow, and that is how they got their name. When activated by an antigen, B cells develop into plasma cells which make **antibodies**; that is, specialized proteins called **immunoglobulins**. Antibody molecules become inserted in the plasma membrane of the plasma cells, and they are also released into the blood and lymph where they are responsible for the humoral (*humor* meaning fluid) immune responses. The antibodies bind specifically to the antigens which stimulated their production.
5. A key feature of the immune response is that, once an organism has been exposed to a particular antigen, it exhibits immune memory. That is, when an organism encounters a foreign antigen,

the immune system mounts a response by committing cells to making antibodies against that foreign antigen. The next time the same antigen is encountered, the organism "remembers," and sufficient antibodies are synthesized rapidly to respond to the new invasion.

6. The establishment of immunity against a particular antigen results from **clonal selection**. This is a process whereby cells that have antibodies specific for the antigen displayed on their surfaces are stimulated to proliferate and secrete that antibody. During development each lymphocyte becomes committed to react with a particular antigen, even though the cell has never "seen" the antigen. For the humoral response system, there is a population of B cells, *each of which can recognize a single antigen*. A particular B cell recognizes an antigen because the B cell has made antibody molecules, which are attached to the outer membrane of the cell and which act as receptor molecules. When an antigen encounters a B cell that has the appropriate antibody receptor capable of binding to the antigen, that B cell is stimulated selectively to proliferate. This produces a clonal population of plasma cells, each of which makes and secretes the identical antibody. It is important to note that *any given cell makes only one specific kind of antibody towards one specific antigen*. The actual immune response, though, may involve the binding of many different antibodies to an array of antigens on such invaders as an infecting virus or bacterium. This binding mediates a variety of other mechanisms which inactivate the invading antigen.

ANTIBODY MOLECULES. All antibody molecules made by a given plasma cell are identical; that is, they have the same protein chains and bind the same antigen. Because there are millions of B cells in the whole organism, millions of different antibody types can be produced, each with a different amino acid sequence and a different antigen-binding specificity.

As a group, antibodies are proteins called *immunoglobulins* (Igs). A stylized antibody (immunoglobulin) molecule of the type IgG is shown in Figure 17.24a, and a model of an antibody molecule based on X-ray crystallography is shown in Figure 17.24b. Both figures show the molecule's two short polypeptide chains, called *light (L) chains*, and two long polypeptide chains, called *heavy (H) chains*. (All antibody molecules also have carbohydrates attached to the regions of H chains not involved in binding with L chains.) The two H chains are held together by disulfide (–S–S–) bonds, and an L chain is bonded to each H chain by disulfide bonds. Other disulfide bonds within each L and H chain cause the chains to fold up into their characteristic shapes.

The overall structure resembles a Y, with the two arms containing the two antigen-binding sites. The two L chains in each Ig molecule are identical, as are the two H chains, so the two antigen-binding sites are identical. The hinge region (see Figure 17.24a) allows the two arms to move in space, making it easier for the antibody to bind an antigen. Also, one arm can then bind an antigen on, say, one virus, while the other arm binds the same antigen on a different virus. Such crosslinking of antibody molecules in solution helps inactivate infecting agents.

In mammals there are five major classes of antibodies: IgA, IgD, IgE, IgG, and IgM. They have different H chains—α (alpha), δ (delta), ϵ (epsilon), γ (gamma), μ (mu), respectively. Two types of L chains are found: κ (kappa) and λ (lambda). Both L chain types are found in all Ig classes, but a given antibody molecule will have either two identical κ chains or two identical λ chains. A complete discussion of the functions of the five Ig classes is beyond the scope of this text. For our purposes, we need to be aware that the most abundant class of immunoglobulin in the blood is IgG, and that IgM plays an important role in the early stages of an antibody response to a previously unrecognized antigen. We will focus on these two antibody classes from now on.

Each polypeptide chain in an antibody is organized into domains of about 110 amino acids (see Figure 17.24a). Each L chain (κ or λ) has two domains, and the H chain of IgG (the γ chain) has four domains, while the IgM's H chain (the μ chain) has five domains. The N-terminal domains of the H and L chains have highly variable amino acid sequences that constitute the antigen-binding sites. These domains, representing in IgG the N-terminal half of the L chain and the N-terminal quarter of the H chain, are termed the *variable*, or V, regions. The V regions are symbolized generically as V_L (for the light chain) and V_H (for the heavy chain). The V_L and V_H regions comprise the antigen-binding sites (see Figure 17.24a). The amino acid sequence of the rest of the L chain is constant for all antibodies (with the same L chain type, i.e., μ or λ) and is termed C_L. Similarly, the amino acid sequence of the rest of the H chain is constant and is termed C_H. For IgG, there are three approximately equal domains of C_H called C_H1, C_H2, and C_H3 (see Figure 17.24a). For IgM, there are four approximately equal domains of C_H called C_H1, C_H2, C_H3, and C_H4. Thus, the production of antibody molecules involves synthesizing polypeptide chains, one part of which varies from molecule to molecule, and the other part of which is constant. How this occurs is discussed in the following section.

~ FIGURE 17.24

IgG antibody molecule. (a) Diagram showing the two heavy and two light chains, and the antigen-binding sites. The heavy and light chains are held together by disulfide bonds; (b) Model of IgG antibody molecule.

a)

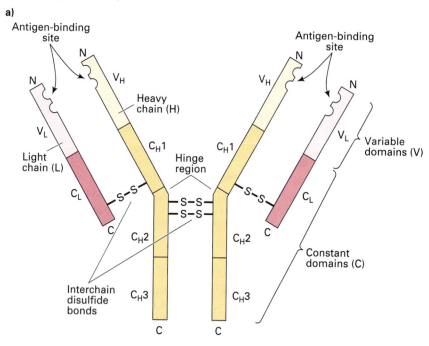

b)

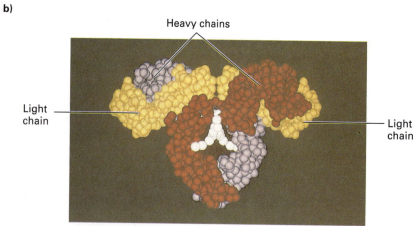

ASSEMBLY OF ANTIBODY GENES FROM GENE SEGMENTS DURING B CELL DEVELOPMENT. It is thought that a mammal may produce from 10^6 to 10^8 different antibodies. Since each antibody molecule consists of one kind of L chain and one kind of H chain, these antibodies theoretically would require 10^3 to 10^4 different L chains and 10^3 to 10^4 different H chains, if L and H chains paired randomly. The dilemma is that the human genome is thought to contain perhaps only 10^5 genes. Therefore, how can the observed diversity be generated? The answer is that variability in L and H chains results from particular DNA rearrangements that occur during B cell development. These rearrangements involve the joining of

different gene segments to form a gene that is transcribed to produce an Ig chain; the process is called *somatic recombination* and the many permutations of recombinations that are possible are responsible largely for the diversity of antibody structure. The process will now be illustrated for mouse immunoglobulin chains.

Light Chain Gene Recombination. In mouse germline DNA, there are a number of gene segments that encode the κ light chain (Figure 17.25). There are three types of segments:

1. Each of the 350 or so L-Vκ segments consists of a leader sequence (L) and a sequence, Vκ, which

~ FIGURE 17.25

Production of the kappa (κ) light chain gene in mouse by recombination of V, J, and C gene segments during development. The rearrangement shown is only one of many possible recombinations.

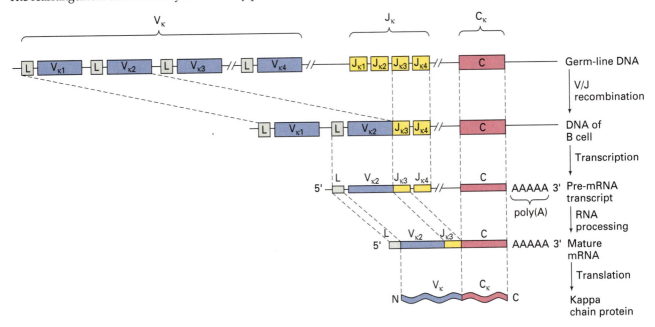

varies from segment to segment. Each Vκ segment encodes most of the amino acids of the light chain variable domain. The leader sequence also encodes a special sequence called a signal sequence (see Chapter 14) that is required for secretion of the Ig molecule; this signal sequence is subsequently removed and is not part of the functional antibody molecule.

2. A C_κ segment specifies the constant domain of the kappa L chain; and

3. Four J_κ segments are joining segments used to join V_κ and C_κ segments in the production of a functional κ light chain gene.

In the pre–B cell, the L-V_κ, J_κ and C_κ segments, in that order, are widely separated on the chromosome. Each L-V_κ segment is about 400 nucleotide pairs long; L-V_κ segments are tandemly arranged on the chromosome with about 7 kb of DNA between them. The J_κ segments are about 30 nucleotide pairs long, and are tandemly arranged about 20 kb apart. A DNA segment of about 2 to 4 kb separates the J_κ segments and the single C_κ segment. (This DNA segment does not code for amino acids in the resulting immunoglobulin polypeptide but is an example of an *intron*, a sequence found between amino acid coding regions in a gene and in the transcript of the gene. It is subsequently removed from the RNA before it is translated. Introns and intron removal are described in more detail in Chapter 13.) As the B cell develops, a particular L-V_κ segment becomes associated with one of the J_κ segments and with the C_κ

segment (see Figure 17.25). In the example L-$V_{\kappa 2}$ has recombined next to $J_{\kappa 3}$. Transcription of this new DNA arrangement produces the primary RNA transcript. As is typical of eukaryotic mRNAs, a poly(A) tail is added to the end of the transcript post-transcriptionally. Recall that a poly(A) tail plays a role in the stability of the resulting mRNA. Removal of the intron from the primary RNA transcript produces the mature mRNA, which has the organization L-$V_{\kappa 2}J_{\kappa 3}C_\kappa$; translation and leader removal produces the κ light chain that the B cell is committed to make.

In the mouse there are about 350 L-V_κ gene segments, four functional J_κ segments, and one C_κ gene segment. Thus, the number of possible κ chain variable regions that can be produced by this mechanism is $350 \times 4 = 1,400$. Further diversity results from imprecise joining of the V_κ and J_κ gene segments. That is, during the joining process a few nucleotide pairs from V_κ and a few nucleotide pairs from J_κ are lost from the DNA at the $V_\kappa J_\kappa$ joint, generating significant diversity in sequence at that point. Thus, diversity of κ light chains results from (1) variability in the sequences of the multiple V_κ gene segment; (2) variability in the sequences of the four J_κ gene segments; and (3) variability in the number of nucleotide pairs deleted at V_κ-J_κ joints.

A similar mechanism exists for mouse λ light chain gene assembly. In this case there are only two L-V_λ gene segments, and four C_λ gene segments, each with its own J_λ gene segment. Thus, fewer λ variable regions can be produced than is the case for κ chains.

~ **FIGURE 17.26**

Production of heavy chain genes in mouse by recombination of V, D, J, and C gene segments during development. Depending upon the C_H segment used, the resulting antibody molecule will be IgM, IgD, IgG, IgE, or IgA. Shown here is the assembly of an IgG heavy chain. This rearrangement is only one of the many thousands possible.

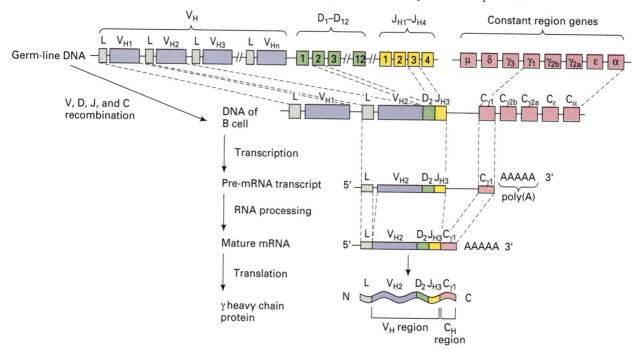

Heavy Chain Gene Recombination. The immunoglobulin heavy chain gene is also encoded by V_H, J_H, and C_H segments. In this case additional diversity is provided by another gene segment, D (diversity), which is located between the V_H segments and the J_H segments (Figure 17.26). As the figure shows, in the germ line of mouse DNA there is a tandem array of L-V_H segments, then a spacer, then 12 D segments, then a spacer, and then 4 J_H segments. After another spacer, the constant region gene segments are arranged in a cluster which, in mouse, has the order μ, δ, γ (four different sequences for four different, but similar, IgG H chain constant domains), ε, and α for the H chain constant domains of IgM, IgD, IgG, IgE, and IgA, respectively. As in L chain gene rearrangements, further antibody diversity results from imprecise joining of the gene segments that comprise the chain's variable region. In Figure 17.26, an IgG heavy chain is assembled.

KEYNOTE

Antibodies are specialized proteins called immunoglobulins, which bind specifically to antigens. Immunity against a particular antigen results from clonal selection, in which cells already making the required antibody are stimulated to prolif-

erate by the specific antigen. Antibody molecules consist of two light (L) chains and two heavy (H) chains. The amino acid sequence of one domain of each type of chain is variable; this variation is responsible for the different antigen-binding site on different antibody molecules. The other domain(s) of each chain are constant in amino acid sequence. In germ-line DNA, the coding regions for immunoglobulin chains are scattered in tandem arrays of gene segments. Thus, for light chains, there are many variable region (V) gene segments, a few joining (J) gene segments, and one constant region (C) gene segment. During development, somatic recombination occurs to bring particular gene segments together into a functional L chain gene. A large number of different L chain genes result from the many possible ways in which the gene segments can recombine. Similar rearrangements occur for H chain genes, but with the addition of several D (diversity) segments that are between V and J, which increase the possible diversity of H chain genes.

GENETIC REGULATION OF DEVELOPMENT IN *DROSOPHILA*

For many years, the fruit fly *Drosophila melanogaster* has been a prominent research organism for geneticists. In the several decades of classical genetics research with *Drosophila*, many mutations affecting development were identified. More recently, other organisms have become model systems for relating gene activity to developmental processes. In the nematode worm *Caenorhabditis elegans (C. elegans)*, sex determination is similar to the *Drosophila* system, the genome is relatively small (8×10^7 bp), genetic crosses and selfings can be done readily, and the body of the worm is transparent (enabling all cells to be seen under the microscope). In addition, development of *C. elegans* is programmed in an invariant way—that is, each adult has a set number of cells, each of which is traceable back to the zygote along particular cell pathways or *cell lineages.* In the zebrafish *(Brachydanio rerio)*, geneticists can easily observe several developmental steps in the transparent embryos. Genetic crosses can be made, large numbers of fish can be bred, and techniques have been developed to genetically screen the organism in an attempt to identify all genes that affect embryogenesis. In the plant *Arabidopsis thaliana*, individuals are small, making it possible to do genetic crosses and analyze large numbers of progeny. Genetic screens have isolated many mutations that affect the development, for example, of flowers and roots.

Nevertheless, *Drosophila* remains a system in which significant progress has been made. Many more developmental mutants have been isolated following extensive genetic screens so that virtually all areas of *Drosophila* development can be studied genetically and molecularly in detail. The discoveries made from such studies have become even more important as discoveries in other systems (e.g., nematode, mouse) indicate that the genes discovered in *Drosophila* have counterparts in all higher organisms, including humans. This implies, of course, that the same mechanisms that control development in *Drosophila* are used in higher organisms as well. We focus on the genetics of *Drosophila* development in this section, providing a brief overview of what is known.

Drosophila Developmental Stages

The production of an adult *Drosophila* from a fertilized egg involves a well-ordered sequence of developmentally programmed events under strict genetic control (Figure 17.27). About 24 hours after fertilization, a *Drosophila* egg hatches into a larva, which undergoes three molts, after which it is called a pupa.

~ **FIGURE 17.27**

Development of an adult *Drosophila* from a fertilized egg.

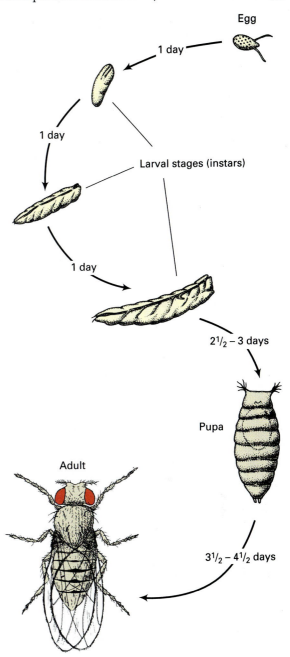

The pupa metamorphoses into an adult fly. The whole process from egg to adult fly takes about 10 to 12 days at 25°C.

Embryonic Development

Development commences with a single fertilized egg, giving rise to cells that have different developmental fates. What follows is a brief discussion of the information that has been obtained about the relationship between the developmental events in the egg and the determination of adult body parts.

Before a mature egg is fertilized, particular molecular gradients are established within it. The posterior end is indicated by the presence of a region called the *polar cytoplasm* (Figure 17.28a). At fertilization, the two parental nuclei are roughly centrally located in the egg. The two nuclei fuse to produce a 2N zygote nucleus (Figure 17.28b). For the first nine divisions, only the nuclei divide in a common cytoplasm—cytokinesis does not occur—to produce what is called a multinucleate *syncytium* (Figure 17.28c). (After seven divisions, some nuclei migrate into the polar cytoplasm, where they become precursors to germ-line cells.) Next, the nuclei migrate and divide to form a layer at the surface of the egg, producing the *syncytial blastoderm* (Figure 17.28d). After four more divisions, membranes form around the nuclei to produce somatic cells, about 4,000 of which make up the *cellular blastoderm* (Figure 17.28e).

Subsequent development of body structures depends upon two processes (Figure 17.29):

1. Gradients of molecules are produced along the anterior-posterior axis and the dorsal-ventral axis of the egg. Somehow, a nucleus is aware of its position with reference to the molecular concentration in the two gradients.
2. Regions are determined in the embryo that correspond to adult body segments. In the cellular blastoderm stage, where the regions are not well defined, they are called *parasegments*. In subsequent stages, where they are visible, they are called *segments*, forming a striped pattern along the anterior-posterior axis of the embryo. The embryonic segments give rise to the segments of the adult fly.

Genes involved in regulating *Drosophila* development are defined by mutations that have a lethal phenotype early in development or that result in the development of abnormal structures (e.g., embryos with abnormal striping, or two anterior ends, etc.).

How does the polarity of the *Drosophila* egg specify the segments of the adult fly body? Some of the large amount of information known will be presented here. Three major classes of developmental genes are involved: maternal effect genes, segmentation genes, and homeotic genes.

Maternal effect genes are expressed by the mother during oogenesis. The key genes identified are *bicoid*, *nanos*, and *torso*, which regulate the formation of the anterior, posterior, and terminal structures, respectively. Mutations in the *bicoid* gene result in the conversion of head and thorax to abdomen, and terminal posterior

~ **FIGURE 17.28**

Embryonic development in *Drosophila*. (a) The fertilized egg, with the two parental nuclei. Polar cytoplasm indicates the posterior end. (b) The two parental nuclei fuse to produce a diploid zygote nucleus. (c) The nucleus undergoes nine divisions in a common cytoplasm to produce a multinucleate syncytium. (d) Nuclei migrate and divide, producing a layer at the periphery of the egg. This stage is the syncytial blastoderm. (e) Nuclei divide four times, a membrane then forms around each to produce the somatic cells of the cellular blastoderm.

a) **Fertilized egg with two parental nuclei**

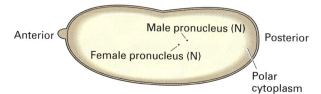

b) **Parental nuclei fuse and produce a diploid zygote nucleus**

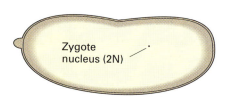

c) **The nucleus divides for nine divisions in a common cytoplasm to give a multinucleate syncytium**

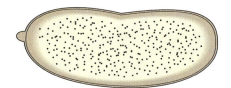

d) **Syncytial blastoderm nuclei migrate and divide, producing a layer at the periphery of the egg**

e) **Cellular blastoderm. Nuclei divide four times; membranes form around them and produce somatic cells**

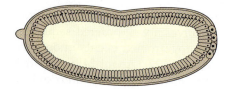

~ FIGURE 17.29

Drosophila development results from gradients in the egg that define parasegments in the cellular blastoderm, and segments in the embryo and adult. The adult segment organization directly reflects the segment pattern of the embryo.

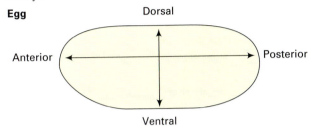

Egg

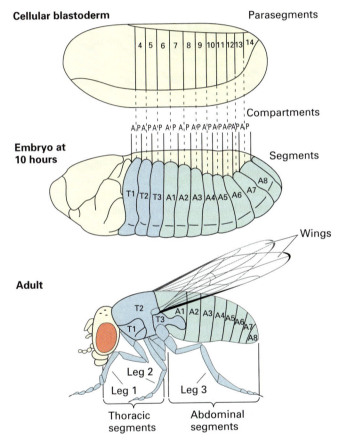

Cellular blastoderm

Embryo at 10 hours

Adult

Similarly, *nanos* mRNAs are localized to the posterior end of the egg. Translation of those mRNAs produces the NANOS protein, which forms a posterior-to-anterior gradient and acts as a morphogen to direct abdomen formation. Mutations in *nanos* result in a no-abdomen phenotype.

The *torso* maternal-effect gene is responsible for the formation of the terminal structures at the anterior and posterior ends of the embryo. Mutations in *torso* result in a no-terminals phenotype. The *torso* gene is transcribed and translated during oogenesis unlike *bicoid* and *nanos*, which are not translated until after fertilization. The TORSO protein is distributed throughout the egg but becomes activated only at the two termini.

Next, the embryo is subdivided into regions through the action of *segmentation genes*, about 25 of which have been identified. Mutations in segmentation genes alter the number of segments or their internal organization but do not affect the overall organizational polarity of the egg. The segmentation genes are subclassified on the basis of their mutant phenotypes (Figure 17.30). Mutations in *gap genes* (e.g., *Krüppel, hunchback*) result in the deletion of several adjacent segments; mutations in *pair rule genes* (e.g., *even-skipped, fushi tarazu*) result in the deletion of the same part of the pattern in every other segment; and mutations in *segment polarity genes* (e.g., *gooseberry, engrailed*) have portions of segments replaced by mirror images of adjacent half segments.

What is the role of segmentation genes in specifying regions of the embryo? Gap genes are activated or repressed by maternal effect genes. The gap gene products lead to an organization of the embryo into broad regions, each of which covers areas that will later develop into several distinct segments. Critical to this broad definition of regions is expression of the *hunchback* gene, which is stimulated by the BICOID protein and inhibited by the NANOS. As a result, a gradient of HUNCHBACK protein is generated with its highest concentration at the anterior end of the embryo. Next, through the transcription regulating action of the gap genes, the pair rule genes are expressed, leading to a division of the embryo into a number of regions, each of which contains a pair of parasegments (see Figure 17.29). The transcription factors encoded by the pair rule genes regulate the expression of the segment polarity genes, which determines regions that will become the segments that are seen in larvae and adults.

Once the segmentation pattern has been determined, a major class of genes called the *homeotic* (structure-determining) *genes* specifies the identity of each segment with respect to the body part that will develop at metamorphosis. Homeotic mutants cause a segment to develop into a different body part from that normally specified.

structures are present at both ends. We now know that the wild-type *bicoid* gene encodes a protein that is a morphogen; that is, it controls development. The *bicoid* gene is transcribed in the mother during oogenesis and the resulting mRNA is deposited into the oocyte where it becomes localized to the anterior pole. Translation of the mRNA occurs after egg deposition, after the mRNA becomes polyadenylated in the cytoplasm. The BICOID protein produced forms a gradient with its highest concentration at the anterior end of the egg, fading to nothing in the posterior third of the egg.

~ **FIGURE 17.30**

Functions of segmentation genes as defined by mutations.

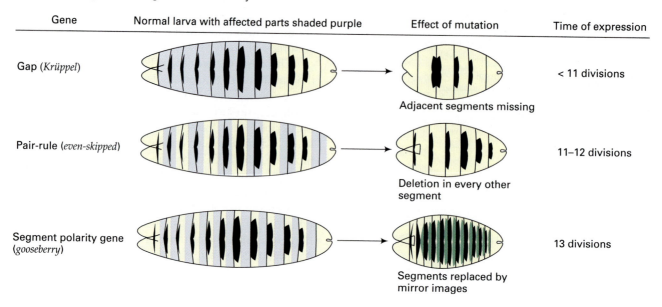

KEYNOTE

Development of *Drosophila* body structures results from gradients along the posterior-anterior and dorsal-ventral axes of the egg and from the subsequent determination of regions in the embryo that directly correspond to adult body segments. As defined by mutations, genes control *Drosophila* development in a temporal regulatory cascade. First, maternal effect genes specify the gradients in the egg, then segmentation genes (gap genes, pair rule genes, and segment polarity genes) determine the segments of the embryo and adult, and homeotic genes next specify the identity of the segments.

Imaginal Discs

Two types of cells are specified by cellular blastoderm cells: (1) those that will produce larval tissues; and (2) those that will develop into the adult tissues and organs. For the latter, certain groups of undifferentiated cells form larval structures called **imaginal discs** (the Latin *imago* means adult), each of which differentiates into a specific part of the adult fly. (Imaginal discs are characteristic of metamorphic insects.) That is, a number of *Drosophila* adult structures develop from imaginal discs, which are determined early in larval development. These adult structures include mouth parts, antennae, eyes, wings, halteres, legs,

and the external genitalia. Imaginal disc cells remain in an embryonic state throughout larval development, even though larval cells differentiate around them. Other structures, such as the nervous system and gut, do not develop from imaginal discs. A determined state that is so programmed is very stable, although rare changes do occur. For instance, there is a clear distinction between the stage when a phenotype is determined and the stage when the differentiation processes that give rise to the phenotype occur.

Thus larvae contain two groups of cells: Cells of one group are involved exclusively with larval development and function, while cells of the other group are grouped into the clusters of cells that comprise imaginal discs. Imaginal disc cells remain in an embryonic state throughout larval development, even though larval cells differentiate around them. Each disc eventually differentiates into a particular adult structure. Figure 17.31 shows the positions of some imaginal discs in a mature larva and the adult structures that develop from them.

From the genetic point of view imaginal discs are excellent subjects for study because each disc develops during the first larval stage, and when it consists of about 20 to 50 cells, it is already programmed to specify its given adult structure—its fate is determined. From then on the number of cells in each disc increases by mitotic division, until by the end of the larval stages, there are many thousand cells per disc.

The nature of the disc determination process is unknown, although at a simple level it probably involves a programming of those genes that are accessible to hormone (ecdysone)-stimulated activation during the pupal stage. Evidence suggests that the

~ **FIGURE 17.31**

Locations of imaginal discs in a mature *Drosophila* larva and the adult structures derived from each disc.

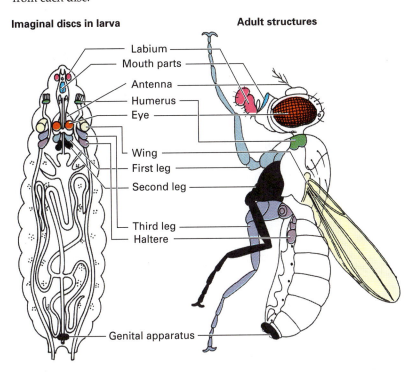

Imaginal discs in larva   **Adult structures**

Labium
Mouth parts
Antenna
Humerus
Eye
Wing
First leg
Second leg
Third leg
Haltere
Genital apparatus

determination process is a remarkably stable event; some mechanism apparently keeps all the cells in a particular disk programmed in the same way. This evidence comes from transplantation experiments in which discs or parts of discs are transplanted from a larva into the abdomen of an adult fly. In the adult abdomen the disc gets larger as its cells grow and divide, but the disc does not differentiate into its adult structure because the adult abdomen lacks the necessary hormones to induce such differentiation. After the disc has grown, it can be reisolated from the abdomen and part of it transplanted into the abdomen of another adult. Continued growth, isolation, and transplantation to a new host (called *serial transplantation*) at least 150 times does not change the programming of the disc. That is, at the end of the series of transplantations, when the disc is transplanted into a larval host (in which the necessary hormones for differentiation exist), the disc will still, with very rare exceptions, develop into the same adult structure as the original disc would have.

TRANSDETERMINATION. Rarely, the determined state of an imaginal disc does change, as the transplantation studies of Ernst Hadorn showed. The disc does not totally dedifferentiate but switches to another determined path in a process called **transdetermination**. For example, an eye disc may become altered so that it specifies a wing structure, or an antenna disc may become altered so that it specifies a leg structure. Significantly, the *entire disc changes its determined state*; no studies have reported, for instance, that half a disc becomes transdetermined while the other half does not. The results of such transdetermination would be easy to detect by visual scrutiny of adult flies since they would be hybrid structures. Figure 17.32, which shows the pathways of transdetermination in *Drosophila* that have been detected by transplantation experiments such as those of Hadorn, indicates that certain discs transdetermine only into certain other disc types. Leg discs, for example, transdetermine into wing discs, but the opposite transdetermination occurs far less frequently.

ᴋEYNOTE

A number of *Drosophila* adult structures develop from imaginal discs, which are determined early in larval development. Rarely, the determined state of an imaginal disc changes so that a different adult structure is produced; this is called transdetermination.

~ **FIGURE 17.32**

Pathways of transdetermination in *Drosophila* disc cells, as derived from transplantation experiments. Colored areas indicate the adult structures that develop from serially transplanted disc material. Arrows indicate the observed transdetermination changes.

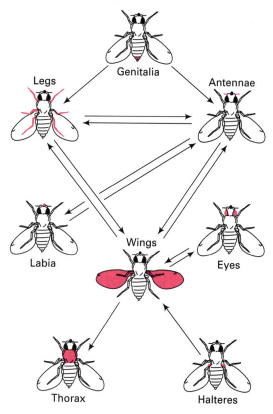

Homeotic Genes

Once the basic segmentation pattern has been laid down, the homeotic genes give a specific developmental identity to each of the segments. The homeotic genes have been defined by mutations that affect the development of the fly. That is, **homeotic mutations** alter the identity of particular segments, transforming them into copies of other segments. The principal pioneer of genetic studies of homeotic mutants is Edward Lewis, and the more recent molecular analysis has been done in many laboratories, including those of Thomas Kaufman, Walter Gehring, W. McGinnis, Matthew Scott, and Welcome Bender.

Lewis's pioneering studies were on a cluster of homeotic genes called the *bithorax* complex (*BX-C*). *BX-C* determines the posterior identity of the fly, namely thoracic segment T3 and abdominal segments A1-A8. *BX-C* contains three complementation groups called *Ultrabithorax* (*Ubx*), *abdominal-A* (*abd-A*), and *Abdominal-B* (*Abd-B*), each of which constitutes one protein-coding transcription unit. Mutations in these homeotic genes are often lethal, so the fly typically does not survive past embryogenesis. Some nonlethal mutations have been characterized, however, which allow an adult fly to develop. Figure 17.33 shows the abnormal adult structures that can result from *bithorax* mutations. A diagram showing the segments of a normal adult fly is in Figure 17.33a: note that the wings are located on segment Thorax 2 (T2), while the pair of halteres (rudimentary wings used as "balancers" in flight) are on segment T3. A photograph of a normal adult fly clearly showing the wings and halteres is presented in Figure 17.33b. Figure 17.33c shows one type of developmental abnormality that can result from nonlethal homeotic mutations in *BX-C*: shown is a fly that is homozygous for three separate mutations in the *Ubx* gene, *abx*, *bx3*, and *pbx*. Collectively these mutations transform segment T3 into an adult structure similar to T2. The transformed segment exhibits a fully developed set of wings. The fly lacks halteres, however, because no normal T3 segment is present.

Another well-studied group of mutations defines another large cluster of homeotic genes called the *Antennapedia* complex (*ANT-C*). *ANT-C* determines the anterior identity of the fly, namely the head and thoracic segments T1 and T2. *ANT-C* contains at least four genes, called *Deformed* (*Dfd*), *fushi tarazu* (*ftz*), *Sex combs reduced* (*Scr*), and *Antennapedia* (*Antp*). Most *ANT-C* mutations are lethal. Among the nonlethal mutations is a group of mutations in *Antp* that result in leg parts instead of an antenna growing out of the cells near the eye during the development of the eye disc (Figure 17.34a and b). Note that the leg has a normal structure, but it is obviously positioned in an abnormal location. A different mutation in *Antp*, called *Aristapedia*, has a different effect: only the distal part of the antenna, the arista, is transformed into a leg (Figure 17.34c).

The homeotic genes *ANT-C* and *BX-C*, therefore, encode products that are involved in controlling the normal development of the relevant adult fly structures.

The *Antennapedia* complex (*ANT-C*) and the *bithorax* complex (*BX-C*) have been cloned. Both complexes are very large. In *ANT-C*, for example, the *Antp* gene

~ FIGURE 17.33

~ **FIGURE 17.33**

(a) Drawing of a normal fly. T = thoracic segment. A = abdominal segment. The haltere (rudimentary wing) is on T3 (see Figure 17.32). (b) Photograph of a normal fly with a single set of wings. (c) Photograph of a fly homozygous for three mutant alleles (*bx3*, *abx*, and *pbx*) that results in the transformation of segment T3 into a structure like T2: namely, a segment with a pair of wings. These flies therefore have two sets of wings, but no halteres.

a)

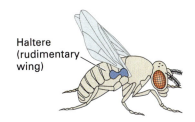

Haltere (rudimentary wing)

b)

c)

is 103-kb long, with many introns; this gene encodes a mature mRNA of only a few kilobase pairs. *BX-C* covers more than 300 kb of DNA and contains only three protein-coding regions amounting to about 50 kb of that DNA: *Ubx*, *abd-A*, and *Abd-B* (Figure 17.35). At least *Ubx* and *abd-A* have introns. Other RNA prod-

ucts are known to be transcribed from *BX-C*, but they do not code for proteins, and they have no known functions. The other 250 kb of DNA in the complex is not transcribed and is believed to consist of regulatory regions of significant size and complexity. The functions of these regulatory regions are to control the expression of the protein-coding genes.

Because the *ANT-C* and *BX-C* protein-coding genes have similar functions but are located in different places in the genome, Lewis predicted that the genes would have related structures. Analysis of the DNA sequences for the genes revealed the presence of similar sequences of about 180 bp that has been named the **homeobox**. The homeobox is part of the protein-coding sequence of each gene, and the corresponding 60-amino-acid part of each protein is called the **homeodomain**.

Homeoboxes have been found in over 20 *Drosophila* genes, most of which regulate development. All homeodomain-containing proteins are DNA-binding proteins. The homeodomain of such proteins binds strongly to an 8-bp consensus recognition sequence upstream of all genes controlled as a unit by the homeotic gene encoding the homeodomain-containing protein. Helix-turn-helix motifs are used in the DNA-binding property of homeodomains. Thus, homeodomain-containing proteins play a role in transcriptional regulation through interaction with specific DNA sequences.

The complete set of homeotic genes and complexes in *Drosophila*—generically the *Hom* genes—consists of *lab, pb, Dfd, Scr, Antp, Ubx, abdA,* and *AbdB*. Most interestingly, these complexes are arranged in the same order along the chromosome as they are expressed along the antero-posterior body axis: this is known as the *colinearity rule*. Homeotic gene complexes are found also in all major animal phyla with the exception of sponges and coelenterates. The homeobox sequences in the *Hom* genes are highly conserved, indicating common function in the wide range of organisms involved. As in *Drosophila*, the homeotic genes of vertebrates—the *Hox* genes—follow the colinearity rule. In mammals, for example, there are four clusters of homeotic genes designated *HoxA–D*. Each cluster is thought to have originated by duplication of a primordial gene cluster followed by subsequent evolutionary divergence. The patterns of *Hox* gene expression, the effects of mutations, and embryological analyses all indicate that the vertebrate genes have homeotic effects similar to *Drosophila* homeotic genes. Further, the studies indicate that the *Hox* genes specify the vertebrate body plan.

~ FIGURE 17.34

(a) Scanning electron micrograph (*left*) and drawing (*right*) of the antennal area of a wild-type fly; (b) Scanning electron micrograph (*left*) and drawing (*right*) of the antennal area of the homeotic mutant of *Drosophila*, *Antennapedia*, in which the antenna is transformed into a leg; (c) Scanning electron micrograph (*left*) and drawing (*right*) of the homeotic mutant of *Drosophila*, *Aristapedia*, in which the arista is transformed into a leg.

a) Normal

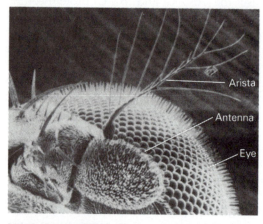

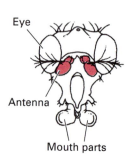

b) Antennapedia

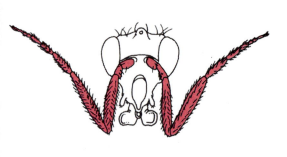

c) Aristapedia

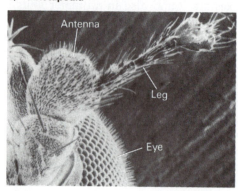

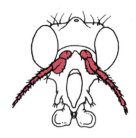

~ FIGURE 17.35

Organization of the *bithorax* complex (BX-C). The DNA spanned by this complex is 300 kb long. The transcription units for *Ubx*, *abdA*, and *AbdB* are shown below the DNA: the exons are shown by colored blocks and the introns by bent, dotted lines. All three genes are transcribed from right to left. Shown above the DNA are regulatory mutants that affect the development of different fly segments.

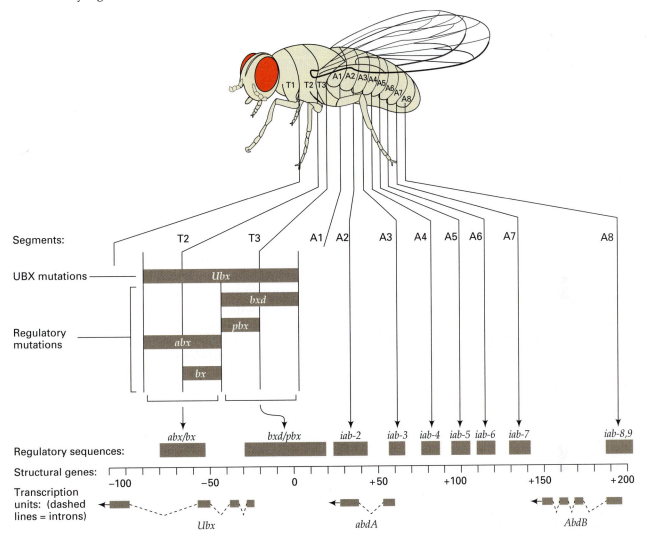

KEYNOTE

Homeotic genes in *Drosophila* determine the developmental identity of a segment, once the basic segmentation pattern has been laid down. Homeotic genes have in common similar DNA sequences called homeoboxes. The homeobox is part of the protein-coding sequence of each gene and the corresponding approximately 60-amino-acid part of the protein is called the homeodomain. Homeoboxes have been found in developmental genes in other organisms, and homeodomains are thought to play a role in regulating transcription by binding to specific DNA sequences.

SUMMARY

In this chapter we have considered a number of examples of gene regulation in eukaryotes. The general picture is one of much greater complexity than prokaryotes. Genes are not organized into operons, since genes of related function are often scattered around the genome; nonetheless, genes are regulated coordinately. The chromosome organization in eukaryotes also makes for greater complexity in regulating gene expression.

Three main topics were discussed in the chapter: the levels of control of gene expression, gene regulation during development and differentiation, and genetic regulation of development in *Drosophila*. Given the information available, each of these topics could be the subject of its own chapter.

Levels of Control of Gene Expression

Gene expression in eukaryotes is regulated at a number of distinct levels. Regulatory systems have been found for: (1) the control of transcription; (2) of precursor-RNA processing; (3) the transport of the mature RNA out of the nucleus; (4) the translation of the mRNAs; (5) the degradation of the mature RNAs; and (6) the degradation of the protein products.

The control of transcription itself involves a number of elements. At the DNA level, transcription of protein-coding genes involves the interaction of transcription factors with promoter elements, and of regulatory proteins with both promoter elements and enhancer elements. Depending on the element and the protein that binds to it, the effect on transcription may be positive or negative. Activation of transcription through protein binding at the distant enhancer elements is thought to involve a looping of the DNA, brought about by interaction of the regulatory proteins bound at the enhancer and promoter elements. While unique regulatory proteins certainly bind to promoter elements and to enhancer elements, some regulatory proteins are shared by the two, indicating that both types of regulatory elements affect transcription by a similar mechanism.

At the chromosome level, the regulation of transcription must deal with the interaction of histones and nonhistones with the DNA. Generally, we can view the eukaryote chromosome as repressed for transcription by this interaction, so activation of transcription can be seen as a derepression phenomenon that results from a loosening of the DNA-protein structure in the region of a gene being activated. Relevant to this, we know that transcriptionally active chromatin is more sensitive to digestion with DNase I than transcriptionally inactive chromatin. Moreover, upstream of active genes are DNase-hypersensitive sites, which are highly sensitive to DNase I digestion. These sites probably correspond to the binding sites for RNA polymerase and regulatory proteins.

Histones play an important role in gene repression by assembling nucleosomes on TATA boxes in promoter regions. Gene activation involves regulatory proteins binding to enhancers and then disrupting the nucleosomes on the TATA boxes. This allows regulatory proteins and transcription factors to bind to promoters to initiate transcription.

Gene activation in many eukaryotes is accompanied in many instances by a decrease in the level of DNA methylation, although the relationship, if any, to chromosome organization changes is not clear.

Gene expression can be regulated at the level of RNA processing. This type of regulation operates to determine the production of mature RNA molecules from precursor-RNA molecules. Two regulatory events that exemplify this level of control are choice of poly(A) site and choice of splice site. In both cases different types of mRNAs are produced, depending upon the choice made. Examples are known of both types of regulation in developmental systems. For example, different classes of immunoglobulin molecules can result from poly(A) site selection, and choice of splice site plays a key role in the *Drosophila* sex determination system.

mRNA transport from the nucleus to the cytoplasm is another important control point in the regulation of gene expression in eukaryotes. A spliceosome retention model has been proposed for this regulatory system. In this model, spliceosome assembly on a precursor-mRNA competes with nuclear export of the molecule. It is argued that, during intron removal, the spliceosome that is complexed with the precursor-mRNA prevents the RNA molecule from leaving the nucleus. Then, when all introns have been removed and the spliceosome has dissociated from the now-mature mRNA, the free mRNA molecule can interact with the nuclear pore and move to the cytoplasm.

Gene expression is also regulated by mRNA translation control and by mRNA degradation control. The latter is believed to be a major control point in the regulation of gene expression, as evidenced by the wide range of mRNA stabilities found within organisms. It is clear that nucleases are ultimately responsible for the degradation of the RNAs, and the signals for the differential mRNA stabilities seem to be a property of the structural features of individual mRNAs. For example, a group of AU-rich sequences in the 3'-untranslated regions of some short-lived mRNAs is responsible for their instability, although how this signal is read by the cell and transmitted to the cellular factors and enzymes that must be directly involved in the mRNA degradation remains under investigation.

Lastly, we discussed the immune response, the structure of antibody molecules, and generation of antibody diversity in mammals. Here a particularly interesting process of chromosomal rearrangements is involved during cell development to bring together parts of the coding regions for the light and heavy polypeptide chains of antibody molecules into the genes that are transcribed. Specifically, each light chain and each heavy chain has a variable (V) domain and a constant (C) domain. For the light chain, in germ-line DNA there are many V domain gene segments tandemly arranged, a few joining (J) gene segments, and one C gene segment. During development, somatic recombination occurs to bring together particular V, J, and C gene segments to form a functional L chain gene. A similar process occurs for assembling a functional H chain gene, in this case with the addition of several D (diversity) segments between V and J. In each cell, a different light chain gene and heavy chain gene are assembled during development, and this is the basis for the huge number of different types of antibodies that can be produced in mammals.

Gene Regulation During Development and Differentiation

This topic is a vast one and brings together material in the domains of the geneticist, the cell biologist, the embryologist, the developmental biologist, and the molecular biologist. Classical and recent experiments have shown that development and differentiation result from differential gene activity of a genome that contains a constant amount of DNA, rather than from a programmed loss of genetic information. Development and differentiation, then, must involve regulation of gene expression, using the levels of control we have just discussed. Adding to the complexity is the fact that we must consider the regulation of a large number of genes for each developmental process, and communication between differentiating tissues, as well as systems for timing the activation and repression of genes during those important events. For example, the structure and function of a cell are often determined early, even though the manifestations of this determination process are not seen until later in development. Such early determination events may involve some preprogramming of genes that will be turned on later. Neither the nature of these determination events, nor the mechanism of timing in developmental processes are well understood in vertebrates, although progress is being made in understanding the genetic regulation of development in certain organisms such as *Drosophila*.

In sum, a great deal of information has been learned in the past decade or so about gene regulation in eukaryotes. We have merely scratched the surface in this chapter. Thousands of researchers are currently working to elaborate the molecular details of gene regulation in model systems. Much of our advancing knowledge has been made possible by the application of recombinant DNA and related technologies, and we can look forward to significant increases in our understanding of eukaryotic gene regulation by the turn of the century.

ANALYTICAL APPROACHES FOR SOLVING GENETICS PROBLEMS

Q17.1 A region of the yeast chromosome specifies three enzyme activities in the histidine biosynthesis pathway; these activities are synthesized coordinately. How would you distinguish among the following three models?

a. Three genes are not organized into an operon. They code for three discrete mRNAs that are translated into three distinct enzymes.

b. Three genes are arranged in an operon. The operon is transcribed to produce a single polygenic mRNA, whose translation produces three distinct enzymes.

c. One gene (a supergene) is transcribed to produce a single mRNA, whose translation produces a single polypeptide with three different enzyme activities.

A17.1 A key feature of an operon (model b) is that a contiguously arranged set of genes is transcribed onto a single polygenic mRNA. Thus a nonsense mutation in a structural gene will result in the loss of not only the enzyme activity coded for by that gene but also the enzyme activities coded for by the structural genes that are more distant from the promoter (see Chapter 16). If there is a supergene coding for a single polypeptide with three different enzyme activities (model c), then nonsense mutations will cause effects similar to polar effects in polygenic mRNAs; that is, a nonsense mutation in the coding region for the first enzyme activity will cause a loss of that activity as well as of the two other enzyme activities. However, if there are three genes that are closely linked but each with its promoter (model a), transcription will produce three distinct mRNAs, so a nonsense mutation will affect only the gene in which it is located and no other gene. Therefore, if nonsense mutations are shown to have polar effects, then either model b or model c is correct, but model a cannot be correct. If nonsense mutations do not have polar effects, no matter in which gene they are located, then model a must be correct.

Characterization of the enzyme activities coded for by the three genes would enable us to distinguish between models b and c. That is, in the operon model b, three distinct polypeptides would be produced. These polypeptides could be isolated and purified individually by using standard techniques.

Note that it would be particularly important to make sure that inhibitors of protein-degrading enzymes, proteases, are present during cell disruption so that if there is a trifunctional polypeptide present, it is not cleaved by the proteases to produce three separable enzyme activities.

Thus if the operon model b is correct, we could show that there are three distinct polypeptides, each exhibiting only one of the enzyme activities—that is, three polypeptides and three enzyme activities. On the other hand, if the supergene model c is correct, it should only be possible to isolate a large polypeptide with all three enzyme activities (again assuming careful isolation and purification procedures); no polypeptides with only one of the enzyme activities should exist.

QUESTIONS AND PROBLEMS

17.1 Promoters, enhancers, transcription factors, and regulatory proteins that are active for one gene typically share structural similarities with these elements in other genes. Nonetheless, the transcriptional control of a gene can be exquisitely specific: it will be specifically transcribed in some tissues at very defined times. Explore how this specificity arises by addressing the following questions.
a. Distinguish between the functions of promoters and enhancers in transcriptional regulation.

b. What structural features are found in proteins that bind these DNA elements?
c. Can an enhancer bound by a regulatory protein stimulate as well as suppress transcription? If so, how?
d. Given that several different genes may contain the same types of promoter and enhancer elements, and a number of transcription factors contain the same structural features, how is transcriptional specificity generated?

*17.2 Eukaryotic organisms have a large number of copies (usually more than a hundred) of the genes that code for ribosomal RNA, yet they have only one copy of each gene that codes for each ribosomal protein. Explain why.

*17.3 The human α, β, γ, δ, ϵ and ζ globin genes are transcriptionally active at various stages of development. Fill in the following table, indicating whether the globin gene in question is sensitive (S) or resistant (R) to DNase I digestion at the developmental stages listed.

	TISSUE		
GLOBIN GENE	EMBRYONIC YOLK SAC	FETAL SPLEEN	ADULT BONE MARROW
α			
β			
γ			
δ			
ζ			
ϵ			

17.4 In *Drosophila*, mutations at four genes known as the *male-specific lethals (msls)* cause male-specific lethality during the larval stages. Analysis has shown that male-specific lethality is caused by defects in the process of dosage compensation (the process by which the single X chromosome of a male is transcribed at twice the rate of either of a female's two X chromosomes; see Chapter 3, pp. 78–79). Examination of polytene chromosomes in normal males and females shows that the X chromosome is less tightly condensed in males than in females, that the X chromosome of the male is enriched in an acetylated form of histone H4, and that the protein products of each of the four *msl* loci are primarily associated with the X chromosome of the male, but not that of the female. Examination of polytene chromosomes in each of the four types of mutant *msl* males reveals that their X chromosome remains about as tightly condensed as an X chromosome in fe-males, and that the absence of any single *MSL*-protein in a male results in the other three *MSL*-proteins not associating with the X chromosome.

a. Based on your understanding of the relationship between chromatin structure and gene transcription, how might a decrease in chromatin condensation be associated with dosage compensation?

b. Based on the data presented above, what potential roles might the protein products of the *msl* loci play in dosage compensation?

c. What evidence is there to suggest that the protein products of the *msl* loci act as a complex?

17.5 A cloned DNA sequence was used to probe a Southern blot. There were two DNA samples on the blot, one from white blood cells and the other from a liver biopsy of the same individual. Both samples had been digested with *Hpa*II. The probe bound to a single 2.2 kb band in the white blood cell DNA, but bound to two bands (1.5 and 0.7 kb) in the liver DNA.

a. Is this difference likely to be due to a somatic mutation in a *Hpa*II site? Explain.

b. How would it affect your answer if you knew that white blood cell and liver DNA from this individual both showed the 2 band pattern when digested with *Msp*I?

*17.6 Both fragile X syndrome and Huntington disease are caused by trinucleotide repeat expansion. Individuals with fragile X syndrome have at least 200 CGG repeats at the 5'-end of the *FMR-1* gene. Individuals with Huntington disease have at least 36 CAG repeats within the protein-coding region of the huntingtin gene.

a. How is gene expression affected by these repeat expansions?

b. Based on your answer to (a), why is the fragile X syndrome recessive, while Huntington disease is dominant?

c. Why is the number of trinucleotide repeats needed to cause the phenotype different for each disease?

17.7 Both peptide and steroid hormones can affect gene regulation of a targeted population of cells.

a. What is a hormone?

b. Distinguish between the mechanisms by which a peptide and a steroid hormone affect gene expression.

c. What role do each of the following have in a physiological response to a peptide or a steroid hormone?
   i. steroid hormone receptor (SHR)
   ii. steroid hormone response element (HRE)
   iii. second messenger
   iv. cAMP and adenylate cyclase

*17.8 The following figure shows the effect of the hormone estrogen on ovalbumin synthesis in the oviduct of 4-day-old chicks. Chicks were given daily injections of estrogen ("Primary Stimulation") and then after 10 days the injections were stopped. Two weeks after withdraw-

al (25 days), the injections were resumed ("Secondary Stimulation").

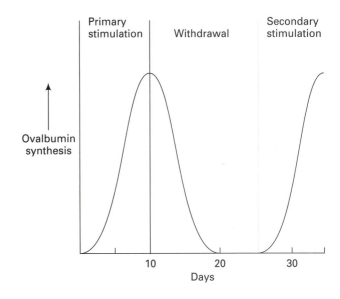

Provide possible explanations of these data.

17.9 Although the primary transcript of a gene may be identical in two different cell types, the translated mRNAs can be quite different. Consequently, in different tissues, distinct protein products can be produced from the same gene. Discuss two different mechanisms by which the production of mature mRNAs can be regulated to this end; give a specific example for each mechanism.

17.10 Although many mRNAs are present in the cytoplasm of unfertilized vertebrate and invertebrate embryos, the rate of protein synthesis is very low. After fertilization, the rate of protein synthesis increases dramatically without new mRNA transcription.

a. What differences are seen in the length of poly(A) tails between inactive, stored mRNAs and actively translated mRNAs?

b. What role does *cytoplasmic* polyadenylation have in this process?

c. What signals are present in mRNAs that control polyadenylation and deadenylation?

d. In what way is deadenylation also important for controlling mRNA degradation?

*17.11 Although eukaryotes lack operons like those found in prokaryotes, the exceptional conserved organization of the *ChAT/VAChT* locus is reminiscent of a prokaryotic operon. *ChAT* is the gene for the enzyme choline acetyltransferase, which synthesizes acetylcholine, a neurotransmitter released by one neuron in order to signal another neuron. *VAChT* is the gene for the vesicular acetylcholine transporter protein, which packages acetylcholine into vesicles prior to its release from a neuron. Both *ChAT* and *VAChT* are expressed in the same neuron.

Part of the *VAChT* gene is nested within the first intron of the *ChAT* gene, and the two genes share a common regulatory region and a first exon. The structure of a primary mRNA and two processed mRNA transcripts produced by this locus are diagrammed below. The common regulatory region important for transcription of the locus in neurons is shown in the DNA, black rectangles in RNA represent exons, lines connecting the exons represent spliced intronic regions, and AUG indicates the translation start codons within the *ChAT* and *VAChT* mRNAs. Polyadenylation sites are not shown.

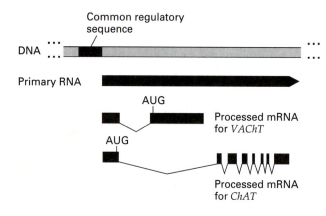

a. In what ways is the organization of the *VAChT/ChAT* locus reminiscent of a bacterial operon?
b. Why is the organization of this locus not structurally equivalent to a bacterial operon?
c. Based on the transcript structures that are shown, what modes of regulation might be used to obtain two different protein products from the single primary mRNA?

17.12 Distinguish between the terms *development* and *differentiation*.

17.13 What is totipotency? Give an example of the evidence for the existence of this phenomenon.

*17.14 In Woody Allen's 1973 film *Sleeper*, the aging leader of a futuristic totalitarian society has been dismembered in a bomb attack. The government wishes to clone the leader from his only remaining intact body part, a nose. The characters Miles and Luna thwart the cloning by abducting the nose and flattening it under a steam roller.
a. In light of the 1996 cloning of the sheep Dolly, how should the cloning have proceeded, if Miles and Luna had not intervened?
b. If methods akin to those used for Dolly had been successful, in what *genetic* ways would the cloned leader be unlike the original?

c. Suppose that, instead of a nose, only mature B cells (B lymphocytes of the immune system) were available. What *genetic* deficits would you expect in the "new leader"?
d. In the set of experiments used to clone Dolly, six additional live lambs were obtained. Why is the production of Dolly more significant than the production of the other lambs?
e. If the cloning of the leader had succeeded, can you make any prediction about whether the "cloned leader" would be interested in perpetuating the totalitarian state?

17.15 Discuss some of the evidence for differential gene activity during development.

*17.16 The enzyme lactate dehydrogenase (LDH) consists of four polypeptides (a tetramer). Two genes are known to specify two polypeptides, A and B, which combine in all possible ways (A_4, A_3B, A_2B_2, AB_3, and B_4) to produce five LDH isozymes. If, instead, LDH consisted of three polypeptides (i.e., it was a trimer), how many possible isozymes would be produced by various combinations of polypeptides A and B?

17.17 Discuss the expression of human hemoglobin genes during development.

17.18 Discuss the organization of the hemoglobin genes in the human genome. Is there any correlation with the temporal expression of the genes during development?

17.19 In humans, β-thalassemia is a disease caused by failure to produce sufficient β-globin chains. In many cases, the mutation causing the disease is a deletion of all or part of the β-globin structural gene. Individuals homozygous for certain of the β-thalassemia mutations are able to survive because their bone marrow cells produce γ-globin chains. The γ-globin chains combine with α-globin chains to produce fetal hemoglobin. In these people, fetal hemoglobin is produced by the bone marrow cells throughout life, whereas normally it is produced in the fetal liver. Use your knowledge about gene regulation during development to suggest a mechanism by which this expression of γ-globin might occur in β-thalassemia.

17.20 What are the polytene chromosomes? Discuss the molecular nature of the puffs that occur in polytene chromosomes during development.

17.21 Puffs of regions of the polytene chromosomes in salivary glands of *Drosophila* are surrounded by RNA molecules. How would you show that this RNA is single-stranded and not double-stranded?

*17.22 In experiment A, ^{3}H-thymidine (a radioactive precursor of DNA) is injected into larvae of *Chironomus*, and the polytene chromosomes of the salivary glands are later examined by autoradiography. The radioactivity is seen to be distributed evenly throughout the polytene chromosomes. In experiment B, ^{3}H-uridine (a radioactive precursor of RNA) is injected into the larvae, and the polytene chromosomes are examined. The radioactivity is first found only around puffs; later, radioactivity is also found in the cytoplasm. In experiment C, actinomycin D (an inhibitor of transcription) is injected into larvae and then ^{3}H-uridine is injected. No radioactivity is found associated with the polytene chromosomes, and few puffs are seen. Those puffs that are present are much smaller than the puffs found in experiments A and B. Interpret these results.

*17.23 The following figure shows the percentage of ribosomes found in polysomes in unfertilized sea urchin oocytes (0 h) and at various times after fertilization:

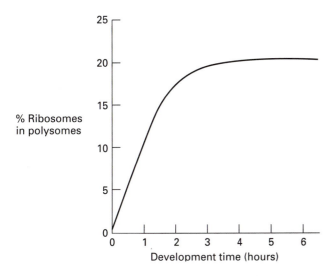

In the unfertilized egg, less than 1 percent of ribosomes are present in polysomes, while at 2 h post-fertilization, about 20 percent of ribosomes are present in polysomes. It is known that no new mRNA is made during the time period shown. How may the data be interpreted?

17.24 The mammalian genome contains about 10^5 genes. Mammals can produce about 10^6 to 10^8 different antibodies. Explain how it is possible for both of the above sentences to be true.

*17.25 Ashley and Connie are identical twins. Ashley has blood type A, and therefore must have anti-B antibodies in her serum. (For the purpose of this problem, ignore the effects of possible new mutations.)
a. What is Connie's blood type?
b. If Ashley's genotype is I^A/I^A, what is Connie's blood type genotype?
c. If Ashley has anti-B antibodies, what type of antibodies would Connie have?
d. Would Ashley's and Connie's anti-B antibodies have identical polypeptide sequences? Explain.

e. Would Ashley's and Connie's β-globin chain have the identical polypeptide sequence? Explain.

17.26 Antibody molecules (Ig) are composed of four polypeptide chains (two of one light chain type and two of one heavy chain type) held together by disulfide bonds.
a. If for the light chain there were 300 different V_κ and four J_κ segments, how many different light chain combinations would be possible?
b. If for the heavy chain there were 200 V_H segments, 12 D segments, and 4 J_H segments, how many heavy chain combinations would be possible?
c. Given the information in parts (a) and (b), what would be the number of possible types of IgG molecules (L + H chain combinations)?

17.27 Define *imaginal disc*, *homeotic mutant*, and *transdetermination*.

*17.28 Imagine that you observed the following mutants (*a–e*) in *Drosophila*. Based on the characteristics given, assign each of the mutants to one of the following categories: maternal gene, segmentation gene, or homeotic gene.
a. Mutant *a*: In homozygotes phenotype is normal, except wings are oriented backward.
b. Mutant *b*: Homozygous females are normal but produce larvae that have a head at each end and no distal ends. Homozygous males produce normal offspring (assuming the mate is not a homozygous female).
c. Mutant *c*: Homozygotes have very short abdomens, which are missing segments A2 through A4.
d. Mutant *d*: Affected flies have wings growing out of their heads in place of eyes.
e. Mutant *e*: Homozygotes have shortened thoracic regions and lack the second and third pair of legs.

*17.29 If actinomycin D, an antibiotic that inhibits RNA synthesis, is added to newly fertilized frog eggs, there is no significant effect on protein synthesis in the eggs. Similar experiments have shown that actinomycin D has little effect on protein synthesis in embryos up until the gastrula stage. After the gastrula stage, however, protein synthesis is significantly inhibited by actinomycin D, and the embryo does not develop any further. Interpret these results.

*17.30 It is possible to excise small pieces of early embryos of the frog, transplant them to older embryos, and follow the course of development of the transplanted material as the older embryo develops. A piece of tissue is excised from a region of the late blastula or early gastrula that would later develop into an eye and is transplanted to three different regions of an older embryo host (see part a of Figure 17.A). If the tissue is transplanted to the head region of the host, it will form eye, brain, and other material characteristic of the head region. If the tissue is transplanted to other regions of the host, it will form organs and tissues characteristic of those regions in normal development (e.g., ear, kidney, etc.). In contrast, if tissue destined to be an eye is excised from a neurula and transplanted into an older embryo host to exactly the same places as used for the blastula/gastrula transplants, in every case the transplanted tissue differentiates into an eye (see part b of Figure 17.A). Explain these results.

17.31 What features of the genes that control body segmentation have been conserved during evolution? How has this been shown?

FIGURE 17.A

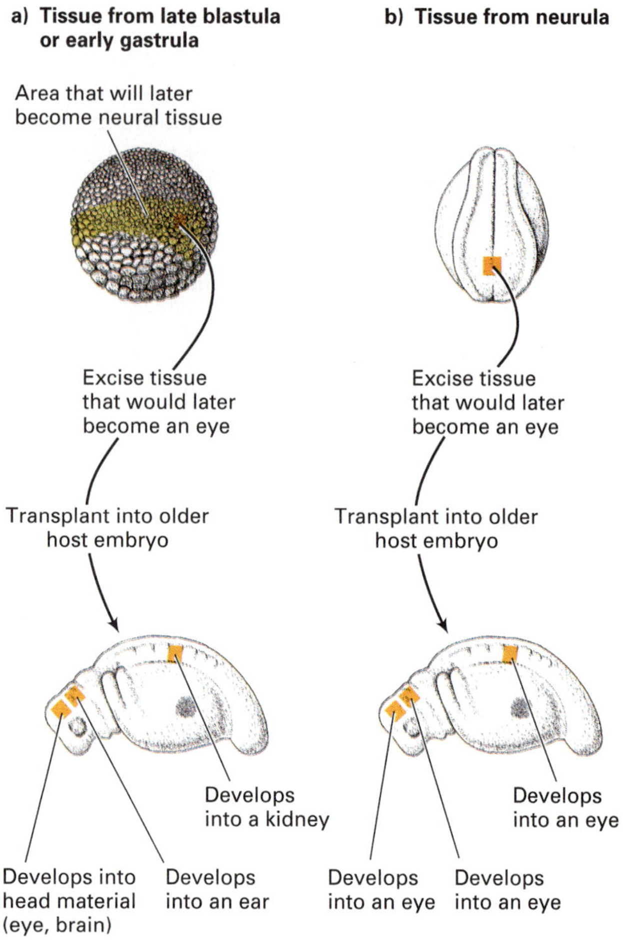

a) **Tissue from late blastula or early gastrula**

Area that will later become neural tissue

Excise tissue that would later become an eye

Transplant into older host embryo

Develops into head material (eye, brain)

Develops into an ear

Develops into a kidney

b) **Tissue from neurula**

Excise tissue that would later become an eye

Transplant into older host embryo

Develops into an eye

Develops into an eye

Develops into an eye

18 GENETICS OF CANCER

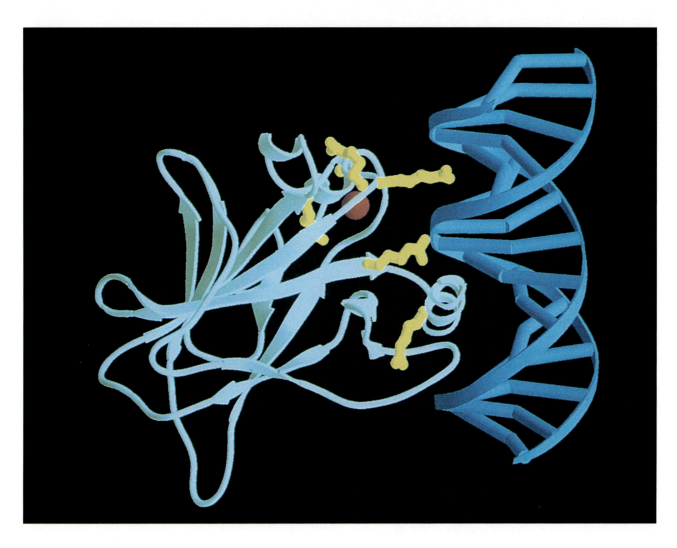

RELATIONSHIP OF THE CELL CYCLE TO CANCER

THE TWO-HIT MUTATION MODEL FOR CANCER

GENES AND CANCER
 Oncogenes
 Tumor Suppressor Genes
 Mutator Genes

THE MULTISTEP NATURE OF CANCER

CHEMICALS AND RADIATION AS CARCINOGENS
 Chemical Carcinogens
 Radiation

PRINCIPAL POINTS

~ Progression of a normal eukaryotic cell through the cell cycle is tightly controlled by a number of molecular factors. Healthy cells grow and divide only when the balance of stimulatory and inhibitory signals received from outside the cell favor cell proliferation. A cancerous cell does not respond to the usual signals and reproduces without constraints.

~ The two-hit mutation model for cancer states that two mutational events are critical for cancer to develop, one in each allele of a cancer-causing gene. In familial (hereditary) cancers one mutation is inherited, predisposing the person to cancer; the other mutation occurs later in somatic cells. In sporadic (nonhereditary) cancers both mutations occur in the somatic cells. This simple two-hit model applies for very few cancers; other cancers involve mutations in many genes.

~ Three classes of genes have been shown to be mutated frequently in cancer. These are oncogenes, tumor suppressor genes, and mutator genes. The products of oncogenes normally stimulate cell proliferation, the products of tumor suppressor genes normally inhibit cell proliferation, and the products of mutator genes are involved in DNA replication and repair. Mutant forms of these three types of genes all have the potential to contribute to the transformation of a cell to a tumorous state.

~ Retroviruses are RNA viruses that replicate via a DNA intermediate. All RNA tumor viruses are retroviruses, but not all retroviruses cause cancer. When a retrovirus infects a cell, the RNA genome is released from the viral particle, and through the action of reverse transcriptase, a cDNA copy of the genome—called the proviral DNA—is synthesized. The proviral DNA integrates into the genome of the host cell. Then, using host transcriptional machinery, viral genes are transcribed, and full-length viral RNAs are produced. Progeny viruses assembled within the cell exit the cell and can infect other cells.

~ When tumor induction occurs after retrovirus infection, it is because of the activity of a viral oncogene (v-*onc*) in that retroviral genome. Retroviruses carrying an oncogene are known as transducing retroviruses. Normal animal cells contain genes with DNA sequences that are similar to those of the viral oncogenes. These cellular genes, called proto-oncogenes, play important roles in regulating cell division and differentiation. When proto-oncogenes are mutated, they induce tumor formation. The mutant phenotype is dominant in that only one allele needs to be mutated for tumors to be induced. In its mutated state a proto-oncogene is called a cellular oncogene (c-*onc*). The viral oncogenes are copies of the cellular proto-oncogenes that the retrovirus has picked up and modified during its life cycle.

~ Tumor suppressor genes, like proto-oncogenes, are involved in the regulation of cell growth and division. Whereas the normal products of proto-oncogenes have a stimulatory role in those processes, the normal products of tumor suppressor genes have inhibitory roles. Therefore, when both alleles of a tumor suppressor gene are inactivated or lost, the inhibitory activity is lost, and unprogrammed cell proliferation can occur. Inactivation of tumor suppressor genes is involved in the development of a wide variety of human cancers, including breast, colon, and lung cancer.

~ The development of most cancers involves the accumulation of mutations in a number of genes over a significant period of life. This multistep nature of cancer typically involves mutational events that activate oncogenes and inactivate tumor suppressor genes, thereby breaking down the multiple mechanisms that regulate growth and differentiation.

~ Various types of radiation and many chemicals increase the frequency with which cells become cancerous. These agents are known as carcinogens. Practically all carcinogens act by causing changes in the genome of the cell. In the case of chemical carcinogens, a few act directly on the genome, but most act indirectly by being converted to active derivatives by cellular enzymes. All carcinogenic forms of radiation act directly.

In Chapter 17 we learned about some of the genetically controlled processes involved in development and differentiation. The picture we have is that, during development, specific tissues and organs arise by genetically programmed cell division and differentiation. Occasionally, dividing and differentiating cells deviate from their normal genetic program and give rise to tissue masses called *tumors*, or *neoplasms* ("new growth"). Figure 18.1 shows a mammogram indicating the presence of a tumor. The process by which a cell loses its ability to remain constrained in its growth properties is called **transformation** (not to be confused with transformation of a cell by uptake of DNA). If the transformed cells stay together in a single mass, the tumor is said to be *benign*. Benign tumors are usually not life threatening, and their surgical removal generally results in a complete cure. Exceptions include many brain tumors, which are life threatening because they impinge on essential cells. If the cells of a tumor can invade and disrupt surrounding tissues, the tumor is said to be *malignant* and is identified as a **cancer**. Cells from malignant tumors can also break off and move through the blood system or lymphatic system, forming new tumors at other locations in the body. The spreading of malignant tumor cells throughout the body is called **metastasis**. Figure 18.2 shows a metastasized tumor—a melanosarcoma—in a human liver. Malignancy can result in death because of damage to critical organs, starvation, secondary infection, metabolic problems, second malignancies, and/or hemorrhage.

The initiation of tumors in an organism is called **oncogenesis** (*onkos,* "mass" or "bulk"; *genesis,* "birth"). There are many different causes of cancer, such as spontaneous genetic changes (e.g., spontaneous gene mutations or chromosome mutations) and exposure to mutagens, radiation, or cancer-inducing viruses *(tumor viruses).* There is also hereditary predisposition to cancer. In this chapter we focus on the genetic basis of tumors and cancers.

~ FIGURE 18.2

Light micrograph of human liver, showing a metastasized tumor, in this case a melanosarcoma (in red).

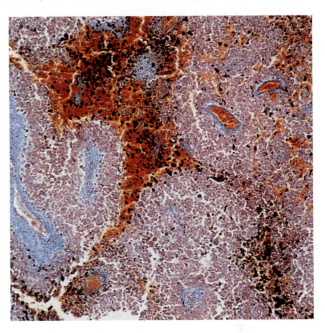

RELATIONSHIP OF THE CELL CYCLE TO CANCER

During development a tissue is produced by cell proliferation. Over a series of divisions, progeny cells begin to express genes that are specific for the tissue, a process called cell differentiation. Cell differentiation is also associated with the progressive loss of the ability of cells to proliferate: the most highly differentiated cell—that is, the one that is fully functional in the tissue—can no longer divide. Such cells are known as *terminally differentiated* cells. They have a finite life span in the tissue and are replaced with younger cells produced by division of *stem cells,* a small fraction of cells in the tissue that are capable of *self-renewal.* To understand neoplastic diseases, both benign and malignant ones, we must realize that the linkage of growth with differentiation of any tissue is not necessary. That is, cells *can* divide without undergoing terminal differentiation. For example, in malignant neoplasms most daughter cells of any replicative event fail to express fully the genetic programs that regulate terminal differentiation.

In Chapter 3 (pp. 51–52) and Chapter 12 (p. 361) we discussed the cell cycle, that is, the cycle of cell growth, mitosis, and cell division in eukaryotes. Recall that the cell cycle in proliferating somatic cells consists of the mitotic phase (M) and an interphase between divisions. Interphase itself consists of three

~ FIGURE 18.1

A mammogram showing the presence of a tumor.

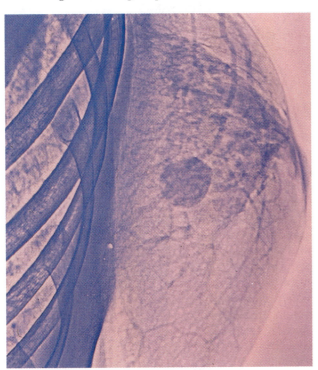

~ **FIGURE 18.3**

General events for regulation of cell division in normal cells. (a) When a growth factor binds to its cell membrane receptor, it acts as a signal to stimulate cell growth. To do that, the signal is transduced into the cell and relayed to the nucleus, activating the expression of a gene or genes that encode a protein or proteins required for the stimulation of cell division. (b) When a growth-inhibiting factor binds to its cell membrane receptor, it acts as a signal to inhibit cell growth. In this case the signal is transduced into the cell and relayed to the nucleus, activating the expression of a gene or genes that encode a protein or proteins required for the inhibition of cell division.

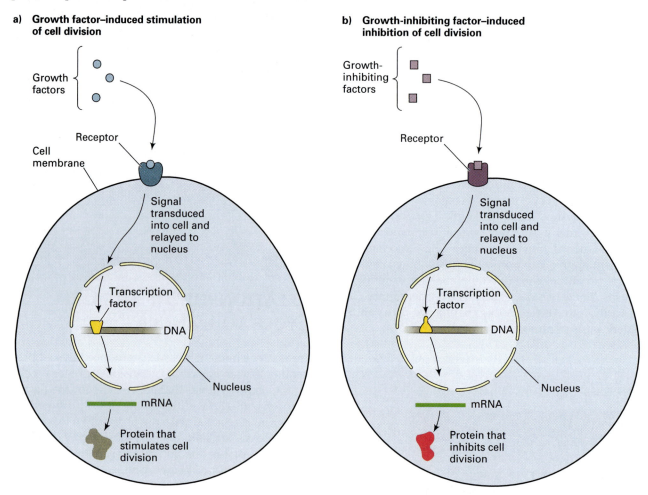

a) **Growth factor–induced stimulation of cell division**

b) **Growth-inhibiting factor–induced inhibition of cell division**

stages: G_1, S, and G_2. During G_1 the cell monitors its size and the environment and, if conditions are appropriate, commits to DNA replication (in the S phase) and the completion of the cycle (G_2 and M). Cells in G_1 that have not committed to DNA replication may enter a special quiescent state called G_0 (G zero), where they might stay for long periods (days to years) before reentering the cell cycle and continuing proliferation.

How is progression of a normal cell through the cell cycle controlled? Several molecular factors have been shown to be involved in such control, as evidenced by their abilities under experimental conditions to influence the progress of cells through division. Perhaps the most important of such factors are the proteins that operate through interaction with receptors embedded in the plasma membrane. When these factors bind to the cell surface receptor, a signal is transmitted into the cytoplasm, initiating a chain of reactions that exerts a regulatory effect on cell division. Depending upon the protein factor, the signal may be to stimulate or to inhibit cell division. *Growth factors* turn on stimulatory pathways for cell division (Figure 18.3a), and *growth-inhibiting factors* turn on inhibitory pathways for cell division (Figure 18.3b). In either event the intracellular chain of reactions involved ultimately causes changes in gene expression. For growth factors, genes that encode proteins needed for the cell division process are turned on; for growth-inhibiting factors, genes that encode proteins with inhibitory effects on cell division are turned on. Normal healthy cells give rise to progeny cells only when the balance of stimulatory and inhibitory signals from outside the cell tips toward favoring cell division. A neoplastic cell, however, has lost control of the cell division process and reproduces without constraints.

𝒦EYNOTE

Progression of a normal eukaryotic cell through the cell cycle is tightly controlled by a number of molecular factors. Healthy cells grow and divide only when the balance of stimulatory and inhibitory signals received from outside the cell favor cell proliferation. A cancerous cell does not respond to the usual signals and reproduces without constraints.

THE TWO-HIT MUTATION MODEL FOR CANCER

All cancers are genetic disorders in the sense that they are caused by changes in DNA that are stably inherited by progeny cells. Our current understanding is that a cancer results from an accumulation of genetic mutations in particular classes of genes in a cell over a period of time. (We will have more to say about this later in the chapter.) This escalating genetic damage causes a progressive loss in the ability of the cell to respond to growth regulatory signals so that eventually the cell will divide uncontrollably, thereby giving rise to a neoplasm. Research over the past twenty years or so has led to the identification of a number of the particular genes related to the onset of cancer.

Anecdotal evidence that genes have a role in cancer came from the observation that there was a high incidence of particular cancers in some human families. Cancers that "run" in families are known as *familial (hereditary) cancers*; cancers that do not appear to be inherited are known as *sporadic* (or *nonhereditary*) *cancers*. Sporadic cancers are more frequent than familial cancers.

One model for the relation of mutations to cancer—the *two-hit model*—came from the study of the onset of retinoblastoma, a childhood cancer of the eye (Figure 18.4). Retinoblastoma occurs from birth to age four years and is the most common eye tumor in children. If discovered early enough, over 90 percent of the eye tumors can be permanently destroyed, usually by gamma radiation. There are two forms of retinoblastoma. In *sporadic retinoblastoma* (60 percent of cases) the development of an eye tumor is a spontaneous event in a patient from a family with no history of the disease. In these cases a *unilateral tumor*

~ FIGURE 18.4

An eye tumor in a patient with retinoblastoma.

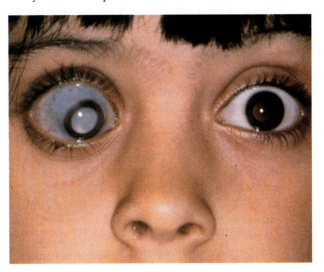

will develop; that is, the tumor is in one eye only. In *hereditary retinoblastoma* (40 percent of cases) the susceptibility to develop the eye tumors is inherited. Patients with this form of retinoblastoma typically develop multiple eye tumors involving both eyes (*bilateral tumors*), usually at an earlier age than is the case for unilateral tumor formation in sporadic retinoblastoma patients. Moreover, the siblings and offspring of hereditary retinoblastoma patients often develop the same type of tumor. Data from pedigree analysis are consistent with a single gene being responsible for retinoblastoma.

In 1971 Alfred Knudson proposed the following to explain the occurrence of hereditary and sporadic forms of retinoblastoma: "Retinoblastoma is a cancer caused by two mutational events. . . . one mutation is inherited via the germinal cells and the second occurs in somatic cells. In the nonhereditary form, both mutations occur in somatic cells."[1]

Knudson's two-hit mutational model, as exemplified by retinoblastoma, is illustrated in Figure 18.5. The assumption is that the retinoblastoma gene (RB^+) is the critical locus, and cancer will develop if both normal alleles of this gene are changed to the RB mutant allele. For sporadic retinoblastoma (Figure 18.5a) Knudson hypothesized that a child is born with two wild-type copies of the retinoblastoma gene (genotype RB^+/RB^+), and both mutations must then occur in the same retinoblast cell. Since the chance of having two such mutational events in the same cell is very low, sporadic retinoblastoma patients would be

[1]Knudson, A. G., Jr. 1971. Mutation and cancer: statistical study of retinoblastoma. *Proc. Natl. Acad. Sci. USA* 68:820–823.

~ FIGURE 18.7

The Rous sarcoma virus (RSV) RNA genome and a suggested mechanism for the integration of the proviral DNA into the host (chicken) chromosome. (a) RSV genome RNA. (b) RSV proviral DNA produced by reverse transcriptase. (c) Circularization of the proviral DNA. (d) Staggered nicks are made in viral and cellular DNAs. (e) By recombination, the viral ends become joined to the ends of the cell's DNA. (f) The single-stranded gaps are filled in, and a complete, double-stranded, integrated RSV provirus results.

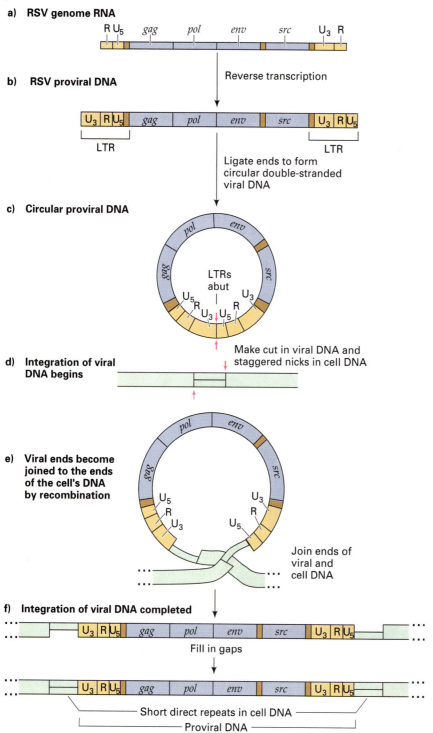

Life Cycle of Retroviruses. In 1910 F. Peyton Rous showed that when pieces of a particular kind of chicken tumor called a *sarcoma* (cancer of the connective tissue or muscle cells) were transplanted into other chickens from the same stock, the second group of chickens also developed sarcomas. Rous showed further that cell-free filtrates of the sarcoma inoculated into chickens resulted in tumor development. These data were interpreted to mean that some kinds of cancer might result from infection with a "filter-passing microbe" or a "virus." We now know that Rous's results (published in 1911) are explained by the activities of the retrovirus *Rous sarcoma virus* (RSV). The RNA genome organization of RSV is shown in Figure 18.7a.

As illustrated in Figure 18.7, when an encapsulated retrovirus such as RSV infects a cell, the RNA genome is released from the viral particle and serves as a template for the synthesis of a double-stranded DNA copy of the retrovirus (the *proviral DNA*). This process, known as reverse transcription, is catalyzed by reverse transcriptase, an enzyme brought into the cell as part of the virus particle. The proviral DNA then integrates into the host chromosome. Once integrated, the proviral DNA is transcribed by the host RNA polymerase II to produce the various viral mRNAs that encode for the individual viral proteins. Progeny RNA genomes are produced by transcription of the entire, integrated viral DNA and packaged into new viral particles that exit the cell and can infect other cells.

Typical retroviruses require three protein-coding genes for the virus life cycle: *gag*, *pol*, and *env*. The *gag* gene encodes a precursor protein that, when cleaved, produces virus particle proteins. The *pol* gene encodes a precursor protein that is cleaved to produce reverse transcriptase and an enzyme needed for the integration of the proviral DNA into the host cell chromosome. The *env* gene encodes the precursor to the envelope glycoprotein.

The ends of all retroviral RNA genomes consist of the sequences R and U_5 (at the left in Figure 18.7a) and U_3 and R (at the right in Figure 18.7a). When proviral DNA is produced by reverse transcriptase,[2] the end sequences are duplicated to produce long terminal repeats (LTRs; see Figure 18.7b). The LTRs contain many of the transcription regulatory signals for the viral sequence (e.g., for initiation of transcription and 3' cleavage and polyadenylation of mRNAs; Figure 18.8).

The first step in the integration of the proviral DNA into the chromosome of the host is ligation of the ends of the linear molecule to produce a circular, double-stranded molecule (Figure 18.7c). This brings two LTRs next to each other. Staggered nicks are made in both the viral and cellular DNAs (Figure 18.7d). By recombination the viral ends become joined to the ends of the cellular DNA; at this point integration has occurred (Figure 18.7e). Lastly, the single-stranded gaps are filled in. The integration of retrovirus proviral DNA results in a duplication of DNA at the target site, producing short, direct repeats in the host cell DNA flanking the provirus.

In addition to life-cycle genes, some retroviruses also carry an oncogene that gives them the ability to transform the cells they infect; these are the *oncogenic retroviruses*. In the case of RSV the oncogene is called *src* (see Figure 18.7a), and, like other retroviral oncogenes, it is not involved in the viral life cycle. Different retroviruses carry different oncogenes. Oncogenic retroviruses may or may not have the capability to replicate. Retroviruses without oncogenes

[2]Reverse transcriptase, unlike most replication DNA polymerases, does not have 3'-to-5' exonuclease activity. Since 3'-to-5' exonuclease activity is involved in proofreading, reverse transcriptases have no capacity to correct errors introduced in the RNA-to-DNA synthesis process. Thus, significant numbers of mutations may be introduced in the DNA product by reverse transcriptase's normal polymerization activities.

~ FIGURE 18.8

Features of LTRs. Most retroviruses (and many "transposable elements" of *Drosophila* and yeast) have LTRs with these features. (The LTRs have been drawn disproportionately large in order to show their features.)

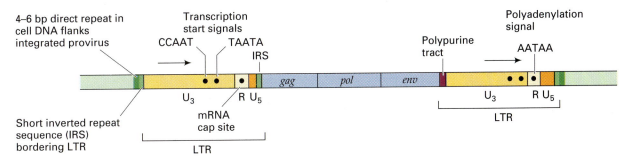

~ FIGURE 18.9

Life cycle of a nononcogenic retrovirus.

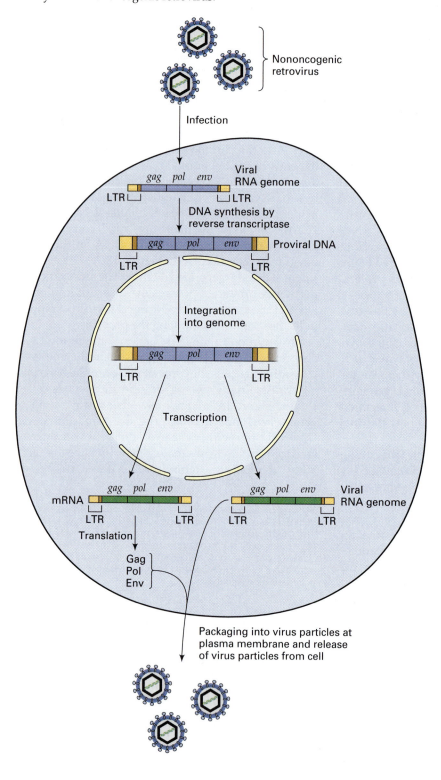

direct their own life cycle but do not change the growth properties of the cells they infect; these are *nononcogenic retroviruses*. Figure 18.9 shows the life cycle of a nononcogenic retrovirus.

In the context of the life cycle of retroviruses, it is appropriate to discuss briefly HIV-1, the causative agent of AIDS, even though this virus does not cause cancer. The retrovirus HIV-1 has a bullet-shaped capsid and is surrounded by a viral envelope in which are embedded viral-encoded gp120 glycoproteins (Figure 18.10a). The HIV-1 genome contains complete *gag*, *pol*, and *env* genes, so HIV can self-propagate. In addition HIV contains several other genes that are not oncogenes but that help control gene expression (Figure 18.10b). For example, *tat* encodes a protein that regulates transcription of the *gag* and *pol* genes and the translation of the resulting mRNA.

The gp120 glycoprotein of the HIV-1 envelope is the main basis for the infection of cells by the virus. This glycoprotein is recognized by the CD4 receptor that is found on the surfaces of a cell type of the immune system called *helper T cells* (a type of T lymphocyte). Other cell types without the CD4 receptor can also bind the HIV-1 virus, apparently because they have other receptors that recognize the gp120 glycoprotein. These cells include the macrophages, which are also involved in the immune response, as well as non-immune system cells such as glial cells of the brain and regulatory cells of the intestinal lining. Once the virus has bound to a receptor on the cell sur-

~ FIGURE 18.10

(a) Schematic drawing of a cross-section through an HIV particle. Note that the capsid of this particular retrovirus is "bullet" shaped. (b) Organization of the HIV genome.

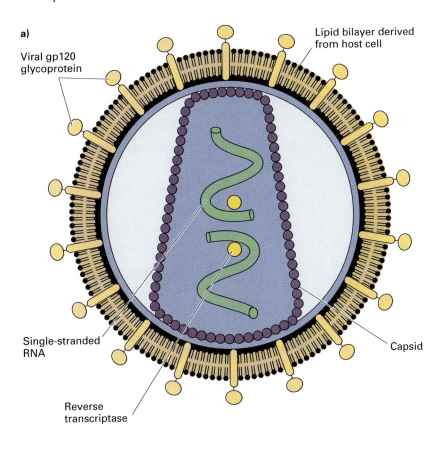

a)

Viral gp120 glycoprotein

Lipid bilayer derived from host cell

Single-stranded RNA

Reverse transcriptase

Capsid

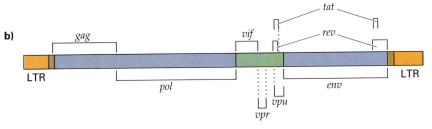

b)

gag

vif

tat

rev

LTR

pol

vpr

vpu

env

LTR

face, the viral particles enter the cell, most likely by fusion of the viral envelope with the plasma membrane of the cell. By this membrane fusion the viral capsid is released into the cell. Next, the viral protein coat is lost, and the viral life cycle is initiated, starting with reverse transcription of the viral RNA into DNA.

Through normal viral replication HIV-1 causes the death of the cell it infects. Thus, by repeated infection and viral replication a steady destruction of helper T cells and of other cells infected by the virus takes place. The decrease in the population of helper T cells effectively disables the responses of macrophages and B cells in the immune response, and the immune system becomes progressively less functional. As a result a person infected with HIV-1 is unable to combat infections by pathogens such as bacteria, viruses, and fungi and becomes susceptible to numerous types of cancers. AIDS patients die most frequently from the infections. To date at least 19 million people worldwide have been infected with HIV-1.

KEYNOTE

Retroviruses are RNA viruses that replicate via a DNA intermediate. All RNA tumor viruses are retroviruses, but not all retroviruses cause cancer. When a retrovirus infects a cell, the RNA genome is released from the viral particle, and through the action of reverse transcriptase, a cDNA copy of the genome— called the proviral DNA—is synthesized. The proviral DNA integrates into the genome of the host cell. Then, using host transcriptional machinery, viral genes are transcribed, and full-length viral RNAs are produced. Progeny viruses assembled within the cell exit the cell and can infect other cells.

Viral Oncogenes. In the case of RSV, tumor induction results from the activity of a particular **viral oncogene** in the retroviral genome. The discovery that the product of a single gene was both necessary and sufficient for tumor induction and formation was a significant breakthrough. Viral oncogenes (which are generically called v-*oncs*) are responsible for many different cancers. Only retroviruses that contain a v-*onc* gene are tumor viruses. The v-*onc* genes are named for the tumor that the virus causes, with the prefix "v" to indicate that the gene is of viral origin. Thus, the v-*onc* gene of RSV is v-*src*. Bacteriophages that have picked up cellular genes are said to transduce the genes to other cells, so such retroviruses are called **transducing retroviruses** because they have

picked up an oncogene from the genome of the cell. (We learn how this happens later.) Table 18.1 lists some transducing retroviruses and their viral oncogenes. Retroviruses that do not carry oncogenes are called *nontransducing retroviruses*.

Cells infected by RSV rapidly transform into the cancerous state because of the activity of the v-*src* gene. Since RSV contains all the genes necessary for viral replication (i.e., *gag*, *env*, and *pol*), an RSV-transformed cell produces progeny RSV particles. In this ability RSV is an exception because all other transducing retroviruses are defective in some of their viral replication genes (Figure 18.11): they can trans-

~ **FIGURE 18.11**

Structures of four defective transducing viruses (not to scale). (a) Avian myeloblastosis virus (AMV) contains the v-*myb* oncogene, which replaces the 3' end of *pol* and most of *env*. (b) Avian defective leukemia virus (DLV) contains the v-*myc* oncogene, which replaces the 3' end of *gag*, all of *pol*, and the 5' end of *env*. (c) Feline sarcoma virus (FeSV) contains the v-*fes* oncogene, which replaces the 3' end of *gag* and all of *pol* and *env*. (d) Abelson murine leukemia virus (AbMLV) contains the v-*abl* oncogene, which replaces the 3' end of *gag* and all of *pol* and *env*.

a) **Avian myeloblastosis virus (AMV) genomic RNA**

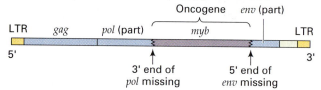

b) **Avian defective leukemia virus (DLV) genomic RNA**

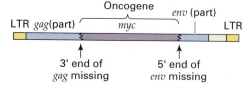

c) **Feline sarcoma virus (FeSV) genomic RNA**

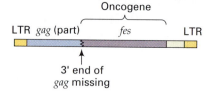

d) **Abelson murine leukemia virus (AbMLV) genomic RNA**

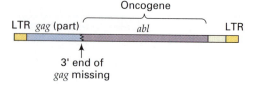

~ **TABLE 18.1**

Some Transducing Retroviruses and Their Viral Oncogenes

ONCOGENE	RETROVIRUS ISOLATE	V-ONC ORIGIN	V-ONC PROTEIN	TYPE OF CANCER
src	Rous sarcoma virus (RSV)	Chicken	pp60^{src}	Sarcoma
yes	Y73-ASV	Chicken	P90$^{gag-yes}$	Sarcoma
abl	Abelson murine leukemia virus (MLV)	Mouse	P90-P160$^{gag-abl}$	Pre–B cell leukemia
erbA	Avian erythroblastosis virus (AEV)	Chicken	P75$^{gag-erbA}$	Erythroblastosis and sarcoma
erbB	Avian erythroblastosis virus (AEV)	Chicken	gp65^{erbB}	Erythroblastosis and sarcoma
fgr	Gardner-Rasheed FeSV	Cat	P70$^{gag-actin-fgr}$	Sarcoma
fms	McDonough (SM)-FeSV	Cat	gp180$^{gag-fms}$	Sarcoma
fos	FBJ (Finkel-Biskis-Jinkins)-MSV	Mouse	pp55^{fos}	Osteosarcoma
mos	Moloney MSV	Mouse	P37$^{env-mos}$	Sarcoma
sis	Simian sarcoma virus (SSV)	Monkey	P28$^{env-sis}$	Sarcoma
	Parodi-Irgens FeSV	Cat	P76$^{gag-sis}$	Sarcoma
myc	MC29	Chicken	P100$^{gag-myc}$	Sarcoma, carcinoma, and myelocytoma
myb	Avian myeloblastosis virus (AMV)	Chicken	p45^{myb}	Myeloblastosis
	AMV-E26	Chicken	P135$^{gag-myb-ets}$	Myeloblastosis and erythroblastosis
rel	Reticuloendotheliosis (REV)	Turkey	p64^{rel}	Reticuloendotheliosis
kit	HZ4-FeSV	Cat	P80$^{gag-kit}$	Sarcoma
raf	3611-MSV	Mouse	P75$^{gag-raf}$	Sarcoma
H-ras	Harvey MSV (Ha-MSV)	Rat	pp21^{ras}	Sarcoma and erythroleukemia
	RaSV	Rat	P29$^{gag-ras}$	Sarcoma?
K-ras	Kirsten MSV (Ki-MSV)	Rat	pp21^{ras}	Sarcoma and erythroleukemia

form cells but are unable to produce progeny viruses because of the absence of one or more genes needed for virus reproduction. These defective retroviruses can produce progeny viral particles (i.e., they can be rescued), however, if cells containing them are also infected with a normal virus (a *helper virus*) that can supply the missing gene products (Figure 18.12).

Cellular Proto-Oncogenes. In the mid-1970s J. Michael Bishop and Harold Varmus, along with a number of other researchers, demonstrated that normal animal cells contain genes with DNA sequences very closely related to the viral oncogenes. (Bishop and Varmus received the Nobel Prize in 1989 for this research.) In the early 1980s R. A. Weinberg and M. Wigler showed independently that a variety of human tumor cells contain oncogenes. These genes, when introduced into other cells growing in culture, transformed those cells into cancer cells. The human oncogenes were found to be very similar to viral oncogenes that had been characterized earlier, even though viruses did not induce the human cancers involved. Moreover, the human oncogenes were shown also to be closely related to genes found in normally growing cells.

In short, most human and other animal oncogenes are mutant forms of normal cellular genes. Such genes in their normal state are called **proto-oncogenes**. Proto-oncogenes have important roles in regulating cell division and differentiation. When proto-oncogenes become mutated such that they induce tumor formation, they are called oncogenes (*onc*s). As we have seen, if they are carried by a virus, oncogenes are known as v-*onc*s. If they reside in the

~ FIGURE 18.12

Transformation of cells by viruses produced by a cell making a defective transforming virus and a helper virus. *Top:* When a cell is infected by both a defective transforming virus and a nondefective helper virus, the result is a transformed cell that produces both types of virus. *Middle:* When a cell is infected by only a defective transforming virus, transformed virus-nonproducer cells are generated. Transforming viruses can be rescued from these cells by infection with a helper virus. *Bottom:* When a cell is infected by only a helper virus, a nontransformed, virus-producing cell results.

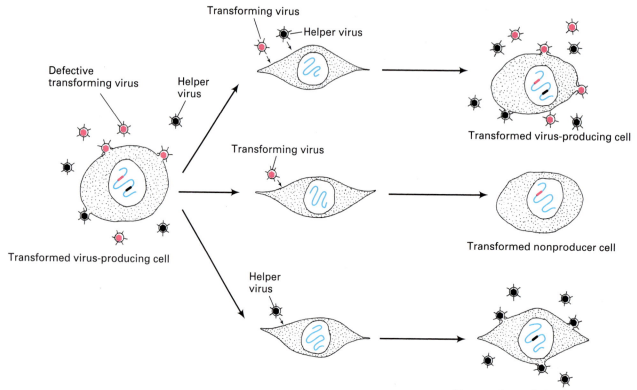

host chromosome, oncogenes are called **cellular oncogenes**, or c-*oncs*. (For a particular oncogene, the *onc* is replaced by the three-letter sequence of the related viral oncogene, e.g., v-*src* and c-*src*.) A transducing retrovirus, then, carries a significantly altered form of a cellular proto-oncogene (now a v-*onc*). When the transducing retrovirus infects a normal cell, the hitchhiking oncogene transforms the cell into a cancer cell. As we will see, normal cells can also become transformed into cancer cells even if a tumor virus does not infect them if the proto-oncogene is converted into a cellular oncogene.

One significant difference between a cellular proto-oncogene and its viral oncogene counterpart is that most proto-oncogenes contain introns that are not present in the corresponding v-*onc*. For example, the chicken *src* proto-oncogene is over 7 kb long, with twelve exons separated by introns (Figure 18.13a). In the RSV RNA genome the v-*src* oncogene is 1.7 kb

long with no introns. Genomic RNA produced by the full-length transcription of the proviral DNA is packaged into virus particles. Three types of mRNA are produced from the RSV proviral DNA: a full-length transcript and two mRNAs generated by differential splicing of the full-length transcript (Figure 18.13b). mRNA transcription starts at a promoter in the left U_3 sequence, and the addition of the poly(A) tail is signaled by a sequence in the right R sequence.

Formation of Transducing Retroviruses. Retroviral DNA (the provirus) integrates randomly into cellular DNA. Sometimes there occurs a genetic rearrangement by which the transcriptional unit of the provirus connects to nearby cellular genes, often by a deletion event involving the loss of some or all of the *gag*, *pol*, and *env* genes. In this way viral RNA contains all or parts of a cellular gene. All viral progeny then carry the cellular

~ FIGURE 18.13

(a) At the top is shown the molecular organization of the chicken *src* proto-oncogene. The gene contains twelve exons (shown as purple boxes). Below is the molecular organization of the Rous sarcoma virus RNA genome to indicate the relationship of nucleotide sequences in the cellular *src* proto-oncogene and v-*src*; v-*src* was generated mostly by intron removal from the cellular *src* proto-oncogene. (b) mRNAs produced by transcription of RSV proviral DNA genome.

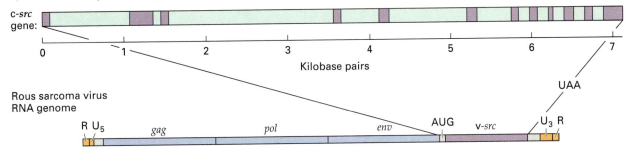

a) **Chicken c-src proto-oncogene**

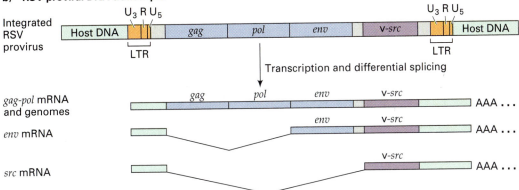

b) **RSV proviral DNA transcripts**

gene and, under the influence of viral promoters in the LTR, express the cellular protein in infected cells. If the cellular gene picked up was a proto-oncogene, the modified retrovirus will be oncogenic.

A molecular model for the formation of a transducing retrovirus is shown in Figure 18.14. In this case, after the retrovirus integrates near a cellular proto-oncogene, a deletion that fuses retroviral sequences with proto-oncogene sequences may occur. Then an mRNA is produced from the retrovirus-oncogene fusion by transcription (and splicing if introns are present in the proto-oncogene). If this mRNA is packaged into a virus particle along with a normal retrovirus genome, reverse transcriptase can produce a new defective transducing virus by switching RNA templates during synthesis of a cDNA strand.

Molecular analysis has shown that cellular DNA contained sequences homologous to viral oncogenes. Southern blot analysis (see Chapter 15, pp. 470–472) using labeled probes revealed that DNA from tumors

induced by an oncogenic retrovirus contained the proviral DNA. Surprisingly, DNA from normal (non-tumorous) cells used as a control also contained sequences detected by the probe. That the viral gene was evolutionarily a transduced cellular gene was shown by several lines of evidence, including the following. (1) The chromosomal position of each cellular proto-oncogene is conserved in eukaryotic cells, but the structure and location of the corresponding v-*onc* in the viral genome are variable. (2) Cellular homologs of v-*onc* are evolutionarily conserved throughout the biological world (e.g., from yeast to man), but v-*onc*s are not conserved between viral strains. (3) Transduction has been achieved experimentally in cell cultures as well as in the whole animal.

PROTEIN PRODUCTS OF PROTO-ONCOGENES.
About one hundred oncogenes have been identified through a number of experiments. To understand

~ **FIGURE 18.14**

Model for the formation of a transducing retrovirus. First a wild-type provirus integrates near a cellular proto-oncogene. Next a hypothetical deletion-fusion event fuses a proto-oncogene exon into the *gag* region of the retrovirus to produce a *gag-oncogene* (*gag-onc*) fusion. The LTR in this fusion directs the synthesis of a transcript that undergoes splicing to produce an mRNA for the *gag-onc* fusion protein. If the cell is also infected with a wild-type (helper) retrovirus, a wild-type RNA and a *gag-onc* RNA can be packaged into one virus particle as part of the virus life cycle. In a virus particle containing the two RNAs, reverse transcriptase can start copying the *gag-onc* RNA, then switch to copying the wild-type RNA to produce a new defective transducing retrovirus.

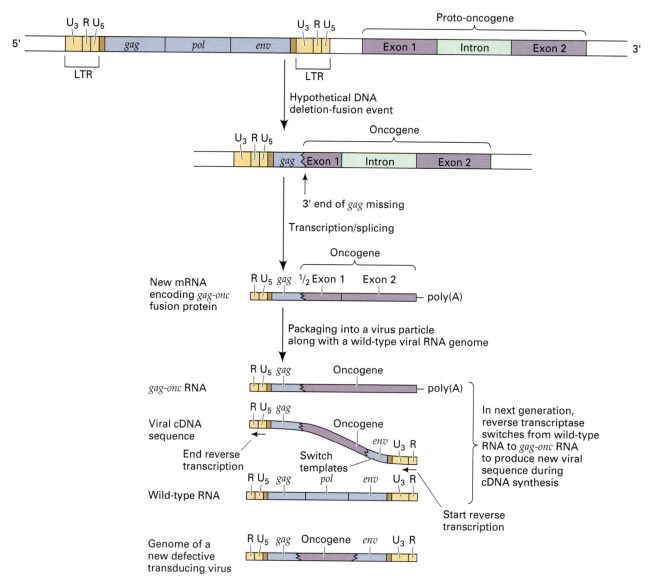

how *v-oncs* cause neoplasia, it is necessary to know the nature and function of the proteins encoded by proto-oncogenes. Based on DNA sequence similarities and similarities in amino acid sequences of the protein products, proto-oncogenes fall into several distinct classes, each with a characteristic type of protein product, as outlined in Table 18.2.

In each case the proto-oncogene product is involved in the positive control of cell growth and division; that is, the products are stimulatory to

~ TABLE 18.2

Examples of the Functions of Oncogene Products

Growth factors	
sis	PDGF B-chain growth factor
int-2	FGF-related growth factor

Receptor and nonreceptor protein-tyrosine and protein-serine/threonine kinases	
src	Membrane-associated nonreceptor protein-tyrosine kinase
fgr	Membrane-associated nonreceptor protein-tyrosine kinase
fps/fes	Nonreceptor protein-tyrosine kinase
kit	Truncated stem cell receptor protein-tyrosine kinase
pim-1	Cytoplasmic protein-serine kinase
mos	Cytoplasmic protein-serine kinase (cytostatic factor)

Receptors lacking protein kinase activity	
mas	Angiotensin receptor

Membrane-associated G proteins activated by surface receptors	
H-*ras*	Membrane-associated GTP-binding/GTPase
K-*ras*	Membrane-associated GTP-binding/GTPase
gsp	Mutant-activated form of G α

Cytoplasmic regulators	
crk	SH-2/3 protein that binds to (and regulates?) phosphotyrosine-containing proteins

Nuclear transcription factors (gene regulators)	
myc	Sequence-specific DNA-binding protein
fos	Combines with c-*jun* product to form AP-1 transcription factor
jun	Sequence-specific DNA-binding protein; part of AP-1
erbA	Dominant negative mutant thyroxine (T3) receptor
ski	Transcription factor?

growth. In the following sections we illustrate the function of protein products of proto-oncogenes by considering just two examples: growth factors and protein kinases.

Growth Factors. The effect of oncogenes on cell growth and division led to an early hypothesis that proto-oncogenes might be regulatory genes involved with the control of cell multiplication during differentiation. Evidence supporting this hypothesis came when the product of the viral oncogene v-*sis* was shown to be identical to part of platelet-derived growth factor (PDGF, a factor found in blood platelets in mammals), which is released after tissue damage. PDGF affects only one type of cell, fibroblasts, causing them to grow and divide. The fibroblasts are part of the wound-healing system. PDGF itself consists of two polypeptides encoded by the proto-oncogene *sis*. The causal link between PDGF and tumor induction was demonstrated in an experiment in which the cloned PDGF gene was introduced into a cell that normally does not make PDGF (i.e., a fibroblast): that cell was transformed into a cancer cell.

We can generalize and say that some cancer cells can result from the excessive or untimely synthesis of growth factors in cells that do not normally produce the factors. Introduction of an altered growth factor gene such as a v-*onc* or mutation of a c-*onc* can cause tumor development.

Protein Kinases. Many proto-oncogenes appear to encode protein kinases, enzymes that catalyze the addition of phosphate groups to target proteins, thereby modifying the proteins' function. The *src* gene product, for example, has been shown to be a nonreceptor protein kinase and has been termed pp60*src*. The viral protein, pp60v-*src*, and the protein encoded by the cellular oncogene, pp60c-*src*, differ in only a few amino acids, and both proteins bind to the inner surface of the plasma membrane.

What is particularly interesting about the *src* protein kinases is that both versions add a phosphate group to the amino acid tyrosine; that is, they are tyrosine protein kinases. Before this discovery, the protein kinases that had been characterized had all been shown to add phosphates only to the amino acids serine or threonine. Because protein phosphorylation was known to be important in effecting a multitude of metabolic changes in cells, the *src* discovery was exciting. It suggested a possible explanation of how the *src* and other tyrosine protein kinase–coding oncogenes might transform a normal cell into a metabolically different cancer cell. For example, a large class of proteins, including the receptors for growth factors, uses protein phosphorylation to transmit its signals through the membrane. Thus, the action of protein kinases appears also to be linked to growth factors and their activities.

CHANGING CELLULAR PROTO-ONCOGENES INTO ONCOGENES. As we have discussed, the products of proto-oncogenes are involved in the positive regulation of cell growth and division. In normal cells expression of proto-oncogenes is tightly controlled so that cell growth and division occur only as appropriate for the cell type involved. However, when proto-oncogenes are changed into oncogenes, the tight control can be lost, and unscheduled cell proliferation can take place. The results of sequencing a number of known oncogenes indicate that many have differences from the normal proto-oncogenes. The general types of changes that have been found are point mutations (base pair substitutions), deletion, gene amplification, and chromosomal translocation.

Point Mutation. Point mutations in the coding region of a gene or in the controlling sequences (promoter, regulatory elements, enhancers) can change a proto-oncogene into an oncogene by causing an increase in either the activity of the gene product or the expression of the gene, leading in turn to an increase in the amount of gene product. For example, the *ras* genes are a family of genes encoding membrane-associated G proteins. A single point mutation, generally in codon 12, 13, or 61, results in a mutant protein that can transform normal cells into malignant cells. These types of mutations are found in malignant cells from patients with bladder cancer, lung cancer, colon cancer, and leukemia. The effect of the mutations is to cause the G proteins to lose their ability to be regulated so that constitutive growth signals (growth signals that are always being produced) are transmitted into the cell. As a result, unregulated cell proliferation can commence.

Deletion. Deletions of part of the coding region or of part of the controlling sequences of a proto-oncogene have been found frequently in oncogenes. The *myc* oncogene, for example, can arise from its proto-oncogene by deletion. The normal gene consists of three exons and two introns; in some commonly found *myc* oncogenes the first exon and most of the first intron are deleted. Transcription is then controlled from sequences in exon 2, which can function as a promoter. The *myc* proto-oncogene encodes a nuclear transcription factor that positively regulates genes involved in cell proliferation. Thus, the deletions in the oncogene forms have brought about a change in the amount or activity of the remaining *myc* protein chain that activates those genes.

Gene Amplification. Some tumors have multiple (sometimes hundreds of) copies of proto-oncogenes. These have probably occurred by a random overreplication of small segments of the genomic DNA. For example, multiple copies of *ras* are found in mouse adrenocortical tumors. In general, extra copies of the proto-oncogene in the cell result in an increased amount of gene product, thereby inducing or con-

tributing to unscheduled cell proliferation. The amplification of proto-oncogenes may not play much of a role in carcinogenesis until the late stages of tumor development.

Chromosomal Translocation. Chromosomal translocations in human tumor cells are rather common, and some are specific for certain tumor types. We have already discussed specific examples of this and the oncogenes involved: chronic myelogenous leukemia (CML) and the Philadelphia chromosome produced by a reciprocal translocation involving chromosomes 9 and 22 (Chapter 7, pp. 201–203, and Figure 7.14) and Burkitt's lymphoma resulting from a reciprocal translocation involving chromosomes 8 and 14 (Chapter 7, pp. 203–204, and Figure 7.15).

CANCER INDUCTION BY RETROVIRUSES. With our new knowledge of how oncogenes can cause unscheduled cell proliferation, we can return to consider retroviruses and discuss their role in carcinogenesis. Retroviruses are common causes of cancer in animals, although only one case is known for humans.

We have seen that a retrovirus can cause cancer if it is a transducing retrovirus and the v-*onc* it carries is expressed. In this case transcription of the v-*onc* takes place under the control of retroviral promoters. Another way in which a retrovirus can cause cancer is if the proviral DNA integrates near a proto-oncogene. As we have discussed, proto-oncogene expression is tightly regulated during cell growth and differentiation. When a retrovirus integrates near a proto-oncogene, however, expression of the proto-oncogene can come under the control of the promoter and enhancer sequences in the retroviral LTR. These retroviral sequences do not respond to the environmental signals that normally regulate proto-oncogene expression, so overexpression of the proto-oncogene is induced. This results, in turn, in transforming the cell to the tumorous state. The process of proto-oncogene activation is called *insertional mutagenesis*. This process occurs rarely in animals and does not occur in humans.

KEYNOTE

When tumor induction occurs after retrovirus infection, it is because of the activity of a viral oncogene (v-*onc*) in that retroviral genome. Retroviruses carrying an oncogene are known as transducing retroviruses. Normal animal cells contain genes with

DNA sequences that are similar to those of the viral oncogenes. These cellular genes, called proto-oncogenes, play important roles in regulating cell division and differentiation. When proto-oncogenes are mutated, they induce tumor formation. The mutant phenotype is dominant in that only one allele needs to be mutated for tumors to be induced. In their mutated state, proto-oncogenes are called cellular oncogenes (c-*onc*s). The viral oncogenes are modified copies of the cellular proto-oncogenes that have been picked up by the retrovirus.

DNA TUMOR VIRUSES.   DNA tumor viruses are oncogenic—they induce cell proliferation—but they do not carry oncogenes like those in RNA tumor viruses. Thus, their mechanism for transforming cells is completely different from that of RNA tumor viruses. DNA tumor viruses transform cells to the cancerous state through the action of a gene or genes that are essential parts of the viral genome. Examples of DNA tumor viruses are found among five of six major families of DNA viruses—papovaviruses, hepatitis B viruses, herpes viruses, adenoviruses, and pox viruses.

DNA tumor viruses normally progress through their life cycles without transforming the cell to a cancerous state. Typically, the virus produces a viral protein that activates DNA replication in the host cell. Then, through the use of host proteins, the viral genome is replicated and transcribed, ultimately producing a large number of progeny viruses and killing the cell. The released viruses can then infect other cells. Rarely, the viral DNA is not replicated and becomes integrated into the host cell genome. If the viral protein that activates DNA replication of the host cell is now synthesized, this protein transforms the cell to the cancerous state by stimulating the quiescent host cell to proliferate; that is, it causes the cell to move from the G_0 phase to the S phase of the cell cycle.

One example of tumors produced by DNA tumor viruses is found in the papovavirus family. This family includes the many known papillomaviruses, some of which cause benign tumors such as skin and venereal warts in humans. Other human papillomaviruses (*HPV-16*, *HPV-18*, or both) cause cervical cancer, which is a leading cause of cancer deaths among women worldwide. The key viral genes involved in this transformation are *E6* and *E7*. The protein prod-

ucts of these two genes cause a change in the levels of cellular proteins that are important in regulating cell growth and division (see Figure 18.16c).

Tumor Suppressor Genes

Oncogenes play a major role in the development of cancer. As we have discussed, the normal counterparts of oncogenes, proto-oncogenes, encode products that have a stimulatory role in cell growth and division. In the late 1960s Henry Harris fused normal rodent cells with cancer cells and observed that some of the resultant hybrid cells did not form tumors, but established a normal growth pattern. Harris hypothesized that the normal cells contained gene products that had the ability to suppress the uncontrolled cell proliferation characteristic of cancer cells. The genes involved were called **tumor suppressor genes**.

Further evidence for the existence of tumor suppressor genes came from data indicating that in certain cancers specific chromosome regions were deleted from both homologs. Logically, if the loss of function of particular genes is correlated with tumor development, then the normal alleles of those genes must suppress tumor formation. In other words the normal products of tumor suppressor genes have an inhibitory role in cell growth and division. Thus, when tumor suppressor genes are inactivated, the inhibitory activity is lost, and unprogrammed cell proliferation can commence. Inactivation of tumor suppressor genes has been linked to the development of a wide variety of human cancers, including breast, colon, and lung cancer. In essence, then, tumor suppressor genes are the opposites of proto-oncogenes. Figure 18.15 compares the effects of tumor suppressor gene and proto-oncogene mutations. In this section we discuss how tumor suppressor genes were found and then look at some specific tumor suppressor genes to exemplify their roles in cancer.

FINDING TUMOR SUPPRESSOR GENES.   In contrast to mutations that activate oncogenes, mutations of tumor suppressor genes are recessive; that is, cell proliferation can be affected only if both alleles are inactivated. Thus, oncogenes can be identified in the laboratory because they stimulate the growth of cells in culture, but that is not the case for tumor suppressor genes. Introducing tumor suppressor genes into cells in culture either results in no change or kills the cell. The isolation of individual tumor suppressor genes was made readily possible only with the development of positional cloning (see Chapter 15, pp. 481–486). In this method genetic variations are searched for in cancer cells or in cells of patients with inherited cancer predisposition. Such genetic

~ FIGURE 18.15

Comparison of the effects of tumor suppressor gene and proto-oncogene mutations. (a) Mutations in both alleles of a tumor suppressor gene are needed for the cell to lose growth control. (b) A mutation in only one allele of a proto-oncogene, converting it to an oncogene, is needed for the cell to lose growth control.

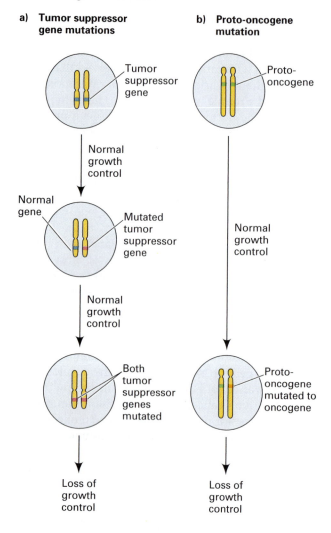

variations indicate the existence of mutant genes in those cells, and that enables researchers to home in on the mutations and hence clone and study the genes. Several tumor suppressor genes have been isolated in this way, and the functions of the proteins they encode are being studied in order to understand their role in the cancers with which they are involved. Table 18.3 lists some of the known tumor suppressor genes in humans. The products of tumor suppressor genes are found throughout the cell.

THE RETINOBLASTOMA TUMOR SUPPRESSOR GENE, _RB_. Retinoblastoma was introduced earlier

in this chapter in the context of Knudson's two-hit mutation model for cancer. Recall that retinoblastoma will develop if both alleles of the _RB_ gene are inactivated. In hereditary retinoblastoma one mutant allele of _RB_ is inherited, so a mutation in the other allele will lead to the development of tumors. In sporadic retinoblastoma both alleles are normal when inherited, and both must mutate or be deleted for tumors to develop. Although the protein encoded by the _RB_ gene is found in a number of cell types, inheritance of a single mutant allele in the germ line results in predisposition to only a narrow subset of cancers, namely, retinoblastoma and osteosarcoma (bone cancer). In contrast, cancers that have resulted entirely from somatic mutations may also involve mutant _RB_ alleles, for example, some lung cancers and some breast cancers.

Genetics. The _RB_ gene is a tumor suppressor gene. Its locus is at chromosome location 13q14.1–q14.2.

The _RB_ gene was cloned in 1986. It spans 180 kb of DNA and encodes a 110-kDa nuclear phosphoprotein (a phosphorylated protein)—pRB—that is involved in regulating cell growth. Studies of the _RB_ gene in patients with retinoblastoma have revealed a number of different mutations that can lead to loss of gene function, including point mutations and deletions. Most _RB_ mutations result in a truncated and unstable pRB. In approximately 5 percent of retinoblastoma patients, the genetic abnormality can be detected by karyotype analysis. The remainder are more difficult to detect, even by molecular analysis. Interestingly, in hereditary retinoblastoma the second _RB_ mutation often results in an identically mutated allele to the inherited one. The interpretation is that the chromosomal region with the wild-type _RB_ allele is replaced by a duplicated copy of the homologous chromosome region that carries the mutant allele. This may occur through mitotic recombination, chromosomal nondisjunction, or gene conversion.

Cell Biology. Recall that the cell cycle involves progression through the stages G_1-S-G_2-M (Chapter 3, pp. 51–52, and Figure 3.5). pRB plays a major role in the cell cycle by regulating the passage of cells from G_1 to S, a transition that is important to the cell because it commits it to progressing through the rest of the cell cycle. In G_1 phase in a normal cell or in a cell heterozygous for an _RB_ mutation, unphosphorylated pRB binds to a complex of E2F and DP1 transcription factors (Figure 18.16a). As long as this status is maintained, the cells will remain in G_1 or enter the quiescent state (called the G_0 phase). If progression through the cell cycle is signaled, phosphorylation of pRB by a cyclin/cyclin dependent kinase (Cdk) com-

~ TABLE 18.3

Some Known or Candidate Tumor Suppressor Genes

GENE	CANCER TYPE	PRODUCT LOCATION	MODE OF ACTION	HEREDITARY SYNDROME	CHROMOSOME LOCATION
APC	Colon carcinoma	Cytoplasm?	Cell adhesion molecule	Hereditary adenomatous polyposis	5q21–q22
BRCA1	Breast cancer	Nucleus	Transcription factor	Breast cancer and ovarian cancer	17q21
BRCA2	Breast cancer	Nucleus	Transcription factor?	Breast cancer	13q12–q13
DCC	Colon carcinoma	Membrane	Cell adhesion molecule	Involved in colorectal cancer	18q21.3
NF1	Neurofibromas	Cytoplasm	GTPase-activator	Neurofibromatosis type 1	17q11.2
NF2	Schwannomas and meningiomas	Inner membrane?	Links membrane to skeleton?	Neurofibromatosis type 2	22q12.2
p16	Melanoma	Nucleus	Transcription factor	Melanoma	9p21
p53	Colon cancer; many others	Nucleus	Transcription factor	Li-Fraumeni syndrome	17p13.1
RB	Retinoblastoma	Nucleus	Transcription factor	Retinoblastoma	13q14.1–q14.2
VHL	Kidney carcinoma	Membrane?	Transcription elongation factor	von Hippel-Lindau disease	3p26–p25
WT1	Nephroblastoma	Nucleus	Transcription factor	Wilms tumor	11p13

Adapted with permission from J. Marx, *Science* 261 (1993): 1385–1387. Copyright © 1993 American Association for the Advancement of Science.

plex (cyclins are proteins involved in control of the cell cycle) occurs, which makes pRB no longer able to bind to E2F. The released E2F molecules bind to genes with binding sites for this transcription factor, and those genes, whose activities are required for entry into S phase, are turned on. Progression of the cell into S is then assured. Once a cell has completed mitosis, pRB is dephosphorylated.

In a cell with two mutant *RB* alleles, pRB is usually truncated and unstable and does not bind to E2F/DP1, which is then free to activate genes for transition to S phase (Figure 18.16b). As a result, cell division takes place in an unprogrammed fashion. Interestingly, several different DNA tumor viruses (e.g., adenovirus, SV40) exert their tumorigenic effects in part by a process in which proteins encoded by their oncogenes form complexes with pRB in the cell, thereby blocking its ability to bind to E2F and inactivating the suppressive function of the protein (Figure 18.16c). In other words these tumor viruses transform cells to the neoplastic state by inactivating a mechanism that inhibits cellular growth.

Despite our present knowledge of the cellular activities of pRB, we do not yet know how children with *RB* mutations develop retinoblastoma. It may be through the role of pRB in development rather than its role in the cell cycle specifically. That is, cells programmed for terminal differentiation may depend critically on pRB in order to establish a state of permanent nonproliferation. Even if this model proves to be correct, the causal relationship of *RB* mutations with retinoblastoma will remain to be determined in detail.

THE *P53* TUMOR SUPPRESSOR GENE. The tumor suppressor gene *p53* is so named because it encodes a protein of molecular weight 53 kDa; that protein is called p53. When both alleles are mutated, *p53* may be involved in the development of perhaps 50 percent of all human cancers, including breast, brain, liver, lung, colorectal, bladder, and blood cancer. This does not mean that *p53* causes 50 percent of human cancers, but that among the usually several genetic changes found in those cancers are mutations in *p53*.

Genetics. The *p53* gene, isolated by positional cloning (see Chapter 15, pp. 481–486), is at chromosome location 17p13.1. As with the *RB* gene, both alleles of *p53* must be inactivated in order for loss of tumor suppression to occur. Individuals who inherit one mutant copy of *p53* develop Li-Fraumeni

of DNA and consists of twenty-four exons; it is transcribed in numerous tissues, including breast and ovary, to produce a 7.8-kb mRNA that is translated to produce a 190-kDa protein with 1,863 amino acids. One region of the protein appears to be a transcription activation domain, suggesting that the protein plays a role in regulating gene expression. Many different mutations in *BRCA1* have been identified in breast cancers. More than 75 percent of the mutations are frameshift mutations or nonsense mutations resulting in shorter than normal proteins.

To localize other genes that confer susceptibility to breast cancer, an international consortium of research groups did linkage analysis of a number of families with many cases of early-onset breast cancers that were not linked to *BRCA1*. From this study a second breast cancer susceptibility gene, *BRCA2*, was mapped to chromosome location 13q12–q13. Unlike *BRCA1*, *BRCA2* does not have an associated high risk of ovarian cancer. The gene has been molecularly cloned and is composed of twenty-seven exons distributed over approximately 70 kb of DNA. The protein predicted to be encoded by the *BRCA2* gene has 3,418 amino acids and some similarity to the *BRCA1*-encoded protein but no similarity to other known proteins.

KEYNOTE

Tumor suppressor genes, like proto-oncogenes, are involved in the regulation of cell growth and division. Whereas the normal products of proto-oncogenes have a stimulatory role in those processes, the normal products of tumor suppressor genes have inhibitory roles. Therefore, when both alleles of a tumor suppressor gene are inactivated or lost, the inhibitory activity is lost, and unprogrammed cell proliferation can occur. Inactivation of tumor suppressor genes is involved in the development of a wide variety of human cancers, including breast, colon, and lung cancer.

Mutator Genes

A **mutator gene** is any gene that, when mutant, increases the spontaneous mutation frequencies of other genes. In a cell the normal (unmutated) forms of mutator genes are involved in such important activities as DNA replication and DNA repair. Mutations of these genes can significantly affect those processes and can make the cell error prone so that it accumu-

lates mutations. For an illustration of how a mutation in a mutator gene can result in cancer, we consider hereditary nonpolyposis colon cancer (HNPCC).

HNPCC is an autosomal dominant genetic disease in which there is an early onset of colorectal cancer. Unlike hereditary (or familial) adenomatous polyposis (FAP) (see p. 609), no adenomas (benign tumors or polyps) are seen in HNPCC, hence its name. HNPCC accounts for perhaps 5–15 percent of colorectal cancers.

As with any genetic disease, researchers were eager to isolate and characterize the gene or genes responsible. Using a number of different approaches to gene mapping, they identified four human genes that when mutated confer a phenotype of hereditary predisposition to HNPCC. Like the hereditary retinoblastoma mutation, when one mutation is inherited through the germ line, tumor formation requires only one mutational event to inactivate the remaining normal allele. Thus, owing to the high probability of such an event, HNPCC appears dominant in pedigrees.

The four human genes identified are *hMSH2*, *hMLH1*, *hPMS1*, and *hPMS2*. The first two together are responsible for about 90 percent of all HNPCC cases, with the others accounting for 5 percent of cases each. DNA sequencing analysis revealed the exciting fact that all four genes were homologous to *E. coli* and yeast genes known to be involved in DNA repair. For example, *hMSH2* is homologous to *E. coli mutS*, and the other three genes have homologies to *E. coli mutL*. The *E. coli* genes have well-characterized roles in mismatch repair, a process for correcting mismatched base pairs left after DNA replication. (Mismatch repair is described in detail in Chapter 19, pp. 639–640.) The yeast genes have similar functions. In other words the human, yeast, and *E. coli* genes described here are all mutator genes because they are involved in DNA repair systems. Mutations in these genes make the DNA replication error prone, and mutation rates increase significantly compared with normal cells. That the human *hMSH2* gene is indeed a mutator gene was confirmed by an experiment in which an *hMSH2* cDNA was cloned into an *E. coli* plasmid and expressed in *E. coli*. The result was a tenfold increase in the accumulation of mutations. DNA-based assays are available for all four genes, allowing carriers to be detected through analysis of blood samples.

THE MULTISTEP NATURE OF CANCER

The development of most cancers is a stepwise process involving an accumulation of mutations in a number of genes. The statistics of increasing occurrence of cancers in humans as they age suggest that perhaps six or seven independent mutations are needed over several decades of life in order for cancer to be induced. The multiple mutational events typically involve both activation of oncogenes and inactivation of tumor suppressor genes, with a resulting breakdown of the multiple cellular mechanisms that regulate growth and differentiation.

As an example, Figure 18.18 illustrates Bert Vogelstein's molecular model of multiple mutations leading to hereditary FAP, a form of colorectal cancer. This model derives from molecular analyses of discrete phases of tumor development as the cancer evolves through histologically identifiable phases. Patients with FAP inherit the loss of a chromosome 5 tumor suppressor gene called *APC* (adenomatous polyposis coli). The same gene can be lost early in carcinogenesis in sporadic tumors. Once both alleles of *APC* are lost in a colon cell, increased cell growth will result. If hypomethylation (decreased methylation) of the DNA occurs, a benign tumor called an *adenoma class I* (a small polyp from the colon or rectum epithelium) can develop. Then, if a mutation converts the chromosome 12 *ras* proto-oncogene into an oncogene, the cells can progress to a larger benign tumor known as *adenoma class II* (a larger polyp). Next, if both copies of the chromosome 18 tumor suppressor gene *DCC* (deleted in colon cancer) are lost, the cells progress into *adenoma class III* (large benign polyps). Deletion of both copies of the chromosome 17 tumor suppressor gene *p53* results in the progression to a carcinoma (a cancer); with other gene losses the cancer metastasizes. Note that this is only one path whereby adenomatous polyposis can occur; others are possible. However, in all paths observed deletions of *APC* and mutations of *ras* usually occur earlier in carcinogenesis than do deletions of *DCC* and *p53*. Progressive changes in function of oncogenes and tumor suppressor genes are also thought to occur for other cancers.

KEYNOTE

The development of most cancers involves an accumulation of mutations in a number of genes over a significant period of life. This multistep nature of cancer typically involves mutational events that activate oncogenes and inactivate tumor suppressor genes, thereby breaking down the multiple mechanisms that regulate growth and differentiation.

~ **FIGURE 18.18**

A multistep molecular event model for the development of hereditary adenomatous polyposis (FAP), a colorectal cancer.

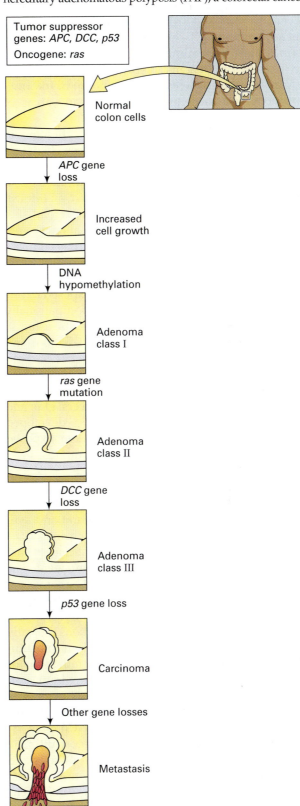

Tumor suppressor genes: *APC, DCC, p53*
Oncogene: *ras*

Normal colon cells

APC gene loss

Increased cell growth

DNA hypomethylation

Adenoma class I

ras gene mutation

Adenoma class II

DCC gene loss

Adenoma class III

p53 gene loss

Carcinoma

Other gene losses

Metastasis

CHEMICALS AND RADIATION AS CARCINOGENS

Several natural and artificial agents increase the frequency with which cells become cancerous. These agents, mostly chemicals and types of radiation, are known as **carcinogens**. Because of the obvious human relevance, there is a vast amount of information about carcinogenesis spanning many areas of biology; only an overview will be given here.

Although we have focussed much attention in this chapter on viruses as causes of cancer, chemicals are responsible for more human cancers than are viruses. Chemical carcinogenesis was discovered in the eighteenth century by Sir Percival Pott, an English surgeon who correlated the incidence of scrotal skin cancer in some of his patients to occupational exposure to coal soot when they had worked as chimney sweeps as children. From the beginning of industrial development in the eighteenth century to the present day, workers in many fields have been exposed to carcinogenic agents and have developed occupationally related cancers. For example, radiologists using X rays and radium (sources of ionizing radiation) and farmers exposed to the sun's ultraviolet (UV) light (nonionizing radiation) have developed skin cancer, asbestos and insulation workers exposed to asbestos have developed bronchial and lung cancers, glue and varnish workers using benzene have developed leukemia (blood cancer), and PVC workers exposed to vinyl chloride have developed liver cancer.

Chemical Carcinogens

Chemical carcinogens include both natural and synthetic chemicals. Two major classes of chemical carcinogens are recognized. *Direct-acting carcinogens* are chemicals that bind to DNA and act as mutagens. The second class is procarcinogens that must be converted metabolically to active derivatives; virtually all of these so-called *ultimate carcinogens* bind to DNA and act as mutagens. In both cases the mutations typically are point mutations that involve base-pair substitutions producing missense or nonsense mutations or base-pair additions or deletions producing frameshift mutations. (The mutagenicity of ultimate carcinogens can be demonstrated in a number of screening tests, including the Ames test described in Chapter 19, pp. 636–637.) Thus, direct-acting and most ultimate carcinogens bring about transformation of cells and the formation of tumors through binding to, and causing changes in, the DNA. Direct-acting carcinogens include alkylating agents such as some anticancer therapeutic chemicals. Examples of procarcinogens are polycyclic aromatic hydrocarbons, azo dyes and

natural metabolites (such as aflatoxin produced from fungal contamination of food), and nitrosamines (produced by nitrites in food). Most chemical carcinogens are procarcinogens.

The metabolic conversion of procarcinogens to ultimate carcinogens is carried out by normal cellular enzymes that function in a variety of pathways that involve, for example, hydrolysis, oxidation, and reduction. If a procarcinogen interacts with the active site of one of the enzymes, then it, too, can be modified by the enzyme activity—being hydrolyzed, oxidized, reduced, and so on—to give rise to the derivative ultimate carcinogen.

Chemical carcinogens are responsible for most cancer deaths in the United States, with the top two causes of cancer—tobacco smoke and diet—being responsible for 50–60 percent of those deaths. Since these two factors are environmental in nature, our risk of cancer can be significantly affected by our life habits. We will discuss only tobacco smoke here.

Overall, smoking—mostly of cigarettes—is responsible for 30 percent of cancer deaths, making tobacco smoke (or rather the chemicals in the smoke) the most lethal carcinogen. Tobacco smoke can cause a number of types of cancers, including lung, upper respiratory tract, esophagus, stomach, and liver cancer. The risk of developing cancer as a result of smoking tobacco is influenced by such factors as the amount of tobacco smoked, its tar content, and how long the person has been smoking. Thus, the younger a person is when he or she begins smoking, the greater the risk of developing cancer later in life. Moreover, there is good evidence that secondhand smoke increases the risk of cancer, making tobacco smoke an environmental concern for all individuals. One of the main types of carcinogens found in tobacco smoke is the polycyclic hydrocarbons. Once converted in the cell to their ultimate carcinogen derivatives, they react with negatively charged molecules, such as DNA, and this can result in mutations.

Radiation

We may be able to avoid or minimize environmental exposure to chemical carcinogens, but avoiding exposure to radiation in its various forms is more difficult. We are exposed to radiation, for example, from the sun, cellular telephones, radioactive radon gas, electric power lines, and some household appliances. Only about 2 percent of all cancer deaths are caused by radiation, and most of the cancers involved are the highly aggressive melanoma skin cancers that can be induced by exposure to the sun's UV light. Ionizing radiation, such as that emitted, for example, from X-ray machines, atomic bombs, decay of some radioactive materials, and radon gas, has the potential to be carcinogenic, although the risk to the general public is generally very low. Ionizing radiation most commonly causes leukemia and thyroid cancer.

Radiation carcinogens all act directly, causing mutations in DNA. The mutagenic effects of ultraviolet light and X rays are discussed in Chapter 19, pp. 629–630. We discuss ultraviolet light as a carcinogen in more detail in the following paragraphs.

UV light is emitted by the sun, along with visible light and infrared radiation. The UV light that reaches Earth is classified into two types, based on its wavelength: ultraviolet A (UVA, spanning 320–400 nm) and ultraviolet B (UVB, spanning 290–320 nm). The intensity of UVA and UVB reaching an individual on Earth depends on a number of factors, including time of day, altitude, and materials in the atmosphere such as dust and other particles. Generally, the ambient level of UVA is one to three orders of magnitude higher than UVB.

UV light causes several forms of skin cancer, the most dangerous of which are directly related to long-term exposure to UV light radiation. Both UVA and UVB play a role in carcinogenesis. Sunburn is mainly caused by UVB, which also induces skin cancer because the radiation in the wavelength range of UVB is mutagenic. These mutagenic effects are discussed in more detail in Chapter 19, p. 630. UVA plays a role in skin cancer by acting to increase the carcinogenic effects of UVB.

The risks of UV light–induced skin cancers can be minimized by reducing one's exposure to the sun (or UV light in any form) and by applying an effective sunscreen when out in the sun. The effectiveness of a sunscreen is indicated in terms of its sun protection factor (SPF). The SPF value tells how many times your natural sunburn protection against UVB the product will provide. However, one must be careful in interpreting what the different values mean. For example, a sunscreen with SPF 15 will block 93 percent of UVB, and a sunscreen with SPF 50 will block 98 percent of UVB. Most sunscreens provide limited or no protection against UVA, which can lead to harm if using a sunscreen with high SPF encourages long exposure to the sun.

Fortunately, many skin cancers are easy to detect and may be removed surgically.

KEYNOTE

Various types of radiation and many chemicals increase the frequency with which cells become cancerous. These agents are known as carcinogens. All carcinogens act by causing changes in the genome of the cell. A few chemical carcinogens act directly on the genome; the majority act indirectly. The latter are metabolically converted by cellular enzymes to ultimate carcinogens that bind to DNA and cause mutations. All carcinogenic forms of radiation act directly.

SUMMARY

To understand the development of cancer (neoplasia), it is necessary to understand how normal cell division is controlled. We know that a normal eukaryotic cell moves through the cell cycle in steps that are tightly controlled by a number of molecular factors. Healthy cells grow and divide only when the balance of stimulatory and inhibitory signals received from outside favor cell proliferation. A cancerous cell does not respond to the usual signals and reproduces without constraints.

Three classes of genes have been shown to be mutated frequently in cancer: oncogenes, tumor suppressor genes, and mutator genes. The products of oncogenes stimulate cell proliferation, the products of tumor suppressor genes inhibit cell proliferation, and the products of mutator genes are involved in the replication and repair of DNA. Mutant forms of these three types of genes all have the potential to contribute to the transformation of a cell to a tumorous state.

Some forms of cancer are caused by tumor viruses. Both DNA and RNA tumor viruses are known. DNA tumor viruses transform cells to the cancerous state through the action of a gene or genes that are essential parts of the viral genome. This is different from the transformation of cells by RNA tumor viruses. All RNA tumor viruses are retroviruses—RNA viruses that replicate via a DNA intermediate—but not all retroviruses cause cancer. When a retrovirus infects a cell, the RNA genome is released from the viral particle, and through the action of reverse transcriptase, a cDNA copy of the genome—called the proviral DNA—is synthesized. The proviral DNA integrates into the host cell's genome, and viral genes are transcribed and full-length viral RNAs are produced using host transcriptional machinery. Progeny viruses assembled within the cell exit the cell and can infect other cells.

The tumor-causing RNA retroviruses contain cancer-inducing genes termed oncogenes. A very significant discovery was that both normal cells and non-viral-induced cancer cells contain sequences that are related to the viral oncogenes. The current model is that tumor-causing retroviruses have picked up normal cellular genes—called proto-oncogenes—while simultaneously losing some of their genetic information. Cellular proto-oncogenes in normal cells function in various ways to regulate cell differentiation. However, in the retrovirus these genes have been modified or their expression regulated differently so that the oncogene protein product, now synthesized under viral control, is both qualitatively and quantitatively changed. Further, because the retrovirus can affect a number of cell types, the oncogene product is now expressed in cells that normally do not contain that particular growth-stimulating factor. The oncogene products, which include growth factors and protein kinases, are directly responsible for the transformation of cells to the cancerous state. It is also possible for cellular proto-oncogenes to mutate, resulting in a stimulatory effect on cell division. In their mutated state proto-oncogenes are called cellular oncogenes. Since only one allele of a proto-oncogene needs to be mutated to cause changes in cell growth and division, the mutations are dominant mutations.

Tumor suppressor genes, like proto-oncogenes, are involved in the regulation of cell growth and division. Whereas the normal products of proto-oncogenes have a stimulatory role in those processes, the normal products of tumor suppressor genes have inhibitory roles. Therefore, when both alleles of a tumor suppressor gene are inactivated or lost, the inhibitory activity is lost, and unprogrammed cell proliferation can occur. Thus, tumor suppressor mutations are recessive in that two mutant alleles are necessary for a phenotypic change. Inactivation of tumor suppressor genes is involved in the development of a wide variety of human cancers, including breast, colon, and lung cancer.

In 1971 a model for cancer development was proposed to explain the disease retinoblastoma, which we now know results from mutations in a tumor suppressor gene. This two-hit mutation model states that two mutational events are needed for cancer to develop, one in each allele of a critical cancer-causing gene. In familial (hereditary) cancers one mutation is inherited, and the other mutation occurs later in somatic cells. In essence the inheritance of one gene mutation predisposes a person to cancer. In sporadic (nonhereditary) cancers both mutations occur in the somatic cells. We know now that this simple two-hit model applies for only a few cancers. The development of most cancers involves an accumulation of mutations in a number of genes over a significant period of life. This multistep nature of cancer typically involves mutational events that activate oncogenes and inactivate tumor suppressor genes, thereby breaking down the multiple mechanisms that regulate growth and differentiation. Mutations of mutator genes can also contribute to the development of cancer because the normal maintenance of the integrity of the genome through accurate DNA replication and efficient DNA repair is adversely affected, making the cell error prone so that it accumulates mutations.

Various types of radiation and many chemicals—collectively known as carcinogens—have been shown to increase the frequency with which cells become cancerous. Practically all carcinogens act by causing changes in the genome of the cell. A few chemical carcinogens act directly on the genome; the majority act indirectly by being converted by cellular enzymes to active derivatives called ultimate carcinogens. Through an understanding of what carcinogens exist in the environment, we can position ourselves to minimize their effects on us and thereby perhaps decrease our risk of cancer.

ANALYTICAL APPROACHES FOR SOLVING GENETICS PROBLEMS

Q18.1 An investigator has found a retrovirus capable of infecting human nerve cells. This is a complete virus, capable of reproducing itself, and it contains no oncogenes. People who are infected suffer a debilitating encephalitis. The investigator has shown that when he infects nerve cells in culture with the complete virus, the nerve cells are killed as the virus reproduces. But if he infects cultured nerve cells with a virus in which he has created deletions in the *env* or *gag* genes, no cell death occurs. The investigator is interested in finding ways to bring about nerve cell growth or regeneration in people who have suffered nerve damage. For example, in a patient with a severed spinal cord, nerve regeneration might relieve paralysis. The investigator has cloned the human nerve growth factor gene and wants to insert it into the genome of his retrovirus from which he has deleted parts of the *env* and *gag* genes. He would then use the engineered retrovirus to infect cultured nerve cells. Adult nerve cells do not normally produce large amounts of nerve growth factor. If he is successful in inducing growth in them without causing any cell death, he would like to move on to clinical trials on injured patients. When the investigator applied for grant support to do this work, his application was denied on grounds that there were inadequate safeguards in the plan. Why might this work be dangerous? What comparisons can you draw between the virus the investigator wants to create and, for example, Avian myeloblastosis virus (see Figure 18.11)?

A18.1 In engineering the retrovirus in the way he plans, the investigator would probably be creating a new cancer virus in which the cloned nerve growth factor gene would be the oncogene. It is, of course, an advantage that the engineered virus would not be able to reproduce itself, but we know that many "wild" cancer viruses are also defective and reproduce with the help of other viruses. If the engineered virus were to infect cells carrying other viruses (for example, wild-type versions of itself) that could supply the *env* and *gag* functions, the new virus could be reproduced and spread. Presumably, infection of normal nerve cells *in vivo* by the engineered retrovirus would sometimes result in abnormally high levels of nerve growth factor, and thus perhaps in the production of nervous system cancers.

In Avian myeloblastosis virus the *pol* and *env* genes are partially deleted. Thus, like our investigator's virus, AMV needs a helper virus to reproduce. In AMV the *myb* oncogene has been inserted; it encodes a nuclear protein presumably involved in control of gene expression. In our new virus the oncogene would be the cloned nerve growth factor gene.[3]

Questions and Problems

*18.1 What is the difference between a hereditary and a sporadic cancer?

18.2 Individuals with hereditary retinoblastoma are heterozygous for a mutation in the *RB* gene; however, their cancerous cells often have two identically mutated *RB* alleles. Describe three different mechanisms by which the normal *RB* allele can be lost. Illustrate your answer with diagrams.

18.3 Distinguish between a transducing retrovirus and a nontransducing retrovirus.

*18.4 In what ways is the mechanism of cell transformation by transducing retroviruses fundamentally different from transformation by DNA tumor viruses? Even though the mechanisms are different, how are both able to cause neoplastic growth?

*18.5 Although there has been a substantial increase in our understanding of the genetic basis for cancer, the vast majority of cases of many types of cancer are not hereditary.

a. How might studying a hereditary form of a cancer provide insight into a similar, more frequent sporadic form?

b. The incidence of cancer in several members of an extended family might reasonably raise the concern as to whether there is a genetic predisposition for cancer in members of the family. What does the term "genetic predisposition" mean? What might be the basis of a genetic predisposition to a cancer that appears as a dominant trait? What issues must be addressed before concluding that a genetic predisposition for a specific type of cancer exists in a particular family?

18.6 Cellular proto-oncogenes and viral oncogenes are related in sequence, but they are not identical. What fundamental difference is there between the two?

18.7 Material that has been biopsied from tumors is useful for discerning both the type of tumor and the stage to which a tumor has progressed. It has been known for a long time that biopsied tissues with more differentiated cellular phenotypes are associated with less advanced tumors. Explain this finding in terms of the multistep nature of cancer.

18.8 An autopsy of a cat that died from feline sarcoma revealed neoplastic cells in the muscle and bone marrow, but not in the brain, liver, or kidney. To gather evidence for the hypothesis that the virus FeSV (Figure 18.11c) contributed to the cancer, Southern blot analysis (see Chapter 15, pp. 470–472) was performed on DNA isolated from these tissues and on a cDNA clone of the FeSV viral genome. The DNA was digested with the enzyme *Hin*dIII, separated by size on an agarose gel, and transferred to a membrane. The resulting Southern blot was hybridized with a ^{32}P-labeled probe made from a 1.0 kb *Hin*dIII fragment of the feline *fes* protooncogene cDNA. The autoradiogram revealed a 3.4 kb band in each lane, with an additional 1.2 kb band in the lanes with muscle and bone marrow DNA. Only a 1.2 kb band was seen in the lane loaded with *Hin*dIII-cut FeSV cDNA. Explain these results, including the size of the bands seen. Do these results support the proposed hypothesis?

18.9 The sequences of proto-oncogenes are highly conserved among a large number of animal species. What hypothesis can you make about the functions of the proto-oncogenes?

18.10 Explain why HIV-1, the causative agent of AIDS, is considered a non-oncogenic retrovirus even though numerous types of cancers are frequently seen in AIDS patients.

18.11 Give two ways in which cancer can be induced by a retrovirus.

---

[3]A propos this question, it is interesting to note that now that the safety of retroviral vectors has been established, they are being used successfully to treat certain human diseases.

*18.12 Proto-oncogenes produce a diverse set of gene products.

a. What types of gene products are made by proto-oncogenes? Do these gene products share any common feature(s)?

b. Which of the following mutations might result in an oncogene?

    i. A deletion of the entire coding region of a proto-oncogene

    ii. A deletion of a silencer that lies 5' to the coding region

    iii. A deletion of an enhancer that lies 3' to the coding region

    iv. A deletion of a 3' splice-site acceptor region

    v. The introduction of a premature stop codon

    vi. A point mutation

    vii. A translocation that places the coding region near a constitutively transcribed gene

    viii. A translocation that places the gene near constitutive heterochromatin

*18.13 You have a culture of normal cells and a culture of cells dividing uncontrollably (isolated from a tumor). Experimentally, how might you determine whether uncontrolled growth was the result of an oncogene or a mutated pair of tumor suppressor alleles?

18.14 What are the four main ways in which a proto-oncogene can be changed into an oncogene?

*18.15 After a retrovirus that does not carry an oncogene infects a particular cell, northern blots indicate that transcription of a particular proto-oncogene became elevated approximately thirteenfold compared with uninfected control cells. Propose a hypothesis to explain this result.

18.16 Explain how progression through the cell cycle is regulated by the phosphorylation of the retinoblastoma protein pRB. What phenotype(s) might you expect in cells where

a. pRB was constitutively phosphorylated?

b. pRB was never phosphorylated?

c. a severely truncated pRB protein was produced that was unable to be phosphorylated?

d. a normal pRB protein was produced at higher than normal levels?

e. a normal pRB protein was produced at lower than normal levels?

18.17 Mutations in the *p53* gene appear to be a major factor in the development of human cancer.

a. Explain what the normal cellular functions of the *p53* gene product are and how alterations in these functions can lead to cancer.

b. Suppose cells in a cancerous growth are shown to have a genetic alteration that results in diminished *p53* gene function. Why can we not immediately conclude that the mutation has *caused* the cancer? How would the effect of the mutation be viewed in light of the current, multistep model of cancer?

*18.18 Some tumor types very frequently have been associated with specific chromosomal translocations. In some cases, these translocations are found as the only cytogenetic abnormality. In each case examined to date, the chromosome breaks that occur result in a chimeric gene that encodes a fusion protein. A partial list of tumor-specific chromosomal translocations in bone and soft tissue tumors is given below.

TUMOR TYPE	TRANSLOCATION	CHARACTERISTIC	GENES
Ewings sarcoma	t(11;22) (q24;q12)	Malignant	FLI1, EWS
	t(21;22) (q22;q12)		ERD, EWS
	t(7;21) (p22;q12)		ATV1, EWS
Soft tissue clear cell carcinoma	t(12;22) (q13;q12)	Malignant	ATF1, EWS
Myxoid chrondrosarcoma	t(9;22) (q22-31;q12)	Malignant	CHN, EWS
Desmoplastic small round cell tumor	t(11;22) (p13;q12)	Malignant	WT1, EWS
Synovial sarcoma	t(X;18) (p11.2;q11.2)	Malignant	SSX1, SSX2, SYT
Lipomas	t(var;12) (var;q13-15)	Benign	var, HMGI-C
Leiomomas	t(12;14) (q13-15)	Benign	HMGI-C, ?

a. What conclusions can you draw from the fact that in some cases, these translocations are found as the only cytogenetic abnormality?

b. How might the formation of a chimeric protein result in tumor formation?

c. Based on the data presented here, can you infer whether the genes near the breakpoints of these translocations are tumor supressor genes or proto-oncogenes? If so, which?

d. Can you speculate how multiple translocations involving the *EWS* gene result in different sarcomas?

e. It is often difficult to diagnose individual sarcoma types based solely on tissue biopsy and clinical symptoms. How might the cloning of the genes involved in translocation breakpoints associated with specific tumors have a practical value in the improvement of tumor diagnosis and management?

18.19 What is apoptosis? Why is the cell death associated with apoptosis desirable and how is it regulated?

*18.20 What mechanisms ensure that cells with heavily damaged DNA are unable to replicate?

18.21 How do radiation and chemical carcinogens induce cancers?

19 GENE MUTATION

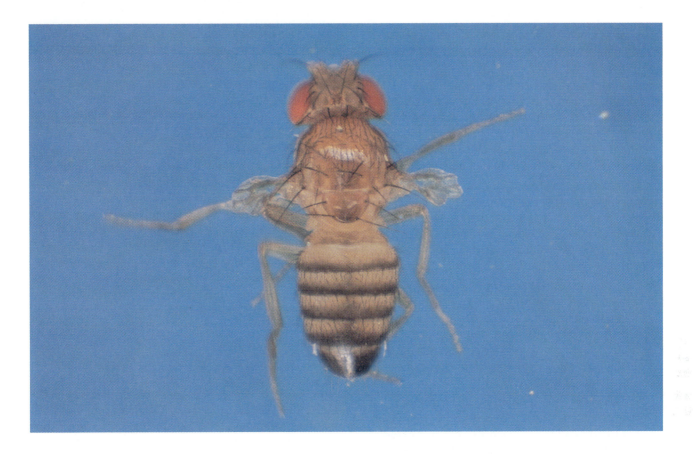

PRINCIPAL POINTS

~ Changes in heritable traits result from random mutation rather than by adaptation to environmental influences.

~ Mutation is the process by which a change in DNA base pairs or a change in the chromosomes is produced. A mutation, then, is the DNA base-pair change or chromosome change resulting from the mutation process. Mutations may occur spontaneously or may be induced experimentally by the application of mutagens.

~ Mutations at the level of the chromosome are called chromosomal mutations (see Chapter 7). Mutations in the sequences of genes at the level of the base pair are called gene mutations. Gene mutations may occur, for example, by a substitution of one base pair for another or by the addition or deletion of one or more base pairs.

~ The consequences of a gene mutation to an organism depend upon a number of factors, especially the extent to which the amino acid coding information is changed. For example, missense mutations cause the substitution of one amino acid for another, and nonsense mutations cause premature termination of synthesis of the polypeptide.

~ The effects of a gene mutation can be reversed either by reversion of the gene sequence to its original state, or by a mutation at a site distinct from that of the original mutation. The latter is called a suppressor mutation. Suppressor mutations that occur within the same gene as the original mutation are intragenic suppressors. They act either by altering a different nucleotide in the same codon affected by the original mutation, or by altering a nucleotide in a different codon. Suppressor mutations that occur in a different gene (the suppressor gene) from the original mutation are called intergenic suppressors. Often, suppression by this class of suppressors involves a tRNA with an altered anticodon.

~ Radiation may cause genetic damage by breaking chromosomes, by producing chemicals that interact with DNA, or by causing unusual bonds between DNA bases. Mutations result if the genetic damage is not repaired.

~ Gene mutations may also be caused by exposure to certain chemicals, called chemical mutagens. Again, mutations result if the genetic damage caused by the mutagen is not repaired. A variety of chemical mutagens is known, and they act in various ways. Base analogs physically replace the proper bases during DNA replication and then shift chemical form so that their base-pairing properties change. Base modifiers cause chemical changes to existing bases, thereby altering their base-pairing properties. Base-pair substitution mutations are the result of treatment with base analogs and base modifiers. Intercalating agents are inserted between existing adjacent bases during replication resulting in single base-pair additions or deletions, that is, frameshift mutations.

~ Both prokaryotes and eukaryotes have a number of repair systems that deal with different kinds of DNA damage. All of the systems use enzymes to make the correction. Without such repair systems, mutations would accumulate and be lethal to the cell or organism. Not all DNA damage is repaired; hence, mutations do appear, but at relatively low frequencies. At high doses of mutagens, repair systems are unable to correct all of the damage, and cell death can result.

~ Geneticists have made great progress in understanding how cellular processes take place by studying mutants that have defects in those processes. A number of screening procedures have been developed to enrich selectively for mutants of interest from a population of mutagenized cells or organisms, and to detect mutations generated by specific molecular targeting methods.

*E*arlier in this text we learned that genetic material is stably passed on from mother cell to daughter cell, and from generation to generation through the faithfulness of the DNA replication process. The genetic material can be changed, however, through a number of ways. Examples include spontaneous changes, errors in the replication process, or the action of particular chemicals or radiation. On the broad scale, there are two types of changes to the genetic material: changes involving whole chromosomes (the topic of Chapter 7) and changes affecting one or a few base pairs. All organisms have mechanisms for repairing base-pair changes, but not all base-pair changes are corrected by the repair systems. The changes that are not repaired are base-pair *mutations*.

Mutations of base pairs can occur anywhere in the genome of an organism. Since not all of the genome of an organism consists of genes, a base-pair mutation will have no phenotypic consequences to

~ FIGURE 19.1

Concept of a mutation in the protein-coding region of a gene. (Note that not all mutations lead to altered proteins and that not all mutations are in protein-coding regions.)

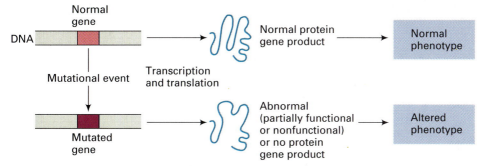

the organism unless it occurs within a gene or in the sequences regulating the gene. Thus, the mutations that have been of particular interest to geneticists are *gene mutations*, that is, those that affect the function of genes. Further, since cell function is the result of protein function, most attention has been focused on mutations affecting the expression of protein-coding genes. Figure 19.1 illustrates in general how a gene mutation can alter a phenotype by changing the function of a protein. We have encountered numerous examples throughout this text of mutations affecting protein-coding genes, including white-flowered pea plants, miniature-winged fruit flies, hemophilia in humans, albino mice, *lacZ* mutants of *E. coli*, and so on. Indeed, it is through the comparison of mutant strains with wild-type (normal) strains that geneticists have been able to obtain an understanding of normal cell function and have advanced our knowledge of the genetics of many organisms.

In this chapter we focus on some of the mechanisms that bring about changes in the DNA at the base-pair level, on some of the repair systems that can repair genetic damage, and on some of the methods used to select for genetic mutants. We will concentrate on gene mutations affecting protein-coding genes and the phenotypic consequences of those changes. As we learn about the specifics of gene mutations, we must be aware that mutations are a major source of genetic variation in a species. Mutations, then, are important elements of the evolutionary process. And, as we have seen throughout our studies thus far, the study of mutants by geneticists has informed us about many biological processes. These comparative investigations of wild type and mutants may take place at a number of levels, but most frequently they involve physiological comparisons and investigations of biochemical and/or molecular events. The information gained from studying the altered process enables geneticists to extrapolate the function of the normal gene product involved.

ADAPTATION VERSUS MUTATION

In the early part of this century, some geneticists supported the theory that variation among organisms resulted from *adaptation* rather than mutation; that is, the environment induced an inheritable change. This theory is essentially Lamarckism, the doctrine of the inheritance of acquired characteristics. Some observations with bacteria fueled the controversy. Bacteria such as *E. coli* exhibit a number of characteristic traits such as the ability or inability to ferment certain sugars, sensitivity or resistance to antibiotics, sensitivity or resistance to infection by phages, etc. Wild-type *E. coli*, for example, is sensitive to the virulent bacteriophage T1, and subcultures challenged with T1 remain sensitive. However, if a culture of wild-type *E. coli* started from a single cell is plated in the presence of an excess of phage T1, most of the bacteria are killed, but a very few survive and produce colonies on the plate. These colonies contain cells that are resistant to T1 because the phage can no longer adsorb to the bacterial surface. All descendants of these resistant cells are also resistant, indicating the heritable nature of the resistance trait. Supporters of the adaptation theory argued that the resistance trait arose as a result of the presence of the T1 phage in the environment. In the opposite camp, supporters of the mutation theory argued that mutations occur randomly so that at any one time in a population of cells some cells have undergone a mutation that makes them resistant to T1 (for this example) even though they have never been exposed to T1. When T1 is subsequently added, the T1-resistant bacteria are selected for.

~ FIGURE 19.2

Representation of a dividing population of T1-phage-sensitive wild-type *E. coli*. At generation 4, T1 phage is added. (a) If one cell mutates to resistance to T1 phage infection at generation 3, then 2 of the 16 cells at generation 4 will be resistant to T1; (b) If one cell mutates to resistance to T1 phage infection at generation 1, then 8 of the 16 cells at generation 4 will be resistant to T1.

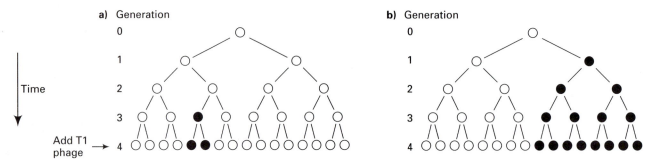

The acquisition of T1 resistance was used by Luria and Delbrück in 1943 to prove that the mutation mechanism was correct and the adaptation mechanism was incorrect. The test they used is known as the *fluctuation test*, which we will now describe.

Consider a dividing population of wild-type *E. coli* that started with a single cell (Figure 19.2). After four generations there will be 16 cells. Let us assume that phage T1 is added at generation 4. If the adaptation theory is correct, then a certain proportion of the generation 4 cells will be induced at that time to become resistant to T1. Most importantly, *that proportion will be the same for all duplicate cultures because adaptation would not commence until T1 was added*. However, if the mutation theory is correct, then the number of generation 4 cells that are resistant to T1 will depend on when in the culturing process the random mutational event occurred that confers resistance to T1. If the mutational event occurs in generation 3 in our example, then 2 of the 16 generation 4 cells will be T1-resistant (Figure 19.2a). However, if the mutational event occurs instead at generation 1, then 8 of the 16 generation 4 cells will be T1-resistant (Figure 19.2b). The key point is that, if the mutation theory is correct, there should be a *fluctuation in the number of T1-resistant cells in generation 4 because the mutation to T1 resistance occurred randomly in the population and did not require the presence of T1*.

Luria and Delbrück initiated a number of independent, 0.2-mL cultures of *E. coli* with the same number of cells. After allowing the cultures to grow in parallel for the same time, a sample from each was taken, plated with phage T1, and the number of phage-resistant colonies was counted. The results of one such experiment are shown in Table 19.1. The results showed a large range in the number of resis-

tant colonies among the duplicate cultures. This high degree of fluctuation in the number of resistant bacteria was taken as proof that resistance was due to random mutation rather than adaptation.

KEYNOTE

Identical cultures of bacteria produced a large range in the number of phage-resistant colonies when phage was added. This was taken as evidence that

~ TABLE 19.1

Comparison of the Number of T1-Phage–Resistant Bacteria in Identical Samples Taken from a Number of Independent Cultures

SAMPLE NUMBER	NUMBER OF T1-RESISTANT COLONIES
1	1
2	0
3	0
4	7
5	0
6	303
7	0
8	0
9	3
10	48
11	1
12	4
Average per sample:	30

heritable traits result from random mutation rather than by adaptation as a result of environmental influence.

MUTATIONS DEFINED

Mutation is the process by which a DNA base-pair change or a chromosome change is produced. A mutation, then, is the DNA base-pair change or chromosome change resulting from the mutation process. Thus, a mutation may be the result of any detectable change that affects DNA's chemical or physical constitution, its replication, its phenotypic function, or the sequence of one or more DNA base pairs (base pairs may, for example, be added, deleted, substituted, reversed in order or inverted, or moved to new positions). Mutations of the DNA are discussed in this chapter, while mutations of a chromosome set are discussed in the next chapter.

A mutation can be transmitted to daughter cells and even to succeeding generations, thereby giving rise to mutant cells or mutant individuals. If a mutant cell gives rise only to somatic cells (in multicellular organisms), a mutant spot or area is produced, but the mutant characteristic is not passed on to the succeeding generation. This type of mutation is called a **somatic mutation**. However, mutations in the germ line of sexually reproducing organisms may be transmitted by the gametes to the next generation, producing an individual with the mutation in both its somatic and germ line cells. Such mutations are called **germ-line mutations**. A somatic mutation affects the individual in which it happens, while a germ-line mutation affects individuals of the subsequent generations.

Types of Mutations

A change in the organization of a chromosome or chromosomes is called a **chromosomal mutation**, also called a **chromosomal aberration** (see Chapter 7). A mutation in a gene sequence is called a **gene mutation**, and it can involve any one of a number of alterations of the DNA sequence of the gene, including base-pair substitutions and additions or deletions of one or more base pairs. Those gene mutations that affect a single base pair of DNA are called **point mutations**.

Mutations can occur spontaneously, but they can also be induced experimentally by the application of a **mutagen**, any physical or chemical agent that signifi-

cantly increases the frequency of mutational events above the spontaneous mutation rate. Mutations that result from treatment with mutagens are called **induced mutations**; naturally occurring mutations are **spontaneous mutations**. There are no qualitative differences between spontaneous and induced mutations. The manifestation of a mutant phenotype is typically the result of a change in DNA that results in the altered function or production of a protein.

With the above background we can now move to the definitions of several terms. While some of the terms apply generally to mutations throughout the genome, some relate specifically to mutations of genes. These terms are illustrated in Figure 19.3. Keep in mind that since the tertiary structure of a polypeptide is a function of its primary amino acid sequence (which is coded for by a gene), a polypeptide synthesized by a mutant strain may be structurally different from the wild-type polypeptide. If so, the mutant polypeptide may be partially functional, nonfunctional, or not produced.

A **base-pair substitution mutation** (point mutation) is a change in a gene such that one base pair is replaced by another base pair; e.g., AT to GC.

A **transition mutation** (Figure 19.3a) is a specific type of base-pair substitution mutation involving a change from one purine-pyrimidine base pair to the other purine-pyrimidine base pair. The four types of transition mutations are AT to GC, GC to AT, TA to CG, and CG to TA.

A **transversion mutation** (Figure 19.3b) is another specific type of base-pair substitution mutation, this one involving a change from a purine-pyrimidine base pair to a pyrimidine-purine base pair. The four types of transversion mutations are AT to TA, GC to CG, AT to CG, and GC to TA.

Mutations can also be defined according to their effects on amino acid sequences in proteins; the following types of mutations can be caused by base-pair substitutions.

A **missense mutation** (Figure 19.3c) is a gene mutation in which a base-pair change in the DNA causes a change in an mRNA codon so that a different amino acid is inserted into the polypeptide in place of the one specified by the wild-type codon, resulting in an altered phenotype. In Figure 19.3c, an AT-to-GC transition mutation changes the DNA from $\frac{5'\text{-AAA-}3'}{3'\text{-TTT-}5'}$ to $\frac{5'\text{-GAA-}3'}{3'\text{-CTT-}5'}$ and this changes the mRNA codon from 5'-AAA-3' (lysine) to 5'-GAA-3' (glutamic acid).

For example, in humans, a single nucleotide pair change in codon 6 of the β-globin gene leads to an amino acid substitution in the β-hemoglobin chain. If

~ **FIGURE 19.3**

Types of base-pair substitution mutations.

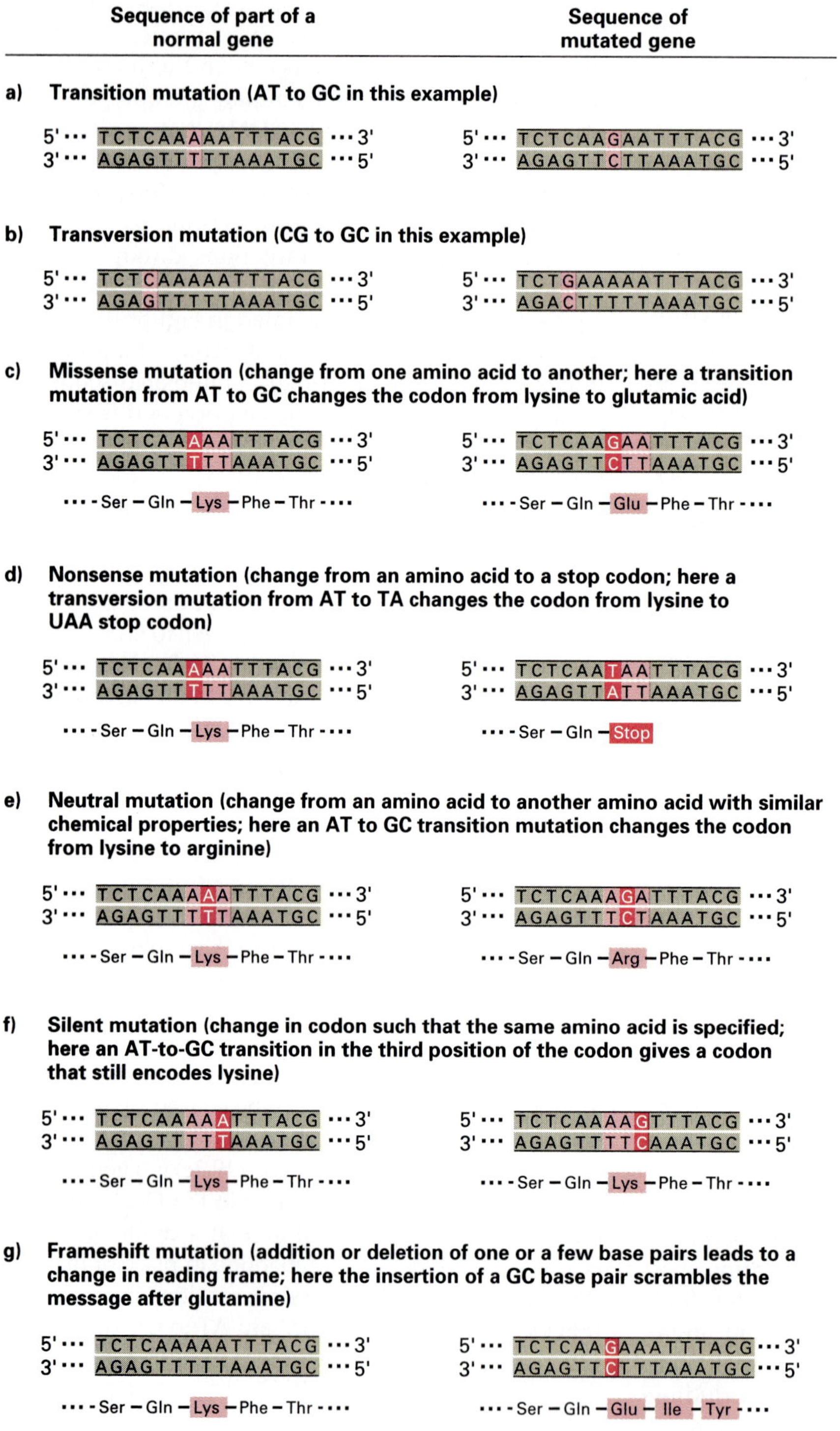

| Sequence of part of a
normal gene | Sequence of
mutated gene |

a)   Transition mutation (AT to GC in this example)

5'··· TCTCAAAAATTTACG ···3'   5'··· TCTCAAGAATTTACG ···3'
3'··· AGAGTTTTTAAATGC ···5'   3'··· AGAGTTCTTAAATGC ···5'

b)   Transversion mutation (CG to GC in this example)

5'··· TCTCAAAAATTTACG ···3'   5'··· TCTGAAAAATTTACG ···3'
3'··· AGAGTTTTTAAATGC ···5'   3'··· AGACTTTTTAAATGC ···5'

c)   Missense mutation (change from one amino acid to another; here a transition mutation from AT to GC changes the codon from lysine to glutamic acid)

5'··· TCTCAAAAATTTACG ···3'   5'··· TCTCAAGAATTTACG ···3'
3'··· AGAGTTTTTAAATGC ···5'   3'··· AGAGTTCTTAAATGC ···5'

···-Ser—Gln—Lys—Phe—Thr-···   ···-Ser—Gln—Glu—Phe—Thr-···

d)   Nonsense mutation (change from an amino acid to a stop codon; here a transversion mutation from AT to TA changes the codon from lysine to UAA stop codon)

5'··· TCTCAAAAATTTACG ···3'   5'··· TCTCAATAATTTACG ···3'
3'··· AGAGTTTTTAAATGC ···5'   3'··· AGAGTTATTAAATGC ···5'

···-Ser—Gln—Lys—Phe—Thr-···   ···-Ser—Gln—Stop

e)   Neutral mutation (change from an amino acid to another amino acid with similar chemical properties; here an AT to GC transition mutation changes the codon from lysine to arginine)

5'··· TCTCAAAAATTTACG ···3'   5'··· TCTCAAAGATTTACG ···3'
3'··· AGAGTTTTTAAATGC ···5'   3'··· AGAGTTTCTAAATGC ···5'

···-Ser—Gln—Lys—Phe—Thr-···   ···-Ser—Gln—Arg—Phe—Thr-···

f)   Silent mutation (change in codon such that the same amino acid is specified; here an AT-to-GC transition in the third position of the codon gives a codon that still encodes lysine)

5'··· TCTCAAAAATTTACG ···3'   5'··· TCTCAAAAGTTTACG ···3'
3'··· AGAGTTTTTAAATGC ···5'   3'··· AGAGTTTTCAAATGC ···5'

···-Ser—Gln—Lys—Phe—Thr-···   ···-Ser—Gln—Lys—Phe—Thr-···

g)   Frameshift mutation (addition or deletion of one or a few base pairs leads to a change in reading frame; here the insertion of a GC base pair scrambles the message after glutamine)

5'··· TCTCAAAAATTTACG ···3'   5'··· TCTCAAGAAATTTACG ···3'
3'··· AGAGTTTTTAAATGC ···5'   3'··· AGAGTTCTTTAAATGC ···5'

···-Ser—Gln—Lys—Phe—Thr-···   ···-Ser—Gln—Glu—Ile—Tyr-···

the individual is homozygous for this mutation, he or she will have sickle-cell anemia.

A **nonsense mutation** (Figure 19.3d) is a base-pair change in the DNA that results in the change of an mRNA codon from one that specifies an amino acid to a chain-terminating (nonsense) codon (UAG, UAA, or UGA). For example, in Figure 19.3d, an AT-to-TA transversion mutation changes the DNA from

5'-AAA-3'   5'-TAA-3'
3'-TTT-5' to 3'-ATT-5'   and this changes the mRNA codon from 5'-AAA-3' (lysine) to 5'-UAA-3', which is a nonsense codon. Because a nonsense mutation gives rise to chain termination at an incorrect place in a polypeptide, the mutation prematurely ends the polypeptide (Figure 19.4). Instead of complete polypeptides, polypeptide fragments (usually nonfunctional) are released from the ribosomes.

A **neutral mutation** (Figure 19.3e) is a base-pair change in a gene that changes a codon in the mRNA such that the resulting amino acid substitution produces no detectable change in the function of the protein translated from that message. A neutral mutation is a subset of missense mutations and is the case where the new codon codes for a different amino acid that is chemically equivalent to the original and hence does not affect the protein's function. In Figure 19.3e, an AT-to-GC transition mutation changes the codon from 5'-AAA-3' to 5'-AGA-3' which substitutes arginine for lysine. Both arginine and lysine are basic amino acids and are sufficiently similar in properties so that the protein's function may well not be altered significantly.

A **silent mutation** (Figure 19.3f) is also a subset of missense mutations and is the case where a base-pair change in a gene that alters a codon in the mRNA such that the *same* amino acid is inserted in the protein. The protein in this case obviously has wild-type function. In Figure 19.3f, a silent mutation results from an AT-to-GC transition mutation which changes the codon from 5'-AAA-3' to 5'-AAG-3', both of which specify lysine.

A **frameshift mutation** (Figure 19.3g) results from the addition or deletion of one or more base pairs in a gene. An addition or deletion of one base pair, for example, shifts the mRNA's reading frame by one base so that incorrect amino acids are added to the polypeptide chain after the mutation site. Often, frameshift mutations generate new codons resulting in a shortened protein, or they result in read-through of the normal stop codon, resulting in longer than normal proteins. In any case, a frameshift mutation usually results in a nonfunctional protein. In Figure 19.3g, an insertion of a GC base pair scrambles the message after the codon specifying glutamine.

~ FIGURE 19.4

A nonsense mutation and its consequence to translation.

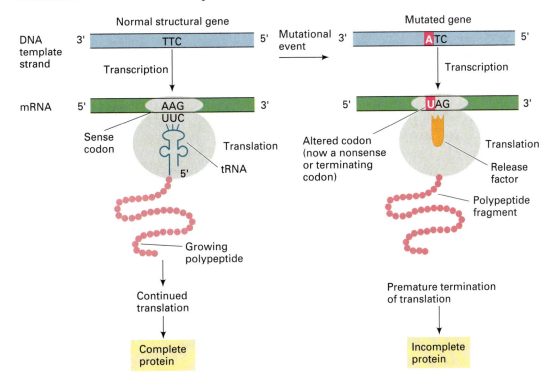

KEYNOTE

Mutation is the process by which DNA base-pair change or a chromosome change occurs. A mutation, then, is the DNA base-pair change or chromosome change resulting from the mutation process. Mutations may occur spontaneously or may be induced experimentally by the application of mutagens. Mutations that affect a single base pair of DNA are called base-pair substitution mutations or point mutations. Mutations in the sequences of genes are called gene mutations.

Reverse Mutations and Suppressor Mutations

Point mutations generally fall into two classes in terms of their effects on the phenotype in comparison to the wild type. **Forward mutations** are mutations that cause the genotype to change from wild type to mutant, and **reverse mutations** (or **reversions** or **back mutations**) are mutations that cause the genotype to change from mutant to wild type.

A gene reversion is a mutational event that causes a change from a mutant phenotype to wild-type or partially wild-type function. Reversion of a nonsense mutation, for instance, occurs when a base-pair change results in a change of the mRNA nonsense codon to a codon for an amino acid. If this gene reversion is back to the wild-type amino acid, the mutation is a **true reversion**. If the reversion is to some other amino acid, the mutation is a *partial reversion,* and complete function may be restored. Reversion of missense mutations can occur in the same way.

The effects of a mutation may be diminished or abolished by a **suppressor mutation**, that is, a mutation at a different site from the original mutation (also called a secondary or **second-site mutation**). *A suppressor mutation does not result in a reversal of the original mutation*; instead, it masks or compensates for the effects of the initial mutation.

There are two major classes of suppressor mutations: those which occur within the same gene as the original mutations but at a different site (called **intragenic** [*intra* = within] **suppressors**); and those occurring in a different gene (called **intergenic** [*inter* = between] **suppressors**). Both intragenic and intergenic suppressors operate to allow production of

functional or partially functional copies of the protein which were initially rendered inactive by the original deleterious mutation. Thus, function can be restored only when both the original mutation and the suppressor mutation are present together in the same cell.

Intragenic suppressors act in one of two ways: by altering a different nucleotide in the same codon in which the original mutation occurred, or by altering a nucleotide in a different codon. Table 19.2 illustrates the first situation. Here, a DNA sequence of three base pairs in the wild type specifies the mRNA codon 5'-CGU-3', which is read as arginine. The original (first) mutation is a CG-to-AT transversion at the first base pair, resulting in the mRNA codon 5'-AGU-3', which specifies serine. The suppressor (second) mutation is a TA-to-AT transversion at the third position, giving the mRNA codon 5'-AGA-3', which is an arginine codon. Thus, with both mutations, the protein will be completely functional in cells. This type of intragenic suppressor can also correct frameshift mutations that result when an extra nucleotide is added or a nucleotide is deleted. Suppression here, for example, occurs when a nucleotide is removed from the same codon in which an earlier mutation inserted a nucleotide. In the second type of intragenic suppressor, a second mutation alters a nucleotide in a different codon. Most often, this type of second mutation suppresses a frameshift mutation; that is, adding a nucleotide near a prior nucleotide deletion to restore reading frame, or deleting a nucleotide near an addition.

Intergenic suppression is the suppression of a mutational defect by a second mutation in another gene. Genes that cause suppression of mutations in other genes are called **suppressor genes**. Many intergenic suppressors work by changing the way the mRNA encoded by the mutant gene is read. Each

~ TABLE 19.2

Example of an Intragenic Suppressor Mutation Altering a Nucleotide in the Same Codon as the Original Mutation

	DNA SEQUENCE	mRNA CODON SPECIFIED	AMINO ACID CODED FOR
Wild-type condition	5'-CGT-3' 3'-GCA-5'	5'-CGU-3'	Arg
Original mutation	5'-AGT-3' 3'-TCA-5'	5'-AGU-3'	Ser
After suppressor mutation	5'-AGA-3' 3'-TCT-5'	5'-AGA-3'	Arg

suppressor gene can suppress the effects of only one type of nonsense, missense, or frameshift mutation; hence, suppressor genes can suppress only a small proportion of the point mutations that theoretically can occur within a gene. On the other hand, a given suppressor gene will suppress all mutations for which it is specific, whatever gene the mutation is in.

The suppressors of nonsense mutations have been well characterized, particularly in *E. coli* and yeast. The suppressor genes in this case often are mutant tRNA genes. That is, particular tRNA genes can mutate so that (in contrast to what occurs with wild-type tRNAs) their anticodons recognize a chain-terminating codon and put an amino acid into the chain. Thus, instead of polypeptide chain synthesis being stopped prematurely as a result of a nonsense mutation, the altered (suppressor) tRNA inserts an amino acid at that position, and full or partial function of the polypeptide may be restored. There are three classes of nonsense suppressors: one for each of the nonsense codons UAG, UAA, and UGA. If, for example, a gene for a tyrosine tRNA (which has the anticodon 3'-AUG-5') is mutated so that the tRNA has

the anticodon 3'-AUC-5', the mutated suppressor tRNA (which will still carry tyrosine) will read the nonsense codon 5'-UAG-3' (Figure 19.5). So, instead of chain termination occurring, tyrosine is inserted at that point in the polypeptide. How functional the complete protein will be will then depend on the effects of the inserted tyrosine in the protein. If it is an important part of the protein, then the incorrect amino acid may not restore function to a significant degree. If it is in a less crucial area, the protein may have some or complete function.

But we now have a dilemma. If we have changed this particular class of tRNA.Tyr so that its anticodon can now read a nonsense codon, it cannot read the original codon that specifies the amino acid it carries. Thus nonsense suppressor tRNAs are typically produced by mutation of tRNA genes that are redundant in the genome; that is, for which several different genes all specifying the same tRNA base sequence exist. Therefore if there is a mutation in one of the redundant genes (usually one that codes for minor amounts of tRNA) so that the tRNA will read, say, UAG, the other genes specifying the same tRNA pro-

~ FIGURE 19.5

Mechanism for action of an intergenic nonsense suppressor mutation that results from mutation of a tRNA gene. In this example, a tRNA.Tyr gene has mutated so that the tRNA's anticodon is changed from 3'-AUG-5' to 3'-AUC-5', which can read a UAG nonsense codon, inserting tyrosine in the polypeptide chain at that codon.

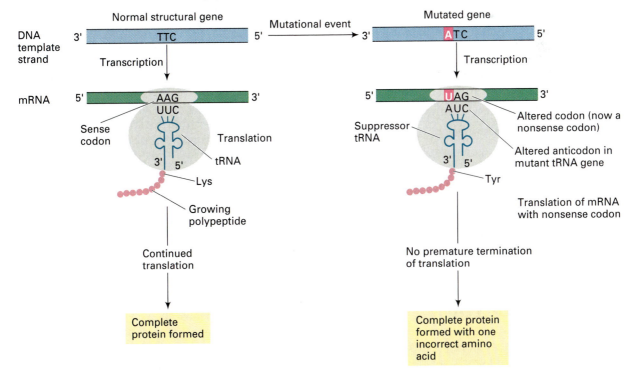

duce a tRNA molecule that will read the normal Tyr codon.

We can view nonsense suppression as a competition between the release factor binding to the nonsense codon and the suppressor tRNA. UAG and UGA suppressor tRNAs do well in this competition, since they succeed in reading over one-half of the nonsense codons. However, UAA suppressor tRNAs are only 1 to 5 percent efficient. Given these figures, suppression of the normal chain-terminating codons—which are frequently UAG and UGA—might also occur, producing longer than normal proteins. Such proteins are known as *read-through proteins*. However, for reasons that are not always clear, UAG and UGA suppressor tRNAs do not produce large numbers of read-through proteins. In some cases, this is because two different stop codons are present in tandem (e.g., UAGUGA).

Lastly, suppressors of +1 frameshift mutations are known. These suppressors have mutations in tRNA genes so that there are four bases in the anticodon rather than three.

KEYNOTE

A suppressor mutation is a mutation at a second site that completely or partially restores a function lost because of a primary mutation at another site. Intragenic suppressors are suppressor mutations that occur within the same gene as the original mutation but at a different site. They act either by altering a different nucleotide in the same codon as that in which the original mutation occurred, or by altering a nucleotide in a different codon. Intergenic suppressors are suppressor mutations that occur in a different gene—a suppressor gene—than that with the original mutation. Intergenic suppressors are known for nonsense, missense, and frameshift mutations, and typically involve a tRNA with an altered anticodon. Thus, nonsense mutations are suppressed by mutated tRNAs which now can read the chain-terminating codon and insert an amino acid at the mutation site.

CAUSES OF MUTATION

Mutations can occur spontaneously or they can be induced. Spontaneous mutations are mutations that occur naturally; that is, without the use of chemical or physical mutagenic agents. They may occur, for example, by errors in cellular processes (such as DNA replication), or by the action of mutagens in the environment (such as UV irradiation in sunlight). Induced mutations are mutations that occur as a result of treatment with known chemical or physical mutagens.

Spontaneous Mutations

All types of point mutations can occur spontaneously. For a long while geneticists thought that spontaneous mutations were produced by mutagens indigenous to the environment, such as radiation and chemicals. However, evidence indicates that although the rate at which spontaneous mutations appear is extremely low, that rate is nonetheless too high to be accounted for by indigenous mutagens alone.

Two different terms are often used to give a quantitative measure of the occurrence of mutations. **Mutation rate** presents the probability of a particular kind of mutation as a function of time, e.g., number per nucleotide pair per generation, or number per gene per generation. **Mutation frequency**, on the other hand, is the number of occurrences of a particular kind of mutation expressed as the proportion of cells or individuals in a population, e.g., number per 100,000 organisms or number per 1 million gametes.

In *Drosophila*, for example, the spontaneous mutation rate for individual genes is about 10^{-4} to 10^{-5} per gene per generation. In humans the rate varies between 10^{-4} and 4×10^{-6} per gene per generation. For eukaryotes in general, the spontaneous mutation rate is 10^{-4} to 10^{-6} per gene per generation and for bacteria and phages, the rate is 10^{-5} to 10^{-7} per gene per generation. (The spontaneous mutation frequencies at specific loci for various organisms are presented in Table 22.5, p. 738.) That the spontaneous mutation rate is affected by the genetic constitution of the organism is shown by the fact that male and female *Drosophila* of the same strain have identical mutation rates, while different strains may exhibit different mutation rates. Further, it is important to note that the rates and frequency values represent the mutations that become fixed in DNA. Most spontaneous errors are corrected by cellular repair systems; only a few remain uncorrected as permanent changes.

Spontaneous mutations can result from any one of a number of events, including errors in DNA replication and spontaneous chemical changes in DNA. Spontaneous mutations can also result from the movement of what are called transposable genetic elements, which is discussed in Chapter 20.

~ FIGURE 19.6

Examples of mismatched bases in DNA involving tautomeric forms of the bases. (a) Mismatched bases resulting from rare forms of pyrimidines; (b) Mismatched bases resulting from rare forms of purines. (From *Science of Genetics*, 6th ed. by George W. Burns and Paul J. Bottino. Copyright © 1989. Adapted by permission of Prentice-Hall, Inc., Upper Saddle River, NJ.)

a)

Rare form of cytosine (C*) Adenine

Rare form of thymine (T*) Guanine

b)

Cytosine Rare form of adenine (A*)

Thymine Rare form of guanine (G*)

DNA REPLICATION ERRORS. Both point mutations and small additions and deletions can result from DNA replication errors.

Base-Pair Substitution Mutations. Base-pair substitution mutations can occur if mismatched base pairs occur during DNA replication. How can mismatched base pairs occur? A mechanism that has been generally accepted involves changes of the bases between alternate chemical forms, called **tautomers**. The change in the chemical form of a base is called a **tautomeric shift**. In its rare form, a base can form different hydrogen bonds and this can lead to mismatched base pairs. For the pyrimidines, a rare form of cytosine can pair with adenine, and a rare form of thymine can pair with guanine (Figure 19.6a), and for the purines, a rare form of adenine can pair with cytosine, and a rare form of guanine can pair with thymine (Figure 19.6b).

Figure 19.7 illustrates how a GC to AT transition mutation can be produced as a result of a tautomeric shift. When the parental DNA (Figure 19.7a) is replicating, a guanine could shift to its rare state on a parental strand (top strand, Figure 19.7b). As a result, a T will be inserted on the new DNA strand giving a mismatched GT base pair after replication is complet-

ed (Figure 19.7c). The rare form of guanine in the GT pair is likely to change back to the normal form so that, in the next round of DNA replication, it acts as a normal G. This will give a progeny DNA molecule with a GC base pair like the original parental DNA (Figure 19.7d). The T in the mismatched GT pair specifies an AT in the other progeny DNA molecule (Figure 19.7d); this is the GC to AT transition mutation. Note that many more mutations would result from tautomeric shifts than are actually observed if it were not for the proofreading activity of DNA polymerases. This activity recognizes many of the mismatches (such as GT), excises them, and replaces them with the correct base pair (see Chapter 12). However, replication of an uncorrected mismatch "fixes" the mutation in the cell. For example, once GT replicates to produce a GC and an AT in the two progeny DNAs, the AT mutation is a normal base pair and will not be recognized as "wrong" by repair enzymes.

However, there are critics of the tautomerism model for spontaneous base-pair substitution mutations and they have proposed a more plausible model for the origin of such mutations. The basis for the criticism is that the assumptions used in calculating the relative proportions of tautomeric forms of the bases in DNA are probably incorrect. Specifically, measure-

~ FIGURE 19.9

Spontaneous generation of addition and deletion mutants by DNA looping-out errors during replication.

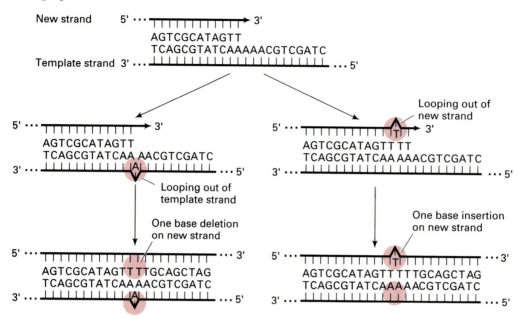

~ FIGURE 19.10

Depurination (loss of a purine) from a single strand of DNA. The sugar-phosphate backbone is unbroken.

Adenine

Adenine released from DNA

~ FIGURE 19.11

Deamination of cytosine to uracil.

NH₂ → Deamination → O

Cytosine   Uracil

~ FIGURE 19.12

Deamination of 5-methylcytosine (5ᵐC) to thymine.

5-methylcytosine (5ᵐC)   Thymine (T)

are corrected by repair mechanisms but 5ᵐC deamination mutations are not, locations of 5ᵐC in the genome often appear as *mutational hot spots*; that is, nucleotides where a higher than average frequency of mutation occurs (Figure 19.13).

Induced Mutations

Since the rate of spontaneous mutation is so low, geneticists use mutagens to increase mutation frequency so that a significant number of organisms have mutations in the gene being studied. Two classes of mutagens are used—radiation and chemical—both of which involve specific mechanisms of action.

RADIATION.   Both X rays and ultraviolet light (UV) are used to induce mutations. X rays are an example of ionizing radiation. Ionizing radiation can penetrate tissues, hence the use of X rays as a diagnostic tool. Collision of ionizing radiation with atoms in its path gives rise to ions and reactive chemical radicals that

can break chemical bonds, including those in DNA. That is, the products of ionizing radiation can induce chromosome breakages, chromosome rearrangements, and damage to DNA, e.g., point mutations. In fact, ionizing radiation is the leading cause of gross chromosomal mutations in humans.

High doses of ionizing radiation kill cells, hence its use in treating some forms of cancer. At certain low levels of ionizing radiation, point mutations are commonly produced; at these levels, there is a linear relationship between the rate of point mutations and radiation dosage. In other words, with twice the amount of ionizing radiation, twice the number of point mutations are induced. Importantly, for many organisms, including humans, the effects of ionizing radiation doses are cumulative. That is, if a particular dose of radiation results in a certain number of point mutations, the same number of point mutations will be induced whether the radiation dose is received over a short period of time, or over a long period of time. Clearly, then, people should be concerned about

~ FIGURE 19.13

Distribution of spontaneous GC to AT transition mutations to stop codons in the *lac* repressor gene of *E. coli*, showing the hot spots for mutation at 5ᵐC nucleotides.

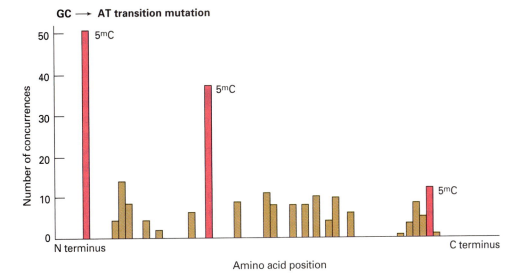

GC ⟶ AT transition mutation

DNA in its rare state and then switches to the normal state during replication. Thus 5BU-induced mutations can be reverted by a second treatment of 5BU.

The base analog 2-aminopurine (2AP) acts as a mutagen in essentially the same way as 5BU does, and like 5BU, 2AP exists in normal and rare states (Figure 19.16a and b). In its normal state 2AP resembles adenine but has an amino group at a different position on the purine ring than does adenine; in its normal state it base-pairs with thymine. In its rare state 2AP resembles guanine and base-pairs with cytosine. As with 5BU, 2AP induces transition mutations, which can be reverted by a second application of 2AP. The mutation will be either AT to GC or GC to AT, depending on its state during its initial incorporation into the DNA and on its state during replication. Note that since the mutations involved are transitions in both cases, 5BU can revert mutations induced by 2AP and vice versa.

Not all base analogs are mutagens. One of the approved drugs given to AIDS patients, AZT (azidothymidine), is an analog of thymidine but it is not a mutagen, because it does not result in base-pair changes. AZT works as an AIDS treatment as follows. The AIDS virus, HIV-I (human immunodeficiency virus-I), is a retrovirus. That is, its genome is RNA, but when the virus enters a cell, the viral enzyme reverse transcriptase (see Chapter 15) makes a DNA copy of the RNA (i.e., a cDNA). That DNA can then be incorporated into the cell's genomic DNA, from where it can direct new viral synthesis. During the reverse transcriptase step, AZT (as a triphosphate derivative) is incorporated into the growing DNA chain as a thymidine analog. However, AZT has an azido group (N_3) on the 3' carbon of its deoxyribose sugar and the missing OH group prevents 5'-to-3'

growth of the DNA chain. This chain termination process is similar to that for the dideoxynucleotides used in DNA sequencing reaction (see Chapter 15, pp. 473-476). AZT is *not* a good substrate for cellular DNA polymerases, however, so host DNA synthesis is not affected. Thus, AZT acts as a selective poison by inhibiting the production of the viral cDNA. This blocks new viral synthesis because a cDNA must be synthesized and incorporated into the host genome in order to code for viral components.

Base-Modifying Agents.   Unlike base analogs, which actually replace bases in the DNA and require DNA replication for their incorporation, a number of chemicals act as mutagens by directly modifying the chemical structure and properties of the bases. Figure 19.17 shows the action of three types of mutagens that work in this way: a deaminating agent, a hydroxylating agent, and an alkylating agent.

Nitrous acid, HNO_2 (Figure 19.17a), is a deaminating agent that removes amino groups ($-NH_2$) from the bases guanine, cytosine, and adenine. Treatment of guanine with nitrous acid produces xanthine, but since this purine base has the same pairing properties as guanine, no mutation results (Figure 19.17a, part 1). However, when cytosine is treated with nitrous acid, uracil (which pairs with adenine) is produced (Figure 19.17a, part 2). The deamination of cytosine by nitrous acid, then, produces a CG-to-TA transition mutation during replication. Likewise, nitrous acid modifies adenine to produce hypoxanthine, a base that pairs with cytosine rather than thymine, thus resulting in an AT-to-GC transition mutation (Figure 19.17a, part 3). A nitrous acid-induced mutation can be reverted by a second treatment with nitrous acid.

~ **FIGURE 19.16**

Base-pairing properties of the normal and rare states of the base-analog mutagen 2-aminopurine (2AP). (a) In its normal state 2AP pairs with thymine; (b) In its rare state 2AP pairs with cytosine.

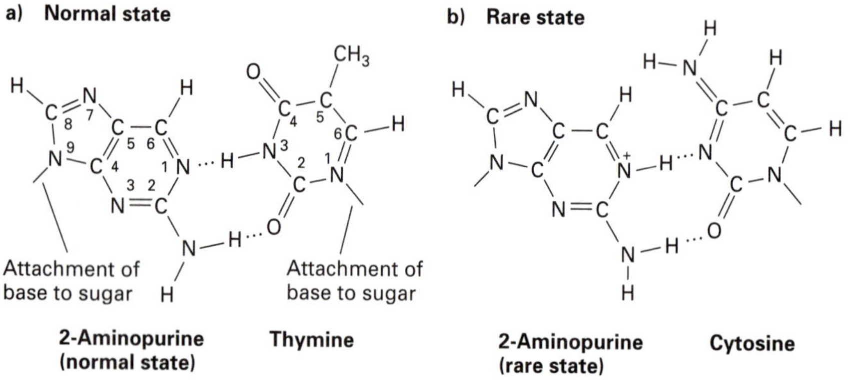

a) **Normal state**

b) **Rare state**

2-Aminopurine (normal state)   Thymine

2-Aminopurine (rare state)   Cytosine

~ FIGURE 19.17

Action of three base-modifying agents. (a) Nitrous acid (HNO_2) modifies (1) guanine, (2) cytosine, and (3) adenine. The cytosine and adenine modifications result in mutations, while the guanine modification does not. (b) Hydroxylamine (NH_2OH) reacts only with cytosine. (c) Methylmethane sulfonate (MMS), an alkylating agent, alkylates guanine. (Note: dr = deoxyribose.)

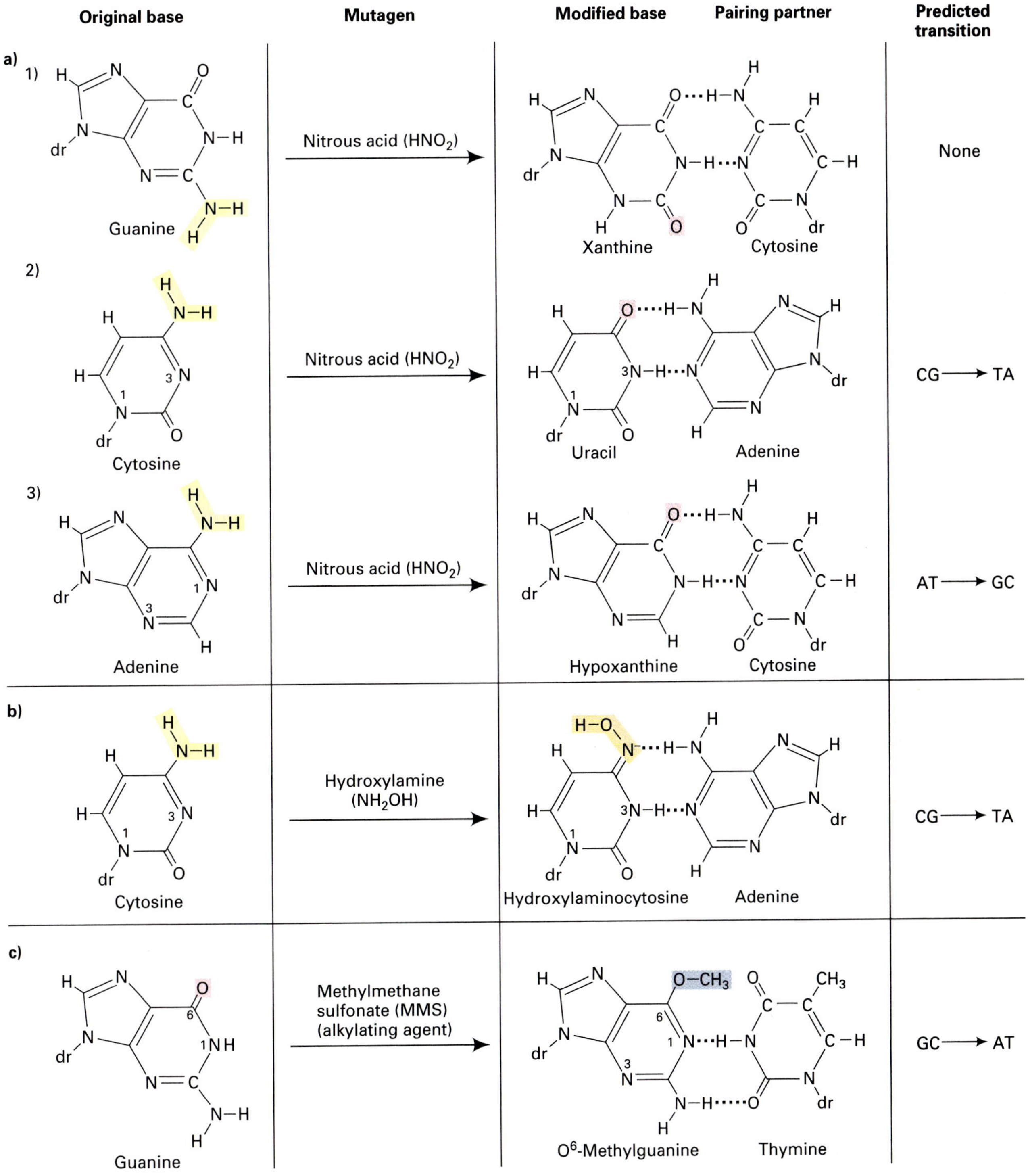

~ FIGURE 19.19

Simplified outline of a site-specific *in vitro* mutagenesis method to introduce a point mutation at a specific site in DNA.

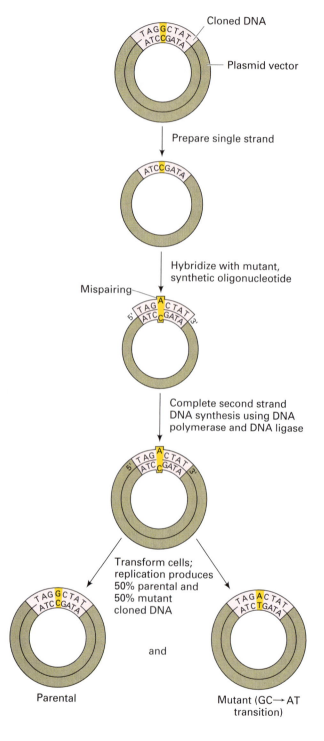

Cloned DNA

Plasmid vector

Prepare single strand

Hybridize with mutant, synthetic oligonucleotide

Mispairing

Complete second strand DNA synthesis using DNA polymerase and DNA ligase

Transform cells; replication produces 50% parental and 50% mutant cloned DNA

and

Parental

Mutant (GC→AT transition)

of various types. With methods such as these, a researcher can, for example, mutate a gene of unknown function, introduce the mutant gene into a cell and investigate the resulting phenotypic changes, thereby gaining insight into the normal gene's function. This is opposite from the traditional genetics approach of working from a gene's phenotype back to the gene responsible for the phenotype. Site-specific mutagenesis has also been used to generate specific changes in key regulatory sequences to investigate which base pairs are important and which are not. Our understanding of promoter and enhancer elements has come, in part, from site-specific mutagenesis of sequences upstream or downstream from a gene.

THE AMES TEST: A SCREEN FOR POTENTIAL MUTAGENS

All the mutagens (radiation and chemicals) we have examined so far are those commonly used to induce mutations during genetic research. Scientists are aware that many chemicals in our environment can have mutagenic effects. And, since some chemicals cause mutations that result in cancerous growth, there is a strong interest in testing new chemicals for their ability to induce mutations.

In the early 1970s a rapid test was developed by Bruce Ames. The **Ames test** uses the bacterium *Salmonella typhimurium* as the test organism. Approximately 10^9 cells of tester bacteria that are auxotrophic for histidine are spread on a culture plate that lacks the amino acid histidine. (Recall that an auxotrophic mutant is unable to make a particular molecule that is essential for growth and it requires a particular nutrient in order to grow. Histidine [*his*] auxotrophs, then, require the presence of histidine in the growth medium in order to grow; normal [*his*⁺] individuals do not.) Typically the tester bacteria are a mixture of several *his* strains that allow detection of both point mutations and frameshift mutations in the test. The bacterial strains are also mutant for the production of a lipopolysaccharide coat that would otherwise limit the entry of chemicals into the cells. Further, the strains contain plasmid-borne mutant genes that produce enzymes involved in error-prone SOS repair that are more efficient than the enzymes in the wild type. Because SOS repair itself causes other mutations (see pp. 640–642), most kinds of DNA damage in the mutant strains are converted to mutations rather than being repaired. Finally, because in humans and other mammals many chemicals are converted to carcinogens in the liver and other tissues by

Other protocols are also commonly used for *in vitro* mutagenesis. For example, there are many approaches for using PCR to make specific mutations

enzymatic detoxification pathways, the culture medium contains a small amount of rat liver homogenate to simulate the effects of those mammalian enzymes. Control plates are also set up that lack the chemical under test.

After incubation for about two days at 37°C, *his⁺* colonies are present on both plates. The colonies on the control plate represent revertants that have arisen by spontaneous mutation. Thus, whether the potential mutagen is indeed a mutagen is demonstrated by a significantly higher number of *his⁺* revertants on the experimental plate, because those colonies represent both induced and spontaneous revertants.

The Ames test is so straightforward to perform that it is in routine use in many laboratories around the world. To date, the Ames test has identified a large number of mutagens, including many environmental chemicals (such as hair dye additives, vinyl chloride, and particular food colorings) which include both synthetic and natural compounds.

The Ames test can show whether or not a chemical is a mutagen, but it cannot indicate whether or not the chemical is a carcinogen (i.e., induces a cancer). That is, carcinogenesis (the induction of cancers) does not necessarily occur because of mutation. In fact, recent data indicate that mitogenesis (induced cell division) plays a dominant role in carcinogenesis. Some carcinogens, then, are *not* mutagens, but cause cancers by inducing mitogenesis.

Several other methods are available to test new or old environmental chemicals for carcinogenic effects. One method is to test a chemical directly for carcinogenicity by setting up experiments with inbred lines of mice or rats. The inbred lines are genetically identical so that a statistical analysis of treated versus untreated animals for the incidence of cancers provides information about the probability that a chemical is carcinogenic. Although these studies are expensive and time-consuming, they are done routinely by research laboratories, often under contract to federal agencies or to private companies. Using animal testing, about half of the natural chemicals (those that are normally present in such things as food) and synthetic chemicals (e.g., pesticides, drugs, and food additives) tested have been shown to be carcinogens.

DNA REPAIR MECHANISMS

Spontaneous and induced mutations constitute damage to the DNA of a cell or an organism. Especially with high doses of mutagens, the mutational damage can be considerable. Both prokaryotic and eukaryotic cells have a number of repair systems to deal with damage to DNA. All of the systems use enzymes to make the correction. Some of the systems directly correct the lesion while others first excise the lesion creating a single-stranded gap and then synthesize new DNA for the resulting gap. If the repair systems are unable to correct all of the lesions, the result is a mutant cell (or organism) or, if too many mutations remain, death of the cell. Clearly, DNA repair systems are very important for the survival of the cell. The fact that the repair systems are not 100 percent efficient makes it possible to isolate mutants for study.

Direct Correction of Mutational Lesions

REPAIR BY DNA POLYMERASE PROOFREADING.

In bacterial genes, the frequency of base-pair substitutions varies from 10^{-7} to 10^{-11} errors per replication event. However, DNA polymerase makes errors in inserting nucleotides while it is synthesizing the new DNA strand perhaps at a frequency of 10^{-5}. The discrepancy between the two values is accounted for by proofreading activity of the polymerase itself (see Chapter 12). That is, most bacterial DNA polymerases, in addition to having 5'-to-3' polymerizing activity, also have a 3'-to-5' exonuclease activity. The latter is necessary to ensure accurate DNA replication. When an incorrect nucleotide is inserted, the error is most often (but not always) detected by the polymerase, perhaps by the fact that the mismatched base pair results in a bulge in the double helix. Or, since the incorrect base cannot form a hydrogen bond with the complementary base, perhaps the polymerase will not add a nucleotide to the growing 3'-OH end unless that nucleotide is properly hydrogen-bonded. Thus, DNA synthesis stalls and cannot proceed until the wrong nucleotide is removed and the proper one put in its place. The incorrect nucleotide is removed by the 3'-to-5' exonuclease activity as the polymerase reverses along the template strand. The polymerase then moves forward again, resuming its 5'-to-3' DNA synthesis activity.

The importance of the 3'-to-5' exonuclease activity of DNA polymerase for maintaining a low mutation rate is shown nicely by the existence of *mutator* mutations in *E. coli*. Strains carrying mutator mutations show a much higher than normal mutation frequency for all genes. These mutations have been shown to affect proteins whose normal functions are required for accurate DNA replication. Relevant to this discussion is the *mutD* mutator gene of *E. coli*, which results in an altered ε (epsilon) subunit of DNA polymerase III, the primary replication enzyme of *E. coli*. *mutD* causes a defect in 3'-to-5' proofreading activity, so that many incorrectly inserted nucleotides are unrepaired.

While 3'-to-5' exonuclease activity is a property of bacterial DNA polymerases, no such enzyme activities are found in eukaryotic DNA polymerases. Proofreading does take place in eukaryotes, although the proofreading activity is not located on the DNA polymerase itself. Presumably it is located on other proteins.

PHOTOREACTIVATION OF UV-INDUCED PYRIMIDINE DIMERS. Direct correction of lesions can occur also in the repair of UV-light-induced thymine (or other pyrimidine) dimers. By this process of **photoreactivation** or **light repair** (Figure 19.20), the dimers are reverted directly to the original form by exposure to visible light in the wavelength range of 320 to 370 nm. Photoreactivation is catalyzed by an enzyme called *photolyase* (encoded by the *phr* gene), which, when activated by a photon of light, splits the dimers apart. Strains with mutations in the *phr* gene are defective in light repair. Photolyase has been found in prokaryotes and in lower eukaryotes, but not in humans. Presumably, it functions by searching along the double helix, seeking the bulges that result when thymine dimers are present. Photolyases are apparently very effective since few thymine dimers (and, hence, mutations) are left after photoreactivation. Photolyase's action can be avoided in experimental situations by keeping irradiated cells in the dark.

~ **FIGURE 19.20**

Repair of a thymine dimer by photoreactivation.

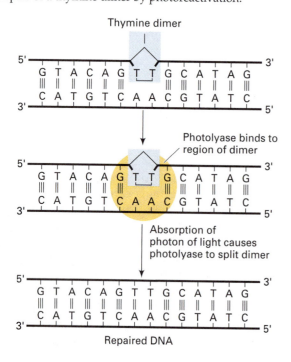

Thymine dimer

Photolyase binds to region of dimer

Absorption of photon of light causes photolyase to split dimer

Repaired DNA

REPAIR OF ALKYLATION DAMAGE. Earlier in this chapter, we discussed the use of base-modifying agents as mutagens (see Figure 19.17). Some of these mutagens—the alkylating agents—transfer alkyl groups (usually methyl or ethyl groups) to reactive sites on the bases or phosphates. A particularly reactive site is the oxygen of carbon 6 in the guanine, as shown in Figure 19.17c. The alkylated guanine pairs with thymine and produces a GC-to-AT transition when the DNA is replicated. Alkylation damage like this can be removed by specific DNA repair enzymes. Note that in this repair system the modified base is not removed from the DNA. In this particular case, an enzyme encoded by the *ada* gene, called O[6]-*methylguanine methyltransferase*, recognizes the O[6]-methylguanine in the DNA and removes the methyl group, thereby changing it back to its original form.

Repair Involving Excision of Base Pairs

EXCISION REPAIR. In 1964, the team of R. P. Boyce and P. Howard-Flanders and that of R. Setlow and W. Carrier discovered a new DNA repair system. They isolated some UV-sensitive mutants of *E. coli*, which, after UV irradiation, exhibited a higher-than-normal rate of induced mutation in the dark. These mutants were called *uvrA* mutants (*uvr* means "UV repair"). The *uvrA* mutants can repair dimers only with the input of light; that is, they have a normal photoreactivation repair system. The investigators hypothesized that there must be another repair system, one that did not require light for function. They called this the **dark repair** or **excision repair** system. Since wild-type organisms can repair dimers in the dark, the wild-type genes are labeled *uvrA*[+].

The excision repair system in *E. coli* corrects not only pyrimidine dimers but also other serious damage-induced distortions of the DNA helix. The system works as diagrammed in Figure 19.21. The DNA helix distortions are recognized by the UvrABC endonuclease, a multisubunit enzyme encoded by the three genes *uvrA*, *uvrB*, and *uvrC*. This enzyme makes one cut in the damaged DNA strand eight nucleotides to the 5' side of the damage (e.g., the dimer) and four nucleotides to the 3' side of the damage. The cuts release a 12-nucleotide stretch of single-stranded DNA containing the damaged base(s). The 12-nucleotide gap is filled by the 5'-to-3' polymerizing activity of DNA polymerase I, and sealed by DNA ligase. The excision repair system is found in most organisms that have been studied, and its mechanism of action is thought to be essentially like the one found in *E. coli*. Occasionally, errors are introduced in the repair synthesis of the DNA, and such errors are another source of mutations resulting from UV radia-

~ FIGURE 19.21

~ FIGURE 19.21

Excision repair of pyrimidine dimer and other damage-induced distortions of DNA initiated by the UvrABC endonuclease.

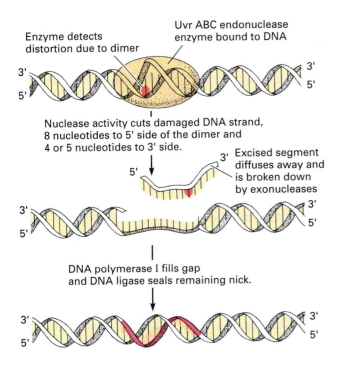

~ FIGURE 19.22

Excision repair of damaged bases involving the action of glycosylase. Glycosylase detects the damaged base and catalyzes its removal. A few adjacent bases are removed by AP endonuclease, and the resulting single-stranded gap is repaired by DNA polymerase I and DNA ligase.

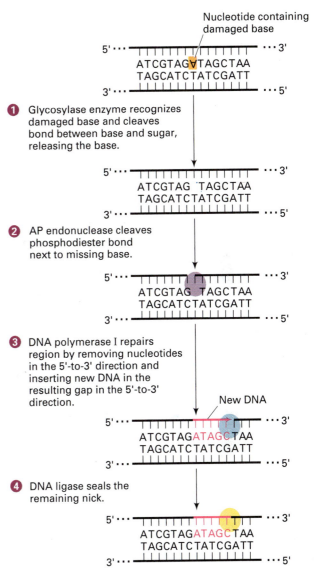

tion. The most common cause of the errors is incorrect pairing of new nucleotides with those in the template strand.

Similar excision repair systems are known in eukaryotes. While the details of these systems have not been elaborated fully, it is known that about 12 genes encode proteins involved in excision repair in yeast and mammalian systems.

REPAIR BY GLYCOSYLASES. Damaged bases can be excised by other means. For example, cells contain a glycosylase enzyme that can detect an individual unnatural base and catalyze its removal from the deoxyribose sugar to which it is attached (Figure 19.22, part 1). This catalytic activity leaves a gap in the DNA where the base was removed. This hole is called an *AP site* (for apurinic, where there is no A or G, or apyrimidinic, where there is no C or T). Such holes can also occur through natural, spontaneous losses of bases, discussed earlier in the chapter. The enzyme *AP endonuclease* recognizes that there is a hole, and cuts the DNA backbone on the 3' side beside the missing base (Figure 19.22, part 2). This leaves a primer end from which DNA polymerase I (in *E. coli*) initiates repair synthesis, using its 5'-3' exonuclease activity to remove a few nucleotides ahead of the missing

base, and then using its 5'-3' DNA polymerizing activity to fill in the gap (Figure 19.22, part 3). DNA ligase seals the remaining nick in the backbone (Figure 19.22, part 4). The activity of DNA polymerase I in this repair system, in which the enzyme simultaneously makes new DNA and removes nucleotides ahead of the growing DNA chain, is called *nick translation*.

REPAIR BY MISMATCH REPAIR. As should be apparent by now, the replication process does not exhibit perfect fidelity in inserting nucleotides in the new chain, so a number of error correction systems have evolved to reduce the frequency of errors (mutations) in the genome. While proofreading by DNA polymerase is an efficient way of correcting many errors soon after they are made, a significant number of errors still remain uncorrected after replication has been completed. Such errors usually involve mismatched base pairs which, in the next round of replication, become fixed as mutations.

Many mismatched base pairs left after DNA replication may be corrected by another system of repair called *mismatch repair*. In *E. coli* the products of three genes, *mutS*, *mutL*, and *mutH*, are involved in the initial stages of mismatch repair (Figure 19.23). The first step in the repair process is the binding of the *mutS*-encoded protein, MutS, to the mismatch, which may be a single base pair mismatch, or small base-pair additions or deletions. (The excision repair system cannot handle this kind of DNA change.) The next step depends upon the repair system recognizing which DNA strand is the newly synthesized strand and which is the template strand (with the correct nucleotide[s]). How does the system know which base in the mismatched pair is the correct one and which is the erroneous one? The answer is that the two strands are distinguished by methylation of a nucleotide in a specific DNA sequence. In *E. coli*, this sequence is GATC, in which the A is usually methylated by the action of *dam*-methylase, an enzyme encoded by the *dam* gene. The GATC sequence is palindromic: that is, the same sequence is read 5'-to-3' on both DNA strands (see Chapter 15). Thus, both A nucleotides in the DNA segment are methylated. However, after replication, the parental DNA strand has a methylated A nucleotide in the GATC sequence, while the A nucleotide in the GATC of the *newly replicated DNA strand* is not methylated until a short time after its synthesis. Therefore, for a short while after replication, the parental strand has a methylated GATC sequence, while the new strand has an unmethylated GATC sequence, called *hemimethylation*. In mismatch repair, the MutS protein bound to the mismatch forms a complex with the *mutL⁻* and *mutH⁻* encoded proteins, MutL and MutH, to bring the unmethylated GATC sequence located about 1 to 2 kb away from the replication site close to the mismatch. The MutH protein then nicks the unmethylated DNA strand at the GATC site, the mismatch is removed by an exonuclease, and the gap is repaired by DNA polymerase III and ligase.

Mismatch repair also takes place in eukaryotes, as evidenced by the presence of genes in yeast and humans that are homologous to *E. coli* mismatch repair genes according to DNA sequencing analysis. It is not clear how the new DNA strand is distinguished from the parental DNA strand in the repair process

because there is little or no methylation of yeast genomic DNA. In humans, four genes named *hMSH2*, *hMLH1*, *hPMS1*, and *hPMS2* have been identified: *hMSH2* is homologous to *E. coli mutS*, and the other three genes have homologies to *E. coli mutL*. The genes are known as *mutator genes* because loss of function of such a gene results in increased accumulation of mutations in the genome. Mutations in any one of the four human mismatch repair genes confers a phenotype of hereditary predisposition to a form of colon cancer called hereditary nonpolyposis colon cancer (HNPCC). HNPCC and the role of mutator genes in cancer are described in Chapter 18, pp. 614–615.

SOS RESPONSE. As we have learned, various mutagens can damage one or more bases in DNA so that specific base pairing cannot occur. Such damage is particularly serious for DNA replication, because the replication enzymes have difficulty at bases in the template DNA strand that cannot specify a complementary base pair on the new strand, so that gaps are often left after the replication fork has passed by. Uncorrected, DNA damage of this kind can therefore be lethal. In *E. coli*, DNA damage induces a complex system in what is called the *SOS response*. (The name SOS comes from the fact that the system is induced as an emergency response to mutational damage.) The *SOS response* operates to allow the cell to survive otherwise lethal events, although often at the expense of generating new mutations.

The SOS response has been studied best in *E. coli*, where the synthesis of many of the enzymes involved in the repair of DNA damage is regulated by the SOS system. Two genes are key to controlling the SOS system: *lexA* and *recA*. That is, *E. coli* cells with mutant *recA* and *lexA* genes have their SOS response permanently turned on. How does the SOS response work? In the uninduced state—when there is no DNA damage—the *lexA*-encoded protein, LexA, functions as a repressor to prevent transcription of about 17 genes whose protein products are involved in the repair of various kinds of DNA damage, including excision repair and the repair of gaps (Figure 19.24a). All of the genes involved have in common a 20-nucleotide regulatory sequence called the SOS box. The *recA*-encoded protein, RecA, is a regulator of the induction of the SOS response. (RecA also is an enzyme involved in recombination of homologous sequences in *E. coli*.) Specifically, when there has been sufficient DNA damage, somehow the RecA protein becomes activated, perhaps by binding to single-stranded DNA (Figure 19.24b). Activated RecA stimulates the LexA protein to cleave itself (recall that activated RecA can also stimulate the self-cleavage of the lamb-

~ **FIGURE 19.23**

Mechanism of mismatch correction repair. The mismatch correction enzyme recognizes which strand the base mismatch is on by reading the methylation state of a nearby GATC sequence. If the sequence is unmethylated, a segment of that DNA strand containing the mismatch is excised and new DNA is inserted.

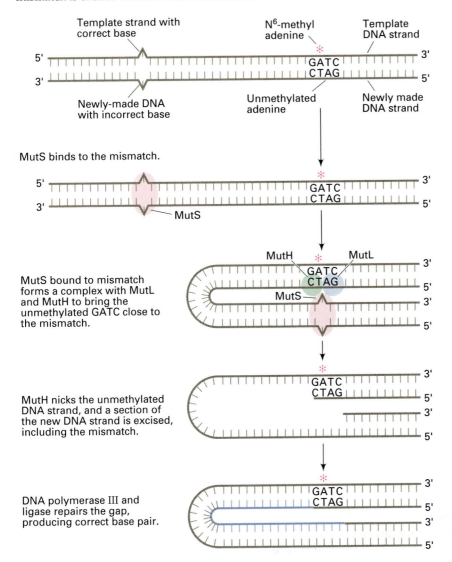

da repressor: see Chapter 16, p. 527), and this relieves the repression of the DNA repair genes. As a result, the DNA repair genes are transcribed, and DNA repair proceeds. After the DNA damage is dealt with, RecA again becomes inactivated, and newly synthesized LexA protein again acts to repress the DNA repair genes.

The SOS response is, itself, a mutagenic system. Originally it was thought that the SOS repair system inserted a base at a damaged DNA site, even though no information was available about which bases were originally present. This process, normally prevented in DNA synthesis, was named *error-prone repair* or *error-prone bypass synthesis*, and was known to involve other genes, including *umuC* and *umuD*. Thus, if the base was the wrong base, then a mutation would have been produced. However, a new model for the formation of mutations in UV-irradiated DNA was recently proposed based on studies of the kinetics of growth and mutagenesis of UV-irradiated DNA in phages that were undergoing SOS repair. The model is as follows: As stated previously, UV irradiation produces pyrimidine dimers in the DNA; that is, T^T, T^C, C^T, and C^C. During repair by the SOS system the T^T dimers are copied quickly and faithfully onto the new DNA segment to give

~ FIGURE 19.24

Outline of the SOS response. (a) The SOS system in the uninduced state; (b) The SOS system when induced by DNA damage. The details of the SOS response are given in the text.

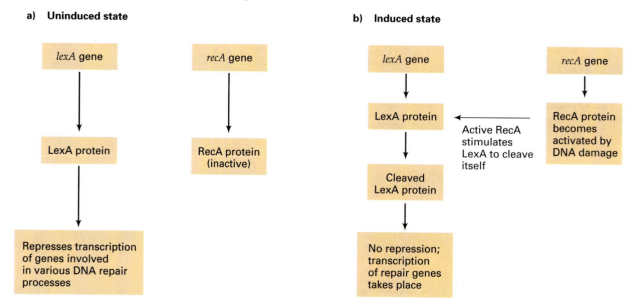

a) **Uninduced state**

b) **Induced state**

A A, the correct nucleotide sequence. However, for dimers involving C nucleotides, the SOS repair system becomes stalled at the dimer because of the mismatch. This delay is time enough for a C in a dimer to become deaminated to a U (uracil) (see Figure 19.11), which would still remain dimerized to its partner. The U then specifies an A in the new DNA synthesized, and in the next generation the original C-G site will be a T-A site; that is, a transition mutation will have been produced. In fact, CG-to-TA transitions are the common result of UV mutagenesis. In essence, this is an *error-free bypass synthesis* since the template is faithfully copied into the new DNA strand. The protein products of the *umuC* and *umuD* genes, UmuC and UmuD, are hypothesized to bind to DNA polymerase III and allow it to insert adenine nucleotides across from the nucleotides involved in the dimer.

𝒦EYNOTE

Mutations constitute damage to the DNA. Both prokaryotes and eukaryotes have a number of repair systems that deal with different kinds of DNA damage. All of the systems use enzymes to make the correction. Without such repair systems lesions would accumulate and be lethal to the cell

or organism. Not all lesions are repaired; hence, mutations do appear, but at relatively low frequencies. At high doses of mutagens, repair systems are unable to correct all of the damage, and cell death may result.

Human Genetic Diseases Resulting from DNA Replication and Repair Errors

Human cells may carry some naturally occurring genetic diseases in which the defect has been attributed to defects in DNA replication or repair. Some of these mutants are listed in Table 19.3. One well-known disease of this type is *xeroderma pigmentosum* (Figure 19.25), which is caused by homozygosity for a recessive mutation. People with this lethal affliction are photosensitive, and portions of their skin that have been exposed to light show intense pigmentation, freckling, and warty growths that may become malignant. The function affected in these people is excision repair of damage caused by ultraviolet light, X rays, or gamma radiation or by chemical treatment. Thus individuals with xeroderma pigmentosum are unable to repair radiation damage to DNA and eventually die, often as a result of malignancies that arise from the damage. Since the disease is inherited, the defect must result from a mutation in a gene coding for a protein involved in the repair of DNA damage.

~ TABLE 19.3

Some Examples of Naturally Occurring Human Cell Mutants That Are Defective in DNA Replication or Repair

Disease and Mode of Inheritance	Symptoms	Functions Affected	Chromosome Location[b]
Xeroderma pigmentosum (XP)—autosomal recessive	Sensitivity to sunlight with skin freckling and cancerous growths on skin; lethal at relatively early age as a result of the malignancies	Repair of DNA damaged by UV irradiation or chemicals	9q34.1
Ataxia-telangiectasia (AT)—autosomal recessive	Muscle coordination defect; propensity for respiratory infection; progressive spinal muscular atrophy in significant proportion of patients in second or third decade of life; marked hypersensitivity to ionizing radiation; cancer prone; high frequency of chromosome breaks leading to translocations and inversions	Repair replication of DNA	11q22.3
Fanconi's anemia (FA)—autosomal recessive	Aplastic anemia[a]; pigmentary changes in skin; malformations of heart, kidney, and limbs; leukemia is a fatal complication; genital abnormalities common in males; spontaneous chromosome breakage	Repair replication of DNA, UV-induced pyrimidine dimers and chemical adducts not excised from DNA; a repair exonuclease, DNA ligase, and transport of DNA repair enzymes have been hypothesized to be defective in FA patients	16q24.3
Bloom's syndrome (BS)—autosomal recessive	Pre- and postnatal growth deficiency; sun-sensitive skin disorder; predisposition to malignancies; chromosome instability; diabetes mellitus frequently develops in second or third decade of life	Elongation of DNA chains intermediate in replication—candidate gene is homologous to *E. coli* helicase Q	15q26.1
Cockayne syndrome (CS)—autosomal recessive	Dwarfism; precociously senile appearance; optic atrophy; deafness; sensitivity to sunlight; mental retardation; disproportionately long limbs; knee contractures produce bowlegged appearance; early death	Precise molecular defect is unknown, but may involve transcription-coupled repair	5
Hereditary nonpolyposis colon cancer (HNPCC)—autosomal dominant	Inherited predisposition to non-polyp-forming colorectal cancer	Defect in mismatch repair develops when the remaining wild-type allele of the inherited mutant allele becomes mutated; homozygosity for mutations in any one of four genes (*hMSH2*, *hMLH1*, *hPMS1* and *hPMS2*, known as mutator genes) has been shown to give rise to HNPCC	2p22–p21

[a]Individuals with aplastic anemia make no, or very few, red blood cells.
[b]If multiple complementation groups exist, the location of the most common defect is given.

~ FIGURE 19.25

An individual with xeroderma pigmentosum.

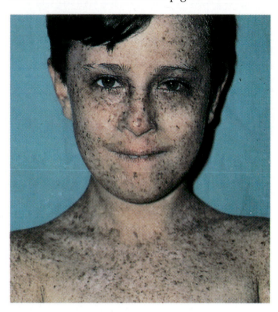

SCREENING PROCEDURES FOR THE ISOLATION OF MUTANTS

Geneticists have made great progress over the years in understanding how normal processes take place by studying mutants that have defects in those processes. Researchers have used mutagens to induce mutations at a greater rate than the rate at which spontaneous mutations occur. As we have mentioned, however, mutagens are not directed in their action toward particular genes. Instead, they change base pairs at random, without regard to the positions of the base pairs in the genetic material. Thus a mutagen's effect can only be studied if the mutation affects the function of a gene governing a phenotype researchers can observe. Mutations in regulatory or protein-coding genes usually result in a phenotypic change, unless the mutation is a neutral or silent one.

Once mutations have occurred, they must be detected if they are to be studied. Mutations of haploid organisms are readily detectable because there is only one copy of the genome. Mutations of diploid organisms, however, may be more difficult to detect. Let us consider mutations of *Drosophila*, an experimental organism in which genetic crosses can be made as desired. In such an organism, dominant mutations will be readily detectable. Recessive mutations are less readily detectable. Sex-linked recessive mutations can be detected because they are expressed in half of the sons of a mutated, heterozygous female. Autosomal recessive mutations can only be detected

if the mutation is homozygous. To simplify detection of different types of mutations, special tester strains of *Drosophila* have been constructed to which strains carrying potentially new mutations are crossed. Analysis of the progeny quickly establishes whether a new mutation has been made. The discussion of these tester strains is beyond the scope of this text.

The detection of mutations in humans is much more difficult than in *Drosophila* because geneticists cannot make controlled crosses. Dominant mutations can be readily detected, of course, but other types of mutations may be revealed only by pedigree analysis or by direct biochemical or molecular probing. Pedigree analysis can detect sex-linked recessive mutations by the mother-son transmission patterns described above. Individuals will be homozygous for a rare autosomal recessive mutations very infrequently, however, so few pedigrees will have individuals exhibiting such a trait.

Thus, the detection of mutations is not necessarily a simple matter. Fortunately, for some organisms of genetic interest, particularly microorganisms, screening procedures have been developed to help geneticists obtain particular mutants of interest from among a heterogeneous mixture in a mutagenized population. What follows are descriptions of a few screening procedures for the isolation of particular types of mutations.

Visible Mutations

Visible mutations affect the morphology or physical appearance of an organism (Figure 19.26). Examples of visible mutants are eye-color or wing-shape mutants of *Drosophila,* coat-color mutants of animals (e.g., albino organisms), colony-size mutants of yeast, and plaque morphology mutants of bacteriophages. Since visible mutations, by definition, are readily apparent, screening is done by inspection.

Nutritional Mutations

An *auxotrophic mutation* (also called a *nutritional* or *biochemical mutation*) affects an organism's ability to make a particular molecule essential for growth. Auxotrophic mutations are most readily detected in microorganisms such as *E. coli*, *yeast*, *Neurospora*, or some unicellular algae that grow on simple and defined growth media from which they synthesize the enzymes used to make all the molecules essential to their growth. A number of screening procedures can isolate auxotrophic mutants, and some of them will be described now.

In 1952 Esther and Joshua Lederberg developed a procedure to screen for auxotrophic mutants of any microorganism that grows in discrete colonies on solid

~ **FIGURE 19.26**

Examples of visible mutants and their wild-type counterparts. (a) White-eyed (mutant) (left) and red-eyed (wild-type) (right) *Drosophila*; (b) Albino (mutant) (left) and agouti (wild type) (right) mice; (c) Field of white-flowered, pink-flowered, and red-flowered snapdragons (flower color is determined by a codominant pair of alleles).

a)

b)

c)

medium. This procedure is called **replica plating**; Figure 19.27 diagrams how replica plating can be used to isolate arginine auxotrophs of a microorganism.

In the replica-plating technique, samples from a culture of a colony-forming organism or cell type that has or has not been mutagenized are plated onto a medium containing the nutrients appropriate for the mutants desired. For example, if we wish to isolate arginine auxotrophs, we would plate the culture on a master plate of minimal medium + arginine. On this medium, wild type and arginine auxotrophs will grow, but no other auxotrophs will grow. After incubation, colonies will be seen on the plate. Each colony consists of a clone of the original cell that landed in that position on the medium; that is, the cells in the colony are genetically identical copies of the cell that initiated the colony. The pattern of the colonies is transferred onto sterile velveteen cloth. Replicas of the original colony pattern on the cloth are then made by gently pressing new plates onto the velveteen. If the new plate contains minimal medium, the wild type colonies will be able to grow, but the arginine auxotrophs will not. So by comparing the patterns on the original minimal medium + arginine master plate with those on the minimal medium replica plate, researchers can readily identify the potential arginine auxotrophs. They can then be picked from the original master plate and cultured for further study.

A useful replica-plating technique to screen for a range of mutant types employs an antibiotic to kill many nonmutant cells but not many or any mutant cells. For example, the antibiotic nystatin (named after New York state) may be used to "enrich" for mutants of yeast; that is, to increase the proportion of mutant to nonmutant cells in the population. Nystatin is an effective killer of growing yeast cells.

Suppose we wish to enrich selectively for adenine auxotrophic (*ade*) mutants of yeast. We incubate the mutagenized population of cells in a culture medium that lacks adenine but contains nystatin. Under these conditions, wild-type cells grow and are killed by the nystatin, while the *ade* (adenine-requiring) cells that either do not grow or grow very slowly are preferentially spared. If the exposure to the antibiotic is carefully limited, the population of surviving cells has a much higher proportion of *ade* mutant cells than is found in the original, untreated population of cells. This method can be modified to screen for a number of other classes of mutants. An essentially identical method with penicillin as the antibiotic may be used with *E. coli*.

Conditional Mutations

The products of many genes—DNA polymerase and RNA polymerase, for example—are important for the

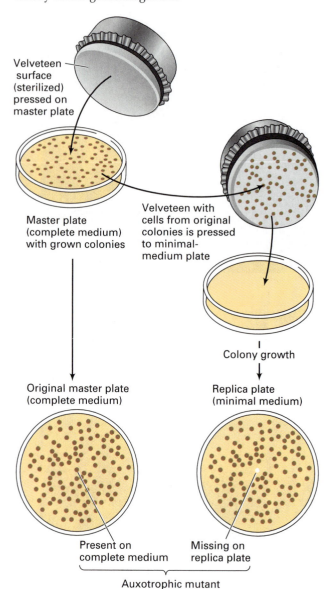

~ FIGURE 19.27

Replica-plating technique to screen for mutant strains of a colony-forming microorganism.

Velveteen surface (sterilized) pressed on master plate

Master plate (complete medium) with grown colonies

Velveteen with cells from original colonies is pressed to minimal-medium plate

Colony growth

Original master plate (complete medium)

Replica plate (minimal medium)

Present on complete medium

Missing on replica plate

Auxotrophic mutant

growth and division of cells, and most mutations in these genes result in a lethal phenotype. The structure and function of this class of genes may be studied by inducing *conditional mutations* in the genes. A common type of conditional mutation to study is a heat-sensitive mutation that is characterized by normal function at the normal growth temperature and no or severely impaired function at a higher temperature. In yeast, for instance, normal growth temperature is 23°C; heat-sensitive mutations typically are isolated at 36°C. Heat sensitivity typically results from a missense mutation causing a change in the amino acid sequence of a protein such that, at the higher temperature, the protein assumes a shape that is nonfunctional.

Essentially the same procedures are used to screen for heat-sensitive mutations of microorganisms as for auxotrophic mutations. For example, replica plating can be used to screen for temperature-sensitive mutants when the replica plate is incubated at a higher temperature than the master plate. That is, such mutants will grow on the master plate but not on the replica plate. Nystatin selection could also be used to enrich for heat-sensitive mutations of yeast that do not grow at 36°C.

Modern Molecular Screens

Increasingly, the isolation of new mutants relies not on screening a population of randomly mutagenized organisms, but on using molecular targeting approaches to introduce the desired mutation. Screening is then crucial to detect those organisms in which that mutation has been achieved. Here we consider one mutational approach used to obtain information about gene function in humans. Because we cannot perform mutational studies with humans, researchers often attempt to mimic human mutations in mice. Such mouse models of human mutations are very valuable for furthering our understanding of the gene involved and may, in the case of disease genes, move us toward diagnosis and a cure.

If we have a cloned human gene, we can easily clone the equivalent mouse gene because the two genes show a high degree of homology. The cloned mouse gene can then be mutagenized *in vitro*—for example, by deleting part of it—to render it completely nonfunctional. Using an appropriate vector, we can microinject the mutated gene into mouse embryos where, in some cells, it will replace one of the resident wild-type genes. The embryos are implanted into a female mouse and progeny mice are screened molecularly (through such techniques as PCR, restriction analysis, or DNA probing) to identify those carrying the mutation; such mice will be heterozygous +/− for the mutation. Mice carrying the mutation are called *knockout mice* because the normal gene has been knocked out by the engineered gene. By interbreeding +/− mice, knockout −/− progeny can be produced with both alleles mutated. Again, molecular screens are used to identify these progeny (assuming they are viable) and they can be used for study. Since the mutated genes used are completely nonfunctional, these mice will show a null phenotype for the gene—that is, a phenotype resulting from a complete absence of gene product.

Many types of knockout mice, both heterozygotes and homozygotes, have been produced to date. One example is knockout mice for the tumor suppressor gene *p53* (see Chapter 18, pp. 609–612). Recall that, in humans, *p53* mutations are responsible for a large proportion of human cancers. The *p53* (−/−) knockout mice show rapid development of spontaneous tumors and a number of other mutant phenotypes. Knockout mice have also been made for the cystic fibrosis (CF) gene encoding cystic fibrosis transmembrane conductance regulator (CFTR) (see Chapter 9, pp. 279–281, and Chapter 15, pp. 481–486). The symptoms are very similar to those of human CF patients, notably major mucous membrane defects in the pharynx, lungs, intestine, and colon, due to defects in chloride ion transport. CF is a lethal disease in humans, with the median age of death in the 30s. In mice, 40 percent of CF gene knockouts die within one week of birth from intestinal obstruction.

The knockout approach for generating specific gene mutations is applicable for any organism that can be transformed by cloned DNA and whose resident genes can be replaced by the introduced DNA. In all applications, molecular screening is done to confirm the presence of the mutations.

KEYNOTE

Geneticists have made great progress in understanding how cellular processes take place by studying mutants that have defects in those processes. With microorganisms, a number of screening procedures enrich for mutants of interest from a heterogeneous mixture of cells in a mutagenized population of cells. More modern approaches involve making specific gene mutations called knockouts, using cloned genes that have been mutated *in vitro*. Molecular screens are used to detect heterozygous and homozygous mutant knockout organisms.

SUMMARY

In this chapter we have seen that genetic damage can occur to the DNA spontaneously, through replication errors, or through treatment with radiation or chemical mutagens. If the genetic damage is not repaired mutations will result, and if there has been too much damage, cell death may result.

Mutations occur spontaneously at a low rate. The mutation rate can be increased through the use of mutagens like irradiation and certain chemicals. Typically, mutagens are used by researchers so that a mutant of interest is more likely to be found in a population of cells. Chemical mutagens work in a number of different ways, such as by acting as base analogs, by modifying bases, or by intercalating into the DNA. The latter results in frameshift mutations, while the others result in base-pair substitution mutations. Bruce Ames has devised a test—the Ames test—which has shown that a number of chemicals (e.g., an environmental or commercial chemical) have the potential to cause mutations in humans. A large number of potential human carcinogens have been found in this way.

Cells possess a number of repair mechanisms that function to correct at least some damage to DNA. These repair mechanisms include: (1) repair by DNA polymerase proofreading, in which a base-pair mismatch in DNA being synthesized is immediately repaired by 3'-to-5' excision; (2) photoreactivation of pyrimidine dimers induced by UV light; (3) excision repair, in which pyrimidine dimers and other DNA damage that distorts the DNA helix are excised and replaced with new DNA; (4) repair of damaged bases by glycosylases and AP endonuclease; and (5) repair by mismatch correction (in which

the methylation state of a DNA sequence signals which DNA strand is newly synthesized so that the mismatched base on that strand is corrected). Any DNA damage that is not repaired may result in a mutation and may have the potential to be lethal to the cell. The collective array of repair enzymes, then, serves to reduce mutation rates for spontaneous errors by several orders of magnitude. Such repair mechanisms cannot cope, however, with the extensive amount of DNA damage that arises from the use of chemical mutagens or UV irradiation, so that many mutations typically result.

Lastly, we considered some examples of how a mutagenized population of cells can be screened for particular mutants of interest, e.g., auxotrophs and conditional mutations. Over the decades, spontaneous and induced mutants have proved invaluable for the genetic analysis of biological function. Recently, techniques have been developed for mutating cloned genes *in vitro* and replacing resident normal genes to produce knockout mutant organisms. Molecular screens are used to detect heterozygous and homozygous knockouts.

ANALYTICAL APPROACHES FOR SOLVING GENETICS PROBLEMS

Q19.1 Five strains of *E. coli* containing mutations that affect the tryptophan synthetase A polypeptide have been isolated. Figure 19.A shows the changes produced in the protein itself in the indicated mutant strains. In addition, *A23* can be further mutated to insert Ile, Thr, Ser, or the wild-type Gly into position 210.

In the following questions, assume that only a single base change can occur at each step.

a. Using the genetic code (see Figure 14.9, p. 427), explain how the two mutations *A23* and *A46* can result in two different amino acids being inserted at position 210. Give the nucleotide sequence of the wild-type gene at that position and the two mutants.

b. Can mutants *A23* and *A46* recombine? Why or why not? If recombination can occur, what would be the result?

c. From what you can infer of the nucleotide sequence in the wild-type gene, indicate, for the codons specifying amino acids 48, 210, 233, and 234, whether or not a nonsense mutant could be generated by a single nucleotide substitution in the gene.

A19.1a. There are no simple ways to answer questions like this one. The best approach is to scrutinize the genetic code dictionary and use a pencil and paper to try to define the codon changes that are compatible with all the data. The number of amino acid changes

FIGURE 19.A

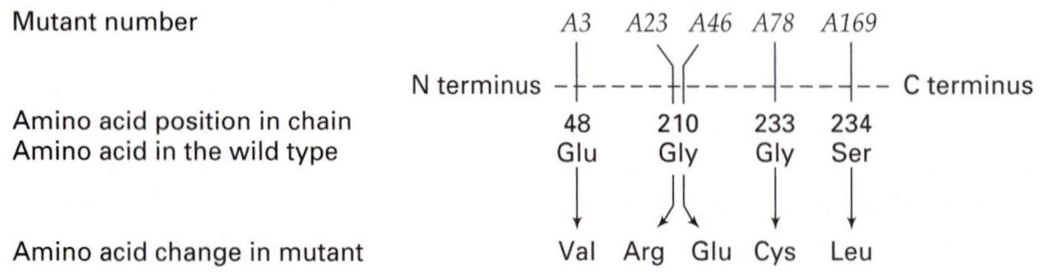

Mutant number	A3	A23	A46	A78	A169

N terminus –|– – – – –||– – – – –|– – –|– – C terminus

Amino acid position in chain	48	210	233	234
Amino acid in the wild type	Glu	Gly	Gly	Ser
Amino acid change in mutant	Val	Arg Glu	Cys	Leu

in position 210 of the polypeptide is helpful in this case. The wild-type amino acid is Gly and the codons for Gly are GGU, GGC, GGA, and GGG. The *A23* mutant has Arg at position 210, and the arginine codons are AGA, AGG, CGU, CGC, CGA, and CGG. Clearly, any Arg codon could be generated by a single base change. Thus, we have to look at the amino acids at 210 generated by further mutations of *A23*. In the case of Ile, the codons are AUU, AUC, and AUA. The *only* way to get from Gly to Arg in one base change and then to Ile in a subsequent single base change is from GGA (Gly) → AGA (Arg) → AUA (Ile). Is this compatible with the other mutational changes from *A23*? There are four possible Thr codons, ACU, ACC, ACA, and ACG, so a mutation from AGA (Arg) to ACA (Thr) would fit. There are six possible Ser codons, UCU, UCC, UCA, UCG, AGU, and AGC, so a mutation from AGA to either AGU or AGC would fit.

Considering the *A46* mutant, the possible codons for Glu are GAA and GAG. Given that the wild-type codon is GGA (Glu), the only possible single base change to give Glu is if the Glu codon in the mutant is GAA. So, the answer to the question is that the wild-type sequence at position 210 is GGA, the sequence in the *A23* mutant is AGA, and the sequence in the *A46* mutant is GAA. In other words, the *A23* and *A46* mutations are in different bases of the codon.

b. The answer to this question follows from the answer deduced in part a. Mutants *A23* and *A46* can recombine since the mutations in the two mutant strains are in different base pairs. The results of a single recombination event (at the DNA level) between the first and second base of the codon in AGA × GAA are a wild-type GGA codon (Gly) and a double mutant AAA codon (Lys). Recombination can also occur between the second and third base of the codon, but the products are AGA and GAA, that is, identical to the parents.

c. Amino acid 48 had a Glu-to-Val change. This change must have involved GAA to GUA or GAG to GUG. In either case the Glu codon can mutate with a single base-pair change to a nonsense codon, that is, UAA or UAG, respectively.

Amino acid 210 in the wild type has a GGA codon, as we have already discussed. This gene could mutate to the UGA nonsense codon with a single base-pair change.

Amino acid 233 had a Gly-to-Cys change. This change must have involved either GGU to UGU or GGC to UGC. In either case the Gly codon cannot mutate to a nonsense codon with one base-pair change.

Amino acid 234 had a Ser-to-Leu change. This change was either UCA to UUA or UCG to UUG. If the Ser codon was UCA, it could be changed to UGA

in one step. But if the Ser codon was UCG, it cannot change to a nonsense codon in one step.

Q19.2 The chemically induced mutations, *a*, *b*, and *c* show specific reversion patterns when subjected to treatment by the following mutagens: 2-aminopurine (AP), 5-bromouracil (BU), proflavin (PRO), and hydroxylamine (HA). The reversion patterns are shown in the following table.

MUTATION	MUTAGENS TESTED IN REVERSION STUDIES			
	AP	BU	PRO	HA
a	−	−	+	−
b	+	+	−	+
c	+	+	−	−

(Note: + indicates many reversions to wild type were found; − indicates no reversions, or very few, to wild type were found.)

For each original mutation (i.e., *a*⁺ to *a*, *b*⁺ to *b*, etc.), indicate the probable base-pair change (i.e., AT to GC, deletion of GC, etc.) and the mutagen that was most probably used to induce the original change.

A19.2 This question tests knowledge of the base-pair changes that can be induced by the various mutagens used.

Mutagen AP is a base-analog mutagen that induces mainly AT-to-GC changes and can cause GC-to-AT changes also. Both of these changes are transitions. Thus AP-induced mutations can be reverted by AP.

Base-analog mutagen BU induces mainly GC-to-AT changes and can cause AT-to-GC changes, so BU-induced mutations can be reverted by BU. Both of the BU-induced changes are transitions.

Proflavin causes single base-pair deletions or additions with no specificity. Proflavin-induced changes can be reverted by a second treatment with proflavin, as we discovered in the discussion of the experiments used to prove the three-letter basis of the genetic code (Chapter 14, pp. 423–425).

Mutagen HA modifies cytosine, causing one-way transitions from GC to AT, and so HA-induced mutations cannot be reverted by HA.

With the above mutagen specificities in mind, we can answer the questions for each mutation in turn:

Mutation *a*⁺ to *a*: The *a* mutation was reverted only by proflavin, indicating that it was a deletion or an addition (i.e., a frameshift mutation). Therefore the original mutation was induced by an intercalating agent such as proflavin, since that is the only class of mutagen that can cause an addition or a deletion.

Mutation *b*⁺ to *b*: The *b* mutation was reverted by AP, BU, or HA. A key here is that HA only causes GC-to-AT changes. Therefore *b* must be GC, and the original *b*⁺ must have been an AT. Thus the mutational change of *b*⁺ to *b* must have been caused by treatment with AP or BU, since these are the two mutagens in the list that are capable of inducing that change.

Mutation c^+ to c: The c mutation was reverted only by AP and BU. Since it could not be reverted by HA, c must be an AT and c^+ a GC. The mutational change from c^+ to c therefore involved a GC-to-AT transition and could have resulted from treatment with AP, BU, or HA.

QUESTIONS AND PROBLEMS

*19.1   Mutations are (choose the correct answer):
a. Caused by genetic recombination.
b. Heritable changes in genetic information.
c. Caused by faulty transcription of the genetic code.
d. Usually but not always beneficial to the development of the individuals in which they occur.

19.2   Answer true or false: Mutations occur more frequently if there is a need for them.

*19.3   Which of the following is *not* a class of mutation?
a. Frameshift
b. Missense
c. Transition
d. Transversion
e. None of the above (i.e., all are classes of mutation)

*19.4   Ultraviolet light usually causes mutations by a mechanism involving (choose the correct answer):
a. One-strand breakage in DNA
b. Light-induced change of thymine to alkylated guanine
c. Induction of thymine dimers and their persistence or imperfect repair
d. Inversion of DNA segments
e. Deletion of DNA segments
f. All of the above

19.5   For the middle region of a particular polypeptide chain, the normal amino acid sequence and the amino acid sequence of several mutants were determined, as shown below (. . . indicates additional, unspecified amino acids). For each mutant, say what DNA level change has occurred, whether the change is a base-pair substitution mutation (transversion or transition, missense or nonsense) or a frameshift mutation, and in which codon the mutation occurred. (Refer to the codon dictionary in Figure 14.9, p. 427.)

					CODON				
	1	2	3	4	5	6	7	8	9
a. Normal: . . .	Phe	Leu	Pro	Thr	Val	Thr	Thr	Arg	Trp
b. Mutant 1: . . .	Phe	Leu	His	His	Gly	Asp	Asp	Thr	Val
c. Mutant 2: . . .	Phe	Leu	Pro	Thr	Met	Thr	Thr	Arg	Trp
d. Mutant 3: . . .	Phe	Leu	Pro	Thr	Val	Thr	Thr	Arg	
e. Mutant 4: . . .	Phe	Pro	Pro	Arg					
f. Mutant 5: . . .	Phe	Leu	Pro	Ser	Val	Thr	Thr	Arg	Trp

*19.6   In mutant strain X of *E. coli*, a leucine tRNA which recognizes the codon 5'-CUG-3' in normal cells has been altered so that it now recognizes the codon 5'-GUG-3'. A missense mutation, which affects amino acid 10 of a particular protein, is suppressed in mutant X cells.
a. What are the anticodons of the two Leu tRNAs, and what mutational event has occurred in mutant X cells?
b. What amino acid would normally be present at position 10 of the protein (without the missense mutation)?
c. What amino acid would be put in at position 10 if the missense mutation is not suppressed (i.e., in normal cells)?
d. What amino acid is inserted at position 10 if the missense mutation is suppressed (i.e., in mutant X cells)?

19.7   A researcher using a model eukaryotic experimental system has identified a temperature-sensitive mutation, $rpIIA^{ts}$, in a gene that encodes a protein subunit of RNA polymerase II. This mutation results from a missense mutation. It has a recessive lethal phenotype at the higher, restrictive temperature, but allows for growth at the lower, permissive temperature. To identify genes whose products interact with the subunit of RNA polymerase II, the researcher designs a screen to isolate mutations that will act as dominant suppressors of the temperature-sensitive recessive lethality.
a. Explain how a new mutation in an interacting protein could suppress the temperature-sensitive lethality of the original mutation.
b. In addition to mutations in interacting proteins, what other type of suppressor mutations might be found?
c. Outline how the researcher might select for the new suppressor mutations.
d. Do you expect the frequency of suppressor mutations to be similar to, much greater than, or much less than the frequency of new mutations at a typical eukaryotic gene?
e. How might this approach be used generally to identify genes whose products interact in controlling transcription?

19.8   In any kind of chemotherapy, the object is to find a means to kill the invading pathogen or cancer cell without killing the cells of the host. To do this successfully, one must find and exploit biological differences between target organisms and host cells. Explain why AZT is effective in terminating viral cDNA synthesis but does not interfere with host DNA replication.

*19.9   The mutant *lacZ-1* was induced by treating *E. coli* cells with acridine, while *lacZ-2* was induced with 5BU. What kinds of mutants are these likely to be? Explain. How could you confirm your predictions by studying the structure of the β-galactosidase in these cells?

*19.10a.The sequence of nucleotides in an mRNA is:

5'-AUGACCCAUUGGUCUCGUUAG-3'

Assuming that ribosomes could translate this mRNA, how many amino acids long would you expect the polypeptide chain made with this messenger to be?

b. Hydroxylamine is a mutagen that results in the replacement of an AT base pair for a GC base pair in the DNA; that is, it induces a transition mutation. When applied to the organism that made the mRNA molecule shown in part a, a strain was isolated in which a mutation occurred at the 11th position of the DNA that coded for the mRNA. How many amino acids long would you expect the polypeptide made by this mutant to be? Why?

19.11 In a series of 94,075 babies born in a particular hospital in Copenhagen, 10 were achondroplastic dwarfs (this is an autosomal dominant condition). Two of these 10 had an achondroplastic parent. The other 8 achondroplastic babies each had two normal parents. What is the apparent mutation rate at the achondroplasia locus?

*19.12 Three of the codons in the genetic code are chain-terminating codons for which no naturally occurring tRNAs exist. Just like any other codons in the DNA, though, these codons can change as a result of base-pair changes in the DNA. Confining yourself to single base-pair changes at a time, determine which amino acids could be inserted in a polypeptide by mutation of these chain-terminating codons: (a) UAG; (b) UAA; (c) UGA. (The genetic code is listed in Figure 14.9, p. 427.)

19.13 The amino acid substitutions in the following figure occur in the α and β chains of human hemoglobin. Those amino acids connected by lines are related by single nucleotide changes. Propose the most likely codon or codons for each of the numbered amino acids. (Refer to the genetic code listed in Figure 14.9, p. 427.)

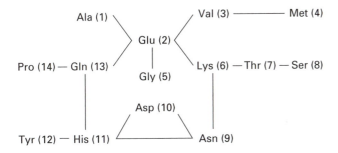

*19.14 Yanofsky studied the tryptophan synthetase of *E. coli* in an attempt to identify the base sequence specifying this protein. The wild type gave a protein with a glycine in position 38. Yanofsky isolated two *trp* mutants, *A23* and *A46*. Mutant *A23* had Arg instead of Gly at position 38, and mutant *A46* had Glu at position 38. Mutant *A23* was plated on minimal medium, and four spontaneous revertants to prototrophy were obtained. The tryptophan synthetase from each of four

revertants was isolated, and the amino acids at position 38 were identified. Revertant 1 had Ile, revertant 2 had Thr, revertant 3 had Ser, and revertant 4 had Gly. In a similar fashion, three revertants from *A46* were recovered, and the tryptophan synthetase from each was isolated and studied. At position 38 revertant 1 had Gly, revertant 2 had Ala, and revertant 3 had Val. A summary of these data is given in the figure. Using the genetic code in Figure 14.9 (p. 427), deduce the codons for the wild type, for the mutants *A23* and *A46*, and for the revertants, and place each designation in the space provided in the following figure.

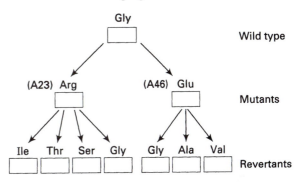

19.15 Consider an enzyme chewase from a theoretical microorganism. In the wild-type cell the chewase has the following sequence of amino acids at positions 39 to 47 (reading from the amino end) in the polypeptide chain:

-Met-Phe-Ala-Asn-His-Lys-Ser-Val-Gly-
  39 40 41 42 43 44 45 46 47

A mutant of the organism was obtained; it lacks chewase activity. The mutant was induced by a mutagen known to cause single base-pair insertions or deletions. Instead of making the complete chewase chain, the mutant makes a short polypeptide chain only 45 amino acids long. The first 38 amino acids are in the same sequence as the first 38 of the normal chewase, but the last 7 amino acids are as follows:

-Met-Leu-Leu-Thr-Ile-Arg-Val
  39 40 41 42 43 44 45

A partial revertant of the mutant was induced by treating it with the same mutagen. The revertant makes a partly active chewase, which differs from the wild-type enzyme only in the following region:

-Met-Leu-Leu-Thr-Ile-Arg-Gly-Val-Gly-
  39 40 41 42 43 44 45 46 47

Using the genetic code given in Figure 14.9, deduce the nucleotide sequences for the mRNA molecules that specify this region of the protein in each of the three strains.

19.16 DNA polymerases from different organisms differ in their fidelity of nucleotide insertion; however, even the "best" DNA polymerases make mistakes. These are usually mismatches. If such mismatches are not corrected, they can become fixed as mutations after the next round of replication.

a. How does DNA polymerase attempt to correct mismatches during DNA replication?

b. What mechanism is used to repair such mismatches if they escape detection by DNA polymerase?

c. How is the "mismatched" base in the newly synthesized strand distinguished from the "correct base" in the template strand?

19.17 Two mechanisms in *E. coli* were described for the repair of DNA damage (thymine dimer formation) after exposure to ultraviolet light: photoreactivation and excision (dark) repair. Compare and contrast these mechanisms, indicating how each achieves repair.

19.18 What chemical reaction do glycosylases carry out and what role do they have in the removal of damaged DNA bases?

19.19 DNA damage by mutagens has very serious consequences for DNA replication. Without specific base pairing, the replication enzymes cannot specify a complementary strand, and gaps are left after the passing of a replication fork.

a. What response has *E. coli* developed to large amounts of DNA damage by mutagens? How is this response coordinately controlled?

b. Why is this response itself a mutagenic system?

c. What effects would loss-of-function mutations in *recA* or *lexA* have on this response?

***19.20** After a culture of *E. coli* cells was treated with the chemical 5-bromouracil, it was noted that the frequency of mutants was much higher than normal. Mutant colonies were then isolated, grown, and treated with nitrous acid; some of the mutant strains reverted to wild type.

a. In terms of the Watson-Crick model, diagram a series of steps by which 5BU may have produced the mutants.

b. Assuming the revertants were not caused by suppressor mutations, indicate the steps by which nitrous acid may have produced the back mutations.

19.21 A single, very hypothetical strand of DNA is composed of this base sequence:

5'-T-HX-U-A-G-BU-enol-2AP-C-BU-X-2AP-imino-3'

In the sequence above, A indicates adenine, T indicates thymine, G indicates guanine, C denotes cytosine, U denotes uracil, BU is 5-bromouracil, 2AP is 2-aminopurine, BU-enol is a tautomer of 5BU, 2AP-imino is a rare tautomer of 2AP, HX is hypoxanthine, and X is xanthine, 5' and 3' are the numbers of the free, OH-containing carbons on the deoxyribose part of the terminal nucleotides.

a. Opposite the bases of the hypothetical strand, and using the shorthand of the base sequence, indicate the sequence of bases on a complementary strand of DNA.

b. Indicate the direction of replication of the new strand by drawing an arrow next to the new strand of DNA from part a.

c. When postmeiotic germ cells of a higher organism are exposed to a chemical mutagen before fertilization, the resulting offspring expressing an induced mutation are almost always mosaics for wild-type and mutant tissue. Give at least one reason that in the progenies of treated individuals these mosaics are found and not the so-called complete or whole-body mutants.

The following information applies to Problems 19.22 through 19.26. A solution of single-stranded DNA is used as the template in a series of reaction mixtures. It has the following base sequence:

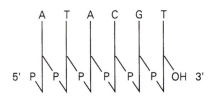

where A = adenine, G = guanine, C = cytosine, T = thymine, H = hypoxanthine, and HNO_2 = nitrous acid. For Problems 19.22 through 19.26, use the shorthand system in the figure and draw the products expected from the reaction mixtures. Assume that a primer is available in each case.

19.22 The DNA template + DNA polymerase + dATP + dGTP + dCTP + dTTP + Mg^{2+}.

***19.23** The DNA template + DNA polymerase + dATP + dGMP + dCTP + dTTP + Mg^{2+}.

19.24 The DNA template + DNA polymerase + dATP + dHTP + dGMP + dTTP + Mg^{2+}.

***19.25** The DNA template is pretreated with HNO_2 + DNA polymerase + dATP + dGTP + dCTP + dTTP + Mg^{2+}.

19.26 The DNA template + DNA polymerase + dATP + dGMP + dHTP + dCTP + dTTP + Mg^{2+}.

19.27 A strong experimental approach to determining the mode of action of mutagens is to examine the revertibility of the products of one mutagen by other mutagens. The following table represents collected data on revertibility of various mutagens on *rII* mutations in phage T2; + indicates majority of mutants reverted, − indicates virtually no reversion; BU = 5-bromouracil, AP = 2-aminopurine, NA = nitrous acid, and HA = hydroxylamine. Fill in the empty spaces.

MUTATION INDUCED BY	PROPORTION OF MUTATIONS REVERTED BY				BASE-PAIR SUBSTITUTION INFERRED
	BU	AP	NA	HA	
BU	+	–	–	–	_____
AP	–	–	+	–	_____
NA	+	+	–	+	_____
HA	–	–	+	–	$GC \rightarrow AT$

19.28a. Nitrous acid deaminates adenine to form hypoxanthine, which forms two hydrogen bonds with cytosine during DNA replication. Following treatment with nitrous acid, a mutant is recovered which contains a protein with an amino acid substitution: the amino acid valine (Val) is located in a position occupied by methionine (Met) in nonmutant organisms. What is the simplest explanation for this observation?

b. Hydroxylamine adds a hydroxyl (OH) group to cytosine, causing it to pair with adenine. Could mutant organisms like those in part (a) be back mutated (returned to normal) using hydroxylamine? Explain.

*19.29 A protein contains the amino acid proline (Pro) at one site. Treatment with nitrous acid, which deaminates C to make it U, produces two different mutants. One mutant has a substitution of serine (Ser) and the other has a substitution of leucine (Leu) at the site.

Treatment of the two mutants produces new mutant strains, each with phenylalanine (Phe) at the site. Treatment of these new Phe-carrying mutants produces no change. The results are summarized in the following figure:

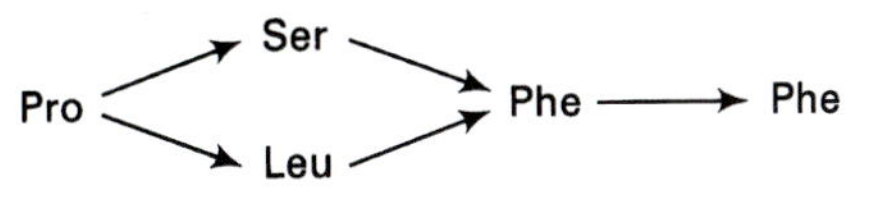

Using the appropriate codons, show how it is possible for nitrous acid to produce these changes, and why further treatment has no influence. (Assume only single nucleotide changes occur at each step.)

*19.30 Three *ara* mutants of *E. coli* were induced by mutagen X. The ability of other mutagens to cause the reverse change (*ara* to *ara*⁺) was tested, with the results shown in the following table.

Assume all *ara*⁺ cells are true revertants. What base changes were probably involved in forming the three original mutations? What kind(s) of mutations are caused by mutagen X?

19.31 As genes have been cloned for a number of human diseases caused by defects in DNA repair and replication, striking evolutionary parallels have been found between human and bacterial DNA repair systems. Discuss the features of DNA repair systems that appear to be shared in these two organisms.

*19.32 In the past few years, a considerable number of knockout mice that lack specific gene functions have been bred. In addition to the knockouts for *p53* and the *CFTR* genes described in this chapter, knockouts for many different *Hox* genes (see Chapter 17), lysosomal hexosaminidase (the enzyme missing in Tay-Sachs disease) and many other enzymes have been made.

a. Why is the ability to make knockout mice important for the treatment and cure of human disease?

b. Consider a situation in which a dominant mutation causes a human disease such as Huntington disease. A hypothetical researcher clones the disease gene and discovers that a novel, abnormal protein is responsible for the disease phenotype. Would you expect the phenotype of a knockout mouse to be similar to the disease phenotype? Why or why not?

Frequency of *ara*⁺ Cells Among Total Cells After Treatment

MUTANT	MUTAGEN				
	NONE	BU	AP	HA	FRAMESHIFT
ara-1	1.5×10^{-8}	5×10^{-5}	1.3×10^{-4}	1.3×10^{-8}	1.6×10^{-8}
ara-2	2×10^{-7}	2×10^{-4}	6×10^{-5}	3×10^{-5}	1.6×10^{-7}
ara-3	6×10^{-7}	10^{-5}	9×10^{-6}	5×10^{-6}	6.5×10^{-7}

20 TRANSPOSABLE ELEMENTS

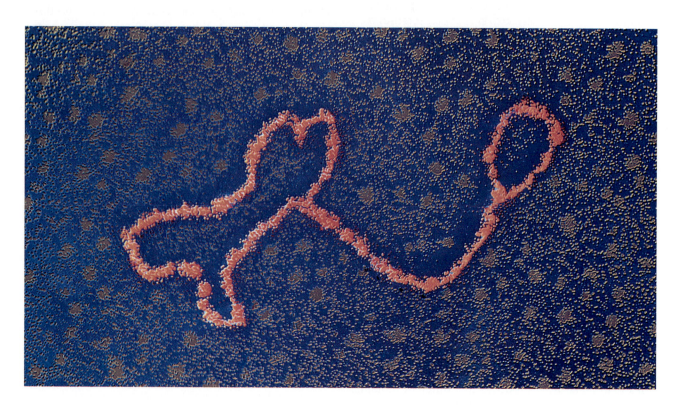

TRANSPOSABLE ELEMENTS IN PROKARYOTES
 Insertion Sequences
 Transposons
 IS Elements and Transposons in Plasmids
 Bacteriophage Mu
TRANSPOSABLE ELEMENTS IN EUKARYOTES
 Transposons in Plants
 Ty Elements in Yeast
 Drosophila Transposons
 Human Retrotransposons

PRINCIPAL POINTS

~ Transposable elements are DNA segments that can insert themselves at one or more sites in a genome. The presence of transposable elements in a cell is usually detected by the changes they bring about in the expression and activities of the genes at or near the chromosomal sites into which they integrate.

~ In bacteria three major types of transposable elements are insertion sequence (IS) elements, transposons (Tn), and some bacteriophages, such as Mu. Insertion sequence elements and transposons have repeated sequences at their ends and encode proteins that are responsible for their transposition. Transposons also have genes that encode other functions, such as drug resistance. When ISs and Tns integrate into the genome, a short host sequence at the target site is duplicated, giving rise to directly repeated sequences flanking the integrated element.

~ Transposable elements in eukaryotes resemble bacterial transposons in general structure and transposition

properties. Depending on the transposon, it may transpose to new sites while leaving a copy behind in the original site or it may excise itself from the chromosome. Transposons integrate at a target site by a precise mechanism so that the integrated elements are flanked at the insertion site by a short duplication of target site DNA. Many plant transposons occur in families, the autonomous elements of which are able to direct their own transposition, and the nonautonomous elements of which are able to transpose only when activated by an autonomous element in the same genome. While most transposons move using a DNA-to-DNA mechanism, some eukaryotic transposons move via an RNA intermediate (using a transposon-encoded reverse transcriptase). Such transposons resemble retroviruses in genome organization and other properties.

The classic picture of genes is one in which the genes are at fixed loci on a chromosome. In Chapter 7, we learned that segments of chromosomes can change their orientation or position in the genome as a result of chromosome mutations. These changes inform us about ways in which genomes might have evolved. For example, a chromosome segment containing a gene might become duplicated. Not only does this increase genome size, but the two copies may, in turn, mutate independently to produce two distinct genes with related function. Moreover, through translocation, genes can be moved to different places in the genome. Genomic rearrangements can occur in other ways. That is, certain genetic elements in the chromosomes of both prokaryotes and eukaryotes have the capacity to move from one location to another in the genome. These mobile genetic elements have been given a number of names in the literature, including controlling elements, jumping genes, insertion sequences, and transposons. We shall use a generic term that has become fairly widely accepted, **transposable elements**, since this term reflects the *transposition* ("change in position") events associated with these elements. Undoubtedly, transposable elements have contributed to the evolution of the genomes of both prokaryotes and eukaryotes through the chromosome rearrangements they cause.

In structure and function transposable elements are similar in both prokaryotes and eukaryotes as well

as their viruses. In prokaryotes, transposable elements can move to new positions on the same chromosome (since there is only one) or onto plasmids or phage chromosomes, while in eukaryotes, transposable elements have the opportunity to move to new positions within the same chromosome or to a different chromosome. In both prokaryotes and eukaryotes, transposable elements are identified by the changes they cause; for example, they can produce mutations by inserting into genes, they can affect gene expression by inserting into gene regulatory sequences, and they can produce various kinds of chromosome mutations. In fact, the effects of transposable elements have been established through genetic, cytological, molecular, and recombinant DNA procedures. But what is the biological function of transposable elements? No one knows for sure, but they have probably been important in genome evolution. It has been argued that a sizable proportion of the genomes of some eukaryotic organisms, including humans, is made up of repetitive DNA that is probably derived from various transposable elements.

TRANSPOSABLE ELEMENTS IN PROKARYOTES

There are three types of transposable elements in prokaryotes: insertion sequence (IS) elements, transposons (Tn), and certain bacteriophages.

Insertion Sequences

An **insertion sequence (IS)**, or **IS element**, is the simplest transposable element found in prokaryotes. An IS element contains only genes required for mobilizing the element and inserting the element into a chromosome at a new location. IS elements are normal constituents of bacterial chromosomes and plasmids.

DISCOVERY OF IS ELEMENTS. The IS elements were first identified in *E. coli* as a result of their effects on the expression of a set of three genes whose products are needed to metabolize the sugar galactose as a carbon source. A certain set of mutations affecting the expression of these genes did not have properties typical of the classes of gene mutations discussed in Chapter 19. Careful analysis of this unusual set of mutations revealed that the mutant phenotypes resulted from the insertion of an approximately 800-base-pair (bp) DNA segment into a gene. This particular DNA segment is now called insertion sequence 1, or IS*1* (Figure 20.1).

PROPERTIES OF IS ELEMENTS. IS*1* is one of a family of genetic elements capable of moving around the genome. It integrates into the chromosome at locations with which it has no homology, thereby distinguishing it from recombination. This event is an example of a *transposition event*.

There are a number of IS elements that have been identified in *E. coli*, including IS*1*, IS*2*, and IS*10R*, each present in 0 to 30 copies per genome, and each with a characteristic length and unique nucleotide sequence (Table 20.1). IS*1* (see Figure 20.1), for example, is 768 bp long, and is present in 4 to 19 copies on the *E. coli* chromosome. IS*2* is present in 0 to 12 copies on the *E. coli* chromosome and in one copy on the *F* plasmid, and IS*10* is found in a class of plasmids called *R* plasmids that can replicate in *E. coli*. Among prokaryotes as a whole, the IS elements range in size from 768 bp to more than 5,000 bp and are normal cell constituents; that is, they are found in most cells.

~ FIGURE 20.1

The insertion sequence (IS) transposable element, IS*1*. The IS element has inverted repeat (IR) sequences at the ends.

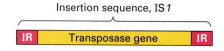

Insertion sequence, IS*1*

Altogether, IS elements constitute approximately 0.3 percent of the cell's genome.

All IS elements that have been sequenced end with perfect or nearly perfect inverted terminal repeats (IRs) of between 9 and 41 bp. This means that essentially the same sequence is found at each end of an IS but in opposite orientations. The inverted repeats of IS*1*, for example, consist of 23 bp of not quite identical sequence (Figure 20.2).

IS TRANSPOSITION. Since IS elements integrate at random points along the chromosome, they often cause mutations by disrupting the coding sequence of a gene or by disrupting a gene's regulatory region. Promoters within the IS elements themselves may also have effects by altering the expression of nearby genes. Additionally, the presence of an IS element in the chromosome can cause chromosome mutations such as deletions and inversions in the adjacent DNA. Finally, deletion and insertion events can also occur as a result of crossing-over between duplicated IS elements in the genome (as in Figure 20.10, p. 663).

When transposition of an IS element takes place, a copy of the IS element inserts into a new chromosome location while the original IS element remains in place. That is, transposition requires the precise replication of the original IS element, using the replication enzymes of the host cell. The actual transposition also requires an enzyme encoded by the IS element called **transposase**. The IR sequences are essential for the transposition process; that is, those sequences are recognized by transposase to initiate transposition. The frequency of transposition is char-

~ TABLE 20.1

Properties of Some IS Elements of *E. coli*

INSERTION SEQUENCE	LENGTH (bp)	INVERTED TERMINAL REPEATING SEQUENCE (IR) (bp)	TARGET SITE DUPLICATIONS (bp)
IS*1*	768	23	9
IS*2*	1,327	32/41	5
IS*10*	1,329	22	9

~ FIGURE 20.2

23-bp inverted terminal repeats (IR) of IS1.

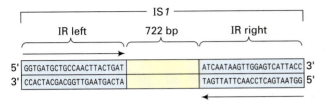

acteristic of each IS element, ranging between 10^{-5} and 10^{-7} per generation.

IS elements insert into the chromosome at sites with which they have no sequence homology. Genetic recombination between nonhomologous sequences is called *illegitimate recombination.* The sites into which IS elements insert are called *target sites.* The process of IS insertion into a chromosome is shown in Figure 20.3. First, a staggered cut is made in the target site and the IS element is then inserted, becoming joined to the

single-stranded ends. The gaps are filled in by DNA polymerase and DNA ligase, producing an integrated IS element with two direct repeats of the target site sequence flanking the IS element. "Direct" in this case means that the two sequences are repeated in the same orientation (see Figure 20.3). The direct repeats are called *target site duplications.* The sizes of target site duplications vary with the IS element, but tend to be small (4 to 13 bp; e.g., see Table 20.1).

Integration of an IS element is not completely random, but there are no general rules. Some IS elements show preference for certain regions (i.e., integration hotspots), while others integrate only at particular sequences.

All copies of a given IS element have the same sequence, including that of the inverted terminal repeats. Mutations that affect the inverted terminal repeat sequences of IS elements affect transposition, indicating that the inverted terminal repeat sequences are the key sequences recognized by transposase during a transposition event.

~ FIGURE 20.3

Schematic of the integration of an IS element into chromosomal DNA. As a result of the integration event, the target site becomes duplicated to produce direct target repeats. Thus, the integrated IS element is characterized by its inverted repeat (IR) sequences flanked by direct target site duplications. Integration involves making staggered cuts in the host target site. After insertion of the IS, the gaps that result are filled in with DNA polymerase and DNA ligase. (Note: The base sequences given for the IR are for illustration only and are not the actual sequences found, either in length or sequences.)

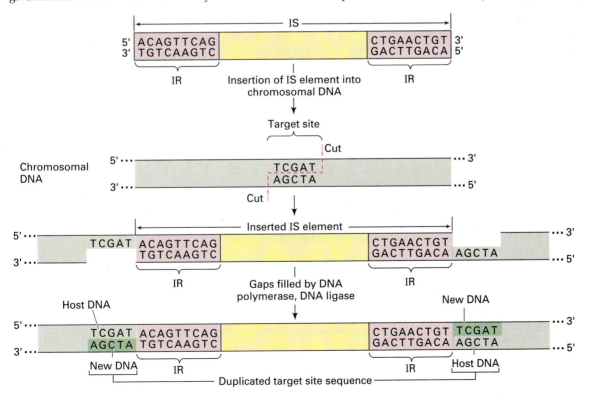

Transposons

A **transposon (Tn)** is more complex than an IS element. A transposon is a mobile DNA segment that, like an IS element, contains genes for the insertion of the DNA segment into the chromosome and for the mobilization of the element to other locations on the chromosome. The properties of some *E. coli* transposons are summarized in Table 20.2.

There are two types of prokaryotic transposons: *composite transposons* and *noncomposite transposons*. *Composite transposons* are complex transposons with a central region containing genes (e.g., drug-resistance genes) flanked on both sides by IS elements (also called *IS modules*). Composite transposons may be thousands of base pairs long. The IS elements are both of the same type and are called IS-*L* (for "left") and IS-*R* (for "right"). Depending upon the transposon, IS-*L* and IS-*R* may be in the same or inverted orientation relative to each other. Because the ISs themselves have terminal inverted repeats, the composite transposons also have terminal inverted repeats.

Figure 20.4 diagrams the structure of the composite transposon Tn*10* to illustrate the general features of such transposons. Nancy Kleckner has done much of the work to show the structure of Tn*10* and how it transposes. The Tn*10* transposon is 9,300 bp long and consists of 6,500 bp of central, nonrepeating DNA containing the tetracycline resistance gene flanked at each end with a 1,400-bp IS element. These IS elements are designated IS*10L* and IS*10R* and are arranged in an inverted orientation. Cells containing Tn*10* are resistant to tetracycline because of the tetracycline resistance gene contained within the central DNA sequence.

Transposition of composite transposons occurs because of the function of the IS elements they contain. One or both IS elements supplies the transposase. The inverted repeats of the IS elements at the

~ FIGURE 20.4

Structure of the composite transposon Tn*10*. The general features of composite transposons are evident: a central region carrying a gene or genes such as for drug resistance flanked by either direct or inverted IS elements. The IS elements themselves each have terminal inverted repeats.

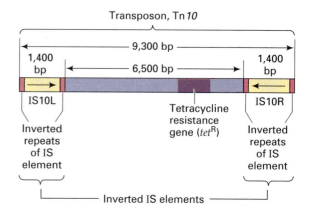

two ends of the transposon are recognized by transposase to initiate transposition (as with transposition of IS elements). Transposition of Tn10 is rare, occurring once in 10^7 cell generations. This is the case because less than one transposase molecule per cell generation is made by Tn*10*. Like IS elements, composite transposons produce target site duplications after transposition. For example, Tn*10* produces a 9-bp target site duplication (see Table 20.2).

Noncomposite transposons, like composite transposons, contain genes such as those for drug resistance. Unlike composite transposons, they do not terminate with IS elements. However, they do have the repeated sequences at their ends that are required for transposition. Tn*3* is a noncomposite transposon (Table 20.2 and Figure 20.5). Tn*3* has 38-bp inverted terminal repeats and contains three genes in its central region. One of those genes, *bla*, encodes β-lacta-

~ TABLE 20.2

Properties of Some Transposons Found in *E. coli*

TRANSPOSON[a]	GENE MARKER(S)	LENGTH (bp)	TERMINAL REPEAT SEQUENCES	IS ORIENTATION	TARGET SITE DUPLICATION
Tn*3*	Resistance to ampicillin	4,957	38 bp; not an IS element	—	5 bp
Tn*9*	Resistance to chloramphenicol	2,638	IS*1L*, IS*1R*	Direct	9 bp
Tn*10*	Resistance to tetracycline	9,300	IS*10L*, IS*10R*	Inverted	9 bp

[a]Tn*3* is a noncomposite transposon and Tn*9* and Tn*10* are both composite transposons.

~ FIGURE 20.5

Structure of the noncomposite transposon Tn3. Tn3 has genes in the central region for three enzymes: *bla* encodes β-lactamase (destroys antibiotics like penicillin and ampicillin); *tnpA* encodes transposase; and *tnpB* encodes resolvase. Transposase and resolvase are involved in the transposition process. Tn3 has short inverted terminal repeats that are unrelated to IS elements.

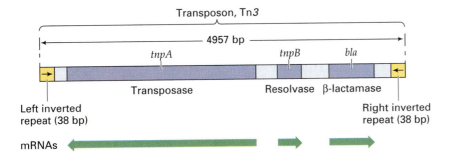

mase which breaks down ampicillin and therefore makes cells containing Tn3 resistant to ampicillin. The other two genes, *tnpA* and *tnpB*, encode the enzymes *transposase* and *resolvase* that are needed for transposition of Tn3. Transposase catalyzes insertion of the Tn into new sites, and resolvase is an enzyme involved in the particular recombinational events associated with transposition. Resolvase is not found in all transposons. The genes for transposition are in the central region for noncomposite transposons, while they are in the terminal IS elements for composite transposons. Like composite transposons, noncomposite transposons cause target site duplications when they move. Tn3, for example, produces a 5-bp target site duplication (Figure 20.6).

A number of detailed models have been generated for transposition of transposons. Figure 20.7 shows a *cointegration* model involving, in this particular case, the transposition of a transposon from one genome to another; for example, from a plasmid to a bacterial chromosome or vice versa. Similar events

can occur between two locations on the same chromosome. First, the donor DNA containing the transposable element fuses with the recipient DNA. Because of the way this occurs, the transposable element becomes duplicated with one copy located at each junction between donor and recipient DNA. This fused product is called a *cointegrate*. Next, the cointegrate is resolved into two products, each with one copy of the transposable element. Since the transposable element becomes duplicated, the process is called *replicative transposition*. Tn3, and related noncomposite transposons, move by replicative transposition.

A second type of transposition mechanism involves the movement of a transposable element from one location to another on the same or different DNA *without* replication of the element. This mechanism is called *conservative (nonreplicative) transposition* or *simple insertion*. In other words, the element is lost from the original position when it transposes. Tn10, for example, transposes by conservative transposition.

~ FIGURE 20.6

DNA sequence of a target site of Tn3: the important 5-nucleotide segment for this transposon is highlighted. After Tn3 inserts, the same 5-nucleotide sequence is found at each end.

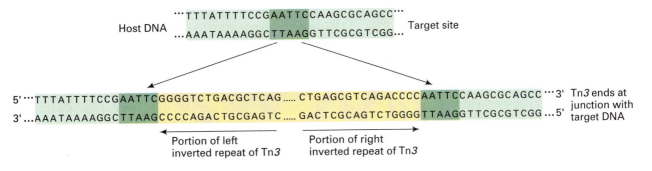

~ FIGURE 20.7

Cointegration model for transposition of a transposable element by replicative transposition. A donor DNA with a transposable element fuses with a recipient DNA. During the fusion, the transposable element is duplicated so that the product is a cointegrate molecule with one transposable element at each junction between donor and recipient DNA. The cointegrate is resolved into two molecules, each with one copy of the transposable element.

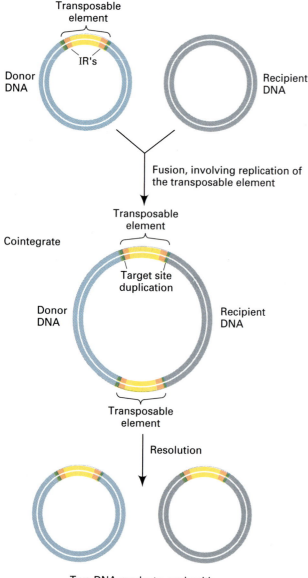

As with the movement of IS elements, transposition of transposons can cause mutations. Insertion of a transposon into the reading frame of a gene will disrupt it, causing a loss of function of that gene. Insertion into a gene's controlling region can cause

changes in the level of expression of the gene depending upon the promoter elements in the transposon and how they are oriented with respect to the gene. Deletion and insertion events also occur as a result of the gene activities of the transposons, and from crossing-over between duplicated transposons in the genome.

IS Elements and Transposons in Plasmids

When we studied bacterial genetics in Chapter 8, we learned that the transfer of genetic material between conjugating *E. coli* is the result of the function of the fertility factor *F*. The *F* factor, a circular double-stranded DNA molecule, is one example of a bacterial **plasmid**, an extrachromosomal genetic element capable of self-replication. Plasmids such as *F* that are also capable of integrating into the bacterial chromosome are called **episomes**.

Figure 20.8a shows the genetic organization of the *E. coli F* factor. It consists of 94,500 bp of DNA that code for a variety of proteins. The important genetic elements are:

1. *tra* (for "transfer") genes, required for the conjugal transfer of the DNA from a donor bacterium to a recipient bacterium
2. genes that encode proteins required for the plasmid's replication
3. four IS elements: two copies of IS3, one of IS2, and one of an insertion sequence element called γδ (gamma-delta)

It is because the *E. coli* chromosome has copies of these four insertion sequences at various positions that the *F* factor can integrate into the *E. coli* chromosome at different sites and in different orientations. That is, *F* factor integration occurs by conventional genetic recombination between the homologous sequences of the insertion elements. Whether separately replicating or integrated into the *E. coli* chromosome, the *tra* genes in the *F* factor direct the conjugal transfer functions.

Another class of plasmids that has medical significance is the *R* plasmid group, which was discovered in Japan in the 1950s. This discovery came about as a result of research into a cure for dysentery, an intestinal disease that is the result of infection by the pathogenic bacterium *Shigella*. The usual treatment of bacterial infections is the administration of antibiotics such as penicillin or ampicillin. However, the *Shigella* strain that occurred in dysentery patients in the Japanese hospitals was found to be resistant not only to penicillin but also to tetracycline, sulfanilamide, chloramphenicol, and streptomycin—all antibiotics

~ **FIGURE 20.8**

Organizational maps: (a) The *E. coli* F factor. The map shows the locations of genes required for transfer of the *F* factor during conjugation and for the replication of *F*; also shown are the insertion elements responsible for *F*'s integration into the bacterial chromosome; (b) A typical *R* plasmid. The map shows the region of homology with the *F* factor (the orange area), the area needed for DNA replication, and the sites of three antibiotic-resistance genes (*amp*, *kn*, and *tet*). These genes are flanked by insertion sequences and hence are true transposons.

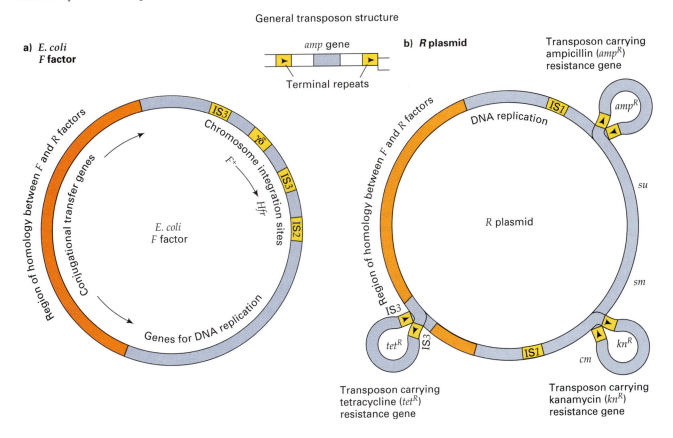

that are usually effective in killing bacteria or stopping their growth.

Scientists discovered that the multiple-resistance phenotype was transmissible to other nonresistant strains of *Shigella* as well as to other bacteria commonly found in the human intestine. Subsequently, they found that the genes responsible for the drug resistances were carried on *R* plasmids, which can promote the transfer of genes between bacteria by conjugation, just as the *F* factor can.

Figure 20.8b shows the genetic organization of one type of *R* plasmid. One segment of an *R* plasmid that is homologous to a segment in the *F* factor is the part needed for the conjugal transfer of genes. That segment and the plasmid-specific genes for DNA replication constitute what is called the *RTF* (resistance transfer factor) region. The rest of the *R* plasmid differs from type to type and includes the antibiotic-resistance genes or other types of genes of medical significance, such as resistance to heavy-metal ions.

The resistance genes in *R* plasmids are, in fact, transposons; that is, each resistance gene is located between flanking, directly repeated segments such as one of the IS modules. Thus, each transposon with its resistance gene in the *R* plasmid can be inserted into new locations on other plasmids or on the bacterial chromosome, while at the same time leaving behind a copy of itself in the original position. Different *R* plasmids contain various combinations of transposons carrying drug-resistance genes. In an individual to whom drugs are being administered, any bacterium containing an *R* plasmid that is resistant to the drugs will be able to survive and propagate, whereas bacteria without resistance genes will be killed. Thus, there is selection pressure for new *R* plasmids to be produced by transposition of transposons that include genes for drug-resistance, especially when a person is exposed to drug therapy. These new *R* plasmids can then be disseminated through populations of the same and other bacterial species by conjugation mediated by the *RTF* region of the plasmids.

Bacteriophage Mu

Mu is a temperate bacteriophage that infects *E. coli*. Recall that *temperate bacteriophages* (e.g., λ) can go through the lytic cycle or enter the lysogenic phase (see Chapter 8). Mu is also a transposon, and it can cause mutations when it transposes. In fact, the name Mu stands for *mutator*.

In the phage particle, the Mu genome is a 37-kb linear piece of DNA consisting mostly of phage DNA, with unequal lengths of host DNA at the two ends (Figure 20.9a). The G segment is a region that can invert; some Mu genomes have the G segment in one orientation, while the others have it in the opposite orientation. When Mu infects *E. coli* and enters the lysogenic state, the Mu genome integrates into the host chromosome by conservative transposition to produce the integrated prophage DNA, flanked by a 5-bp direct repeat of the host target site sequence (Figure 20.9b). During integration, the flanking host DNA that was present in the particle is lost. In the lysogenic state, a phage-encoded repressor prevents most Mu gene expression (recall our discussion of the λ repressor in Chapters 8 and 16), and the Mu prophage replicates when the *E. coli* chromosome replicates.

Mu's lytic cycle is different from that of λ. The Mu genome integrates in a different way than for lysogeny and remains integrated throughout the lytic phase.

Replication of the Mu genome occurs by replicative transposition. The details of Mu transposition remain to be determined.

Transposition by Mu causes mutations of various kinds. Apart from simple insertions, Mu can also cause deletions, inversions, and translocations. For example, by homologous recombination between two identical copies, Mu (and other multicopy transposons) can cause deletions (Figure 20.10a) and inversions (Figure 20.10b). Deletion occurs if the two Mu prophages or transposons are in the same orientation in the chromosome, while inversion occurs if the two are in opposite orientations.

𝒦EYNOTE

Transposable elements are unique DNA segments that can insert themselves at one or more sites in a genome. The presence of transposable elements in a cell is usually detected by the changes they bring about in the expression and activities of the genes at or near the chromosomal sites into which they integrate. In prokaryotes the three major types of transposable elements are insertion sequence (IS) elements, transposons (Tn), and bacteriophages such as Mu.

~ **FIGURE 20.9**

Temperate bacteriophage Mu genome shown (a) in phage particles and (b) integrated into the *E. coli* chromosome as a prophage.

a) Phage DNA present in virus particles

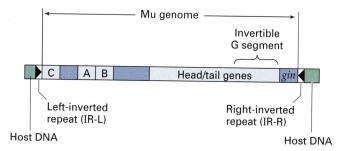

b) Prophage DNA

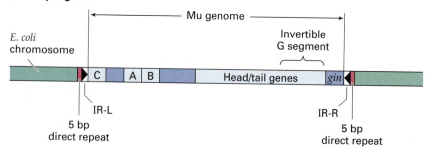

~ FIGURE 20.10

Production of deletion or inversion by homologous recombination between two Mu genomes or two transposons (for orientation purposes the ends of the DNA segment are labeled A and B). (a) For a deletion to occur, the two elements are in the same orientation; (b) For an inversion to occur, the two elements are in opposite orientation.

a) Deletion by recombination between two Mu genomes or transposons in same orientation

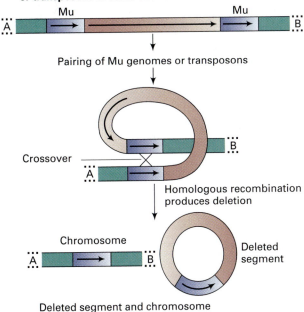

b) Inversion by recombination between two Mu genomes or transposons in opposite orientations

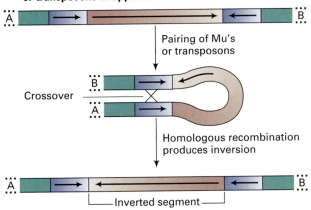

TRANSPOSABLE ELEMENTS IN EUKARYOTES

In the 1930s, Marcus Rhoades observed an interesting phenomenon in Black Mexican sweet corn. That is, when allele *a* of the *A1* locus is homozygous, no purple anthocyanin pigment is produced, so the aleurone layer of the seed is colorless. However, in the presence of *Dt* (dotted), a dominant mutant allele of a locus on another chromosome, a very different phenotype results. Plants of genotype *a/a Dt/–* produce seeds with dots of purple color as if allele *a* had mutated to the dominant wild-type allele *A*1. Moreover, the number of doses of the *Dt* allele affected the number of dots in the triploid aleurone tissue: one dose gave an average of 7.2 dots per seed, two doses gave 22.2, and three doses gave 121.9. Rhoades interpreted his results to indicate that the *Dt* was a **mutator gene**, that is, a mutant gene that increases the spontaneous mutation frequencies of other genes. Indeed, this system was used as a classical example of mutator genes for many years.

In the 1940s and 1950s, Barbara McClintock did a series of elegant genetic experiments with corn (*Zea mays*) that led her to hypothesize the existence of what she called "controlling elements," which modify or suppress gene activity in corn and which are mobile in the genome. Decades later, the controlling elements she studied were shown to be transposable elements. Similarly, more recent work has shown that Rhoade's mutator gene *Dt* is also a transposable element. For her original work in deducing the presence of transposable elements, Barbara McClintock was awarded the Nobel Prize in 1983. A fascinating and moving biographical sketch of Barbara McClintock is given in Box 20.1.

Transposable elements have been identified in many eukaryotes, and they have been studied mostly in yeast, *Drosophila*, corn, and humans. In general, their structure and function are very similar to those of prokaryotic transposable elements. Eukaryotic transposable elements can integrate into chromosomes at a number of sites. Thus, such elements may be able to affect the function of virtually any gene, turning it on or off, depending on the element involved and how it integrates into the gene. The integration events themselves, like those of most prokaryotic transposable elements, involve nonhomologous recombination. Many of the eukaryotic transposable elements carry genes. However, while some of the genes must code for enzymes required for transposition, in most cases the functions of the genes are unknown.

Transposons in Plants

Some of the transposable elements found in plants are presented in Table 20.3 on p. 667. Like the transposons we have already discussed, plant transposons have inverted repeated (IR) sequences at their ends and generate short direct repeats of the target site DNA when they integrate.

BOX 20.1
BARBARA MCCLINTOCK
(1902–1992)

Barbara McClintock's remarkable life spanned the history of genetics in the twentieth century. Barbara McClintock was born in Hartford, Connecticut, to Sara Handy McClintock, an accomplished pianist as well as a poet and painter, and Thomas Henry McClintock, a physician. Both parents were quite unconventional in their attitudes toward child rearing: they were interested in what their children would and could be, rather than what they should be.

During her high school years Barbara discovered science, and she loved to learn and figure things out. After high school Barbara attended Cornell University, where she flourished both socially and intellectually. She enjoyed her social life, but her comfort with solitude and the tremendous joy she experienced in knowing, learning, and understanding were to be the defining themes of her life. The decisions she made during her university years were consistent with her adamant individuality and self-containment. In Barbara's junior year, after a particularly exciting undergraduate course in genetics, her professor invited her to take a graduate course in genetics. After that she was treated much like a graduate student; by the time she had finished her undergraduate course work, there was no question in her mind: she had to continue her studies of genetics.

At Cornell, genetics was taught in the plant-breeding department, which at the time did not take female graduate students. To circumvent this obstacle, McClintock registered in the botany department with a major in cytology and a minor in genetics and zoology. She began to work as a paid assistant to Lowell Randolph, a cytologist. McClintock and Randolph did not get along well and soon dissolved their working relationship, but as McClintock's colleague and lifelong friend Marcus Rhoades later wrote, "Their brief association was momentous because it led to the birth of maize cytogenetics." McClintock discovered that the metaphase or late prophase chromosomes in the first microspore mitosis were far better for cytological discrimination than were root tip

Barbara McClintock in 1947.

chromosomes. In a few weeks she prepared detailed drawings of the maize chromosomes, which she published in *Science*.

This was McClintock's first major contribution to maize genetics and laid the groundwork for a veritable explosion of discoveries that connected the behavior of chromosomes to the genetic properties of an organism, defining the new field of cytogenetics. McClintock was awarded a Ph.D. in 1927 and appointed an instructor at Cornell, where she continued to work with maize. The Cornell maize genetics group was small, including Professor R. A. Emerson, the founder of maize genetics, McClintock, George Beadle, C. R. Burnham, Marcus Rhoades, and Lowell Randolph, together with a few graduate students. By all accounts McClintock was the intellectual driving force of this talented group.

In 1929, a new graduate student, Harriet Creighton, joined the group and was guided by McClintock. Their work showed, for the first time, that genetic recombination is a reflection of the physical exchange of chromosome segments (see Chapter 5, pp. 139–140). A paper on the work by Creighton and McClintock, published in 1931, was perhaps McClintock's first seminal contribution to the science of genetics.

Although McClintock's fame was growing, she had no permanent position. Cornell had no female professors in fields other than home economics, so her prospects were dismal. She had already attained international recognition, but as a woman she had little hope of securing a permanent academic position at a major research university. R. A. Emerson obtained a grant from the Rockefeller Foundation to support her work for two years, allowing her to continue to work independently. McClintock was discouraged and resentful of the disparity between her prospects and those of her male counterparts. Her extraordinary talents and accomplishments were widely appreciated, but she was also seen as "difficult" by many of her colleagues, in large part because of her quick mind and intolerance of second-rate work and thinking.

In 1936, Lewis Stadler convinced the University of Missouri to offer McClintock an assistant professorship. She accepted the position and began to follow the behavior of maize chromosomes that had been broken by X-irradiation. However, soon after her arrival at Missouri she understood that hers was a special appointment. She found herself excluded from regular academic activities, including faculty meetings. In 1941, she took a leave of absence from Missouri and departed with no intention of returning. She wrote to her friend Marcus Rhoades, who was planning to go to Cold Spring Harbor for the summer to grow his corn. An invitation for McClintock was arranged through Milislav Demerec (member and later the director of the genetics department of the Carnegie Institution of Washington, then the dominant research laboratory at Cold Spring Harbor), who offered her a year's research appointment. Though hesitant to commit herself, McClintock accepted. When Demerec later offered her an appointment as a permanent member of the research staff, McClintock accepted, still unsure whether she would stay. Her dislike of making commitments was a given: she insisted that she would never have become a scientist in today's world of grants because she could not have committed herself to a written research plan. It was the unexpected that fascinated her, and she was always ready to pursue an observation that didn't fit. Nevertheless, McClintock did stay at Carnegie until 1967.

At Carnegie McClintock continued her studies on the behavior of broken chromosomes. In 1944, she was elected to the National Academy of Sciences and in 1945 to the presidency of the Genetics Society of America. In these same two years McClintock reported observing "an interesting type of chromosomal behavior" involving the repeated loss of one of the broken chromosomes from cells during development. What struck her as odd was that in this particular stock it was always chromosome 9 that broke, and it always broke at the same place. McClintock called the unstable chromosome site *Dissociation* or *Ds* because "the most readily recognizable consequence of its actions is this dissociation." She quickly established that the *Ds* locus would "undergo dissociation mutations only when a particular dominant factor is present." She named this factor *Activator* (*Ac*) because it activated chromosome breakage at *Ds*. She also reached the extraordinary conclusion that *Ac* was not only required for *Ds*-mediated chromosome breakage, but also could destabilize previously stable mutations. But more than that, and unprecedented, the chromosome-breaking *Ds* locus could "change its position in the chromosome," a phenomenon she called *transposition*. Moreover, she had evidence that the *Ac* locus was required for transposition of *Ds* and that, like the *Ds* locus, the *Ac* locus was mobile.

Within several years McClintock had established beyond a doubt that both the *Ac* and *Ds* loci were not only capable of changing their positions on the genetic map, but also of inserting into loci to cause unstable mutations. She presented a paper on her work at the Cold Spring Harbor Symposium of 1951. The reaction to her presentation ranged from perplexed to hostile. Later she published several papers in refereed journals, but from the paucity of reprint requests, she inferred on the part of the larger biological community an equally cool reaction to the astonishing news that genes could move.

McClintock's work had taken her far outside the scientific mainstream, and in a profound sense she had lost her ability to communicate with her colleagues. By her own admission McClintock had neither a gift for written exposition nor a talent for explaining complex phenomena in simple terms. But there are more important factors: the very notion that genes can move was in deep contradiction to the assumption of the regular relationships among genes that underlies the construction of linkage maps and the physical mapping of genes onto chromosomes. The concept that genetic elements can move would undoubtedly have met with resistance regardless of author and presentation.

McClintock was deeply frustrated by her failure to communicate, but her fascination with the unfolding story of transposition was sufficient to keep her working at the highest level of physical and mental intensity she could sustain. By the time of her formal retirement, she had accumulated a rich store of knowledge about the genetic behavior of two markedly different transposable element families. And beginning about the time her active fieldwork ended, transposable genetic elements began to surface in one experimental organism after another.

These later discoveries came in an altogether different age. In the two decades between McClintock's original genetic discovery of transposition and its rediscovery, genetics had undergone as profound a change as the cytogenetic revolution that had occurred in the second and third decades of the century. The genetic material had been identified as DNA, the manner in which information is encoded in the genes had been deciphered, and methods had been devised to isolate and study individual genes. Genes were no longer abstract entities known only by the consequences of their alteration or loss; they were real bits of nucleic acids that could be isolated, visualized, subtly altered, and reintroduced into living organisms.

By the time the maize transposable elements were cloned and their molecular analysis initiated, the importance of McClintock's discovery of transposition was widely recognized, and her public recognition was growing. For example, she received the National Medal of Science in 1970, she was named Prize Fellow laureate of the

MacArthur Foundation and she received the Lasker Basic Medical Research Award in 1981, and in 1982 she shared the Horwitz Prize. Finally, in 1983, thirty-five years after publication of the first evidence for transposition, McClintock was awarded the Nobel Prize for physiology or medicine.

McClintock was sure she would die at ninety, and a few months after her ninetieth birthday she was gone, drifting away from life gently, as a leaf from an autumn tree. What Barbara McClintock was and what she left behind are eloquently expressed in a few short lines written many years earlier by her friend and champion, Marcus Rhoades, whose death preceded hers by a few months:

One of the remarkable things about Barbara McClintock's surpassingly beautiful investigations is that they came solely from her own labors. Without technical help of any kind she has by virtue of her boundless energy, her complete devotion to science, her originality and ingenuity, and her quick and high intelligence made a series of significant discoveries unparalleled in the history of cytogenetics. A skilled experimentalist, a master at interpreting cytological detail, a brilliant theoretician, she has had an illuminating and pervasive role in the development of cytology and genetics.

Adapted by permission of Nina Fedoroff and by courtesy of the National Academy of Sciences, Washington, DC.

GENERAL PROPERTIES OF PLANT TRANSPOSONS. When a plant transposon inserts into a chromosome, the consequence depends upon the properties of the transposon. Typically, the effects range from activation or repression of adjacent genes, to chromosome mutations such as duplications, deletions, inversions, translocations, or chromosome breakage. That is, as with bacterial IS elements and transposons, generally, transposition of transposons into genes causes mutations. Disruption of a gene typically results in a *null mutation*, that is, a mutation where the function of a gene is reduced to zero. If a transposon moves into a gene's promoter, the efficiency of that promoter can be decreased or obliterated. Alternatively, the transposon may provide promoter function itself, and lead to an increase in gene expression.

Geneticists have identified several families of transposons (controlling elements) in corn. Each family consists of a characteristic array of transposons with respect to numbers, types, and locations. There are two forms of transposons in each family: *autonomous elements*, which can transpose by themselves; and *nonautonomous elements*, which cannot transpose by themselves because they lack the gene for transposition—they require the presence of an autonomous element to supply the missing functions. Often the nonautonomous element is a deletion derivative of the autonomous element in the family. When either element is inserted into a gene, a mutant allele of that gene is produced. When an autonomous element is inserted into a host gene, the resulting mutant allele is *unstable* because the element can

~ **TABLE 20.3**

Examples of Transposable Elements in Plants

TRANSPOSABLE ELEMENT	PROPERTIES
Ac	In corn, *Ac* (*Activator*) is the autonomous element of the *Ac/Ds* family.
Ds	In corn, *Ds* is a nonautonomous element whose transposition is dependent on the presence of an active *Ac* element. *Ds* (*Dissociation*) refers to the ability of particular *Ds* elements to induce chromosome breakage at their insertion site.
Mu1	*Mu1* (*Mutator 1*) is an autonomous transposable element in corn originally identified in strains with a very high frequency of spontaneous mutations.
Tam	Transposable element in *Antirrhinum majus*
Tgm	Transposable element in *Glycine max*

excise and transpose to a new location. The frequency of transposition out of a gene is higher than the spontaneous reversion frequency for a regular point mutation; hence, the allele produced by an autonomous element is referred to as a *mutable allele*.

By contrast, mutant alleles resulting from the insertion of a nonautonomous element in a gene are *stable*, since the element is unable to transpose out of the locus by itself. However, if the autonomous element of its family is also present in the genome or is introduced, that element can provide the enzymes needed for transposition, and the nonautonomous element will now exhibit properties of autonomous elements with respect to transposition around the genome.

MCCLINTOCK'S DISCOVERY OF TRANSPOSONS IN PLANTS. A number of different genes must function together for the synthesis of red anthocyanin pigment, which gives the corn kernel a purple color. Classical genetic experiments had shown that mutation of any one of these genes causes a kernel to be unpigmented. In her classic experiments, Barbara McClintock studied kernels that, rather than being purple or white, exhibited spots of purple pigment on an otherwise white kernel (Figure 20.11). She knew that the phenotype was the result of an unstable mutation. From her careful genetic and cytological studies she came to the conclusion that the spotted phenotype was not the result of any conventional kind of mutation (such as a point mutation) but rather was due to a controlling element, which we now know is a transposon.

The explanation for the spotted kernels is as follows: If the corn plant carries a wild-type *C* gene, the kernel will be purple; *c* (colorless) mutations block purple pigment production, so the kernel is colorless. During kernel development, revertants of the mutation occur, leading to a spot of purple pigment. The genetic nature of the reversion is supported by the

~ **FIGURE 20.11**

Corn kernels, some of which show spots of pigment produced by cells in which a transposable genetic element had transposed out of a pigment-producing gene, thereby allowing function of that gene to be restored. The cells in the white areas of the kernel lack pigment because a pigment-producing gene continues to be inactivated by the presence of a transposable element within that gene.

fact that descendants of the cell which underwent the reversion also can produce the pigment. The earlier in development the reversion occurs, the larger is the purple spot. McClintock determined that the original *c* (colorless) mutation resulted from a "mobile controlling element" (in modern terms, a transposon), called *Ds* for "dissociation," being inserted into the *C* gene (Figure 20.12a and b). We now know this to take place by a transposition event of a nonautonomous transposon. Another mobile controlling element, called *Ac* for "activator" (which we now know is an autonomous element) is required for transposition of *Ds* into the gene. *Ac* can also result in *Ds* transposing (excising perfectly) out of the *c* gene, giving a wild-type revertant; i.e., a purple spot (Figure 20.12c). The *Ac-Ds* system is discussed below.

The remarkable fact of McClintock's conclusion was that at the time there was no precedent for the existence of transposable genetic elements—indeed, the genome was thought to be very static with regard to gene locations. Only much more recently have transposable genetic elements been widely identified and studied, and only in 1983 was direct evidence obtained for the movable genetic elements proposed by McClintock.

THE *Ac-Ds* CONTROLLING ELEMENTS IN CORN.

Transposition events in corn are often followed by visible phenotypic changes when the transposon inserts near a gene. Such changes are best seen in heterozygotes when the dominant allele is inactivated by the transposon, resulting in a cell with a recessive phenotype. Mitotic progeny of that cell remain in the same location, giving rise to a tissue spot or sector with a clone of cells that exhibit the recessive phenotype. As we have discussed, tissue sectors are readily seen in kernels as colored patches in a background of a different color.

The *Ac-Ds* family of controlling elements has been studied in detail. McClintock's mapping studies had shown that *Ac* and *Ds* were not conventional genetic loci because they could move to new locations

~ FIGURE 20.12

Kernel color in corn and transposon effects. (a) Purple kernels result from the active *C* gene. (b) Colorless kernels can result when the *Ac* transposon activates *Ds* transposition and *Ds* inserts into *C*, producing a mutation. (c) Spotted kernels result by reversion of the *c* mutation during kernel development when *Ac* activates *Ds* transposition out of the *C* gene.

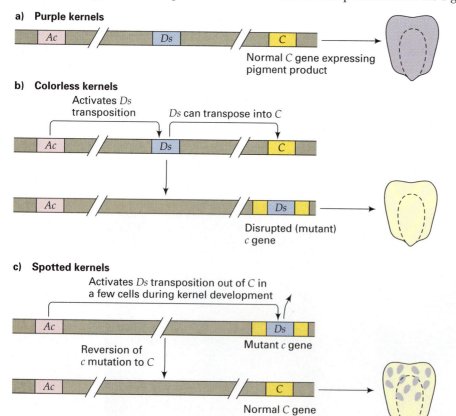

a) **Purple kernels**

Normal *C* gene expressing pigment product

b) **Colorless kernels**

Activates *Ds* transposition

Ds can transpose into *C*

Disrupted (mutant) *c* gene

c) **Spotted kernels**

Activates *Ds* transposition out of *C* in a few cells during kernel development

Mutant *c* gene

Reversion of *c* mutation to *C*

Normal *C* gene

in the genome; that is, they could transpose and cause mutations by inserting into other loci. *Ac* is the autonomous element of the family, and, hence, mutations caused by *Ac* are unstable. *Ds* is the nonautonomous element of the family. *Ds* mutations are stable if only *Ds* is present; they are unstable in the presence of an *Ac* element. In the mid-1950s McClintock demonstrated that at least some *Ds* elements were derived from *Ac* elements.

Ac is 4,563 bp long, with 11-bp imperfect terminal inverted repeats (IRs) (Figure 20.13a). *Ac* contains a single transcription unit that comprises most of the element's length. A single 3.5-kb mRNA is produced, which encodes the 807–amino acid transposase enzyme. Upon insertion into a chromosome site, an 8-bp direct duplication of the target site is generated.

Ds does not transpose in the absence of *Ac* and remains as a stable insertion in the chromosome. When an *Ac* element is present, it activates *Ds*, causing it to transpose to a new site or to break the chromosome in which it is located.

Ds elements are heterogeneous in length and sequence; a number of them are diagrammed in Figure 20.13b. All *Ds* elements have the same terminal IRs as *Ac* elements, and many of these *Ds* elements have been generated from *Ac* by deletion of various lengths (e.g., *Ds9*, *Ds2d1*, *Ds2d2*, *Ds6*; Figure

~ FIGURE 20.13

The structure of the *Ac* autonomous transposable element of corn, and of several *Ds* nonautonomous elements derived from *Ac*. (a) The *Ac* element is 4,563 bp long and has imperfect 11-bp inverted terminal repeats. *Ac* has a single transcription unit, comprising most of the element, and going from left to right. The five exons of the resulting RNA are shown in green within the *Ac* element itself. (b) Several *Ds* elements are shown in the same way as for the *Ac* element. They are derived from *Ac* by internal deletions of various lengths.

a) Activator element (*Ac*)

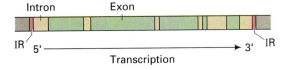

b) Dissociation elements (*Ds*)

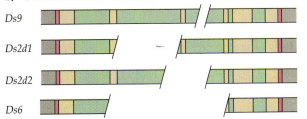

20.13b) or by more complex sequence rearrangements. The deletions and complex sequence rearrangements are responsible for the transposition-defective phenotype of the various *Ds* elements.

Genetic evidence indicates that both the timing and frequency of transposition of the *Ac* and *Ds* elements, as well as the genetic rearrangements associated with the elements, are developmentally regulated. This is a unique property of corn transposons. Transposition of the *Ac* element does not itself involve replication of the element, so that it does not leave a copy of itself in the original chromosomal location when it transposes: that is, *Ac* exhibits conservative transposition. *Ac* transposes only during chromosome replication (Figure 20.14). Consider a chromosome with one copy of *Ac* at a site called the *donor site*. When the chromosome region containing *Ac* replicates, two copies of *Ac* result, one on each progeny chromatid. There are two possible results of *Ac* transposition, depending on whether transposition occurs to a replicated or unreplicated chromosome site.

First, let us consider transposition to a replicated chromosome site (Figure 20.14a). If one of the two *Ac* elements transposes at this time, an empty donor site is left on one chromatid, while an *Ac* element remains in the homologous donor site on the other chromatid. The transposing *Ac* element inserts into a new, already replicated recipient site, which is often on the same chromosome. In Figure 20.14a the site is shown on the same chromatid as the parental *Ac* element. Thus, in the case of transposition to an already-replicated site, there is no net increase in the number of *Ac* elements.

Figure 20.14b shows transposition of one *Ac* element to an unreplicated chromosome site. As with the first case, one of the two *Ac* elements transposes, leaving an empty donor site on one chromatid and an *Ac* element in the homologous donor site on the other chromatid. But now the transposing element inserts into a nearby recipient site that has yet to be replicated. When that region of the chromosome replicates, the result will be a copy of the transposed *Ac* element on both chromatids, in addition to the one original copy of the *Ac* element at the donor site on one chromatid. Thus, in the case of transposition to an unreplicated recipient site, there is a net increase in the number of *Ac* elements.

Transposition of most *Ds* elements occurs in the same way as *Ac* transposition, with functions for transposition supplied by an *Ac* element in the genome.

MENDEL'S WRINKLED PEAS. In Chapter 2 we discussed the wrinkled-pea trait that Mendel used in his

~ FIGURE 20.14

The *Ac* transposition mechanism. (a) Transposition to an already replicated recipient site results in no net increase in the number of *Ac* elements in the genome; (b) Transposition to an unreplicated recipient site results in a net increase in the number of *Ac* elements when the region of the chromosome containing the transposed element is subsequently replicated.

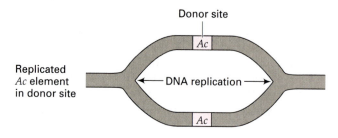

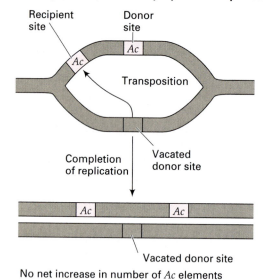
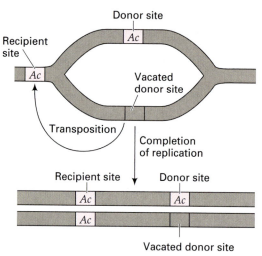

a) **Transposition to an already-replicated recipient site**

No net increase in number of *Ac* elements

b) **Transposition to an unreplicated recipient site**

Net increase in number of *Ac* elements

genetics experiments. In *RR* (wild-type) peas, starch grains are large and simple, while in *rr* (wrinkled) peas they are small and deeply fissured. *RR* seeds contain larger amounts of starch, and lower levels of sucrose than *rr* seeds. The sucrose difference leads to a higher water content and larger size of developing *rr* seeds. However, when the seeds mature, the *rr* seeds lose a larger proportion of their volume and this leads to the wrinkled phenotype.

At the biochemical level, the *rr* genotype results in a complete absence of one form of starch-branching enzyme (SBEI) in developing embryos. SBEI is important in determining the starch content of embryos so that in *rr* plants starch content is reduced. The SBEI gene involved has been cloned and shows 100 percent cosegregation with the *r* locus in genetic experiments, indicating that the *r* gene encodes the form of SBEI associated with wrinkled peas. Molecular analysis has shown that *rr* plant lines descended from those that Mendel used in his experiments have a 0.8-kb

transposon inserted into the *R* gene, resulting in the *r* allele. The particular transposon is very similar to the *Ac/Ds* family of transposons we have already discussed.

*K*EYNOTE

The mechanism of transposition of plant transposons is quite similar to transposition of bacterial IS elements or transposons. Transposons integrate at a target site by a precise mechanism so that the integrated elements are flanked at the insertion site by a short duplication of target site DNA of characteristic length. Many plant transposons occur in families, the autonomous elements of which are able to direct their own transposition, and the nonautonomous elements of which are able to transpose only when activated by an autonomous element in the same genome. Most nonautonomous elements are derived from autonomous elements by internal deletions or complex sequence rearrangements.

Ty Elements in Yeast

Ty STRUCTURE AND PROPERTIES.

All *Ty* transposable elements of yeast have a number of structural properties in common with bacterial transposons. That is, they have terminal repeated sequences, they integrate at sites with which they have no homology, and they generate a target site duplication (of 5 bp) upon insertion.

A *Ty* element is diagrammed in Figure 20.15. The element is about 5.9 kb long and includes two 334-bp-long, directly repeated sequences called long terminal repeats (LTR) or deltas (δ), one at each end of the element. The deltas consist of about 70 percent AT, and each contains a promoter and sequences recognized by transposing enzymes. The *Ty* elements encode a single, 5,700-nucleotide mRNA that begins at the promoter in the delta at the 5' end of the element (see Figure 20.15). The mRNA transcript contains two open reading frames (ORFs); that is, regions with a start codon in frame with a chain-terminating codon, indicating that two proteins could be produced from the mRNA. The two regions have been designated *TyA* and *TyB*. The number of copies of this element varies between strains, although on the average a strain will contain about 35.

Because *Ty* has no genetic marker associated with it, it is much more difficult to follow its movements around the genome than it is to follow the drug-resis-

tance markers of bacterial transposons. In at least 12 cases in which *Ty* element movements have been followed, a 5-bp terminal repeat of the yeast target DNA has been generated upon integration of the element. Since there is a delta at each end of a *Ty* element, a frequent event is recombination between two deltas by homologous recombination (Figure 20.16). This produces a released *Ty* element that has one delta and leaves a delta behind in the yeast chromosome. Studies have shown that there are at least a hundred copies of delta per yeast cell. These delta "droppings" are incapable of transposition, but because they contain transcription initiation signals, they can affect the expression of the genes near their location. Since there are multiple *Ty* elements in a cell, recombination can also occur between *Ty* elements located at different positions in the genome. The result can be translocations, deletions, or inversions.

Ty ELEMENTS AND RETROVIRUSES.

Retroviruses are single-stranded RNA viruses that replicate via double-stranded DNA intermediates. That is, when a retrovirus infects a cell, its RNA genome is copied by viral reverse transcriptase to produce a double-stranded DNA. The DNA integrates into the host's chromosome where it can be transcribed to produce progeny RNA viral genomes and viral protein mRNAs.

Analysis of yeast *Ty* elements revealed a number of similarities in their organization to that of retroviruses. This analysis led to the hypothesis that the *Ty*

~ FIGURE 20.16

Excision of a *Ty* element by homologous recombination between directly repeated deltas, resulting in a delta being left behind at the original *Ty* insertion site.

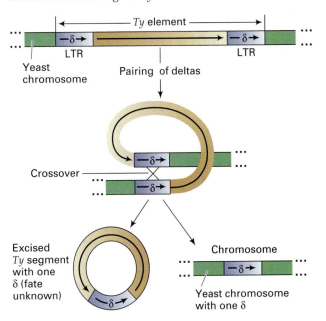

~ FIGURE 20.15

The *Ty*-transposable element of yeast. ORF = open reading frame. LTR = long terminal repeat, called "delta."

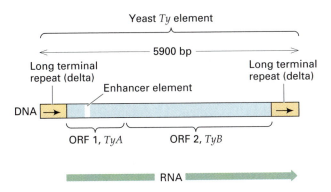

elements move to new chromosomal locations using the same mechanisms as those involved in provirus production and integration of retroviruses. That is, rather than transposing DNA to DNA as is the case with bacterial transposons and most eukaryotic transposons, *Ty* elements were hypothesized to transpose by making an RNA copy of the integrated DNA sequence and then by creating a new *Ty* element by reverse transcription. The new element would then integrate at a new chromosome location.

Evidence substantiating the hypothesis was obtained through experiments using *Ty* elements that had been constructed using recombinant DNA techniques; these *Ty* elements had special features that enabled their transposition to be monitored easily. One compelling piece of evidence came from experiments in which an intron was placed into the *Ty* element by recombinant DNA techniques (there is none in normal *Ty* elements) and the element was monitored from the beginning through the transposition event. At the new location, the *Ty* element no longer had the intron sequence, a fact that was consistent with the notion that transposition occurred via an RNA intermediate. The intron had been removed by the usual splicing processes that function to remove introns from pre-mRNAs.

It is now accepted that *Ty* elements transpose via an RNA intermediate. Subsequently it has been demonstrated that *Ty* elements encode a reverse transcriptase. Moreover, *Ty* viruslike particles (*Ty*-VLPs) containing *Ty* RNA and reverse transcriptase activity have been identified. Because of their similarity with retroviruses in this regard, *Ty* elements have been referred to as *retrotransposons*.

Drosophila Transposons

A number of classes of transposons have been identified in *Drosophila*. In this organism it is estimated that about 15 percent of the genome is mobile.

THE *COPIA* RETROTRANSPOSON.   Figure 20.17 diagrams the structure of a *copia* transposable genetic element of *Drosophila*. There are several families of *copia*-like elements. The members of each family are highly conserved and are located at about 5 to 100 widely scattered sites in the genome. The *copia* elements code for abundant mRNAs found in *Drosophila*. All *copia* elements found in *Drosophila* are capable of transposition, although there are differences among *Drosophila* strains in the number and distribution of these elements.

The structure of the *copia* element is similar to that of the *Ty* elements of yeast. Directly repeated termini (LTRs) of 276 bp are found at either end of a 5,000-bp segment of DNA. Inverted repeats of 17 bp are located at the ends of each LTR. Like yeast *Ty* elements, *copia* elements transpose via an RNA intermediate, using a reverse transcriptase catalyzed process. Viruslike particles (VLPs) have also been identified with *copia*. Integration of a *copia* element into the genome, like other transposons, results in a duplication of a short target site sequence, in this case 3 to 6 bp long.

HYBRID DYSGENESIS AND *P* ELEMENTS.   **Hybrid dysgenesis** is the appearance of a series of defects, including mutations, chromosomal aberrations, and sterility, when certain strains of *Drosophila melanogaster* are crossed. For example, hybrid dysgenesis occurs when female laboratory strains are mated to males from natural populations (Figure 20.18a). The laboratory strain is said to be of the *M cytotype* (maternal contributing cell type) and the naturally occurring strain is said to be of the *P cytotype* (paternal contributing cell type), thus the cross is M♀ × P♂. In the reciprocal P♀ × M♂ cross, hybrid dysgenesis does not occur and all progeny are fertile (Figure 20.18b).

Hybrid dysgenesis primarily affects germ line cells. That is, F$_1$ hybrids of a M♀ × P♂ cross have normal somatic tissues, but gonads do not develop, and the flies are sterile. Hybrid dysgenesis results when chromosomes of the P male parent become exposed to the cytoplasm derived from the M female parent. Cytoplasm from P females does not induce hybrid dysgenesis, so progeny of P♀ × P♂ or P♀ × M♂ are fully fertile.

The model for the induction of hybrid dysgenesis in the M♀ × P♂ crosses is as follows. The haploid genome of the P male has about 40 copies of a family of transposons called *P elements*. The M strain has no *P* elements. *P* elements vary in length from 500 to 2,900 bp and each has 31-bp inverted terminal repeats. The shorter *P* elements (which are nonautonomous elements, like *Ds* in corn) are derived from the longest *P* element by internal deletions. The longest *P* elements are autonomous elements like *Ac* in corn; that is, they encode a transposase, which can catalyze their own transposition and the transposition of the shorter *P* elements (Figure 20.19). Insertion

Structure of the transposable element *copia*, a retrotransposon found in *Drosophila melanogaster*.

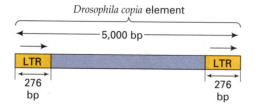

Drosophila copia element

~ **FIGURE 20.18**

Hybrid dysgenesis, exemplified by the production of sterile flies, results from (a) a cross of M♀ × P♂, but not from (b) a cross of P♀ × M♂.

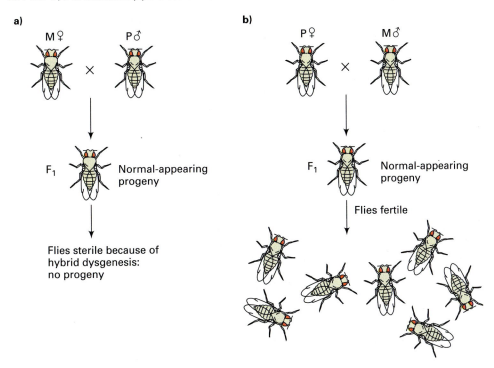

a)

M♀   P♂

×

F₁   Normal-appearing progeny

Flies sterile because of hybrid dysgenesis: no progeny

b)

P♀   M♂

×

F₁   Normal-appearing progeny

Flies fertile

of a *P* element into a new site results in a direct repeat of the target site. In the hybrid flies produced from the M♀ × P♂ crosses, the chromosomes have *P* elements inserted at new sites. This occurs because *P* element transposition has become activated, and such activation occurs only in the germ line.

P elements are thought to encode a repressor protein that prevents the transcription of the transposase gene, therefore preventing transposition of *P* elements. Thus, in the P cytotype, the *P* elements are sta-

ble in the chromosomes. However, in hybrids from M♀ × P♂, the cytoplasm of the egg is derived from the M cytotype, which lacks *P* repressors. Transposase is produced, activating *P* element transposition, which leads to the hybrid dysgenesis phenomenon. This transposase has been characterized biochemically and molecularly. It is an 87 kDa DNA-binding protein. The protein binds to *P* element DNA located adjacent to the inverted repeats, but does not interact with the repeats themselves.

~ **FIGURE 20.19**

Structure of the autonomous *P* transposable element found in *Drosophila melanogaster.*

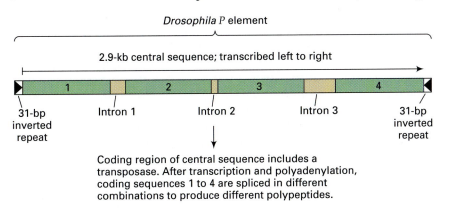

Drosophila P element

2.9-kb central sequence; transcribed left to right

| 1 | | 2 | | 3 | | 4 |

31-bp inverted repeat   Intron 1   Intron 2   Intron 3   31-bp inverted repeat

Coding region of central sequence includes a transposase. After transcription and polyadenylation, coding sequences 1 to 4 are spliced in different combinations to produce different polypeptides.

Apart from their role in hybrid dysgenesis, *P* elements are very important vectors for transferring genes into the germ line of *Drosophila* embryos, allowing genetic manipulation of the organism. Figure 20.20 illustrates an experiment by G. M. Rubin and A. G. Spradling in which the wild-type *rosy+* (*ry+*) gene was introduced into a strain homozygous for a mutant *rosy* allele. The wild-type *rosy* gene was introduced into the middle of a *P* element by recombinant DNA techniques and cloned in a plasmid. The recombinant plasmids were then microinjected into *rosy* embryos in the regions that would become the germ-line cells. *P* element–encoded transposase then catalyzed the movement of the *P* element, along with the wild-type *rosy* gene it contained, to the *Drosophila* genome in some of the germ-line cells. When the flies resulting from these embryos produced gametes, they contained the wild-type *rosy* gene and, hence, descendants of these individuals had normal eye color.

KEYNOTE

Transposable genetic elements in eukaryotes are typically transposons. Transposons can transpose to new sites while leaving a copy behind in the original site, or they may excise themselves from the chromosome. When the excision is imperfect, deletions can occur, and by various recombination events other chromosomal rearrangements such as inversions and duplications may occur. While most transposons move by using a DNA-to-DNA mechanism, some eukaryotic transposons such as yeast *Ty* elements and *Drosophila copia* elements transpose via an RNA intermediate (using a transposon encoded reverse transcriptase), thereby resembling retroviruses.

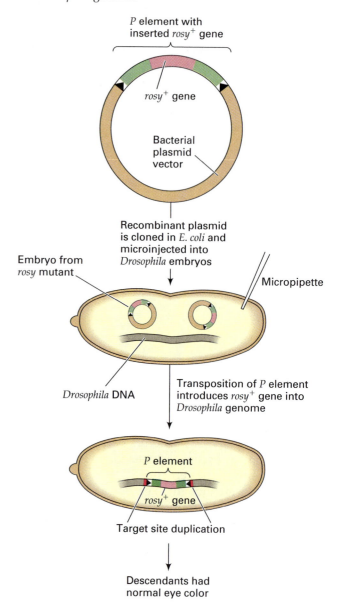

~ **FIGURE 20.20**

Illustration of the use of *P* elements to introduce genes into the *Drosophila* genome.

Human Retrotransposons

We have learned that yeast *Ty* elements and *Drosophila copia* elements are both retrotransposons, moving around the genome via RNA intermediates. There is good evidence to indicate that retrotransposons are present also in mammalian genomes.

In Chapter 11 we discussed the different repetitive classes of DNA sequences found in the genome. Of relevance here are the SINEs (short interspersed sequences) and LINEs (long interspersed sequences) found in the moderately repetitive class of sequences. SINEs are 100- to 300-bp repeated sequences interspersed between unique-sequence DNA 1,000 to 2,000 bp in length. LINEs are repeated sequences greater than 5,000 bp in length interspersed among unique-sequence DNA of up to 35,000 bp or more in length. Both SINEs and LINEs occur in families where family members are related by sequence.

In humans a very abundant SINEs family is the *Alu family*. The repeated sequence in this family is about 300 bp long and is repeated between 300,000 and 500,000 times in the genome, amounting to up to

3 percent of the total genomic DNA. The name for the family comes for the fact that the sequence contains a restriction site for the enzyme *Alu*I ("Al-you-one"). Over evolutionary time, members of the family have diverged so the sequences of the individuals in the family are related, but not identical.

Sequence analysis of members of the Alu family indicated that each Alu sequence is flanked by direct repeats of 7 to 20 bp. Since integrated transposons and retrotransposons that we have discussed are flanked by direct repeats resulting from the integration event, it is hypothesized that Alu sequences are transposable elements. Even though most of the moderately repetitive DNA in the genome is not transcribed, there is evidence that at least some members of the Alu family can be transcribed. Thus, a model is that transcriptionally active Alu sequences are actually retrotransposons that move via an RNA intermediate, much like the yeast *Ty* element and the *Drosophila copia* element.

Evidence that Alu sequences can transpose has come from the study of a young male patient with neurofibromatosis, a genetic disease caused by an autosomal dominant mutation. Individuals with neurofibromatosis develop tumorlike growths (neurofibromas) over the body (see Chapter 4, p. 119). DNA analysis showed that an Alu sequence was present in one of the introns of the neurofibromatosis gene of the patient. Analysis of RNA transcripts from this gene showed that they were longer than those from normal individuals or from other neurofibromatosis patients with different mutational lesions. The presence of the Alu sequence in the intron disrupts the processing of the transcript, causing one exon to be lost completely from the mature mRNA. As a result, the protein encoded is 800 amino acids shorter than normal and is nonfunctional. Neither parent has neurofibromatosis and neither has an Alu sequence in the neurofibromatosis gene. As was mentioned earlier, individual members of the Alu family are not identical in sequence. This made it possible to track down the same Alu sequence in the patient's parents. From this analysis, it was concluded that an Alu sequence mostly likely inserted into the neurofibromatosis gene by retrotransposition in the germ line of the father. Since the movable Alu sequences (called *Alu elements*) do not encode any of the enzymes presumably needed for retrotransposition, those functions must be provided in some other, as yet unknown, way.

One mammalian LINEs family, LINEs-1 (also called *L1 elements*), is also thought to have the means to transpose via an RNA intermediate; i.e., they are retrotransposons. In humans, there are 50,000 to 100,000 copies of the L1 element, comprising about 5 percent of the genome. The maximum length of L1 elements is 6,500 bp, although only about 3,500 of them in the genome are of that full length, the rest having various length internal deletions (much like corn *Ds* elements). The full-length L1 elements contain a large open reading frame that has homology with known reverse transcriptases. When the yeast *Ty* element reverse transcriptase gene was replaced with the putative reverse transcriptase gene from L1, the *Ty* element was able to transpose. Point mutations introduced into the sequence abolished the enzyme activity indicating that the L1 sequence can indeed make a functional reverse transcriptase. Thus, the current thinking is that L1 elements are a class of ubiquitous retrotransposons. These elements do not have LTRs, so they are not closely related to the retrotransposons we have already discussed. Interestingly, in 1991 two unrelated cases of hemophilia in children were shown to result from insertions of an L1 element into the factor VIII gene, the product of which is required for normal blood clotting. Molecular analysis showed that the insertion was not present in either set of parents, leading to the conclusion that the L1 element had newly transposed. More broadly, these results show that L1 elements in humans can indeed transpose and that they can cause disease by insertional mutagenesis.

SUMMARY

Bacteria and eukaryotic cells contain a variety of transposable elements that have the property of moving from one site to another in the genome. In bacteria there are three types of transposable elements: insertion sequence (IS) elements, transposons (Tn), and certain bacteriophages. The simplest type of transposable element is an IS element. An IS element typically consists of inverted terminal repeat sequences flanking a coding region, the products of which provide transposition activity. Tn elements are more complex in that they contain other genes. There are two types of prokaryotic transposons. Composite transposons consist of a central region flanked on both sides by IS elements. The central region contains genes, such as those for drug resistance. The IS elements contain the genes encoding the proteins required for transposition. Noncomposite transposons consist of a central region containing genes (e.g., for drug resistance), but they do not end with IS elements. Instead, short repeated sequences are found at their ends that are required for transposition. In these transposons, the transposition functions are encoded by genes in the central region.

Transposable elements in eukaryotes resemble bacterial transposons in general structure and transposition properties. Corn transposons often occur as families, each family containing an autonomous element (an element capable of transposing by itself) and one or more nonautonomous elements (elements that can only transpose if the autonomous element of the family is also present in the genome). Interestingly, and perhaps uniquely, the timing and frequency of transposition of corn transposons is developmentally regulated. Some eukaryotic transposons, such as yeast *Ty* elements, *Drosophila copia* elements, human (and maybe other mammalian) SINEs and LINEs family members transpose via an RNA intermediate (using a transposon-encoded reverse transcriptase in the case of *Ty*, *copia*, and LINEs). These types of transposons resemble retroviruses in genome organization and other properties, and hence have been called retrotransposons.

The presence of bacterial transposons (IS and Tn elements) and eukaryotic transposons in a cell is usually detected by the changes they bring about in the expression and activities of the genes at or near the chromosomal sites into which they integrate. Gene expression may be increased or decreased if the element inserts into a promoter or other regulatory sequence, mutant alleles of a gene can be produced if an element inserts within the coding sequence of the gene, and various chromosome rearrangements or chromosome breakage events can occur as a result of the mechanics of transposition. In addition, transcription of genes can be turned on next to a transposon because of the action of promoters in the transposons themselves. Often, mutant alleles produced by insertion of a transposable element are unstable, since they revert when the transposable element undergoes a new transposition event.

ANALYTICAL APPROACHES FOR SOLVING GENETICS PROBLEMS

Q20.1 Imagine that you are a corn geneticist and are interested in a gene you call *zma,* which is involved in formation of the tiny hairlike structures on the upper surfaces of leaves. You have a cDNA clone of this gene. In a particular strain of corn that contains many copies of *Ac* and *Ds,* but no other transposable elements, you observe a mutation of the *zma* gene. You want to figure out whether this mutation does or does not involve the insertion of a transposable element into the *zma* gene. How would you proceed? Suggest at least two approaches, and say how your expectations for an inserted transposable element would differ from your expectations for an ordinary gene mutation.

A20.1 One approach would be to make a detailed examination of leaf surfaces in mutant plants. Since there are many copies of *Ac* in the strain, if a transposable element has inserted into *zma* it should be able to leave again, so that the mutation of *zma* would be unstable. The leaf surfaces should then show a patchy distribution of regions with and regions without the hairlike structures. A simple point mutation would be expected to be relatively stable.

A second approach would be to digest the DNA from mutant plants and DNA from normal plants with a particular restriction endonuclease, run the digested DNAs on a gel and prepare a Southern blot, and probe the blot using the cDNA. If a transposable element has inserted into the *zma* gene in the mutant plants, then the probe should bind to different molecular weight fragments in mutant as compared to normal DNAs. This would not be the case if a simple point mutation had occurred.

QUESTIONS AND PROBLEMS

20.1 Distinguish between prokaryotic insertion elements and transposons. How do composite transposons differ from noncomposite transposons?

20.2 What are the properties in common between bacterial and eukaryotic transposable elements?

20.3 An IS element became inserted into the *lacZ* gene of *E. coli.* Later, a small deletion occurred in this gene, which removed forty base pairs, starting to the left of the IS element. Ten *lacZ* base pairs were removed, including the left copy of the target site, and the thirty leftmost base pairs of the IS element were removed. What will be the consequence of this deletion?

20.4 Although the detailed mechanisms by which transposable elements transpose differ considerably, some features underlying transposition are shared. Examine the shared and different features by answering the following questions.

a. Use an example to illustrate different transposition mechanisms that require:

  i. DNA replication of the element

  ii. no DNA replication of the element

  iii. an RNA intermediate

b. What evidence is there that the inverted or direct terminal repeat sequences found in transposable elements are essential for transposition?

c. Do all transposable elements generate a target site duplication following insertion?

20.5 In addition to single gene mutations caused by the insertion of transposable elements, the frequency of chromosomal aberrations such as deletions or inversions can be increased when transposable elements are present. How?

*20.6 An understanding of the molecular structures of *Ac* and *Ds* elements, together with an understanding of the basis for their transposition, has allowed for a clearer, molecular-based interpretation of Barbara McClintock's observations. Ponder the significance of this on the acceptance of her work and use your understanding of transposition of *Ac* and *Ds* elements in corn to propose an explanation for the following results.

P:	A × C	A × D
F$_1$:	all purple	all purple
F$_2$:	3 purple : 1 colorless	3 purple : 1 colorless
P:	B × C	B × D
F$_1$:	all purple	all purple
F$_2$:	3 purple : 1 spotted*	3 purple : 1 colorless

*spotted = kernels with purple spots in a colorless background

a. Two different true-breeding strains of corn with colorless kernels A and B are crossed with each of two different true-breeding strains with purple kernels C and D. The F$_1$ from each cross is selfed, with the following results:

b. When strain B is crossed to a large number of other true-breeding purple strains, those that show a 3 purple : 1 spotted phenotype in the F$_2$ can be grouped into three categories according to the degree of spotting. Some show relatively large spots distributed in sectors within the kernels, some show smaller spots distributed throughout the kernels, and some show a very fine speckling pattern.

20.7 A geneticist was studying glucose metabolism in yeast, and had deduced both the normal structure of the enzyme glucose-6-phosphatase (G6Pase) and the DNA sequence of its coding region. She had been using a wild-type strain called A to study another enzyme for many generations, when she noticed a morphologically peculiar mutant had arisen from one of the strain A cultures. She grew the mutant up into a large stock and found that the defect in this mutant involved a markedly reduced G6Pase activity. She isolated the G6Pase protein from these mutant cells and found it was present in normal amounts but had an abnormal structure. The N-terminal 70 percent of the protein was normal. The C-terminal 30 percent was present but altered in sequence by a frame shift reflecting the insertion of 1 base pair, and the N-terminal 70 percent and the C-terminal 30 percent were separated by 111 new amino acids unrelated to normal G6Pase. These amino acids represented predominantly the AT rich codons (Phe, Leu, Asn, Lys, Ile, Tyr). There were also two extra amino acids added at the C-terminal end. Explain these results.

*20.8 Consider two theoretical yeast transposons, A and B. Each contains an intron. Each transposes to a new location in the yeast genome and then is examined for the presence of the intron. In the new locations, you find that A has no intron, but B does. What can you conclude about the mechanisms of transposon movement for A and B from these facts?

20.9 When certain strains of *Drosophila melanogaster* are crossed, hybrid dysgenesis can result in mutations, chromosomal aberrations, and sterility. A curious feature of hybrid dysgenesis is that it has been seen when females from certain laboratory strains (M cytotype) are mated to males from wild populations (P cytotype), but not when males from the same laboratory strains are mated to females from wild populations.

a. What occurs during the development of the F$_1$ hybrids to cause hybrid dysgenesis?

b. Why does it occur only in F$_1$ hybrids from M females × P males, and not in F$_1$ hybrids from P females × M males?

c. Why does hybrid dysgenesis only affect the F$_1$ germ line, and not F$_1$ somatic cells?

d. Some strains of P cytotype males are more potent at causing hybrid dysgenesis than others. Why might this be the case?

e. How might you test your explanation in (c)? Assume that you have a plasmid vector with an intact *P* element cloned into it.

20.10 In the experiment described in Figure 20.20, Rubin and Spradling demonstrated that they could introduce a wild-type *ry*$^+$ (*rosy*$^+$) gene into a strain homozygous for a mutant *ry* allele by using *P* element–mediated gene transformation.

a. If a transformed male was mated to an M cytotype female, would you expect the *P* element construct used by Rubin and Spradling to transpose to another site in the genome of the F$_1$?

b. If a transformed male was mated to a P cytotype female, would you expect the *P* element construct used by Rubin and Spradling to transpose to another site in the genome of the F$_1$?

c. Crosses of transformed flies to well-defined laboratory strains are more often useful for experimentation than crosses to animals from wild populations. Based on your answers to (a) and (b), what problems may arise from using transformants obtained by the approach in Figure 20.20?

20.11 After the discovery that *P* elements could be used to develop transformation vectors in *Drosophila melanogaster*, attempts were made to use them for the development of germ-line transformation in several different insect species. Recently, Savakis and his colleagues have succeeded in using a different transposable element found in *Drosophila*, the *Minos* element, to develop germ-line transformation for both *Drosophila melanogaster* and the medfly, *Ceratitus capitata*, a major agricultural pest present in Mediterranean climates.

a. What is the value to developing a transformation vector for an insect pest?

b. What basic information about the *Minos* element would need to be gathered before it could be used for germ-line transformation?

21 EXTRANUCLEAR GENETICS

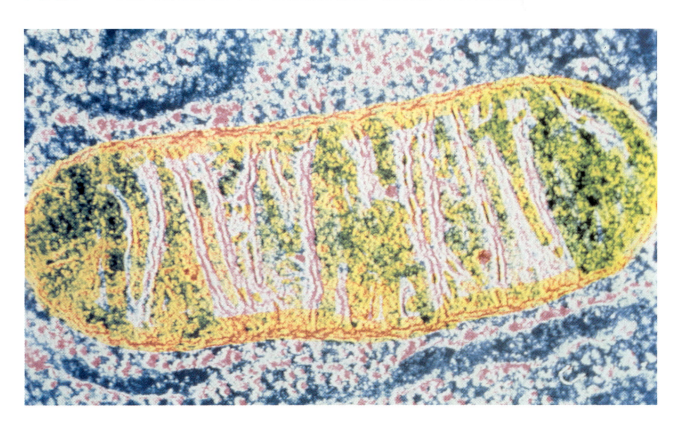

ORGANIZATION OF EXTRANUCLEAR GENOMES
  Mitochondrial Genome
  Chloroplast Genome
  RNA Editing
  Origin of Mitochondria and Chloroplasts
RULES OF EXTRANUCLEAR INHERITANCE
EXAMPLES OF EXTRANUCLEAR INHERITANCE
  Leaf Variegation in the Higher Plant *Mirabilis jalapa*
  The [*poky*] Mutant of *Neurospora*
  Yeast *petite* Mutants
  Extranuclear Genetics of *Chlamydomonas*
  Human Genetic Diseases and Mitochondrial DNA Defects
  Exceptions to Maternal Inheritance
  Infectious Heredity—Killer Yeast
MATERNAL EFFECT
GENOMIC IMPRINTING

PRINCIPAL POINTS

~ Both mitochondria and chloroplasts contain their own DNA genomes. The DNA in most species' mitochondria and all chloroplasts is circular, double-stranded, and supercoiled. The mitochondrial DNA of some species is linear. Typically mitochondria and chloroplasts contain several nucleoid regions in which the DNA is located, and each nucleoid contains several copies of the DNA molecule.

~ The mitochondrial and chloroplast genomes contain genes for the rRNA components of the ribosomes of these organelles, for many (if not all) of the tRNAs used in organellar protein synthesis, and for a few proteins that remain in the organelles and perform functions specific to the organelles. All other proteins are nuclear-encoded, synthesized on cytoplasmic ribosomes, and transported into the organelles. The genetic code is different from that found in nuclear protein-coding genes in the mitochondria of some organisms. Mitochondrial DNA analysis is used to study genetic relationships between individuals because of the maternal inheritance of mitochondria and polymorphisms of mitochondrial DNA.

~ The inheritance of mitochondrial and chloroplast genes follows rules different from those for nuclear genes: no meiotic segregation is involved, uniparental (usually maternal) inheritance is typically exhibited, extranuclear genes are non-mappable to the known nuclear linkage groups, and a phenotype resulting from an extranuclear mutation persists after nuclear substitution.

~ Examples of extranuclearly inherited gene mutations, involving genes in mitochondrial DNA or chloroplast DNA, include leaf variegation in four o'clock plants, certain slow-growing mutants in fungi, and certain antibiotic resistance or dependence traits in *Chlamydomonas*.

~ Not all cases of extranuclear inheritance result from genes on mitochondrial DNA or chloroplast DNA. Many other examples in eukaryotes result from infectious heredity, in which symbiotic, cytoplasmically located bacteria or viruses are transmitted when cytoplasms mix.

~ Maternal effect is defined as a phenotype in an individual that is established by the maternal nuclear genome, as the result of mRNA and/or proteins that are deposited in the oocyte prior to fertilization. These inclusions direct early development of the embryo. Maternal effect is different from extranuclear inheritance. The maternal inheritance pattern of extranuclear genes occurs because the zygote obtains most of its cellular organelles (containing the extranuclear genes) from the female parent.

~ While most genes are expressed in a way that is independent of parental origin, the expression of certain genes is determined by whether the gene is inherited from the female or male parent. This phenomenon is called genomic imprinting. Attention is being focussed on methylation patterns as a basis for imprinting.

*I*n our discussion of eukaryotic genetics up to now, we have analyzed the structure and expression of the genes located on chromosomes in the nucleus and have defined rules for the segregation of nuclear genes. The nucleus is not the only place in the cell in which DNA is located, however. Outside the nucleus, DNA is found in two principal cytoplasmic organelles, the mitochondrion (found in both animals and plants) and the chloroplast (found only in green plants). The genes in these mitochondrial and chloroplast genomes have become known as extrachromo-somal genes, cytoplasmic genes, non-Mendelian genes, organellar genes, or extranuclear genes.

Although *extranuclear* is used in the discussions that follow, the term *non-Mendelian* is also informative because extranuclear genes do not follow the rules of Mendelian inheritance like nuclear genes. Since the cytoplasm is inherited from the mother, the inheritance of such cytoplasmic factors is strictly maternal. Recently, the application of modern molecular biology techniques has led to rapid advances in knowledge about the organization of extranuclear genomes.

In this chapter we first examine some of these advances and then discuss the inheritance patterns of extranuclear genes.

ORGANIZATION OF EXTRANUCLEAR GENOMES

Mitochondrial Genome

Mitochondria, organelles found in the cytoplasm of all aerobic animal and plant cells, are the principal sources of energy in the cell (see Figure 1.7, p. 8). They contain the enzymes of the Krebs cycle, carry out oxidative phosphorylation, and are involved in fatty acid biosynthesis. The presence of DNA in mitochondria has been shown by a variety of techniques, including electron microscopy, autoradiography, and the isolation from mitochondria of a DNA species with a buoyant density different from that of nuclear DNA by virtue of different GC percentages (Figure 21.1). The application of recombinant DNA techniques is leading to detailed restriction maps of mitochondrial DNA (mtDNA) from many species, and the complete DNA sequences of a number of mtDNA molecules are now known. Some of the properties of mitochondrial DNA, or mtDNA, will now be described.

STRUCTURE. Many mitochondrial (mt) genomes are circular, double-stranded, supercoiled DNA molecules. Linear mitochondrial genomes are found in some protozoa and some fungi. In many cases the GC content of mtDNA differs greatly from that of the nuclear DNA, which allows mtDNA to be isolated by CsCl equilibrium density gradient centrifugation. No structural proteins are associated with the mtDNA. Figure 21.2 presents an electron micrograph of a circular mtDNA molecule.

The gene content is very similar among mitochondrial genomes from different species in both number and function. Despite that, the size of the genome varies tremendously from organism to organism. In animals the circular mitochondrial genome is less than 20 kb. For example, the sizes of three mammalian mtDNA genomes determined by DNA sequencing are: human—16,569 bp; mouse—16,275 bp; and cow—16,338 bp. In the frog *Xenopus laevis* the mtDNA is about 18,400 bp, which is the same value found for mtDNA from *Drosophila melanogaster*. The complete sequences of these two species' mtDNAs are not yet known. By contrast, the circular yeast mitochondrial genome is about 80 kb (80,000 bp). The mitochondrial genomes of plants are even larger, ranging from 100,000 to 2 million base

pairs. For example, the mtDNA of pea is 110,000 bp and that of corn is 570,000 bp.

Despite the differences in the amount of DNA, mitochondria from all organisms contain about the same amount of unique-sequence DNA; that is, DNA that codes for functional products. Therefore, despite the disparity in mitochondrial size, the main difference between animal, plant, and fungal mitochondria is that essentially the entire mitochondrial genomes of animals encode products, whereas the mitochondrial genomes of fungi and plants include a lot of DNA that does not code for products, in addition to the coding sequences.

While the informational content of the mitochondrial genome is very small in comparison to the nuclear genome, the relative amount of mitochondrial DNA is actually quite large. Within mitochondria are *nucleoid regions* (similar to those of bacterial cells), each

~ FIGURE 21.1

Nuclear and mitochondrial DNA. Mitochondrial DNA can be detected under the electron microscope as small circles and can often be separated from nuclear DNA by CsCl density gradient centrifugation, since the two DNAs differ in buoyant density.

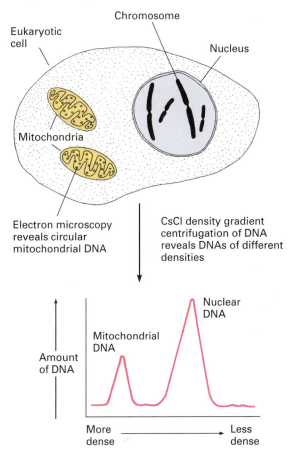

~ FIGURE 21.2

Electron micrograph of mitochondrial DNA.

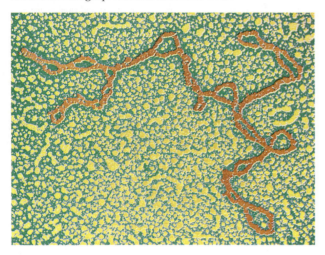

~ FIGURE 21.3

Electron micrograph of the D loop structure found in replicating mitochondrial DNA; a small D loop is shown.

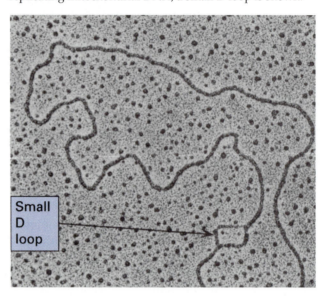

Small D loop

of which contains several copies of the mitochondrial chromosome. Yeast has between 4 and 5 mtDNA molecules per nucleoid, and each mitochondrion has 10 to 30 nucleoids. Since each yeast cell has between 1 and 45 mitochondria per cell, there are between 40 and 6,750 mtDNA molecules per cell. Each mtDNA molecule is 80 kb so there is between 3,200 and 540,000 kb of mitochondrial DNA in the cell, compared with 17,500 kb of genomic DNA in the haploid nucleus.

REPLICATION. Replication of mtDNA is semi-conservative, uses DNA polymerases specific to the mitochondrion, and involves no proofreading. As for replication of nuclear and other DNA, RNA primers are synthesized for initiation. The mitochondrial DNA replication process occurs throughout the cell cycle, without any preference for the S phase of the cell cycle, which is when nuclear DNA replicates. Some differences appear in the details of mtDNA replication among eukaryotes, and the displacement loop (D loop) model (deduced from observing animal mitochondria *in vivo*) is useful as a general scheme. Figure 21.3 shows an electron micrograph of an early stage of D loop replication; Figure 21.4 diagrams the model in more detail. The model is as follows:

~ FIGURE 21.4

Model for mitochondrial DNA replication that involves the formation of a D loop structure.

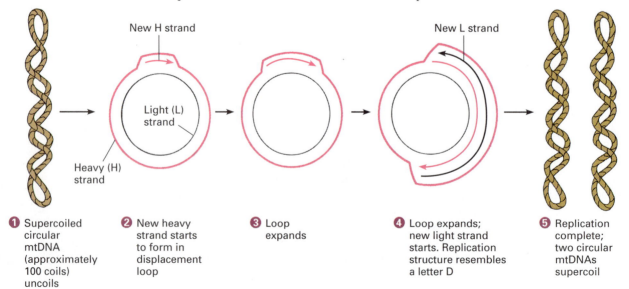

New H strand

New L strand

Light (L) strand

Heavy (H) strand

❶ Supercoiled circular mtDNA (approximately 100 coils) uncoils

❷ New heavy strand starts to form in displacement loop

❸ Loop expands

❹ Loop expands; new light strand starts. Replication structure resembles a letter D

❺ Replication complete; two circular mtDNAs supercoil

In most animals the two strands of mtDNA have different densities, so they are called the H (heavy) and L (light) strands. The D loop model of replication shows relative asynchrony in DNA replication for the two complementary H and L strands. In the D loop model, the synthesis of the new H strand is started at a replication origin for the H strand and forms a D loop structure. As the new H strand extends to about halfway around the molecule, initiation of synthesis of the new L strand at a second replication origin takes place. Both strands are completed by continuous replication. Lastly, the circular DNAs are each converted to a supercoiled form, with approximately one hundred superhelical twists.

How does duplication of the mitochondrion itself occur? The two models proposed were (1) *de novo* synthesis, in which new mitochondria are assembled from simpler components; and (2) growth and division, in which new mitochondria are produced by division of preexisting mitochondria. Experimental evidence obtained by David Luck in 1963 indicates that the second model is correct; that is, mitochondria grow and divide.

GENE ORGANIZATION OF MTDNA. Our present-day understanding of the gene organization of mtDNA derives from DNA sequencing experiments, restriction mapping, and analysis of the mRNA, rRNA, and tRNA transcripts from mitochondria. As mentioned previously, the complete sequences of a number of mtDNA genomes are now known. Analysis of the sequences enables the precise organization of genes and intergene regions to be determined. Figure 21.5 presents the map of the genes of human mtDNA.

~ **FIGURE 21.5**

Map of the genes of human mitochondrial DNA. The outer circle shows the genes transcribed from the H (heavy) strand, and the inner circle shows the genes transcribed from the L (light) strand. The origins of replication (*ori*) for the H and L strands are indicated. tRNA genes are shown in purple and protein-coding genes in yellow. Code: ATPase 6 and 8: components of the mitochondrial ATPase complex. COI, COII, and COIII: cytochrome c oxidase subunits. cyt b: cytochrome b. ND1-6: NADH dehydrogenase components (previously URF1-6, where URF = unidentified reading frame).

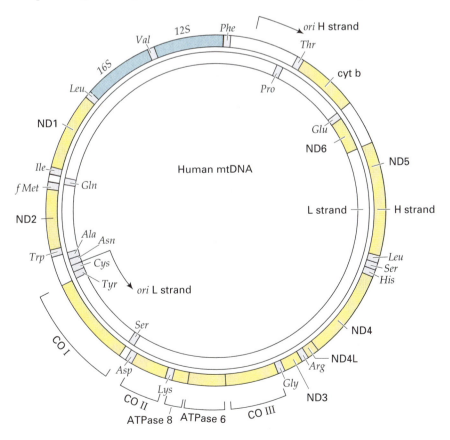

In general, mitochondrial DNA contains information for a number of mitochondrial components, such as tRNAs, rRNAs, and some of the polypeptide subunits of the proteins cytochrome oxidase, NADH-dehydrogenase and ATPase. Other components found in the mitochondria are encoded by nuclear genes and must be imported into the mitochondria. These components include the DNA polymerase and other proteins for mtDNA replication, RNA polymerase and other proteins for transcription, ribosomal proteins for ribosome assembly, protein factors for translation, the aminoacyl tRNA synthetases, and the other polypeptide subunits for the three proteins mentioned above.

Messenger RNAs synthesized within the mitochondria remain in the organelle and are translated by mitochondrial ribosomes which are assembled within the organelle. Like cytoplasmic ribosomes and bacterial ribosomes, mitochondrial ribosomes consist of two subunits. In humans, the 60S mitochondrial ribosomes consist of 45S and 35S subunits. There are only two rRNAs in the mitochondrial ribosome: 16S rRNA in the large subunit and 12S rRNA in the small subunit. There is one gene for each rRNA in the mitochondrial genome. Figure 21.6 diagrams the organization of the two rRNA genes and illustrates the variation in the organization of rRNA genes in selected mitochondrial genomes. In animals, as exemplified by HeLa (human) cells, the genes for the small (12S) and large (16S) rRNAs are very closely linked to one another on the H strand. In the spacer separating them, one tRNA gene is found. By contrast, in fungi the two genes are separated by a large amount of DNA: 6 kb in *Neurospora* and about 30 kb in yeast. A number of tRNA genes are located in these spacer regions. The rRNA genes of *Neurospora* and yeast both contain introns (see Figure 21.6). In the protozoan *Tetrahymena* there is an extra copy of the large rRNA gene, but not of the small rRNA gene. This extra copy has presumably arisen by a duplication event and is not commonly found in other mitochondrial genomes.

The ribosomal proteins found in mitochondrial ribosomes generally are encoded by nuclear genes and transported into the mitochondria from the cytoplasm. In a few organisms, one or more mitochondrial ribosomal proteins are encoded by mitochondrial genes. Mitochondrial ribosomal proteins are different from the proteins of cytoplasmic ribosomes.

The mitochondrial protein-coding genes are found in both strands. Their positions were identified in two ways. One method involved searching the DNA sequence for possible start and stop signals in the same reading frame (an *open reading frame*, or ORF, also called *unidentified reading frame*, or URF). The signals had to be separated by significant lengths; the sequences in between had to be able to code for

~ FIGURE 21.6

Organization of mitochondrial rRNA and mitochondrial tRNA genes in HeLa cells and in a number of organisms.

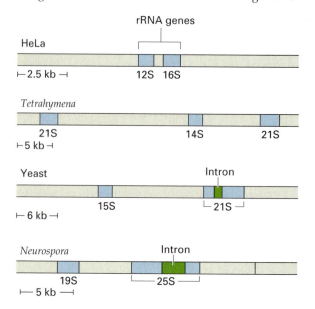

amino acids. This search was accompanied by using computer programs designed for this purpose. The second method (used primarily by G. Attardi and his group) was to align the 5' and 3' end proximal sequences of mitochondrial mRNAs with the corresponding mitochondrial DNA sequence. All of the human mitochondrial ORFs have been assigned a function.

TRANSCRIPTION OF MtDNA. The transcription of mammalian mtDNA is unusual. Rather than each gene being transcribed independently, both strands of the entire mitochondrial genome are transcribed into two single RNA strands. The origins for both strands are near the DNA replication origin for the H strand (see Figure 21.5). In addition, some shorter transcripts are made of the H strand that stop at the far end of the 16S rRNA gene. All the transcripts are made by an RNA polymerase encoded by a nuclear gene. The exact mechanism of transcription initiation is not known, although the promoter elements are known to be in short sequences flanking the transcription start sites.

How are the large RNA transcripts processed to produce the mature mRNAs, rRNAs, and tRNAs? Notice in the organization map for human mtDNA shown in Figure 21.5 that most of the genes encoding the rRNAs and the mRNAs are punctuated by tRNA genes. A current model for processing has the tRNAs playing a crucial role. That is, the tRNA sequences fold up into characteristic cloverleaf shapes which are recognized by specific enzymes that cut the tRNAs out of the transcript. Since there are essentially no gaps between genes in animal mtDNA, the removal of the tRNAs liberates essentially complete mRNAs and rRNAs. The processed transcripts are then modified to produce the mature RNAs. That is, a poly(A) tail is added to the 3' ends of the mRNAs, and CCA is added to the 3' ends of the tRNAs. Mitochondrial mRNAs have no 5' cap.

Interestingly, the DNA sequences for some of the mitochondrial mRNAs do not encode complete chain-terminating codons. Instead, the processed transcripts end with either U or UA. The subsequent addition of a poly(A) tail completes the missing part(s) of a UAA stop codon.

The transcription and RNA processing mechanisms just described are characteristic only of animal mitochondria. The much larger mitochondrial genomes of yeast and plants, for example, do not have tRNA genes separating other genes, and the gaps between genes are large. In these systems, transcription termination is signalled by other, non-tRNA sequences in the DNA. In yeast, the gaps between the genes contain long AT-rich sequences that have no obvious biological function. Introns are never found in animal mitochondrial genomes, but they are found in yeast, plants, and some other organisms.

TRANSLATION IN THE MITOCHONDRIA. The mRNAs in animal mitochondria have no 5' cap and the start codon is very near the 5' end of the molecule so that there is virtually no 5' leader sequence. As a result, the initiation of translation must be quite different from that for cytoplasmic mRNAs, where both of those features are present. In other words, mitochondrial ribosomes in animals must bind to the mRNAs and orient themselves to start translation in a unique way. Again, yeast and plant mitochondrial mRNAs are more "traditional" in sequence; although there is no 5' cap, there is a 5' leader so translation initiation may be like that in the cytoplasm; that is, at the first AUG.

The process of protein synthesis in the mitochondrion is, in some ways, analogous to the process of protein synthesis in bacteria. In all mitochondria, like *E. coli*, a mitochondrial tRNA.fMet is used in the initiation of protein synthesis. Special mitochondrial initiation factors (IFs), elongation factors (EFs), and release factors (RFs) distinct from cytoplasmic factors are used for the translation process. The initiation factors (IFs) are also very similar to those in bacteria. In a number of cases mitochondrial and bacterial components are interchangeable in an *in vitro*, protein-synthesizing system. For example, *Neurospora crassa* ribosomes are able to recognize, bind, and translocate *E. coli* fMet-tRNA.fMet in response to AUG, and *E. coli* IFs can substitute for mt IFs in catalyzing the initiation of protein synthesis.

Another piece of evidence for the similarity between *E. coli* and mt ribosomes is that the latter are sensitive to most inhibitors of bacterial ribosome function, including streptomycin, spectinomycin, neomycin, and chloramphenicol. Also, mt ribosomes are generally insensitive to antibiotics or other agents to which cytoplasmic ribosomes are sensitive, such as cycloheximide. By the selective use of antibiotics, the site of synthesis of proteins found in mitochondria can be investigated. For example, those mt proteins that are synthesized in the presence of cycloheximide, an inhibitor of cytoplasmic ribosomes, must be made on mt ribosomes. Conversely, those mt proteins made in the presence of chloramphenicol, an inhibitor of mt ribosomes, must be made on cytoplasmic ribosomes. Using this approach, scientists have discovered that four of the seven subunits of the mitochondrial enzyme cytochrome oxidase in yeast mitochondria are made in the cytoplasm, and the other three are made in the mitochondria (see Figure 21.7).

For protein synthesis, only plant mitochondria use the universal nuclear genetic code. Mitochondria of other organisms have differences from that universal code, although there is no one pattern for the differences. For example, the differences in the human mitochondrial code from the universal nuclear code are as follows (Figure 21.8):

1. AUA and AUG both encode methionine instead of only AUG in the nuclear code. In the nuclear code, AUA specifies isoleucine.
2. UGA in mitochondria specifies tryptophan while in the nucleus UGA is a stop codon.
3. AGA and AGG are stop codons in the mitochondria and arginine codons in the nucleus.

Interestingly, the mitochondrial genetic codes are not the same in all organisms. For example, Table 21.1 shows the points of difference between human and yeast mitochondrial genetic codes. Further, differences in the nuclear genetic code have also been found in ciliated protozoans. In other words, the genetic code is not quite as universal as was originally thought.

There are also differences in mitochondria with respect to the tRNAs and how they read the mRNAs. We learned in Chapter 14 about base-pairing wobble, in which one tRNA may read more than one codon (see Table 14.1, p. 428). Recall that there are many instances in which the first two bases in the codon are the same and the same amino acid is specified, whether the third base is U or C (e.g., UUU or UUC = Phe), A or G (e.g., AAA or AAG = Lys), or U, C, A, or G (e.g., GUU, GUC, GUA, or GUG = Val). The term *family box* is used for groups of codons of this kind: UUU or UUC are a family box for phenylalanine

~ FIGURE 21.7

Synthesis of the multisubunit protein cytochrome oxidase takes place on both cytoplasmic and mitochondrial ribosomes.

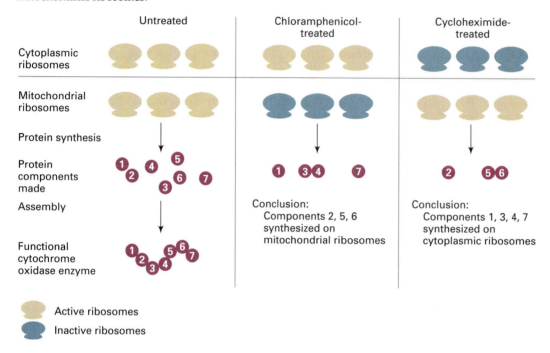

Active ribosomes

Inactive ribosomes

(Phe), for example. In the cytoplasm, no tRNA can read more than three of these related codons, even with wobble. Even when the same amino acid is specified by the first two bases no matter what the third base is (a four-codon family box), the four codons need *two* tRNAs in order for them to be read (see Figure 21.8's "universal code" columns). Many of the mitochondrial tRNAs, however, apparently have the ability to read all four members in a four-codon family box. Such tRNAs have U (uracil) in the anticodon at the position corresponding to the third letter of the codon. In other words, wobble is "more liberal" in the mitochondria, so some tRNAs are able to recognize four different codons. In the "human mitochondria" columns in Figure 21.8, one tRNA reads each of the codons in a shaded box. As a result of the extended wobble, only 22 tRNA genes are needed in mammalian mitochondria to read all 60 sense codons. (This correlates with the 22 tRNA genes identified in human mtDNA through the use of a computer program designed to search the DNA sequence for polynucleotide stretches that could form the classical cloverleaf, two-dimensional structure.) In contrast, with wobble, a theoretical minimum of 32 tRNAs is needed to read the universal code. In *E. coli*, where all the tRNAs have been identified, 40 different tRNAs are needed to read all 61 sense codons (see Figure 21.8).

Structurally, a number of features thought to be invariable in cytoplasmic tRNAs in terms of biological function vary in mitochondrial tRNAs. For example, one mitochondrial serine tRNA has one complete loop missing, yet it is still functional.

MITOCHONDRIAL RIBOSOMES. Mitochondrial ribosomes are structurally diverse among organisms, with all of them having a higher protein/RNA ratio

~ TABLE 21.1

Differences Between Human and Yeast Mitochondrial Genetic Codes

| | | AMINO ACID | |
| | | MITOCHONDRIAL CODE | |
CODON[a]	NUCLEAR CODE	MAMMAL	YEAST
UGA	Termination	Tryptophan	Tryptophan
AUA	Isoleucine	Methionine	Isoleucine
CUN[b]	Leucine	Leucine	Threonine
AGG, AGA	Arginine	Termination	Arginine
CGN[b]	Arginine	Arginine	Termination?

[a]All sequences read 5' to 3'.
[b]N = any one of the four bases A, G, U, and C.

~ FIGURE 21.8

Comparison of the universal (Univ) and human mitochondrial (Human mito) genetic codes. Each shaded box indicates the amino acid encoded by the codons involved. Codons that have different properties in the human mitochondrial code vs. the universal code are shown as white letters in red boxes. The numbers in parentheses below each amino acid indicate the number of different tRNAs that are used to read the codons involved in the box. For the universal code, the values are for *E. coli* tRNAs. For example, for the Pro (proline) box, three tRNAs are needed to read the CCU, CCC, CCA, and CCG codons in *E. coli*, and only one tRNA is needed to read the same codons in the human mitochondrial code.

First letter	U	Univ.	Human mito.	C	Univ.	Human mito.	A	Univ.	Human mito.	G	Univ.	Human mito.	Third letter
U	UUU	Phe (1)	Phe (1)	UCU	Ser (3)	Ser (1)	UAU	Tyr (1)	Tyr (1)	UGU	Cys (1)	Cys (1)	U
	UUC			UCC			UAC			UGC			C
	UUA	Leu (2)	Leu (1)	UCA			UAA	Stop	Stop	UGA	Stop	**Trp (1)**	A
	UUG			UCG			UAG	Stop	Stop	UGG	Trp (1)	Trp	G
C	CUU	Leu (3)	Leu (1)	CCU	Pro (3)	Pro (1)	CAU	His (1)	His (1)	CGU	Arg (2)	Arg (1)	U
	CUC			CCC			CAC			CGC			C
	CUA			CCA			CAA	Gln (2)	Gln (1)	CGA			A
	CUG			CCG			CAG			CGG			G
A	AUU	Ile (2)	Ile (1)	ACU	Thr (3)	Thr (1)	AAU	Asn (1)	Asn (1)	AGU	Ser (1)	Ser (1)	U
	AUC			ACC			AAC			AGC			C
	AUA		**Met (1)**	ACA			AAA	Lys (1)	Lys (1)	AGA	Arg (1)	**Stop**	A
	AUG	Met (1)	Met	ACG			AAG			AGG		**Stop**	G
G	GUU	Val (2)	Val (1)	GCU	Ala (2)	Ala (1)	GAU	Asp (1)	Asp (1)	GGU	Gly (3)	Gly (1)	U
	GUC			GCC			GAC			GGC			C
	GUA			GCA			GAA	Glu (1)	Glu (1)	CGA			A
	GUG			GCG			GAG			GGG			G

than *E. coli* ribosomes. Table 21.2 shows examples of structural characteristics of mitochondrial ribosomes in humans and yeast. In animals, mitochondrial ribosomes are about the same size or perhaps slightly larger than *E. coli* ribosomes in terms of molecular weight and volume. However, owing to the high protein/RNA content, animal ribosomes have a relatively low buoyant density of 55S to 60S. In fungi such as yeast and *Neurospora*, the mt ribosomes have a sedimentation coefficient of 70S to 75S.

We can make some rough generalizations about the rRNA content of mt ribosomes. In most cases the ribosomes lack the 5S and 5.8S rRNA components characteristic of cytoplasmic ribosomes. Higher plants do have a 5S rRNA molecule in mt ribosomes but it is distinct from that molecule found in cytoplasmic ribosomes. In other words, most mt ribosomes have two unequal-sized subunits, with a single rRNA molecule in each subunit. As we have already discussed, in animal mitochondria these rRNA molecules have sedimentation coefficients of 12S and 16S, which are significantly smaller than the large rRNAs of *E. coli* ribosomes.

The number of ribosomal proteins found in mt ribosomes is not well defined. Yeast appears to have 60 to 70 proteins in the mt ribosomes, while animal

~ TABLE 21.2

Relative Size (in S Values) of Mitochondrial (mt) Ribosomes and Cytoplasmic (cyto) Ribosomes in Humans and Yeast Compared with *E. coli* Ribosomes

COMPONENT	HUMAN (HeLa) mt	HUMAN (HeLa) cyto	YEAST mt	YEAST cyto	*E. coli*
RIBOSOMES	60S	74S	75S	80S	70S
Large subunit	45S	60S	53S	60S	50S
Small subunit	35S	40S	35S	40S	30S
RIBOSOMAL RNA					
Large ribosomal subunit	16S	28S	21S	26S	23S
		5.8S		5.8S	
		5S		5S	
Small ribosomal subunit	12S	18S	15S	18S	16S

mitochondrial ribosomes have between 70 and 100 ribosomal proteins.

The proteins of mitochondrial ribosomes appear to be totally distinct from the proteins found in cytoplasmic ribosomes, and with one possible exception in both *Neurospora* and in yeast, mitochondrial proteins are encoded by nuclear genes. Hence, these proteins are synthesized on cytoplasmic ribosomes and subsequently transported into the mitochondria. This process points again to the remarkable abilities of cells to direct the products of protein synthesis to the specific sites where they are needed. We know that cytoplasmic ribosomes in eukaryotes contain about 65 to 70 ribosomal proteins, and lower eukaryotes appear to have 60 to 70 ribosomal proteins in mitochondrial ribosomes. About half of them subsequently migrate into the nucleus to be used in the assembly of cytoplasmic ribosomes, while the remainder enter the mitochondria, where they are used in the assembly of mitochondrial ribosomes and of transcripts of the two rRNA genes located on the mitochondrial chromosome. But we have much to learn about migration between cellular compartments.

INVESTIGATING GENETIC RELATIONSHIPS BY MtDNA ANALYSIS. In Chapter 15 we discussed some types of DNA analysis that can be done using molecular techniques. A number of such cases involved polymorphisms in the DNA sequences. In human mtDNA one 400-bp region is highly polymorphic and this, along with the fact that the vast majority of mitochondria is maternally inherited, means that maternal lineages are practically unique. Thus, maternal-line relations between individuals can be investigated by using PCR to analyze mtDNA for polymorphisms.

An example of the use of mtDNA analysis involves the last tsar and tsarina of Russia and their children. During the Bolshevik Revolution of 1917, Tsar Nicholas Romanov II was overthrown and exiled, and in 1918 the tsar and his family were executed by Bolshevik guards. Rumors persisted, however, that one of the tsar's daughters, Princess Anastasia, escaped the execution. In 1922 a woman came forward in Berlin claiming to be Anastasia. In 1928, using the name Anna Anderson, she came to the United States, where she lived until her death in 1984. Though she claimed until she died that she was Anastasia, there was insufficient information available to prove or disprove her claim during her lifetime. In 1993, mtDNA analysis was done on bones found two years earlier in a shallow grave in the Russian town to which the Romanovs had been exiled. The DNA samples were compared to a blood sample provided by Prince Philip, Duke of Edinburgh, who is the grand nephew of the Tsarina Alexandra. (Prince Philip's grandmother was Princess Victoria, Alexandra's sister.) The results showed unequivocally that the bones were the remains of the tsarina and three of their five children. That is, their mtDNA patterns perfectly matched the mtDNA of Prince Philip, indicating they all belonged to the same maternal lineage. The bones of the tsar were identified in a similar way by matching mtDNA patterns with those of two living relatives. Soon afterward, mtDNA analysis proved that Anna Anderson was not Anastasia, since her mtDNA pattern did not match that of Prince Philip. It is not clear whether any of the three children was Anastasia, although a Russian government commission has stated that there is "definite proof" that one of the skeletons is that of Anastasia.

The case of the Romanovs is an example in which mtDNA analysis was a powerful tool for analyzing maternal lineages in humans. Mitochondrial DNA analysis is being used for studying genetic relationships in many other organisms as well (see Chapter 22, pp. 763). Mitochondrial DNA analysis is also being used in conservation biology studies to assess the extent of genetic variability in natural populations. One such study is being done with the threatened Yellowstone National Park grizzly bear as a model population for many other endangered species of predators.

Chloroplast Genome

Chloroplasts are cellular organelles found only in green plants, photosynthetic protists, and cyanobacteria (blue-green algae), and are the site of photosynthesis in the cells containing them. Chloroplasts are characterized by a double membrane surrounding an internal, chlorophyll-containing lamellar structure embedded in a protein-rich stroma (see Figure 1.8, p. 8). Like mitochondria, chloroplasts contain their own genomes, although we do not know as much about the chloroplast (cp) genome as we do about the mitochondrial genome.

STRUCTURE AND REPLICATION.

In many respects the structure of the chloroplast genome is similar to the structure of mitochondrial genomes. In all cases the DNA is double-stranded, circular, devoid of structural proteins, and supercoiled. An electron micrograph of cpDNA is shown in Figure 21.9a. In many cases the GC content of cpDNA differs greatly from that of the nuclear and mitochondrial DNA, which allows cpDNA to be isolated by CsCl equilibrium density gradient centrifugation. Figure 21.9b shows the result of such an experiment using total cellular DNA from the unicellular alga *Chlamydomonas* (Figure 21.10). Two discrete bands of DNA are seen (apart from the two marker bands) and indicate the presence of molecular species with different base-pair compositions. Since the cpDNA has a GC content of 36 percent compared with 64 percent for nuclear DNA and 71 percent for mtDNA, the cpDNA produces a *satellite band* that can be isolated and studied further. The relatively large amount of nuclear DNA results in a fairly broad peak that does not permit resolution of the mtDNA peak.

Chloroplast DNA is much larger than animal mtDNA, with a size between 80 kb and 600 kb. The DNA sequences of the chloroplast genomes of a few organisms have been completely determined. For example, the tobacco genome is 155,844 bp, the rice genome is 134,525 bp, and the liverwort *Marchantia polymorpha* genome is 121,024 bp. In other plants, the cpDNA genome size is approximated from restriction mapping and other methods of analysis. For example, *Chlamydomonas* cpDNA is 195 kb, *Euglena* cpDNA is 130-152 kb, *Ginkgo* (a gymnosperm) cpDNA is 158 kb, pea cpDNA is 120 kb, and tomato cpDNA is 158 kb. The chloroplast genome contains a lot of noncoding DNA sequences.

The number of copies of cpDNA per chloroplast varies from species to species. In all cases there are multiple copies per chloroplast, and these copies are found in nucleoid regions that are also present in multiple copies. For example, leaf cells of the garden beet have between 4 and 8 cpDNA molecules per nucleoid, from 4 to 18 nucleoids per chloroplast, and about 40 chloroplasts per cell, giving almost 6,000 cpDNA molecules per cell. In *Chlamydomonas* the one chloroplast in a cell generally contains between 500 and 1,500 cpDNA molecules.

It is not yet known exactly how replication of cpDNA occurs. Chloroplasts themselves grow and divide in essentially the same way as do mitochondria.

GENE ORGANIZATION OF CPDNA.

The chloroplast genome contains genes for all the chloroplast rRNAs (16S, 23S, 4.5S, and 5S) and tRNAs, and genes for some, but not all, of the proteins required by their genetic apparatus (e.g., ribosomal proteins, RNA polymerase subunits, translation factors, etc.) or for photosynthesis. Introns are found in some, but not all, of the protein-coding and tRNA genes in cpDNA. All of the mRNAs transcribed from chloroplast genes are translated by chloroplast ribosomes. Other proteins found in the chloroplast are encoded by nuclear genes. For purposes of comparison, the organization of the chloroplast genome in rice is presented in Figure 21.11.

Characteristically the chloroplast genome contains two copies of each of the rRNA genes. (By contrast, only one copy of each mitochondrial rRNA gene is found in mtDNA.) The two sets of chloroplast rRNA genes are located in two identical 10- to 25-kb repeated sequences located in the genome in an inverted orientation. The inverted repeats are designated IR_A and IR_B in Figure 21.11. Other genes are found in the repeated sequence so they are also duplicated in the chloroplast genome. The location of these repeats defines a short single copy (SSC) region and a long single copy (LSC) region of the chloroplast genome. Recombination between the repeats can produce an inversion of the SSC region relative to the LSC region.

Both tobacco and rice have 30 tRNA genes, while *Marchantia* has 32 tRNA genes. Almost 100 ORFs (open reading frames) have been identified by computer analysis of the cpDNA sequences. Approximately 60 of

~ FIGURE 21.9

Chloroplast (cp) DNA: (a) Electron micrograph of a chloroplast DNA molecule; (b) Example of the results of a CsCl equilibrium density gradient centrifugation experiment done with total DNA extracted from the unicellular alga *Chlamydomonas*. The large amount of nuclear DNA results in a broad peak that obscures the mtDNA peak.

a)

b)   **Satellite cpDNA band in CsCl gradient**

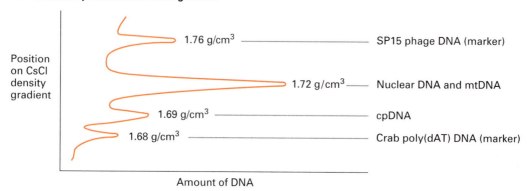

Position on CsCl density gradient

1.76 g/cm³ ——————————— SP15 phage DNA (marker)

1.72 g/cm³ ——— Nuclear DNA and mtDNA

1.69 g/cm³ ——————— cpDNA

1.68 g/cm³ ——————————— Crab poly(dAT) DNA (marker)

Amount of DNA

these ORFs have been correlated with known functional genes, and the others remain to be defined. The ORFs are quite highly conserved among the plants whose sequences are presently known, although some ORFs conserved between tobacco and liverwort are absent in rice. Structural changes are evident in the rice cpDNA compared with tobacco cpDNA, notably an increased length of the inverted repeats such that some genes are duplicated in rice that are present in single copies in tobacco and liverwort, and a series of inversions in the LSC sequence.

CHLOROPLAST RIBOSOMES AND PROTEIN SYNTHESIS.

Chloroplast protein synthesis uses ribosomes that are completely distinct from mitochondri-

~ FIGURE 21.10

Chlamydomonas.

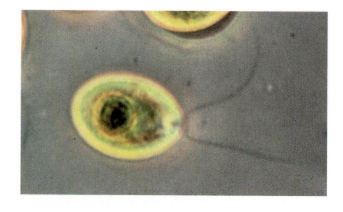

~ FIGURE 21.11

Organizations of the chloroplast genome of rice (*Oryza sativa*) (From "Complete Sequence of the Rice (*Oryza sativa*) Chloroplast . . ." by Hiratsuka et al. in *Molecular and General Genetics*, Vol. 217, 1989. Reprinted by permission of Springer-Verlag, New York, Inc.)

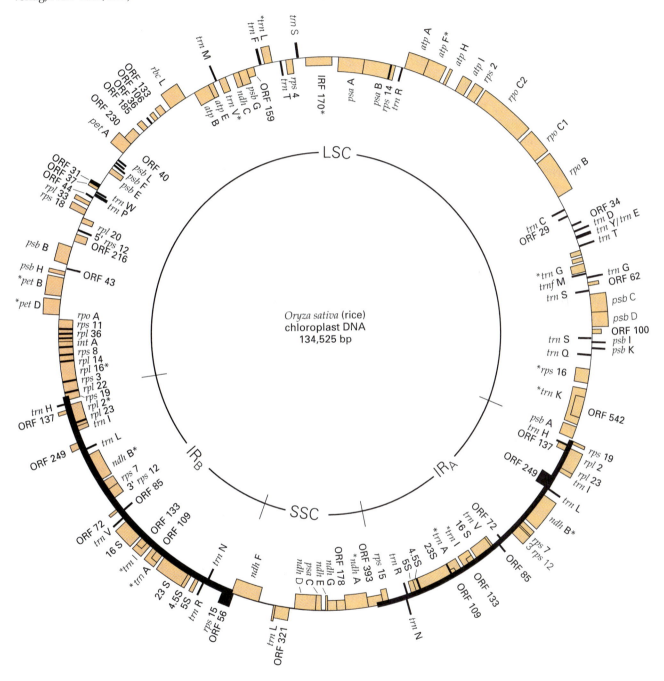

al ribosomes and cytoplasmic ribosomes. They have a sedimentation coefficient of 70S and consist of two unequal-sized subunits, 50S and 30S—thus they are similar in sedimentation coefficients to prokaryotic ribosomes.

The large subunit of the chloroplast ribosome contains one copy each of 23S, 5S, and 4.5S rRNAs. The small subunit of the chloroplast ribosome con-

tains one copy of a 16S rRNA. The number of ribosomal proteins in each ribosomal subunit is less well defined, although we do know that some are nuclear encoded and some are chloroplast encoded.

Protein synthesis in chloroplasts is similar to the process in prokaryotes. Formylmethionyl tRNA is used to initiate all proteins, and the formylation reaction is catalyzed by a transformylase localized in the chloroplast. The chloroplast uses its own initiation factors (IFs), elongation factors (EFs), and release factors (RFs) that are distinct from those of the cytoplasmic protein synthesis system. The universal genetic code is used in chloroplast protein synthesis.

Like mitochondrial ribosomes, chloroplast ribosomes are resistant to cycloheximide, an inhibitor of cytoplasmic ribosomes, but are sensitive to virtually all inhibitors known to block prokaryotic protein synthesis. Using selective antibiotics in essentially the same way as we described for the synthesis of multisubunit mitochondrial proteins (see Figure 21.7), the synthesis of chloroplast proteins has been examined. One important chloroplast protein is ribulose bisphosphate decarboxylase, the first enzyme used in the pathway for the fixation of carbon dioxide in the photosynthetic process. This enzyme is a major protein and is found in the chloroplasts of all plant tissues. Since it constitutes about 50 percent of the protein found in green-plant tissue, it is the most prevalent protein in the world. Ribulose bisphosphate decarboxylase contains eight polypeptides, four identical small ones and four identical large ones. The antibiotic approach as well as a genetic approach have shown that the small polypeptide is coded by a nuclear gene, while the large polypeptide is coded by a chloroplast gene.

The ribosomes themselves are found both free and membrane-bound in the chloroplast. The free ribosomes appear to make the large subunit of the ribulose bisphosphate decarboxylase enzyme, while ribosomes bound to the membrane are presumed to make the hydrophobic proteins that are important in photosynthesis. Hydrophobic proteins are located in the membranes of the chloroplast.

KEYNOTE

Both mitochondria and chloroplasts contain their own DNA genomes. The DNA in most species' mitochondria and all chloroplasts is circular, double-stranded, and supercoiled. The mitochondrial DNA of some species is linear. The mitochondrial and chloroplast genomes contain genes for the rRNA components of the ribosomes of these organelles, for many (if not all) of the tRNAs used in organellar protein synthesis, and for a few proteins that remain in the organelles and perform functions specific to the organelles. All other proteins are nuclear-encoded, synthesized on cytoplasmic ribosomes, and imported into the organelles. At least in the mitochondria of some organisms, the genetic code is different from that found in nuclear protein-coding genes. Because humans and many other organisms receive most of their mitochondria from their mothers, maternal lineages are practically unique. Thus, maternal line relations between individuals can be investigated by mtDNA analysis.

RNA Editing

We are already aware that the production of a mature nuclear-encoded eukaryotic mRNA involves a number of cotranscriptional and posttranscriptional modifications, including 5' capping, 3' polyadenylation, and splicing to remove introns. The synthesis of mitochondrial and chloroplast mRNAs does not involve 5' capping, but may involve polyadenylation and possibly splicing. In the mid 1980s a new phenomenon came to light in the production of certain mRNAs in the mitochondria of some protozoa. This phenomenon is called **RNA editing** and involves the posttranscriptional insertion and/or deletion of nucleotides in a mRNA molecule. For example, a conserved protein encoded by mtDNA is subunit III of cytochrome oxidase (COIII). The sequences of the *COIII* gene and its mRNA transcripts were compared in the protozoans *Trypanosome brucei* (*Tb*, for short), *Crithidia fasiculata* (*Cf*, for short), and *Leishmania tarentolae* (*Lt*, for short) (Figure 21.12). While the mRNA sequences are highly conserved among the three organisms, only the *Cf* and *Lt* mtDNA sequences are colinear with the mRNAs. Strikingly, the *Tb* gene has a sequence which cannot produce the mRNA it apparently encodes. Careful comparison of the DNA and mRNA sequences indicates that the differences between the two can be accounted for by U nucleotides in the mRNA not encoded in the DNA and T nucleotides in the DNA not found in the transcript. The model is that the transcript of the *Tb COIII* gene is edited once it is made to add U nucleotides in the appropriate places and remove the U nucleotides encoded by the T nucleotides in the DNA. As Figure 21.12 shows, we are not talking about one or two changes, but extensive insertions of U nucleotides, and the deletions of several templated T nucleotides. The magnitude of the changes is even more apparent when the whole sequence is examined—over 50 percent of the mature mRNA consists of posttranscriptionally added U nucleotides! This RNA editing must be accurate in order to reconstitute the appropriate reading frame for translation. A discussion of the models proposed to explain this type of RNA editing is beyond the scope of this text. A special RNA molecule, called a

∼ FIGURE 21.12

Comparison of the DNA sequences of the cytochrome oxidase subunit III gene (*COIII*) in the protozoans *Trypanosome brucei* (*Tb*), *Crithridia fasiculata* (*Cf*), and *Leishmania tarentolae* (*Lt*), aligned with the conserved mRNA for *Tb*. The lowercase *u*'s are the U nucleotides added to the transcript by RNA editing. The template Ts in *Tb* DNA that are not in the RNA transcript are highlighted in blue.

Region of *COIII* gene transcript

Tb DNA	G GTTTTTGG AGG G GTTTTG G G A A GA GAG	
Tb RNA	uuGuGUUUUUGGuuuAGGuuuuuuuuGuuG UUGuuGuuuuGuAuuAuGAuuGAGu	
Cf DNA	TTTTTATTTTGATTTCGTTTTTTTTTATG TGTATTATTTGTGCTTTGATCCGCT	
LT DNA	TTTTTATTTTGATTTCGTTTTTTTTTATG TGTTTATTTATGTTATGAGTAGGA	
Tb Protein	Leu Cys Phe Trp Phe Arg Phe Phe Cys Cys Cys Cys Phe Val Leu Trp Leu Ser	

guide RNA (gRNA), is involved in the process. The gRNA pairs with the mRNA transcript and is thought to be responsible for cleaving the transcript, templating the missing U nucleotides, and ligating the transcript back together again.

Since the discovery of RNA editing in *T. brucei*, other types of RNA editing have been described. In the mitochondria of the slime mold *Physarum polycephalum*, single C nucleotides are added posttranscriptionally at many positions of several transcripts. In higher plants, the sequences of most mitochondrial transcripts are edited by C to U changes and to some extent by U to C changes. C-to-U editing is also involved in the production of an AUG initiation codon from an ACG codon in some chloroplast mRNAs in some higher plants. Lastly, C-to-U editing occurs in the mRNA transcribed from the nuclear gene encoding the cytoplasmic apolipoprotein B in mammals. This editing results in tissue-specific generation of a stop codon.

In sum, more and more examples of RNA editing are being discovered and described. How broadly this phenomenon exists currently cannot be predicted. Possibly the RNA editing process itself is an additional regulatory control point.

Origin of Mitochondria and Chloroplasts

How might mitochondria and chloroplasts have arisen? The widely accepted **endosymbiont hypothesis** is that mitochondria and chloroplasts originated as free-living prokaryotes that invaded primitive eukaryotic cells and established a mutually beneficial (symbiotic) relationship. According to this hypothesis, eukaryotic cells started out as anaerobic organisms that lacked mitochondria and chloroplasts. At some point in their evolution, about a billion (10^9) years ago, a eukaryotic cell established a symbiotic relationship with a purple nonsulfur photosynthetic bacterium. Over time, the oxidative phosphorylation activi-

ties of the bacterium became used for the benefits of the eukaryotic cell, and, in the presence of oxygen, photosynthetic activity was lost. Eventually, the eukaryotic cell became dependent upon the intracellular bacterium for survival and the mitochondrion was formed. The chloroplasts of plants and algae are hypothesized to have occurred simultaneously (or perhaps later) by the ingestion of an oxygen-producing photosynthetic bacterium (a cyanobacterium) by a eukaryotic cell.

As we have just discussed, many proteins found in mitochondria and chloroplasts are encoded by nuclear genes. Thus, further evolution of mitochondria and chloroplasts must have involved the extensive transfer of genes from the organelles to the nuclear DNA.

RULES OF EXTRANUCLEAR INHERITANCE

Since the pattern of inheritance shown by genes located in organelles differs strikingly from the pattern shown by nuclear genes, the term **non-Mendelian inheritance** is appropriate to use when we are discussing extranuclear genes. In fact, if results are obtained from genetic crosses that do not conform to predictions based on the inheritance of nuclear genes, this is a good reason to suspect extranuclear inheritance.

Here are the four main characteristics of *extranuclear inheritance*:

1. In higher eukaryotes, the results of reciprocal crosses involving extranuclear genes are not the same as reciprocal crosses involving nuclear genes. (Recall from Chapters 2 and 3 that, in a reciprocal cross, the sexes of the parents are reversed in each case. For example, if A and B represent contrasting genotypes, A♀ × B♂ and A♂ × B♀ would represent a pair of reciprocal crosses.)

Extranuclear genes usually exhibit the phenomenon of **uniparental inheritance** from generation to generation. In uniparental inheritance, all progeny (both males and females) have the phenotype of only one parent. (The fact that both males and females resemble one of the parents distinguishes uniparental inheritance of an extranuclear gene from that of a sex-linked nuclear gene.) Typically, for higher eukaryotes it is the mother's phenotype that is expressed exclusively, a phenomenon called **maternal inheritance**. Maternal inheritance occurs because the amount of cytoplasm in the female gamete, the egg, usually greatly exceeds that in the male gamete, the sperm. Therefore, the zygote receives most of its cytoplasm (containing the extranuclear genes in the organelles; i.e., the mitochondria and, where applicable, the chloroplasts) from the female parent and a negligible amount from the male parent.

In contrast, the results of reciprocal crosses between a wild-type and a mutant strain are identical if the genes are located on nuclear chromosomes. One exception for nuclear genes occurs when X-linked genes are involved (see Chapter 3), but even then the results are distinct from those for extranuclear inheritance.

2. Extranuclear genes cannot be mapped to the chromosomes in the nucleus. If the nuclear chromosomes of an organism are well mapped, any new mutations of nuclear genes can be mapped by standard genetic mapping crosses. If a new mutation does not show linkage to any of the nuclear genes, it is probably an allele of an extranuclear gene.

3. Ratios typical of Mendelian segregation are not found. For nuclear gene mutations the rules of Mendelian segregation predict that the two alternative phenotypes will segregate in predicted ratios, depending on linkages and dominance. Such segregation patterns are not characteristic of extranuclear genes—the uniparental inheritance pattern characteristic of extranuclear genes is clearly a deviation from Mendelian segregation.

4. Extranuclear inheritance is not affected by substituting a nucleus with a different genotype. When a particular phenotype persists after the nucleus is replaced with one with a different genotype, this indicates that the phenotype is likely to be controlled by an extranuclear genome.

Keynote

The inheritance of extranuclear genes follows rules different from those for nuclear genes. In particular,

no meiotic segregation is involved, generally uniparental (and often maternal) inheritance is seen, extranuclear genes are not mappable to the known nuclear-linkage groups, and the phenotype persists even after nuclear substitution.

Examples of Extranuclear Inheritance

In this section we discuss the properties of a selected number of mutations in extranuclear chromosomes in order to illustrate the principles of extranuclear inheritance.

Leaf Variegation in the Higher Plant *Mirabilis jalapa*

One of the first exceptions to Mendelian inheritance was demonstrated in 1909 through the work of Carl Correns, a plant geneticist and one of the three rediscoverers of Mendelian principles of inheritance. Correns had been studying the inheritance of a number of variegated-leaf forms in a number of flowering plants. Many of these phenotypes were controlled by genes that showed typical Mendelian inheritance. One phenotype that did not follow the expected pattern was the yellowish white patches of a variegated strain of *Mirabilis jalapa* (also called the four o'clock, or the marvel of Peru; Figure 21.13). This strain, called *albomaculata*, also has occasional shoots that are wholly green or wholly yellow-white (Figure 21.14).

All types of shoots (variegated, green, white) give rise to flowers, so crosses can be made by taking pollen from one type of flower and fertilizing a flower on the same or different type of shoot. Table 21.3 summarizes the results of these intercrosses. The flowers on green shoots give only green progeny, regardless of whether the pollen is from green, white, or variegated shoots. Flowers on white shoots give only white progeny, regardless of the source of the pollen. (However, because the whiteness indicates the absence of chlorophyll and hence an inability to carry out photosynthesis, the white progeny die soon after seed germination.) Finally, flowers on variegated shoots all produce three types of progeny—completely green, completely white, and variegated—regardless of the type of the pollen. No pattern is seen for the relative proportions of these three types. In subsequent generations maternal inheritance is always seen in the same patterns just described.

~ FIGURE 21.13

The four o'clock, *Mirabilis jalapa*.

~ FIGURE 21.14

Leaf variegation in *Mirabilis jalapa*. Shoots that are all green, all white, and variegated are found on the same plant, and flowers may form on any of these shoots.

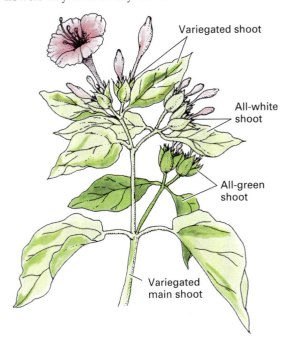

In sum, these breeding experiments with *M. jalapa* show results that indicate maternal inheritance. That is, the *progeny phenotype in each case was the same as that of the maternal parent* (i.e., the color of the progeny shoots resembled the color of the parental flower), which is indicative of maternal inheritance. Moreover, the results of reciprocal crosses differed from the expected patterns, and there was a lack of constant proportions of the different phenotypic classes in the segregating progeny. These last two properties are also characteristic of traits showing extranuclear inheritance and are not expected for traits showing Mendelian inheritance.

The basis for the green color of higher plants is the presence of the green pigment chlorophyll in large numbers of chloroplasts. Green shoots in *Mirabilis* have a normal complement of chloroplasts. White shoots have abnormal, colorless chloroplasts called leukoplasts. Leukoplasts lack chlorophyll and hence are incapable of carrying out photosynthesis.

The simplest explanation for the inheritance of leaf color in the *albomaculata* strain of *Mirabilis jalapa* is that the abnormal chloroplasts are defective as a result of a mutant gene in the cpDNA. During plant growth the two types of organelles, chloroplasts and

~ TABLE 21.3

Results of Crosses of Variegated Plants of *Mirabilis jalapa*

PHENOTYPE OF BRANCH-BEARING ♀ EGG PARENT	PHENOTYPE OF BRANCH-BEARING ♂ (POLLEN) PARENT	PHENOTYPE OF PROGENY
White	White	White
	Green	White
	Variegated	White
Green	White	Green
	Green	Green
	Variegated	Green
Variegated	White	Variegated, green, or white
	Green	Variegated, green, or white
	Variegated	Variegated, green, or white

leukoplasts, segregate so that a particular cell and its progeny cells may receive only chloroplasts (leading to green tissues), only leukoplasts (leading to white tissues), or a mixture of chloroplasts and leukoplasts (leading to variegation). This model is shown in Figure 21.15.

In a variegated plant, then, the white shoots derive from cells in which, through segregation, only

leukoplasts are present. Green shoots are similarly derived from cells containing only chloroplasts. Variegated shoots are generated from cells that contain chloroplasts and leukoplasts so, when they segregate later, patches of white tissue are produced on the shoots and leaves. Let us take this model one step further. The flowers on a green shoot have only chloroplasts, and so through maternal inheritance these

~ **FIGURE 21.15**

Model for the inheritance of leaf color in the four o'clock, *Mirabilis jalapa*.

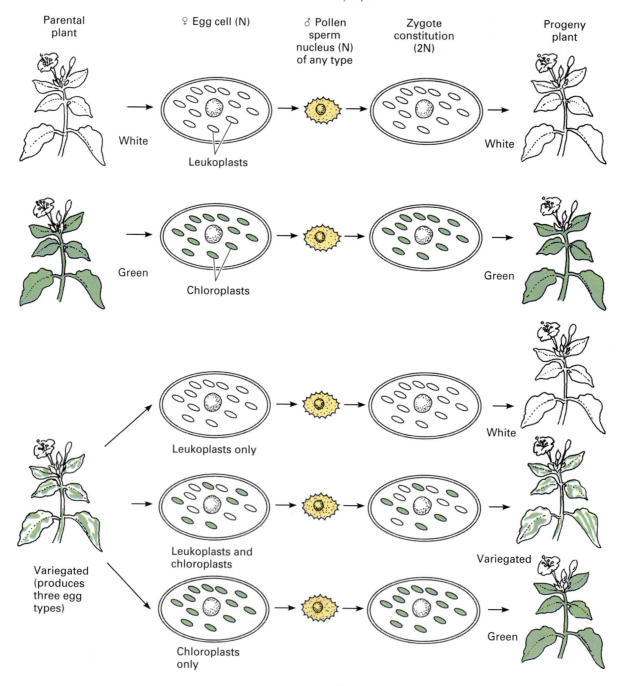

chloroplasts form the basis of the phenotype of the next generation. Similar, logical arguments can be made for the flowers on white or variegated shoots.

This simple model has three assumptions. One is that the pollen contributes essentially no cytoplasmic information (that is, no chloroplasts or leukoplasts) to the egg. This is a reasonable assumption since the egg is much larger than the pollen. Thus in the zygote the extranuclear genetic determinants come from the egg. The second assumption is that the chloroplast genome replicates autonomously and, by growth and division of plastids (the general term for photosynthetic organelles such as chloroplasts), the wild-type and mutant cpDNA molecules have the potential to segregate randomly to the new plastids so that pure plastid lines can be generated from an original mixed line. The third assumption is that segregation of plastids to daughter cells is random, so that some daughters receive chloroplasts, some receive leukoplasts, and some receive mixtures.

$\mathcal{K}$EYNOTE

The leaf color phenotypes in a variegated strain of the four o'clock, *Mirabilis jalapa*, show maternal inheritance. The abnormal chloroplasts in white tissue are the result of a mutant gene in the cpDNA, and the observed inheritance patterns follow the segregation of cpDNA.

The [*poky*] Mutant of *Neurospora*

The [*poky*] mutant of *Neurospora*, which involves a change in mtDNA, illustrates a number of the classical expectations of extranuclear inheritance. (The square brackets around the mutant symbols indicate an extranuclear gene.) The [*poky*] mutant is also called [*mi-1*], where *mi* represents *maternal inheritance*. The phenotype of the [*poky*] mutant is a much slower growth than wild type, either on solid medium or in liquid medium.

Neurospora crassa is an obligate aerobe; that is, it requires oxygen in order to grow and survive, so mitochondrial functions are essential for its growth. Biochemical analysis showed that the [*poky*] mutant is defective in aerobic respiration as a result of changes in the cytochrome complement of the mitochondria. The three principal cytochromes are a + a_3, b, and c. Compared with the wild type, [*poky*] lacks cytochromes a + a_3 and b; [*poky*] also has a greater amount of cytochrome c. The change in the cytochrome spectrum affects the ability of the mitochondria to generate sufficient ATP to support rapid growth, so slow growth results.

In *Neurospora* the sexual phase of the life cycle is initiated following a fusion of nuclei from mating type *A* and *a* parents. A sexual cross can be made in one of two ways: either by putting both parents on the crossing medium simultaneously or by inoculating the medium with one strain, and after three or four days at 25°C, adding the other parent. In the latter case the first parent on the medium produces all the *protoperithecia*, the bodies that will give rise to the true fruiting bodies in which are the asci with the sexual spores.

Compared with the conidia, the asexual spores, the protoperithecia have a tremendous amount of cytoplasm and hence can be considered the female parent in much the same way as an egg of a plant or an animal is the female parent. Now by adding conidia of a strain of the opposite mating type to the crossing medium, we have what is called a *controlled cross* in which one strain acts as the female and the other as the male parent. Using a strain to produce the protoperithecia as the female parent and conidia of another strain as a male parent, geneticists can make reciprocal crosses to determine whether any trait shows extranuclear inheritance. This experiment is now illustrated for the [*poky*] mutant.

Mary and Herschel Mitchell did reciprocal crosses between [*poky*] and the wild type, with the following results:

[*poky*] ♀ × wild type ♂ → all [*poky*]
wild type ♀ × [*poky*] ♂ → all wild type

In other words, all the progeny show the same phenotype as the maternal parent, indicating maternal inheritance as a characteristic for the [*poky*] mutation.

This analysis can be made more refined by using tetrad analysis (Chapter 6) to follow the phenotype more closely. Recall that in *Neurospora* the eight products of a meiosis and subsequent mitosis are retained in linear order within the ascus. The eight ascospores can be removed from the asci, and strains germinated from the spores can be analyzed for the particular phenotype being followed. By doing tetrad analysis for the [*poky*] ♀ × wild type ♂ cross, we find an 8:0 ratio of [*poky*]:wild-type progeny (Figure 21.16a) and for the wild type ♀ × [*poky*] ♂ cross, we get a 0:8 ratio of [*poky*] wild-type progeny with regard to the growth phenotype (Figure 21.16b). At the same time nuclear genes show the 4:4 segregation expected of Mendelian inheritance. Again, the simplest explanation for these

~ FIGURE 21.16

Results of reciprocal crosses of [*poky*] and normal (wild-type) *Neurospora*. (a) [*poky*] ♀ × normal ♂; (b) normal ♀ × [*poky*] ♂.

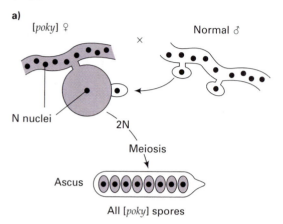

a)

[*poky*] ♀   Normal ♂

×

N nuclei

2N

Meiosis

Ascus

All [*poky*] spores

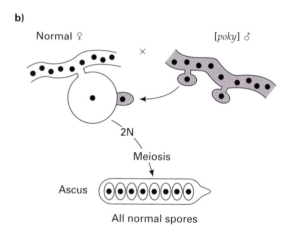

b)

Normal ♀   [*poky*] ♂

×

2N

Meiosis

Ascus

All normal spores

results is that [*poky*] is determined by an extranuclear gene (located in this case in the mitochondria) that exhibits a pattern of maternal inheritance.

The cytochrome deficiency phenotype in [*poky*] results from a defect in mitochondrial protein synthesis. The [*poky*] mutant has been shown to be grossly deficient in 19S rRNA, the rRNA component of the small mitochondrial ribosomal subunit, which leads to a decreased amount of small ribosomal subunits in the organelle and, hence, a deficiency in protein synthesis capability. The molecular basis of the 19S rRNA deficiency in [*poky*] has been traced to a 4-bp deletion in the promoter for that rRNA.

𝒦EYNOTE

The slow-growing [*poky*] mutant of *Neurospora crassa* shows maternal inheritance and deficiencies for some mitochondrial cytochromes. The molecu-

lar defect in [*poky*] is a deletion in the promoter for the rRNA gene of the small mitochondrial ribosomal subunit.

Yeast *petite* Mutants

Yeast grows as single cells. Thus, on solid media, yeast forms discrete colonies consisting of many thousands of individual cells clustered together. Yeast can grow with or without oxygen. In the absence of oxygen yeast obtains energy for cell growth and cellular metabolism through fermentation metabolism, in which the mitochondria are not involved.

CHARACTERISTICS OF *PETITE* MUTANTS.   In the late 1940s, Boris Ephrussi and his colleagues studied the growth of yeast cells on a solid medium that allowed growth to occur either by aerobic respiration or by fermentation. Occasionally, they noticed a colony that was much smaller than wild-type colonies. Since Ephrussi was French, the small colonies were called *petites* (French for "small"), and the wild-type colonies were called *grandes* (French for "big"). Ephrussi found that *petite* colonies were small not because the cells were small but because the growth rate of the mutant *petite* strain is significantly slower than that of the wild type. Thus there are fewer cells in the *petite* colonies.

Biochemical analysis shows that the *petites* are essentially incapable of carrying out aerobic respiration; they must obtain their energy primarily from fermentation, which is a relatively inefficient process.

In an unmutagenized population of cells between 0.1 and 1 percent of the cells spontaneously become *petite*. In the presence of an intercalating agent (a chemical that can wedge itself between adjacent base pairs in DNA: see Chapter 19) such as ethidium bromide, 100 percent of the cells become *petite*! The yeast *petite* system is particularly useful in studies of extranuclear inheritance because there is an abundance of *petite* mutants and because yeast cells that lack mitochondrial functions can still survive and grow. Such mutations to respiratory deficiencies in yeast are automatically conditional mutants. On a medium that can support fermentation and aerobic growth, the *petites* grow more slowly than the *grandes*, whereas on a medium that supports only aerobic respiration, *petites* are unable to grow.

NUCLEAR, NEUTRAL, AND SUPPRESSIVE *PETITE* MUTANTS.   Let us now turn to the genetics of *petite* mutants. Some *petites* have as their basis a mutation in the nuclear genome—a characteristic that is not surprising since some subunits of some mitochondrial proteins are encoded by nuclear genes. The mutations in these *nuclear petites* (also called *segregational petites*) are called *pet⁻*. When a *pet⁻* mutant is crossed with the wild type (*pet⁺*), the diploid is *pet⁺/pet⁻*, which produces *grande* colonies. When this *pet⁺/pet⁻* cell goes through meiosis, each resulting tetrad shows a 2:2 segregation of *grande* (*pet⁺*) : *petite* (*pet⁻*) phenotype. This result is typical of Mendelian inheritance, so we will consider these *petites* no further, except to say that nuclear *petites* occur much less frequently than extranuclear *petites*.

Two other classes of *petites*, the *neutral petites* and *suppressive petites*, exhibit the traits of extranuclear inheritance. Figure 21.17 shows the inheritance pattern of *neutral petites* (symbolized [*rho⁻N*]). When a *neutral petite* is crossed with normal wild-type cells ([*rho⁺N*]), the resulting [*rho⁺N*]/[*rho⁻N*] diploids all produce *grande* colonies. When these diploids go through meiosis, all resulting meiotic tetrads show a 0:4 ratio of *petite*:*grande* (wild type), while at the same time nuclear markers segregate 2:2. (The name *neutral*, then, refers to that fact that this class of *petites* does not affect the wild type.) This result is a classical example of uniparental inheritance, in which all progeny have the phenotype of only one parent. *This phenomenon is not maternal inheritance, however, since the two haploid cells that fuse to produce the diploid are the same size and contribute equally to the cytoplasm.*

What is the nature of the *neutral petite* mutation? As we have said, *petites* can be induced readily by treatment with intercalating agents like ethidium bromide. With prolonged treatment the majority of *petites* produced are of the *neutral* type. The mitochondrial genome is implicated because cytochromes are altered, because there is evidence for extranuclear inheritance, and because the mitochondria are the only other site in the yeast cell in which genetic material is found.

An examination of the mitochondrial genetic material in the *neutral petites* reveals a remarkable characteristic: Essentially 99 to 100 percent of the mtDNA is missing. Not surprisingly, then, the *neutral petites* are unable to perform mitochondrial functions. They survive, however, because fermentation processes are localized in the cytoplasm. In genetic crosses with wild-type yeasts normal (wild-type) mitochondria form a population from which new, normal (wild-type) mitochondria are produced in all progeny, and hence the *petite* trait is lost after one generation.

The second class of *petites* that shows extranuclear inheritance is the *suppressive petites* (symbolized [*rho⁻S*]). The *suppressive petites* are different from the *neutrals* because they *do* have an effect on the wild type. Most *petite* mutants are of the suppressive type. Like the neutrals and the nuclear *petites*, the [*rho⁻S*] petites are deficient in mitochondrial protein synthesis.

The inheritance pattern of *suppressive petites* is different from those of nuclear and *neutral petites* (Figure 21.18). When a [*rho⁺*]/[*rho⁻S*] diploid is formed, it has respiratory properties intermediate between those of normal and *petite* strains. If this diploid is allowed to divide mitotically a number of times, the diploid progeny population will have up to 99 percent *petites* (see Figure 21.18). (The name *suppressive*, then, refers to that fact that this class of *petites* overcomes the wild type so that a respiratory-deficient phenotype results.) Sporulation of any of the *petites* in that population produces tetrads with a 4:0 ratio of *petite*:*grande* (wild type), while at the same time nuclear markers segregate 2:2 (see Figure 21.18). Sporulation of any of the few wild-type diploids in the population produces tetrads with a 0:4 ratio of *petite*:*grande* strains, and a 2:2 nuclear marker segregation ratio (see Figure 21.18).

The *suppressive petite* mutations result from changes in the mtDNA. They start out as deletions of part of the mtDNA, and then, by some correction mechanism, sequences that are not deleted become duplicated until the normal amount of mtDNA is restored. In some *suppressive petites*, over 90 percent of the base pairs are ATs as a result of this process. During the correction events rearrangements of the mtDNA sometimes occur. Since the protein-coding genes in the mitochondrial genome are widely scattered, these deletions and rearrangements lead to deficiencies in the enzymes involved in aerobic respiration, and a *petite* colony results. *Suppressive petites* are proposed to have a suppressive effect over normal mitochondria in one of two ways:

1. *Suppressive* mitochondria could replicate faster than normal mitochondria and simply overrun them in the cell. The variations seen in suppressiveness between strains would then reflect the relative competitiveness of *suppressive* mitochondria in replication.
2. *Suppressive* and normal mitochondria could fuse and recombination between rearranged *suppressive* mtDNA and normal mtDNA could severely alter the latter's gene organization and produce a *petite* phenotype.

ᴋEYNOTE

Yeast *petite* mutants grow slowly and have various deficiencies in mitochondrial functions as a result of alterations in mitochondrial DNA. The *petite* mutants show extranuclear inheritance, with particular patterns of inheritance varying with the type of *petite* involved.

~ **FIGURE 21.17**

Extrachromosomal inheritance of neutral *petite* in yeast.

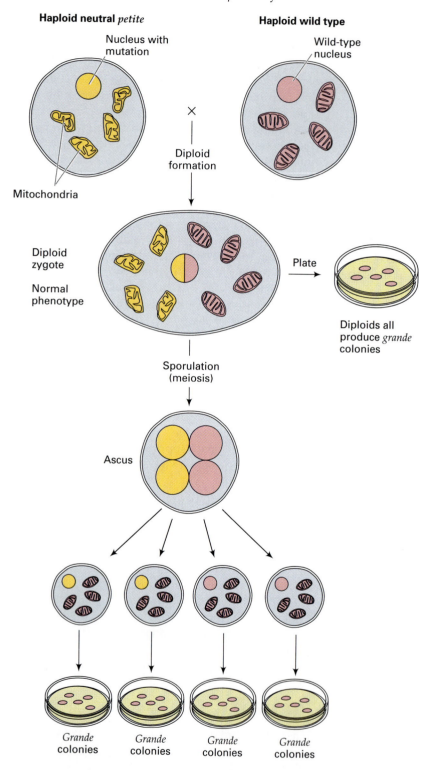

Sum: Nuclear gene segregation: 2 neutral *petite* mutants: 2 wild type
        Extranuclear segregation: 0 neutral *petite* mutants: 4 wild type

~ **FIGURE 21.18**

Extrachromosomal inheritance of suppressive *petite* in yeast.

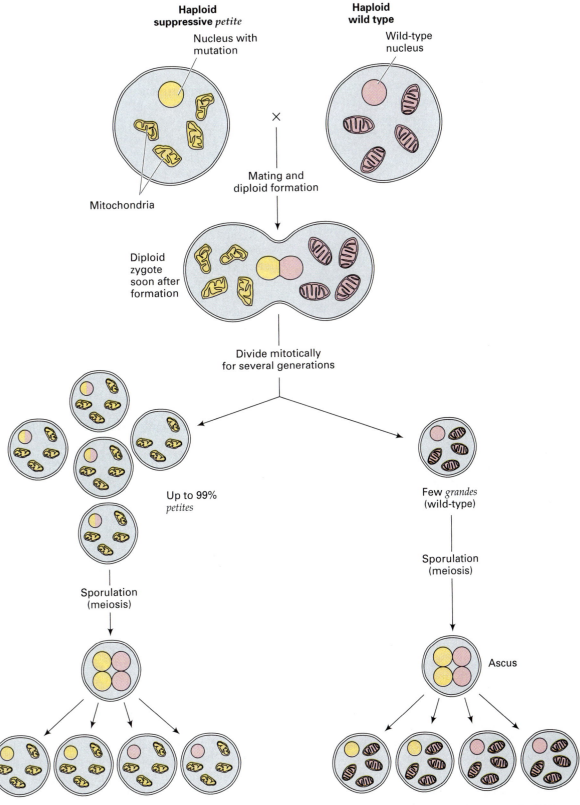

Haploid suppressive *petite*
Nucleus with mutation

Haploid wild type
Wild-type nucleus

Mitochondria

Mating and diploid formation

Diploid zygote soon after formation

Divide mitotically for several generations

Up to 99% *petites*

Few *grandes* (wild-type)

Sporulation (meiosis)

Sporulation (meiosis)

Ascus

Nuclear gene segregation: 2 mutants: 2 wild type
Extranuclear segregation: 4 *petites*: 0 wild type

Nuclear gene segregation: 2 mutants: 2 wild type
Extranuclear segregation: 0 *petites*: 4 wild type

Extranuclear Genetics of *Chlamydomonas*

One of the most thorough analyses of chloroplast genetics was begun by Ruth Sager and her colleagues in their work with the unicellular alga *Chlamydomonas reinhardtii*. This motile, haploid organism has two flagella and a single chloroplast that contains many copies of cpDNA. There are two mating types, *mt*⁺ and *mt*⁻. In a process known as syngamous mating, a zygote is formed by fusion of two equal-sized cells (which, therefore, contribute an equal amount of cytoplasm), one of each mating type. A thick-walled cyst develops around the zygote. After meiosis, four haploid progeny cells are produced, and since mating type is determined by a nuclear gene, a 2:2 segregation of *mt*⁺:*mt*⁻ mating-types results.

From a systematic study of traits affecting this organism, researchers identified a number of traits that showed the expected patterns for extranuclear inheritance. Some of these extranuclear traits were shown to involve resistance to or dependence on antibiotics that affect chloroplast protein synthesis by changing ribosome structure and/or function.

One trait that is inherited in an extranuclear manner confers erythromycin resistance ([*ery*ʳ]) on the organism. Wild-type *Chlamydomonas* cells are erythromycin-sensitive ([*ery*ˢ]). If we perform a cross of *mt*⁺ [*ery*ʳ] × *mt*⁻ [*ery*ˢ], about 95 percent of the offspring are erythromycin-resistant phenotypes (Figure 21.19a). This result is a classic instance of uniparental inheritance, an attribute we have come to expect of extranuclear traits. The reciprocal cross, *mt*⁻ [*ery*ʳ] × *mt*⁺ [*ery*ˢ] (Figure 21.19b) also shows uniparental inheritance about 95 percent of the time, although here the erythromycin-sensitive phenotype is inherited. Thus even though both parents contribute equal amounts of cytoplasm to the zygote, the progeny *almost always* resemble the *mt*⁺ parent with regard to chloroplast-controlled phenotypes. Somehow, preferential segregation of one chloroplast type occurs in this organism, or perhaps one parental type is inactivated preferentially.

A number of *Chlamydomonas* mutants in addition to *ery*ʳ show uniparental inheritance, with progeny of crosses *always* resembling the phenotype of the *mt*⁺ parent. These mutations are in genes in the chloroplast genome. Some of the evidence is as follows. The density of the chloroplast DNA is different from that of nuclear DNA and from that of mitochondrial DNA, so it is possible to study cpDNA selectively. By using density labels, it is possible to make the cpDNA of the *mt*⁺ and *mt*⁻ parents different in density (recall the Meselson and Stahl experiment, Chapter 12, pp. 345–348). When *mt*⁺/*mt*⁻ zygotes are examined for their cpDNA, the cpDNA from the *mt*⁻ parent has always disappeared. This loss of cpDNA from the *mt*⁻ parent clearly parallels the loss of uniparental genes (such as the erythromycin genes) of the *mt*⁻ parent in the progeny of the genetic crosses.

As we indicated above, only about 95 percent of the zygotes in the crosses involving chloroplast genes show uniparental inheritance. The other 5 percent of the zygotes show extranuclear traits from both parents. Such zygotes are said to show **biparental inheritance**, indicating that both types of chloroplast chromosomes are present and active in these zygotes. The genetic condition of these zygotes is defined as a **cytohet**, the term deriving from "*cyto*plasmically *het*erozygous." In many instances the extranuclear traits of biparental zygotes segregate into pure types (i.e., either erythromycin-sensitive or erythromycin-resistant) on successive mitotic divisions. This phenomenon presumably involves the segregation of the different chloroplast chromosomes, and hence the different chloroplasts, into pure types. This situation parallels the situation we discussed for *Mirabilis*. Since biparental zygotes are occasionally produced, geneticists are able to make crosses between strains carrying different extranuclear traits that are chloroplast-controlled. Occasionally, a biparental zygote does not segregate the two traits; instead, both traits continue to be expressed in subsequent generations. The explanation is that a recombination event takes place between the two types of cpDNA so that a recombinant chromosome carrying both alleles originally carried by the two parents is produced. From such segregation data it is possible to construct genetic maps of the chloroplast genome.

𝒦EYNOTE

A number of genes have been identified and mapped in the *Chlamydomonas* chloroplast genome. These genes are inherited in an extranuclear manner.

Human Genetic Diseases and Mitochondrial DNA Defects

A number of human genetic diseases have been shown to be the result of mtDNA gene mutations. These diseases show maternal inheritance as would be expected for defects involving mitochondria. The following are some brief, selected examples of these diseases.

~ FIGURE 21.19

Uniparental inheritance in *Chlamydomonas*. (a) From a cross of *mt⁺* [*ery^r*] × *mt⁻* [*ery^s*], 95 percent of the zygotes give tetrads that segregate 2:2 for the nuclear mating type genes, and 4:0 for the extranuclear gene carried by the *mt⁺* parent (here, [*ery^r*]); (b) From the reciprocal cross of a *mt⁻* [*ery^r*] × *mt⁺* [*ery^s*], 95 percent of the zygotes give tetrads segregating 2 *mt⁺*:2 *mt⁻* and 0 [*ery^r*]:4 [*ery^s*], again showing uniparental inheritance for the extranuclear trait of the *mt⁺* parent.

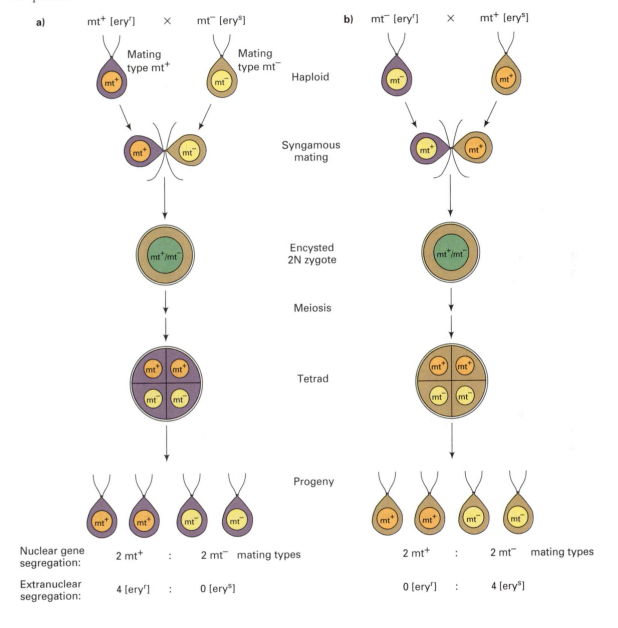

Leber's hereditary optic neuropathy (LHON): This disease affects mid-life adults resulting in complete or partial blindness from optic nerve degeneration. Eighteen missense mutations in the mitochondrial genes for the proteins ND1, ND2, ND4, ND5, ND6, cytb, COI, COIII, and ATPase 6 (see Figure 21.5) lead to LHON. Those proteins are included in mitochondrial electron transport chain enzyme complexes. The electron transport chain drives cellular ATP pro-

duction by oxidative phosphorylation. It appears that death of the optic nerve in LHON is a common result of oxidative phosphorylation defects, here brought about by the inhibition of the electron transport chain.

Kearns-Sayre syndrome: People with this syndrome have encephalomyopathy, a brain disease. The cause of the syndrome is large deletions at various positions in the mtDNA. One model is that each deletion removes one or more tRNA genes whose products are essential for mitochondrial protein synthesis. In some unknown way, this leads to development of the syndrome.

Myoclonic Epilepsy and Ragged-Red Fiber (MERRF) disease: People with this disease exhibit dementia, deafness, and seizures. The mitochondria of these individuals are abnormal in appearance. The disease is caused by a single nucleotide substitution in the gene for tRNA.Lys. The mutated tRNA has a detrimental effect on mitochondrial protein synthesis, and somehow this gives rise to the various phenotypes of the disease.

In most diseases resulting from mtDNA defects, the cells of affected individuals have a mixture of mutant and normal mitochondria. This condition is known as **heteroplasmy**. Characteristically the proportions of the two mitochondrial types vary from tissue to tissue and from individual to individual within a pedigree. The severity of the disease symptoms correlates approximately with the relative amount of mutant mitochondria.

Exceptions to Maternal Inheritance

Strict maternal inheritance has been considered to be the case for extranuclear mutations in animals and plants where the female gamete contributes the majority of the cytoplasm to the zygote. However, exceptions are coming to light. The following are some examples.

1. Exploiting DNA sequence differences between mtDNAs of two inbred lines of mice, researchers have used PCR (polymerase chain reaction: see Chapter 15, pp. 476–477) to demonstrate that paternally inherited mtDNA molecules are present at a frequency of 10^{-4} relative to maternal mtDNA molecules. This heteroplasmy of paternal and maternal mitochondria has potentially significant evolutionary implications. That is, it has been generally considered that maternal and paternal mtDNA remain distinct because of the strict maternal inheritance of mitochondria. However, if heteroplasmy can occur, then there is a likelihood of genetic recombination between maternally derived and paternally derived mtDNA molecules. Such recombination will lead to significant diversity of mtDNA in an individual. The extent to which this phenomenon occurs is presently unknown but the fact that it exists at all makes it necessary to be cautious about conclusions made using a purely maternal inheritance of mtDNA.

2. In most angiosperms (flowering plants), the plastids are inherited only from the maternal parent. In some angiosperms, however, plastids are in- herited at high frequency from both parents, or mostly from the paternal parent. For example, bi-parental inheritance of plastids is seen in the evening primrose, *Oenothera*. Paternal inheritance of plastids is the rule in conifers, which are gymnosperms.

Infectious Heredity—Killer Yeast

The examples of extranuclear inheritance we have discussed so far have all resulted from mutations in the mtDNA or cpDNA. There are other examples of eukaryotic extranuclear inheritance that are due to the presence of cytoplasmic bacteria or viruses coexisting with the eukaryotes in a symbiotic relationship. One example is the killer phenomenon in yeast. Some yeast strains secrete a killer toxin, which will kill sensitive strains of yeast. (Killer strains are immune to their own toxin.) The killer phenomenon results from the presence in the cell's cytoplasm of two types of viruses, L and M (Figure 21.20a). Neither appear to cause deleterious effects on the host cell.

The L virus consists of a protein capsid within which is one 4.6-kb double-stranded (ds) RNA genome called L-dsRNA. L-dsRNA encodes the capsid proteins for both L and M viruses and the viral polymerase required for viral RNA replication. So, since all viral particles are encoded by an L-dsRNA, M viruses are only found in cells if L viruses are also present. The M virus consists of a virus particle encoded by L-dsRNA, and two identical copies of a 1.8-kb double-stranded RNA genome called M-dsRNA. M-dsRNA encodes the killer toxin protein, which is secreted from the cell. The same protein confers immunity upon the killer cell.

Sensitive yeast cells are cells that can be killed by the M-encoded killer toxin; there are two types (Figure 21.20b). One type has only L viruses, and the other has neither L nor M viruses. In both of these types, no immunity function is produced because killer toxin is not made.

Unlike most viruses, the yeast L and M viruses are not found outside of the cell, so sensitive yeast cells cannot be infected by viruses that invade from outside. Rather, virus transmission from yeast to yeast occurs whenever there is cytoplasmic mixing, most commonly when two yeast cells mate. All progeny of the mating will inherit copies of the viruses in the parental cells, illustrating an infectious mechanism of cytoplasmic inheritance. For example, if a killer yeast mates with a sensitive yeast that lacks viruses (obviously before killing it), the resulting diploid will be a killer because of the presence of both L and M viruses from the killer yeast. When ascospores are produced, both virus types will be dis-

~ FIGURE 21.20

The killer phenomenon in yeast. (a) Killer yeast contains two virus types, L and M, each of which contains a double-stranded RNA genome. L-dsRNA encodes both virus particles and the replication enzyme required for L and M virus replication. M-dsRNA encodes the killer toxin; (b) Sensitive yeast, which can be killed by killer toxin, either have L viruses but no M viruses, or have neither virus type.

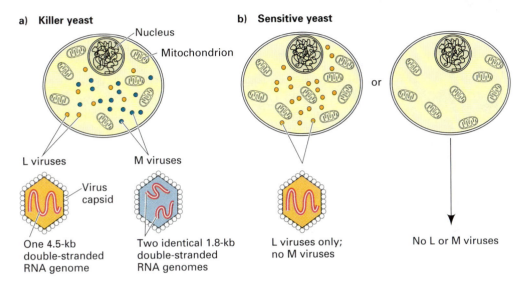

a) Killer yeast

Nucleus
Mitochondrion
L viruses
M viruses
Virus capsid
One 4.5-kb double-stranded RNA genome
Two identical 1.8-kb double-stranded RNA genomes

b) Sensitive yeast

or

L viruses only; no M viruses

No L or M viruses

---

tributed to them. Thus, the ascospores will all give rise to cells that are killers. The killer phenomenon in yeast, then, is a case of extranuclear, uniparental inheritance resulting from infectious heredity of virus particles.

KEYNOTE

Not all cases of extranuclear inheritance result from genes on mtDNA or cpDNA. Many other examples in eukaryotes result from infectious heredity in which symbiotic, cytoplasmically located bacteria or viruses are transmitted when cytoplasms mix. The killer phenomenon in yeast is such an example: It results from the infectious heredity of cytoplasmically located viruses.

MATERNAL EFFECT

The maternal inheritance pattern of extranuclear genes is distinct from the phenomenon of **maternal effect**, which is defined as the phenotype in an individual that is established by the maternal nuclear genome, as the result of mRNA and/or proteins that are deposited in the oocyte prior to fertilization.

These inclusions direct early development of the embryo. That is, in maternal inheritance the progeny always have the maternal phenotype, whereas in maternal effect the progeny always have the phenotype specified by the maternal nuclear *genotype*. *Maternal effect does not involve any extranuclear genes and is discussed here to make the distinction from extranuclear inheritance clear.*

Maternal effect is seen, for instance, in the inheritance of the direction of coiling of the shell of the snail *Limnaea peregra* (Figure 21.21). The shell-coiling trait is determined by a single pair of *nuclear* alleles, D for coiling to the right (dextral coiling) and d for coiling to the left (sinistral coiling). The D allele is completely dominant to the d allele and the shell-coiling phenotype is *always determined by the genotype of the mother*. The latter is shown by the results of reciprocal crosses between a true-breeding, dextral-coiling, and a sinistral-coiling snail (Figure 21.22). All the F_1s have the same genotype since a nuclear gene is involved, yet the *phenotype* is different for the reciprocal crosses.

In the cross of a dextral (D/D) female with a sinistral (d/d) male (Figure 21.22a), the F1s are all D/d in genotype and dextral in phenotype. Selfing the F_1 produces F_2s with a 1:2:1 ratio of D/D, D/d, and d/d genotypes. *All* of the F_2s are dextral, even the d/d snails whose genotype would seem to indicate sinistral phenotype. Here is our first encounter with maternal effect; the d/d snails have a coiling phenotype not specified by the genotype they have, but one

~ FIGURE 21.21

The snail, *Limnaea peregra*.

specified by the genotype of their mother (*D/d*). Selfing the F₂ snails gives F₃ progeny, 3/4 of which are dextral and 1/4 of which are sinistral. The latter are the *d/d* progeny of the F₂ *d/d* snails; these F₃ snails are sinistral because their phenotype reflects their mother's (the F₂) genotype.

Similar results are seen in the reciprocal cross of a sinistral (*d/d*) female with a dextral (*D/D*) male (Figure 21.22b). The F₁s are all *D/d* in genotype yet they are sinistral in phenotype because the mother was *d/d* in genotype. Selfing the F₁ produces F₂s all of which are dextral for the same reason as the reciprocal cross already described. The genotypes and phenotypes of the F₂ and F₃ generations are the same as for the reciprocal cross (see Figure 21.22b) and for the same reasons.

The above results do *not* fit our criteria for extranuclear inheritance. That is, if the coil direction phenotype were controlled by an extranuclear gene, the progeny would always exhibit the *phenotype* of the mother, owing to maternal inheritance. Here, the coiling phenotype is governed directly by the nuclear *genotype* of the mother and is an example of maternal effect. But what is the basis for the coiling? Cytological analysis of developing eggs has shown that the orientation of the mitotic spindle in the first mitotic division following fertilization controls the direction of coiling. It has been hypothesized that the mother encodes products that are deposited in the oocyte and that direct the orientation of the mitosis

spindle and therefore of direction of cell cleavage. Thus, a mother of genotype *D/−* will deposit gene products that specify a dextral (right-handed) coiling, and a mother of genotype *d/d* will deposit gene products that specify a sinistral (left-handed) coiling. Support for this hypothesis has come from the experiments of G. Freeman and J. Lundelius. They injected cytoplasm from dextrally coiling snails into the eggs of *d/d* mothers (who would normally specify sinistrally coiling progeny) and found that the resulting embryos coiled dextrally. In the reciprocal experiment, cytoplasm from sinistrally coiling snails injected into *D/−* mothers had no effect on the resulting embryos; that is, they were still dextrally coiling. The interpretation is that: (1) the *D* allele specifies a product that is deposited in egg and causes the next-generation embryos to coil dextrally; and (2) the *d* allele produces a defective or no product so that *d/d* snails produce embryos that coil in the sinistral direction by default.

We encountered examples of maternal effect earlier in our discussion of the genetic control of *Drosophila* development (Chapter 17, pp. 570–571). The class of genes called *maternal effect genes* are nuclear genes that are expressed by the mother during oogenesis. The products of these genes are deposited in the egg and function to specify the gradients in the egg that control spatial organization in early development. These genes were identified by studying the properties of mutants which did not develop normally. For example, mothers homozygous for a mutated maternal effect gene, *bicoid* (*bcd*), produce mutant embryos that have no heads or thoraxes, only abdomens. On the basis of that result, the *bcd* gene was deduced to play in important role in determining normal anterior development. Relevant to our discussion here is that the *phenotype* (mutant) of the progeny embryos reflects the *genotype* (*bcd/bcd*) and not the phenotype (normal) of the mothers; that is, maternal effect is involved. The role of the *bcd* protein in development was discussed in Chapter 17 (pp. 570–571).

KEYNOTE

Maternal effect is different from extranuclear inheritance. The maternal inheritance pattern of extranuclear genes occurs because the zygote contains most of its organelles (containing the extranuclear genes) from the female parent, whereas in the maternal effect the trait inherited is controlled by the maternal *nuclear* genotype before the fertilization of the egg and does not involve extranuclear genes.

~ FIGURE 21.22

Inheritance of the direction of shell coiling in the snail *Limnaea peregra* is an example of maternal effect. (a) Cross between true-breeding dextral-coiling female (*D/D*) and sinistral-coiling male (*d/d*); (b) Cross between sinistral-coiling female (*d/d*) and dextral-coiling male (*D/D*).

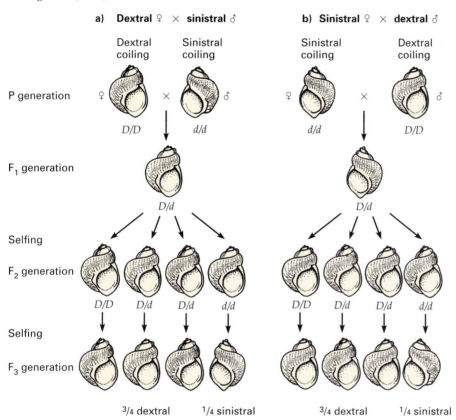

GENOMIC IMPRINTING

An implicit assumption of Mendelian genetics is that the expression of a gene is totally independent of whether it is of maternal or paternal origin. Most genes align with this assumption. However, there are instances where the expression of certain genes is determined by whether the gene is inherited from the female or male parent, a phenomenon called **genomic imprinting** or **parental imprinting**. As for maternal effect, genomic imprinting does not involve any extranuclear genes and is discussed here because of the maternal versus paternal influence on gene expression.

One example of genomic imprinting came from studies in mice. A transgenic mouse line (see Chapter 19, p. 647) was constructed using an autosomal transgene containing an oncogene whose expression could easily be assayed for. If the transgene was inherited from the male parent, it was expressed in the heart and in no other tissue. If the transgene was donated from the female parent, the transgene was not expressed at all. This pattern of expression clearly fits our definition of genomic imprinting. Molecular analysis indicated that the expression pattern correlated precisely with the methylation state of the transgene. (Recall from Chapter 17 that decreasing the methylation state of a gene can correlate with increased expression of the gene.) That is, the transgene is methylated in the female parent and is inherited from the female in a methylated state; it is not transcribed. The transgene is also methylated in the male parent but is demethylated during gametogenesis in the male, so it is inherited from the male in an unmethylated state and is active. Methylation is increasingly being implicated in other examples of genomic imprinting.

Some human genetic diseases appear to result from genomic imprinting, for example, the *Prader-Willi* and *Angelman syndromes*. The frequency of Prader-Willi syndrome (PWS) is about 1 in 25,000 births. PWS patients are typically small and weak at birth, and they exhibit a number of symptoms including retardation and poor feeding due to diminished swallowing and sucking reflexes. The feeding difficulties improve by the age of 6 months and, from about 12 months onward, a pattern of uncontrollable eating develops, resulting in obesity and attendant psychological problems. Adolescents have poor motor skills and insatiable hunger. Adults are short relative to their family members and often develop a form of diabetes because of the eating disorder. PWS patients rarely live beyond 30 years unless strict weight-control regimes are maintained to control the diabetes.

PWS is caused by deletion or disruption of a gene or several genes in region 15q11–q13 of chromosome 15. Pedigree analysis has shown that in 70–80 percent of cases examined, the deletion/disruption event occurred in the father and that genomic imprinting plays a role. That is, in a PWS child, the activities of some genes in region 15q11–q13 on the maternal chromosome 15 are normally suppressed by methylation as a result of genomic imprinting. The paternally inherited alleles are necessary for normal development, but because of the gene deletion/disruption event in the father, those genes are inactive and the PW phenotype results.

Angelman syndrome (AS) is characterized by a number of symptoms including severe motor and intellectual retardation, smaller-than-normal head size, jerky limb movements, hyperactivity, and frequent unprovoked laughter. In about 50 percent of AS patients, a deletion of region 15q11–q13 is seen. This is the same region affected in PWS patients. Indeed, it seems that AS can be caused in much the same way as PWS, except that in AS maternally inherited alleles of the genes involved are needed for normal development. That is, the paternally inherited alleles are inactive because of methylation brought about by the genomic imprinting phenomenon, which causes AS to develop if the maternally inherited alleles are deleted or disrupted. Support for this conclusion is provided by the development of AS as a result of inheritance of a nondeletion genetic muation in 15q11–q13 from the mother, but not from the father.

Genomic imprinting may also affect the symptoms of other human genetic diseases. Fragile X syndrome (Chapter 7, pp. 205–206) is one such example in which methylation is clearly responsible for the imprinting.

KEYNOTE

While most genes are expressed in a way that is totally independent of parental origin, the expression of certain genes is determined by whether the gene is inherited from the female or male parent. This phenomenon is called genomic imprinting. Attention is being focussed on methylation patterns as a basis for imprinting.

SUMMARY

Both mitochondria and chloroplasts contain DNA, the length of which varies from organism to organism. The genomes of both organelles are naked, usually circular, double-stranded DNAs. The two genomes contain genes that are not duplicated in the nuclear genome; hence, the organelle genes are contributing different information for the function of the cell. The organellar genes encode the rRNA components of the ribosomes (that are assembled and function in the organelles), and for many (if not all) of the tRNAs used in organellar protein synthesis. Some of the organelle proteins are encoded by the organelle; the remainder are nuclear-encoded. Nuclear-encoded proteins found in organelles are synthesized on cytoplasmic ribosomes, then imported into the appropriate organelles. These proteins include the ribosomal proteins for the organellar ribosomes and, in the case of mitochondria, component proteins of cytochromes.

The existence of polymorphic sequences in mtDNA and the fact that humans and many other organisms exhibit maternal inheritance of mitochondria mean that maternal lineages are practically unique. Thus, the genetic relationships between individuals can readily be investigated by mtDNA analysis. This method is used in many areas of biology, including conservation biology. There are many examples of mitochondrial and chloroplast mutants. The inheritance of these extranuclear genes follows rules different from those for nuclear genes, and this is how extranuclear genes were originally identified. For extranuclear genes, no meiotic segregation is involved, uniparental (and usually maternal) inheritance is exhibited, a trait resulting from an extranuclear mutation persists after nuclear substitution, and extranuclear genes cannot be mapped to the known nuclear linkage groups. Not all cases of extranuclear inheritance result from genes on mtDNA or cpDNA. Many other examples result from infectious heredity, in which cytoplasmically located bacteria or viruses are transmitted when cytoplasms mix.

A specific inheritance pattern sometimes observed is the phenomenon in which the mother specifically affects the phenotype of the offspring. This phenomenon, called the maternal effect, is defined as the phenotype in an individual that is established by the maternal nuclear genome, as the result of mRNA and/or proteins that are deposited in the oocyte prior to fertilization. These inclusions direct early development of the embryo. Maternal effect is different from extranuclear inheritance. That is, the maternal inheritance pattern of extranuclear genes occurs because the zygote receives most of its organelles (containing the extranuclear genes) from the female parent, whereas in the maternal effect the trait inherited is controlled by the maternal *nuclear* genotype before the fertilization of the egg and does *not* involve extranuclear genes.

Finally, we discussed genomic imprinting, the phenomenon in which the expression of a gene is determined by whether the gene is inherited from the female or male parent as opposed to being expressed independently of parental origin. This phenomenon does not involve extranuclear genes, but it does involve an influence on gene expression of one or other parent. Methylation has been shown to be responsible for imprinting in some cases.

ANALYTICAL APPROACHES TO SOLVING GENETICS PROBLEMS

Q21.1 In *Neurospora*, strains of [*poky*] have been isolated that have reverted to nearly wild-type growth, although they still retain the abnormal metabolism and cytochrome pattern characteristic of [*poky*]. These strains are called *fast*-[*poky*]. A cross of a *fast*-[*poky*] female with a wild-type male gives a 1:1 segregation of [*poky*]:*fast*-[*poky*] ascospores in all asci. Interpret these results, and predict the results you would expect from the reciprocal of the stated cross.

A21.1 The [*poky*] mutation shows maternal inheritance, so when it is used as a female parent, all the progeny ascospores will carry the [*poky*] determinant. Conventionally, extranuclear genes are designated within square brackets so in this case the cross with respect to the [*poky*] determinant was [*poky*] × [*N*], where *N* signifies normal cytoplasm. All progeny are [*poky*]. The asci show a 1:1 segregation for the [*poky*] and *fast*-[*poky*] phenotypes. This ratio is characteristic of a nuclear gene segregating in the cross. Therefore the simplest explanation is that the factor that causes [*poky*] strains to be *fast*-[*poky*] is a nuclear gene mutation; we can call it *F* (its actual designated name). The *F* gene segregates in meiosis, as do all nuclear genes. The cross can be rewritten as [*poky*]*F* ♀ × [*N*]+ ♂, which gives a 1:1 segregation of [*poky*]+ and [*poky*]*F*. The former is [*poky*] and the latter is *fast*-[*poky*].

With these results behind us, the reciprocal cross may be diagrammed as [*N*]+ ♀ × [*poky*]*F* ♂. All the progeny spores from this cross are [*N*], half of them being *F* and half of them being +. If *F* has no effect on normal cytoplasm, then these two classes of spores would be phenotypically indistinguishable, which is the case.

Q21.2 Four slow-growing mutant strains of *Neurospora crassa*, coded *a*, *b*, *c*, and *d*, were isolated. All have an abnormal system of respiratory mitochondrial enzymes. The inheritance patterns of these mutants were tested in controlled crosses with the wild type, with the following results:

PROTO-PERITHECIAL (FEMALE) PARENT		CONIDIAL (MALE) PARENT	PROGENY (ASCOSPORES)	
			WILD TYPE	SLOW GROWING
Wild type	×	*a*	847	0
a	×	Wild type	0	659
Wild type	×	*b*	1,113	0
b	×	Wild type	0	2,071
Wild type	×	*c*	596	590
Wild type	×	*d*	1,050	1,035

Give a genetic interpretation of these results.

A21.2 This question asks us to consider the expected transmission patterns for nuclear genes and for extranuclear genes. The nuclear genes will have a 1:1 segregation in the offspring, since this organism is a haploid organism and hence should exhibit no differences in the segregation patterns, whichever strain is the maternal parent. On the other hand, a distinguishing characteristic of extranuclear genes is a difference in the results of reciprocal crosses. In *Neurospora* this characteristic is usually manifested by all progeny having the phenotype of the maternal parent. With these ideas in mind we can analyze each mutant in turn.

Mutant *a* shows a clear difference in its segregation in reciprocal crosses and is, in fact, a classic case of maternal inheritance. The interpretation here is that the gene is extranuclear; hence, the gene must be in the mitochondrion. The [*poky*] mutant described in this chapter shows this type of inheritance pattern.

By the same reasoning used to analyze mutant *a*, the mutation in strain *b* must also be extranuclear.

Mutants *c* and *d* segregate 1:1, indicating that the mutations involved are in the nuclear genome. In these cases we need not consider the reciprocal cross since there is no evidence for maternal inheritance. In fact, the actual mutations that are the basis for this question cause sterility, so the reciprocal cross cannot be done. We can confirm that the mutations are in the nuclear genome by doing mapping experiments, using known nuclear markers. Evidence of linkage to such markers would confirm that the mutations are not extranuclear.

QUESTIONS AND PROBLEMS

21.1 Compare and contrast the structure of the nuclear genome, the mitochondrial genome, and the chloroplast genome.

*21.2 Imagine you have discovered a new genus of yeast. In the course of your studies on this organism, you isolate DNA and subject it to CsCl density gradient centrifugation. You observe a major peak at a density of 1.75 g/cm^3 and a minor peak at a density of 1.70 g/cm^3. How could you determine whether the minor peak represents organellar (presumably mitochondrial) DNA, as opposed to a relatively AT-rich repeated sequence in the nuclear genome?

21.3 In what ways is mitochondrial DNA replication unlike DNA replication in nuclear chromosomes? In what ways are they similar?

21.4 How do mitochondria reproduce? What is the evidence for the method you describe?

21.5 What genes are present in the human mitochondrial genome?

21.6 What conclusions can you draw from the fact that most nuclear-encoded mRNAs and all mitochondrial mRNAs have a poly(A) tail at the 3' end?

*21.7 A substantial body of evidence indicates that defects in mitochondrial energy production may contribute to the neuronal cell death seen in a number of late-onset neurodegenerative diseases, including Alzheimer's disease, Parkinson's disease, Huntington disease, and amyotrophic lateral sclerosis (ALS, Lou Gehrig disease). Some, but not all, of these diseases have been associated with mutations in the nuclear genome. One experimental system that has been developed to evaluate the contributions of the mitochondrial genome to these diseases uses a cytoplasmic hybrid known as a "cybid." Cybids are made by repopulating a tissue-culture cell line that has been made mitochondria-deficient with mitochondria from the cytoplasm of a human platelet cell. The cybids thus have nuclear DNA from the tissue-culture cell, and mitochondrial DNA from the human platelet cell.

The mitochondrial protein cytochrome oxidase has subunits encoded by both nuclear and mitochondrial genes. Patients with Alzheimer's disease have been reported to have lower levels of cytochrome oxidase than age-matched controls.

a. What is the evidence that cytochrome oxidase has subunits encoded by both nuclear and mitochondrial genes?
b. Given the means to assay cytochrome oxidase activity, how would you investigate whether the decreased levels of cytochrome oxidase activity in Alzheimer's patients could be ascribed to nuclear or mitochondrial genetic defects? What controls would you create?
c. If you are able to demonstrate that the mitochondrial contribution to cytochrome oxidase is responsible for lowered cytochrome oxidase activity, can you conclude that each of the mitochondria of an affected individual has an identical defect?

*21.8 Discuss the differences between the universal genetic code of the nuclear genes of most eukaryotes and the code found in human mitochondria. Is there any advantage to the mitochondrial code?

21.9 When the DNA sequences for most of the mRNAs in human mitochondria are examined, no nonsense codons are found at their termini. Instead, either U or UA is found. Explain this result.

21.10 Compare and contrast the cytoplasmic, mitochondrial, and chloroplast protein-synthesizing systems.

21.11 Compare and contrast the organization of the ribosomal RNA genes in mitochondria and in chloroplasts.

*21.12 What features of extranuclear inheritance distinguish it from the inheritance of nuclear genes?

21.13 Distinguish between maternal effect and extranuclear inheritance.

21.14 Reciprocal crosses between two types of the evening primrose, *Oenothera hookeri* and *Oenothera muricata*, produce the following effects on the plastids:

O. hookeri female × *O. muricata* male → yellow plastids
O. muricata female × *O. hookeri* male → green plastids

Explain the difference between these results, noting that the chromosome constitution is the same in both types.

*21.15 A series of crosses are performed with a recessive mutation in *Drosophila* called *tudor*. Homozygous *tudor* animals appear normal and can be generated from the cross of two heterozygotes, but a true-breeding *tudor* strain cannot be maintained. When homozygous *tudor* males are crossed to homozygous *tudor* females, both of whom appear to be phenotypically normal, a normal-appearing F_1 is produced. However, when F_1 males are crossed to wild-type females, or when F_1 females are crossed to wild-type males, no progeny are ever produced. The same results are seen in the F_1 progeny of homozygous *tudor* females crossed to wild-type males. The F_1 progeny of homozygous *tudor* males crossed to wild-type females appear normal and they are capable of issuing progeny when mated either with each other or with wild-type animals.

a. How would you classify the *tudor* mutation? Why?

b. What might cause the *tudor* phenotype?

*21.16 A form of male sterility in corn is maternally inherited. Plants of a male-sterile line crossed with normal pollen give male-sterile plants. Some lines of corn carry a dominant, so-called restorer (*Rf*) gene, which restores pollen fertility in male-sterile lines.

a. If a male-sterile plant is crossed with pollen from a plant homozygous for gene *Rf*, what will be the genotype and phenotype of the F_1?

b. If the F_1 plants of part a are used as females in a testcross with pollen from a normal plant (*rf/rf*), what would be the result? Give genotypes and phenotypes, and designate the type of cytoplasm.

*21.17 In *Neurospora* a chromosomal gene *F* suppresses the slow-growth characteristic of the [*poky*] phenotype and makes a [*poky*] culture into a *fast*-[*poky*] culture, which still has abnormal cytochromes. Gene *F* in combination with normal cytoplasm has no detectable effect. (Hint: Since both nuclear and extranuclear genes have to be considered, it will be convenient to use symbols to distinguish the two. Thus cytoplasmic genes will be designated in square brackets; e.g., [*N*] for normal cytoplasm, [*poky*] for poky.)

a. A cross in which *fast*-[*poky*] is used as the female (protoperithecial) parent and a normal wild-type strain is used as the male parent gives half [*poky*] and half *fast*-[*poky*] progeny ascospores. What is the genetic interpretation of these results?

b. What would be the result of the reciprocal cross of the cross described in part a, that is, normal female × *fast*-[*poky*] male?

21.18 Distinguish between nuclear (segregational), neutral, and suppressive *petite* mutants of yeast.

*21.19 In yeast a haploid nuclear (segregational) *petite* is crossed with a neutral *petite*. Assuming that both strains have no other abnormal phenotypes, what proportion of the progeny ascospores are expected to be *petite* in phenotype if the diploid zygote undergoes meiosis?

21.20 When grown on a medium containing acriflavin, a yeast culture produces a large number of very small

(*tiny*) cells that grow very slowly. How would you determine whether the slow-growth phenotype was the result of a cytoplasmic factor or a nuclear gene?

21.21 *Drosophila melanogaster* has a sex-linked, recessive, mutant gene called *maroon-like* (*ma-l*). Homozygous *ma-l* females or hemizygous *ma-l* males have light-colored eyes, owing to the absence of the active enzyme xanthine dehydrogenase, which is involved in the synthesis of eye pigments. When heterozygous $ma\text{-}l^+/ma\text{-}l$ females are crossed with *ma-l* males, all the offspring are phenotypically wild type. However, half the female offspring from this cross, when crossed back to *ma-l* males, give all *ma-l* progeny. The other half of the females, when crossed to *ma-l* males, give all phenotypically wild-type progeny. What is the explanation for these results?

*21.22 When females of a particular mutant strain of *Drosophila melanogaster* are crossed to wild-type males, all the viable progeny flies are females. Hypothetically, this result could be the consequence of either a sex-linked, male-specific lethal mutation or a maternally inherited factor that is lethal to males. What crosses would you perform in order to distinguish between these alternatives?

21.23 Reciprocal crosses between two *Drosophila* species, *D. melanogaster* and *D. simulans*, produce the following results:

melanogaster ♀ × *simulans* ♂ → females only
simulans ♀ × *melanogaster* ♂ → males, with few or no females

Propose a possible explanation for these results.

21.24 Some *Drosophila* flies are very sensitive to carbon dioxide—they become anesthetized when it is administered to them. The sensitive flies have a cytoplasmic particle called *sigma* that has many properties of a virus. Resistant flies lack *sigma*. The sensitivity to carbon dioxide shows strictly maternal inheritance. What would be the outcome of the following two crosses: (a) sensitive female × resistant male and (b) sensitive male × resistant female?

*21.25 A few years ago the political situation in Chile was such that very many young adults were kidnapped, tortured, and killed by government agents. When abducted young women had young children or were pregnant, those children were often taken and given to government supporters to raise as their own. Now that the political situation has changed, grandparents of stolen children are trying to locate and reclaim their grandchildren. Imagine that you are a judge in a trial centering on the custody of a child. Mr. and Mrs. Escobar believe Carlos Mendoza is the son of their abducted, murdered daughter. If this is true, then Mr.

and Mrs. Sanchez are the paternal grandparents of the child, as their son (also abducted and murdered) was the husband of the Escobars' daughter. Mr. and Mrs. Mendoza claim Carlos is their natural child. The attorney for the Escobar and Sanchez couples informs you that scientists have discovered a series of RFLPs in human mitochondrial DNA. He tells you his clients are eager to be tested, and ask that you order that Mr. and Mrs. Mendoza and Carlos be tested also.

a. Can mitochondrial RFLP data be helpful in this case? In what way?

b. Do all seven parties need to be tested? If not, who actually needs to be tested in this case? Explain your choices.

c. Assume the critical people have been tested, and you have received the results. How would the results determine your decision?

21.26   The analysis of mitochondrial DNA has been very useful in assessing the history of specific human populations. For example, a 9-bp deletion in the small intergenic region between the genes for cytochrome oxidase subunit II and tRNA.Lys (see Figure 21.5) has been a very informative marker to trace the origins of Polynesians. The deletion is widely distributed across Southeast Asia and the Pacific, and is present in 80–100 percent of individuals in the different populations within Polynesia. One of the most polymorphic regions of the mitochondrial genome is found in the region between the genes for tRNA.Phe and tRNA.Thr. In Asians with the 9-bp deletion, a specific set of DNA sequence polymorphisms in this region is found. Using the 9-bp deletion and the DNA sequence polymorphisms as markers, comparative analysis of Asian populations has found a genetic "trail" of mitochondrial DNA variation. The trail begins in Taiwan, winds through the Philippines and Indonesia, proceeds along the coast of New Guinea, and then moves into Polynesia. Based on an estimated rate of mutation in the tRNA.Phe to tRNA.Thr region, this expansion of mitochondrial DNA variants is thought to be about 6,000 years old. This is in agreement with linguistic and archeological evidence that associates Polynesian origins with the spread of the Austronesian language family out of Taiwan between 6,000 and 8,000 years ago.

a. Why are these types of mitochondrial DNA polymorphisms such good markers for tracing human migration patterns?

b. Why is it important to correlate findings from mitochondrial DNA polymorphisms with other (non-DNA) assessment methods?

c. Why might sequences in the tRNA.Phe to tRNA.Thr region be more polymorphic than other sequences in the mitochondrial genome?

d. The 9-bp deletion has also been found in human populations in Africa. What different explanations are possible for this, and how might these explanations be evaluated?

21.27   The pedigree in the figure below shows a family in which an inherited disease called Leber's optic atrophy is segregating. This condition causes blindness in adulthood. Studies have recently shown that the mutant gene causing Leber's optic atrophy is located in the mitochondrial genome.

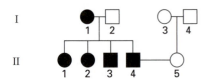

a. Assuming II-4 marries a normal person, what proportion of his offspring should inherit Leber's optic atrophy?

b. What proportion of the sons of II-2 should be affected?

c. What proportion of the daughters of II-2 should be affected?

*21.28   The inheritance of the direction of shell coiling in the snail *Limnaea peregra* has been studied extensively. A snail produced by a cross between two individuals has a shell with a right-hand twist (dextral-coiling). This snail produces only left-hand (sinistral) progeny on selfing. What are the genotypes of the F₁ snail and its parents?

*21.29   Beckwith-Wiedemann syndrome (BWS) is characterized by fetal malformation and cancer. BWS has been mapped to chromosome region 11p15.5 by its association with balanced chromosomal rearrangements. This region contains an embryonic tumor suppressor gene and a number of other genes, including *KVLQT1*. Mutations in *KVLQT1* cause a dominantly inherited heart defect that results in cardiac arrhythmia and sudden death. When a wide variety of childhood and adult tumors are examined for gene expression in region 11p15.5, a preferential loss of expression of the paternal alleles is seen. Most normal fetal tissues only express the maternal allele, but heart tissue typically expresses maternal and paternal alleles of *KVLQT1*.

a. What evidence would be necessary to show that genomic imprinting may play a role in Beckwith-Wiedemann syndrome? From the description above, does such evidence exist?

b. What is the molecular basis for imprinting? How is it detected?

c. What kind of evidence would be necessary to show that imprinting is involved with the dominantly inherited heart defect associated with the *KVLQT1* gene?

d. What similarities, if any, do you see in the imprinting of genes in this region with those for the Prader-Willi and Angelman syndromes discussed in this chapter (pp. 707–708)?

21.30 The rate of trinucleotide repeat expansion at an allele with an unstable trinucleotide repeat depends on whether the allele is transmitted maternally or paternally. For example, in fragile X syndrome, the maternal transmission of a non-disease allele with 150–200 CGG repeats results in a disease allele with over 200 CGG repeats. The paternal transmission of a similar-sized allele has little effect on the number of CGG repeats. In other diseases, alleles with the longest trinucleotide repeat expansion are transmitted only by affected fathers.

Although the age of disease onset usually correlates with the length of trinucleotide repeat expansion, the age of onset for some diseases can differ in individuals with identical trinucleotide repeat expansions. This is intriguing when one considers that the extent of DNA methylation at some genes can vary with age. Based on what you know about imprinting and trinucleotide repeat expansion, speculate on what potential role(s) imprinting might have on the rate of trinucleotide repeat expansion and how it might contribute to the timing of disease onset.

22 POPULATION GENETICS

PRINCIPAL POINTS

~ The genetic structure of a population is determined by the total of all alleles (the gene pool). In the case of diploid, sexually interbreeding individuals, the structure is also characterized by the distribution of alleles into genotypes. Thus, the genetic structure is described in terms of both allelic and genotypic frequencies. Except for rare mutations, individuals are born and die with the same set of alleles; what changes genetically over time (evolves) is the genetic structure of a group of individuals, reproductively connected in a Mendelian population.

~ There is usually a great deal of genetic variation among individuals within populations.

~ The genetic structure of a species can vary both geographically and temporally.

~ Principles of population genetics are necessary to explain the evolution of species differences and adaptations. A change in the genetic structure of a population over time is equivalent to an evolutionary change.

~ Principles of population genetics can be applied to the management of rare and endangered species. Genetic diversity is best maintained by establishing a population with adequate founders, expanding the population rapidly, avoiding inbreeding, and maintaining equal sex ratio and equal family size.

~ The Hardy-Weinberg law states that in a large, randomly mating population, free from evolutionary processes, the allelic frequencies do not change, and the genotypic frequencies stabilize after one generation in the proportions p^2, $2pq$, and q^2, where p and q equal the allelic frequencies of the population.

~ The classical, balance, and neutral mutation models generate testable hypotheses and are used to explain how much genetic variation should exist within natural populations and what processes are responsible for the observed variation.

~ Mutation, genetic drift, migration, and natural selection are processes that can alter allelic frequencies of a population.

~ Recurrent mutation changes allelic frequencies, and at equilibrium the relative rates of forward and reverse mutations determine allelic frequencies of a population in the absence of other processes.

~ Genetic drift is random change in allelic frequencies due to chance. Genetic drift produces genetic change within populations, genetic differentiation among populations, and loss of genetic variation within populations.

~ Migration, also termed gene flow, involves movement of alleles among populations. Migration can alter the allelic frequencies of a population, and it tends to reduce genetic divergence among populations.

~ Natural selection is differential reproduction of genotypes. It is measured by Darwinian fitness, which is the relative reproductive ability of genotypes. Natural selection can produce a number of different effects on the gene pool of a population.

~ Nonrandom mating affects the genotypic frequencies of a population. Inbreeding leads to an increase in homozygosity, and outbreeding leads to an increase in heterozygosity.

~ Rates of evolution can be measured by comparing DNA or RNA sequences. Different genes and different parts of the same gene tend to evolve at different rates.

~ In eukaryotic organisms, genes frequently occur in multiple copies with identical or similar sequences. A group of such genes is termed a multiple gene family.

~ The mitochondrial DNA of some organisms evolves at a faster rate than the nuclear DNA.

~ Concerted evolution is a process that maintains sequence uniformity among multiple copies of the same sequence within a species.

~ Evolutionary relationships among organisms can be revealed by the study of DNA and RNA sequences.

The science of genetics can be broadly divided into four major subdisciplines: transmission genetics (also called classical genetics), molecular genetics, population genetics, and quantitative genetics. Each of these four areas focuses on a different aspect of heredity. **Transmission genetics** is primarily concerned with genetic processes that occur within individuals, and how genes are passed from one individual to another. Thus, the unit of study for transmission genetics is the *individual*. In **molecular genetics**, we are largely interested in the molecular nature of heredity — how genetic information is encoded within the DNA and how biochemical processes of the cell translate the genetic information into influencing the phenotype. Consequently, in molecular genetics we focus on the *cell*. **Population genetics**, the subject of this chapter, is the field of genetics that studies heredity in groups of individuals for traits that are determined by one, or only a few genes. **Quantitative genetics**, the subject of the next chapter,

also considers the heredity of traits in groups of individuals, but the traits of concern are determined by many genes simultaneously. Because these latter two fields are based on Mendelian principles, but taken to the level of the group, they are very amenable to mathematical treatment. In fact these areas provide us with the oldest and richest examples of the success of mathematical theory in biology. The impetus for the development of these areas came after the rediscovery of Mendel's work and the obviously great implications it had for Darwinian theory. In fact, the fusion of Mendelian theory with Darwinian theory is called the neo-Darwinian synthesis and was championed by Sir Ronald Fisher, Sewall Wright, and J. B. S. Haldane (Figure 22.1). The neo-Darwinian synthesis saved natural selection from obscurity, and is the foundation of a large part of modern biology.

Population geneticists investigate the patterns of genetic variation found among individuals within groups (the **genetic structure** of populations) and

~ **FIGURE 22.1**

Sir Ronald Fisher (a), Sewall Wright (b), and J. B. S. Haldane (c) are considered the major architects of neo-Darwinian theory.

a)   b)   c)

how these patterns vary geographically and evolve over time. In this discipline, our perspective shifts away from the individual and the cell, and focuses instead upon a **Mendelian population**. A Mendelian population is a group of *interbreeding* individuals, who share a common set of genes. The genes shared by the individuals of a Mendelian population are called the **gene pool**. To understand the genetics of the evolutionary process, we study the gene pool of a Mendelian population, rather than the genotypes of its individual members. An understanding of the genetic structure of a population is also a key to our understanding of the importance of genetic resources and the importance of genes for the conservation of species and biodiversity.

Questions frequently studied by population geneticists include the following:

1. How much genetic variation is found in natural populations, and what processes control the amount of variation observed?
2. What evolutionary processes shape the genetic structures of populations?
3. What processes are responsible for producing genetic divergence among populations?
4. How do biological characteristics of a population, such as breeding system, fecundity, and age structure, influence the gene pool of the population?

To answer these questions, population geneticists frequently develop mathematical models and equations to describe what happens to the gene pool of a population under various conditions. As a result of regularity of the meiotic process in gamete formation and the ensuing mechanisms of gamete fusion during sexual reproduction, a great deal of predictability concerning the genetic structure of a population can be modeled. An example is the set of equations that describes the influence of random mating on the allele and genotypic frequencies of an infinitely large population, a model called the **Hardy-Weinberg law**, which we discuss later in the chapter. At first, students often have difficulty grasping the significance of models such as these, because the models are frequently simple and require numerous assumptions that are unlikely to be met by organisms in the real world. Beginning with simple models is useful, however, because with such models we can examine what happens to the genetic structure of a population when we deliberately violate one assumption after another and then in combination. Once we understand the results of the simple models, we can incorporate more realistic conditions into the equations. In the case of the Hardy-Weinberg law, the assumptions of infinite-

ly large size and random mating may seem unrealistic, but these assumptions are necessary at first for simplifying the mathematical analysis. We cannot begin to understand the impact of nonrandom mating and limited population size on allelic frequencies until we first know what happens under the simpler conditions of random mating and large population size.

GENETIC STRUCTURE OF POPULATIONS

Genotypic Frequencies

To study the genetic structure of a Mendelian population, population geneticists must first describe the gene pool of the group quantitatively. This is done by calculating genotypic frequencies and allelic frequencies within the population. A frequency is a proportion, and always ranges between 0 and 1. If 43 percent of the people in a group have red hair, the frequency of red hair in the group is 0.43. To calculate the **genotypic frequencies** at a specific locus, we count the number of individuals with one particular genotype and divide this number by the total number of individuals in the population. We do this for each of the genotypes at the locus. The sum of the genotypic frequencies should be 1. Consider a locus that determines the pattern of spots in the scarlet tiger moth, *Panaxia dominula* (Figure 22.2). Three genotypes are present in most populations, and each genotype produces a different phenotype. E. B. Ford collected moths at one locality in England and found the following numbers of genotypes: 452 *BB*, 43 *Bb*, and 2 *bb*, out of a total of 497 moths. The genotypic frequencies (where f = frequency of) are therefore:

$$
\begin{aligned}
f(BB) &= 452/497 = 0.909 \\
f(Bb) &= 43/497 = 0.087 \\
f(bb) &= 2/497 = \underline{0.004} \\
&\quad\text{Total} \quad \underline{1.000}
\end{aligned}
$$

Allelic Frequencies

Although genotypic frequencies at a single gene locus are useful for examining the effects of certain evolutionary processes on a population, in most cases population geneticists use frequencies of alleles to describe the gene pool. The use of **allelic frequencies** offers several advantages over the genotypic frequencies. First, in sexually reproducing organisms, genotypes break down to alleles when gametes are formed, and alleles, not genotypes, are passed from

~ **FIGURE 22.2**

Panaxia dominula, the scarlet tiger moth. The top two moths are normal homozygotes (*BB*), those in the middle two rows are heterozygotes (*Bb*), and the bottom moth is the rare homozygote (*bb*).

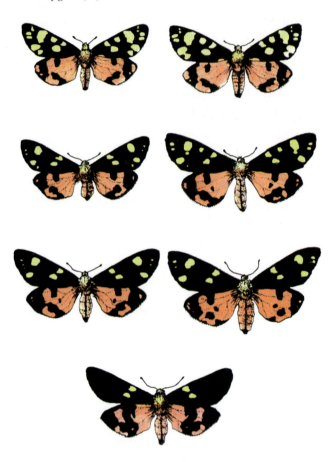

one generation to the next. Consequently, only alleles have continuity over time, and the gene pool evolves through changes in the frequencies of alleles. Furthermore, there are always fewer alleles than genotypes, and so the gene pool can be described with fewer parameters when allelic frequencies are used. For example, if there are three alleles segregating at a particular locus, six genotypes will form and frequencies must be calculated for each to describe the gene pool.

Allelic frequencies may be calculated in either of two ways: from the observed numbers of different genotypes at a particular locus, or from the genotypic frequencies. First, we can calculate the allelic frequencies directly from the *numbers* of genotypes. In this

method, we count the number of alleles of one type at a particular locus and divide it by the total number of alleles at that locus in the population:

$$\text{allelic frequency} = \frac{\text{Number of copies of a given allele in the population}}{\text{Sum of all alleles in the population}}$$

As an example, imagine a population of 1,000 diploid individuals with 353 *AA*, 494 *Aa*, and 153 *aa* individuals. Each *AA* individual has two *A* alleles, while each *Aa* heterozygote possesses only a single *A* allele. Therefore, the number of *A* alleles in the population is (2 × the number of *AA* homozygotes) + (the number of *Aa* heterozygotes), or (2 × 353) + 494 = 1,200. Because every diploid individual has two alleles, the total number of alleles in the population will be twice the number of individuals, or 2 × 1,000. Using the equation given above, the allelic frequency is 1,200/2,000 = 0.60. When two alleles are present at a locus, we can use the following formula for calculating allelic frequencies:

$$p = f(A) = \frac{(2 \times \text{number of } AA \text{ homozygotes}) + (\text{number of } Aa \text{ heterozygotes})}{(2 \times \text{total number of individuals})}$$

The second method of calculating allelic frequencies goes through the step of first calculating genotypic frequencies as demonstrated previously. In this example $f(AA) = 0.353$; $f(Aa) = 0.494$; and $f(aa) = 0.153$. From these genotypic frequencies we calculate the allele frequencies as follows:

$$p = f(A) = (\text{frequency of the } AA \text{ homozygote}) + (\tfrac{1}{2} \times \text{frequency of the } Aa \text{ heterozygote})$$

$$q = f(a) = (\text{frequency of the } aa \text{ homozygote}) + (\tfrac{1}{2} \times \text{frequency of the } Aa \text{ heterozygote})$$

The frequencies of two alleles, $f(A)$ and $f(a)$, are commonly symbolized as p and q. The allelic frequencies for a locus, like the genotypic frequencies, should always add up to 1. This is because in a one gene locus model that has only two alleles, 100 percent (i.e., the frequency = 1) of the alleles are accounted for by the sum of the percentages of the two alleles. Therefore, once p is calculated, q can be easily obtained by subtraction: $1 - p = q$.

ALLELIC FREQUENCIES WITH MULTIPLE ALLELES.

Suppose we have three alleles—A^1, A^2, and A^3—at a locus, and we want to determine the allelic frequencies. Here, we employ the same rule that we used with two alleles: we add up the number of alleles of each type and divide by the total number of alleles in the population:

$$p = f(A^1) = \frac{(2 \times A^1A^1) + (A^1A^2) + (A^1A^3)}{(2 \times \text{total number of individuals})}$$

$$q = f(A^2) = \frac{(2 \times A^2A^2) + (A^1A^2) + (A^2A^3)}{(2 \times \text{total number of individuals})}$$

$$r = f(A^3) = \frac{(2 \times A^3A^3) + (A^1A^3) + (A^2A^3)}{(2 \times \text{total number of individuals})}$$

To illustrate the calculation of allelic frequencies when more than two alleles are present, we will use data from a study on genetic variation in milkweed beetles. Walter Eanes and his coworkers examined allelic frequencies at a locus that codes for the enzyme phosphoglucomutase (PGM). Three alleles were found at this locus; each allele codes for a different molecular variant of the enzyme. In one population, the following numbers of genotypes were collected:

$$
\begin{array}{rcr}
AA & = & 4 \\
AB & = & 41 \\
BB & = & 84 \\
AC & = & 25 \\
BC & = & 88 \\
CC & = & 32 \\
\hline
\text{Total} & = & 274
\end{array}
$$

The frequencies of the alleles are calculated as follows:

$$f(A) = p = \frac{(2 \times 4) + 41 + 25}{(2 \times 274)} = 0.135$$

$$f(B) = q = \frac{(2 \times 84) + 41 + 88}{(2 \times 274)} = 0.542$$

$$f(C) = r = \frac{(2 \times 32) + 88 + 25}{(2 \times 274)} = 0.323$$

As seen in these calculations, we add twice the number of homozygotes that possess the allele and one times each of the heterozygotes that have the allele. We then divide by twice the number of individuals in the population, which represents the total

number of alleles present. In the top part of the equation, notice that, for each allelic frequency, we do not add all the heterozygotes because some of the heterozygotes do not have the allele; for example, in calculating the allelic frequency of A, we do not add the number of BC heterozygotes in the top part of the equation. BC individuals do not have an A allele. We can use the same procedure for calculating allelic frequencies when four or more alleles are present.

The second method for calculating allelic frequencies from *genotypic frequencies* can also be used here. This calculation may be quicker if we have already determined the frequencies of the genotypes. The frequency of the homozygote is added to half of the heterozygote frequency because half of the heterozygote's alleles are A and half are a. If three alleles (A^1, A^2, and A^3) are present in the population, the allelic frequencies are

$$p = f(A^1) = f(A^1A^1) + \tfrac{1}{2} f(A^1A^2) + \tfrac{1}{2} f(A^1A^3)$$

$$q = f(A^2) = f(A^2A^2) + \tfrac{1}{2} f(A^1A^2) + \tfrac{1}{2} f(A^2A^3)$$

$$r = f(A^3) = f(A^3A^3) + \tfrac{1}{2} f(A^1A^3) + \tfrac{1}{2} f(A^2A^3)$$

Although calculating allelic frequencies from genotypic frequencies may be quicker than calculating them directly from the numbers of genotypes, more rounding error will occur. As a result, calculations from direct counts are usually preferred. Calculating genotype frequencies and allelic frequencies are illustrated for a one-gene locus, three-allele example in Box 22.1.

ALLELIC FREQUENCIES AT AN X-LINKED LOCUS.

Calculation of allelic frequencies at an X-linked locus is slightly more complicated, because males have only a single X-linked allele. However, we can use the same rules we used for autosomal loci. Remember that each homozygous female carries two X-linked alleles; heterozygous females have only one of that particular allele, and all males may have only a single X-linked allele. To determine the number of alleles at an X-linked locus, we multiply the number of homozygous females by 2, then add the number of heterozygous females and the number of hemizygous males. We next divide by the total number of alleles in the population. When determining the total number of alleles, we add twice the number of females (because each female has two X-linked alleles) to the number of males (who have a single allele at X-linked loci). Using this reasoning, the frequencies of two

BOX 22.1

Sample Calculation of Genotypic and Allelic Frequencies for Hemoglobin Variants Among Nigerians Where Multiple Alleles Are Present

Hemoglobin Genotypes

AA	AS	SS	AC	SC	CC	Total
2,017	783	4	173	14	11	3,002

Calculation of Genotypic Frequencies

$$\text{Genotypic frequency} = \frac{\text{Number of individuals with the genotype}}{\text{Total number of individuals}}$$

$$f(SS) = \frac{4}{3,002} = 0.0013 \qquad f(AA) = \frac{2,017}{3,002} = 0.672 \qquad f(AC) = \frac{173}{3,002} = 0.058$$

$$f(AS) = \frac{783}{3,002} = 0.261 \qquad f(SC) = \frac{14}{3,002} = 0.0047 \qquad f(CC) = \frac{11}{3,002} = 0.0037$$

Calculation of Allelic Frequencies from the Number of Individuals with a Particular Genotype

$$\text{Allelic frequency} = \frac{\text{Number of copies of a given allele in the population}}{\text{Sum of all alleles in the population}}$$

$$f(S) = \frac{(2 \times \text{number of } SS \text{ individuals}) + (\text{number of } AS \text{ individuals}) + (\text{number of } SC \text{ individuals})}{2 \times \text{total number of individuals}}$$

$$f(S) = \frac{(2 \times 4) + 783 + 14}{(2 \times 3,002)} = \frac{805}{6,004} = 0.134$$

$$f(A) = \frac{(2 \times \text{number of } AA \text{ individuals}) + (\text{number of } AS \text{ individuals}) + (\text{number of } AC \text{ individuals})}{(2 \times \text{total number of individuals})}$$

$$f(A) = \frac{(2 \times 2,017) + 783 + 173}{(2 \times 3,002)} = \frac{4,990}{6,004} = 0.831$$

$$f(C) = \frac{(2 \times \text{number of } CC \text{ individuals}) + (\text{number of } AC \text{ individuals}) + (\text{number of } SC \text{ individuals})}{(2 \times \text{total number of individuals})}$$

$$f(C) = \frac{(2 \times 11) + 173 + 14}{(2 \times 3,002)} = \frac{209}{6,004} = 0.035$$

Calculation of Allelic Frequencies from the Frequencies of Particular Genotypes

$f(S) = f(SS) + \frac{1}{2} f(AS) + \frac{1}{2} f(SC)$

$f(S) = 0.0013 + \frac{1}{2} \times (0.261) + \frac{1}{2} \times .0047 = 0.134$

$f(A) = f(AA) + \frac{1}{2} f(AS) + \frac{1}{2} f(AC)$

$f(A) = 0.672 + \frac{1}{2} \times 0.261 + \frac{1}{2} \times 0.058 = 0.831$

$f(C) = f(CC) + \frac{1}{2} f(SC) + \frac{1}{2} f(AC)$

$f(C) = 0.0037 + \frac{1}{2} \times 0.0047 + \frac{1}{2} \times 0.058 = 0.035$

alleles at an X-linked locus (X^A and X^a) are determined with the following equations:

$$p = f(X^A) = \frac{(2 \times X^A X^A \text{ females}) + (X^A X^a \text{ females}) + (X^A Y \text{ males})}{(2 \times \text{number of females}) + (\text{number of males})}$$

$$q = f(X^a) = \frac{(2 \times X^a X^a \text{ females}) + (X^A X^a \text{ females}) + (X^a Y \text{ males})}{(2 \times \text{number of females}) + (\text{number of males})}$$

Allelic frequencies at an X-linked locus can be determined from the genotypic frequencies by

$$p = f(X^A) = f(X^A X^A) + \tfrac{1}{2} f(X^A X^a) + f(X^A Y)$$

$$q = f(X^a) = f(X^a X^a) + \tfrac{1}{2} f(X^A X^a) + f(X^a Y)$$

Students should strive to understand the logic behind these calculations, not just memorize the formulas. If you fully understand the basis of the calculations, you will not need to remember the exact equations and will be able to determine allelic frequencies for any situation.

$\mathcal{K}$EYNOTE

The genetic structure of a population is determined by the total of all alleles (the gene pool). In the case of diploid, sexually interbreeding individuals, the structure is also characterized by the distribution of alleles into genotypes. The genetic structure can be described in terms of allelic and genotypic frequencies. Except for rare mutations, individuals are born and die with the same set of alleles; what changes genetically over time (evolves) is the hereditary makeup of a group of individuals, reproductively connected in a Mendelian population.

THE HARDY-WEINBERG LAW

The Hardy-Weinberg law is the most important principle in population genetics, for it offers a simple explanation for how the Mendelian principles that result from meiosis and sexual reproduction influence allelic and genotypic frequencies of a population. It provides us with a "null model" which we can use to generate testable hypotheses concerning the effects of a variety of processes on genetic structure of populations. The Hardy-Weinberg law is named after the

two individuals who independently discovered it in the early 1900s (Box 22.2). We begin our discussion of the Hardy-Weinberg law by simply stating what it tells us about the gene pool of a population. We then explore the implications of this principle, and briefly discuss how the Hardy-Weinberg law is derived. Finally, we present some applications of the Hardy-Weinberg law and test a population to determine if the genotypes are in Hardy-Weinberg proportions.

The Hardy-Weinberg law is divided into three parts—a set of assumptions and two major results. A simple statement of the law follows:

Part 1: In an infinitely large, randomly mating population, free from mutation, migration, and natural selection (note that there are five assumptions);

Part 2: The frequencies of the alleles do not change over time; and

Part 3: As long as mating is random, the genotypic frequencies will remain in the proportions p^2 (frequency of AA), $2pq$ (frequency of Aa), and q^2 (frequency of aa), where p is the allelic frequency of A and q is the allelic frequency of a. The sum of the genotypic frequencies should be equal to 1 (that is, $p^2 + 2pq + q^2 = 1$).

In short, the Hardy-Weinberg law explains what happens to the allelic and genotypic frequencies of a population as the alleles are passed from generation to generation in the absence of evolutionarily relevant processes. In other words, if the assumptions listed in part 1 are met, alleles would be expected to combine into genotypes based on simple laws of probability and the population is in Hardy-Weinberg equilibrium. Thus, genotype frequencies can be predicted from allele frequencies.

Assumptions of the Hardy-Weinberg Law

Part 1 of the Hardy-Weinberg law presents certain conditions, or assumptions, which must be present for the law to apply. First, the law indicates that the population must be infinitely large. If a population is limited in size, chance deviations from expected ratios can cause changes in allelic frequency, a phenomenon called **genetic drift**. It is true that the assumption of infinite size in part 1 is unrealistic — no population has an infinite number of individuals. But large populations will look very similar to populations that are mathematically infinitely large. (We will discuss this phenomenon later, when we examine genetic drift in more detail.) At this point, we merely want to stress that populations need not be infinitely large for the Hardy-Weinberg law to hold true.

A second condition of the Hardy-Weinberg law is that mating must be random. **Random mating** refers

BOX 22.2

Hardy, Weinberg, and the History of Their Contribution to Population Genetics

Godfrey H. Hardy (1877–1947), a mathematician at Cambridge University, often met R. C. Punnett, the Mendelian geneticist, at the faculty club. One day in 1908 Punnett told Hardy of a problem in genetics that he attributed to a strong critic of Mendelism, G. U. Yule (Yule later denied having raised the problem). Supposedly, Yule said that if the allele for short fingers (brachydactyly) was dominant (which it is) and its allele for normal-length fingers was recessive, then short fingers ought to become more common with each generation. In time virtually everyone in Britain should have short fingers. Punnett believed the argument was incorrect, but he could not prove it.

Hardy was able to write a few equations showing that, given any particular frequency of alleles for short fingers and alleles for normal fingers in a population, the relative number of people with short fingers and people with normal fingers will stay the same generation after generation providing no natural selection is involved that favors one phenotype or the other in producing offspring. Hardy published a short paper describing the relationship between genotypes and phenotypes in populations, and within a few weeks a paper was published by Wilhelm Weinberg (1862–1937), a German physician of Stuttgart, that clearly stated the same relationship. The Hardy-Weinberg law signaled the beginning of modern population genetics.

To be complete, we should note that in 1903 the American geneticist W. E. Castle of Harvard University was the first to recognize the relationship between allele and genotypic frequencies, but it was Hardy and Weinberg who clearly described the relationship in mathematical terms. Thus, the law is sometimes referred to as the Castle-Hardy-Weinberg law.

---

to matings between genotypes occurring in proportion to the frequencies of the genotypes in the population. More specifically, the probability of a mating between two genotypes is equal to the product of the two genotypic frequencies.

To illustrate random mating, consider the *M-N* blood types in humans, discussed in Chapter 4. The *M-N* blood type is due to an antigen on the surface of a red blood cell, similar to the ABO antigens, except that incompatibility in the *M-N* system does not cause problems during blood transfusion. The *M-N* blood type is determined by one locus with two codominant alleles, L^M and L^N. In a population of Eskimos, the frequencies of the three *M-N* genotypes are $L^M/L^M = 0.835$, $L^M/L^N = 0.156$, and $L^N/L^N = 0.009$. If Eskimos interbreed randomly, the probability of a mating between an L^M/L^M male and an L^M/L^M female is equal to the frequency of L^M/L^M times the frequency of $L^M/L^M = 0.835 \times 0.835 = 0.697$. Similarly, the probabilities of other possible matings are equal to the products of the genotypic frequencies when mating is random.

The requirement of random matings for the Hardy-Weinberg law is often misinterpreted. Many students assume, incorrectly, that the population must be interbreeding randomly for all traits for the Hardy-Weinberg law to hold. If this were true, human

populations would never obey the Hardy-Weinberg law, because humans do not mate randomly. Humans mate preferentially for height, IQ, skin color, socioeconomic status, and other traits. However, while mating may be nonrandom for some traits, most humans still mate randomly for the *M-N* blood types; few of us even know what our *M-N* blood types are. The principles of the Hardy-Weinberg law apply to any locus for which random mating occurs, even if mating is nonrandom for other loci.

Finally, for the Hardy-Weinberg law to work, the population must be free from mutation, migration, and natural selection (described in detail below). In other words the gene pool must be closed to the addition or subtraction of alleles, and we are interested in how allelic frequencies are related to genotypic frequencies solely on the basis of meiosis and sexual reproduction. Therefore, the influence of other evolutionarily relevant processes must be excluded. Later we will discuss these other evolutionary processes and their effect on the gene pool of a population. This condition (that no evolutionary processes act upon the population) applies only to the locus in question — a population may be subject to evolutionary processes acting on some genes, while still meeting the Hardy-Weinberg assumptions at other loci.

Predictions of the Hardy-Weinberg Law

If the conditions of the Hardy-Weinberg law are met, the population will be in genetic equilibrium, and two results are expected. First, the frequencies of the alleles will not change from one generation to the next, and therefore the gene pool is not evolving at this locus. Second, the genotypic frequencies will be in the proportions p^2, $2pq$, and q^2 after one generation of random mating. Also the genotypic frequencies will remain constant in these proportions as long as all the conditions required by the Hardy-Weinberg law continue to be met. When the genotypes are in these proportions, the population is said to be in Hardy-Weinberg equilibrium. An important use of the Hardy-Weinberg law is that it provides a mechanism for determining the genotypic frequencies from the allelic frequencies when the population is in equilibrium. This in turn provides a way of evaluating which assumptions are being violated when the expected theoretical distribution of genotypes does not match an empirically determined distribution.

To summarize, the Hardy-Weinberg law makes several predictions about the allelic frequencies and the genotypic frequencies of a population when certain conditions are satisfied. The necessary conditions are that the population is large, randomly mating, and free from mutation, migration, and natural selection. When these conditions are met, the Hardy-Weinberg law indicates that allelic frequencies will not change and genotypic frequencies will be determined by the allelic frequencies, occurring in the proportions p^2, $2pq$, and q^2.

Derivation of the Hardy-Weinberg Law

The Hardy-Weinberg law states that when a population is in equilibrium, the genotypic frequencies will be in the proportions p^2, $2pq$, and q^2. To understand the basis of these frequencies at equilibrium, consider a hypothetical population in which the frequency of allele A is p and the frequency of allele a is q. In producing gametes, each genotype passes on both alleles that it possesses with equal frequency; therefore, the frequencies of A and a in the gametes are also p and q. If one thinks of the gametes as being in a pool, the random formation of zygotes involves reaching into the pool and drawing two gametes at random. The genotypes that then form after repeatedly drawing two gametes at a time will then be in frequencies that are predicted by the probabilities of drawing the particular allele bearing gametes (see product rule, Chapter 2, page 28). Table 22.1 shows the combinations of gametes when mating is random. This table illustrates

the relationship between the allelic frequencies and the genotypic frequencies, which forms the basis of the Hardy-Weinberg law. We see that when gametes pair randomly, the genotypes will occur in the proportions p^2 (AA), $2pq$ (Aa), and q^2 (aa). These genotypic proportions result from the expansion of the square of the allelic frequencies $(p + q)^2 = p^2 + 2pq + q^2$, and the genotypes reach these proportions after one generation of random mating.

The Hardy-Weinberg law also states that allelic and genotypic frequencies remain constant generation after generation if the population remains large, randomly mating, and free from mutation, migration, and natural selection (i.e., evolutionary processes). This result can also be understood by considering a hypothetical, randomly mating population, as illustrated in Table 22.2. In Table 22.2, all possible matings are given. By definition, random mating means that the frequency of mating between two genotypes will be equal to the product of the genotypic frequencies. For example, the frequency of an $AA \times AA$ mating will be equal to p^2 (the frequency of AA) $\times p^2$ (the frequency of AA) $= p^4$. The frequencies of the offspring produced from each mating are also presented in Table 22.2.

We see that the sum of the probabilities of $AA \times Aa$ (or $2p^3q$) and $Aa \times AA$ (or $2p^3q$) matings is $4p^3q$, and we know from Mendelian principles that these crosses produce $1/2$ AA and $1/2$ Aa offspring. Therefore, the probability of obtaining AA offspring from these matings is $4p^3q \times 1/2 = 2p^3q$. The frequencies of offspring produced by each type of mating are presented in the body of the table. At the bottom of the table, the total frequency for each genotype is obtained by addition. As we can see, after random mating the genotypic frequencies are still p^2, $2pq$, and q^2 and the allelic frequencies remain at p and q. The frequencies of the population can thus be represented in the zygotic and gametic stages as follows:

Zygotes	Gametes
$p^2\ AA + 2pq\ Aa + q^2\ aa$	$p\ A + q\ a$

~ TABLE 22.1

Possible Combinations of A and a Gametes from Gametic Pools for a Population

Gametes		$\male$	
		$p\ A$	$q\ a$
$\female$	$p\ A$	$p^2\ AA$	$pq\ Aa$
	$q\ a$	$pq\ Aa$	$q^2\ aa$

In sum, $p^2\ AA + 2pq\ Aa + q^2\ aa = 1.00$

~ **TABLE 22.2**

Algebraic Proof of Genetic Equilibrium in a Randomly Mating Population for One Gene Locus with Two Alleles

TYPE OF MATING ♀ ♂	MATING FREQUENCY	Offspring Frequencies Contributed to the Next Generation by a Particular Mating		
		AA	Aa	aa
$p^2\,AA \times p^2\,AA$	p^4	p^4	—	—
$p^2\,AA \times 2\,pq\,Aa$ [a] $2\,pq\,Aa \times p^2\,AA$	$4\,p^3q$	$2\,p^3q$	$2\,p^3q$	—
$p^2\,AA \times q^2\,aa$ $q^2\,aa \times p^2\,AA$	$2\,p^2q^2$	—	$2\,p^2q^2$	—
$2\,pq\,Aa \times 2\,pq\,Aa$	$4\,p^2q^2$	p^2q^2	$2\,p^2q^2$	p^2q^2
$2\,pq\,Aa \times q^2\,aa$ $q^2\,aa \times 2\,pq\,Aa$	$4\,pq^3$	—	$2\,pq^3$	$2\,pq^3$
$q^2\,aa \times q^2\,aa$	q^4	—	—	q^4
Totals	$(p^2 + 2\,pq + q^2)^2 = 1$	$p^2(p^2 + 2\,pq + q^2) = p^2$	$2\,pq(p^2 + 2\,pq + q^2) = 2\,pq$	$q^2(p^2 + 2\,pq + q^2) = q^2$

Genotype frequencies = $(p + q)^2 = p^2 + 2\,pq + q^2 = 1$ in each generation afterward.

Gene (allele) frequencies = $p(A) + q(a) = 1$ in each generation afterward.

[a]For example, matings between AA and Aa will occur at $p^2 \times 2\,pq = 2\,p^3q$ for $AA \times Aa$ and at $p^2 \times 2\,pq = 2\,p^3q$ for $Aa \times AA$ for a total of $4\,p^3q$. Two progeny types, AA and Aa, result in equal proportions from these matings. Therefore, offspring frequencies are $2\,p^3q$ (i.e., $1/2 \times 4\,p^3q$) for AA and for Aa.

Each generation of zygotes produces A and a gametes in proportions p and q. The gametes unite to form AA, Aa, and aa zygotes in the proportions p^2, $2pq$, and q^2, and the cycle is repeated indefinitely as long as the assumptions of the Hardy-Weinberg law hold. This short proof gives the theoretical basis for the Hardy-Weinberg law.

The Hardy-Weinberg law indicates that at equilibrium, the genotypic frequencies depend upon the frequencies of the alleles. This relationship between allelic frequencies and genotypic frequencies for a locus with two alleles is represented in Figure 22.3. Several aspects of this relationship should be noted: (1) The maximum frequency of the heterozygote is 0.5, and this maximum value occurs only when the frequencies of A and a are both 0.5; (2) if allelic frequencies are between 0.33 and 0.66, the heterozygote is the most numerous genotype; (3) when the frequency of one allele is low, the homozygote for that allele is the rarest of the genotypes.

This point is also illustrated by the distribution of genetic diseases in humans, which are frequently rare and recessive. For a rare recessive trait, the frequency of the gene causing the trait will be much higher than the frequency of the trait itself, because most of the rare alleles are in non-affected heterozygotes (i.e., carriers). Albinism, for example, is a rare recessive condition in humans. One form of albinism is *tyrosinase-*

~ **FIGURE 22.3**

Relationship of the frequencies of the genotypes AA, Aa, and aa, to the frequencies of alleles A and a (in values of p [top abscissa] and q [bottom abscissa], respectively) in populations that meet the assumptions of the Hardy-Weinberg law. Any single population will be defined by a single vertical line such as $p = 0.3$ and $q = 0.7$.

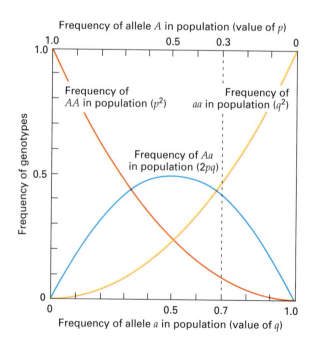

negative albinism (see Chapter 9, pp. 272–274). In this type of albinism, affected individuals have no tyrosinase activity, which is required for normal production of pigment. Among North American whites, the frequency of tyrosinase-negative albinism is roughly 1 in 40,000, or 0.000025. Since albinism is a recessive condition, the genotype of affected individuals is *aa*. If we assume that the population meets the assumptions of the Hardy-Weinberg law, the frequency of the *aa* genotypes equals q^2. If $q^2 = 0.000025$, then $q = 0.005$ and $p = 1 - q = 0.995$. The heterozygote frequency is therefore $2pq = 2 \times 0.995 \times 0.005 = 0.00995$ (almost 1 percent). Thus, while the frequency of albinos is low (1 in 40,000), individuals heterozygous for albinism are much more common (almost 1 in 100). Heterozygotes for recessive traits can be common, even when the trait is rare.

KEYNOTE

The Hardy-Weinberg law describes what happens to allelic and genotypic frequencies of a large population, when gametes fuse randomly and there is no mutation, migration, or natural selection. If these conditions are met, allelic frequencies do not change from generation to generation, and the genotypic frequencies stabilize after one generation in the proportions p^2, $2pq$, and q^2, where p and q equal the frequencies of the alleles in the population.

Extensions of the Hardy-Weinberg Law to Loci with More than Two Alleles

When two alleles are present at a locus, the Hardy-Weinberg law tells us that at equilibrium the frequencies of the genotypes will be p^2, $2pq$, and q^2, which is the square of the allelic frequencies $(p + q)^2$. This is a simple binomial expansion and this principle of probability theory can be extended to any number of alleles that are sampled two at a time into a diploid zygote. For example, if three alleles are present (e.g., alleles *A*, *B*, and *C*) with frequencies equal to p, q, and r, the frequencies of the genotypes at equilibrium are also given by the square of the allelic frequencies:

$$(p + q + r)^2 = p^2 (AA) + 2pq (AB) + q^2 (BB) + 2pr (AC) + 2qr (BC) + r^2 (CC)$$

In the blue mussel found along the Atlantic coast of North America, three alleles are common at a locus coding for the enzyme leucine aminopeptidase (LAP). For a population of mussels inhabiting Long Island Sound (discussed later—see Figure 22.6), R. K. Koehn and colleagues determined that the frequencies of the three alleles were as follows:

ALLELE	FREQUENCY
LAP^{98}	$p = 0.52$
LAP^{96}	$q = 0.31$
LAP^{94}	$r = 0.17$

If the population were in Hardy-Weinberg equilibrium, the expected genotypic frequencies would be

GENOTYPE	EXPECTED FREQUENCY			
LAP^{98}/LAP^{98}	p^2	$= (0.52)^2$	$=$	0.27
LAP^{98}/LAP^{96}	$2pq$	$= 2(0.52)(0.31)$	$=$	0.32
LAP^{96}/LAP^{96}	q^2	$= (0.31)^2$	$=$	0.10
LAP^{96}/LAP^{94}	$2qr$	$= 2(0.31)(0.17)$	$=$	0.10
LAP^{94}/LAP^{98}	$2pr$	$= 2(0.52)(0.17)$	$=$	0.18
LAP^{94}/LAP^{94}	r^2	$= (0.17)^2$	$=$	0.03

The square of the allelic frequencies can be used in the same way to estimate the expected frequencies of the genotypes when four or more alleles are present at a locus.

Extensions of the Hardy-Weinberg Law to Sex-linked Alleles

If alleles are X-linked, females may be homozygous or heterozygous, but males carry only a single allele for each X-linked locus. For X-linked alleles in females, the Hardy-Weinberg frequencies are the same as those for autosomal loci: p^2 ($X^A X^A$), $2pq$ ($X^A X^B$), and q^2 ($X^B X^B$). In males, however, the frequencies of the genotypes will be p ($X^A Y$) and q ($X^B Y$), the same as the frequencies of the alleles in the population. For this reason, recessive X-linked traits are more frequent among males than among females. To illustrate this concept, consider red-green color blindness, which is an X-linked recessive trait. The frequency of the color-blind allele varies among human ethnic groups; the frequency among African Americans is 0.039. At equilibrium, the expected frequency of color-blind males in this group is $q = 0.039$, but the frequency of color-blind females is only $q^2 = (0.039)^2 = 0.0015$.

When random mating occurs within a population, the equilibrium genotypic frequencies are reached in one generation. However, if the alleles are X-linked and the sexes differ in allelic frequency, the equilibrium frequencies are approached over several generations. This is because males receive their X chromosome from their mother only, while females receive an X chromosome from both the mother and the father. Consequently, the frequency of an X-linked allele in males will be the same as the frequency of that allele in their mothers, whereas the frequency in females will be the average of that in mothers and fathers. With random mating, the allelic frequencies in the two sexes oscillate back and forth each generation, and the difference in allelic frequency between the sexes is reduced by half each generation, as shown in Figure 22.4. Once the allelic frequencies of the males and females are equal, the frequencies of the genotypes will be in Hardy-Weinberg proportions after one more generation of random mating.

Testing for Hardy-Weinberg Proportions

The Hardy-Weinberg law can be used as a null hypothesis to which the genetic structure of any particular population can be compared. If the observed genetic structure does not match the expected structure based on the law we can begin to ask about which of the assumptions are being violated. To determine whether the genotypes of a population are in Hardy-Weinberg proportions, we first compute p and q from the observed frequencies of the genotypes. (Note: it is important not to take the square roots of the homozygote frequencies to obtain allele frequencies because to do that already assumes that the population is in equilibrium. Thus, allele frequencies should be calculated from the sum of homozygote frequencies plus one half the heterozygote frequencies—for a more complete discussion, see p. 718.) Once we have obtained these allelic frequencies, we can calculate the expected genotypic frequencies

$(p^2, 2pq,$ and $q^2)$ and compare these frequencies with the actual observed frequencies of the genotypes using a chi-square test (see Chapter 5). The chi-square test gives us the probability that the difference between what we observed and what we expect under the Hardy-Weinberg law is due to chance.

To illustrate this procedure, consider a locus that codes for transferrin (a blood protein) in the red-backed vole, *Clethrionomys gapperi*. Three genotypes are found at the transferrin locus: *MM*, *MJ*, and *JJ*. In a population of voles trapped in the Northwest Territories of Canada in 1976, 12 *MM* individuals, 53 *MJ* individuals, and 12 *JJ* individuals were found. To determine if the genotypes are in Hardy-Weinberg proportions, we first calculate the allelic frequencies for the population using our familiar formula:

$$p = \frac{(2 \times \text{number of homozygotes}) + (\text{number of heterozygotes})}{(2 \times \text{total number of individuals})}$$

Therefore:

$$p = f(M) = \frac{(2 \times 12) + (53)}{(2 \times 77)} = 0.50$$

$$q = 1 - p = 0.50$$

Using p and q calculated from the observed genotypes, we can now compute the expected Hardy-Weinberg proportions for the genotypes: $f(MM) = p^2 = (0.50)^2 = 0.25$; $f(MJ) = 2pq = 2(0.50)(0.50) = 0.50$; and $f(JJ) = q^2 = (0.50)^2 = 0.25$. However, for the chi-square test, actual numbers of individuals are needed, not the proportions. To obtain the expected numbers, we simply multiply each expected proportion times the total number of individuals counted (N), as shown below.

	EXPECTED	OBSERVED
$f(MM) = p^2 \times N$		
$= 0.25 \times 77 =$	19.3	12
$f(MJ) = 2pq \times N$		
$= 0.50 \times 77 =$	38.5	53
$f(JJ) = q^2 \times N$		
$= 0.25 \times 77 =$	19.3	12

With observed and expected numbers, we can compute a chi-square value to determine the probability that the differences between observed and expected numbers could be the result of chance. The chi-square is computed using the same formula that we employed for analyzing genetic crosses; that is, d, the deviation, is calculated for each class as (observed − expected); d^2, the deviation squared, is divided by the expected number e for each class; and chi-square (χ^2) is computed as the sum of all d^2/e values. For this

~ FIGURE 22.4

Representation of the gradual approach to equilibrium of an X-linked gene with an initial frequency of 1.0 in females and 0 in males.

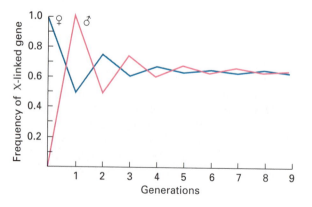

example, $\chi^2 = 10.98$. We now need to find this value in the chi-square table (see Table 2.5, p. 37) under the appropriate degrees of freedom. This step is not as straightforward as in our previous χ^2 analyses. In those examples the number of degrees of freedom was the number of classes in the sample minus 1. Here, however, while there are three classes, there is only one degree of freedom, because the frequencies of alleles in a population have no theoretically expected values. Thus p must be estimated from the observations themselves. So one degree of freedom is lost for every parameter (p in this case) that must be calculated from the data. Another degree of freedom is lost because, for the fixed number of individuals, once all but one of the classes have been determined, the last class has no degree of freedom and is set automatically. Therefore with three classes (MM, MJ, and JJ), two degrees of freedom are lost, leaving one degree of freedom.

In the chi-square table under the column for one degree of freedom, the chi-square value of 10.98 indicates a P value less than 0.05. Thus the probability that the differences between the observed and expected values is due to chance is very low. That is, the observed numbers of genotypes do not fit the expected numbers under Hardy-Weinberg law.

Using the Hardy-Weinberg Law to Estimate Allelic Frequencies

An important application of the Hardy-Weinberg law is the calculation of allelic frequencies when one or more alleles is recessive. For example, we have seen that albinism in humans results from an autosomal recessive gene. Normally, this trait is rare, but among the Hopi Indians of Arizona, albinism is remarkably common. Charles M. Woolf and Frank C. Dukepoo visited the Hopi villages in 1969 and observed 26 cases of albinism in a total population of about 6,000 Hopis (Figure 22.5). This gave a frequency for the trait of 26/6,000, or 0.0043, which is much higher than the frequency of albinism in most populations. Although we have calculated the frequency of the trait, we cannot directly determine the frequency of the gene for albinism because we cannot distinguish between heterozygous individuals and those individuals homozygous for the normal allele. Recall that our computation of the allelic frequency involves counting the number of alleles:

$$p = \frac{(2 \times \text{number of homozygotes}) + (\text{number of heterozygotes})}{2N}$$

But since heterozygotes for a recessive trait such as albinism cannot be identified, this is impossible. Nevertheless, we can determine the allelic frequency from the Hardy-Weinberg law if we assume that the population is in equilibrium. At equilibrium, the frequency of the homozygous recessive genotype is q^2. For albinism among the Hopis, $q^2 = 0.0043$, and q can be obtained by taking the square root of the frequency of the trait. Therefore, $q = \sqrt{0.0043} = 0.065$, and $p = 1 - q = 0.935$. Following the Hardy-Weinberg law, the frequency of heterozygotes in the population will be $2pq = 2 \times 0.935 \times 0.065 = 0.122$. Thus, one out of eight Hopis, on the average, carries an allele for albinism!

We should not forget that this method of calculating allelic frequency rests upon the assumptions of the Hardy-Weinberg law. If the conditions of the Hardy-Weinberg law do not apply, then our estimate of allelic frequency will be inaccurate. Also, once we calculate allelic frequencies with these assumptions, we cannot then test the population to determine if the genotypic frequencies are in the Hardy-Weinberg expected proportions. To do so would involve circular reasoning, for we assumed Hardy-Weinberg proportions in the first place to calculate the allelic frequencies. To repeat an important point made earlier, the Hardy-Weinberg law provides us with a "null model" that explains the genetic structure of populations in the absence of a variety of complicating processes. As such it is a valuable tool for exploring deviations from the "null model" which are frequently observed in real populations. Before we explore how the model can be used to discern the causes of deviations of observed populations from theoretically

~ FIGURE 22.5

Three Hopi girls, photographed about 1900. The middle child has albinism. Albinism, an autosomal recessive disorder, occurs in high frequency among the Hopi Indians of Arizona.

expected populations, we will look at the genetic structure of some real populations.

GENETIC VARIATION IN SPACE AND TIME

The genetic structure of populations can vary in space and time. This means that the frequencies and distri-bution of alleles can vary in samples of the same species in different areas or samples from the same area collected at different times. Figure 22.6 shows how the frequencies of 3 alleles at the locus for the enzyme leucine aminopeptidase gradually change in a geographic series of samples of blue mussels that inhabit Long Island Sound. Most populations of plants and animals that have widespread geographic distributions show differences of this sort in the allele frequencies of their component populations. In addi-

~ **FIGURE 22.6**

Geographic variation in frequencies of the alleles LAP^{94}, LAP^{96}, and LAP^{98} of the locus coding for the enzyme leucine amino peptidase (LAP) in the blue mussel.

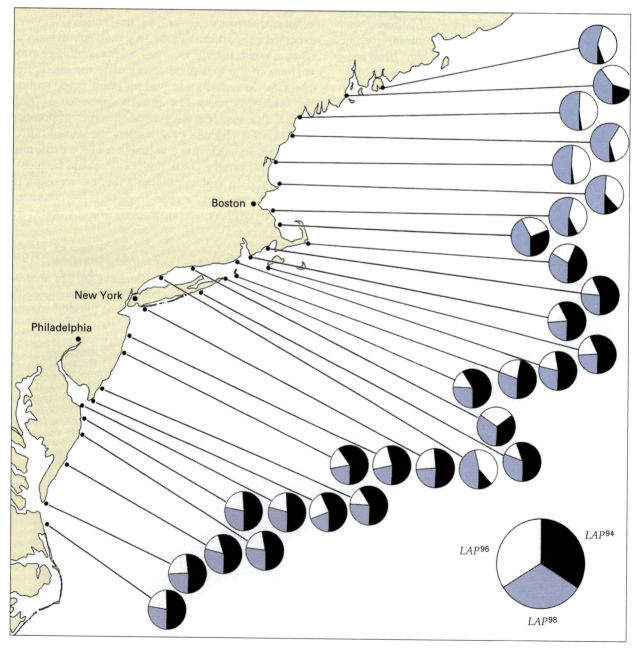

~ **FIGURE 22.7**

Temporal variation in the locus coding for the enzyme esterase 4F in the prairie vole, *Microtus ochrogaster*. The four populations are close to each other and near Lawrence, Kansas.

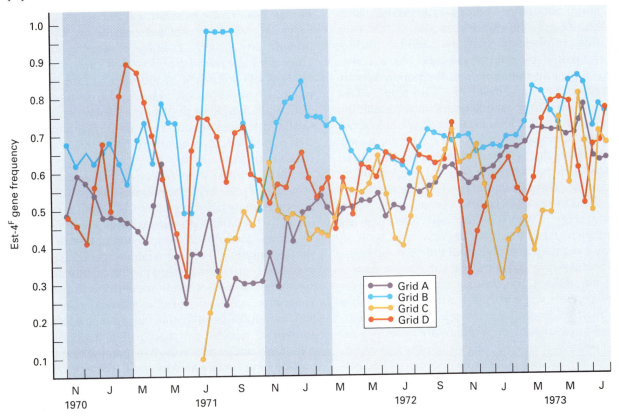

tion, the genetic structure of a population can change over time such as shown in Figure 22.7.

The fact that genetic structure varies geographically means that not all populations of a particular species are genetically the same. Thus, in terms of the future evolution of a species and conservation of the genetic resources of a species, attention has to be paid to the fact that there is a spatial component to genetic variation. The fact that genetic structure of a population can change over time implies that populations in different locations may change differently and thus lead to further variation in space and the evolution of novel genetic structure. A closer look at the genetic structure of populations and what processes cause structure to change will help us understand how genetic differences evolve and how we may best conserve the great diversity of genetic resources that exist.

GENETIC VARIATION IN NATURAL POPULATIONS

One of the most significant questions addressed in population genetics is how much genetic variation exists within natural populations. The genetic variation within populations is important for several reasons. First, it determines the potential for evolutionary change and adaptation. The amount of variation also provides us with important clues about the relative importance of various evolutionary processes, since some processes increase variation while others decrease it. The manner in which new species arise may depend upon the amount of genetic variation harbored within populations. In addition, the ability of a population to persist over time can be influenced by how much genetic variation it has to draw upon should environments change. For all these reasons, population geneticists are interested in measuring genetic variation, attempting to understand the evolutionary processes that affect it, and understanding the effects of human environmental disturbance that may alter it.

Models of Genetic Variation

During the 1940s and 1950s, population geneticists developed several opposing hypotheses concerning the amount of genetic variation within natural populations. These hypotheses, or models of genetic variation, have become known as the classical model, the balance model, and the neutral-mutation model.

The classical model emerged primarily from the work of laboratory geneticists, who proposed that most natural populations possess little genetic variation. According to the **classical model**, within each population one allele functions best, and this allele is strongly favored by natural selection. As a consequence, almost all individuals in the population are homozygous for this "best" or "wild-type" allele. New alleles arise from time to time through mutation, but almost all are deleterious and are kept in low frequency by selection. Once in a long while, a new mutation arises that is better than the wild-type allele. This new allele increases the survival and reproduction of the individuals that carry it, so the frequency of the allele increases over time because of its selective advantage. Eventually the new allele reaches high frequency and becomes the new wild type. In this way a population evolves, but little genetic variation is found within the population at any one time.

As we will see, the classical model is no longer tenable because the gene pool of a population has been found to consist of many alleles at each locus; therefore individuals in populations are heterozygous at numerous loci. To account for the presence of this variation, the balance model was developed. The **balance model** proposes that natural selection actively maintains genetic variation within a population through balancing selection. Balancing selection is natural selection that maintains a balance between alleles, preventing any single allele from reaching high frequency. One form of balancing selection is heterozygote superiority, in which the heterozygote has higher fitness than either homozygote.

An alternative explanation for the existence of large amounts of genetic variation in populations is provided by the proponents of the **neutral mutation model**. They maintain that recurrent mutation and random changes in allele frequencies are sufficient to explain the large amounts of genetic variation that exist within populations and it is not necessary to invoke natural selection. While the classical model is untenable it still has proponents in the popular literature. Among population geneticists, however, there are strong advocates of both the balance and neutral models. This tension is one of the exciting areas of current research in the field.

Measuring Genetic Variation with Protein Electrophoresis

For many years, population geneticists were constrained in establishing how much variation existed within natural populations. Naturalists recognized that plants and animals in nature frequently differ in

phenotype (Figure 22.8), but the genetic basis of most traits is too complex to assign specific genotypes to individuals. A few traits and alleles that behaved in a Mendelian fashion, such as spot patterns in butterflies and shell color in snails, gave observable genetic variation, but these isolated cases were too few to provide any general estimate of genetic variation. Then cytological evidence was used such as the morphology of chromosomes and their banding patterns especially in the salivary glands of fruit flies. It was difficult at this point to distinguish between the classical and balance models because population geneticists required data on genotypes at many loci of many individuals from multiple species.

In 1966, population geneticists began to apply the principle of protein electrophoresis to the study of natural populations. Electrophoresis is a biochemical technique that separates proteins with different molecular structures. (See Box 22.3 for a description of protein electrophoresis.) For population geneticists, electrophoresis provides a technique for quickly determining the genotypes of many individuals at many loci. This procedure is now used to examine genetic variation in hundreds of plant and animal species. The amount of genetic variation within a population is commonly measured with two parameters, the **proportion of polymorphic loci** and **heterozygosity**.

A polymorphic locus is any locus that has more than one allele present within a population. The proportion of polymorphic loci (P) is calculated by dividing the number of polymorphic loci by the total number of loci examined. For example, suppose we found that of 33 loci in a population of green frogs, 18 were polymorphic. The proportion of polymorphic loci would be $18/33 = 0.55$. Heterozygosity (H) is the proportion of an individual's loci that are heterozygous. Suppose we analyzed the genotypes of green frogs from one population at a locus coding for esterase and found that the frequency of heterozygotes was 0.09. Heterozygosity for this locus would be 0.09. We would average this heterozygosity with those for other loci and obtain an estimate of heterozygosity for the population. The expected heterozygosity is one-third of the proportion of polymorphic loci.

Table 22.3 (p. 733) presents estimates of the proportion of polymorphic loci and heterozygosity for many species that have been surveyed with electrophoresis. The results of these studies are unambiguous—most species possess large amounts of genetic variation in their proteins, and the classical model is clearly wrong. Actually, the technique of standard electrophoresis misses a large proportion of the genetic variation that is present, because only genetic variants that cause a change in the movement of the protein on a gel will be observed. Therefore, the true amount of genetic variation is even greater than that revealed by this technique.

~ FIGURE 22.8

Extensive phenotypic variation exists in most natural populations, as apparent in the color patterns of the Cuban tree snail, a species treasured for its varicolored beauty.

Studies of electrophoretic variation in natural populations showed that the classical model was incorrect, because populations possess large amounts of genetic variation. However, this did not prove that the balance model was correct. The balance model predicts that much genetic variation is found within natural populations, and this turns out to be the case.

But the balance model also proposes that large amounts of genetic variation are maintained by natural selection, and this fact was not conclusively demonstrated by electrophoretic studies.

At this point, the nature of the controversy changed. Previously, the central question was how much genetic variation exists. Now, emphasis shifted

BOX 22.3

Analysis of Genetic Variation with Protein Electrophoresis

Gel electrophoresis has been widely used to study genetic variation in natural populations. This process is a biochemical technique that separates large molecules on the basis of size, shape, and charge. To examine genetic variation in proteins with electrophoresis, separate tissue samples, taken from a number of individuals, are ground up, releasing the proteins into an aqueous solution. The individual solutions are then inserted into a gel made of starch, agar, polyacrylamide, or some other porous substance. Electrodes are placed at the ends of the gel, as shown in Box Figure 22.1a, and a direct electrical current is supplied, setting up an electrical field across the gel.

Because proteins are charged molecules, they will migrate within the electrical field. Molecules with different shape, size, and/or charge will migrate at different rates and will separate from one another within the gel over time. If two individuals have different genotypes at a locus coding for a protein, they will produce slightly different molecular forms of the protein, which can sometimes be separated and identified with electrophoresis.

After a current has been supplied to the gel for several hours and the proteins allowed to separate, the current is turned off and the proteins visualized. The gel now contains a large number of different enzymes and proteins; to identify the product of a single locus, one must stain for a specific enzyme or protein. This is frequently accomplished by having the enzyme in the gel carry out a specific biochemical reaction. The substrate for the reaction is added to the gel, along with a dye that changes color when the reaction takes place. A colored band will appear on the gel wherever the enzyme is located.

If two individuals have different molecular forms of the enzyme, indicating differences in genotypes, bands for those individuals will appear at different locations on the gel. By examining the pattern of bands produced, it is also possible to determine whether an individual is homozygous or heterozygous for a particular allele. A diagram of the banding pattern is shown in Box Figure 22.1b. Histochemical stains are available for dozens of enzymes, so the genotypes of many individuals may be quickly obtained for studying variation at a number of loci.

~ BOX FIGURE 22.1

The technique of protein electrophoresis used to measure genetic variation in natural populations. (a) Solutions of proteins from different individuals are inserted into slots in the gel, and a direct current is supplied to separate the proteins. (b) After electrophoresis, enzyme-specific staining reactions are used to reveal the proteins. The pattern of bands on the gel indicates the genotype of each individual.

a)

Electrode
Gel   Sample slot   Contact wick
Electrode
Buffer tank

b)

Gel
Colored bands showing positions of enzymes
Sample slot
Solution with substrate and salt
Staining box

~ **TABLE 22.3**

Genic Variation in Some Major Groups of Animals and in Plants

Group	Number of Species or Forms	Mean Number of Loci Examined per Species	Mean Proportion of Loci	
			Polymorphic per Population	Heterozygous per Individual
Insects				
Drosophila	28	24	0.529 ± 0.030[a]	0.150 ± 0.010
Others	4	28	0.531	0.151
Haplodiploid wasps	6	15	0.243 ± 0.039	0.062 ± 0.007
Marine invertebrates	9	26	0.587 ± 0.084	0.147 ± 0.019
Snails				
Land	5	18	0.437	0.150
Marine	5	17	0.175	0.083
Fish	14	21	0.306 ± 0.047	0.078 ± 0.012
Amphibians	11	22	0.336 ± 0.034	0.082 ± 0.008
Reptiles	9	21	0.231 ± 0.032	0.047 ± 0.008
Birds	4	19	0.145	0.042
Rodents	26	26	0.202 ± 0.015	0.054 ± 0.005
Large Mammals[b]	4	40	0.233	0.037
Plants[c]	8	8	0.464 ± 0.064	0.170 ± 0.031

[a] Values are mean ± standard error.
[b] Human, chimpanzee, pigtailed macaque, and southern elephant seal.
[c] Predominantly outcrossing species; mean gene diversity is 0.233 ± 0.029.

to the question of what maintains the extensive variation observed. A new model arose to replace the classical model. This model, called the **neutral-mutation model**, acknowledges the presence of extensive genetic variation in proteins but proposes that this variation is neutral with regard to natural selection. This does not mean that the proteins detected by electrophoresis have no function, but rather that the different genotypes are physiologically equivalent. Therefore, natural selection does not act on the neutral alleles, and random processes such as mutation and genetic drift shape the patterns of genetic variation that we see in natural populations. The neutral-mutation model proposes that variation at some loci does affect fitness, and natural selection eliminates variation at these loci.

Distinguishing between the balance model and the neutral-mutation model has been difficult. Population geneticists still argue over the relative merits of the two, and no clear consensus has emerged as to which model of genetic variation is correct. In reality, both models may be partly correct; at some loci genetic variation may be essentially neutral, while at others, it may be acted upon by natural selection.

KEYNOTE

Two models were developed to explain the amount of variation in natural populations. The classical model predicted that little variation would occur, and the balance model proposed that much genetic variation would be maintained in natural populations by natural selection. Through the use of protein electrophoresis, population geneticists eventually demonstrated that natural populations harbor much genetic variation, thus disproving the classical model. However, the processes responsible for keeping this variation within populations are still controversial. The neutral-mutation model proposes that the genetic variation detected by elec-

trophoresis is neutral with regard to natural selection, whereas the balance model proposes that this variation is favored by natural selection.

Measuring Genetic Variation with RFLPs and DNA Sequencing

As technology improves, techniques in molecular genetics increasingly provide the means to examine directly nucleotide sequence differences in the DNA and to determine unambiguously the amount of genetic variation present in natural populations. One such technique employs restriction enzymes, which were discussed in Chapter 15, for detecting genetic variation. You will recall that restriction enzymes make double-stranded cuts in DNA at specific base sequences.

Most restriction enzymes recognize a sequence of four bases or six bases; see Table 15.1 (p. 450). For example, the restriction enzyme *Bam*HI recognizes the sequence $\frac{\text{GGATCC}}{\text{CCTAGG}}$; whenever this sequence appears in the DNA, *Bam*HI will cut the DNA. The resulting fragments can be separated by agarose gel electrophoresis and can be observed by staining the DNA or by using probes for specific genes (see Chapter 15).

Suppose that two individuals differ in one or more nucleotides at a particular DNA sequence and that the differences occur at a site recognized by a restriction enzyme (Figure 22.9). One individual has a DNA molecule with the restriction site, but the other individual does not, because the sequence of DNA nucleotides differ. If the DNA from these two individuals is mixed with the restriction enzyme and the resulting fragments are separated on a gel, the two individuals produce different patterns of fragments,

~ FIGURE 22.9

DNA from individual 1 and individual 2 differ in one nucleotide, found within the sequence recognized by the restriction enzyme *Bam*HI. Individual 1's DNA contains the *Bam*HI restriction sequence and is cleaved by the enzyme. Individual 2's DNA lacks the *Bam*HI restriction sequence and is not cleaved by the enzyme. When placed on an agarose gel and separated by electrophoresis, the DNAs from 1 and 2 produce different patterns on the gel. This variation is called a restriction fragment length polymorphism.

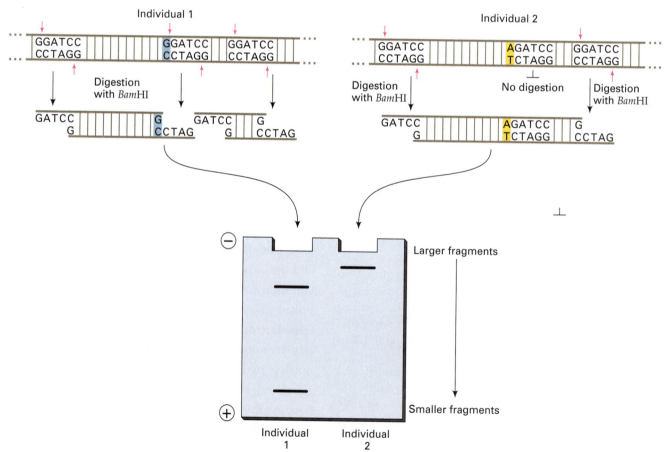

as shown in Figure 22.9. The different patterns on the gel are termed restriction fragment length polymorphisms, or RFLPs (see Chapter 15, p. 480). They indicate that the DNA sequences of the two individuals differ. RFLPs are inherited in the same way that alleles coding for other traits are inherited, only the RFLPs do not produce any outward phenotypes; their phenotypes are the fragment patterns produced on a gel when the DNA is cut by the restriction enzyme. RFLPs can be used as genetic markers for mapping genes. They can also provide information about how DNA sequences differ among individuals. Such differences involve only a small part of the DNA, specifically those few nucleotides recognized by the restriction enzyme. However, if we assume that restriction sites occur randomly in the DNA, which is not an unreasonable assumption since the sites are not

expressed as traits, the presence or absence of restriction sites can be used to estimate the overall differences in sequence.

To illustrate the use of RFLPs for estimating genetic variation, suppose we isolate DNA from 5 wild mice, cut the DNA with the restriction enzyme *Bam*HI, and separate the fragments with agarose gel electrophoresis. We then transfer the DNA to a membrane filter, using the Southern blot technique (see Chapter 15, p. 473), and add a probe that will detect the gene for β-globin. A typical set of restriction patterns that might be obtained is shown in Figure 22.10. Remember that each mouse carries two copies of the β-globin gene, one on each homologous chromosome. Thus, a mouse could be +/+ (the restriction site is present on both chromosomes), +/− (the restriction site is present on one chromosome and is absent on

~ **FIGURE 22.10**

Restriction patterns from five mice. The patterns differ in the presence (+) or absence (−) of a particular restriction site (middle one of the three shown). Each mouse has two homologous chromosomes, each of which potentially carries the restriction site. Thus a mouse may be +/+ (has the restriction site on both chromosomes), +/− (has the restriction site on one chromosome), or −/− (has the restriction site on neither chromosome). When the restriction site is present, the DNA is broken into two fragments after digestion with the restriction enzyme and separation with electrophoresis.

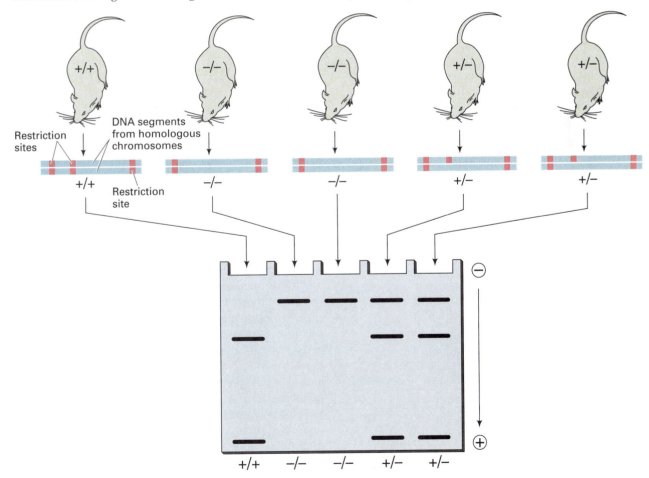

the other), or −/− (the restriction site is absent on both chromosomes). For the ten chromosomes present among these particular five mice, four have the restriction site and six do not.

To calculate the expected heterozygosity in nucleotide sequence, we use the formula:

$$H_{nuc} = \frac{n(\Sigma c_i) - \Sigma c_i^2}{j(\Sigma c_i)(n-1)}$$

In this equation, n equals the number of homologous DNA molecules examined. In our example we looked at five mice, each with two homologous chromosomes, so $n = 10$. The quantity j equals the number of nucleotides in the restriction site; in our example this is six, because $BamHI$ recognizes a six-base sequence. For each restriction site i, c_i represents the number of molecules in the sample that were cleaved at that restriction site. In our example, we examined only a single restriction site, so we have a single value of c_i which is four. If additional restriction enzymes were used, we would have a value of c_i for each site recognized by an enzyme. The symbol Σc_i is the total number of cuts at all cleavage sites in all the chromosomes. Since we examined only a single restriction site, and that site was cut in four of the chromosomes, $\Sigma c_i = 4$. Thus we obtain

$$H_{nuc} = \frac{10(4) - (4)^2}{6(4)(10-1)} = \frac{40 - 16}{24(9)} = \frac{24}{216} = 0.11$$

The use of this equation assumes that each RFLP results from a single nucleotide difference.

Nucleotide heterozygosity has been studied for a number of different organisms through the use of restriction enzymes. A few examples are shown in Table 22.4. Nucleotide heterozygosity typically varies from 0.002 to 0.02 in eukaryotic organisms. This means that an individual is heterozygous (contains different nucleotides on the two homologous chromosomes) at about one in every 50 to 500 nucleotides. Another way to interpret nucleotide heterozygosity is

that two randomly chosen chromosomes from a population will differ at about one in every 50 to 500 nucleotides.

One disadvantage of using RFLPs for examining genetic variation is that this method reveals variation at only a small subset of the nucleotides that make up a gene. With RFLP analysis, we are taking a small sample of the nucleotides (those recognized by the restriction enzyme) and using this sample to estimate the overall level of variation. DNA sequencing, which was described in Chapter 15 (pp. 474–476), provides a method for detecting all nucleotide differences that exist among a set of DNA molecules. For example, Martin Krietman sequenced 11 copies (obtained from different fruit flies) of a 2,659–base pair segment of the alcohol dehydrogenase gene in *Drosophila melanogaster*. Among the 11 copies, he found different nucleotides at 43 positions within the 2,659–base pair segment. Furthermore, only 3 of the 11 copies were identical at all nucleotides examined—thus, there were 8 different alleles (at the nucleotide level) among the 11 copies of this gene! This, along with other issues raised in Chapter 15 (such as DNA typing [= DNA fingerprinting] and VNTRs) suggests that populations harbor a tremendous amount of genetic variation in their DNA sequences.

CHANGES IN GENETIC STRUCTURE OF POPULATIONS

We have seen that meiosis and gamete fusion do not, by themselves, generate changes in the allelic frequencies of a population. When the population is large, randomly mating, and free from mutation, migration, and natural selection, no evolution occurs. The genetic structure of the population is in equilibrium. For many populations, however, the conditions required by the Hardy-Weinberg law do not hold. Populations are frequently small, mating may be nonrandom, and mutation, migration, and natural selection may be occurring. In these circumstances, allelic frequencies do change, and the gene pool of the population evolves in response to the interplay of these processes. In the following sections we discuss the role of four evolutionary processes—mutation, genetic drift, migration, and natural selection—in changing the allele frequencies of a population. We also discuss the effects of nonrandom mating on genotype frequencies. We first consider violations of the Hardy-Weinberg equilibrium assumptions one by one. We then consider several cases where two assumptions are violated simultaneously. Be aware, however, that in real populations all of the assump-

∼ TABLE 22.4

Estimates of Nucleotide Heterozygosity for DNA Sequences

DNA SEQUENCES	ORGANISM	H_{nuc}
β-Globin genes	Humans	0.002
Growth hormone gene	Humans	0.002
Alcohol dehydrogenase gene	Fruit fly	0.006
Mitochondrial DNA	Humans	0.004
H4 gene region	Sea urchin	0.019

tions can be violated simultaneously, but our understanding is best enhanced by considering simpler situations.

Mutation

One process that can potentially alter the frequencies of alleles within a population is mutation. As we discussed in Chapter 19, gene mutations consist of heritable changes in the DNA that occur within a locus. Usually a mutation converts one allelic form of a gene to another. The rate at which mutations arise is generally low, but varies among loci and among species (Table 22.5). Certain genes modify overall mutation rates, and many environmental factors, such as chemicals, radiation, and infectious agents, may increase the number of mutations.

Evolution is a two-step process: first, genetic variation arises; then, different alleles increase or decrease in frequency in response to evolutionary processes. Mutation is potentially important in both steps.

Ultimately, mutation is the source of all new genetic variation; new combinations of alleles may arise through recombination, but new alleles only occur as a result of mutation. Thus, mutation provides the raw genetic material upon which evolution acts. Most mutations will be detrimental and will be eliminated from the population. A few mutations, however, will convey some advantage to the individuals that possess them and will spread through the population. Whether a mutation is detrimental or advantageous depends upon the specific environment, and if the environment changes, previously harmful or neutral mutations may become beneficial. For example, after the widespread use of the insecticide DDT, insects with mutations that conferred resistance to DDT were capable of surviving and reproducing; because of this advantage, the mutations spread and many insect populations quickly evolved resistance to DDT. Mutations are of fundamental importance to the process of evolution, since they provide the genetic variation upon which other evolutionary processes act.

Mutations also have the potential to affect evolution by contributing to the second step in the evolutionary process—by changing the frequency of alleles within a population. Consider a population that consists of 50 individuals, all with the genotype AA. The frequency of $A(p)$ is 1.00 $[(2 \times 50)/100]$. If one A allele mutates to a, the population now consists of 49 AA individuals and 1 Aa individual. The frequency of p is now $[(2 \times 49) + 1]/100 = 0.99$. When another mutation occurs, the frequency of A drops to 0.98. If these mutations continue to occur at a low but steady rate over long periods of time, the frequency of A will

eventually decline to zero, and the frequency of a will reach a value of 1.00.

The mutation of A to a is referred to as a *forward mutation*. For most genes, mutations also occur in the reverse direction; a may mutate to A. These mutations are called *reverse mutations,* and they typically occur at a lower rate than forward mutations. The forward mutation rate — the rate at which A mutates to a $(A \rightarrow a)$—is symbolized with u; the reverse mutation rate — the rate at which a mutates to A $(A \rightarrow a)$— is symbolized with v. Consider a hypothetical population in which the frequency of A is p and the frequency of a is q. We assume that the population is large and that no selection occurs on the alleles. Each generation, a proportion u of all A alleles mutates to a. The actual number mutating depends upon both u and the frequency of A alleles. For example, suppose the population consists of 100,000 alleles. If u equals 10^{-4}, one out of every 10,000 A alleles mutates to a. When $p = 1.00$, all 100,000 alleles in the population are A, and free to mutate to a, so $10^{-4} \times 100,000 = 10$ A alleles should mutate to a. However, if $p = 0.10$, only 10,000 alleles are A and free to mutate to a. Therefore, with a mutation rate of 10^{-4} only 1 of the A alleles will undergo mutation. The decrease in the frequency of A resulting from mutation of $A \rightarrow a$ is equal to up; the increase in frequency resulting from $A \leftarrow a$ is equal to vq. As a result of mutation, the amount that A decreases in one generation is equal to the increase in A alleles due to reverse mutations minus the decrease in A alleles due to forward mutations. Given that we have a forward mutation rate increasing the frequency of a and a reverse mutation rate decreasing the frequency of a, it is intuitively easy to see that eventually, the population achieves equilibrium, in which the number of alleles undergoing forward mutation is exactly equal to the number of alleles undergoing reverse mutation. At this point, no further change in allelic frequency occurs, in spite of the fact that forward and reverse mutations continue to take place. With some simple algebra population genetics theorists have shown that the equilibrium frequency for a is

$$\hat{q} = \frac{u}{u + v}$$

and the equilibrium value for p is

$$\hat{p} = \frac{v}{u + v}$$

Consider a population in which the initial allelic frequencies are $p = 0.9$ and $q = 0.1$ and the forward and reverse mutation rates are $u = 5 \times 10^{-5}$ and $v = 2 \times 10^{-5}$, respectively (These values are similar to forward and

~ TABLE 22.5

Spontaneous Mutation Frequencies at Specific Loci for Various Organisms[a]

ORGANISM	TRAIT	MUTATION PER 100,000 GAMETES[b]
T2 Bacteriophage (virus)	To rapid lysis ($r^+ \rightarrow r$)	7
	To new host range ($h^+ \rightarrow h$)	0.001
E. coli K12 (bacterium)	To streptomycin resistance	0.00004
	To phage T1 resistance	0.003
	To leucine independence	0.00007
	To arginine independence	0.0004
	To tryptophan independence	0.006
	To arabinose dependence	0.2
Salmonella typhimurium (bacterium)	To threonine resistance	0.41
	To histidine dependence	0.2
	To tryptophan independence	0.005
Diplococcus pneumoniae (bacterium)	To penicillin resistance	0.01
Neurospora crassa	To adenine independence	0.0008–0.029
	To inositol independence	0.001–0.010
	(One inos allele, JH5202)	1.5
Drosophila melanogaster males	y^+ to yellow	12
	bw^+ to brown	3
	e^+ to ebony	2
	ey^+ to eyeless	6
Corn	Wx to waxy	0.00
	Sh to shrunken	0.12
	C to colorless	0.23
	Su to sugary	0.24
	Pr to purple	1.10
	I to i	10.60
	R^r to r^r	49.20
Mouse	a^+ to nonagouti	2.97
	b^+ to brown	0.39
	c^+ to albino	1.02
	d^+ to dilute	1.25
	ln^+ to leaden	0.80
	Reverse mutations for above genes	0.27
Chinese hamster somatic cell tissue culture	To azaguanine resistance	0.0015
	To glutamine independence	0.014
Humans	Achondroplasia	0.6–1.3
	Aniridia	0.3–0.5
	Dystrophia myotonica	0.8–1.1
	Epiloia	0.4–1
	Huntington disease	0.5
	Intestinal polyposis	1.3
	Neurofibromatosis	5–10
	Osteogenesis imperfecta	0.7–1.3
	Pelger's anomaly	1.7–2.7
	Retinoblastoma	0.5–1.2

[a]Mutations to independence for nutritional substances are from the auxotrophic condition (e.g., leu) to the prototrophic condition (e.g., leu+).
[b]Mutation frequency estimates of viruses, bacteria, Neurospora, and Chinese hamster somatic cells are based on particle or cell counts rather than gametes.

Source: From Genetics, 3d ed. by Monroe W. Strickberger. Copyright © 1985. Adapted by permission of Prentice-Hall, Inc., Upper Saddle River, NJ.

reverse mutation rates observed for many genes). In the first generation the change in allelic frequency is

$$\Delta p = vq - up$$
$$= (2 \times 10^{-5} \times 0.1) - (5 \times 10^{-5} \times 0.9)$$
$$\Delta p = -0.000043$$

The frequency of A decreases by only four-thousandths of 1 percent. At equilibrium, the frequency of the a allele, $\hat{q}$, equals

$$\hat{q} = \frac{u}{u + v}$$

$$\hat{q} = \frac{5 \times 10^{-5}}{(5 \times 10^{-5}) + (2 \times 10^{-5})} = 0.714$$

If no other processes act on a population, after many generations the alleles will reach equilibrium. Therefore, mutation rates determine the allelic frequencies of the population in the absence of other evolutionary processes. However, because mutation rates are so low, the change in allelic frequency due to mutation pressure is exceedingly slow. Furthermore, as allelic frequencies approach their equilibrium values, the change in frequency becomes smaller and smaller. Suppose that the mutation rate of $A \rightarrow a$ is $u = 10^{-5}$; suppose also that no reverse mutation occurs. With no reverse mutation, the equilibrium frequency of A is 0.0. If the frequency of A is initially 1.00, 1,000 generations are required to change the frequency of A from 1.00 to 0.99. To change the frequency from 0.50 to 0.49, 2,000 generations are required, and to change it from 0.1 to 0.09, 10,000 generations are necessary. If some reverse mutation occurs, the rate of change is even slower.

In practice, mutation by itself changes the allelic frequencies at such a slow rate that populations are rarely in mutational equilibrium. Other processes have more profound effects on allelic frequencies, and mutation alone rarely determines the allelic frequencies of a population. For example, achondroplastic dwarfism is an autosomal dominant trait in humans that arises through recurrent mutation. However, the frequency of this disorder in human populations is determined by an interaction of mutation pressure and natural selection.

𝒦EYNOTE

When we study what happens when we violate the assumption of the Hardy-Weinberg equilibrium of the absence of mutation, we see that mutation is the only way in which novel genetic material can come to exist within a species. The larger a population the more potential there is for a novel mutation to arise. In addition, mutations that occur repeatedly can alter the allelic frequencies of a population over time, provided that other evolutionary processes are not active. Eventually, a mutational equilibrium is reached, in which the allelic frequencies of the population remain constant in spite of continuing mutation; the allelic frequencies at this equilibrium are a function of the forward and reverse mutation rates.

Genetic Drift

Another major assumption of the Hardy-Weinberg law is that the population is infinitely large. Real populations are not infinite in size, but frequently they are large enough that expected ratios are realized and chance factors have small effects on allelic frequencies. Some populations are small, however, and in these groups chance factors may produce large changes in allelic frequencies. Random change in allelic frequency due to chance is called **genetic drift**, or simply *drift* for short. Sewall Wright (see Figure 22.1b), a brilliant population geneticist who laid much of the theoretical foundation of the discipline, championed the importance of genetic drift in the 1930s, so sometimes genetic drift is called the *Sewall Wright effect* in his honor.

CHANCE CHANGES IN ALLELIC FREQUENCY. Changes in allelic frequency resulting from random events can have important evolutionary implications in small populations. In addition, such changes can have important consequences for the conservation of a rare or endangered species. To see how chance can play a big role in altering the genetic structure of a population, imagine a small group of humans inhabiting a South Pacific island. Suppose that this population consists of only ten individuals, five of whom have green eyes and five of whom have brown eyes. For this example, we assume that eye color is determined by a single locus (actually a number of genes control eye color) and that the allele for green eyes is recessive to brown (BB and Bb codes for brown eyes and bb codes for green). The frequency of the allele for green eyes is 0.6 in the island population. A typhoon strikes the island, killing 50 percent of the population; five of the inhabitants perish in the storm. Those five individuals who die all have brown eyes. Eye color in no way affects the probability of surviving; the fact that only those with green eyes survive is strictly the result of chance. After the typhoon, the allelic frequency for green eyes is 1.0. Evolution has occurred in this population—the frequency of the green-eye allele has changed from 0.6 to 1.0, simply as a result of chance.

Now, imagine the same scenario, but this time with a population of 1,000 individuals. As before, 50 percent of the population has green eyes and 50 percent has brown eyes. A typhoon strikes the island and kills half the population. How likely is it that, just by chance, all 500 people who perish will have brown eyes? In a population of 1,000 individuals, the probability of this occurring by chance is extremely remote. This example illustrates an important characteristic of genetic drift — chance factors are likely to produce significant changes in allelic frequencies only in small populations.

Random factors producing mortality in natural populations, such as the typhoon in the above example, is only one of several ways in which genetic drift arises. Chance deviations from expected ratios of gametes and zygotes also produce genetic drift. We have seen the importance of chance deviations from expected ratios in the genetic crosses we studied in earlier chapters. For example, when we cross a heterozygote with a homozygote ($Aa \times aa$), we expect 50 percent of the progeny to be heterozygous and 50 percent to be homozygous. We do not expect to get exactly 50 percent every time, however, and if the number of progeny is small, the observed ratio may differ considerably from the expected. Recall that the Hardy-Weinberg law is based upon random mating and expected ratios of progeny resulting from each type of mating (see Table 22.2). If the actual number of progeny differs from the expected ratio due to chance, genotypes may not be in Hardy-Weinberg proportions. As a result, changes in allelic frequencies may occur.

Chance deviations from expected proportions arise from a general phenomenon called **sampling error**. Imagine that a population produces an infinitely large pool of gametes, with alleles in the proportions p and q. If random mating occurs and all the gametes unite to form zygotes, the proportions of the genotypes will be equal to p^2, $2pq$, and q^2, and the frequencies of the alleles in these zygotes will remain p and q. If the number of progeny is limited, however, the gametes that unite to form the progeny constitute a sample from the infinite pool of potential gametes. Just by chance, or by "error," this sample may deviate from the larger pool; the smaller the sample, the larger the potential deviation.

Flipping a coin is analogous to the situation in which sampling error occurs. When we flip a coin, we expect 50 percent heads and 50 percent tails. If we flip the coin 1,000 times, we will get very close to that expected fifty-fifty ratio. But, if we flip the coin only four times, we would not be surprised if by chance we obtain 3 heads and 1 tail, or even all tails. When the sample—in this case the number of flips—is small, the sampling error can be large. All genetic drift arises from such sampling error.

MEASURING GENETIC DRIFT. Genetic drift is random, and thus we cannot predict what the allelic frequencies will be after drift has occurred. However, since sampling error is related to the size of the population, we can make predictions about the magnitude of genetic drift. Ecologists often measure population size by counting the number of individuals, but not all individuals contribute gametes to the next generation. To determine the magnitude of genetic drift, we must know the **effective population** size, which equals the equivalent number of adults contributing gametes to the next generation. If the sexes are equal in number and all individuals have an equal probability of producing offspring, the effective population size equals the number of breeding adults in the population. However, when males and females are not present in equal numbers, the effective population size is

$$N_e = \frac{4 \times N_f \times N_m}{N_f + N_m}$$

where N_f equals the number of breeding females and N_m equals the number of breeding males.

Students often have difficulty understanding why this equation must be used—why the effective population size is not simply the number of breeding adults. The reason is that males, as a group, contribute half of all genes to the next generation and females, as a group, contribute the other half. Therefore, in a population of 70 females and 2 males, the two males are not genetically equivalent to two females; each male contributes $1/2 \times 1/2 = 0.25$ of the genes to the next generation, whereas each female contributes $1/2 \times 1/70 = 0.007$ of all genes. The small number of males disproportionately influences what alleles are present in the next generation. Using the above equation, the effective population size is $N_e = (4 \times 70 \times 2)/(70 + 2) = 7.8$, or approximately 8 breeding adults. What this means is that in a population of 70 females and 2 males, genetic drift will occur as if the population had only four breeding males and four breeding females. Therefore, genetic drift will have a much greater effect in this population than in one with 72 breeding adults equally divided between males and females.

Other factors, such as differential production of offspring, fluctuating population size, and overlapping generations can further reduce the effective population size. Considering these complications, it is quite difficult to measure effective population size accurately.

The amount of variation among populations resulting from genetic drift is measured by the **variance[1] of allelic frequency**, which equals

$$s_p{}^2 = \frac{pq}{2N_e}$$

where N_e equals the effective population size and p and q equal the allelic frequencies. A more useful measure is the **standard error[1]** of allelic frequency, which is the square root of the variance of allelic frequency:

$$s_p = \sqrt{\frac{pq}{2N_e}}$$

The standard error can be used to calculate the 95 percent confidence limits of allelic frequency, which indicate the expected range of p in 95 percent of such populations. The 95 percent confidence limits equal *approximately* $p + 2s_p$. Suppose, for example, that $p = 0.8$ in a population with N_e equal to 50. The standard error in allelic frequency is $s_p = \sqrt{(pq/2N_e)} = 0.04$. The 95 percent confidence limits for p are therefore $p + 2s_p = 0.72 \leq p \leq 0.88$. To interpret the 95 percent confidence limits, imagine that 100 populations with N_e of 50 have p initially equal to 0.8. Genetic drift may cause allelic frequencies in some populations to change; the 95 percent confidence limits tell us that in the next generation, 95 of the original 100 populations should have p within the range of 0.72 to 0.88. Therefore, if we observe a change in p greater than this, say from 0.8 to 0.68, we know that the probability that this change will occur by genetic drift is less than 0.05. Most likely we would conclude that some process other than genetic drift contributed to the observed change in allelic frequency.

CAUSES OF GENETIC DRIFT. All genetic drift arises from sampling error, but there are several ways in which sampling error occurs in natural populations. First, as already discussed genetic drift arises when population size remains continuously small over many generations. Undoubtedly, this situation is fre-

quent, particularly where populations occupy marginal habitats, or when competition for resources limits population growth. In such populations, genetic drift plays an important role in the evolution of allelic frequencies. Many species are spread out over a large geographic range. This can result in a species consisting of numerous populations of small size, each undergoing drift independently. In addition, human intervention such as the clear cutting of forests can result in the fragmentation of previously large continuous populations into small subdivided ones, again, each showing genetic drift.

The effect of genetic drift arising from small population size is seen in a classic laboratory experiment conducted by Buri with *Drosophila melanogaster*. Buri examined the frequency of two alleles, bw^{75} and bw, at a locus that determines eye color in the fruit flies. He set up 107 experimental populations, and the initial frequency of bw^{75} was 0.5 in each. The flies in each population interbred randomly, and in each generation Buri randomly selected 8 males and 8 females to be the parents for the next generation. Thus, effective population size was always 16 individuals. The distribution of allelic frequencies in these 107 populations is presented in Figure 22.11. Notice that the allelic frequencies in the early generations were clumped around 0.5, but genetic drift caused the frequencies in the populations to spread out or diverge over time. By generation 19, the frequency of bw^{75} was 0 or 1 in most populations. What is most elegant about this experiment is that it closely matched the theoretically expected effects of drift demonstrated by Fisher and Wright (Figure 22.12) and speaks to the power of strictly theoretical approaches in making predictions about the real world.

Another way in which genetic drift arises is through **founder effect**. Founder effect occurs when a population is initially established by a small number of breeding individuals. Although the population may subsequently grow in size and later consist of a large number of individuals, the gene pool of the population is derived from the genes present in the original founders. Chance may play a significant role in determining which genes were present among the founders, and this has a profound effect upon the gene pool of subsequent generations. Founder effects have frequently been used to explain the subsequent evolution of new species, but their importance to the process of species formation is currently under intense study and debate.

Many excellent examples of founder effect come from the study of human populations. Consider the inhabitants of Tristan da Cunha, a small, isolated island in the South Atlantic. This island was first per-

---

[1]A more thorough discussion of statistical concepts, including variance and standard error, is presented in the following chapter, pp. 776–784.

~ FIGURE 22.11

Results of Buri's study of genetic drift in 107 populations of *Drosophila melanogaster*. Shown are the distributions of the frequency of the bw^{75} allele among the populations in 19 consecutive generations. Each population consisted of 16 individuals.

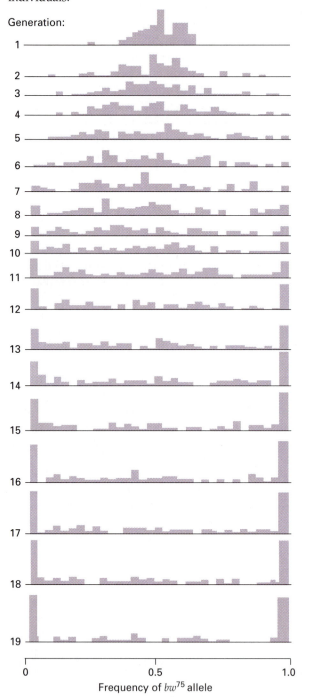

Generation:

Frequency of bw^{75} allele

~ FIGURE 22.12

Theoretically predicted results of the Fisher-Wright model for the fate of neutral alleles in a sample of populations where each population contains 16 individuals and the initial allele frequency is 0.5.

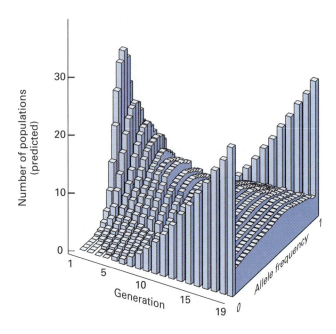

from the distant island of St. Helena, but for the most part the island remained a genetic isolate. In 1961, a volcano on Tristan da Cunha erupted, and the population of almost 300 inhabitants was evacuated to England. During their two-year stay in England, geneticists studied the islanders and reconstructed the genetic history of the population. These studies revealed that the gene pool of Tristan da Cunha was strongly influenced by genetic drift.

Three forms of genetic drift occurred in the evolution of the island's population. First, founder effect took place at the initial settlement. By 1855, the population of Tristan da Cunha consisted of about 100 individuals, but 26 percent of the genes of the population in 1855 were contributed by William Glass and his wife. Even in 1961, these original two settlers contributed 14 percent of all the genes in the 300 individuals of the population. The particular genes that Glass and other original founders carried heavily influenced the subsequent gene pool of the population. Second, population size remained small throughout the history of the settlement, and sampling error continually occurred.

manently settled by William Glass, a Scotsman, and his family in 1817. (Several earlier attempts at settlement failed.) They were joined by a few additional settlers, some shipwrecked sailors, and a few women

A third form of sampling error, called **bottleneck effect**, also played an important role in the population of Tristan da Cunha. Bottleneck effect is a form of genetic drift that occurs when a population is drastically reduced in size. During such a population reduction, some genes may be lost from the gene pool as a result of chance. Recall our earlier example of the population consisting of 10 individuals inhabiting a South Pacific island. When a typhoon struck the island, the population size was reduced to five, and by chance, all individuals with brown eyes perished in the storm, changing the frequency of green eyes from 0.6 to 1.0. This is an example of bottleneck effect. Bottleneck effect can be viewed as a type of founder effect, since the population is refounded by those few individuals that survive the reduction.

Two severe bottlenecks occurred in the history of Tristan da Cunha. The first took place around 1856 and was precipitated by two events: the death of William Glass and the arrival of a missionary who encouraged the inhabitants to leave the island. At this time many islanders emigrated to America and to South Africa, and the population dropped from 103 individuals at the end of 1855 to 33 in 1857. A second bottleneck occurred in 1885. The island of Tristan da Cunha has no natural harbor, and the islanders intercepted passing ships for trade by rowing out in small boats. On November 28, 1885, 15 of the adult males on the island put out in a small boat to make contact with a passing ship. In full view of the entire island community, the boat capsized and all 15 men drowned. Following this disaster, only four adult males were left on the island, one of whom was insane and two of whom were old. Many of the widows and their families left the island during the next few years, and the population size dropped from 106 to 59. Both bottlenecks had a major effect on the gene pool of the population. All the genes contributed by several settlers were lost, and the relative contributions of others were altered by these events. Thus, the gene pool of Tristan da Cunha has been influenced by

genetic drift in the form of founder effect, small population size, and bottleneck effect.

As we shall see later, when we discuss migration, gene flow among populations increases the effective population size and reduces the effects of genetic drift. Small breeding units that lack gene flow are genetically isolated from other groups and often experience considerable genetic drift, even though surrounded by much larger populations. A good example is a religious sect, known as the Dunkers, found in eastern Pennsylvania. Between 1719 and 1729, fifty Dunker families emigrated from Germany and settled in the United States. Since that time, the Dunkers have remained an isolated group, rarely marrying outside of the sect, and the number of individuals in their communities has always been relatively small.

During the 1950s geneticists studied one of the original Dunker communities in Franklin County, Pennsylvania. At the time of the study, this population had about 300 members, and the population size had remained relatively constant for many generations. The investigators found that some of the allelic frequencies in the Dunkers were very different from the frequencies found among the general population of the United States. The Pennsylvania frequencies were also different from the frequencies of the West German population from which the Dunkers descended. Table 22.6 presents some of the allelic frequencies at the ABO blood group locus. The ABO allele frequencies among the Dunkers are not the same as those in either the United States population or the West German population. Nor are the Dunker frequencies intermediate between the United States and German frequencies. (Intermediate frequencies might be expected if intermixing of Dunkers and Americans had occurred.) The most likely explanation for the unique Dunker allelic frequencies observed is that genetic drift has produced random change in the gene pool. Founder effect probably occurred when the original 50 families emigrated

~ TABLE 22.6

Frequencies of Alleles Controlling the ABO Blood Group System in Three Human Populations

POPULATION	ALLELE FREQUENCIES			PHENOTYPE (BLOOD GROUP) FREQUENCIES			
	I^A	I^B	i	A	B	AB	O
Dunker	0.38	0.03	0.59	0.593	0.036	0.023	0.348
United States	0.26	0.04	0.70	0.431	0.058	0.021	0.490
West Germany	0.29	0.07	0.64	0.455	0.095	0.041	0.410

from Germany, and genetic drift has most likely continued to influence allelic frequencies each generation since 1729, because the population size has remained small.

Effects of genetic drift. Genetic drift produces changes in allelic frequencies, and these changes have several effects on the genetic structure of populations. First, genetic drift causes the allelic frequencies of a population to change over time. This is illustrated in Figure 22.13. The different lines represent allelic frequencies in several populations over a number of generations. Although all populations begin with an allelic frequency equal to 0.50, the frequencies in each population change over time as a result of sampling error. In each generation, the allelic frequency may increase or decrease, and over time the frequencies wander randomly or drift (hence the name "genetic drift"). Sometimes, within 30 generations, just by chance, the allelic frequency reaches a value of 0.0 or 1.0. At this point, one allele is lost from the population and the population is said to be *fixed* for the remaining allele in a one gene locus two allele example. Once an allele has reached fixation, no further change in allelic frequency can occur, unless the other allele is reintroduced through mutation or migration. The probability of fixation in a population increases with time, as shown theoretically by Motoo Kimura in Figure 22.14. If the initial allelic frequencies

~ Figure 22.14

The average time to fixation or loss of an allele from a population as a function of population size and initial allele frequency as predicted by Kimura. For example if the initial allele frequency is 0.3 and the population size was 10 it would take just under 2.5×10 generations on average for the allele to be lost or fixed in the population.

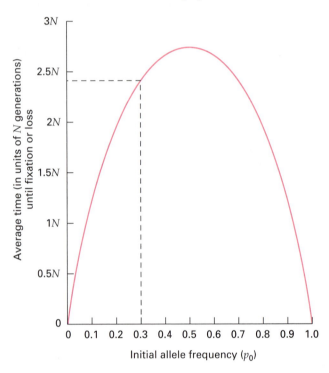

~ Figure 22.13

The effect of genetic drift on the frequency (q) of allele A^2 in four populations. Each population begins with q equal to 0.5, and the effective populations size for each is 20. The mean frequency of allele A^2 for the four replicates is indicated by the red line. These results were obtained by a computer simulation.

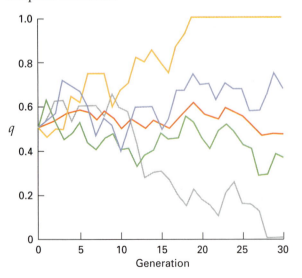

are equal, which allele becomes fixed is strictly random. If, on the other hand, initial allelic frequencies are not equal, the rare allele is more likely to be lost. During this process of genetic drift and fixation, the number of heterozygotes in the population also decreases, and after fixation, the population heterozygosity is zero. As heterozygosity decreases and alleles become fixed, populations lose genetic variation; thus, the second effect of genetic drift is a reduction in genetic variation within populations.

Since genetic drift causes random change in allelic frequency, the allelic frequencies in separate, individual populations will not change in the same direction. Therefore, populations diverge in their allelic frequencies through genetic drift. This is illustrated in Figures 22.11, 22.12, and 22.13; in all figures all the populations begin with p and q equal to 0.5. After a few generations, the allelic frequencies of the populations diverge, and this divergence increases over generations. The maximum divergence in allelic frequencies is reached when all populations are fixed for one or the other allele. If allelic frequencies are initially equal to 0.5, approximately half of the populations will be fixed for one allele, and half will be fixed for the other.

Because genetic drift is greater in small populations and leads to genetic divergence, we expect more variance in allelic frequency among small populations than among large populations. Such a relationship has been observed in studies of natural populations. R. K. Selander, for example, studied genetic variation in populations of the house mouse inhabiting barns in Texas. Through systematic trapping, he was able to estimate population size, and using electrophoresis he examined the variance in allelic frequency at two loci, a locus coding for the enzyme esterase (*Est-3*) and a locus coding for hemoglobin (*Hbb*). Selander found that the variance in allelic frequency among small populations was several times larger than that among large populations (Table 22.7), an observation consistent with our understanding of how genetic drift leads to population divergence.

While full development is beyond the scope of this treatment, it is important to note that Kimura's result (see Figure 22.14) also forms the basis of the **neutral theory of molecular evolution.** A new mutation that is not selectively disadvantageous, but neutral with respect to natural selection will initially be at a low frequency. Therefore it will most likely be lost from the population as a result of genetic drift. Occasionally, however, by chance alone, the new mutation will drift to fixation and the gene locus will have evolved. Given that mutation is a recurring event a gene locus will accumulate differences over time by chance alone. In this way the genes of two related lineages, such as humans and other primates, can be compared and used to estimate the date since they last shared a common ancestor. In this way it is estimated that humans and chimpanzees last shared a common ancestor six million years ago. This area is also currently under great debate and study.

KEYNOTE

Genetic drift, or chance changes in allelic frequency due to sampling error, can have important evolutionary and survival implications for small populations. Genetic drift leads to loss of genetic variation within populations, genetic divergence among populations, and random fluctuation in the allelic frequencies of a population over time. Genetic drift can also explain how molecules in different species accumulate differences on a seemingly regular basis, forming the basis for the neutral theory of molecular evolution.

Migration

One of the assumptions of the Hardy-Weinberg law is that the population is closed and not influenced by other populations. Many populations are not completely isolated, however, and exchange genes with other populations of the same species. Individuals migrating into a population may introduce new alleles to the gene pool and alter the frequencies of existing alleles. Thus **migration** has the potential to disrupt Hardy-Weinberg equilibrium and may influence the evolution of allelic frequencies within populations.

The term *migration* usually implies movement of organisms. In population genetics, however, we are interested in the movement of genes, which may or may not occur when organisms move. Movement of genes takes place only when organisms or gametes migrate, and contribute their genes to the gene pool of the recipient population. This process is also referred to as **gene flow**.

~ TABLE 22.7

Variance in Allelic Frequency Among Populations as a Function of Population Size

TYPE OF POPULATION	NUMBER OF POPULATIONS	MEAN ALLELIC FREQUENCY		VARIANCE IN ALLELIC FREQUENCY	
		Est-3[b]	Hbb[s]	Est-3[b]	Hbb[s]
Small (N < 50)	29	0.418	0.849	0.051	0.188
Large (N > 50)	13	0.372	0.843	0.013	0.008

Gene flow has two major effects on a population. First, it introduces new alleles to the population. Because mutation is generally a rare event, a specific mutant allele may arise in one population and not in another. Gene flow spreads unique alleles to other populations and, like mutation, is a source of genetic variation for the population. Second, when the allelic frequencies of migrants and the recipient population differ, gene flow changes the allelic frequencies within the recipient population. Through exchange of genes, different populations remain similar, and thus, migration is a homogenizing force that tends to prevent populations from accumulating genetic differences among them.

To illustrate the effect of migration on allelic frequencies, we will consider a simple model in which gene flow occurs in only one direction, from population I to population II. Suppose that the frequency of allele A in population I (p_I) is 0.8 and the frequency of A in population II (p_{II}) is 0.5. Each generation some individuals migrate from population I to population II, and these migrants are a random sample of the genotypes in population I. After migration, population II actually consists of two groups of individuals: the migrants with $p_I = 0.8$, and the residents with $p_{II} = 0.5$. The migrants now make up a proportion of population II, which we will designate m. The frequency of A in population II after migration (p'_{II}) is

$$p'_{II} = mp_I + (1 - m)p_{II}$$

We see that the frequency of A after migration is determined by the proportion of A alleles in the two groups that now comprise population II. The first component, mp_I, represents the A alleles in the migrants—we multiply the proportion of the population that consists of migrants (m) by the allelic frequency of the migrants (p). The second component represents the A alleles in the residents, and equals the proportion of the population consisting of residents ($1 - m$) multiplied by the allelic frequency in the residents (p_{II}). Adding these two components together gives us the allelic frequency of A in population II after migration. This model of gene flow is diagrammed in Figure 22.15.

The change in allelic frequency in population II as a result of migration (Δp) equals the original frequency of A subtracted from the frequency of A after migration:

$$\Delta p = p'_{II} - p_{II}$$

In the previous equation, we found that p'_{II} equaled $mp_I + (1 - m)p_{II}$, so the change in allelic frequency can be written as

$$\Delta p = mp_I + (1 - m)p_{II} - p_{II}$$

∼ FIGURE 22.15

Theoretical model illustrating the effect of migration on the gene pool of a population. After migration, population II consists of two groups, the migrants with allelic frequency of p_I and the original residents with allelic frequency p_{II}.

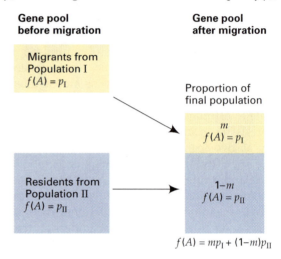

Multiplying $(1 - m)$ by p_{II} in the above equation, we obtain

$$\Delta p = mp_I + p_{II} - mp_{II} - p_{II}$$
$$\Delta p = mp_I - mp_{II}$$
$$\Delta p = m(p_I - p_{II})$$

This final equation indicates that the change in allelic frequency from migration depends upon two factors: the proportion of the migrants in the final population and the difference in allelic frequency between the migrants and the residents. If no differences exist in the allelic frequency of migrants and residents ($p_I - p_{II} = 0$), then we can see that the change in allelic frequency is zero. Populations must differ in their allelic frequencies in order for migration to affect the makeup of the gene pool. With continued migration, p_I and p_{II} become increasingly similar, and, as a result, the change in allelic frequency due to migration decreases. Eventually, allelic frequencies in the two populations will be equal, and no further change will occur. This is only true, however, when other factors besides migration do not influence allelic frequencies.

An additional point to be noted is that migration among populations tends to increase the effective size of the populations. As we have seen, small population size leads to genetic drift, and genetic drift causes populations to diverge. Migration, on the other hand, reduces divergence among populations, effectively increasing the size of the individual populations. Extensive gene flow has been shown to occur, for example, among populations of the monarch butterfly (Figure 22.16). Even a small amount of gene flow can

~ FIGURE 22.16

Extensive gene flow occurs among populations of the monarch butterfly (a). The butterflies overwinter in Mexico (b) and then migrate north during spring and summer to breeding grounds as far away as northern Canada. Extensive gene flow occurs during the migration period with the result that monarch populations display relatively little genetic divergence.

a)

b)

reduce the effect of genetic drift. Calculations have shown that a single migrant moving between two populations every other generation will prevent the two populations from becoming fixed for different alleles.

The effects of gene flow have important ramifications not only for the evolution of species but for the conservation of species as well. As discussed earlier, many species that have wide geographic ranges show variation in genetic structure over the species range. Part of the natural genetic structure of a species could include population subdivision where populations are loosely connected to each other by gene flow. Since gene flow has large consequences for the maintenance of genetic diversity, this feature of population genetic structure needs to be taken into account by those interested in conserving genetic structure.

KEYNOTE

Migration of individuals into a population may alter the makeup of the population gene pool, if the genes carried by the migrants differ from those of the resident population. Migration, also termed gene flow, tends to reduce genetic divergence among populations and increases the effective size of the population. The amount of migration among populations of the same species determines how much genetic substructuring exists and whether different populations of the same species become quite different from each other genetically.

Natural Selection

We have now examined three major evolutionary processes capable of changing allelic frequencies and producing evolution—mutation, genetic drift, and migration. These processes alter the gene pool of a population, and they certainly influence the evolution of a species. However, mutation, migration, and genetic drift do not result in adaptation. Adaptation is the process by which traits evolve that make organisms more suited to their immediate environment; these traits increase the organism's chances of surviving and reproducing. Adaptation is responsible for the many extraordinary traits seen in nature — wings that enable a hummingbird to fly backward, leaves of the pitcher plant that capture and devour insects, brains that allow humans to speak, read, and love. These biological features and countless other exquisite traits are the product of adaptation (Figure 22.17). Genetic drift, mutation, and migration all influence the pattern and process of adaptation, but adaptation arises chiefly from natural selection. **Natural selection** is the dominant force in the evolution of many traits and has shaped much of the phenotypic variation observed in nature.

Charles Darwin and Alfred Russel Wallace (Figure 22.18) independently developed the concept of natural selection in the middle of the nineteenth century, although some earlier naturalists had similar ideas. In 1858, Darwin and Wallace's theory was presented to the Linnaean Society of London and was enthusiastically received by other scientists. Darwin pursued the theory of evolution further than Wallace, amassing hundreds of observations to support it, and publishing his ideas in a book entitled *On the Origin of Species*. For his innumerable contributions to our understanding of

~ FIGURE 22.17

Natural selection produces organisms that are finely adapted to their environment, such as this lizard with cryptic coloration, which allows it to blend in with its natural surroundings.

natural selection, Darwin is frequently regarded as the "father" of evolutionary theory.

Natural selection can be defined as differential reproduction of genotypes. It simply means that individuals with certain genes produce more offspring than others; therefore those genes increase in frequency in the next generation, as further discussed in Chapter 23. Through natural selection, traits that contribute to survival and reproduction increase over time. In this way organisms adapt to their environment.

SELECTION IN NATURAL POPULATIONS. A classic example of selection in natural populations is the evolution of melanic (dark) forms of moths in association with industrial pollution, a phenomenon known as "industrial melanism." Melanic phenotypes have appeared in a number of different species of moths found in the industrial regions of Europe, North America, and England. One of the best studied cases involves the peppered moth, *Biston betularia*. The common phenotype of this species, termed the *typical* form, is a greyish white color with black mottling over the body and wings.

Prior to 1848, all peppered moths collected in England possessed this *typical* phenotype, but in 1848, a single black moth was collected near Manchester, England. This new phenotype, called *carbonaria*, presumably arose by mutation, and rapidly increased in frequency around Manchester and in other industrial regions. By 1900, the *carbonaria* phenotype had reached a frequency of more than 90 percent in several populations. High frequencies of *carbonaria* appeared to be associated with industrial regions, whereas the *typical* phenotype remained common in more rural districts. Laboratory studies by a number of investigators, including E. B. Ford and R. Goldschmidt, demonstrated that the *carbonaria* phenotype was dominant to the *typical* phenotype. A third phenotype was also discovered, which was somewhat intermediate to *typical* and *carbonaria*; this phenotype, *insularia*, was produced by a dominant allele at a different locus.

H. B. D. Kettlewell investigated color polymorphism in the peppered moth, demonstrating that the

~ FIGURE 22.18

Charles Darwin (a) and Alfred Russel Wallace (b) should be given equal credit for developing the theory of evolution through natural selection.

a)   b)

increase in the *carbonaria* phenotype occurred as a result of strong selection against the *typical* form in polluted woods. Peppered moths are nocturnal; during the day they rest on the trunks of lichen-covered trees. Birds frequently prey upon the moths during the day, but because the lichens that cover the trees are naturally grey in color, the *typical* form of the peppered moth is well camouflaged against this background (Figure 22.19a). In industrial areas, however, extensive pollution beginning with the industrial revolution in the mid-nineteenth century had killed most of the lichens and covered the tree trunks with black soot. Against this black background, the *typical* phenotype was quite conspicuous and was readily consumed by birds. In contrast, the *carbonaria* form was well camouflaged against the blackened trees and had a higher rate of survival than the *typical* phenotype in polluted areas (Figure 22.19b). Because *carbonaria* survived better in polluted woods, more *carbonaria* genes were transmitted to the next generation; thus, the *carbonaria* phenotype increased in frequency in industrial areas. In rural areas, where pollution was absent, the *carbonaria* phenotype was conspicuous and the *typical* form was camouflaged; in these regions the frequency of the *typical* form remained high.

Kettlewell demonstrated that selection affected the frequencies of the two phenotypes by conducting a series of mark-and-recapture experiments involving dark and light moths in smoky, industrial Birmingham, England, and in nonindustrialized Dorset. As predicted, the *typical* phenotype was favored in Dorset, and *carbonaria* was favored in Birmingham.

FITNESS AND COEFFICIENT OF SELECTION. Darwin described natural selection primarily in terms of survival, and even today, many nonbiologists think of natural selection in terms of a struggle for existence. However, what is most important in the process of natural selection is the relative number of genes that are contributed to future generations. Certainly the ability to survive is important, but survival alone will not ensure that genes are passed on; reproduction must also occur. Therefore, we measure natural selection by assessing reproduction. Natural selection is measured in terms of **Darwinian fitness**, which is defined as the relative reproductive ability of a genotype.

Darwinian fitness is often symbolized as W and also called the adaptive value of a genotype. Since it is a measure of the relative reproductive ability, population geneticists usually assign an adaptive value of 1 to a genotype that produces the most offspring. The fitnesses of the other genotypes are assigned relative to this. For example, suppose that the genotype G^1G^1 on the average produces 8 offspring, G^1G^2 produces an average of 4 offspring, and G^2G^2 produces an average of 2 offspring. The G^1G^1 genotype has the highest reproductive output, so its fitness is 1 ($W_{11} = 1.0$). Genotype G^1G^2 produces on the average 4 offspring for the 8 produced by the most fit genotype, so the fitness of G^1G^2 (W_{12}) is 4/8 = 0.5. Similarly, G^2G^2 produces 2 offspring for the 8 produced by G^1G^1, so the

---

~ FIGURE 22.19

Biston betularia, the peppered moth, and its dark form *carbonaria* (a) on the trunk of a lichened tree in the unpolluted countryside, and (b) on the trunk of a tree with dark bark. On the lichened tree, the dark form of the moth is readily seen, whereas the light form is well camouflaged. On the dark tree, the dark form of the moth is well camouflaged.

a)   b)

~ TABLE 22.8

Computation of Fitness Values and Selection Coefficients of Three Genotypes

	GENOTYPES		
	G^1G^1	G^1G^2	G^2G^2
Number of breeding adults in one generation	16	10	20
Number of offspring produced by all adults of the genotype in the next generation	128	40	40
Average number of offspring produced per breeding adult	$128/16 = 8$	$40/10 = 4$	$40/20 = 2$
Fitness W (relative number of offspring produced)	$8/8 = 1$	$4/8 = 0.5$	$2/8 = 0.25$
Selection coefficient ($s = 1 - \overline{W}$)	$1 - 1 = 0$	$1 - 0.5 = 0.5$	$1 - 0.25 = 0.75$

fitness of G^2G^2 (W_{22}) is $2/8 = 0.25$. Table 22.8 illustrates the calculation of relative fitness values.

Assigning fitness values to genotypes has to be done with great care and often a variety of assumptions are made. For example, equating fitness to the number of offspring produced may result in errors. For example, David Lack found that starling nests that had more eggs in them wound up successfully rearing fewer chicks. Assigning higher fitness values to those birds that laid more eggs would have been incorrect. The fitness associated with a genotype is difficult to pin down by looking at a snapshot of a part of the organism's life. A genotype that may yield high performance at one part of a life might result in low performance at another. This phenomenon has been termed **antagonistic pleiotropy**. Again, measuring fitness is difficult at best and caution is needed to avoid subjective (and erroneous) assignment of fitness values. Nevertheless, Darwinian fitness tells us how well a genotype is doing in terms of natural selection. A related measure is the **selection coefficient**, which is a measure of the relative intensity of selection against a genotype. The selection coefficient is symbolized by s and equals $1 - W$. In our example, the selection coefficients for G^1G^1 are $s = 0$; for G^1G^2, $s = 0.5$; for G^2G^2, $s = 0.75$.

EFFECT OF SELECTION ON ALLELIC FREQUENCIES. Natural selection produces a number of different effects. At times, natural selection eliminates genetic variation, and at other times it maintains variation; it can change allelic frequencies or prevent allelic frequencies from changing; it can produce genetic divergence among populations or maintain genetic uniformity. Which of these effects occurs

depends primarily on the relative fitness of the genotypes and on the frequencies of the alleles in the population.

The change in allelic frequency that results from natural selection can be calculated by constructing a table such as Table 22.9. This "table method" can be used for any type of single-locus trait, whether the trait is dominant, codominant, recessive, or overdominant. To use the table method we begin by listing the genotypes (A^1A^1, A^1A^2, and A^2A^2) and their initial frequencies. If random mating has just taken place, the genotypes are in Hardy-Weinberg proportions and the initial frequencies are p^2, $2pq$, and q^2. We then list the fitnesses for each of the genotypes, W_{11}, W_{12}, and W_{22}. Now, suppose that selection occurs and only some of the genotypes survive. The contribution of each genotype to the next generation will be equal to the initial frequency of the genotype multiplied by its fitness. For A^1A^1 this will be $p^2 \times W_{11}$. Notice that the contributions of the three genotypes do not add up to one. We calculate the relative contributions of each genotype by dividing each by the mean fitness of the population. The *mean fitness of the population* equals $p^2W_{11} + 2pqW_{12} + q^2W_{22} = \overline{W}$. This gives us the relative frequencies of the genotypes after selection. We then calculate the new allelic frequency (p') from the genotypes after selection, using our familiar formula, $p' =$ (frequency of A^1A^1) + (1/2 × frequency of A^1A^2). Finally, the change in allelic frequency resulting from selection equals $p' - p$. A sample calculation using some actual allelic frequencies and fitness values is presented in Table 22.10.

The wide range of generally unappreciated effects of natural selection discussed above can now be understood in terms of the adaptive values in Table 22.9. Again remembering that we will arbitrarily set

~ TABLE 22.9

General Method of Determining Change in Allelic Frequency Due to Natural Selection

	GENOTYPES		
	A^1A^1	A^1A^2	A^2A^2
Initial genotypic frequencies	p^2	$2pq$	q^2
Fitness[a]	W_{11}	W_{12}	W_{22}
Frequency after selection	p^2W_{11}	$2pqW_{12}$	q^2W_{22}
Relative genotypic frequency after selection	$P' = \dfrac{p^2W_{11}}{\overline{W}^b}$	$H' = \dfrac{2pqW_{12}}{\overline{W}}$	$Q' = \dfrac{q^2W_{22}}{\overline{W}}$

Allelic frequency after selection $= p' = P' + 1/2(H')$
$$q' = 1 - p'.$$
Change in allelic frequency due to selection $= \Delta p = p' - p$.

[a]For simplicity, fitness in this example is considered to be the probability of survival. Change in allelic frequency due to differences in the number of offspring produced by the genotypes is calculated in the same manner.

[b]$\overline{W} = p^2W_{11} + 2pqW_{12} + q^2W_{22}$

~ TABLE 22.10

General Method of Determining Change in Allelic Frequency Due to Natural Selection When Initial Allelic Frequencies Are $p = 0.6$ and $q = 0.4$

	GENOTYPES		
	A^1A^1	A^1A^2	A^2A^2
Initial genotypic frequencies	p^2 $(0.6)^2 = 0.36$	$2pq$ $2(0.6)(0.4) = 0.48$	q^2 $(0.4)^2 = 0.16$
Fitness	$W_{11} = 0$	$W_{12} = 0.4$	$W_{22} = 1$
Frequency after selection	$p^2W_{11} =$ $(0.36)(0) = 0$	$2pqW_{12} =$ $(0.48)(0.4) =$ 0.19	$q^2W_{22} =$ $(0.16)(1)$ $= 0.16$
Relative genotypic frequency after selection	$P' = \dfrac{p^2W_{11}}{\overline{W}^a}$ $P' = 0/0.35 = 0$	$H' = \dfrac{2pqW_{12}}{\overline{W}}$ $H' = 0.19/0.35$ $= 0.54$	$Q' = \dfrac{q^2W_{22}}{\overline{W}}$ $Q' = 0.16/0.35$ $= 0.46$

Allelic frequency after selection $p' = P' + 1/2(H')$
$$p' = 0 + 1/2(0.54) = 0.27$$
$$q' = 1 - p' = 1 - 0.27 = 0.73$$
Change in allelic frequency due to selection $= \Delta p = p' - p$
$$\Delta p = 0.27 - 0.6 = -0.33$$

[a]$\overline{W} = p^2W_{11} + 2pqW_{12} + q^2W_{22}$
$\overline{W} = 0 + 0.19 + 0.16$
$\overline{W} = 0.35$

the genotype with highest fitness to 1.0, we get a variety of classes of natural selection each with its own effects by permuting all possible relationships among fitness values for the genotypes. These include:

1. $W_{11} = W_{12} = W_{22} = 1.0$. All adaptive values are equal and there is no selection.
2. $W_{11} = W_{12} < 1.0$ and $W_{22} = 1.0$. The heterozygote has an adaptive value equal to a homozygote, but less than the best adaptive value of the other homozygote. Natural selection is operating against a dominant allele.
3. $W_{22} < 1.0$ and $W_{12} = W_{11} = 1.0$. The heterozygote along with a homozygote has the best adaptive value which is greater than that of the other homozygote. Natural selection is operating against a recessive allele.
4. $W_{11} < W_{12} < 1.0$ and $W_{22} = 1.0$. The heterozygote has an intermediate adaptive value. Natural selection is operating without effects of dominance.
5. W_{11}, $W_{22} < 1.0$ and $W_{12} = 1.0$. The heterozygote has the best adaptive value and the two homozygotes have a lower adaptive value that may or may not be the same. Natural selection is favoring the heterozygote.
6. $W_{12} < W_{11}$, $W_{22} = 1.0$. The heterozygote has lower fitness than both the homozygotes. Only one of the homozygotes needs to have an adaptive value equal to 1.0. Natural selection is favoring the homozygotes.

Each of the five cases of natural selection results in a characteristic pattern of change in the genetic structure of a population. Cases two, three, and four are all a type of natural selection called directional selection and result in the elimination or at least great reduction in one of the alleles. Case five is very different and results in no evolutionary change once a stable equilibrium has been reached. Case six results in what looks like a directional change in allele frequency, but the allele that is selected against depends on the initial allele frequency. We will now consider how several of the above cases affect the genetic structure of a population.

SELECTION AGAINST A RECESSIVE TRAIT. Cases two, three, and four listed above are all similar in that they result in a directed change in the allele frequency of a population. This directional effect is the one that is most often associated with natural selection. The case of the peppered moth discussed earlier would fall into this category. We will discuss one of these cases in more detail because many important traits and most new mutations are recessive and have reduced fitness. When a trait is completely recessive

(case 3 above), both the heterozygote and the dominant homozygote have a fitness of 1, while the recessive homozygote has reduced fitness, as shown below.

Genotype	Fitness
AA	1
Aa	1
aa	$1 - s$

If the genotypes are initially in Hardy-Weinberg proportions, the contribution of each genotype to the next generation will be the frequency times the fitness.

AA	$p^2 \times 1 = p^2$
Aa	$2pq \times 1 = 2pq$
aa	$q^2 \times (1 - s) = q^2 - sq^2$

The mean fitness of the population is $p^2 + 2pq + q^2 - sq^2$. Since $p^2 + 2pq + q^2 = 1$, the mean fitness becomes $1 - sq^2$, and the normalized genotypic frequencies after selection are

AA	$\dfrac{p^2}{1 - sq^2}$
Aa	$\dfrac{2pq}{1 - sq^2}$
aa	$\dfrac{q^2 - sq^2}{1 - sq^2}$

To obtain q', the frequency after selection, we add the frequency of the aa homozygote and half the frequency of the heterozygote.

$$q' = \frac{q^2 - sq^2}{1 - sq^2} + \frac{1}{2} \times \frac{2pq}{1 - sq^2}$$

$$= \frac{q^2 - sq^2}{1 - sq^2} + \frac{pq}{1 - sq^2}$$

$$= \frac{q^2 - sq^2 + pq}{1 - sq^2}$$

$$= \frac{q^2 + pq - sq^2}{1 - sq^2}$$

$$= \frac{q(q + p) - sq^2}{1 - sq^2}$$

Since $(q + p) = 1$,

$$q' = \frac{q - sq^2}{1 - sq^2}$$

Therefore, the change in the frequency of a after one generation of selection is

$$\Delta q = q' - q$$

$$= \frac{q - sq^2}{1 - sq^2} - q$$

$$= \frac{q - sq^2}{1 - sq^2} - \frac{q(1 - sq^2)}{(1 - sq^2)}$$

$$= \frac{q - sq^2 - q(1 - sq^2)}{1 - sq^2}$$

$$= \frac{q - sq^2 - q + sq^3}{1 - sq^2}$$

$$= \frac{-sq^2 + sq^3}{1 - sq^2}$$

$$= \frac{-sq^2(1 - q)}{1 - sq^2}$$

So $\Delta q = -spq^2/(1 - sq^2)$ because $1 - q = p$. When $\Delta q = 0$, no further change occurs in allelic frequencies. Notice that there is a negative sign in the equation to the left of spq^2; because the values of s, p, and q are always positive or zero, Δq is negative or zero. Thus, the value of q will decrease with selection.

Selection also depends on the actual frequencies of the allele in the population. This is because the relative proportions of Aa and aa individuals at various frequencies of allele a influence how effectively selection can reduce a detrimental recessive trait. When the frequency of a recessive allele is relatively high, many homozygous recessive individuals are present in the population and will have low fitness, causing a large change in the allelic frequency. When the allelic frequency is low, however, the homozygous recessive genotype is rare, and little change in allelic frequency occurs.

Table 22.11 shows the magnitude of change in allelic frequency each generation in three populations with different initial allelic frequencies. Population 1 begins with allelic frequency q equal to 0.9, population 2 begins with q equal to 0.5, and population 3 begins with q equal to 0.1. In this example the homozygous recessive genotype (aa) has a fitness of 0 and the other two genotypes (AA and Aa) have a fitness of 1 (recessive lethal condition). When the frequency of q is high, as in population 1, the change in allelic frequency is large; in the first generation, q drops from 0.9 to 0.47. However, when q is small, as in population 3, the change in q is much less; here q drops from 0.1 to 0.091 in the first generation. So, as q becomes smaller, the change in q becomes less. Because of this diminishing change in frequency, it is virtually impossible to eliminate a recessive trait from the population entirely. This is also understood if one realizes that the final recessive alleles in a population will almost always find themselves in the heterozygote condition. This hiding of recessive alleles in the heterozygote is called a **protected polymorphism**. This result only applies, however, to completely recessive traits; if the fitness of the heterozygote is also reduced (Case 4 above), the change in allelic frequency will be more rapid, because now selection also acts against the heterozygote in addition to the homozygote. The effect of dominance on changes in allelic frequency as a result of selection is illustrated in Figure 22.20.

We have gone through a lengthy discussion of the effects of selection on a recessive trait to illustrate

~ TABLE 22.11

Effectiveness of Selection Against a Recessive Lethal Genotype at Different Initial Frequencies

| | GENOTYPE ($q^2\,aa$) AND ALLELE ($q\,a$) FREQUENCIES | | | | | |
| | POPULATION 1 | | POPULATION 2 | | POPULATION 3 | |
GENERATION	q^2	q	q^2	q	q^2	q
0	0.810	0.900	0.250	0.500	0.010	0.100
1	0.224	0.474	0.112	0.333	0.008	0.091
2	0.103	0.321	0.063	0.250	0.007	0.083
3	0.059	0.243	0.040	0.200	0.006	0.077
4	0.038	0.195	0.028	0.167	0.005	0.071
5	0.027	0.163	0.020	0.143	0.004	0.066
.						
.						
.						
10			0.007	0.08		

~ **FIGURE 22.20**

Comparison of changes in frequency of an allele with different types of dominance, holding selection coefficient constant. A_1 is the favored allele.

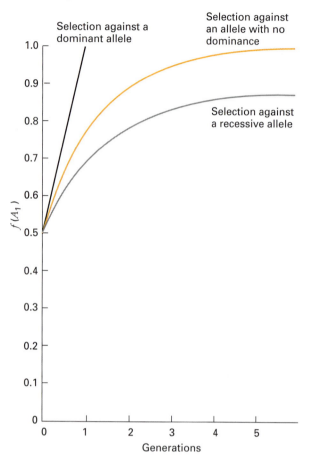

how the formula for change in allelic frequency can be derived from our general table method of allelic frequency change under selection. Similar derivations can be carried out for dominant traits and codominant traits (Cases two and four above). We will not present those derivations here, but the appropriate formulas for calculating changes in allelic frequency under different types of dominance are presented in Table 22.12. Remember, however, it is possible to calculate changes in allelic frequency for any type of trait using the table method.

HETEROZYGOTE SUPERIORITY. Natural selection does not always result in a directional change in allele frequency and a decrease in genetic variation. Some forms of selection, in fact, result in the maintenance of genetic variation and form the backbone of the balanced model of genetic variation discussed earlier. The simplest type of balancing selection is called **heterosis**, **overdominance**, or **heterozygote superiority**. An equilibrium of allelic frequencies arises when the heterozygote has higher fitness than either of the homozygotes. In this case (Case 5), both alleles are maintained in the population, because both are favored in the heterozygote genotype. Allelic frequencies will change as a result of selection until the equilibrium point is reached and then will remain stable. The allelic frequencies at which the population reaches equilibrium depend upon the relative fitnesses of the two homozygotes. If the selection coefficient of AA is s and the selection coefficient

~ **TABLE 22.12**

Formulas for Calculating Change in Allelic Frequency After One Generation of Selection

TYPE OF SELECTION	FITNESSES OF GENOTYPES			CALCULATION OF CHANGE IN ALLELIC FREQUENCY
	A^1A^1	A^1A^2	A^2A^2	
Selection against recessive homozygote	1	1	$1 - s$	$\Delta q = \dfrac{-spq^2}{1 - sq^2}$
Selection against a dominant allele	$1 - s$	$1 - s$	1	$\Delta p = \dfrac{-spq^2}{1 - s + sq^2}$
Selection with no dominance	1	$(1 - s/2)$	$1 - s$	$\Delta q = \dfrac{-spq/2}{1 - sq}$
Selection which favors the heterozygote (overdominance)	$1 - s$	1	$1 - t$	$\Delta q = \dfrac{pq(sp - tq)}{1 - sp^2 - tq^2}$
Selection against the heterozygote	1	$1 - s$	1	$\Delta q = \dfrac{spq(q - p)}{1 - 2spq}$
General	W_{11}	W_{12}	W_{22}	$\Delta q = \dfrac{pq[p(W_{12} - W_{11}) + q(W_{22} - W_{12})]}{\overline{W}}$ [a]

[a]Note: For calculation of $\overline{W}$ see Table 22.9.

of *aa* is *t*, it can be shown algebraically that at equilibrium

$$\hat{p} = f(A) = t/(s + t)$$

and

$$\hat{q} = f(a) = s/(s + t)$$

Notice that if selection against both homozygotes is the same (i.e., *s* = *t*) then the equilibrium allele frequency is 0.5. As selection against the homozygotes becomes less symmetrical the equilibrium allele frequency shifts in the direction of the most fit homozygote.

The most famous example of heterozygote superiority operating in nature is provided by human sickle-cell anemia. Sickle-cell anemia results from a mutation in the gene coding for beta hemoglobin. In some populations, there are three hemoglobin genotypes: *Hb-A/Hb-A*, *Hb-A/Hb-S*, and *Hb-S/Hb-S*. Individuals with the *Hb-A/Hb-A* genotype have completely normal red blood cells; *Hb-S/Hb-S* individuals have sickle-cell anemia; and *Hb-A/Hb-S* individuals

have sickle-cell trait, a mild form of sickle-cell anemia. In an environment in which malaria is common, the heterozygotes are at a selective advantage over the two homozygotes. Apparently, the abnormal hemoglobin mixture in the heterozygotes provides an unfavorable environment for the growth or maintenance of the malarial parasite in the red cell. The heterozygotes therefore have greater resistance to malaria and thus higher fitness than *Hb-A/Hb-A* individuals. The *Hb-S/Hb-S* individuals are at a serious selective disadvantage because they have sickle-cell anemia. As a result, in malaria-infested areas in which the *Hb-S* gene is also found, an equilibrium state is established in which a significant number of *Hb-S* alleles are found in the heterozygotes because of the selective advantage of this genotype. The distributions of malaria and the *Hb-S* allele are illustrated in Figure 22.21. Thus, despite the seriousness of this genetic disease, natural selection cannot eliminate this detrimental allele from the population because the allele has beneficial effects in the heterozygote state.

~ **FIGURE 22.21**

The distribution of malaria caused by the parasite *Plasmodium falciparum* coincides with distribution of the *Hb-S* allele for sickle-cell anemia. The frequency of *Hb-S* is high in areas where malaria is common, because *Hb-A/Hb-S* heterozygotes are resistant to malarial infection.

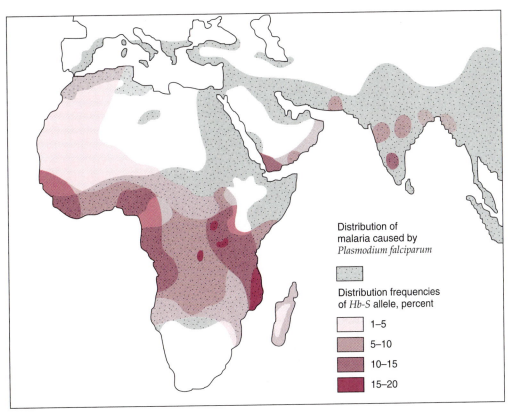

Distribution of
malaria caused by
Plasmodium falciparum

Distribution frequencies
of *Hb-S* allele, percent

1–5

5–10

10–15

15–20

KEYNOTE

Natural selection involves differential reproduction of genotypes and is measured in terms of Darwinian fitness, the relative reproductive contribution of a genotype. The effects of selection depend on the relative fitnesses of the different genotypes. Directional selection results in the directional change in allele frequency with the disfavored allele being eliminated from the population in the cases where it is dominant or codominant, but persisting in the population at low frequencies if it is invisible in the heterozygote. In either case directional selection decreases the amount of genetic variation in a population. Balancing selection, exemplified here as heterozygote superiority, results in the maintenance of genetic variation in the population.

Simultaneous Effects of Mutation and Selection

So far we have considered violations of the assumptions of the Hardy-Weinberg law one at a time. We saw that mutation, migration, small population size, and natural selection all result in changing the genetic structure of a population away from equilibrium. Population genetics theory has been extended to accommodate violations of several assumptions simultaneously; much has been learned by taking this approach. Here, as an example, we will just consider the simultaneous effects of mutation and natural selection.

As we have seen, natural selection can do many things, including reducing the frequency of a deleterious recessive allele. As the frequency of the allele becomes low, the change in frequency diminishes with each generation. When the allele is rare, the change in frequency is very slight. Opposing this reduction in the allele's frequency due to selection is mutation pressure, which will continually produce new alleles and tend to increase the frequency. Eventually a balance, or equilibrium, is reached, in which the input of new alleles by recurrent mutation is exactly counterbalanced by the loss of alleles through natural selection. When equilibrium is obtained, the allele's frequency remains stable, in spite of the fact that selection and mutation continue,

unless the equilibrium is perturbed by some other process.

Consider a population in which selection occurs against a deleterious recessive allele, a. As we saw on pp. 752–753, the amount a will change in one generation (Δq) as a result of selection is

$$\Delta q = -spq^2/(1 - sq^2)$$

For a rare recessive allele, q^2 will be near 0 and the denominator in this equation, $1 - sq^2$, will be approximately 1, so that the decrease in frequency caused by selection is given by

$$\Delta q = -spq^2$$

At the same time, the frequency of the a allele increases as a result of mutation from A to a. Provided the frequency of a is low, the reverse mutation of a to A essentially can be ignored. Equilibrium between selection and mutation occurs when the decrease in allelic frequency produced by selection is the same as the increase produced by mutation:

$$spq^2 = up$$

We can predict the frequency of a at equilibrium (q) by rearranging this equation:

$$sq^2 = u$$
$$q^2 = u/s$$

and

$$\hat{q} = \sqrt{u/s}$$

If the recessive homozygote is lethal ($s = 1$), the equation becomes

$$\hat{q} = \sqrt{u}$$

As an example of the balance between mutation and selection, consider a recessive gene for which the mutation rate is 10^{-6} and s is 0.1. At equilibrium, the frequency of the gene will be $\hat{q} = \sqrt{10^{-6}/0.1}$. Most recessive deleterious traits remain within a population at low frequency because of equilibrium between mutation and selection.

For a dominant allele A, the frequency at equilibrium ($\hat{p}$) is

$$\hat{p} = u/s$$

If the mutation rate is 10^{-6} and s is 0.1, as in the above example, the frequency of the dominant gene at equilibrium will be $10^{-6}/0.1 = 0.00001$, which is considerably less than the equilibrium frequency for a recessive allele with the same fitness and mutation rate. This is because selection cannot act on a recessive allele in the heterozygote state, whereas both the homozygote and the heterozygote for a dominant allele have reduced fitness. For this reason, detrimental dominant alleles generally are less common than recessive ones.

Nonrandom Mating

A fundamental assumption of the Hardy-Weinberg law is that members of the population mate randomly. But, many populations do not mate randomly for some traits, and when nonrandom mating occurs, the genotypes will not exist in the proportions predicted by the Hardy-Weinberg law.

One form of nonrandom mating is **positive assortative mating**, which occurs when individuals with similar phenotypes mate preferentially. Positive assortative mating is common in natural populations. For example, humans mate assortatively for height; tall men and tall women marry more frequently and short men and short women marry more frequently than would be expected on a random basis. **Negative assortative mating** occurs when phenotypically dissimilar individuals mate more often than randomly chosen individuals. If humans exhibited negative assortative mating for height, tall men and short women would marry preferentially and short men and tall women would marry preferentially. Neither positive nor negative assortative mating affects the allelic frequencies of a population, but they may influence the genotypic frequencies if the phenotypes for which assortative mating occurs are genetically determined.

Two other departures from random mating are **inbreeding** and **outbreeding**. Inbreeding involves preferential mating between close relatives, and thus, inbreeding is really positive assortative mating for relatedness. Outbreeding is preferential mating between nonrelated individuals and is a form of negative assortative mating for relatedness. Inbreeding is often measured in terms of the coefficient of inbreeding (F). The greater the value of F the greater the reduction in heterozygosity relative to that expected from the Hardy-Weinberg expectation. Thus,

$$F = \frac{\text{expected heterozygosity} - \text{observed heterozygosity}}{\text{expected heterozygosity}}$$

In random mating, $F = 0$, as observed heterozygosity and expected heterozygosity are equal. Regular systems of inbreeding exist, however, such as self-fertilization, sib mating, and mating between first cousins. After one generation of mating in such systems the value of F would be 0.5, 0.25, and 0.06, respectively.

The most extreme case of inbreeding is self-fertilization, which occurs in many plants and a few animals, such as some snails. The effects of self-fertilization are illustrated in Table 22.13. Assume that we begin with a population consisting entirely of *Aa* heterozygotes, and that all individuals in this population reproduce by self-fertilization. After one generation of self-fertilization, the progeny will consist of 1/4 *AA*, 1/2 *Aa*, and 1/4 *aa*. Now only half of the population consists of heterozygotes. When this generation undergoes self-fertilization, the *AA* homozygotes will produce only *AA* progeny, and the *aa* homozygotes will produce only *aa* progeny. When the heterozygotes reproduce, however, only half of their progeny will be heterozygous like the parents, and the other half will be homozygous (1/4 *AA* and 1/4 *aa*). This means that in each generation of self-fertilization, the percentage of heterozygotes decreases by 50 percent. After a large number of generations, there will be no heterozygotes and the population will be divided equally between the two homozygous genotypes. Note that the popu-

~ TABLE 22.13

Relative Genotype Distributions Resulting from Self-fertilization over Several Generations Starting with an *Aa* Individual

GENERATION	FREQUENCIES OF GENOTYPES		
	AA	*Aa*	*aa*
0	0	1	0
1	1/4	1/2	1/4
2	1/4 + 1/8 = 3/8	1/4	1/4 + 1/8 = 3/8
3	3/8 + 1/16 = 7/16	1/8	3/8 + 1/16 = 7/16
4	7/16 + 1/32 = 15/32	1/16	7/16 + 1/32 = 15/32
5	15/32 + 1/64 = 31/64	1/32	15/32 + 1/64 = 31/64
n	$[1 - (1/2)^n]/2$	$(1/2)^n$	$[1 - (1/2)^n]/2$
∞	1/2	0	1/2

lation was in Hardy-Weinberg proportions in the first generation after self-fertilization, but after further rounds the proportion of homozygotes is greater than that predicted by the Hardy-Weinberg law.

It should be noted that inbreeding has very similar effects to genetic drift in small populations. In both cases heterozygosity decreases and homozygosity increases. In the case of inbreeding in large populations, however, allele frequencies stay the same while homozygosity increases, whereas in the case of drift allele frequency changes while homozygosity increases.

The result of continued self-fertilization is to increase homozygosity at the expense of heterozygosity. The frequencies of alleles *A* and *a* remain constant, while the frequencies of the three genotypes change significantly. When less intensive inbreeding occurs, similar, but less pronounced, effects occur. On the other hand, outbreeding increases heterozygosity.

KEYNOTE

Inbreeding involves preferential mating between close relatives, and outbreeding is preferential mating between unrelated individuals. Continued inbreeding increases homozygosity within a population, whereas outbreeding tends to increase heterozygosity.

SUMMARY OF THE EFFECTS OF EVOLUTIONARY PROCESSES ON THE GENETIC STRUCTURE OF A POPULATION

Let us now review the major effects of the different evolutionary processes on (1) changes in allelic frequency within a population, (2) genetic divergence among populations, and (3) increases and decreases in genetic variation within populations.

Changes in Allelic Frequency Within a Population

Mutation, migration, genetic drift, and selection all have the potential to change the allelic frequencies of a population over time. Mutation, however, usually occurs at such a low rate that the change resulting from mutation pressure alone is frequently negligible. Genetic drift will produce substantial changes in allelic frequency when population size is small. Furthermore, mutation, migration, and selection may lead to equilibria, where these processes continue to act, but the allelic frequencies no longer change. Nonrandom mating does not change allelic frequencies, but it does affect the genotypic frequencies of a population: inbreeding leads to increases in homozygosity and outbreeding produces an excess of heterozygotes.

Genetic Divergence Among Populations

Several evolutionary processes lead to genetic divergence among populations. Because genetic drift is a random process, allelic frequencies in different populations may drift in different directions, so genetic drift can produce genetic divergence among populations. Migration among populations has just the opposite effect, increasing effective population size and equalizing allelic frequency differences among populations. If the population size is small, different mutations may arise in different populations, and, therefore, mutation may contribute to population differentiation. Natural selection can increase genetic differences among populations by favoring different alleles in different populations, or it can prevent divergence by keeping allelic frequencies among populations uniform. Nonrandom mating will not, by itself, generate genetic differences among populations, although it may contribute to the effects of other processes by increasing or decreasing effective population size.

Increases and Decreases in Genetic Variation Within Populations

Migration and mutation tend to increase genetic variation within populations by introducing new alleles to the gene pool. Genetic drift produces the opposite effect, decreasing genetic variation within small populations through loss of alleles. Because inbreeding leads to increases in homozygosity, it also diminishes genetic variation within populations; outbreeding, on the other hand, increases genetic variation by increasing heterozygosity. Natural selection may increase or decrease genetic variation; if one particular allele is favored, other alleles decrease in frequency and may be eliminated from the population by selection. Alternatively, natural selection may increase genetic variation within populations through overdominance and other forms of balancing selection.

In practice, these evolutionary processes never act in isolation but combine and interact in complex ways. In most natural populations, the combined effects of these processes and their interaction determine the pattern of genetic variation observed in the gene pool over time.

SUMMARY OF THE EFFECTS OF EVOLUTIONARY PROCESSES ON THE CONSERVATION OF GENETIC RESOURCES

Over the past several decades considerable attention has been drawn to the problems of extinction and biodiversity. It is estimated that there are approximately 2 million known species and as many as 30 million that are yet to be described. Species are becoming extinct at a very rapid and increasing rate with some not overly pessimistic estimates putting the value in the thousands each year. Many of the genetic principles and processes that have been discussed in this chapter relate to this conservation problem. As we have seen, populations have genetic structure; conserving this structure may warrant special attention. For example, techniques of **population viability analysis** are concerned with estimating how large a population has to be in order to keep it from going extinct for a particular period of time with a certain degree of certainty. If one wants to ensure that a population has the potential to evolve over long periods of time, an adequate gene pool must be maintained. Clearly, the genetic structure of a population and how genetic variation within populations affects the probability of extinction will require a great deal of study. The problem is particularly acute for rare and already endangered species. The effects of unintentional inbreeding of species in zoos and game management programs are diminishing as population genetic principles are being employed to manage genetic structures of populations more carefully. For example, we have seen that inbreeding, genetic drift, and selection can all decrease genetic variation, and populations may need to be maintained at certain genetic effective sizes to ensure that ample amounts of variation remain. In addition, migration between populations and how a population is subdivided geographically has been shown to have potentially large consequences on how much genetic variation is retained in a particular population or species; geographic structure in relation to genetic structure will also need to be explored further.

MOLECULAR GENETIC TECHNIQUES AND EVOLUTION

As discussed earlier, population geneticists have begun to apply molecular genetic techniques to studies of genetic variation within populations and to questions about the molecular basis of evolution. By using restriction mapping and DNA sequencing methods (see also Chapter 15), biologists can now examine evolution at the most basic genetic level, the level of the DNA. These studies have not altered the basic principles of population genetics, but they have provided a more complete and detailed picture of how natural processes produce evolutionary change.

DNA Sequence Variation

An important question in the study of evolution is how the patterns and rates of evolution differ among genes and among different parts of the same gene. This work extends work of the 1970s and 1980s that considered the rates of change of amino acid sequences in proteins of different organisms. The neutral theory of molecular evolution that was discussed earlier in this chapter was very influential in this area, and together with empirical data on the evolution of differences in proteins, gave rise to the idea that molecules can be used as "molecular clocks" by which divergence times of lineages such as humans and chimpanzees can be estimated. These methods also can be applied to DNA sequence differences. Rates of evolution of given DNA sequences can be measured by comparing sequences in two different organisms that diverged from a common ancestor at some known time in the past. We assume that the common ancestor had a single DNA sequence. Then, after the two organisms diverged from this ancestor, their DNA sequences underwent independent evolutionary changes, producing the differences we see in their sequences today. For example, it is believed that most major groups of mammals diverged from a common ancestor some 65 million years ago. Suppose we examined a particular DNA sequence, perhaps the gene for growth hormone, in mice and humans; we might find that the sequences for this gene in mouse and human differ in 20 nucleotides. These 20 nucleotides must have changed during the past 65 million years since the split between the mouse and human evolutionary lines. To calculate the rate of evolutionary change in this gene, we first compute the number of nucleotide substitutions (changes) that occurred to produce the 20 nucleotide differences we observe today. There are a number of different mathematical methods for estimating this number, which is usually expressed as nucleotide substitutions per nucleotide site, so that the rate of evolution is independent of the length of the sequences compared. To obtain rates of change, we then divide our value of nucleotide substitutions per nucleotide site by the number of years of evolutionary time that separate the two organisms. The rate of change obtained is then expressed as substitutions per nucleotide site per year. In our example with the growth hormone gene, we might find that the rate of change was 4×10^{-9} substitutions per site per year.

Studies of nucleotide sequences in numerous genes have revealed that different parts of genes evolve at different rates. Recall from our discussion of molecular genetics that a typical eukaryotic gene is made up of some nucleotides that specify the amino acid sequence of a protein (coding sequences) and other nucleotides that do not code for amino acids in a protein (noncoding sequences). Noncoding sequences include introns, leader regions, and trailer regions which are transcribed but not translated, and 5' and 3' flanking sequences that are not transcribed. Additional noncoding sequences include pseudogenes, which are nucleotide sequences that no longer produce a functional gene product because of mutations. Even within the coding regions of a functional gene, not all nucleotide substitutions produce a corresponding change in the amino acid sequence of a protein. In particular, many mutations occurring at the third position of the codon have no effect on the amino acid sequence of the protein, because such mutations produce a synonymous codon—one that codes for the same amino acid (see Chapter 14).

Different regions of the gene are apparently subject to different evolutionary processes, and evolve at different rates. Table 22.14 shows relative rates of change in different parts of mammalian genes. Within functional genes, notice that the highest rate of change involves synonymous changes in the coding sequences. The rate of synonymous nucleotide change is about five times greater than the observed rate of nonsynonymous changes. Synonymous changes do not alter the amino acid sequence of the protein. Thus, the high rate of evolutionary change seen there is not unexpected, because these changes do not affect a protein's functioning. Synonymous and nonsynonymous mutations probably arise with equal frequency. However, nonsynonymous changes that arise within coding sequences are usually detrimental to fitness and are eliminated by natural selection, whereas synonymous mutations are rarely detrimental, so they are tolerated.

You will notice in Table 22.14 that high rates of evolutionary change also occur in the 3' flanking

regions of functional genes. Like synonymous changes, sequences in the 3' flanking regions have no known effect on the amino acid sequence and usually have little effect on gene expression; most mutations that occur here will be tolerated by natural selection. Rates of change in introns are also high, but not as high as the synonymous changes and those in the 3' flanking regions. Although the sequences in the introns do not code for proteins, the intron must be properly spliced out for the mRNA to be translated into a functional protein. A few sequences within the intron are critical for proper splicing; these include the consensus sequences at the 5' and 3' ends of the intron and the sequence at the branch point (see Chapter 13, pp. 397–398). As a result, not all changes in introns will be tolerated, so the overall rate of evolution is a bit lower than that seen in synonymous coding sequences and in the 3' flanking region.

Lower rates of evolutionary change are seen in the 5' flanking region. Although this region is neither transcribed nor translated, it does contain the promoter for the gene; thus, sequences in the 5' flanking region are important for gene expression. Important consensus sequences, such as the TATA box, are found here. Mutations in consensus sequences may prevent the gene from being transcribed and thus will have detrimental effects on the fitness of the organism; natural selection eliminates these mutations and keeps change in this region low.

Next in evolutionary rate are the leader and trailer regions (see Table 22.14), which have somewhat lower rates than the 5' flanking region. Although leaders and trailers are not translated, they are tran-

~ TABLE 22.14

Relative Rates of Evolutionary Change in DNA Sequences of Mammalian Genes

SEQUENCE	NUCLEOTIDE SUBSTITUTIONS PER SITE PER YEAR ($\times 10^{-9}$)
Functional Genes	
5' Flanking region	2.36
Leader	1.74
Coding sequence—synonymous	4.65
Coding sequence—nonsynonymous	0.88
Intron	3.70
Trailer	1.88
3' Flanking region	4.46
Pseudogenes	4.85

scribed, and they provide important signals for processing the mRNA and for attachment of the ribosome to the mRNA. Nucleotide substitutions in these regions are therefore limited. The lowest rate of evolution is seen in the nonsynonymous coding sequences. Alteration of these nucleotides changes the amino acid sequence of the protein, and most mutations that occur here are eliminated by natural selection.

One final thing to note in Table 22.14 is that the highest rate of evolution is seen in nonfunctional pseudogenes. Among human globin pseudogenes, for example, the rate of nucleotide change is approximately 10 times that observed in the coding sequences of functional globin genes. The high rate of evolution observed in these sequences occurs because pseudogenes no longer code for proteins; since changes in these genes do not affect an individual's fitness, changes are not eliminated by natural selection. In summary, we observe what seems to make intuitive sense: the highest rates of evolutionary change occur in those sequences that have the least effect on function.

An interesting observation from the study of nucleotide sequences is that although the rate of evolution for synonymous nucleotide changes is much higher than that of nonsynonymous changes, rates of synonymous changes are often not quite as high as those observed in pseudogenes. This observation suggests that synonymous mutations are not completely neutral with respect to each other; natural selection may favor some synonymous codons over others. This hypothesis is reinforced by the finding that synonymous codons are not used equally throughout the coding sequences. For example, six different codons specify the amino acid leucine (UUA, UUG, CUU, CUC, CUA, and CUG), but 60 percent of the leucine codons in bacteria are CUG, and in yeast, 80 percent are UUG. Since the synonymous codons specify the same amino acid, selection must be favoring some synonymous codons over others. Remember that some synonymous codons pair with different tRNAs that carry the same amino acid. Therefore, a mutation to a synonymous codon does not change the amino acid, but it may change the tRNA. Studies of the different tRNAs reveal that within a cell, the amounts of the isoacceptor tRNAs (different tRNAs that accept the same amino acid) differ, and the most abundant tRNAs are those that pair with the most frequently used codons. Selection may favor one synonymous codon over another because the tRNA for the codon is more abundant, and translation of that codon is more efficient. Alternatively, the bonding energy between codon and anticodon of synonymous codons may differ because different bases are paired; this difference in bonding energy may be subject to natural selection.

It should be kept in mind that while differences in evolutionary rate among parts of a gene exist and can often be related to the purported function of the particular region, when we average rates over an entire gene we find that different functional genes evolve at different rates (Table 22.15). For example, the rate of nonsynonymous nucleotide substitution in the prolactin gene of mammals is over 300 times higher than that found in the histone H4 gene of mammals, an observation which parallels that made on primary protein structure (i.e., amino acid sequence). Mitochondrial DNA on the other hand has been found to be more valuable as a molecular clock due to its more uniform distribution and regular rate of accumulation of nucleotide sequence differences.

DNA Length Polymorphisms

In addition to evolution of nucleotide sequences through nucleotide substitution, variation frequently occurs in the number of nucleotides found within a gene. These variations are called DNA length polymorphisms, and they arise through deletions and insertions of relatively short stretches of nucleotides. For example, DNA length polymorphisms have been observed in the alcohol dehydrogenase gene of *Drosophila melanogaster*. In a study mentioned earlier (see p. 736), Martin Krietman sequenced 11 copies of this gene in fruit flies. In addition to extensive variation in nucleotide sequence, Krietman found six insertions and deletions in the 11 copies of the gene he examined. All of these were confined to introns and flanking regions of the DNA—none were found within exons. Insertions and deletions within exons

~ **TABLE 22.15**

Relative Rates of Evolutionary Change in DNA Sequences of Different Mammalian Genes[a]

GENE	SYNONYMOUS RATE	NONSYNONYMOUS RATE
Histone H4	0.004	1.43
Insulin	0.16	5.41
Prolactin	1.29	5.59
α-globin	0.56	3.94
β-globin	0.87	2.96
Albumin	0.92	6.72
α-fetoprotein	1.21	4.90

[a]All rates are in nucleotide substitutions per site per year $\times 10^{-9}$.

usually alter the reading frame, so they will be selected against. As a result, insertions and deletions are most commonly found in noncoding regions of the DNA. However, some insertions and deletions have been found in the coding regions of certain genes. Another class of DNA length polymorphisms involves variation in the number of copies of a particular gene. For example, among individual fruit flies, the number of copies of ribosomal genes varies extensively. The number of copies of transposons also varies extensively among individuals and is responsible for some DNA length polymorphisms.

Evolution of Multigene Families Through Gene Duplication

In eukaryotic organisms, we frequently find multiple copies of genes, all having identical or similar sequences. A group of such genes is termed a **multigene family**, which is defined as a set of related genes that have evolved from some ancestral gene through the process of gene duplication. Members of a gene family may be clustered together or dispersed on different chromosomes.

An example of a multigene family is the globin gene family, which consists of the genes that code for the polypeptide chains making up the hemoglobin molecule. The organization of this multigene family in humans was discussed in Chapter 17. The globin multigene family in humans is comprised of seven α-like genes found on chromosome 16 and six β-like genes found on chromosome 11. Globin genes are also found in other animals, and globin-like genes are even found in plants, suggesting that this is a very ancient gene family. Almost all functional globin genes in animal species have the same general structure, consisting of three exons separated by two introns. However, the numbers of globin genes and their order varies among species, as is shown for the β-like genes in Figure 22.22. Because all globin genes have similarities in structure and sequence, it appears that an ancestral globin gene (perhaps most like the present day myoglobin gene) duplicated and diverged to produce an ancestral α-like gene and an ancestral β-like gene. These two genes then underwent repeated duplications, giving rise to the various α-like and β-like genes found in vertebrates today. Repeated gene duplication, such as that giving rise to the globin gene family, appears to be a frequent evolutionary occurrence. Indeed, the number of copies of globin genes varies in some human populations. For example, most humans have two α-globin genes on chromosome 16, as shown in Figure 17.21 (p. 562). However, some individuals have a single α-globin gene on chromosome 16; other individuals have three or even four copies of the α-globin gene on one of their chromosomes. These observations indicate that duplication and deletion of genes in a multigene family are constant, ongoing processes. Gene duplications and deletions often arise as a result of misalignment of sequences during crossing over, a process termed *unequal crossing over* (see Chapter 7). Duplications may also arise through transposition (see Chapter 20). Following gene duplication, the separate copies of a gene may undergo changes in sequence. In some cases, mutations arise that render a copy of the gene nonfunctional, creating a pseudogene. In other cases, the change of nucleotide sequence may lead to different functions for the protein product of a gene.

~ **FIGURE 22.22**

Organization of the globin gene families in several mammalian species.

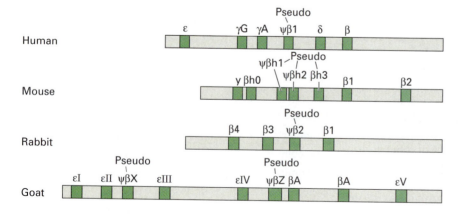

Evolution in Mitochondrial DNA Sequences

In Chapter 20 we discussed the mitochondrial genome. We saw that animal mitochondrial DNA (mtDNA) is comprised of approximately 15,000 base pairs and that this DNA encodes 2 rRNAs, 22 to 23 tRNAs, and 10 to 12 proteins. A number of recent studies have examined sequence variation in mtDNA. These studies reveal that mtDNA evolution differs from evolution typically observed in nuclear DNA. For example, nucleotide sequences within animal mtDNA evolve at a faster rate than coding sequences in nuclear genes. In fact, the mtDNA of mammals undergoes evolutionary change at a rate that is 5 to 10 times faster than that typically observed in mammalian nuclear DNA. However, the rapid rate of evolution observed in animal mtDNA is not universal for all eukaryotes. For unknown reasons, plant mtDNA appears to evolve at an even slower rate than plant nuclear DNA.

It is not entirely clear why animal mtDNA undergoes rapid evolutionary change, but the reason may be related to a higher mutation rate in mtDNA sequences. Higher mutation rates could occur in mtDNA because DNA repair mechanisms are lacking or because the DNA polymerase involved in mtDNA replication is more prone to errors. Another possible reason for the higher rate of evolution in animal mtDNA is that the selection pressure that normally eliminates many mutations in nuclear genes is relaxed in the mitochondria. In other words, changes in the proteins, tRNAs, and rRNAs encoded by the mtDNA sequences might be less detrimental to individual fitness than similar changes in the proteins, tRNAs, and rRNAs encoded by nuclear genes.

Whatever the cause of accelerated evolution of mtDNA, the fact that these sequences change rapidly makes them ideal for assessing evolutionary relationships among groups of closely related organisms. For example, analyses of mtDNA sequences have been used to investigate the relationships among humans, chimpanzees, gorillas, orangutans, and gibbons, a group whose precise evolutionary relationships have been controversial.

Mitochondrial DNA also differs from nuclear DNA in that all mtDNA is inherited clonally from the mother; mitochondria are located in the cytoplasm and only the egg cell contributes cytoplasm to a zygote. As a consequence, mtDNA does not undergo meiosis, and all offspring should be identical to the maternal genotype for mtDNA sequences (the offspring are clones for mtDNA genes). This pattern of inheritance allows matriarchal lineages (descendants

from one female) to be traced, and provides a means for examining family structure in some populations. This in conjunction with the rapid and regular rate of accumulation of nucleotide sequence differences have allowed mtDNA to become a valuable tool for comparing closely related lineages. An example of geographic variation in mtDNA sequences is shown in Figure 22.23.

Concerted Evolution

One of the surprising findings from studies of DNA sequence variation is the observation that some molecular force, or processes, maintain uniformity of sequence in multiple copies of a gene. This phenomenon has been termed **concerted evolution** or **molecular drive**. As we discussed in previous chapters, some genes in eukaryotes exist in multiple copies. Ribosomal RNA genes in complex organisms, for example, typically exist in hundreds or thousands of copies. Undoubtedly, these multiple copies arose through duplication. Following duplication, individual copies of a gene might be expected to acquire mutations and diverge. Selection might limit mutations in coding regions, but if many copies exist, we would expect some divergence to occur, especially in the noncoding sequences. Contrary to this expectation, numerous studies have revealed that nucleotide sequences in the different copies of a gene are fre-

~ **FIGURE 22.23**

The geographic distribution of mitochondrial DNA phenotypes in the pocket gopher, *Geomys pinetis*, in the southeastern United States, showing geographic separation of mitochondrial lineages. *Bam*HI M and *Bam*HI N are different restriction patterns revealed when the mitochondrial DNA is cleaved by the enzyme *Bam*HI.

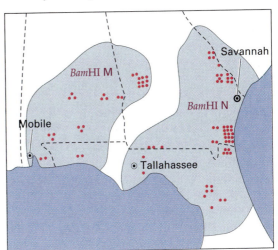

quently quite homogeneous. Furthermore, the non-coding sequences are also homogeneous, which suggests that natural selection is not responsible. When the same genes are examined in a second, closely related species, that group's sequences are also homogeneous, but are frequently different from the homogeneous sequence found in the first species.

These observations have led to the conclusion that some molecular process continually enforces uniformity among multiple copies of the same sequence within a species. At the same time, the process allows for rapid differentiation among species. The mechanism of concerted evolution is not fully understood, but concerted evolution has important consequences for how genes evolve, and it represents an evolutionary force that was unknown before the application of modern molecular techniques to population genetics.

Evolutionary Relationships Revealed by RNA and Nuclear DNA Sequences

Within the past few years, molecular genetics has provided powerful tools for deciphering the evolutionary history of life. Because evolution is defined as genetic change, genetic relationships are of primary importance in the construction of evolutionary trees. However, until recently, it was impossible to examine the genes directly. In the past, evolutionary biologists were forced to rely entirely on comparison of phenotypes to infer genetic similarities and differences. They assumed that if the phenotypes were similar, the genes coding for the phenotypes were similar; if the phenotypes were different, the genes were different. Thus, phenotypes were used for evolutionary studies. Originally the phenotypes examined consisted largely of gross anatomical features. Later, behavioral, ultrastructural, and biochemical characteristics were also studied. Comparisons of such traits were successfully used to construct evolutionary trees for many groups of plants and animals, and, indeed, are still the basis of most evolutionary studies today.

However, relying on the study of such traits has limitations. Sometimes, similar phenotypes can evolve in organisms that are distantly related. For example, if a naive biologist tried to construct an evolutionary tree on the basis of whether wings were present or absent, he might place birds, bats, and insects in the same evolutionary group, since all have wings. In this particular case, it is fairly obvious that these three organisms are not closely related—they differ in many features besides presence of wings, and the wings themselves are very different. But this extreme example shows that phenotypes can sometimes be misleading about evolutionary relationships, and

phenotypic similarities do not necessarily reflect genetic similarities.

Another problem with relying on a comparison of phenotypes is that not all organisms have a number of easily studied phenotypic features. For example, the study of evolutionary relationships among bacteria has always been problematic, because bacteria have few obvious traits that correlate with the degree of genetic relatedness. A third problem arises when we try to compare distantly related organisms. How do we compare, for example, bacteria and mammals, which have very few traits in common?

We have seen how DNA sequencing methods and analysis of restriction fragment length polymorphisms provide information about DNA sequences. DNA sequences provide the most accurate and reliable information upon which to infer evolutionary relationships. They allow direct comparison of the genetic differences among organisms, they are easily quantified, and all organisms have them (all organisms have at least some genes in common, such as tRNA genes, rRNA genes, and genes for a few proteins). Because of these considerable advantages, many evolutionary biologists have turned to DNA sequences for assessing evolutionary relationships and for constructing evolutionary trees.

One case where sequence data has provided new information about evolutionary relationships is in our understanding of the primary divisions of life. Many years ago, biologists divided all of life into two major groups, the plants and the animals. As more organisms were discovered and their features examined in more detail, this simple dichotomy became unworkable. It was later recognized that organisms can be divided into prokaryotes and eukaryotes on the basis of cell structure. More recently, several primary divisions of life have been recognized, such as the five kingdoms (prokaryotes, protista, plants, fungi, and animals) proposed by Whittaker.

In recent years, RNA and DNA sequences have been used to uncover the primary lines of evolutionary history among all organisms. In one study, Norman Pace and his colleagues constructed an evolutionary tree of life based upon the sequences found in the 16S-like rRNA, which all organisms possess. As illustrated in Figure 22.24, their evolutionary tree revealed three major evolutionary groups: the Bacteria (the traditional prokaryotes), the Eukarya (eukaryotes), and the Archaea (a relatively little known group of bacteria). Bacteria and Archaea, although both prokaryotic, were found to be as different genetically as Bacteria and Eukarya. The deep evolutionary differences that separate the Bacteria and the Archaea were not obvious on the basis of phenotype; this became clear only after their nucleotide sequences

~ FIGURE 22.24

An evolutionary tree of life revealed by comparison of 16S rRNA sequences. (Reprinted with permission from N. Pace, "A Molecular View of Microbial Diversity in the Biosphere" in *Science* 276 (1997):735. Copyright © 1997 American Association for the Advancement of Science.)

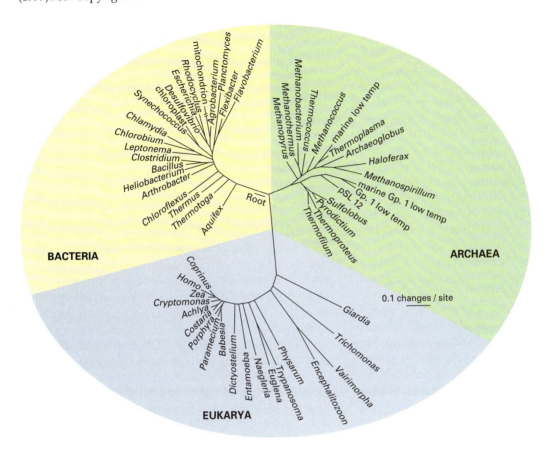

were compared. Sequences of other genes, including 5S rRNAs, large rRNAs, and genes for some basic proteins, support the idea that three major evolutionary groups exist among living organisms.

Another field in which DNA sequences are being used to study evolutionary relationships is human evolution. In contrast to the extensive variation that is observed in size, body shape, facial features, skin color, etc., genetic differences among human populations are relatively small. For example, analysis of mtDNA sequences shows that the mean difference in sequence between two human populations is about 0.33 percent. Other primates exhibit much larger differences. For example, the two subspecies of orangutan differ by as much as 5 percent. This indicates that all human groups are closely related. Nevertheless, some genetic differences do occur among different human groups. Surprisingly, the greatest differences are not found between populations located on different continents, but are seen among human populations residing in Africa. All other human populations show fewer differences than we find among the African populations. Many experts interpret these findings to mean that humans experienced their origin and early evolutionary diver-

gence in Africa. After a number of genetically differentiated populations had evolved in Africa, it is hypothesized that a small group of humans may have migrated out of Africa and given rise to all other human populations. This hypothesis has been termed the "out of Africa" theory. Sequence data from both mitochondrial DNA and nuclear genes are consistent with this hypothesis. Although the out-of-Africa theory is not universally accepted, DNA sequence data is playing an increasingly important role in the study of human evolution and indeed in the study of the evolution of many lineages.

SUMMARY

Population genetics is the study of the genetic structure of populations and species and how the structure changes or evolves over time. The gene pool of a population is the total of all genes within a Mendelian population, and it is described in terms of the allelic and genotypic frequencies. The Hardy-Weinberg law describes what happens to allelic and genotypic frequencies of a large, randomly mating population free from evolutionary processes; when these conditions are met, allelic frequencies do not change, and genotypic frequencies stabilize after one generation in the proportions p^2, $2pq$, q^2, where p and q equal the allelic frequencies of the population.

The classical, balance, and neutral mutation models have generated testable hypotheses which help explain how much genetic variation should exist within natural populations and what processes are responsible for the variation observed. Protein electrophoresis showed that most populations of plants and animals contain large amounts of genetic variation, proving that the classical model was wrong. However, attempts to determine whether most genetic variation is maintained by natural selection (the balance model) or by the neutral processes of genetic drift and mutation (the neutral mutation model) have provided no clear answer, tending to indicate that the evolution of genetic structure proceeds as a result of many processes sometimes acting simultaneously and sometimes in sequence.

Mutation, genetic drift, migration, and natural selection are processes that can alter allelic frequencies of a population. Recurrent mutation changes the allelic frequencies of a population; the relative rates of forward and reverse mutation will determine allelic frequencies of a population in the absence of other processes. Genetic drift, chance change in allelic frequencies due to small effective population size, leads to a loss of genetic variation within a population, genetic divergence among populations, and change

of allelic frequency within a population. Migration tends to reduce genetic divergence among populations and increases effective population size. Natural selection is differential reproduction of genotypes. The relative reproductive contribution of genotypes is measured in terms of Darwinian fitness. The effects of natural selection depend upon the fitnesses of the genotypes, the degree of dominance, and the frequencies of the alleles in the population. Nonrandom mating affects the effective population size and genotypic frequencies of a population; the allelic frequencies are unaffected. One type of nonrandom mating, inbreeding, leads to an increase in homozygosity.

New techniques of molecular genetics, including analysis of restriction fragment length polymorphisms and RNA and DNA sequences, have supported prior insights obtained from analyses of proteins with respect to evolutionary processes. Different parts of a gene are found to evolve at different rates; those parts of the gene that have the least effect on fitness appear to evolve at the highest rates. In addition to changes in nucleotide sequences, molecular evolution involves variation in DNA length polymorphisms. Multigene families evolve through repeated duplication of genes, followed by genetic divergence of their sequences. Mitochondrial DNA of animals appears to evolve at a faster and more uniform rate than nuclear DNA. RNA and DNA sequences can be used for inferring evolutionary rates and relationships among organisms.

ANALYTICAL APPROACHES FOR SOLVING GENETICS PROBLEMS

Q22.1 In a population of 2,000 gaboon vipers, a genetic difference with respect to venom exists at a single locus. The alleles are incompletely dominant. The population shows 100 individuals homozygous for the t allele (genotype tt, nonpoisonous), 800 heterozygous (genotype Tt, mildly poisonous), and 1,100 homozygous for the T allele (genotype TT, deadly poisonous).

a. What is the frequency of the t allele in the population?

b. Are the genotypes in Hardy-Weinberg equilibrium?

A22.1 This question addresses the basics of calculating allelic frequencies and relating them to the genotype frequencies expected of a population in Hardy-Weinberg equilibrium.

a. The t frequency can be calculated from the information given, since the trait is an incompletely dominant one. There are 2,000 individuals in the popula-

tion under study, meaning a total of 4,000 alleles at the T/t locus. The number of t alleles is given by

$$(2 \times tt \text{ homozygotes}) + (1 \times Tt \text{ heterozygotes})$$
$$= (2 \times 100) + (1 \times 800) = 1,000$$

This calculation is straightforward, since both alleles in the nonpoisonous snakes are t, while only one of the two alleles in the mildly poisonous snakes is t. Since the total number of alleles under study is 4,000, the frequency of t alleles is $1,000/4,000 = 0.25$. This system is a two-allele system, so the frequency of T must be 0.75.

b. For the genotypes to be in Hardy-Weinberg equilibrium, the distribution must be p^2 TT + $2pq$ Tt + q^2 tt genotypes, where p is the frequency of the T allele and q is the frequency of the t allele. In part a we established that the frequency of T is 0.75 and the frequency of t is 0.25. Therefore $p = 0.75$ and $q = 0.25$. Using these values, we can determine the expected genotype frequencies if this population is in Hardy-Weinberg equilibrium:

$$(0.75)^2\ TT + 2(0.75)(0.25)\ Tt + (0.25)^2\ tt$$

This expression gives 0.5625 TT + 0.3750 Tt + 0.0625 tt. Thus with 2,000 individuals in the population we would expect 1,125 TT, 750 Tt, and 125 tt. These values are close to the values given in the question, suggesting that the population is indeed in genetic equilibrium.

To check this result, we should perform a chi-square analysis (see Chapter 5), using the given numbers (not frequencies) of the three genotypes as the observed numbers and the calculated numbers as the expected numbers. The chi-square analysis is as follows, where $d =$ (observed − expected):

Genotype	Observed	Expected	d	d^2	d^2/e
TT	1,100	1,125	−25	625	0.556
Tt	800	750	+50	2,500	3.334
tt	100	125	−25	625	5.000
Totals	2,000	2,000	0		8.890

Thus the chi-square value (i.e., the sum of all the d^2/e values) is 8.89. For the reasons discussed in the text for a similar example, there is only one degree of freedom. Looking up the chi-square value in the chi-square table (Table 2.5 p. 37), we find a P value of approximately 0.0025. So about 25 times out of 10,000 we would expect chance deviations of the magnitude observed. In other words, our hypothesis that the population is in Hardy-Weinberg equilibrium is not substantiated. In this case our guess that it was in equilibrium was inaccurate. Nonetheless, the population is not greatly removed from an equilibrium state.

Q22.2 About one normal allele in 30,000 mutates to the X-linked recessive allele for hemophilia in each human generation. Assume for the purposes of this problem that one h allele in 300,000 mutates to the normal alternative in each generation. (Note that in reality it is difficult to measure the reverse mutation of a human recessive allele that is essentially lethal, like the allele for hemophilia.) The mutation frequencies are indicated in the following equation:

$$h^+ \xrightarrow[\xleftarrow{\;v\;}]{\;u\;} h$$

where $u = 10v$. What allelic frequencies would prevail at equilibrium under mutation pressures alone in these circumstances?

A22.2 This question seeks to test our understanding of the effects of mutation on allelic frequencies. In the chapter we discussed the consequences of mutation pressure. The conclusion was that if A mutates to a at n times the frequency that a mutates back to A, then at equilibrium the value of q will be $\hat{q} = u/(u + v)$ or $\hat{q} = nv/(n + 1)v$. Applying this general derivation to this particular problem, we simply use the values given. We are told that the forward mutation rate is 10 times the reverse mutation rate, or $u = 10v$. At equilibrium the value of q will be $\hat{q} = u/(u + v)$. Since $u = 10v$, this equation becomes $\hat{q} = 10v/11v$, so $q = 10/11$, or 0.909. Therefore, at equilibrium brought about by mutation pressures the frequency of h (the hemophilia allele) will be 0.909, and the frequency of h^+ (the normal allele) will be $\hat{p}$, that is, $(1 − \hat{q}) = (1 − 0.909) = 0.091$.

QUESTIONS AND PROBLEMS

*22.1 In the European land snail, *Cepaea nemoralis*, multiple alleles at a single locus determine shell color. The allele for brown (C^B) is dominant to the allele for pink (C^P) and to the allele for yellow (C^Y). Pink is recessive to brown, but is dominant to yellow, and yellow is recessive to pink and brown. Thus, the dominance hierarchy among these alleles is $C^B > C^P > C^Y$. In one population of *Cepaea*, the following color phenotypes were recorded:

Brown	236
Pink	231
Yellow	33
Total	500

Assuming that this population is in Hardy-Weinberg equilibrium (large, randomly mating, and free from evolutionary processes), calculate the frequencies of the C^B, C^Y, and C^P alleles.

22.2 Three alleles are found at a locus coding for malate dehydrogenase (MDH) in the spotted chorus frog. Chorus frogs were collected from a breeding pond, and each frog's genotype at the MDH locus was determined with electrophoresis. The following numbers of genotypes were found:

M^1M^1	8
M^1M^2	35
M^2M^2	20
M^1M^3	53
M^2M^3	76
M^3M^3	62
Total	254

a. Calculate the frequencies of the M^1, M^2, and M^3 alleles in this population.

b. Using a chi-square test, determine whether the MDH genotypes in this population are in Hardy-Weinberg proportions.

22.3 In a large interbreeding population 81 percent of the individuals are homozygous for a recessive character. In the absence of mutation or selection, what percentage of the next generation would be homozygous recessives? Homozygous dominants? Heterozygotes?

*22.4 Let A and a represent dominant and recessive alleles whose respective frequencies are p and q in a given interbreeding population at equilibrium (with $p + q = 1$).

a. If 16 percent of the individuals in the population have recessive phenotypes, what percentage of the total number of recessive genes exist in the heterozygous condition?
b. If 1.0 percent of the individuals were homozygous recessive, what percentage of the recessive genes would occur in heterozygotes?

*22.5 A population has eight times as many heterozygotes as homozygous recessives. What is the frequency of the recessive gene?

22.6 In a large population of range cattle the following ratios are observed: 49 percent red (RR), 42 percent roan (Rr), and 9 percent white (rr).
a. What percentage of the gametes that give rise to the next generation of cattle in this population will contain allele R?
b. In another cattle population only 1 percent of the animals are white and 99 percent are either red or roan. What is the percentage of r alleles in this case?

22.7 In a gene pool the alleles A and a have initial frequencies of p and q, respectively. Prove that the allelic frequencies and zygotic frequencies do not change from generation to generation as long as there is no selection, mutation, or migration, the population is large, and the individuals mate at random.

*22.8 The S-s antigen system in humans is controlled by two codominant alleles, S and s. In a group of 3,146 individuals the following genotypic frequencies were found: 188 SS, 717 Ss, and 2,241 ss.
a. Calculate the frequency of the S and s alleles.
b. Determine whether the genotypic frequencies conform to the Hardy-Weinberg equilibrium by using the chi-square test.

22.9 Refer to Problem 22.8. A third allele is sometimes found at the S locus. This allele S^u is recessive to both the S and the s alleles and can only be detected in the homozygous state. If the frequencies of the alleles S, s, and S^u are p, q, and r, respectively, what would be the expected frequencies of the phenotypes S–, Ss, s–, and $S^u S^u$?

22.10 In a large interbreeding human population 60 percent of individuals belong to blood group O (genotype i/i). Assuming negligible mutation and no selective advantage of one blood type over another, what percentage of the grandchildren of the present population will be type O?

*22.11 A selectively neutral, recessive character appears in 0.40 of the males and in 0.16 of the females in a randomly interbreeding population. What is the gene's frequency? How many females are heterozygous for it? How many males are heterozygous for it?

22.12 Suppose you found two distinguishable types of individuals in wild populations of some organism in the following frequencies:

	TYPE 1	TYPE 2
Females	99%	1%
Males	90%	10%

The difference is known to be inherited. What is its genetic basis?

*22.13 Red-green color blindness is due to a sex-linked recessive gene. About 64 women out of 10,000 are color-blind. What proportion of men would be expected to show the trait if mating is random?

22.14 About 8 percent of the men in a population are red-green color-blind (owing to a sex-linked recessive gene). Answer the following questions, assuming random mating in the population, with respect to color blindness.
a. What percentage of women would be expected to be color-blind?
b. What percentage of women would be expected to be heterozygous?
c. What percentage of men would be expected to have normal vision two generations later?

22.15 List some of the basic differences in the classical, balance, and neutral-mutation models of genetic variation.

*22.16 Two alleles of a locus, A and a, can be interconverted by mutation:

$$A \xrightarrow{u} \xleftarrow{v} a$$

and u is a mutation rate of 6.0×10^{-7}, and v is a mutation rate of 6.0×10^{-8}. What will be the frequencies of A and a at mutational equilibrium, assuming no selective difference, no migration, and no random fluctuation caused by genetic drift?

22.17a. Calculate the effective population size (N_e) for a breeding population of 50 adult males and 50 adult females.
b. Calculate the effective population size (N_e) for a breeding population of 60 adult males and 40 adult females.

c. Calculate the effective population size (N_e) for a breeding population of 10 adult males and 90 adult females.

d. Calculate the effective population size (N_e) for a breeding population of 2 adult males and 98 adult females.

22.18 In a population of 40 adult males and 40 adult females, the frequency of allele A is 0.6 and the frequency of allele a is 0.4.

a. Calculate the 95 percent confidence limits of the allelic frequency for A.

b. Another population with the same allelic frequencies consists of only 4 adult males and 4 adult females. Calculate the 95 percent confidence limits of the allelic frequency for A in this population.

c. What are the 95 percent confidence limits of A if the population consists of 76 females and 4 males?

22.19 The land snail *Cepaea nemoralis* is native to Europe but has been accidentally introduced into North America at several localities. These introductions occurred when a few snails were inadvertently transported on plants, building supplies, soil, or other cargo. The snails subsequently multiplied and established large, viable populations in North America.

Assume that today the average size of *Cepaea* populations found in North America is equal to the average size of *Cepaea* populations in Europe. What predictions can you make about the amounts of genetic variation present in European and North American populations of *Cepaea*? Explain your reasoning.

*22.20 A population of 80 adult squirrels resides on campus, and the frequency of the *Est¹* allele among these squirrels is 0.70. Another population of squirrels is found in a nearby woods, and there, the frequency of the *Est¹* allele is 0.5. During a severe winter, 20 of the squirrels from the woods population migrate to campus in search of food and join the campus population. What will be the allelic frequency of *Est¹* in the campus population after migration?

22.21 Upon sampling three populations and determining genotypes, you find the following three genotype distributions. What would each of these distributions imply with regard to selective advantages of population structure?

POPULATION	AA	Aa	aa
1	0.04	0.32	0.64
2	0.12	0.87	0.01
3	0.45	0.10	0.45

22.22 The frequency of two adaptively neutral alleles in a large population is 70 percent A : 30 percent a. The population is wiped out by an epidemic, leaving only four individuals, who produce many offspring. What is the probability that the population several years later will be 100 percent AA? (Assume no mutations.)

*22.23 A completely recessive gene, owing to changed environmental circumstances, becomes lethal in a certain population. It was previously neutral, and its frequency was 0.5.

a. What was the genotype distribution when the recessive genotype was not selected against?

b. What will be the allelic frequency after one generation in the altered environment?

c. What will be the allelic frequency after two generations?

22.24 Human individuals homozygous for a certain recessive autosomal gene die before reaching reproductive age. In spite of this removal of all affected individuals, there is no indication that homozygotes occur less frequently in succeeding generations. To what might you attribute the constant rate of appearance of recessives?

*22.25 A completely recessive gene (Q^1) has a frequency of 0.7 in a large population, and the Q^1Q^1 homozygote has a relative fitness of 0.6.

a. What will be the frequency of Q^1 after one generation of selection?

b. If there is no dominance at this locus (the fitness of the heterozygote is intermediate to the fitnesses of the homozygotes), what will the allelic frequency be after one generation of selection?

c. If Q^1 is dominant, what will the allelic frequency be after one generation of selection?

22.26 As discussed earlier in this chapter, the gene for sickle-cell anemia exhibits overdominance. An individual who is an *Hb-A/Hb-S* heterozygote has increased resistance to malaria and therefore has greater fitness than the *Hb-A/Hb-A* homozygote, who is susceptible to malaria, and the *Hb-S/Hb-S* homozygote, who has sickle-cell anemia. Suppose that the fitness values of the genotypes in Africa are as presented below:

$$Hb\text{-}A/Hb\text{-}A = 0.88$$
$$Hb\text{-}A/Hb\text{-}S = 1.00$$
$$Hb\text{-}S/Hb\text{-}S = 0.14$$

Give the expected equilibrium frequencies of the sickle-cell gene (*Hb-S*).

*22.27 Achondroplasia, a type of dwarfism in humans, is caused by an autosomal dominant gene. The mutation rate for achondroplasia is about 5×10^{-5} and the fitness of achondroplastic dwarfs has been estimated to be about 0.2, compared with unaffected individuals. What is the equilibrium frequency of the achondroplasia gene based on this mutation rate and fitness value?

22.28 To answer the following questions, consider the spontaneous mutation frequencies tabulated in Table 22.5.

a. In humans, why is the frequency of forward mutations to neurofibromatosis an order of magnitude larger than that for the other human diseases?

b. In *E. coli*, why is the frequency of mutations to arabinose dependence two to four orders of magnitude larger than the frequency of mutations to leucine, arginine, or tryptophan independence?

c. What factors influence the spontaneous mutation frequency for a specific trait?

22.29 The frequencies of the L^M and L^N blood group alleles are the same in each of the populations I, II, and III, but the genotypes' frequencies are not the same, as shown below. Which of the populations is most likely to show each of the following characteristics: random mating, inbreeding, genetic drift. Explain your answers.

	$M - M$	$M - N$	$N - N$
I	0.50	0.40	0.10
II	0.49	0.42	0.09
III	0.45	0.50	0.05

*22.30 DNA was collected from 100 people randomly sampled from a given human population and was digested with the restriction enzyme *Bam*HI, the fragments were separated by electrophoresis, and then transferred to a membrane filter using the Southern blot technique. The blots were probed with a particular cloned sequence. Three different patterns of hybridization were seen on the blots. Some DNA samples (56 of them) showed a single band of 6.3 kb, others (6) showed a single band at 4.1 kb, and yet others (38) showed both the 6.3 and the 4.1 kb bands.

a. Interpret these results in terms of *Bam*HI sites.

b. What are the frequencies of the restriction site alleles?

c. Does this population appear to be in Hardy-Weinberg equilibrium for the relevant restriction site(s)?

22.31 DNA was isolated from 10 nine-banded armadillos and cut with the restriction enzyme *Hind*III. *Hind*III recognizes the six-base sequence 5'- AAGCTT - 3' 3'- TTCGAA - 5'. The DNA fragments that resulted from the restriction reaction were separated with agarose electrophoresis and transferred to nitrocellulose using Southern blotting. A labeled probe for the β-hemoglobin gene was added, which resulted in the following set of restriction patterns. Note: +/+ indicates that the restriction site was present on both chromosomes of the individual, +/− indicates that the restriction site was present on one chromosome and absent on one chromosome of the individual, and −/− indicates that the restriction site was absent on both chromosomes of the individual:

+/+ −/− +/− +/− −/− +/− +/+ +/− −/− +/+
——   ——   ——   ——   ——   ——   ——   ——   ——   ——

——   ——   ——   ——   ——   ——

——   ——   ——   ——   ——   ——

Calculate the expected heterozygosity in nucleotide sequence.

*22.32 Fifty tiger salamanders from one pond in west Texas were examined for genetic variation by using the technique of protein electrophoresis. The genotype of each salamander was determined for five loci (AmPep, ADH, PGM, MDH, and LDH-1). No variation was found at AmPep, ADH, and LDH-1; in other words, all individuals were homozygous for the same allele at these loci. The following numbers of genotypes were observed at the MDH and PGM loci.

MDH GENOTYPES	NUMBER OF INDIVIDUALS	PGM GENOTYPES	NUMBER OF INDIVIDUALS
AA	11	DD	35
AB	35	DE	10
BB	4	EE	5

Calculate the proportion of polymorphic loci and the heterozygosity for this population.

*22.33 The success of a population depends in part on its reproductive rate, which may be affected by low genetic variability. Vyse and his colleagues have been interested in the conservation of small populations of grizzly bears, studying 304 members of 30 grizzly bear family groups in a population in northwestern Alaska. They have identified a set of polymorphic loci with these alleles, allele frequencies, and observed heterozygosities (obs. het.):

Locus *G1A*		Locus *G10X*		Locus *G10C*		Locus *G10L*	
Obs. het.: 0.776		Obs. het.: 0.783		Obs. het.: 0.770		Obs. het.: 0.651	
Allele	Freq.	Allele	Freq.	Allele	Freq.	Allele	Freq.
A194	0.398	X137	0.395	C105	0.355	L155	0.487
A184	0.240	X135	0.211	C103	0.257	L157	0.276
A192	0.211	X141	0.211	C111	0.240	L161	0.128
A180	0.086	X133	0.102	C113	0.092	L159	0.089
A190	0.036	X131	0.053	C107	0.043	L171	0.013
A200	0.016	X129	0.030	C101	0.010	L163	0.007
A186	0.007				C109	0.003		
A188	0.006							

The genotypes of a mother bear and her three cubs are shown here.

MOTHER	CUB #1	CUB #2	CUB #3
A184, A192	A184, A194	A184, A192	A184, A194
X135, X137	X135, X137	X133, X135	X137, X141
C105, C113	C105, C111	C105, C105	C111, C113
L155, L159	L155, L157	L159, L161	L155, L155

a. How do the observed heterozygosities compare to the expected heterozygosities? On the basis of this information, can you tell whether this grizzly bear population is in Hardy-Weinberg equilibrium?

b. What can you infer about the paternity of the mother's three cubs? How might paternity information affect the genetic variability and the effective population size in this population of grizzly bears?

22.34 What factors cause genetic drift?

*22.35 What are the primary effects of the following evolutionary processes on the gene and genotypic frequencies of a population?

a. mutation
b. migration
c. genetic drift
d. inbreeding

22.36 Explain how overdominance leads to an increased frequency of sickle-cell anemia in areas where malaria is widespread.

22.37 Since 1968, Pinter has studied the population dynamics of the montane vole, a small rodent in the Grand Teton mountains in Wyoming. For more than twenty-five years, severe periodic fluctuations in population density have been negatively correlated with precipitation levels: vole density sharply declines every few years when spring precipitation is extremely high.

a. Propose several hypotheses concerning the genetic structure of the population of montane voles in two separate sampling sites if
   i. there is negligible migration of voles between sampling sites
   ii. there is substantial migration of voles between sampling sites

b. How would you gather data to evaluate these hypotheses?

*22.38 Suppose we examine the rates of nucleotide substitution in two 300-nucleotide sequences of DNA isolated from humans. In the first sequence (sequence A), we find a nucleotide substitution rate of 4.88×10^{-9} substitutions per site per year. The substitution rate is the same for synonymous and nonsynonymous substitutions. In the second sequence (sequence B), we find a synonymous substitution rate of 4.66×10^{-9} substitutions per site per year and a nonsynonymous substitution rate of 0.70×10^{-9} substitutions per site per year. Referring to

Table 22.14 (p. 760), what might you conclude about the possible functions of sequence A and sequence B?

22.39 What are some of the characteristics of mitochondrial DNA evolution in animals?

22.40 Since the non-recombining portion of the relatively small, gene-sparse Y chromosome is inherited solely through males, it provides a record of the mutational events that have occurred along male lineages through evolution. It is thus the male counterpart of maternally inherited mitochondrial DNA. Unlike mitochondrial DNA, however, there is a relative paucity of DNA sequence polymorphisms on the human Y chromosome: in the ethnically diverse subjects studied, only one base substitution was found in 1,400 bp of DNA sequenced from 12 males, only 3 base substitutions were found in 15,680 bp of DNA sequenced in 5 males, and no variation was found in a 729 bp intron in the *ZFY* gene sequenced in 38 males.

a. Why was the lack of DNA sequence polymorphism surprising?

b. How might this finding be explained?

22.41 What is concerted evolution?

22.42 What are some of the advantages of using DNA sequences to infer evolutionary relationships?

*22.43 Multiple, geographically isolated populations of tortoises exist on the Galapagos islands off the coast of Ecuador. Several populations are endangered, partly because of hunting and partly because of illegal capture and trade. In principle, a significant number of tortoises in captivity (either in zoos or "private collections") could be used to repopulate some of the endangered populations.

a. Suppose you are interested in returning a particular captive tortoise to its native subpopulation, but you have no record of its capture. How could you determine its original subpopulation?

b. A researcher planned to characterize one subpopulation of tortoises. In her first field season, she tagged and collected blood samples from all animals in the subpopulation. When she returns a year later, she could not locate two animals. She learns that two untagged, smuggled tortoises are being held by U.S. customs officials. How can she assess whether the smuggled animals are from her field site?

23 QUANTITATIVE GENETICS

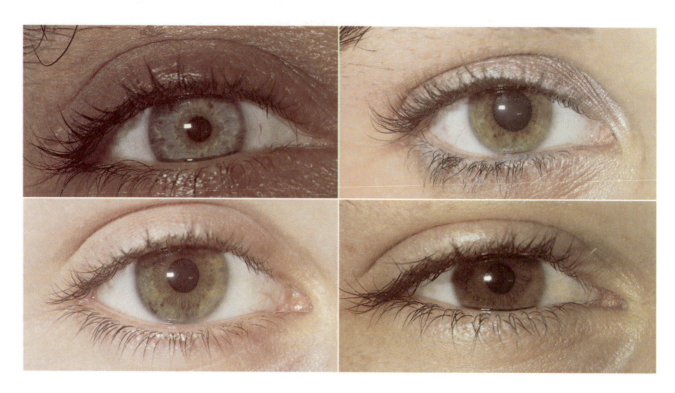

PRINCIPAL POINTS

~ Discontinuous traits exhibit only a few distinct phenotypes. Continuous traits display a range of phenotypes.

~ Continuous traits have many phenotypes because they are encoded by many genotypes (are polygenic) and/or because environmental factors cause each genotype to produce a range of phenotypes.

~ Continuous traits are studied by using samples of populations and statistical concepts such as the mean, variance, and correlation of characters, combined with statistical techniques such as analysis of variance and regression analysis.

~ The polygene or multiple-gene hypothesis of inheritance proposes that quantitative traits are determined by a number of genes, whose effects add together to determine the phenotype.

~ Variation among individuals can be partitioned into genetic and environmental components. However, genotypes may behave differently in different environments and thus caution must be exercised when designing and interpreting experiments that measure genetic and environmental contributions to phenotypic variation.

~ The broad-sense heritability of a trait is the proportion of the phenotypic variance that results from genetic differences among individuals. The narrow-sense heritability is the proportion of the phenotypic variance that results from additive genetic variance. Both of these measures are dependent on a particular population in a particular environment.

~ The amount that a trait changes in one generation as a result of selection for the trait is called the response to selection. The magnitude of the response to selection depends upon the selection differential and the narrow-sense heritability.

~ Genetic correlations arise when two traits are influenced by the same genes. When a trait is selected, genetically correlated traits will also exhibit a response to selection.

---

*I*n the previous chapter, we saw that there is a great deal of genetic variation among individuals. The amount of variation, and how is it distributed among individuals, provides the population with genetic structure. In this chapter on quantitative genetics we shift attention from the purely genetic and consider the "phenotypic structure" of a population and especially the relationship between the genetic structure and the phenotypic structure. Knowledge of the genetic structure, and the factors that preserve it and change it, is fundamental to our understanding of the evolutionary process and management of rare and endangered species. This focus on genetic structure is a logical extension of the earlier chapters that covered transmission genetics and Mendelian traits and molecular aspects of gene structure, function, and expression. By isolating mutants that affect the phenotype and by comparing the mutants with the wild type, geneticists were able to come to grips with Mendel's laws and later describe the molecular basis for the mutant phenotype. The mutations used in these early studies and later in molecular studies, and in fact most of the traits we have studied up to this point, have been character-ized by the presence of a few distinct phenotypes. The effects of variant alleles and a single gene locus were observable at the level of the organism so the phenotype could be used as a quick assay for the genotype. The seed coats of pea plants, for example, were either grey or white, the seed pods were green or yellow, and the plants were tall or short. In each trait the different phenotypes were distinct, and each phenotype was easily separated from all other phenotypes. Traits such as these, with only a few distinct phenotypes, are called **discontinuous traits**. Some additional examples of discontinuous traits are the ABO blood types, coat colors in mice, the prototrophic versus auxotrophic mutants of bacteria, and the presence or absence of extra digits on the hand. The phenotypes of a discontinuous trait can be described in qualitative terms, and all individuals can be placed into a few phenotypic categories, as shown in Figure 23.1.

For discontinuous traits, a simple relationship usually exists between the genotype and the phenotype. In most cases, each genotype produces only a single phenotype, and frequently each phenotype results from a single genotype. When dominance or epistasis occurs, the same phenotype may be pro-

~ FIGURE 23.1

Discontinuous distribution of shell color in the snail *Cepaea nemoralus* from a population in England.

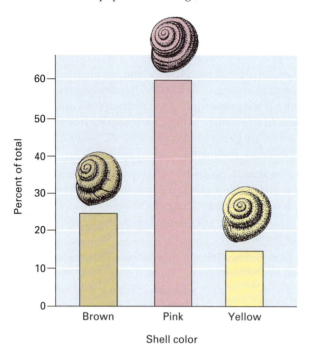

~ FIGURE 23.2

Distribution of birth weight of babies (males + females) born to teenagers in Portland, Oregon in 1992.

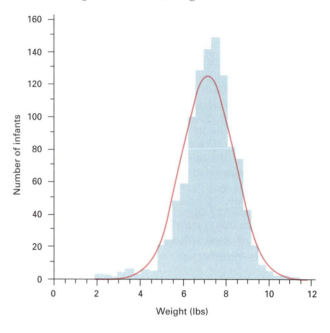

duced by several different genotypes, but the relationship between the genes and the trait remains simple. Because of this simple relationship, genotypes can be inferred by studying the phenotypes of parents and offspring, and in this way Mendel was able to work out the basic principles of heredity.

We saw in Chapter 4 that the relationship between the genotype and the phenotype is not so simple. Variable **penetrance** and **expressivity**, as well as **pleiotropy** and **epistasis,** can be quite common. In addition, single genotypes can give rise to a range of phenotypes as the genotype interacts with variable environments during development to give rise to a **norm of reaction**. As a result of these and other factors, there are not many traits with phenotypes that fall into a few distinct categories. Many traits (probably most), such as birth weight and adult height in humans, protein content in corn, and number of eggs laid by *Drosophila*, exhibit a wide range of possible phenotypes. Traits such as these, with a continuous distribution of phenotypes, are called **continuous traits**. The distribution of a continuous trait—birth weight in humans—is illustrated in Figure 23.2. Since the phenotypes of continuous traits must be described in quantitative terms, such traits are also known as quantitative traits; and the study of the inheritance of quantitative traits comprises the field of **quantitative genetics**.

Numerous traits have a continuous distribution; quantitative genetics therefore plays an important role in our understanding of evolution, conservation, and other areas of applied biology. In agriculture, for example, crop yield, rate of weight gain, milk production, and fat content are all continuous traits that are studied with the techniques of quantitative genetics. In psychology, methods of quantitative genetics are frequently employed for the study of complex behavioral traits, such as IQ, learning ability, and personality. Geneticists also use these methods to study other continuous traits found in humans, such as blood pressure, antibody titer, fingerprint pattern, and birth weight (see Figure 23.2). The traditional techniques of transmission genetics and the newer techniques of molecular genetics provide little information about the inheritance of these complex traits; different, more quantitative methods are required.

THE NATURE OF CONTINUOUS TRAITS

Biologists began developing statistical techniques for the study of continuous traits during the latter part of the nineteenth century, even before they were aware of Mendel's principles of heredity. Francis Galton and his associate Karl Pearson studied a number of continuous traits in humans, such as height, weight, and mental traits. They demonstrated that many traits of parents and their offspring are statistically associated. From this result they were able to infer that these traits are inherited, but they were not successful in determining how genetic transmission occurs. After the rediscovery of Mendel's work, considerable controversy arose over whether continuous traits also follow Mendel's principles, or whether they are inherited in some different fashion. Around 1903, Wilhelm Johannsen conducted a series of experiments on the inheritance of seed weight of beans, and he demonstrated that continuous variation is partly genetic and partly environmental. Several years later, Herman Nilsson-Ehle, working with wheat, proposed that continuous traits are determined by multiple genes, each of which segregates according to Mendelian principles. This was followed by Sir Ronald Fisher's theoretical demonstration that the basic mathematics of Mendelian and population genetics theory that we learned in the previous chapter can be extended to traits that are determined by multiple gene loci.

Why Some Traits Have Continuous Phenotypes

Continuous traits by definition have a continuous range of phenotypes. To understand the inheritance of continuous traits, we must first determine why some traits have many phenotypes.

Multiple phenotypes of a trait arise in several ways. Frequently a range of phenotypes occurs because numerous genotypes exist among the individuals of a group—this happens when the trait is influenced by a large number of loci. For example, when a single locus with two alleles determines a trait, three genotypes are present: AA, Aa, and aa. With two loci, each with two alleles, the number of genotypes is $3^2 = 9$ ($AA\ BB$, $Aa\ BB$, $AA\ Bb$, $Aa\ Bb$, $AA\ bb$, $aa\ BB$, $aa\ Bb$, $Aa\ bb$, and $aa\ bb$). In general, the number of genotypes is 3^n, where n equals the number of loci with two alleles; if more than two alleles are present at a locus, the number of genotypes is even greater. As the number of loci influencing a trait increases, the number of genotypes quickly becomes large. Traits that are encoded by many loci are referred to as **polygenic traits**. If each genotype in a polygenic trait encodes a separate phenotype, many phenotypes will be present. And because many phenotypes are present and the differences between phenotypes are slight, the trait appears to be continuous.

More frequently, several genotypes of a polygenic trait produce the same phenotype. For example, with dominance, only two phenotypes are produced by the three genotypes at a locus (AA and Aa produce one phenotype and aa produces a different phenotype). If dominance occurs among the alleles at each of three loci, the number of genotypes is $3^3 = 27$, but the number of phenotypes is only $2^3 = 8$. When epistatic interactions occur among the alleles at different loci, as discussed in Chapter 4, several genotypes may code for the same phenotype. In many polygenic traits, multiple genotypes code for the same phenotype, and the relationship between genotype and phenotype is obscured.

A second reason that a trait may have a range of phenotypes is that environmental factors also affect the trait. When environmental factors exert an influence on the phenotype, each genotype is capable of producing a range of phenotypes (the **norm of reaction**). Which phenotype is expressed depends both on the genotype and on the specific environment in which the genotype is found. For most continuous traits, both multiple genotypes and environmental factors influence the phenotype; such a trait is **multifactorial**.

When multiple genes and environmental factors influence a trait, one does not find the simple relationship between genotype and phenotype that exists in discontinuous traits. Therefore, (1) the rules of inheritance that we learned from transmission genetics, (2) the mechanisms of gene function that we learned from molecular genetics, and (3) the principles of genetic structure and function that we learned from population genetics are necessary for an understanding of how genes influence the phenotype, but are not sufficient for our understanding of how genes influence continuous traits. In order to understand the genetic basis of these traits and their inheritance, we must employ special concepts and analytical procedures.

KEYNOTE

Discontinuous traits exhibit only a few distinct phenotypes and can be described in qualitative terms. Continuous traits, on the other hand, display a spectrum of phenotypes and must be described in quantitative terms. The relationship between the genotype and the phenotype is complex. One genotype may give rise to a range of phenotypes, and many genotypes can give rise to the same phenotype. Numerous phenotypes are present in a continuous trait because the trait may be encoded by many loci, producing many genotypes, and because environmental factors may cause each genotype to

produce a range of phenotypes. Understanding how variation among individuals in a particular trait is determined during development is the major underlying theme of quantitative genetics.

Questions Studied in Quantitative Genetics

Not only are the methods used in quantitative genetics different from those we have previously studied, but the fundamental nature of the questions asked is also different. In transmission genetics, we frequently determined the probability of inheriting a discontinuous trait. With most continuous traits, however, numerous genes are involved and no simple relationship exists between the genotype and the phenotype; thus, we cannot make precise predictions about the probability of inheriting a continuous trait as we did for simple discontinuous traits. For the same reason, it is difficult to identify the DNA sequences that code for the trait, so molecular studies of continuous traits are usually impossible. But continuous traits are common and therefore require special attention. What follows is a set of questions frequently studied by quantitative geneticists.

1. To what degree does the observed variation in phenotype result from differences in genotype and to what degree does this variation reflect the influence of different environments? In our study of discontinuous traits, this question assumed little importance, because the differences in phenotype were assumed to reflect differences in genes.
2. How many genes determine the phenotype of a trait? When only a few loci are involved and the trait is discontinuous, the number of loci involved can often be determined by examining the phenotypic ratios in genetic crosses. With complex, continuous traits, however, determining the number of loci involved is more difficult.
3. Are the contributions of the determining genes equal? Or do a few genes have major effects on the trait and other genes only modify the phenotype slightly?
4. To what degree do alleles at the different loci interact with one another? Are the effects of alleles additive?
5. When selection occurs for a particular phenotype, how rapidly does the trait change? Do other traits change at the same time?
6. What is the best method for selecting and mating individuals so as to produce desired phenotypes in the progeny?

STATISTICAL TOOLS

As indicated above, one of the fundamental questions that is addressed in the study of quantitative traits is how much of the variation that exists among individuals in populations is genetically determined and how much is environmentally induced. It is paradigmatic that genes do not express themselves outside of an environmental context, and similarly without genes there would be nothing to express. Thus, at the heart of the field of quantitative genetics (the only field of science that addresses this question explicitly) is the perennial question of **nature versus nurture** or genes versus environment. Notice the traditional phrasing of the problem which pits one against the other, implying that they are mutually exclusive. This poor phrasing has resulted in much bad science and needless argumentation. Nevertheless, in quantitative genetic terms we can phrase the problem in terms of variation: how much of the variation in some aspect of the phenotype (V_P) is due to genetic variation (V_G) and how much is due to environmental variation (V_E), or

$$V_P = V_G + V_E.$$

In order to work this equation we have to learn how to measure variation in phenotype and how to partition the variation into genetic and environmental components. For this we need to understand some statistical methodology, much of which was developed specifically to deal with these genetic issues.

Samples and Populations

Suppose we want to describe some aspect of a trait in a large group of individuals. For example, we might be interested in the average birth weight of infants born in New York City during the year 1987. Since thousands of babies are born in New York City every year, collecting information on each baby's weight might not be practical. An alternative method would be to collect information on a subset of the group, say birth weights on 100 infants born in New York City during 1987, and then use the average obtained on this subset as an estimate of the average for the entire city. Biologists and other scientists commonly employ this sampling procedure in data collection, and statistics are necessary for analyzing such data. The group of ultimate interest (in our example, all infants born in New York City during 1987) is called the **population**, and the subset used to give us information about the population (our set of 100 babies) is called a **sample**. For a sample to give us confidence in information about the population, it must be large enough so that chance differences between the sample and the population are not misleading. If our sample consisted of only a single baby, and that infant was unusually large, then our estimate of the average birth weight of all babies would not be very accurate. The sample must also be a random subset of the population. If all the babies in our sample came from Hope Hospital for Premature Infants, then we would grossly underestimate the true birth weight of the population. While this might seem obvious, a great many errors in judgment are made because data were not collected randomly. Figure 23.2 shows a distribution of birth weight in humans.

$\mathcal{K}$EYNOTE

To describe and study a large group of individuals, scientists frequently examine a subset of the group. This subset is called a sample, and the sample provides information about the larger group, which is termed the population. The sample must be of reasonable size and it must be a random subset of the larger group to provide accurate information about the population.

Distributions

When we studied discontinuous traits, we were able to describe the phenotypes found among a group of individuals by stating the proportion of individuals falling into each phenotypic class. As we discussed earlier, continuous traits exhibit a range of phenotypes, and describing the phenotypes found within a group of individuals is more complicated. One means of summarizing the phenotypes of a continuous trait is with a **frequency distribution**, which is a description of the population in terms of the proportion of individuals that fall within a certain range of phenotypes.

To make a frequency distribution, classes are constructed to consist of individuals falling within a specified range of the phenotype, and the number of individuals in each class is counted. Table 23.1 pre-

~ **FIGURE 23.3**

Frequency histogram for bean weight in *Phaseolus vulgaris* plotted from the data in Table 23.1. A normal curve has been fitted to the data and is superimposed on the frequency histogram.

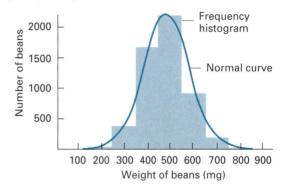

sents a frequency distribution constructed from the data in Johannsen's study of the inheritance of weight in the dwarf bean, *Phaseolus vulgaris*. As shown in the table, Johannsen weighed 5,494 beans from the F_2 progeny of a cross and classified them into nine groups or classes, each of which covered a 100-milligram (mg) range of weight. A frequency distribution such as this can be displayed graphically by plotting the phenotypes in a frequency histogram, as shown in Figure 23.3 for Johannsen's beans. In the histogram, the phenotypic classes are indicated along the horizontal axis and the number present in each class is plotted on the vertical axis. If a curve is drawn tracing the outline of the histogram, the curve assumes a shape that is characteristic of the frequency distribution. Several different types of distributions are illustrated in Figure 23.4.

Many continuous phenotypes exhibit a symmetrical, bell-shaped distribution similar to the one shown in Figure 23.4a and superimposed over the data in Figure 23.3. This type of distribution is called a **normal distribution**. The normal distribution is a theoretical distribution that has specific properties and is produced when a large number of independent factors influence the measurement. We will make use of these properties later. Since many continuous traits are multifactorial (influenced by multiple genes and

~ **TABLE 23.1**

Weight of 5,494 F_2 Beans (Seeds of *Phaseolus vulgaris*) Observed by Johannsen in 1903

Weight (mg) (Midpoint of range)	50–150 (100)	150–250 (200)	250–350 (300)	350–450 (400)	450–550 (500)	550–650 (600)	650–750 (700)	750–850 (800)	850–950 (900)
Number of beans	5	38	370	1,676	2,255	928	187	33	2

~ FIGURE 23.4

Three different types of distributions. (a) A normal distribution of percent sucrose in sugar beets; (b) A skewed distribution representing coat color in guinea pigs; (c) A bimodal distribution of the size of female rattlesnakes.

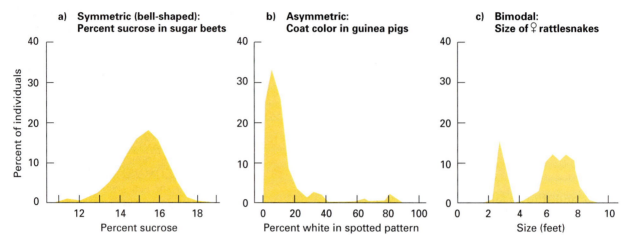

multiple environmental factors), observing a nearly normal distribution for these traits is not surprising. Two other types of distribution are illustrated in Figures 23.4b and 23.4c.

The Mean

A frequency distribution of a phenotypic trait can be summarized in the form of two convenient statistics, the mean and the variance. The **mean**, which is also known as the average, gives us information about where the center of the distribution of the phenotypes in a sample is located along a continuous range of possibilities. For example, the frequency distribution of body sizes of elephants and mice might look the same, but they are located in very different locations along a continuum of possible sizes. The mean of a sample is calculated by simply adding up all the individual measurements (x_1, x_2, x_3, etc.) and dividing by the number of measurements we added (n). We can represent this calculation with the following formula:

$$\text{Mean} = \bar{x} = \frac{x_1 + x_2 + x_3 + \ldots x_n}{n}$$

This equation can be abbreviated by using the symbol Σ, which means the summation, and x_i, which means the ith (or individual) value of x:

$$\text{Mean} = \bar{x} = \frac{\Sigma x_i}{n}$$

When data are taken directly from a frequency distribution, such as the distribution of bean weights seen in Table 23.1, several individuals will belong to the same phenotypic class. In this case, the mean can be estimated by first multiplying the midpoint of each class, m_i (in this case the middle weight in each class) by the number of individuals in that class, or f_i. These products are then added up and divided by the total number of measurements:

$$\bar{x} = \frac{\Sigma f_i m_i}{n}$$

Table 23.2 includes a sample calculation of the mean (and other statistics, see below) body lengths of 10 spotted salamanders from Penobscot County, Maine.

The mean is frequently used in quantitative genetics to characterize the phenotypes of a group of individuals. For example, in an early study of continuous variation, Edward M. East examined the inheritance of flower length in several strains of the tobacco plant. He crossed a short-flowered strain of tobacco with a long-flowered strain. Within each strain, however, flower length varied some, so East reported that the mean phenotype of the short strain was 40.4 mm and the mean phenotype of the long strain was 93.1 mm. The F_1 progeny, which consisted of 173 plants, had a mean flower length of 63.5 mm. In this situation, the mean provides a convenient way to quickly characterize the phenotypes of parents and offspring.

The Variance and the Standard Deviation

A second statistic that provides key information about a distribution is the **variance**. The variance is a measure of how much the individual measurements spread out around the mean—how variable the measurements are. Two distributions may have the same mean but very different variances, as shown in Figure

~ TABLE 23.2

Sample Calculations of the Mean, Variance, and Standard Deviation for Body Length of 10 Spotted Salamanders from Penobscot County, Maine

BODY LENGTH (x_i) (mm)	$(x_i - \bar{x})$	$(x_i - \bar{x})^2$
65	$(65 - 57.1) = 7.9$	$7.9^2 = 62.41$
54	$(54 - 57.1) = -3.1$	$-3.1^2 = 9.61$
56	$(56 - 57.1) = -1.1$	$-1.1^2 = 1.2$
60	$(60 - 57.1) = 2.9$	$2.9^2 = 8.41$
56	$(56 - 57.1) = -1.1$	$1.1^2 = 1.21$
55	$(55 - 57.1) = -2.1$	$2.1^2 = 4.41$
53	$(53 - 57.1) = -4.1$	$-4.1^2 = 16.81$
55	$(55 - 57.1) = -2.1$	$-2.1^2 = 4.41$
58	$(58 - 57.1) = 0.9$	$0.9^2 = 0.81$
59	$(59 - 57.1) = 1.9$	$1.9^2 = 3.61$
$\Sigma x_i = 571$		$\Sigma(x_i - \bar{x})^2 = 112.9$

Mean $= \bar{x} = \dfrac{\Sigma x_i}{n} = \dfrac{571}{10} = 57.1$

Variance $= s_x^2 = \dfrac{\Sigma(x_i - \bar{x})^2}{n - 1} = \dfrac{112.9}{9} = 12.54$

Standard deviation $= s_x = \sqrt{12.54} = 3.54$

23.5. The variance, symbolized as s^2, is defined as the average squared deviation from the mean.

$$\text{Variance} = s^2 = \frac{\Sigma(x_i - \bar{x})^2}{n - 1}$$

It can be calculated by first subtracting the mean from each individual measurement. Each value obtained from this subtraction is squared and all the squared values are added up. The sum of this calculation is then divided by the number of original measurements minus one. (For mathematical reasons, which we will not discuss here, the variance is obtained by dividing $n - 1$ instead of n.)

Another statistic, closely related to the variance, is the **standard deviation,** which is simply the square root of the variance:

$$\text{Standard deviation} = s = \sqrt{s^2}$$

The standard deviation is often preferred to the variance, because the standard deviation is in the same units as the original measurements, while the variance is in the units squared. Sample calculations for the variance and the standard deviation are presented in Table 23.2. A broad curve implies considerable variability in the quantity measured and a correspondingly large standard deviation. A narrow curve, in contrast, indicates relatively little variability

in the quantity measured and a correspondingly small standard deviation.

The importance of the normal distribution can now become clear. Once we know the mean and the standard deviation, a theoretical normal distribution is completely specified. It always has the shape indi-

~ FIGURE 23.5

Graphs showing three distributions with the same mean but different variances.

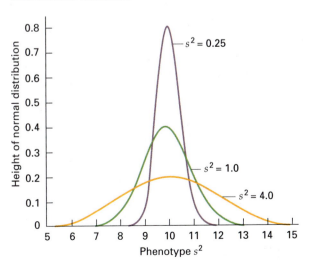

~ FIGURE 23.6

Normal distribution curve showing the proportions of the data in the distribution that are included within certain multiples of the standard deviation.

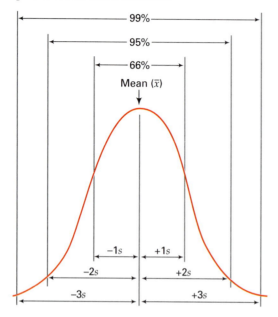

cated in Figure 23.6, where 66 percent of the individual observations have values within one standard deviation above or below (± 1s) the mean of the distribution, about 95 percent of the values fall within two standard deviations (± 2s) of the mean, and over 99 percent fall within three standard deviations (± 3s). We can infer many things about our data and experiments using these objectively determined percentages. We can also use a statistical technique called the **analysis of variance** developed by Sir Ronald Fisher (see Figure 22.1a) to help partition variance into component parts (see section on analysis of variance below).

The variance and the standard deviation can provide us with valuable information about the phenotypes of a group of individuals. In our discussion of the mean, we saw how East used the mean to describe flower length of parents and offspring in crosses of the tobacco plant. When East crossed a strain of tobacco with short flowers to a strain with long flowers, the F_1 offspring had a mean flower length of 63.5 mm, which was intermediate to the phenotypes of the parents. When he intercrossed the F_1, the mean flower length of the F_2 offspring was 68.8 mm, approximately the same as the mean phenotype of the F_1. However, the F_2 progeny differed from the F_1 in an important attribute that is not apparent if we only examine the means of the phenotype—the F_2 were more variable in phenotype than the F_1. The variance in the flower

length of the F_2 was 42.4 mm², whereas the variance in the F_1 was only 8.6 mm². This finding indicated that more genotypes were present among the F_2 progeny than in the F_1. Thus, the mean and the variance are both necessary for fully describing the distribution of phenotypes among a group of individuals.

Correlation

A difficulty encountered when thinking about the phenotype is that it is somewhat artificial to pick out traits and study them in isolation. Organisms are composites of a multitude of traits. Some of these traits like height and weight may actually be two members of a more general trait called size. Genes and environmental factors that affect the development of size may affect both traits. Genes that affect height may have pleiotropic effects on weight. In other words, two or more traits are often associated or *correlated*. This means that if one variable changes, the other is also likely to change. For example, head size and body size are correlated in most animals—individuals with large bodies have correspondingly large heads, and those with small bodies tend to have small heads. The **correlation coefficient** is a statistic that measures the strength of the association between two variables. Suppose we have two variables, x and y (x might equal body length and y might equal head width), and we wish to calculate the correlation between them. We begin by obtaining the covariance of x and y. The **covariance** is computed by taking the first x value and subtracting it from the mean of x and then multiplying the result by the deviation of the first y value from the mean of y. This is done for each pair of x and y values and the products are added together. The sum is then divided by $n - 1$ to give the covariance of x and y, where n equals the number of xy pairs:

$$\text{cov}_{xy} = \frac{\Sigma(x_i - \bar{x})(y_i - \bar{y})}{n - 1}$$

An algebraically equivalent equation, which is easier to compute, is

$$\text{cov}_{xy} = \frac{\Sigma x_i y_i - \frac{1}{n}(\Sigma x_i \Sigma y_i)}{n - 1}$$

where $\Sigma x_i y_i$ is the sum of each value of x multiplied by each corresponding value of y, Σx_i is the sum of all x values, and Σy_i is the sum of all y values. The correlation coefficient r can then be obtained by dividing the covariance by the product of the standard deviations of x and y.

$$\text{Correlation coefficient} = r = \frac{\text{cov}_{xy}}{s_x s_y}$$

~ TABLE 23.3

Sample Calculation of the Correlation Coefficient for Body Length and Head Width of Tiger Salamanders

BODY LENGTH (mm)			HEAD WIDTH (mm)			
x_i	$x_i - \bar{x}$	$(x_i - \bar{x})^2$	y_i	$y_i - \bar{y}$	$(y_i - \bar{y})^2$	$x_i y_i$
72.00	−7.92	62.67	17.00	−0.75	0.56	1224
62.00	−17.92	321.01	14.00	−3.75	14.06	868
86.00	6.08	37.01	20.00	2.25	5.06	1720
76.00	−3.92	15.34	14.00	−3.75	14.06	1064
64.00	−15.92	253.34	15.00	2.75	7.56	960
82.00	2.08	4.34	20.00	2.25	5.06	1640
71.00	−8.92	79.51	15.00	−2.75	7.56	1065
96.00	16.08	258.67	21.00	3.25	10.56	2016
87.00	7.08	50.17	19.00	1.25	1.56	1653
103.00	23.08	532.84	23.00	5.25	27.56	2369
86.00	6.08	37.01	18.00	0.25	0.06	1548
74.00	−5.92	35.01	17.00	−0.75	0.56	1258
$\Sigma x_i =$ 959.00		$\Sigma (x_i - \bar{x})^2 =$ 1,686.92	$\Sigma y_i =$ 213.00		$\Sigma (y_i - \bar{y})^2 =$ 94.25	$\Sigma x_i y_i =$ 17,385

$\bar{x} = \Sigma x_i / n = 959/12 = 79.92$

$\bar{y} = \Sigma y_i / n = 213/12 = 17.75$

Variance of $x = s_x^2 = \Sigma (x_i - \bar{x})^2 / n - 1 = 1,686.92/11 = 153.35$

Standard deviation of $\bar{x} = s_x = \sqrt{s_x^2} = \sqrt{153.35} = 12.38$

Variance of $y = s_y^2 = \Sigma (y_i - \bar{x})^2 / n - 1 = 94.25/11 = 8.57$

Standard deviation of $y = s_y = \sqrt{s^2} = \sqrt{8.57} = 2.93$

Covariance $= \text{cov}_{xy} = (\Sigma x_i y_i - 1/n(\Sigma x_i \Sigma y_i))/n - 1$

$\text{cov}_{xy} = \dfrac{(17,385 - 1/12(959 \times 213))}{12 - 1}$

$\text{cov}_{xy} = 32.97$

Correlation coefficient $= r = \text{cov}_{xy}/(s_x s_y) = 32.97/(12.38 \times 2.93)$

$r = 0.91$

where s_x equals the standard deviation of x, and s_y equals the standard deviation of y. Table 23.3 gives a sample calculation of the correlation coefficient between two variables.

The correlation coefficient can range from −1 to +1. The sign of the correlation coefficient, whether it is positive or negative, indicates the direction of the correlation. If the correlation coefficient is positive, then an increase in one variable tends to be associated with an increase in the other variable. If seed size and seed number are positively correlated in sunflowers, for example, plants with larger seeds also tend to produce more seeds. Positive correlations are illustrated in Figure 23.7b, c, d, and f. A negative correlation coefficient indicates that an increase in one variable is associated with a decrease in the other. If seed size and seed number are negatively correlated, plants with large seeds tend to produce fewer seeds

on the average than plants with smaller seeds. Figure 23.7e represents a negative correlation. The absolute value of the correlation coefficient (its magnitude if the sign is ignored) provides information about the strength of the association. When the correlation coefficient is close to −1 or to +1, the correlation is strong, meaning that a change in one variable is almost always associated with a corresponding change in the other variable. For example, the x and y variables in Figure 23.7f are strongly associated, and have a correlation coefficient of 0.9. On the other hand, a correlation coefficient near 0 indicates a weak relationship between the variables, as is illustrated in Figure 23.7b.

Several important points about correlation coefficients warrant emphasis. First, a correlation between variables means only that the variables are associated; correlation *does not imply that a cause-effect relationship*

Scatter diagrams showing the correlation of x and y variables. Diagrams b, c, d, and f show positive correlations, whereas diagram e shows a negative correlation. The absolute value of the correlation coefficient (r) indicates the strength of association. For example, diagram f illustrates a relatively strong correlation and diagram b illustrates a relatively weak correlation. In diagram a, the x and y variables are not correlated.

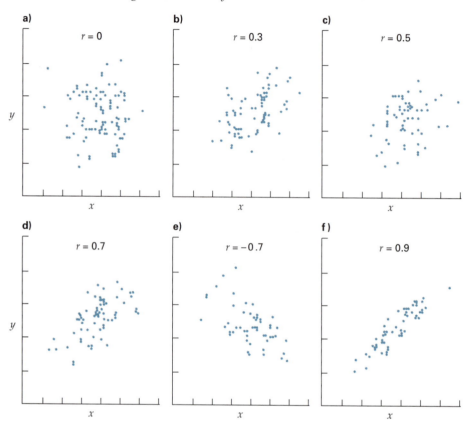

exists. The classic example of a noncausal correlation between two variables is the positive correlation that exists between number of Baptist ministers and liquor consumption in cities with population size over 10,000. One should not conclude from this correlation that Baptist ministers are consuming all the alcohol. Alcohol consumption and number of Baptist ministers are associated because both are positively correlated with a third factor, population size; larger cities contain more Baptist ministers and also have higher alcohol consumption. Assuming that two factors are causally related because they are correlated may lead to erroneous conclusions.

Another important point is that *correlation is not the same thing as identity*. Correlation only means that a change in one variable is associated with a corresponding change in the other variable. Two variables can be highly correlated, and yet have very different values. For example, the height of today's college-age males is correlated with the height of their fathers; tall fathers tend to produce tall sons and short fathers tend to produce short sons. This correlation results from the fact that genes influence human height. However, most college-age males today are taller than their fathers, probably because better diet and health care have increased the average height of all individuals in recent years. Thus, fathers and sons exhibit a correlation in height, but they are not the same height.

KEYNOTE

The correlation coefficient is a measure of how strongly two variables are associated. A positive correlation coefficient indicates that the two variables change in the same direction; an increase in one variable is usually associated with a corresponding increase in the other variable. When the

correlation coefficient is negative, the variables are inversely related; an increase in one variable is most often associated with a decrease in the other. The absolute value of the correlation coefficient provides information about the strength of the association. Correlation does not imply that a cause-effect relationship exists between the two variables.

Regression

The correlation coefficient tells us about the strength of association between variables and indicates whether the relationship is positive or negative, but it provides little information about the precise relationship between the variables. Often we are interested in knowing how much of a change in one variable is associated with a given change in another variable. For example, if we know the IQ of parents can we predict the IQ of offspring? Or, returning to our example of the correlation between heights of father and son, we might ask: If a father is 6 feet tall, what is the most likely height of his son? To answer this question, **regression** analysis is used.

The relationship between two variables can be expressed in the form of a **regression line**, as shown in Figure 23.8 for the relationship between heights of father and son. Each point on the graph represents the actual height of a father (value on the horizontal or x axis) and the height of his son (value on the vertical or y axis). The regression line is a mathematically computed line that represents the best fit of a line to the points; what this means is that the squared vertical distance from the points to the regression line is minimized. The regression line can be represented with the equation

$$y = a + bx$$

where x and y represent the values of the two variables (in Figure 23.8 the heights of father and son, respectively), b represents the **slope** (also called the **regression coefficient**), and a is the **y intercept**. The slope can be calculated from the covariance of x and y and the variance of x in the following manner:

$$\text{Slope} = b = \frac{\text{cov}_{xy}}{s_x^2}$$

The slope indicates how much of an increase in the variable on the y axis is associated with a unit increase in the variable on the x axis. For example, a slope of 0.5 for the regression of father and son height

~ FIGURE 23.8

Regression of son's height on father's height. Each point represents a pair of data for the height of a father and his son. The regression equation is $y = 36.05 + 0.49x$.

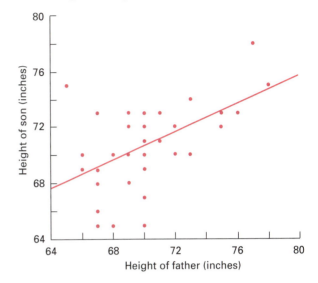

would mean that for each 1-inch increase in height of a father, the expected height of the son would increase 0.5 inches. The y intercept is the expected value of y when x is zero (the point at which the regression line crosses the y axis). Examples of regression lines with different slopes are presented in Figure 23.9. Regression analysis is one method that is commonly used for measuring the extent to which variation in a trait is genetically determined, as is described later in the section on heritability.

Analysis of Variance

One last statistical technique that we will mention briefly is **analysis of variance**. Analysis of variance, often called simply **ANOVA**, is a powerful statistical procedure for determining if differences in means are significant and for dividing the variance into components. We might be interested in knowing, for example, if males with the XYY karyotype (see Chapter 3) differ in height from males with a normal XY karyotype. We would proceed by first calculating the mean height of a sample of XYY males and the mean height of a sample of XY males. Suppose we found the mean height of our sample of XYY males was 74 inches, while the mean height of our sample of XY males was 70. The means are different, but our result could be due to chance differences between the samples, rather than to some factor associated with the extra Y chromosome.

~ FIGURE 23.9

Regression lines with different slopes. The slope indicates how much of a change in the y variable is associated with a change in the x variable.

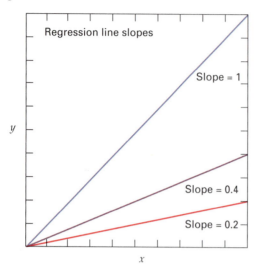

Analysis of variance can provide us with the probability that the difference in means of two samples results from chance. For example, it might indicate that there is less than a 1 percent probability that the difference we observed in the mean heights resulted from chance. We would probably conclude, then, that the difference in mean heights of XYY males and XY males is not due to chance differences in our samples, but results from some significant factor associated with the difference in chromosomes (i.e., karyotype). The analysis of variance can also be used to determine how much of the variation in height is associated with the difference in karyotype. We might find that the difference in the karyotypes is associated with 40 percent of the overall variation in height among the individuals in our samples. Factors other than difference in the number of Y chromosomes (perhaps genes on other chromosomes, diet, health care, etc.) would be responsible for the other 60 percent of the variation.

The calculations involved in analysis of variance are beyond the scope of this book, but the concept of breaking down the variance into components—called partitioning the variance—is important in quantitative genetics. For example, frequently we are interested in knowing how much of the variation in a trait is associated with genetic differences among individuals and how much of the variation is associated with environmental factors. Suppose that we wanted to increase milk production in a herd of dairy cattle. We might use analysis of variance to determine how much of the variation in milk production among the cows results

from environmental differences and how much arises because of genetic differences. If much of the variation is genetic, we could increase milk production by selective breeding. On the other hand, if most of the variation is environmental, selective breeding will do little to increase milk production, and our efforts would be better directed toward providing the optimum environment for high production. Analysis of variance is often used in this type of problem.

POLYGENIC INHERITANCE

In the examples of polygenic inheritance described in the following sections, geneticists did not know at first how these traits were inherited, although it was apparent that their pattern of inheritance differed from that of discontinuous traits.

Inheritance of Ear Length in Corn

An organism that has been the subject of genetic and cytological studies for many years is corn, *Zea mays*. Ear length is one of the traits of the corn plant that came under scrutiny in a classic study that demonstrated quantitative inheritance for that trait. In this study, reported in 1913, Rollins Emerson and E. East started their experiments with two pure-breeding strains of corn, each of which displayed little variation in ear length. The two varieties were Black Mexican sweet corn (which had short ears of mean length 6.63 cm) and Tom Thumb popcorn (which had long ears of mean length 16.80 cm).

Emerson and East crossed the two strains and then interbred the F_1 plants. Figure 23.10 presents the results in both photographs and histograms. Note that the F_1s have a mean ear length of 12.12 cm, which is approximately intermediate between the mean ear lengths of the two parental lines. The parental plants are pure breeding, so we can assume that each is homozygous for whatever genes control the lengths of their ears. Since the two parental plants differ in ear length, though, each must be genetically different. When two pure-breeding strains are crossed, the F_1 plants will be heterozygous for all genes, and all plants should have the same genotype. Therefore the range of ear length phenotypes seen in the F_1 plants must be due to factors other than genetic differences; these other factors are probably environmental, since it is impossible to grow plants in exactly identical conditions.

In the F_2 the mean ear length of 12.89 cm is about the same as the mean for the F_1 population, but the F_2 population has a much larger variation around the mean than the F_1 population has. This variation is easy to see in Figure 23.10b; it can also be shown by

~ FIGURE 23.10

Inheritance of ear length in corn: (a) Representative corn ears from the parental, F_1, and F_2 generations from an experiment in which two pure-breeding corn strains that differ in ear length were crossed and then the F_1s interbred. (b) Histograms of the distributions of ear length (in centimeters) of ears of corn from the experiment represented in part (a); the vertical axes represent the percentages of the different populations found at each ear length.

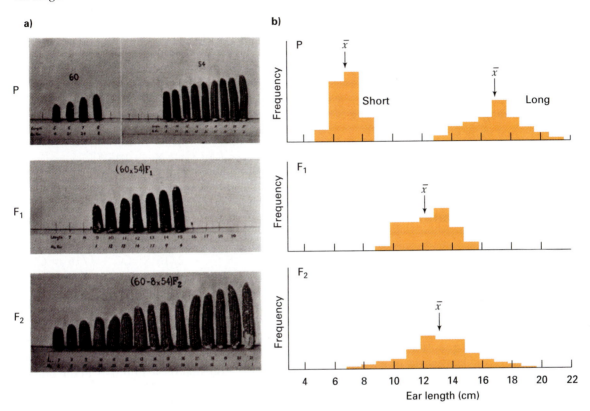

calculating the standard deviation s. The standard deviation of the long-eared parent is 1.887, and that of the short-eared parent is 0.816. In the F_1, $s = 1.519$, and in the F_2, $s = 2.252$. These numbers confirm that the F_2 has greater variability, something we could conclude by looking at the data.

Is this variation the result of the effects of environmental factors? Certainly, if the environment was responsible for variation in the parental and the F_1 generations, we have every reason to believe that it would have a similar effect on the F_2. However, we have no reason to suppose that the environment would have a greater influence on the F_2 than on the other two generations, and thus there must be another explanation for the greater variation in ear length in the F_2 generation. A more reasonable hypothesis is that the increased variability of the F_2 results from the presence of greater genetic variation in the F_2.

Setting aside the environmental influence for the moment, the data reveal four observations that apply generally to quantitative-inheritance studies similar to this one:

1. The mean value of the quantitative trait in the F_1 is approximately intermediate between the means of the two true-breeding parental lines.
2. The mean value for the trait in the F_2 is approximately equal to the mean for the F_1 population.
3. The F_2 shows more variability around the mean than the F_1 does.
4. The extreme values for the quantitative trait in the F_2 extend further into the distribution of the two parental values than do the extreme values of the F_1.

Can the data presented be explained in terms of standard single gene locus Mendelian genetic principles that govern the inheritance of discontinuous traits? No, the data cannot support such an interpretation. That is, if a single gene were responsible for the two phenotypes of the original parents (AA = homozygous, long; aa = homozygous, short), then the F_1 data could be explained if we assume incomplete dominance. However, crossing the F_1 heterozygote (Aa) should produce a 1:2:1 ratio of AA, Aa, and aa, or long, intermediate, and short phenotypes. Clearly, the data do not fall into such discrete classes.

KEYNOTE

For a quantitative trait, the F_1 progeny of a cross between two phenotypically distinct, pure-breeding parents has a phenotype intermediate between the parental phenotypes. The F_2 shows more variability than the F_1, with a mean phenotype close to that of the F_1. The extreme phenotypes of the F_2 extend well beyond the range of the F_1 and into the ranges of the two parental values.

Polygene Hypothesis for Quantitative Inheritance

The simplest explanation for the data obtained from Emerson and East's experiments on corn ear length and from other experiments with quantitative traits is that quantitative traits are controlled by not one but by many genes. This explanation, called the **polygene** or **multiple-gene hypothesis for quantitative inheritance**, is regarded as one of the landmarks of genetic thought.

The polygene hypothesis can be traced back to 1909 and the classic work of Hermann Nilsson-Ehle, who studied the color of wheat kernels. Like Mendel and like Emerson and East, Nilsson-Ehle started by crossing true-breeding lines of plants with red kernels and plants with white kernels. The F_1 had grains that were all the same shade of an intermediate color between red and white. At this point he could not rule out incomplete dominance as the basis for the F_1 results. However, when he intercrossed the F_1s, a number of the F_2 progeny showed a ratio of approximately 15 red (all shades) : 1 white kernels, clearly a deviation from a 3:1 ratio expected for a monohybrid cross. He recognized four discrete shades of red, in addition to white, among the progeny. He counted the relative number of each class and found a 1:4:6:4:1 phenotypic ratio of wheat with dark red, medium red, intermediate red, light red, and white kernels. Note that 1/16 of the F_2 has a kernel phenotype as extreme as the original red parent, and 1/16 has a kernel phenotype as extreme as the original white parent.

How can the data be interpreted in genetic terms? Recall from Chapter 4 (Analytical Approaches for Solving Genetics Problems, question 4.3c, p. 128) that

a 15:1 ratio of two alternative characteristics resulted from the interaction of the products of two genes, each of which affects the same trait; the two genes are known as duplicate genes. The explanation for the 15:1 ratio was that two allelic pairs are involved in determining the phenotypes segregating in the cross. Since several of the F_2 populations from the wheat crosses exhibited a 15 red : 1 white ratio, we can apply that explanation to the kernel trait.

Let us hypothesize that there are two pairs of independently segregating alleles that control the production of red pigment: alleles R (red) and C (crimson) result in red pigment and alleles r and c result in the lack of pigment. Nilsson-Ehle's parental cross and the F_1 genotypes can then be shown as follows:

P $RR\ CC$ × $rr\ cc$
       (dark red)   (white)
F_1 $Rr\ Cc$
       (intermediate red)

When the F_1 is interbred, the distribution of genotypes in the F_2 is that typical of dihybrid inheritance, that is, $1/16\ RR\ CC + 2/16\ Rr\ CC + 1/16\ rr\ CC + 2/16\ RR\ Cc + 4/16\ Rr\ Cc + 2/16\ rr\ Cc + 1/16\ RR\ cc + 2/16\ Rr\ cc + 1/16\ rr\ cc$. If R and C are dominant to r and c, the 9:3:3:1 phenotypic ratio characteristic of dihybrid inheritance will result. For the wheat kernel color phenotype, then, dominance is not the simple answer, since the observed phenotypic ratio approximates 1:4:6:4:1. Note that these numbers in the phenotypic ratio are the same as the coefficients in the **binomial expansion** of $(a + b)^4$. The following calculation demonstrates how we can arrive at the coefficients and their associated terms of this expansion. In essence, we multiply $(a + b)$ by $(a + b)$, then multiply the product by $(a + b)$, and so on:

$$
\begin{array}{ll}
& a + b \\
\times & a + b \\
\hline
= & a^2 + ab \\
& \quad\ \ + ab + b^2 \\
\hline
= & a^2 + 2ab + b^2 \qquad \text{—this is } (a+b)^2 \\
\times & a + b \\
\hline
= & a^3 + 2a^2b + ab^2 \\
& \quad\ \ + a^2b + 2ab^2 + b^3 \\
\hline
= & a^3 + 3a^2b + 3ab^2 + b^3 \qquad \text{—this is } (a+b)^3 \\
\times & a + b \\
\hline
= & a^4 + 3a^3b + 3a^2b^2 + ab^3 \\
& \quad\ \ + a^3b + 3a^2b^2 + 3ab^3 + b^4 \\
\hline
= & a^4 + 4a^3b + 6a^2b^2 + 4ab^3 + b^4 \quad \text{—this is } (a+b)^4 \\
\end{array}
$$

and so on.

Returning to the wheat kernel color phenotypic distribution, a simple explanation, then, is that each dose of a gene controlling pigment production allows the synthesis of a certain amount of pigment. Therefore the intensity of red coloration is a function of the number of R or C alleles in the genotype; $RR\,CC$ (term a^4 in the expansion) would be dark red and $rr\,cc$ (b^4 in the expansion) would be white. Table 23.4 summarizes this situation with regard to the five phenotypic classes observed by Nilsson-Ehle. In other words, the genes represented by capital letters code for products that add to the phenotypic characteristic; for example, each allele, R or C, causes more red pigment to be added to the wheat kernel color phenotype. Alleles that contribute to the phenotype are called **contributing alleles.** The alleles that do not have any effect on the phenotypes of the quantitative trait, such as the r and c alleles in the wheat kernel color trait, are called **noncontributing alleles.** Thus the inheritance of red kernel color in wheat is an example of a multiple gene or polygene series of as many as four contributing alleles.

We must be cautious in interpreting the genetic basis of this particular quantitative trait, though. Some F_2 populations show only three phenotypic classes with a 3:1 ratio of red to white, while other F_2 populations show a 63:1 ratio of red to white, with discrete classes of color between the dark red and the white. These results indicate that the genetic basis for the quantitative trait can vary with the strain of wheat involved. The 3:1 case could be explained by a single-gene system with two contributing alleles, while the 63:1 case could indicate a polygene series with six contributing alleles. The number of discrete classes in the latter case would be seven, with the proportion of each class following the coefficients in the binomial expansion of $(a + b)^6$, that is, 1:6:15:20:15:6:1.

The multiple-gene hypothesis that fits the wheat kernel color example so well has been applied to other examples of quantitative inheritance, including ear length in corn. In its basic form the multiple-gene hypothesis proposes that a number of the attributes of quantitative inheritance can be explained on the basis of the action and segregation of a number of allelic pairs that each have a small but additive effect on the phenotype. These allelic pairs with small effects are called **polygenes.** For the most part the multiple-gene hypothesis is satisfactory as a working hypothesis for interpreting many quantitative traits. The whole picture of quantitative traits is very complicated, though, and there are still many gaps in our understanding of quantitative inheritance. In addition, we have a lot to learn about the molecular aspects of quantitative traits. For example, while the proposal that a number of alleles each function to produce a particular amount of pigment is an attractive hypothesis, what does that hypothesis mean at the molecular level? Is some regulation of product output exerted at the translational level or in a biochemical pathway? In a large polygenic series, how many biochemical pathways are controlled? Thus quantitative inheritance provides an explanation for the inheritance of continuous traits that is compatible with Mendel's laws, but many aspects of quantitative inheritance are still unknown.

$\mathcal{K}$EYNOTE

Quantitative traits are based on genes in a multiple-gene series, or polygenes. The multiple-gene hypothesis assumes that contributing and noncontributing alleles in the series operate so that as the number of contributing alleles increases, there is an additive (or occasionally multiplicative) effect on the phenotype.

Determining the Number and Location of Polygenes for a Quantitative Trait

As the previous example shows, a quantitative trait need not be determined by a large number of genes. Even a few gene loci, where the differences among individuals with the same genotype can be greater than the average difference between genotypes, will give rise to continuously distributed traits. One of the outstanding questions in genetics is: How does the genetic structure of the population interact with the environment to determine the phenotypic structure of

~ TABLE 23.4

Genetic Explanation for the Number and Proportions of F_2 Phenotypes for the Quantitative Trait Red Kernel Color in Wheat

GENOTYPE	NUMBER OF CONTRBUTING ALLELES FOR RED	PHENOTYPE	FRACTION OF F_2
$RR\,CC$	4	Dark red	1/16
$RR\,Cc$ or $Rr\,CC$	3	Medium red	4/16
$RR\,cc$ or $rr\,CC$ or $Rr\,Cc$	2	Intermediate red	6/16
$rr\,Cc$ or $Rr\,cc$	1	Light red	4/16
$rr\,cc$	0	White	1/16

the population? One important way of getting at this question is to try to determine how many genes generally influence the variation in quantitative characters.

As you might imagine, it is not an easy task to estimate the number of genes controlling a quantitative trait. If a large number of genes are involved, many different phenotypes may be present, and the progeny are not easily placed into distinct phenotypic classes. The fewer the genes the easier the task, and artificial breeding programs, where the changes in variance among inbred, hybrid, and backcrossed lines are calculated, can be used to make these estimates. In this way it has been estimated that skin color in humans is determined by a minimum of 4 to 6 genes; fruit weight in tomatoes is determined by 7 to 11 genes; and oil content in corn by a minimum of 17 genes.

Where discrete phenotypic classes can be identified, we can estimate the number of genes involved. In Nilsson-Ehle's studies of the inheritance of red kernel color in wheat, for example, all F_2 populations were not identical. While crosses of some red-kerneled varieties gave F_2 progeny that fell into five phenotypic classes, others gave seven phenotypic classes. The latter crosses showed an approximately 1:6:15:20:15:6:1 ratio of different shades, ranging from dark red to white. Since the relative proportions of the phenotypic classes correspond to the coefficients in the binomial expansion of $(a + b)^6$, it is likely that three pairs of alleles controlled kernel color in this particular cross.

In general, the relationship between the number of independently segregating allelic pairs in a polygene series and the number of quantitative-trait phenotypes in the F_2 is as shown in Table 23.5. When one pair of alleles controls a trait, then 1/4 of the F_2 should show one phenotypic extreme (e.g., the darkest red) and 1/4 should show the other phenotypic extreme (e.g., white). The general formula $(1/2)^{2n}$ $(= (1/4)^n)$ describes the predicted probability for an extreme phenotype when n is the number of pairs of segregating alleles involved. For two genes, and therefore two pairs of alleles, the fraction of the F_2 population with an extreme character is $(1/2)^4 = 1/16$. For three genes (three pairs of alleles) the fraction is 1/64, as in the wheat kernel example. As the number of genes increases, the fraction of the F_2 with an extreme phenotype decreases, and it does so very rapidly. For five genes the fraction is 1/1,024, and for ten genes it is 1/1,048,576.

The number of genotypic classes also increases rapidly as the number of pairs of alleles increases. The beginnings of this increase are shown in Table 23.5. Three pairs of alleles yield 27 possible genotypic classes in the F_2; five pairs of alleles yield 243 genotypes; and ten pairs of alleles yield 59,049 genotypes. In general, the number of genotypic classes in the F_2 is $(3)^n$, where n is the number of pairs of alleles involved.

As the number of pairs of alleles in a multiple-gene series increases, the F_2 population rapidly assumes a continuum of phenotypic variation for which distinctions between classes are impossible. In short, when the number of pairs of alleles is five or more, it is difficult for us to be sure of the number of pairs of alleles involved in controlling a quantitative effect. Other methods are available for determining the number of pairs of alleles; they have as their basis more sophisticated mathematical treatments. Even with these methods, however, we still encounter significant complications caused by the environment,

~ TABLE 23.5

Probability Information for a Quantitative Character Controlled by a Polygenic Series in Which Independently Segregating Allelic Pairs Have Equivalent, Additive Effects

NUMBER OF ALLELIC PAIRS	NUMBER OF SEGREGATING ALLELES	FRACTION OF POPULATION SHOWING AN EXTREME EXPRESSION OF THE CHARACTER	NUMBER OF GENOTYPES IN F_2	NUMBER OF PHENOTYPES IN F_2	F_2 PHENOTYPIC RATIOS ARE THE COEFFICIENTS IN THE BINOMIAL EXPANSION
1	2	$(1/2)^2 = 1/4$	$(3)^1 = 3$	3	$(a + b)^2$
2	4	$(1/2)^4 = 1/16$	$(3)^2 = 9$	5	$(a + b)^4$
3	6	$(1/2)^6 = 1/64$	$(3)^3 = 27$	7	$(a + b)^6$
4	8	$(1/2)^8 = 1/256$	$(3)^4 = 81$	9	$(a + b)^8$
n	$2n$	$(1/2)^{2n} = (1/4)^n$	$(3)^n$	$2n + 1$	$(a + b)^{2n}$

different dominance relationships of the genes in the series, and linkage.

In addition, traditional and molecular methods for finding genes that influence traits can be used in the case of quantitative traits. As was discussed in Chapter 5 we can use traditional methods of linkage analysis. We can also use molecular linkage methods as discussed in Chapter 6 (pp. 182–185). The task of finding multiple gene loci that influence a quantitative trait can be very tedious. Nevertheless, the potential rewards of discovering whether in fact most quantitative traits are controlled by a relatively small number of genes with large effects versus many genes with small effects is going to be well worth the effort.

HERITABILITY

Heritability is the proportion of a population's phenotypic variation that is attributable to genetic factors. The term has frequently been misused and it is often tossed about without firm scientific basis. For example, when individuals in a family resemble each other in some aspect of the phenotype, be it stature or intelligence, a genetic basis is often assumed to be responsible for the similarity. But, consider that all the resemblance among family members could just as easily be due to their shared environment as opposed to their shared genes. Carefully planned quantitative genetic experiments are the only way to get at the correct answer. The concept of heritability is used to examine the relative contributions of genes and environment to variation in a specific trait. As we have seen, continuous traits are frequently influenced by multiple genes and by environmental factors. For example, multifactorial traits such as weight of cattle, number of eggs laid by chickens, and the amount of fleece produced by sheep are important for breeding programs and agricultural management. Many ecologically important traits, such as variation in body size, fecundity, and developmental rate are also multifactorial, and the genetic contribution to this variation is important for understanding how natural populations evolve. The extent to which genetic and environmental factors contribute to human variation in traits like blood pressure and birth weight is important for health care. Also, the extent to which genes affect human behaviors, such as alcoholism and criminality, may be useful in establishing social policy, but utmost care needs to be taken because of the ease of misuse of supposedly scientific information.

The following section deals with two types of heritability: **broad-sense heritability** and **narrow-sense heritability**. To assess heritability, we must first measure the variation in the trait and then we must partition that variance into components attributable to different causes.

Components of the Phenotypic Variance

As noted earlier (p. 776), the **phenotypic variance,** represented by V_P, is a measure of the variability of a trait. It is calculated by computing the variance of the trait for a group of individuals, as was outlined in this chapter's section on statistics. Differences among individuals arise from several factors. It is paradigmatic that all differences among individuals arise as differences in the developmental process. The developmental process is in turn driven by physiology, which is directed by genetic information and influenced by the environment. Therefore, we can partition the phenotypic variance into several components attributable to different sources. First, some of the phenotypic variation may arise because of genetic differences among individuals (different genotypes within the group). This contribution to the phenotypic variation is called **genetic variance** and is represented by the symbol V_G. Figure 23.11a shows a situation where all the variation is due to two genotypes and none is due to the environment. As noted, additional variation often results from environmental differences among the individuals; in other words, different environments experienced by individuals may contribute to the differences in their phenotypes. The **environmental variance** is symbolized by V_E, and by definition, it includes any nongenetic source of variation. Temperature, nutrition, and parental care are examples of obvious environmental factors that may cause differences during development among individuals. Figure 23.11b shows a situation where all the variation is due to the environment, and Figure 23.11c shows the situation where both genes and environment contribute to the variation. Thus, we have the basic nature-nurture equation that was discussed earlier:

$$V_P = V_G + V_E$$

One hundred percent of the variation among individuals is accounted for by genetic and environmental influences; however, the partitioning of phenotypic variance is more complicated than this. The sum of the genetically caused variance and environmentally caused variance may not add up to the total phenotypic variance. This is because the genetically caused variance and environmentally caused variance may covary and another term (COV_{GxE}) is needed. For example, let's say milk production in cows is influenced by genes, but it is also influenced by the

~ FIGURE 23.11

Hypothetical example of the effects of genes and environments on plant height. (a) Plant height variation is influenced by the two genotypes with individuals of the G2 type having greater height than individuals of the G1 type. Plant height is independent of the temperature in which the plants are raised. (b) All the variation is environmental and the two genotypes are indistinguishable. (c) Both the genotypes and the environments exert an additive influence on variation in height. (d) Both genotype and environment exert an influence but the relative effects of the two genotype are dependent on the environment. This is an example of genotype by environment interaction.

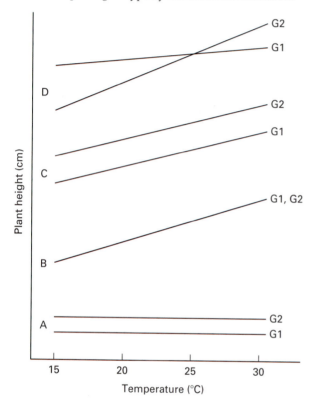

when the genotypes are moved to a warm climate, genotype *Aa* is now 60 cm, as compared with genotype *AA*, which is only 50 cm in height. This relationship is shown in Figure 23.11d. In this example, both genotypes grow taller in the warm environment, thus there is an environmental effect on variance. There is also a genetic effect, but it is not the kind that results in consistent resemblance among relatives. The genetic effect depends on the environment. The relative performance of the genotypes switches in the two environments. Therefore, both environmental differences (temperature) and genetic differences (genotypes) contribute to the phenotypic variance. However, the effects of genotype and environment cannot simply be added together. An additional component that accounts for how genotype and environment interact must be considered, and this is V_{GxE}.

The phenotypic variance, composed of differences arising from genetic variation, environmental variation, genetic-environmental covariation, and genetic-environmental interaction, can be represented by the following equation:

$$V_P = V_G + V_E + COV_{GxE} + V_{GxE}$$

The relative contributions of these four factors to the phenotypic variance depend upon the genetic composition of the population, the specific environment, and the manner in which the genes interact with the environment.

KEYNOTE

Variation among individuals can be partitioned into genetic and environmental components. The fact that genotypes might not be distributed randomly across environments, and that genotypes may behave differently in different environments, provides a caution that the results of an experiment determining the relative importance of genetic and environmental factors may depend in non-obvious ways on the environment in which the experiment is performed.

amount of feed a farmer provides. The farmer knows his cows and provides the offspring of good milking cows more feed and poor milking cows less feed. In this way the variance in milk production is increased beyond that which would be expected on the basis of genes and environment operating independently.

In addition, we may want to know if we can expect offspring to resemble their parents. Just knowing that there is a genetic component to the variation does not answer this question. A reason for this is that there is another source of phenotypic variance that is called genetic-environmental interaction, represented by V_{GxE}. Genetic-environmental interaction occurs when the relative effects of the genotypes differ among environments. For example, in a cold environment, genotype *AA* of a plant may be 40 cm in height and genotype *Aa* may be 35 cm in height. However,

The complications continue, however, in that the genetic variance V_G can be further subdivided into components arising from different types of interactions between genes. Some of the genetic variance occurs as a result of the average effects of the different alleles on the phenotype. For example, an allele *g* may, on the average, contribute 2 centimeters in height to a plant, and the allele *G* may contribute 4 centimeters. In this case, the *gg* homozygote would contribute 2 + 2 = 4 centimeters in height, the *Gg* heterozygote would contribute 2 + 4 = 6 centimeters in height, and the *GG* homozygote would contribute 4 + 4 = 8 centimeters in height. To determine the genetic contribution to height, we would then add the effects of alleles at this locus to the effects of alleles at other loci which might influence the phenotype. Genes such as these are said to have additive effects, and this type of genetic variation is called **additive genetic variance**. You may recall that the genes studied by Nilsson-Ehle, which determine kernel color in wheat, are strictly additive in this way. Some alleles contribute to the pigment of the kernel and others do not; the added effects of all the individual contributing alleles determine the phenotype of the kernel. Thus, the genotypes *AA bb*, *aa BB*, and *Aa Bb* all produce the same phenotype, since each genotype has two contributing alleles. The phenotypic variance arising from the additive effects of genes is the additive genetic variance and is symbolized by V_A.

Some genes may exhibit dominance, and this comprises another source of genetic variance, the **dominance variance** (V_D). Dominance occurs when one allele masks the expression of the other allele at the locus. If dominance is present, the individual effects of the alleles are not strictly additive; we must also consider how alleles at a locus interact. In the presence of dominance, the heterozygote *Gg* would contribute 8 centimeters in height to the phenotype, the same amount as the *GG* homozygote. Thus a population consisting of both genotypes would have genetic variation but there would be no corresponding phenotypic variation. As the degree of dominance diminishes the genotypic differences more clearly become phenotypic differences as the dominance variance turns into **additive genetic variance**.

Finally, epistatic interactions may occur among alleles at different loci. Recall that when epistasis exists, alleles at different loci interact to determine the phenotype. Thus, we might have three genotypes at one gene locus but their penetrance could be affected by variation at other gene loci. The presence of epistasis adds another source of genetic variation, called epistatic or **interaction variance** (V_I). So we can partition the genetic variance as follows:

$$V_G = V_A + V_D + V_I$$

and the total phenotypic variance up to this point can then be summarized as

$$V_P = V_A + V_D + V_I + V_E + \text{COV}_{GxE} + V_{GxE}$$

Just as we partitioned the genetic component of variance into three subcomponents, the environmental component of variance can also be partitioned. For example, individuals in a population may be exposed to varying temperature or nutritional environments

during development. These environmental effects result in somewhat irreversible differences among individuals and are called **general environmental effects** (V_{Eg}). So for example, an individual that is raised in a nutritionally deprived region may have a smaller body size. Some environmental variation results in immediate changes in the phenotype such as skin pigment differences upon exposure to sun. These can often be reversible and are called **special environmental effects** (V_{Es}). A special category of environmental effects called **maternal effects** (V_{Em}) are so common that they deserve special mention. For example, variation in the size of humans at birth will have genetic and environmental components. The genetic component would be due to the specific genotypic differences among the infants. The environment of the infants up to that time is the uterus of their mothers and this would constitute a maternal effect. Clearly, the study of how much variation in birth weights of infants is due to a maternal effect is important in the study of risk factors associated with birth weights. Thus, at this point our more completely partitioned nature versus nurture equation is as follows:

$$V_P = V_A + V_D + V_I + V_{Eg} + V_{Es} + V_{Em} + \text{COV}_{GxE} + V_{GxE}$$

Partitioning the phenotypic variance into these components is useful for thinking about the contribution of different factors to the variation in phenotype. It is very difficult to design experiments that can analyze all of these components simultaneously and assumptions about some usually have to be made. For example, it is often assumed that there is no genotype by environment interaction (V_{GxE}), but the well-trained geneticist would always remember that the results of such an experiment must be presented with appropriate caution.

Broad-Sense and Narrow-Sense Heritability

Again, geneticists frequently partition the phenotypic variance of a trait to determine the extent to which variation among individuals results from genetic differences. Thus, they are interested in how much of the phenotypic variance V_P can be attributed to genetic variance V_G. This quantity, the proportion of the phenotypic variance that consists of genetic variance, is called the broad-sense heritability and is expressed as follows:

$$\text{Broad-sense heritability} = h_B^2 = \frac{V_G}{V_P}$$

The heritability of a trait can range from 0 to 1. A broad-sense heritability of 0 indicates that none of the variation in phenotype among individuals results from genetic differences. A heritability of 0.5 means that 50 percent of the phenotypic variation arises from genetic differences among individuals, and a heritability of 1 would suggest that all the phenotypic variance is genetically based. Broad-sense heritability includes genetic variation from all types of genes. It ignores the fact that the genetic variance may be of the additive, dominance, or interactive sort. It also assumes that genotype by environment interaction (V_{GxE}) is not important. Thus, its usefulness is questionable.

More frequently, we are more interested in the proportion of the phenotypic variation that results only from additive genetic effects. This is because the additive genetic component of variation allows us to predict the phenotype of the offspring from the phenotypes of the parents. They are responsible for resemblance among relatives when there isn't a common environmental cause. To understand the reason for this, consider a cross involving a trait that results from the effect of alleles at a single locus. One parent is A^1A^1 and 10 cm tall; the other parent is A^2A^2 and is 20 cm tall. All the offspring from this cross (the F_1) will be A^1A^2. If the alleles are additive and contribute equally to height, the offspring should be 15 cm tall, exactly intermediate between the parents. However, if A^2 is dominant, all the offspring will be 20 cm, resembling only one of the parents. Thus, if dominant genes are important, the offspring will not be intermediate to the parental phenotypes. In a similar fashion, epistatic genes will not always contribute to the resemblance between parents and offspring.

Because the additive genetic variance allows one to make accurate predictions about the resemblance between offspring and parents, quantitative geneticists frequently determine the proportion of the phenotypic variance that results from additive genetic variance, a quantity referred to as the **narrow-sense heritability**. The additive genetic variance is also that variation that responds to selection in a predictable way, and thus the narrow-sense heritability provides information about how a trait will evolve. The narrow-sense heritability is represented mathematically as

$$\text{Narrow-sense heritability} = h_N^2 = \frac{V_A}{V_P}$$

Understanding Heritability

Despite their utility, heritability estimates have a number of significant limitations. Unfortunately, these limitations are often ignored. As a result, heritability is one of the most misunderstood and widely abused concepts in all of genetics. Before we discuss how heritability is determined, it is important that we list some of the important qualifications and limitations of heritability.

1. Broad-sense heritability does not indicate the extent to which a trait is genetic. What heritability does measure is the *proportion of the phenotypic variance* among individuals in a population that results from genetic differences. The "proportion of the phenotypic variance among individuals in a population that results from genetic differences" may seem like the same thing as "the extent to which a trait is genetic," but these two statements actually mean quite different things. Genes often influence the development of a trait, and thus the trait may be said to be genetic. However, the differences in phenotype among individuals may not be genetic at all. Let us consider an example. Our ability to learn about the game of football is dependent upon the proper development of our nervous system, which is clearly controlled by genes. Thus we could say that knowledge of the game of football is a genetically influenced trait. However, the differences we see among individuals in their knowledge of football is usually not due to genes but to different environments and personal interests. Since broad-sense heritability measures the proportion of the phenotypic variation among individuals that results from genetic differences, broad-sense heritability for knowledge of football would be zero, despite the fact that genes do influence our ability to have knowledge of football.

   Consider another more general situation. Suppose that genes are critical in determining the phenotypes of a trait. If all individuals in a population have identical genes at the loci that control the trait, then the genetic variance is zero ($V_G = 0$). Because broad-sense heritability equals the proportion of the phenotypic variance that results from genetic variance ($h_B^2 = V_G/V_P$), and V_G is zero, the heritability must be zero. Although the heritability in this case is zero, it would be incorrect to assume that genes play no role in the development of the trait. Similarly, a high heritability does not negate the importance of environmental factors influencing a trait; a high heritability might simply mean that the environmental factors that influence the trait are relatively uniform among the individuals studied.

2. Heritability does not indicate what proportion of an individual's phenotype is genetic. Since it is based on the variance, which can only be calculated on a group of individuals, heritability is characteristic of a population. An individual does not have heritability—a population does.

3. Heritability is not fixed for a trait. Thus, there is no universal heritability for a trait like human stature. Rather, the heritability value for a trait depends upon the genetic makeup and the spe-

cific environment of the population. We could calculate the heritability of stature among Hopi Indians living in Arizona, for example, but heritability calculated for other populations and other environments might be very different.

To illustrate this point, suppose that we calculated broad sense heritability for adult height on individuals living in a small New England town and obtained a value of 0.7. This would indicate that 70 percent of the variation in adult height among these individuals is due to genetic variation. Heritability for other populations might not be the same. Residents of San Francisco, for example, might be more ethnically heterogeneous than the inhabitants of the small New England town; therefore the San Francisco population would have more genetic variation for height. Let us assume that the environmental variance of the two populations was about the same. If the environmental variances of the two populations were similar, but the genetic variance was greater in San Francisco, then heritability calculated for the height of San Francisco residents also would be greater.

Genes are not the only factor that influences height in humans; diet, an environmental effect, is also a major determinant of height. Since most individuals in our small New England town probably receive an adequate diet, at least in terms of calories, the environmental variance for height would not be large. In a developing nation, however, some individuals might receive adequate nutrition, while the diet of others might be severely deficient. Because greater differences in diet exist, in such a nation the environmental variance for height would be larger, and as a result, the heritability of height would be less. Thus heritability calculated for human height might differ substantially for residents of the small New England town, residents of San Francisco, and residents of a developing country.

These examples illustrate that heritability can be applied only to a specific group of individuals in a specific environment. If the genetic composition of the group is different, or the environment is different, the original heritability value is no longer valid. Changing groups or environments does not alter the way in which genes affect the trait, but it may change the amount of genetic and environmental variance for the trait, which would then alter the heritability.

4. Even if heritability is high in each of two populations, and the populations differ markedly in a particular trait, one cannot assume that the populations are genetically different. For example, suppose that you obtain some genetically vari-

able mice and divide them into two groups. You feed one group a nutritionally rich diet, and you are careful to provide each mouse with exactly the same amount of food, space, water, and other environmental necessities. The mice grow to a large size because of the rich diet. When you measure heritability for adult body weight, you obtain a high value of 0.93. The high heritability is not surprising since the mice were genetically variable and environmental differences were kept at a minimum. The second group of mice comes from the same genetic stock, but you feed them an impoverished diet, lacking in calories and essential nutrients; again, each mouse gets exactly the same amount of food, space, water, and necessities. Because of the poor diet, the mice of this second group are all smaller than those in the first group. When you calculate heritability for adult weight in the small mice, you again obtain a high value of 0.93, because the mice were genetically variable and the environmental differences were kept to a minimum. Because the heritability of body weight is high in both groups and the mice of the two groups differ in adult weight, some people might suggest that the mice are genetically different in body size. Yet, any claim that the mice of the two groups differ genetically is clearly wrong—both groups came from the same stock. The important point is that heritability cannot be used to draw conclusions about the nature of differences between populations. If we draw an analogy between this example of body size in mice to book reading ability in humans, we see how easy it is to misapply quantitative genetic approaches to socially loaded human issues. Let's say we had two groups of humans and determined that variation in book reading ability within each group had a high heritability. One group was raised in a book-rich environment and most individuals could read well. The other group was raised in a book-poor environment and individuals could only read poorly. What conclusions would you draw concerning the genetic differences between the two populations? Can book reading ability be enhanced by social intervention programs, or is it hopeless because book-reading ability is "genetic"?

5. Traits shared by members of the same family do not necessarily have high heritability. When members of the same family share a trait, the trait is said to be **familial**. Familial traits as mentioned earlier may arise because family members share genes or because they are exposed to the same environmental factors. Thus, familiality is not the same as heritability.

͞KEYNOTE

Broad-sense heritability of a trait represents the proportion of the phenotypic variance that results from genetic differences among individuals. The narrow-sense heritability is more restricted—it measures the proportion of the phenotypic variance that results from additive genetic variance. Narrow-sense heritability allows quantitative geneticists to make predictions about the resemblance between parents and offspring, and it represents that part of the phenotypic variance that responds to natural or artificial selection in a predictable manner.

How Heritability Is Calculated

A number of different methods are available for calculating heritability. Many of the methods involve comparing related and unrelated individuals, or comparing individuals with different degrees of relatedness. If genes are important in determining the phenotypic variance, then closely related individuals should be more similar in phenotype, since they have more genes in common. Alternatively, if environmental factors are responsible for determining differences in the trait, then related individuals should be no more similar in phenotype than unrelated individuals. An important point to remember is that *the related individuals studied must not share a more common environment*. We assume that if related individuals are

more similar in phenotype, it is because they share similar genes. If related individuals also share a more common environment than unrelated individuals, separating the effects of genes and environment is much more difficult and frequently impossible. The absence of common environmental factors among related individuals can often be achieved in domestic plants and animals, and common environments may not exist among family members in the wild; however, this is very difficult to obtain in humans, where family structure and extended parental care create common environments for many related individuals. Some of the methods used to calculate heritability include comparison of parents and offspring, comparison of full and half sibs, comparison of identical and nonidentical twins, and response-to-selection experiments.

HERITABILITY FROM PARENT-OFFSPRING REGRESSION. If the additive genetic component of variation is important in determining the differences among individuals, then we expect that offspring should resemble their parents. To quantify the degree to which genes influence a trait, we must first measure the phenotypes of parents and offspring in a series of families, and then statistically analyze the relationship between their phenotypes. Correlation and regression are appropriate for this type of problem.

We can represent the relationship between offspring phenotype and parental phenotype by plotting the mean phenotype of the parents against the mean phenotype of the offspring, as shown in Figure 23.12. Each point on the graph represents one family. If the points are randomly scattered across the plot, as in Figure 23.12c, then no relationship exists between the

~ **FIGURE 23.12**

Three hypothetical regressions of mean parental wing length on mean offspring wing length in *Drosophila*. In each case, the slope of the regression line (*b*) equals the narrow-sense heritability (h_N^2). (See text for explanation)

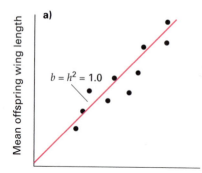
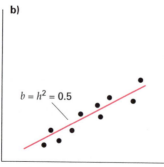
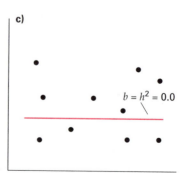

Mean parent wing length

traits of parents and offspring. We would conclude that additive genetic differences are not important in determining the phenotypic variability and that the heritability is low. On the other hand, if a definite relationship exists between the phenotypes of parents and offspring, as shown in Figure 23.12a and 23.12b, then additive genetic variance is more important and heritability is high (assuming that no common environmental effects between parents and offspring influence the trait).

In a plot of the parent and offspring phenotypes, the slope of the regression line can provide us with information about the magnitude of the heritability (see section of this chapter on statistics). If the slope is 0, as shown in Figure 23.12c, then the narrow-sense heritability h_N^2 is zero. When the slope of the parent-offspring regression is 1, as in Figure 23.12a, the mean offspring phenotype is exactly intermediate to the phenotype of the two parents, and genes with additive effects determine all the phenotypic differences. If the slope is less than 1 but greater than zero, as in Figure 23.12b, both additive genes and nonadditive factors (genes with dominance, genes with epistasis, and environmental factors) affect the phenotypic vari-

ation. It is possible to show mathematically that when the mean phenotype of the offspring is regressed against the mean phenotype of the parents, the narrow-sense heritability (h_N^2) equals the slope of the regression line (b):

$$h_N^2 = b \text{ (for the regression of mean offspring}$$
$$\text{phenotype and mean parental phenotype)}$$

When the mean phenotype of the offspring is regressed against the phenotype of only one parent, the narrow-sense heritability is twice the slope:

$$h_N^2 = 2b \text{ (for the regression of mean offspring}$$
$$\text{phenotype and one parental phenotype)}$$

This is because an offspring shares only half its genes with one of its parents. Similarly, the factor by which the slope must be multiplied to obtain heritability increases with increased distance between relatives.

Heritability values for a number of traits in different species are given in Table 23.6. These heritability values are based upon various populations and have been determined using a variety of methods, including the parent-offspring method. Estimates of heri-

~ TABLE 23.6

Heritability Values for Some Traits in Humans, Domesticated Animals, and Natural Populations[a]

ORGANISM	TRAIT	HERITABILITY
Humans	Stature	0.65
	Serum immunoglobulin (IgG) level	0.45
Cattle	Milk yield	0.35
	Butterfat content	0.40
	Body weight	0.65
Pigs	Back-fat thickness	0.70
	Litter size	0.05
Poultry	Egg weight	0.50
	Egg production (to 72 weeks)	0.10
	Body weight (at 32 weeks)	0.55
Mice	Body weight	0.35
Drosophila	Abdominal bristle number	0.50
Jewelweed	Germination time	0.29
Milkweed bugs	Wing length (females)	0.87
	Fecundity (females)	0.50
Spring peeper (frog)	Size at metamorphosis	0.69
Wood frog	Development rate (mountain population)	0.31
	Size at metamorphosis (mountain population)	0.62

[a]The estimates given in this table apply to particular populations in particular environments; heritability values for other individuals may differ.

tability are rarely precise, and most measured heritability values have large standard errors. This lack of precision is reflected in the fact that heritabilities calculated for the same trait in the same organism often vary widely. Heritability values calculated for human traits must be viewed with special caution, given the difficulties of separating genetic and environmental influences.

RESPONSE TO SELECTION

Two fields of study in which quantitative genetics has played a particularly important role are plant and animal breeding and evolutionary biology. Both fields are concerned with genetic change within groups of organisms: in the case of plant and animal breeding, genetic change can lead to improvement in yield, hardiness, size, and other agriculturally important qualities; in the case of evolutionary biology, genetic change occurs in natural populations as a result of the processes discussed in Chapter 22. **Evolution** can be defined as genetic change that takes place over time within a group of organisms. Therefore, both evolutionary biologists and plant and animal breeders are interested in the process of evolution, and both use the methods of quantitative genetics to predict the rate and magnitude of genetic change.

The essential element of natural selection is that individuals with certain genotypes leave more offspring than others. In this way, groups of individuals change or evolve over time and become better adapted to their particular environment. Humans bring about evolution in domestic plants and animals through the similar process of **artificial selection**. In artificial selection, humans, not nature, select the individuals that are to survive and reproduce. If the selected traits have a genetic basis, then the genetic structure of the selected population will change over time and evolve, just as traits in natural populations evolve as a result of natural selection. Artificial selection can be a powerful tool in bringing about rapid evolutionary change, as evidenced by the extensive variation observed among domesticated plants and animals. For example, all breeds of domestic dogs are derived from one species that was domesticated some 10,000 years ago. The large number of breeds that exist today, encompassing a tremendous variety of sizes, shapes, colors, and even behaviors, has been produced by artificial selection and selective breeding (see Figure 2.1, p. 19).

Both the process of natural selection, as described by Charles Darwin, and artificial selection, practiced by plant and animal breeders, depend upon the presence of genetic variation. Only if genetic variation is present within a population of individuals can that population change genetically and evolve. Furthermore, the amount and the type of genetic variation present is extremely important in determining how fast evolution will occur. Both evolutionary biologists and breeders are interested in the question of how much genetic variation for a particular trait exists within a population. As we have seen, quantitative genetics is often employed to answer this question.

Estimating the Response to Selection

When natural or artificial selection is imposed upon a phenotype, the phenotype will change from one generation to the next, provided that genetic variation underlying the trait is present in the population. The amount that the phenotype changes in one generation is termed the **selection response**. To illustrate the concept of selection response, suppose a geneticist wishes to produce a strain of *Drosophila melanogaster* with large body size. In order to increase body size in fruit flies, the geneticist would first examine flies from a genetically diverse population and would measure the body size of these *unselected flies*. Suppose that our geneticist found the mean body weight of the unselected flies to be 1.3 mg. After determining the mean body weight of this population, the geneticist would select flies that were endowed with large bodies (assume that the mean body weight of these selected flies was 3.0 mg). He would then place the large, selected flies in a separate culture vial and allow them to interbreed. After the F_1 offspring of these selected parents emerged, the geneticist would measure the body weights of the F_1 flies and compare them with the body weights of the original, unselected population.

What our geneticist has done in this procedure is to apply selection for large body size to the population of fruit flies. If genetic variation underlies the variation in body size of the original population, the offspring of the selected flies will resemble their parents and the mean body size of the F_1 generation will be greater than the mean body size of the original population. If the F_1 flies have a mean body weight of 2.0 mg, which is considerably larger than the mean body weight of 1.3 mg observed in the original, unselected population, a response to selection has occurred.

The amount of change that occurs in one generation, or the selection response, is dependent upon two things: the narrow-sense heritability and the **selection differential**. The selection differential is defined as the difference between the mean phenotype of the selected parents and the mean phenotype of the unselected population. In our example of body size in fruit flies, the original population had a mean weight of 1.3

mg, and the mean weight of the selected parents was 3.0 mg, so the selection differential is 3.0 mg – 1.3 mg = 1.7 mg. The selection response is related to the selection differential and the heritability by the following formula:

$$\text{Selection response} = \text{narrow-sense heritability} \times \text{selection differential}$$

When the geneticist applied artificial selection to body size in fruit flies, the difference in the mean body weight of the F_1 flies and the original population was 2.0 mg – 1.3 mg = 0.7 mg, which is the response to selection. We now have values for two of the three parameters in the above equation: the selection response (0.7 mg) and the selection differential (1.7 mg). By rearranging the formula for the selection response, we can solve for the narrow-sense heritability.

$$\text{Narrow-sense heritability} = h_N^2 = \text{selection response}/\text{selection differential}$$

$$h_N^2 = 0.7 \text{ mg}/1.7 \text{ mg} = 0.41$$

Measuring the response to selection provides another means for determining the narrow-sense heritability, and heritabilities for many traits are determined in this way.

A trait will continue to respond to selection, generation after generation, as long as genetic variation for the trait remains within the population. The results from an actual, long-term selection experiment on phototaxis in *Drosophila pseudoobscura* are presented in Figure 23.13. Phototaxis is a behavioral response to light. In this study, flies were scored for the number of times the fly moved toward light in a total of 15 light–dark choices. Two different response-to-selection experiments were carried out. In one, attraction to light was selected, and in the other, avoidance of light was selected. As can be seen in Figure 23.13, the fruit flies responded to selection for positive and negative phototactic behavior for a number of generations. Eventually, however, the response to selection tapered off, and finally no further directional change in phototactic behavior occurred. One possible reason for this lack of response in later generations is that no more genetic variation for phototactic behavior existed within the population. In other words, all flies at this point were homozygous for all the alleles affecting the behavior. If this were the case, phototactic behavior could not undergo further evolution in this population unless input of additional genetic variation occurred. More often, some variation still exists for the trait, even after the selection response levels off, but the population fails to respond to selection because the genes for the selected trait have detrimental effects on other traits. These detrimental effects occur because of genetic correlations, which are discussed in the next section.

KEYNOTE

The amount that a trait changes in one generation as a result of selection for the trait is called the selection response or the response to selection. The magnitude of selection response depends on both the intensity of selection, what is termed the selection differential, and the narrow-sense heritability.

Genetic Correlations

The phenotypes of two or more traits may be associated or correlated; this means that the traits do not vary independently. For example, fair skin, blond hair, and blue eyes are often found together in the same individual. The association is not perfect—we sometimes see individuals with dark hair, fair skin, and blue eyes—but the traits are found together with enough regularity for us to say that they are correlated. The **phenotypic correlation** between two traits can be computed by measuring two phenotypes on a

~ FIGURE 23.13

Selection for phototaxis in *Drosophila pseudoobscura*. The upper graph is the line selected for avoidance of light. The lower graph is the line selected for attraction to light. The phototactic score is the number of times the fly moved toward the light out of a total of 15 light-dark choices.

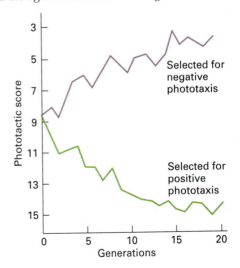

number of individuals and then calculating a correlation coefficient for the two traits (see the section on statistics for a discussion of correlation coefficients). One reason for a phenotypic correlation among traits is that the traits are influenced by a common set of genes. Indeed, this is the most likely reason for the association among hair color, eye color, and skin color in humans. Rarely do genes affect only a single trait. More commonly, each gene influences a number of traits, and this is particularly true for the polygenes that influence continuous traits. When genes affect multiple phenotypes, we say they are *pleiotropic*, a concept we introduced in Chapter 9. If a gene is pleiotropic, it simultaneously affects two or more traits, and thus the phenotypes of those two traits will be correlated. For example, the genes that affect growth rates in humans influence both weight and height, so these two phenotypes tend to be correlated. When pleiotropy is present (i.e., some of the genes influencing two traits are the same), we say that a **genetic correlation** exists for the traits.

Pleiotropy is not the only cause of phenotypic correlations among traits. Environmental factors may also influence several traits simultaneously and may cause nonrandom associations between phenotypes. For example, adding fertilizer to the soil often causes plants both to grow taller and to produce more flowers. If we measured plant height and counted the number of flowers on a group of responsive plants, some of which received fertilizer and some of which did not, we would find that the two traits are correlated; those plants receiving fertilizer would be tall and would have many flowers, while those without fertilizer would be short and have few flowers. However, this phenotypic correlation does not result from any genetic correlation (pleiotropy), but from the common effect of the environmental factor, the fertilizer, on both traits.

Genetic correlations may be positive or negative. A positive correlation means that genes causing an increase in the magnitude of one trait bring about a simultaneous increase in the magnitude of the other. In chickens, body weight and egg weight have a positive genetic correlation. If breeders select for heavier chickens, thereby favoring the genes for large body size, the size of the chickens will increase and the mean weight of the eggs produced by these chickens will also increase. This increase in egg weight occurs because the genes that produce heavier chickens have a similar effect on egg weight.

Other traits exhibit negative genetic correlations. In this case, genes that cause an increase in one trait tend to produce a corresponding decrease in another trait. Egg weight and egg number in chickens have a negative genetic correlation. When breeders select for chickens that produce larger eggs, the average egg size increases, but the number of eggs laid by each chicken decreases.

Genetic correlations are important for predicting the changes in phenotype that result from selection. If two traits are genetically correlated, they evolve together. Any genetic change in one trait occurring as a result of selection will simultaneously alter the other trait, since both are influenced by the same genes. Such correlation often presents problems for plant and animal breeders. As an example, milk yield and butterfat content have a negative genetic correlation in cattle. The same genes that cause an increase in milk production bring about a decrease in butterfat content of the milk. Thus, when breeders select for increased milk yield, the amount of milk produced by the cows may go up, but the butterfat content will decrease at the same time. Negative correlations among traits often place practical constraints on the ability of breeders to select for desirable traits. Knowing about the presence of genetic correlations before undertaking an expensive breeding program is essential in order to avoid the production of associated undesirable traits in the selected stock.

An organism's ability to adapt to a particular environment is also strongly influenced by genetic correlations among traits, and thus genetic correlations are of considerable interest to evolutionary biologists. As an illustration, consider two traits in tadpoles: developmental rate and size at metamorphosis. Most tadpoles are found in small ponds and pools, where fish (potential predators) are absent and food is abundant. A major liability in using this type of aquatic habitat is that ponds often dry up, frequently before the tadpoles have developed sufficiently to metamorphose into frogs and leave the water. One might expect, then, that natural selection would favor a maximum rate of development in tadpoles, so that the tadpoles could quickly metamorphose into frogs. However, many species of tadpoles fail to develop at maximum rates, contrary to what we might expect to evolve under natural selection. One reason for the slow rate of development is that a negative genetic correlation may exist between developmental rate and body size at metamorphosis. Genes that accelerate development also tend to cause metamorphosis at a smaller size, at least in some populations. Thus, selection for fast metamorphosis will also produce smaller frogs, and size is extremely important in determining the survival of young frogs. Small frogs tend to lose water more rapidly in the terrestrial environment, are more likely to be eaten by predators, and have more difficulty finding sufficient food. The negative genetic correlation between developmental rate and body size at metamorphosis places constraints on the frogs' ability to evolve rapid development and on their ability to evolve large body size at metamorphosis. Knowing about such genetic correlations is important for understanding how animals adapt or fail to adapt to a particular environment. Table 23.7 presents some genetic correlations that have been detected in studies of quantitative genetics.

*K*EYNOTE

Genetic correlations arise when two traits are influenced by the same genes. When a trait is selected, any genetically correlated traits will also exhibit a selection response. Thus, the evolution of a population, the outcome of an artificial breeding program, or the ability of a population to respond evolutionarily and avoid extinction, depends on the simultaneous integration and correlation of many aspects of the phenotype.

SUMMARY

Quantitative genetics is the field of genetics that studies the inheritance of continuous, or quantitative traits—those traits with a range of phenotypes. Continuous traits usually result from the influence of multiple genes and environmental factors. Statistics such as the mean, variance, and standard deviation can be used to describe continuous traits; correlation, regression, and analysis of variance are statistical procedures that are used for studying continuous traits.

Polygenic traits are caused by genes at multiple loci, each of which follows the principles of Mendelian inheritance. The multiple gene hypothesis assumes that the effects of the alleles at a locus are small and additive, but at present this is at best a working model and much more needs to be known.

The broad-sense heritability is the proportion of the phenotypic variance in a population that is due to genetic differences. The narrow-sense heritability indicates the proportion of the phenotypic variance that results from additive genetic variance.

The response to selection is the amount a trait changes in one generation as a result of selection; it depends upon the narrow-sense heritability and the selection differential. Genetic correlation occurs when

~ TABLE 23.7

Genetic Correlations among Traits in Humans, Domesticated Animals, and Natural Populations[a]

ORGANISM	TRAITS	GENETIC CORRELATION
Humans	IgG, IgM	0.07
Cattle	Butterfat content, milk yield	−0.38
Pigs	Weight gain, back-fat thickness	0.13
	Weight gain, efficiency	0.69
Chickens	Egg weight, egg production	−0.31
	Body weight, egg weight	0.42
	Body weight, egg production	−0.17
Mice	Body weight, tail length	0.29
Jewelweed	Seed weight, germination time	−0.81
Milkweed bugs	Wing length, fecundity	−0.57
Wood frogs	Developmental rate, size at metamorphosis	−0.86
Drosophila	Early life fecundity, resistance to starvation	−0.91

[a]The estimates given in this table apply to particular populations in particular environments; genetic correlations for other individuals may differ.

two traits are influenced by the same genes. When a trait responds to selection, any genetically correlated trait will also exhibit a selection response. This will greatly influence the direction and speed of both natural and artificial selection to change the genetic and phenotypic structures of populations. Evolutionary responses that result from the effects of natural selection depend on heritability in the narrow-sense.

The field of quantitative genetics strives to clarify the relationship between a complex genetic architecture and complex phenotypic architecture. Are most continuous traits controlled by many genes with small effects or are they controlled by relatively few genes with major effects? How important are epistatic interactions and pleiotropic effects? These are not easy questions to answer. But, in the answers will lie extremely important findings that will impact the way in which we view the relationship between the genotype and phenotype, which in turn impacts our understanding of developmental biology and evolution.

ANALYTICAL APPROACHES FOR SOLVING GENETICS PROBLEMS

Q23.1   Assume that genes A, B, C, and D are members of a multiple-gene series that control a quantitative trait. Each of these genes has a duplicate, cumulative effect in that each contributes 3 cm of height to the organism when it is present. Each gene assorts independently. In addition, gene L is always present in the homozygous state, and the LL genotype contributes a constant 40 cm of height. The alleles a, b, c, and d do not contribute anything to the height of the organism. If we ignore height variation caused by environmental factors, an organism with genotype $AA\ BB\ CC\ DD\ LL$ would be 64 cm high, and one with genotype $aa\ bb\ cc\ dd\ LL$ would be 40 cm. A cross is made of $AA\ bb\ CC\ DD\ LL \times aa\ BB\ cc\ DD\ LL$ and is carried into the F_2 by selfing of the F_1.

a.   How does the size of the F_1 individuals compare with the size of each of the parents?

b.   Compare the mean of the F_1 with the mean of the F_2, and comment on your findings.

c.   What proportion of the F_2 population would show the same height as the $AA\ bb\ CC\ DD\ LL$ parent?

d.   What proportion of the F_2 population would show the same height as the $aa\ BB\ cc\ DD\ LL$ parent?

e.   What proportion of the F_2 population would breed true for the height shown by the $aa\ BB\ cc\ DD\ LL$ parent?

f.   What proportion of the F_2 population would breed true for the height characteristic of F_1 individuals?

A23.1   This question explores our understanding of the basic genetics involved in a multiple-gene series that in this case controls a quantitative trait. The approach we will take is essentially the same as the approach used with a series of independently assorting genes that control distinctly different traits. That is, we make predictions on the basis of genotypes and relate the results to phenotypes, or we make predictions on the basis of phenotypes and relate the results to genotypes.

a.   Each allele represented by a capital letter contributes 3 cm of height to the base height of 40 cm, which is controlled by the ever present LL homozygosity. Therefore the $AA\ bb\ CC\ DD\ LL$ parent, which has six capital-letter alleles from the A-through-D, multiple-gene series, is $40 + (6 \times 3) = 58$ cm high. Similarly, the $aa\ BB\ cc\ DD\ LL$ parent has four capital-letter alleles and therefore is $40 + 12 = 52$ cm high. The F_1 from a cross between these two individuals would be heterozygous for the A, B, and C loci and homozygous D and L, that is, $Aa\ Bb\ Cc\ DD\ LL$. This progeny has five capital-letter alleles apart from LL and hence is $40 + 15 = 55$ cm high.

b.   The F_2 is derived from a self of the $Aa\ Bb\ Cc\ DD\ LL$ F_1. All the F_2 individuals will be $DD\ LL$, making them at least $40 + 6 = 46$ cm high. Now we must deal with the heterozygosity at the other three loci. What we need to calculate is the relative proportions of individuals with all the various possible numbers of capital-letter alleles. This calculation is equivalent to determining the relative distribution of three independently assorting traits, each showing incomplete dominance. In other words, we must calculate directly the relative frequencies of all possible genotypes for the three loci and collect those with no, one, two, three, four, five, and six capital-letter alleles. Thus the probability of getting an individual with two capital-letter alleles for each locus is $1/4$, the probability of getting an individual with one capital-letter allele for each locus is $1/2$, and the probability of getting an individual with no capital-letter alleles for each locus is $1/4$. So the probability of getting an F_2 individual with six capital-letter alleles for the A, B, and C loci is $(1/4)^3 = 1/64$, and the same probability is obtained for an individual with no capital-letter alleles. This analysis gives us a clue about how we should consider all the possible combinations of genotypes that have the other numbers of capital-letter alleles. That is, the simplest approach is to compute the coefficients in the binomial expansion of $(a + b)^6$, as explained on p. 786. The expansion gives a 1:6:15:20:15:6:1 distribution of zero, one, two, three, four, five, and six capital-letter alleles, respectively. Now since each capital-letter allele in the A, B, and C set contributes 3 cm of height over the 46-cm height given by the $DD\ LL$ genotype common to all, then the F_2 individuals would fall into the following distribution:

NUMBER OF CAPITAL-LETTER ALLELES	HEIGHT ADDED TO BASIC HEIGHT OF 46 CM FOR COMMON $DD\ LL$ GENOTYPE (cm)	HEIGHT OF INDIVIDUALS (cm)	FREQUENCY
6	18	64	1
5	15	61	6
4	12	58	15
3	9	55	20
2	6	52	15
1	3	49	6
0	0	46	1

The distribution is clearly symmetrical, giving an average of 55 cm, the same height shown in F_1 individuals.

c. The $AA\ bb\ CC\ DD\ LL$ parent was 58 cm, so we can read the proportion of F_2 individuals that show this same height directly from the table in part b. The answer is 15/64.

d. The $aa\ BB\ cc\ DD\ LL$ parent was 52 cm, and from the table in part b the proportion of F_2 individuals that show this same height is 15/64.

e. We are asked to determine the proportion of the F_2 population that would breed true for the height shown by the $aa\ BB\ cc\ DD\ LL$ parent, which was 52 cm. To breed true, the organism must be homozygous. We have also established that $DD\ LL$ is a constant genotype for the F_2 individuals, giving a basic height of 46 cm. Therefore for a height of 52 cm, two additional, active, capital-letter alleles must be present, apart from those at the D and L loci. With the requirement for homozygosity, there are only three genotypes that give a 52-cm height; they are $AA\ bb\ cc\ DD\ LL$, $aa\ BB\ cc\ DD\ LL$, and $aa\ bb\ CC\ DD\ LL$. The probability of each combination occurring in the F_2 is 1/64, so the answer to the problem is 1/64 + 1/64 + 1/64 = 3/64. (Note that the individual probability for each genotype may be calculated. That is, probability of AA = 1/4, probability of bb 1/4, probability of cc = 1/4, probability of $DD\ LL$ = 1, giving an overall probability for $AA\ bb\ cc\ DD\ LL$ of 1/64.)

f. We are asked to determine the proportion of the F_2 population that would breed true for the height characteristic of F_1 individuals. Again, the basic height given by $DD\ LL$ is 46 cm. The F_1 height is 55 cm, so three capital-letter alleles must be present in addition to $DD\ LL$ to give that height, since (3×3) cm = 9 cm, and 9 cm + 46 cm = 55 cm. However, since an individual must be homozygous to be true breeding, the answer to this question is none, since 3 is an odd number, meaning that at least one locus must be heterozygous in order to get the 55-cm height.

Q23.2 Five field mice collected in Texas had weights of 15.5 g, 10.3 g, 11.7 g, 17.9 g, and 14.1 g. Five mice collected in Michigan had weights of 20.2 g, 21.1 g, 20.4 g, 22.0 g, and 19.7 g. Calculate the mean weight and the variance in weight for mice from Texas and for mice from Michigan.

A23.2 To answer this question, we use the formulas given in the chapter section on statistics. The formula for the mean is

$$\bar{x} = \frac{\Sigma x_i}{n}$$

The symbol Σ means to add, and the x_i represents all the individual values. We begin by summing up all the weights of the mice from Texas:

$$\Sigma x_i = 15.5 + 10.3 + 11.7 + 17.9 + 14.1 = 69.5$$

Next, we divide this summation by n, which represents the number of values added together. In this case, we added together five weights, so $n = 5$. The mean for the Texas mice is therefore

$$\frac{\Sigma x_i}{n} = \frac{69.5}{5} = 13.9$$

To calculate the variance in weight among the Texas mice, we utilize the formula

$$s^2 = \frac{\Sigma(x_i - \bar{x})^2}{n - 1}$$

We must take each individual weight and subtract it from the mean weight of the group. Each value obtained from this subtraction is then squared, and all squared values are added up, as shown below.

1.5 − 13.9 = 1.6	$(1.6)^2 =$	2.56
10.3 − 13.9 = −3.6	$(−3.6)^2 =$	12.96
11.7 − 13.9 = 2.2	$(−2.2)^2 =$	4.84
17.9 − 13.9 = 4.0	$(4.0)^2 =$	16.0
14.1 − 13.9 = 0.2	$(0.2)^2 =$	0.04
		36.4

The sum of all the squared values is 36.4. All that remains for us to do is to divide this sum by $n - 1$, which is $5 - 1 = 4$.

$$s^2 = \frac{\Sigma(x_i - \bar{x})^2}{n-1} = \frac{36.4}{4} = 9.1$$

The mean and the variance for the Texas mice are therefore 13.9 and 9.1.

We now repeat these steps for the mice from Michigan.

$$\Sigma x_i = 20.2 + 21.2 + 20.4 + 22.0 + 19.7 = 103.5$$

$$\frac{\Sigma x_i}{n} = \frac{36.4}{4} = 9.1$$

$$s^2 = \frac{\Sigma(x_i - \bar{x})^2}{n-1}$$

$20.2 - 20.7 = -0.5$	$(-0.5)^2 = 0.25$
$21.2 - 20.7 = 0.5$	$(0.5)^2 = 0.25$
$20.4 - 20.7 = -0.3$	$(-0.3)^2 = 0.9$
$22.0 - 20.7 = 1.3$	$(1.3)^2 = 1.69$
$19.7 - 20.7 = -1.0$	$(-1.0)^2 = 1.0$
	4.09

$$s^2 = \frac{\Sigma(x_i - \bar{x})^2}{n-1} = \frac{4.09}{4} = 1.23$$

The mean and the variance for the Michigan mice are 20.79 and 1.23.

We conclude that the Michigan mice are considerably heavier than the Texas mice and the Michigan mice also exhibit less variance in weight.

QUESTIONS AND PROBLEMS

*23.1 The following measurements of head width and wing length were made on a series of steamer-ducks:

SPECIMEN	HEAD WIDTH (cm)	WING LENGTH (cm)
1	2.75	30.3
2	3.20	36.2
3	2.86	31.4
4	3.24	35.7
5	3.16	33.4
6	3.32	34.8
7	2.52	27.2
8	4.16	52.7

a. Calculate the mean and the standard deviation of head width and of wing length for these eight birds.
b. Calculate the correlation coefficient for the relationship between head width and wing length in this series of ducks.
c. What conclusions can you make about the association between head width and wing length in steamer-ducks?

23.2 Answer the following questions:
a. In a family of six children, what is the probability that three will be girls and three will be boys?
b. In a family of five children, what is the probability that one will be a boy and four will be girls?
c. What is the probability that in a family of six children, all will be boys?

*23.3 In flipping a coin, there is a 50 percent chance of obtaining heads and a 50 percent chance of obtaining tails on each flip. If you flip a coin 10 times, what is the probability of obtaining exactly 5 heads and 5 tails?

*23.4 The F_1 generation from a cross of two pure-breeding parents that differ in a size character is usually no more variable than the parents. Explain.

23.5 If two pure-breeding strains, differing in a size trait, are crossed, is it possible for F_2 individuals to have phenotypes that are more extreme than either grandparent (i.e., be larger than the largest or smaller than the smallest in the parental generation)? Explain.

23.6 Two pairs of genes with two alleles each, A/a and B/b, determine plant height additively in a population. The homozygote $AA\ BB$ is 50 cm tall, the homozygote $aa\ bb$ is 30 cm tall.
a. What is the F_1 height in a cross between the two homozygous stocks?
b. What genotypes in the F_2 will show a height of 40 cm after an $F_1 \times F_1$ cross?
c. What will be the F_2 frequency of the 40-cm plants?

*23.7 Three independently segregating genes (A, B, C), each with two alleles, determine height in a plant. Each capital-letter allele adds 2 cm to a base height of 2 cm.
a. What are the heights expected in the F_1 progeny of a cross between homozygous strains $AA\ BB\ CC$ (14 cm) $\times aa\ bb\ cc$ (2 cm)?
b. What is the distribution of heights (frequency and phenotype) expected in an $F_1 \times F_1$ cross?
c. What proportion of F_2 plants will have heights equal to the heights of the original two parental strains?
d. What proportion of the F_2 will breed true for the height shown by the F_1?

23.8 Repeat Problem 23.7, but assume that each capital-letter allele acts to double the existing height; for example, $Aa\ bb\ cc = 4$ cm, $AA\ bb\ cc = 8$ cm, $AA\ Bb\ cc = 16$ cm, and so on.

23.9 Assume three equally and additively contributing pairs of alleles control flower length in nasturtiums. A completely homozygous plant with 10-mm flowers is crossed to a completely homozygous plant with 30-mm flowers. F_1 plants all have flowers about 20 mm long. F_2 plants show a range of lengths from 10 to 30 mm, with about 1/64 of the F_2 having 10-mm flowers and 1/64 having 30-mm flowers. What distribution of flower length would you expect to see in the offspring of a cross between an F_1 plant and the 30-mm parent?

*23.10 In a particular experiment the mean internode length in spikes of the barley variety *asplund* was found to be 2.12 mm. In the variety *abed binder* the mean internode length was found to be 3.17 mm. The mean of the F_1 of a cross between the two varieties was approximately 2.7 mm. The F_2 gave a continuous range of variation from one parental extreme to the other. Analysis of the F_3 generation showed that in the F_2 8 out of the total 125 individuals were of the *asplund* type, giving a mean of 2.19. mm. Eight other individuals were similar to the parent *abed binder*, giving a mean internode length of 3.24 mm. Is the internode length in spikes of barley a discontinuous or a quantitative trait? Why?

23.11 From the information given in Problem 23.10, determine how many gene pairs involved in the determination of internode length are segregating in the F_2.

23.12 Assume that the difference between a type of oats yielding about 4 g per plant and a type yielding 10 g is the result of three equal and cumulative multiple-gene pairs *AA BB CC*. If you cross the type yielding 4 g with the type yielding 10 g, what will be the phenotypes of the F_1 and the F_2? What will be their distribution?

*23.13 Assume that in squashes the difference in fruit weight between a 3-lb type and a 6-lb type is due to three allelic pairs, *A/a*, *B/b*, and *C/c*. Each capital-letter allele contributes a half pound to the weight of the squash. From a cross of a 3-lb plant (*aa bb cc*) with a 6-lb plant (*AA BB CC*), what will be the phenotypes of the F_1 and the F_2? What will be their distribution?

23.14 Refer to the assumptions stated in Problem 23.13. Determine the range in fruit weight of the offspring in the following squash crosses: (a) *AA Bb CC × aa Bb Cc*; (b) *AA bb Cc × Aa BB cc*; (c) *aa BB cc × AA BB cc*.

*23.15 Assume that the difference between a corn plant 10 dm (decimeters) high and one 26 dm high is due to four pairs of equal and cumulative multiple alleles, with the 26-dm plants being *AA BB CC DD* and the 10-dm plants being *aa bb cc dd*.
a. What will be the size and genotype of an F_1 from a cross between these two true-breeding types?
b. Determine the limits of height variation in the offspring from the following crosses:
   (**1**) *Aa BB cc dd × Aa bb Cc dd*;
   (**2**) *aa BB cc dd × Aa Bb Cc dd*;
   (**3**) *AA BB Cc DD × aa BB cc Dd*;
   (**4**) *Aa Bb Cc Dd × Aa bb Cc Dd*.

23.16 Refer to the assumptions given in Problem 23.15. But for this problem two 14-dm corn plants, when

crossed, give nothing but 14-dm offspring (case A). Two other 14-dm plants give one 18-dm, four 16-dm, six 14-dm, four 12-dm, and one 10-dm offspring (case B). Two other 14-dm plants, when crossed, give one 16-dm, two 14-dm, and one 12-dm offspring (case C). What genotypes for each of these 14-dm parents (cases A, B, and C) would explain these results? Would it be possible to get a plant taller than 48 dm by selection in any of these families?

*23.17 Pigmentation in the imaginary river-bottom dweller *Mucus yuccas* is a quantitative character controlled by a set of five independently segregating polygenes with two alleles each: *A/a*, *B/b*, *C/c*, *D/d*, and *E/e*. Pigment is deposited at three different levels, dependent on the threshold of gene products produced by the capital-lettered alleles. Greyish-brown pigmentation is seen if at least four capital alleles are present, light-tan pigmentation is seen if two or three capital alleles are present, and whitish-blue pigmentation is seen if these thresholds are not met. If an *AA BB CC DD EE* animal is crossed to a true-breeding *aa bb cc dd ee* animal and the progeny are selfed, what kinds of phenotypes are expected in the F_1 and F_2?

*23.18 Alzheimer's disease (AD) is the leading cause of dementia in elderly individuals. Evidence that genetic alterations are involved in AD comes from three sources: the incidence of AD in first-degree relatives, the incidence in pairs of twins, and pedigree analysis. There is a 24–50 percent risk of AD by age 90 in first-degree relatives of AD individuals, a 40–50 percent risk of AD in the identical (monozygotic) twin of an AD individual, and a 10–50 percent risk of AD in the fraternal (dizygotic) twin of an AD individual. AD individuals in a subset of families showing AD have an alteration in the *APP* (amyloid protein) gene on chromosome 21. AD individuals in another subset of AD families have a particular allele (*E4*) at the *APOE* (apolipoprotein E) gene on chromosome 19. Individuals homozygous for the *E4* allele have increased risk of AD and earlier disease onset than heterozygotes. Population studies have shown that 40–50 percent of AD cases are associated with alterations in the *APOE* gene, but less than 1 percent of AD cases are associated with mutations in the *APP* gene.
a. In what sense might AD be considered a polygenic trait?
b. If AD has a genetic basis, why are identical twins not equally affected?

23.19 Since monozygotic twins share all of their genetic material and dizygotic twins share, on average, half of their genetic material, twin studies can sometimes be useful for evaluating the genetic contribution to a trait. Consider the following two instances.

An intelligence quotient (IQ) assesses intellectual performance on a standardized test that involves reasoning ability, memory, and knowledge of an individual's language and culture. IQ scores are transformed so that the population mean score is 100 and 95 percent of the individuals have scores in the range between 70 and 130. Observations in the United States and England found that monozygotic twins had an average difference of 6 IQ points, dizygotic twins had an average difference of 11 points, and random pairs of individuals had an average difference of 21 points.

In a large sample of pairs of twins in the United States where one twin was a smoker, 83 percent of monozygotic twins both smoked, while 62 percent of dizygotic twins both smoked.

From these data, can you infer the genetic determination of IQ or smoking?

23.20   A quantitative geneticist determines the following variance components for leaf width in a population of wildflowers growing along a roadside in Kentucky:

Additive genetic variance (V_A)	= 4.2
Dominance genetic variance (V_D)	= 1.6
Interaction genetic variance (V_I)	= 0.3
Environmental variance (V_E)	= 2.7
Genetic-environmental variance (V_{GE})	= 0.0

a. Calculate the broad-sense heritability and the narrow-sense heritability for leaf width in this population of wildflowers.
b. What do the heritabilities obtained in part a indicate about the genetic nature of leaf width variation in this plant?

*23.21   Assume all genetic variance affecting seed weight in beans is genetically determined and is additive. From a population where the mean seed weight was 0.88 g, a farmer selected two seeds, each weighing 1.02 g. He planted these and crossed the resulting plants to each other, then collected and weighed their seeds. The mean weight of their seeds was 0.96 g. What is the narrow-sense heritability of seed weight?

*23.22   Members of the inbred rat strain SHR are salt sensitive: they respond to a high salt environment by developing hypertension. Members of a different inbred rat strain, TIS, are not salt sensitive. Imagine you placed a population consisting only of SHR rats in an environment that was variable in regard to distribution of salt, so that some rats would be exposed to more salt than others. What would be the heritability of blood pressure in this population?

*23.23   In Kansas a farmer is growing a variety of wheat called TK138. He calculates the narrow-sense heritability for yield (the amount of wheat produced per acre) and finds that the heritability of yield for TK138 is 0.95. The next year he visits a farm in Poland and observes that a

Russian variety of wheat, UG334, growing there has only about 40 percent as much yield as TK138 grown on his farm in Kansas. Since he found the heritability of yield in his wheat to be very high, he concludes that the American variety of wheat (TK138) is genetically superior to the Russian variety (UG334), and he tells the Polish farmers that they can increase their yield by using TK138. What is wrong with his conclusion?

23.24   Dermatoglyphics are the patterns of the ridged skin found on the fingertips, toes, palms, and soles. (Fingerprints are dermatoglyphics.) Classification of dermatoglyphics is frequently based on the number of triradii; a triradius is a point from which three ridge systems separate at angles of 120°. The number of triradii on all ten fingers was counted for each member of several families, and the results are tabulated below.

FAMILY	MEAN NUMBER OF TRIRADII IN THE PARENTS	MEAN NUMBER OF TRIRADII IN THE OFFSPRING
I	14.5	12.5
II	8.5	10.0
III	13.5	12.5
IV	9.0	7.0
V	10.0	9.0
VI	9.5	9.5
VII	11.5	11.0
VIII	9.5	9.5
IX	15.0	17.5
X	10.0	10.0

a. Calculate the narrow-sense heritability for the number of triradii by the regression of the mean phenotype of the parents against the mean phenotype of the offspring.
b. What does your calculated heritability value indicate about the relative contributions of genetic variation and environmental variation to the differences observed in number of triradii?

*23.25   A scientist wishes to determine the narrow-sense heritability of tail length in mice. He measures tail length among the mice of a population and finds a mean tail length of 9.7 cm. He then selects the ten mice in the population with the longest tails; mean tail length in these selected mice is 14.3 cm. He interbreeds the mice with the long tails and examines tail length in their progeny. The mean tail length in the F_1 progeny of the selected mice is 13 cm.

Calculate the selection differential, the response to selection, and the narrow-sense heritability for tail length in these mice.

23.26   Suppose that the narrow-sense heritability of wool length in a breed of sheep is 0.92, and the narrow-sense heritability of body size is 0.87. The genetic correlation between wool length and body size is –0.84. If a breeder selects for sheep with longer wool, what will be the most likely effects on wool length and on body size?

*23.27 The heights of ten college-age males and the heights of their fathers are presented below.

HEIGHT OF SON (INCHES)	HEIGHT OF FATHER (INCHES)
70	70
72	76
71	72
64	70
66	70
70	68
74	78
70	74
73	69

a. Calculate the mean and the variance of height for the sons and do the same for the fathers.
b. Calculate the correlation coefficient for the relationship between the height of father and height of son.
c. Determine the narrow-sense heritability of height in this group by regression of the son's height on the height of father.

*23.28 The narrow-sense heritability of egg weight in a particular flock of chickens is 0.60. A farmer selects for increased egg weight in this flock. The difference in the mean egg weight of the unselected chickens and the selected chickens is 10 g. How much should egg weight increase in the offspring of the selected chickens?

23.29 Members of a strain of White Leghorn chickens are selectively crossed to produce two lines, A and B, that show improved egg production. The progeny from a cross of lines A and B are used for commercial egg production. The selection strategy is shown in Figure 23.A. The mean number of eggs produced in the first egg-production year and the mean egg-weight (in grams) from hens at an age of 240 days is given for animals at each step of the selection procedure.

a. What is the narrow-sense heritability for the traits at each selection step?
b. Why does the response of the traits to selection change during the selection process?
c. What percentage increase in production is obtained when lines A and B are crossed?
d. With the possible exception of dairy cattle, commercial livestock are hybrids produced by crossing breeds, lines, or strains already selected for a set of desirable traits. Why?

*23.30 The variances tabulated below were determined for measurements of body length, antenna bristle number, and egg production in a moth species. Which of these characters would be most rapidly changed by natural selection? Which character would be least affected by natural selection?

VARIANCE	BODY LENGTH	ANTENNA BRISTLE NUMBER	EGG PRODUC-TION
Phenotypic (V_P)	798	342	145
Additive (V_A)	132	21	21
Dominance (V_D)	122	126	24
Interaction (V_I)	118	136	34
Genetic-environmental ($V_{G \times E}$)	81	23	21
Maternal effects (V_{Em})	345	36	45

FIGURE 23.A

Parental White Leghorn Strain (196 eggs, 51 g)

$\rightarrow$ Select Set A (279 eggs, 57 g) $\rightarrow$ Select Set B (310 eggs, 53 g)
    $\downarrow$ $\downarrow$
    F_1 (208 eggs, 54 g) F_1 (217 eggs, 52 gm)
    $\rightarrow$ Select F_1 (271 eggs, 60 g) $\rightarrow$ Select F_1 (310 eggs, 55 g)
        $\downarrow$ $\downarrow$
        F_2 (214 eggs, 55 g) F_2 (224 eggs, 53 g)
        $\rightarrow$ Select F_2 (292 eggs, 61 g) $\rightarrow$ Select F_2 (315 eggs, 57 g)
            $\downarrow$ $\downarrow$
            Line A (218 eggs, 56 g)   Line B (230 eggs, 54 g)
Line A $\times$ Line B $\rightarrow$ Hybrid used for commercial production (262 eggs, 57 g)

GLOSSARY

acrocentric chromosome A chromosome with a centromere near the end such that it has one long arm plus a stalk and a satellite.

additive genetic variance (VA) Genetic variance that arises from the additive effects of genes on the phenotype.

adenine (A) A purine base found in RNA and DNA; in double-stranded DNA adenine pairs with the pyrimidine thymine.

allele One of two or more alternative forms of a single gene locus. Different alleles of a gene each have a unique nucleotide sequence, and their activities are all concerned with the same biochemical and developmental process, although their individual phenotypes may differ.

allelic frequencies The frequencies of alleles at a locus occurring among individuals in a population.

allelomorph (allele) A term coined by William Bateson; literally means "alternative form"; later shortened by others to *allele*.

allopolyploidy Polyploidy involving two or more genetically distinct sets of chromosomes.

alternation of generations The two distinct reproductive phases of green plants in which stages alternate between haploid cells and diploid cells (gametophyte cells and sporophyte cells).

Ames test A test developed by Bruce Ames in the early 1970s that investigates new or old environmental chemicals for carcinogenic effects. It uses the bacterium *Salmonella typhimurium* as a test organism for mutagenicity of compounds.

amino acids The building blocks of polypeptides. There are 20 different amino acids.

aminoacyl-tRNA A tRNA molecule covalently bound to an amino acid. This complex brings the amino acid to the ribosome so that it can be used in polypeptide synthesis.

aminoacyl-tRNA synthetase An enzyme that catalyzes the addition of a specific amino acid to a tRNA molecule. Since there are 20 amino acids, there are also 20 synthetases.

amniocentesis A procedure in which a sample of amniotic sac fluid is withdrawn from the amniotic sac of a developing fetus and cells are cultured and examined for chromosomal abnormalities.

analysis of variance (ANOVA) A series of statistical procedures for examining differences in means and for partitioning variance.

anaphase The stage in mitosis or meiosis during which the sister chromatids (mitosis) or homologous chromosomes (meiosis) separate and migrate toward the opposite poles of the cell.

anaphase II The second stage of meiosis during which the centromeres (and therefore the chromatids) are pulled to the opposite poles of the spindle. The separated chromatids are now referred to as chromosomes in their own right.

aneuploidy The abnormal condition in which one or more whole chromosomes of a normal set of chromosomes either are missing or are present in more than the usual number of copies. Aneuploidy also refers to the abnormal condition in which a part or parts of a chromosome or chromosomes are duplicated or deleted.

antibody A protein molecule that recognizes and binds to a foreign substance introduced into the organism.

anticodon A three-nucleotide sequence that pairs with a codon in mRNA by complementary base pairing.

antigen Any large molecule that stimulates the production of specific antibodies or that binds specifically to an antibody.

applied research Research done with an eye towards making products that can be commercialized, or at least made available to humankind or practical benefit.

artificial selection Human determination as to which individuals will survive and reproduce. If the selected traits have a genetic basis, they will change and evolve.

asexual (vegetative) reproduction Reproduction in which a new individual develops either from a single cell or from a group of cells in the absence of any sexual process.

attenuation A regulatory mechanism in certain bacterial biosynthetic operons that controls gene expression by causing RNA polymerase to terminate transcription.

autonomously replicating sequences (ARS elements) Specific sequences (e.g., in baker's yeast, *Saccharomyces cerevisiae*) that, when included as part of an extrachromosomal, circular DNA molecule, confer on that molecule the ability to replicate autonomously.

autopolyploidy Polyploidy involving more than two chromosome sets of the same species.

autosome A chromosome other than a sex chromosome.

auxotroph A mutant strain of a given organism that is unable to synthesize a molecule required for growth and therefore must have the molecule supplied in the growth medium in order for it to grow.

auxotrophic mutation (nutritional, biochemical) A mutation that affects an organism's ability to make a particular molecule essential for growth.

bacteria Spherical, rod-shaped or spiral-shaped, single-cellular or multi-cellular, filamentous prokaryotic organisms.

balance model A hypothesis of genetic variation proposing that balancing selection maintains large amounts of genetic variation within populations.

Barr body A highly condensed mass of chromatin found in the nuclei of normal females, but not in the nuclei of normal male cells. It represents a cytologically condensed and inactivated X chromosome.

base analog A chemical whose molecular structure is extremely similar to the bases normally found in DNA.

base-pair substitution mutation A change in a gene such that one base pair is replaced by another base pair; for instance, an AT is replaced by a GC pair.

basic research Research done to further knowledge for knowledge's sake.

bidirectional replication The DNA synthesis that takes place in both directions away from the origin of the replication point.

binomial distribution The theoretical frequency distribution of events that have two possible outcomes.

biochemical mutation *See* **auxotrophic mutation**.

biparental inheritance Plant zygotes that show traits indicating chloroplast chromosomes from both parents are present and active.

bivalent A pair of homologous, synapsed chromosomes during the first meiotic division.

bottleneck effect A form of genetic drift that occurs when a population is drastically reduced in size. Some genes may be lost from the gene pool as a result of chance.

branch-point sequence The consensus sequence in mammalian cells, YNCURA, (where Y is a pyrimidine, R is a purine, and N is any base), to which the free 5' end of the intron loops and binds to the A nucleotide in the sequence during intron splicing.

broad-sense heritability A quantity representing the proportion of the phenotypic variance that consists of genetic variance.

C value The amount of DNA found in the haploid set of chromosomes.

CAAT box One of the eukaryotic promoter elements found in approximately 80 base pairs upstream of the initiation site, but it can function at a number of other locations and in either orientation with respect to the start point. The consensus sequence is 5'-GGCCAATCT-3'.

cancer Diseases characterized by the uncontrolled and abnormal division of eukaryotic cells and by the spread of the disease (metastasis) to disparate sites in the organism.

5'- capping The addition of a methylated guanine nucleotide (a "cap") to the 5' end of a premessenger RNA molecule; the cap is retained on the mature mRNA molecule.

catabolite repression (glucose effect) The inactivation of an inducible bacterial operon in the presence of glucose even though the operon's inducer is present.

cDNA DNA copies made from an RNA template catalyzed by the enzyme reverse transcriptase.

cDNA library The collection of molecular clones that contains cDNA copies of the entire mRNA population of a cell. *See also* **cDNA**.

cell cycle The cyclical process of growth and cellular reproduction in unicellular and multicellular eukaryotes. The cycle includes nuclear division, or mitosis, and cell division, or cytokinesis.

cell division A process whereby one cell divides to produce two cells.

cell-free, protein-synthesizing system A system, isolated from cells, that contains ribosomes, mRNA, tRNAs with amino acids attached, and all the necessary protein factors for the in vitro synthesis of polypeptides.

cellular oncogene (c-*onc*) The genes, present in a functional state in cancerous cells, that are responsible for the cancerous state.

centi-Morgan (cM) The map unit. It is sometimes called a centi-Morgan in honor of T. H. Morgan.

centromere (kinetochore) A specialized region of a chromosome seen as a constriction under the microscope. This region is important in the activities of the chromosomes during cellular division.

chain-terminating codon One of three codons for which no normal tRNA molecule exists with an appropriate anticodon. A nonsense codon in an mRNA specifies the termination of polypeptide synthesis.

character An observable phenotypic feature of the developing or fully developed organism that is the result of gene action.

charged tRNA The product of an amino acid added to a tRNA.

charging The act of adding the amino acid to the tRNA.

chiasma A cross-shaped structure formed during crossing-over and visible during the diplonema stage of meiosis.

chiasma interference (chromosomal interference) The physical interference caused by the breaking and rejoining of chromatids that reduces the probability of more than one crossing-over event occurring near another one in one part of the meiotic tetrad.

chi-square (χ^2) test A statistical procedure that determines what constitutes a significant difference between observed results and results expected on the basis of a particular hypothesis; a goodness-of-fit test.

chloroplast The cellular organelle found only in green plants that is the site of photosynthesis in the cells containing it.

chorionic villus sampling A procedure in which a sample of chorionic villus tissue of a developing fetus is examined for chromosomal abnormalities.

chromatid One of the two visibly distinct longitudinal subunits of all replicated chromosomes that becomes visible between early prophase and metaphase of mitosis.

chromatin The piece of DNA-protein complex that is studied and analyzed. Each chromatin fragment reflects the general features of chromosomes but not the specifics of any individual chromosome.

chromosomal interference *See* **chiasma interference**.

chromosome The genetic material of the cell, complexed with protein and organized into a number of linear structures. It literally means "colored body," because the threadlike structures are visible under the microscope only after they are stained with dyes.

chromosomal mutation The variation from the wild-type condition in either chromosome number or chromosome structure.

chromosome aberration *See* **chromosomal mutation**.

chromosome theory of inheritance The theory that the chromosomes are the carriers of the genes. The first clear formulation of the theory was made by both Sutton and Boveri, who independently recognized that the transmission of chromosomes from one generation to the next closely paralleled the pattern of transmission of genes from one generation to the next.

chromosome walking A process to identify adjacent clones in a genomic library. In chromosome walking, a piece of DNA is used to probe a genomic library to find an overlapping clone; then a piece of that clone is used as a probe to screen the library again for an overlapping clone; and so on.

cis-dominance The phenomenon of a gene or DNA sequence controlling only genes that are on the same contiguous piece of DNA.

cis-trans (complementation) test A test developed by E. Lewis, used to determine whether two different mutations are within the same cistron (gene).

classical model A hypothesis of genetic variation proposing that natural populations contain little genetic variation as a result of strong selection for one allele.

clonal selection A process whereby cells that already have antibodies specific to an antigen on their surfaces are stimulated to proliferate and secrete that antibody.

cloning The generation of many copies of a DNA molecule (e.g., a recombinant DNA molecule) by replication in a suitable host.

cloning vector (cloning vehicle) A double-stranded DNA molecule that is able to replicate autonomously in a host cell and with which a DNA fragment (or fragments) can be bonded to form a recombinant DNA molecule for cloning.

coding sequence The part of an mRNA molecule that specifies the amino acid sequence of a polypeptide during translation.

codominance The situation in which the heterozygote exhibits the phenotypes of both homozygotes.

codon A group of three adjacent nucleotides in an mRNA molecule that specifies either one amino acid in a polypeptide chain or the termination of polypeptide synthesis.

coefficient of coincidence A number that expresses the extent of chiasma interference throughout a genetic map; the ratio of observed double-crossover frequency to expected double-crossover frequency. Interference is equal to 1 minus the coefficient of coincidence.

combinatorial gene regulation Transcriptional control (i.e., whether a gene is active or inactive) achieved by combining relatively few regulatory proteins (negative and positive) binding to particular DNA sequences.

complementary-base pairing The hydrogen bonding between a particular purine and a particular pyrimidine in double-stranded nucleic acid molecules (DNA–DNA, DNA–RNA, or RNA–RNA). The major specific pairings are guanine with cytosine and adenine with thymine or uracil.

complementary DNA *See* **cDNA.**

complementation test *See* **cis-trans test.**

complete dominance The case in which one allele is dominant to the other so that at the phenotypic level the heterozygote is essentially indistinguishable from the homozygous dominant.

complete recessiveness The situation in which an allele is phenotypically expressed only when it is homozygous.

concerted evolution (molecular drive) A poorly understood evolutionary process that produces uniformity of sequence in multiple copies of a gene.

conditional mutant A mutant organism that is normal under one set of conditions but becomes seriously impaired or dies under other conditions.

conjugation A process having a unidirectional transfer of genetic information through direct cellular contact between a donor ("male") and a recipient ("female") bacterial cell.

consensus sequence The sequence indicating which nucleotide is found most frequently at each position.

conservative model A DNA replication scheme in which the two parental strands of DNA remain together and serve as a template for the synthesis of a new daughter double helix.

constitutive gene A gene whose products are essential to the normal functioning of the cell, no matter what the life-supporting environmental conditions are. These genes are always active in growing cells.

constitutive heterochromatin Condensed chromatin that is always genetically inactive, and is found at homologous sites on chromosome pairs.

continuous trait *See* **quantitative (continuous) traits.**

contributing allele An allele that contributes to the phenotype.

controlling site A specific sequence of nucleotide pairs adjacent to the gene where the transcription of a gene occurs in response to a particular molecular event.

coordinate induction The simultaneous transcription and translation of two or more genes brought about by the presence of an inducer.

core enzyme The portion of the *E. coli* RNA polymerase that is the active enzyme and can be written as $\alpha_2\beta\beta'\omega$.

correlation coefficient A statistic that measures the strength of the association between two variables.

cosmids Cloning vectors derived from plasmids that are capable of cloning large fragments of DNA. In addition to the features of plasmid cloning vectors (i.e., origin of replication and selectable marker(s) for growth in bacteria), cosmids contain phage lambda *cos* sites which permit recombinant DNA molecules that are constructed to be packaged into the lambda phage head in vitro.

cotransduction The simultaneous transduction of two or more bacterial genes; a good indication that the bacterial genes are closely linked.

cotranslational transport The movement of a protein into the ER simultaneously with its synthesis.

coupling An arrangement in which the two wild-type alleles are on one homologous chromosome and the two recessive mutant alleles are on the other.

covariance A statistic used to calculate the correlation coefficient between two variables. The covariance is calculated by taking the sum of $(x - \bar{x})(y - \bar{y})$ over all pairs of values for the variables x and y, where $\bar{x}$ is the mean of the x values and $\bar{y}$ is the mean of all y values.

crisscross inheritance A type of gene transmission passed from a male parent to a female child to a male grandchild.

cross *See* **cross-fertilization.**

cross-fertilization (cross) A term used for the fusion of male and female gametes from different individuals; the bringing together of genetic material from different individuals for the purpose of genetic recombination.

crossing-over A term introduced by Morgan and E. Cattell, in 1912, to describe the process of reciprocal chromosomal interchange by which recombinants arise.

cytohet The genetic condition of plant zygotes that display biparental inheritance; the term is derived from "*cyto*plasmically *het*erozygous."

cytokinesis A term that refers to the division of the cytoplasm. The two new nuclei compartmentalize into separate daughter cells, and the mitotic cell division process is completed.

cytological marker A cytologically distinguishable feature of chromosomes.

cytosine (C) A pyrimidine base found in RNA and DNA. In double-stranded DNA, cytosine pairs with the purine guanine.

dark repair *See* **excision repair.**

Darwinian fitness The relative reproductive ability of a genotype.

degeneracy A multiple coding; more than one codon per amino acid.

degradation control Regulation of the RNA breakdown rate in the cytoplasm.

deleted Refers to when a chromosome breaks off spontaneously and is lost.

deletion (deficiency) A chromosomal mutation resulting in the loss of a segment of the genetic material and the genetic information contained therein from a chromosome.

deoxyribonuclease (DNase) An enzyme that catalyzes the degradation of DNA to nucleotides.

deoxyribonucleic acid (DNA) A polymeric molecule consisting of deoxyribonucleotide building blocks that in a double-stranded, double-helical form is the genetic material of most organisms.

deoxyribonucleotide The basic building block of DNA, consisting of a sugar (deoxyribose), a base, and a phosphate.

deoxyribose The pentose (five-carbon) sugar found in DNA.

development The process of regulated growth that results from the interaction of the genome with cytoplasm and the environment. It involves a programmed sequence of phenotypic events that are typically irreversible.

diakinesis The stage that follows diplonema and during which the four chromatids of each tetrad are most condensed and the chiasmata often terminalize.

dicentric bridge *See* **dicentric chromosome.**

dicentric chromosome A chromosome with two centromeres. For example, as a result of the crossover between genes B and C in the inversion loop, one recombinant chromatid becomes stretched across the cell as the two centromeres begin to migrate, forming a dicentric bridge.

dideoxy nucleotide A modified nucleotide that has a 3'-H on the deoxyribose sugar rather than a 3'-OH. If a dideoxy nucleoside triphosphate (ddNTP) is used in a DNA synthesis reaction, the ddNTP can be incorporated into the growing chain. However, no further DNA synthesis can then occur because no phosphodiester bond can be formed with an incoming DNA precursor.

dideoxy (Sanger) sequencing A method of rapid sequencing of DNA molecules developed by Fred Sanger. This technique incorporates the use of dideoxy nucleotides in a DNA polymerase-catalyzed DNA synthesis reaction.

differentiation An aspect of development that involves the formation of different types of cells, tissues, and organs through the processes of specific regulation of gene expression.

dihybrid cross A cross between two dihybrids of the same type. Individuals that are heterozygous for two pairs of alleles at two different loci are called dihybrid.

dioecious A term referring to plant species that have both male and female sex organs on different individuals.

diploid (2N) A eukaryotic cell with two sets of chromosomes.

diplonema The second stage of prophase I in which the chromosomes begin to repel one another and tend to move apart.

discontinuous DNA replication A DNA replication involving the synthesis of short DNA segments, which are subsequently linked to form a long polynucleotide chain.

discontinuous trait A heritable trait in which the mutant phenotype is sharply distinct from the alternative, wild-type phenotype.

disjunction The process in anaphase during which sister chromatid pairs undergo separation.

dispersive model A DNA replication scheme in which the parental double helix is cleaved into double-stranded DNA segments that act as templates for the synthesis of new double-stranded DNA segments. Somehow, the segments reassemble into complete DNA double helices, with parental and progeny DNA segments interspersed.

DNA *See* **deoxyribonucleic acid**.

DNA fingerprinting *See* **DNA typing**.

DNA helicase An enzyme that catalyzes the unwinding of the DNA double helix during replication in *E. coli*; product of the *rep* gene.

DNA ligase (polynucleotide ligase) An enzyme that catalyzes the formation of a covalent bond between free single-stranded ends of DNA molecules during DNA replication and DNA repair.

DNA polymerase An enzyme that catalyzes the synthesis of DNA.

DNA polymerase I An *E. coli* enzyme that catalyzes DNA synthesis, originally called the Kornberg enzyme.

DNA primase The enzyme in DNA replication that catalyzes the synthesis of a short nucleic acid primer.

DNA typing The use of restriction fragment length polymorphisms DNA analysis to identify an individual.

docking protein An integral protein membrane of the endoplasmic reticulum (ER) to which the nascent polypeptide-signal recognition particle (SRP)-ribosome complex binds to facilitate the binding of the polypeptide's signal sequence and associated ribosome to the ER.

dominance variance (represented by V_D) Genetic variance that arises from the dominance effects of genes.

dominant An allele or phenotype that is expressed in either the homozygous or the heterozygous state.

dominant lethal allele Allele that will exhibit a lethal phenotype when present in the heterozygous condition.

dosage compensation A mechanism in mammals which compensates for X chromosomes in excess of the normal complement. *See also* **Barr body**.

double crossover Two crossovers occurring in a particular region of a chromosome in a meiosis.

double fertilization An event found only in the life cycle of flowering plants. It is the fusing of the sperm cell with the two nuclei of the gametophyte's central cell to form the cell that will become the endosperm of the seed.

Down syndrome *See* **trisomy-21**.

duplication A chromosomal mutation that results in the doubling of a segment of a chromosome.

effective population size The effective number of adults contributing gametes to the next generation.

effector A small molecule involved in the control of expression of many regulated genes.

endosymbiont hypothesis The hypothesis that mitochondria and chloroplasts originated as free-living prokaryotes that invaded primitive eukaryotic cells and established a mutually beneficial (symbiotic) relationship.

enhancer sequence (enhancer element) In eukaryotes, a type of DNA sequence element having a strong, positive effect on transcription by RNA polymerase II.

environmental sex determination The process by which the environment plays a major role in determining the sex of an organism.

environmental variance (represented by V_E) Any nongenetic source of phenotypic variation among individuals.

episome An autonomously replicating plasmid (a circular, double-stranded DNA molecule) that is capable of integrating into the host cell's chromosome.

epistasis A form of gene interaction in which one gene interferes with the phenotypic expression of another nonallelic gene so that the phenotype is governed by the former gene and not by the latter gene when both genes are present in the genotype.

erythroblasts Red blood cell precursors.

essential genes Genes that when mutated can result in a lethal phenotype.

euchromatin Chromatin that is condensed during division but becomes uncoiled during interphase.

eukaryote A term that literally means "true nucleus." Eukaryotes are organisms that have cells in which the genetic material is located in a membrane-bound nucleus. Eukaryotes can be unicellular or multicellular.

euploid The condition in which an organism or cell has one complete set of chromosomes, or an exact multiple of complete sets.

excision repair (dark repair) An enzyme-catalyzed, light-independent process of repair of ultraviolet-light–induced thymine dimers in DNA that involves removal of the dimers and synthesis of a new piece of DNA complementary to the undamaged strand.

exon The part of an mRNA molecule that specifies the amino acid sequence of a polypeptide during translation. *See also* **coding sequence**.

expressivity The degree to which a particular genotype is expressed in the phenotype.

F_1 generation The first filial generation produced by crossing two parental strains.

F_2 generation The second filial generation produced by selfing the F_1.

F-pili (sex pili) Hairlike cell surface components produced by cells containing the F factor, which allow the physical union of F^+ and F^- cells or *Hfr* and F^- cells to take place.

facultative heterochromatin Chromatin that may become condensed throughout the cell cycle and may contain genes that are inactivated when the chromatin becomes condensed.

familial trait A trait shared by members of a family.

fine-structure mapping A high-resolution mapping of allelic sites within a gene.

first filial generation *See also* **F_1 generation**; the offspring that result from the first experimental crossing of animals or plants.

first law *See* **principle of segregation**.

fitness *See* **Darwinian fitness**

formylmethionine (fMet) A specially modified amino acid involving the addition of a formyl group to the amino group of methionine. It is the first amino acid incorporated into a polypeptide chain in prokaryotes and in eukaryotic cellular organelles.

forward mutation A mutational change from a wild-type allele to a mutant allele.

founder effect A phenomenon that occurs when the isolate effect is exhibited by a small breeding unit that has formed by migration of a small number of individuals from a large population.

frameshift mutation A mutational addition or deletion of a base pair in a gene that disrupts the normal reading frame of an mRNA, which is read in groups of three bases.

frequency distribution A means of summarizing the phenotypes of a continuous trait whereby the population is described in terms of the proportion of individuals that have each phenotype.

gametes Mature reproductive cells that are specialized for sexual fusion. Each gamete is haploid and fuses with a cell of similar origin but of opposite sex to produce a diploid zygote.

gametogenesis The formation of male and female gametes by meiosis.

gametophyte The haploid sexual generation in the life cycle of plants that produces the gametes.

GC box A eukaryotic promoter element with the consensus sequence 5'-GGGCGGG-3' that can be found in either orientation upstream of the transcription initiation site. The GC boxes appear to help the RNA polymerase near the transcription start point.

gene (Mendelian factor) The determinant of a characteristic of an organism. Genetic information is coded in the DNA, which is responsible for species and individual variation. A gene's nucleotide sequence specifies a polypeptide or RNA and is subject to mutational alteration.

gene flow The movement of genes that takes place when organisms migrate and then reproduce, contributing their genes to the gene pool of the recipient population.

gene locus *See* **locus**.

gene marker *See* **genetic marker**.

gene mutation A heritable alteration of the genetic material, usually from one allelic form to another.

gene pool The total genetic information encoded in the total genes in a breeding population existing at a given time.

generalized transduction A type of transduction in which any gene may be transferred between bacteria.

gene redundancy A situation in which tRNA genes occur two or more times in the *E. coli* chromosome.

gene regulatory elements Base-pair sequences associated with a gene, which are involved in the regulation of gene expression.

gene segregation *See* **principle of segregation (first law)**.

genetic code The base-pair information that specifies the amino acid sequence of a polypeptide.

genetic correlation An association between the genes that determine two traits.

genetic counseling The procedures whereby the risks of prospective parents having a child who expresses a genetic disease are evaluated and explained to them. The genetic counselor typically makes predictions about the probabilities of particular traits (deleterious or not) occurring among children of a couple.

genetic drift Any change in gene frequency due to chance in a population.

genetic engineering The alteration of the genetic constitution of cells or individuals by directed and selective modification, insertion, or deletion of an individual gene or genes. In some cases, novel gene combinations are made by joining DNA fragments from different organisms.

genetic map (linkage map) A representation of the genetic distance separating nonallelic gene loci in a linkage structure.

genetic mapping The use of genetic crosses to locate genes on chromosomes relative to one another.

genetic marker (gene marker) Any genetically controlled phenotypic difference used in genetic analysis, particularly in the detection of genetic recombination events.

genetic recombination A process by which parents with different genetic characters give rise to progeny so that genes in which the parents differed are associated in new combinations. For example, from *A B* and *a b* the recombinants *A b* and *a B* are produced.

genetics The science of heredity that involves the structure and function of genes and the way genes are passed from one generation to the next.

genetic variance (represented by V_G) Genetic sources of phenotypic variation among individuals of a population; includes dominance genetic variance, additive genetic variance, and epistatic genetic variance.

genome The total amount of genetic material in a chromosome set; in eukaryotes, this is the amount of genetic material in the haploid set of chromosomes of the organism.

genomic imprinting Phenomenon in which the expression of certain genes is determined by whether the gene is inherited from the female or male parent.

genomic library The collection of molecular clones that contains at least one copy of every DNA sequence in the genome.

genotype The complete genetic makeup of an organism.

genotypic frequencies The frequencies or percentages of different genotypes found within a population.

genotypic sex determination The process by which the sex chromosomes play a decisive role in the inheritance and determination of sex.

germ-line mutations Mutations in the germline of sexually reproducing organisms may be transmitted by the gametes to the next generation, giving rise to an individual with the mutant state in both its somatic and germ-line cells.

glucose effect *See* **catabolite repression**.

Goldberg-Hogness box (TATA box, or TATA element) Found approximately at position–30 from the transcription initiation site. The Goldberg-Hogness sequence is considered to be the likely eukaryotic promoter sequence. The consensus sequence for the Goldberg-Hogness box is TATAAAAA.

group I-intron self-splicing *See* **self-splicing**.

guanine (G) Purine base found in RNA and DNA. In double-stranded DNA, guanine pairs with the pyrimidine cytosine.

haploid (N) A cell or an individual with one copy of each nuclear chromosome.

Hardy-Weinberg law (Hardy-Weinberg equilibrium, Hardy-Weinberg law of genetic equilibrium) An extension of Mendel's laws of inheritance that describes the expected relationship between gene frequencies in natural populations and the frequencies of individuals of various genotypes in the same populations.

harlequin chromosomes 5-bromodeoxyuridine (5-BUdR), a thymidine analog, is incorporated into DNA during replication. When both DNA strands contain 5-BUdR, the chromatid stains less intensely than when only one DNA strand contains the analog. When cells are grown in the presence of 5-BUdR for two replication cycles, the two sister chromatids stained differentially are called harlequin chromosomes.

helix-destabilizing proteins *See* **single-strand DNA-binding (SSB) proteins.**

hemizygous The condition of X-linked genes in males. Males that have an X chromosome with an allele for a particular gene but do not have another allele of that gene in the gene complement are hemizygous.

hereditary trait A characteristic under control of the genes that is transmitted from one generation to another.

heritability The proportion of phenotypic variation in a population attributable to genetic factors.

hermaphroditic For animals (e.g., nematode), the species in which each individual has both testes and ovaries; in plants, the species that have both stamens and pistils on the same flower.

heterochromatin Chromatin that remains condensed throughout the cell cycle and is genetically inactive.

heterogametic sex The sex that has sex chromosomes of different types (e.g., XY) and therefore produces two kinds of gametes with respect to the sex chromosomes.

heterogeneous nuclear RNA (hnRNA) The RNA molecules of various sizes that exist in a large population in the nucleus. Some of the RNA molecules are precursors to mature mRNAs.

heterokaryon A cell or collection of cells (as in a mycelium) possessing genetically different nuclei (regardless of their number) in a common cytoplasm.

heterosis The phenomenon in which the heterozygous genotypes with respect to one or more characters are superior in comparison with the corresponding homozygous genotypes in terms of growth, survival, phenotypic expression, and fertility.

heterozygosity The proportion of individuals heterozygous at a locus; the state of being heterozygous; *see also* **heterozygous.**

heterozygote advantage *See* **overdominance.**

heterozygous A term describing a diploid organism having different alleles of one or more genes and therefore producing gametes of different genotypes.

Hfr (high-frequency recombination) A male cell in *E. coli* with the F factor integrated into the bacterial chromosome. When the F factor promotes conjugation with a female (F–) cell, bacterial genes are transferred to the female cell with high frequency.

highly repetitive sequence A DNA sequence that is repeated between 10^5 and 10^7 times in the genome.

histone One of a class of basic proteins that are complexed with DNA in chromosomes and that play a major role in determining the structure of eukaryotic nuclear chromosomes.

homeobox A 180-bp consensus sequence found in the protein-coding sequences of genes that regulate development.

homeodomain The 60-amino-acid part of proteins that corresponds to the homeobox sequence of genes. All homeodomain-containing proteins appear to be located in the nucleus.

homeotic mutations Mutations that alter the identity of particular segments, transforming them into copies of other segments.

homogametic sex The gender in the species, most often the female, that produces only the X sex chromosome.

homolog Each individual member of a pair of homologous chromosomes.

homologous chromosomes The members of a chromosome pair that are identical in the arrangement of genes they contain and in their visible structure.

homozygosity The state of being homozygous; *see also* **homozygous.**

homozygous A term describing a diploid organism having the same alleles at one or more genes and therefore producing gametes of identical genotypes.

homozygous dominant A diploid organism that has the same dominant allele for a given gene locus on both members of a homologous pair of chromosomes.

homozygous recessive A diploid organism that has the same recessive allele for a given gene locus on both members of a homologous pair of chromosomes.

human genome project A project to obtain the sequence of the complete 3 billion (3×10^9) nucleotide pairs of the human genome, and to map all of the estimated 50,000 to 100,000 human genes.

hybrid dysgenesis The appearance of a series of defects, including mutations, chromosomal aberrations, and sterility, when certain strains of *Drosophila melanogaster* are crossed.

hypersensitive sites (hypersensitive regions) Sites in the regions of DNA around transcriptionally active genes that are highly sensitive to digestion by DNase I.

hypothetico-deductive method of investigation Research method involving making observations, forming hypotheses to explain the observations, making experimental predictions based on the hypotheses, and, finally, testing the predictions. The last step produces new observations and so a cycle is set up leading to a refinement of the hypotheses and perhaps eventually to the establishment of a law or an accepted principle.

imaginal discs In the *Drosophila* blastoderm, undifferentiated cells that will develop into adult tissue and organs.

immunoglobulins Specialized proteins (antibodies) secreted by B cells that circulate in the blood and lymph and that are responsible for humoral immune responses.

inbreeding Preferential mating between close relatives.

incomplete (partial) dominance The condition resulting when one allele is not completely dominant to another allele so that the heterozygote has a phenotype between that shown in individuals homozygous for either individual allele involved. An example of partial dominance is the frizzle chicken.

induced mutation A mutation that results from treatment with mutagens.

inducer A chemical or environmental agent for bacterial operons that brings about the transcription of an operon.

induction The synthesis of a gene product (or products) in response to the action of an inducer, that is, a chemical or environmental agent.

insertion sequence (IS) element The simplest transposable genetic element found in prokaryotes. It is a mobile segment of DNA that contains genes required for the process of insertion of the DNA segment into a chromosome and for the mobilization of the element to different locations.

interaction variance (represented by V_I) Genetic variance that arises from epistatic interactions among genes.

intergenic suppressor A mutation whose effect is to suppress the phenotypic consequences of another mutation in a gene distinct from the gene in which the suppressor mutation is located.

internal control region (ICR) Promoter sequence, recognized by RNA polymerase III, that is located within the gene sequence, for instance, in tRNA genes and 5S rRNA genes of eukaryotes.

intervening sequence (ivs) *See* **intron.**

intragenic suppressors A mutation whose effect is to suppress the phenotypic consequences of another mutation within the same gene in which the suppressor mutation is located.

intron A nucleotide sequence in eukaryotes that must be excised from a structural gene transcript in order to convert the transcript into a mature messenger RNA molecule containing only coding sequences that can be translated into the amino acid sequence of a polypeptide.

inversion A chromosomal mutation that results when a segment of a chromosome is excised and then reintegrated in an orientation 180° from the original orientation.

IS element *See* **insertion sequence (IS) element.**

karyotype A complete set of all the metaphase chromatid pairs in a cell (literally, "nucleus type").

kinetochore *See* **centromere.**

Klinefelter syndrome A human clinical syndrome that results from disomy for the X chromosome in a male, which results in a 47,XXY male. Many of the affected males are mentally deficient, have underdeveloped testes, and are taller than average.

lagging strand In DNA replication, the DNA strand that is synthesized discontinuously in the 5' to 3' direction away from the replication fork.

leader sequence One of three main parts of the mRNA molecule. The leader sequence is located at the 5' end of the mRNA molecule and contains the coded information that the ribosome and special proteins read to tell it where to begin the synthesis of the polypeptide.

leading strand In DNA replication, the DNA strand synthesized continuously in the 5' to 3' direction toward the replication fork.

leptonema The stage during meiosis in prophase I at which the chromosomes have begun to coil and are visible.

lethal allele An allele that results in the death of an organism.

light repair *See* **photoreactivation.**

LINES (long interspersed repeated sequences) The dispersed families of repeated sequences in mammals that are several thousand base pairs in length and occur > 20,000 times in the genome.

linkage A term describing genes located on the same chromosome.

linkage map *See* **genetic map.**

linked genes Genes that are located on the same chromosome.

linker *See* **restriction site linker.**

locus (*plural*, loci) The position of a gene on a genetic map; the specific place on a chromosome where a gene is located.

lod score method The lod (logarithm of **od**ds) score method is a statistical analysis, usually performed by computer programs, based on data from pedigrees. It is used to test for linkage between two loci in humans.

lyonization A mechanism in mammals that allows them to compensate for X chromosomes in excess of the normal complement. The excess X chromosomes are cytologically condensed and inactivated, and they do not play a role in much of the development of the individual. The name derives from the discoverer of the phenomenon, Mary Lyon.

lysogenic A term describing a bacterium that contains a temperate phage in the prophage state. The bacterium is said to be lysogenic for that phage. On induction phage reproduction is initiated, progeny phages are produced, and the bacterial cell lyses.

lysogenic pathway A path, besides the lytic cycle, that a phage can follow. The chromosome does not replicate; instead, it inserts itself physically into a specific region of the host cell's chromosome in a way that is essentially the same as *F* factor integration.

lysogeny The phenomenon of the insertion of a temperate phage chromosome into a bacterial chromosome, where it replicates when the bacterial chromosome replicates. In this state the phage genome is repressed and is said to be in the prophage state.

lytic cycle A type of phage life cycle in which the phage takes over the bacterium and directs its growth and reproductive activities to express the phage's genes and to produce progeny phages.

macromolecule A large molecule (such as DNA, RNA, and proteins) that has a molecular weight of at least a few thousand daltons.

mapping functions Mathematical formulas that are used to correct the observed recombination values for the incidence of multiple crossovers.

map unit (mu) A unit of measurement used for the distance between two gene pairs on a genetic map. A crossover frequency of 1 percent between two genes equals 1 map unit. *See also* **centi-Morgan.**

maternal effect The phenotype in an individual that is established by the maternal nuclear genome, as the result of mRNA and/or proteins that are deposited in the oocyte prior to fertilization. These inclusions direct early development in the embryo.

maternal inheritance A phenomenon in which the mother's phenotype is expressed exclusively.

mating types A genic system in which two sexes are morphologically indistinguishable but carry different alleles and will mate.

Maxam-Gilbert sequencing A method of rapid sequencing of DNA molecules developed by Allan Maxam and Walter Gilbert. The technique uses specific chemical reactions to break DNA at specific nucleotides. The DNA is first radiolabeled with ^{32}P at the 5' or 3' end of a chain. Second, the DNA is chemically modified and cleaved at various points along the backbone. Third, the DNA fragments are analyzed with polyacrylamide gel electrophoresis.

mean The average of a set of numbers, calculated by adding all the values represented and dividing by the number of values.

megasporogenesis The formation in flowering plants of megaspores and the production of the embryo sac (the female gametophyte).

meiosis Two successive nuclear divisions of a diploid nucleus that result in the formation of haploid gametes or of meiospores having one-half the genetic material of the original cell.

meiosis I The first meiotic division that results in the reduction of the number of chromosomes. This division consists of four stages: prophase I, metaphase I, anaphase I, and telophase I.

meiosis II The second meiotic division, resulting in the separation of the chromatids.

Mendelian factor *See* **gene.**

Mendelian population An interbreeding group of individuals sharing a common gene pool; the basic unit of study in population genetics.

messenger RNA (mRNA) The RNA molecule that contains the coded information for the amino acid sequence of a protein.

metacentric chromosome A chromosome that has the centromere approximately in the center of the chromosome.

metaphase A stage in mitosis or meiosis in which chromosomes become aligned along the equatorial plane of the spindle.

metaphase II The second stage of meiosis during which the centromeres line up on the equator of the second-division spindles (in each of two daughter cells formed from meiosis I).

metaphase plate The plane where the chromosomes become aligned during metaphase.

metastasis The spreading of malignant tumor cells throughout the body so that tumors develop at new sites.

microsporogenesis The formation in flowering plants of microspores in the anthers and the production of the male gametophyte (pollen), normally from diploid microsporocytes.

migration Movement of organisms from one location to another.

missense mutation A gene mutation in which a base-pair change in the DNA causes a change in an mRNA codon, with the result that a different amino acid is inserted into the polypeptide in place of one specified by the wild-type codon.

mitochondria The organelles found in the cytoplasm of all aerobic animal and plant cells; the principal sources of energy in the cell.

mitosis The process of nuclear division in haploid or diploid cells producing daughter nuclei that contain identical chromosome complements and that are genetically identical to one another and to the parent nucleus from which they arose.

mitotic crossing-over (mitotic recombination) A genetic recombination that occurs following the rare pairing of homologs during mitosis of a diploid cell.

moderately repetitive sequence A DNA sequence that is reiterated from a few to as many as 10^3 to 10^5 times in the genome.

molecular cloning *See* **cloning.**

molecular drive *See* **concerted evolution.**

molecular genetics A subdivision of the science of genetics involving how genetic information is encoded within the DNA and how biochemical processes of the cell translate the genetic information into the phenotype.

monoecious A term referring to plants in which male and female gametes are produced in the same individual.

monohybrid cross A cross between two individuals that are both heterozygous for the same pair of alleles (e.g., *Aa × Aa*). By extension, the term also refers to crosses involving the purebreeding parents that differ with respect to the alleles of one locus (e.g., *AA × aa*).

monoploidy An aberrant, aneuploid state in a normally diploid cell or organism in which one chromosome is missing, leaving one chromosome with no homolog.

monosomy An aberrant, aneuploid state in a normally diploid cell or organism in which one chromosome is missing, leaving one chromosome with no homolog.

mRNA splicing A process whereby an intervening sequence between two coding sequences in an RNA molecule is excised and the coding sequences ligated (spliced) together.

multifactorial trait A trait influenced by multiple genes and environmental factors.

multigene family A set of related genes that have evolved from some ancestral gene through the process of gene duplication.

multiple alleles Many alternative forms of a single gene.

multiple cloning site *See* **polylinker.**

multiple crossovers More than one crossover occurring in a particular region of a chromosome in a meiosis.

mutagen Any physical or chemical agent that significantly increases the frequency of mutational events above a spontaneous mutation rate.

mutant allele Any alternative to the wild-type allele of a gene. Mutant alleles may be dominant or recessive to wild-type alleles.

mutation Any detectable and heritable change in the genetic material not caused by genetic recombination.

mutation frequency The number of occurrences of a particular kind of mutation in a population of cells or individuals.

mutation rate The probability of a particular kind of mutation as a function of time.

mutator gene A mutant gene that increases the spontaneous mutation frequencies of other genes.

narrow-sense heritability The proportion of the phenotypic variance that results from additive genetic variance.

natural selection Differential reproduction of genotypes.

negative assortative mating A mating that occurs between dissimilar individuals more often than it does between randomly chosen individuals.

neutral mutation A base-pair change in the gene that changes a codon in the mRNA so that there is no change in the function of the protein translated from that message.

neutral-mutation hypothesis A hypothesis that replaced the classical model by acknowledging the presence of extensive genetic variation in proteins, but proposing that this variation is neutral with regard to natural selection.

nitrogenous base A nitrogen-containing base that, along with a pentose sugar and a phosphate, is one of the three parts of a nucleotide, the building block of RNA and DNA.

noncontributing alleles The alleles that do not have any effect on the phenotype of the quantitative trait.

nondisjunction (primary disjunction) A failure of homologous chromosomes or sister chromatids to separate at anaphase.

nonhistone A type of acidic or neutral protein found in chromatin.

nonhomologous chromosomes The chromosomes containing dissimilar genes that do not pair during meiosis.

non-Mendelian inheritance (cytoplasmic inheritance) The inheritance of characters determined by genes not located on the nuclear chromosomes but on mitochondrial or chloroplast chromosomes. Such genes show inheritance patterns distinctly different from those of nuclear genes.

nonparental-ditype (NPD) One of three types of tetrads possible when two genes are segregating in a cross. The NPD tetrad contains four nuclei, all of which have recombinant (nonparental) genotypes, that is, two of each possible type.

nonsense codon *See* **chain-terminating codon.**

nonsense mutation A gene mutation in which a base-pair change in the DNA causes a change in an mRNA codon from an amino acid-coding codon to a chain-terminating (nonsense) codon. As a result, polypeptide chain synthesis is terminated prematurely and is therefore either nonfunctional or, at best, partially functional.

nontranscribed spacer (NTS) sequence Sequences, which are not transcribed, found between transcription units in rDNA. Important sequences that control transcription of the rDNA are within the NTS.

normal distribution A probability distribution in statistics, graphically displayed as a bell-shaped curve.

norm of reaction The extent to which the phenotype produced by a genotype varies with the environment.

northern blot analysis A similar technique to Southern blotting except that RNA rather than DNA is separated and transferred to a filter for hybridization with a probe.

nuclease An enzyme that catalyzes the degradation of a nucleic acid by breaking phosphodiester bonds. Nucleases specific for DNA are termed deoxyribonucleases (DNases), and nucleases specific for RNA are termed ribonucleases (RNases).

nucleofilament A fiber seen in chromatin. It is approximately 10 nm in diameter and consists of DNA wrapped around nucleosome cores.

nucleoid Central region in a bacterial cell in which the chromosome is compacted.

nucleolus An organelle within the eukaryotic nucleus; the site of transcription of the ribosomal RNA genes and assembly of the ribosomal subunits.

nucleoside phosphate *See* **nucleotide.**

nucleosome The basic structural unit of eukaryotic nuclear chromosomes, consisting of two molecules each of the four core histones (H2A, H2B, H3, and H4, the histone octamer), a single molecule of the linker histone H1, and about 180 bp of DNA.

nucleotide A monomeric molecule of RNA and DNA that consists of three distinct parts: a pentose (ribose in RNA, deoxyribose in DNA), a nitrogenous base, and a phosphate group.

nucleus A discrete structure within the cell that is bounded by a nuclear membrane. It contains most of the genetic material of the cell.

nullisomy The aberrant, aneuploid state in a normally diploid cell or organism in which there is a loss of one pair of homologous chromosomes.

nutritional mutation *See* **auxotrophic mutation.**

Okazaki fragments The relatively short, single-stranded DNA fragments in discontinuous DNA replication that are synthesized during DNA replication and that are subsequently covalently joined to make a continuous strand.

oligomers (oligo = few) Short DNA molecules.

oncogene A gene whose action promotes cell proliferation. Oncogenes are altered forms of proto-oncogenes.

oncogenesis Tumor (cancer) initiation in an organism.

one gene–one enzyme hypothesis The hypothesis, based on Beadle and Tatum's studies in biochemical genetics, that each gene controls the synthesis of one enzyme.

one gene–one polypeptide hypothesis Updated version of the one gene–one enzyme hypothesis, which states that each gene controls the synthesis of a polypeptide chain.

oogenesis The development in the gonad of the female germ cell (egg cell) of animals.

open reading frame In a segment of DNA, a potential protein-coding sequence identified by an initiator codon in frame with a chain-terminating codon.

operator The controlling site, that is adjacent to a promoter, that is responsible for controlling the transcription of genes that are contiguous to the promoter.

operon A cluster of genes whose expressions are regulated together by operator-regulator protein interactions, plus the operator region itself and the promoter.

ordered tetrads A structure resulting from meiosis in which the four meiotic products are in an order reflecting exactly the orientation of the four chromatids at the metaphase plate in meiosis I.

origin A specific site on the chromosome at which the double helix denatures into single strands and continues to unwind as the replication fork(s) migrates.

origin of replication A specific DNA sequence that is required for the initiation of DNA replication in prokaryotes.

outbreeding Preferential mating between nonrelated individuals.

overdominance (heterozygote advantage) Condition in which the heterozygote has higher fitness than either of the homozygotes.

ovum A mature egg cell. In the second meiotic division, the secondary oocyte produces two haploid cells; the large cell rapidly matures into the ovum.

pachynema The stage in meiosis (mid-prophase I) during which the homologous pairs of chromosomes exchange chromosome regions.

paracentric inversion An inversion in which the inverted segment occurs on one chromosome arm and does not include the centromere.

parasexual system A system that achieves genetic recombination by means other than the regular alternation of meiosis and fertilization.

parental-ditype (PD) One of three types of tetrads possible when two genes are segregating in a cross. The PD tetrad contains four nuclei, all of which are parental genotypes, with two of one parent and two of the other parent.

parental genotypes (parental classes, parentals) Individuals among progeny of crosses that have combinations of genetic markers like one or other of the parents in the parental generation.

parental imprinting *See* **genomic imprinting.**

partial dominance *See* **incomplete (partial) dominance.**

particulate factors The term Mendel used to describe the factors that carried hereditary information and were transmitted from parents to progeny through the gametes. We now know these factors by the name *genes*.

pedigree analysis A family tree investigation that involves the careful compilation of phenotypic records of the family over several generations.

penetrance The frequency with which a dominant or homozygous recessive gene manifests itself in the phenotype of an individual.

pentose sugar A 5-carbon sugar that, along with a nitrogenous base and a phosphate group, is one of the three parts of a nucleotide, the building block of RNA and DNA.

peptide bond A covalent bond in a polypeptide chain that joins the α-carboxyl group of one amino acid to the α-amino group of the adjacent amino acid.

peptidyl transferase The enzyme that catalyzes the formation of the peptide bond in protein synthesis.

pericentric inversion An inversion in which the inverted segment includes the parts of both chromosome arms and therefore includes the centromere.

P generation The parental generation, i.e., the immediate parents of an F_1.

phage lysate The progeny phages released following lysis of phage-infected bacteria.

phage vector A phage that carries pieces of bacterial DNA between bacterial strains in the process of transduction.

phenocopy An abnormal individual resulting from special environmental conditions. It mimics a similar phenotype caused by gene mutation.

phenotype The observable properties of an organism that are produced by the genotype and its interaction with the environment.

phenotypic correlation An association between two traits.

phenotypic variance (represented by V_p) A measure of a trait's variability.

phosphate group A component, along with a pentose sugar and a nitrogenous base, of a nucleotide, the building block of RNA and DNA. Because phosphate groups are acidic in nature, DNA and RNA are called nucleic acids.

phosphodiester bond A covalent bond in RNA and DNA between a sugar and a phosphate. Phosphodiester bonds form the repeating sugar-phosphate array of the backbone of DNA and RNA.

photoreactivation (light repair) One way by which thymine dimers can be repaired. The dimers are reverted directly to the original form by exposure to visible light in the wavelength range 320–370 nm.

physical map Map of physically identifiable regions or markers on genomic DNA, constructed without genetic recombination analysis.

pistil The female reproductive organ in a flowering plant that typically consists of the stigma, the style, and the ovary.

plaque A round, clear area in a lawn of bacteria on solid medium that results from the lysis of cells by repeated cycles of phage lytic growth.

plasma membrane Lipid bilayer that surrounds the cytoplasm of both animal and plant cells.

plasmid An extrachromosomal genetic element consisting of double-stranded DNA that replicates autonomously from the host chromosome.

pleiotropy Multiple phenotypic effects resulting from a single mutant gene.

point mutants Organisms whose phenotypes result from an alteration of a single nucleotide pair.

point mutation A mutation caused by a substitution of one base pair for another.

polarity A term referring to a bacterial operon that codes for a polygenic mRNA. It is the phenomenon whereby certain nonsense mutations not only result in the loss of activity of the enzyme encoded by the gene in which they are located but also reduce significantly or abolish the synthesis of enzymes coded by structural genes on the operatordistal side of the mutation. The mutations are called *polar mutations*.

poly(A) polymerase The enzyme that catalyzes the production of the 3' poly(A) tail.

poly(A) site The 3' end of mRNA to which 50 to 250 adenine nucleotides are added as part of mRNA posttranscriptional modification.

poly(A) tail A sequence of 50 to 250 adenine nucleotides that is added as a posttranscriptional modification at the 3' ends of most eukaryotic mRNAs.

polygene (multiple-gene) hypothesis for quantitative inheritance The hypothesis that quantitative traits are controlled by many genes.

polygenic mRNA (polycistronic mRNA) A single mRNA transcript in prokaryotic operons of two or more adjacent structural genes that specifies the amino acid sequences of the corresponding polypeptides.

polygenic traits Traits encoded by many loci.

polylinker (multiple cloning site) A region of clustered unique restriction sites in a cloning vector.

polymerase chain reaction (PCR) A method used to replicate defined DNA sequences selectively and repeatedly from a DNA mixture.

polynucleotide A linear sequence of nucleotides in DNA or RNA.

polypeptide A polymeric, covalently bonded linear arrangement of amino acids joined by peptide bonds.

polyploidy The condition of a cell or organism that has more than its normal number of sets of chromosomes.

polyribosome (polysome) The complex between an mRNA molecule and all the ribosomes that are translating it simultaneously.

polytene chromosome A special type of chromosome representing a bundle of numerous chromatids that have arisen by repeated cycles of replication of single chromatids without nuclear division. This type of chromosome is characteristic of various tissues of Diptera.

population A group of interbreeding individuals that share a set of genes.

population genetics A branch of genetics that describes in mathematical terms the consequences of Mendelian inheritance on the population level.

positional cloning The isolation of a gene associated with a genetic disease on the basis of its approximate chromosomal position.

position effect A change in the phenotypic effect of one or more genes as a result of a change in their position in the genome.

positive assortative mating A mating that occurs more frequently between individuals who are phenotypically similar than it does among randomly chosen individuals.

posttranslational transport Transport in which synthesis of the protein is completed before import into the organelle takes place.

precursor mRNA (primary transcripts; pre-mRNA) The initial transcript of a gene that is modified and/or processed to produce the mature, functional mRNA molecule. In eukaryotes, for example, the transcript is modified at both the 5' and the 3' ends, and in a number of cases RNA sequences that do not code for amino acids are present and must be excised.

precursor RNA molecule (primary transcripts; pre-RNA) The initial transcript whose processing may involve the addition and/or removal of bases, the chemical modification of some bases, or the cleavage of sequences from the precursors.

precursor rRNA (pre-rRNA) A primary transcript of adjacent rRNA genes (16S, 23S, and 5S rRNA genes in prokaryotes; 18S, 5.8S, and 28S rRNA genes in eukaryotes) plus flanking and spacer DNA that must be processed to release the mature rRNA molecules.

precursor tRNA (pre-tRNA) A primary transcript of a tRNA gene whose bases must be extensively modified and that must be processed to remove extra RNA sequences in order to produce the mature tRNA molecule. In some cases, the primary transcript may contain the sequences of two or more tRNA molecules.

Pribnow box A part of the promoter sequence in prokaryotic genomes that is located at about 10 base pairs upstream from the transcription starting point. The consensus sequence for the Pribnow box is TATAAT. The Pribnow box is often referred to as the TATA box.

primary nondisjunction (nondisjunction) A rare event in which sister chromatids (in mitosis) or chromosomes contained in pairing configurations (in meiosis) fail to be distributed to opposite poles.

primary transcripts *See* **precursor RNA molecules.**

primer *See* **RNA primer.**

primosome A complex of *E. coli* primase, helicase, and perhaps other polypeptides that together become functional in catalyzing the initiation of DNA synthesis.

principle of independent assortment (second law) The law that the factors (genes) for different traits assort independently of one another. In other words, genes on different chromosomes behave independently in the production of gametes.

principle of segregation (first law) The law that two members of a gene pair (alleles) segregate (separate) from each other during the formation of gametes. As a result, one-half the gametes carry one allele and the other half carry the other allele.

probability The ratio of the number of times a particular event occurs to the number of trials during which the event could have happened.

proband In human genetics, an affected person, with whom the study of a character in a family begins. (*See also* **propositus; proposita.**)

product rule The rule that the probability of two independent events occurring simultaneously is the product of each of their probabilities.

prokaryote A cellular organism whose genetic material is not located within a membrane-bound nucleus (cf. eukaryote).

promoter elements (modules) Consensus sequences found in the promoter region of the transcription initiation site. The elements are the TATA box (or Goldberg-Hogness box), CAT element, and the GC element.

promoter site (promoter sequence, promoter) A specific regulatory nucleotide sequence in the DNA to which RNA polymerase binds for the initiation of transcription.

proofreading In DNA synthesis, the process of recognizing a basepair error during the polymerization events and correcting it. Proofreading is a property of the DNA polymerase in prokaryotic cells.

prophage A temperate bacteriophage integrated into the chromosome of a lysogenic bacterium. It replicates with the replication of the host cell's chromosome.

prophase The first stage in mitosis or meiosis during which the chromosomes (already replicated) condense and become visible under the microscope.

prophase I The first stage of meiosis. There are several stages of prophase I, including leptonema, zygonema, pachynema, diplonema, and diakinesis.

prophase II The second stage of meiosis during which there is chromosome contraction.

proportion of polymorphic loci A ratio calculated by determining the number of polymorphic loci and dividing by the total number of loci examined.

proposita In human genetics, an affected female person, with whom the study of a character in a family begins. (*See also* **proband.**)

propositus In human genetics, an affected male person, with whom the study of a character in a family begins. (*See also* **proband.**)

protein One of a group of high-molecular weight, nitrogen-containing organic compounds of complex shape and composition.

protein degradation control Regulation of the protein degradation rate.

proto-oncogenes A gene that in normal cells functions to control the normal proliferation of cells, and that when mutated or changed in any other way becomes an oncogene.

prototroph A strain that is a wild type for all nutritional requirement genes and thus requires no supplements in its growth medium.

pseudodominance The unexpected expression of a recessive trait, caused by the absence of a dominant allele.

Punnett square A matrix that describes all the possible gametic fusions that will give rise to the zygotes that will produce the next generation.

pure-breeding *See* **true-breeding (pure-breeding) strain.**

purine A type of nitrogenous base. In DNA and RNA the purines are adenine and guanine.

pyrimidine A type of nitrogenous base. Cytosine is a pyrimidine in DNA and RNA; thymine is a pyrimidine in DNA; and uracil is a pyrimidine in RNA.

Q banding A staining technique in which metaphase chromosomes are stained with quinacrine mustard to produce temporary fluorescent Q bands on the chromosomes.

quantitative genetics Study of the inheritance of quantitative traits.

quantitative (continuous) traits Traits that show a continuous variation in phenotype over a range.

random mating Matings between genotypes occurring in proportion to the frequencies of the genotypes in the population.

rDNA repeat units The tandem arrays of rRNA genes, 18S-5.8S-28S, repeated many times along the chromosome.

recessive An allele or phenotype that is expressed only in the homozygous state.

recessive lethal allele An allele that causes lethality when it is homozygous.

reciprocal cross A cross of males and females of one trait with males and females of another trait. In the garden pea example

a reciprocal cross for smooth and wrinkled seeds is smooth female × wrinkled male and wrinkled female × smooth male.

recombinant chromosome A chromosome that emerges from meiosis with a combination of genes different from a parental combination of genes.

recombinant DNA molecule A new type of DNA sequence that has been constructed or engineered in the test tube from two or more distinct DNA sequences.

recombinant DNA technology A collection of experimental procedures that allow molecular biologists to splice a DNA fragment from one organism into DNA from another organism and to clone the new recombinant DNA molecule. It includes the development and application of particular molecular techniques, such as biotechnology or genetic engineering. This technology is important, for example, in the production of antibiotics, hormones, and other medical agents used in the diagnosis and treatment of certain genetic diseases.

recombinants The individuals or cells that have nonparental combinations of genes as a result of the processes of genetic recombination.

recombination *See* **genetic recombination.**

regression A statistical analysis assessing the association between two variables.

regression line A mathematically computed line that represents the best fit of a line to the points.

regulated gene A gene whose activity is controlled in response to the needs of a cell or organism.

regulatory factors Proteins active in the activation or repression of transcription of the gene.

release factors *See* **termination factors.**

replica plating The procedure for transferring the pattern of colonies from a master plate to a new plate. In this procedure, a velveteen pad on a cylinder is pressed lightly onto the surface of the master plate, thereby picking up a few cells from each colony to inoculate onto the new plate.

replication bubble Opposing replication forks found with the local denaturing of DNA during replication.

replication fork A Y-shaped structure formed when a double-stranded DNA molecule unwinds to expose the two single-stranded template strands for DNA replication.

replication machine (replisome) The complex formed by the close association of the key proteins used during DNA replication.

replicon (replication unit) The stretch of DNA in eukaryotes from the origin of replication to the two termini of replication on each side of the origin.

repressor *See* **repressor gene.**

repressor gene A regulatory gene whose product is a protein that controls the transcriptional activity of a particular operon.

repressor molecule The protein product of a repressor gene.

repulsion An arrangement in which each homologous chromosome carries the wild-type allele of one gene and the mutant allele of the other one.

restriction endonucleases (restriction enzymes) Enzymes important for analyzing DNA and for constructing recombinant DNA molecules because of their ability to cleave double-stranded DNA molecules at specific nucleotide pair sequences.

restriction enzymes *See* **restriction endonucleases.**

restriction fragment length polymorphisms (RFLPs) The different restriction maps that result from different patterns of distribution of restriction sites. They are detected by the presence of restriction fragments of different lengths on gels.

restriction map A genetic map of DNA showing the relative positions of restriction enzyme cleavage sites.

restriction site linker (linker) A relatively short, double-stranded oligodeoxyribonucleotide about 8 to 12 nucleotide pairs long that is synthesized by chemical means and that contains the cleavage site for a specific restriction enzyme within its sequence.

retroviruses Single-stranded RNA viruses that replicate via double-stranded DNA intermediates. The DNA integrates into the host's chromosome where it can be transcribed.

reverse genetics *See* **positional cloning.**

reverse mutation (reversion) A mutational change from a mutant allele back to a wild-type allele.

reverse transcriptase An enzyme (an RNA-dependent DNA polymerase) that makes a complementary DNA copy of an mRNA strand.

ribonuclease (RNase) An enzyme that catalyzes the degradation of RNA to nucleotides.

ribonucleic acid (RNA) A usually single-stranded polymeric molecule consisting of ribonucleotide building blocks. RNA is chemically very similar to DNA. The three major types of RNA in cells are ribosomal RNA (rRNA), transfer RNA (tRNA), and messenger RNA (mRNA), each of which performs an essential role in protein synthesis (translation). In some viruses, RNA is the genetic material.

ribonucleotide The basic building block of RNA consisting of a sugar (ribose), a base, and a phosphate.

ribose The pentose sugar component of the nucleotide building block of RNA.

ribosomal DNA (rDNA) The regions of the DNA that contain the genes for the rRNAs in prokaryotes and eukaryotes.

ribosomal proteins The proteins that along with rRNA molecules comprise the ribosomes of prokaryotes and eukaryotes.

ribosomal RNA (rRNA) The RNA molecules of discrete sizes that along with ribosomal proteins comprise ribosomes of prokaryotes and eukaryotes.

ribosome A complex cellular particle composed of ribosomal protein and rRNA molecules that is the site of amino acid polymerization during protein synthesis.

ribosome-binding site The nucleotide sequence on an mRNA molecule on which the ribosome becomes oriented in the correct reading frame for the initiation of protein synthesis.

R looping (R loops) A technique developed by M. Thomas, R. White, and R. Davis in which molecules of double-stranded DNA are incubated at temperatures below their denaturing temperature to open up short stretches of the DNA double helix so that single-stranded RNA molecules can begin to form DNA/RNA hybrids where the two are complementary. The DNA/RNA hybrid forms an R loop by displacing a single-stranded section of DNA.

RNA *See* **ribonucleic acid.**

RNA editing Posttranscriptional insertion and/or deletion of nucleotides in an mRNA molecule.

RNA ligase An enzyme that splices together the RNA pieces once the intervening sequence is removed from the pre-tRNA.

RNA polymerase An enzyme that catalyzes the synthesis of RNA molecules from a DNA template in a process called *transcription*.

RNA polymerase I An enzyme in eukaryotes located in the nucleolus that catalyzes the transcription of the 18S, 5.8S, and 28S rRNA genes.

RNA polymerase II An enzyme in eukaryotes found only in the nucleoplasm of the nucleus. It catalyzes the transcription of mRNA-coding genes.

RNA polymerase III An enzyme in eukaryotes found only in the nucleoplasm. It catalyzes the transcription of the tRNA and 5S rRNA genes.

RNA primer A preexisting polynucleotide chain in DNA replication to which new nucleotides can be added.

RNA processing control The second level of control of gene expression in eukaryotes. This level involves regulating the production of mature RNA molecules from precursor-RNA molecules.

RNA synthesis *See* **transcription.**

Robertsonian translocation A type of nonreciprocal translocation in which the long arms of two nonhomologous acrocentric chromosomes become attached to a single centromere.

sample The subset used to give information about a population. It must be of reasonable size and it must be a random subset of the larger group in order to provide accurate information about the population.

sampling error The phenomenon in which chance deviations from expected proportions arise in small samples.

satellite DNA The DNA that forms a band in an equilibrium density band that is distinct from the band constituting the majority of the genomic DNA as a result of a different buoyant density.

secondary nondisjunction A nondisjunction of the Xs in the progeny of females that were produced by a primary nondisjunction.

secondary oocyte A large cell produced by the primary oocyte. In the ovaries of female animals the diploid primary oocyte goes through meiosis I and unequal cytokinesis to produce two cells; the large cell is called the secondary oocyte.

second law *See* **principle of independent assortment.**

second-site mutation *See* **suppressor mutation.**

selection coefficient A measure of the relative intensity of selection against a genotype.

selection differential In natural and artificial selection, the difference between the mean phenotype of the selected parents and the mean phenotype of the unselected population.

selection response The amount that a phenotype changes in one generation when selection is applied to a group of individuals.

self-fertilization (selfing) The union of male and female gametes from the same individual.

self-splicing The excision of introns from some precursor RNA molecules that occurs by a protein-independent reaction in some organisms.

semiconservative replication model A DNA replication scheme in which each daughter molecule retains one of the parental strands.

semidiscontinuous Concerning DNA replication, when one new strand is synthesized continuously and the other discontinuously. *See also* **discontinuous DNA replication.**

sense codon A sense codon, as opposed to a nonsense codon, in an mRNA molecule specifies an amino acid in the corresponding polypeptide.

sequence tagged site (STS) A short segment of DNA that defines a unique position in the human genome; an STS is usually detected by the **polymerase chain reaction (PCR).**

sex chromosome A chromosome in eukaryotic organisms that is represented differently in the two sexes. In many organisms, one sex possesses a pair of visibly different chromosomes. One is an X chromosome, and the other is a Y chromosome. Commonly, the XX sex is female and the XY sex is male.

sex-influenced traits The traits that appear in both sexes but either the frequency of occurrence in the two sexes is different or there is a different relationship between genotype and phenotype.

sex-limited trait A genetically controlled character that is phenotypically exhibited in only one of the two sexes.

sex-linked *See* **X-linked.**

sexual reproduction The reproduction involving the fusion of haploid gametes produced by meiosis.

shuttle vector A cloning vector that can replicate in two or more host organisms. Shuttle vectors are used for experiments in which recombinant DNA is to be introduced into organisms other than *E. coli.*

signal hypothesis The hypothesis that the secretion of proteins from a cell occurs through the binding of a hydrophobic amino terminal extension to the membrane and the subsequent removal and degradation of the extension in the cisternal space of the endoplasmic reticulum.

signal peptidase An enzyme in the cisternal space of the ER that catalyzes removal of the signal sequence from the polypeptide.

signal recognition particle (SRP) In eukaryotes, a complex of a small RNA molecule with six proteins, which can temporarily halt protein synthesis by recognizing the signal sequence of a nascent polypeptide destined to be translocated through the ER, binding to it, and thereby blocking further translation of the mRNA.

signal sequence The hydrophobic, amino terminal extension found on proteins that are secreted from a cell. The amino terminus (extension) is removed and degraded in the cisternal space of the endoplasmic reticulum.

silencer *See* **silencer element.**

silencer element In eukaryotes, a transcriptional regulatory element that decreases RNA transcription rather than stimulating it like other enhancer elements.

silent mutation A mutational change resulting in a protein with a wild-type function because of an unchanged amino acid sequence.

simple telomeric sequences Simple, tandemly repeated DNA sequences at, or very close to, the extreme ends of the chromosomal DNA molecules.

SINEs (short interspersed repeated sequences) One class of interspersed and highly repeated sequences that consists of dispersed families with unit lengths of fewer than 500 base pairs and repeated for as many as hundreds of thousands of copies in the genome.

single-strand DNA binding (SSB) proteins (helix-destabilizing proteins) Proteins that help the DNA unwinding process by stabilizing the single-stranded DNA.

sister chromatid A chromatid derived from replication of one chromosome during interphase of the cell cycle.

slope (regression coefficient) The change in one variable (y) associated with a unit increase in another variable (x).

small nuclear ribonucleoprotein particles (snRNPs) The complexes formed by small nuclear RNAs and proteins in which the processing of pre-mRNA molecules occurs.

small nuclear RNA (snRNA) Found only in eukaryotes, one of four major classes of RNA molecules produced by transcription. snRNAs are used in the processing of pre-mRNA molecules.

somatic cell hybridization The fusion of two genetically different somatic cells of the same or different species to generate a somatic hybrid for genetic analysis.

somatic mutation A mutation in a cell that produces a mutant spot or area, but the mutant characteristic is not passed on to the succeeding generation.

Southern blot technique A technique invented by E. M. Southern and used in analyzing genes and gene transcripts, in which DNA fragments are transferred from a gel to a nitrocellulose filter.

spacer sequences Transcribed sequences that are found between, and flanking, coding RNA sequences. Spacer sequences are removed during processing of pre-rRNA and pre-tRNA to produce mature molecules.

specialized transducing phage A temperate bacteriophage that can transduce only a certain section of the bacterial chromosome.

specialized transduction A type of transduction in which only specific genes are transferred.

spermatogenesis Development of the male animal germ cell within the male gonad.

sperm cells (spermatozoa) The male gametes; the spermatozoa produced by the testes in male animals.

spliceosomes The splicing complexes formed by the association of several snRNPs bound to the pre-mRNA.

spontaneous mutations The mutations that occur without the use of chemical or physical mutagenic agents.

sporophyte The haploid, asexual generation in the life cycle of plants that produces haploid spores by meiosis.

stamen The male reproductive organ in a flowering plant that usually consists of a stalk, called a filament, bearing a pollen-producing anther.

standard deviation The square root of the variance. It measures the extent to which each measurement in the data set differs from the mean value and is used as a measure of the extent of variability in a population.

standard error of gene frequency A measure of the amount of variation among the gene frequencies of populations. It is the square root of the variance of gene frequency.

steroid response element (REs) The DNA sequence to which steroid hormones will bind to activate a gene.

stop codon *See* **chain-terminating codon.**

structural gene A gene that codes for an mRNA molecule and hence for a polypeptide chain.

submetacentric chromosome A chromosome that has the centromere nearer one end than the other. Such chromosomes appear J-shaped at anaphase.

sum rule The rule that the probability of either one of two mutually exclusive events occurring is the sum of their individual probabilities.

suppressor gene A gene that causes suppression of mutations in other genes.

suppressor mutation A mutation at a second site that totally or partially restores a function lost because of a primary mutation at another site.

Svedberg units (S values) The conversions for sedimentation rates in sucrose density centrifugation. Svedberg units are used as a rough indication of relative sizes of the components being analyzed.

synapsis The intimate association of homologous chromosomes brought about by the formation of a zipperlike structure along the length of the chromatids called the **synaptonemal complex.**

synaptonemal complex A complex structure spanning the region between meiotically paired (synapsed) chromosomes that is concerned with crossing-over rather than with chromosome pairing.

synkaryon A fusion nucleus produced following the fusion of cells with genetically different nuclei.

syntenic The genes that are localized to a particular chromosome by using an experimental approach (literally "together thread"; the term is similar to *linked*).

TATA element *See* **Goldberg-Hogness box.**

tautomeric shift The change in the chemical form of a DNA (or RNA) base.

tautomers Alternate chemical forms in which DNA (or RNA) bases are able to exist.

telocentric chromosome A chromosome that has the centromere more or less at one end.

telomere-associated sequences Repeated, complex DNA sequences extending from the molecular gene of chromosomal DNA, suspected to mediate many of the telomere-specific interactions.

telophase A stage during which the migration of the daughter chromosomes to the two poles is completed.

telophase II The last stage of meiosis II during which a nuclear membrane forms around each set of chromosomes, and cytokinesis takes place.

template strand The unwound single strand of DNA on which new strands are made (following complementary base pairing rules).

termination factors (release factors; RF) The specific proteins in polypeptide synthesis (translation) that read the chain termination codons and then initiate a series of specific events to terminate polypeptide synthesis.

terminator *See* **transcription terminator sequence**.

testcross A cross of an individual of unknown genotype, usually expressing the dominant phenotype, with a homozygous recessive individual in order to determine the genotype of the individual.

testis-determining factor Gene product in placental mammals that sets the switch toward male sexual differentiation.

tetrad analysis Genetic analysis of all the products of a single meiotic event. Tetrad analysis is possible in those organisms in which the four products of a single nucleus that has undergone meiosis are grouped together in a single structure.

tetrasomy The aberrant, aneuploid state in a normally diploid cell or organism in which an extra chromosome pair results in the presence of four copies of one chromosome type and two copies of every other chromosome type.

tetratype (T) One of the three types of tetrads possible when two genes are segregating in a cross. The T tetrad contains two parental and two recombinant nuclei, one of each parental type and one of each recombinant type.

three-point testcross A test involving three genes within a relatively short section of the chromosome. It is used to map genes for their order in the chromosome and for the distance between them.

thymine (T) A pyrimidine base found in DNA but not in RNA. In double-stranded DNA, thymine pairs with adenine.

topoisomerases A class of enzymes that catalyze the supercoiling of DNA.

totipotency The capacity of a nucleus to direct events through all the stages in development and therefore produce a normal adult.

trailer sequence The sequence of the mRNA molecule beginning at the end of the amino acid-coding sequence and ending at the 3' end of the mRNA. The trailer sequence is not translated and varies in length from molecule to molecule.

transconjugants In bacteria, the recipients inheriting donor DNA in the process of conjugation.

transcription The transfer of information from a double-stranded DNA molecule to a single-stranded RNA molecule. It is also called *RNA synthesis*.

transcriptional control The first level of control of gene expression in eukaryotes. This level involves regulating whether or not a gene is to be transcribed and the rate at which transcripts are produced.

transcription factors (TFs) Specific proteins that are required for the initiation of transcription by each of the three eukaryotic RNA polymerases. Each polymerase uses its own set of TFs.

transcription terminator sequence (terminator) A transcription regulatory sequence located at the distal end of a gene that signals the termination of transcription.

transdetermination During development, a process whereby an imaginal disc does not totally dedifferentiate but switches to another determined path.

trans-dominant The phenomenon of a gene or DNA sequence controlling genes that are on a different piece (strand) of DNA.

transducing phage The phage that is the vehicle by which genetic material is shuttled between bacteria.

transducing retroviruses Retroviruses that have picked up an oncogene from the cellular genome.

transductants In bacteria, the recipients inheriting donor DNA in the process of transduction.

transduction A process by which bacteriophages mediate the transfer of bacterial genetic information from one bacterium (the donor) to another (the recipient); a process whereby pieces of bacterial DNA are carried between bacterial strains by a phage.

transfer RNA (tRNA) One of the four classes of RNA molecules produced by transcription and involved in protein synthesis; molecules that bring amino acids to the ribosome, where they are matched to the transcribed message on the mRNA.

transformant The genetic recombinant generated by the transformation process.

transformation (a) A process in which genetic information is transferred by means of extracellular pieces of DNA in bacteria. (b) The failure of cells to remain constrained in their growth properties and give rise to tumors.

transgene A gene introduced into the genome of an organism by genetic manipulation in order to alter its genotype.

transgenic organism An organism that has had its genotype altered by the introduction of a gene into its genome by genetic manipulation.

transition mutation A specific type of base-pair substitution mutation that involves a change in the DNA from one purine-pyrimidine base pair to the other purine-pyrimidine base pair at a particular site (e.g., AT to GC).

transit peptidase The enzyme that removes transit sequences from proteins transported into organelles.

transit sequences The extra sequences at the N-terminal ends of proteins that are necessary and sufficient for posttranslational transport into organelles.

translation (protein synthesis) The conversion in the cell of the mRNA base sequence information into an amino acid sequence of a polypeptide.

translational control The regulation of protein synthesis by ribosome synthesis among mRNAs.

translocation (transposition) (a) A chromosomal mutation involving a change in position of a chromosome segment (or segments) and the gene sequences it contains. (b) In polypeptide synthesis, translocation is the movement of the ribosome, one codon at a time, along the mRNA toward the 3' end.

transmission genetics (classical genetics) A subdivision of the science of genetics primarily dealing with how genes are passed from one individual to another.

transport control Regulating the number of transcripts that exit the nucleus to the cytoplasm.

transposable element A genetic element of chromosomes of both prokaryotes and eukaryotes that has the capacity to mobilize itself and move from one location to another in the genome.

transposase An enzyme encoded by the IS element of a transposon that catalyzes transposition activity of a transposable element.

transposon (Tn) A mobile DNA segment that contains genes for the insertion of the DNA segment into the chromosome and for mobilization of the element to other locations on the chromosomes.

transversion mutation A specific type of base-pair substitution mutation that involves a change in the DNA from a purine-pyrimidine base pair to a pyrimidine-purine base pair at the same site (e.g., AT to TA or GC to TA).

trihybrid cross A cross between individuals of the same type that are heterozygous for three pairs of alleles at three different loci.

trisomy An aberrant, aneuploid state in a normally diploid cell or organism in which there are three copies of a particular chromosome instead of two copies.

trisomy-21 A human clinical condition characterized by various abnormalities. It is caused by the presence of an extra copy of chromosome 21.

true-breeding (pure-breeding) strain A strain allowed to self-fertilize for many generations to ensure that the traits to be studied are inherited and unchanging.

true reversion A point mutation from mutant back to wild type in which the change codes for the original amino acid of the wild type.

tumor viruses Viruses that induce cells to dedifferentiate and to divide to produce a tumor.

Turner syndrome A human clinical syndrome that results from monosomy for the X chromosome in the female, which gives a 45,X female. These females fail to develop secondary sexual characteristics, tend to be short, have weblike necks, have poorly developed breasts, are usually infertile, and exhibit mental deficiencies.

twin spots Two adjacent cell groups that differ in genotype and phenotype. They result from mitotic crossing-over within the somatic cells of a heterozygous individual.

uniparental inheritance A phenomenon, usually exhibited by extranuclear genes, in which all progeny have the phenotype of only one parent.

unique (single-copy) sequence A class of DNA sequences that has one to a few copies per genome.

upstream activator sequences (UASs) In yeast, elements that are functionally similar to enhancers in other eukaryotes. UASs can function in either orientation and at variable distances upstream of the promoter.

uracil (U) A pyrimidine base found in RNA but not in DNA.

variance A statistical measure of how values vary from the mean.

variance of gene frequency The variance in the frequency of an allele among a group of populations.

vegetative reproduction *See* **asexual reproduction.**

viral oncogene A viral gene that transforms a cell it infects to a cancerous state. *See also* **cellular oncogene; oncogenesis.**

virulent phage A phage like T4, which always follows the lytic cycle when it infects bacteria.

visible mutation A mutation that affects the morphology or physical appearance of an organism.

wild type A strain, organism, or gene of the type that is designated as the standard for the organism with respect to genotype and phenotype.

wild-type allele The allele designated as the standard ("normal") for a strain of organism.

wobble hypothesis A theory proposed by Francis Crick which proposes that the base at the 5' end of the anticodon (3' end of the codon) is not as constrained as the other two bases. This feature allows for less exact base pairing so that the 5' end of the anticodon can potentially pair with one of three different bases at the 3' end of the codon.

X chromosome A sex chromosome present in two copies in the homogametic sex and in one copy in the heterogametic sex.

X chromosome-autosome balance system A genotypic sex determination system. The main factor in sex determination is the ratio between the numbers of X chromosomes and autosomes. Sex is determined at the time of fertilization, and sex differences are assumed to be due to the action during development of two sets of genes located in the X chromosomes and in the autosomes.

X chromosome nondisjunction An event occurring when the two X chromosomes fail to separate in meiosis so that eggs are produced either with two X chromosomes or with no X chromosomes, instead of the usual one X chromosome.

X-linked Referring to genes located on the X chromosome.

X-linked dominant trait A trait due to a dominant mutant gene carried on the X chromosome.

X-linked recessive trait A trait due to a recessive mutant gene carried on the X chromosome.

Y chromosome A sex chromosome that when present is found in one copy in the heterogametic sex, along with an X chromosome, and is not present in the homogametic sex. Not all organisms with sex chromosomes have a Y chromosome.

yeast artificial chromosome (YAC) A cloning vector in which DNA fragments several hundred kilobase pairs long can be cloned in yeast. A YAC is a linear vector with a yeast telomere at each end, a centromere, a sequence for autonomous replication in yeast, a selectable marker for yeast, and a polylinker.

Y-intercept In a regression analysis, the value of y when x is zero.

Y-linked (holandric ["wholly male"]) trait A trait due to a mutant gene carried on the Y chromosome but with no counterpart on the X.

zygonema The stage during meiosis in prophase I at which homologous chromosomes begin to pair in a highly specific way (like a zipper).

zygote The cell produced by the fusion of the male and female gametes.

SUGGESTED READING

CHAPTER 1: GENETICS: AN INTRODUCTION

Sturtevant, A. H. 1965. *A history of genetics.* New York: Harper & Row.

CHAPTER 2: MENDELIAN GENETICS

Bateson, W. 1909. *Mendel's principles of heredity.* Cambridge: Cambridge University Press.

Mendel, G. 1866. Experiments in plant hybridization (translation). In *Classic papers in genetics,* J. A. Peters, ed. 1959. Englewood Cliffs, NJ: Prentice-Hall.

Peters, J. A., ed. 1959. *Classic papers in genetics.* Englewood Cliffs, NJ: Prentice-Hall.

Sandler, I., and Sandler, L. 1985. A conceptual ambiguity that contributed to the neglect of Mendel's paper. *Hist. Phil. Life Sci.* 7:3–70.

Tschermak-Seysenegg, E. von. 1951. The rediscovery of Mendel's work. *J. Hered.* 42:163–171.

CHAPTER 3: CHROMOSOMAL BASIS OF INHERITANCE, SEX DETERMINATION, AND SEX LINKAGE

Barr, M. L. 1960. Sexual dimorphism in interphase nuclei. *Am. J. Hum. Genet.* 12:118–127.

Bogan, J. S., and Page, D. C. 1994. Ovary? Testis?—A mammalian dilemma. *Cell* 76:603–607.

Boggs, R. T., Gregor, P., Idriss, S., Belote, J. M., and McKeown, M. 1987. Regulation of sexual differentiation in *Drosophila melanogaster* via alternative splicing of RNA from the transformer. *Cell* 50:739–747.

Bridges, C. B. 1916. Nondisjunction as a proof of the chromosome theory of heredity. *Genetics* 1:1–52, 107–163.

———. 1925. Sex in relation to chromosomes and genes. *Am. Natur.* 59:127–137.

Capel, B. 1995. New bedfellows in the mammalian sex-determination affair. *Trends Genet.* 11:161–163.

Dice, L. R. 1946. Symbols for human pedigree charts. *J. Hered.* 37:11–15.

Disteche, C. M. 1995. Escape from X inactivation in human and mouse. *Trends Genet.* 11:17–22.

Egel, R. 1995. The synaptonemal complex and the distribution of meiotic recombination events. *Trends Genet.* 11:206–208.

Eicher, E. M., and Washburn, L. L. 1986. Genetic control of primary sex determination in mice. *Annu. Rev. Genet.* 20:327–360.

Ellis, N., and Goodfellow, P. N. 1989. The mammalian pseudoautosomal region. *Trends Genet.* 5:406–410.

Farabee, W. C. 1905. Inheritance of digital malformations in man. *Papers Peabody Museum Amer. Arch. Ethnol. (Harvard Univ.)* 3:65–78.

Fuller, M. T., and Wilson, P. G. 1992. Force and counterforce in the mitotic spindle. *Cell* 71:547–550.

Haqq, C. M., King, C.-Y., Ukiyama, E., Falsafi, S., Haqq, T. N., Donahoe, P. K., and Weiss, M. A. 1994. Molecular basis of mammalian sexual determination: activation of Müllerian inhibiting substance gene expression by SRY. *Science* 266:1494–1500.

Hawley, R. S., and Arbel, T. 1993. Yeast genetics and the fall of the classical view of meiosis. *Cell* 72:301–303.

Hodgkin, J. 1987. Sex determination and dosage compensation in *Caenorhabditis elegans. Annu. Rev. Genet.* 21:133–154.

———. 1989. *Drosophila* sex determination: A cascade of regulated splicing. *Cell* 56:905–906.

———. 1993. Molecular cloning and duplication of the nematode sex-determining gene *tra-1. Genetics* 133:543–560.

Jiménez, R., Sánchez, A., Burgos, M., and Díaz de la Guardia, R. 1996. Puzzling out the genetics of mammalian sex determination. *Trends Genet.* 12:164–166.

Kay, G. F., Barton, S. C., Surani, M. A., and Rastan, S. 1994. Imprinting and X chromosome counting mechanisms determine *Xist* expression in early mouse development. *Cell* 77:639–650.

Koopman, P., Gubbay, J., Vivian, N., Goodfellow, P., and Lovell-Badge, R. 1991. Male development of chromosomally female mice transgenic for *Sry. Nature* 351:117–121.

Lee, J. T., Strauss, W. M., Dausman, J. A., and Jaenisch, R. 1996. A 450 kb transgene displays properties of the mammalian X-inactivation center. *Cell* 86:83–94.

Lyon, M. F. 1962. Sex chromatin and gene action in the mammalian X-chromosome. *Am. J. Hum. Genet.* 14:135–148.

McClung, C. E. 1902. The accessory chromosome—sex determinant? *Biol. Bull.* 3:43–84.

McElreavy, K., Vilain, E., Abbas, N., Costa, J.-M., Souleyreau, N., Kucheria, K., Boucekkine, C., Thibaud, E., Brauner, R., Flamant, F., and Fellous, M. 1992. XY sex reversal associated with a deletion 5' to the *SRY* "HMG box" in the testis-determining region. *Proc. Natl. Acad. Sci. USA* 89:11016–11020.

McKusick, V. A. 1965. The royal hemophilia. *Sci. Am.* 213:88–95.

Migeon, B. R. 1994. X-chromosome inactivation: molecular mechanisms and genetic consequences. *Trends Genet.* 10:230–235.

Morgan, L. V. 1922. Non criss-cross inheritance in *Drosophila melanogaster. Biol. Bull.* 42:267–274.

Morgan, T. H. 1910. Sex-limited inheritance in *Drosophila. Science* 32:120–122.

———. 1911. An attempt to analyze the constitution of the chromosomes on the basis of sex-limited inheritance in *Drosophila. J. Exp. Zool.* 11:365–414.

Page, D. C. 1985. Sex-reversal: Deletion mapping of the male-determining function of the human Y chromosome. *Cold Spring Harbor Symp. Quant. Biol.* 51:229–235.

Page, D. C., de la Chapelle, A., and Weissenbach, J. 1985. Chromosome Y-specific DNA in related human XX males. *Nature* 315:224–226.

Page, D. C., Mosher, R., Simpson, E. M., Fisher, E. M. C., Mardon, G., Pollack, J., McGillivray, B., de la Chapelle, A., and Brown, L. G. 1987. The sex-determining region of the human Y chromosome encodes a finger protein. *Cell* 51:1091–1104.

Palmer, M. S., Sinclair, A. H., Berta, P., Ellis, N. A., Goodfellow, P. N., Abbas, N. E., and Fellous, M. 1990. Genetic evidence that ZFY is not the testis-determining factor. *Nature* 342:937–939.

Penny, G. D., Kay, G. F., Sheardown, S. A., Rastan, S., and Brockdorff, N. 1996. Requirement for *Xist* in X chromosome inactivation. *Nature* 379:131–137.

Rivera-Pomar, R., and Jäckle, H. 1996. From gradients to stripes in *Drosophila* embryogenesis: filling in the gaps. *Trends Genet.* 12:478–483.

Sinclair, A. H., Berta, P., Palmer, M. S., Hawkins, J. R., Griffiths, B. L., Smith, M. J., Foster, J. W., Frischauf, A.-M., Lovell-Badge, R., and Goodfellow, P. N. 1990. A gene from the human sex-determining region encodes a protein with homology to a conserved DNA-binding motif. *Nature* 346:240–244.

Staehelin, L. A., and Hepler, P. K. 1996. Cytokinesis in higher plants. *Cell* 84:821–824.

Stern, C., Centerwall, W. P., and Sarkar, Q. S. 1964. New data on the problem of Y-linkage of hairy pinnae. *Am. J. Hum. Genet.* 16:455–471.

Sutton, W. S. 1903. The chromosomes in heredity. *Biol. Bull.* 4:231–251.

Willard, H. F. 1996. X chromosome inactivation, *XIST*, and pursuit of the X-inactivation center. *Cell* 86:5–7.

Wilson, E. B. 1905. The chromosomes in relation to the determination of sex in insects. *Science* 22:500–502.

CHAPTER 4: EXTENSIONS OF MENDELIAN GENETIC ANALYSIS

Bultman, S. J., Michaud, E. J., and Woychik, R. P. 1992. Molecular characterization of the mouse *agouti* locus. *Cell* 71:1195–1204.

Ginsburg, V. 1972. Enzymatic basis for blood groups. *Methods Enzymol.* 36:131–149.

Huntington's Disease Collaborative Research Group. 1993. A novel gene containing a trinucleotide repeat that is expanded and unstable on Huntington's disease chromosomes. *Cell* 72:971–983.

Landauer, W. 1948. Hereditary abnormalities and their chemically induced phenocopies. *Growth Symposium* 12:171–200.

Landsteiner, K., and Levine, P. 1927. Further observations on individual differences of human blood. *Proc. Soc. Exp. Biol. Med.* 24:941–942.

Siracusa, L. D. 1994. The *agouti* gene: turned on to yellow. *Trends Genet.* 10:423–428.

CHAPTER 5: GENETIC MAPPING IN EUKARYOTES I

Bateson, W., Saunders, E. R., and Punnett, R. G. 1905. Experimental studies in the physiology of heredity. *Rep. Evol. Committee R. Soc.* II:1–55, 80–99.

Blixt, S. 1975. Why didn't Mendel find linkage? *Nature* 256:206.

Creighton, H. S., and McClintock, B. 1931. A correlation of cytological and genetical crossing-over in *Zea mays. Proc. Natl. Acad. Sci. USA.* 17:492–497.

Morgan, T. H. 1910. Sex-limited inheritance in *Drosophila. Science* 32:120–122.

———. 1910. The method of inheritance of two sex-limited characters in the same animal. *Proc. Soc. Exp. Biol. Med.* 8:17.

———. 1911. An attempt to analyze the constitution of the chromosomes on the basis of sex-limited inheritance in *Drosophila. J. Exp. Zool.* 11:365–414.

———. 1911. Random segregation versus coupling in Mendelian inheritance. *Science* 34:384.

Morgan, T. H., Sturtevant, A. H., Müller, H. J., and Bridges, C. B. 1915. *The mechanism of Mendelian heredity.* New York: Henry Holt.

Sturtevant, A. H. 1913. The linear arrangement of six sex-linked factors in *Drosophila* as shown by their mode of association. *J. Exp. Zool.* 14:43–59.

Sutton, W. S. 1903. The chromosomes in heredity. *Biol. Bull.* 4:231–251.

CHAPTER 6: GENETIC MAPPING IN EUKARYOTES II

Barratt, R. W., Newmeyer, D., Perkins, D. D., and Garnjobst, L. 1954. Map construction in *Neurospora crassa. Adv. Genet.* 6:1–93.

Chaleff, R. S., and Carlson, P. S. 1974. Somatic cell genetics of higher plants. *Annu. Rev. Genet.* 8:267–278.

Dib, C., and many other authors. 1996. A comprehensive genetic map of the human genome based on 5,264 microsatellites. *Nature* 380:152–154.

Emery, A. E. H. 1976. *Methodology in Medical Genetics: An Introduction to Statistical Methods.* Churchill Livingstone, Edinburgh, Scotland.

Ephrussi, B., and Weiss, M. C. 1969. Hybrid somatic cells. *Sci. Am.* 220:26–35.

Fincham, J. R. S., Day, P. R., and Radford, A. 1979. *Fungal genetics.* 3rd ed. Oxford: Blackwell Scientific.

Kao, F., Jones, C., and Puck, T. T. 1976. Genetics of somatic mammalian cells: Genetic, immunologic, and biochemical analysis with Chinese hamster cell hybrids containing selected human chromosomes. *Proc. Natl. Acad. Sci. USA* 73:193–197.

McKusick, V. A. 1971. The mapping of human chromosomes. *Sci. Am.* 224:104–113.

Pontecorvo, G., and Kafer, E. 1958. Genetic analysis based on mitotic recombination. *Adv. Genet.* 9:71–104.

Pritchard, R. H. 1955. The linear arrangement of a series of alleles of *Aspergillus nidulans. Heredity* 9:343–371.

Ried, T., Baldini, A., Rand, T. C., and Ward, D. C. 1992. Simultaneous visualization of seven different DNA probes by *in situ* hybridization using combinatorial fluorescence and digital imaging microscopy. *Proc. Natl. Acad. Sci. USA* 89:1388–1392.

Ruddle, F. H., and Kucherlapati, R. S. 1974. Hybrid cells and human genes. *Sci. Am.* 231:36–44.

Stern, C. 1936. Somatic crossing-over and segregation in *Drosophila melanogaster. Genetics* 21:625–730.

CHAPTER 7: CHROMOSOMAL MUTATIONS

Ada, G. L., and Nossal, G. 1987. The clonal selection theory. *Sci. Am.* 257(2):62–69.

Auerbach, C. 1976. *Mutation research.* London: Chapman and Hall.

Barr, M. L., and Bertram, E. G. 1949. A morphological distinction between neurones of the male and female, and the behavior of the nucleolar satellite during accelerated nucleoprotein synthesis. *Nature* 163:676–677.

Blackwell, T. K., and Alt, F. W. 1989. Mechanism and developmental program of immunoglobulin gene rearrangements in mammals. *Annu. Rev. Genet.* 23:605–636.

Bloom, A. D. 1972. Induced chromosome aberrations in man. *Adv. Hum. Genet.* 3:99–153.

Borst, P., and Greaves, D. R. 1987. Programmed gene rearrangements altering gene expression. *Science* 235:658–667.

Caskey, C. T., Pizzuti, A., Fu, Y.-H., Fenwick, R. G., and Nelson, D. L. 1992. Triplet repeat mutations in human disease. *Science* 256:784–789.

Dalla-Favera, R., Martinotti, S., Gallo, R., Erickson, J., and Croce, C. 1983. Translocation and rearrangements of the *c-myc* oncogene locus in human undifferentiated B-cell lymphomas. *Science* 219:963–997.

DeKlein, A., van Kessel, A. G., Grosveld, G., Bartram, C. R., Hagemeijer, A., Bootsma, D., Spurr, N. K., Heisterkamp, N., Groffen, J., and Stephenson, J. R. 1982. A cellular oncogene is translocated to the Philadelphia chromosome in chronic myelocytic leukemia. *Nature* 300:765–767.

Donelson, J. E. 1989. DNA rearrangements and antigenic variation in African Trypanosomes. In D. E. Berg and M. M. Howe, eds., *Mobile DNA*, pp. 763–782. Washington, DC: American Society for Microbiology.

Haber, J. E. 1992. Mating-type gene switching in *Saccharomyces cerevisiae*. *Trends Genet.* 8:446–452.

Huntington's Disease Collaborative Research Group. 1993. A novel gene containing a trinucleotide repeat that is expanded and unstable in Huntington's disease chromosome. *Cell* 72:971–983.

Kremer, E., Pritchard, M., Lynch, M., Yu, S., Holman, K., Baker, E., Warren, S. T., Schlessinger, D., Sutherland, G. R., and Richards, R. I. 1991. Mapping of DNA instability at the fragile X to a trinucleotide repeat sequence p(CGG)*n*. *Science* 252:1711–1714.

Lyon, M. F. 1961. Gene action in the X-chromosomes of the mouse (*Mus. musculus L*). *Nature* 190:372–373.

Nasmyth, K. A. 1982. Molecular genetics of yeast mating type. *Annu. Rev. Genet.* 16:439–500.

Penrose, L. S., and Smith, G. F. 1966. *Down's anomaly*. Boston: Little, Brown.

Richards, R. I., and Sutherland, G. R. 1992. Dynamic mutations: A new class of mutations causing human disease. *Cell* 70:709–712.

———. 1992. Fragile X syndrome: The molecular picture comes into focus. *Trends Genet.* 8:249–255.

Ried, T., Baldini, A., Rand, T. C., and Ward, D. C. 1992. Simultaneous visualization of seven different DNA probes by *in situ* hybridization using combinatorial fluorescence and digital imaging microscopy. *Proc. Natl. Acad. Sci. USA* 89:1388–1392.

Robertson, B. D., and Meyer, T. F. 1992. Genetic variation in pathogenic bacteria. *Trends Genet.* 8:422–427.

Rowley, J. D. 1973. A new consistent chromosomal abnormality in chronic myelogenous leukemia identified by quinacrine fluorescence and Giemsa staining. *Nature* 243:290–293.

Shaw, M. W. 1962. Familial mongolism. *Cytogenetics* 1:141–179.

Siomi, H., Siomi, M. C., Nussbaum, R. L., and Dreyfuss, G. 1993. The protein product of the fragile X gene, *FMR1*, has characteristics of an RNA-binding protein. *Cell* 74:291–298.

Tarleton, J. C., and Saul, R. A. 1993. Molecular genetic advances in fragile X syndrome. *J. Pediatrics* 122:169–185.

Van de Putte, P., and Goosen, N. 1992. DNA inversions in phages and bacteria. *Trends Genet.* 8:457–462.

Van der Ploeg, L. H. T., Gottesdiener, K., and Lee, M. G.-S. 1992. Antigenic variation in African trypanosomes. *Trends Genet.* 8:452–457.

Verkerk, A. J. M. H., Piertti, M., Sutcliff, J. S., Fu, Y.-H., Kuhl, D. P. A., Pizzuti, A., Reiner, O., Richards, S., Victoria, M. F., Zhang, F., Eussen, B. E., van Ommen, G.-J. B., Blonden, L. A. J., Riggins, G. J., Chastain, J. L., Kunst, C. B., Galjaard, H., Caskey, C. T., Nelson, D. L., Oostra, B. A., and Warrent, S. T. 1991. Identification of a gene (*FMR–1*) containing a CGG repeat coincident with a breakpoint cluster region exhibiting length variation in fragile X syndrome. *Cell* 65:905–914.

CHAPTER 8: GENETIC RECOMBINATION IN BACTERIA AND BACTERIOPHAGES

Archer, L. J. 1973. *Bacterial transformation*. New York: Academic Press.

Benzer, S. 1959. On the topology of the genetic fine structure. *Proc. Natl. Acad. Sci. USA* 45:1607–1620.

———. 1961. On the topography of the genetic fine structure. *Proc. Natl. Acad. Sci. USA* 47:403–415.

———. 1962. The fine structure of the gene. *Sci. Am.* 206:70–84.

Campbell, A. 1969. *Episomes*. New York: Harper & Row.

Curtiss, R. 1969. Bacterial conjugation. *Annu. Rev. Microbiol.* 23:69–136.

Ellis, E. L., and Delbruck, M. 1939. The growth of bacteriophage. *J. Gen. Physiol.* 22:365–384.

Fincham, J. 1966. *Genetic complementation*. New York: W. A. Benjamin.

Hayes, W. 1968. *The genetics of bacteria and their viruses*, 2nd ed. New York: Wiley.

Hershey, A. D., and Rotman, R. 1949. Genetic recombination between host-range and plaque-type mutants of bacteriophage in single bacterial cells. *Genetics* 34:44–71.

Hotchkiss, R. D., and Gabor, M. 1970. Bacterial transformation with special reference to recombination processes. *Annu. Rev. Genet.* 4:193–224.

Jacob, F., and Wollman, E. L. 1951. *Sexuality and the genetics of bacteria*. New York: Academic Press.

Ravin, A. W. 1961. The genetics of transformation. *Adv. Genet.* 10:61–163.

Susman, M. 1970. General bacterial genetics. *Annu. Rev. Genet.* 4:135–176.

Vielmetter, W., Bonhoeffer, F., and Schutte, A. 1968. Genetic evidence for transfer of a single DNA strand during bacterial conjugation. *J. Mol. Biol.* 37:81–86.

Wollman, E. L., Jacob, F., and Hayes, W. 1962. Conjugation and genetic recombination in *E. coli* K-12. *Cold Spring Harbor Symp. Quant. Biol.* 21:141–162.

Zinder, N., and Lederberg, J. L. 1952. Genetic exchange in *Salmonella*. *J. Bacteriol.* 64:679–699.

CHAPTER 9: THE BEGINNINGS OF MOLECULAR GENETICS: GENE FUNCTION

Beadle, G. W., and Ephrussi, B. 1937. Development of eye colors in *Drosophila*: Diffusible substances and their interrelationships. *Genetics* 22:76–86.

Beadle, G. W., and Tatum, E. L. 1942. Genetic control of biochemical reactions in *Neurospora*. *Proc. Natl. Acad. Sci. USA* 27:499–506.

Collins, F. 1992. Cystic fibrosis: Molecular biology and therapeutic implications. *Science* 256:774–779.

Doggett, N. A., Cheng, J.-F., Smith, C. L., and Cantor, C. R. 1989. The Huntington's disease locus is most likely within 325 kilobases of the chromosome 4p telomere. *Proc. Natl. Acad. Sci. USA* 86:10011–10014.

Galjaard, H. 1986. Biochemical diagnosis of genetic diseases. *Experientia* 42:1075–1085.

Garrod, A. E. 1909. *Inborn errors of metabolism*. New York: Oxford University Press.

Gilbert, F., Kucherlapati, R., Creagan, R. P., Murnane, M. J., Darlington, G. J., and Ruddle, F. H. 1975. Tay-Sachs' and Sandhoff's diseases: The assignment of genes for hexosaminidase A and B to individual human chromosomes. *Proc. Natl. Acad. Sci. USA* 72:263–267.

Gusella, J. F., Wexler, N. S., Conneally, P. M., Naylor, S. L., Anderson, M. A., Tanzi, R. E., Watkins, P. C., Ottina, K., Wallace, M. R., Sakaguchi, A. Y., Young, A. B., Shoulson, I., Bonilla, E., and Martin, J. B. 1993. A polymorphic DNA marker genetically linked to Huntington's disease. *Nature* 306:234–238.

Guttler, F., and Woo, S. L. C. 1986. Molecular genetics of PKU. *J. Inherited Metab. Dis.* 9 Suppl.1:58–68.

Harris, H. 1975. *The principles of human biochemical genetics.* Amsterdam: North-Holland.

Ingram, V. M. 1963. *The hemoglobins in genetics and evolution.* New York: Columbia University Press.

McIntosh, I., and Cutting, G. R. 1992. Cystic fibrosis transmembrane conductance regulator and the etiology and pathogenesis of cystic fibrosis. *FASEB J.* 6:2775–2782.

Maniatis, T., Fritsch, E. F., Lauer, J., and Lawn, R. M. 1980. The molecular genetics of human hemoglobins. *Annu. Rev. Genet.* 14:145–178.

Motulsky, A. G. 1964. Hereditary red cell traits and malaria. *Am. J. Trop. Med. Hyg.* 13:147–158.

———. 1973. Frequency of sickling disorders in U.S. blacks. *N. Engl. J. Med.* 288:31–33.

Neel, J. V. 1949. The inheritance of sickle-cell anemia. *Science* 110:64–66.

Pauling, L., Itano, H. A., Singer, S. J., and Wells, J. C. 1949. Sickle-cell anemia, a molecular disease. *Science* 110:543–548.

Riordan, J. R., Rommens, J. M., Kerem, B., Alon, N., Rozmahel, R., Grzelczak, Z., Zielenski, J., Lok, S., Plavsic, N., Chou, J. L., Drumm, M. L., Ianuzzi, M. C., Collins, F. S., and Tsui, L.-C. 1989. Identification of the cystic fibrosis gene: Cloning and characterization of complementary DNA. *Science* 245:1066–1073.

Rommens, J. M., Ianuzzi, M. C., Kerem, B., Drumm, M. L., Melmer, G., Dean, M., Rozmahel, R., Cole, J. L., Kennedy, D., Hidaka, N., Zsiga, M., Buchwald, M., Riordan, J. R., Tsui, L.-C., and Collins, F. S. 1989. Identification of the cystic fibrosis gene: Chromosome walking and jumping. *Science* 245:1059–1065.

Scriver, C. R., and Clow, C. L. 1980. Phenylketonuria and other phenylalanine hydroxylation mutants in man. *Annu. Rev. Genet.* 14:179–202.

Srb, A. M., and Horowitz, N. H. 1944. The ornithine cycle in *Neurospora* and its genetic control. *J. Biol. Chem.* 154:129–139.

Stout, J. T., and Caskey, C. T. 1988. The Lesch-Nyhan syndrome: Clinical, molecular and genetic aspects. *Trends Genet.* 4:175–178.

Woo, S. L. C., Lidsky, A. S., Guttler, F., Chandra, T., and Robson, K. J. H. 1983. Cloned human phenylalanine hydroxylase gene allows prenatal diagnosis and carrier detection of classical phenylketonuria. *Nature* 300:151–155.

CHAPTER 10: THE STRUCTURE OF GENETIC MATERIAL

Andrews, R., Halligan, N. L., and Halligan, B. D. 1993. Nonamer binding protein induces a bend in the immunoglobulin gene recombinational signal sequence. *Biochem. Biophys. Res. Commun.* 193:139–145.

Arndt-Jovin, D. J., Udvardy, A., Garner, M. M., Ritter, S., and Jovin, T. M. 1993. Z-DNA binding and inhibition by GTP of *Drosophila* topoisomerase II. *Biochemistry* 32:4862–4872.

Avery, O. T., MacLeod, C. M., and McCarty, M. 1944. Studies on the chemical nature of the substance inducing transformation of pneumococcal types. Induction of transformation by a deoxyribonucleic acid fraction isolated from pneumococcus type III. *J. Exp. Med.* 79:137–158.

Chargaff, E. 1951. Structure and function of nucleic acids as cell constituents. *Fed. Proc.* 10:654–659.

Dickerson, R. E. 1983. The DNA helix and how it is read. *Sci. Am.* 249 (December):94–111.

Fraenkel-Conrat, H., and Singer, B. 1957. Virus reconstitution: Combination of protein and nucleic acid from different strains. *Biochim. Biophys. Acta* 24:540–548.

Franklin, R. E., and Gosling, R. 1953. Molecular configuration of sodium thymonucleate. *Nature* 171:740–741.

Geis, I. 1983. Visualizing the anatomy of A, B, and Z-DNAs. *J. Biomol. Struct. Dynam.* 1:581–591.

Gierer, A., and Schramm, G. 1956. Infectivity of ribonucleic acid from tobacco mosaic virus. *Nature* 177:702–703.

Griffith, F. 1928. The significance of pneumococcal types. *J. Hyg.* (Lond.) 27:113–159.

Gruskin, E. A., and Rich, A. 1993. B-DNA to Z-DNA structural transitions in the SV40 enhancer: Stabilization of Z-DNA in negatively supercoiled DNA minicircles. *Biochemistry* 32:2167–2176.

Hershey, A. D., and Chase, M. 1952. Independent functions of viral protein and nucleic acid in growth and bacteriophage. *J. Gen. Physiol.* 36:39–56.

Jaworski, A., Hsieh, W.-T., Blaho, J. A., Larson, J. E., and Wells, R. D. 1988. Left-handed DNA in vivo. *Science* 238:773–777.

Krishna, P., Kennedy, B. P., van de Sande, J. H., and McGhee, J. D. 1988. Yolk proteins from nematodes, chickens, and frogs bind strongly and preferentially to left-handed Z-DNA. *J. Biol. Chem.* 263:19066–19070.

Nardulli, A. M., Greene, G. L., and Sharpiro, D. J. 1993. Human estrogen receptor bound to an estrogen response element bends DNA. *Mol. Endocrinol.* 7:331–340.

Pauling, L., and Corey, R. B. 1956. Specific hydrogen-bond formation between pyrimidines and purines in deoxyribonucleic acids. *Arch. Biochem. Biophys.* 65:164–181.

Rich, A., Nordheim, A., and Wang, A. H.-J. 1984. The chemistry and biology of left-handed Z-DNA. *Annu. Rev. Biochem.* 53:791–846.

Structures of DNA. 1982. *Cold Spring Harbor Symp. Quant. Biol.* 47. Cold Spring Harbor, NY: Cold Spring Harbor Laboratory.

van der Vliet, P. C., and Verrijzer, C. P. 1993. Bending of DNA by transcription factors. *BioEssays* 15:25–32.

Wang, A. H.-J., Quigley, G. J., Kolpak, F. J., Crawford, J. L., van Boom, J. H., van der Marel, G., and Rich, A. 1979. Molecular structure of a left-handed double helical DNA fragment at atomic resolution. *Nature* 282:680–686.

Wang, J. C. 1982. DNA topoisomerases. *Sci. Am.* 247:94–109.

Watson, J. D. 1968. *The double helix.* New York:Atheneum.

Watson, J. D., and Crick, F. H. C. 1953. Genetical implications of the structure of deoxyribonucleic acid. *Nature* 171:964–969.

———. 1953. Molecular structure of nucleic acids. A structure for deoxyribose nucleic acid. *Nature* 171:737–738.

Wilkins, M. H. F., Stokes, A. R., and Wilson, H. R. 1953. Molecular structure of deoxypentose nucleic acids. *Nature* 171:738–740.

Wing, R. M., Drew, H. R., Takano, T., Broka, C., Tanaka, S., Itakura, K., and Dickerson, R. E. 1980. Crystal structure analysis of a complete turn of B-DNA. *Nature* 287:755–758.

Withers, B. E., and Dunbar, J. C. 1993. The endonuclease isoschizomers, SmaI and XmaI, bend DNA in opposite orientations. *Nucl. Acids Res.* 21:2571–2577.

CHAPTER 11: THE ORGANIZATION OF DNA IN CHROMOSOMES

Amati, B. B., and Gasser, S. M. 1988. Chromosomal ARS and CEN elements bind specifically to the yeast nuclear scaffold. *Cell* 54:967–978.

Blackburn, E. H. 1994. Telomeres: No end in sight. *Cell* 77:621–623.

Blackburn, E. H., and Szostak, J. W. 1984. The molecular structure of centromeres and telomeres. *Annu. Rev. Biochem.* 53:163–194.

Bloom, K. S., Amaya, E., Carbon, J., Clarke, L., Hill, A., and Yeh, E. 1984. Chromatin conformation of yeast centromeres. *J. Cell Biol.* 99:1559–1568.

Britten, R. J., and Kohne, D. E. 1968. Repeated sequences in DNA. *Science* 161:529–540.

Burlingame, R. W., Love, W. E., Wang, B.-C., Hamlin, R., Xuang, N.-H., and Moudranakis, E. N. 1985. Crystallographic structure of the octameric histone core of the nucleosome at a resolution of 33 Å. *Science* 228:546–553.

Cai, M., and Davis, R. W. 1990. Yeast centromere binding protein CBF1 of the helix-loop-helix protein family is required for chromosome stability and methionine prototrophy. *Cell* 61:437–446.

Carbon, J. 1984. Yeast centromeres: Structure and function. *Cell* 37:351–353.

Clarke, L. 1990. Centromeres of budding and fission yeasts. *Trends Genet.* 6:150–154.

Comings, D. 1978. Mechanisms of chromosome banding and implications for chromosome structure. *Annu. Rev. Genet.* 12:25–46.

D'Ambrosio, E., Waitzikin, S. D., Whitney, F. R., Salemme, A., and Furano, A. V. 1985. Structure of the highly repeated, long interspersed DNA family (LINE or L1Rn) of the rat. *Mol. Cell. Biol.* 6:411–424.

Freifelder, D. 1978. *The DNA molecule. Structure and properties.* San Francisco: Freeman.

Gellert, M. 1981. DNA topoisomerases. *Annu. Rev. Biochem.* 50:879–910.

Grosschedl, R., Giese, K., and Pagel, J. 1994. HMG domain proteins: architectural elements in the assembly of nucleoprotein structures. *Trends Genet.* 10:94–100.

Jelinek, W. R., and Schmid, C. W. 1982. Repetitive sequences in eukaryotic DNA and their expression. *Annu. Rev. Biochem.* 51:813–844.

Korenberg, J. R., and Rykowski, M. C. 1988. Human genome organization: Alu, LINES, and the molecular structure of metaphase chromosome bands. *Cell* 53:391–400.

Kornberg, R. D. 1977. Structure of chromatin. *Annu. Rev. Biochem.* 46:931–954.

Kornberg, R. D., and Klug, A. 1981. The nucleosome. *Sci. Am.* 244 (2):52–64.

Levis, R. W., Ganesan, R., Houtchens, K., Tolar, L. A., and Sheen, F. 1993. Transposons in place of telomeric repeats at a Drosophila telomere. *Cell* 75:1083–1093.

Lewin, B. 1980. *Gene expression,* 2nd ed., vol. 2, *Eucaryotic chromosomes.* New York: Wiley.

Long, E. O., and Dawid, I. B. 1980. Repeated genes in eukaryotes. *Annu. Rev. Biochem.* 49:727–764.

Ludérus, M. E. E., den Blaauwen, J. L., de Smit, O. J. B., Compton, D. A., and van Driel, R. 1994. Binding of matrix attachment regions to lamin polymers involves single-stranded regions and the minor groove. *Mol. Cell. Biol.* 14:6297–6305.

MacHattie, L. A., Ritchie, D. A., and Thomas, C. A. 1967. Terminal repetition in permuted T2 bacteriophage DNA molecules. *J. Mol. Biol.* 23:355–363.

Marmur, J., Rownd, R., and Schildkraut, C. L. 1963. Denaturation and renaturation of deoxyribonucleic acid. *Prog. Nucleic Acid Res. Mol. Biol.* 1:231–300.

Mason, J. M. , and Biessmann, A. 1995. The unusual telomeres of *Drosophila. Trends Genet.* 11:58–62.

Mirkovich, J., Mirault, M.-E., and Laemmli, U. K. 1984. Organization of the higher-order chromatin loop: Specific DNA attachment sites on nuclear scaffold. *Cell* 39:223–232.

Morse, R. H., and Simpson, R. T. 1988. DNA in the nucleosome. *Cell* 54:285–287.

Moyzis, R. K. 1991. The human telomere. *Sci. Amer.* 265 (Aug):48–55.

Moyzis, R. K., Buckingham, J. M., Cram, L. S., Dani, M., Deaven, L. L., Jones, M. D., Meyne, J., Ratliff, R. L., and Wu, J.-R. 1988. A highly conserved repetitive DNA sequence $(TTAGGG)_n$, present at the telomeres of human chromosomes. *Proc. Natl. Acad. Sci. USA* 85:6622–6626.

Olins, A. L., Carlson, R. D., and Olins, D. E. 1975. Visualization of chromatin substructure: nu-bodies. *J. Cell Biol.* 64:528–537.

Pettijohn, D. E. 1988. Histone-like proteins and bacterial chromosome structure. *J. Biol. Chem.* 263:12793–12796.

Pluta, A. F., Mackay, A. M., Ainsztein, A. M., Goldberg, I. G., and Earnshaw, W. C. 1995. The centromere: hub of chromosomal activities. *Science* 270:1591–1594.

Pruss, D., Bartholomew, B., Persinger, J., Hayes, J., Arents, G., Moudrianakis, E. N., and Wolfe, A. P. 1996. An asymmetric model for the nucleosome: a binding site for linker histones inside the DNA gyres. *Science* 274:614–617.

Richards, E. J., and Ausubel, F. M. 1988. Isolation of a higher eukaryotic telomere from *Arabidopsis thaliana. Cell* 53:127–136.

Roth, S. Y., and Allis, C. D. 1996. Histone acetylation and chromatin assembly: A single escort, multiple dances? *Cell* 87:5–8.

Singer, M. F. 1982. Highly repeated sequences in mammalian genomes. *Int. Rev. Cytol.* 76:67–112.

———. 1982. SINEs and LINEs: Highly repeated short and long interspersed sequences in mammalian genomes. *Cell* 28:133–134.

Singer, M. F., and Skowronski, J. 1985. Making sense out of LINES: Long interspersed repeat sequences in mammalian genomes. *Trends Biochem. Sci.* (March):119–121.

Sinsheimer, R. L. 1959. A single-stranded deoxyribonucleic acid from bacteriophage ΦX174. *J. Mol. Biol.* 1:43–53.

Streisinger, G., Edgar, R. S., and Denhardt, G. H. 1964. Chromosome structure in phage T4. I. Circularity of the linkage map. *Proc. Natl. Acad. Sci. USA* 5:775–779.

Thomas, C. A., and MacHattie, L. A. 1967. The anatomy of viral DNA molecules. *Annu. Rev. Biochem.* 36:485–518.

van Holde, K. E. 1988. *Chromatin.* New York: Springer-Verlag.

van Holde, K., and Zlatanova, J. 1996. What determines the folding of the chromatin fiber? *Proc. Natl. Acad. Sci. USA.* 93:10548–10555.

Williamson, J. R., Raghuraman, M. K., and Cech, T. R. 1989. Monovalent cation-induced structure of telomeric DNA: The G-quartet model. *Cell* 59:871–880.

Woodcock, C. L. F., Frado, L.-L. Y., and Rattner, J. B. 1984. The higher-order structure of chromatin: Evidence for a helical ribbon arrangement. *J. Cell Biol.* 99:42–52.

Worcel, A. 1978. Molecular architecture of the chromatin fiber. *Cold Spring Harbor Symp. Quant. Biol.* 42:313–324.

Worcel, A., and Benyajati, C. 1977. Higher order coiling of DNA in chromatin. *Cell* 12:83–100.

Worcel, A., and Burgi, E. 1972. On the structure of the folded chromosome of *Escherichia coli. J. Mol. Biol.* 71:127–147.

Zimmerman, S. B. 1982. The three-dimensional structure of DNA. *Annu. Rev. Biochem.* 51:395–427.

CHAPTER 12: DNA REPLICATION AND RECOMBINATION

Baker, T. A., and Wickner, S. H. 1992. Genetics and enzymology of DNA replication in *Escherichia coli. Annu. Rev. Genet.* 26:447–477.

Beese, L. S., Derbyshire, V., and Steitz, T. A. 1993. Structure of DNA polymerase I Klenow fragment bound to duplex DNA. *Science* 260:352–355.

Bell, S. P., and Stillman, B. 1992. ATP-dependent recognition of eukaryotic origins of DNA replication by a multiprotein complex. *Nature* 357:128–134.

Benbow, R. M., Zhao, J., and Larson, D. D. 1992. On the nature of origins of DNA replication in eukaryotes. *BioEssays* 14:661–670.

Biswas, S. B., and Biswas, E. E. 1990. ARS binding factor I of the yeast *Saccharomyces cerevisiae* binds to sequences in telomeric and nontelomeric autonomously replicating sequences. *Mol. Cell. Biol.* 10:810–815.

Blackburn, E. H. 1991. Structure and function of telomeres. *Nature* 350:569–573.

———. 1992. Telomerases. *Annu. Rev. Biochem.* 61:113–129.

Budd, M. E., and Campbell, J. L. 1993. DNA polymerases δ and ε are required for chromosomal replication in *Saccharomyces cerevisiae*. *Mol. Cell. Biol.* 13:496–505.

Coverley, D., and Laskey, R. A. 1994. Regulation of eukaryotic DNA replication. *Annu. Rev. Biochem.* 63:745–776.

Cox, M. M., and Lehman, I. R. 1987. Enzymes of general recombination. *Annu. Rev. Biochem.* 56:229–262.

Cozzarelli, N. R. 1980. DNA gyrase and the supercoiling of DNA. *Science* 207:953–960.

DeLucia, P., and Cairns, J. 1969. Isolation of an *E. coli* strain with a mutation affecting DNA polymerase. *Nature* 224:1164–1166.

DePamphilis, M. S. 1988. Transcriptional elements as components of eukaryotic origins of DNA replication. *Cell* 52:635–638.

Diller, J. D., and Raghuraman, M. K. 1994. Eukaryotic replication origins: control in space and time. *Trends Biochem.* 19:320–325.

Fangman, W. L., and Brewer, B. J. 1992. A question of time: Replication origins of eukaryotic chromosomes. *Cell* 71:363–366.

Gilbert, W., and Dressler, D. 1968. DNA replication: The rolling circle model. *Cold Spring Harbor Symp. Quant. Biol.* 33:473–484.

Gilley, D., and Blackburn, E. H. 1996. Specific RNA residue interactions required for enzymatic functions of Tetrahymena telomerase. *Mol. Cell. Biol.* 16:66–75.

Greider, C. W. 1990. Telomeres, telomerase, and senescence. *BioEssays* 12:363–369.

Hamlin, J. L. 1992. Mammalian origins of replication. *BioEssays* 14:651–659.

Holliday, R. 1964. A mechanism for gene conversion in fungi. *Genet. Res.* 5:282–304.

Huberman, J. A. 1987. Eukaryotic DNA replication: A complex picture partially clarified. *Cell* 48:7–8.

Huberman, J. A. 1995. Prokaryotic and eukaryotic replicons. *Cell* 82:535–542.

Huberman, J. A., and Riggs, A. D. 1968. On the mechanism of DNA replication in mammalian chromosomes. *J. Mol. Biol.* 32:327–341.

Kornberg, A. 1960. Biologic synthesis of deoxyribonucleic acid. *Science* 131:1503–1508.

Kornberg, A., and Baker, T. A. 1992. *DNA replication,* 2nd ed. New York: Freeman.

Lendvay, T. S., Morris, D. K., Sah, J., Balasubramanian, B., and Lundblad, V. 1996. Senescence mutants of *Saccharomyces cerevisiae* with a defect in telomere replication identify three additional *EST* genes. *Genetics* 144:1399–1412.

Marians, K. J. 1992. Prokaryotic DNA replication. *Annu. Rev. Biochem.* 61:673–719.

Mason, J. M., and Biessmann, A. 1995. The unusual telomeres of *Drosophila. Trends Genet.* 11:58–62.

Masters, M., and Broda, P. 1971. Evidence for the bidirectional replication of the *E. coli* chromosome. *Nature New Biol.* 232:137–140.

Meselson, M., and Radding, C. M. 1975. A general model for genetic recombination. *Proc. Natl. Acad. Sci. USA* 72:358–361.

Meselson, M., and Stahl, F. W. 1958. The replication of DNA in *Escherichia coli. Proc. Natl. Acad. Sci. USA* 44:671–682.

Modrich, P. 1987. DNA mismatch correction. *Annu. Rev. Biochem.* 56:435–466.

Nasmyth, K. 1996. At the heart of the budding yeast cell cycle. *Trends Genet.* 12:405–412.

Ogawa, T., Baker, T. A., van der Ende, A., and Kornberg, A. 1985. Initiation of enzymatic replication at the origin of the *Escherichia coli* chromosome: Contributions of RNA polymerase and primase. *Proc. Natl. Acad. Sci. USA* 82:3562–3566.

Ogawa, T., and Okazaki, T. 1980. Discontinuous DNA replication. *Annu. Rev. Biochem.* 49:424–457.

Okazaki, R. T., Okazaki, K., Sakobe, K., Sugimoto, K., and Sugino, A. 1968. Mechanism of DNA chain growth. I. Possible discontinuity and unusual secondary structure of newly synthesized chains. *Proc. Natl. Acad. Sci. USA* 59:598–605.

Pardee, A. B., Dubrow, R., Hamlin, J. L., and Kleitzien, R. F. 1978. Animal cell cycle. *Annu. Rev. Biochem.* 47:715–750.

Radding, C. 1982. Homologous pairing and strand exchange in genetic recombination. *Annu. Rev. Genet.* 16:405–437.

Recombination at the DNA level, vol. 49. 1984. Cold Spring Harbor Symposium on Quantitative Biology. Cold Spring Harbor, NY. Cold Spring Harbor Laboratory.

Reynolds, A. E., McCarroll, E. M., Newlon, C. S., and Fangman, W. L. 1989. Time of replication of ARS elements along yeast chromosome III. *Mol. Cell. Biol.* 9:4488–4494.

Runge, K. W., and Zakian, V. A. 1996. *TEL2,* an essential gene required for telomere length regulation and telomere position effect in *Saccharomyces cerevisiae. Mol. Cell. Biol.* 16:3094–3105.

Shippen-Lentz, D., and Blackburn, E. H. 1990. Functional evidence for an RNA template in telomerase. *Science* 247:546–552.

Simchen, G. 1978. Cell cycle mutants. *Annu. Rev. Genet.* 12:161–191.

Steiner, B. R., Hidaka, K., and Futcher, B. 1996. Association of the Est1 protein with telomerase activity in yeast. *Proc. Natl. Acad. Sci. USA* 93:2817–2821.

Stillman, B. 1994. Smart machines at the DNA replication fork. *Cell* 78:725–728.

Stukenberg, P. T., Turner, J., and O'Donnell, M. 1994. An explanation for lagging strand replication: polymerase hopping among DNA sliding clamps. *Cell* 78:877–887.

Szostak, J., Orr-Weaver, T., Rothstein, R., and Stahl, F. 1983. The double-strand break repair model for recombination. *Cell* 33:25–35.

Taylor, J. H. 1970. The structure and duplication of chromosomes. In *Genetic organization,* E. Caspari and A. Ravin, eds., vol. 1, pp. 163–221. New York: Academic Press.

Van der Ende, A., Baker, T. A., Ogawa, T., and Kornberg, A. 1985. Initiation of enzymatic replication at the origin of the *Escherichia coli* chromosome: Primase as the sole priming enzyme. *Proc. Natl. Acad. Sci. USA* 82:3954–3958.

Weissbach, A. 1977. Eukaryotic DNA polymerases. *Annu. Rev. Biochem.* 46:25–47.

Wellinger, R. J., Ethier, K., Labrecque, P., and Zakian, V. A. 1996. Evidence for a new step in telomere maintenance. *Cell* 85:423–433.

West, S. C. 1992. Enzymes and molecular mechanisms of genetic recombination. *Annu. Rev. Biochem.* 61:603–640.

Zyskind, J. W., and Smith, D. W. 1986. The bacterial origin of replication, *oriC. Cell* 46:489–490.

CHAPTER 13: TRANSCRIPTION, RNA MOLECULES, AND RNA PROCESSING

Atchison, M. L. 1988. Enhancers: Mechanisms of action and cell specificity. *Annu. Rev. Cell Biol.* 4:127–153.

Baker, S. M., and Platt, T. 1986. Pol I transcription: Which comes first, the end or the beginning? *Cell* 47:839–840.

Banerjee, A. K. 1980. 5'-terminal cap structure in eucaryotic messenger ribonucleic acids. *Microbiol. Rev.* 44:175–205.

Belfort, M. 1989. Bacteriophage introns: Parasites within parasites? *Trends Genet.* 5:209–213.

Bell, S. P., Pikaard, C. P., Reeder, R. H., and Tjian, R. 1989. Molecular mechanisms governing species-specific transcription of ribosomal RNA. *Cell* 59:489–497.

Bogenhagen, D. F., Sakonju, S., and Brown, D. D. 1980. A control region in the center of the 5S RNA gene directs specific initiation of transcription II: The 3' border of the region. *Cell* 19:27–35.

Brand, A. H., Breeden, L., Abraham, J., Sternglanz, R., and Nasmyth, K. 1987. Characterization of a "silencer" in yeast: A DNA sequence with properties opposite to those of a transcriptional enhancer. *Cell* 41:41–48.

Breathnach, R., and Chambon, P. 1981. Organization and expression of eucaryotic split genes coding for proteins. *Annu. Rev. Biochem.* 50:349–383.

Breathnach, R., Mandel, J. L., and Chambon, P. 1977. Ovalbumin gene is split in chicken DNA. *Nature* 270:314–318.

Breitbart, R. E., Andreadis, A., and Nadal-Ginard, B. 1987. Alternative splicing: A ubiquitous mechanism for the generation of multiple protein isoforms from single genes. *Annu. Rev. Biochem.* 56:467–495.

Brennan, C. A., Dombroski, A. J., and Platt, T. 1987. Transcription termination factor rho is an RNA-DNA helicase. *Cell* 48:945–952.

Brody, E., and Abelson, J. 1985. The "spliceosome": Yeast premessenger RNA associates with a 40S complex in a splicing-dependent reaction. *Science* 228:963–967.

Buratowski, S. 1994. The basics of basal transcription by RNA polymerase II. *Cell* 77:1–3.

Busby, S., and Ebright, R. H. 1994. Promoter structure, promoter recognition, and transcription activation in prokaryotes. *Cell* 79:743–746.

Cech, T. R. 1983. RNA splicing: Three themes with variations. *Cell* 34:713–716.

———. 1985. Self-splicing RNA: Implications for evolution. *Int. Rev. Cytol.* 93:3–22.

———. 1986. Ribosomal RNA gene expression in *Tetrahymena*: Transcription and RNA splicing. *Mol. biol. ciliated protozoa*, pp. 203–225. New York: Academic Press.

———. 1986. The generality of self-splicing RNA: Relationship to nuclear mRNA splicing. *Cell* 44:207–210.

Chambliss, G., Craven, G. R., Davies, J., Davis, K., Kahan, L., and Nomura, M., eds. 1980. *Ribosomes: Structure, function, and genetics.* Baltimore: University Park Press.

Choi, Y. D., Grabowski, P. J., Sharp, P. A., and Dreyfuss, G. 1986. Heterogeneous nuclear ribonucleoproteins: Role in RNA splicing. *Science* 231:1534–1539.

Chu, F. K., Maley, G. F., West, D. K., Belfort, M., and Maley, F. 1986. Characterization of the intron in the phage T4 thymidylate synthase gene and evidence for its self-excision from the primary transcript. *Cell* 45:157–166.

Clark, D. J., and Felsenfeld, G. 1992. A nucleosome core is transferred out of the path of a transcribing polymerase. *Cell* 71:11–22.

Crick, F. H. C. 1979. Split genes and RNA splicing. *Science* 204:264–271.

Dynan, W. S. 1989. Modularity in promoters and enhancers. *Cell* 58:1–4.

Eick, D., Wedel, A., and Heumann, H. 1994. From initiation to elongation: Comparison of transcription by prokaryotic and eukaryotic RNA polymerases. *Trends Genet.* 10:292–296.

Geiduschek, E. P., and Tocchini-Valentini, G. P. 1988. Transcription by RNA polymerase III. *Annu. Rev. Biochem.* 57:873–914.

Gesteland, R. F., and Atkins, J. F., eds. 1993. *The RNA world.* Cold Spring Harbor, NY: Cold Spring Harbor Laboratory Press.

Goodrich, J. A., Cutler, G., and Tjian, R. 1996. Contacts in context: Promoter specificity and macromolecular interactions in transcription. *Cell* 84:825–830.

Grabowski, P. J., Seiler, S. R., and Sharp, P. A. 1985. A multicomponent complex is involved in the splicing of messenger RNA precursors. *Cell* 42:355–367.

Green, M. R., 1986. Pre-mRNA splicing. *Annu. Rev. Genet.* 20:671–708.

———. 1989. Pre-mRNA processing and mRNA nuclear export. *Curr. Opinion Cell Biol.* 1:519–525.

———. 1991. Biochemical mechanisms of constitutive and regulated pre-mRNA splicing. *Annu. Rev. Cell Biol.* 7:559–599.

Greenblatt, J. 1991. RNA polymerase-associated transcription factors. *Trends Biochem. Sci.* 11:408–411.

Guarente, L. 1988. UASs and enhancers: Common mechanism of transcriptional activation in yeast and mammals. *Cell* 52:303–305.

Guarente, L., and Birmingham-McDonogh, O. 1992. Conservation and evolution of transcriptional mechanisms in eukaryotes. *Trends Genet.* 8:27–32.

Guthrie, C. 1986. Finding functions for small nuclear RNAs in yeast. *Trends Biochem. Sci.* (October):430–434.

———. 1992. Messenger RNA splicing in yeast: Clues to why the spliceosome is a ribonucleoprotein. *Science* 253:157–163.

Guthrie, C., and Patterson, B. 1988. Spliceosomal snRNAs. *Annu. Rev. Genet.* 22:387–419.

Halle, J.-P., and Meisterernst, M. 1996. Gene expression: Increasing evidence for a transcriptosome. *Trends Genet.* 12:161–163.

Helman, J. D., and Chamberlin, M. J. 1988. Structure and function of bacterial sigma factors. *Annu. Rev. Biochem.* 57:839–872.

Horowitz, D. S., and Krainer, A. R. 1994. Mechanisms for selecting 5' splice sites in mammalian pre-mRNA splicing. *Trends Genet.* 10:100–105.

Jacob, S. T. 1986. Transcription of eukaryotic ribosomal RNA genes. *Mol. Cell. Biochem.* 70:11–20.

Jeffreys, A. J., and Flavell, R. A. 1977. The rabbit beta-globin gene contains a large insert in the coding sequence. *Cell* 12:1097–1108.

Kassavetis, G. A., Braun, B. R., Nguyen, L. H., and Geiduschek, E. P. 1990. *S. cerevisiae* TFIIIB is the transcription initiation factor proper of RNA polymerase III, while TFIIIA and TFIIIC are assembly factors. *Cell* 60:235–245.

Katagiri, F., and Chua, N.-H. 1992. Plant transcription factors: Present knowledge and future challenges. *Trends Genet.* 8:22–27.

Kustu, S., North, A. K., and Weiss, D. S. 1991. Prokaryotic transcriptional enhancers and enhancer-binding proteins. *Trends Biochem. Sci.* 16:397–402.

Labhart, P., and Reeder, R. H. 1987. Ribosomal precursor 3' end formation requires a conserved element upstream of the promoter. *Cell* 50:51–57.

Lang, W. H., Morrow, B. E., Ju, Q., Warner, J. R., and Reeder, R. H. 1994. A model for transcription termination by RNA polymerase I. *Cell* 79:527–534.

McKnight, S. L., and Kingsbury, R. 1982. Transcriptional control signals of a eukaryotic protein-coding gene. *Science* 217:316–324.

McStay, B., and Reeder, R. H. 1986. A termination site for Xenopus RNA polymerase I also acts as an element of an adjacent promoter. *Cell* 47:913–920.

Marmur, J., Greenspan, C. M., Palecek, E., Kahan, F. M., Levine, J., and Mandel, M. 1963. Specificity of the complementary RNA formed by *Bacillus subtilis* infected with bacteriophage SP8. *Cold Spring Harbor Symp. Quant. Biol.* 28:191–199.

Moore, P. B. 1988. The ribosome returns. *Nature* 331:223–227.

Murphy, S., Moorefield, B., and Pieler, T. 1989. Common mechanisms of promoter recognition by RNA polymerases II and III. *Trends Genet.* 5:122–126.

Nilsen, T. W. 1994. RNA-RNA interactions in the spliceosome: Unraveling the ties that bind. *Cell* 78:1–4.

Nomura, M. 1973. Assembly of bacterial ribosomes. *Science* 179:864–873.

Nomura, M., and Morgan, E. A. 1977. Genetics of bacterial ribosomes. *Annu. Rev. Genet.* 11:297–347.

Nomura, M., Morgan, E. A., and Jaskunas, S. R. 1977. Genetics of bacterial ribosomes. *Annu. Rev. Genet.* 11:297–347.

O'Hare, K. 1995. mRNA 3' ends in focus. *Trends Genet.* 11:253–257.

Pabo, C. O., and Sauer, R. T. 1992. Transcription factors: Structural families and principles of DNA recognition. *Annu. Rev. Biochem.* 61:1053–1093.

Padgett, R. A., Grabowski, P. J., Konarska, M. M., and Sharp, P. A. 1985. Splicing messenger RNA precursors: Branch sites and lariat RNAs. *Trends Biochem. Sci.* (April):154–157.

Pruss, G. J., and Drlica, K. 1989. DNA supercoiling and prokaryotic transcription. *Cell* 56:521–523.

Reeder, R. H. 1989. Regulatory elements of the generic ribosomal gene. *Curr. Opinion Cell Biol.* 1:466–474.

Roeder, R. G. 1991. The complexities of eukaryotic transcription initiation: Regulation of preinitiation complex assembly. *Trends Biochem. Sci.* 16:402–408.

Rogers, J. H. 1989. How were introns inserted into nuclear genes? *Trends Genet.* 5:213–216.

Schmidt, F. J. 1985. RNA splicing in prokaryotes: Bacteriophage T4 leads the way. *Cell* 41:339–340.

Sharp, P. A. 1985. On the origin of RNA splicing and introns. *Cell* 42:397–400.

Sharp, P. A. 1994. Split genes and RNA splicing. Nobel lecture. *Cell* 77:805–815.

Smith, C. W. J., Porro, E. B., Patton, J. G., and Nadal-Ginard, B. 1989. Scanning from an independently specified branch point defines the 3' splice site of mammalian introns. *Nature* 342:243–247.

Sollner-Webb, B. 1988. Surprises in polymerase III transcription. *Cell* 52:153–154.

Srivastava, A. K., and Schlessinger, D. 1990. Mechanism and regulation of bacterial ribosomal RNA processing. *Annu. Rev. Microbiol.* 44:105–129.

Stragier, P. 1991. Dances with sigmas. *EMBO J.* 10:3559–3566.

Struhl, K. 1996. Chromatin structure and RNA polymerase II connection: Implications for transcription. *Cell* 84:179–182.

Symons, R. H. 1992. Small catalytic RNAs. *Annu. Rev. Biochem.* 61:641–671.

Thompson, C. C., and McKnight, S. L. 1992. Anatomy of an enhancer. *Trends Genet.* 8:232–236.

Tilghman, S. M., Curis, P. J., Tiemeier, D. C., Leder, P., and Weissman, C. 1978. The intervening sequence of a mouse β-globin gene is transcribed within the 15S β-globin mRNA precursor. *Proc. Natl. Acad. Sci. USA* 75:1309–1313.

Tilghman, S. M., Tiemeier, D. C., Seidman, J. G., Peterlin, B. M., Sullivan, M., Maizel, J. V., and Leder, P. 1978. Intervening sequence of DNA identified in the structural portion of a mouse beta-globin gene. *Proc. Natl. Acad. Sci. USA* 78:725–729.

Van der Sande, C. A. F. M., Kulkens, T., Kramer, A. B., de Wijs, I. J., van Heerikhuizen, H., Klootwijk, J., and Planta, R. J. 1989. Termination of transcription by yeast RNA polymerase I. *Nucleic Acids Res.* 17:9127–9146.

Voss, S. D., Schlokat, U., and Gruss, P. 1986. The role of enhancers in the regulation of cell-type-specific transcriptional control. *Trends Biochem. Sci.* (July):287–289.

Wahle, E., and Keller, W. 1992. The biochemistry of 3'-end cleavage and polyadenylation of messenger RNA precursors. *Annu. Rev. Biochem.* 61:419–440.

Weinstock, R., Sweet, R., Weiss, M., Cedar, H., and Axel, R. 1978. Intragenic DNA spacers interrupt the ovalbumin gene. *Proc. Natl. Acad. Sci. USA* 75:1299–1303.

Weis, L., and Reinberg, D. 1992. Transcription by RNA polymerase II: Initiator-directed formation of transcription-competent complexes. *FASEB J.* 6:3300–3309.

White, R. J., and Jackson, S. P. 1992. The TATA-binding protein: A central role in transcription by RNA polymerases I, II, and III. *Trends Genet.* 8:284–288.

Wolffe, 1994. Transcription: In tune with the histones. *Cell* 77:13–16.

Zaug, A. J., and Cech, T. R. 1986. The intervening sequence RNA of *Tetrahymena* is an enzyme. *Science* 231:470–475.

CHAPTER 14: THE GENETIC CODE AND THE TRANSLATION OF THE GENETIC MESSAGE

Bachmair, A., Finley, D., and Varshavsky, A. 1986. In vivo half-life of a protein is a function of its amino-terminal residue. *Science* 234:179–186.

Blobel, G., and Dobberstein, B. 1975. Transfer of proteins across membranes. I. Presence of proteolytically processed and unprocessed nascent immunoglobulin light chains on membrane-bound ribosomes of murine myeloma. *J. Cell Biol.* 67:835–851.

Brenner, S., Jacob, F., and Meselson, M. 1961. An unstable intermediate carrying information from genes to ribosomes for protein synthesis. *Nature* 190:576–581.

Burgess, T. L., and Kelly, R. B. 1987. Constitutive and regulated secretion of proteins. *Annu. Rev. Cell Biol.* 3:243–293.

Cold Spring Harbor Symposia for Quantitative Biology. 1966: *The genetic code*, vol. 31. Cold Spring Harbor, NY: Cold Spring Harbor Laboratory.

Colman, A., and Robinson, C. 1986. Protein import into organelles: Hierarchical targeting signals. *Cell* 46:321–322.

Crick, F. H. C. 1966. Codon-anticodon pairing: The wobble hypothesis. *J. Mol. Biol.* 19:548–555.

Crick, F. H. C., Barnett, L., Brenner, S., and Watts-Tobin, R. J. 1961. General nature of the genetic code for proteins. *Nature* 192:1227–1232.

Dahlberg, A. 1989. The functional role of ribosomal RNA in protein synthesis. *Cell* 57:525–529.

Dingwall, C. 1985. The accumulation of proteins in the nucleus. *Trends Biochem. Sci.* (February):64–66.

Dingwall, C., and Laskey, R. A. 1986. Protein import into the cell nucleus. *Annu. Rev. Cell Biol.* 2:367–390.

Garen, A. 1968. Sense and nonsense in the genetic code. *Science* 160:149–159.

Griffiths, G., and Simons, K. 1986. The trans Golgi network: Sorting at the exit site of the Golgi complex. *Science* 234:438–442.

Horowitz, S., and Gorovsky, M. A. 1985. An unusual genetic code in nuclear genes of *Tetrahymena*. *Proc. Natl. Acad. Sci. USA* 82:2452–2455.

Jackson, R. J., and Standart, N. 1990. Do the poly(A) tail and 3' untranslated region control mRNA translation? *Cell* 62:15–24.

Khorana, H. G. 1966–67. Polynucleotide synthesis and the genetic code. *Harvey Lectures* 62:79–105.

Kozak, M. 1983. Comparison of initiation of protein synthesis in procaryotes, eucaryotes, and organelles. *Microbiol. Rev.* 47:145.

———. 1989. Context effects and inefficient initiation at non-AUG codons in eucaryotic cell-free translation systems. *Mol. Cell. Biol.* 9:5073–5080.

Lingappa, V. R. 1991. More than just a channel: Provocative new features of protein traffic across the ER membrane. *Cell* 65:527–530.

McCarthy, J. E. G., and R. Brimacombe. 1994. Prokaryotic translation: The interactive pathway leading to initiation. *Trends Genet.* 10:402–407.

Meyer, D. I. 1982. The signal hypothesis—A working model. *Trends Biochem. Sci.* 7:320–321.

Moore, P. B. 1988. The ribosome returns. *Nature* 331:223–227.

Morgan, A. R., Wells, R. D., and Khorana, H. G. 1966. Studies on polynucleotides. LIX. Further codon assignments from amino acid incorporation directed by ribopolynucleotides containing repeating trinucleotide sequences. *Proc. Natl. Acad. Sci. USA* 56:1899–1906.

Nierhaus, K. H. 1990. The allosteric three-site model for the ribosomal elongation cycle: Features and future. *Biochemistry* 29:4997–5008.

Nirenberg, M., and Leder, P. 1964. RNA code words and protein synthesis. *Science* 145:1399–1407.

Nirenberg, M., and Matthaei, J. H. 1961. The dependence of cell-free protein synthesis in *E. coli* upon naturally occurring or synthetic polyribonucleotides. *Proc. Natl. Acad. Sci. USA* 47:1588–1602.

Noller, H. F., Hoffarth, V., and Zimniak, L. 1992. Unusual resistance of peptidyl transferase to protein extraction procedures. *Science* 256:1416–1419.

Pfeffer, S. R., and Rothman, J. E. 1987. Biosynthetic protein transport and sorting by the endoplasmic reticulum and Golgi. *Annu. Rev. Biochem.* 56:829–852.

Rogers, S., Wells, R., and Rechsteiner, M. 1986. Amino acid sequences common to rapidly degraded proteins: The PEST hypothesis. *Science* 234:364–368.

Ryan, K. R., and Jensen, R. E. 1995. Protein translocation across mitochondrial membranes: What a long, strange trip it is. *Cell* 83:517–519.

Schekman, R. 1985. Protein localization and membrane traffic in yeast. *Annu. Rev. Cell Biol.* 1:115–143.

Schmidt, G. W., and Mishkind, M. L. 1986. The transport of proteins into chloroplasts. *Annu. Rev. Biochem.* 55:879–912.

Schnell, D. J. 1995. Shedding light on the chloroplast protein import machinery. *Cell* 83:521–524.

Shine, J., and Delgarno, L. 1974. The 3'-terminal sequence of *Escherichia coli* 16S ribosomal RNA: Complementarity to nonsense triplet and ribosome binding sites. *Proc. Natl. Acad. Sci. USA* 71:1342–1346.

Silver, P. A. 1991. How proteins enter the nucleus. *Cell* 64:489–497.

Verner, K., and Schatz, G. 1988. Protein translocation across membranes. *Science* 241:1307–1313.

Walter, P., Gilmore, R., and Blobel, G. 1984. Protein translocation across the endoplasmic reticulum. *Cell* 38:5–8.

Walter, P., and Lingappa, V. 1986. Mechanism of protein translocation across the endoplasmic reticulum membrane. *Annu. Rev. Cell Biol.* 2:499–516.

Watson, J. D. 1963. The involvement of RNA in the synthesis of proteins. *Science* 140:17–26.

Zheng, N., and Gierasch, L. M. 1996. Signal sequences: The same yet different. *Cell* 86:849–852.

CHAPTER 15: RECOMBINANT DNA TECHNOLOGY AND THE MANIPULATION OF DNA

Anderson, W. F. 1992. Human gene therapy. *Science* 256:808–813.

Antonarakis, S. E. 1989. Diagnosis of genetic disorders at the DNA level. *N. Eng. J. Med.* 320:153–163.

Arber, W. 1965. Host-controlled modification of bacteriophage. *Annu. Rev. Microbiol.* 19:365–378.

Arber, W., and Dussoix, D. 1962. Host specificity of DNA produced by *Escherichia coli* I. Host controlled modification of bacteriophage lambda. *J. Mol. Biol.* 5:18–36.

Arnheim, N., and Erlich, H. 1992. Polymerase chain reaction strategy. *Annu. Rev. Biochem.* 61:131–156.

Boyer, H. W. 1971. DNA restriction and modification mechanisms in bacteria. *Annu. Rev. Microbiol.* 25:153–176.

Bult, C. J., and 39 other authors. 1996. Complete genome sequence of the methanogenic archaeon, *Methanococcus jannaschii*. *Science* 273:1058–1073.

Collins, F. 1992. Cystic fibrosis: Molecular biology and therapeutic implications. *Science* 256:774–779.

Culver, K. V., and Blaese, R. M. 1994. Gene therapy for cancer. *Trends Genet.* 10:174–178.

Danna, K., and Nathans, D. 1971. Specific cleavage of simian virus 40 DNA by restriction endonuclease of *Haemophilus influenzae*. *Proc. Natl. Acad. Sci. USA* 68:2913–2917.

Dib, C., Fauré, S., Fizames, C., Samson, D., Drouot, N., Vignal, A., Millasseau, P., Marc, S., Hazan, J., Seboun, E., Lathrop, M., Gyapay, G., Morissette, J., and Weissenbach, J. 1996. A comprehensive genetic map of the human genome based on 5,264 microsatellites. *Nature* 380:152–154.

Dussoix, D., and Arber, W. 1962. Host specificity of DNA produced by *Escherichia coli*. II. Control over acceptance of DNA from infecting phage lambda. *J. Mol. Biol.* 5:37–49.

Eisenstein, B. I. 1990. The polymerase chain reaction: A new method of using molecular genetics for medical diagnosis. *N. Eng. J. Med.* 322:178–183.

Erlich, H. A., and Arnheim, N. 1992. Genetic analysis using the polymerase chain reaction. *Annu. Rev. Genet.* 26:479–506.

Feinberg, A. P., and Vogelstein, B. 1983. A technique for radiolabeling DNA restriction endonuclease fragments to high specific activity. *Anal. Biochem.* 132:6–13.

———. 1984. Addendum: A technique for radiolabeling DNA restriction endonuclease fragments to high specific activity. *Anal. Biochem.* 137:266–267.

Foote, S., Vollrath, D., Hilton, A., and Page, D. C. 1992. The human Y chromosome: Overlapping DNA clones spanning the euchromatic region. *Science* 258:60–66.

Gilliam, T. C., Tanzi, R. E., Haines, J. L., Bonner, T. I., Faryniarz, A. G., Hobbs, W. J., MacDonald, M. E., Cheng, S. V., Folstein, S. E., Conneally, P. M., Wexler, N. S., and Gusella, J. F. 1987. Localization of the Huntington's disease gene to a small segment of chromosome 4 flanked by *D4S10* and the telomere. *Cell* 50:565–571.

Green, E. D., and Olson, M. V. 1990. Chromosomal region of the cystic fibrosis gene in yeast artificial chromosomes: A model for human genome mapping. *Science* 250:94–98.

Harris, J. D., and Lemoine, N. R. 1996. Strategies for targeted gene therapy. *Trends Genet.* 12:400–405.

Huntington's Disease Collaborative Research Group. 1993. A novel gene containing a trinucleotide repeat that is expanded and unstable on Huntington's disease chromosomes. *Cell* 72:971–983.

Kay, M. A., and Woo, S. L. C. 1994. Gene therapy for metabolic disorders. *Trends Genet.* 10:253–257.

Kerem, B.-S., Rommens, J. M., Buchanan, J. A., Markiewicz, D., Cox, T. K., Chakravarti, A., Buchwald, M., and Tsui, L.-C. 1989. Identification of the cystic fibrosis gene: Genetic analysis. *Science* 245:1073–1080.

Klee, H., Horsch, R., and Rogers, S. 1987. *Agrobacterium*-mediated plant transformation and its further applications to plant biology. *Annu. Rev. Plant Physiol.* 38:467–486.

Knowlton, R. G., Cohen-Haguenauer, O., Van Cong, N., Frézal, J., Brown, V. A., Barker, D., Braman, J. C., Schumm, J. W., Tsui, L.-C., Buchwald, M., and Donis-Keller, H. 1985. A polymorphic DNA marker linked to cystic fibrosis is located on chromosome 7. *Nature* 318:380–385.

Koenig, M., Hoffman, E. P., Bertelson, C. J., Monaco, A. P., Feener, C., and Kunkel, L. M. 1987. Complete cloning of the Duchenne muscular dystrophy (DMD) cDNA and preliminary genomic organization of the DMD gene in normal and affected individuals. *Cell* 50:509–517.

Luria, S. E. 1953. Host-induced modification of viruses. *Cold Spring Harbor Symp. Quant. Biol.* 18:237–244.

Mandel, J.-L., Monaco, A. P., Nelson, D. L., Schlessinger, D., and Willard, H. 1992. Genome analysis and the human X chromosome. *Science* 258:103–109.

Maxam, A. M., and Gilbert, W. 1977. A new method for sequencing DNA. *Proc. Natl. Acad. Sci. USA* 74:560–564.

Morgan, R. A., and Anderson, W. F. 1993. Human gene therapy. *Annu. Rev. Biochem.* 62:191–217.

Mulligan, R. C. 1993. The basic science of gene therapy. *Science* 260:926–932.

Mullis, K. B. 1990. The unusual origin of the polymerase chain reaction. *Sci. Am.* (April):56–65.

Mullis, K. B., and Faloona, F. A. 1987. Specific synthesis of DNA *in vitro* via a polymerase-catalyzed chain reaction. *Meth. Enzymol.* 155:335–350.

Murray, J. M., Davies, K. E., Harper, P. S., Meredith, L., Mueller, C. R., and Williamson, R. 1982. Linkage relationship of a cloned DNA sequence on the short arm of the X chromosome to Duchenne muscular dystrophy. *Nature* 300:69–71.

Pääbo, S. 1993. Ancient DNA. *Sci. Amer.* (November):86–92.

Pendick, D. 1992. Better than the real thing. *Science News* 142:376–377.

Riles, L., Dutchik, J. E., Baktha, A., McCauley, B. K., Thayer, E. C., Leckie, M. P., Braden, V. V., Depke, J. E., and Olson, M. V. 1993. Physical maps of the six smallest chromosomes of *Saccharomyces cerevisiae* at a resolution of 2.6 kilobase pairs. *Genetics* 134:81–150.

Riordan, J. R., Rommens, J. M., Kerem, B., Alon, N., Rozmahel, R., Grzelczak, Z., Zielenski, J., Lok, S., Plavsic, N., Chou, J. L., Drumm, M. L., Ianuzzi, M. C., Collins, F. S., and Tsui, L.-C. 1989. Identification of the cystic fibrosis gene: Cloning and characterization of complementary DNA. *Science* 245:1066–1073.

Rommens, J. M., Ianuzzi, M. C., Kerem, B., Drumm, M. L., Melmer, G., Dean, M., Rozmahel, R., Cole, J. L., Kennedy, D., Hidaka, N., Zsiga, M., Buchwald, M., Riordan, J. R., Tsui, L.-C., and Collins, F. S. 1989. Identification of the cystic fibrosis gene: Chromosome walking and jumping. *Science* 245:1059–1065.

Sambrook, J., Fritsch, E. F., and Maniatis, T. 1989. *Molecular cloning: A laboratory manual*, 2nd ed. Cold Spring Harbor, NY: Cold Spring Harbor Laboratory.

Sanger, F., and Coulson, A. R. 1975. A rapid method for determining sequences in DNA by primed synthesis with DNA polymerase. *J. Mol. Biol.* 94:441–448.

Schlessinger, D. 1990. Yeast artificial chromosomes: Tools for mapping and analysis of complex genomes. *Trends Genet.* 6:248–258.

Southern, E. M. 1975. Detection of specific sequences among DNA fragments separated by gel electrophoresis. *J. Mol. Biol.* 98:503–517.

Vollrath, D., Foote, S., Hilton, A., Brown, L. G., Beer-Romero, P., Bogan, J. S., and Page, D. C. 1992. The human Y chromosome: A 43-interval map based on naturally occurring deletions. *Science* 258:52–59.

Wainwright, B. J. 1993. The isolation of disease genes by positional cloning. *Medical J. Australia* 159:170–174.

Watson, J. D., Gilman, M., Witkowski, J., and Zoller, M. 1992. *Recombinant DNA*, 2nd ed. New York: Scientific American Books, Freeman.

White, R., and Lalouel, J.-M. 1988. Chromosome mapping with DNA markers. *Sci. Am.* 258(February):40–48.

White, T. J., Arnheim, N., and Erlich, H. A. 1989. The polymerase chain reaction. *Trends Genet.* 5:185–188.

Wicking, C., and Williamson, B. 1991. From linked marker to gene. *Trends Genet.* 7:288–293.

CHAPTER 16: REGULATION OF GENE EXPRESSION IN BACTERIA AND BACTERIOPHAGES

Aloni, Y., and Hay, N. 1985. Attenuation may regulate gene expression in animal viruses and cells. *CRC Crit. Rev. Biochem.* 18:327–383.

Bertrand, K., Korn, L., Lee, F., Platt, T., Squires, C. L., Squires, C., and Yanofsky, C. 1975. New features of the structure and regulation of the tryptophan operon of *Escherichia coli*. *Science* 189:22–26.

Bertrand, K., and Yanofsky, C. 1976. Regulation of transcription termination in the leader region of the tryptophan operon of *Escherichia coli* involves tryptophan as its metabolic product. *J. Mol. Biol.* 103:339–349.

Dickson, R. C., Abelson, J., Barnes, W. M., and Reznikoff, W. S. 1975. Genetic regulation: The *lac* control region. *Science* 187:27–35.

Fisher, R. F., Das, A., Kolter, R., Winkler, M. E., and Yanofsky, C. 1985. Analysis of the requirements for transcription pausing in the tryptophan operon. *J. Mol. Biol.* 182:397–409.

Gilbert, W., Maizels, N., and Maxam, A. 1974. Sequences of controlling regions of the lactose operon. *Cold Spring Harbor Symp. Quant. Biol.* 38:845–855.

Gilbert, W., and Muller-Hill, B. 1966. Isolation of the *lac* repressor. *Proc. Natl. Acad. Sci. USA* 56:1891–1898.

Jacob, F. 1965. Genetic mapping of the elements of the lactose region of *Escherichia coli*. *Biochem. Biophys. Res. Commun.* 18:693–701.

Jacob, F., and Monod, J. 1961. Genetic regulatory mechanisms in the synthesis of proteins. *J. Mol. Biol.* 3:318–356.

Lee, F., and Yanofsky, C. 1977. Transcription termination at the trp operon attenuators of *Escherichia coli* and *Salmonella typhimurium*: RNA secondary structure and regulation of termination. *Proc. Natl. Acad. Sci. USA* 74:4365–4369.

Lewis, M., Chang, G., Horton, N. C., Kercher, M. A., Pace, H. C., Schumacher, M. A., Brennan, R. G., and Lu, P. 1996. Crystal structure of the lactose operon repressor and its complexes with DNA and inducer. *Science* 271:1247–1254.

Maizels, N. 1974. *E. coli* lactose operon ribosome binding site. *Nature (New Biol.)* 249:647–649.

Matthews, K. S. 1996. The whole lactose repressor. *Science* 271:1245–1246.

Niu, W., Kim, Y., Tau, G., Heyduk, T., and Ebright, R. H. 1996. Transcription activation at class II CAP-dependent promoters: Two interactions between CAP and RNA polymerase. *Cell* 87:1123–1134.

Pabo, C. O., Sauer, R. T., Sturtevant, J. M., and Ptashne, M. 1979. The λ repressor contains two domains. *Proc. Natl. Acad. Sci. USA* 76:1608–4612.

Ptashne, M. 1967. Isolation of the λ phage repressor. *Proc. Natl. Acad. Sci. USA* 57:306–313.

———. 1984. Repressors. *Trends Biochem. Sci.* 9:142–145.

———. 1992. *A genetic switch*, 2nd ed. Oxford: Cell Press and Blackwell Scientific Publications.

Ptashne, M., and Gilbert, W. 1970. Genetic repressors. *Sci. Am.* 222 (June):36–44.

Winkler, M. E., and Yanofsky, C. 1981. Pausing of RNA polymerase during in vitro transcription of the tryptophan operon leader region. *Biochemistry* 20:3738–3744.

Yanofsky, C. 1981. Attenuation in the control of expression of bacterial operons. *Nature* 289:751–758.

———. 1987. Operon-specific control by transcription attenuation. *Trends Genet.* 3:356–360.

Yanofsky, C., and Kolter, R. 1982. Attenuation in amino acid biosynthetic operons. *Annu. Rev. Genet.* 16:113–134.

CHAPTER 17: REGULATION OF GENE EXPRESSION AND DEVELOPMENT IN EUKARYOTES

Albrecht, E. B., and Salz, H. K. 1993. The *Drosophila* sex determination gene *snf is* utilized for the establishment of the female-specific splicing pattern of *Sex-lethal*. *Genetics* 134:801–807.

Ashburner, M. 1990. Puffs, genes, and hormones revisited. *Cell* 61:1–3.

Atwater, J. A., Wisdom, R., and Verma, I. M. 1990. Regulated mRNA stability. *Annu. Rev. Genet.* 24:519–541.

Bachvarova, R. F. 1992. A maternal tail of poly(A): The long and the short of it. *Cell* 69:895–897.

Beachy, P. A. 1990. A molecular view of the *Ultrabithorax* homeotic gene of *Drosophila*. *Trends Genet.* 6:46–51.

Beato, M., Herrlich, P., and Schützm, G. 1995. Steroid hormone receptors: Many actors in search of a plot. *Cell* 83:851–857.

Beelman, C. A., and Parker, R. 1995. Degradation of mRNA in eukaryotes. *Cell* 81:179–182.

Beelman, C. A., Stevens, A., Caponigro, G., LaGrandeur, T. E., Hatfield, L., Fortner, D. M., and Parker, R. 1996. An essential component of the decapping enzyme required for normal rates of mRNA turnover. *Nature* 382:642–646.

Beerman, W., and Clever, U. 1964. Chromosome puffs. *Sci. Am.* 210:50–58.

Blackwell, T. K. and F. W. 1988. Immunoglobin genes. In *Molecular immunology*, B. D. Homes and D. M. Glover, eds., pp. 1–60. Washington, DC: IRL Press.

Boggs, R. T., Gregor, P., Idriss, S., Belote, J. M., and McKeown, M. 1987. Regulation of sexual differentiation in *D. melanogaster* via alternative splicing of RNA from the *transformer gene*. *Cell* 50:739–747.

Bonner, J. J., and Pardue, M. L. 1976. Ecdysone-stimulated RNA synthesis in imaginal discs of *Drosophila melanogaster*. *Chromosoma* 58:87–99.

Cedar, H. 1988. DNA methylation and gene activity. *Cell* 53:3–4.

Chen, C.-Y. A., and Shyu, A.-B. 1995. AU-rich elements: characterization and importance of mRNA degradation. *Trends Biochem. Sci.* 20:465–470.

Davidson, E. H. 1976. *Genetic activity in early development*, 2nd ed. New York: Academic Press.

Davis, M. M., Calame, K., Early, P. W., Livant, D. L., Joho, R., Weissman, I. L., and Hood, L. 1980. An immunoglobulin heavy chain gene is formed by at least two recombinational events. *Nature* 283:733–739.

Davis, M. M., Kim, S. K., and Hood, L. 1980. Immunoglobulin class switching: Developmentally regulated DNA rearrangements during differentiation. *Cell* 22:1–2.

De Robertis, E. M., and Gurdon, J. B. 1977. Gene activation in somatic nuclei after injection into amphibian oocytes. *Proc. Natl. Acad. Sci. USA* 74:2470–2474.

Doerfler, W. 1983. DNA methylation and gene activity. *Annu. Rev. Biochem.* 52:93–124.

Efstratiadis, A., Posakony, J. W., Maniatis, T., Lawn, R. M., O'Connell, C., Spritz, R. A., DeRiel, J. K., Forget, B. G., Weissman, S. M., Slighton, J. L., Blechtl, A. E., Smithies, O., Baralle, F. E., Shoulders, C. C., and Proudfoot, N. J. 1980. The structure and evolution of the human β-globin gene family. *Cell* 21:653–668.

Evans, R. M., and Hollenberg, S. M. 1988. Zinc fingers: Gilt by association. *Cell* 52:1–3.

Gehring, W. 1979. Developmental genetics of *Drosophila*. *Annu. Rev. Genet.* 10:209–252.

Gellert, M. 1992. V(D)J recombination gets a break. *Trends Genet.* 8:408–412.

Green, M. R. 1989. Pre-mRNA processing and mRNA nuclear export. *Curr. Opinion Cell Biol.* 1:519–525.

Gross, D. S., and Garrard, W. T. 1987. Poising chromatin for transcription. *Trends Biochem. Sci.* (August):293–297.

———. 1988. Nuclease hypersensitive sites in chromatin. *Annu. Rev. Biochem.* 57:159–197.

Grunstein, M. 1992. Histones as regulators of genes. *Sci. Am.* 267 (October):68–74B.

Gurdon, J. B. 1968. Transplanted nuclei and cell differentiation. *Sci. Am.* 219:24–35.

Gurdon, J. B., Laskey, R. A., and Reeves, R. 1975. The developmental capacity of nuclei transplanted from keratinized skin cells of adult frogs. *J. Embryol. Exp. Morph.* 34:93–112.

Hadorn, E. 1968. Transdetermination in cells. *Sci. Am.* 219:110–120.

Hanna-Rose, W., and Hansen, U. 1996. Active repression mechanisms of eukaryotic transcription repressors. *Trends Genet.* 12:229–234.

Hershko, A., and Ciechanover, A. 1992. The ubiquitin system for protein degradation. *Annu. Rev. Biochem.* 61:761–807.

Hochstrasser, M. 1996. Protein degradation or regulation: Ub the judge. *Cell* 84:813–815.

Hodgkin, J. 1989. *Drosophila* sex determination: A cascade of regulated splicing. *Cell* 56:905–906.

Horabin, J. I., and Schedl, P. 1993. Regulated splicing of the *Drosophila Sex-lethal* male exon involves a blockage mechanism. *Mol. Cell. Biol.* 13:1408–1414.

Johnston, M., Flick, J. S, and Pexton, T. 1994. Multiple mechanisms provide rapid and stringent repression of *GAL* gene expression in *Saccharomyces cerevisiae*. *Mol. Cell. Biol.* 14:3834–3841.

Karlsson, S., and Nienhuis, A. W. 1985. Development regulation of human globin genes. *Annu. Rev. Biochem.* 54:1071–1078.

Keyes, L. N., Cline, T. W., and Schedl, P. 1992. The primary sex determination signal of *Drosophila* acts at the level of transcription. *Cell* 68:933–943.

Landschulz, W. H., Johnson, P. F., and McKnight, S. L. 1988. The leucine zipper: A hypothetical structure common to a new class of DNA binding protein. *Science* 240:1759–1763.

Lucas, P. C., and Granner, D. K. 1992. Hormone response domains in gene transcription. *Annu. Rev. Biochem.* 61:1131–1173.

Oettinger, M. A., Schatz, D. G., Gorka, C., and Baltimore, D. 1990. *RAG-1* and *RAG-2*, adjacent genes that synergistically activate V(D)J recombination. *Science* 248:1517–1522.

O'Malley, B. W., and Schrader, W. T. 1976. The receptors of steroid hormones. *Sci. Am.* 234:32–43.

Pabo, C. O., and Sauer, R. T. 1992. Transcription factors: Structural families and principles of DNA recognition. *Annu. Rev. Biochem.* 61:1053–1093.

Pankratz, M. J., and Jäckle, H. 1990. Making stripes in the *Drosophila* embryo. *Trends Genet.* 6:287–292.

Paranjape, S. M., Kamakaka, R. T., and Kadonaga, J. T. 1994. Role of chromatin structure in the regulation of transcription by RNA polymerase II. *Annu. Rev. Biochem.* 63:265–297.

Parthun, M. R., and Jaehning, J. A. 1992. A transcriptionally active form of GAL4 is phosphorylated and associated with GAL80. *Mol. Cell. Biol.* 12:4981–4987.

Postlethwait, J. H., and Schneiderman, H. A. 1973. Developmental genetics of *Drosophila* imaginal discs. *Annu. Rev. Genet.* 7:381–433.

Prescott, D. M. 1992. The unusual organization and processing of genomic DNA in hypotrichous ciliates. *Trends Genet.* 8:439–445.

Ptashne, M. 1989. How gene activators work. *Sci. Am.* 243 (January): 41–47.

Rhodes, D., and Klug, A. 1993. Zinc fingers. *Sci. Am.* 259 (February):56–65.

Rogers, J. O., Early, H., Carter, C., Calame, K., Bond., M., Hood, L., and Wall, R. 1980. Two mRNAs with different 3' ends encode membrane-bound and secreted forms of immunoglobin chain. *Cell* 20:303–312.

Ross, J. 1996. Control of messenger RNA stability in higher eukaryotes. *Trends Genet.* 12:171–175.

Scott, M. P., Tamkun, J. W., and Hartzell III, G. W. 1989. The structure and function of the homeodomain. *Biochim. Biophys. Acta* 989:25–48.

Studitsky, V. M., Clark, D. J., and Felsenfeld, G. 1994. A histone octamer can step around a transcribing polymerase without leaving the template. *Cell* 76:371–382.

Takagaki, Y., Seipelt, R. L., Peterson, M. L., and Manley, J. L. 1996. The polyadenylation factor CstF–64 regulates alternative processing of IgM heavy chain pre-mRNA during B cell differentiation. *Cell* 87:941–952.

Tobler, H., Etter, A., and Müller, F. 1992. Chromatin diminution in nematode development. *Trends Genet.* 8:427–432.

Tsai, M.-J., and O'Malley, B. W. 1994. Molecular mechanisms of action of steroid/thyroid receptor superfamily members. *Annu. Rev. Biochem.* 63:451–486.

Varshavsky, A. 1992. The N-end rule. *Cell* 69:725–735.

Varshavksy, A. 1996. The N-end rule: Functions, mysteries, uses. *Proc. Natl. Acad. Sci. USA* 93:12142–12149.

Verdine, G. L. 1994. The flip side of DNA methylation. *Cell* 76:197–200.

Wang, T. Y., Kostraba, N. C., and Newman, R. S. 1976. Selective transcription of DNA mediated by nonhistone proteins. *Prog. Nucleic Acid Res. Mol. Biol.* 19:447–462.

Wilmut, I., Schnieke, A. E., McWhir, J., Kind, A. J., and Campbell, K. H. S. 1997. Viable offspring derived from fetal and adult mammalian cells. *Nature* 385:810–813.

Wolffe, A. P. 1994. Transcription: In tune with the histones. *Cell* 77:13–16.

Wolffe, A. P., and Pruss, D. 1996. Targeting chromatin disruption: transcription regulators that acetylate histones. *Cell* 84:817–819.

CHAPTER 18: GENETICS OF CANCER

Baltimore, D. 1985. Retroviruses and retrotransposons: The role of reverse transcription in shaping the eukaryotic genome. *Cell* 40:481–482.

Bishop, J. M. 1983. Cancer genes come of age. *Cell* 32:1018–1020.

———. 1983. Cellular oncogenes and retroviruses. *Annu. Rev. Biochem.* 52:301–354.

———. 1987. The molecular genetics of cancer. *Science* 235:305–311.

Cavenee, W. K., and White, R. L. 1995. The genetic basis of cancer. *Sci. Amer.* (March): 72–79.

Cleaver, J. E. 1994. It was a very good year for DNA repair. *Cell* 76:1–4.

Fishel, R., Lescoe, M. K., Rao, M. R. S., Copeland, N. G., Jenkins, N. A., Garber, J., Kane, M., and Kolodner, R. 1994. The human mutator gene homolog MSH2 and its association with hereditary nonpolyposis colon cancer. *Cell* 75:1027–1038.

Hartwell, L. H., and Kastan, M. B. 1994. Cell cycle control and cancer. *Science* 266:1821–1828.

Jiricny, J. 1994. Colon cancer and DNA repair: Have mismatches met their match? *Trends Genet.* 10:164–168.

Kamb, A. 1995. Cell-cycle regulators and cancer. *Trends Genet.* 11:136–140.

Kingston, R. E., Baldwin, A. S., and Sharp, P. A. 1985. Transcription control by oncogenes. *Cell* 41:3–5.

Leach, F. S., and many other authors. 1993. Mutations of a *mutS* homolog in hereditary nonpolyposis colorectal cancer. *Cell* 75:1215–1225.

Mancini, M. A., Shan, B., Nickerson, J. A., Penman, S., and Lee, W.-H. 1994. The retinoblastoma gene product is a cell cycle-dependent, nuclear matrix-associated protein. *Proc. Natl. Acad. Sci. USA* 91:418–422.

Rabbitts, T. H. 1994. Chromosomal translocations in human cancer. *Nature* 372:143–149.

Ratner, L., Gallo, R. C., and Wong-Staal, F. 1985. Cloning of human oncogenes. In *Recombinant DNA research and virus,* edited by Y. Becker. Boston: Martinus Nijhoff.

Ratner, L., Josephs, S. F., and Wong-Staal, F. 1985. Oncogenes: Their role in neoplastic transformation. *Annu. Rev. Microbiol.* 39:419–449.

Rebbeck, T. R., Couch, F. J., Kant, J., Calzone, K., DeShano, M., Peng, Y., Chen, K., Garber, J. E., and Weber, B. L. 1996. Genetic heterogeneity in hereditary breast cancer: Role of *BRCA1* and *BRCA2*. *Am. J. Hum. Genet.* 59:547–553.

Weinberg, R. A., 1995. The retinoblastoma protein and cell cycle protein. *Cell* 81:323–330.

Wooster, R., and many other authors. 1994. Localization of a breast cancer susceptibility gene, *BRCA2*, to chromosome 13q12–13. *Science* 265:2088–2090.

———. 1995. Identification of the breast cancer susceptibility gene *BRCA2*. *Nature* 378:789–792.

Wooster, R., and Stratton, M. R. 1995. Breast cancer susceptibility: A complex disease unravels. *Trends Genet.* 11:3–5.

CHAPTER 19: GENETIC MUTATION

Ames, B. N., Durston, W. E., Yamasaki, E. and Lee, F. 1973. Carcinogens are mutagens: A simple test system combining liver homogenates for activation and bacteria for detection. *Proc. Natl. Acad. Sci. USA* 70:2281–2285.

Ames, B. N., and Gold, L. S. 1990. Too many rodent carcinogens: Mitogenesis increases mutagenesis. *Science* 249:970–971.

Boyce, R. P., and Howard-Flanders, P. 1964. Release of ultraviolet light-induced thymine dimers from DNA in *E. coli* K12. *Proc. Natl. Acad. Sci. USA* 51:293–300.

Cleaver, J. E. 1994. It was a very good year for DNA repair. *Cell* 76:1–4.

Devoret, R. 1979. Bacterial tests for potential carcinogens. *Sci. Am.* 241:40–49.

Fishel, R., Lescoe, M. K., Rao, M. R. S., Copeland, N. G., Jenkins, N. A., Garber, J., Kane, M., and Kolodner, R. 1993. The human mutator gene homolog *MSH2* and its association with hereditary nonpolyposis colon cancer. *Cell* 75:1027–1038.

Lederberg, J., and Lederberg, E. M. 1952. Replica plating and indirect selection of bacterial mutants. *J. Bacteriol.* 63:399–406.

Luria, S. E., and Delbrück, M. 1943. Mutations of bacteria from virus sensitivity to virus resistance. *Genetics* 28:491–511.

Modrich, P. 1987. DNA mismatch correction. *Annu. Rev. Biochem.* 56:435–466.

Morgan, A. R. 1993. Base mismatches and mutagenesis: How important is tautomerism? *Trends Biochem. Sci.* 18:160–163.

Radman, M., and Wagner, R. 1986. Mismatch repair in *Escherichia coli*. *Annu. Rev. Genet.* 20:523–538.

Setlow, R. B., and Carrier, W. L. 1964. The disappearance of thymine dimers from DNA: An error-correcting mechanism. *Proc. Natl. Acad. Sci. USA* 51:226–231.

Tessman, I., Liu, S.-K., and Kennedy, A. 1992. Mechanism of SOS mutagenesis of UV-irradiated DNA: Mostly error-free processing of deaminated cytosine. *Proc. Natl. Acad. Sci. USA* 89:1159–1163.

CHAPTER 20: TRANSPOSABLE ELEMENTS

Adams, S. E., Mellor, J., Gull, K., Sim, R. B., Tuite, M. F., Kingsman, S. M., and Kingsman, A. J. 1987. The functions and relationships of Ty-VLP proteins in yeast reflect those of mammalian retroviral proteins. *Cell* 49:111–119.

Berg, D. E., and Howe, M. M. 1989. *Mobile DNA*. Washington, DC: American Society of Microbiology

Bhattacharyya, M. K., Smith, A. M., Ellis, T. H. N., Hedley, C., and Martin, C. 1990. The wrinkled-seed character of pea described by Mendel is caused by a transposon-like insertion in a gene encoding starch-branching enzyme. *Cell* 60:115–122.

Boeke, J. D., and Corces, V. G. 1989. Transcription and reverse transcription of retrotransposons. *Annu. Rev. Microbiol.* 43:403–434.

Boeke, J. D., Garfinkel, D. J., Styles, C. A., and Fink, G. R. 1985. Ty elements transpose through an RNA intermediate. *Cell* 40:491–500.

Bucheton, A. 1990. I transposable elements and I-R hybrid dysgenesis in *Drosophila*. *Trends Genet.* 6:16–21.

Cohen, S. N., and Shapiro, J. A. 1980. Transposable genetic elements. *Sci. Am.* 242:40–49.

Engles, W. R. 1983. The P family of transposable elements in *Drosophila*. *Annu. Rev. Genet.* 17:315–344.

Federoff, N. V. 1989. About maize transposable elements and development. *Cell* 56:181–191.

Foster, T. J., Davis, M. A., Roberts, D. E., Takashita, K., and Kleckner, N. 1981. Genetic organization of transposon Tn10. *Cell* 23:201–213.

Garfinkel, D. J., Boeke, J. D., and Fink, G. R. 1985. Ty element transposition: Reverse transcriptase and virus-like particles. *Cell* 42:507–517.

Iida, S., Meyer, J., and Arber, W. 1983. Prokaryotic IS elements. In *Mobile genetic elements*, J. A. Shapiro, ed., pp. 159–221. New York: Academic Press.

Kingsman, A. J., and Kingsman, S. M. 1988. Ty: A retroelement moving forward. *Cell* 53:333–335.

Kleckner, N. 1981. Transposable elements in prokaryotes. *Annu. Rev. Genet.* 15:341–404.

McClintock, B. 1939. The behavior in successive nuclear divisions of a chromosome broken at meiosis. *Proc. Natl. Acad. Sci. USA* 25:405–416.

———. 1950. The origin and behavior of mutable loci in maize. *Proc. Natl. Acad. Sci. USA* 36:344–355.

———. 1951. Chromosome organization and genic expression. *Cold Spring Harbor Symp. Quant. Biol.* 16:13–47.

———. 1953. Induction of instability at selected loci in maize. *Genetics* 38:579–599.

———. 1956. Controlling elements and the gene. *Cold Spring Harbor Symp. Quant. Biol.* 21:197–216.

———. 1961. Some parallels between gene control systems in maize and in bacteria. *Am. Naturalist* 95:265–277.

———. 1965. The control of gene action in maize. *Brookhaven Symp. Biol.* 18:162 ff.

———. 1984. The significance of responses of the genome to challenge. Nobel lecture. *Science* 226:792–801.

Spradling, A. C., and Rubin, G. M. 1981. *Drosophila* genome organization: Conserved and dynamic aspects. *Annu. Rev. Genet.* 15:219–264.

Varmus, H. E. 1982. Form and function of retroviral proviruses. *Science* 216:812–821.

CHAPTER 21: EXTRANUCLEAR GENETICS

Ashwell, M., and Work, T. S. 1970. The biogenesis of mitochondria. *Annu. Rev. Biochem.* 39:251–290.

Birky, C. W. 1978. Transmission genetics of mitochondria and chloroplasts. *Annu. Rev. Genet.* 12:471–512.

Brown, M. D., Voljavec, A. S., Lott, M. T., MacDonald, I., and Wallace, D. C. 1992. Leber's hereditary optic neuropathy: A model for mitochondrial neurodegenerative diseases. *FASEB J.* 6:2791–2799.

Cattaneo, R. 1991. Different types of messenger RNA editing. *Annu. Rev. Genet.* 25:71–88.

Chiu, W.-L., and Sears, B. B. 1993. Plastome-genome interactions affect plastid transmission in Oenothera. *Genetics* 133:989–997.

Clayton, D. A. 1982. Replication of animal mitochondrial DNA. *Cell* 28:693–705.

Ephrussi, B. 1953. *Nucleo-cytoplasmic relations in microorganisms.* New York: Oxford University Press.

Freeman, G., and Lundelius, J. W. 1982. The developmental genetics of dextrality and sinistrality in the gastropod *Limnaea peregra*. *Wilhelm Roux Arch. Dev. Biol.* 191:69–83.

Gellissen, G., and Michaelis, G. 1987. Gene transfer: Mitochondria to nucleus. *Ann. N.Y Acad. Sci.* 503:391–401.

Grivell, L. 1983. Mitochondrial DNA. *Sci. Am.* 225(March):78–89.

Gyllensten, U., Wharton, D., Josefsson, A., and Wilson, A. C. 1991. Paternal inheritance of mitochondrial DNA in mice. *Nature* 352:255–257.

Herbert, A. 1996. RNA editing, introns and evolution. *Trends Genet.* 12:6–9.

Hiratsuka, J., Shimada, H., Whittier, R., Ishibashi, T., Sakamoto, M., Mori, M., Kondo, C., Honji, Y., Sun, C.-R., Meng, B.-Y., Li, Y-Q., Kanno, A., Nishizawa, Y., Hirai, A., Shinozaki, K., and Sugiura, M. 1989. The complete sequence of the rice (*Oryza saliva*) chloroplast genome: Intermolecular recombination between distinct tRNA genes accounts for a major plastid DNA inversion during the evolution of cereals. *Mol. Gen. Genet.* 217:185–194.

Hoch, B., Maier, R. M., Appel, K., Igloi, G. L., and Kössel, H. 1991. Editing of chloroplast mRNA by creation of an initiation codon. *Nature* 353:178–180.

Kirk, J. T. O. 1971. Chloroplast structure and biogenesis. *Annu. Rev. Biochem.* 40:161–196.

Lander, E. S., and Lodish, H. 1990. Mitochondrial diseases: Gene mapping and gene therapy. *Cell* 61:925–926.

Lee, S. B., and Taylor, J. W. 1992. Uniparental inheritance and replacement of mitochondrial DNA in *Neurospora tetrasperma*. *Genetics* 134:1063–1075.

Levings, C. S. 1983. The plant mitochondrial genome and its mutants. *Cell* 32:659–661.

Lu, B., and Hanson, M. R. 1992. A single nuclear gene specifies the abundance and extent of RNA editing of a plant mitochondrial transcript. *Nucl. Acids Res.* 20:5699–5703.

Matzke, M., and Matzke, A. J. M. 1993. Genomic imprinting in plants: Parental effects and *trans*-inactivation phenomena. *Annu. Rev. Plant. Physiol. Plant Mol. Biol.* 44:53–76.

Ohyama, K., Fukuzawa, H., Kohchi, T., Sano, T., Sano, S., Shirai, H., Umesono, K., Shiki, Y., Takeuchi, M., Chang, Z., Aota, S.-I., Inokuchi, H., and Ozeki, H. 1988. Structure and organization of *Marchantia polymorpha* chloroplast genome. I. Cloning and gene identification. *J. Mol. Biol.* 203:281–298.

Reik, W. 1989. Genomic imprinting and genetic disorders in man. *Trends Genet.* 5:331–336.

Rochaix, J. D. 1978. Restriction endonuclease map of the chloroplast DNA of *Chlamydomonas reinhardi*. *J. Mol. Biol.* 126:597–617.

Schuster, W., and Brennicke, A. 1994. The plant mitochondrial genome: Physical structure, information content, RNA editing, and gene migration to the nucleus. *Annu. Rev. Plant. Physiol. Plant Mol. Biol.* 45:61–78.

Scott, J. 1995. A place in the world for RNA editing. *Cell* 81:833–836.

Simpson, L., and Thiemann, O. H. 1995. Sense from nonsense: RNA editing in mitochondria of kinetoplastid protozoa and slime molds. *Cell* 81:837–840.

Swain, J. L., Stewart, T. A., and Leder, P. 1987. Parental legacy determines methylation and expression of an autosomal transgene: A molecular mechanism for parental imprinting. *Cell* 50:719–727.

Turmel, M., Lemieux, B., and Lemieux, C. 1988. The chloroplast genome of the green alga *Chlamydomonas moewusii*: Localization of protein-coding genes and transcriptionally active regions. *Mol. Gen. Genet.* 214:412 –419.

Umesono, K., and Ozeki, H. 1987. Chloroplast gene organization in plants. *Trends Genet.* 3:281–287.

Van Winkle-Swift, K. P., and Birky, C. W. 1978. The nonreciprocality of organelle gene recombination in *Chlamydomonas reinhardi* and *Saccharomyces cerevisiae*. *Mol. Gen. Genet.* 166:193–209.

Wallace, D. C. 1992. Diseases of the mitochondrial DNA. *Annu. Rev. Biochem.* 61:1175–1212.

Yang, X., and Griffiths, A. J. F. 1992. Male transmission of linear plasmids and mitochondrial DNA in the fungus *Neurospora*. *Genetics* 134:1055–1062.

CHAPTER 22: POPULATION GENETICS

Avise, J. C. 1986. Mitochondrial DNA and the evolutionary genetics of higher animals. *Phil. Trans. Roy. Soc. Lond.*, Ser. B 321:325–342.

Avise, J., Giblin-Davidson, C. C., Laerm, J., Patton, J. C., and Lansman, R. A. 1979. Mitochondrial DNA clones and matriarchal phylogeny within and among geographic populations of the pocket gopher *Geomys pinetis*. *Proc. Natl. Acad. Sci. USA* 76:6694–6698.

Ayala, F. J., ed. 1976. *Molecular evolution*. Sunderland, MA: Sinauer.

Buri, P. 1956. Gene frequency in small populations of mutant *Drosophila*. *Evolution* 10:367–402.

Clarke, C. A., and Sheppard, P. M. 1966. A local survey of the distribution of industrial melanic forms in the moth *Biston betularia* and estimates of the selective values of these in an industrial environment. *Proc. R. Soc. Lond. [Biol.]* 165:424–439.

Crow, J. F. 1986. *Basic concepts in population, quantitative, and evolutionary genetics*. New York: Freeman.

Darwin, C. 1860. *On the origin of species by means of natural selection, or the preservation of favoured races in the struggle for life*. New York: Appleton.

Dobzhansky, T. 1951. *Genetics and the origin of species*, 3rd ed. New York: Columbia University Press.

Fisher, R. A. 1930. *The genetical theory of natural selection*. Oxford: Clarendon Press.

Ford, E. B. 1971. *Ecological genetics*, 3rd ed. London: Chapman and Hall.

Glass, B., Sacks, M. S., Jahn, E. F., and Hess, C. 1952. Genetic drift in a religious isolate: An analysis of the causes of variation in blood group and other gene frequencies in a small population. *Amer. Natur.* 86:145–159.

Gray, M. W. 1989. Origin and evolution of mitochondrial DNA. *Annu. Rev. Cell Biol.* 5:25–50.

Hardy, G. H. 1908. Mendelian proportions in a mixed population. *Science* 28:49–50.

Hartl, D. L., and Clark, A. G. 1988. *Principles of population genetics*, 2nd ed. Sunderland, MA: Sinauer.

Hedrick, P. H. 1983. *Genetics of populations*. Boston: Science Books International.

Hillis, D. M., and Moritz, C. 1990. *Molecular systematics*. Sunderland, MA: Sinauer.

Hillis, D. M., Moritz, C., Porter, C. A., and Baker, R. J. 1991. Evidence for biased gene conversion in concerted evolution of ribosomal DNA. *Science* 251:308–309.

Kettlewell, H. B. D. 1961. The phenomenon of industrial melanism in the Lepidoptera. *Annu. Rev. Entomol.* 6:245–262.

Koehn, R., and Hilbish, T. J. 1987. The adaptive importance of genetic variation. *Amer. Sci.* 75:134–141.

Kreitman, M. 1983. Nucleotide polymorphism at the alcohol dehydrogenase locus of *Drosophila melanogaster*. *Nature* 304:412–417.

Lehman, N., Eisenhawer, A., Hansen, K., Mech, L. D., Peterson, R. O., Gogan, P. J., and Wayne, R. K. 1991. Introgression of coyote mitochondrial DNA into sympatric North American gray wolf populations. *Evolution* 45:104–119.

Lewontin, R. C. 1974. *The genetic basis of evolutionary change*. New York: Columbia University Press.

Lewontin, R. C. 1985. Population genetics. *Annu. Rev. Genet.* 19:81–102.

Lewontin, R. C., Moore, J. A., Provine, W. B., and Wallace, B. 1981. *Dobzhansky's genetics of natural populations I-XLIII*. New York: Columbia University Press.

Li, W-H. 1991. *Fundamentals of molecular evolution*. Sunderland, MA: Sinauer.

Li, W-H., Luo, C. C., and Wu, C. I. 1985. Evolution of DNA sequences. *Molecular evolutionary genetics*, R. J. MacIntyre, ed., pp. 1–94. New York: Plenum.

MacIntyre, R. J., ed. 1985. *Molecular evolutionary genetics*. New York: Plenum.

Maniatis, T., Fritsch, E. F., Lauer, L., and Lawn, R. M. 1980. The molecular genetics of human hemoglobin. *Annu. Rev. Genet.* 14:145–178.

Maynard Smith, J. 1989. *Evolutionary genetics*. Oxford: Oxford University Press.

Nei, M. 1987. *Molecular evolutionary genetics*. New York: Columbia University Press.

Nei, M., and Koehn, R. K. 1983. *Evolution of genes and proteins*. Sunderland, MA: Sinauer.

Pace, N. R., Olsen, G. J., and Woese, C. R. 1986. Ribosomal RNA phylogeny and the primary lines of evolutionary descent. *Cell* 45:325–326.

Schonewald-Cox, C. M., Chambers, S. M., MacBryde, B., and Thomas, L. 1983. *Genetics and conservation: A reference for managing wild animal and plant populations*. Menlo Park, CA: Benjamin/Cummings.

Selander, R. K., and Kaufman, D. W. 1975. Self-fertilization and genetic population structure in a colonizing land snail. *Proc. Natl. Acad. Sci. USA* 70:1186–1190.

Soulé, M. E., ed. 1986. *Conservation biology: The science of scarcity and diversity.* Sunderland, MA: Sinauer.

Speiss, E. 1977. *Genes in populations.* New York: Wiley.

Stringer, C. B. 1990. The emergence of modern humans. *Sci. Am.* (December):98–104.

Stringer, C. B., and Andrews, P. 1988. Genetic and fossil evidence for the origin of modern humans. *Science* 239:1263–1268.

Woese, C. R. 1981. Archaebacteria. *Sci. Am.* 244:98–122.

CHAPTER 23: QUANTITATIVE GENETICS

Blumer, M. G. 1980. *The mathematical theory of quantitative genetics.* Oxford: Clarendon Press.

Darwin, C. 1860. *On the origin of species by means of natural selection, or the preservation of favoured races in the struggle for life.* New York: Appleton.

Dobzhansky, T., and Pavlovsky, O. 1969. Artificial and natural selection for two behavioral traits in *Drosophila pseudoobscura. Proc. Natl. Acad. Sci. USA.* 62:75–80.

East, E. M. 1910. A Mendelian interpretation of variation that is apparently continuous. *Am. Natur.* 44:65–82.

———. 1916. Studies on size inheritance in *Nicotiana. Genetics* 1:164–176.

East, E. M., and Jones, D. F. 1919. *Inbreeding and outbreeding.* Philadelphia: Lippincott.

Falconer, D. S. 1989. *Introduction to quantitative genetics.* New York: Wiley.

Hill, W. G., ed. 1984. *Quantitative genetics,* Parts I and II. New York, Van Nostrand Reinhold.

Lander, E. S., and Botstein, D. 1989. Mapping Mendelian factors umderlying quantitative traits using RFLP linkage maps. *Genetics* 121:185–199.

Mather, K. 1943. Polygenic inheritance and natural selection. *Biol. Rev.* 18:32–64.

Nilsson-Ehle, H. 1909. Kreuzungsuntersuchungen an Hafer und Weizen. *Lunds Univ. Aarskr. N. F. Atd.,* Ser. 2, 5 (2):1–122.

Paterson, A. H., Lander, E. S., Hewitt, J. D., Person, S., Lincoln, S. E., and Tanksley, S. D. 1988. Resolution of quantitative traits into Mendelian factors by using a complete RFLP linkage map. *Nature* 335:721–726.

Selander, R. K., and Kaufman, D. W. 1975. Self-fertilization and genetic population structure in a colonizing land snail. *Proc. Natl. Acad. Sci. USA* 70:1186–1190.

Sokal, R. R., and Rohlf, F. J. 1981. *Biometry,* 2nd ed. San Francisco: Freeman.

Sokal, R. R., and Thoday, J. M., eds. 1979. *Quantitative genetic variation.* New York: Academic Press.

Thoday, J. M. 1961. Location of polygenes. *Nature* 191:368–370.

Thompson, J. N., Jr., and Thompson, J. M., eds. 1979. *Quantitative genetic variation.* New York: Academic Press.

Weir, B. S., Eisen, E. J., Goodman, M. M., and Namkoong, G., eds. 1988. *Proceedings of the second international conference on quantitative genetics.* Sunderland, MA: Sinauer.

Solutions to Selected Questions and Problems

Chapter 2

2.1 a. red

b. 3 red, 1 yellow

c. all red

d. ½ red, ½ yellow

2.3 The F_2 genotypic ratio (if C is colored, c is colorless) is ¼ CC : ½ Cc : ¼ cc. If we consider just the colored plants, there is a 1:2 ratio of CC homozygotes to Cc heterozygotes. Therefore, if a colored plant is picked at random, the probability that it is CC is ⅓ (i.e., ⅓ of the *colored* plants are homozygous), and the probability that it is Cc is ⅔. Only if a Cc plant is selfed will more than one phenotypic class be found among its progeny; therefore, the answer is ⅔.

2.4 a. Parents are Rr (rough) and rr (smooth); F_1 are Rr (rough) and rr (smooth).

b. $Rr \times Rr \to$ ¼ rough, ¼ smooth

2.6 Progeny ratio approximates 3:1, and so the parent is heterozygous. Of the dominant progeny, there is a 1:2 ratio of homozygous to heterozygous, so ⅓ will breed true.

2.7 Black is dominant to brown. If B is the allele for black and b for brown, then female X is Bb and female Y is BB. The male is bb.

2.10

PARENTS		PROGENY		FEMALE PARENT
female $\times$	male	grey	white	GENOTYPE?
a. grey $\times$	white	81	82	Gg
b. grey $\times$	grey	118	39	Gg
c. grey $\times$	white	74	0	GG
d. grey $\times$	grey	90	0	GG or Gg

2.11 a. To obtain a white babbit in a cross between F_1 babbits, both babbits must be Bb in genotype, and a bb offspring must be produced by these parents. Among the black F_1 progeny, there is a ⅓ chance of picking a BB individual and a ⅔ chance of picking a Bb individual. The chance of two Bb parents giving a bb offspring is ¼. Thus:

P(white offspring) = P(both F_1 babbits are Bb and a bb offspring is produced)

= P(both F_1 babbits are Bb) $\times$ P(bb offspring)

= (⅔ × ⅔) × (¼)

= ⅑

b. If he crosses an F_1 male (Bb or BB) to the parental female (Bb), two types of crosses are possible. The crosses and probabilities are (1) Bb (F_1 male) × Bb (parental female); $P = $ ⅔ × 1 = ⅔ and (2) BB (F_1 male) × Bb (parental female); $P = $ ⅓ × 1 = ⅓. Only the first cross can produce white progeny, ¼ of the time. Thus, the chance that this strategy will yield white progeny is:

$P = $ ⅔ (chance of $Bb \times Bb$ cross) × ¼ (chance of bb offspring)

= ⅙

c. The best strategy is as follows. Remate the initial two black babbits (both are Bb) to obtain a white male offspring ($P = $ ¼ [white bb] × ½ [male] = ⅛). Retain this male and breed it back to its mother. This cross would be $Bb \times bb$ and give ½ white (bb) and ½ black (Bb) offspring. The progeny of this cross could be used to develop a "breeding colony" consisting of black (Bb) females and white (bb) males. These would consistently produce half white and half black offspring.

2.13 Try fitting the data to a model in which catnip sensitivity/insensitivity is controlled by a pair of alleles at one gene. Hypothesize that, because sensitivity is seen in all of the progeny of the initial mating between catnip-sensitive Cleopatra and catnip-insensitive Antony, sensitivity is dominant. If S is the sensitive allele and s the insensitive allele, then the initial cross was $S- \times ss$, and the progeny are Ss. If two of the Ss kittens mate, one would expect 3 Ss (sensitive) : 1 ss (insensitive) kittens. In the mating with Augustus, the cross would be $Ss \times ss$ and should give a 1 Ss (sensitive) : 1 ss (insensitive) progeny ratio. The observed progeny ratios are not far off from these expectations.

An alternative hypothesis is that sensitivity (s) is recessive, and insensitivity (S) is dominant. For Antony and Cleopatra to have sensitive (ss) offspring, they would need to be Ss and ss, respectively. When two of their ss progeny mate, only sensitive, ss offspring should be produced. This is not observed, so this hypothesis does not explain the data.

2.15 a. $WW\,Dd \times ww\,dd$

b. $Ww\,dd \times Ww\,dd$

c. $ww\,DD \times WW\,dd$

d. $Ww\,Dd \times Ww\,dd$

e. $Ww\,Dd \times Ww\,dd$

2.16 The cross is $Aa\,Bb\,Cc \times Aa\,Bb\,Cc$.

a. Considering the A gene alone, the probability of an offspring showing the A trait from $Aa \times Aa$ is ¾. Similarly, the probability of showing the B trait from $Bb \times Bb$ is ¾ and the C trait from $Cc \times Cc$ is ¾. Therefore, the probability of a given progeny being phenotypically ABC (using the product rule) is ¾ × ¾ × ¾ = 27/64.

b. Considering the A gene, the probability of an AA offspring from $Aa \times Aa$ is ¼. The same probability is the case for a BB offspring and for a CC offspring. Therefore, the probability of an $AA\,BB\,CC$ offspring (from the product rule) is ¼ × ¼ × ¼ = 1/64.

2.20 a. The F_1 has the genotype $Aa\,Bb\,Cc^h$ and is all agouti, black. The F_2 is 27/64 agouti, black; 9/64 agouti, black, Himalayan; 9/64 agouti, brown; 9/64 black; 3/64 agouti, brown, Himalayan; 3/64 black, Himalayan; 3/64 brown; 1/64 brown, Himalayan.

b. F_2 animals that are black and agouti (and either Himalayan or have full-body pigmentation) have the genotypes $A- B- C-$ and $A- B- c^hc^h$. Among the $A-$ animals, ⅔ are Aa. Among the $B-$ animals, ⅓ are BB. Among the $C-$ and c^hc^h animals, ½ are Cc^h. Thus, among all of the black, agouti F_2, ⅔ × ⅓ × ½ = ⅑ are $Aa\,BB\,Cc^h$. If one limits the animals under consideration to those that are black, agouti, *and not Himalayan* (i.e., $A- B- C-$), then ⅔ of the $C-$ animals are Cc^h, and the proportion of $Aa\,BB\,Cc^h$ animals is ⅔ × ⅓ × ⅔ = 4/27.

c. From the cross $Aa\,Bb\,Cc^h \times Aa\,Bb\,Cc^h$, ¼ of the progeny will be bb and show brown pigment. This will be the case regardless of whether the animals are pigmented over their entire bodies or are Himalayan. Thus, ¼ of the Himalayan mice will show brown pigment.

d. From the cross $Aa\,Bb\,Cc^h \times Aa\,Bb\,Cc^h$, ¾ of the progeny will be $B-$ and show black pigment. This will be the case regardless of whether the animals are agouti or nonagouti. Thus, ¾ of the agouti mice will show black pigment.

2.24 Mating type C is determined only by the genotype *aa bb*. Thus, "C" must be genotype *aa bb*. Crosses of the other strains to "C," then, are testcrosses and can tell us the genotypes of the strains. Therefore, "A" is *Aa Bb*, "B" is *aa Bb*, and "D" is *Aa bb*.

2.25 a. The cross is *W/w R/r × W r*. Workers are females, and in the progeny of the cross they are: ¼ each *W/W R/r* (black, use wax) : ¼ *W/w R/r* (black, use wax) : ¼ *W/W r/r* (black, use resin) : ¼ *W/w r/r* (black, use resin). In sum, all workers are black-eyed, while ½ use wax and ½ use resin.

b. ¼ each *W R, W r, w R, w r*.

c. Here the question is that the female in question is heterozygous, not that her progeny will be wingless. Madonna is *C/c*, so there is a ¼ chance that her granddaughter will be heterozygous and that wingless males will be found in the hive (i.e., ½ that her daughter is heterozygous × ½ that the daughter of the daughter is heterozygous).

d. For the F₄ generation great-great-granddaughter to be heterozygous is ½ × ½ × ½ × ½ = ¹⁄₁₆.

2.26 a. Mother must be heterozygous *Aa* in order to have children who have the trait.

b. Father is homozygous *aa* because he expresses the trait.

c. The cross is *Aa × aa*, so children will be *aa* if they have the trait (II.2 and II.5) and *Aa* if they do not have the trait (II.1, II.3, and II.4).

d. From the cross *Aa × aa*, the prediction is that ½ of the progeny will be *Aa* (normal) and ½ will be *aa* (expressing the trait). There are five children, two of whom have the trait and three of whom are normal. Thus, the ratio fits as well as it could for five children.

2.29 An uncertainty exists as to whether the brother of the man's wife's paternal grandmother had Gaucher's disease. If this distant relative had the disease, a disease allele might have been passed on to the man's wife. Therefore, in a "worst-case scenario," consider this distant relative to have had the disease. Under this scenario, the pedigree is:

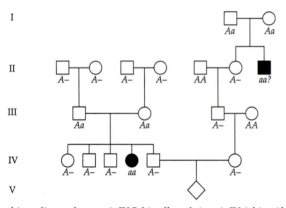

In this pedigree, the man is IV-5, his affected sister is IV-4, his wife is IV-6, and the brother of his wife's paternal grandmother is II-7. II-7 is affected but her parents are not, so the disease must be a recessive trait, and both of her parents must be heterozygous.

To calculate the chance that, if the couple has a child, the child will be affected, calculate the probability that V-1 can receive an *a* allele from each parent. For each parent, this is ½ the chance a parent is *Aa*.

The parents of IV-5 must be *Aa*, and IV-5 is *A–*, so there is a ⅔ chance that IV-5 is *Aa*. Thus, there is a ½ × ⅔ = ⅓ chance that IV-5 will pass the *a* allele to V-1.

The probability that IV-6 is *Aa* is more complex, and is:
Prob(III-3 was *Aa* and III-3 passed the *a* allele on to IV-6)
= Prob([II-6 was *Aa* and II-6 passed *a* to III-3] *and* III-3 passed *a* to IV-6)
= ([⅔ × ½] × ½) = ⅙

Thus, there is a ⅙ × ½ = ¹⁄₁₂ chance that IV-6 will pass the *a* allele to V-1. In this worst-case scenario, the chance that both parents will pass on an *a* allele and have an affected child is ¹⁄₁₂ × ⅓ = ¹⁄₃₆.

If the brother of the wife's paternal grandmother did not have the disease, IV-6 would be *AA*, and the child would be *A–* and phenotypically normal.

CHAPTER 3

3.1 c

3.3 c

3.6 a. Yes, if a sexual mating system exists in that species. In that case, two haploid cells can fuse to produce a diploid cell, which can then go through meiosis to produce haploid progeny. The fungi *Neurospora crassa* and *Saccharomyces cerevisiae* exemplify this positioning of meiosis in the life cycle.

b. No, because a diploid cell cannot be formed and meiosis occurs only starting with a diploid cell.

3.8 c

3.10 a. metaphase
b. anaphase

3.14 a. The chance that a gamete would have a particular maternal chromosome is ½. Therefore, the chance of obtaining a gamete with all three maternal chromosomes is $(½)^3 = ⅛$.

b. The set of gametes with some maternal and paternal chromosomes is composed of all gametes *except* those that have only maternal or only paternal chromosomes. That is, *P*(gamete with both maternal and paternal chromosomes) = 1 – *P*(gamete with only maternal or only paternal chromosomes). From part (a), the chance of a gamete having chromosomes from only one parent is ⅛. Using the sum rule, *P*(gamete with both maternal and paternal chromosomes) = 1 – (⅛ + ⅛) = ¾.

3.15 One of the long chromosomes and the short chromosome might be members of a heteromorphic pair—i.e., X and Y chromosomes, respectively.

3.17 False. Owing to the randomness of independent assortment and to crossing over, both of which characterize meiosis, the probability of any two sperm cells being genetically identical is extremely remote.

3.19 a. 17 + 26 = 43 chromosomes.

b. Similar chromosomes pair in meiosis. The pairing pattern seen in the hybrid indicates that some of the chromosomes in the arctic and red foxes share evolutionary similarity, but others do not. Unpaired chromosomes will not segregate in an orderly manner, giving rise to unbalanced meiotic products with either extra or missing chromosomes. This can lead to sterility for two reasons. First, meiotic products that are missing chromosomes may not have genes necessary to form gametes. Second, even if gametes are able to form, a zygote generated from them will not have the chromosome set from the hybrid, the red, or the arctic fox. The zygote will have missing or extra genes, causing it to be inviable.

3.20 The probability of a given homolog going to one particular pole is ½. The probability of all five paternal chromosomes going to the same pole is $(½)^5 = ¹⁄₃₂$. The same answer applies for all maternal chromosomes going to one pole.

3.21

	MOTHER'S		FATHER'S	
	MOTHER	FATHER	MOTHER	FATHER
X chromosome	Yes	Yes	No	No
Y chromosome	No	No	No	Yes

3.24 Fathers always give their X chromosome to their daughters, so the woman must be heterozygous for the color-blindness trait and is *c⁺c*. Her husband received his X chromosome from his mother and has normal color vision, so he is *c⁺Y*. The cross is therefore *c⁺c × c⁺Y*. All daughters will receive the paternal X bearing the *c⁺* allele and have normal color vision. Sons will receive the maternal X, so half will be cY and be color-blind, and half will be c^+Y and have normal color vision.

3.26 If c is the red-green color-blindness allele and a is the albino allele, the parents' genotypes are $c^+c^+\,aa$ ♀ and $cY\,a^+a^+$ ♂. All children will be normal visioned (c^+c ♀ and c^+Y ♂) and normally pigmented (a^+a, both sexes).

3.27 The parentals are $ww\,vg^+vg^+$ ♀ and $w^+Y\,vgvg$ ♂.

a. The F_1 males are all $wY\,vg^+vg$, white eyes, long wings. The F_1 females are all $w^+w\,vg^+vg$, red eyes, long wings.

b. The F_2 females are $\frac{3}{8}$ red, long; $\frac{3}{8}$ white, long; $\frac{1}{8}$ red, vestigial; $\frac{1}{8}$ white, vestigial. The same ratios of the respective phenotypes apply for the males.

c. The cross of F_1 male with the parental female is $wY\,vg^+vg \times ww\,vg^+vg^+$. All progeny have white eyes and long wings. The cross of an F_1 female with the parental male is $w^+w\,vg^+vg \times w^+Y\,vgvg$. Female progeny: All have red eyes, half have long wings, and half have vestigial wings. Male progeny: $\frac{1}{4}$ red, long; $\frac{1}{4}$ red, vestigial; $\frac{1}{4}$ white, long; $\frac{1}{4}$ white, vestigial.

3.31 The simplest hypothesis is that brown-colored teeth are determined by a sex-linked dominant mutant allele. Man A was BY and his wife was bb. All sons will be bY normals and cannot pass on the trait. All the daughters receive the X chromosome from their father, and so they are Bb with brown enamel. Half their sons will have brown teeth because half their sons receive the X chromosome with the B mutant allele.

3.33 a. To have produced a child with cystic fibrosis, both parents must have been heterozygous for the autosomal recessive mutant gene. Therefore, the probability of their next child having cystic fibrosis is $\frac{1}{4}$ (i.e., $\frac{1}{4}$ of the progeny from $Aa \times Aa$ will be aa).

b. Nonaffected children can be either AA or Aa. From a cross $Aa \times Aa$, there will be a progeny ratio of $1\,AA : 2\,Aa : 1\,aa$. Therefore, $\frac{2}{3}$ of the nonaffected children will be Aa heterozygotes.

3.36 The answer is given in the following Punnett squares. Note: Nondisjunction that occurs in individuals with a normal set of chromosomes is designated as primary nondisjunction. When it occurs in the progeny of such individuals, it is known as secondary nondisjunction.

a.

Original cross: $ww \times w^+Y$

EGGS	SPERM	
	w^+	Y
ww	www^+ Usually dies	wwY
0	w^+0 Sterile, red ♂	$Y0$ Dies

b.

Backcross: wwY ♀ $\times w^+\,Y$ ♂

EGGS		SPERM	
		w^+	Y
Normal X segregation	wY	w^+wY Red ♀	wYY White ♂
	w	w^+w Red ♀	wY White ♂
Secondary nondisjunction	ww	w^+ww Triplo-X Usually dies	wwY White ♀
	Y	w^+Y Red ♂	YY Dies

3.40

	PEDIGREE A	PEDIGREE B	PEDIGREE C
Autosomal recessive	Yes	Yes	Yes
Autosomal dominant	Yes	Yes	No
X-linked recessive	Yes	Yes	No
X-linked dominant	No	No	No

3.42 a. Y-linked inheritance can be excluded because females are affected. X-linked recessive inheritance can also be excluded because an affected mother (I.2) has a normal son (II.5). Autosomal recessive inheritance can also be excluded because in such a case, two affected parents such as II.1 and II.2 could only have affected offspring, which they do not.

b. The two remaining mechanisms of inheritance are X-linked dominant and autosomal dominant. Genotypes can be written to satisfy both mechanisms of inheritance. Of these two, X-linked dominant inheritance may be more likely because II.6 and II.7 have only affected daughters, indicating crisscross inheritance. If the trait were autosomal dominant, one would expect half of the daughters and half of the sons to be affected.

3.45 a is false. The father passes on his X to his daughters, so *none* of the sons will be affected.

b is false. Because the disease is rare, the mother would be heterozygous and only 50% of her daughters would receive the X with the dominant mutant allele.

c is true. All daughters receive the father's X chromosome.

d is false. Only 50% of her sons would receive the X with the dominant mutant allele.

3.47 a. True. Two affected individuals will always have affected children ($aa \times aa$ can give only aa offspring).

b. False. An autosomal trait is inherited independent of sex type.

c. Need not be true. The trait could be masked by normal dominant alleles through many generations before two heterozygotes marry and produce affected, homozygous offspring.

d. Could be true. If the trait is rare, an unaffected individual marrying into the pedigree is likely to be homozygous for a normal allele. The trait is recessive, and the children receive the dominant, normal allele from the unaffected parent, so the children will be normal. This answer would not be true if the unaffected individual was heterozygous. In this case, half of the children would be affected.

3.49 Because hemophilia is an X-linked trait, the most likely explanation is that random inactivation of X chromosomes (lyonization, see p. 79) produces individuals with different proportions of cells with the normal allele. Thus, if some women had only 40% of their cells with an active h^+ allele, and 60% with the h (hemophilia) allele, but other women had 60% of their cells with an active h^+ allele, and 40% with the h allele, there would be a significant difference in the amount of clotting factor these two individuals would make.

CHAPTER 4

4.2 Six possible genotypes: w/w, $w/w1$, $w/w2$, $w1/w1$, $w1/w2$, $w2/w2$.

4.6 The woman's genotype is I^A/I^B, and the man's genotype is I^A/i.

a. $\frac{1}{2} \times \frac{1}{2} = \frac{1}{4}$.

b. Zero. A blood group O baby is not possible.

c. $\frac{1}{2}$ (probability of male) $\times \frac{1}{4}$ (probability of AB) $\times \frac{1}{2}$ (probability of male) $\times \frac{1}{4}$ (probability of B) $= \frac{1}{64}$ probability that all four conditions will be fulfilled.

4.8 Blood type O, because genotype is i/i.

4.10 Half will be C^R/C^W and hence will resemble the parents.

4.12 a. The cross is $F/F\,G^N/G^N \times f/f\,G^O/G^O$. The F_1 is $F/f\,G^O/G^N$, which is fuzzy with round leaf glands.

b. Interbreeding the F_1 gives $\frac{3}{16}$ fuzzy, oval-glanded; $\frac{6}{16}$ fuzzy, round-glanded; $\frac{3}{16}$ fuzzy, no-glanded; $\frac{1}{16}$ smooth, oval-glanded; $\frac{2}{16}$ smooth, round-glanded; $\frac{1}{16}$ smooth, no-glanded.

c. Cross is $F/f\,G^O/G^N \times f/f\,G^O/G^O$. The progeny are ¼ fuzzy, oval-glanded; ¼ fuzzy, round-glanded; ¼ smooth, oval-glanded; ¼ smooth, round-glanded.

4.18 To show no segregation among the progeny, the chosen plant must be homozygous. The genotypes comprising the $\frac{9}{16}$ colored plants are: $1\,A/A\,B/B : 2\,A/a\,B/B : 2\,A/A\,B/b : 4\,A/a\,B/b$. Only one of these genotypes is homozygous; the answer is ⅑.

4.19 There are two independently assorting genes involved. Because of epistasis, only two phenotypes are seen in the F_2; that is, $A/-\,B/-$ genotypes are runner, and the other three genotypes are bunch. Thus, the original cross was $A/A\,b/b \times a/a\,B/B$, giving an F_1 of $A/a\,B/b$.

4.20 a. The cross is $A/a\,B/b \times A/a\,B/b$, which gives $\frac{9}{16}\,A/-\,B/-$: $\frac{3}{16}\,A/-\,b/b : \frac{3}{16}\,a/a\,B/- : \frac{1}{16}\,a/a\,b/b$. The $A/-\,B/-$ rabbits are not deaf because they produce both substances needed for hearing. The other three genotypic classes result in deafness because one or the other or both of the enzymes needed for hearing are not produced. Thus, the phenotypic ratio is 9 hearing rabbits : 7 deaf rabbits.

b. Epistasis. In this case, it is duplicate recessive epistasis.

c. The cross is $a/a\,B/b \times A/a\,B/b$, which gives $\frac{3}{8}\,A/a\,B/-$: $\frac{1}{8}\,A/a\,b/b : \frac{3}{8}\,a/a\,B/- : \frac{1}{8}\,a/a\,b/b$. Only the $A/a\,B/-$ rabbits can hear; the rest are deaf. Thus, the phenotypic ratio is 3 hearing rabbits : 5 deaf rabbits.

4.23 a. If A^Y governs yellow and a^+ governs nonyellow (agouti), then A^Y/A^Y are lethal, A^Y/a^+ are yellow, and y/y are agouti. Let c^+ determine colored coat and c determine albino. The parental genotypes, then, are $A^Y/a^+\,c^+/c$ (yellow) and $A^Y/a^+\,c/c$ (white).

b. The proportion is 2 yellow : 1 agouti : 1 albino. None of the yellows breed true because they are all heterozygous, with homozygous A^Y/A^Y individuals being lethal.

4.25 a. $Y/Y\,R/R$ (crimson) $\times y/y\,r/r$ (white) gives $Y/y\,R/r$ F_1 plants, which have magenta-rose flowers. Selfing the F_1 gives an F_2 as follows: $\frac{1}{16}$ crimson ($Y/Y\,R/R$), $\frac{2}{16}$ orange-red ($Y/Y\,R/r$), $\frac{1}{16}$ yellow ($Y/Y\,r/r$), $\frac{2}{16}$ magenta ($Y/y\,R/R$), $\frac{4}{16}$ magenta-rose (Y/y R/r), $\frac{2}{16}$ pale yellow ($Y/y\,r/r$), and $\frac{4}{16}$ white ($y/y\,R/R$, $y/y\,R/r$, and $y/y\,r/r$). Progeny of the F_1 backcrossed to the crimson parent are ¼ crimson, ¼ orange-red, ¼ magenta, and ¼ magenta-rose.

b. $Y/Y\,R/r \times Y/y\,r/r$ gives ¼ orange-red ($Y/Y\,R/r$), ¼ yellow ($Y/Y\,r/r$), ¼ magenta-rose ($Y/r\,R/r$), ¼ pale yellow (Y/y r/r).

c. $Y/y\,r/r$ (yellow) $\times y/y\,R/r$ (white) gives ½ magenta-rose ($Y/y\,R/r$) and ½ pale yellow ($Y/y\,r/r$).

4.27 a. The simplest approach is to calculate the proportion of progeny that will be black and then subtract that answer from 1. The black progeny have the genotype $A/-\,B/-\,C/-$, and the proportion of these progeny is $(\frac{3}{4})^3$. Therefore, the proportion of colorless progeny is $1 - (\frac{3}{4})^3 = 1 - \frac{27}{64} = \frac{37}{64}$.

b. Black is produced only when $c/c\,A/-\,B/-$ results, and this offspring occurs with the frequency $(\frac{1}{4})(\frac{3}{4})(\frac{3}{4}) = \frac{9}{64}$ black, which gives $\frac{55}{64}$ colorless.

4.32 In males, H/H and H/h are horned, and h/h is hornless; in females, H/H is horned, H/h and h/h are hornless. The cross is an $H/H\,W/W$ male $\times h/h$ w/w female. The F_1 is $H/h\,W/w$, which gives horned white males and hornless white females. Interbreeding the F_1 gives the following F_2:

	MALE	FEMALE
$\frac{3}{16}\,H/H\,W/-$	horned, white	horned, white
$\frac{6}{16}\,H/h\,W/-$	horned, white	hornless, white
$\frac{3}{16}\,h/h\,W/-$	hornless, white	hornless, white
$\frac{1}{16}\,H/H\,w/w$	horned, black	horned, black
$\frac{2}{16}\,H/h\,w/w$	horned, black	hornless, black
$\frac{1}{16}\,h/h\,w/w$	hornless, black	hornless, black

In sum, the ratios are $\frac{9}{16}$ horned white : $\frac{3}{16}$ hornless white : $\frac{3}{16}$ horned black : $\frac{1}{16}$ hornless black males and $\frac{3}{16}$ horned white : $\frac{9}{16}$ hornless white : $\frac{1}{16}$ horned black : $\frac{3}{16}$ hornless black females.

4.34 Ewe A is $H/h\,w/w$; B is $H/h\,W/w$ or $h/h\,W/w$; C is H/H w/w; D is $H/h\,W/w$; the ram is $H/h\,W/w$.
4.35 c

CHAPTER 5

5.2 From the χ^2 test, $\chi^2 = 16.10$; P is less than 0.01 at three degrees of freedom. This test reveals that the two genes do not fit a 1:1:1:1 ratio. It does not say why. Linkage might seem reasonable until it is realized that the minority classes are not reciprocal classes (both carry the aa phenotype). If the segregation at each locus is considered, however, the $B/-$: b/b ratio is about 1:1 (203:197), but the $A/-$: a/a ratio is not (240:160). The departure, then, specifically results from a deficiency of a/a individuals. This departure should be confirmed in other crosses that test the segregation at locus A. In corn, further evidence would be that the a/a deficiency might show up as a class of ungerminated seeds or of seedlings that die early.

5.4 There are 158 progeny rabbits. The recombinant classes are the English plus Angora and the non-English plus short-haired, and so the map distance between the genes is $(11 + 6)/158 \times 100\%$ = 10.8% = 10.8 mu.

5.6 You were lucky in that a double mutant hatched. Based on the presence of one double mutant, you can eliminate linkage between vg and m. This is because there is no crossing-over in the *Drosophila* male, and there is no way to produce a recombinant $vg\,m$ gamete in the male F_1. The crosses performed here were

$$P \qquad \frac{vg^+}{vg^+}\,\frac{m}{m}\ \text{females} \times \frac{vg}{vg}\,\frac{m^+}{m^+}\ \text{males}$$

$$\text{All } F_1 \text{ are}\quad \frac{vg^+}{vg}\,\frac{m^+}{m}\ \text{(wild type)}$$

The results of this cross told you that m is *not* X-linked.

Therefore, the F_1 cross was a dihybrid cross, and we expect a 9:3:3:1 F_2 ($vg^+\,m^+$ [wild type] : $vg\,m^+ : vg^+\,m : vg\,m$) ratio. In the small sample, by chance, a double mutant appeared. This tells us that vg and m are segregating independently. If vg and m had been on the same chromosome, the crosses performed would have been

$$P \qquad \frac{vg^+\,m}{vg^+\,m} \times \frac{vg\,m^+}{vg\,m^+}$$

$$F_1 \qquad \frac{vg^+\,m}{vg\,m^+} \qquad \text{(All wild type)}$$

$F_1 \times F_1 \to F_2$ phenotypes are in the following chart:

		♂ gametes	
		$vg^+\,m$	$vg\,m^+$
♀ gametes	$vg^+\,m$	maroon	wild type
	$vg\,m^+$	wild type	vestigial
	$vg^+\,m^+$	wild type	wild type
	$vg\,m$	maroon	vestigial

The phenotypic ratio would be 4:2:2:0 (wild:vestigial:maroon:vestigial-maroon).

There is no way to get a double mutant from this cross *regardless* of the distance between the two linked genes (0–100 percent recombination). So if the two genes were linked one would never find a double mutant progeny in the F_2. Since a double mutant did hatch, this was enough to allow the preliminary (but correct) conclusion that m was not on chromosome 2. Although m could be on chromosome 3 or 4, this experiment does not distinguish between these two possibilities.

5.7

Note that the map distances are not strictly additive—e.g., $a - d = 38$ and $a - c = 8$, but $d - e = 45$—because of the effects of multiple crossovers, as described in the chapter.

5.11 45% a b^+, 45% a^+ b, 5% a^+ b^+, 5% a b

5.12 a. $0.035 + 0.465 = 0.50$

b. All the daughters have a^+ b^+ phenotype.

5.15 a. The genotype of the F_1 is A B/a b. The gametes produced by the F_1 are 40% A B; 40% a b; 10% A b; 10% a B. From a testcross of the F_1 with a b/a b, we get 40% A B/a b; 40% a b/a b; 10% A b/a b; 10% a B/a b.

b. The F_1 genotype is A B/a B. The gametes produced by the F_1 are 40% A b; 40% a B; 10% A B; 10% a b. From a testcross of the F_1 with a b/a b, we get 40% A b/a b; 40% a B/a b; 10% A B/a b; 10% a b/a b. This question illustrates that map distance is computed between sites in the chromosome, and it does not matter whether the genes are in coupling or in repulsion; the percentage of recombinants will be the same, even though the phenotypic classes constituting the recombinants differ.

5.17 Each chromosome pair segregates independently. We can compute the relative proportions of gametes produced for each homologous pair of chromosomes separately from the known map distances (P = parental; R = recombinant):

P		R		P		R		P		R	
Ab	0.4	Ab	0.1	CD	0.45	Cd	0.05	EF	0.35	Ef	0.15
ab	0.4	aB	0.1	cd	0.45	cD	0.05	ef	0.35	eF	0.15

To answer the question, simply multiply the probabilities of getting the particular gamete from the F_1 multiple heterozygote.

a. A B C D E F = $0.4 \times 0.45 \times 0.35 = 0.063$ (6.3%)

b. A B C d e f = $0.4 \times 0.05 \times 0.35 = 0.007$ (9.7%)

c. A b c D E f = $0.1 \times 0.05 \times 0.15 = 0.00075$ (0.075%)

d. a B C d e f = $0.1 \times 0.05 \times 0.35 = 0.00175$ (0.175%)

e. a b c D e F = $0.4 \times 0.05 \times 0.15 = 0.003$ (0.3%)

5.18 a. 47.5% each of DP h and dp h; 2.5% each of Dp h and dp h.

b. 23.75% each of DP H, Dp h, dP H, and dP h; 1.25% each of DP H, DP h, dp H, and dp h.

5.23 a. F m W and f M w will be the least frequent (because of double crossovers).

b. M f W and m F w will be the least frequent.

c. M w F and m W f will be the least frequent.

5.25 a. By doing a three-point mapping analysis as described in the chapter, we find that the order of genes in the chromosome is dp-b-hk, with 35.5 mu between dp and b, and 5.4 mu between b and hk.

b. (1) The frequency of observed double crossover is 1.4%, the frequency of expected double crossovers is 1.4%, and the frequency of expected double crossovers is 1.9%. The coefficient of coincidence, therefore, is $1.4/1.9 = 0.73$. (2) The interference value is given by 1 – coefficient of coincidence, 0.27.

5.28 a. a^+ c b/a c^+ b^+

b. a^+ c^+ b^+/Y

c.

![map: a — c — b; 5 and 10]

5.31 The two X chromosomes in the female have the genotypes a + and + b. The lethal must be on the + b chromosomes of the female parent because there are far fewer parental + b chromosomes in the progeny males than a + chromosomes. Therefore, we can diagram the cross as:

All female progeny of the cross will be wild type. At this point, however, we do not know the order of the three genes on the chromosome.

You would expect 1,000 males normally, but you get only 499. The remainder can be considered the lethal (l) progeny. This fact allows you to predict the entire set of eight genotypes expected among the zygotes of the cross. This is done by knowing that a + is really + +, where the last + sign is the wild-type allele of the lethal. Thus, each genotype seen is accompanied in this cross by one carrying the l allele. Writing them out and recognizing the equality of reciprocal classes, we have:

405 a + +	and	405 + b l
44 + b +	and	44 a + +
48 + + +	and	48 a b l
2 a b +	and	2 + + l

This is:
810 parentals (a + + and + b l)
88 single crossovers, one region (+ b + and a + +)
96 single crossovers, other region (+ + + and a b l)
4 double crossovers (a b + and + + l)
Traditional methods lead us to the map:

![map: a — b — l; 10 and 9.2]

5.32 The starting point is to use the coefficient of coincidence value to calculate the observed number of double crossovers. Once that number is in hand, the known map unit values are used with the formula for map distance between genes in three-point testcrosses to calculate the observed number of single crossovers for the two gene distances. The answer is:

Both males and females are distributed as:

a + b	372
+ c +	373
a c +	67
+ + b	68
a + +	57
+ c b	58
a c b	2
+ + +	3

5.35 a. a, b, c, and d are linked on the same chromosome, because the percentage of recombinations is less than 50. e is on a separate chromosome because it segregates independently from the four other genes. The map is constructed by calculating the recombination frequency for all possible pairs of a, b, c, and d. The map distances allow an order to be determined. The map is shown below.

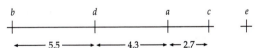

b. To get an a^+ b^+ c^+ d^+ e^+ fly, there must be a crossover between b and d, d and a, and a and c. ½ of the progeny from those crossovers are a^+ b^+ c^+ d^+, and the other ½ are a b c d. ½ of all progeny are e^+. The answer is 1.6×10^{-5}, that is, $0.055 \times 0.043 \times 0.0027 \times 0.5$ (for the half of the progeny produced by the triple crossover that have all the wild-type alleles) $\times 0.5$ (to give the proportion that are e^+).

5.38 First, use the chi-square test to evaluate the hypothesis that there is no relationship between chestnut coat color and "class." If, by using the chi-square test, we can reject this hypothesis, further investigation into the potential linkage of the coat color gene and a class gene would be warranted. In the statement of the problem, two assumptions concerning the mates of Sharpen Up are given. In the initial chi-square test proposed here, these are the major assumptions. Any other assumptions that are needed follow from the hypothesis of the chi-square test. However, if one begins to evaluate linkage of the chestnut gene and a class gene, a number of further assumptions will be needed.

Given the hypothesis that there is no relationship between class and chestnut coat color, we can assume further that the likelihood of obtaining a "classy" horse in a mating with Sharpen Up is uniform with regard to coat color. Then, since 83 classy horses were produced from a total of 367 + 260 = 627 progeny, the chance of obtaining a classy horse from a mating with Sharpen Up is 83/627 = 0.1324, independent of coat color. To perform the chi-square test, we need to determine the expected numbers of classy chestnut and classy bay horses and compare this to the observed number of classy chestnut and classy bay horses.

The number of chestnut horses expected in the progeny of Sharpen Up can be determined using assumptions 1 and 2 as stated in the problem. Under assumption 1, three types of crosses occurred equally frequently: homozygous bay × chestnut, heterozygous bay/chestnut × chestnut, and chestnut × chestnut. Chestnut is recessive to bay, so a homozygous bay × chestnut cross will produce no chestnut progeny, a heterozygous bay × chestnut cross will produce 50 percent chestnut progeny, and a chestnut × chestnut cross will produce 100 percent chestnut progeny. Thus, if each of these crosses occurred equally frequently, 50 percent, or 0.50 × 627 = 314, of the progeny are expected to be chestnut under assumption 1. (In this answer, fractions of a horse will be rounded up or down to the nearest "whole" horse.) The remaining 627 − 314 = 313 progeny were bay.

We have calculated (using the assumption that the frequency of classy horses is uniform) that the frequency of classy horses is 13.24 percent. Then, under assumption 1, we would expect 314 × 0.1324 = 42 chestnut classy progeny and 41 bay classy progeny. The observed numbers of classy horses were 45 chestnut and 38 bay. For these values, $\chi^2 = ([45 − 42]^2/42 + [38 − 41]^2/41) = 0.43$, df = 1, $0.50 < P < 0.70$. Under assumption 1, then, we would accept as being possible the hypothesis that chestnut coat color and class are unrelated.

Under assumption 2, two types of crosses occurred equally frequently: heterozygous bay/chestnut × chestnut and chestnut × chestnut. The heterozygous bay/chestnut × chestnut cross will produce 50 percent chestnut progeny, and the chestnut × chestnut cross will produce 100 percent chestnut progeny. Thus, if both types of crosses occur equally frequently, 75 percent, or 0.75 × 627 = 470, of the progeny are expected to be chestnut under assumption 2. The remaining 627 − 470 = 157 progeny are bay.

If the frequency of classy horses is 13.24 percent, then under assumption 2, we would expect 470 × 0.1324 = 62 chestnut classy progeny and 21 bay classy progeny. The observed numbers of classy horses were 45 chestnut and 38 bay. For these values, $\chi^2 = ([45 − 62]^2/62 + [38 − 21]^2/21) = 18.4$, df = 1, $P < 0.001$. Under assumption 2, then, we would reject as being unlikely the hypothesis that the chestnut coat color gene and class are unrelated. It would be reasonable to consider the hypothesis that a gene closely linked to chestnut might contribute to class.

Notice that the evidence for a relationship between chestnut coat and class hinges on knowing what alleles at the chestnut/bay gene were present in the mares bred to Sharpen Up. This information is available (although not in this problem). Additional assumptions that might be required to test specifically for linkage to a class gene are assumptions about the number of alleles in the population of horses, the dominance relationship(s) between them, and whether they are in repulsion or in coupling with the chestnut allele.

CHAPTER 6

6.4

6.6

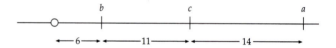

6.9 The two genes can be considered to assort independently because the number of parental ditype (20) is approximately equal to the number of nonparental ditype (18).

6.12 a. From applying the tetrad analysis formula,

$$\text{map distance} = \frac{\frac{1}{2}T + \text{NPD}}{\text{Total}} \times 100\%$$

the *a-b* distance is 19.6 map units, the *b-c* distance is 11 map units, and the *a-c* distance is 14 map units. Thus, the gene order is *a-c-b*.

b. Centromere distances are calculated from the formula (% second division segregation tetrads)/2.

Because we have all gene-gene distances, all we need to know is whether *a* or *b* is nearer the centromere. *a* is 20.7 map units from the centromere, and *b* is 6 mu. Thus, the map looks like this:

```
            b              c              a
   o--------|--------------|--------------|-----
     ←6→   ←---11---→   ←----14----→
```

6.13 Map distance is ½ × % of second division segregation asci, which in this case is 5 map units.

6.15 The cross is w/Y (white male) × w^{ch}/w^{ch} (cherry female), so the female progeny are heterozygous at the white gene and are w/wch. When a single mitotic crossover occurs between a gene and its centromere in a heterozygote, two daughter cells that are each homozygous for a different allele can be produced (see Figure 6.10a, p. 181). If mitotic recombination occurred in a w/wch individual during the development of the eye, and crossovers occurred between the white gene and the centromere (which is most of the X chromosome), w/w and w^{ch}/wch cells would be produced in a w/wch background.

The alleles at the white gene control the amount of pigment deposited in cells of the eye and show partial dominance (see Table 4.2, p. 102). In phenotypically white w/w homozygotes, 0.44 percent of wild-type pigment levels are found; in cherry-colored w^{ch}/wch homozygotes, 4.1 percent of wild-type pigment levels are found. It is reasonable to expect that w/wch heterozygotes will have pink eyes and have about 2 percent of wild-type pigment levels. If mitotic recombination occurs, twin spots will be seen: in a pink background (w/w^{ch}), there will be neighboring white (w/w) and cherry (w^{ch}/w^{ch}) twin spots. The cells in the spots have descended from a cell in which there was a mitotic recombination event.

CHAPTER 7

7.1 a. pericentric inversion (DₒE F inverted)

b. nonreciprocal translocation (BC moved to other arm)

c. tandem duplication (EF duplicated)

d. reverse tandem duplication (EF duplicated)

e. deletion (C deleted)

7.2 See text, pp. 197–199.

7.5 a.

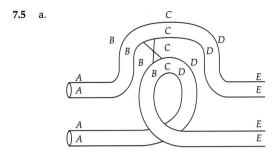

b. From part a, the crossover between B and C results in the following:

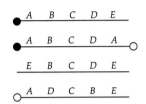

c. Paracentric inversion, because the centromere is not included in the inverted DNA segment.

7.8 a → c → e → d
　　　　　　　↘ b

That is, the following sequence occurred:
a.　A B C D E F G H I
↓　　　　⌐————⌐ Inversion
c.　A B F E D C G H I
↓　　　　　⌐————⌐ Inversion
e.　A B F E H G C D I
d and b are derived from e by separate inversion events:
e.　A B F E H G C D I
↓　　　⌐————⌐ Inversion
d.　A B F C G H E D I
and
e.　A B F E H G C D I
↓　⌐————⌐ Inversion
b.　H E F B A G C D I

7.9 a. Parents of Rec(8) individuals are heterozygous for a pericentric inversion with breakpoints at 8p23.1 and 8q22.1, so their 8q-duplication and 8p-deletion most likely arose from a single crossover within the pericentric inversion. Such an event is diagrammed in Figure 7.11 (p. 200).

b. As shown in Figure 7.11 (p. 200), if a single crossover occurs between two nonsister chromatids in an inversion heterozygote, two of the products will have the noncrossover chromosomes (one with the normal-ordered chromosome and one with the inversion chromosome), and two of the products will be duplication/deletion products. Here, one of these products (that having a duplication for 8q22.1 to 8q-ter and a deletion for 8p23.1 to 8p-ter) will contribute to a viable zygote with Rec(8) syndrome. The other product (that having a deletion for 8p22.1 to 8p-ter and a duplication for 8p23.1 to 8p-ter) is not mentioned in the problem. Perhaps if it is contributed to a zygote, the zygote will not survive. In this case, $\frac{1}{3}$ of the surviving zygotes will have Rec(8) syndrome, the remaining $\frac{2}{3}$ will be normal, and $\frac{1}{2}$ of the normal zygotes will have the chromosome 8 inversion.

c. The phenotypes of Rec(8) individuals could vary for one or a combination of reasons. (1) There could be variation in the precise location of the inversion breakpoints in the parents of Rec(8) individuals, leading to differences in the genes that are duplicated and deleted in Rec(8) individuals. That is, there may be in the population several different chromosome 8 inversions that have breakpoints in the vicinity of 8q22.1 and 8p23.1. When single crossovers occur in inversion heterozygotes, the progeny would have different degrees of aneuploidy. This could result in different symptoms. In addition, slight variation in the precise location of the breakpoints could lead to gene activation or gene inactivation in different regions of chromosome 8. (2) The genetic background in which the duplications and deletions are found could vary. That is, the addition or subtraction of a subset of genes could have different effects, depending on the genetic interactions between the genes present in three copies or one copy and other genes in the genome. For example, if there were genetic interactions between a key duplicated/deleted gene and another gene in the genome, allelic differences in either gene could strongly affect the phenotype seen. In this case, genes inherited from the father that are different from those inherited from the mother and grandmother could contribute to the phenotype. (3) There could be environmental effects that exacerbate the effects of the deleted and duplicated region. These effects are not uniform in the population of Rec(8) individuals, so these could contribute to the observed phenotypic variability. Environment is used in a broad sense here to include the conditions present during fetal development because many of the symptoms associated with Rec(8) syndrome are developmental abnormalities. (4) There may be other, cytologically invisible mutations associated with the Rec(8) individuals. The presence or absence of additional mutations could strongly affect the phenotype.

d. The child has the chromosome 8 inversion, but not the duplication/deletion chromosome that results from a single crossover in an inversion heterozygote, so she is an atypical Rec(8) individual. There are several explanations for why some of her symptoms overlap with those of Rec(8) syndrome. She may have an additional mutation near one of the Rec(8) breakpoints, in a region that is duplicated or deleted in Rec(8) syndrome, or in a gene that interacts with genes in the duplicated or deleted regions. Alternatively, the inversion may disrupt the function of a gene or genes at one or both breakpoints, and the inversion may normally be an asymptomatic condition. In this case, the inversion chromosome (in her mother and grandmother) would bear a recessive mutation. If she had a new allelic mutation, or her paternally contributed chromosome had an allelic mutation, she would be affected. This could also explain why she has only some of the symptoms of Rec(8) syndrome; she would have fewer genes affected than most Rec(8) individuals.

Small deletions would be cytologically invisible, as would point mutations. Thus, the explanations given could not be evaluated by karyotype analysis. FISH, STS, and/or DNA sequence analyses (see Chapter 6, pp. 184–185) could be used to evaluate the integrity of the chromosomal regions near the breakpoints.

7.10 a. Paracentric inversion.

b. Single crossover within the inversion loop:

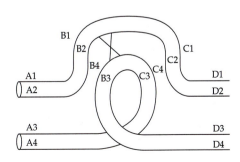

c. Four-strand double crossover within the inversion loop:

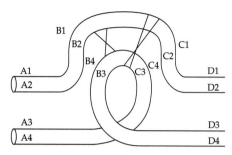

7.12 a. Mr. Lambert is heterozygous for a pericentric inversion of chromosome 6. One of the breakpoints is within the fourth light band up from the centromere, and the other is in the sixth dark band below the centromere. Mrs. Lambert's chromosomes are normal.

b. When Mr. Lambert's chromosome 6s paired, they formed an inversion loop that included the centromere. Crossing over occurred within the loop and gave rise to the partially duplicated, partially deficient 6, which the child received.

c. The child's abnormalities stem from having three copies of some, and only one copy of other, chromosome 6 regions. The top part of the short arm is duplicated, and there is a deficiency of the distal part of the long arm in this case.

d. The inversion appears to cover more than half of the length of chromosome 6, so crossing over will occur in this region in the majority of meioses. In the minority of meioses where crossing over has occurred outside the loop or where it has occurred within the loop but the child receives an uncrossed-over chromatid, the child can be normal. There is significant risk for abnormality, so monitoring of fetal chromosomes should be done.

7.15

```
        F  |  F
        E  |  E
        D  |  D
A B C ___|  |___ O N M
```

```
A B C ___|  |___ O N M
        P  |  P
        Q  |  Q
        R  |  R
```

7.17 a. Mr. Denton has normal chromosomes. Mrs. Denton is heterozygous for a balanced reciprocal translocation between chromosomes 6 and 12. Most of the short arm of chromosome 6 has been reciprocally translocated onto the long arm of chromosome 12. The breakpoints appear to be in the thick dark band just above the centromere of 6 and in the third dark band below the centromere of 12.

b. The child received a normal 6 and a normal 12 from his father. In meiosis in Mrs. Denton, the 6s and 12s paired and formed a cross-like figure in prophase I, segregation of adjacent nonhomologous centromeres to the same pole occurred, and the child received a gamete containing a normal 6 and one of the translocation chromosomes. (See Figure 7.13.)

c. The child received a normal chromosome 6 and a normal chromosome 12 from Mr. Denton. The child is not phenotypically normal because of partial trisomy (it has three copies of part of the short arm of chromosome 6) and partial monosomy (it has only one copy of most of the long arm of chromosome 12).

d. Given that segregation of adjacent homologous centromeres to the same pole will be relatively rare in this case, there should be about a 50% chance that a given conception will be chromosomally unbalanced. Of the 50% balanced ones, 50% would be translocation heterozygotes.

e. Prenatal monitoring of fetal chromosomes could be done, followed by therapeutic abortion of chromosomally unbalanced fetuses.

7.21 a. 45
  b. 47
  c. 23
  d. 69
  e. 48

7.22 b. An individual with three instead of two chromosomes is said to be *trisomic*.

7.23 a. Mother. The color-blind Turner syndrome child must have received her X chromosome from her father because the father carries the only color-blind allele. Therefore, the mother must have produced an egg with no X chromosomes.

b. Father. Because the Turner syndrome child does not have color blindness, she must have received her X chromosome from the homozygous normal mother. Therefore, the father must have produced a sperm with no X or Y chromosome.

7.24 This problem considers what happens when a chromosome is lost at the very first mitotic division (and only at that division).

a. The cross is $y/y +/+$ (female) $\times +/Y\ pal/pal$ (male), with progeny $y/+\ pal/+$ (daughters) and $y/Y\ pal/+$ (sons). The paternally contributed X is found only in the $y/+\ pal/+$ daughter, so we need to consider only the consequence of its loss in daughters. If a paternally contributed X chromosome (+) is lost during the first mitotic division in a $y/+\ pal/+$ zygote, one daughter cell will lose an X chromosome and be $y\ pal/+$. The other daughter cell will have two X chromosomes and be $y/+\ pal/+$. The cell with two X chromosomes would be female (XX) and produce nonyellow cells ($y/+$), and the cell with one X chromosome would be male (XO) and produce yellow cells (y). The animal will be a mosaic with cells of two sex types that are marked by yellow (male) or grey (female) cuticle.

b. The cross is $+/+\ eye/eye$ (female) $\times pal/pal +/+$ (male), with progeny $pal/+\ eye/+$ (daughters and sons). The paternally contributed fourth chromosome is +. If it is lost during the first mitotic division in a $pal/+\ eye/+$ zygote, one daughter cell will lose a fourth chromosome and be $pal/+\ eye$. The other daughter cell will have two fourth chromosomes and be normal. The animal will be a mosaic with some cells that are haploid for the fourth chromosome and some cells that are diploid for the fourth chromosome. If a patch of haplo-4 cells contributes to the formation of an eye during development, the eye will be reduced in size.

c. The cross is $+/+\ e/e$ (female) $\times pal/pal +/+$ (male), with progeny $pal/+\ e/+$. The paternally contributed third chromosome is +. If it is lost during the first mitotic division in a $pal/+\ e/+$ zygote, one daughter cell will lose a third chromosome and be $pal/+\ e$. This cell is inviable and so will not be recovered in the organism, should the organism survive. Consequently, if the organism survives, it will be phenotypically normal ($pal/+\ e/+$).

7.28 a. *AA aa*

b. If we label the four alleles *A1*, *A2*, *a1*, and *a2*, there are six possible gamete types: *A1 A2*, *A1 a1*, *A1 a2*, *A2 a1*, *A2 a2*, and *a1 a2*, i.e., $\frac{1}{6}$ *AA*, $\frac{4}{6}$ *Aa*, and $\frac{1}{6}$ *aa*. The possible gamete pairings are as follows:

	$\frac{1}{6}$ *AA*	$\frac{4}{6}$ *Aa*	$\frac{1}{6}$ *aa*
$\frac{1}{6}$ *AA*	$\frac{1}{36}$ *AAAA*	$\frac{4}{36}$ *AAAa*	$\frac{1}{36}$ *AAaa*
$\frac{4}{6}$ *Aa*	$\frac{4}{36}$ *AAaa*	$\frac{16}{36}$ *AAaa*	$\frac{4}{36}$ *Aaaa*
$\frac{1}{6}$ *aa*	$\frac{1}{36}$ *AAaa*	$\frac{4}{36}$ *Aaaa*	$\frac{1}{36}$ *aaaa*

Phenotypically this gives $\frac{35}{36}$ *A* : $\frac{1}{36}$ *a*

7.32 $(7N + 10N) \times 2 = 34$ (4N)

CHAPTER 8

8.1 Strain A is *thr leu+* and B is *thr+ leu*. Transformed B should be *thr+ leu+*, and its presence should be detected on a medium containing neither threonine nor leucine.

8.2 The whole chromosome would have to be transferred in order for the recipient to become a donor in an *Hfr × F-* cross; that is, the F factor in the *Hfr* strain is transferred to the *F-* cell last. This transfer takes approximately 100 min, and usually the conjugal unions break apart before then.

8.3

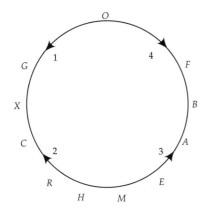

8.5 a. To select initially for *c+ str^R* recombinants, plate the progeny on minimal medium without compound C, but supplemented with streptomycin and compounds A, B, D, E, F, G, and H. To assess the complete genotype of the *c+ str^R* recombinants, replica plate them onto different minimal media supplemented with streptomycin and all but two of the compounds (compound C and one other). For example, to test if a *c+ str^R* colony was also *a+*, replica plate it onto a medium that lacks compound A but is supplemented with streptomycin and B, D, E, F, G, and H. If the colony is able to grow on this medium, it will be *a+ c+ str^R*. If it is unable to grow, it will be *a c+ str^R*.

b. Strain #1. No *c+* recombinants are ever obtained, so strain #1 is unable to transfer *c+*. This means it is (1) *F-*; (2) *Hfr*, but with the F factor inserted either far from *c+* or close to it but in an orientation so that genes are transferred in a direction opposite to *c+*; or (3) *F'*, but with *c+* in the bacterial chromosome. It should not be *F+* because then, at a very low frequency, some *c+* recombinants would be obtained.

Strain #2. Since *c+* recombinants are obtained at 6 min and *g+*, *h+*, *a+*, and *b+* recombinants are obtained in subsequent time intervals, strain #2 is *Hfr*. The genes are transferred in the following order: *c+*, *g+*, *h+*, *a+*, and *b+*. From the times of their transfer, the map position of the genes is origin (0) – *c+* (6) – *g+* (8) – *h+* (11) – *a+* (14) – *b+* (16). The location of genes *d+*, *e+*, and *f+* cannot be precisely determined. They are not transferred in an *Hfr × F-* cross, so they are either far away from the F factor insertion site or close to it but near the fertility genes, which are only rarely transferred by an *Hfr* strain. When the recombinants obtained from the strain #2 × F- mating at the 16-minute time period are crossed to an *amp^R F-* strain, *c+* is not transferred. If these recombinants cannot conjugate with *F-*, this indicates that although strain #2 is fertile, it did not transfer a complete F factor. It therefore must be *Hfr*.

Strain #3. Strain #3 transfers *c+* within one minute and *g+* by three minutes. From analysis of the strain #2 × F- cross, we knew that these genes are two minutes apart. The data support this conclusion. Since no other recombinants are obtained, no other genes are transferred. This suggests that strain #3 is *F'* and that the segment of DNA containing *c+* and *g+* is in the *F'* factor. If this is the case, the complete F factor will be transferred in a strain #3 × F- cross if the mating is allowed to proceed long enough. This is observed: *c+* recombinants

from the strain #3 × F- cross obtained at sixteen minutes are able to transfer *c+* to an *F- amp^R* strain. Therefore, strain #3 is *F'*.

c. Information known with certainty is shown in the following diagram. The locations of genes in strains #1 and #3 are inferred from crosses with strain #2. The locations of genes *d+*, *e+*, and *f+* are unknown.

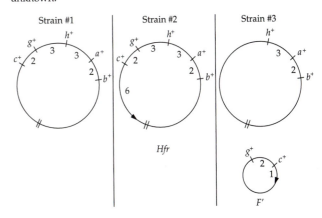

8.8 a. GT
b. ST
c. ST
d. GT
e. GT
f. ST
g. ST
h. B
i. N

8.10 The correct order is *aroA-cmlB-pyrD*.

8.12 Working from the principle that a relatively high cotransduction frequency means that the genes are relatively close, the order of the genes is *supD-cheA-cheB-eda*.

8.13 0.07 mu. The plaques produced on *K12(λ)* are wild-type *r+* phages, and in the undiluted lysate there are 470 × 5 per milliliter (because 0.2 mL was plated), or 2350/mL. The *r+* phages were generated by recombination between the two *rII* mutations. The other product of the recombination event is the double mutant, and it does not grow on *K12(λ)*. Therefore, the true number of recombinants in the population is actually equivalent to twice the number of *r+* phages because for every wild-type phage produced, there ought to be a doubly mutant recombinant produced. Thus, there are 4,700 recombinants/mL. The total number of phages in the lysate is 627 × (dilution factor) × (1 mL divided by the sample size plated) per milliliter, or 672 × 1,000 × 10 = 6,720,000/mL. The map distance between the mutations is (4,700/6,720,000) × 100% = 0.07% = 0.07 mu.

8.15 There are two answers that are compatible with the data:

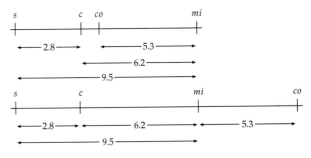

8.17 0.5% recombination (the numbers of plaques counted are so small that this value is a rather rough approximation).

8.19

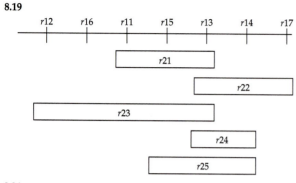

8.21

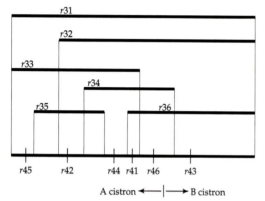

A cistron ◄────|────► B cistron

8.23 a. 3 genes

b. *A*, *D*, and *F* are in one; *B* and *G* are in the second; and *C* and *E* are in the third.

CHAPTER 9

9.2 The double homozygote should have PKU but not AKU. The PKU block should prevent most homogentisic acid from being formed, so that it could not accumulate to high levels.

9.4 a. The simplest approach is to calculate the proportion of the F_2 that are colored and to subtract that answer from one. In this case, the noncolorless are the brown or black progeny. The proportion of progeny that make at least the brown pigment is given by the probability of having the following genotype: $a^+/- b^+/- c^+/- (d^+/d^+, d^+/d,$ or $d/d)$. The answer is $\frac{3}{4} \times \frac{3}{4} \times \frac{3}{4} = \frac{27}{64}$. Therefore, the proportion of colorless is $1 - \frac{27}{64} = \frac{37}{64}$.

b. The brown progeny have the following genotype: $a^+/- b^+/- c^+/- d/d$. The probability of getting individuals with this genotype is $\frac{3}{4} \times \frac{3}{4} \times \frac{3}{4} \times \frac{1}{4} = \frac{27}{256}$.

9.8 a. Half have white eyes (the sons), and half have fire red eyes (the daughters).

b. All have fire red eyes.

c. All have brown eyes.

d. All are w^+; ¼ are $bw^+/- st^+/-$, red; ¼ are $bw^+/- st/st$, scarlet; ¼ are $bw/bw\ st^+/-$, brown; ¼ are $bw/bw\ st/st$, the color of 3-hydroxykynurenine plus the color of the precursor to biopterin, or colorless.

9.9 Wild-type T4 will produce progeny phages at all three temperatures. Let us suppose that model (1) is correct. If cells infected with the double mutant are first incubated at 17°C and then shifted to 42°C, progeny phages will be produced, and the cells will lyse. The explanation is as follows: the first step, *A* to *B*, is controlled by a gene whose product is heat-sensitive. At 17°C, the enzyme works, and *A* is converted to *B*, but *B* cannot then be converted to mature phages because that step is cold-sensitive. When the temperature is raised to 42°C, the *A*-to-*B* step is blocked, but the accumulated *B* can be converted to mature phages because the enzyme involved with

that step is cold-sensitive and functional at the high temperature. If model (2) is the correct pathway, then progeny phages should be produced in a 42°–to–17°C temperature shift, but not vice versa. In general, two gene product functions can be ordered by this method whenever one temperature shift allows phage production and the reciprocal shift does not, according to the following rules: (1) if a low-to-high temperature results in phages but a high-to-low temperature does not, the *hs* step precedes the *cs* step (model 1); (2) if a high-to-low temperature results in phages but a low-to-high temperature does not, the *cs* step precedes the *hs* step (model 2).

9.11 A mutant cell can take up a substance and grow on it only if that substance occurs in the metabolic pathway at a point after the mutant's own block. Since *trpE* can be fed by *C, D, F* and *A*, *E* is blocked earlier in the pathway than the others, *E* is also earlier than *B* because *B* can feed *F*. Thus, anthranilate synthetase is the first enzyme in the pathway. *F* can feed *C* and *C* can feed *D*. *A* and *B* both feed *F*, so the order of the steps is *E, D, C, F, [AB]*. After anthranilate synthetase, the enzymes are, in order, PRA transferase, PRA isomerase, IGP synthetase, and tryptophan synthetase.

9.13 a. *c d*⁺ and *c*⁺ *d*

b. The genes are not linked because parental ditype (PD) and nonparental ditype (NPD) tetrads occur in equal frequencies.

c. The pathway is Y to X to Z, with *d* blocking the synthesis of Y and *c* blocking the synthesis of X and Y.

9.16 A single *p*⁺ allele provides 50 percent of the enzyme activity seen in a *p*⁺/*p*⁺ homozygote. Since *p*⁺ is dominant (i.e., *P*– *CC* plants are purple), this appears to be enough activity to give a wild-type phenotype. If a plant with less than 50 percent of normal activity does not synthesize enough purple pigment for a wild-type phenotype (e.g., 25 percent of normal activity gives a light purple flower), and a plant with more than 100 percent of normal activity produces noticeably darker purple pigmentation, the following phenotypes should be seen.

Genotype	Percent of +/+ Activity	Percent of +/+ Activity When Mixed 50:50 with +/+ Extract	(A) Homozygote Phenotype	(B) Heterozygote Phenotype	(C) Hemizygote Phenotype	(D) Allele Classification
p^+/p^+	100	100	purple	purple	purple	wild type
p^1/p^1	20	60	light purple	purple	very light purple	hypomorph
p^2/p^2	0	50	white	purple	white	amorph
p^3/p^3	300	200	very dark purple	dark purple	dark purple	hypermorph
p^4/p^4	0	5	white	very light purple	white	antimorph
p^5/p^5	0ᵃ	50ᵇ	red	reddish purple	red	neomorph

ᵃProduces red, not purple, pigment.
ᵇProduces red and purple pigments.

9.18 Baby Joan, whose blood type is O, must belong to the Smith family. Mrs. Jones, being of blood type AB, could not have an O baby. Therefore, Baby Jane must be hers.

9.19 a. Normal parents have affected offspring, so the disease appears to be recessive. However, patients with 50 percent of GSS activity have a mild form of the disease; therefore, an individual who was heterozygous (mutant/+) for a complete loss-of-function GSS mutation might show mild symptoms. Thus, in a population, the disease may show variable penetrance. The expressivity of the disease in individuals will depend on the nature of their GSS mutation. The alleles discussed here appear to be recessive, but a complete loss-of-function allele may show partial dominance.

b. Patient 1, with 9 percent of normal GSS activity, has a more severe form of the disease. Patient 2, with 50 percent of normal GSS activity, has a less severe form of the disease. Thus, increased disease severity is associated with less GSS enzyme activity.

c. The two amino acid substitutions may disrupt different regions of the enzyme's structure (consider the effect of various amino acid substitutions on hemoglobin function that are discussed in the text, pp. 277–278). Amino acids vary in their polarity and charge, so different amino acid substitutions within the same structural region could have varying chemical effects on protein structure. This too could lead to disparate levels of enzymatic function. (For a discussion of the chemical dissimilarities between amino acids, see Chapter 14).

d. By analogy with the disease PKU discussed in the text, 5-oxoproline is produced only when a precursor to glutathione accumulates in large amounts due to a block in a biosynthetic pathway. When GSS levels are 9 percent of normal, this occurs. When GSS levels are 50 percent of normal, sufficient GSS enzyme is present to complete the pathway sufficiently to prevent high levels of 5-oxoproline.

e. The mutations are allelic, because both the severe and the mild forms of the disease are associated with alterations in the same polypeptide that is a component of the GSS enzyme. (Although the data in this problem suggest that the GSS enzyme is composed of a single polypeptide, they do not exclude the possibility that GSS has multiple polypeptide subunits.)

f. If GSS is normally found in fetal fibroblasts, one could, in principle, measure GSS activity in fibroblasts obtained via amniocentesis. The GSS enzyme level in cells from at-risk fetuses could be compared to that in normal control samples to predict disease due to inadequate GSS levels. Some variation in GSS level might be seen, depending on the allele(s) present. More than one mutation is present in the population, so it is important to devise a functional test that assesses GSS activity, rather than a test that identifies a single mutant allele based on the nature of its DNA mutation.

CHAPTER 10

10.4 d

10.5 a, b, and c. C, N and H are all present both in proteins and in DNA, so label would be located both within and on the surface of the host cell.

10.8 3' and 5' carbons are connected by a phosphodiester bond.

10.13 The evidence for only two base pair combinations in DNA, that is, A-T and G-C, comes from quantitative measurements of the four bases in double-stranded DNA isolated from a wide variety of organisms. In all cases, the amount of A was shown to equal the amount of T, and the amount of G was shown to equal the amount of C. Moreover, different DNAs exhibit different base ratios (stated as the %GC), so although A = T and G = C, in most organisms, (A + T) does not equal (G + C). The simplest hypothesis, therefore, is that there are two base pairs in DNA, A-T and G-C, and the proportion of the two base pairs varies from organism to organism.

10.15 a. 3' TCAATGGACTAGCAT 5'

b. 3'AAGAGTTCTTAAGGT 5'

10.16 The adenine-thymine base pair is held together by two hydrogen bonds, but the guanine-cytosine base pair is held together by three hydrogen bonds. Thus, the guanine-cytosine base pair requires more energy to break it and is the harder pair to break apart.

10.18 b, c, and d

10.21 C = 17% therefore G = 17% and the % GC = 34. Therefore, the % AT = 66 and the % A = 66/2 = 33.

10.30 Evidence that DNA is flexible comes from studies of DNA in solution and studies of DNA-protein interactions. In solution, DNA undergoes shape changes in response to ionic and temperature conditions: its flexibility is enhanced by the presence of certain base-

pair sequences. When bound by certain proteins, DNA has been shown to bend. Such proteins include restriction enzymes that cleave DNA at specific base sequences, enzymes that mediate DNA recombination, and proteins that bind to regulatory DNA sequences to control gene expression (transcription).

Based on the kinds of evidence that shows DNA is a flexible molecule, its flexibility appears to be important for its packaging in chromosomes, for the completion of enzymatic reactions in which DNA is a substrate (e.g., recombination), and for expression of the genetic information that it holds.

10.32 a. 200,000

b. 10,000

c. 3.4×10^4 nm

10.34 The probability any group of four bases would be GUUA is $(0.3)(0.25)(0.25)(0.2) = 0.00375$. A molecule 10^6 nucleotides long contains very nearly 10^6 groups of four bases. (The first group of four is bases 1, 2, 3, and 4, the second group of four is bases 2, 3, 4, and 5, etc.) Thus, the number of occurrences of GUUA should be about $(0.00375)(10^6) = 3,750$.

CHAPTER 11

11.1 a.

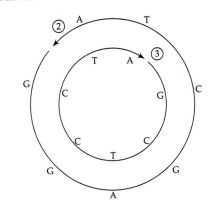

b.

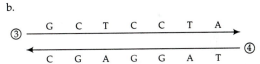

11.3 a. T4 chromosomes are circularly permuted, terminally redundant molecules. Upon denaturation and renaturation, there would be a few perfectly matched, double-stranded molecules, but most molecules would have single-stranded ends. Molecules with very long single-stranded ends might form complex, partially circularized structures.

b. T7 chromosomes have a unique sequence that is the same for all molecules. Therefore, all molecules would be perfectly matched.

c. λ chromosomes are linear; each chromosome has the same sequence, and there are single-stranded "sticky ends" that permit the molecules to form circles. Therefore, all molecules would have single-stranded ends or be circularized (to nicked circles).

11.5 a. $\frac{9}{24} = 0.375$. Since S is 37.5% of the cycle, 37.5% of cells should be in S at any instant, and these are the ones that would incorporate label.

b. A little over 4 h. Such cells would have to take up some ^{3}H (a few minutes), pass through G_2 (4 h), and pass through mitotic prophase (several minutes).

c. Both. Each chromatid is a double-stranded DNA molecule containing one old and one newly synthesized DNA strand.

d. A little more than 13 h. Such cells would have to pass through nearly all of S, all of G_2, and through prophase.

11.9

DNA TYPE	CHROMATIN TYPE
Barr body (inactivated DNA)	Facultative heterochromatin
Centromere	Constitutive heterochromatin
Telomere	Constitutive heterochromatin
Most expressed genes	Euchromatin

11.13 a. D, G, J

b. A, E, F, H

c. I

11.16 a. The belt forms a right-handed helix. Although you wrapped the belt around the can axis in a counterclockwise direction from your orientation (looking down at the can), the belt was winding up and around the side of the can in a clockwise direction from its orientation. While the belt is wrapped around the can, curve the fingers of your right hand over the belt and use your index finger to trace the direction of the belt's spiral. Your right index finger will trace the spiral upward, the same direction your thumb points when you wrap your hand around the can. Therefore, the belt has formed a right-handed helix.

b. Three

c. Three. The number of helical turns is unchanged, although the twist in the belt is.

d. The belt appears more twisted because the pitch of the helix was altered, and the edges of the belt (positioned much like the complementary base pairs of a double helix) are twisted more tightly.

e. When it is twisted around the can, the length of the belt decreases by about 70–80 percent, depending on the initial length of the belt and the diameter of the can.

f. Yes. As the DNA of linear chromosomes is wrapped around histones to form the 10-nm nucleofilament, it becomes supercoiled. In much the same manner as you must add twists to the belt in order for it to lie flat on the surface of the can, supercoils must be introduced into the DNA for it to wrap around the histones.

g. Topoisomerases increase or reduce the level of negative supercoiling in DNA. In order for linear DNA to be packaged, negative supercoils must be added to it.

11.8 See pp. 333–335. Each of the known *Saccharomyces cerevisiae* centromeres has a common core centromere region, which consists of three sequence domains. All three domains are important for centromere function. The centromeres of other eukaryotic organisms that have been characterized are quite different from the *Saccharomyces* centromeres.

11.23 a. These findings support the current view that telomeres are specialized chromosome structures with two distinct structural components: simple telomeric sequences and telomere-associated sequences. They show that functional genes do not reside in the telomeric region, consistent with the view that telomeres are heterochromatic and have special, protective functions in chromosomes. They also add significantly to our knowledge of the structure of telomeric and near-telomeric regions. For example, they document the considerable distance over which the telomere-associated sequences are found, about 36 kb, and give a sense of the number, size, and density of genes in the region near this telomere.

b. If the rearrangement is broken in the α-globin gene, it could alter the protein produced, resulting in a mutation. A rearrangement near the gene, but not in it, might result in the region becoming heterochromatic and so silence the gene (much like the heterochromatization of the X chromosome into a Barr body silences genes on the X).

c. At least in this region, Alu sequences are apparently found more often in AT-rich areas. These areas do not seem to be as gene rich as adjacent GC-rich areas. Thus, this class of moderately repetitive sequences, as well as the genes in this area, appear to have a nonrandom distribution.

CHAPTER 12

12.3 Key: $^{15}N - ^{15}N$ DNA = HH; $^{15}N - ^{14}N$ DNA = HL; $^{14}N - ^{14}N$ DNA = LL

a. Generation 1: all HL; 2: ½ HL, ½ LL; 3: ¼ HL, ¾ LL; 4: ⅛ HL, ⅞ LL; 6: ¹⁄₃₂ HL, ³¹⁄₃₂ LL; 8: ¹⁄₁₂₈ HL, ¹²⁷⁄₁₂₈ LL

b. Generation 1: ½ HH, ½ LL; 2: ¼ HH, ¾ LL; 3: ⅛ HH, ⅞ LL; 4: ¹⁄₁₆ HH, ¹⁵⁄₁₆ LL; 6: ¹⁄₆₄ HH, ⁶³⁄₆₄ LL; 8: ¹⁄₂₅₆ HH, ²⁵⁵⁄₂₅₆ LL

12.6 a. That DNA replication is semiconservative does not *a priori* require DNA replication to be semidiscontinuous. For example, if both of the two old strands were completely unwound and replication were initiated from the 3' end of each, it could proceed continuously in a 5'-to-3' direction along both strands. Alternatively, if DNA polymerase were able to synthesize DNA in both 3'-to-5' and 5'-to-3' directions, DNA replication could proceed continuously on both DNA strands.

b. That DNA replication is semidiscontinuous does ensure that it is semiconservative. In the semidiscontinuous model, each old, separated strand serves as a template for a new strand. This is the essence of the semiconservative model.

c. That DNA polymerase synthesizes just one new strand from each "old" single-stranded template and can synthesize in only one direction (5'-to-3') ensures that replication is semiconservative.

12.7 See text, pp. 356–359. DNA polymerase; intact, high-molecular-weight DNA; dATP, dGTP, dTTP, and dCTP; and magnesium ions are needed.

12.8 Two principal lines of evidence show that the Kornberg enzyme is not the enzyme involved in *E. coli* chromosome duplication *in vivo*: First, a mutant that is deficient in the Kornberg enzyme nonetheless grows, replicates the DNA, and divides. Second, other mutants that are temperature-sensitive for DNA synthesis and do not replicate the DNA or divide have Kornberg enzyme activity at the nonpermissive temperature.

12.13 A primer strand is a nucleic acid sequence that is extended by DNA synthesis activities. A template strand directs the base sequence of the DNA strand being made; for example, an A on the template causes a T to be inserted on the new chain, and so on.

12.14 a. One base pair is 0.34 nm, and the chromosome is 1,100 μm. Thus, the number of base pairs is $(1,100/0.34) \times 1,000 = 3.24 \times 10^6$.

b. There are ten base pairs per turn in a normal DNA double helix; therefore, it has a total of 3.24×10^5 turns.

c. 3.24×10^5 turns and 60 min for unidirectional synthesis; therefore $(3.24 \times 10^5/60$ turns per minute $= 5,400$ revolutions per minute.

d.

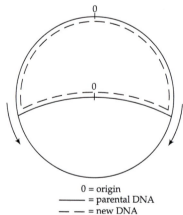

0 = origin
——— = parental DNA
- - - = new DNA

12.19 See text, pp. 354–357.

12.24 See text, pp. 348–351 and 365–366 and Tables 12.1 and 12.4.

12.27 Assuming these cells have the typical 4 h G_2 period, you would add ^3H thymidine to the medium, wait 4.5 h, and prepare a slide of metaphase chromosomes. Autoradiography would then be done on this chromosome preparation. Regions of chromosomes displaying silver grains are then the late-replicating regions. Cells at earlier stages in the S period when they began to take up ^{3}H will not have had sufficient time to reach metaphase.

12.32 See text, pp. 372–373. Hybrid DNA resulting from branch migration may contain mismatched sequences. Mismatched sequences are recognized by special enzymes that remove a short segment of one DNA strand and then fill in the gap. The mismatch repair does not recognize which is the correct sequence.

12.34 In each case, there is evidence that the segregation of one of the alleles in the tetrad has resulted from gene conversion caused by mismatch repair of heteroduplex DNA.

For the *a1 a2+* × *a1+ a2* cross, the tetrad shows 2:2 segregation of *a1+*:*a1* but 3:1 segregation of *a2+*:*a2*, indicating gene conversion of an *a2* allele to its wild-type counterpart. Similarly, the *a1 a3+* × *a1+ a3* cross shows 2:2 segregation of *a3+*:*a3* and 3:1 segregation of *a1+*:*a1* resulting from gene conversion of an *a1* allele to *a1+*. In the *a2 a3+* × *a2+ a3* cross, the *a2* allele segregates in a Mendelian fashion while the *a3* allele segregates 3:1 *a3+*:*a3* as a result of gene conversion of one *a3* allele to *a3+*.

Chapter 13

13.1 The DNA contains deoxyribose and thymine, whereas RNA contains ribose and uracil, respectively. Also, DNA is usually double-stranded, but RNA is usually single-stranded.

13.5 See text, pp. 383–388, 390–391, 401–408, 410–412.

13.7 RNA polymerase I transcribes the major rRNA genes that code for 18S, 5.8S, and 28S rRNAs; RNA polymerase II transcribes the protein-coding genes to produce mRNA molecules; and RNA polymerase III transcribes the 5S rRNA genes, the tRNA genes, and the genes for small nuclear RNA (snRNA) molecules.

In the cell, the 18S, 5.8S, 28S, and 5S rRNAs are structural components of ribosomes. The mRNAs are translated to produce proteins, the tRNAs bring amino acids to the ribosome to donate to the growing polypeptide chain during protein synthesis, and at least some of the snRNAs are involved in RNA processing events.

13.10 An enhancer element is defined as a DNA sequence that somehow, without regard to its position relative to the gene or its orientation in the DNA, increases the amount of DNA synthesized from the gene it controls (discussed in the chapter on pp. 389–390).

13.13 The three classes of RNA are mRNA (p. 391–394), tRNA (pp. 408–409), and rRNA (pp. 400–401).

13.18 For mRNA: 5' capping and 3' polyadenylation in eukaryotes. For tRNA: modification of a number of the bases, removal of 5' leader and 3' trailer sequences (if present), and addition of a three-nucleotide sequence (5'-CCA-3') at the 3' end of the tRNA. For rRNA in prokaryotes a pre-rRNA molecule is processed to the mature 16S, 23S, and 5S rRNAs. In eukaryotes, a precursor molecule is processed to the 18S, 5.8S, and 28S rRNA molecules; 5S rRNA is made elsewhere. The large rRNAs in both types of organisms are methylated. Additionally, the same rRNAs in eukaryotes are pseudouridylated.

13.20 a. This would prevent splicing out of any sequences corresponding to introns and would result in abnormal sequences of most proteins. This should be lethal.

b. If splicing is not affected, this should have no phenotypic consequence.

c. This would prevent the splicing out of intron 2 material. Since the 5' cut could still be made, it is likely this mutation would lead to the absence of functional mRNA and the absence of β-globin. (This should produce a phenotype similar to that seen when the β-globin gene is deleted. In that case, there is anemia compensated by increased production of fetal hemoglobin. The condition is called β-thalassemia.)

13.23 a. This would be possible because the 5S genes occur in clusters apart from the other rRNA genes.

b. This would not be possible because the 18S gene copies are interspersed among 5.8S and 28S genes.

c. This would be possible. These three occur in tandem arrays.

d. This would not be possible because the 5S genes are located separately within the chromosomes.

13.25 a. 3

b. 1, 2, 3, 4

c. 3

d. 1, 2

e. 1

f. 4

g. 3

h. 3

i. a

Chapter 14

14.1 b. mRNA

14.2 c. Linear, folded chains of amino acids

14.11 a. 4 A : 6 C

AAA = $(^4/_{10})(^4/_{10})(^4/_{10})$ = 0.064, or 6.4% Lys
AAC = $(^4/_{10})(^4/_{10})(^6/_{10})$ = 0.096, or 9.6% Asn
ACA = $(^4/_{10})(^6/_{10})(^4/_{10})$ = 0.096, or 9.6% Thr
CAA = $(^6/_{10})(^4/_{10})(^4/_{10})$ = 0.096, or 9.6% Gln
CCC = $(^6/_{10})(^6/_{10})(^6/_{10})$ = 0.216, or 21.6% Pro
CCA = $(^6/_{10})(^6/_{10})(^4/_{10})$ = 0.144, or 14.4% Pro
CAC = $(^6/_{10})(^4/_{10})(^6/_{10})$ = 0.144, or 14.4% His
ACC = $(^4/_{10})(^6/_{10})(^6/_{10})$ = 0.144, or 14.4% Thr

In sum, 6.4% Lys, 9.6% Asn, 9.6% Gln, 36.0% Pro, 24.0% Thr, and 14.4% His.

b. 4 G : 1 C

GGG = $(^4/_5)(^4/_5)(^4/_5)$ = 0.512, or 51.2% Gly
GGC = $(^4/_5)(^4/_5)(^1/_5)$ = 0.128, or 12.8% Gly
CCG = $(^4/_5)(^1/_5)(^4/_5)$ = 0.128, or 12.8% Ala
CGG = $(^1/_5)(^4/_5)(^4/_5)$ = 0.128, or 12.8% Arg
CCC = $(^1/_5)(^1/_5)(^1/_5)$ = 0.008, or 0.8% Pro
CCG = $(^1/_5)(^1/_5)(^4/_5)$ = 0.032, or 3.2% Arg
GCC = $(^4/_5)(^1/_5)(^1/_5)$ = 0.032, or 3.2% Ala

In sum, 64.0% Gly, 16.0% Ala, 16.0% Arg, and 4.0% Pro.

c. 1 A : 3 U : 1 C; the same logic is followed here, using $^1/_5$ as the fraction for A

AAA = 0.008, or 0.8% Lys
AAU = 0.024, or 2.4% Asn
AUA = 0.024, or 2.4% Ile
UAA = 0.024, or 2.4% Chain terminating
AUU = 0.072, or 7.2% Ile
UAU = 0.072, or 7.2% Tyr
UUA = 0.072, or 7.2% Leu
UUU = 0.216, or 21.6% Phe
AAC = 0.008, or 0.8% Asn
ACA = 0.008, or 0.8% Thr
CAA = 0.008, or 0.8% Gln
ACC = 0.008, or 0.8% Thr
CAC = 0.008, or 0.8% His
CCA = 0.008, or 0.8% Pro
CCC = 0.008, or 0.8% Pro
UUC = 0.072, or 7.2% Phe
UCU = 0.072, or 7.2% Ser
CUU = 0.072, or 7.2% Leu
UCC = 0.024, or 2.4% Ser
CUC = 0.024, or 2.4% Leu
CCU = 0.024, or 2.4% Pro

UCA = 0.024, or 2.4% Ser
UAC = 0.024, or 2.4% Tyr
CUA = 0.024, or 2.4% Leu
CAU = 0.024, or 2.4% His
AUC = 0.024, or 2.4% Ile
ACU = 0.024, or 2.4% Thr

In sum, 0.8% Lys, 3.2% Asn, 12.0% Ile, 2.4% Chain terminating, 9.6% Tyr, 19.2% Leu, 28.8% Phe, 4.0% Thr, 0.8% Gln, 3.2% His, 4.0% Pro, and 12.0% Ser. The likelihood is that the chain would not be long because of the chance of the chain-terminating codon.

   d. 1 A : 1 U : 1 G : 1 C; all 64 codons will be generated. The probability of each codon is $\frac{1}{64}$, so there is a $\frac{3}{64}$ chance of the codon being a chain-terminating codon. With those exceptions, the relative proportion of amino acid incorporation is directly dependent on the codon degeneracy for each amino acid, and that can be determined by inspecting the code word dictionary

14.12

Word Size	Number of combinations
a. 5	$2^5 = 32$
b. 3	$3^3 = 27$
c. 2	$5^2 = 25$

(The minimum word sizes must uniquely designate twenty amino acids.)

14.14 a. AAA ATA AAA ATA etc.
   b. TTT TAT TTT TAT etc.
   c. AAA for Phe and AUA for Tyr

14.16 No chain-terminating codons can be produced from only As and Gs. But the stop codon UAA can be made from As and Us. Therefore, the population A proteins will be longer than those from population B. Most of the population B proteins will be free in solution rather than attached to ribosomes.

14.18 Met-Val-Ser-Ser-Pro-Ile-Gly-Ala-Ala-Ile-Ser . . . (In fact, either of the Ile residues might be replaced by Met in a particular molecule.) The normal tRNA recognizes the AUC codon and therefore carries Ile. The mutant tRNA will recognize the codon AUG and insert Ile there (although the normal tRNA-Met will compete for these sites). The N terminal Met will not be replaced by Ile because it requires the special tRNA-fMet for initiation to occur.

14.20 Normal: Met-Phe-Ser-Asn-Tyr- . . . -Met-Gly-Trp-Val.
   Mutant *a*: Met-Phe-Ser-Asn
   Mutant *b*: Starts at later AUG to give: Met-Gly-Trp-Val
   Mutant *c*: Met-Phe-Ser-Asn-Tyr- . . . -Met-Gly-Trp-Val
   Mutant *d*: Met-Phe-Ser-Lys-Tyr- . . . -Met-Gly-Trp-Val
   Mutant *e*: Met-Phe-Ser-Asn-Ser- . . . -Trp-Gly-Gly-Trp . . .
(no stop codon; protein continues)
   Mutant *f*: Met-Phe-Ser-Asn-Tyr- . . . -Met-Gly-Trp-Val-Trp . . .
(no stop codon; protein continues)

14.24 a. The pre-mRNA must be substantially processed by RNA splicing (removal of introns) and polyadenylation to a smaller mature mRNA. (As is discussed in Chapter 15, this gene has twenty-four exons!)

   b. A 1,480 amino acid protein requires $1,480 \times 3 = 4,440$ bases of protein-coding sequence. This leaves $6,500 - 4,440 = 2,060$ bases of 5' untranslated leader and 3' untranslated trailer sequence in the mature mRNA, about 32 percent.

   c. The ΔF508 mutation could be caused by a DNA deletion for the three base pairs encoding the mRNA codon for phenylalanine. This codon is UUY (Y = U or C), and the DNA sequence of the non-template strand is 5'-TTY-3'. The segment of DNA containing these bases could be deleted in the appropriate region of the gene.

   d. If positioned at random and solely within the coding region of a gene (i.e., not in 3' or 5' untranslated sequences or in intronic sequences), a deletion of three base pairs results either in an mRNA missing a single codon or an mRNA missing bases from two adjacent codons. If three of the six bases from two adjacent codons were

deleted, the remaining three bases would form a single codon. In this case, an incorrect amino acid might be inserted into the polypeptide at the site of the left codon, and the amino acid encoded by the right codon would be deleted. If the 3' base of the left codon were deleted, it would be replaced by the 3' base of the right codon. The code is degenerate and wobble occurs in the 3' base, so this type of deletion might not alter the amino acid specified by the left codon. The adjacent amino acid would still be deleted, however.

CHAPTER 15

15.1 At any one position in the DNA, there are four possibilities for the base pair: A-T, T-A, G-C, and C-G. Therefore, the length of the base-pair sequence that the enzyme recognizes is given by the power to which four must be raised to equal (or approximately equal) the average size of the DNA fragment produced by enzyme digestion. The answer in this case is 6; that is, 4 to the power of 6 = 4,096. If the enzyme instead recognized a four base-pair sequence, the average size of the DNA fragment would be 4 to the power of 4 = 256.

15.7 Genomic libraries were discussed on pp. 458–460, and cloning in a lambda vector by replacing a central section of the λ chromosome with foreign DNA on pp. 454 and 456. The steps to make a yeast genomic library are:
   1. Isolate high-molecular-weight DNA from yeast nuclei.
   2. Perform a partial digest of the DNA with a restriction enzyme that cuts frequently (e.g., *Sau*3A), and isolate DNA fragments of the correct size range for cloning in the λ vector by sucrose gradient centrifugation.
   3. Remove the central section of an appropriate λ vector by digestion with *Bam*HI.
   4. Ligate the left and right λ arms to the yeast DNA fragments. The *Sau*3A and *Bam*HI sticky ends are complementary (see pp. 459–460).
   5. Package the recombinant DNA molecules in vitro into λ phage particles.
   6. Infect *E. coli* cells with the λ phage population, and collect progeny phages produced by cell lysis. These phages represent the yeast genomic library.

15.9 a. The number of clones that must be screened to have a specified probability of finding at least one copy of a DNA sequence represented in a genomic library is calculated using the formula $N = (\ln[1 - p])/(\ln[1 - f])$. Here, N is the number of recombinant DNA molecules that must be screened, p is the chance of finding at least one positive clone, and f is the fractional proportion of the genome in a single recombinant DNA molecule. Here, $p = 0.95$, and f varies: for i, $f = 7/(3 \times 10^6)$; for ii, $f = 15/(3 \times 10^6)$; for iii, $f = 40/(3 \times 10^6)$; for iv, $f = 350/(3 \times 10^6)$. The number of clones required for each vector is (i) 1.3×10^6 plasmids, (ii) 6.0×10^5 phages, (iii) 2.2×10^5 cosmids, and (iv) 2.6×10^4 YACs.

   b. Screening libraries with larger average inserts requires one to screen fewer clones. This advantage must be evaluated relative to the added difficulty of analyzing the larger inserts of positive clones. For example, restriction mapping 350 kb is substantially more difficult than restriction mapping 15 kb. A single gene ranges in size from hundreds of base pairs to hundreds of kilobase pairs, so an essential question to consider is how the cloned sequences will be analyzed and used after they are identified.

   c. Genomic clones provide for the analysis of gene structure: intron/exon boundaries, transcriptional control regions, and polyadenylation sites. This analysis is important for evaluating how the expression of a gene is controlled. Mutations in regulatory regions can affect the expression of a gene, so analysis of these regions is important for understanding the molecular basis of a mutation.

15.12 Use cDNA. Human genomic DNA contains introns. mRNA transcribed off genomic DNA needs to be processed before it can be translated. Bacteria do not contain these processing enzymes. So in the bacterium, even if the human mRNA were translated, the protein it would code for would not be insulin. cDNA is the complementary copy of a functional mRNA molecule. So when this is the template, its mRNA transcript will be functional and when translated, human pro-insulin will be synthesized.

15.14 a. Both cDNAs hybridize to the same bands on a genomic Southern blot, so they are copies of mRNAs transcribed from the same sequences. Therefore, they are likely from the same gene.

b. Different bands on the northern blot indicate that the primary mRNA for this gene may be processed differently in brain and liver tissue. For example, it is possible that the 0.8-kb size difference between the two bands reflects a 0.8-kb intron that is spliced out in brain tissue that is not spliced out in liver tissue.

c. The two cDNAs are copies of mRNAs found in two different tissues. The northern blot indicates that there are some differences between the mRNAs in the different tissues. Thus, it is not surprising that the restriction maps are not identical. Note that the ends of the restriction maps are identical (the same EcoRI-HindIII and BamHI fragments), but the internal regions are not (the brain cDNA lacks the 0.5-kb EcoRI-HindIII fragment and some of each adjoining fragment).

d. The genomic Southern blot gives an indication of the gene organization at the DNA level, but the cDNA maps indicate the structure of the mRNA transcript(s). When the cDNA is used to probe genomic DNA sequences, it will hybridize to any sequences that are transcribed. Restriction sites in the genome do not delineate where the transcribed regions are, so the probe can hybridize to genomic DNA fragments that are only partly transcribed. That is, the probe can hybridize to transcribed sequences that are "connected to" nontranscribed sequences. Thus, the large (7.8-, 7.4-, 6.1-, and 3.6-kb) bands reflect the parts of the cDNA that hybridize to genomic DNA fragments that are only partly transcribed. They are the same size fragments that appear in the liver cDNA, so the smaller fragments (2.0, 1.4, and 1.3 kb) represent fragments that are entirely transcribed. A possible gene organization based on these data is illustrated in the figure.

Possible Gene Structure

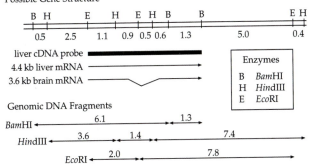

15.17 The restriction map is:

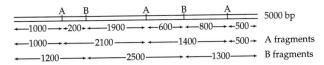

The map is built by considering the large fragments first. For example, the 2,500-bp B fragment is cut with A to produce 1,900-bp and 600-bp fragments. The 2,100-bp A fragment is cut with B to produce the same 1,900-bp fragment and a 200-bp fragment. Thus, the 200-bp and 600-bp fragments must be on opposite sides of the 1,900-bp fragment. The map is extended in a step-by-step fashion by next considering other cuts that produced 200-bp fragments, 600-bp fragments, and so on.

15.18 The sequencing gel banding pattern is shown in the following figure:

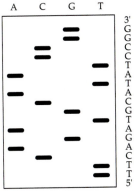

15.21 I-1 is heterozygous for the RFLP and the disease. Inspection of the blot figure shows three different "haplotypes" (DNA types) are segregating in this family. Let us name them A, B, and C in decreasing order of size. Let us use d for the normal allele of the disease gene, and D for the allele causing the disease. I-1 is then Dd and AB, but we don't know whether he is DA/dB or DB/dA, so we cannot tell which of his offspring are recombinant and which are nonrecombinant. I-2 is dB/dB, so I-1 must be DA/dB in order to produce dB/dB progeny, assuming linkage. Given the genotypes of I-1 and I-2, the genotypes of II-2 and II-5 must both be DA/dB, and II-3 and II-4 must both be dB/dB. Thus, all four generation II progeny of I-1 and I-2 are nonrecombinant. II-2 (DA/dB) pairs with II-1 (dC/dC) and has six offspring. III-1 is dB/dC, III-2 is DA/dC, III-4 is DA/dC, III-5 is dB/dC, III-6 is dA/dC, and III-7 is DA/dC. Of these offspring, only III-6 is recombinant because it is not DA.

The pairing of II-5 (DA/dB) and II-6 (dA/dA) produces six offspring, of which only III-12 (dA/dA) and III-13 (DB/dA) are recombinants. That is, III-12 has received the recombinant chromosome dA, and III-13 has received the recombinant chromosomes DB.

The pairing of III-2 (DA/dC) with III-3 (dB/dB) produces three nonrecombinant progeny: IV-1 is DA/dB, IV-2 is DA/dB, and IV-3 is dC/dB. The pairing of III-13 (DB/dA) with III-14 (dC/dC) produces four offspring, of which only IV-6 is recombinant (dB/dC).

Overall there are nineteen individuals whose recombinant or nonrecombinant status can be ascertained. Of these, four (III-6, III-12, III-13, and IV-6) are recombinant. $^4/_{19} = 0.21$, which is less than the 0.5 expected from independent assortment. But is it significantly less? To find out, we must do the χ^2 test. Independent assortment predicts $^{19}/_2 = 9.5$ recombinants and the same number of nonrecombinants. The value of χ^2 is thus 6.368, with one degree of freedom. The χ^2 table (Chapter 2, p. 37) indicates that this difference is significant, so we can conclude the two loci are linked.

15.23 Chromosomes bearing CF mutations have a shorter fragment than chromosomes bearing wild-type alleles. Both parents (M and P lanes) have two bands, indicating that each has a normal and mutant chromosome. The fetus (F lane) also has two bands, indicating that it, too, is heterozygous. It has a normal gene, so it will be normal.

15.24 Let us use the symbol "A" to represent the heavier molecular weight band detected by probe A and "a" to represent the lighter molecular weight band. Similarly, let us use "B" to symbolize the heavier molecular weight band detected by probe B and "b" to symbolize the doublet of lighter bands. Then the genotype of strain J can be designated as Ab and that of strain K as aB.

First, the data indicate the region homologous to probe A is linked to that homologous to probe B. This is reflected in the large excess of parental ditype asci (50) over nonparental ditypes (2). Therefore, we need to map only one centromere.

The data show that the region homologous to probe A is on the opposite side of the centromere from the region homologous to probe B. This is reflected in the fact that the two single-exchange classes (32 asci and 12 asci) show first-division segregation for the

A region only (12) or for the B region only (32), and you get second-division segregation for both the A and B regions only in the double-exchange asci (4, 2, and 2).

Considering the distance from region A to the centromere, we see that A shows second-division segregation in 20 of the 100 asci. Gene-centromere distance is given by this formula:

$$\frac{\text{\% second-division segregation asci}}{2}$$

Here, then, the gene-centromere distance is 20%/2 = 10 mu. To calculate this distance another way, the 20 asci contain 80 meiotic products, and half of these have had an exchange between A and the centromere. Thus, 40 of a total of 400 meiotic products have experienced an exchange between A and the centromere. 40/400 = 0.1, so the A region is 10 mu from the centromere.

The B region shows second-division segregation in 40 asci, out of 100 assayed, giving a gene-centromere distance of 40%/2 = 20 mu. Calculating gene-centromere distance another way, the 40 asci contain 80 meiotic products that have inherited chromosomes that have crossed over between B and the centromere. 80/400 = 0.2, so the B region is 20 mu from the centromere.

Given that A is 10 mu from the centromere and B is 20 mu from the centromere, the one remaining fact to determine is whether A and B are on the same side of a centromere or on opposite sides. If they were on the same side, the prediction would be that they are about 10 mu apart. If they were on the opposite sides of the centromere, the prediction would be that they are about 30 mu apart. Map distance using tetrad analysis is given by the formula

$$\frac{\frac{1}{2}T + NPD}{\text{Total}} \times 100$$

$$= \frac{\frac{1}{2}(48) + 2}{100} \times 100$$

$$= 26 \text{ mu}$$

Thus, our map is

CHAPTER 16

16.3 A constitutive phenotype can be the result of either an *lacI⁻* or an *lacOᶜ* mutation.

16.4 a. A missense mutation results in partial or complete loss of β-galactosidase activity, but there would be no loss of permeases and transacetylase activities.

b. A nonsense mutation is likely to have polar effects unless the mutation is very close to the normal chain-terminating codon for β-galactosidase. Thus, permease and transacetylase activities would be lost in addition to the loss of β-galactosidase activity if nonsense mutation occurred near the 5′ end of the *lacz* gene.

16.6 *lacI⁺ lacOᶜ lacZ⁺ lacY⁻/lacI⁺ lacO⁺ lacZ⁻ lacY⁺*
(It cannot be ruled out that one of the repressor genes is *lacI⁻*.)

16.8 Answer is in Table 16.A (p. S-17).

16.11 The CAP, in a complex with cAMP, is required to facilitate RNA polymerase binding to the *lac* promoter. The RNA polymerase binding occurs only in the absence of glucose and only if the operator is not occupied by a repressor (i.e., if lactose is absent). A mutation in the CAP gene, then, would render the *lac* operon incapable of expression because RNA polymerase would not be able to recognize the promoter.

16.13 See pp. 517–521. For a wild-type *trp* operon, the absence of tryptophan results in antitermination; that is, the structural genes are transcribed, and the tryptophan biosynthetic enzymes are then made. This occurs because the absence of tryptophan results in the absence of, or at least a very low level of, Trp-tRNA.Trp, which, in

turn, causes the ribosome translating the leader sequence to stall at the Trp codons. With the ribosome stalled at that location, the RNA being synthesized just ahead of the ribosome by RNA polymerase assumes a secondary structure that favors continued transcription of the structural genes by the polymerase. If the two Trp codons were mutated to stop codons, then the mutant operon would function in the same way as the wild-type operon in the absence of tryptophan because the ribosome would stall in the same place, and the anti-termination would be produced, resulting in transcription of the structural genes.

For a wild-type *trp* operon, the presence of tryptophan turns the transcription of the structural genes off. This happens because the presence of tryptophan leads to the accumulation of Trp-tRNA.Trp, which allows the ribosome to read the two Trp codons and stall at the normal stop codon for the leader sequence. When stalled in that position, the antitermination signal cannot form in the RNA being synthesized; instead, a termination signal is formed, resulting in the termination of transcription. In a mutant *trp* operon with two stop codons instead of the Trp codons, the ribosome will still stall at the premature stop codons even in the presence of tryptophan (and therefore Trp-tRNA.Trp). And, as discussed previously, this results in an antitermination signal so that the structural genes will be transcribed.

In sum, in both the presence and the absence of tryptophan, the mutant *trp* operon will show no attenuation; that is, the structural genes will be transcribed in both cases, and the tryptophan biosynthetic enzymes will be synthesized.

16.17 The *cI* gene codes for repressor protein that functions to keep the lytic functions of the phage repressed when lambda is in the lysogenic state. A *cI* mutant strain would be unable to establish lysogeny, and thus it would also follow the lytic pathway.

16.21 a. Proteins that interact with DNA without site specificity are likely to have configurations of amino acids that interact with structural features shared among all DNA molecules, such as the negatively charged sugar-phosphate backbone. For example, the histone proteins have high percentages of the positively charged amino acids lysine and arginine and interact without site specificity via ionic interactions with the negatively charged DNA backbone.

b. Proteins that interact with DNA with site specificity have amino acid configurations that interact with particular DNA nucleotides. For example, the R-groups on interacting amino acids could be positioned to form hydrogen bonds with different atoms in the nucleotide bases. When enough hydrogen bonds form between a particular set of amino acids in the protein and a set of bases in a region of DNA, the affinity of the protein for the region is high enough to result in a site-specific interaction.

c. The specificity of Cro and the λ repressor interactions with the λ operator sites O_{R1}, O_{R2}, and O_{R3} must arise from interactions between a subset of amino acids in these proteins and particular DNA nucleotides at the operator sites. Although some of the affinity of one of these proteins for a particular operator site may derive from interactions between the sugar-phosphate backbone of these nucleotides and amino acids of the protein, this does not explain the site specificity. The affinity of a protein for an operator site must also derive from interactions between the amino acids of the protein and specific bases in nucleotides at the operator site. The ability of the protein to bind at multiple sites probably reflects that the operators have only slightly different DNA sequences.

To account for different affinities of the λ repressor and Cro for each operator site, consider the following. Each DNA base that interacts with an amino acid in these proteins contributes to the strength of the interaction between the protein and the operator site. If hydrogen bonds form between some bases and a set of amino acids, the base sequence able to form the largest total number of hydrogen bonds will have the highest affinity for the protein. Because the amino acid sequences in the λ repressor and Cro differ, their relative affinities for each operator site differ.

TABLE 16.A

GENOTYPE	INDUCER ABSENT β-GALACTOSIDASE	INDUCER ABSENT PERMEASE	INDUCER PRESENT β-GALACTOSIDASE	INDUCER PRESENT PERMEASE
a. I^+ P^+ O^+ Z^+ Y^+	−	−	+	+
b. I^+ P^+ O^+ Z^- Y^+	−	−	−	+
c. I^+ P^+ O^+ Z^+ Y^-	−	−	+	−
d. I^- P^+ O^+ Z^+ Y^+	+	+	+	+
e. I^s P^+ O^+ Z^+ Y^+	−	−	−	−
f. I^+ P^+ O^c Z^+ Y^+	+	+	+	+
g. I^s P^+ O^c Z^+ Y^+	+	+	+	+
h. I^+ P^+ O^c Z^+ Y^-	+	−	+	−
i. I^{-d} P^+ O^+ Z^+ Y^+	+	+	+	+
j. $\dfrac{I^-\ P^+\ O^+\ Z^+\ Y^+}{I^+\ P^+\ O^+\ Z^-\ Y^-}$	−	−	+	+
k. $\dfrac{I^-\ P^+\ O^+\ Z^+\ Y^-}{I^+\ P^+\ O^+\ Z^-\ Y^+}$	−	−	+	+
l. $\dfrac{I^s\ P^+\ O^+\ Z^+\ Y^-}{I^+\ P^+\ O^+\ Z^-\ Y^+}$	−	−	−	−
m. $\dfrac{I^+\ P^+\ O^c\ Z^+\ Y^+}{I^+\ P^+\ O^+\ Z^-\ Y^-}$	−	+	+	+
n. $\dfrac{I^{-d}\ P^+\ O^c\ Z^+\ Y^-}{I^+\ P^+\ O^+\ Z^-\ Y^+}$	+	−	+	+
o. $\dfrac{I^s\ P^+\ O^+\ Z^+\ Y^+}{I^+\ P^+\ O^c\ Z^+\ Y^+}$	+	+	+	+
p. $\dfrac{I^{-d}\ P^+\ O^+\ Z^+\ Y^-}{I^+\ P^+\ O^+\ Z^-\ Y^+}$	+	+	+	+
q. $\dfrac{I^+\ P^-\ O^c\ Z^+\ Y^-}{I^+\ P^+\ O^+\ Z^+\ Y^-}$	−	−	+	+
r. $\dfrac{I^+\ P^-\ O^+\ Z^+\ Y^-}{I^+\ P^+\ O^c\ Z^-\ Z^+}$	−	+	−	+
s. $\dfrac{I^-\ P^-\ O^+\ Z^+\ Y^+}{I^+\ P^+\ O^+\ Z^-\ Y^-}$	−	−	−	−
t. $\dfrac{I^-\ P^+\ O^+\ Z^+\ Y^-}{I^+\ P^-\ O^+\ Z^-\ Y^+}$	−	−	+	−

16.23 a. Prokaryotic organisms cannot individually control their immediate environment and do not have the capacity to store nutrients, so it is essential for them to adapt rapidly to even subtle environmental changes. At operons such as the tryptophan operon, the control of transcription initiation can be repressed about seventyfold by the binding of an active repressor (tryptophan bound to the aporepressor) to its operator. Attenuation can lead to a further eightfold to tenfold repression. Each mechanism is sensitive to the concentration of tryptophan in the cell. Together, the different mechanisms allow for a wider range of responses than is possible from either mechanism alone.

b. At the tryptophan operon in *E. coli*, transcription can be terminated by attenuation if sufficient tryptophan is already present in the cell. If cells are starved for tryptophan, attenuation will not occur. As the ribosome attempts to translate the leader transcript, it stalls at tandem Trp codons while awaiting Trp-tRNA.Trp. An antitermination stem-loop structure forms (the 2:3 pairing in Figure 16.17), which allows RNA polymerase to continue past the attenuator and transcribe the structural genes. In contrast to this, transcription antitermination mediated by the N protein of lambda results when the N protein allows transcription to proceed past a transcription termination signal so that additional genes in an operon can be transcribed. At the tryptophan operon, the antiterminator signal is a particular RNA secondary structure; in lambda, an antiterminator protein allows transcription readthrough.

c. (i) If the 2:3 pairing in Figure 16.17 were destabilized by mutation, antitermination would be decreased and attenuation increased. The cellular phenotype would most likely be *trp*− because fewer or no full-length transcripts would be produced and little or no synthetic enzyme would be made.

(ii) The N protein is a regulator of early gene function and is required before lambda chooses between a lytic and a lysogenic pathway. If the N protein does not function, transcription will not proceed leftward past N or rightward past *cro* (see Figure 16.21, p. 525). In this event, *cII* would not be transcribed. Since *cII* is needed to activate transcription at *cI* and produce the lambda repressor, lysogeny is not possible. Since *cII* is also needed for the production of the DNA replication proteins O and P, as well as for the production of Q, a protein needed to turn on the late genes needed for cell lysis, lytic growth is not possible either. Thus, an *N*− mutation is pleiotropic (has multiple effects) and will prevent lambda from reproducing.

CHAPTER 17

17.2 The final product of the rRNA genes is an rRNA molecule. Hence, a large number of genes are required to produce the large number of rRNA molecules required for ribosome biosynthesis. Ribosomal proteins, in contrast, are the end products of the translation of mRNAs, which can be "read" over and over to produce the large number of ribosomal protein molecules required for ribosome biosynthesis.

17.3

GLOBIN GENE	EMBRYONIC YOLK SAC	FETAL SPLEEN	ADULT BONE MARROW
α	R	S	S
β	R	R	S
γ	R	S	R
δ	R	R	S
ζ	S	R	R
ε	S	R	R

17.6 a. In the fragile X syndrome, the expanded CGG repeat results in hypermethylation and transcriptional silencing. In Huntington disease, the expanded CAG repeat results in the inclusion of a polyglutamine stretch within the huntingtin protein, which causes it to have a novel, abnormal function.

b. A heterozygote with a CGG repeat expansion near one copy of the *FMR-1* gene will still have one normal copy of the *FMR-1* gene. The normal gene can produce a normal product, even if the other is silenced, so fragile X syndrome is recessive. In contrast, a novel, abnormal protein is produced by the CAG expansion in the disease allele in Huntington disease. The disease phenotype is due to the presence of the abnormal protein, so the disease trait is dominant.

c. Transcriptional silencing may require significant amounts of hypermethylation and so require more CGG repeats for an effect to be seen. In contrast, protein function may be altered by a relatively smaller sized polyglutamine expansion.

17.8 The data show clearly that the synthesis of ovalbumin is dependent upon the presence of the hormone estrogen. The data do not indicate at what level estrogen works. Theoretically, it could act to increase transcription of the ovalbumin gene, to stabilize the ovalbumin mRNA, to stabilize the ovalbumin protein, or to stimulate transport of the ovalbumin mRNA out of the nucleus. Experiments in which the levels of the ovalbumin mRNA were measured have shown that the production of ovalbumin is primarily regulated at the level of transcription.

17.11 a. In bacterial operons, a common regulatory region controls the production of single mRNA from which multiple protein products are translated. These products function in a related biochemical pathway. Here, two proteins that are involved in the synthesis and packaging of acetylcholine are both produced from a common primary mRNA transcript.

b. Unlike the proteins translated from an mRNA synthesized from a bacterial operon, the protein products produced at the VAChT/ChAT locus are not translated sequentially from the same mRNA. Here, the primary mRNA appears to be alternatively processed to produce two distinct mature mRNAs. These mRNAs are translated starting at different points, producing various proteins.

c. At least two mechanisms are involved in the production of the different ChAT and VAChT proteins: alternative mRNA processing and alternative translation initiation. After the first exon, an alternative 3' splice site is used in the two different mRNAs. In addition, different AUG start codons are employed.

17.14 a. Based on the work of Wilmut and his colleagues, the nose cells would first be dissociated and grown in tissue culture. The cells would be induced into a quiescent state (the G_0 phase of the cell cycle) by reducing the concentration of growth serum in the medium. Then they would be fused with enucleated oocytes from a donor female and allowed to grow and divide by mitosis to produce embryos. The embryos would be implanted into a surrogate female.

b. Although the nuclear genome would generally be identical to that in the original nose cell, cytoplasmic organelles would presumably derive from those in the enucleated oocyte. Hence, the mitochondrial DNA would not derive from the original leader. In addition, telomeres in an older individual are "shorter," so one might expect the telomeres in the cloned leader to be those of an older individual.

c. In mature B cells, DNA rearrangements at the heavy- and light-chain immunoglobulin genes have occurred. One would expect the cloned leader to be immunocompromised because he would be unable to make the wide spectrum of antibodies present in a normal individual.

d. The production of Dolly was significant because it demonstrated the apparent complete totipotency of a nucleus from a mature mammary epithelium cell. The other six lambs developed from the transplanted nuclei of embryonic or fetal cells, which one might expect to be less determined and have a higher degree of totipotency due to their younger developmental age.

e. There is no way to predict the psychological profile of the cloned leader based on his genetic identity. Even identical twins, who are genetically more identical than such a clone, do not always share behavioral traits.

17.16 four: A_3, A_2B, AB_2, B_3

17.22 Experiment A results in all the DNA becoming radioactively labeled, and so because DNA is a fundamental and major component of polytene chromosomes, radioactivity is evident throughout the chromosomes. Experiment B results in radioactive labeling of RNA molecules. Because radioactivity is first found only around puffs and later in the cytoplasm, we can hypothesize that the puffs are sites of transcriptional activity. Initially, radioactivity is found in RNA that is being synthesized, and the later appearance of radioactivity in the cytoplasm reflects the completed RNA molecules that have left the puffs and are being translated in the cytoplasm. Experiment C provides additional support for the hypothesis that transcriptional activity is associated with puffs. That is, the inhibition of RNA transcription by actinomycin D blocks the appearance of ^{3}H-uridine (which would be in RNA) at puffs. In fact, puffs are much smaller, indicating that the puffing process is intimately associated with the onset of transcriptional activity for the gene(s) in that region of the chromosomes.

17.23 Preexisting mRNAs (stored in the oocyte) are recruited into polysomes as development begins following fertilization.

17.25 a. Connie must also have blood type A.

b. Connie must also have the blood type genotype I^A/I^A.

c. Connie has anti-B antibodies.

d. No, Ashley's and Connie's anti-B antibodies would not have identical polypeptide sequences. During development, their IgG DNAs were rearranged independently, so their IgG mRNAs would be different. (In addition, the splicing process would further diversify their functional mRNAs.) So their IgG antibodies would be different; that is, they would have different polypeptide sequences.

e. Yes. Their genes for β-globin would be identical, so the sequences of their β-globin chains would be identical (as would be those of all other sequences except for the immunoglobulins).

17.28

MUTANT	CLASS
a	segmentation gene (segment polarity)
b	maternal gene (anterior-posterior gradient)
c	segmentation gene (gap)
d	homeotic (eye to wing transdetermination)
e	segmentation gene (gap)

17.29 Preexisting mRNA that was by the mother and deposited in the egg prior to fertilization is translated up until the gastrula stage. After gastrulation, new mRNA synthesis is necessary for production of proteins needed for subsequent embryo development.

17.30 The tissue taken from the blastula/gastrula has not yet been committed to its final differentiated state in terms of its genetic programming; that is, it is not yet determined. Thus, when it is transplanted into the host, the determined tissues surrounding the transplant in the host communicate with the transplanted tissue and cause it to be determined in the same way as they are; for example, tissue transplanted to a future head area will differentiate into head material, and so on. In contrast, tissues in the neurula stage are now determined as to their final tissue state once development is complete. Thus, tissue transplanted from a neurula to an older embryo cannot be influenced by the determined, surrounding tissues and will develop into the tissue type for which it is determined, in this case, an eye.

CHAPTER 18

18.1 Hereditary cancer is associated with the inheritance of a germ-line mutation; sporadic cancer is not. Consequently, hereditary cancer "runs" in families. For some cancers, both hereditary and sporadic forms exist, with the hereditary form being much less frequent. For example, retinoblastoma occurs when both normal alleles of the tumor suppressor gene *RB* are inactivated. In hereditary retinoblastoma, a mutated, inactive allele is transmitted via the germ line. Retinoblastoma occurs in cells of an *RB/+* heterozygote when an additional somatic mutation occurs. In the sporadic form of the disease, retinoblastoma occurs when both alleles are inactivated somatically.

18.4 Tumor growth induced by transforming retroviruses results either from the activity of a single viral oncogene or the activation of a proto-oncogene due to the nearby integration of the proviral DNA. The oncogene can cause abnormal cellular proliferation via the variety of mechanisms discussed in the text. The expression of a proto-oncogene, normally tightly regulated during cell growth and development, can be altered if it comes under the control of the promoter and enhancer sequences in the retroviral LTR.

DNA tumor viruses do not carry oncogenes. They transform cells through the action of a gene or genes within their genomes. For example, in a rare event, the DNA virus can be integrated into the host genome, and the DNA replication of the host cell may be stimulated by a viral protein that activates viral DNA replication. This would cause the cell to move from the G_0 to the S phase of the cell cycle.

For both transducing retroviruses and DNA tumor viruses, an abnormally expressed protein(s) leads to the activation of the cell from G_0 to S, and abnormal cell growth occurs.

18.5 a. Studies of hereditary forms of cancer have led to insights into the fundamental cellular processes affected by cancer. For example, substantial insights into the important role of DNA repair and the relationship between the control of the cell cycle and DNA repair have come from analyses of the genes responsible for hereditary forms of human colorectal cancer. For breast cancer, studying the normal functions of the *BRCA1* and *BRCA2* genes promises to provide substantial insights into breast and ovarian cancer.

b. Genetic predisposition for cancer refers to the presence of an inherited mutation that, with additional somatic mutations during the lifespan of an individual, can lead to cancer. For diseases such as retinoblastoma, a genetic predisposition has been associated with the inheritance of a recessive allele of the *RB* tumor suppressor gene. Retinoblastoma occurs in *RB/+* individuals when the normal allele is mutated in somatic cells and the pRB protein no longer functions. Because somatic mutation is likely, the disease appears dominant in pedigrees.

Although there is a substantial understanding of the genetic basis for cancer and the genetic abnormalities present in somatic cancerous cells, there are also substantial environmental risk factors for specific cancers. Environmental risk factors must be investigated thoroughly when a pedigree is evaluated for showing a genetic predisposition for cancer.

18.12 a. Proto-oncogenes encode a diverse set of gene products that include growth factors, receptor and nonreceptor protein kinases, receptors lacking protein kinase activity, membrane-associated GTP-binding proteins, cytoplasmic regulators involved in intracellular signaling, and nuclear transcription factors. These gene products all function in intercellular and intracellular circuits that regulate cell division and differentiation.

b. In general, mutations that activate a proto-oncogene convert it into an oncogene. Since (i), (iii), and (viii) cause a decrease in gene expression, they are unlikely to result in an oncogene. Since (ii) and (vii) could activate gene expression, they could result in an oncogene. Mutations (iv), (v), and (vi) cannot be predicted with certainty. The deletion of a 3' splice-site acceptor would alter the mature mRNA and possibly the protein produced and may or may not affect the protein's function and regulation. Similarly, it is difficult to predict the effect of a nonspecific point mutation or a premature stop codon. The text presents examples where these types of muta-

tions have caused the activation of a proto-oncogene and resulted in an oncogene.

18.13 Experimentally fuse cells from the two cell lines and then test the resultant hybrids for their ability to form tumors. If the uncontrolled growth of the tumor cell line was due to a mutated pair of tumor suppressor alleles, the normal alleles present in the normal cell line would "rescue" the tumor cell line defect. The hybrid line would grow normally and be unable to form a tumor. If the uncontrolled growth of the tumor cell line was due to an oncogene, the oncogene would also be present in the hybrid cell line. The hybrid line would grow uncontrollably and form a tumor.

18.15 One hypothesis is that the proviral DNA has integrated near the proto-oncogene, and the expression of the proto-oncogene has come under the control of promoter and enhancer sequences in the retroviral LTR. This could be assessed by performing a whole-genome Southern blot analysis to determine whether the organization of the genomic DNA sequences near the proto-oncogene has been altered.

18.18 a. The fact that some translocations are found as the only cytogenetic abnormality in certain cancers most likely means that they represent a key event in tumor formation. It does not necessarily mean that they are the primary cause of the tumor or the first of many mutational events.

b. A chimeric fusion protein may have functional properties different than either of the two proteins from which it derives. If it results in the activation of a proto-oncogene product into a protein that has oncogenic properties or if it results in the inactivation of a tumor suppressor gene product, it could play a key role in the genetic cascade of events leading to tumor formation.

c. Before drawing conclusions as to whether these chromosomal aberrations inactivate the function of tumor suppressor genes or activate quiescent proto-oncogenes, one needs additional molecular information on the effects of the translocation breakpoints on specific transcripts. Finding that the translocation breakpoints result in a lack of gene transcription or in transcripts that encode nonfunctional products would support the hypothesis that the translocation inactivated a tumor suppressor gene. Finding that the translocation breakpoints result in activation of gene transcription or in the production of an active fusion protein would support the hypothesis that the translocation activated a previously quiescent proto-oncogene.

d. One hypothesis is that the various fusion proteins that result from different translocations involving the *EWS* gene somehow result in the transcription activation of different proto-oncogenes and this leads to the different sarcomas that are seen. (Sarcomas are cancers found in tissues that include, among others, muscle, bone, fat, and blood vessels.)

e. If translocation breakpoints are conserved within a tumor type, molecular-based diagnostics can be developed to identify the breakpoints relatively quickly from a tissue biopsy. For example, if the genes at the breakpoints have been cloned, PCR methods can be used to address whether the gene is intact or disrupted, using the DNA from cells of a tumor biopsy. Primers can be designed to amplify different segments of the normal gene. Then PCR reactions containing these primers and either normal, control DNA or tumor-cell DNA can be set up to determine if each segment of a candidate gene is intact (a PCR product of the expected size is obtained) or disrupted (no PCR product will be obtained because the gene has been rearranged).

Such molecular analyses would provide fast, accurate tumor diagnosis. If the several tumor types respond differentially to various regimens of therapeutic intervention, then a more rapid, unequivocal diagnosis of a particular tumor type should allow for the earlier prescription of a more optimized regime of therapeutic intervention. In addition, understanding the nature of the normal gene products of the affected genes may allow for the development of sarcoma-specific therapies.

18.20 Apoptosis is programmed, or suicidal, cell death. Cells targeted for apoptosis are those that have large amounts of DNA damage and so are at a greater risk for neoplastic transformation. Apoptosis is regulated by p53, among other proteins. In cells with large amounts of DNA damage, p53 is stabilized. This leads to the activation of *WAF1*, whose product, p21, causes cells to arrest in G_1 by binding to cyclin/Cdk complexes and blocking kinase activity needed for progression from G_1 to S. If the DNA damage is unable to be repaired, apoptosis will be induced by p53.

CHAPTER 19

19.1 b. Mutations are heritable changes in genetic information.

19.3 e. None of these (i.e., all are classes of mutations).

19.4 c. Ultraviolet light usually causes the induction of thymine dimers and their persistence or imperfect repair.

19.6 a. The normal anticodon was 5'-CAG-3', and the mutant one is 5'-CAC-3'. The mutational event was a CG to GC transversion.

b. Presumably Leu

c. Val

d. Leu

19.9 Acridine is an intercalating agent and so can be expected to induce frame shift mutations. 5BU is incorporated in place of T, but is relatively likely to be read as C by DNA polymerase because of a shift from its normal to its rare chemical state. Thus, 5BU-induced mutations would be expected to be point mutations, usually TA to CG transitions. If these expectations are realized, *lacZ-1* would probably contain a single amino acid difference from the normal β-galactosidase, although it could be a truncated normal protein due to a nonsense point mutation. *lacZ-2* should have a completely altered amino acid sequence after some point and might also be truncated.

19.10 a. Six.

b. Three. The mutation results in replacement of the UGG codon in the mRNA by UAG, which is a chain termination codon.

19.12 a. UAG: CAG Gln; AAG Lys; GAA Glu; UUG Leu; UCG Ser; UGG Trp; UAU Tyr; UAC Tyr; UAA chain terminating.

b. UAA: CAA Gln; AAA Lys; GAA Glu; UUA Leu; UCA Ser; UGA chain terminating; UAU Tyr; UAC Tyr; UAG chain terminating.

c. UGA: CGA Arg; AGA Arg; GGA Gly; UUA Leu; UCA Ser; UAA chain terminating; UGU Cys; UGC Cys; UGG Trp.

19.14

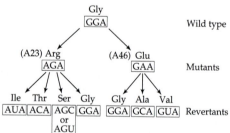

19.20 a. 5BU in its normal form is a T analog; in its rare form, it resembles C. The mutation is an AT-to-GC transition.

$$A-T \longrightarrow A-5BU \longrightarrow G-5BU \longrightarrow G-C$$

b. Nitrous acid can deaminate C to U, and so the reversion is to the original AT pair.

19.23 The reaction stops when it needs dGTP; only dGMP is present.

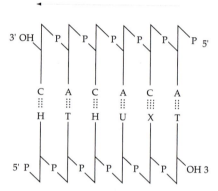

19.25 HNO_2 converts A to H, C to U, and G to X.

19.29 At the DNA level, nitrous acid deaminates C to make it U. As a result, CG base pairs are changed to TA base pairs. At the RNA level, this means that C nucleotides can be changed to U nucleotides.

According to the genetic code dictionary, Pro is specified by CCX (where X is U, C, G, or A); Ser is specified by UCX or AGU or AGC; Leu is specified by CUX or UUA or UUG; and Phe is specified by UUU or UUC.

To go from Pro to Ser to Phe requires changing the first two C nucleotides of the CCX codon to two U nucleotides of the UUU or UUC codon. This rules out the AGU and AGC codons for Ser. Since the Phe is unaffected by further nitrous acid treatment, the Phe codon must not contain a C (a CG at the DNA level); thus, the Phe codon must be UUU, and this means the Ser codon must be UCU and the Pro codon must be CCU.

Then, to go from Pro to Leu to Phe starting with CCU requires that the Leu codon must be CUU.

In sum, the changes that are compatible with the given data are:

```
                Ser
                UCU
         ↗            ↘
Pro                      Phe      Phe
CCU                      UUU  →   UUU
         ↘            ↗
                Leu
                CUU
```

19.30 *ara⁺* to *ara-1*: This mutation is CG to AT for it is reverted by base analogs but not by HA or the frameshift mutagen. *ara⁺* to *ara-2*: This mutation is AT to GC because it is reverted by base analogs and HA but not by frameshift mutagen. *ara⁺* to *ara-3*: This mutation is AT to GC for the same reasons as given for the second mutant. Mutagen X causes transition mutations in both directions because mutants are revertible by base analogs, some are revertible by HA, and none (if this is a representative sample) by frameshift mutagens.

19.32 a. The production of knockout mice allows for the development of an animal model system to investigate human disease. It allows for greater investigation into the consequences of not having a particular gene function and also allows for the development of therapeutic approaches in a model system that closely parallels the human disease.

b. Knockout mice lack a particular gene function, so the phenotype of a knockout mouse will not necessarily mimic the phenotype of a dominant mutation that has an additional, abnormal function. To design a mouse model in this instance, one must add a copy of a dominant allele to the normal genetic background or replace a normal gene with the dominant allele. Only in this way would the aberrant protein product of the gene be expressed and would the phenotype it causes be observed.

CHAPTER 20

20.6 a. As is described in Figure 20.12 (p. 668), the spotted kernel phenotype arises from the transposition of *Ds* elements out of the *C* gene during kernel development. This means that the mutation in colorless strain B is caused by an insertion of a *Ds* element in the *C* gene. It is able to transpose because the purple strain C has *Ac* elements that can activate *Ds* transposition.

Spotted kernels are not seen in any of the other crosses, so transposition of *Ds* elements is not occurring in them. Since spotted kernels are not seen in the A × C cross, the colorless phenotype in strain A is not caused by the insertion of a *Ds* element. Spotted kernels are not seen in the B × D cross, so the D strain lacks *Ac* elements. For these reasons, the crosses A × C, A × D, and B × D all show only expected Mendelian patterns of inheritance.

b. The degree of spotting will depend on the number of intact *Ac* elements in the true-breeding purple strain. A greater number of intact *Ac* elements will result in a greater amount of transposase and so cause both an increase in the excision of the *Ds* element from the *C* gene and a greater degree of chromosome breakage at the *C* gene. The timing and frequency of transposition of the *Ac* and *Ds* elements, as well as the genetic rearrangements associated with the elements, are developmentally controlled. Transposition later in the development of the kernel will produce smaller spots.

20.8 Since a transposition results in the loss of the intron, it may be hypothesized that transposition in A occurs via an RNA intermediate. That is, intron removal occurs only at the RNA level.

The lack of intron removal during B transposition suggests that in B there is a DNA-DNA transposition mechanism, without any RNA intermediate.

CHAPTER 21

21.2 There are various possibilities. You could isolate the minor peak and examine it using the electron microscope. If the molecules in this peak are circular, they are unlikely to be nuclear fragments. You could grow the yeast in the presence of an intercalating agent such as acridine and see whether this treatment causes the minor peak species to disappear. If it does, the minor peak is organellar in origin. You could isolate the minor peak, label it (by nick translation, for example), and hybridize this labeled DNA to DNA within suitably prepared yeast cells. Then, using the electron microscope, you could determine whether the label is found over the nucleus or the mitochondria. Finally you could isolate the minor peak DNA and study its homology to other yeast mitochondria.

21.7 a. Selective use of antibiotics that inhibit translation by mitochondrial vs. cytoplasmic ribosomes has shown that components 2, 5, and 6 of cytochrome oxidase are synthesized on mitochondrial ribosomes and components 1, 3, 4, and 7 are synthesized on cytoplasmic ribosomes. See Figure 21.7 (p. 683) and the accompanying text on pp. 683–684.

b. Compare cytochrome oxidase activity in cybids made with platelets from diseased individuals and in cybids made with platelets from age-matched control individuals. It is important to assess a number of different enzyme activities associated with mitochondrial proteins to ensure that the deficits in cytochrome oxidase were specific.

c. As discussed in the text (pp. 700–702), the cells of individuals with diseases resulting from mitochondrial DNA defects have a mixture of mutant and normal mitochondria. That is, they show heteroplasmy. Thus, assays in cybids are measurements of the enzyme activity present in a population of mitochondria in a cell. It would be unlikely that each of the mitochondria of an affected individual has an identical defect.

21.8 The two codes are different in codon designations. The mitochondrial code has more extensive wobble so that fewer tRNAs are needed to read all possible sense codons. As a consequence, fewer mitochondrial genes are necessary. The advantage is that fewer tRNAs are needed, and hence fewer tRNA genes need be present than for cytoplasmic tRNAs.

21.12 The features of extranuclear inheritance are differences in reciprocal-cross results (not related to sex), nonmappability to known nuclear chromosomes, Mendelian segregation not followed, and the indifference to nuclear substitution.

21.15 a. The *tudor* mutation is a maternal effect mutation. Homozygous *tudor* mothers give rise to sterile progeny, regardless of their mate.

b. The "grandchildless" phenotype results from the absence of some maternally packaged component in the egg needed for the development of the F_1's germ line.

21.16 a. If normal cytoplasm is [N] and male-sterile cytoplasm is [Ms], then the F_1 genotype is [Ms] *Rf/rf*, and the phenotype is male-fertile.

b. The cross is [Ms] *Rf/rf* ♀ × [N] *rf/rf* ♂, giving 50 percent [Ms] *Rf/rf* and 50 percent [Ms] *rf/rf* progeny. Thus, the phenotypes are ½ male-fertile and ½ male-sterile.

21.17 a. Parents: [poky] *F* ♀ × [N] + ♂; progeny: [poky] *F* and [poky] + in equal numbers. Because standard [poky] {[poky]+) is found among the offspring, gene *F* must not effect a permanent alteration of [poky] cytoplasm. The 1:1 ratio of [poky] to *fast*-[poky] indicates that all progeny have the [poky] mitochondrial genotype (by maternal inheritance) and that the *F* gene must be a nuclear gene segregating according to Mendelian principles. Thus, the [poky] progeny are [poky]+, and the *fast*-[poky] are [poky]F.

b. Parents: [N]+ × [poky]F; progeny: [N]+ and [N]F in equal numbers. These two types are phenotypically indistinguishable.

21.19 ½ petite, ½ wild-type (*grande*)

21.22 The first possibility is that the results are the consequences of a sex-linked lethal gene. The females would be homozygous for a dominant gene *L* that is lethal in males but not in females. In this case, mating the F_1 females of an *L/L* × +/Y cross to +/Y males should give a sex ratio of 2 females : 1 male in the progeny flies.

The second possibility is that the trait is cytoplasmically transmitted via the egg and is lethal to males. In this case, the same F_1 females should continue to have only female progeny when mated with +/Y males.

21.25 a. Carlos Mendoza will have inherited his mitochondrial DNA from his mother, and she will have inherited it from her mother. If Mrs. Mendoza and Mrs. Escobar have different mitochondrial RFLPs, then it can be determined which of them contributed mitochondria to Carlos.

b. None of the potential grandfathers need to be tested, because they will not have given any mitochondria to Carlos. In addition, there is no point in testing Mrs. Sanchez. She may have given mitochondria to Carlos' father, but the father did not pass them on to Carlos.

c. If Mrs. Mendoza and Mrs. Escobar do not differ in RFLP, the data will not be helpful. If they do differ, and if Carlos matches Mrs. Mendoza, the case should be dismissed. If Carlos matches Mrs. Escobar, then the Escobar and Sanchez couples are indeed the grandparents, and the Mendozas have claimed a stolen child.

21.28 The parental snails were D/d female and $d/-$ male. The F_1 snail is d/d. (Given the F_1 genotype, the male can be either homozygous d or heterozygous, but the determination cannot be made from the data given.)

21.29 a. It would be necessary to show that individuals with Beckwith-Wiedemann syndrome have the loss of expression of a parental allele that contributes to the onset of BWS. From what is described in the question, there is a tumor suppressor gene that, if not expressed correctly, could result in tumor formation. In addition, the region appears to be subject to imprinting because of the expression pattern of genes in the region in tumors and the expression pattern of *KVLQT1* in some tissues of normal individuals. The loss of expression of one tumor suppressor allele could give rise to tumor formation due to the sole expression of a mutant allele.

b. The molecular basis for imprinting is a decrease in the expression of a specific allele due to the heritable methylation of that allele in one parent. To demonstrate imprinting, one must identify which of a paternally or maternally inherited allele is transcribed. This can be done if the parental alleles are polymorphic and the differences between them can be detected by analysis of their mRNA or protein products.

c. It would be necessary to demonstrate that only one of the two parental alleles is expressed in the heart of diseased individuals. For example, if imprinting occurred abnormally, the levels of mRNA for the *KVLQT1* gene could be lowered, and this might lead to the disease state.

d. In both situations, a region of a chromosome appears to be subject to imprinting, and multiple genes are involved. This suggests that there may be signals for methylation in certain regions, and if these signals are disrupted, abnormal imprinting can result in disease.

CHAPTER 22

22.2 Brown = C^BC^B, C^BC^P, C^BC^Y $p^2 + 2pq + 2pr = 236/500 = 0.472$
Pink = C^PC^P, C^PC^Y, $= q^2 + 2qr = 231/500 = 0.462$
Yellow = $C^YC^Y = r^2 = 33/500 = 0.066$

$$f(C^YC^Y) = r^2$$
$$\sqrt{f(C^YC^Y)} = qr^2$$
$$\sqrt{33/500} = r = \sqrt{0.066} = 0.26$$
$$f(C^PC^P + C^PC^Y + C^YC^Y) = q^2 + 2qr + r^2$$
$$f(C^PC^P + C^PC^Y + C^YC^Y) = (q + r)^2$$
$$\frac{231 + 33}{500} = (q + r)^2$$
$$0.528 = (q + r)^2$$
$$\sqrt{0.528} = q + r$$
$$0.727 = q + r$$
$$r = 0.26$$
$$0.727 - r = q$$
$$q = 0.467$$
$$p = 1 - q - r$$
$$= 1 - 0.467 - 0.26 = 0.273$$
So, $f(C^B) = p = 0.273$
$f(C^Y) = r = 0.26$
$f(C^P) = q = 0.467$

22.4 a. $\sqrt{0.16} = 0.4 = 40\%$ = frequency of recessive alleles; $1 - 0.4 = 0.6 = 60\%$ = frequency of dominant alleles; $2pq = (2)(0.4)(0.6) = 0.48$ = probability of heterozygous diploids. Then $(0.48)/[(2 \times 0.16) + 0.48] = 0.48/0.80 = 60\%$ of recessive alleles are heterozygous.

b. If $q^2 = 1\% = 0.01$, then $q = 0.1$, $p = 0.9$, and $2pq$ 0.18 heterozygous diploids. Therefore, $(0.18)/[(2 \times 0.01) + 0.18] = 0.18/0.20 = 90\%$ of recessive alleles in heterozygotes.

22.5 $2pq/q^2 = 8$, and so $2p = 8q$; then $2(1 - q) = 8q$, and $2 = 10q$, or $q = 0.2$

22.8 a. Let p = the frequency of S and q = the frequency of s. Then

$$p = \frac{2(188)\ SS + 717\ Ss}{2(3146)} = \frac{1093}{6292} = 0.1737$$

$$q = \frac{717\ Ss + 2\ (2241)\ ss}{2(3146)} = \frac{5199}{6292} = 0.8263$$

b.

CLASS	OBSERVED	EXPECTED	d	d^2/e
SS	188	94.9	+93.1	91.235
Ss	717	903.1	−186.1	38.361
ss	2241	2147.9	+93.1	4.032
	3146	3145.9	0	133.628

There is only one degree of freedom because the three genotypic classes are completely specified by two gene frequencies, namely, p and q. Thus, the number of degrees of freedom = number of genes – 1. The χ^2 value for this example is 133.628, which, for one degree of freedom, gives a P value less than 0.0001. Therefore, the distribution of genotypes differs significantly from the Hardy-Weinberg equilibrium.

22.11 Since the frequency of the trait is different in males and females, the character might be caused by a sex-linked recessive gene. If the frequency of this gene is q, females would occur with the character at a frequency of q^2, and males with the character would occur at a frequency of q. The frequency of males is $q = 0.4$, and thus we may predict that the frequency of females would be $(0.4)2 = 0.16$ if this is a sex-linked gene. This result fits the observed data. Therefore, the frequency of heterozygous females is $2pq = 2(0.6)(0.4) = 0.48$. For sex-linked genes, no heterozygous males exist.

22.13 64/10,000 are color blind, i.e., $0.0064 = q^2$, and so $q = 0.08$ = probability of color-blind male.

22.16

$$q = \frac{u}{u + v} = \frac{6 \times 10^{-7}}{(6 \times 10^{-7}) + (6 \times 10^{-8})}$$

$$= \frac{6 \times 10^{-7}}{(6 \times 10^{-7}) + (0.6 + 10^{-7})} = \frac{6}{6.6} = 0.91$$

$$p = 1 - q = 1 - 0.91 = 0.09$$

Thus, the frequencies are 0.008 AA, 0.16 Aa, and 0.828 aa.

22.20 $p'_{II} = mp_I + (1 - m)p_{II}$
$p'_{II} = (0.20)(0.5) + (1 - 0.20)(0.70) = 0.66$

22.23 a. When selectively neutral, the genes distribute themselves according to the Hardy-Weinberg law, so 0.25 are AA, 0.05 are Aa, and 0.25 are aa.

b. $q = 0.33$

c. $q = 0.25$

22.25 a. $q = 0.63$

b. $q = 0.64$

c. $q = 0.66$

22.27 $q = u/s = (5 \times 10^{-5})/0.8 = 0.0000625$

22.30 a. The data fit the idea that a single *Bam*HI site varies. The probe is homologous to a region wholly within the 4.1-kb piece bounded on one end by the variable *Bam*HI site and on the other end by a constant site. When the variable site is present, the hybridized fragment is 4.1 kb. When the variable site is absent, the fragment extends to the next constant *Bam*HI site, and is 6.7 kb long. People with only 4.1- or only 6.7-kb bands are homozygotes; people with both are heterozygotes.

b. The "+" allele of the variable site is present in $2(6) + 38 = 50$ chromosomes, and the "−" allele is present in $2(56) + 38 = 130$ chromosomes. Thus, $f(+)$ is 0.25 and $f(-)$ is 0.75.

c. If the population is in Hardy-Weinberg equilibrium, we would expect $(0.25)^2$ or 0.0625 of the sample to show only the 4.1-kb band. This would be 6.25 individuals. We observed 6. We expect $(0.75)^2$ or 0.5625 to be homozygous for the 6.7-kb band, which is 56.25 individuals. We saw 56. Finally, we would expect $2(0.25)(0.75)$ or 0.375 to be heterozygotes, or 37.5 individuals. We observed 38. The observed numbers are so close to the expected that a χ^2 test is unnecessary.

22.32 To calculate the percent polymorphic loci, we divide the number of loci with more than one allele (two) by the total number of loci examined (five).

$$\frac{2}{5} = 0.40$$

Heterozygosity is calculated by averaging the frequency of heterozygotes for each locus. The frequency of heterozygotes for the *AmPep*, *ADH*, and *LDH-1* loci is zero. Thirty-five out of fifty individuals are heterozygous at the *MDH* locus and ten out of fifty individuals are heterozygous at the *PGM* locus. Thus,

$$\frac{0 + 0 + 0 + \frac{35}{50} + \frac{10}{50}}{5} = \frac{0.9}{5} = 0.18$$

22.33 a. The expected heterozygosity is 1 – (frequency of expected homozygotes). If the frequency of alleles in the population is p_1, p_2, p_3, ... , p_n, the expected frequency of homozygotes is $p_1^2 + p_2^2 + p_3^2 + ... + p_n^2$. For locus *G1A*, the expected frequency of homozygotes is $(0.398)^2 + (0.240)^2 + (0.211)^2 + (0.086)^2 + (0.036)^2 + (0.016)^2 + (0.007)^2 + (0.006)^2 = 0.270$, and the expected heterozygosity is $1 - 0.270 = 0.730$. The expected heterozygosities for the other loci are *G10X*, 0.741; *G10C*, 0.740; and *G10L*, 0.662. These are approximately the observed frequencies of heterozygosities. The numbers and types of different heterozygotes are not given, so it is not possible to employ the chi-square test to evaluate directly if the population is in Hardy-Weinberg equilibrium. The population appears to be close to Hardy-Weinberg equilibrium.

b. The three cubs of the mother show evidence of multiple paternity. For each of the loci *G10X* and *G10L*, three alleles present in the cubs must have been contributed paternally (*G10X: X133, X135* or *X137, X141; G10L: L155, L157, L161*). This could have happened only if the cubs were sired by at least two different fathers. Multiple paternity within one set of cubs would tend to increase the genetic variability in the population because it would allow a larger number of males to contribute gametes seen in the next generation. $N_e = (4 \times N_f \times N_m)/(N_f + N_m)$, so a larger N_m will tend to increase the effective population size.

22.35 a. Mutation leads to change in gene frequencies within a population if no other forces are acting and introduces new genetic variation. If population size is small, mutation may lead to genetic differentiation among populations.

b. Migration increases population size and genetic variation within populations and equalizes gene frequencies among populations.

c. Genetic drift reduces genetic variation within populations, leads to genetic change over time, increases genetic differences among populations, and increases homozygosity within populations.

d. Inbreeding increases homozygosity within populations and decreases genetic variation.

22.38 Comparison with the rates of nucleotide substitution in Table 22.14 indicates that sequence A has as high a rate as that typically observed in mammalian pseudogenes. In addition, the rates of synonymous and nonsynonymous substitutions are the same. These observations suggest that sequence A is either a pseudogene or is a sequence that provides no function, because high rates of substitution are observed when sequences are functionless.

Sequence B has a relatively low rate of nonsynonymous substitution but a relatively high rate of synonymous substitution. This is the pattern we expect when a sequence codes for a protein; thus, sequence B probably encodes a protein.

22.43 In both instances, protein electrophoresis, RFLP analyses, and DNA sequence analysis of specific genes could be used to analyze the genotype of the captured individuals and members of each island population. In (a), the captured individual should be returned to the subpopulation with which it shows the least genetic variation. In (b), evaluate the genotype of the two missing tortoises using DNA from the previously collected blood samples, and compare these genotypes to the genotypes of the two captured tortoises. If a captured animal was taken from the field site, its genotype will exactly match a genotype obtained from one of the blood samples.

CHAPTER 23

23.1 a. Head width: mean = 3.15, standard deviation = 0.49, wing length: mean = 35.21, standard deviation = 7.68

b. Correlation coefficient = $r = 0.973$

c. Head width and wing length display a strong positive association, meaning that ducks with wider heads tend to have longer wings.

23.3 252/1,024

23.4 Each pure-breeding parent is homozygous for the genes (however many there are) controlling the size character, and hence each parent is homogeneous in type. A cross of two pure-breeding strains will generate an F_1 heterozygous for those loci controlling the size trait. Since the F_1 is genetically homogeneous (all heterozygotes), it shows no greater variability than the parents.

23.7 a. 8 cm

b. 1 (2 cm) : 6 (4 cm) : 15 (6 cm) : 20 (8 cm) : 15 (10 cm) : 6 (12 cm) : 1 (14 cm)

c. Proportion of F_2 of *AA BB CC* is $(\frac{1}{4})^3$; proportion of F_2 of *aa bb cc* is $(\frac{1}{4})^3$; proportion of F_2 with heights equal to one or other is $(\frac{1}{4})^3 + (\frac{1}{4})^3 = \frac{2}{64}$.

d. The F_1 height of 8 cm can be produced only by maintaining heterozygosity at no less than one gene pair (e.g., *AA Bb cc*). Thus, although 20 of 64 F_2 plants are expected to be 8 cm tall, none of these will breed true for this height.

23.10 It is a quantitative trait. Variation appears to be continuous over a range rather than falling into three discrete classes. Also, the pattern in which the F_1 mean falls between the parental means and in which the F_2 individuals show a continuous range of variation from one parental extreme to the other is typical of quantitative inheritance.

23.13 The cross is *aa bb cc* (3 lb) $\times$ *AA BB CC* (6 lb), which gives an F_1 that is *Aa Bb Cc* and which weighs 4.5 lb. The distribution of phenotypes in the F_2 is 1 (3 lb) : 6 (3.5 lb) : 15 (4 lb) : 20 (4.5 lb) : 15 (5 lb) : 6 (5.5 lb) : 1 (6 lb).

23.15 a. The progeny are *Aa Bb Cc Dd*, which are 18 dm high.

b. (1) The minimum number of capital-letter alleles is one, and the maximum number is four, giving a height range of 12 to 18 dm. (2) The minimum number is one, and the maximum number is four, giving a height range of 12 to 18 dm. (3) The minimum number is four, and the maximum number is six, giving a height range of 18 to 22 dm. (4) The minimum number is zero, and the maximum number is seven, giving a height range of 10 to 24 dm.

23.17 Let n = total number of alleles, s = number of capital alleles, t = number of lowercase alleles, a = chance of obtaining a capital allele, b = chance of obtaining a lowercase allele, and $x!$ = $(x)(x-1)(x-2) \ldots (1)$, with $0!$ = 1. Then the chance p of obtaining progeny with a specified number of each type of allele is given by:

$$p(s,t) = \frac{n!}{s!t!} a^s b^t$$

$$p(0,10) = \frac{10!}{0!10!} \left(\frac{1}{2}\right)^0 \left(\frac{1}{2}\right)^{10} = \frac{1}{1{,}024} \left.\vphantom{\begin{array}{c}1\\1\end{array}}\right\} \frac{11}{1{,}024} \text{ bluish-white}$$

$$p(1,9) = \frac{10!}{1!9!} \left(\frac{1}{2}\right)^1 \left(\frac{1}{2}\right)^9 = \frac{10}{1{,}024}$$

$$p(2,8) = \frac{10!}{2!8!} \left(\frac{1}{2}\right)^2 \left(\frac{1}{2}\right)^8 = \frac{45}{1{,}024} \left.\vphantom{\begin{array}{c}1\\1\end{array}}\right\} \frac{165}{1{,}024} \text{ light tan}$$

$$p(3,7) = \frac{10!}{3!7!} \left(\frac{1}{2}\right)^3 \left(\frac{1}{2}\right)^7 = \frac{120}{1{,}024}$$

$$1 - \frac{11 + 165}{1{,}024} = \frac{848}{1{,}024} \text{ greyish-brown}$$

23.18 Some proportion of cases of Alzheimer's disease appear to be attributable to genetic factors. Multiple genes that increase the risk for AD have been identified, some of which seem to act in a dose-dependent manner. Thus, a number of different genes could contribute to the onset of AD, with some having a greater contribution than others.

b. Consider two explanations. (1) If AD can be caused by either environmental agents or mutation, the presence of AD in both twins could be due to the presence of an abnormal allele in both or exposure of both twins to adverse environmental conditions. The presence of AD in only one twin may be due to exposure of only one twin to the causative environmental agent(s). (2) The presence of a mutation may only increase the risk of disease and not determine its occurrence. The penetrance of an allele may be affected by the environment.

23.21 The selection differential is 0.14 g. The selection response is 0.08 g. $h^2 = 0.08/0.14 = 0.57$.

23.22 Zero because there would be no genetic variability in the population to affect blood pressure. Blood pressure would vary only because of salt exposure.

23.23 Heritability is specific to a particular population and to a specific environment and cannot be used to draw conclusions about the basis of populational differences. Because the environments of the farms in Kansas and in Poland differ and because the two wheat varieties differ in their genetic makeup, the heritability of yield calculated in Kansas cannot be applied to the wheat grown in Poland. Furthermore, the yield of TK138 would most likely be different in Poland and might even be less than the yield of the Russian variety when grown in Poland.

23.25 Selection differential = 14.3 − 9.7 = 4.6

Selection response = 13 − 9.7 = 3.3

Narrow-sense heritability = selection response/selection differential = 3.3/4.6 = 0.72

23.27 a. Fathers: mean = 71.9, variance = 11.6

Sons: mean = 70, variance = 10.25

b. Correlation = 0.49

c. Slope = b = 0.46

Narrow-sense heritability = 2(b) = 2(0.46) = 0.92

23.28 Selection response = narrow-sense heritability × selection differential.

Selection response = 0.60 × 10 g = 6 g

23.30 The variance that responds to selection is the additive genetic variance. The narrow-sense heritability for each of the traits is V_A/V_P and can be used to predict how a trait will respond to selection. This is 0.165 for body length, 0.061 for antennal bristle number, and 0.144 for egg production. Thus, body length will respond most to selection, and antenna bristle number will respond least to selection.

CREDITS

TEXT AND ILLUSTRATION CREDITS

Figure 1.6: From Bruce Alberts et al., *Molecular Biology of the Cell*. Copyright ©1989 by Bruce Alberts, Dennis Bray, Julian Lewis, Martin Raff, Keith Roberts, and James D. Watson. Reprinted by permission of Garland Publishing Company Inc.

Figure 3.15: From *Biological Science*, 4th ed. by William T. Keeton and James L. Gould with Carol Grant Gould. Copyright ©1986, 1980, 1979, 1978, 1972, 1967 by W. W. Norton & Company, Inc. Reprinted by permission of W. W. Norton & Company, Inc.

Table 4.2: Data from D. J. Nolte, "The Eye-Pigmentary System of *Drosophila*" in *Heredity* 13 (1959): 219–281.

Figure 5.5: Figure adapted from *Genetics*, 2nd ed. by Ursula Goodenough. Copyright ©1978 by Holt Rinehart and Winston, Inc. Reproduced by permission of the publisher.

Figure 5.20: From *Introduction to Genetics* by A. H. Sturtevant and G. W. Beadle, 1962. Reprinted by permission of Dover Publishers, Inc.

Page 167: Text quotation from *Daily Racing Form*, Sunday October 23, 1994.

Figure 6.12: From "The Genetics Revolution" in *Time*, January 17, 1994. Copyright ©1994 by Time Inc. Reprinted by permission.

Figure 7.14: Reprinted by permission of the publisher from "High Resolution Chromosomes of the +(9:22) Leukemias" by O. Prakash and J. J. Yunis in *Cancer and Cytogenetics II*, 1984. Copyright ©1984 by Elsevier Science Inc.

Figure 7.16: From *General Genetics* by A. M. Srb, R. D. Owen, and R. S. Edgar. Copyright ©1965 by W. H. Freeman and Company. Used with permission.

Figure 8.31: From W. H. Wood and H. R. Revel, "The Genome of Bacteriophage T4" in *Bacteriological Review* 40 (1976): 847–868. Reprinted by permission.

Figure 9.9: Copyright ©Irving Geis.

Table 11.2: Data from P. L. Altman and D. S. Dittmer (eds.), *Biology Data Book*, 2nd ed., Vol. 1 (Bethesda, MD: Federation of American Societies for Experimental Biology, 1972)

Figure 11.4: From *Genes IV* by Benjamin Lewin. Published by Oxford University Press and Cell Press, 1990. Copyright ©1990 Cell Press. Reprinted by permission.

Table 11.4: Reproduced with permission, from the *Annual Review of Biochemistry*, Vol. 44. Copyright ©1975 by Annual Reviews Inc.

Figure 11.9: From D. M. Freifelder, *Essentials of Molecular Biology*, 2nd ed. Copyright ©1985 by Jones and Bartlett. Boston: Jones and Bartlett Publishers. Reprinted with permission.

Figure 11.14: From R. J. Britten and E. Davidson, "Distribution of Genome Size in Animals" in *Quarterly Review of Biology* 46 (1971): 111. Reprinted by permission.

Figure 11.21: From Bruce Alberts et al., *Molecular Biology of the Cell*. Copyright ©1989 by Bruce Alberts, Dennis Bray, Julian Lewis, Martin Raff, Keith Roberts, and James D. Watson. Reprinted by permission of Garland Publishing Company Inc.

Figure 11.25: From *Principles of Cell Biology* by Lewis Kleinsmith and Valerie M. Kish. Copyright ©1988 by HarperCollins Publishers, Inc. Reprinted by permission of Addison Wesley Longman, Inc.

Figures 12.4, 12.8: From Bruce Alberts et al., *Molecular Biology of the Cell*. Copyright ©1989 by Bruce Alberts, Dennis Bray, Julian Lewis, Martin Raff, Keith Roberts, and James D. Watson. Reprinted by permission of Garland Publishing Company Inc.

Figure 12.6: From J. D. Watson, et al. *Molecular Biology of the Gene*, Vols. I and II, 4th ed. Copyright ©1965, 1970, 1976, 1987 by James D. Watson. Reprinted by permission of Addison Wesley Longman, Inc.

Figure 12.14: Adapted from *Biology Concepts and Connections*, 2nd ed. by Campbell et al., Figure 8.10, p. 136. Copyright ©1997 by The Benjamin/Cummings Publishing Company.

Figure 13.5: From Charles Yanofsky, "Attenuation in the Control of Expression of Bacterial Operons" in *Nature* 289 (1981): 751. Copyright ©1981 by Macmillan Journals Ltd. Reprinted by permission.

Figure 13.12: Adapted from N. Proudfoot "Ending the Message Is Not So Simple" in

Cell 87 (1996): 779–781. Copyright ©1996 by Cell Press. Reprinted by permission.

Figure 13.24: From *Genes IV* by Benjamin Lewin. Published by Oxford University Press and Cell Press, 1990. Copyright ©1990 Cell Press. Reprinted by permission.

Figure 14.4c, d: Copyright ©Irving Geis.

Figure 14.8: From J. D. Watson, et al. *Molecular Biology of the Gene*, Vols. I and II, 4th ed. Copyright ©1965, 1970, 1976, 1987 by James D. Watson. Reprinted by permission of Addison Wesley Longman, Inc.

Table 15.1: Adapted from R. J. Roberts in *CRC Critical Reviews in Biochemistry* 4 (1976): 123. Copyright ©1976 CRC Press, Inc. Boca Raton, Florida. Reprinted with permission.

Figure 15.29: From *Genetics* by Robert F. Weaver and Philip W. Hedrick. Copyright ©1989 by Wm. C. Brown Publishers. Reprinted by permission.

Figure 16.19: From Charles Yanofsky, "Attenuation in the Control of Expression of Bacterial Operons" in *Nature* 289 (1981). Copyright ©1981 by Macmillan Journals Ltd. Reprinted by permission.

Table 17.1: From L. Chan and B. W. O'Malley in *Annals of Internal Medicine* 89 (1978): 649. Reprinted by permission.

Figure 17.12: From Bruce Alberts et al., *Molecular Biology of the Cell*. Copyright ©1989 by Bruce Alberts, Dennis Bray, Julian Lewis, Martin Raff, Keith Roberts, and James D. Watson. Reprinted by permission of Garland Publishing Company Inc.

Figure 17.22: Adapted from Painter in *Journal of Heredity* 25 (1934): 465–467. Reprinted by permission of Oxford University Press, London.

Figure 17.29: From *Genes IV* by Benjamin Lewin. Published by Oxford University Press and Cell Press, 1990. Copyright ©1990 by Cell Press. Reprinted by permission.

Figure 17.32: Illustration by Bunji Tagawa in "Transdetermination in Cells", *Scientific American*, November 1968, p. 116.

Table 18.1: Adapted from *RNA Tumor Viruses*, 2nd ed. by R. Weiss, et al. (eds.). (New York: Cold Spring Harbor Laboratory,

1985.), Supplement Chapter 9, Table 9S.1, pp. 252–253.

Table 18.2: Adapted from T. Hunter, "Cooperation Between Oncogenes" in *Cell* 64 (1991): 249–270. Copyright ©1991 Cell Press. Reprinted by permission.

Figure 18.13: Based on data from Takeya and H. Manfusa in *Cell* 32 (1983): 881–890.

Figure 18.14: From J. D. Watson, et al. *Molecular Biology of the Gene*, Vols. I and II, 4th ed. Copyright ©1965, 1970, 1976, 1987 by James D. Watson. Reprinted by permission of Addison Wesley Longman, Inc.

Table 19.1: Data from S. E. Luria and M. Delbruck in *Genetics* 28 (1943): 491–511.

Figure 19.13: From J. H. Miller and K. B. Low, "Distribution of Spontaneous GC and AC Transition Mutations to Stop Codons in the *lac* Gene of *E. coli*" in *Cell* 37 (1984): 675–682. Copyright ©1984 by Cell Press. Reprinted by permission.

Box 20.1: Abridged version of a memoir on Barbara McClintock by Nina Federoff for the Biological Memoir Series of the National Academy of Sciences, Vol. 68. Adapted by permission of Nina Fedoroff and by courtesy of the National Academy of Sciences, Washington, DC.

Table 20.3: With permission, from the *Annual Review Genetics*, Volume 20, Copyright ©1986, by Annual Reviews Inc.

Figure 20.15: Figure adapted by permission of Gerald B. Fink.

Figure 21.9b: From R. Sager and D. Lane, "Republication of Chloroplast DNA in Zygotes of Chlamydomonas" in *Federation Proceedings of the Federation of American Societies for Experimental Biology* 38 (1969): 347.

Figure 21.12: From N. Maizels and Winer, "RNA Editing" in *Nature* 334 (1988): 469. Copyright ©1988 by Macmillan Magazines Limited. Reprinted by permission.

Figures 21.14, 21.15: Reprinted by permission from *New Britton and Brown Illustrated Flora of the Northeastern United States and Adjacent Canada* by Henry Allen Gleason et al. Copyright ©1952 by the New York Botanical Garden.

Box 22.1: Data from F. B. Livingstone, 1973. Data on the Abnormal Hemoglobins and Glucose-6 Phosphate Dehydrogenase Deficiency in Human Populations of 1967–1973 in *Contributions in Human Biology No. 1*, Museum of Anthropology, University of Michigan.

Figure 22.2: From *Ecological Genetics* by E. B. Ford, 1975. Reprinted by permission of Chapman and Hall Ltd.

Figure 22.3: From P. Buri in *Evolution* 10 (1956): 367. Reprinted by permission of the Society for the Study of Evolution.

Table 22.3: From R. K. Selander, "Genetic Variation in Natural Populations" in *Molecular Evolution* by F. J. Ayala, 1976. Reprinted by permission of Sinauer Associates Inc.

Figure 22.6: From R. K. Koehn et al. in *Evolution* 30 (1976): 6. Reprinted by permission of the Society for the Study of Evolution.

Figure 22.7: From M. S. Gaines et al. in *Evolution* 32 (1978): 729. Reprinted by permission of the Society for the Study of Evolution.

Table 22.7: Data from R. K. Selander, "Behavior and Genetic Variation in Natural Populations (*Mus musculus*)" in *American Zoologist* 10 (1970): 53–66.

Figure 22.11: From P. Buri in *Evolution* 10 (1956): 367. Reprinted by permission of the Society for the Study of Evolution.

Figures 22.12, 14: From *Principles of Population Genetics* by J. L Hartl and J. G. Clark, 1989. Reprinted by permission of Sinauer Associates Inc.

Figure 22.13: From Philip Hedrick, *Genetics of Population,* 1983. Copyright ©1983 Jones and Bartlett Publishers. (Boston: Jones and Bartlett Publishers, Inc.) Reprinted with permission.

Table 22.14: From W. Li, C. Luo, and C. Wu, "Evolution of DNA Sequences" in *Molecular Evolutionary Genetics* 2 (1985): 150–174 by R. J. MacIntyre, ed. Reprinted by permission of Plenum Publishing Corporation.

Table 22.15: Data from W. Li et al. in *Molecular Biology Evolution* 2 (1985): 250–174.

Figure 22.21: Adapted from A. C. Allison, "Abnormal Hemoglobin and Erythrovute Enzyme-Deficiency Traits" in *Generic Variation in Human Population* by G. A. Harrison ed. (Oxford: Elsevier Science, 1961).

Figure 23.2: Data from Multnomah County Department of Public Health.

Figure 23.4: From *Evolution and the Genetics of Populations*, Vol. 1 by Sewall Wright. Copyright © by The University of Chicago Press. All rights reserved. Reprinted by permission.

Table 23.6: D. S. Falconer, *Introduction to Quantitative Genetics,* 2nd ed., 1981. New York: Longman. T. Mitchell-Olds, *Evolution* 40 (1986): 107–116. J. O. Palmer and H. Dingle, *Evolution* 40 (1986): 767–777. J. Travis et al., *Evolution* 41 (1987): 145–156. K. A. Berven, *Evolution* 41 (1987): 1088–1097.

Table 23.7: D. S. Falconer, *Introduction to Quantitative Genetics,* 2nd ed., 1981. New York: Longman. T. Mitchell-Olds, *Evolution* 40 (1986): 107–116. J. O. Palmer and H. Dingle, *Evolution* 40 (1986): 767–777. J. Travis et al., *Evolution* 41 (1987): 145–156. K. A. Berven, *Evolution* 41 (1987): 1088–1097. P. M.

Service and M. R. Rose, *Evolution* 39 (1985): 943–944.

Figure 23.10: From *An Introduction to Genetics* by A. H. Sturtevant and G. W. Beadle, 1962. Reprinted by permission of Dover Publishers, Inc.

Figure 23.13: Data from *Introduction to Quantitative Genetics,* 2nd ed. by D. S. Falconer, 1981. Reprinted by permission of Addison Wesley Longman, Inc., London.

PHOTOGRAPH CREDITS

Chapter 1   Opener: ©Howard Hughes Medical Institute/Peter Arnold, Inc. 1.3a: Calgene Fresh Inc. 1.3b: ©SIU/Visuals Unlimited. 1.4a: ©J. Forsdyke/Gene Cox/SPL/Photo Researchers, Inc. 1.4b: ©Grant Heilman/Grant Heilman Photography. 1.4c: Courtesy of John Sulston, Medical Research Council/Laboratory of Molecular Biology. 1.4d: ©Dr. Jeremy Burges/SPL/Photo Researchers, Inc. 1.4g: Peter J. Russell. 1.4h: ©David M. Phillips/Visuals Unlimited. 1.4i: ©K. Aufderheide/Visuals Unlimited. 1.4j: ©Cabisco/Visuals Unlimited. 1.4k: ©John Colwell/Grant Heilman Photography. 1.4l: ©Larry Lefever/Grant Heilman Photography. 1.4m: ©Hans Reinhard/Bruce Coleman, Inc. 1.9: ©Custom Medical Stock Photo. 1.11: ©Biophoto Associates/SS/Photo Researchers, Inc. 1.12 (left): ©Jane Grushow/Grant Heilman Photography. 1.12 (right): ©H. Reinhard/Okapia/Photo Researchers, Inc. 1.16: Courtesy of Dr. James A. Lake, *Scientific American*, 84–97, August 1981.

Chapter 2   Opener: ©Nigel Cattlin/Holt Studios Int./Earth Scenes. 2.1a–e: ©Sharon Eide. 2.3a: ©Chick Harrity/U.S. News & World Report. 2.3b: Harvey Stein. 2.4: ©The Granger Collection, New York. 2.18a: ©G. Whiteley/Photo Researchers, Inc. 2.18b: ©Biophoto Associates/Photo Researchers, Inc. 2.21a (left): ©Davila/ Retna Ltd. 2.21a (right): ©Gary Gershoff/ Retna Ltd. 2.22a: From the *Journal of Heredity* 23 (September 1932): 345.

Chapter 3   Opener: ©Dr. Gerald Schatten/SPL/Photo Researchers, Inc. 3.6a (left), 3.6b (left), 3.6c (left), 3.6d (left), and 3.6e (left): ©Ed Reschke/Peter Arnold, Inc. 3.6a (right), 3.6b (right), 3.6c (right), 3.6d (right), and 3.6e (right): ©Michael Abbey/SS/Photo Researchers, Inc. 3.9a: ©G. F. Bahr/Armed Forces Institute. 3.9b: ©K. G. Murti/ Visuals Unlimited. 3.9c: J. R. Paulson and U. K. Laemmli, *Cell* 12 (1977): 817–828. ©1977. M.I.T. Photo Courtesy of Dr. U. K. Laemmli. 3.11c: ©David Scharf/ Peter Arnold, Inc. 3.13: Courtesy of M. Westergaard and D. von Wettstein. 3.31, 3.32: Digamber S. Borgaonkar, Ph.D. 3.33a: ©George Wilder/Visuals Unlimited. 3.33b:

©M. Abbey 1981/Photo Researchers, Inc. 3.34a, c: Courtesy of John Sulston, Medical Research Council/Laboratory of Molecular Biology. 3.37a: Courtesy of Dr. Carol Witkop, University of Minnesota. 3.39: Courtesy of the American Society of Human Genetics. From Stern, Centerwall, and Sakar, "New Data on the Problem of Hairy Pinnae," *American Journal of Human Genetics* 15 (1964): 74–75.

Chapter 4 Opener: ©Jean Paul Ferrero/Ardea London, Ltd./Ardea Photographics. 4.5a: ©Fritz Prenzel/Animals Animals. 4.5b: International American Albino Association, Inc. 4.5c: ©Larry Lefever/Grant Heilman Photography. 4.7a: ©Barbara J. Wright/Animals Animals. 4.7b: ©Larry Lefever/Grant Heilman Photography. 4.7c: ©C. Prescott Allen/Animals Animals. 4.7d: ©Larry Lefever/Grant Heilman Photography. 4.11b: ©Hans Reinhard/OKAPIA/Photo Researchers, Inc. 4.11c: ©E. R. Degginger. 4.19: From "The True History of the Elephant Man" by Michael Howell and Peter Ford, *British Medical Journal*, 1890. ©1980 Allison and Busby Ltd., London, England. 4.21: James H. Tonsgard, M.D., Wyler Children's Hospital, University of Chicago Medical Center. 4.22 (top): ©Carolina Biological Supply/Phototake. 4.22 (bottom): ©Photo Researchers, Inc. 4.25: John Blasdel for Scott Foresman. 4.26a: ©Paul Conklin.

Chapter 5 Opener: 6.1b: ©Dr. Jeremy Burgess/SPL/Photo Researchers, Inc.

Chapter 6 Opener: ©A. B. Dowsett/SPL/Photo Researchers, Inc. 6.11: ©Dr. T. Reid and Dr. D. Ward/Peter Arnold, Inc.

Chapter 7 Opener: ©Department of Energy/Photo Researchers, Inc. 7.4a: Dr. Liard Jackson, Thomas Jefferson Medical College. 7.4b: C. Weinkove and R. McDonald, *S Afr Med J* 43 (1969): 318 from *Syndromes of the Head and Neck*, 3rd ed., by Robert Golin, M. Michael Cohen, and L. Stefan Levin, Oxford University Press, 1990. 7.10a: Courtesy of Grant L. Pyrah, Department of Biology, Southwest Missouri State University. 7.17a: Courtesy of Christine J. Harrison, from the *Journal of Medical Genetics*, Vol. 20, pp. 280–285, 1983. 7.20b: ©M. Coleman/Visuals Unlimited. 7.23, 7.24: Dr. Liard Jackson, Thomas Jefferson Medical College.

Chapter 8 Opener: ©Dr. Dennis Kunkel/Phototake. 8.1: ©Custom Medical Stock Photo. 8.2: ©Michael Gabridge/Custom Medical Stock Photo. 8.7: Micrograph courtesy of David P. Allison, Biology Division, Oak Ridge National Laboratory. 8.13a, b: Dr. Harold W. Fisher. 8.18: ©Bruce Iverson. 8.20: Courtesy of Gunther S. Stent, University of California. 8.21: Courtesy of Dr. D. P. Snustad, Department of Genetics and

Cell Biology, College of Biological Sciences, University of Minnesota.

Chapter 9 9.8a: ©Bill Longcore/Science Source/Photo Researchers, Inc. 9.8b: ©Jackie Lewis/Royal Free Hospital/Science Photo Library/Photo Researchers, Inc.

Chapters 9 and 10 Openers: ©Ken Eward/Science Source/Photo Researchers, Inc.

Chapter 10 10.1: ©Manfred Kage/Peter Arnold, Inc. 10.12: Cold Spring Harbor Laboratory. 10.13: Courtesy of Professor M. H. F. Willkins, Biophysics Department, King's College, London. 10.16a–c: Richard Pastor, Courtesy of the Food and Drug Administration. 10.17a–c: ©Irving Geiss.

Chapter 11 Opener: ©Jan Hinsch/Science Photo Library/Photo Researchers, Inc. 11.2: Jack Griffith. 11.3: ©Dr. Gopal Murti/Photo Researchers, Inc. 11.8a: Courtesy of Dr. Harold W. Fisher, University of Rhode Island. 11.8b: From W. J. Thomas and R. W. Horne, "The Structure of Bacteriophage phiX174," *Virology* 15 (1961): 2–7. ©Academic Press. 11.8c: Courtesy of David Dressler, Department of Neurobiology, Harvard Medical School. 11.11: H. Lisco, from Bloom and Fawcett, Eugenia Fawcett, *Textbook of Histology*, ©D. W. Fawcett. 11.12: ©J. F. Gennaro/Photo Researchers, Inc. 11.13: Published in Sumner et al., *Nature New Biology* 232 (1971): 31–32, Fig. 2. Courtesy of Dr. A. T. Sumner, MRC Human Genetics Unit, Western General Hospital, Edinburgh, Scotland. 11.16: Jack Griffith. 11.17: Barbara Hamakalo. 11.19: ©J. R. Paulsen, U. K. Laemmli, and D. W. Fawcett/Visuals Unlimited.

Chapter 12 Opener: ©Clive Freeman/The Royal Institution/Science Photo Library/Photo Researchers, Inc. 12.11a: Courtesy of N. P. Salzman, National Institutes of Health. 12.15, 12.16a: Courtesy of Dr. Sheldon Wolff, Courtesy University of California, San Francisco.

Chapter 13 Opener: ©Professor Oscar L. Miller/Science Photo Library/Photo Researchers, Inc. 13.10: Courtesy of O. L. Miller Jr., Lewis and Clark Professor of Biology, University of Virginia. 13.13: Courtesy of Philip Leder, National Institutes of Health, from Tilghman, Curtis, Tiemeier, Leder, and Weisman, *Proceedings of the National Academy of Sciences*, 75 (1979): 1309. 13.17: Dr. James A. Lake, *Journal of Molecular Biology*, 15 (1976): 131–159. 13.21: ©O. L. Miller/B. R. Beatty/D. W. Fawcett/Visuals Unlimited. 13.27b: Tripos Associates. 13.A (top): Courtesy of Professor Pierre Chamboni, Louis Pasteur University, Strasbourg, France.

Chapter 14 Opener: ©Harstock Medical Art Company/Custom Medical Stock Photo. 14.18: E. Kiseleva, from Don W. Faw-

cett, *Textbook of Histology*, 12th ed., ©Chapman and Hall, 1994.

Chapter 15 Opener: ©Jean-Claude Revy/Phototake. 15.22: Pharmacia Biotech Inc.

Chapter 16 Opener: ©Ken Eward/Science Source/Photo Researchers, Inc. 16.10: Courtesy of Mitchell Lewis et al., *Science* 271: 1247–1254. ©1996.

Chapter 17 Opener: Courtesy of Stephen Paddock, James Langeland, Peter DeVries and Sean B. Carroll of the Howard Hughes Medical Institute at the University of Wisconsin (*Bio-Techniques*, January 1993). 17.3a: Courtesy of Michael Pique, The Scripps Research Institute, La Jolla, CA. Based on an NMR structure by Peter E. Wright, The Scripps Research Institute. 17.3b: ©Leonard Lessin/Peter Arnold, Inc. 17.3c: ©Ken Eward/Science Source/Photo Researchers, Inc. 17.4: Courtesy of Dr. Harold Weintraub, Fred Hutchison Cancer Research Center, Seattle. Used by permission. 17.24b: ©Richard Feldmann/Phototake. 17.33b, c: Courtesy of Prof. E. B. Lewis, California Institute of Technology. 17.34: From David Suzuki et al., *Introduction to Genetic Analysis*, p. 485. ©1986 W. H. Freeman and Company. Used with permission.

Chapter 18 Opener: Courtesy of Y. Cho et al., *Science* 265: 346–355. ©1994 by AAAS, kindly provided by N. P. Pavletich. 18.1: ©SIU/Visuals Unlimited. 18.2: ©Astrid & Hanns-Frieder Michler/Science Photo Library/Photo Researchers, Inc. 18.4: ©Custom Medical Stock.

Chapter 19 Opener: ©Runk/Schoenberger/Grant Heilman Photography. 19.25: ©Ken Greer/Visuals Unlimited. 19.26a (left): ©CBSC/Phototake. 19.26a (right): ©Jean-Claude Revy/Phototake. 19.26b (left): ©Hans Reinhard/Bruce Coleman, Inc. 19.26b (right): ©Jane Burton/Bruce Coleman, Inc. 19.26c: Derek Fell.

Chapter 20 Opener: ©Professor Stanley Cohen/Science Photo Library/Photo Researchers, Inc. 20.11: Virginia Walbot, Stanford University. Page 664: ©Associated Press/Worldwide Photos.

Chapter 21 Opener: ©Science Photo Library/Custom Medical Stock. 21.2: ©CNRI/Science Photo Library/Photo Researchers, Inc. 21.3: Professor David A. Clayton. 21.10: ©M. I. Walker/Science Source/Photo Researchers, Inc. 21.13: ©E. R. Degginger. 21.21: ©Kim Taylor/Bruce Coleman, Inc.

Chapter 22 Opener: ©Dieter and Mary Plage/Bruce Coleman, Inc. 22.1a: ©Corbis-Bettmann. 22.1b: Courtesy of Hildegard Adler. 22.1c: ©UPI/Corbis-Bettmann. 22.5: Neg. No. ANTHRO 118/Field Museum of Natural History, Chicago. 22.8: Chip Clark.

22.16a: ©Kevin Byron/Bruce Coleman, Inc. 22.16b: ©W. Perry Conway/Tom Stack and Associates. 22.17: ©Cliff B. Frith/Bruce Coleman, Inc. 22.18a: ©ARCHIV/Photo Researchers, Inc. 22.18b: ©The Granger Collection, New York. 22.19: ©Breck Kent.

Chapter 23 Opener: ©Camera M.D. Studios.

TRADEMARK ACKNOWLEDGMENTS

Flavr Savr is a trademark of Calgene, Inc.

NutraSweet is a registered trademark of NutraSweet Company, Inc.

Roundup is a registered trademark of Monsanto Company (De Corp).

Roundup Ready soybeans is a registered trademark of Monsanto Company (De Corp).

Index

Page numbers in *italics* indicate material in figures and tables.